Main groups — Transition metals — Main groups

1 1A	2 2A	3 3B	4 4B	5 5B	6 6B	7 7B	8 8B	9 8B	10 8B	11 1B	12 2B	13 3A	14 4A	15 5A	16 6A	17 7A	18 8A
1 **H** 1.00794																	2 **He** 4.00260
3 **Li** 6.941	4 **Be** 9.01218											5 **B** 10.81	6 **C** 12.011	7 **N** 14.0067	8 **O** 15.9994	9 **F** 18.998403	10 **Ne** 20.1797
11 **Na** 22.98977	12 **Mg** 24.305											13 **Al** 26.98154	14 **Si** 28.0855	15 **P** 30.97376	16 **S** 32.065	17 **Cl** 35.453	18 **Ar** 39.948
19 **K** 39.0983	20 **Ca** 40.078	21 **Sc** 44.9559	22 **Ti** 47.867	23 **V** 50.9415	24 **Cr** 51.996	25 **Mn** 54.9380	26 **Fe** 55.845	27 **Co** 58.9332	28 **Ni** 58.6934	29 **Cu** 63.546	30 **Zn** 65.38	31 **Ga** 69.723	32 **Ge** 72.64	33 **As** 74.9216	34 **Se** 78.96	35 **Br** 79.904	36 **Kr** 83.798
37 **Rb** 85.4678	38 **Sr** 87.62	39 **Y** 88.9059	40 **Zr** 91.224	41 **Nb** 92.9064	42 **Mo** 95.96	43 **Tc** (98)	44 **Ru** 101.07	45 **Rh** 102.9055	46 **Pd** 106.42	47 **Ag** 107.8682	48 **Cd** 112.41	49 **In** 114.82	50 **Sn** 118.710	51 **Sb** 121.760	52 **Te** 127.60	53 **I** 126.9045	54 **Xe** 131.29
55 **Cs** 132.9055	56 **Ba** 137.33	57 ***La** 138.9055	72 **Hf** 178.49	73 **Ta** 180.9479	74 **W** 183.84	75 **Re** 186.207	76 **Os** 190.23	77 **Ir** 192.217	78 **Pt** 195.08	79 **Au** 196.9666	80 **Hg** 200.59	81 **Tl** 204.3833	82 **Pb** 207.2	83 **Bi** 208.9804	84 **Po** (209)	85 **At** (210)	86 **Rn** (222)
87 **Fr** (223)	88 **Ra** 226.0254	89 **†Ac** 227.0278	104 **Rf** (267)	105 **Db** (268)	106 **Sg** (271)	107 **Bh** (272)	108 **Hs** (270)	109 **Mt** (276)	110 **Ds** (281)	111 **Rg** (280)	112 **Cn** (285)	113 **Uut** (284)	114 **Fl** (289)	115 **Uup** (288)	116 **Lv** (293)	117 **Uus** (294)	118 **Uuo** (294)

*Lanthanide series	58 **Ce** 140.116	59 **Pr** 140.9077	60 **Nd** 144.242	61 **Pm** (145)	62 **Sm** 150.36	63 **Eu** 151.964	64 **Gd** 157.25	65 **Tb** 158.9254	66 **Dy** 162.500	67 **Ho** 164.9303	68 **Er** 167.259	69 **Tm** 168.9342	70 **Yb** 173.05	71 **Lu** 174.9668
†Actinide series	90 **Th** 232.0381	91 **Pa** 231.0359	92 **U** 238.0289	93 **Np** 237.048	94 **Pu** (244)	95 **Am** (243)	96 **Cm** (247)	97 **Bk** (247)	98 **Cf** (251)	99 **Es** (252)	100 **Fm** (257)	101 **Md** (258)	102 **No** (259)	103 **Lr** (262)

Source: Data from *Atomic Weights of the Elements 2009*, IUPAC. See iupac.org/publications/pac/83/2/0359/
Values in parentheses are mass numbers of the longest-lived known isotopes.

Greek Alphabet

A	α	alpha	N	ν	nu
B	β	beta	Ξ	ξ	xi
Γ	γ	gamma	O	o	omicron
Δ	δ	delta	Π	π	pi
E	ε	epsilon	P	ρ	rho
Z	ζ	zeta	Σ	σ	sigma
H	η	eta	T	τ	tau
Θ	θ	theta	Υ	υ	upsilon
I	ι	iota	Φ	ϕ	phi
K	κ	kappa	X	χ	chi
Λ	λ	lambda	Ψ	ψ	psi
M	μ	mu	Ω	ω	omega

Names and Symbols for the Elements

Z	*Symbol*	*Name*	*Z*	*Symbol*	*Name*	*Z*	*Symbol*	*Name*
1	H	Hydrogen	41	Nb	Niobium	81	Tl	Thallium
2	He	Helium	42	Mo	Molybdenum	82	Pb	Lead (Plumbum)
3	Li	Lithium	43	Tc	Technetium	83	Bi	Bismuth
4	Be	Beryllium	44	Ru	Ruthenium	84	Po	Polonium
5	B	Boron	45	Rh	Rhodium	85	At	Astatine
6	C	Carbon	46	Pd	Palladium	86	Rn	Radon
7	N	Nitrogen	47	Ag	Silver (Argentum)	87	Fr	Francium
8	O	Oxygen	48	Cd	Cadmium	88	Ra	Radium
9	F	Fluorine	49	In	Indium	89	Ac	Actinium
10	Ne	Neon	50	Sn	Tin (Stannum)	90	Th	Thorium
11	Na	Sodium (Natrium)	51	Sb	Antimony (Stibium)	91	Pa	Protactinium
12	Mg	Magnesium	52	Te	Tellurium	92	U	Uranium
13	Al	Aluminum	53	I	Iodine	93	Np	Neptunium
14	Si	Silicon	54	Xe	Xenon	94	Pu	Plutonium
15	P	Phosphorus	55	Cs	Cesium	95	Am	Americium
16	S	Sulfur	56	Ba	Barium	96	Cm	Curium
17	Cl	Chlorine	57	La	Lanthanum	97	Bk	Berkelium
18	Ar	Argon	58	Ce	Cerium	98	Cf	Californium
19	K	Potassium (Kalium)	59	Pr	Praseodymium	99	Es	Einsteinium
20	Ca	Calcium	60	Nd	Neodymium	100	Fm	Fermium
21	Sc	Scandium	61	Pm	Promethium	101	Md	Mendelevium
22	Ti	Titanium	62	Sm	Samarium	102	No	Nobelium
23	V	Vanadium	63	Eu	Europium	103	Lr	Lawrencium
24	Cr	Chromium	64	Gd	Gadolinium	104	Rf	Rutherfordium
25	Mn	Manganese	65	Tb	Terbium	105	Db	Dubnium
26	Fe	Iron (Ferrum)	66	Dy	Dysprosium	106	Sg	Seaborgium
27	Co	Cobalt	67	Ho	Holmium	107	Bh	Bohrium
28	Ni	Nickel	68	Er	Erbium	108	Hs	Hassium
29	Cu	Copper (Cuprum)	69	Tm	Thulium	109	Mt	Meitnerium
30	Zn	Zinc	70	Yb	Ytterbium	110	Ds	Darmstadtium
31	Ga	Gallium	71	Lu	Lutetium	111	Rg	Roentgenium
32	Ge	Germanium	72	Hf	Hafnium	112	Cn	Copernicium
33	As	Arsenic	73	Ta	Tantalum	113	Uut	Ununtrium
34	Se	Selenium	74	W	Tungsten (Wolfram)	114	Fl	Flerovium
35	Br	Bromine	75	Re	Rhenium	115	Uup	Ununpentium
36	Kr	Krypton	76	Os	Osmium	116	Lv	Livermorium
37	Rb	Rubidium	77	Ir	Iridium	117	Uus	Ununseptium
38	Sr	Strontium	78	Pt	Platinum	118	Uuo	Ununoctium
39	Y	Yttrium	79	Au	Gold (Aurum)			
40	Zr	Zirconium	80	Hg	Mercury (Hydrargyrum)			

The names in parentheses are the sources of the symbols, but are not used when referring to the elements. Iron, copper, silver, tin, gold, and lead (and sometimes antimony) anions are named by using the name in parentheses. For example, $[Fe(CN)_6]^{3-}$ is called hexacyanoferrate (III).

FIFTH EDITION

Inorganic Chemistry

Gary L. Miessler
St. Olaf College

Paul J. Fischer
Macalester College

Donald A. Tarr
St. Olaf College

PEARSON

Boston Columbus Indianapolis New York San Francisco Upper Saddle River
Amsterdam Cape Town Dubai London Madrid Milan Munich Paris Montréal Toronto
Delhi Mexico City São Paulo Sydney Hong Kong Seoul Singapore Taipei Tokyo

Editor in Chief: Adam Jaworski
Executive Editor: Jeanne Zalesky
Senior Marketing Manager: Jonathan Cottrell
Project Editor: Jessica Moro
Assistant Editor: Coleen Morrison
Editorial Assistant: Lisa Tarabokjia
Marketing Assistant: Nicola Houston
Associate Media Producer: Erin Fleming
Managing Editor, Chemistry and Geosciences: Gina M. Cheselka
Production Project Manager: Edward Thomas
Production Management/Composition: GEX Publishing Services
Illustrations: Imagineering, Inc.
Design Manager: Mark Ong
Interior and Cover Design: Gary Hespenheide
Photo Permissions Manager: Maya Melenchuk
Text Permissions Manager: Joseph Croscup
Text Permissions Research: Electronic Publishing Services, Inc.
Operations Specialist: Jeffrey Sargent
Cover Image Credit: Image of the d_{z^2} orbital of the iron atom within ferrocene, $Fe(C_5H_5)_2$. Courtesy of Gary Miessler.

Credits and acknowledgments borrowed from other sources and reproduced, with permission, in this textbook appear on the appropriate page within the text.

Crystal structures that appear in this text were generated from data obtained from The Cambridge Crystallographic Data Centre. Visualization of the structures was created using Mercury CSD 2.0 and Diamond.

The Cambridge Structural Database: a quarter of a million crystal structures and rising
F. H. Allen, *Acta Cryst.*, B**58**, 380–388, 2002. These data can be obtained free of charge from The Cambridge Crystallographic Data Centre via *www.ccdc.cam.ac.uk/data_request/cif*

Mercury CSD 2.0 - New Features for the Visualization and Investigation of Crystal Structures
C. F. Macrae, I. J. Bruno, J. A. Chisholm, P. R. Edgington, P. McCabe, E. Pidcock, L. Rodriguez-Monge, R. Taylor, J. van de Streek and P. A. Wood, *J. Appl. Cryst.*, 41, 466–470, 2008
[DOI: 10.1107/S0021889807067908] <*dx.doi.org/10.1107/S0021889807067908*>

Diamond - Crystal and Molecular Structure Visualization
Crystal Impact - Dr. H. Putz & Dr. K. Brandenburg GbR, Kreuzherrenstr. 102, 53227 Bonn, Germany
www.crystalimpact.com/diamond.

Library of Congress Cataloging-in-Publication Data
Miessler, Gary L.
Inorganic chemistry. — Fifth edition / Gary L. Miessler, St. Olaf College, Paul J. Fischer, Macalester College.
pages cm
Includes index.
ISBN-13: 978-0-321-81105-9 (student edition)
ISBN-10: 0-321-81105-4 (student edition)
1. Chemistry, Inorganic—Textbooks. I. Fischer, Paul J. II. Title.
QD151.3.M54 2014
546—dc23
2012037305

4 5 6 7 8 9 10—EBM—16 15 14

www.pearsonhighered.com

ISBN-10: 0-321-81105-4
ISBN-13: 978-0-321-81105-9

Brief Contents

Contents

Chapter 11

Coordination Chemistry III: Electronic Spectra 403

Chapter 12

Coordination Chemistry IV: Reactions and Mechanisms 437

Chapter 13

Organometallic Chemistry 475

Preface

The rapid development of inorganic chemistry makes ever more challenging the task of providing a textbook that is contemporary and meets the needs of those who use it. We appreciate the constructive suggestions provided by students, faculty, and reviewers, and have adopted much of this advice, keeping in mind the constraints imposed by space and the scope of the book. The main emphasis in preparing this edition has been to bring it up to date while providing clarity and a variety of helpful features.

New to the Fifth Edition:

- New and expanded discussions have been incorporated in many chapters to reflect topics of contemporary interest: for example, frustrated Lewis pairs (Chapter 6), IUPAC guidelines defining hydrogen bonds (Chapter 6), multiple bonding between Group 13 elements (Chapter 8), graphyne (Chapter 8), developments in noble gas chemistry (Chapter 8), metal–organic frameworks (Chapter 9), pincer ligands (Chapter 9), the magnetochemical series (Chapter 10), photosensitizers (Chapter 11), polyyne and polyene carbon "wires" (Chapter 13), percent buried volume of ligands (Chapter 14), and introductions to C—H bond activation, Pd-catalyzed cross-coupling, and sigma-bond metathesis (Chapter 14).
- To better represent the shapes of molecular orbitals, we are providing new images, generated by molecular modeling software, for most of the orbitals presented in Chapter 5.
- In a similar vein, to more accurately depict the shapes of many molecules, we have generated new images using CIF files from available crystal structure determinations. We hope that readers will find these images a significant improvement over the line drawings and ORTEP images that they replace.
- The discussion of electronegativity in connection with the VSEPR model in Chapter 3 has been expanded, and group electronegativity has been added.
- In response to users' requests, the projection operator approach has been added in the context of molecular orbitals of nonlinear molecules in Chapter 5. Chapter 8 includes more elaboration on Frost diagrams, and additional magnetic susceptibility content has been incorporated into Chapter 10.
- Chapter 6 has been reorganized to highlight contemporary aspects of acid–base chemistry and to include a broader range of measures of relative strengths of acids and bases.
- In Chapter 9 numerous new images have been added to provide more contemporary examples of the geometries of coordination complexes and coordination frameworks.
- The Covalent Bond Classification Method and MLX plots are now introduced in Chapter 13.
- Approximately 15% of end-of-chapter problems are new, with most based on the recent inorganic literature. To further encourage in-depth engagement with the literature, more problems involving extracting and interpreting information from the literature have been included. The total number of problems is more than 580.

- The values of physical constants inside the back cover have been revised to use the most recent values cited on the NIST Web site.
- This edition expands the use of color to better highlight the art and chemistry within the text and to improve readability of tables.

The need to add new material to keep up with the pace of developments in inorganic chemistry while maintaining a reasonable length is challenging, and difficult content decisions must be made. To permit space for increased narrative content while not significantly expanding the length of the book, Appendix B, containing tables of numerical data, has been placed online for free access.

We hope that the text will serve readers well. We will appreciate feedback and advice as we look ahead to edition 6.

SUPPLEMENTS

For the Instructor

ADVANCED CHEMISTRY WEBSITE The new **Advanced Chemistry Series** supports upper-level course work with cutting-edge content delivered by experienced authors and innovative multimedia. We realize chemistry can be a difficult area of study and we want to do all we can to encourage not just completion of course work, but also the building of the foundations of remarkable scholarly and professional success. Pearson Education is honored to be partnering with chemistry instructors and future STEM majors. To learn more about Pearson's **Advanced Chemistry Series**, explore other titles, or access materials to accompany this text and others in the series, please visit *www.pearsonhighered.com/advchemistry.*

For the Student

SOLUTIONS MANUAL (ISBN: 0321814134) by Gary L. Miessler, Paul J. Fischer, and Donald A. Tarr. This manual includes fully worked-out solutions to all end-of-chapter problems in the text.

Dedication and Acknowledgments

We wish to dedicate this textbook to our doctoral research advisors Louis H. Pignolet (Miessler) and John E. Ellis (Fischer) on the occasion of their seventieth birthdays. These chemists have inspired us throughout their careers by their exceptional creativity for chemical synthesis and dedication to the discipline of scholarship. We are grateful to have been trained by these stellar witnesses to the vocation of inorganic chemistry.

We thank Kaitlin Hellie for generating molecular orbital images (Chapter 5), Susan Green for simulating photoelectron spectra (Chapter 5), Zoey Rose Herm for generating images of metal–organic frameworks (Chapter 9), and Laura Avena for assistance with images generated from CIF files. We are also grateful to Sophia Hayes for useful advice on projection operators and Robert Rossi and Gerard Parkin for helpful discussions. We would also like to thank Andrew Mobley (Grinnell College), Dave Finster (Wittenberg University) and Adam Johnson (Harvey Mudd College) for their accuracy review of our text. We appreciate all that Jeanne Zalesky and Coleen Morrison, our editors at Pearson, and Jacki Russell at GEX Publishing Services have contributed.

Finally, we greatly value the helpful suggestions of the reviewers and other faculty listed below and of the many students at St. Olaf College and Macalester College who have pointed out needed improvements. While not all suggestions could be included because of constraints on the scope and length of the text, we are grateful for the many individuals who have offered constructive feedback. All of these ideas improve our teaching of inorganic chemistry and will be considered anew for the next edition.

Reviewers of the Fifth Edition of *Inorganic Chemistry*

Christopher Bradley
Texas Tech University

Stephen Contakes
Westmont College

Mariusz Kozik
Canisius College

Evonne Rezler
FL Atlantic University

Sheila Smith
University of Michigan-Dearborn

Matt Whited
Carleton College

Peter Zhao
East Tennessee State University

Reviewers of Previous Editions of *Inorganic Chemistry*

John Arnold
University of California–Berkeley

Ronald Bailey
Rensselaer Polytechnic University

Robert Balahura
University of Guelph

Craig Barnes
University of Tennessee–Knoxville

Daniel Bedgood
Arizona State University

Simon Bott
University of Houston

Joe Bruno
Wesleyan University

James J. Dechter
University of Central Oklahoma

Nancy Deluca
University of Massachusetts-Lowell

Charles Dismukes
Princeton University

Kate Doan
Kenyon College

Charles Drain
Hunter College

Jim Finholt
Carleton College

Derek P. Gates
University of British Columbia

Daniel Haworth
Marquette University

Stephanie K. Hurst
Northern Arizona University

Michael Johnson
University of Georgia

Jerome Kiester
University of Buffalo

Katrina Miranda
University of Arizona

Michael Moran
West Chester University

Wyatt Murphy
Seton Hall University

Mary-Ann Pearsall
Drew University

Laura Pence
University of Hartford

Greg Peters
University of Memphis

Cortland Pierpont
University of Colorado

Robert Pike
College of William and Mary

Jeffrey Rack
Ohio University

Gregory Robinson
University of Georgia

Lothar Stahl
University of North Dakota

Karen Stephens
Whitworth College

Robert Stockland
Bucknell University

Dennis Strommen
Idaho State University

Patrick Sullivan
Iowa State University

Duane Swank
Pacific Lutheran University

William Tolman
University of Minnesota

Robert Troy
Central Connecticut State University

Edward Vitz
Kutztown University

Richard Watt
University of New Mexico

Tim Zauche
University of Wisconsin–Platteville

Chris Ziegler
University of Akron

Gary L. Miessler
St. Olaf College
Northfield, Minnesota

Paul J. Fischer
Macalester College
St. Paul, Minnesota

Chapter 1

Introduction to Inorganic Chemistry

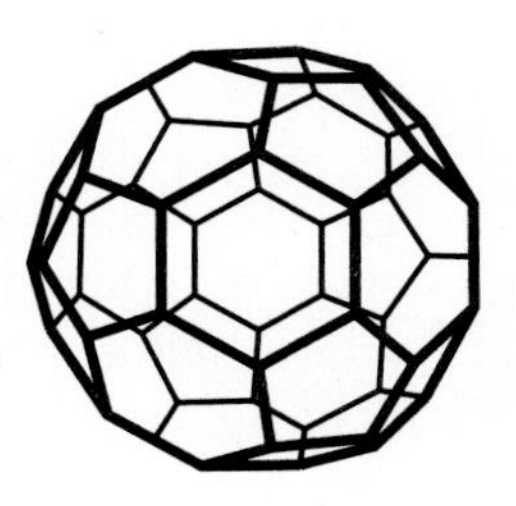

1.1 What Is Inorganic Chemistry?

If organic chemistry is defined as the chemistry of hydrocarbon compounds and their derivatives, inorganic chemistry can be described broadly as the chemistry of "everything else." This includes all the remaining elements in the periodic table, as well as carbon, which plays a major and growing role in inorganic chemistry. The large field of organometallic chemistry bridges both areas by considering compounds containing metal–carbon bonds; it also includes catalysis of many organic reactions. Bioinorganic chemistry bridges biochemistry and inorganic chemistry and has an important focus on medical applications. Environmental chemistry includes the study of both inorganic and organic compounds. In short, the inorganic realm is vast, providing essentially limitless areas for investigation and potential practical applications.

1.2 Contrasts with Organic Chemistry

Some comparisons between organic and inorganic compounds are in order. In both areas, single, double, and triple covalent bonds are found (**Figure 1.1**); for inorganic compounds, these include direct metal—metal bonds and metal—carbon bonds. Although the maximum number of bonds between two carbon atoms is three, there are many compounds that contain quadruple bonds between metal atoms. In addition to the sigma and pi bonds common in organic chemistry, quadruply bonded metal atoms contain a delta (δ) bond (**Figure 1.2**); a combination of one sigma bond, two pi bonds, and one delta bond makes up the quadruple bond. The delta bond is possible in these cases because the metal atoms have *d* orbitals to use in bonding, whereas carbon has only *s* and *p* orbitals energetically accessible for bonding.

Compounds with "fivefold" bonds between transition metals have been reported (**Figure 1.3**), accompanied by debate as to whether these bonds merit the designation "quintuple."

In organic compounds, hydrogen is nearly always bonded to a single carbon. In inorganic compounds, hydrogen is frequently encountered as a bridging atom between two or more other atoms. Bridging hydrogen atoms can also occur in metal cluster compounds, in which hydrogen atoms form bridges across edges or faces of polyhedra of metal atoms. Alkyl groups may also act as bridges in inorganic compounds, a function rarely encountered in organic chemistry except in reaction intermediates. Examples of terminal and bridging hydrogen atoms and alkyl groups in inorganic compounds are in **Figure 1.4**.

Some of the most striking differences between the chemistry of carbon and that of many other elements are in coordination number and geometry. Although carbon is usually limited to a maximum coordination number of four (a maximum of four atoms bonded

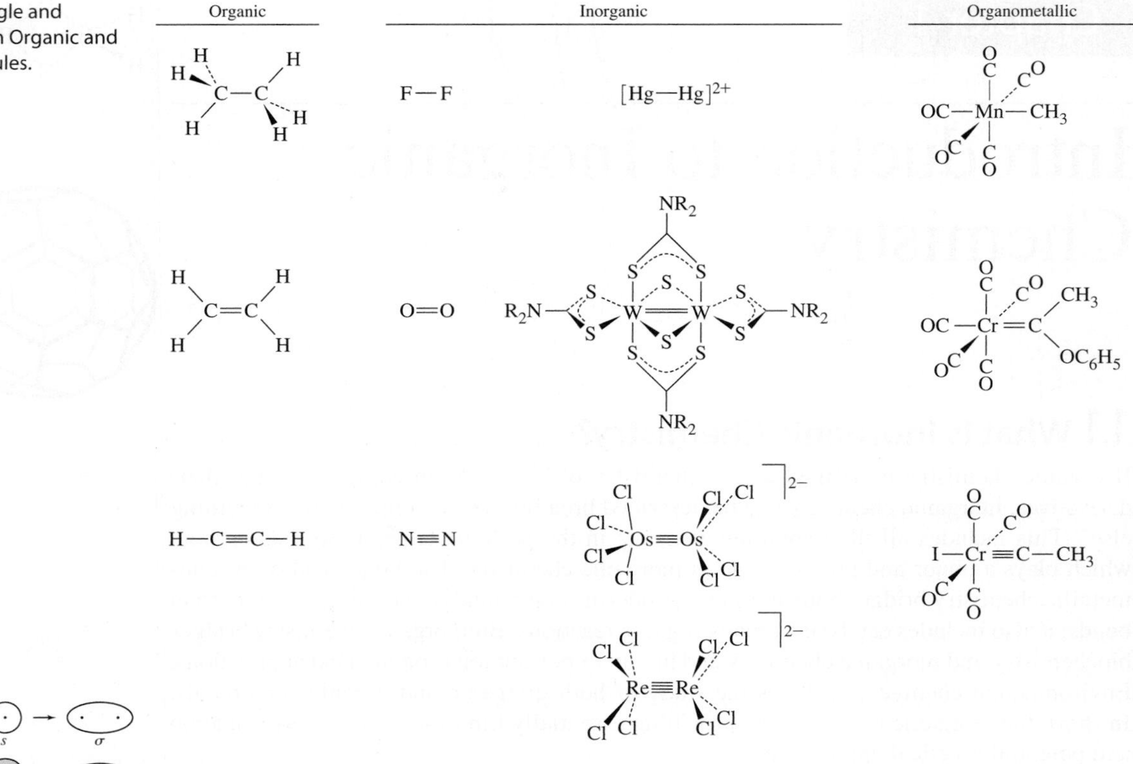

FIGURE 1.1 Single and Multiple Bonds in Organic and Inorganic Molecules.

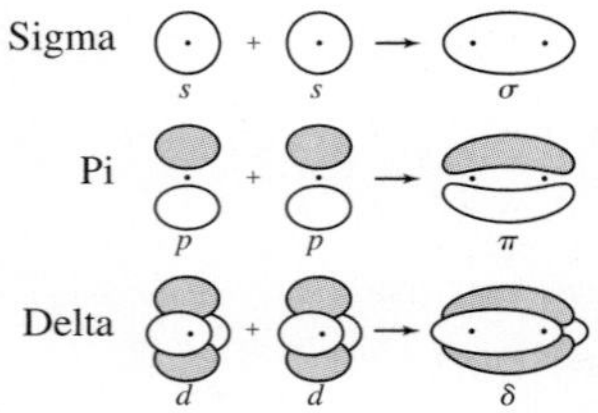

FIGURE 1.2 Examples of Bonding Interactions.

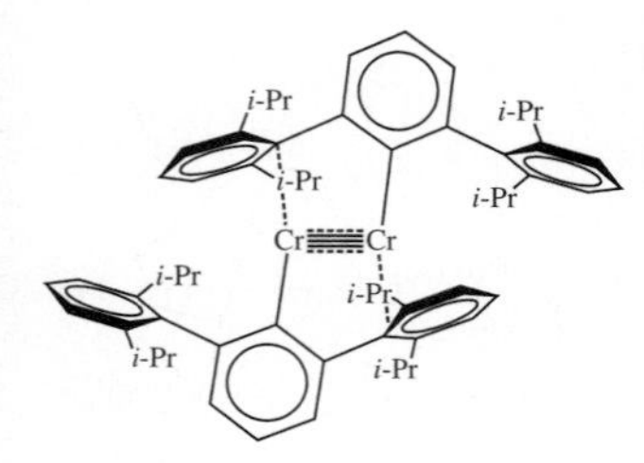

FIGURE 1.3 Example of Fivefold Bonding.

to carbon, as in CH_4), numerous inorganic compounds have central atoms with coordination numbers of five, six, seven, and higher; the most common coordination geometry for transition metals is an octahedral arrangement around a central atom, as shown for $[TiF_6]^{3-}$ (**Figure 1.5**). Furthermore, inorganic compounds present coordination geometries different from those found for carbon. For example, although 4-coordinate carbon is nearly always tetrahedral, both tetrahedral and square-planar shapes occur for 4-coordinate compounds of both metals and nonmetals. When metals are in the center, with anions or neutral molecules (*ligands*) bonded to them (frequently through N, O, or S), these are called *coordination complexes*; when carbon is the element directly bonded to metal atoms or ions, they are also classified as *organometallic* complexes.

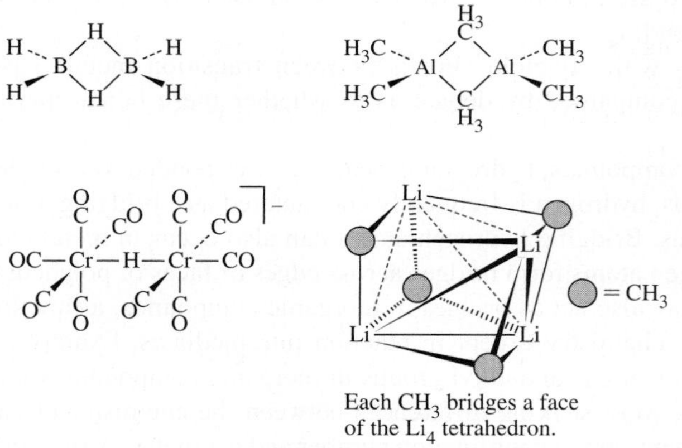

Each CH_3 bridges a face of the Li_4 tetrahedron.

FIGURE 1.4 Examples of Inorganic Compounds Containing Terminal and Bridging Hydrogens and Alkyl Groups.

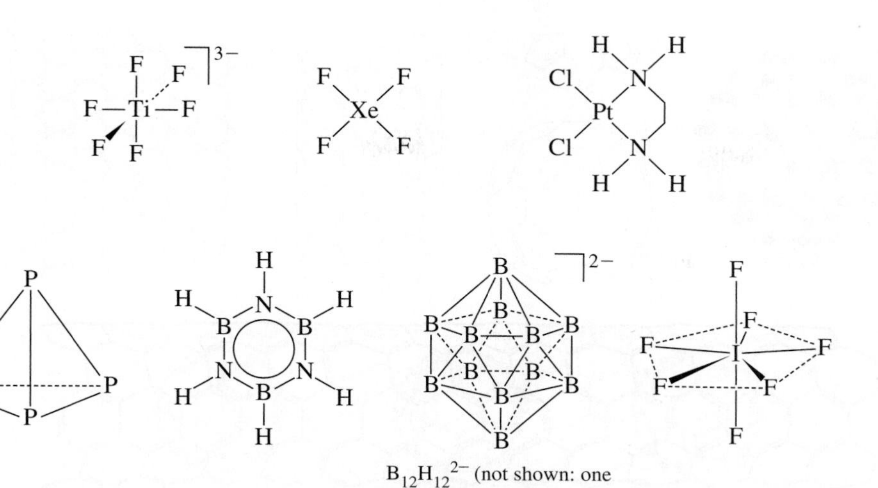

FIGURE 1.5 Examples of Geometries of Inorganic Compounds.

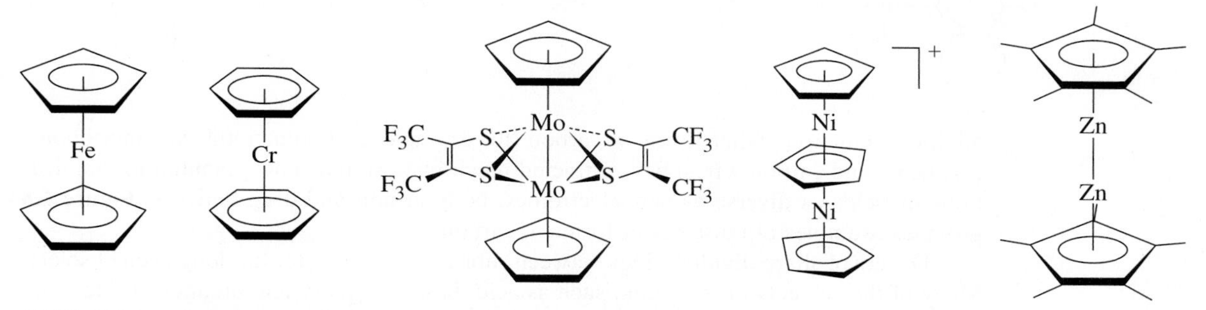

FIGURE 1.6 Inorganic Compounds Containing Pi-Bonded Aromatic Rings.

The tetrahedral geometry usually found in 4-coordinate compounds of carbon also occurs in a different form in some inorganic molecules. Methane contains four hydrogens in a regular tetrahedron around carbon. Elemental phosphorus is tetratomic (P_4) and tetrahedral, but with no central atom. Other elements can also form molecules in which outer atoms surround a central cavity; an example is boron, which forms numerous structures containing icosahedral B_{12} units. Examples of some of the geometries found for inorganic compounds are in Figure 1.5.

Aromatic rings are common in organic chemistry, and aryl groups can also form sigma bonds to metals. However, aromatic rings can also bond to metals in a dramatically different fashion using their pi orbitals, as shown in **Figure 1.6** and in this book's cover illustration. The result is a metal atom bonded above the center of the ring, almost as if suspended in space. In many cases, metal atoms are sandwiched between two aromatic rings. Multiple-decker sandwiches of metals and aromatic rings are also known.

Carbon plays an unusual role in a number of metal *cluster compounds* in which a carbon atom is at the center of a polyhedron of metal atoms. Examples of carbon-centered clusters with five, six, or more surrounding metals are known (**Figure 1.7**). The striking role that carbon plays in these clusters has provided a challenge to theoretical inorganic chemists.

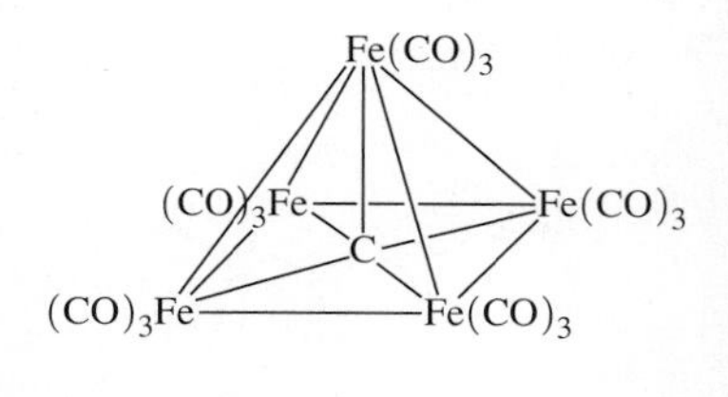

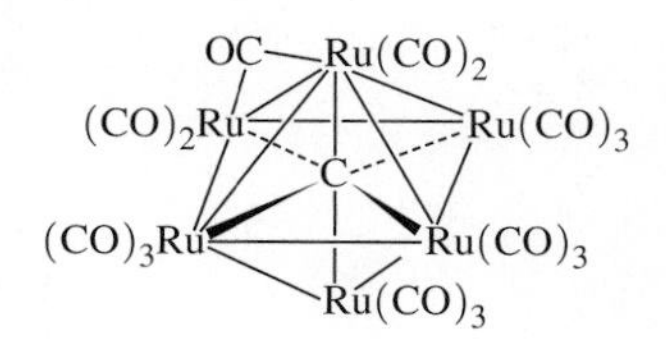

FIGURE 1.7 Carbon-Centered Metal Clusters.

In addition, since the mid-1980s the chemistry of elemental carbon has flourished. This phenomenon began with the discovery of fullerenes, most notably the cluster C_{60}, dubbed "buckminsterfullerene" after the developer of the geodesic dome. Many other fullerenes (buckyballs) are now known and serve as cores of a variety of derivatives. In

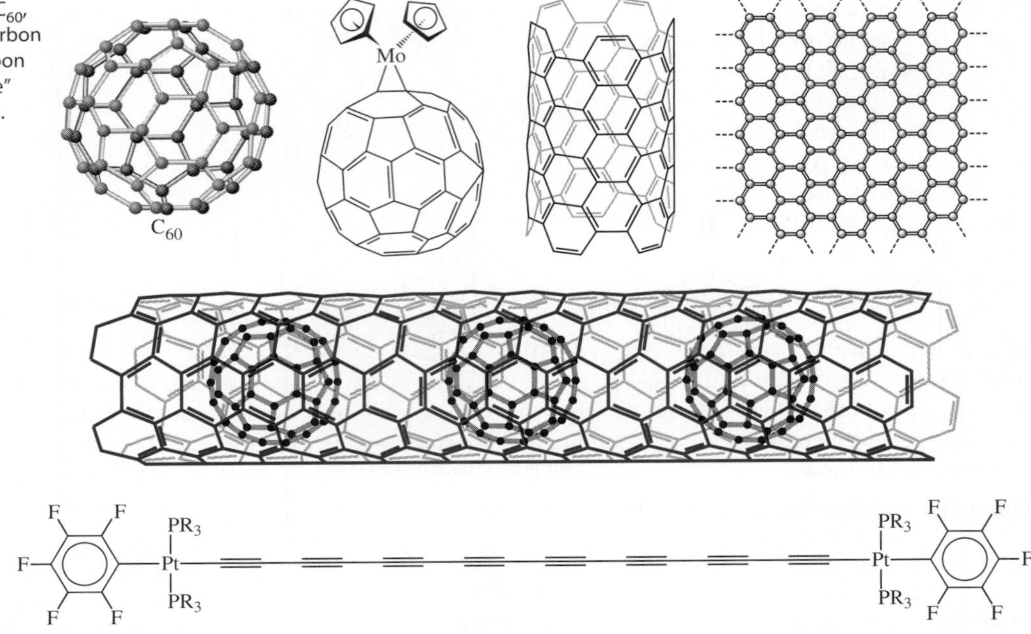

FIGURE 1.8 The Fullerene C_{60}, a Fullerene Compound, a Carbon Nanotube, Graphene, a Carbon Peapod, and a Polyyne "Wire" Connecting Platinum Atoms.

addition, numerous other forms of carbon (for example, carbon nanotubes, nanoribbons, graphene, and carbon wires) have attracted much interest and show potential for applications in fields as diverse as nanoelectronics, body armor, and drug delivery. **Figure 1.8** provides examples of these newer forms of carbon.

The era of sharp dividing lines between subfields in chemistry has long been obsolete. Many of the subjects in this book, such as acid–base chemistry and organometallic reactions, are of vital interest to organic chemists. Other topics such as oxidation–reduction reactions, spectra, and solubility relations interest analytical chemists. Subjects related to structure determination, spectra, conductivity, and theories of bonding appeal to physical chemists. Finally, the use of organometallic catalysts provides a connection to petroleum and polymer chemistry, and coordination compounds such as hemoglobin and metal-containing enzymes provide a similar tie to biochemistry. Many inorganic chemists work with professionals in other fields to apply chemical discoveries to addressing modern challenges in medicine, energy, the environment, materials science, and other fields. In brief, modern inorganic chemistry is not a fragmented field of study, but has numerous interconnections with other fields of science, medicine, technology, and other disciplines.

The remainder of this chapter is devoted to a short history of the origins of inorganic chemistry and perspective on more recent developments, intended to provide a sense of connection to the past and to place some aspects of inorganic chemistry within the context of larger historical events. In later chapters, brief historical context is provided with the same intention.

1.3 The History of Inorganic Chemistry

Even before alchemy became a subject of study, many chemical reactions were used and their products applied to daily life. The first metals used were probably gold and copper, which can be found in the metallic state in nature. Copper can also be readily formed by the reduction of malachite—basic copper carbonate, $Cu_2(CO_3)(OH)_2$—in charcoal fires. Silver, tin, antimony, and lead were also known as early as 3000 BCE. Iron appeared in classical Greece and in other areas around the Mediterranean Sea by 1500 BCE. At about the same time, colored glasses and ceramic glazes were introduced, largely composed of silicon dioxide (SiO_2, the major component of sand) and other metallic oxides, which had been melted and allowed to cool to amorphous solids.

Alchemists were active in China, Egypt, and other centers of civilization early in the first centuries CE. Although much effort went into attempts to "transmute" base metals into gold, alchemists also described many other chemical reactions and operations. Distillation, sublimation, crystallization, and other techniques were developed and used in their studies. Because of the political and social changes of the time, alchemy shifted into the Arab world and later—about 1000 to 1500 CE—reappeared in Europe. Gunpowder was used in Chinese fireworks as early as 1150, and alchemy was also widespread in China and India at that time. Alchemists appeared in art, literature, and science until at least 1600, by which time chemistry was beginning to take shape as a science. Roger Bacon (1214–1294), recognized as one of the first great experimental scientists, also wrote extensively about alchemy.

By the seventeenth century, the common strong acids—nitric, sulfuric, and hydrochloric—were known, and systematic descriptions of common salts and their reactions were being accumulated. As experimental techniques improved, the quantitative study of chemical reactions and the properties of gases became more common, atomic and molecular weights were determined more accurately, and the groundwork was laid for what later became the periodic table of the elements. By 1869, the concepts of atoms and molecules were well established, and it was possible for Mendeleev and Meyer to propose different forms of the periodic table. **Figure 1.9** illustrates Mendeleev's original periodic table.*

The chemical industry, which had been in existence since very early times in the form of factories for purifying salts and for smelting and refining metals, expanded as methods for preparing relatively pure materials became common. In 1896, Becquerel discovered radioactivity, and another area of study was opened. Studies of subatomic particles, spectra, and electricity led to the atomic theory of Bohr in 1913, which was soon modified by the quantum mechanics of Schrödinger and Heisenberg in 1926 and 1927.

Inorganic chemistry as a field of study was extremely important during the early years of the exploration and development of mineral resources. Qualitative analysis methods were

			Ti = 50	Zr = 90	? = 180
			V = 51	Nb = 94	Ta = 182
			Cr = 52	Mo = 96	W = 186
			Mn = 53	Rh = 104.4	Pt = 197.4
			Fe = 56	Ru = 104.2	Ir = 198
			Ni = Co = 59	Pd = 106.6	Os = 199
H = 1			Cu = 63.4	Ag = 108	Hg = 200
	Be = 9.4	Mg = 24	Zn = 65.2	Cd = 112	
	B = 11	Al = 27.4	? = 68	Ur = 116	Au = 197?
	C = 12	Si = 28	? = 70	Sn = 118	
	N = 14	P = 31	As = 75	Sb = 122	Bi = 210?
	O = 16	S = 32	Se = 79.4	Te = 128?	
	F = 19	Cl = 35.5	Br = 80	J = 127	
Li = 7	Na = 23	K = 39	Rb = 85.4	Cs = 133	Tl = 204
		Ca = 40	Sr = 87.6	Ba = 137	Pb = 207
		? = 45	Ce = 92		
		?Er = 56	La = 94		
		?Yt = 60	Di = 95		
		?In = 75.6	Th = 118 ?		

FIGURE 1.9 Mendeleev's 1869 Periodic Table. Two years later, Mendeleev revised his table into a form similar to a modern short-form periodic table, with eight groups across.

*The original table was published in *Zeitschrift für Chemie*, **1869**, *12*, 405. It can be found in English translation, together with a page from the German article, at web.lemoyne.edu/~giunta/mendeleev.html. See M. Laing, *J. Chem. Educ.*, **2008**, *85*, 63 for illustrations of Mendeleev's various versions of the periodic table, including his handwritten draft of the 1869 table.

developed to help identify minerals and, combined with quantitative methods, to assess their purity and value. As the Industrial Revolution progressed, so did the chemical industry. By the early twentieth century, plants for the high volume production of ammonia, nitric acid, sulfuric acid, sodium hydroxide, and many other inorganic chemicals were common.

Early in the twentieth century, Werner and Jørgensen made considerable progress on understanding the coordination chemistry of transition metals and also discovered a number of organometallic compounds. Nevertheless, the popularity of inorganic chemistry as a field of study gradually declined during most of the first half of the century. The need for inorganic chemists to work on military projects during World War II rejuvenated interest in the field. As work was done on many projects (not least of which was the Manhattan Project, in which scientists developed the fission bomb), new areas of research appeared, and new theories were proposed that prompted further experimental work. A great expansion of inorganic chemistry began in the 1940s, sparked by the enthusiasm and ideas generated during World War II.

In the 1950s, an earlier method used to describe the spectra of metal ions surrounded by negatively charged ions in crystals (**crystal field theory**)[1] was extended by the use of molecular orbital theory[2] to develop **ligand field theory** for use in coordination compounds, in which metal ions are surrounded by ions or molecules that donate electron pairs. This theory gave a more complete picture of the bonding in these compounds. The field developed rapidly as a result of this theoretical framework, availability of new instruments, and the generally reawakened interest in inorganic chemistry.

In 1955, Ziegler[3] and Natta[4] discovered organometallic compounds that could catalyze the polymerization of ethylene at lower temperatures and pressures than the common industrial method at that time. In addition, the polyethylene formed was more likely to be made up of linear, rather than branched, molecules and, as a consequence, was stronger and more durable. Other catalysts were soon developed, and their study contributed to the rapid expansion of organometallic chemistry, still a rapidly growing area.

The study of biological materials containing metal atoms has also progressed rapidly. The development of new experimental methods allowed more thorough study of these compounds, and the related theoretical work provided connections to other areas of study. Attempts to make *model* compounds that have chemical and biological activity similar to the natural compounds have also led to many new synthetic techniques. Two of the many biological molecules that contain metals are in **Figure 1.10**. Although these molecules have very different roles, they share similar ring systems.

One current area that bridges organometallic chemistry and bioinorganic chemistry is the conversion of nitrogen to ammonia:

$$N_2 + 3\,H_2 \longrightarrow 2\,NH_3$$

This reaction is one of the most important industrial processes, with over 100 million tons of ammonia produced annually worldwide, primarily for fertilizer. However, in spite of metal oxide catalysts introduced in the Haber–Bosch process in 1913, and improved since then, it is also a reaction that requires temperatures between 350 and 550 °C and from 150–350 atm pressure and that still results in a yield of only 15 percent ammonia. Bacteria, however, manage to fix nitrogen (convert it to ammonia and then to nitrite and nitrate) at 0.8 atm at room temperature in nodules on the roots of legumes. The nitrogenase enzyme that catalyzes this reaction is a complex iron–molybdenum–sulfur protein. The structure of its active sites has been determined by X-ray crystallography.[5] A vigorous area of modern inorganic research is to design reactions that could be carried out on an industrial scale that model the reaction of nitrogenase to generate ammonia under mild conditions. It is estimated that as much as 1 percent of the world's total energy consumption is currently used for the Haber–Bosch process.

Inorganic chemistry also has medical applications. Notable among these is the development of platinum-containing antitumor agents, the first of which was the *cis* isomer of $Pt(NH_3)_2Cl_2$,

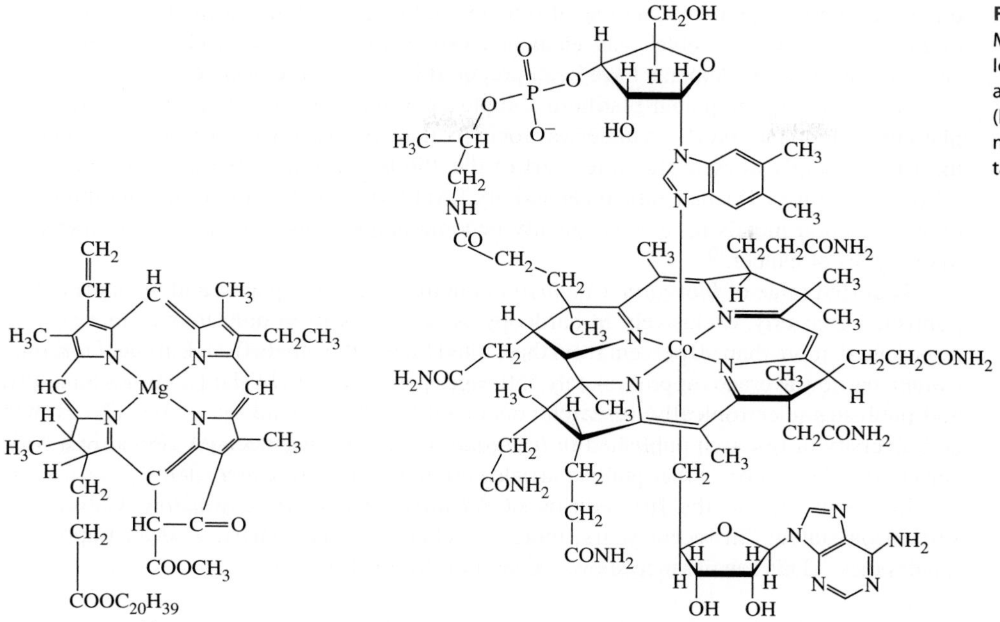

FIGURE 1.10 Biological Molecules Containing Metal Ions. (a) Chlorophyll a, the active agent in photosynthesis. (b) Vitamin B_{12} coenzyme, a naturally occurring organometallic compound.

cisplatin. First approved for clinical use approximately 30 years ago, cisplatin has served as the prototype for a variety of anticancer agents; for example, satraplatin, the first orally available platinum anticancer drug to reach clinical trials.* These two compounds are in **Figure 1.11**.

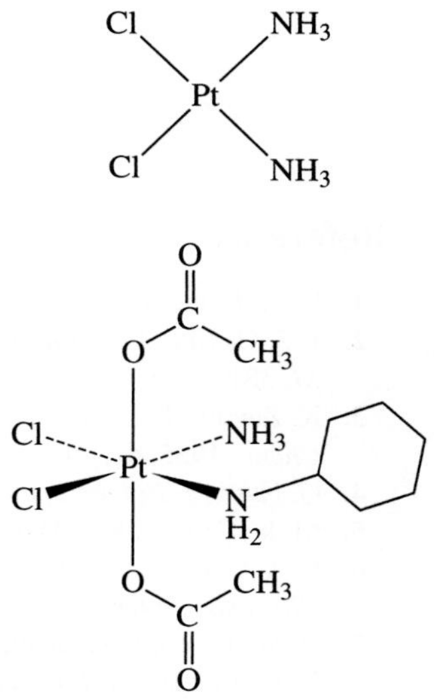

FIGURE 1.11 Cisplatin and Satraplatin.

1.4 Perspective

The premier issue of the journal *Inorganic Chemistry*** was published in February 1962. Much of the focus of that issue was on classic coordination chemistry, with more than half its research papers on synthesis of coordination complexes and their structures and properties. A few papers were on compounds of nonmetals and on organometallic chemistry, then a relatively new field; several were on thermodynamics or spectroscopy. All of these topics have developed considerably in the subsequent half-century, but much of the evolution of inorganic chemistry has been into realms unforeseen in 1962.

The 1962 publication of the first edition of F. A. Cotton and G. Wilkinson's landmark text *Advanced Inorganic Chemistry*[6] provides a convenient reference point for the status of inorganic chemistry at that time. For example, this text cited only the two long-known forms of carbon, diamond and graphite, although it did mention "amorphous forms" attributed to microcrystalline graphite. It would not be until more than two decades later that carbon chemistry would explode with the seminal discovery of C_{60} in 1985 by Kroto, Curl, Smalley, and colleagues,[7] followed by other fullerenes, nanotubes, graphene, and other forms of carbon (Figure 1.8) with the potential to have major impacts on electronics, materials science, medicine, and other realms of science and technology.

As another example, at the beginning of 1962 the elements helium through radon were commonly dubbed "inert" gases, believed to "form no chemically bound compounds" because of the stability of their electron configurations. Later that same year, Bartlett

*For reviews of modes of interaction of cisplatin and related drugs, see P. C. A. Bruijnincx, P. J. Sadler, *Curr. Opin. Chem. Bio.*, **2008**, *12*, 197 and F. Arnesano, G. Natile, *Coord. Chem. Rev.*, **2009**, *253*, 2070.

**The authors of this issue of *Inorganic Chemistry* were a distinguished group, including five recipients of the Priestley Medal, the highest honor conferred by the American Chemical Society, and 1983 Nobel Laureate Henry Taube.

reported the first chemical reactions of xenon with PtF_6, launching the synthetic chemistry of the now-renamed "noble" gas elements, especially xenon and krypton;[8] numerous compounds of these elements have been prepared in succeeding decades.

Numerous square planar platinum complexes were known by 1962; the chemistry of platinum compounds had been underway for more than a century. However, it was not known until Rosenberg's work in the latter part of the 1960s that one of these, *cis*-$Pt(NH_3)_2Cl_2$ (cisplatin, Figure 1.11), had anticancer activity.[9] Antitumor agents containing platinum and other transition metals have subsequently become major tools in treatment regimens for many types of cancer.[10]

That first issue of *Inorganic Chemistry* contained only 188 pages, and the journal was published quarterly, exclusively in hardcopy. Researchers from only four countries were represented, more than 90 percent from the United States, the others from Europe. *Inorganic Chemistry* now averages approximately 550 pages per issue, is published 24 times annually, and publishes (electronically) research conducted broadly around the globe. The growth and diversity of research published in *Inorganic Chemistry* has been paralleled in a wide variety of other journals that publish articles on inorganic and related fields.

In the preface to the first edition of *Advanced Inorganic Chemistry*, Cotton and Wilkinson stated, "in recent years, inorganic chemistry has experienced an impressive renaissance." This renaissance shows no sign of diminishing.

With this brief survey of the marvelously complex field of inorganic chemistry, we now turn to the details in the remainder of this book. The topics included provide a broad introduction to the field. However, even a cursory examination of a chemical library or one of the many inorganic journals shows some important aspects of inorganic chemistry that must be omitted in a textbook of moderate length. The references cited in this text suggest resources for further study, including historical sources, texts, and reference works that provide useful additional material.

References

1. H. A. Bethe, *Ann. Physik*, **1929**, *3*, 133.
2. J. S. Griffith, L. E. Orgel, *Q. Rev. Chem. Soc.*, **1957**, *XI*, 381.
3. K. Ziegler, E. Holzkamp, H. Breil, H. Martin, *Angew. Chem.*, **1955**, *67*, 541.
4. G. Natta, *J. Polym. Sci.*, **1955**, *16*, 143.
5. M. K. Chan, J. Kin, D. C. Rees, *Science*, **1993**, *260*, 792.
6. F. A. Cotton, G. Wilkinson, *Advanced Inorganic Chemistry*, Interscience, John Wiley & Sons, 1962.
7. H. W, Kroto, J. R. Heath, S. C. O'Brien, R. F. Curl, R. E. Smalley, *Nature (London)*, **1985**, *318*, 162.
8. N. Bartlett, D. H. Lohmann, *Proc. Chem. Soc.*, **1962**, 115; N. Bartlett, *Proc. Chem. Soc.*, **1962**, 218.
9. B. Rosenberg, L. VanCamp, J. E. Trosko, V. H. Mansour, *Nature*, **1969**, *222*, 385.
10. C. G. Hartinger, N. Metzler-Nolte, P. J. Dyson, *Organometallics*, **2012**, *31*, 5677 and P. C. A. Bruijnincx, P. J. Sadler, *Adv. Inorg. Chem.*, **2009**, *61*, 1; G. N. Kaluđerović, R. Paschke, *Curr. Med. Chem.*, **2011**, *18*, 4738.

General References

For those who are interested in the historical development of inorganic chemistry focused on metal coordination compounds during the period 1798–1935, copies of key research papers, including translations, are provided in the three-volume set *Classics in Coordination Chemistry*, G. B. Kauffman, ed., Dover Publications, N.Y. 1968, 1976, 1978. Among the many general reference works available, three of the most useful and complete are N. N. Greenwood and A. Earnshaw's *Chemistry of the Elements*, 2nd ed., Butterworth-Heinemann, Oxford, 1997; F. A. Cotton, G. Wilkinson, C. A. Murillo, and M. Bochman's *Advanced Inorganic Chemistry*, 6th ed., John Wiley & Sons, New York, 1999; and A. F. Wells's *Structural Inorganic Chemistry*, 5th ed., Oxford University Press, New York, 1984. An interesting study of inorganic reactions from a different perspective can be found in G. Wulfsberg's *Principles of Descriptive Inorganic Chemistry*, Brooks/Cole, Belmont, CA, 1987.

Chapter 2

Atomic Structure

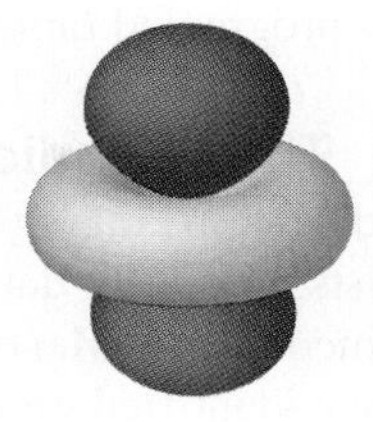

Understanding the structure of the atom has been a fundamental challenge for centuries. It is possible to gain a practical understanding of atomic and molecular structure using only a moderate amount of mathematics rather than the mathematical sophistication of quantum mechanics. This chapter introduces the fundamentals needed to explain atomic structure in qualitative and semiquantitative terms.

2.1 Historical Development of Atomic Theory

Although the Greek philosophers Democritus (460–370 BCE) and Epicurus (341–270 BCE) presented views of nature that included atoms, many centuries passed before experimental studies could establish the quantitative relationships needed for a coherent atomic theory. In 1808, John Dalton published *A New System of Chemical Philosophy*,[1] in which he proposed that

> ... the ultimate particles of all homogeneous bodies are perfectly alike in weight, figure, etc. In other words, every particle of water is like every other particle of water; every particle of hydrogen is like every other particle of hydrogen, etc.[2]

and that atoms combine in simple numerical ratios to form compounds. The terminology he used has since been modified, but he clearly presented the concepts of atoms and molecules, and made quantitative observations of the masses and volumes of substances as they combined to form new substances. For example, in describing the reaction between the gases hydrogen and oxygen to form water Dalton said that

> When two measures of hydrogen and one of oxygen gas are mixed, and fired by the electric spark, the whole is converted into steam, and if the pressure be great, this steam becomes water. It is most probable then that there is the same number of particles in two measures of hydrogen as in one of oxygen.[3]

Because Dalton was not aware of the diatomic nature of the molecules H_2 and O_2, which he assumed to be monatomic H and O, he did not find the correct formula of water, and therefore his surmise about the relative numbers of particles in "measures" of the gases is inconsistent with the modern concept of the mole and the chemical equation $2H_2 + O_2 \rightarrow 2H_2O$.

Only a few years later, Avogadro used data from Gay-Lussac to argue that equal volumes of gas at equal temperatures and pressures contain the same number of molecules, but uncertainties about the nature of sulfur, phosphorus, arsenic, and mercury vapors delayed acceptance of this idea. Widespread confusion about atomic weights and molecular formulas contributed to the delay; in 1861, Kekulé gave 19 different possible formulas for acetic acid![4] In the 1850s, Cannizzaro revived the argument of Avogadro and argued that everyone should use the same set of atomic weights rather than the many different sets then being used. At a meeting in Karlsruhe in 1860, Cannizzaro distributed a pamphlet describing his views.[5] His proposal was eventually accepted, and a consistent set of atomic weights and formulas evolved. In 1869, Mendeleev[6] and Meyer[7] independently proposed periodic tables nearly like those used today, and from that time the development of atomic theory progressed rapidly.

2.1.1 The Periodic Table

The idea of arranging the elements into a periodic table had been considered by many chemists, but either data to support the idea were insufficient or the classification schemes were incomplete. Mendeleev and Meyer organized the elements in order of atomic weight and then identified groups of elements with similar properties. By arranging these groups in rows and columns, and by considering similarities in chemical behavior as well as atomic weight, Mendeleev found vacancies in the table and was able to predict the properties of several elements—gallium, scandium, germanium, and polonium—that had not yet been discovered. When his predictions proved accurate, the concept of a periodic table was quickly accepted (see **Figure 1.11**). The discovery of additional elements not known in Mendeleev's time and the synthesis of heavy elements have led to the modern periodic table, shown inside the front cover of this text.

In the modern periodic table, a horizontal row of elements is called a **period** and a vertical column is a **group**. The traditional designations of groups in the United States differ from those used in Europe. The International Union of Pure and Applied Chemistry (IUPAC) has recommended that the groups be numbered 1 through 18. In this text, we will use primarily the IUPAC group numbers. Some sections of the periodic table have traditional names, as shown in **Figure 2.1**.

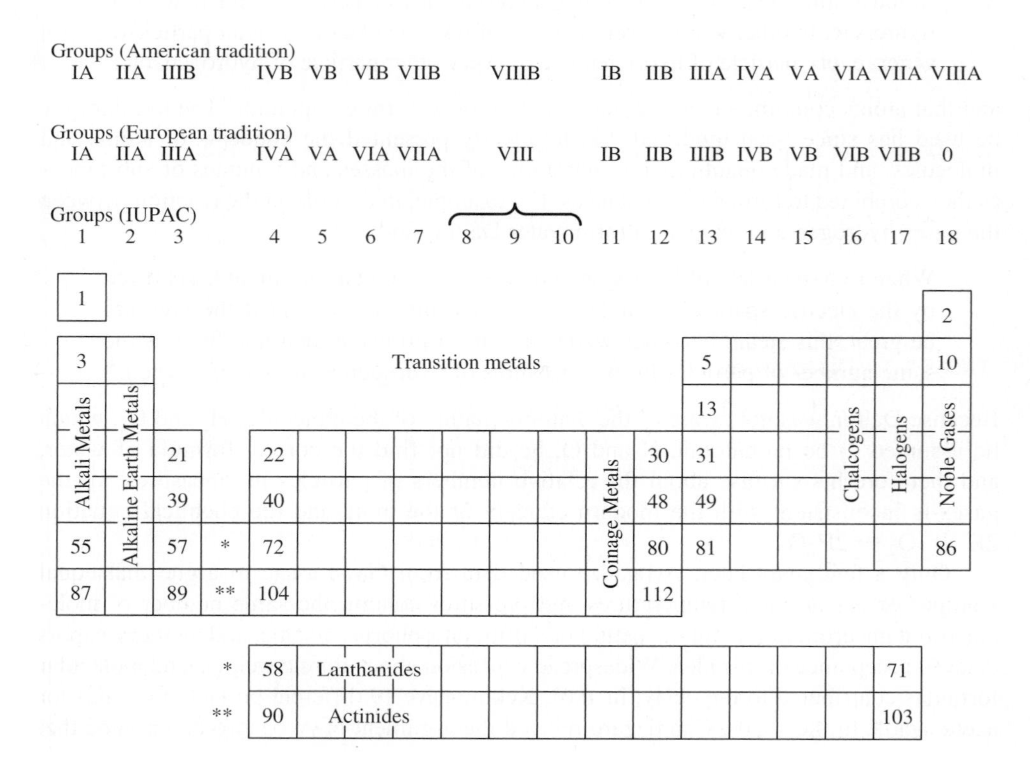

FIGURE 2.1 Numbering Schemes and Names for Parts of the Periodic Table.

2.1.2 Discovery of Subatomic Particles and the Bohr Atom

During the 50 years after the periodic tables of Mendeleev and Meyer were proposed, experimental advances came rapidly. Some of these discoveries are listed in **Table 2.1**.

Parallel discoveries in atomic spectra showed that each element emits light of specific energies when excited by an electric discharge or heat. In 1885, Balmer showed that the energies of visible light emitted by the hydrogen atom are given by the equation

$$E = R_H\left(\frac{1}{2^2} - \frac{1}{n_h^2}\right)$$

where n_h = integer, with $n_h > 2$

R_H = Rydberg constant for hydrogen

$= 1.097 \times 10^7\ \text{m}^{-1} = 2.179 \times 10^{-18}\ \text{J} = 13.61\ \text{eV}$

and the energy of the light emitted is related to the wavelength, frequency, and wavenumber of the light, as given by the equation

$$E = h\nu = \frac{hc}{\lambda} = hc\bar{\nu}$$

where* h = Planck constant $= 6.626 \times 10^{-34}$ J s

ν = frequency of the light, in s^{-1}

c = speed of light $= 2.998 \times 10^8\ \text{m s}^{-1}$

λ = wavelength of the light, frequently in nm

$\bar{\nu}$ = wavenumber of the light, usually in cm^{-1}

In addition to emission of visible light, as described by the Balmer equation, infrared and ultraviolet emissions were also discovered in the spectrum of the hydrogen atom. The energies of these emissions could be described by replacing 2^2 by integers n_l^2 in Balmer's original equation, with the condition that $n_l < n_h$ (l for lower level, h for higher level). These quantities, n, are called **quantum numbers**. (These are the **principal quantum numbers**; other quantum numbers are discussed in **Section 2.2.2**.) The origin of this energy was unknown until Niels Bohr's quantum theory of the atom,[8] first published in 1913 and refined over the following decade. This theory assumed that negatively charged electrons in atoms move in stable circular orbits around the positively charged nucleus with no absorption or emission of energy. However, electrons may absorb light of certain specific energies

TABLE 2.1 Discoveries in Atomic Structure

Year	Discoverer	Discovery
1896	A. H. Becquerel	Discovered radioactivity of uranium
1897	J. J. Thomson	Showed that electrons have a negative charge, with charge/mass $= 1.76 \times 10^{11}$ C/kg
1909	R. A. Millikan	Measured the electronic charge as 1.60×10^{-19} C; therefore, mass of electron $= 9.11 \times 10^{-31}$ kg
1911	E. Rutherford	Established the nuclear model of the atom: a very small, heavy nucleus surrounded by mostly empty space
1913	H. G. J. Moseley	Determined nuclear charges by X-ray emission, establishing atomic numbers as more fundamental than atomic masses

* More accurate values for the constants and energy conversion factors are given inside the back cover of this book.

and be excited to orbits of higher energy; they may also emit light of specific energies and fall to orbits of lower energy. The energy of the light emitted or absorbed can be found, according to the Bohr model of the hydrogen atom, from the equation

$$E = R\left(\frac{1}{n_l^2} - \frac{1}{n_h^2}\right)$$

where $$R = \frac{2\pi^2 \mu Z^2 e^4}{(4\pi\varepsilon_0)^2 h^2}$$

μ = reduced mass of the electron/nucleus combination:

$$\frac{1}{\mu} = \frac{1}{m_e} + \frac{1}{m_{nucleus}}$$

m_e = mass of the electron

$m_{nucleus}$ = mass of the nucleus

Z = charge of the nucleus

e = electronic charge

h = Planck constant

n_h = quantum number describing the higher energy state

n_l = quantum number describing the lower energy state

$4\pi\varepsilon_0$ = permittivity of a vacuum

This equation shows that the Rydberg constant depends on the mass of the nucleus and on various fundamental constants. If the atom is hydrogen, the subscript H is commonly appended to the Rydberg constant (R_H).

Examples of the transitions observed for the hydrogen atom and the energy levels responsible are shown in **Figure 2.2**. As the electrons drop from level n_h to n_l, energy is released in the form of electromagnetic radiation. Conversely, if radiation of the correct energy is absorbed by an atom, electrons are raised from level n_l to level n_h. The inverse-square dependence of energy on n results in energy levels that are far apart in energy at small n and become much closer in energy at larger n. In the upper limit, as n approaches infinity, the energy approaches a limit of zero. Individual electrons can have more energy, but above this point, they are no longer part of the atom; an infinite quantum number means that the nucleus and the electron are separate entities.

EXERCISE 2.1

Determine the energy of the transition from $n_h = 3$ to $n_l = 2$ for the hydrogen atom, in both joules and cm^{-1} (a common unit in spectroscopy, often used as an energy unit, since $\bar{\nu}$ is proportional to E). This transition results in a red line in the visible emission spectrum of hydrogen. (Solutions to the exercises are given in Appendix A.)

When applied to the hydrogen atom, Bohr's theory worked well; however, the theory failed when atoms with two or more electrons were considered. Modifications such as elliptical rather than circular orbits were unsuccessfully introduced in attempts to fit the data to Bohr's theory.[9] The developing experimental science of atomic spectroscopy provided extensive data for testing Bohr's theory and its modifications. In spite of the efforts to "fix" the Bohr theory, the theory ultimately proved unsatisfactory; the energy levels predicted by the Bohr equation above and shown in Figure 2.2 are valid only for the hydrogen atom and

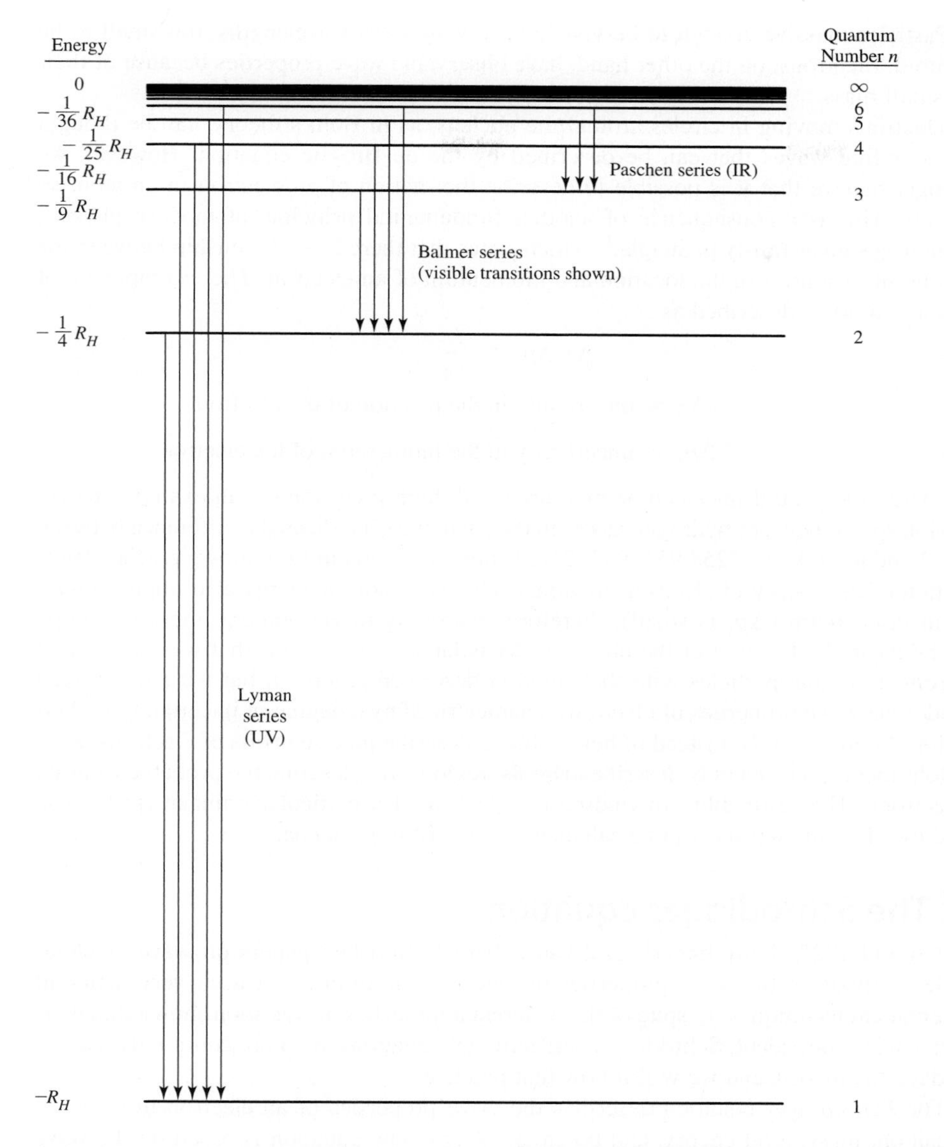

FIGURE 2.2 Hydrogen Atom Energy Levels.

other one-electron situations* such as He^+, Li^{2+}, and Be^{3+}. A fundamental characteristic of the electron—its wave nature—needed to be considered.

The de Broglie equation, proposed in the 1920s,[10] accounted for the electron's wave nature. According to de Broglie, all moving particles have wave properties described by the equation

$$\lambda = \frac{h}{mu}$$

λ = wavelength of the particle
h = Planck constant
m = mass of the particle
u = velocity of the particle

---

* Multiplying R_H by Z^2, the square of the nuclear charge, and adjusting the reduced mass accordingly provides an equation that describes these more exotic one-electron situations.

Particles massive enough to be visible have very short wavelengths, too small to be measured. Electrons, on the other hand, have observable wave properties because of their very small mass.

Electrons moving in circles around the nucleus, as in Bohr's theory, can be thought of as standing waves that can be described by the de Broglie equation. However, we no longer believe that it is possible to describe the motion of an electron in an atom so precisely. This is a consequence of another fundamental principle of modern physics, **Heisenberg's uncertainty principle**,[11] which states that there is a relationship between the inherent uncertainties in the location and momentum of an electron. The x component of this uncertainty is described as

$$\Delta x \, \Delta p_x \geq \frac{h}{4\pi}$$

$$\Delta x = \text{uncertainty in the position of the electron}$$

$$\Delta p_x = \text{uncertainty in the momentum of the electron}$$

The energy of spectral lines can be measured with high precision (as an example, recent emission spectral data of hydrogen atoms in the solar corona indicated a difference between $n_h = 2$ and $n_l = 1$ of 82258.9543992821(23) cm^{-1})![12] This in turn allows precise determination of the energy of electrons in atoms. This precision in energy also implies precision in momentum (Δp_x is small); therefore, according to Heisenberg, there is a large uncertainty in the location of the electron (Δx is large). This means that we cannot treat electrons as simple particles with their motion described precisely, but we must instead consider the wave properties of electrons, characterized by a degree of uncertainty in their location. In other words, instead of being able to describe precise **orbits** of electrons, as in the Bohr theory, we can only describe **orbitals**, regions that describe the probable location of electrons. The **probability** of finding the electron at a particular point in space, also called the **electron density**, can be calculated—at least in principle.

2.2 The Schrödinger Equation

In 1926 and 1927, Schrödinger[13] and Heisenberg[11] published papers on wave mechanics, descriptions of the wave properties of electrons in atoms, that used very different mathematical techniques. In spite of the different approaches, it was soon shown that their theories were equivalent. Schrödinger's differential equations are more commonly used to introduce the theory, and we will follow that practice.

The Schrödinger equation describes the wave properties of an electron in terms of its position, mass, total energy, and potential energy. The equation is based on the **wave function**, Ψ, which describes an electron wave in space; in other words, it describes an atomic orbital. In its simplest notation, the equation is

$$H\Psi = E\Psi$$

$$H = \text{Hamiltonian operator}$$

$$E = \text{energy of the electron}$$

$$\Psi = \text{wave function}$$

The **Hamiltonian operator**, frequently called simply the *Hamiltonian*, includes derivatives that **operate** on the wave function.* When the Hamiltonian is carried out, the result is a constant (the energy) times Ψ. The operation can be performed on any wave function

*An *operator* is an instruction or set of instructions that states what to do with the function that follows it. It may be a simple instruction such as "multiply the following function by 6," or it may be much more complicated than the Hamiltonian. The Hamiltonian operator is sometimes written $\hat{H}$ with the ˆ (hat) symbol designating an operator.

describing an atomic orbital. Different orbitals have different wave functions and different values of E. This is another way of describing quantization in that each orbital, characterized by its own function Ψ, has a characteristic energy.

In the form used for calculating energy levels, the Hamiltonian operator for one-electron systems is

$$H = \frac{-h^2}{8\pi^2 m}\left(\frac{\partial^2}{\partial x^2} + \frac{\partial^2}{\partial y^2} + \frac{\partial^2}{\partial z^2}\right) - \frac{Ze^2}{4\pi\varepsilon_0\sqrt{x^2 + y^2 + z^2}}$$

This part of the operator describes the *kinetic energy* of the electron, its energy of motion.

This part of the operator describes the *potential energy* of the electron, the result of electrostatic attraction between the electron and the nucleus. It is commonly designated as V.

where

h = Planck constant

m = mass of the electron

e = charge of the electron

$\sqrt{x^2 + y^2 + z^2} = r$ = distance from the nucleus

Z = charge of the nucleus

$4\pi\varepsilon_0$ = permittivity of a vacuum

This operator can be applied to a wave function Ψ,

$$\left[\frac{-h^2}{8\pi^2 m}\left(\frac{\partial^2}{\partial x^2} + \frac{\partial^2}{\partial y^2} + \frac{\partial^2}{\partial z^2}\right) + V(x, y, z)\right]\Psi(x, y, z) = E\Psi(x, y, z)$$

where

$$V = \frac{-Ze^2}{4\pi\varepsilon_0 r} = \frac{-Ze^2}{4\pi\varepsilon_0\sqrt{x^2 + y^2 + z^2}}$$

The potential energy V is a result of electrostatic attraction between the electron and the nucleus. Attractive forces, such as those between a positive nucleus and a negative electron, are defined by convention to have a negative potential energy. An electron near the nucleus (small r) is strongly attracted to the nucleus and has a large negative potential energy. Electrons farther from the nucleus have potential energies that are small and negative. For an electron at infinite distance from the nucleus ($r = \infty$), the attraction between the nucleus and the electron is zero, and the potential energy is zero. The hydrogen atom energy level diagram in Figure 2.2 illustrates these concepts.

Because n varies from 1 to ∞, and every atomic orbital is described by a unique Ψ, there is no limit to the number of solutions of the Schrödinger equation for an atom. *Each Ψ describes the wave properties of a given electron in a particular orbital.* The probability of finding an electron at a given point in space is proportional to Ψ^2. A number of conditions are required for a physically realistic solution for Ψ:

1. **1.** The wave function Ψ must be single-valued. — There cannot be two probabilities for an electron at any position in space.
2. **2.** The wave function Ψ and its first derivatives must be continuous. — The probability must be defined at all positions in space and cannot change abruptly from one point to the next.
3. **3.** The wave function Ψ must approach zero as r approaches infinity. — For large distances from the nucleus, the probability must grow smaller and smaller (the atom must be finite).

4. **The integral** $\displaystyle\int_{all\ space} \Psi_A \Psi_A^{\,*}\, d\tau = 1$ — The total probability of an electron being *somewhere* in space = 1. This is called **normalizing** the wave function.*

5. **The integral** $\displaystyle\int_{all\ space} \Psi_A \Psi_B^{\,*}\, d\tau = 0$ — Ψ_A and Ψ_B are wave functions for electrons in different orbitals within the same atom. All orbitals in an atom must be orthogonal to each other. In some cases, this means that the axes of orbitals must be perpendicular, as with the p_x, p_y, and p_z orbitals.

2.2.1 The Particle in a Box

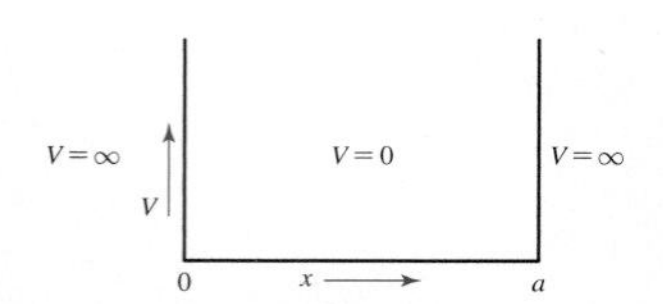

FIGURE 2.3 Potential Energy Well for the Particle in a Box.

A simple example of the wave equation, the particle in a one-dimensional box, shows how these conditions are used. We will give an outline of the method; details are available elsewhere.** The "box" is shown in **Figure 2.3**. The potential energy $V(x)$ inside the box, between $x = 0$ and $x = a$, is defined to be zero. Outside the box, the potential energy is infinite. This means that the particle is completely trapped in the box and would require an infinite amount of energy to leave the box. However, there are no forces acting on it within the box.

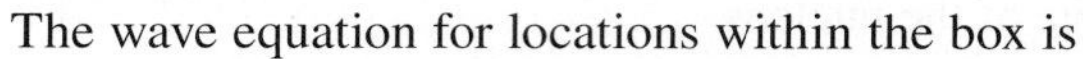

The wave equation for locations within the box is

$$\frac{-h^2}{8\pi^2 m}\left(\frac{\partial^2 \Psi(x)}{\partial x^2}\right) = E\,\Psi(x), \quad \text{because } V(x) = 0$$

Sine and cosine functions have the properties we associate with waves—a well-defined wavelength and amplitude—and we may therefore propose that the wave characteristics of our particle may be described by a combination of sine and cosine functions. A general solution to describe the possible waves in the box would then be

$$\Psi = A \sin rx + B \cos sx$$

where A, B, r, and s are constants. Substitution into the wave equation allows solution for r and s (see Problem 8a at the end of the chapter):

$$r = s = \sqrt{2mE}\,\frac{2\pi}{h}$$

Because Ψ must be continuous and must equal zero at $x < 0$ and $x > a$ (because the particle is confined to the box), Ψ must go to zero at $x = 0$ and $x = a$. Because $\cos sx = 1$ for $x = 0$, Ψ can equal zero in the general solution above only if $B = 0$. This reduces the expression for Ψ to

$$\Psi = A \sin rx$$

At $x = a$, Ψ must also equal zero; therefore, $\sin ra = 0$, which is possible only if ra is an integral multiple of π:

$$ra = \pm n\pi \quad \text{or} \quad r = \frac{\pm n\pi}{a}$$

---

* Because the wave functions may have imaginary values (containing $\sqrt{-1}$), Ψ^* (where Ψ^* designates the complex conjugate of Ψ) is used to make the integral real. In many cases, the wave functions themselves are real, and this integral becomes $\displaystyle\int_{all\ space} \Psi_A^2\, d\tau$.

** G. M. Barrow, *Physical Chemistry*, 6th ed., McGraw-Hill, New York, 1996, pp. 65, 430, calls this the "particle on a line" problem. Other physical chemistry texts also include solutions to this problem.

where n = any integer $\neq 0$.* Because both positive and negative values yield the same results, substituting the positive value for r into the solution for r gives

$$r = \frac{n\pi}{a} = \sqrt{2mE}\,\frac{2\pi}{h}$$

This expression may be solved for E:

$$E = \frac{n^2h^2}{8ma^2}$$

These are the energy levels predicted by the particle-in-a-box model for any particle in a one-dimensional box of length a. The energy levels are quantized according to **quantum numbers** $n = 1, 2, 3, \ldots$

Substituting $r = n\pi/a$ into the wave function gives

$$\Psi = A \sin \frac{n\pi x}{a}$$

And applying the normalizing requirement $\int \Psi\Psi^* \, d\tau = 1$ gives

$$A = \sqrt{\frac{2}{a}}$$

The total solution is then

$$\Psi = \sqrt{\frac{2}{a}} \sin \frac{n\pi x}{a}$$

The resulting wave functions and their squares for the first three states—the ground state ($n = 1$) and first two excited states ($n = 2$ and $n = 3$)—are plotted in **Figure 2.4**.

The squared wave functions are the probability densities; they show one difference between classical and quantum mechanical behavior of an electron in such a box. Classical mechanics predicts that the electron has equal probability of being at any point in the box. The wave nature of the electron gives it varied probabilities at different locations in the box. The greater the square of the electron wave amplitude, the greater the probability of the electron being located at the specified coordinate when at the quantized energy defined by the Ψ.

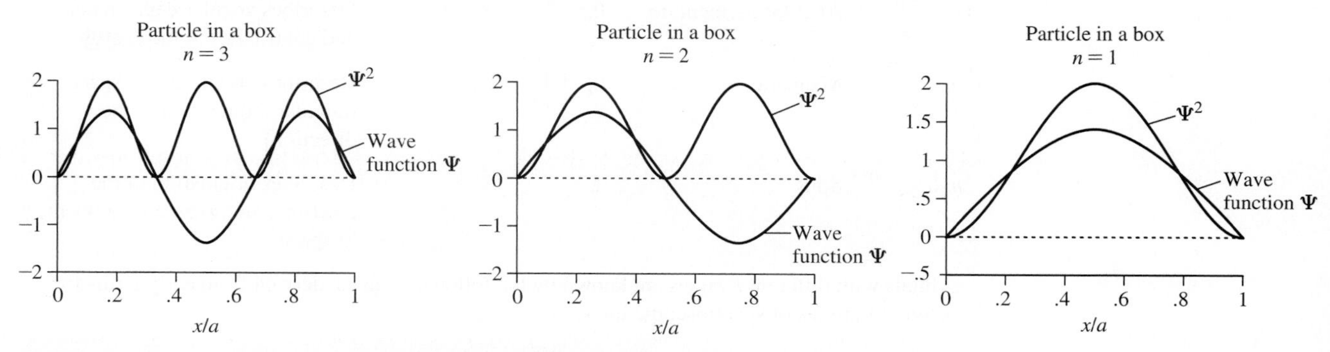

FIGURE 2.4 Wave Functions and Their Squares for the Particle in a Box with $n = 1, 2$, and 3.

* If $n = 0$, then $r = 0$ and $\Psi = 0$ at all points. The probability of finding the particle is $\int \Psi\Psi^* \, dx = 0$; if the particle is an electron, there is then no electron at all.

2.2.2 Quantum Numbers and Atomic Wave Functions

The particle-in-a-box example shows how a wave function operates in one dimension. Mathematically, atomic orbitals are discrete solutions of the three-dimensional Schrödinger equations. The same methods used for the one-dimensional box can be expanded to three dimensions for atoms. These orbital equations include three quantum numbers, n, l, and m_l. A fourth quantum number, m_s, a result of relativistic corrections to the Schrödinger equation, completes the description by accounting for the magnetic moment of the electron. The quantum numbers are summarized in **Table 2.2**. **Tables 2.3** and **2.4** describe wave functions.

The quantum number n is primarily responsible for determining the overall energy of an atomic orbital; the other quantum numbers have smaller effects on the energy. The quantum number l determines the angular momentum and shape of an orbital. The quantum number m_l determines the orientation of the angular momentum vector in a magnetic field, or the position of the orbital in space, as shown in Table 2.3. The quantum number m_s determines the orientation of the electron's magnetic moment in a magnetic field, either in the direction of the field $(+\frac{1}{2})$ or opposed to it $(-\frac{1}{2})$. When no field is present, all m_l values associated with a given n—all three p orbitals or all five d orbitals—have the same energy, and both m_s values have the same energy. Together, the quantum numbers n, l, and m_l define an atomic orbital.

The quantum number m_s describes the electron spin within the orbital. This fourth quantum number is consistent with a famous experimental observation. When a beam of alkali metal atoms (each with a single valence electron) is passed through a magnetic field, the beam splits into two parts; half the atoms are attracted by one magnet pole, and half are attracted by the opposite pole. Because in classical physics spinning charged particles generate magnetic moments, it is common to attribute an electron's magnetic moment to its spin—as if an electron were a tiny bar magnet—with the orientation of the magnetic field vector a function of the spin direction (counterclockwise vs. clockwise). However, the spin of an electron is a purely quantum mechanical property; application of classical mechanics to an electron is inaccurate.

One feature that should be mentioned is the appearance of $i(=\sqrt{-1})$ in the p and d orbital wave equations in Table 2.3. Because it is much more convenient to work with

TABLE 2.2 Quantum Numbers and Their Properties

Symbol	Name	Values	Role
n	Principal	$1, 2, 3, \ldots$	Determines the major part of the energy
l	Angular momentum*	$0, 1, 2, \ldots, n-1$	Describes angular dependence and contributes to the energy
m_l	Magnetic	$0, \pm 1, \pm 2, \ldots, \pm l$	Describes orientation in space (angular momentum in the z direction)
m_s	Spin	$\pm\frac{1}{2}$	Describes orientation of the electron spin (magnetic moment) in space

Orbitals with different l values are known by the following labels, derived from early terms for different families of spectroscopic lines:

l	0	1	2	3	4	5, ...
Label	s	p	d	f	g	continuing alphabetically

* Also called the azimuthal quantum number.

TABLE 2.3 Hydrogen Atom Wave Functions: Angular Functions

Angular Factors					Real Wave Functions			
Related to Angular Momentum				Functions of θ	In Polar Coordinates	In Cartesian Coordinates	Shapes	Label
l	m_l	Φ	Θ		$\Theta\Phi(\theta, \phi)$	$\Theta\Phi(x, y, z)$		
$0(s)$	0	$\frac{1}{\sqrt{2\pi}}$	$\frac{1}{\sqrt{2}}$		$\frac{1}{2\sqrt{\pi}}$	$\frac{1}{2\sqrt{\pi}}$		s
$1(p)$	0	$\frac{1}{\sqrt{2\pi}}$	$\frac{\sqrt{6}}{2}\cos\theta$		$\frac{1}{2}\sqrt{\frac{3}{\pi}}\cos\theta$	$\frac{1}{2}\sqrt{\frac{3}{\pi}}\frac{z}{r}$		p_z
	+1	$\frac{1}{\sqrt{2\pi}}e^{i\phi}$	$\frac{\sqrt{3}}{2}\sin\theta$		$\frac{1}{2}\sqrt{\frac{3}{\pi}}\sin\theta\cos\phi$	$\frac{1}{2}\sqrt{\frac{3}{\pi}}\frac{x}{r}$		p_x
	−1	$\frac{1}{\sqrt{2\pi}}e^{-i\phi}$	$\frac{\sqrt{3}}{2}\sin\theta$		$\frac{1}{2}\sqrt{\frac{3}{\pi}}\sin\theta\sin\phi$	$\frac{1}{2}\sqrt{\frac{3}{\pi}}\frac{y}{r}$		p_y
$2(d)$	0	$\frac{1}{\sqrt{2\pi}}$	$\frac{1}{2}\sqrt{\frac{5}{2}}(3\cos^2\theta - 1)$		$\frac{1}{4}\sqrt{\frac{5}{\pi}}(3\cos^2\theta - 1)$	$\frac{1}{4}\sqrt{\frac{5}{\pi}}\frac{(2z^2 - x^2 - y^2)}{r^2}$		d_{z^2}
	+1	$\frac{1}{\sqrt{2\pi}}e^{i\phi}$	$\frac{\sqrt{15}}{2}\cos\theta\sin\theta$		$\frac{1}{2}\sqrt{\frac{15}{\pi}}\cos\theta\sin\theta\cos\phi$	$\frac{1}{2}\sqrt{\frac{15}{\pi}}\frac{xz}{r^2}$		d_{xz}
	−1	$\frac{1}{\sqrt{2\pi}}e^{-i\phi}$	$\frac{\sqrt{15}}{2}\cos\theta\sin\theta$		$\frac{1}{2}\sqrt{\frac{15}{\pi}}\cos\theta\sin\theta\sin\phi$	$\frac{1}{2}\sqrt{\frac{15}{\pi}}\frac{yz}{r^2}$		d_{yz}
	+2	$\frac{1}{\sqrt{2\pi}}e^{2i\phi}$	$\frac{\sqrt{15}}{4}\sin^2\theta$		$\frac{1}{4}\sqrt{\frac{15}{\pi}}\sin^2\theta\cos 2\phi$	$\frac{1}{4}\sqrt{\frac{15}{\pi}}\frac{(x^2 - y^2)}{r^2}$		$d_{x^2-y^2}$
	−2	$\frac{1}{\sqrt{2\pi}}e^{-2i\phi}$	$\frac{\sqrt{15}}{4}\sin^2\theta$		$\frac{1}{4}\sqrt{\frac{15}{\pi}}\sin^2\theta\sin 2\phi$	$\frac{1}{4}\sqrt{\frac{15}{\pi}}\frac{xy}{r^2}$		d_{xy}

Source: Hydrogen Atom Wave Functions: Angular Functions, *Physical Chemistry*, 5th ed.,Gordon Barrow (c) 1988. McGraw-Hill Companies, Inc.

NOTE: The relations $(e^{i\phi} - e^{-i\phi})/(2i) = \sin\phi$ and $(e^{i\phi} + e^{-i\phi})/2 = \cos\phi$ can be used to convert the exponential imaginary functions to real trigonometric functions, combining the two orbitals with $m_l = \pm 1$ to give two orbitals with $\sin\phi$ and $\cos\phi$. In a similar fashion, the orbitals with $m_l = \pm 2$ result in real functions with $\cos^2\phi$ and $\sin^2\phi$. These functions have then been converted to Cartesian form by using the functions $x = r\sin\theta\cos\phi$, $y = r\sin\theta\sin\phi$, and $z = r\cos\theta$.

real functions than complex functions, we usually take advantage of another property of the wave equation. For differential equations of this type, any linear combination of solutions to the equation—sums or differences of the functions, with each multiplied by any coefficient—is also a solution to the equation. The combinations usually chosen for the p orbitals are the sum and difference of the p orbitals having $m_l = +1$ and –1, normalized by multiplying by the constants $\frac{1}{\sqrt{2}}$ and $\frac{i}{\sqrt{2}}$, respectively:

$$\Psi_{2p_x} = \frac{1}{\sqrt{2}}(\Psi_{+1} + \Psi_{-1}) = \frac{1}{2}\sqrt{\frac{3}{\pi}}\,[R(r)]\,\sin\theta\cos\phi$$

$$\Psi_{2p_y} = \frac{i}{\sqrt{2}}(\Psi_{+1} - \Psi_{-1}) = \frac{1}{2}\sqrt{\frac{3}{\pi}}\,[R(r)]\,\sin\theta\sin\phi$$

TABLE 2.4 Hydrogen Atom Wave Functions: Radial Functions

Radial Functions $R(r)$, with $\sigma = Zr/a_0$			
Orbital	n	l	$R(r)$
$1s$	1	0	$R_{1s} = 2\left[\frac{Z}{a_0}\right]^{3/2} e^{-\sigma}$
$2s$	2	0	$R_{2s} = 2\left[\frac{Z}{2a_0}\right]^{3/2} (2 - \sigma)e^{-\sigma/2}$
$2p$		1	$R_{2p} = \frac{1}{\sqrt{3}}\left[\frac{Z}{2a_0}\right]^{3/2} \sigma e^{-\sigma/2}$
$3s$	3	0	$R_{3s} = \frac{2}{27}\left[\frac{Z}{3a_0}\right]^{3/2} (27 - 18\sigma + 2\sigma^2)e^{-\sigma/3}$
$3p$		1	$R_{3p} = \frac{1}{81\sqrt{3}}\left[\frac{2Z}{a_0}\right]^{3/2} (6 - \sigma)\sigma e^{-\sigma/3}$
$3d$		2	$R_{3d} = \frac{1}{81\sqrt{15}}\left[\frac{2Z}{a_0}\right]^{3/2} \sigma^2 e^{-\sigma/3}$

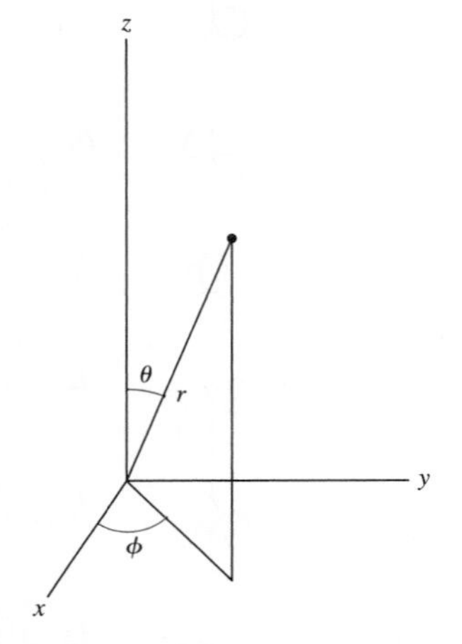

Spherical coordinates

The same procedure used on the d orbital functions for $m_l = \pm 1$ and ± 2 gives the functions in the column headed $\Theta\Phi(\theta, \phi)$ in Table 2.3, which are the familiar d orbitals. The d_{z^2} orbital ($m_l = 0$) actually uses the function $2z^2 - x^2 - y^2$, which we shorten to z^2 for convenience.* These functions are now real functions, so $\Psi = \Psi^*$ and $\Psi\Psi^* = \Psi^2$.

A more detailed look at the Schrödinger equation shows the mathematical origin of atomic orbitals. In three dimensions, Ψ may be expressed in terms of Cartesian coordinates (x, y, z) or in terms of spherical coordinates (r, θ, ϕ). Spherical coordinates, as shown in **Figure 2.5**, are especially useful in that r represents the distance from the nucleus. The spherical coordinate θ is the angle from the z axis, varying from 0 to π, and ϕ is the angle from the x axis, varying from 0 to 2π. Conversion between Cartesian and spherical coordinates is carried out with the following expressions:

$$x = r \sin \theta \cos \phi$$
$$y = r \sin \theta \sin \phi$$
$$z = r \cos \theta$$

In spherical coordinates, the three sides of the volume element are $r\,d\theta$, $r \sin \theta\, d\phi$, and dr. The product of the three sides is $r^2 \sin \theta\, d\theta\, d\phi\, dr$, equivalent to $dx\, dy\, dz$. The volume of the thin shell between r and $r + dr$ is $4\pi r^2\, dr$, which is the integral over ϕ from 0 to π and over θ from 0 to 2π. This integral is useful in describing the electron density as a function of distance from the nucleus.

Ψ can be factored into a radial component and two angular components. The **radial function R** describes electron density at different distances from the nucleus; the **angular functions Θ** and **Φ** describe the shape of the orbital and its orientation in space. The two angular factors are sometimes combined into one factor, called Y:

$$\Psi(r, \theta, \phi) = R(r)\Theta(\theta)\Phi(\phi) = R(r)Y(\theta, \phi)$$

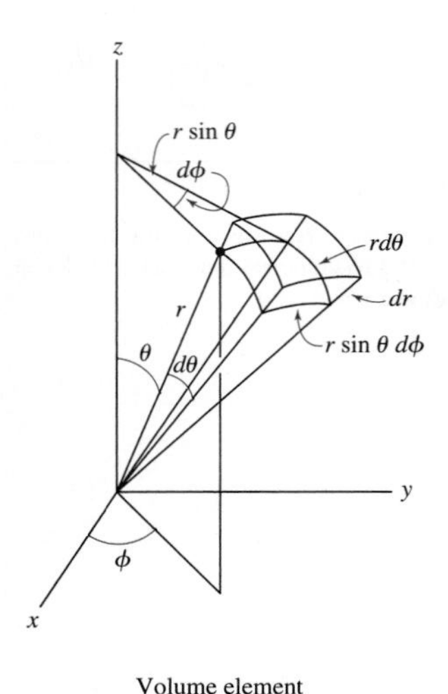

Volume element

FIGURE 2.5 Spherical Coordinates and Volume Element for a Spherical Shell in Spherical Coordinates.

*We should really call this the $d_{2z^2-x^2-y^2}$ orbital!

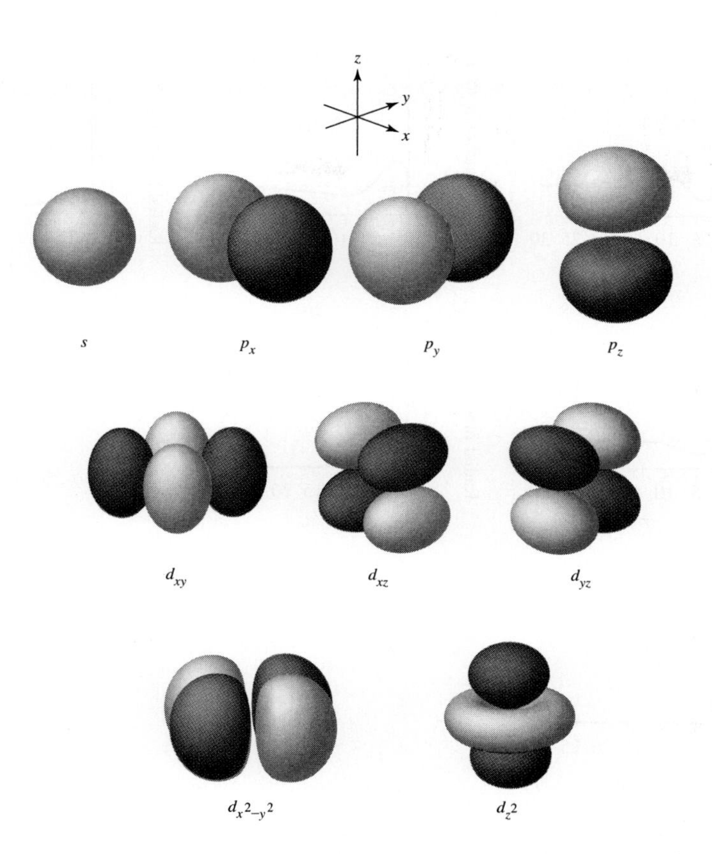

FIGURE 2.6 Selected Atomic Orbitals.

(Selected Atomic Orbitals by Gary O. Spessard and Gary L. Miessler. Reprinted by permission.)

R is a function only of r; Y is a function of θ and ϕ, and it gives the distinctive shapes of s, p, d, and other orbitals. R, Θ and Φ are shown separately in Tables 2.3 and 2.4.

Angular Functions

The angular functions Θ and Φ determine how the probability changes from point to point at a given distance from the center of the atom; in other words, they give the shape of the orbitals and their orientation in space. The angular functions Θ and Φ are determined by the quantum numbers l and m_l. The shapes of s, p, and d orbitals are shown in Table 2.3 and **Figure 2.6**.

In the center of Table 2.3 are the shapes for the Θ portion; when the Φ portion is included, with values of $\phi = 0$ to 2π, the three-dimensional shapes in the far-right column are formed. In the three-dimensional diagrams of orbitals in Table 2.3, the orbital lobes are shaded where the wave function is negative. *The different shadings of the lobes represent different signs of the wave function* Ψ. It is useful to distinguish regions of opposite signs for bonding purposes, as we will see in Chapter 5.

Radial Functions

The radial factor $R(r)$ (Table 2.4) is determined by the quantum numbers n and l, the principal and angular momentum quantum numbers.

The **radial probability function** is $4\pi r^2 R^2$. This function describes the probability of finding the electron at a given distance from the nucleus, summed over all angles, with the $4\pi r^2$ factor the result of integrating over all angles. The radial wave functions and radial probability functions are plotted for the $n = 1$, 2, and 3 orbitals in **Figure 2.7**. Both $R(r)$ and $4\pi r^2 R^2$ are scaled with a_0, the Bohr radius, to give reasonable units on the axes of the

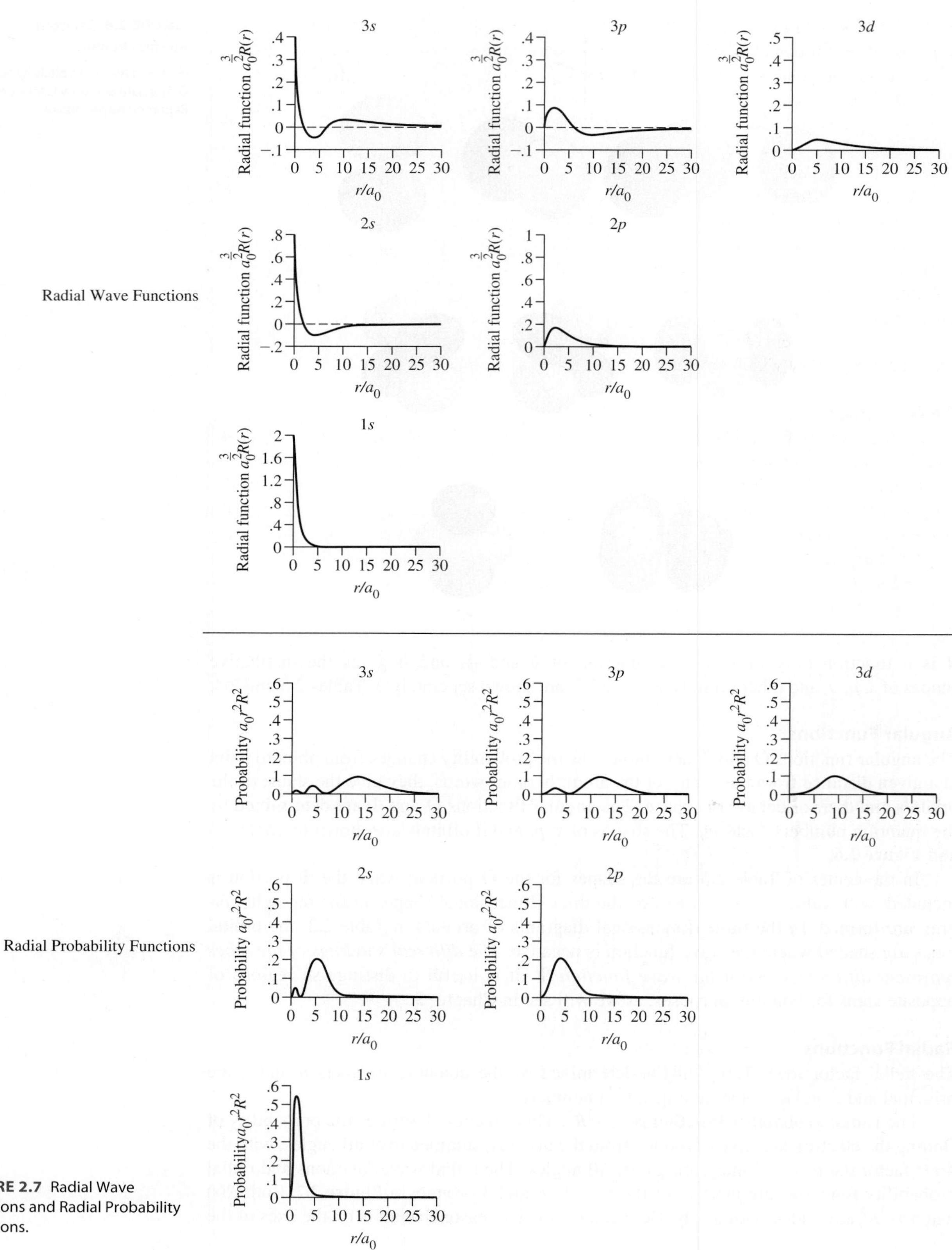

FIGURE 2.7 Radial Wave Functions and Radial Probability Functions.

graphs. The Bohr radius, $a_0 = 52.9$ pm, is a common unit in quantum mechanics. It is the value of r at the maximum of Ψ^2 for a hydrogen 1s orbital (the most probable distance from the hydrogen nucleus for the 1s electron), and it is also the radius of the $n = 1$ orbit according to the Bohr model.

In all the radial probability plots, the electron density, or probability of finding the electron, falls off rapidly beyond its maximum as the distance from the nucleus increases. It falls off most quickly for the 1s orbital; by $r = 5a_0$, the probability is approaching zero. By contrast, the 3d orbital has a maximum at $r = 9a_0$ and does not approach zero until approximately $r = 20a_0$. All the orbitals, including the s orbitals, have zero probability at the center of the nucleus, because $4\pi r^2 R^2 = 0$ at $r = 0$. The radial probability functions are a combination of $4\pi r^2$, which increases rapidly with r, and R^2, which may have maxima and minima, but generally decreases exponentially with r. The product of these two factors gives the characteristic probabilities seen in the plots. Because chemical reactions depend on the shape and extent of orbitals at large distances from the nucleus, the radial probability functions help show which orbitals are most likely to be involved in reactions.

Nodal Surfaces

At large distances from the nucleus, the electron density, or probability of finding the electron, falls off rapidly. The 2s orbital also has a **nodal surface**, a surface with zero electron density, in this case a sphere with $r = 2a_0$ where the probability is zero. Nodes appear naturally as a result of the wave nature of the electron. A node is a surface where the wave function is zero as it changes sign (as at $r = 2a_0$ in the 2s orbital); this requires that $\Psi = 0$, and the probability of finding the electron at any point on the surface is also zero.

If the probability of finding an electron is zero ($\Psi^2 = 0$), Ψ must also be equal to zero. Because

$$\Psi(r, \theta, \phi) = R(r)Y(\theta, \phi)$$

in order for $\Psi = 0$, either $R(r) = 0$ or $Y(\theta, \phi) = 0$. We can therefore determine nodal surfaces by determining under what conditions $R = 0$ or $Y = 0$.

Table 2.5 summarizes the nodes for several orbitals. Note that the total number of nodes in any orbital is $n - 1$ if the conical nodes of some d and f orbitals count as two nodes.*

TABLE 2.5 Nodal Surfaces

Angular Nodes [$Y(\theta, \phi) = 0$]	
Examples (number of angular nodes)	
s orbitals	0
p orbitals	1 plane for each orbital
d orbitals	2 planes for each orbital except d_{z^2}
	1 conical surface for d_{z^2}

Radial Nodes [$R(r) = 0$]					
Examples (number of radial nodes)					
1s	0	2p	0	3d	0
2s	1	3p	1	4d	1
3s	2	4p	2	5d	2

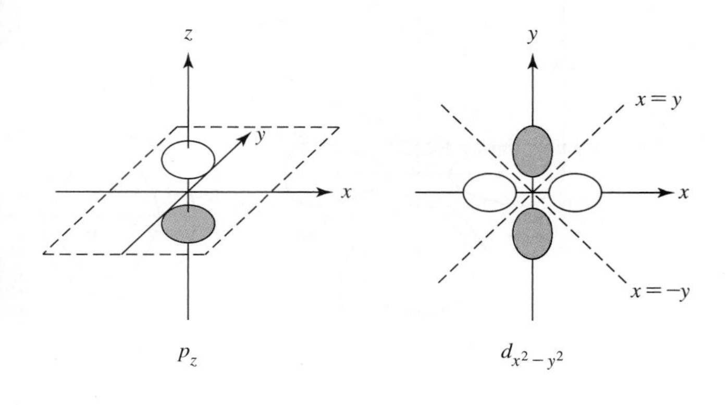

*Mathematically, the nodal surface for the d_{z^2} orbital is one surface, but in this instance, it fits the pattern better if thought of as two nodes.

Angular nodes result when $Y = 0$, and are planar or conical. Angular nodes can be determined in terms of θ and ϕ but may be easier to visualize if Y is expressed in Cartesian (x, y, z) coordinates (see Table 2.3). In addition, the regions where the wave function is positive and where it is negative can be found. This information will be useful in working with molecular orbitals in later chapters. There are l angular nodes in any orbital, with the conical surface in the d_{z^2} orbitals—and other orbitals having conical nodes—counted as two nodes.

Radial nodes (spherical nodes) result when $R = 0$. They give the atom a layered appearance, shown in **Figure 2.8** for the $3s$ and $3p_z$ orbitals. These nodes occur when the radial function changes sign; they are depicted in the radial function graphs by $R(r) = 0$ and in the radial probability graphs by $4\pi r^2 R^2 = 0$. The lowest energy orbitals of each classification ($1s$, $2p$, $3d$, $4f$, etc.) have no radial nodes. The number of radial nodes increases as n increases; the number of radial nodes for a given orbital is always* equal to $n - l - 1$.

Nodal surfaces can be puzzling. For example, a p orbital has a nodal plane through the nucleus. How can an electron be on both sides of a node at the same time without ever having been at the node, at which the probability is zero? One explanation is that the probability does not go quite to zero** on the basis of relativistic arguments.

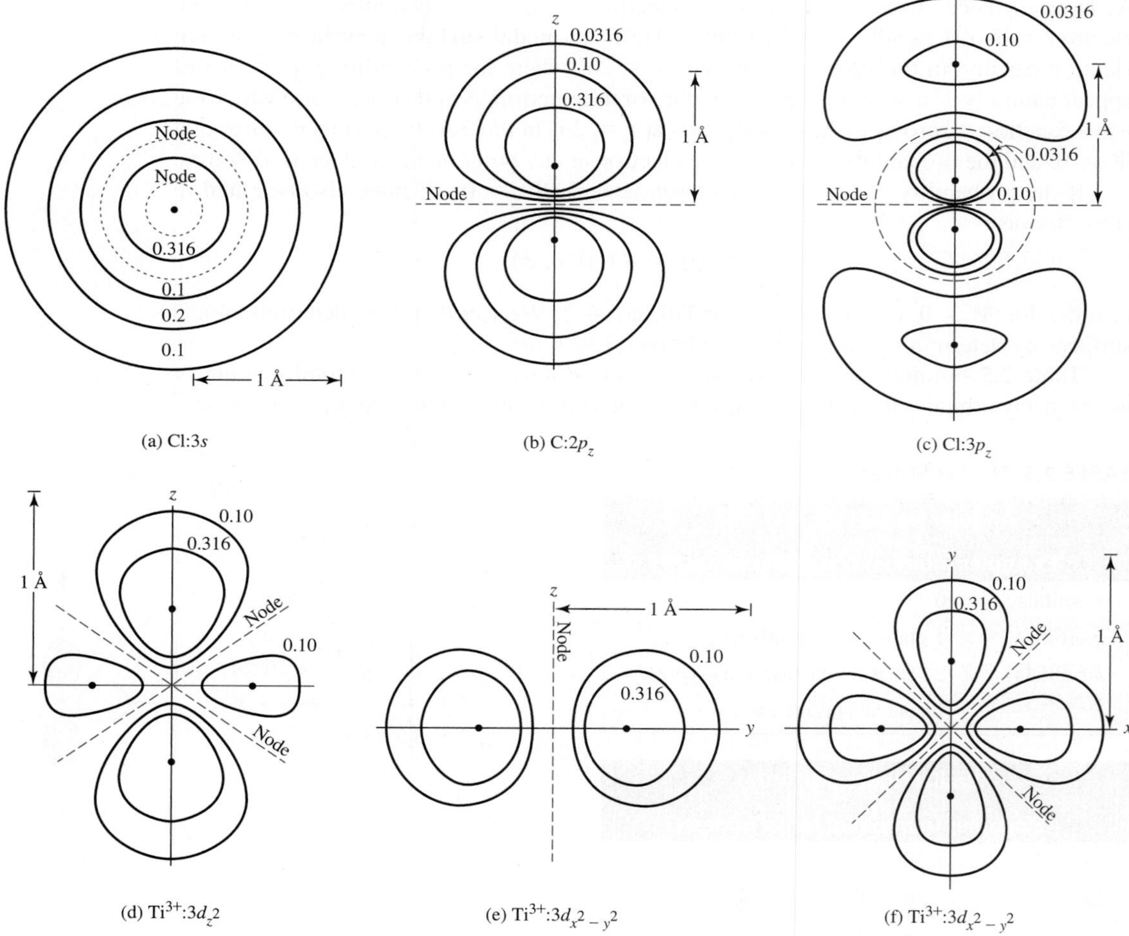

FIGURE 2.8 Constant Electron Density Surfaces for Selected Atomic Orbitals. (a)–(d) The cross-sectional plane is any plane containing the z axis. (e) The cross section is taken through the xz or yz plane. (f) The cross section is taken through the xy plane.

(Figures (b)–(f) Reproduced with permission from E. A. Orgyzlo and G.B. Porter, in *J. Chem. Educ.*, 40, 258. Copyright **1963**. American Chemical Society.)

* Again, counting a conical nodal surface, such as for a d_{z^2} orbital, as two nodes.

** A. Szabo, *J. Chem. Educ.*, **1969**, *46*, 678 explains that the electron probability at a nodal surface has a very small but finite value.

Another explanation is that such a question really has no meaning for an electron behaving as a wave. Recall the particle-in-a-box example. Figure 2.4 shows nodes at $x/a = 0.5$ for $n = 2$ and at $x/a = 0.33$ and 0.67 for $n = 3$. The same diagrams could represent the amplitudes of the motion of vibrating strings at the fundamental frequency ($n = 1$) and multiples of 2 and 3. A plucked violin string vibrates at a specific frequency, and nodes at which the amplitude of vibration is zero are a natural result. Zero amplitude does not mean that the string does not exist at these points but simply that the magnitude of the vibration is zero. An electron wave exists at the node as well as on both sides of a nodal surface, just as a violin string exists at the nodes and on both sides of points having zero amplitude.

Still another explanation, in a lighter vein, was suggested by R. M. Fuoss to one of the authors in a class on bonding. Paraphrased from St. Thomas Aquinas, "Angels are not material beings. Therefore, they can be first in one place and later in another without ever having been in between." If the word "electrons" replaces the word "angels," a semitheological interpretation of nodes would result.

EXAMPLE 2.1

Nodal structure of p_z The angular factor Y is given in Table 2.3 in terms of Cartesian coordinates:

$$Y = \frac{1}{2}\sqrt{\frac{3}{\pi}}\,\frac{z}{r}$$

This orbital is designated p_z because z appears in the Y expression. For an angular node, Y must equal zero, which is true only if $z = 0$. Therefore, $z = 0$ (the xy plane) is an angular nodal surface for the p_z orbital, as shown in Table 2.5 and Figure 2.8. The wave function is positive where $z > 0$ and negative where $z < 0$. In addition, a $2p_z$ orbital has no radial (spherical) nodes, a $3p_z$ orbital has one radial node, and so on.

Nodal structure of $d_{x^2-y^2}$

$$Y = \frac{1}{4}\sqrt{\frac{15}{\pi}}\,\frac{(x^2 - y^2)}{r^2}$$

Here, the expression $x^2 - y^2$ appears in the equation, so the designation is $d_{x^2-y^2}$.

Because there are two solutions to the equation $Y = 0$ (setting $x^2 - y^2 = 0$, the solutions are $x = y$ and $x = -y$), the planes defined by these equations are the angular nodal surfaces. They are planes containing the z axis and making 45° angles with the x and y axes (see Table 2.5). The function is positive where $x > y$ and negative where $x < y$. In addition, a $3d_{x^2-y^2}$ orbital has no radial nodes, a $4d_{x^2-y^2}$ has one radial node, and so on.

EXERCISE 2.2 Describe the angular nodal surfaces for a d_{z^2} orbital, whose angular wave function is

$$Y = \frac{1}{4}\sqrt{\frac{5}{\pi}}\,\frac{(2z^2 - x^2 - y^2)}{r^2}$$

EXERCISE 2.3 Describe the angular nodal surfaces for a d_{xz} orbital, whose angular wave function is

$$Y = \frac{1}{2}\sqrt{\frac{15}{\pi}}\,\frac{xz}{r^2}$$

The result of the calculations is the set of atomic orbitals familiar to chemists. Figure 2.6 shows diagrams of s, p, and d orbitals, and Figure 2.8 shows lines of constant electron density in several orbitals. Different shadings of the orbital lobes in Figure 2.6 indicate different signs of the electron wave amplitude, and the outer surfaces shown enclose 90% of the total electron density of the orbitals. The orbitals we use are the common ones used by chemists; others that are also solutions of the Schrödinger equation can be chosen for special purposes.[14]

Angular functions for *f* orbitals are provided in Appendix B-8. The reader is encouraged to make use of Internet resources that display a wide range of atomic orbitals—including *f*, *g*, and higher orbitals—show radial and angular nodes, and provide additional information.*

2.2.3 The Aufbau Principle

Limitations on the values of the quantum numbers lead to the **aufbau** (German, *Aufbau*, *building up*) **principle**, where the buildup of electrons in atoms results from continually increasing the quantum numbers. The energy level pattern in Figure 2.2 describes electron behavior in a hydrogen atom, where there is only one electron. However, interactions between electrons in polyelectronic atoms require that the order of filling orbitals be specified when more than one electron is in the same atom. In this process, we start with the lowest n, l, and m_l values (1, 0, and 0, respectively) and either of the m_s values (we will arbitrarily use $+\frac{1}{2}$ first). Three rules will then give us the proper order for the remaining electrons, as we increase the quantum numbers in the order m_l, m_s, l, and n.

1. Electrons are placed in orbitals to give the lowest total electronic energy to the atom. This means that the lowest values of n and l are filled first. Because the orbitals within each subshell (p, d, etc.) have the same energy, the orders for values of m_l and m_s are indeterminate.
2. The **Pauli exclusion principle**[15] requires that each electron in an atom have a unique set of quantum numbers. At least one quantum number must be different from those of every other electron. This principle does not come from the Schrödinger equation, but from experimental determination of electronic structures.
3. **Hund's rule of maximum multiplicity**[16] requires that electrons be placed in orbitals to give the maximum total spin possible (the maximum number of parallel spins). Two electrons in the same orbital have a higher energy than two electrons in different orbitals because of electrostatic repulsion (see below); electrons in the same orbital repel each other more than electrons in separate orbitals. Therefore, this rule is a consequence of the lowest possible energy rule (Rule 1). When there are one to six electrons in a p subshell, the required arrangements are those given in **Table 2.6**. (The **spin multiplicity** is the number of unpaired electrons plus 1, or $n + 1$). Any other arrangement of electrons results in fewer unpaired electrons.**

TABLE 2.6 Hund's Rule and Multiplicity

Number of Electrons	Arrangement	Unpaired e^-	Multiplicity
1	$\underline{\uparrow}$ __ __	1	2
2	$\underline{\uparrow}$ $\underline{\uparrow}$ __	2	3
3	$\underline{\uparrow}$ $\underline{\uparrow}$ $\underline{\uparrow}$	3	4
4	$\underline{\uparrow\downarrow}$ $\underline{\uparrow}$ $\underline{\uparrow}$	2	3
5	$\underline{\uparrow\downarrow}$ $\underline{\uparrow\downarrow}$ $\underline{\uparrow}$	1	2
6	$\underline{\uparrow\downarrow}$ $\underline{\uparrow\downarrow}$ $\underline{\uparrow\downarrow}$	0	1

* Two examples are http://www.orbitals.com and http://winter.group.shef.ac.uk/orbitron.

** This is only one of Hund's rules; others are described in Chapter 11.

Hund's rule is a consequence of the energy required for pairing electrons in the same orbital. When two negatively charged electrons occupy the same region of space (same orbital) in an atom, they repel each other, with a **Coulombic energy of repulsion**, Π_c, per pair of electrons. As a result, this repulsive force favors electrons in different orbitals (different regions of space) over electrons in the same orbitals.

In addition, there is an **exchange energy**, Π_e, which arises from purely quantum mechanical considerations. This energy depends on the number of possible exchanges between two electrons with the same energy and the same spin. For example, the electron configuration of a carbon atom is $1s^2 2s^2 2p^2$. The $2p$ electrons can be placed in the p orbitals in three ways:

(1) ↑↓ ____ ____ (2) ↑ ____ ↓ ____ ____ (3) ↑ ____ ↑ ____ ____

Each of these corresponds to a state having a particular energy. State (1) involves Coulombic energy of repulsion, Π_c, because it is the only one that pairs electrons in the same orbital. The energy of this state is higher than that of the other two by Π_c as a result of electron–electron repulsion.

In the first two states, there is only one possible way to arrange the electrons to give the same diagram, because there is only a single electron in each having + or – spin; these electrons can be distinguished from each other on this basis. However, in the third state, the electrons have the same spin and are therefore indistinguishable from each other. Therefore, there are two possible ways in which the electrons can be arranged:

↑1 ____ ↑2 ____ ____ ↑2 ____ ↑1 ____ ____ (one exchange of electrons)

Because there are two possible ways in which the electrons in state (3) can be arranged, we can say that there is *one pair* of possible exchanges between these arrangements, described as *one exchange of parallel electrons*. The energy involved in such an exchange of parallel electrons is designated Π_e; each exchange stabilizes (lowers the energy of) an electronic state, favoring states with more parallel spins (Hund's rule). Therefore, state (3), which is stabilized by one exchange of parallel electrons, is lower in energy than state (2) by Π_e.

The results of considering the effects of Coulombic and exchange energies for the p^2 configuration may be summarized in an energy diagram:

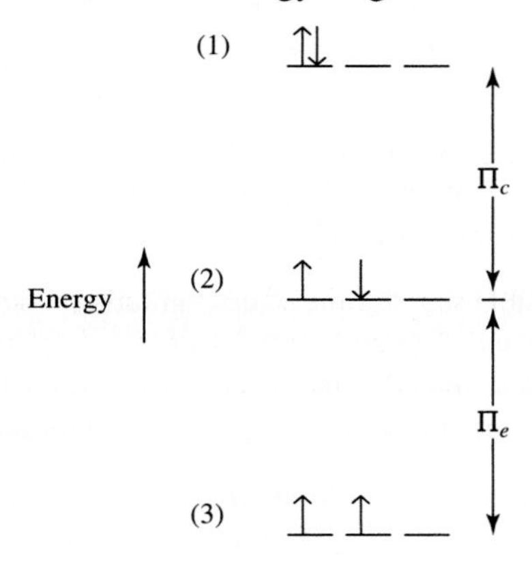

State (3) is the most stable; its electrons are in separate orbitals and have parallel spin; because state (3) has one possible exchange of electrons with parallel spin, it is lower in energy than state (2) by Π_e. State (1) is highest in energy because it has two electrons in the same orbital and is therefore higher in energy than state (2) by Π_c. Neither state (1) nor state (2) is stabilized by exchange interactions (zero Π_e).

In summary:

Coulombic energy of repulsion Π_c is a consequence of repulsion between electrons in the same orbital; the greater the number of such paired electrons, the higher the energy of the state.*

Exchange energy Π_e is a consequence of parallel electron spins in separate orbitals; the greater the number of such parallel spins (and consequently the greater the number of exchanges), the lower the energy of the state.

Both Coulombic and exchange energies must be taken into account when comparing the energies of different electronic states.

EXAMPLE 2.2

Oxygen

With four p electrons, oxygen could have two unpaired electrons ($\underline{\uparrow\downarrow}$ $\underline{\uparrow}$ $\underline{\uparrow}$), or it could have no unpaired electrons ($\underline{\uparrow\downarrow}$ $\underline{\uparrow\downarrow}$ $\underline{\quad}$).

a. Determine the number of electrons that could be exchanged in each case, and find the Coulombic and exchange energies.

$\underline{\uparrow\downarrow}$ $\underline{\uparrow}$ $\underline{\uparrow}$ This configuration has one pair, energy contribution Π_c.

$\underline{\uparrow\downarrow}$ $\underline{\uparrow}$ $\underline{\uparrow}$ One electron with $\downarrow$ spin and no possibility of exchange.

$\underline{\uparrow\downarrow}$ $\underline{\uparrow}$ $\underline{\uparrow}$ Four possible arrangements for electrons with $\uparrow$ spin; three exchange possibilities (1–2, 1–3, 2–3), shown below; energy contribution $3\Pi_e$.

$\underline{\uparrow 1}$ $\underline{\uparrow 2}$ $\underline{\uparrow 3}$ $\quad$ $\underline{\uparrow 2}$ $\underline{\uparrow 1}$ $\underline{\uparrow 3}$ $\quad$ $\underline{\uparrow 3}$ $\underline{\uparrow 2}$ $\underline{\uparrow 1}$ $\quad$ $\underline{\uparrow 1}$ $\underline{\uparrow 3}$ $\underline{\uparrow 2}$

Overall, $3\Pi_e + \Pi_c$.

$\underline{\uparrow\downarrow}$ $\underline{\uparrow\downarrow}$ $\underline{\quad}$ has two pairs in the same orbitals and one exchange possibility for each spin pair.

Overall, $2\Pi_e + 2\Pi_c$.

b. Which state, $\underline{\uparrow\downarrow}$ $\underline{\uparrow}$ $\underline{\uparrow}$, or $\underline{\uparrow\downarrow}$ $\underline{\uparrow\downarrow}$ $\underline{\quad}$, is lower in energy?

The state $\underline{\uparrow\downarrow}$ $\underline{\uparrow}$ $\underline{\uparrow}$ is lower in energy because it has less Coulombic energy of repulsion (Π_c in comparison with $2\Pi_c$) and is stabilized by a greater number of exchanges ($3\Pi_e$ in comparison with $2\Pi_e$).

EXERCISE 2.4 A third possible state for the p^4 configuration would be $\underline{\uparrow\downarrow}$ $\underline{\uparrow}$ $\underline{\downarrow}$. Determine the Coulombic and exchange energies of this state, and compare its energy with the energies of the states determined in the preceding example. Draw a sketch showing the relative energies of these three states for oxygen's p^4 configuration.

EXERCISE 2.5 A nitrogen atom, with three $2p$ electrons, could have three unpaired electrons ($\underline{\uparrow}$ $\underline{\uparrow}$ $\underline{\uparrow}$), or it could have one unpaired electron ($\underline{\uparrow\downarrow}$ $\underline{\uparrow}$ $\underline{\quad}$).

a. Determine the number of electrons that could be exchanged in each case and the Coulombic and exchange energies. Which state would be lower in energy?

*In atoms with more than one electron (polyelectronic atoms), all electrons are subject to some Coulombic repulsion energy, but this contribution is significantly higher for electrons that are paired within atomic orbitals.

b. A third possible state for a $2p^3$ configuration would be $\underline{\uparrow}$ $\underline{\uparrow}$ $\underline{\downarrow}$. Determine its Coulombic and exchange energies, and compare the energy of this state with the energies determined in part **a.**

When the orbitals are **degenerate** (have the same energy), both Coulombic and exchange energies favor unpaired configurations over paired configurations. However, if there is a difference in energy between the levels involved, this difference, together with the Coulombic and exchange energies, determines the final configuration, with the configuration of lowest energy expected as the ground state; energy minimization is the driving force. For atoms, this usually means that one subshell (*s*, *p*, *d*) is filled before another has any electrons. However, this approach is insufficient in some transition elements, because the 4*s* and 3*d* (or the higher corresponding levels) are so close in energy that the sum of the Coulombic and exchange terms is nearly the same as the difference in energy between the 4*s* and 3*d*. Section 2.2.4 considers these cases.

Many schemes have been used to predict the order of filling of atomic orbitals. Klechkowsky's rule states that the order of filling of the orbitals proceeds from the lowest available value for the sum $n + l$. When two combinations have the same value, the one with the smaller value of n is filled first; thus, $4s\,(n + l = 4 + 0)$ fills before $3d\,(n + l = 3 + 2)$. Combined with the other rules, this gives the order of filling of most of the orbitals.*

One of the simplest methods that fits most atoms uses the periodic table organized as in **Figure 2.9**. The electron configurations of hydrogen and helium are clearly $1s^1$ and $1s^2$. After that, the elements in the first two columns on the left (Groups 1 and 2) are filling *s* orbitals, with $l = 0$; those in the six columns on the right (Groups 13 to 18) are filling *p* orbitals, with $l = 1$; and the ten in the middle (the transition elements, Groups 3 to 12) are filling

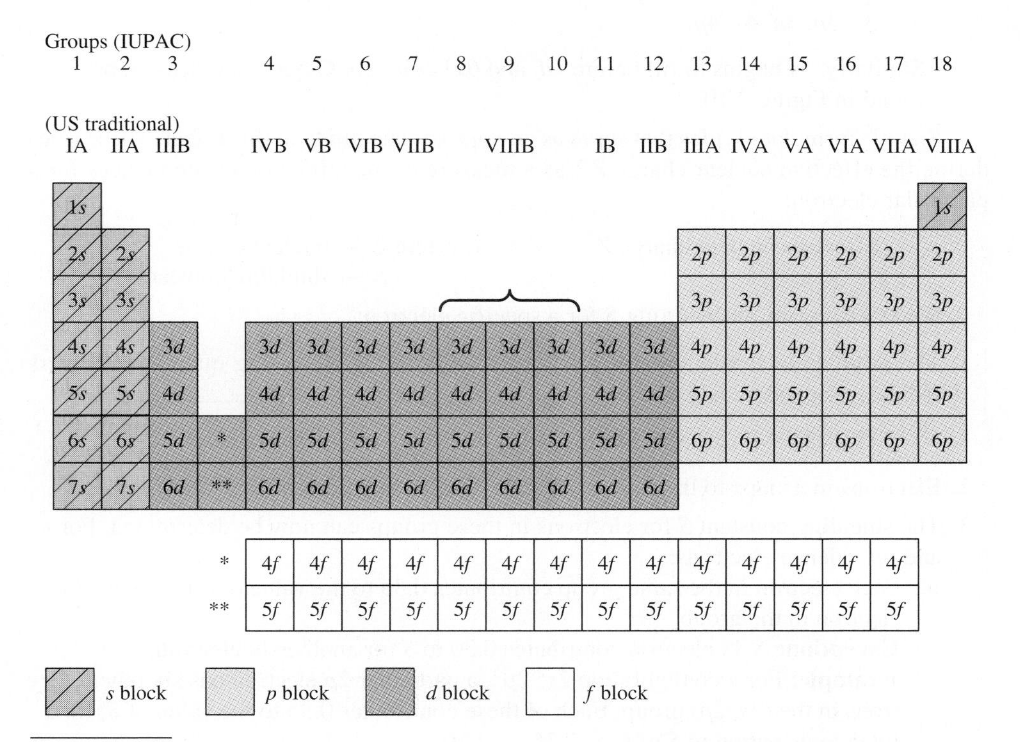

FIGURE 2.9 Atomic Orbital Filling in the Periodic Table.

---

* For recent perspective on electron configurations, energies of atomic orbitals, the periodic system, and related topics, see S-G. Wang and W. H. E. Schwarz, *Angew. Chem. Int. Ed.*, **2009**, *48*, 3404.

d orbitals, with $l = 2$. The lanthanide and actinide series (numbers 58 to 71 and 90 to 103) are filling *f* orbitals, with $l = 3$. These two methods are oversimplifications, as shown in the following paragraphs, but they do fit most atoms and provide starting points for the others.

2.2.4 Shielding

In polyelectronic atoms, energies of specific levels are difficult to predict quantitatively. A useful approach to such predictions uses the concept of shielding: each electron acts as a shield for electrons farther from the nucleus, reducing the attraction between the nucleus and the more distant electrons.

Although the quantum number n is most important in determining the energy, quantum number l must also be included in calculating the energy in atoms having more than one electron. As the atomic number increases, electrons are drawn toward the nucleus, and the orbital energies become more negative. Although the energies decrease with increasing Z, the changes are somewhat irregular because of the shielding of outer electrons by inner electrons. The electron configurations of atoms from the resulting order of orbital filling are shown in **Table 2.7**.

As a result of shielding and other subtle interactions between electrons, exclusive reliance on n to rank orbital energies (higher energy with higher quantum number n), which works for one-electron species, holds only for orbitals with lowest values of n (see **Figure 2.10**) in polyelectronic species. In multielectron atoms (and ions), for higher values of n, as the split in orbital energies with different values of quantum number l becomes comparable in magnitude to the differences in energy caused by n, the simplest order does not hold.

For example, consider the $n = 3$ and $n = 4$ sets in Figure 2.10. For many atoms the 4*s* orbital is lower in energy than the 3*d* orbitals; consequently the order of filling is ...3*s*, 3*p*, 4*s*, 3*d*, 4*p*... rather than the order based strictly on increasing n ...3*s*, 3*p*, 3*d*, 4*s*, 4*p*...

Similarly, 5*s* begins to fill before 4*d*, and 6*s* before 5*d*. Other examples can be found in Figure 2.10.

Slater[17] formulated rules that serve as an approximate guide to this effect. These rules define the effective nuclear charge Z^* as a measure of the attraction of the nucleus for a particular electron:

$$\text{Effective nuclear charge } Z^* = Z - S, \text{ where } Z = \text{nuclear charge}, \quad S = \text{shielding constant}$$

Slater's rules for determining S for a specific electron:*

1. The atom's electronic structure is written in order of increasing quantum numbers n and l, grouped as follows:

   (1*s*) (2*s*, 2*p*) (3*s*, 3*p*) (3*d*) (4*s*, 4*p*) (4*d*) (4*f*) (5*s*, 5*p*) (5*d*) (and so on)

2. Electrons in groups to the right in this list do not shield electrons to their left.
3. The shielding constant S for electrons in these groups can now be determined. For *ns* and *np* valence electrons:
   a. Each electron in the same group contributes 0.35 to the value of S for each other electron in the group.
   Exception: A 1*s* electron contributes 0.30 to S for another 1*s* electron.
   Example: For a configuration $2s^2\,2p^5$, a particular 2*p* electron has six other electrons in the (2*s*, 2*p*) group. Each of these contributes 0.35 to the value of S, for a total contribution to S of $6 \times 0.35 = 2.10$.

---

* Slater's original numbering scheme has been changed for convenience.

TABLE 2.7 Electron Configurations of the Elements

Element	Z	Configuration	Element	Z	Configuration
H	1	$1s^1$	Cs	55	$[Xe]6s^1$
He	2	$1s^2$	Ba	56	$[Xe]6s^2$
Li	3	$[He]2s^1$	La	57	*$[Xe]6s^2 5d^1$
Be	4	$[He]2s^2$	Ce	58	*$[Xe]6s^2 4f^1 5d^1$
B	5	$[He]2s^2 2p^1$	Pr	59	$[Xe]6s^2 4f^3$
C	6	$[He]2s^2 2p^2$	Nd	60	$[Xe]6s^2 4f^4$
N	7	$[He]2s^2 2p^3$	Pm	61	$[Xe]6s^2 4f^5$
O	8	$[He]2s^2 2p^4$	Sm	62	$[Xe]6s^2 4f^6$
F	9	$[He]2s^2 2p^5$	Eu	63	$[Xe]6s^2 4f^7$
Ne	10	$[He]2s^2 2p^6$	Gd	64	*$[Xe]6s^2 4f^7 5d^1$
			Tb	65	$[Xe]6s^2 4f^9$
Na	11	$[Ne]3s^1$	Dy	66	$[Xe]6s^2 4f^{10}$
Mg	12	$[Ne]3s^2$	Ho	67	$[Xe]6s^2 4f^{11}$
Al	13	$[Ne]3s^2 3p^1$	Er	68	$[Xe]6s^2 4f^{12}$
Si	14	$[Ne]3s^2 3p^2$	Tm	69	$[Xe]6s^2 4f^{13}$
P	15	$[Ne]3s^2 3p^3$	Yb	70	$[Xe]6s^2 4f^{14}$
S	16	$[Ne]3s^2 3p^4$	Lu	71	$[Xe]6s^2 4f^{14} 5d^1$
Cl	17	$[Ne]3s^2 3p^5$	Hf	72	$[Xe]6s^2 4f^{14} 5d^2$
Ar	18	$[Ne]3s^2 3p^6$	Ta	73	$[Xe]6s^2 4f^{14} 5d^3$
			W	74	$[Xe]6s^2 4f^{14} 5d^4$
K	19	$[Ar]4s^1$	Re	75	$[Xe]6s^2 4f^{14} 5d^5$
Ca	20	$[Ar]4s^2$	Os	76	$[Xe]6s^2 4f^{14} 5d^6$
Sc	21	$[Ar]4s^2 3d^1$	Ir	77	$[Xe]6s^2 4f^{14} 5d^7$
Ti	22	$[Ar]4s^2 3d^2$	Pt	78	*$[Xe]6s^1 4f^{14} 5d^9$
V	23	$[Ar]4s^2 3d^3$	Au	79	*$[Xe]6s^1 4f^{14} 5d^{10}$
Cr	24	*$[Ar]4s^1 3d^5$	Hg	80	$[Xe]6s^2 4f^{14} 5d^{10}$
Mn	25	$[Ar]4s^2 3d^5$	Tl	81	$[Xe]6s^2 4f^{14} 5d^{10} 6p^1$
Fe	26	$[Ar]4s^2 3d^6$	Pb	82	$[Xe]6s^2 4f^{14} 5d^{10} 6p^2$
Co	27	$[Ar]4s^2 3d^7$	Bi	83	$[Xe]6s^2 4f^{14} 5d^{10} 6p^3$
Ni	28	$[Ar]4s^2 3d^8$	Po	84	$[Xe]6s^2 4f^{14} 5d^{10} 6p^4$
Cu	29	*$[Ar]4s^1 3d^{10}$	At	85	$[Xe]6s^2 4f^{14} 5d^{10} 6p^5$
Zn	30	$[Ar]4s^2 3d^{10}$	Rn	86	$[Xe]6s^2 4f^{14} 5d^{10} 6p^6$
Ga	31	$[Ar]4s^2 3d^{10} 4p^1$			
Ge	32	$[Ar]4s^2 3d^{10} 4p^2$	Fr	87	$[Rn]7s^1$
As	33	$[Ar]4s^2 3d^{10} 4p^3$	Ra	88	$[Rn]7s^2$
Se	34	$[Ar]4s^2 3d^{10} 4p^4$	Ac	89	*$[Rn]7s^2 6d^1$
Br	35	$[Ar]4s^2 3d^{10} 4p^5$	Th	90	*$[Rn]7s^2 6d^2$
Kr	36	$[Ar]4s^2 3d^{10} 4p^6$	Pa	91	*$[Rn]7s^2 5f^2 6d^1$
			U	92	*$[Rn]7s^2 5f^3 6d^1$
Rb	37	$[Kr]5s^1$	Np	93	*$[Rn]7s^2 5f^4 6d^1$
Sr	38	$[Kr]5s^2$	Pu	94	$[Rn]7s^2 5f^6$
			Am	95	$[Rn]7s^2 5f^7$
Y	39	$[Kr]5s^2 4d^1$	Cm	96	*$[Rn]7s^2 5f^7 6d^1$
Zr	40	$[Kr]5s^2 4d^2$	Bk	97	$[Rn]7s^2 5f^9$
Nb	41	*$[Kr]5s^1 4d^4$	Cf	98	*$[Rn]7s^2 5f^9 6d^1$
Mo	42	*$[Kr]5s^1 4d^5$	Es	99	$[Rn]7s^2 5f^{11}$
Tc	43	$[Kr]5s^2 4d^5$	Fm	100	$[Rn]7s^2 5f^{12}$
Ru	44	*$[Kr]5s^1 4d^7$	Md	101	$[Rn]7s^2 5f^{13}$
Rh	45	*$[Kr]5s^1 4d^8$	No	102	$[Rn]7s^2 5f^{14}$
Pd	46	*$[Kr]4d^{10}$	Lr	103	$[Rn]7s^2 5f^{14} 6d^1$
Ag	47	*$[Kr]5s^1 4d^{10}$	Rf	104	$[Rn]7s^2 5f^{14} 6d^2$
Cd	48	$[Kr]5s^2 4d^{10}$	Db	105	$[Rn]7s^2 5f^{14} 6d^3$
In	49	$[Kr]5s^2 4d^{10} 5p^1$	Sg	106	$[Rn]7s^2 5f^{14} 6d^4$
Sn	50	$[Kr]5s^2 4d^{10} 5p^2$	Bh	107	$[Rn]7s^2 5f^{14} 6d^5$
Sb	51	$[Kr]5s^2 4d^{10} 5p^3$	Hs	108	$[Rn]7s^2 5f^{14} 6d^6$
Te	52	$[Kr]5s^2 4d^{10} 5p^4$	Mt	109	$[Rn]7s^2 5f^{14} 6d^7$
I	53	$[Kr]5s^2 4d^{10} 5p^5$	Ds	110	*$[Rn]7s^1 5f^{14} 6d^9$
Xe	54	$[Kr]5s^2 4d^{10} 5p^6$	Rg	111	*$[Rn]7s^1 5f^{14} 6d^{10}$
			Cn[a]	112	$[Rn]7s^2 5f^{14} 6d^{10}$

* Elements with configurations that do not follow the simple order of orbital filling.

[a] Evidence for elements 113–118 has been reviewed by IUPAC; see R. C. Barber, P. J. Karol, H. Nakahara, E. Vardaci, E. W. Vogt, *Pure Appl. Chem.*, **2011**, *83*, 1485. In May 2012, IUPAC officially named element 114 (flerovium, symbol Fl) and element 116 (livermorium, Lv).

Source: Actinide configurations are from J. J. Katz, G. T. Seaborg, and L. R. Morss, *The Chemistry of the Actinide Elements*, 2nd ed., Chapman and Hall, New York and London, 1986. Configurations for elements 100 to 112 are predicted, not experimental.

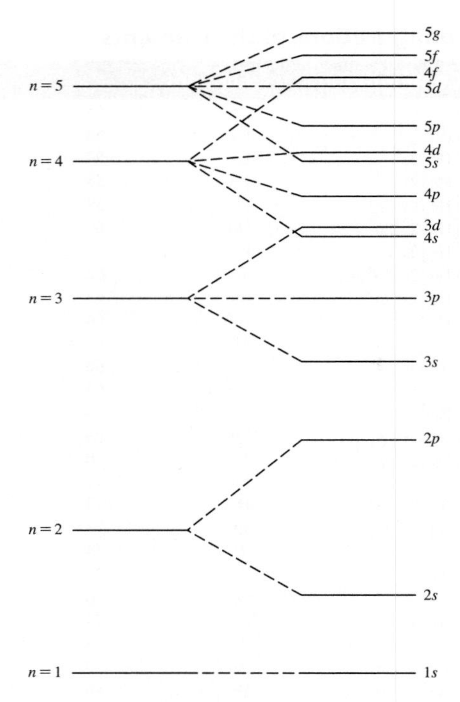

FIGURE 2.10 Energy Level Splitting and Overlap. The differences between the upper levels are exaggerated for easier visualization. This diagram provides unambiguous electron configurations for elements hydrogen to vanadium.

  b. Each electron in $n - 1$ groups contribute 0.85 to S.
  Example: For the 3*s* electron of sodium, there are eight electrons in the (2*s*, 2*p*) group. Each of these electrons contributes 0.85 to the value of S, a total contribution of $8 \times 0.85 = 6.80$.
  c. Each electron in $n - 2$ or lower groups contributes 1.00 to S.

4. For *nd* and *nf* valence electrons:
  a. Each electron in the same group contributes 0.35 to the value of S for each other electron in the group. (Same rule as **3a.**)
  b. Each electron in groups to the left contributes 1.00 to S.

These rules are used to calculate the shielding constant S for valence electrons. Subtracting S from the total nuclear charge Z gives the effective nuclear charge Z^* on the selected electron:

$$Z^* = Z - S$$

Calculations of S and Z^* follow.

EXAMPLE 2.3

Oxygen

Use Slater's rules to calculate the shielding constant and effective nuclear charge of a 2*p* electron.

Rule 1: The electron configuration is written using Slater's groupings, in order:

$$(1s^2)(2s^2, 2p^4)$$

To calculate S for a valence 2*p* electron:

Rule 3a: Each other electron in the $(2s^2, 2p^4)$ group contributes 0.35 to S.
Total contribution $= 5 \times 0.35 = 1.75$

Rule 3b: Each 1*s* electron contributes 0.85 to S.
Total contribution $= 2 \times 0.85 = 1.70$

Total $S = 1.75 + 1.70 = 3.45$
Effective nuclear charge $Z^* = 8 - 3.45 = 4.55$
So rather than feeling the full +8 nuclear charge, a $2p$ electron is calculated to feel a charge of +4.55, or about 57% of the full nuclear charge.

Nickel

Use Slater's rules to calculate the shielding constant and effective nuclear charge of a $3d$ and $4s$ electron.

Rule 1: The electron configuration is written $(1s^2)(2s^2, 2p^6)(3s^2, 3p^6)(3d^8)(4s^2)$

For a $3d$ electron:

Rule 4a: Each other electron in the $(3d^8)$ group contributes 0.35 to S.
Total contribution $= 7 \times 0.35 = 2.45$

Rule 4b: Each electron in groups to the left of $(3d^8)$ contributes 1.00 to S.
Total contribution $= 18 \times 1.00 = 18.00$
Total $S = 2.45 + 18.00 = 20.45$
Effective nuclear charge $Z^* = 28 - 20.45 = 7.55$

For a $4s$ electron:

Rule 3a: The other electron in the $(4s^2)$ group contributes 0.35 to S.

Rule 3b: Each electron in the $(3s^2, 3p^6)(3d^8)$ groups $(n - 1)$ contributes 0.85.
Total contribution $= 16 \times 0.85 = 13.60$

Rule 3c: Each other electron to the left contributes 1.00. Total contribution $=$ $10 \times 1.00 = 10.00$
Total $S = 0.35 + 13.60 + 10.00 = 23.95$
Effective nuclear charge $Z^* = 28 - 23.95 = 4.05$

The effective nuclear charge for the $4s$ electron is considerably smaller than the value for the $3d$ electron. This is equivalent to stating that the $4s$ electron is held less tightly than the $3d$ and should therefore be the first removed in ionization. This is consistent with experimental observations on nickel compounds. Ni^{2+}, the most common oxidation state of nickel, has a configuration of [Ar] $3d^8$, rather than [Ar] $3d^6 4s^2$, corresponding to loss of the $4s$ electrons from nickel atoms. All the transition metal atoms follow this same pattern of losing ns electrons more readily than $(n - 1)d$ electrons.

EXERCISE 2.6 Calculate the effective nuclear charge on a $5s$, $5p$, and $4d$ electron in a tin atom.

EXERCISE 2.7 Calculate the effective nuclear charge on a $7s$, $5f$, and $6d$ electron in a uranium atom.

Justification for Slater's rules comes from the electron probability curves for the orbitals; Slater devised these rules semiempirically using equations modeled after wavefunction equations to fit experimental data for atoms. Slater's approach results in rules that provide useful approximations for the effective nuclear charge an electron in an atom actually experiences after shielding is taken into account. The s and p orbitals have higher probabilities near the nucleus than do d orbitals of the same n, as shown earlier in Figure 2.7. Therefore, the shielding of $3d$ electrons by $(3s, 3p)$ electrons is calculated as 100% effective, a contribution of 1.00. At the same time, shielding of $3s$ or $3p$ electrons by $(2s, 2p)$ electrons is estimated as 85% effective, a contribution of 0.85, because the $3s$ and $3p$ orbitals have regions of significant probability close to the nucleus. Therefore, electrons in these orbitals are not completely shielded by $(2s, 2p)$ electrons.

A complication arises at Cr ($Z = 24$) and Cu ($Z = 29$) in the first transition series and in an increasing number of atoms with higher atomic numbers in the second and third transition series. This effect places an extra electron in the $3d$ level and removes one electron from the $4s$ level. Cr, for example, has a configuration of [Ar] $4s^1\ 3d^5$ rather than [Ar] $4s^2\ 3d^4$. Traditionally, this phenomenon has often been explained as a consequence of the "special stability of half-filled subshells." Half-filled and filled d and f subshells are, in fact, fairly common, as shown in **Figure 2.11**. A more complete explanation considers both the effects of increasing nuclear charge on the energies of the $4s$ and $3d$ levels and the interactions between electrons sharing the same orbital.[18] This approach requires totaling all contributions to the energy of the configuration of electrons, including the Coulombic and exchange energies; results of the complete calculations are consistent with the configurations determined by experimental data.

Slater's rules have been refined to improve their match with experimental data. One relatively simple refinement is based on the ionization energies for the elements hydrogen through xenon, and it provides a calculation procedure similar to that proposed by Slater.[19] A more elaborate method incorporates exponential screening and provides energies that are in closer agreement with experimental values.[20]

Another explanation that is more pictorial and considers electron–electron interactions was proposed by Rich.[21] He explained electronic structures of atoms by considering the difference in energy between the energy of one electron in an orbital and two electrons in the same orbital. Although the orbital itself is usually assumed to have only one energy, the electrostatic repulsion of the two electrons in one orbital adds the electron-pairing energy described in Section 2.2.3 as part of Hund's rule. We can visualize two parallel energy levels, each with electrons of only one spin, separated by the electron-pairing energy, as shown in **Figure 2.12**.

For example, an Sc atom has the valence configuration $4s^2\ 3d^1$. By Rich's approach, the first electron is arbitrarily considered to have $m_s = -\frac{1}{2}$. The second electron, with $m_s = +\frac{1}{2}$, completes the $4s^2$ configuration—but the total energy of these two electrons is greater than twice the energy of the first electron, because of the Coulombic energy of repulsion, Π_c. In Figure 2.12(a) Sc is shown as having three electrons: in ascending order these are $4s$ ($m_s = -\frac{1}{2}$), $4s$ ($m_s = +\frac{1}{2}$), and $3d$ ($m_s = -\frac{1}{2}$). The next element, Ti, also

Na	Mg				Half-filled d					Filled d		Al	Si	P	S	Cl	Ar
K	Ca	Sc $3d^1$	Ti $3d^2$	V $3d^3$	Cr $3d^5$ $4s^1$	Mn $3d^5$ $4s^2$	Fe $3d^6$	Co $3d^7$	Ni $3d^8$	Cu $3d^{10}$ $4s^1$	Zn $3d^{10}$ $4s^2$	Ga	Ge	As	Se	Br	Kr
Rb	Sr	Y $4d^1$	Zr $4d^2$	Nb $4d^4$ $5s^1$	Mo $4d^5$ $5s^1$	Tc $4d^5$ $5s^2$	Ru $4d^7$ $5s^1$	Rh $4d^8$ $5s^1$	Pd $4d^{10}$	Ag $4d^{10}$ $5s^1$	Cd $4d^{10}$ $5s^2$	In	Sn	Sb	Te	I	Xe
Cs	Ba	La $5d^1$	* Hf $4f^{14}$ $5d^2$	Ta $4f^{14}$ $5d^3$	W $4f^{14}$ $5d^4$	Re $5d^5$ $6s^2$	Os $5d^6$	Ir $5d^7$	Pt $5d^9$ $6s^1$	Au $5d^{10}$ $6s^1$	Hg $5d^{10}$ $6s^2$	Tl	Pb	Bi	Po	At	Rn
Fr	Ra	Ac $6d^1$	** Rf $5f^{14}$ $6d^2$	Db $5f^{14}$ $6d^3$	Sg $5f^{14}$ $6d^4$	Bh $5f^{14}$ $6d^5$	Hs $5f^{14}$ $6d^6$	Mt $5f^{14}$ $6d^7$	Ds $6d^9$	Rg $6d^{10}$ $7s^1$	Cn $6d^{10}$ $7s^2$		Uuq		Uuh		Uuo

					Half-filled f								Filled f	
*	Ce $4f^1$ $5d^1$	Pr $4f^3$	Nd $4f^4$	Pm $4f^5$	Sm $4f^6$	Eu $4f^7$	Gd $4f^7$ $5d^1$	Tb $4f^9$	Dy $4f^{10}$	Ho $4f^{11}$	Er $4f^{12}$	Tm $4f^{13}$	Yb $4f^{14}$	Lu $4f^{14}$ $5d^1$
**	Th $6d^2$	Pa $5f^2$ $6d^1$	U $5f^3$ $6d^1$	Np $5f^4$ $6d^1$	Pu $5f^6$	Am $5f^7$	Cm $5f^7$ $6d^1$	Bk $5f^9$	Cf $5f^9$ $6d^1$	Es $5f^{11}$	Fm $5f^{12}$	Md $5f^{13}$	No $5f^{14}$	Lr $5f^{14}$ $6d^1$

FIGURE 2.11 Electron Configurations of Transition Metals, Including Lanthanides and Actinides. Solid lines surrounding elements designate filled (d^{10} or f^{14}) or half-filled (d^5 or f^7) subshells. Dashed lines surrounding elements designate irregularities in sequential orbital filling, also found within some of the solid lines.

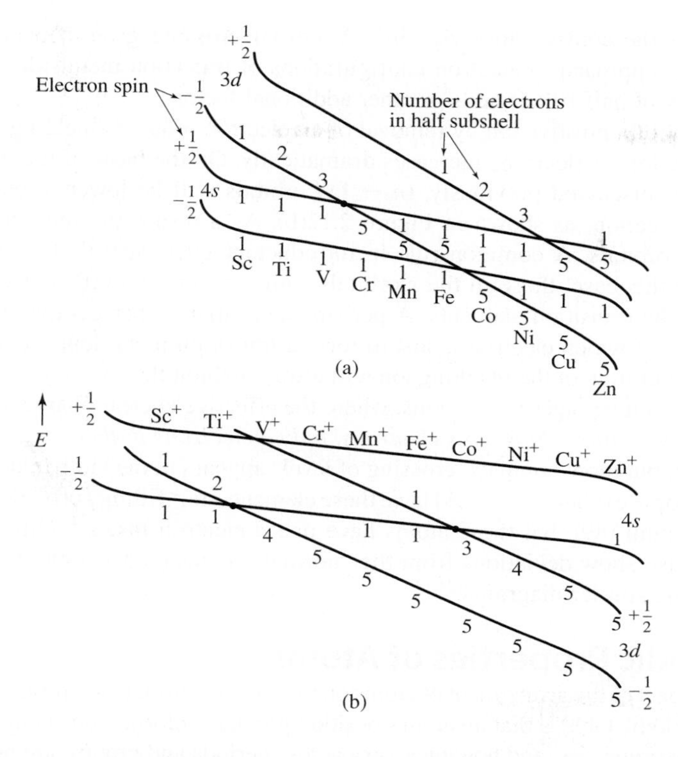

FIGURE 2.12 Schematic Energy Levels for Transition Elements. (a) Schematic interpretation of electron configurations for transition elements in terms of intraorbital repulsion and trends in subshell energies. (b) A similar diagram for ions, showing the shift in the crossover points on removal of an electron. The shift is even more pronounced for metal ions having 2+ or greater charges. As a consequence, transition-metal ions with 2+ or greater charges have no *s* electrons, only *d* electrons in their outer levels. Similar diagrams, although more complex, can be drawn for the heavier transition elements and the lanthanides.

(Rich, R. L., *Periodic Coorelate*, 1st Ed., (c) 1965. Reprinted and Electronically reproduced by permission of Pearson Education Inc, Upper Saddle River, NJ 07458.)

has one 4*s* electron with each spin, then two 3*d* electrons, each with $m_s = -\frac{1}{2}$. The two 3*d* electrons, by Hund's rule, have parallel spin.

As the number of protons in the nucleus increases, the effective nuclear charge for all electrons increases and the energy levels decrease in energy; their electrons become more stable. Figure 2.12 illustrates that the energy of the 3*d* subshell decreases more dramatically relative to 4*s* as one moves across the first transition series; this trend generally holds for $(n - 1)d$ and *ns* orbitals. A rationalization for this trend is that orbitals with shorter most probable distances to the nucleus are stabilized more as *Z* increases relative to orbitals with greater most probable distances. Because the 3*d* orbitals have shorter most probable distances from the nucleus than the 4*s* orbital, the 3*d* orbitals are stabilized more than the 4*s* as the nuclear charge increases.

The effective nuclear charge that an electron experiences generally increases as the most probable distance of the electron from the nucleus decreases; these electrons are less susceptible to shielding by electrons farther from the nucleus (for example, in Slater's rules electrons with greater most probable distances to the electron in question do not contribute at all to *S*). Since the most probable distance from the nucleus increases as *n* increases (Figure 2.7), the 3*d* subshell ultimately stabilizes its electrons more than the 4*s* orbital once *Z* gets sufficiently high. Regardless of the relative orbital energies, the observed electronic configuration is always the one of lowest energy. Electrons fill the lowest available orbitals in order up to their capacity, with the results shown in Figure 2.12 and in Table 2.7.

The schematic diagram in Figure 2.12(a) shows the order in which the levels fill, from bottom to top in energy. For example, Ti has two 4*s* electrons, one in each spin level, and two 3*d* electrons, both with the same spin. Fe has two 4*s* electrons, one in each spin level, five 3*d* electrons with spin $-\frac{1}{2}$, and one 3*d* electron with spin $+\frac{1}{2}$. For vanadium, the first two electrons enter the 4*s*, $-\frac{1}{2}$ and 4*s*, $+\frac{1}{2}$ levels; the next three are all in the 3*d*, $-\frac{1}{2}$ level, and vanadium has the configuration $4s^2\ 3d^3$. The 3*d*, $-\frac{1}{2}$ line crosses the 4*s*, $+\frac{1}{2}$ line between V and Cr. When the six electrons of chromium are filled in from the lowest level, chromium has the configuration $4s^1\ 3d^5$. A similar crossing gives copper its $4s^1\ 3d^{10}$ structure. This approach to electron configurations of transition metals does not depend on the stability of half-filled shells or other additional factors.

Formation of a positive ion by removal of an electron reduces shielding; the effective nuclear charge for all electrons increases dramatically. On the basis of the most probable distance effect discussed previously, $(n-1)d$ orbitals will be lower in energy than ns orbitals in the cation, as shown in Figure 2.12(b). As a result, the remaining electrons occupy the d orbitals. A common rule in introductory chemistry is that electrons with highest n—in this case, those in the s orbitals—are always removed first when ions are formed from the transition elements. A perhaps more mature perspective on this idea is that regardless of which electron is lost to form a transition metal ion, the lowest energy electron configuration of the resulting ion will always exhibit the vacancy in the ns orbital. This effect is even stronger for 2+ ions, where the effective nuclear charge is even higher. *Transition metal cations have no s electrons, only d electrons in their outer levels.*

A similar, but more complex, crossing of levels appears in the lanthanide and actinide series. The simple explanation would have these elements start filling *f* orbitals at lanthanum (57) and actinium (89), but these atoms have one d electron instead. Other elements in these series also show deviations from the "normal" sequence. Rich has explained these situations using similar diagrams.[21]

2.3 Periodic Properties of Atoms

A valuable aspect of the arrangment of atoms on the basis of similar electronic configurations within the periodic table is that an atom's position provides information about its properties. Some of these properties, and how they vary across periods and groups, are now discussed.

2.3.1 Ionization Energy

The ionization energy, also known as the *ionization potential*, is the energy required to remove an electron from a gaseous atom or ion:

$$A^{n+}(g) \longrightarrow A^{(n+1)+}(g) + e^- \qquad \text{ionization energy } (IE) = \Delta U$$

where $n = 0$ (first ionization energy), $n = 1$ (second ionization energy), and so on.

As would be expected from the effects of shielding, the ionization energy varies with different nuclei and different numbers of electrons. Trends for the first ionization energies of the early elements in the periodic table are shown in **Figure 2.13**. The general trend across a period is an increase in ionization energy as the nuclear charge increases. However, the experimental values show a break in the trend in the second period at boron and again at oxygen. Because boron is the first atom to have an electron in a higher energy $2p$ orbital that is shielded somewhat by the $2s$ electrons, boron's $2p$ electron is more easily lost than the $2s$ electrons of beryllium; boron has the lower ionization energy.

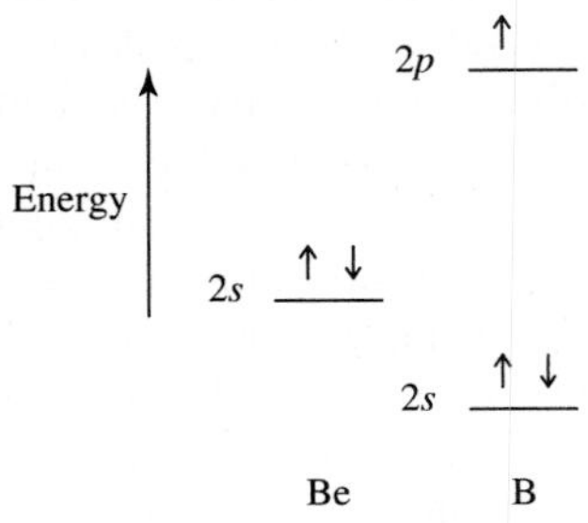

Energy
$2p$ ↑
$2s$ ↑↓
$2s$ ↑↓
Be B

At the fourth $2p$ electron, at oxygen, a similar decrease in ionization energy occurs. Here, the fourth electron shares an orbital with one of the three previous $2p$ electrons

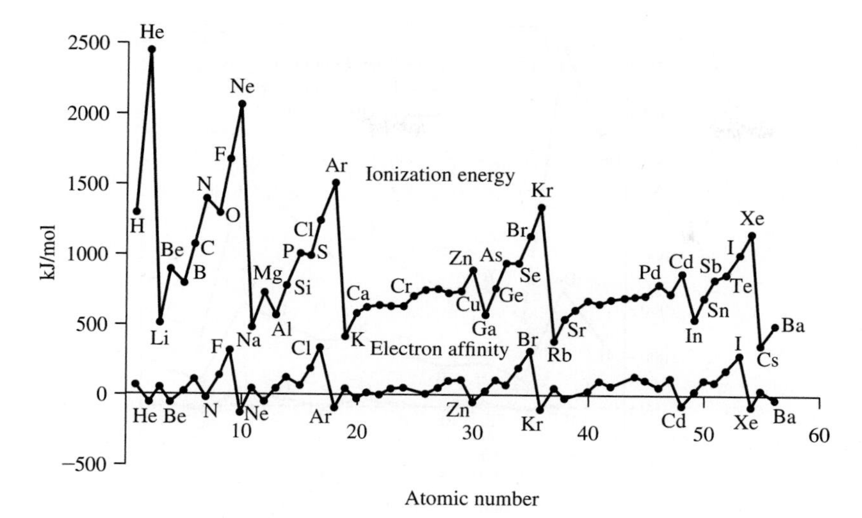

FIGURE 2.13 Ionization Energies and Electron Affinities. Ionization energy = ΔU for $M(g) \longrightarrow M^+(g) + e^-$

(Data from C. E. Moore, *Ionization Potentials and Ionization Limits, National Standards Reference Data Series*, U.S. National Bureau of Standards, Washington, DC, **1970**, NSRDS-NBS 34) Electron affinity = ΔU for $M^-(g) \longrightarrow M(g) + e^-$ Data from H. Hotop and W. C. Lineberger, *J. Phys. Chem. Ref. Data*, **1985**, *14*, 731). Numerical values are in Appendices B-2 and B-3.

($\underline{\uparrow\downarrow}$ $\underline{\uparrow}$ $\underline{\uparrow}$), and the repulsion between the paired electrons (Π_c) reduces the energy necessary to remove an electron from oxygen; oxygen has a lower ionization energy than nitrogen, which has the $2p$ configuration $\underline{\uparrow}$ $\underline{\uparrow}$ $\underline{\uparrow}$.

Similar patterns appear in the other periods, for example Na through Ar and K through Kr, omitting the transition metals. The transition metals have less dramatic differences in ionization energies, with the effects of shielding and increasing nuclear charge more nearly in balance.

Much larger decreases in ionization energy occur at the start of each new period, because the change to the next major quantum number requires that the new *s* electron have a much higher energy. The maxima at the noble gases decrease with increasing *Z*, because the outer electrons are farther from the nucleus in the heavier elements. Overall, the trends are toward higher ionization energy from left to right in the periodic table (the major change) and lower ionization energy from top to bottom (a minor change). The differences described in the previous paragraph are superimposed on these more general changes.

2.3.2 Electron Affinity

Electron affinity can be defined as the energy required to remove an electron from a negative ion:*

$$A^-(g) \longrightarrow A(g) + e^- \qquad \text{electron affinity } (EA) = \Delta U$$

Because of the similarity of this reaction to the ionization for an atom, electron affinity is sometimes described as the *zeroth ionization energy*. This reaction is endothermic (positive ΔU) except for the noble gases and the alkaline earth elements. The pattern of electron affinities with changing *Z*, shown in Figure 2.13, is similar to that of the ionization energies, but for one larger *Z* value (one more electron for each species) and with much smaller absolute numbers. For either of the reactions, removal of the first electron past a noble gas configuration is easy, so the noble gases have the lowest electron affinities. The electron affinities are all much smaller than the corresponding ionization energies, because electron removal from a negative ion (that features more shielding of the nuclear charge) is easier than removal from a neutral atom.

Comparison of the ionization and electron affinity graphs in Figure 2.13 shows similar zigzag patterns, but with the two graphs displaced by one element: for example, electron affinity shows a peak at F and valley at Ne, and ionization energy a peak at Ne and valley at

*Historically, the definition has been $-\Delta U$ for the reverse reaction, adding an electron to the neutral atom. The definition we use avoids the sign change.

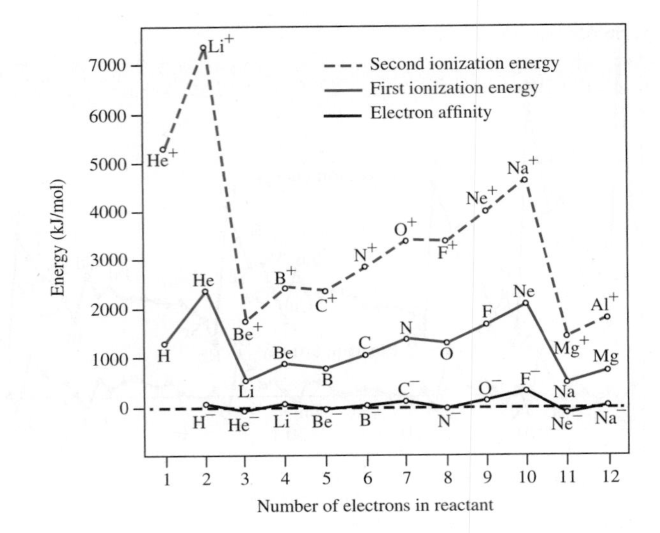

FIGURE 2.14 First and Second Ionization Energies and Electron Affinities

(©2003 Beth Abdella and Gary Miessler, reproduced with permission)

Na. The patterns in these two quantities can more easily be seen by plotting energy against the number of electrons in each reactant, as shown in **Figure 2.14** for electron affinity and first and second ionization energy.

The peaks and valleys match for all three graphs because the electron configurations match—for example, there are peaks at 10 electrons and valleys at 11 electrons. At 10 electrons, all three reactant species (F^-, Ne, and Na^+) have identical $1s^2 2s^2 2p^6$ configurations; these are by definition isoelectronic species. The relatively high energy necessary to remove an electron from these configurations is typical for configurations in which electron shells are complete. The next electron, in an 11-electron configuration, is the first to occupy a higher energy 3*s* orbital and is much more easily lost, providing a valley in each graph, corresponding to removal of an electron from the 11-electron species Ne^-, Na, and Mg^+.

EXERCISE 2.8

Explain why all three graphs in Figure 2.14 have maxima at 4 electrons and minima at 5 electrons.

2.3.3 Covalent and Ionic Radii

The sizes of atoms and ions are also related to the ionization energies and electron affinities. As the nuclear charge increases, the electrons are pulled in toward the center of the atom, and the size of any particular orbital decreases. On the other hand, as the nuclear charge increases, more electrons are added to the atom, and their mutual repulsion keeps the outer orbitals large. The interaction of these two effects, increasing nuclear charge and increasing number of electrons, results in a gradual decrease in atomic size across each period. **Table 2.8** gives nonpolar covalent radii, based on bond distances in nonpolar molecules. There are other measures of atomic size, such as the van der Waals radius, in which collisions with other atoms are used to define the size. It is difficult to obtain consistent data for any such measure, because the polarity, chemical structure, and physical state of molecules change drastically from one compound to another. The numbers shown here are sufficient for a general comparison of different elements.

There are similar challenges in determining the size of ions. Because the stable ions of the different elements have different charges and different numbers of electrons, as well as different crystal structures for their compounds, it is difficult to find a suitable set of numbers for comparison. Earlier data were based on Pauling's approach, in which the ratio of the radii of isoelectronic ions was assumed to be equal to the ratio of their effective

TABLE 2.8 Nonpolar Covalent Radii (pm)

1	2	3	4	5	6	7	8	9	10	11	12	13	14	15	16	17	18
H																	He
32																	31
Li	Be											B	C	N	O	F	Ne
123	89											82	77	75	73	71	69
Na	Mg											Al	Si	P	S	Cl	Ar
154	136											118	111	106	102	99	98
K	Ca	Sc	Ti	V	Cr	Mn	Fe	Co	Ni	Cu	Zn	Ga	Ge	As	Se	Br	Kr
203	174	144	132	122	118	117	117	116	115	117	125	126	122	120	117	114	111
Rb	Sr	Y	Zr	Nb	Mo	Tc	Ru	Rh	Pd	Ag	Cd	In	Sn	Sb	Te	I	Xe
216	191	162	145	134	130	127	125	125	128	134	148	144	140	140	136	133	126
Cs	Ba	La	Hf	Ta	W	Re	Os	Ir	Pt	Au	Hg	Tl	Pb	Bi	Po	At	Ra
235	198	169	144	134	130	128	126	127	130	134	149	148	147	146	(146)	(145)	

Source: Data from R. T. Sanderson, *Inorganic Chemistry*, Reinhold, New York, 1967, p. 74; and E. C. M. Chen, J. G. Dojahn, W. E. Wentworth, *J. Phys. Chem. A*, **1997**, *101*, 3088.

TABLE 2.9 Crystal Radii for Selected Ions

	Z	Element	Radius (pm)
Alkali metal ions	3	Li^+	90
	11	Na^+	116
	19	K^+	152
	37	Rb^+	166
	55	Cs^+	181
Alkaline earth ions	4	Be^{2+}	59
	12	Mg^{2+}	86
	20	Ca^{2+}	114
	38	Sr^{2+}	132
	56	Ba^{2+}	149
Other cations	13	Al^{3+}	68
	30	Zn^{2+}	88
Halide ions	9	F^-	119
	17	Cl^-	167
	35	Br^-	182
	53	I^-	206
Other anions	8	O^{2-}	126
	16	S^{2-}	170

Source: Data from R. D. Shannon, *Acta Crystallogr*. **1976**, *A32*, 751 for six-coordinate ions. A longer list is given in Appendix B-1.

nuclear charges. More recent calculations are based on a number of considerations, including electron density maps from X-ray data that show larger cations and smaller anions than those previously found. Those in **Table 2.9** and Appendix B-1 were called "crystal radii" by Shannon[22] and are generally different from the older values of "ionic radii" by +14 pm for cations and –14 pm for anions, as well as being revised to accommodate more recent measurements. The radii in Table 2.9 and Appendix B-1 can be used for rough estimation of the packing of ions in crystals and other calculations, as long as the "fuzzy" nature of atoms and ions is kept in mind.

Factors that influence ionic size include the coordination number of the ion, the covalent character of the bonding, distortions of regular crystal geometries, and delocalization of electrons (metallic or semiconducting character, described in Chapter 7). The radius of the anion is also influenced by the size and charge of the cation. Conversely, the anion exerts a smaller influence on the radius of the cation.[23] The table in Appendix B-1 shows the effect of coordination number.

The values in **Table 2.10** show that anions are generally larger than cations with similar numbers of electrons. The radius decreases as nuclear charge increases for ions with the same electronic structure, with the charge on cations having a strong effect, for example in the series Na^+, Mg^{2+}, Al^{3+}. Within a group, the ionic radius increases as Z increases because of the larger number of electrons in the ions and, for the same element, the radius decreases with increasing charge on the cation. Examples of these trends are shown in Tables 2.10, **2.11**, and **2.12**.

TABLE 2.10 Crystal Radius and Nuclear Charge

Ion	Protons	Electrons	Radius (pm)
O^{2-}	8	10	126
F^-	9	10	119
Na^+	11	10	116
Mg^{2+}	12	10	86
Al^{3+}	13	10	68

TABLE 2.11 Crystal Radius and Total Number of Electrons

Ion	Protons	Electrons	Radius (pm)
O^{2-}	8	10	126
S^{2-}	16	18	170
Se^{2-}	34	36	184
Te^{2-}	52	54	207

TABLE 2.12 Crystal Radius and Ionic Charge

Ion	Protons	Electrons	Radius (pm)
Ti^{2+}	22	20	100
Ti^{3+}	22	19	81
Ti^{4+}	22	18	75

References

1. John Dalton, *A New System of Chemical Philosophy*, 1808; reprinted with an introduction by Alexander Joseph, Peter Owen Limited, London, 1965.
2. Ibid., p. 113.
3. Ibid., p. 133.
4. J. R. Partington, *A Short History of Chemistry*, 3rd ed., Macmillan, London, 1957; reprinted, 1960, Harper & Row, New York, p. 255.
5. Ibid., pp. 256–258.
6. D. I. Mendeleev, *J. Russ. Phys. Chem. Soc.*, **1869**, *i*, 60.
7. L. Meyer, *Justus Liebigs Ann. Chem.*, **1870**, *Suppl. vii*, 354.
8. N. Bohr, *Philos. Mag.*, **1913**, *26*, 1.
9. G. Herzberg, *Atomic Spectra and Atomic Structure*, 2nd ed., Dover Publications, New York, 1994, p. 18.
10. L. de Broglie, *Philos. Mag.*, **1924**, *47*, 446; *Ann. Phys. Paris,* **1925**, *3*, 22.
11. W. Heisenberg, *Z. Phys.*, **1927**, *43*, 172.
12. Ralchenko, Yu., Kramida, A.E., Reader, J., and NIST ASD Team (2011). *NIST Atomic Spectra Database* (ver. 4.1.0), [Online]. Available: *http://physics.nist.gov/asd* [2012, August 21]. National Institute of Standards and Technology, Gaithersburg, MD.
13. E. Schrödinger, *Ann. Phys. (Leipzig)*, **1926**, *79*, 361, 489, 734; **1926**, *80*, 437; **1926**, *81*, 109; *Naturwissenshaften*, **1926**, *14*, 664; *Phys. Rev.*, **1926**. *28*, 1049.
14. R. E. Powell, *J. Chem. Educ.*, **1968**, *45*, 45.
15. W. Pauli, *Z. Physik*, **1925**, *31*, 765.
16. F. Hund, *Z. Physik*, **1925**, *33*, 345.
17. J.C. Slater. *Phys. Rev.*, **1930**, *36*, 57.
18. L. G. Vanquickenborne, K. Pierloot, D. Devoghel, *J. Chem. Ed.*, 1994, 71, 468
19. J. L. Reed, *J. Chem. Educ.*, **1999**, *76*, 802.
20. W. Eek, S. Nordholm, G. B. Bacskay, *Chem. Educator*, **2006**, *11*, 235.
21. R. L. Rich, *Periodic Correlations*, W. A. Benjamin, Menlo Park, CA, 1965, pp. 9–11.
22. R. D. Shannon, *Acta Crystallogr.*, **1976**, A32, 751.
23. O. Johnson, *Inorg. Chem.*, **1973**, *12*, 780.

General References

Additional information on the history of atomic theory can be found in J. R. Partington, *A Short History of Chemistry*, 3rd ed., Macmillan, London, 1957, reprinted by Harper & Row, New York, 1960, and in the *Journal of Chemical Education*. For an introduction to atomic theory and orbitals, see V. M. S. Gil, *Orbitals in Chemistry: A Modern Guide for Students*, Cambridge University Press, Cambridge, 2000, UK, pp. 1–69. A more thorough treatment of the electronic structure of atoms is in M. Gerloch, *Orbitals, Terms, and States*, John Wiley & Sons, New York, 1986. Many Internet sites provide images of atomic orbitals, their wave equations, nodal behavior, and other characteristics. Two examples are http://www.orbitals.com and http://winter.group.shef.ac.uk/orbitron.

Problems

2.1 Determine the de Broglie wavelength of
- **a.** an electron moving at 1/10 the speed of light.
- **b.** a 400 g Frisbee moving at 10 km/h.
- **c.** an 8.0-pound bowling ball rolling down the lane with a velocity of 2.0 meters per second.
- **d.** a 13.7 g hummingbird flying at a speed of 30.0 miles per hour.

2.2 Using the equation $E = R_H\left(\frac{1}{2^2} - \frac{1}{n_h^2}\right)$, determine the energies and wavelengths of the visible emission bands in the atomic spectrum of hydrogen arising from $n_h = 4, 5$, and 6. (The red line, corresponding to $n_h = 3$, was calculated in Exercise 2.1.)

2.3 The transition from the $n = 7$ to the $n = 2$ level of the hydrogen atom is accompanied by the emission of radiation slightly beyond the range of human perception, in the ultraviolet region. Determine the energy and wavelength.

2.4 Emissions are observed at wavelengths of 383.65 and 379.90 nm for transitions from excited states of the hydrogen atom to the $n = 2$ state. Determine the quantum numbers n_h for these emissions.

2.5 What is the least amount of energy that can be emitted by an excited electron in a hydrogen atom falling from an excited state directly to the $n = 3$ state? What is the quantum number n for the excited state? Humans cannot visually observe the photons emitted in this process. Why not?

2.6 Hydrogen atom emission spectra measured from the solar corona indicated that the $4s$ orbital was 102823.8530211 cm^{-1}, and $3s$ orbital 97492.221701 cm^{-1}, respectively, above the $1s$ ground state. (These energies have tiny uncertainties, and can be treated as exact numbers for the sake of this problem.) Calculate the difference in energy (J) between these levels on the basis of these data, and compare this difference to that obtained by the Balmer equation in Section 2.1.2. How well does the Balmer equation work for hydrogen? (Data from Y. Ralchenko, A. E. Kramida, J. Reader, and NIST ASD Team (2011). *NIST Atomic Spectra Database* (ver. 4.1.0), [Online]. Available: http:// physics.nist.gov/ asd [2012, January 18]. National Institute of Standards and Technology, Gaithersburg, MD.)

2.7 The Rydberg constant equation has two terms that vary depending on the species under consideration, the reduced mass of the electron/nucleus combination and the charge of the nucleus (*Z*).

a. Determine the approximate *ratio* between the Rydberg constants for isoelectronic He^+ (consider the most abundant helium-4 isotope) and H. The masses of the electron, proton, and that of the He^+ nucleus (He^{2+} is an α particle) are given on the inside back cover of this text.

b. Use this ratio to calculate an approximate Rydberg constant (*J*) for He^+.

c. The difference between the He^+ 2*s* and 1*s* orbitals was reported as 329179.76197(20) cm^{-1}. Calculate the He^+ Rydberg constant from this spectral line for comparison to your value from **b.**

(Data from the same reference as Problem 2.6.)

2.8 The details of several steps in the particle-in-a-box model in this chapter have been omitted. Work out the details of the following steps:

a. Show that if $\Psi = A \sin rx + B \cos sx$ (*A*, *B*, *r*, and *s* are constants) is a solution to the wave equation for the one-dimensional box, then

$$r = s = \sqrt{2mE}\left(\frac{2\pi}{h}\right)$$

b. Show that if $\Psi = A \sin rx$, the boundary conditions ($\Psi = 0$ when $x = 0$ and $x = a$) require that $r = \pm\frac{n\pi}{a}$, where $n =$ any integer other than zero.

c. Show that if $r = \pm\frac{n\pi}{a}$, the energy levels of the particle are given by $E = \frac{n^2h^2}{8ma^2}$

d. Show that substituting the value of *r* given in part c into $\Psi = A \sin rx$ and applying the normalizing requirement gives $A = \sqrt{2/a}$.

2.9 For the $3p_z$ and $4d_{xz}$ hydrogen-like atomic orbitals, sketch the following:

a. The radial function *R*

b. The radial probability function $a_0r^2R^2$

c. Contour maps of electron density.

2.10 Repeat the exercise in Problem 2.9, for the 4*s* and $5d_{x^2-y^2}$ orbitals.

2.11 Repeat the exercise in Problem 2.9, for the 5*s* and $4d_{z^2}$ orbitals.

2.12 The $4f_{z(x^2-y^2)}$ orbital has the angular function $Y = (\text{constant})\ z(x^2 - y^2)/r^3$.

a. How many radial nodes does this orbital have?

b. How many angular nodes does it have?

c. Write equations to define the angular nodal surfaces. What shapes are these surfaces?

d. Sketch the shape of the orbital, and show all radial and angular nodes.

2.13 Repeat the exercise in Problem 2.12, for the $5f_{xyz}$ orbital, which has $Y = (\text{constant})\ xyz/r^3$.

2.14 The label for an f_{z^3} orbital, like that for a d_{z^2} orbital, is an abbreviation. The actual angular function for this orbital is $Y = (\text{constant}) \times z(5z^2 - 3r^2)/r^3$. Repeat the exercise in Problem 2.12, for a $4f_{z^3}$ orbital. (*Note*: recall that $r^2 = x^2 + y^2 + z^2$).

2.15 **a.** Determine the possible values for the *l* and m_l quantum numbers for a 5*d* electron, a 4*f* electron, and a 7*g* electron.

b. Determine the possible values for all four quantum numbers for a 3*d* electron.

c. What values of m_l are possible for *f* orbitals?

d. At most, how many electrons can occupy a 4*d* orbital?

2.16 **a.** What are the values of quantum numbers *l* and *n* for a 5*d* electron?

b. At most, how many 4*d* electrons can an atom have? Of these electrons how many, at most, can have $m_s = -\frac{1}{2}$?

c. A 5*f* electron has what value of quantum number *l*? What values of m_l may it have?

d. What values of the quantum number m_l are possible for a subshell having $l = 4$?

2.17 **a.** At most, how many electrons in an atom can have both $n = 5$ and $l = 3$?

b. A 5*d* electron has what possible values of the quantum number m_l?

c. What value of quantum number *l* do *p* orbitals have? For what values of *n* do *p* orbitals occur?

d. What is the quantum number *l* for *g* orbitals? How many orbitals are in a *g* subshell?

2.18 Determine the Coulombic and exchange energies for the following states, and determine which state is favored (has lower energy):

a. ↑ ↑ and ↑↓ ___

b. ↑ ↑ ↑ and ↑↓ ↑ ___

2.19 Two excited states for a d^4 configuration are shown. Which is likely to have lower energy? Explain your choice in terms of Coulombic and exchange energies.

W: ↑ ↑ ↓ ↓ ___

X: ↑ ↑ ↑ ↓ ___

2.20 Two excited states for a d^5 configuration are shown. Which is likely to have lower energy? Why? Explain your choice in terms of Coulombic and exchange energies.

Y: ↑ ↑ ↑ ↓ ↓

Z: ↑ ↑ ↑ ↑ ↓

2.21 What states are possible for a d^3 configuration? Determine the Coulombic and exchange energies for each, and rank the states in terms of relative energy.

2.22 Provide explanations of the following phenomena:

a. The electron configuration of Cr is [Ar] $4s^1 3d^5$ rather than [Ar] $4s^2 3d^4$.

b. The electron configuration of Ti is [Ar] $4s^2 3d^2$, but that of Cr^{2+} is [Ar] $3d^4$.

2.23 Give electron configurations for the following:

a. V

b. Br

c. Ru^{3+}

d. Hg^{2+}

e. Sb

2.24 Predict the electron configurations of the following metal anions:

a. Rb^-

b. Pt^{2-} (See: A. Karbov, J. Nuss, U. Weding, M. Jansen, *Angew. Chem. Int. Ed.*, **2003**, *42*, 4818.)

2.25 Radial probability plots shed insight on issues of shielding and effective nuclear charge. Interpret the radial probability functions in Figure 2.7 to explain why the general order of orbital filling is $n = 1$, followed by $n = 2$, followed by $n = 3$. Interpret the graphs for 3*s*, 3*p*, and 3*d* to rationalize the filling order for these orbitals.

2.26 Briefly explain the following on the basis of electron configurations:

a. Fluorine forms an ion having a charge of 1–.

b. The most common ion formed by zinc has a 2+ charge.

c. The electron configuration of the molybdenum atom is [Kr] $5s^1 4d^5$ rather than [Kr] $5s^2 4d^4$.

2.27 Briefly explain the following on the basis of electron configurations:

a. The most common ion formed by silver has a 1+ charge.

b. Cm has the outer electron configuration $s^2d^1 f^7$ rather than $s^2 f^8$.

c. Sn often forms an ion having a charge of 2+ (the *stannous* ion).

2.28 **a.** Which 2+ ion has two 3*d* electrons? Which has eight 3*d* electrons?

b. Which is the more likely configuration for Mn^{2+}: [Ar] $4s^2 3d^3$ or [Ar] $3d^5$?

2.29 Using Slater's rules, determine *Z** for

a. a 3*p* electron in P, S, Cl, and Ar. Is the calculated value of *Z** consistent with the relative sizes of these atoms?

b. a 2*p* electron in O^{2-}, F^-, Na^+ and Mg^{2+}. Is the calculated value of *Z** consistent with the relative sizes of these ions?

c. a 4*s* and a 3*d* electron of Cu. Which type of electron is more likely to be lost when copper forms a positive ion?

d. a 4*f* electron in Ce, Pr, and Nd. There is a decrease in size, commonly known as the **lanthanide contraction,** with increasing atomic number in the lanthanides. Are your values of *Z** consistent with this trend?

2.30 A sample calculation in this chapter showed that, according to Slater's rules, a 3*d* electron of nickel has a higher effective nuclear charge than a 4*s* electron. Is the same true for early first-row transition metals? Using Slater's rules, calculate *S* and *Z** for 4*s* and 3*d* electrons of Sc and Ti, and comment on the similarities or differences with Ni.

2.31 Ionization energies should depend on the effective nuclear charge that holds the electrons in the atom. Calculate *Z** (Slater's rules) for N, P, and As. Do their ionization energies seem to match these effective nuclear charges? If not, what other factors influence the ionization energies?

2.32 Prepare a diagram such as the one in Figure 2.12(a) for the fifth period in the periodic table, elements Zr through Pd. The configurations in Table 2.7 can be used to determine the crossover points of the lines. Can a diagram be drawn that is completely consistent with the configurations in the table?

2.33 Why are the ionization energies of the alkali metals in the order Li >, Na > K > Rb?

2.34 The second ionization of carbon ($C^+ \longrightarrow C^{2+} + e^-$) and the first ionization of boron ($B \longrightarrow B^+ + e^+$) both fit the reaction $1s^2 2s^2 2p^1 \longrightarrow 1s^2 2s^2 + e^-$. Compare the two ionization energies (24.383 eV and 8.298 eV, respectively) and the effective nuclear charge *Z**. Is this an adequate explanation of the difference in ionization energies? If not, suggest other factors.

2.35 Explain why all three graphs in Figure 2.14 have maxima at 4 electrons and minima at 5 electrons.

2.36 **a.** For a graph of third ionization energy against atomic number, predict the positions of peaks and valleys for elements through atomic number 12. Compare the positions of these peaks and valleys with those for first ionization energies shown in Figure 2.13.

b. How would a graph of third ionization energies against the number of electrons in reactant compare with the other graphs shown in Figure 2.14? Explain briefly.

2.37 The second ionization energy involves removing an electron from a positively charged ion in the gas phase (see preceding problem). How would a graph of second ionization energy vs. atomic number for the elements helium through neon compare with the graph of first ionization energy in Figure 2.13? Be specific in comparing the positions of peaks and valleys.

2.38 In each of the following pairs, pick the element with the higher ionization energy and explain your choice.

a. Fe, Ru

b. P, S

c. K, Br

d. C, N

e. Cd, In

f. Cl, F

2.39 On the basis of electron configurations, explain why
a. sulfur has a lower electron affinity than chlorine.
b. iodine has a lower electron affinity than bromine.
c. boron has a lower ionization energy than beryllium.
d. sulfur has a lower ionization energy than phosphorus.

2.40 **a.** The graph of ionization energy versus atomic number for the elements Na through Ar (Figure 2.13) shows maxima at Mg and P and minima at Al and S. Explain these maxima and minima.
b. The graph of electron affinity versus atomic number for the elements Na through Ar (Figure 2.13) also shows maxima and minima, but shifted by one element in comparison with the ionization energy graph. Why are the maxima and minima shifted in this way?

2.41 The second ionization energy of He is almost exactly four times the ionization energy of H, and the third ionization energy of Li is almost exactly nine times the ionization energy of H:

	IE (MJ mol^{-1})
$H(g) \longrightarrow H^+(g) + e^-$	1.3120
$He^+(g) \longrightarrow He^{2+}(g) + e^-$	5.2504
$Li^{2+}(g) \longrightarrow Li^{3+}(g) + e^-$	11.8149

Explain this trend on the basis of the Bohr equation for energy levels of single-electron systems.

2.42 The size of the transition-metal atoms decreases slightly from left to right in the periodic table. What factors must be considered in explaining this decrease? In particular, why does the size decrease at all, and why is the decrease so gradual?

2.43 Predict the largest and smallest radius in each series, and account for your choices:
a. Se^{2-} Br^- Rb^+ Sr^{2+}
b. Y^{3+} Zr^{4+} Nb^{5+}
c. Co^{4+} Co^{3+} Co^{2+} Co

2.44 Select the best choice, and briefly indicate the reason for each choice:
a. Largest radius: Na^+ Ne F^-
b. Greatest volume: S^{2-} Se^{2-} Te^{2-}
c. Highest ionization energy: Na Mg Al
d. Most energy necessary to remove an electron: Fe Fe^{2+} Fe^{3+}
e. Highest electron affinity: O F Ne

2.45 Select the best choice, and briefly indicate the reason for your choice:
a. Smallest radius: Sc Ti V
b. Greatest volume: S^{2-} Ar Ca^{2+}
c. Lowest ionization energy: K Rb Cs
d. Highest electron affinity: Cl Br I
e. Most energy necessary to remove an electron: Cu Cu^+ Cu^{2+}

2.46 There are a number of Web sites that display atomic orbitals. Use a search engine to find a complete set of the *f* orbitals.
a. How many orbitals are there in one set (for example, a set of 4*f* orbitals)?
b. Describe the angular nodes of the orbitals.
c. Observe what happens to the number of radial nodes as the principal quantum number is increased.
d. Include the URL for the site you used for each, along with sketches or printouts of the orbitals. (Two useful Web sites at this writing are orbitals.com and winter.group.shef.ac.uk/orbitron.)

2.47 Repeat the exercise in Problem 2.46, this time for a set of *g* orbitals.

Chapter 3

Simple Bonding Theory

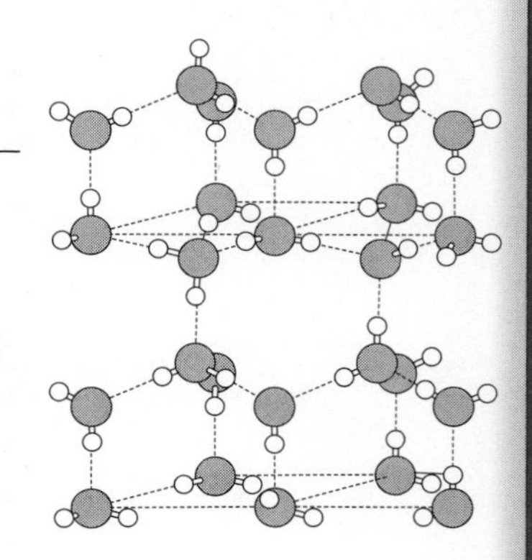

We now turn from the use of quantum mechanics and its description of the atom to an elementary description of molecules. Although most of our discussion of chemical bonding uses the molecular orbital approach, less rigorous methods that provide approximate pictures of the shapes and polarities of molecules are also useful. This chapter provides an overview of Lewis dot structures, valence shell electron-pair repulsion (VSEPR), and related topics. Molecular orbital descriptions of some of the same molecules are presented in Chapter 5 and later chapters; the ideas of this chapter provide a starting point for that more modern treatment.

Ultimately, any description of bonding must be consistent with experimental data on bond lengths, bond angles, and bond strengths. Angles and distances are most frequently determined by diffraction (X-ray crystallography, electron diffraction, neutron diffraction) or spectroscopic (microwave, infrared) methods. For many molecules, there is general agreement on the nature of the bonding, although there are alternative ways to describe it. For others, there is considerable difference of opinion on the best way to describe the bonding. In this chapter and in Chapter 5, we describe some useful qualitative approaches, including some of the opposing views.

3.1 Lewis Electron-Dot Diagrams

Lewis electron-dot diagrams, although oversimplified, provide a good starting point for analyzing the bonding in molecules. Credit for their initial use goes to G. N. Lewis,[1] an American chemist who contributed much to the understanding of thermodynamics and chemical bonding in the early twentieth century. In Lewis diagrams, bonds between two atoms exist when they share one or more pairs of electrons. In addition, some molecules have nonbonding pairs, also called *lone pairs*, of electrons on atoms. These electrons contribute to the shape and reactivity of the molecule but do not directly bond the atoms together. Most Lewis structures are based on the concept that eight **valence electrons**, corresponding to s and p electrons outside the noble gas core, form a particularly stable arrangement, as in the noble gases with s^2p^6 configurations. An exception is hydrogen, which is stable with two valence electrons. Also, some molecules require more than eight electrons around a given central atom, and some molecules require fewer than eight electrons.

Simple molecules such as water follow the **octet rule**, in which eight electrons surround the central atom. Each hydrogen atom shares two electrons with the oxygen, forming the familiar structure with two bonds; the O atom accommodates two bonding pairs and two lone pairs:*

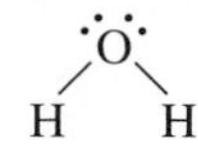

*We will see in Chapter 5 that the treatment of water via molecular orbital theory results in an electronic structure in which each of these electron pairs has a unique energy. This model is supported by spectroscopic evidence, and indicates one limitation of the Lewis model.

Shared electrons are considered to contribute to the electronic requirements of both atoms involved; thus, the electron pairs shared by H and O in the water molecule are counted toward both the 8-electron requirement of oxygen and the 2-electron requirement of hydrogen.

The Lewis model defines double bonds as containing four electrons and triple bonds as containing six electrons:

$$:\ddot{O}=C=\ddot{O}: \qquad H-C\equiv C-H$$

3.1.1 Resonance

In many Lewis structures, the choice of which atoms are connected by multiple bonds is arbitrary. When alternate locations for single bonds and multiple bonds are possible that all afford valid Lewis structures, a structure demonstrating each option should be drawn. For example, three drawings (resonance structures) of CO_3^{2-} are needed (**Figure 3.1**) to show the double bond in each of the three possible C—O positions. In fact, experimental evidence shows that all three C—O bonds are equivalent, with bond lengths (129 pm) between typical C—O double-bond and single-bond distances (116 pm and 143 pm, respectively). All three drawings are necessary to describe the structure, with each drawing contributing equally to describe the bonding in the actual ion. This is called **resonance**; there is more than one possible way in which the valence electrons can be placed in a Lewis structure. Note that in resonance structures, such as those shown for CO_3^{2-} in Figure 3.1, the electrons are arranged differently, but the nuclei remain in fixed positions.

The species CO_3^{2-} and NO_3^- have the same number of electrons (*i.e.*, they are **isoelectronic**) and use the same orbitals for bonding. Their Lewis diagrams are identical except for the identity and formal charge (Section 3.1.3) of the central atom.

When a molecule has several resonance structures, its overall electronic energy is lowered, making it more stable. Just as the energy levels of a particle in a box are lowered by making the box larger, the electronic energy levels of the bonding electrons are lowered when the electrons can occupy a larger space. The molecular orbital description of this effect is presented in Chapter 5.

3.1.2 Higher Electron Counts

When it is impossible to draw a structure consistent with the octet rule because additional valence electrons remain to be assigned after the octet rule is satisfied on all atoms, it is necessary to increase the number of electrons around the central atom. An option limited to elements of the third and higher periods is to use *d* orbitals for this expansion, although theoretical work suggests that expansion beyond the *s* and *p* orbitals is unnecessary for most main group molecules.[2] In most cases, two or four added electrons will complete the bonding, but more can be added if necessary. For example, 10 electrons are required around chlorine in ClF_3 and 12 around sulfur in SF_6 (**Figure 3.2**). The increased number of electrons is often described as an *expanded shell* or an *expanded electron count*. The term **hypervalent** is used to describe central atoms that have electron counts greater than the atom's usual requirement.

There are examples with even more electrons around the central atom, such as IF_7 (14 electrons), $[TaF_8]^{3-}$ (16 electrons), and $[XeF_8]^{2-}$ (18 electrons). There are rarely more than 18 electrons (2 for *s*, 6 for *p*, and 10 for *d* orbitals) around a single atom in the top half of the periodic table, and crowding of the outer atoms usually keeps the number below this, even for much heavier atoms that have *f* orbitals energetically available.

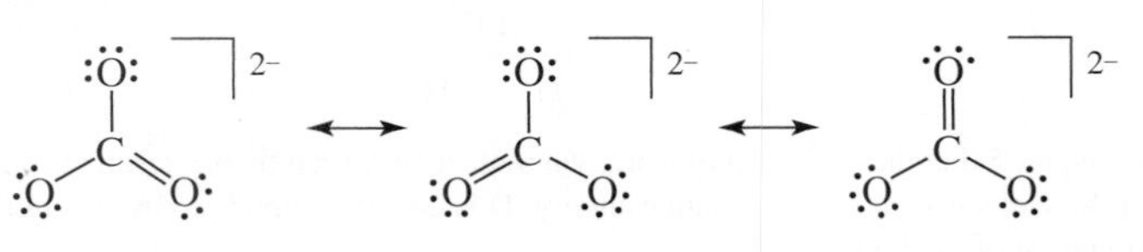

FIGURE 3.1 Lewis Diagrams for CO_3^{2-}.

3.1.3 Formal Charge

Formal charge is the apparent electronic charge of each atom in a molecule, based on the electron-dot structure. Formal charges help assess resonance structures and molecular topology, and they are presented here as a simplified method of describing structures, just as the Bohr model is a simple method of describing electronic configurations in atoms. Both of these methods have limitations, and other approaches are more accurate, but they can be useful as long as their imperfections are kept in mind.

Formal charges can help in eliminating resonance structures expected to contribute very little to the electronic ground state of the molecule, and, in some cases, suggesting multiple bonds beyond those required by the octet rule. It is essential, however, to remember that formal charge is only a tool for assessing Lewis structures, not a measure of any actual charge on the atoms. The number of valence electrons available in a free atom of an element minus the total for that atom in the molecule—determined by counting lone pairs as two electrons and bonding pairs as one electron assigned to each atom—is the formal charge on the atom:

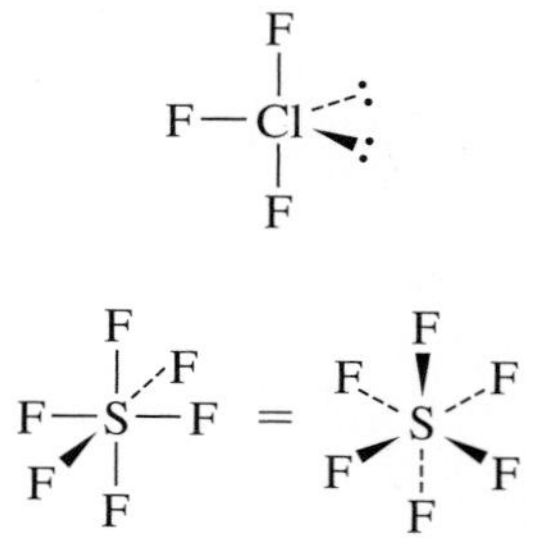

FIGURE 3.2 Structures of ClF_3 and SF_6.

$$\text{Formal charge} = \begin{pmatrix}\text{number of valence}\\\text{electrons in a free}\\\text{atom of the element}\end{pmatrix} - \begin{pmatrix}\text{number of unshared}\\\text{electrons on the atom}\end{pmatrix} - \begin{pmatrix}\text{number of bonds}\\\text{to the atom}\end{pmatrix}$$

In addition,

$$\text{Charge on molecule or ion} = \text{sum of formal charges}$$

Resonance structures that contribute more to the electronic ground state of the species generally (a) have smaller magnitudes of formal charges, (b) place negative formal charges on more electronegative elements (in the upper right-hand part of the periodic table), and (c) have smaller separation of charges. Three examples—SCN^-, OCN^-, and CNO^-—will illustrate the use of formal charges in describing electronic structures.

EXAMPLE 3.1

SCN^-

In the thiocyanate ion, SCN^-, three resonance structures are consistent with the electron-dot method, as shown in **Figure 3.3**. Structure A has only one negative formal charge on the nitrogen atom, the most electronegative atom in the ion. Structure B has a single negative charge on the S, which is less electronegative than N. Structure C has charges of 2− on N and 1+ on S, consistent with the relative electronegativities of these atoms but also has a large magnitude 2− charge and greater charge separation than the other structures. Therefore these structures lead to the prediction that structure A contributes the most to the electronic ground state of SCN^-, structure B contributes an intermediate amount, and any contribution from C is minor in describing the electronic ground state of SCN^-.

The bond lengths in **Table 3.1** are somewhat consistent with this conclusion, with SCN^- bond lengths between those of structures A and B. Protonation of the ion forms HNCS, consistent with a negative charge on N in SCN^-. The bond lengths in HNCS are close to those of double bonds, consistent with the structure H—N=C=S.

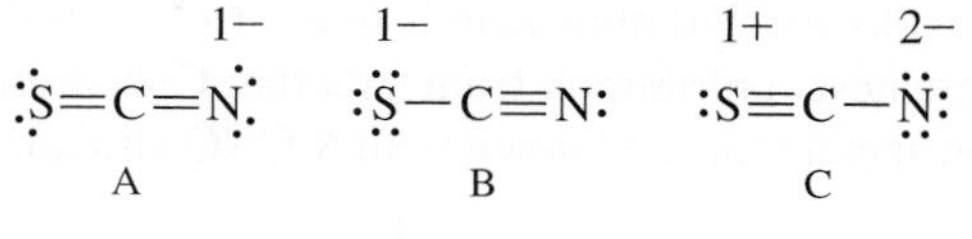

FIGURE 3.3 Resonance Structures of Thiocyanate, SCN^-.

TABLE 3.1 Table of S—C and C—N Bond Lengths (pm)

	S—C	C—N
SCN^- (in NaSCN)	165	118
HNCS	156	122
Single bond	181	147
Double bond	155	128 (approximate)
Triple bond		116

Data from A. F. Wells, *Structural Inorganic Chemistry*, 5th ed., Oxford University Press, New York, 1984, pp. 807, 926, 934–936.

EXAMPLE 3.2

OCN^-

The isoelectronic cyanate ion, OCN^- (**Figure 3.4**), has the same possibilities, but the larger electronegativity of O is expected to make structure B contribute more to the electronic ground state in cyanate relative the contribution of B in thiocyanate. The protonation of cyanate results in two isomers, 97% HNCO and 3% HOCN, consistent with a major contribution of structure A and a small, but significant, contribution from B. The bond lengths in OCN^- and HNCO in **Table 3.2** are reasonably consistent with this analysis. Formal charge arguments provide a good starting point to assess Lewis structures, and reactivity patterns are also useful to gain experimental insight about electron distributions.

A: $\ddot{O}=C=\ddot{N}$ (formal charge 1− on N)

B: $\ddot{O}-C\equiv N$ (formal charge 1− on O)

C: $O\equiv C-\ddot{N}$ (formal charges 1+ on O, 2− on N)

FIGURE 3.4 Resonance Structures of Cyanate, OCN^-.

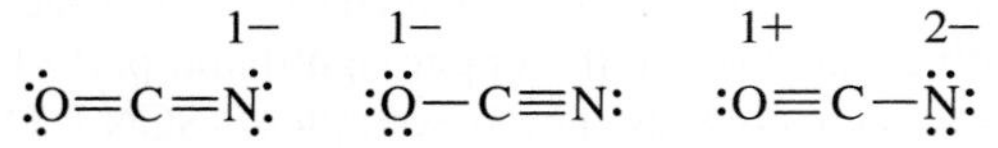

TABLE 3.2 Table of O—C and C—N Bond Lengths (pm)

	O—C	C—N
OCN^-	126	117
HNCO	118	120
Single bond	143	147
Double bond	116 (CO_2)	128 (approximate)
Triple bond	113 (CO)	116

Data from A. F. Wells, *Structural Inorganic Chemistry*, 5th ed., Oxford University Press, New York, 1984, pp. 807, 926, 933–934; S. E. Bradforth, E. H. Kim, E. W. Arnold, D. M. Neumark, *J. Chem. Phys*., **1993**, *98*, 800.

EXAMPLE 3.3

CNO^-

The isomeric fulminate ion, CNO^- (**Figure 3.5**), can be drawn with three similar structures, but the resulting formal charges have larger magnitudes than in OCN^-. Because the order of electronegativities is $C < N < O$, none of these are ideal structures, and it is not surprising that this ion is unstable. The only common fulminate salts are of mercury and silver; both are explosive. Fulminic acid is linear HCNO in the vapor phase, consistent with the greatest contribution from structure C; coordination complexes of CNO^- with transition-metal ions are known with MCNO structures.[3]

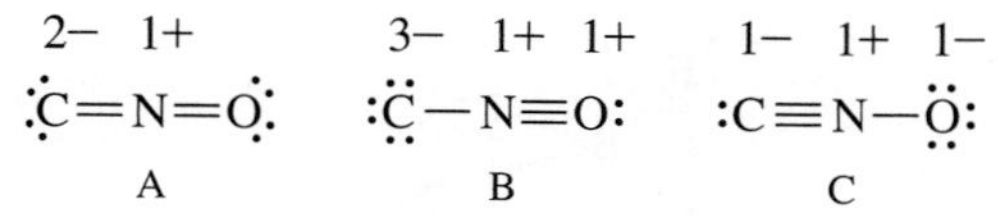

2− 1+	3− 1+ 1+	1− 1+ 1−
:C=N=O:	:C−N≡O:	:C≡N−O:
A	B	C

FIGURE 3.5 Resonance Structures of Fulminate, CNO^-.

EXERCISE 3.1 Use electron-dot diagrams and formal charges to predict the bond order for each bond in POF_3, SOF_4, and SO_3F^-.

Some molecules have satisfactory electron-dot structures with octets but have more reasonable formal charge distributions in their structures with expanded electron counts. In each of the cases in **Figure 3.6**, the actual molecules and ions are consistent with electron counts greater than 8 on the central atom and with a large contribution from the resonance structure that uses multiple bonds to minimize formal charges. The multiple bonds may also influence the shapes of the molecules.

3.1.4 Multiple Bonds in Be and B Compounds

A few molecules—such as BeF_2, $BeCl_2$, and BF_3—seem to require multiple bonds to satisfy the octet rule for Be and B, even though multiple bonds for F and Cl are not generally expected on the basis of the high electronegativities of these halogens. Structures minimizing formal charges for these molecules have only four electrons in the valence shell of Be and six electrons in the valence shell of B, in both cases fewer than the usual octet. The alternative, requiring eight electrons on the central atom, predicts multiple bonds, with BeF_2 analogous to CO_2 and BF_3 analogous to SO_3 (**Figure 3.7**). These structures, however, result in nonideal formal charges (2− on Be and 1+ on F in BeF_2, and 1− on B and 1+ on the double-bonded F in BF_3) on the basis of the usual rules.

<table>
<tr><th></th><th>Octet</th><th></th><th></th><th>Expanded</th><th></th><th></th><th></th></tr>
<tr><th>Molecule</th><th></th><th>Atom</th><th>Formal Charge</th><th></th><th>Atom</th><th>Formal Charge</th><th>Expanded to:</th></tr>
<tr><td>SNF_3</td><td>:N:
|
:F−S−F:
|
:F:</td><td>S
N</td><td>2+
2−</td><td>N
|||
:F−S−F:
|
:F:</td><td>S
N</td><td>0
0</td><td>12</td></tr>
<tr><td>SO_2Cl_2</td><td>:O:
|
:Cl−S−O:
|
:Cl:</td><td>S
O</td><td>2+
1−</td><td>O
||
:Cl−S=O:
|
:Cl:</td><td>S
O</td><td>0
0</td><td>12</td></tr>
<tr><td>XeO_3</td><td>:O:
|
:O−Xe−O:</td><td>Xe
O</td><td>3+
1−</td><td>O
||
:O=Xe=O:</td><td>Xe
O</td><td>0
0</td><td>14</td></tr>
<tr><td>SO_3^{2-}</td><td>:O:
|
:O−S−O:</td><td>S
O</td><td>1+
1−</td><td>O
||
:O−S−O:</td><td>S
O</td><td>0
0,1−</td><td>10</td></tr>
</table>

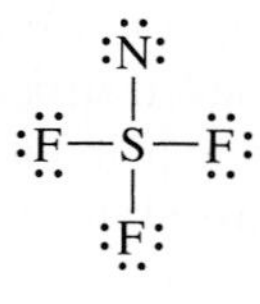

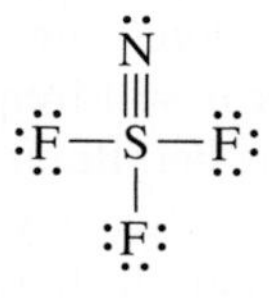

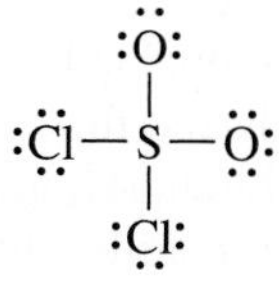

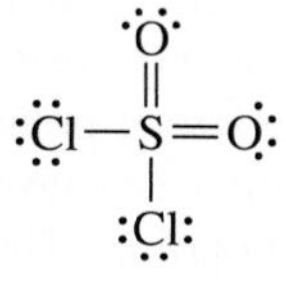

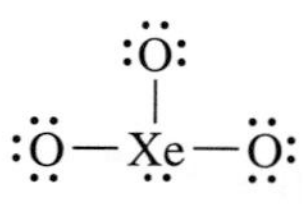

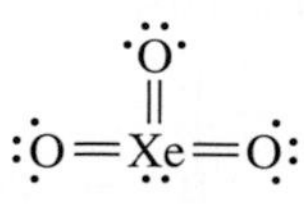

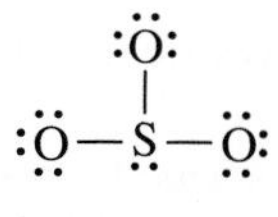

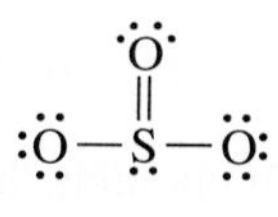

FIGURE 3.6 Formal Charge and Expanded Electron Counts on Central Atom.

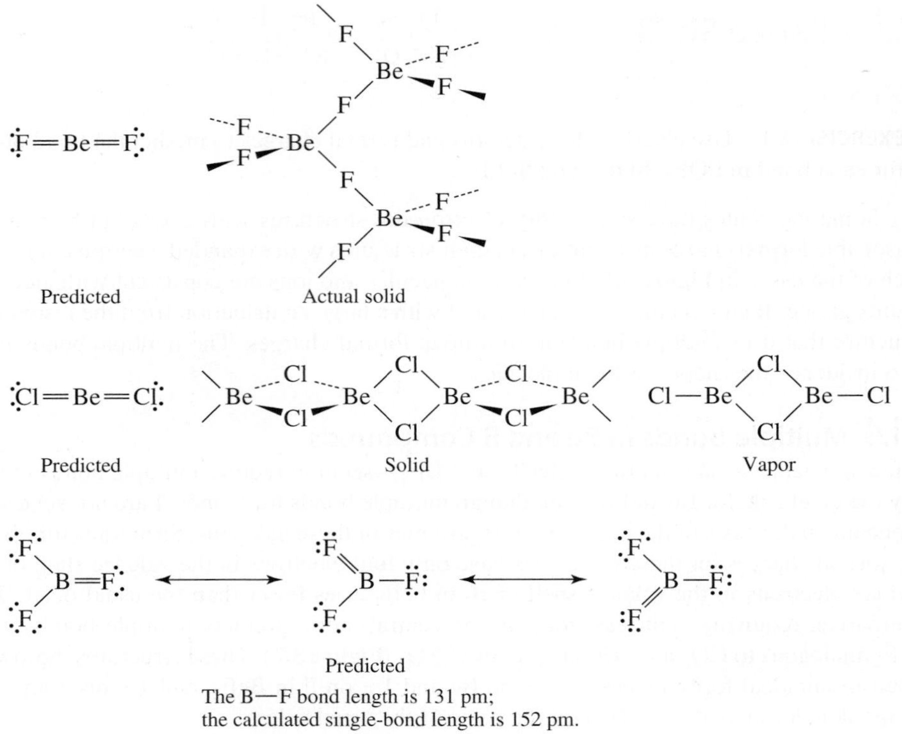

FIGURE 3.7 Structures of BeF_2, $BeCl_2$, and BF_3. (Data from A. F. Wells, *Structural Inorganic Chemistry*, 5th ed., Oxford University Press, Oxford, England, 1984, pp. 412, 1047.)

Predicted   Actual solid

Predicted   Solid   Vapor

Predicted

The B—F bond length is 131 pm;
the calculated single-bond length is 152 pm.

In solid BeF_2, a complex network is formed with a Be atom coordination number of 4 (see Figure 3.7). $BeCl_2$ dimerizes to a 3-coordinate structure in the vapor phase, but the linear monomer is formed at high temperatures. This monomeric structure is unstable due to the electronic deficiency at Be; in the dimer and the network formed in the solid-state, the halogen atoms share lone pairs with the Be atom in an attempt to fill beryllium's valence shell. The monomer is still frequently drawn as a singly bonded structure, with only four electrons around the beryllium and the ability to accept lone pairs of other molecules to relieve its electronic deficiency (Lewis acid behavior, discussed in Chapter 6).

Bond lengths in all the boron trihalides are shorter than expected for single bonds, so the partial double-bond character predicted seems reasonable despite the nonideal formal charges of these resonance forms. While a small amount of double bonding is possible in these molecules, the strong polarity of the B–halogen bonds and the ligand close-packing (LCP) model (Section 3.2.4) have been used to account for the short bonds without the need to invoke multiple bonding. The boron trihalides combine readily with other molecules that can contribute a lone pair of electrons (Lewis bases), forming a roughly tetrahedral structure with four bonds:

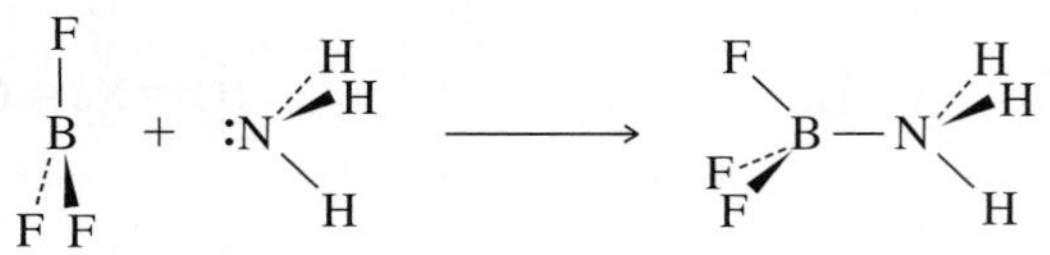

$$BF_3 + :NH_3 \longrightarrow F_3B{-}NH_3$$

Because of this tendency, boron trihalides are frequently drawn with only six electrons around the boron.

Other boron compounds that cannot be adequately described via simple electron-dot structures include hydrides such as B_2H_6, and many more complex molecules. Their structures are discussed in Chapters 8 and 15.

3.2 Valence Shell Electron-Pair Repulsion

Valence shell electron-pair repulsion (VSEPR) is an approach that provides a method for predicting the shape of molecules based on the electron-pair electrostatic repulsion described by Sidgwick and Powell[4] in 1940 and further developed by Gillespie and Nyholm[5] in 1957 and in the succeeding decades. Despite this method's simple approach, based on Lewis electron-dot structures, the VSEPR method in most cases predicts shapes that compare favorably with those determined experimentally. However, this approach at best provides approximate shapes for molecules. The most common method of determining the actual structures is X-ray diffraction, although electron diffraction, neutron diffraction, and many spectroscopic methods are also used.[6] In Chapter 5, we will provide molecular orbital approaches to describe bonding in simple molecules.

The basis of the VSEPR approach is that electrons repel each other because they are negatively charged. Quantum mechanical rules dictate that electrons can be accommodated in the same region of space as bonding pairs or lone pairs, but each pair repels all other pairs. According to the VSEPR model, therefore, molecules adopt geometries such that valence electron pairs position themselves as far from each other as possible to minimize electron–electron repulsions. A molecule can be described by the generic formula AX_mE_n, where *A* is the central atom, *X* stands for any atom or group of atoms surrounding the central atom, and *E* represents a lone pair of electrons. The **steric number*** **(SN = *m* + *n*)** is the total number of positions occupied by atoms or lone pairs around a central atom; lone pairs and bonding pairs both influence the molecular shape.

Carbon dioxide is a molecule with two atoms attached (SN = 2) to the central atom via double bonds. The electrons in each double bond must be between C and O, and the repulsion between these electron groups forces a linear structure on the molecule. Sulfur trioxide has three atoms bound to the sulfur (SN = 3), with equivalent partial double-bond character between sulfur and each oxygen, a conclusion rendered by analysis of its resonance forms. The best positions for the oxygens to minimize electron–electron repulsions in this molecule are at the corners of an equilateral triangle, with O—S—O bond angles of 120°. The multiple bonding does not affect the geometry, because all three bonds are equivalent in terms of bond order.

O=C=O

O=S(=O)=O (trigonal planar SO_3)

The same pattern of finding the Lewis structure and then matching it to a geometry that minimizes the repulsive energy of bonding electrons is followed through steric numbers 4, 5, 6, 7, and 8 where the outer atoms are identical in each molecule, as shown in **Figure 3.8**.

Bond angles and distances are uniform in each of these structures with two, three, four, and six electron pairs. Neither the corresponding 5- nor 7-coordinate structures can have uniform angles and distances, because there are no regular polyhedra with these numbers of vertices. The 5-coordinate molecules have a trigonal bipyramidal structure, with a central triangular plane of three positions plus two other positions above and below the center of the plane. The 7-coordinate molecules have a pentagonal bipyramidal structure, with a pentagonal plane of five positions and positions above and below the center of the plane. The regular square antiprism structure (SN = 8) is like a cube that has had the top face twisted 45° into the antiprism arrangement, as shown in **Figure 3.9**. It has three different bond angles for adjacent fluorines. $[TaF_8]^{3-}$ has square antiprismatic geometry but is distorted from this ideal in the solid.[7]

**The steric number* is also called the *number of electron pair domains*.

Steric Number	Geometry	Examples	Calculated Bond Angles	
2	Linear	CO_2	180°	O=C=O
3	Trigonal (triangular)	SO_3	120°	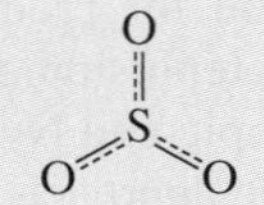
4	Tetrahedral	CH_4	109.5°	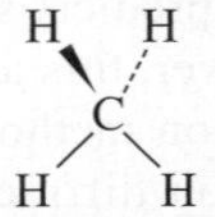
5	Trigonal bipyramidal	PCl_5	120°, 90°	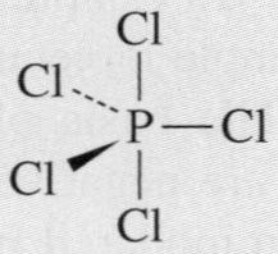
6	Octahedral	SF_6	90°	
7	Pentagonal bipyramidal	IF_7	72°, 90°	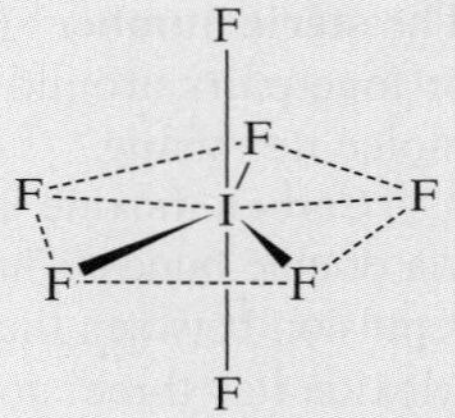
8	Square antiprismatic	$[TaF_8]^{3-}$	70.5°, 99.6°, 109.5°	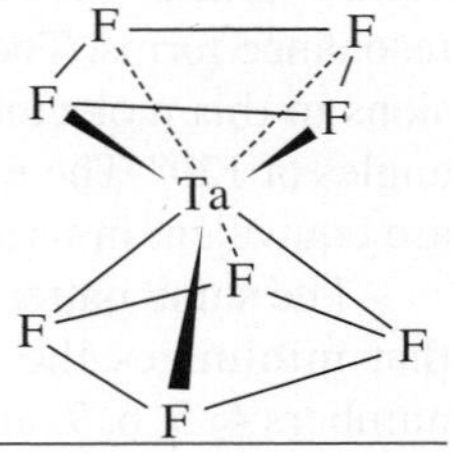

FIGURE 3.8 VSEPR Predictions.

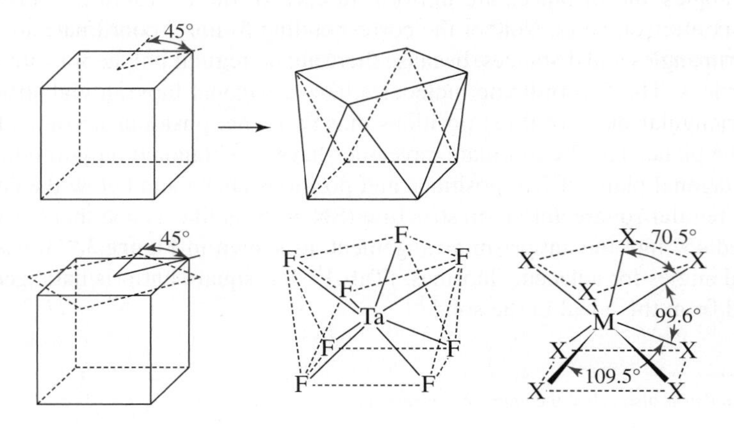

FIGURE 3.9 Conversion of a Cube into a Square Antiprism.

3.2.1 Lone-Pair Repulsion

Bonding models are useful only if their explanations are consistent with experimental data. New theories are continually being suggested and tested. Because we are working with such a wide variety of atoms and molecular structures, a single approach will unlikely work for all of them. Although the fundamental ideas of atomic and molecular structures are relatively simple, their application to complex molecules is not. To a first approximation, lone pairs, single bonds, double bonds, and triple bonds can all be treated similarly when predicting molecular shapes. However, better predictions of overall shapes can be made by considering some important differences between lone pairs and bonding pairs. These methods are sufficient to show the trends and explain the bonding, as in rationalizing why the H—N—H angle in ammonia is smaller than the tetrahedral angle in methane and larger than the H—O—H angle in water.

As a general guideline, the VSEPR model predicts that electron-pair repulsions involving lone pairs (*lp*) are stronger than those involving bonding pairs (*bp*) in the order

lp–*lp* *repulsions* > *lp*–*bp* *repulsions* > *bp*–*bp* *repulsions*

Steric Number = 4

The isoelectronic molecules CH_4, NH_3, and H_2O (**Figure 3.10**) illustrate the effect of lone pairs on molecular shape. Methane has four identical bonds between carbon and each of the hydrogens. When the four pairs of electrons are arranged as far from each other as possible, the result is the familiar tetrahedral shape. The tetrahedron, with all H—C—H angles measuring 109.5°, has four identical bonds.

Ammonia also has four pairs of electrons around the central atom, but three are bonding pairs between N and H, and the fourth is a lone pair on the nitrogen. The nuclei form a trigonal pyramid with the three bonding pairs; the lone pair occupies the fourth region in space resulting in a tetrahedral arrangement of the four electron groups. Because each of the three bonding pairs is attracted by two positively charged nuclei (H and N), these pairs are largely confined to the regions between the H and N atoms. The lone pair, on the other hand, is attracted solely by the nitrogen nucleus; it has no second nucleus to confine it to a small region of space. Consequently, the lone pair tends to spread out and to occupy more space around the nitrogen than the bonding pairs. As a result, the H—N—H angles are 106.6°, nearly 3° smaller than the angles in methane.

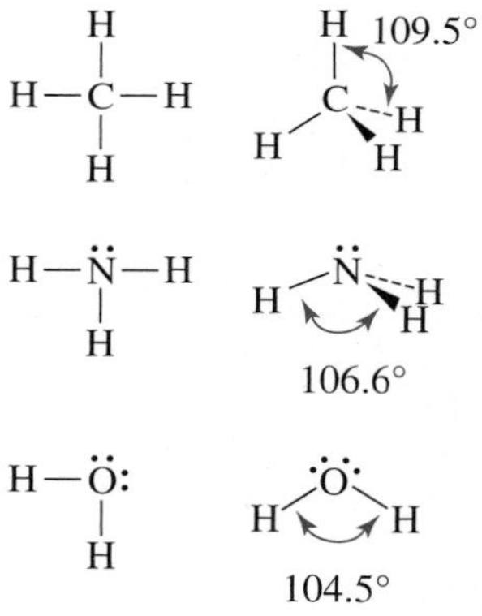

The same principles apply to the water molecule, in which two lone pairs and two bonding pairs repel each other. Again, the electron pairs adopt a nearly tetrahedral arrangement, with the atoms arranged in a V shape. The angle of largest repulsion, between the two lone pairs, cannot be measured. However, the lone pair–bonding pair (*lp–bp*) repulsion is greater than the bonding pair–bonding pair (*bp–bp*) repulsion; as a result, the H—O—H bond angle is only 104.5°, another 2.1° decrease from the ammonia angles. The net result is that we can predict approximate molecular shapes by assigning more space to lone electron pairs; lone pairs are able to spread out and occupy more space since they are attracted to one nucleus rather than two.

FIGURE 3.10 Shapes of Methane, Ammonia, and Water.

Steric Number = 5

For the trigonal bipyramidal geometry, there are two unique locations for electron pairs, axial and equatorial. If there is a single lone pair, for example in SF_4, the lone pair occupies an equatorial position. This position provides the lone pair with the most space and minimizes the interactions between the lone pair and bonding pairs. If the lone pair were axial, it would have three 90° interactions with bonding pairs; in an equatorial position, it has only two such interactions, as shown in **Figure 3.11**. The actual structure is distorted by the lone pair as it spreads out in space and effectively squeezes the rest of the molecule together.

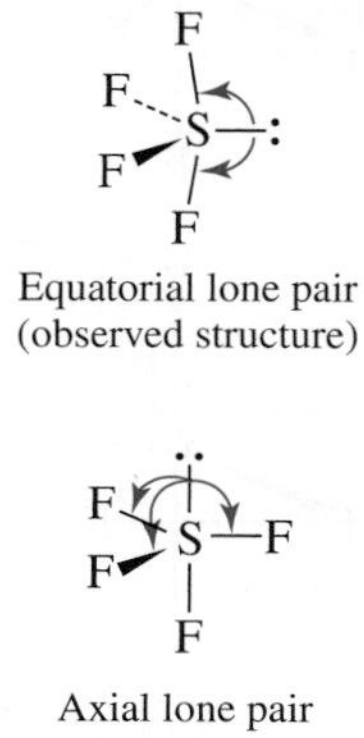

FIGURE 3.11 Possible Structures of SF_4.

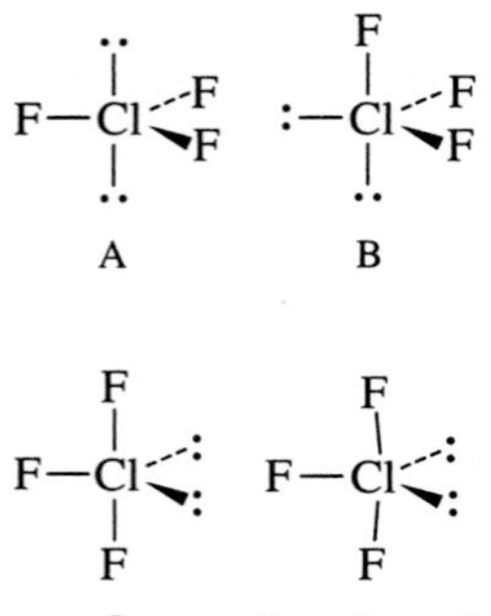

A B C Experimental

FIGURE 3.12 Possible Structures of ClF_3.

ClF_3 provides a second example of the influence of lone pairs in molecules having a steric number of 5. There are three possible structures for ClF_3, as shown in **Figure 3.12**.

In determining the feasibility of different structures, lone pair–lone pair interactions should be considered first, followed by lone pair–bonding pair interactions. These interactions at angles of 90° or less are generally considered destabilizing; larger angles generally render structures more feasible. For example, in ClF_3, structure B can be eliminated quickly because of the 90° *lp*–*lp* angle. The *lp*–*lp* angles are large for A and C, so the choice must come from the *lp*–*bp* and *bp*–*bp* angles. Because the *lp*–*bp* angles are more important, C, which has only four 90° *lp*–*bp* interactions, is favored over A, which has six such interactions. Experiments have confirmed that the structure is based on C, with slight distortions due to the lone pairs. The lone pair–bonding pair repulsion causes the *lp*–*bp* angles to be larger than 90° and the *bp*–*bp* angles to be less than 90° (actually, 87.5°). The Cl—F bond distances show the repulsive effects as well, with the axial fluorines (approximately 90° *lp*–*bp* angles) at 169.8 pm and the equatorial fluorine (in the plane with two lone pairs) at 159.8 pm.[8] Angles involving lone pairs cannot be determined experimentally.

	Angles in Possible Structures			
Interaction	A	B	C	Experimental
lp–*lp*	180°	90°	120°	Cannot be determined
lp–*bp*	6 at 90°	3 at 90°	4 at 90°	Cannot be determined
		2 at 120°	2 at 120°	
bp–*bp*	3 at 120°	2 at 90°	2 at 90°	2 at 87.5°
		1 at 120°		Axial Cl—F 169.8 pm Equatorial Cl—F 159.8 pm

Additional examples of structures with lone pairs are illustrated in **Figure 3.13**. The structures based on a trigonal bipyramidal arrangement of electron pairs around a central atom always place any lone pairs in the equatorial plane, as in SF_4, BrF_3, and XeF_2. The resulting shapes minimize both lone pair–lone pair and lone pair–bonding pair repulsions. The shapes are called *seesaw* (SF_4), *distorted T* (BrF_3), and *linear* (XeF_2).

Steric Numbers = 6 and 7

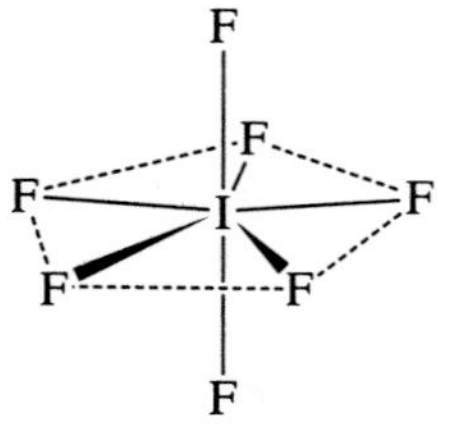

In octahedral structures, all six positions are equivalent. When a single lone pair is present, it typically repels adjacent bonding pairs, reducing bond angles accordingly, as for IF_5 in Figure 3.13. In octahedron-based structures with two lone pairs, lone pair–lone pair repulsion is minimized if these pairs are *trans*, and this is the shape that is adopted. Square planar XeF_4, also shown in Figure 3.13, is an example. Recently XeF_3^-, which would be expected to have a steric number of 6 and three lone pairs, has been reported in the gas phase, but attempts to prepare salts of this ion have been unsuccessful.[9]

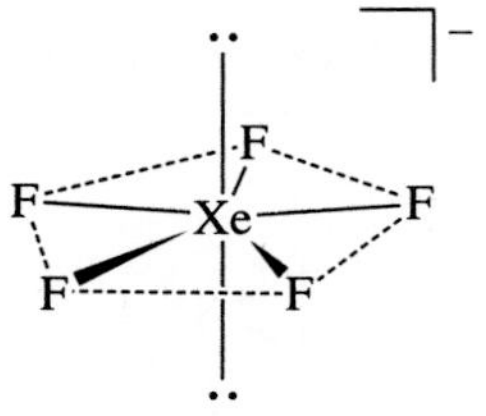

The shape that minimizes electron-pair repulsions for a steric number of 7 is the pentagonal bipyramid, shown in Figure 3.8. IF_7 (in the margin) and TeF_7^{2-} exhibit this shape, with both axial and equatorial fluorines. If a single lone pair is present, in some cases the lone pair causes distortion. The nature of this distortion is not always easy to ascertain; XeF_6 is a classic example.[10] In other cases the structure is octahedral (see Problem 3.26) with the lone pair not stereochemically active.* Two lone pairs minimize their repulsions by adopting axial (*trans*) positions, with the atoms all in the equatorial plane. Two known examples are XeF_5^- (in the margin) and IF_5^{2-}.

*A lone pair that appears in the Lewis-dot structure but has no apparent effect on the molecular geometry is classified as not stereochemically active. The VSEPR model assumes that all lone pairs are stereochemically active and therefore do affect the molecular geometry.

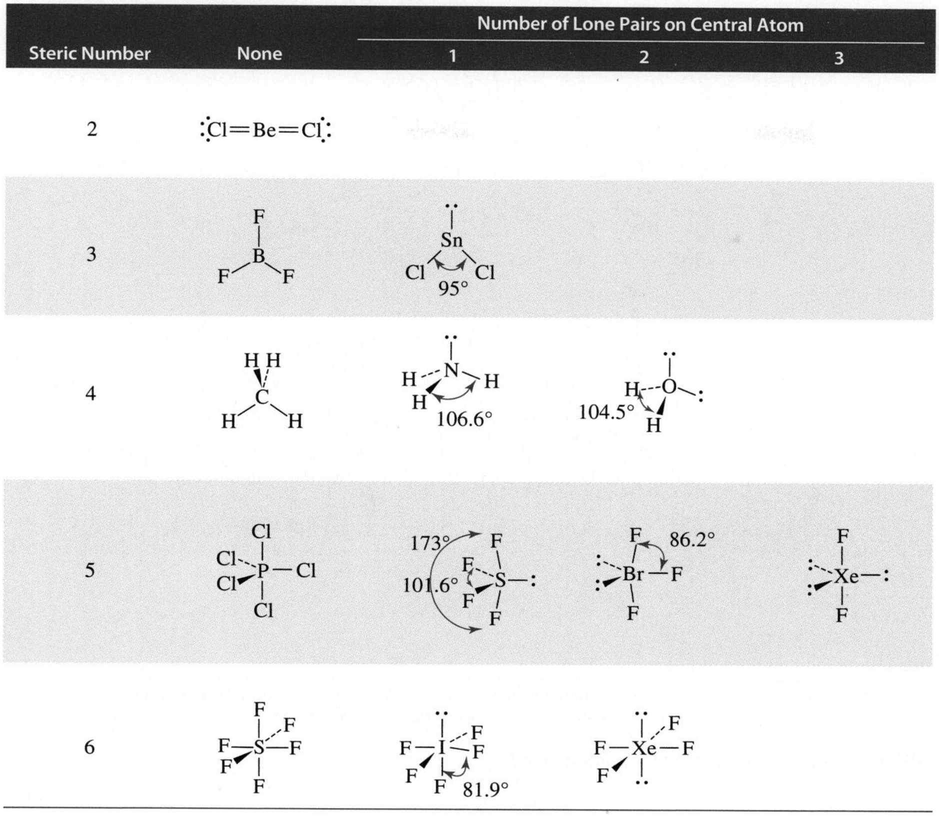

FIGURE 3.13 Structures Containing Lone Pairs.

Steric Number	None	Number of Lone Pairs on Central Atom: 1	2	3
2	:Cl=Be=Cl:			
3	BF_3	$SnCl_2$ (95°)		
4	CH_4	NH_3 (106.6°)	H_2O (104.5°)	
5	PCl_5	SF_4 (173°, 101.6°)	BrF_3 (86.2°)	XeF_2
6	SF_6	IF_5 (81.9°)	XeF_4	

EXAMPLE 3.4

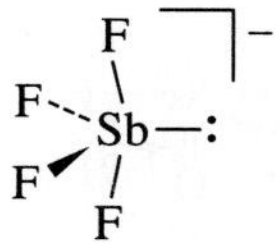

SbF_4^- has a single lone pair on Sb. Its structure is therefore similar to SF_4, with a lone pair occupying an equatorial position. This lone pair causes considerable distortion, giving an F—Sb—F (axial positions) angle of 155° and an F—Sb—F (equatorial) angle of 90°.

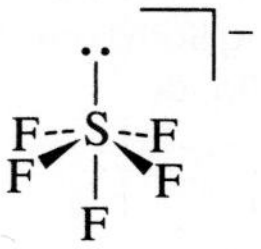

SF_5^- has a single lone pair. Its structure is based on an octahedron, with the ion distorted away from the lone pair, as in IF_5.

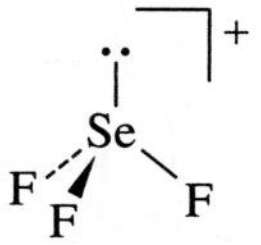

SeF_3^+ has a single lone pair. This lone pair reduces the Se—F bond angle significantly, to 94°.

EXERCISE 3.2 Predict the structures of the following ions. Include a description of distortions from the ideal angles (for example, less than 109.5° because...).

NH_2^- NH_4^+ I_3^- PCl_6^-

3.2.2 Multiple Bonds

The VSEPR model considers double and triple bonds to have slightly greater repulsive effects than single bonds because of the repulsive effect of π electrons that increase the electron density between the bonded atoms beyond that present in a σ bond. For example, the $H_3C—C—CH_3$ angle in $(CH_3)_2C{=}CH_2$ is smaller, and the $H_3C—C{=}CH_2$ angle is larger than the trigonal 120° (**Figure 3.14**).[11]

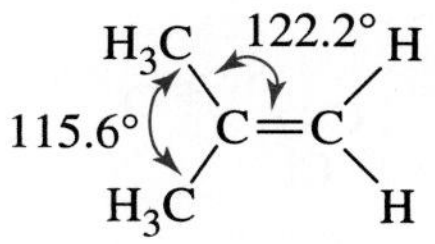

FIGURE 3.14 Bond Angles in $(CH_3)_2C{=}CH_2$.

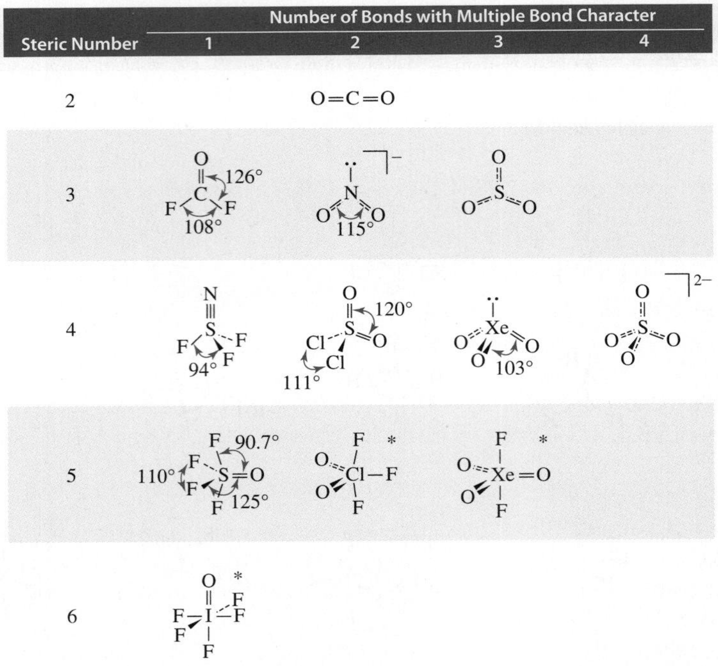

Steric Number	Number of Bonds with Multiple Bond Character 1	2	3	4
2		O=C=O		
3				
4				
5				
6				

* The bond angles of these molecules have not been determined accurately. However, spectroscopic measurements are consistent with the structures shown.

FIGURE 3.15 Structures Containing Multiple Bonds.

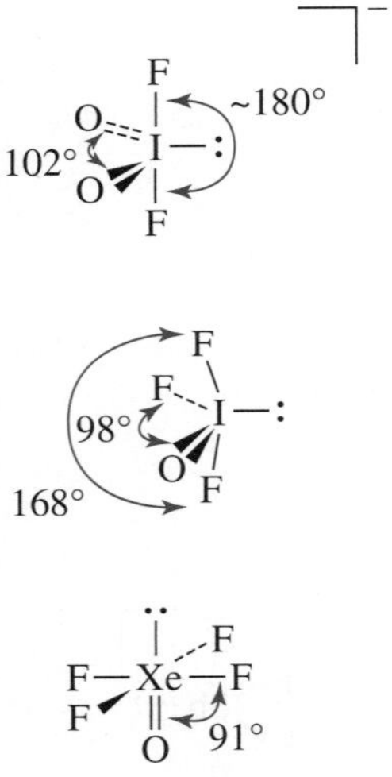

FIGURE 3.16 Structures Containing Both Lone Pairs and Multiple Bonds.

Additional examples of the effect of multiple bonds on molecular geometry are shown in **Figure 3.15**. Comparing Figures 3.13 and 3.15, we see that multiple bonds tend to occupy the same positions as lone pairs. For example, the double bonds to oxygen in SOF_4, ClO_2F_3, and XeO_3F_2 are all equatorial, as are the lone pairs in the matching compounds of steric number 5, SF_4, BrF_3, and XeF_2. Multiple bonds, like lone pairs, also tend to occupy more space than single bonds, causing distortions that squeeze the rest of the molecule together. In molecules that have both lone pairs and multiple bonds, these features may compete for space; examples are shown in **Figure 3.16**. As a generalization, lone pairs often have a greater influence than multiple bonds in dictating molecular geometry.

EXAMPLE 3.5

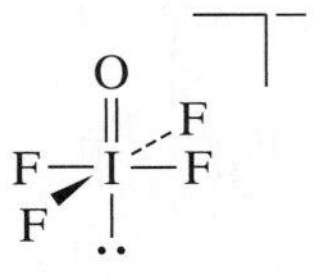

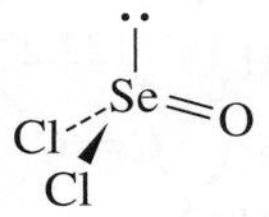

HCP, like HCN, is linear, with a triple bond: H—C≡P:

$\mathbf{IOF_4^-}$ has a single lone pair on the side opposite the oxygen. The lone pair has a slightly greater repulsive effect than the double bond to oxygen, as shown by the average O—I—F angle of 89°. (The extra repulsive character of the I=O bond places it opposite the lone pair.)

$\mathbf{SeOCl_2}$ has both a lone pair and a selenium–oxygen double bond. The lone pair has a greater effect than the double bond; the Cl—Se—Cl angle is reduced to 97° by this effect, and the Cl—Se—O angle is 106°.

EXERCISE 3.3 Predict the structures of the following. Indicate the direction of distortions from the regular structures.

$XeOF_2$ $ClOF_3$ $SOCl_2$

3.2.3 Electronegativity and Atomic Size Effects

Electronegativity is a measure of an atom's ability to attract electrons from a neighboring atom to which it is bonded; it can be viewed as the ability of an atom to win the competition to attract shared electrons. Electronegativity was mentioned earlier as a guide in the use of formal charges. It also can play an important role in determining the arrangement of outer atoms around a central atom and in rationalizing bond angles. The effects of electronegativity and atomic size frequently parallel each other, but in some cases, the sizes of outer atoms and groups may play the more important role.

Electronegativity Scales

Linus Pauling introduced the concept of electronegativity in the 1930s as a means of describing bond energies. Pauling recognized that polar bonds have higher bond energies than nonpolar bonds formed from the same elements. For example, he observed that the bond energy of HCl, 432 kJ/mol, was much higher than the average of the bond energies of H_2 (436 kJ/mol) and Cl_2 (243 kJ/mol).* He related the difference between actual and average bond energies to the difference in electronegativity between the elements involved. He also made adjustments for the sake of convenience, most notably to give the elements C through F equally spaced values of 2.5 through 4.0.** Some early Pauling electronegativity values are in **Table 3.3**. The value of 4.0 for fluorine is still commonly used as a reference point for other electronegativity scales.

More recent values have been derived from other molecular and atomic properties, such as ionization energies and electron affinities. **Table 3.4** summarizes approaches used for a variety of electronegativity scales; examining differences among these is beyond the scope of this text. In most cases the different methods give similar electronegativity values, sometimes with the exception of the transition metals.[12] We choose to use the values reported by Mann, Meek, and Allen (**Table 3.5**) based on configuration energies (CE), the average ionization energies of valence electrons in ground state free atoms. For *s*- and *p*-block elements the configuration energies are defined as follows:[13]

$$\text{CE} = \frac{n\varepsilon_s + m\varepsilon_p}{n + m}$$

where n = number of *s* electrons
m = number of *p* electrons
ε_s, ε_s = experimental 1-electron *s* and *p* energies†

TABLE 3.3 Early Values of Pauling Electronegativities

H 2.1			
C 2.5	N 3.0	O 3.5	F 4.0
Si 1.8	P 2.1	S 2.5	Cl 3.0
Ge 1.8	As 2.0	Se 2.4	Br 2.8

*Values used by Pauling, converted to kJ/mol. L. Pauling, *The Nature of the Chemical Bond*, 3rd ed., 1960, Cornell University Press, Ithaca, NY, p. 81.

Earlier, Pauling had assigned fluorine an electronegativity of 2.00; see L. Pauling, *J. Am. Chem. Soc.*, **1932, *54*, 3570.

†Multiplet averaged values from C. E. Moore, *Ionization Potentials and Ionization Limits Derived From the Analyses of Optical Spectra*, NSRDS-NBS-34, Washington, D.C., 1971; *Atomic Energy Levels*, NSRDS-35, Washington, D.C., 1971, Vol. III.

TABLE 3.4 Electronegativity Scales

Principal Authors	Method of Calculation or Description
Pauling[14]	Bond energies
Mulliken[15]	Average of electron affinity and ionization energy
Allred & Rochow[16]	Electrostatic attraction proportional to Z^*/r^2
Sanderson[17]	Electron densities of atoms
Pearson[18]	Average of electron affinity and ionization energy
Allen[19]	Average energy of valence shell electrons, configuration energies
Jaffé[20]	Orbital electronegativities

TABLE 3.5 Electronegativity (Pauling Units)

1	2	12	13	14	15	16	17	18
H								He
2.300								4.160
Li	Be		B	C	N	O	F	Ne
0.912	1.576		2.051	2.544	3.066	3.610	4.193	4.787
Na	Mg		Al	Si	P	S	Cl	Ar
0.869	1.293		1.613	1.916	2.253	2.589	2.869	3.242
K	Ca	Zn	Ga	Ge	As	Se	Br	Kr
0.734	1.034	1.588	1.756	1.994	2.211	2.424	2.685	2.966
Rb	Sr	Cd	In	Sn	Sb	Te	I	Xe
0.706	0.963	1.521	1.656	1.824	1.984	2.158	2.359	2.582
Cs	Ba	Hg	Tl	Pb	Bi	Po	At	Rn
0.659	0.881	1.765	1.789	1.854	(2.01)	(2.19)	(2.39)	(2.60)

Source: J. B. Mann, T. L. Meek, L. C. Allen, *J. Am. Chem. Soc.*, **2000**, 122, 2780, Table 2.

The configuration energies are multiplied by a constant to give values comparable to the Pauling scale to enable convenient comparison between the scales. A more complete list of electronegativities based on configuration energies is in Appendix B.4.* A graphical representation of electronegativity is in Figure 8.2.

Pauling's calculation of electronegativities from bond energies requires averaging over a number of compounds in an attempt to minimize experimental uncertainties and other minor effects. Methods that use ionization energies and other atomic properties can be calculated more directly. The electronegativities reported here and in Appendix B.4 are suitable for most uses, but the actual values for atoms in different molecules can differ depending on the specific electronic environment of the atoms. The concept of electronegativity varying for a given atom on the basis of its specific bonds within a molecule is usually not introduced in introductory chemistry, but is a consequence of modern electronegativity scales.

It is important to emphasize that all electronegativities are measures of an atom's ability to attract electrons from a neighboring atom *to which it is bonded.* A critique of all electronegativity scales, and particularly Pauling's, is that each scale cannot be successfully applied to all situations; all of these scales have deficiencies on the basis of the specific assumptions used in their development.[21]

*For a recent approach that addresses some of the limitations of the Allen method, see P. Politzer, Z. P. Shields, F. A. Bulat, J. S. Murray, *J. Chem. Theory Comput.*, **2011**, *7*, 377.

With the exception of helium and neon, which have large calculated electronegativities and no known stable compounds, fluorine has the largest value, and electronegativity decreases toward the lower left corner of the periodic table. Although usually classified with Group 1 (IA), hydrogen is quite dissimilar from the alkali metals in its electronegativity, as well as in many other chemical and physical properties. Hydrogen's chemistry is distinctive from all the groups.

Electronegativities of the noble gases can be calculated more easily from ionization energies than from bond energies. Because the noble gases have higher ionization energies than the halogens, calculations suggest that the electronegativities of the noble gases may exceed those of the halogens (Table 3.5).[22] The noble gas atoms are somewhat smaller than the neighboring halogen atoms—for example, Ne is smaller than F—as a consequence of a greater effective nuclear charge. This charge, which is able to attract noble gas electrons strongly toward the nucleus, is also likely to exert a strong attraction on electrons of neighboring atoms; hence, the high electronegativities predicted for the noble gases are reasonable.

Electronegativity and Bond Angles

By the VSEPR approach, trends in many bond angles can be explained by electronegativity. Consider the bond angles in the following molecules:

Molecule	X–P–X Angle (°)	Molecule	X–S–X Angle (°)
PF_3	97.8	OSF_2	92.3
PCl_3	100.3	$OSCl_2$	96.2
PBr_3	101.0	$OSBr_2$	98.2

As the electronegativity of the halogen increases, the halogen exerts a stronger pull on electron pairs it shares with the central atom. This effect reduces the concentration of electrons near the central atom, decreasing somewhat the repulsion between the bonding pairs near the central atom, and allows the lone pair to have more impact in compressing the halogen–central atom–halogen angles. Consequently, the molecules with the most electronegative *outer* atoms, PF_3 and OSF_2, have the smallest angles.

If the central atom remains the same, molecules that have a larger difference in electronegativity values between their central and outer atoms have smaller bond angles. The atom with larger electronegativity draws the shared electrons toward itself and away from the central atom, reducing the repulsive effect of these electrons. The compounds of the halogens in **Table 3.6** show this effect; the compounds containing fluorine have smaller angles than those containing chlorine, which in turn have smaller angles than those containing bromine. The lone pair exerts a relatively larger effect, and forces smaller bond angles, as the electronegativity of the outer atom increases. An alternative explanation for this trend is size: as the size of the outer atom increases in the order $F < Cl < Br$, the bond angle increases. Additional compounds showing the effects of electronegativity on bond angles are also given in Table 3.6.

Similar considerations can be made in situations where the outer atoms remain the same, but the central atom is changed, for example,

Molecule	Bond Angle (°)	Molecule	Bond Angle (°)
H_2O	104.5	NCl_3	106.8
H_2S	92.1	PCl_3	100.3
H_2Se	90.6	$AsCl_3$	98.9

In these cases, as the central atom becomes more electronegative, it pulls electrons in bonding pairs more strongly toward itself, increasing the concentration of electrons near the central atom.

TABLE 3.6 Bond Angles and Lengths

Molecule	Bond Angle (°)	Bond Length (pm)	Molecule	Bond Angle (°)	Bond Length (pm)	Molecule	Bond Angle (°)	Bond Length (pm)	Molecule	Bond Angle (°)	Bond Length (pm)
H_2O	104.5	97	OF_2	103.3	96	OCl_2	110.9	170			
H_2S	92.1	135	SF_2	98.0	159	SCl_2	102.7	201			
H_2Se	90.6	146				$SeCl_2$	99.6	216			
H_2Te	90.2	169				$TeCl_2$	97.0	233			
NH_3	106.6	101.5	NF_3	102.2	137	NCl_3	106.8	175			
PH_3	93.2	142	PF_3	97.8	157	PCl_3	100.3	204	PBr_3	101.0	220
AsH_3	92.1	151.9	AsF_3	95.8	170.6	$AsCl_3$	98.9	217	$AsBr_3$	99.8	236
SbH_3	91.6	170.7	SbF_3	87.3	192	$SbCl_3$	97.2	233	$SbBr_3$	98.2	249

Source: N. N. Greenwood and A. Earnshaw, *Chemistry of the Elements*, 2nd ed., Butterworth-Heinemann, Oxford, 1997, pp. 557, 767; A. F. Wells, *Structural Inorganic Chemistry*, 5th ed., Oxford University Press, Oxford, 1987, pp. 705, 793, 846, and 879; R. J. Gillespie and I. Hargittai, *The VSEPR Model of Molecular Geometry*, Allyn and Bacon, Needham Heights, MA, 1991.

The net effect is that an increase in bonding pair–bonding pair repulsions near the *central* atom increases the bond angles. In these situations the molecule with the most electronegative central atom has the largest bond angles. Additional examples can be found in Table 3.6, where molecules having the same outer atoms, but different central atoms, are shown in the same column.

EXERCISE 3.4

Which molecule has the smallest bond angle in each series?

a. $OSeF_2$ $OSeCl_2$ $OSeBr_2$ (halogen–Se–halogen angle)
b. $SbCl_3$ $SbBr_3$ SbI_3
c. PI_3 AsI_3 SbI_3

Effects of Size

In the examples considered so far, the most electronegative atoms have also been the smallest. For example, the smallest halogen, fluorine, is also the most electronegative. Consequently, we could have predicted the trends in bond angles on the basis of atomic size, with the smallest atoms capable of being crowded together most closely. It is important to also consider situations in which size and electronegativity might have opposite effects, where a smaller outer group is *less* electronegative than a larger group attached to a central atom. For example,

Molecule	C—N—C Angle (°)
$N(CH_3)_3$	110.9
$N(CF_3)_3$	117.9

In this case VSEPR would predict that the more electronegative CF_3 groups would lead to a smaller bond angle because they would withdraw electrons more strongly than CH_3 groups. That the bond angle in $N(CF_3)_3$ is actually 7° larger than in $N(CH_3)_3$ suggests that in this case, size is the more important factor, with the larger CF_3 groups requiring more space. The point at which the size of outer atoms and groups becomes more important than electronegativity can be difficult to predict, but the potential of large outer atoms and groups to affect molecular shape should not be dismissed.

Molecules Having Steric Number = 5

For main group atoms having a steric number of 5, it is instructive to consider the relative bond lengths for axial and equatorial positions. For example, in PCl_5, SF_4, and ClF_3, the central atom–axial distances are longer than the distances to equatorial atoms, as shown in **Figure 3.17**. This effect has been attributed to the greater repulsion of lone and bonding pairs with atoms in axial positions (three 90° interactions) than with atoms in equatorial positions (two 90° interactions).

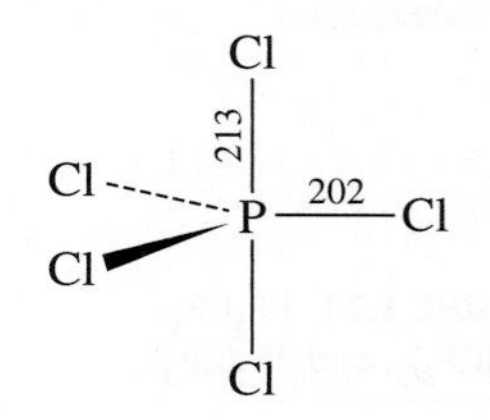

In addition, there is a tendency for less electronegative groups to occupy equatorial positions, similar to lone pairs and multiply bonded atoms. For example, in phosphorus compounds having both fluorine and chlorine atoms, in each case the chlorines occupy equatorial positions (**Figure 3.18**). The same tendency is shown in compounds having formulas PF_4CH_3, $PF_3(CH_3)_2$, and $PF_2(CH_3)_3$, with the less electronegative CH_3 groups also equatorial (**Figure 3.19**). One can envision the electron density of the P—A bond, where A is the less electronegative atom, being concentrated closer to the phosphorus in such cases, leading to a preference for equatorial positions by similar reasoning applied to lone pairs and multiple bonds.

The relative effects on bond angles by less electronegative atoms are, however, typically less than for lone pairs and multiple bonds. For example, the bond angle to equatorial positions opposite the Cl atom in PF_4Cl is only slightly less than 120°, in contrast to the greater reduction in comparable angles in SF_4 and SOF_4 (**Figure 3.20**).

Predicting structures in some cases is challenging. Phosphorus compounds containing both fluorine atoms and CF_3 groups provide an intriguing example. CF_3 is an electron withdrawing group whose electronegativity has been calculated to be comparable to the more electronegative halogen atoms.* Does CF_3 favor equatorial positions more strongly than F? Trigonal bipyramidal phosphorus compounds containing varying numbers of F and CF_3 groups with both axial and equatorial CF_3 groups are known (**Figure 3.21**). When two or three CF_3 groups are present, the orientations are truly a challenge to explain: these groups are axial in $PF_3(CF_3)_2$ but equatorial in $PF_2(CF_3)_3$! In both cases the more symmetrical structure, with identical equatorial groups, is preferred.**

FIGURE 3.17 Bond Distances in PCl_5, SF_4, and ClF_3.

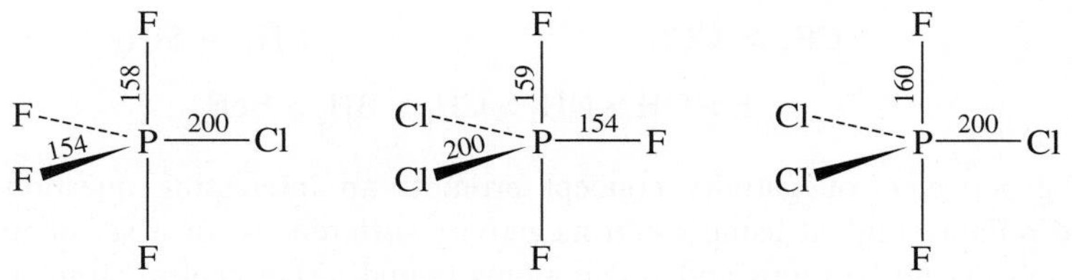

FIGURE 3.18 $PClF_4$, PCl_2F_3, and PCl_3F_2.

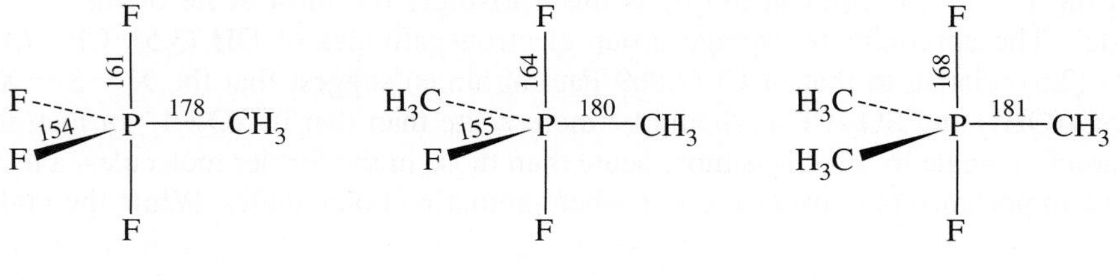

FIGURE 3.19 PF_4CH_3, $PF_3(CH_3)_2$, and $PF_2(CH_3)_3$.

*For an analysis of different approaches to determining the electronegativity of CF_3, see J. E. True, T. D. Thomas, R. W. Winter, G. L. Gard, *Inorg. Chem.*, **2003**, *42*, 4437.

See H. Oberhammer, J. Grobe, D. Le Van, *Inorg. Chem.*, **1982, *21*, 275 for a discussion of these structures.

FIGURE 3.20 Bond angles in PF_4Cl, SF_4, and SOF_4.

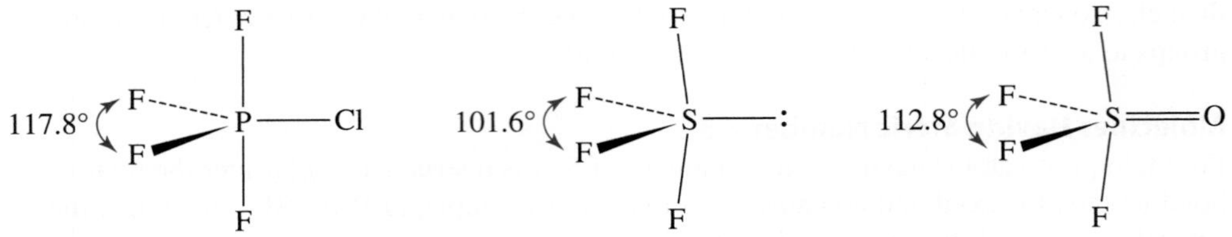

FIGURE 3.21 PF_4CF_3, $PF_3(CF_3)_2$, and $PF_2(CF_3)_3$.

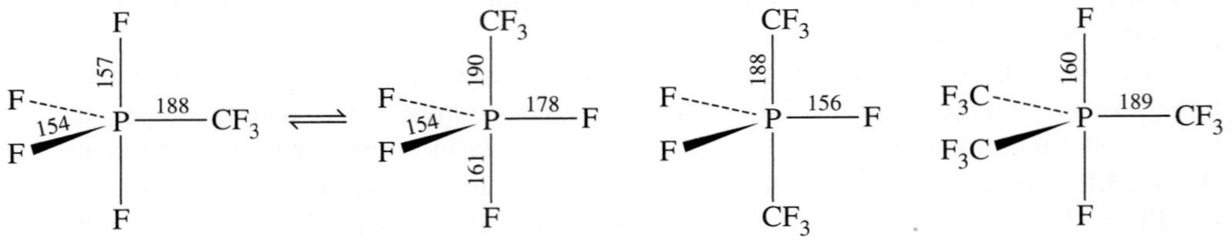

Group Electronegativities

As in the case of individual atoms, numerous approaches have been taken to estimate electron-attracting abilities of groups such as CH_3, CF_3, and OH which may be bonded to central atoms. For example, a CF_3 group would be expected to attract electrons more strongly than a CH_3 group, potentially affecting molecular shapes and reactivities, and CF_3 should therefore be assigned a higher electronegativity than CH_3. Although published values of group electronegativities agree that CF_3 is more electronegative than CH_3, the reported values vary widely, 2.71–3.45 for CF_3 and 2.45–3.05 for CH_3.*

Despite wide variation in proposed group electronegativity values, trends emerge when examining sets of values determined by different calculation methods that follow expectations based on the electronegativities of the component atoms, as in the case of CF_3 and CH_3. Examples in order of decreasing group electronegativity include the following:

$$CF_3 > CHF_2 > CH_2F > CH_3$$

$$CF_3 > CCl_3 \qquad\qquad CH_3 > SiH_3$$

$$F > OH > NH_2 > CH_3 > BH_2 > BeH$$

The group electronegativity concept prompts an interesting question: can bond angles be reliably ranked using electronegativity differences for a set of related molecules containing *both* groups and single atoms bound to the central atom? Consider the molecules in **Figure 3.22**. The S—F bond has the largest electronegativity difference, and the F—S—F bond in SO_2F_2 is unsurprisingly the most acute of the X—S—X angles. The approximate average group electronegativities of OH (3.5), CF_3, (3.1) and CH_3 (2.6) relative to that of Cl (2.869 Pauling units) suggest that the X—S—X angle in $SO_2(OH)_2$ and $SO_2(CF_3)_2$ should be more acute than that in SO_2Cl_2. In fact, the corresponding angle in SO_2Cl_2 is more acute than those in the former molecules, a testament to the importance of considering size when predicting bond angles. While the ranking of

*For comparisons of values of group electronegativities and associated references, see M. D. Moran, J-P. Jones, A. A. Wilson, S. Houle, G. K. S. Prakash, G. A. Olah, N. Vasdev, *Chem. Educator*, **2011**, *16*, 164 and L. D. Garner O'Neale, A. F. Bonamy, T. L. Meek, B. G. Patrick, *J. Mol. Struct. (THEOCHEM)*, **2003**, *639*, 151.

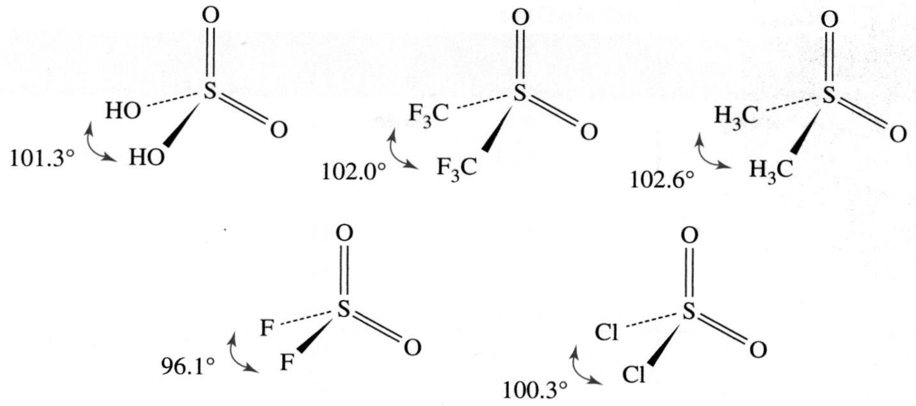

FIGURE 3.22 Bond Angles and Group Electronegativity.

the bond angles for $SO_2(OH)_2 < SO_2(CF_3)_2 < SO_2(CH_3)_2$ is consistent with the group electronegativity order, the small difference in these angles (1.3°) is surprising on the basis of the relatively wide variation (~0.9 units) in group electronegativities. In contrast, the electronegativity difference between F and Cl (1.324 Pauling units) is only slightly greater, yet the result is a rather large 4.2° difference in X—S—X angles. Bond angle prediction clearly depends on multiple factors. Size and possible hydrogen bonding between outer atoms and groups can also affect bond angles and distances.

EXERCISE 3.5

Briefly account for the following observations:

a. The bond angle in NCl_3 is nearly 5 degrees larger than in NF_3.

b. The S–F axial distance in SOF_3 is longer than the S–F equatorial distance.

c. In $Te(CH_3)_2I_2$ the methyl groups are in equatorial, rather than axial, positions.

d. The O—S—O bond angle in $FSO_2(OCH_3)$ is larger than in $FSO_2(CH_3)$.

3.2.4 Ligand Close Packing

The **ligand close-packing (LCP)** model developed by Gillespie[23] uses the distances between outer atoms in molecules as a guide to molecular shapes. For a series of molecules having the same central atom, the *non*bonded distances* between the outer atoms are consistent, but the bond angles and bond lengths change. The results of the LCP approach are in many ways consistent with those of the VSEPR model but focus primarily on the outer atoms rather than on the immediate environment of the central atom.

For example, it was found that in a series of boron compounds, BF_2X and BF_3X, the fluorine–fluorine distance remained nearly constant for a wide variety of X groups, even if the steric number changed from 3 to 4, as shown in **Table 3.7**. Similar results were obtained for a variety of other central atoms: chlorine–chlorine *non*bonded distances were similar in compounds in which the central atom was carbon, oxygen–oxygen nonbonded distances were similar when the central atom was beryllium, and so forth.[24]

*Three dots (· · ·) will be used to designate distances between atoms that are not directly covalently bonded to each other.

TABLE 3.7 Ligand Close-Packing Data

Molecule	Coordination Number of B	B—F Distance (pm)	FBF Angle (°)	F···F Distance (pm)
BF_3	3	130.7	120.0	226
BF_2OH	3	132.3	118.0	227
BF_2NH_2	3	132.5	117.9	227
BF_2Cl	3	131.5	118.1	226
BF_2H	3	131.1	118.3	225
BF_2BF_2	3	131.7	117.2	225
BF_4^-	4	138.2	109.5	226
$BF_3CH_3^-$	4	142.4	105.4	227
$BF_3CF_3^-$	4	139.1	109.9	228
BF_3PH_3	4	137.2	112.1	228
BF_3NMe_3	4	137.2	111.5	229

Source: R. J. Gillespie and P. L. A. Popelier, *Chemical Bonding and Molecular Geometry*, Oxford University Press, New York, 2001, p. 119; Table 5.3, R. J. Gillespie, *Coord. Chem. Rev.*, **2000**, 197, 51.

In the LCP model, ligands (outer atoms) are viewed as exhibiting a specific radius when bonded to a certain central atom.* If the outer atoms pack tightly together, as assumed in this model, the distance between the nuclei of the atoms will then be the sum of these ligand radii. For example, a fluorine atom, when attached to a central boron, has a ligand radius of 113 pm. When two fluorines are attached to a central boron, as in the examples in Table 3.7, the distance between their nuclei will be the sum of the ligand radii, in this case 226 pm. This value matches the F · · · F distances of the examples in the table. Examples of how this approach can be used to describe molecular shapes are presented in the following discussion.

Ligand Close Packing and Bond Distances

The LCP model predicts that nonbonded atom–atom distances in molecules remain approximately the same, even if the bond angles around the central atom are changed. For example, the fluorine–fluorine distances in NF_4^+ and NF_3 are both 212 pm, even though the F—N—F bond angles are significantly different (**Figure 3.23**).

FIGURE 3.23 NF_4^+ and NF_3.

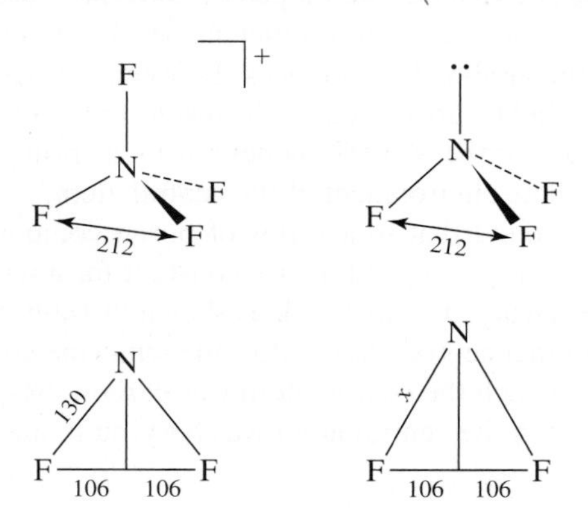

*Values of ligand radii can be found in R. J. Gillespie, E. A. Robinson, *Compt. Rend. Chimie*, **2005**, *8*, 1631.

VSEPR predicts that NF_3 should have the smaller bond angle, and it does: 102.3° in comparison with the tetrahedral angle of 109.5° in NF_4^+. Because the F · · · F distance remains essentially unchanged, the N—F distance in NF_3 must be longer than the 130 pm in NF_4^+. This can be illustrated using simple trigonometry, as also shown Figure 3.23. In NF_3, because

$$\sin\left(\frac{102.3°}{2}\right) = \frac{\frac{1}{2}(\text{F} \cdots \text{F distance})}{x},$$

$$x = \text{N—F bond distance} = \frac{106 \text{ pm}}{\sin 51.15°} = 136 \text{ pm (experimental: 136.5 pm)}$$

As expected, the smaller the F—N—F angle, the longer the N—F bond must be.

In short, the LCP model complements the VSEPR approach; whereas VSEPR predicts that a lone pair will cause a smaller bond angle on the opposite side, LCP predicts that the outer atom · · · outer atom distances should remain essentially unchanged, requiring longer central atom–outer atom distances. An atom that is multiply bonded to a central atom has a similar effect, as shown in the following example.

EXAMPLE 3.6

In PF_4^+ the F · · · F and P—F distances are 238 pm and 145.7 pm, respectively. Predict the P—F distance in POF_3, which has an F—P—F angle of 101.1°.

SOLUTION

The LCP model predicts that the F · · · F distances should be approximately the same in both structures. Sketches similar to those in Figure 3.23 can be drawn to illustrate the angles opposite the double bond:

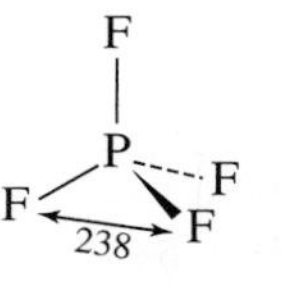

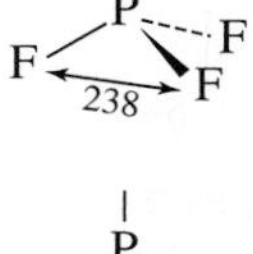

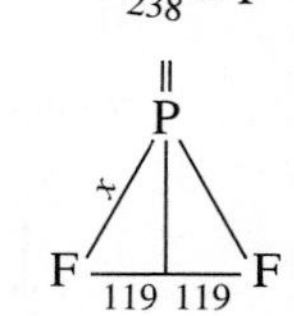

In this situation,

$$x = \text{P—F bond distance} = \frac{119 \text{ pm}}{\sin 50.55°} = 154 \text{ pm (experimental: 152 pm)}$$

(The actual F · · · F distance in POF_3 is 236 pm, slightly shorter than in PF_4^+. If this value is used in the calculation x = 153 pm, a better match for the experimental value.)

In the example, the LCP model predicts the two structures to have approximately the same size trigonal base of fluorines. A longer P—F distance and smaller F—P—F angle are consistent with the VSEPR approach, which would predict a smaller bond angle arising as a consequence of repulsion by the electrons in the double bond.

EXERCISE 3.6 Does this approach work for different steric numbers? BCl_3 has B—Cl and F · · · F distances of 174 and 301 pm. Using the LCP model, predict the B—Cl distance in BCl_4^- and compare with the experimental value of 183 pm.

Gillespie and Popelier have also described several other approaches to molecular geometry, together with their advantages and disadvantages,[25] and in an interview Gillespie has provided additional perspective on the VSEPR, LCP, and related concepts, in addition to interesting historical background.[26]

3.3 Molecular Polarity

When atoms with different electronegativities are bonded, the resulting molecule has polar bonds, with the electrons of the bond concentrated, perhaps very slightly, on the more electronegative atom; the greater the difference in electronegativity, the more polar the bond. As a result, the bonds are dipolar, with relatively positive and negative ends. This polarity can cause specific interactions between molecules, depending on the overall molecular structure.

Experimentally, the polarity of molecules is measured indirectly by measuring the dielectric constant, which is the ratio of the capacitance of a cell filled with the substance to be measured to the capacitance of the same cell with a vacuum between the electrodes. Orientation of polar molecules in the electric field partially cancels out the effect of the field and results in a larger dielectric constant. Measurements at different temperatures allow calculation of the **dipole moment** for the molecule, defined as $\mu = Qr$, where r is the distance between the centers of positive and negative charge and Q is the difference between these charges.*

Dipole moments of diatomic molecules can be calculated directly. In more complex molecules, vector addition of the individual bond dipole moments gives the net molecular dipole moment. However, it is usually not possible to calculate molecular dipoles directly from bond dipoles. Table 3.8 shows experimental and calculated dipole moments of chloromethanes. The values calculated from vectors use C—H and C—Cl bond dipole moments of 1.3×10^{-30} and 4.9×10^{-30} C m, respectively, and tetrahedral bond angles. Clearly, calculating dipole moments is more complex than simply adding the vectors for individual bond moments. However, for many purposes, a qualitative approach is sufficient.

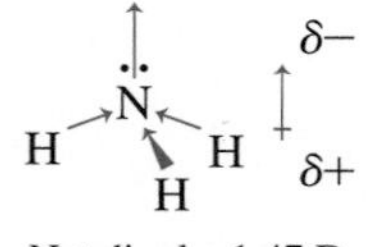

Net dipole, 1.47 D

Net dipole, 1.85 D

Net dipole, 0.23 D

FIGURE 3.24 Bond Dipoles and Molecular Dipoles.

The dipole moments of NH_3, H_2O, and NF_3 (**Figure 3.24**) reveal the often dramatic effect of lone pairs. In ammonia, the averaged N—H bond polarities and the lone pair point in the same direction, resulting in a large dipole moment. Water has an even larger dipole moment; the polarities of the O—H bonds and the two lone pairs result in polarities that reinforce each other. On the other hand, NF_3 has a very small dipole moment, the result of the polarity of the three N—F bonds opposing the electron-rich lone pair. The sum of the three N—F bond moments is larger than the lone pair effect, and the lone pair is the positive end of the molecule. In cases such as those of NF_3 and SO_2 with opposing polarities, the dipole direction is not easily predicted. SO_2 has a large dipole moment (1.63 D), with the polarity of the lone pair outweighing that of the S—O bonds, and the sulfur atom partially negative even though oxygen is more electronegative.

Molecules with dipole moments interact electrostatically with each other and with other polar molecules. When the dipoles are large enough, the molecules orient themselves with the positive end of one molecule toward the negative end of another, and higher melting and boiling points result. Details of the most dramatic effects are given in the discussion of hydrogen bonding later in this chapter and in Chapter 6.

TABLE 3.8 Dipole Moments of Chloromethanes

Molecule	Experimental (D)	Calculated from Vectors (D)
CH_3Cl	1.90	1.77
CH_2Cl_2	1.60	2.008
$CHCl_3$	1.04	1.82

Source: Experimental data, *CRC* Handbook *of Chemistry and Physics*, 92nd ed., Taylor and Francis Group, LLC, 2011–2012, pp. 9–54, 9–55, and 9–59.

*The SI unit for dipole moment is a Coulomb meter (C m), but a commonly used unit is the debye (D). One D = 3.33564×10^{-30} C m.

On the other hand, if the molecule has a highly symmetric structure, or if the polarities of different bonds cancel each other out, the molecule as a whole may have no net dipole moment, even though the individual bonds are quite polar. Tetrahedral molecules such as CH_4 and CCl_4, trigonal molecules and ions such as SO_3, NO_3^-, and CO_3^{2-}, and molecules having identical outer atoms for steric numbers 5 and 6 such as PCl_5 and SF_6, are all nonpolar. The C—H bond has very little polarity, but the bonds in the other molecules and ions are quite polar. In all these cases, the sum of all the polar bonds is zero because of the symmetry of the molecules, as shown in **Figure 3.25**.

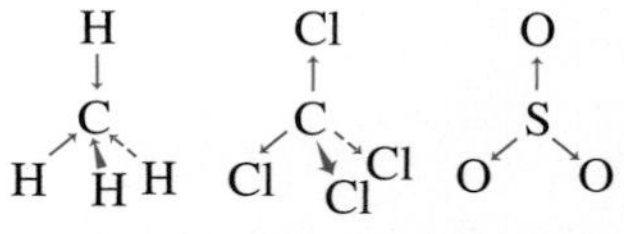

Zero net dipole for all three

FIGURE 3.25 Cancellation of Bond Dipoles Due to Molecular Symmetry.

Even nonpolar molecules participate in intermolecular attractions. Small fluctuations in the electron density in such molecules create small dipoles with extremely short lifetimes. These dipoles in turn attract or repel electrons in adjacent molecules, inducing dipoles in them as well. The result is an overall attraction among molecules. These attractive forces are called **London forces** or **dispersion forces**, and they make liquefaction of the noble gases and nonpolar molecules—such as hydrogen, nitrogen, and carbon dioxide—possible. As a general rule, London forces become more important as the number of electrons in a molecule increases. An increasing electron count generally increases the shielding of nuclear charge, rendering the electron cloud more polarizable and susceptible to perturbation from external dipoles.

3.4 Hydrogen Bonding

Ammonia, water, and hydrogen fluoride all have much higher boiling points than other similar molecules, as shown in **Figure 3.26**. These high boiling points are caused by hydrogen bonds, in which hydrogen atoms bonded to nitrogen, oxygen, or fluorine also form weaker bonds to a lone pair of electrons on another nitrogen, oxygen, or fluorine. Bonds between hydrogen and these strongly electronegative atoms are very polar, with a partial positive charge on the hydrogen. This partially positive hydrogen is strongly attracted to the partially negative nitrogen, oxygen or fluorine of neighboring molecules. In the past, the attractions among these molecules were considered primarily electrostatic in nature, but an alternative molecular orbital approach (described in Chapters 5 and 6) gives a more complete description of this phenomenon. Regardless of the detailed explanation of

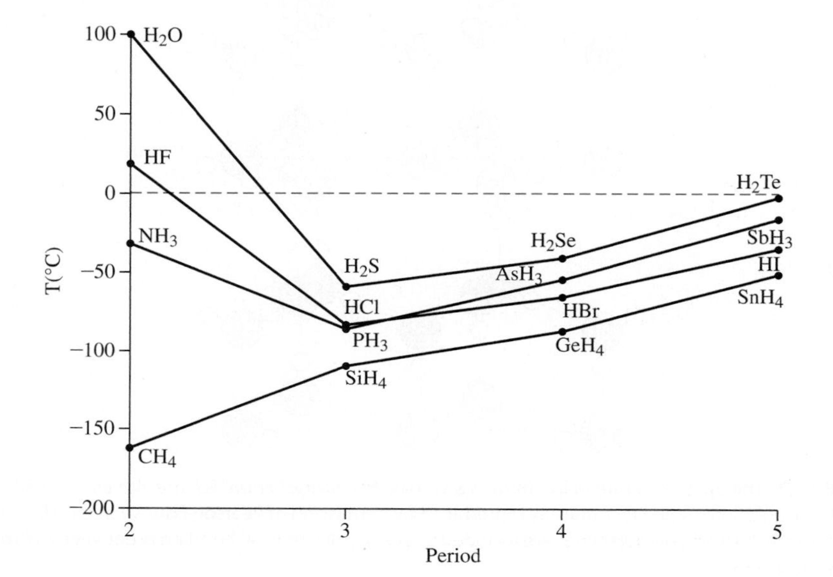

FIGURE 3.26 Boiling Points of Hydrogen Compounds.

the forces involved in hydrogen bonding, the strongly positive hydrogen and the strongly negative lone pairs tend to line up and hold the molecules together. Other atoms with high electronegativity, such as chlorine, can also enable formation of hydrogen bonds in polar molecules such as chloroform, $CHCl_3$. The definition of what atoms can participate in hydrogen bonds has expanded dramatically beyond the traditional atoms of N, O, and F, as described in Chapter 6.

In general, boiling points rise with increasing molecular weight, both because the additional mass requires higher temperature for rapid movement of the molecules and because the larger number of electrons in the heavier molecules provides larger London forces. The difference in temperature between the actual boiling point of water and the extrapolation of the line connecting the boiling points of the heavier analogous compounds is almost 200° C. Ammonia and hydrogen fluoride have similar but smaller differences from the extrapolated values for their families. Water has a much larger effect, because each molecule can have as many as four hydrogen bonds (two through the lone pairs and two through the hydrogen atoms). Hydrogen fluoride can average no more than two hydrogen bonds, because hydrogen fluoride has only one hydrogen.

Water has other unusual properties because of hydrogen bonding. For example, the freezing point of water is much higher than that of similar molecules. An even more striking feature is the decrease in density as water freezes. The tetrahedral structure around each oxygen atom, with two regular bonds to hydrogen and two hydrogen bonds to other molecules, requires a very open structure with large spaces between ice molecules (**Figure 3.27**). This makes the solid less dense than the liquid water surrounding it, so ice floats. Life on Earth would be very different if this were not so. Lakes, rivers, and oceans would freeze from the bottom up, ice cubes would sink, and ice fishing would be impossible. The results are difficult to imagine, but would certainly require a much different biology and geology. The same forces cause coiling of protein (**Figure 3.28**) and polynucleic acid molecules; a combination of hydrogen bonding with other dipolar forces imposes considerable secondary structure on these large molecules. In Figure 3.28(a), hydrogen bonds between carbonyl oxygen atoms and hydrogens attached to nitrogen atoms

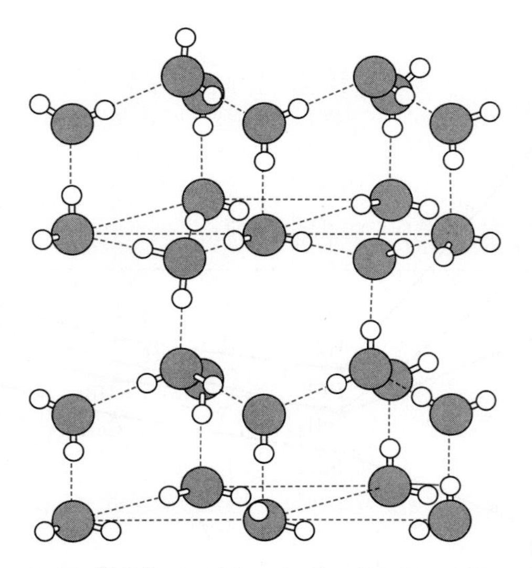

FIGURE 3.27 The open structure of ice. (Brown & Lemay, *Chemistry: Central Science*, 4th Ed., © 1988, pp. 628, 946. Reprinted and Electronically reproduced by permission of Pearson Education Inc, Upper Saddle River, NJ 07458.) The rectangular lines are included to aid visualization; all bonding is between hydrogen and oxygen atoms.

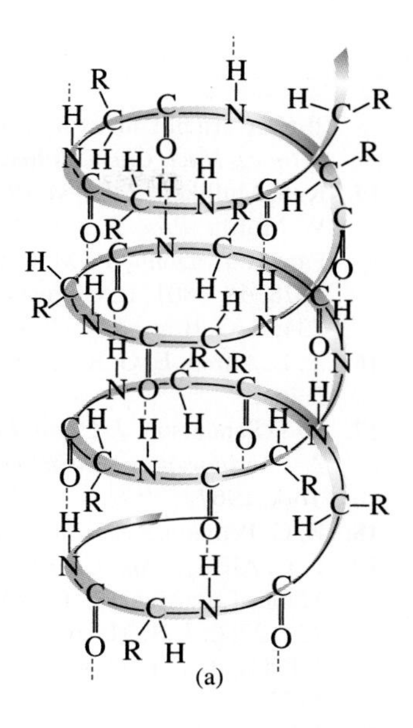

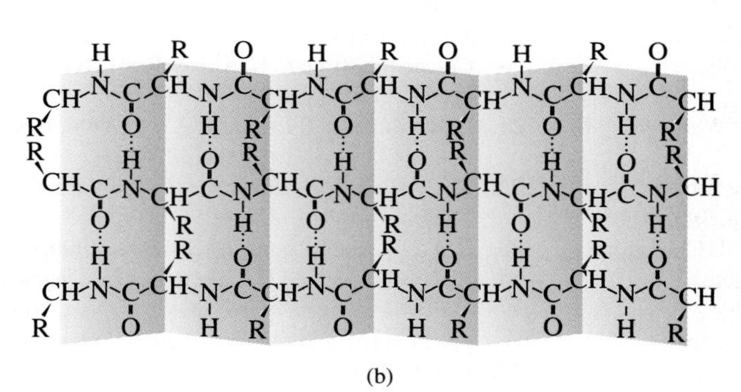

FIGURE 3.28 Hydrogen-Bonded Protein Structures. (a) A protein α helix. Peptide carbonyls and N—H hydrogens on adjacent turns of the helix are hydrogen bonded. (Brown & Lemay, *Chemistry: Central Science*, 4th Ed., © 1988, pp. 628, 946. Reprinted and Electronically reproduced by permission of Pearson Education Inc, Upper Saddle River, NJ 07458.) The pleated sheet arrangement is shown in (b). Each peptide carbonyl group is hydrogen bonded to an N—H hydrogen on an adjacent peptide chain. (Wade, L. G., *Organic Chemistry*, 1st Ed., © 1988. Reprinted with permission and Electronically reproduced by permission of Pearson Education Inc., Upper Saddle River NJ.)

hold the molecule in a helical structure. In Figure 3.28(b), similar hydrogen bonds hold the parallel peptide chains together; the bond angles of the chains result in the pleated appearance of the sheet formed by the peptides. These are two of the many different structures that can be formed from peptides, depending on the side-chain groups R and the surrounding environment.

Another example is a theory of anesthesia by non-hydrogen bonding molecules such as cyclopropane, chloroform, and nitrous oxide, proposed by Pauling.[27] These molecules are of a size and shape that can fit neatly into a hydrogen-bonded water structure with even larger open spaces than ordinary ice. Such structures, with molecules trapped in holes in a solid, are called **clathrates**. Pauling proposed that similar hydrogen-bonded microcrystals form even more readily in nerve tissue because of the presence of other solutes in the tissue. These microcrystals could then interfere with the transmission of nerve impulses. Similar structures of methane and water hold large quantities of methane in the polar ice caps. The amount of methane in such crystals can be so great that they burn if ignited.[28]

More specific interactions involving the sharing of electron pairs between molecules are discussed in connection with acid–base chemistry in Chapter 6.

References

1. G. N. Lewis, *J. Am. Chem. Soc.*, **1916**, *38*, 762; *Valence and the Structure of Atoms and Molecules*, Chemical Catalogue Co., New York, 1923.
2. L. Suidan, J. K. Badenhoop, E. D. Glendening, F. Weinhold, *J. Chem. Educ.*, **1995**, *72*, 583; J. Cioslowski, S. T. Mixon, *Inorg. Chem.*, **1993**, *32*, 3209; E. Magnusson, *J. Am. Chem. Soc.*, **1990**, *112*, 7940.
3. A. G. Sharpe, "Cyanides and Fulminates," in G. Wilkinson, R. D. Gillard, and J. S. McClevert, eds., *Comprehensive Coordination Chemistry*, Vol. 2, Pergamon Press, New York, 1987, pp. 12–14.
4. N. V. Sidgwick, H. M. Powell, *Proc. R. Soc.*, **1940**, *A176*, 153.
5. R. J. Gillespie, R. S. Nyholm, *Q. Rev. Chem. Soc.*, **1957**, *XI*, 339; R. J. Gillespie, *J. Chem. Educ.*, **1970**, *47*, 18; and R. J. Gillespie, *Coord. Chem. Rev.*, **2008**, *252*, 1315. The last reference provides a useful synopsis of 50 years of the VSEPR model.
6. G. M. Barrow, *Physical Chemistry*, 6th ed., McGraw-Hill, New York, 1988, pp. 567–699; R. S. Drago, *Physical Methods for Chemists*, 2nd ed., Saunders College Publishing, Philadelphia, 1977, pp. 689–711.
7. J. L. Hoard, W. J. Martin, M. E. Smith, J. F. Whitney, *J. Am. Chem. Soc.*, **1954**, *76*, 3820.
8. A. F. Wells, *Structural Inorganic Chemistry*, 5th ed., Oxford University Press, New York, 1984, p. 390.
9. N. Vasdev, M. D. Moran, H. M. Tuononen, R. Chirakal, R. J. Suontamo, A. D. Bain, G. J. Schrobilgen, *Inorg. Chem.*, **2010**, *49*, 8997.
10. See, for example, S. Hoyer, T. Emmler, K. Seppelt, *J. Fluorine Chem.*, *2006*, *127*, 1415, and references therein on the various structural modifications of XeF_6.
11. R. J. Gillespie and I. Hargittai, *The VSEPR Model of Molecular Geometry*, Allyn & Bacon, Boston, 1991, p. 77.
12. J. B. Mann, T. L. Meek, E. T. Knight, J. F. Capitani, L. C. Allen, *J. Am. Chem. Soc.*, **2000**, *122*, 5132.
13. L. C. Allen, *J. Am. Chem. Soc.*, **1989**, *111*, 9003.
14. L. Pauling, *The Nature of the Chemical Bond*, 3rd ed., 1960, Cornell University Press, Ithaca, NY; A. L. Allred, *J. Inorg. Nucl. Chem.*, **1961**, *17*, 215.
15. R. S. Mulliken, *J. Chem. Phys.*, **1934**, *2*, 782; **1935**, *3*, 573; W. Moffitt, *Proc. R. Soc. (London)*, **1950**, *A202*, 548; R. G. Parr, R. A. Donnelly, M. Levy, W. E. Palke, *J. Chem. Phys.*, **1978**, *68*, 3801; R. G. Pearson, *Inorg. Chem.*, **1988**, *27*, 734; S. G. Bratsch, *J. Chem. Educ.*, **1988**, *65*, 34, 223.
16. A. L. Allred, E. G. Rochow, *J. Inorg. Nucl. Chem.*, **1958**, *5*,264.
17. R. T. Sanderson, *J. Chem. Educ.*, **1952**, *29*, 539; **1954**, *31*, 2, 238; *Inorganic Chemistry*, Van Nostrand-Reinhold, New York, 1967.
18. R. G. Pearson, *Acc. Chem. Res.*, **1990**, *23*, 1.
19. L. C. Allen, *J. Am. Chem. Soc.*, **1989**, *111*, 9003; J. B. Mann, T. L. Meek, L. C. Allen, *J. Am. Chem. Soc.*, **2000**, *122*, 2780; J. B. Mann, T. L. Meek, E. T. Knight, J. F. Capitani, L. C. Allen, *J. Am. Chem. Soc.*, **2000**, *122*, 5132.
20. J. Hinze, H. H. Jaffé, *J. Am. Chem. Soc.*, **1962**, *84*, 540; *J. Phys. Chem.*, **1963**, *67*, 1501; J. E. Huheey, *Inorganic Chemistry*, 3rd ed., Harper & Row, New York, 1983, pp. 152–156.
21. L. R. Murphy, T. L. Meek, A. L. Allred, L. C. Allen, *J. Am. Chem. Soc.*, **2000**, *122*, 5867.
22. L. C. Allen, J. E. Huheey, *J. Inorg. Nucl. Chem.*, **1980**, *42*, 1523.
23. R. J. Gillespie, *Coord. Chem. Rev.*, **2000,** *197*, 51.
24. R. J. Gillespie and P. L. A. Popelier, *Chemical Bonding and Molecular Geometry*, Oxford, New York, 2001, pp. 113–133 and references cited therein.
25. R. J. Gillespie and P. L. A. Popelier, *Chemical Bonding and Molecular Geometry*, Oxford University Press, New York, 2001, pp. 134–180.
26. L. Cardellini, *J. Chem. Educ.*, **2010**, *87*, 482.
27. L. Pauling, *Science*, **1961**, *134*, 15.
28. L. A. Stern, S. H. Kirby, W. B. Durham, *Science*, **1996**, *273*, 1765 (cover picture), 1843.

General References

Good sources for bond lengths and bond angles are the works of Wells, Greenwood and Earnshaw, and Cotton and Wilkinson cited in Chapter 1. Reviews of electron-dot diagrams and formal charges can be found in most general chemistry texts. One of the best VSEPR references is still the early paper by R. J. Gillespie and R. S. Nyholm, *Q. Rev. Chem. Soc.*, **1957**, *XI*, 339–380. More recent expositions of the theory are in R. J. Gillespie and I. Hargittai, *The VSEPR Model of Molecular Geometry*, Allyn & Bacon, Boston, 1991, and R. J. Gillespie and P. L. A. Popelier, *Chemical Bonding and Molecular Geometry: From Lewis to Electron Densities*, Oxford University Press, New York, 2001. The latter also presents a useful introduction to the Ligand Close-Packing (LCP) model. Gillespie has also provided a retrospective on VSEPR in R. J. Gillespie, *Coord. Chem. Rev.*, **2008**, *252*, 1315, and he and Robinson have compared the VSEPR, LCP, and valence-bond descriptions of molecular geometry in R. J. Gillespie and E. A. Robinson, *Chem. Soc. Rev.*, **2005**, *34*, 396. Additional perspective on VSEPR, LCP, and related topics is provided by an interview with R. J. Gillespie in L. Cardellini, *J. Chem. Ed.*, **2010**, *87*, 482. Molecular orbital arguments for the shapes of many of the same molecules are presented in B. M. Gimarc, *Molecular Structure and Bonding*, Academic Press, New York, 1979, and J. K. Burdett, *Molecular Shapes*, John Wiley & Sons, New York, 1980. Two interesting perspectives on Lewis structures are in G. H. Purser, *J. Chem. Educ.*, **1999**, *76*, 1013 and R. F. See, *J. Chem. Educ.*, **2009**, *86*, 1241.

Problems

3.1 The dimethyldithiocarbamate ion, $[S_2CN(CH_3)_2]^-$, has the following skeletal structure:

S\C—N(CH3)2 /S

a. Give the important resonance structures of this ion, including any formal charges where necessary. Select the resonance structure likely to provide the best description of this ion.

b. Repeat for the dimethylthiocarbamate ion, $[OSCN(CH_3)_2]^-$.

3.2 Several resonance structures are possible for each of the following ions. For each, draw these resonance structures, assign formal charges, and select the resonance structure likely to provide the best description for the ion.

a. Selenocyanate ion, $SeCN^-$

b. Thioformate ion, $[H—C(O)S]^-$

c. Dithiocarbonate, $[S_2CO]^{2-}$ (C is central)

3.3 Draw the resonance structures for the isoelectronic ions NSO^- and SNO^-, and assign formal charges. Which ion is likely to be more stable?

3.4 Three isomers having the formula N_2CO are known: ONCN (nitrosyl cyanide), ONNC (nitrosyl isocyanide), and NOCN (isonitrosyl cyanide). Draw the most important resonance structures of these isomers, and determine the formal charges. Which isomer do you predict to be the most stable (lowest energy) form? (See G. Maier, H. P. Reinsenauer, J. Eckwert, M. Naumann, M. De Marco, *Angew. Chem., Int. Ed.*, **1997**, *36*, 1707.)

3.5 Show the possible resonance structures for nitrous oxide, N_2O (the central atom is nitrogen). Indicate nonzero formal charges where they are present. Which resonance structure gives the best representation of this molecule?

3.6 Nitric acid, which exists as HNO_3 molecules in the absence of water, has the skeletal structure shown. Show the important resonance structures of HNO_3, and designate the formal charges on each atom.

H—O—N(O)O

3.7 L. C. Allen has suggested that a more meaningful formal charge can be obtained by taking into account the electronegativities of the atoms involved. Allen's formula for this type of charge—referred to as the *Lewis–Langmuir (L–L) charge*—of an atom, A, bonded to another atom, B, is

$$\text{L–L charge} = \frac{\text{(US) group}}{\text{number of A}} - \frac{\text{number of unshared}}{\text{electrons on A}} - 2\sum_B \frac{\chi_A}{\chi_A + \chi_B}\left(\begin{array}{c}\text{number of bonds}\\ \text{between A and B}\end{array}\right)$$

where χ_A and χ_B designate the electronegativities. Using this equation, calculate the L–L charges for CO, NO^-, and HF, and compare the results with the corresponding formal charges. Do you think the L–L charges are a better representation of electron distribution? (See L. C. Allen, *J. Am. Chem. Soc.*, **1989**, *111*, 9115; L. D. Garner, T. L. Meek, B. G. Patrick, *THEOCHEM*, **2003**, *620*, 43.)

3.8 Give Lewis dot structures and sketch the shapes of the following:

a. $SeCl_4$
b. I_3^-
c. $PSCl_3$ (P is central)
d. IF_4^-
e. PH_2^-
f. TeF_4^{2-}
g. N_3^-
h. $SeOCl_4$ (Se is central)
i. PH_4^+

3.9 Give Lewis dot structures and sketch the shapes of the following:

a. ICl_2^-
b. H_3PO_3 (one H is bonded to P)
c. BH_4^-
d. $POCl_3$
e. IO_4^-
f. $IO(OH)_5$
g. $SOCl_2$
h. $ClOF_4^-$
i. XeO_2F_2

3.10 Give Lewis dot structures and sketch the shapes of the following:

a. SOF_6 (one F is attached to O)
b. POF_3
c. ClO_2
d. NO_2
e. $S_2O_4^{2-}$ (symmetric, with an S—S bond)
f. N_2H_4 (symmetric, with an N—N bond)
g. $ClOF_2^+$
h. CS_2
i. $XeOF_5^-$

3.11 Explain the trends in bond angles and bond lengths of the following ions:

	X—O (pm)	O—X—O Angle
ClO_3^-	149	107°
BrO_3^-	165	104°
IO_3^-	181	100°

3.12 Select from each set the molecule or ion having the smallest bond angle, and briefly explain your choice:

a. NH_3, PH_3, or AsH_3

b. $OSCl_2$ or OSF_2 (halogen—sulfur—halogen angle)

c. NO_2^- or O_3

d. ClO_3^- or BrO_3^-

3.13 a. Compare the structures of the azide ion, N_3^-, and the ozone molecule, O_3.

b. How would you expect the structure of the ozonide ion, O_3^-, to differ from that of ozone?

3.14 Consider the series OCl_2, $O(CH_3)_2$, and $O(SiH_3)_2$, which have bond angles at the oxygen atom of 110.9°, 111.8°, and 144.1° respectively. Account for this trend.

3.15 Two ions isoelectronic with carbon suboxide, C_3O_2, are N_5^+ and $OCNCO^+$. Whereas C_3O_2 is linear, both N_5^+ and $OCNCO^+$ are bent at the central nitrogen. Suggest an explanation. Also predict which has the smaller outer atom—N—outer atom angle and explain your reasoning. (See I. Bernhardi, T. Drews, K. Seppelt, *Angew. Chem., Int. Ed.*, **1999**, *38*, 2232; K. O. Christe, W. W. Wilson, J. A. Sheehy, J. A. Boatz, *Angew. Chem., Int. Ed.*, **1999**, *38*, 2004.)

3.16 Explain the following:

a. Ethylene, C_2H_4, is a planar molecule, but hydrazine, N_2H_4, is not.

b. ICl_2^- is linear, but NH_2^- is bent.

c. Of the compounds mercury(II) cyanate, $Hg(OCN)_2$, and mercury(II) fulminate, $Hg(CNO)_2$, one is highly explosive, and the other is not.

3.17 Explain the following:

a. PCl_5 is a stable molecule, but NCl_5 is not.

b. SF_4 and SF_6 are known, but OF_4 and OF_6 are not.

3.18 X-ray crystal structures of $ClOF_3$ and $BrOF_3$ have been determined.

a. Would you expect the lone pair on the central halogen to be axial or equatorial in these molecules? Why?

b. Which molecule would you predict to have the smaller $F_{equatorial}$—central atom—oxygen angle? Explain your reasoning. (See A. Ellern, J. A. Boatz, K. O. Christe, T. Drews, K. Seppelt, Z. *Anorg. Allg. Chem.*, **2002**, *628*, 1991.)

3.19 Make the following comparisons about the molecules shown next, and briefly explain your choices.

a. Which molecule has the smaller H_3C—group15 atom—CH_3 angle?

b. Which molecule has the smaller $H_3C—Al—CH_3$ angle?

c. Which molecule has the longer Al—C bond distance?

$(H_3C)_3N—Al(CH_3)_3$ $(H_3C)_3P—Al(CH_3)_3$

3.20 Predict and sketch the structure of the (as yet) hypothetical ion IF_3^{2-}.

3.21 A solution containing the $IO_2F_2^-$ ion reacts slowly with excess fluoride ion to form $IO_2F_3^{2-}$.

a. Sketch the isomers that might be possible matching the formula $IO_2F_3^{2-}$.

b. Of these structures, which do you think is most likely? Why?

c. Propose a formula of a xenon compound or ion isoelectronic with $IO_2F_3^{2-}$.

(See J. P. Mack, J. A. Boatz, M. Gerken, *Inorg. Chem.*, **2008**, *47*, 3243.)

3.22 The $XeOF_3^-$ anion has been reported recently (D. S. Brock and G. J. Schrobilgen, *J. Am. Chem. Soc.*, **2010**, *133*, 6265).

a. Several structures matching this formula are conceivable. Sketch these, select the structure that is most likely, and justify your choice.

b. What is notable about the structure of this ion?

3.23 Predict the structure of $I(CF_3)Cl_2$. Do you expect the CF_3 group to be in an axial or equatorial position? Why? (See R. Minkwitz, M. Merkei, *Inorg. Chem.*, **1999**, *38*, 5041.)

3.24 a. Which has the longer axial P—F distance, $PF_2(CH_3)_3$ or $PF_2(CF_3)_3$? Explain briefly.

b. Al_2O has oxygen as central atom. Predict the approximate bond angle in this molecule and explain your answer.

c. Predict the structure of CAl_4. (See X. Li, L-S. Wang, A. I. Boldyrev, J. Simons, *J. Am. Chem. Soc.*, **1999**, *121*, 6033.)

3.25 The structures of TeF_4 and $TeCl_4$ in the gas phase have been studied by electron diffraction (S. A. Shlykov, N. I. Giricheva, A. V. Titov, M. Szwak, D. Lentz, G. V. Girichev, *Dalton Trans.*, **2010**, *39*, 3245).

a. Would you expect the Te—X (axial) distances in these molecules to be longer or shorter than than Te—X (equatorial) distances? Explain briefly.

b. Which compound would you predict to have the smaller X(axial)—Te—X(axial) angles? The smaller X(equatorial)—Te—X(equatorial) angles? Explain briefly.

3.26 $SeCl_6^{2-}$, $TeCl_6^{2-}$, and ClF_6^- are all octahedral, but SeF_6^{2-} and IF_6^- are distorted, with a lone pair on the central atom apparently influencing the shape. Suggest a reason for the difference in shape of these two groups of ions. (See J. Pilmé, E. A. Robinson, R. J. Gillespie, *Inorg. Chem.*, **2006**, *45*, 6198.)

3.27 When XeF_4 is reacted with a solution of water in CH_3CN solvent, the product $F_2OXeN{\equiv}CCH_3$ is formed. Applying a vacuum to crystals of this product resulted in slow removal of CH_3CN:

$$F_2OXeN{\equiv}CCH_3 \longrightarrow XeOF_2 + CH_3CN$$

Propose structures for $F_2OXeN{\equiv}CCH_3$ and $XeOF_2$. (See D. S. Brock, V. Bilir, H. P. A. Mercier, G. J. Schrobilgen, *J. Am. Chem. Soc.*, **2007**, *129*, 3598.)

3.28 The thiazyl dichloride ion, $NSCl_2^-$, is isoelectronic with thionyl dichloride, $OSCl_2$.

a. Which of these species has the smaller Cl—S—Cl angle? Explain briefly.
b. Which do you predict to have the longer S—Cl bond? Why? (See E. Kessenich, F. Kopp, P. Mayer, A. Schulz, *Angew. Chem., Int. Ed.*, **2001**, *40*, 1904.)

3.29 Sketch the most likely structure of PCl_3Br_2 and explain your reasoning.

3.30 a. Are the CF_3 groups in $PCl_3(CF_3)_2$ more likely axial or equatorial? Explain briefly.
b. Are the axial or equatorial bonds likely to be longer in $SbCl_5$? Explain briefly.

3.31 Of the molecules $ClSO_2CH_3$, $ClSO_2CF_3$, and $ClSO_2CCl_3$, which has the largest X—S—X angle? Explain briefly.

3.32 Of the molecules FSO_2F, $FSO_2(OCH_3)$, and FSO_2CH_3, which has the smallest O—S—O angle? Explain briefly.

3.33 Elemental Se and Te react with 4-tetrafluoropyridyl silver(I) to afford $Se(C_5F_4N)_2$ and $Te(C_5F_4N)_2$. Two independent bent molecules were found for each compound in the solid state with C—Se—C angles of 95.47(12)° and 96.16(13)°, and C—Te—C angles of 90.86(18)° and 91.73(18)°, respectively (Aboulkacem, S.; Naumann, D.; Tyrra, W.; Pantenburg, I. *Organometallics*, **2012**, *31*, 1559).

$$2\ (C_5F_4N)\text{–}Ag + E \xrightarrow[E=Se,\ Te]{} \left(C_5F_4N\right)_2E + 2\ Ag$$

a. Explain why the angles are more acute for the Te compound relative to the Se compound.
b. These angles are approximately 0.8° (Se) and 2.0° (Te) more acute than those in the related pentafluorophenyl (C_6F_5) compounds. The greater compression of these angles in the 4-tetrafluoropyridyl compounds has been postulated on the basis of group electronegativity differences. Explain the logic associated with this hypothesis.

3.34 Which has the smaller F—P—F angle, PF_4^+ or PF_3O? Which has the longer fluorine–fluorine distance? Explain briefly.

3.35 Account for the trend in P—F_{axial} distances in the compounds $PF_4(CH_3)$, $PF_3(CH_3)_2$, and $PF_2(CH_3)_3$. (See Figure 3.19.)

3.36 Although the C—F distances and the F—C—F bond angles differ considerably in $F_2C{=}CF_2$, F_2CO, CF_4, and F_3CO^- (C—F distances 131.9 to 139.2 pm; F—C—F bond angles 101.3° to 109.5°), the F···F distance in all four structures is very nearly the same (215 to 218 pm). Explain, using the LCP model of Gillespie. (See R. J. Gillespie, *Coord. Chem. Rev.*, **2000**, *197*, 51.)

3.37 The Cl···Cl distance in CCl_4 is 289 pm, and the C—Cl bond distance is 171.1 pm. Using the LCP model, calculate the C—Cl distance in Cl_2CO, which has a Cl—C—Cl angle of 111.8°.

3.38 The F—C—F angle in F_2CO, shown here, is 109.5°; the C—F distance is 131.7 pm, and the F···F distance is 215 pm. On the basis of the LCP model, predict the C—F distance in the CF_3^+ ion.

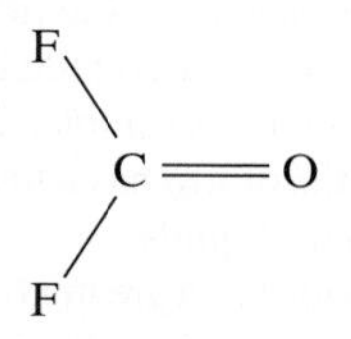

3.39 Compounds in which hydrogen is the outer atom can provide challenges to theories of chemical bonding. Consider the following molecules. Using one or more of the approaches described in this chapter, provide a rationale for HOF having the smallest bond angle in this set.

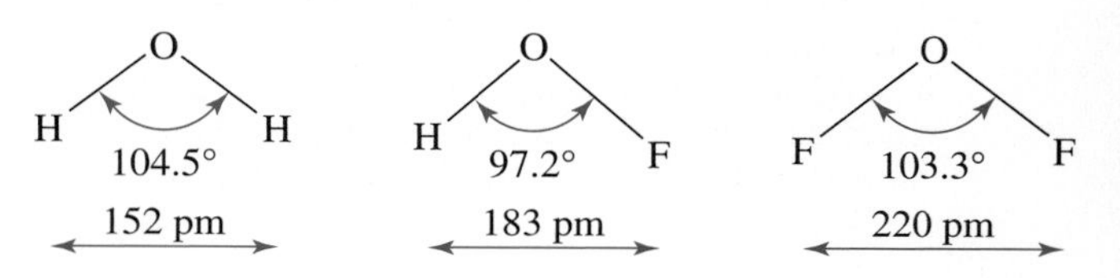

3.40 For each of the following bonds, indicate which atom is more negative, then rank the series in order of polarity.

a. C—N	b. N—O	c. C—I
d. O—Cl	e. P—Br	f. S—Cl

3.41 Give Lewis dot structures and shapes for the following:

a. $VOCl_3$	b. PCl_3	c. SOF_4
d. SO_3	e. ICl_3	f. SF_6
g. IF_7	h. XeO_2F_4	i. CF_2Cl_2
j. P_4O_6		

(P_4O_6 is a closed structure with overall tetrahedral arrangement of phosphorus atoms; an oxygen atom bridges each pair of phosphorus atoms.)

3.42 Give Lewis dot structures and sketch the shapes for the following:

a. PH_3	b. H_2Se	c. SeF_4
d. PF_5	e. IF_5	f. XeO_3
g. BF_2Cl	h. $SnCl_2$	i. KrF_2
j. $IO_2F_5^{2-}$		

3.43 Which of the molecules in Problem 3.41 are polar?

3.44 Which of the molecules in Problem 3.42 are polar?

3.45 Provide explanations for the following:
a. Methanol, CH_3OH, has a much higher boiling point than methyl mercaptan, CH_3SH.
b. Carbon monoxide has slightly higher melting and boiling points than N_2.
c. The *ortho* isomer of hydroxybenzoic acid $[C_6H_4(OH)(CO_2H)]$ has a much lower melting point than the *meta* and *para* isomers.
d. The boiling points of the noble gases increase with atomic number.

e. Acetic acid in the gas phase has a significantly lower pressure (approaching a limit of one half) than predicted by the ideal gas law.

f. Mixtures of acetone and chloroform exhibit significant negative deviations from Raoult's law, which states that the vapor pressure of a volatile liquid is proportional to its mole fraction. For example, an equimolar mixture of acetone and chloroform has a lower vapor pressure than either of the pure liquids.

g. Carbon monoxide has a greater bond-dissociation energy (1072 kJ/mol) than molecular nitrogen (945 kJ/mol).

3.46 Structural data for classical molecules is updated as new experiments are devised to improve precision and accuracy. Consider the trifluoromethyl compounds $E(CF_3)_3$ where E = P, As, Sb. The following C—E—C angles were determined for gas-phase $E(CF_3)_3$ by electron diffraction in the middle of the twentieth century: E = P, 99.6(25)°; E = As, 100.1(35)°; E = Sb, 100.0(35)°. (Berger, R. J. F.; Mitzel, N. W. *J. Mol. Struc.*, **2010**, *978*, 205).

a. Does the trend in these angles seem reasonable? Explain.

b. A more rigorous electron diffraction gas-phase experiment found a C—As—C angle for $As(CF_3)_3$ of 95.4(3)°, that compared favorably to the computationally determined angle of 95.9°. Which of the other two C—E—C angles seems more suspect, and should be reinvestigated using modern techniques? Explain.

Chapter 4

Symmetry and Group Theory

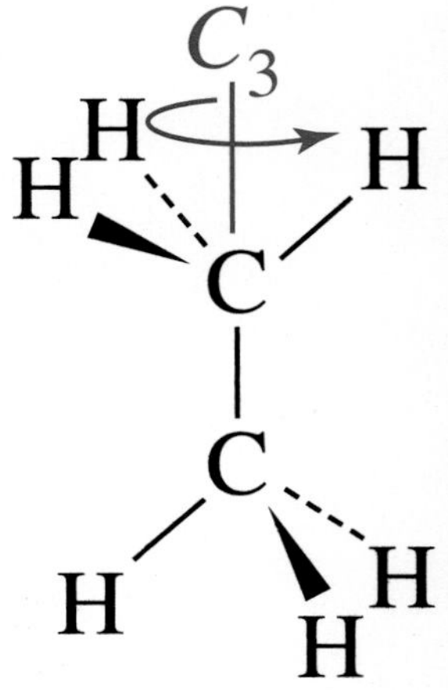

Symmetry is a phenomenon of the natural world, as well as the world of human invention (**Figure 4.1**). In nature, many flowers and plants, snowflakes, insects, certain fruits and vegetables, and a wide variety of microscopic plants and animals exhibit characteristic symmetry. Many engineering achievements have a degree of symmetry that contributes to their esthetic appeal. Examples include cloverleaf intersections, the pyramids of ancient Egypt, and the Eiffel Tower.

Symmetry concepts are extremely useful in chemistry. By analyzing the symmetry of molecules, we can predict infrared spectra, describe orbitals used in bonding, predict optical activity, interpret electronic spectra, and study a number of additional molecular properties. In this chapter, we first define symmetry very specifically in terms of five fundamental symmetry operations. We then describe how molecules can be classified on the basis of the types of symmetry they possess. We conclude with examples of how symmetry can be used to predict optical activity of molecules and to determine the number and types of infrared- and Raman-active molecular vibrations.

In later chapters, symmetry will be a valuable tool in the construction of molecular orbitals (Chapters 5 and 10) and in the interpretation of electronic spectra of coordination compounds (Chapter 11) and vibrational spectra of organometallic compounds (Chapter 13).

A molecular model kit is a useful study aid for this chapter, even for those who can visualize three-dimensional objects easily. We strongly encourage the use of such a kit.*

4.1 Symmetry Elements and Operations

All molecules can be described in terms of their symmetry, even if it is only to say they have none. Molecules or any other objects may contain **symmetry elements** such as mirror planes, axes of rotation, and inversion centers. The actual reflection, rotation, or inversion is called a **symmetry operation**. To contain a given symmetry element, a molecule must have exactly the same appearance after the operation as before. In other words, photographs of the molecule (if such photographs were possible) taken from the same location before and after the symmetry operation would be indistinguishable. If a symmetry operation yields a molecule that can be distinguished from the original in any way, that operation is *not* a symmetry operation of the molecule. The examples in Figures 4.2 through 4.6 illustrate molecular symmetry operations and elements.

The **identity operation** (E) causes no change in the molecule. It is included for mathematical completeness. An identity operation is characteristic of every molecule, even if it has no other symmetry.

*An excellent web site that discusses symmetry with molecular animations has been developed by Dean Johnston, and can be found at symmetry.otterbein.edu

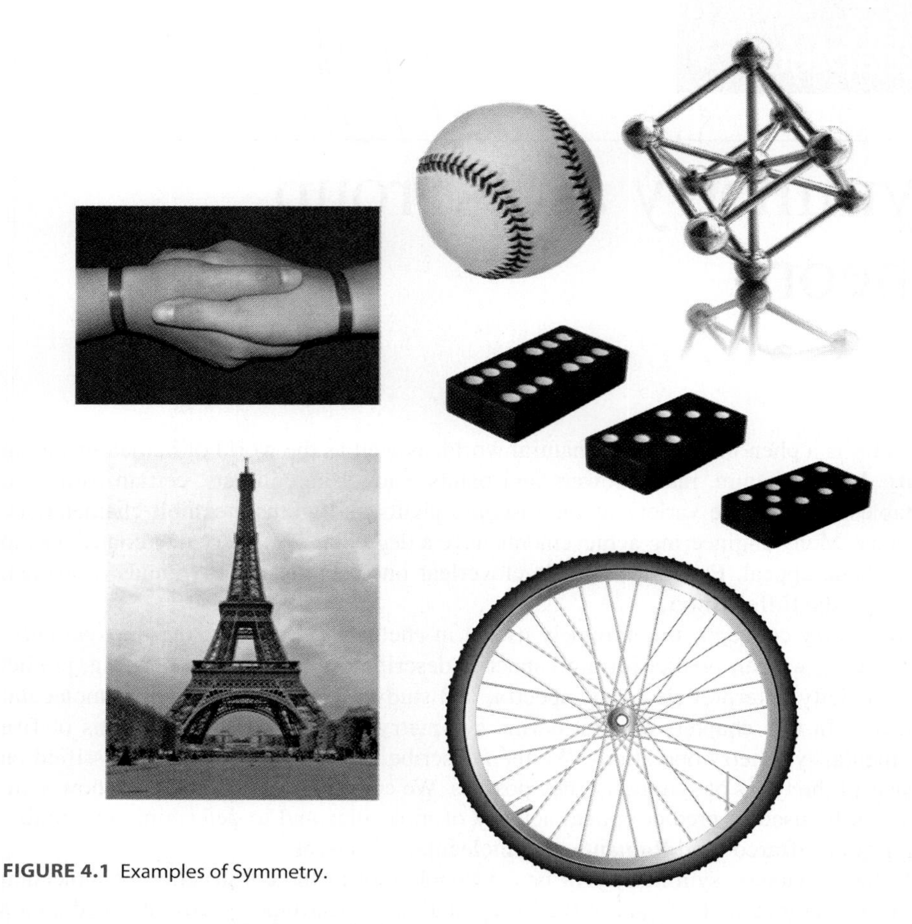

FIGURE 4.1 Examples of Symmetry.

The **rotation operation** (C_n), also called **proper rotation**, is rotation through 360°/n about a rotation axis. We use counterclockwise rotation as a positive rotation. An example of a molecule having a threefold (C_3) axis is $CHCl_3$. The rotation axis is coincident with the C—H bond axis, and the rotation angle is 360°/3 = 120°. Two C_3 operations may be performed consecutively to give a new rotation of 240°. The resulting operation is designated C_3^2 and is also a symmetry operation of the molecule. Three successive C_3 operations are the same as the identity operation ($C_3^3 \equiv E$).

Many molecules and other objects have multiple rotation axes. For example, snowflakes (**Figure 4.2**) exhibit complex shapes that are nearly always hexagonal and approximately planar. The line through the center of the flake perpendicular to the plane of the flake contains a twofold (C_2) axis, a threefold (C_3) axis, and a sixfold (C_6) axis. Rotations by 240° (C_3^2) and 300° (C_6^5) are also symmetry operations of the snowflake.

Rotation Angle	Symmetry Operation
60°	C_6
120°	$C_3 (\equiv C_6^2)$
180°	$C_2 (\equiv C_6^3)$
240°	$C_3^2 (\equiv C_6^4)$
300°	C_6^5
360°	$C_6^6 (\equiv E)$

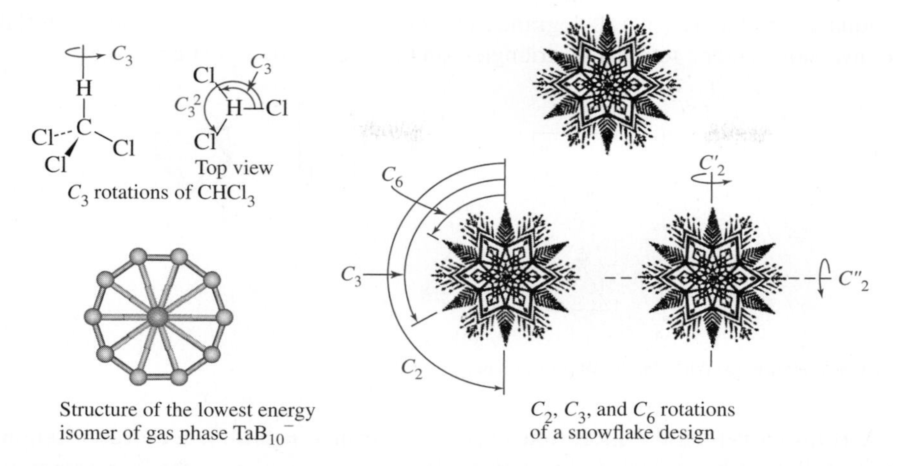

FIGURE 4.2 Rotations.

There are also two sets of three C_2 axes in the plane of the snowflake, one set through opposite points and one through the cut-in regions between the points. One of each of these axes is shown in Figure 4.2. In molecules with more than one rotation axis, the C_n axis having the largest value of n is the **highest order rotation axis** or **principal axis**. The highest order rotation axis for a snowflake is the C_6 axis. (In assigning Cartesian coordinates, the highest order C_n axis is usually chosen as the z axis.) When necessary, the C_2 axes perpendicular to the principal axis are designated with primes; a single prime (C_2') indicates that the axis passes through several atoms of the molecule, whereas a double prime (C_2'') indicates that it passes between the outer atoms.

Finding rotation axes for some three-dimensional figures can be more difficult but the same in principle. Nature can be extraordinary when it comes to symmetry—the lowest energy isomer of gas phase TaB_{10}^- (Figure 4.2) has a ten-fold rotation axis!*

In the **reflection operation** (σ) the molecule contains a mirror plane. If details such as hair style and location of internal organs are ignored, the human body has a left–right mirror plane, as in **Figure 4.3**. Many molecules have mirror planes, although they may not be obvious to identify. The reflection operation exchanges left and right, as if each point had moved perpendicularly through the plane to a position exactly as far from the plane as when it started. Linear objects, such as a round wood pencil, or molecules, such as acetylene and carbon dioxide, have an infinite number of mirror planes that include the center line of the object.

When the plane is perpendicular to the principal axis of rotation, it is called σ_h (horizontal). Other planes, which contain the principal axis of rotation, are labeled σ_v or σ_d.

Inversion (i) is a more complex operation. Each point moves through the center of the molecule to a position opposite the original position and as far from the central point as where it started.** An example of a molecule having a center of inversion is ethane in the staggered conformation; this inversion operation is shown in **Figure 4.4**.

Many molecules that seem at first glance to have an inversion center do not; for example, methane and other tetrahedral molecules lack inversion symmetry. To see this, hold a methane model with two hydrogen atoms in the vertical plane on the right and two hydrogen atoms in the horizontal plane on the left, as in Figure 4.4. Inversion results in two hydrogen atoms in the horizontal plane on the right and two hydrogen atoms in the vertical plane on the left. Inversion is therefore *not* a symmetry operation of methane, because the orientation of the molecule following the i operation differs from the original orientation.

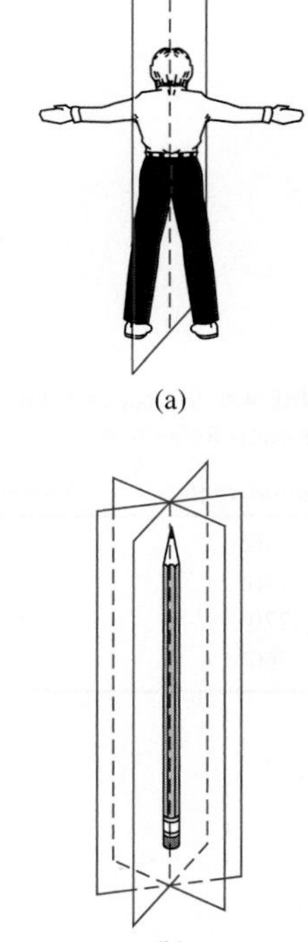

FIGURE 4.3 Reflections.

*The extraordinary observation of this gas phase anion is described in T. R. Galeev, C. Romanescu, W.-L. Li, L.-S. Wang, A. I. Boldyrev, *Angew. Chem., Int. Ed.*, **2012**, *51*, 2101.

**This operation must be distinguished from the inversion of a tetrahedral carbon in a bimolecular reaction, which is more like that of an umbrella in a high wind.

Center of inversion

No center of inversion

FIGURE 4.4 Inversion.

Squares, rectangles, parallelograms, rectangular solids, octahedra, and snowflakes have inversion centers; tetrahedra, triangles, and pentagons do not (**Figure 4.5**).

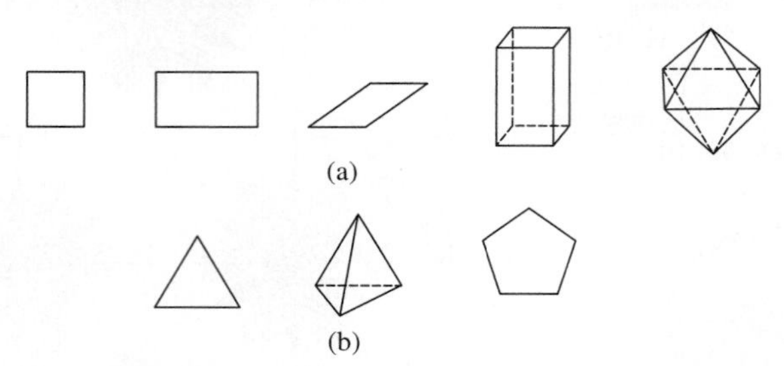

FIGURE 4.5 Figures (a) With and (b) Without Inversion Centers.

A **rotation-reflection operation (S_n)**, or **improper rotation**, requires rotation of $360°/n$, followed by reflection through a plane perpendicular to the axis of rotation. In methane, for example, a line through the carbon and bisecting the angle between two hydrogen atoms on each side is an S_4 axis. There are three such lines, for a total of three S_4 axes; a tetrahedron has six edges, and each of these axes bisects a pair of opposite edges. The operation requires a 90° rotation of the molecule, followed by reflection through the plane perpendicular to the axis of rotation. Two S_n operations in succession generate a $C_{n/2}$ operation. For example, in methane, two S_4 operations are equivalent to a C_2. These operations are shown in **Figure 4.6**, along with a table of C and S equivalences for methane.

Molecules sometimes have an S_n axis that is coincident with a C_n axis. For example, in addition to the rotation axes described previously, snowflakes have $S_2 (\equiv i)$, S_3, and S_6 axes coincident with the C_6 axis. Molecules may also have S_{2n} axes coincident with C_n; methane is an example, with S_4 axes coincident with C_2 axes, as shown in Figure 4.6.

Note that an S_2 operation is the same as inversion, and an S_1 operation is the same as a reflection plane. The i and σ notations are preferred in these cases.* Symmetry elements and operations are summarized in **Table 4.1**.

FIGURE 4.6 Improper Rotation or Rotation-Reflection.

Rotation angle	*Symmetry operation*
90°	S_4
180°	C_2 $(= S_4^2)$
270°	S_4^3
360°	E $(= S_4^4)$

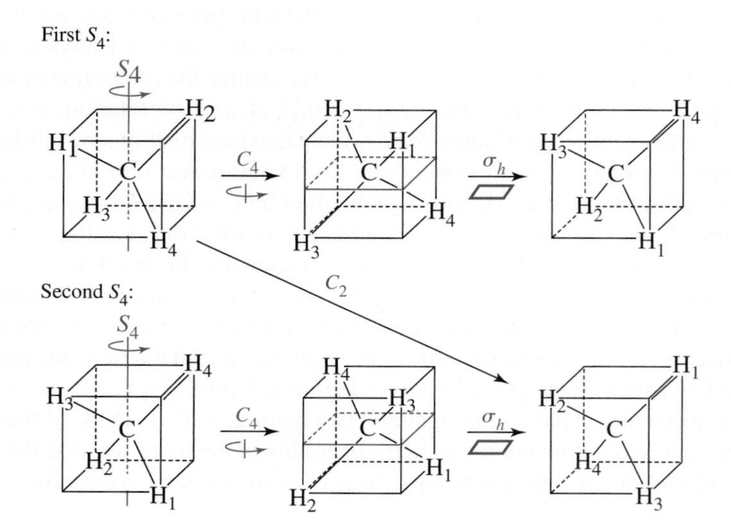

*This preference originates from the group theory requirement of maximizing the number of unique classes of symmetry operations associated with a molecule, to be discussed in Section 4.3.3.

TABLE 4.1 Summary Table of Symmetry Elements and Operations

Symmetry Operation	Symmetry Element	Operation	Examples
Identity, E	None	All atoms unshifted	CHFClBr 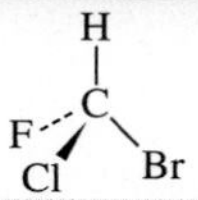
Rotation, C_n	Rotation axis	Rotation by $360°/n$	
C_2			*p*-dichlorobenzene 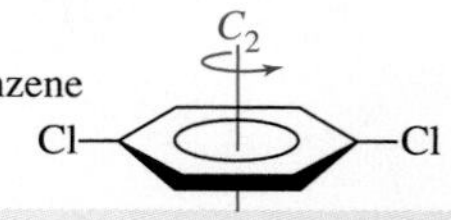
C_3			NH_3 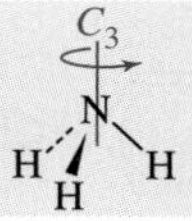
C_4			$[PtCl_4]^{2-}$ 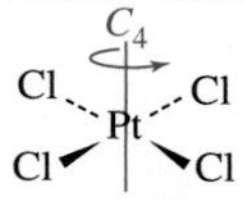
C_5			Cyclopentadienyl group 
C_6			Benzene 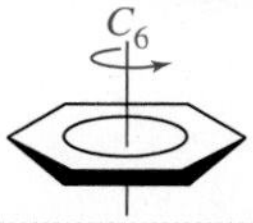
Reflection, σ	Mirror plane	Reflection through a mirror plane	H_2O 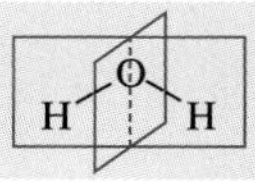
Inversion, i	Inversion center (point)	Inversion through the center	Ferrocene (staggered) 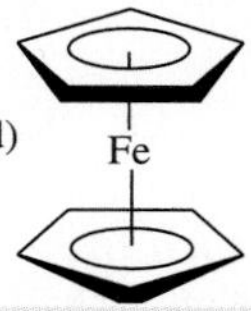
Rotation-reflection, S_n	Rotation-reflection axis (improper axis)	Rotation by $360°/n$, followed by reflection in the plane perpendicular to the rotation axis	
S_4			CH_4 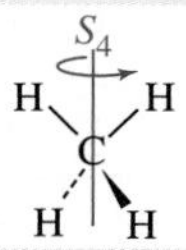
S_6			Ethane (staggered) 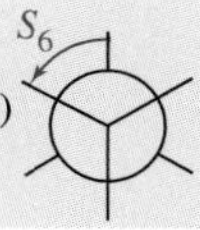
S_{10}			Ferrocene (staggered) 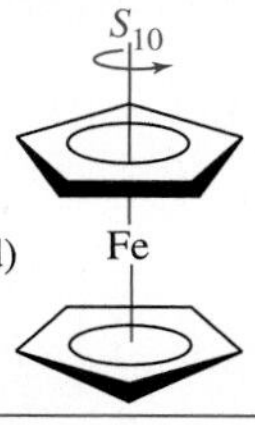

EXAMPLE 4.1

Find all the symmetry elements in the following molecules. Consider only the atoms when assigning symmetry. While lone pairs influence shapes, molecular symmetry is based on the geometry of the atoms.

H_2O

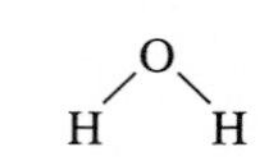

H_2O has two planes of symmetry, one in the plane of the molecule and one perpendicular to the molecular plane, as shown in Table 4.1. It also has a C_2 axis collinear with the intersection of the mirror planes. H_2O has no inversion center.

***p*-Dichlorobenzene**

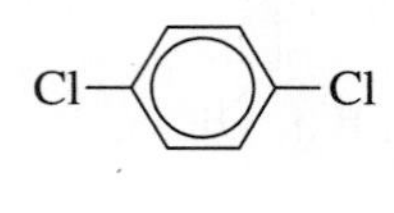

This molecule has three mirror planes: the molecular plane; a plane perpendicular to the molecule, passing through both chlorines; and a plane perpendicular to the first two, bisecting the molecule between the chlorines. It also has three C_2 axes, one perpendicular to the molecular plane (see Table 4.1) and two within the plane, one passing through both chlorines and one perpendicular to the axis passing through the chlorines. Finally, *p*-dichlorobenzene has an inversion center.

Ethane (staggered conformation)

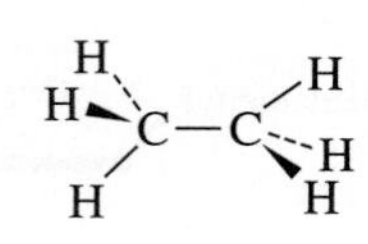

Staggered ethane has three mirror planes, each containing the C—C bond axis and passing through two hydrogens on opposite ends of the molecule. It has a C_3 axis collinear with the carbon–carbon bond and three C_2 axes bisecting the angles between the mirror planes. (Use of a model is helpful for viewing the C_2 axes.) Staggered ethane also has a center of inversion and an S_6 axis collinear with the C_3 axis (see Table 4.1).

EXERCISE 4.1 Using a point in a three dimensional Cartesian coordinate system, show that $S_2 \equiv i$ and $S_1 \equiv \sigma$.

EXERCISE 4.2 Find all the symmetry elements in the following molecules:

NH_3 Cyclohexane (boat conformation) Cyclohexane (chair conformation) XeF_2

4.2 Point Groups

Each molecule has a set of symmetry operations that describes the molecule's overall symmetry. This set of symmetry operations is called the **point group** of the molecule. **Group theory**, the mathematical treatment of the properties of groups, can be used to determine the molecular orbitals, vibrations, and other molecular properties. Navigation of a standard sequence of steps allows systematic deduction of point groups. A flowchart of these steps is shown in **Figure 4.7**.

1. Determine whether the molecule exhibits very low symmetry (C_1, C_s, C_i) or high symmetry (T_d, O_h, $C_{\infty v}$, $D_{\infty h}$ or I_h) as described in **Tables 4.2** and **4.3**.
2. If not, find the rotation axis with the highest n, the highest order C_n (or *principal*) axis, for the molecule.
3. Does the molecule have any C_2 axes perpendicular to the principal C_n axis? If it does, there will be n of such C_2 axes, and the molecule is in the D set of groups. If not, it is in the C or S set.

4. Does the molecule have a mirror plane (σ_h) perpendicular to the principal C_n axis? If so, it is classified as C_{nh} or D_{nh}. If not, continue with Step 5.
5. Does the molecule have any mirror planes that contain the principal C_n axis (σ_v or σ_d)? If so, it is classified as C_{nv} or D_{nd}. If not, but it is in the D set, it is classified as D_n. If the molecule is in the C or S set, continue with Step 6.
6. Is there an S_{2n} axis collinear with the principal C_n axis? If so, it is classified as S_{2n}. If not, the molecule is classified as C_n.

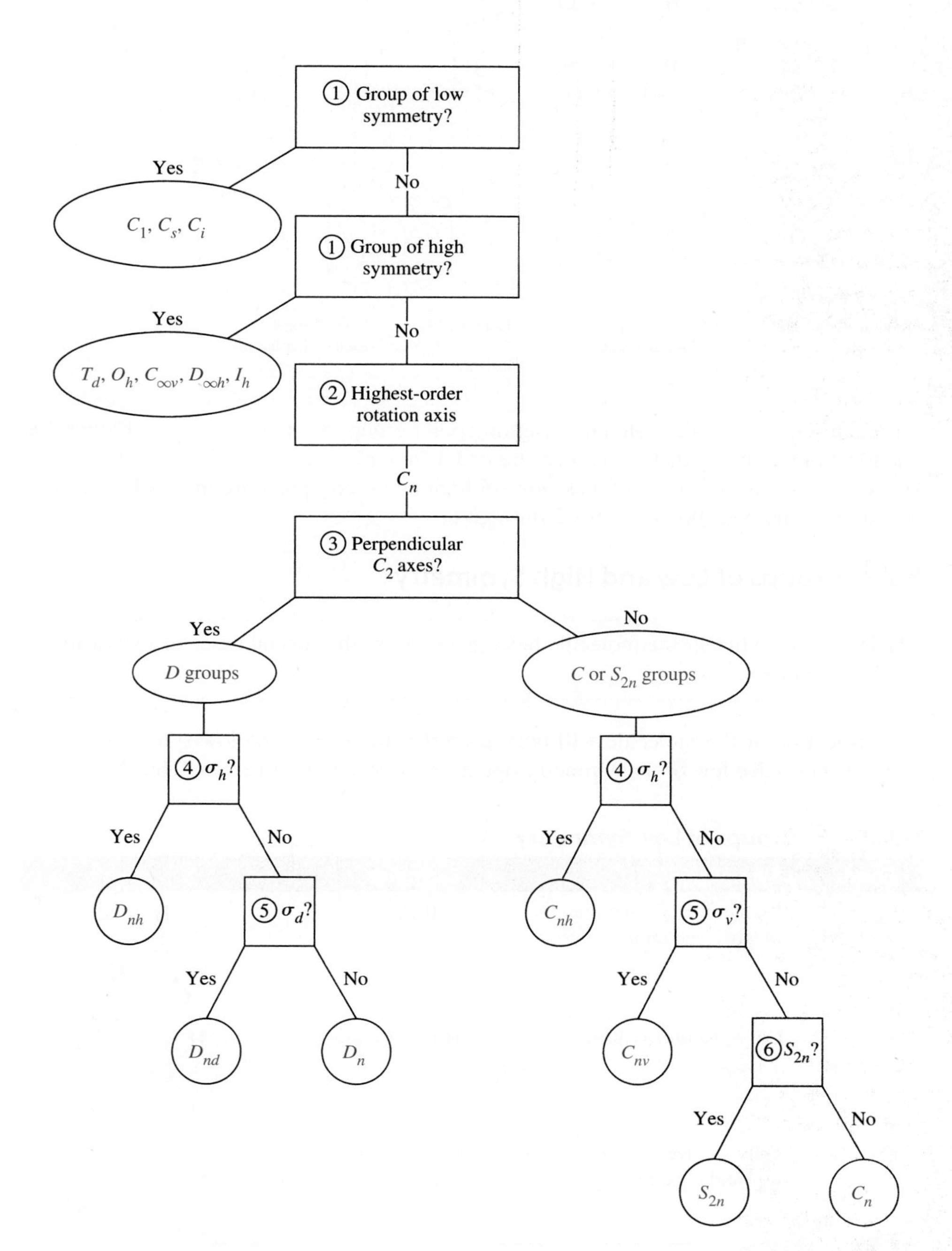

FIGURE 4.7 Diagram of the Point Group Assignment Method.

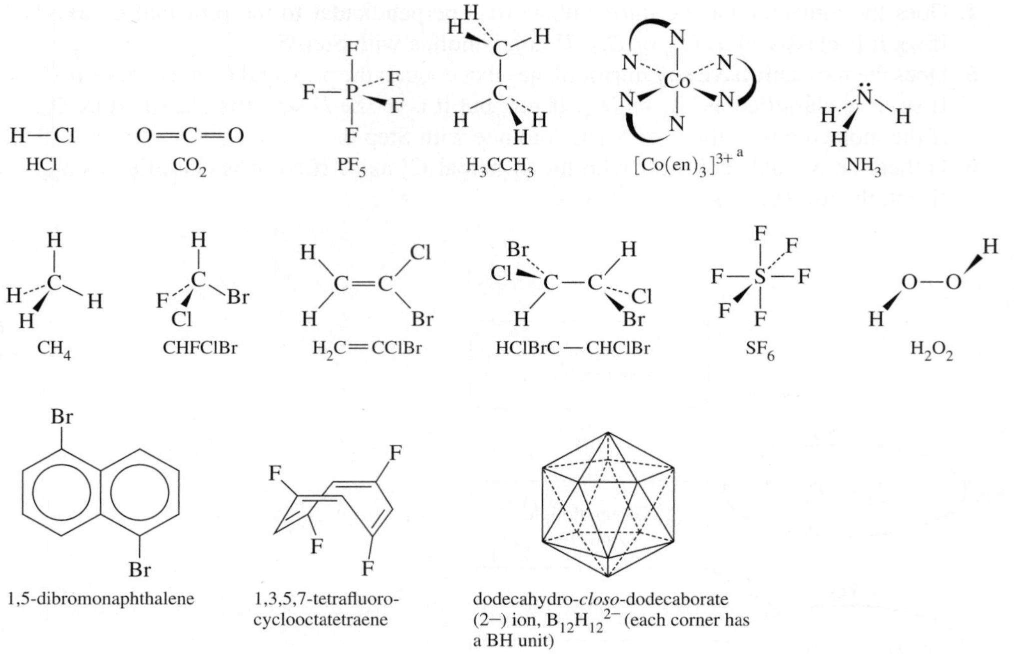

FIGURE 4.8 Molecules to Be Assigned to Point Groups. [a]en = ethylenediamine = $NH_2CH_2CH_2NH_2$, represented by N⌒N.

HCl CO_2 PF_5 H_3CCH_3 $[Co(en)_3]^{3+}$ [a] NH_3

CH_4 CHFClBr $H_2C{=}CClBr$ HClBrC—CHClBr SF_6 H_2O_2

1,5-dibromonaphthalene

1,3,5,7-tetrafluorocyclooctatetraene

dodecahydro-*closo*-dodecaborate (2−) ion, $B_{12}H_{12}^{2-}$ (each corner has a BH unit)

Each step is now illustrated by assigning point groups to the molecules in **Figure 4.8**. The low- and high-symmetry cases are treated differently because of their special nature. Molecules that are not in one of these low- or high-symmetry point groups can be assigned to a point group by following steps 2 through 6.

4.2.1 Groups of Low and High Symmetry

1. Determine whether the molecule belongs to one of the special cases of low or high symmetry.

Inspection of the molecule will determine if it fits one of the low-symmetry cases. These groups have few or no symmetry operations and are described in Table 4.2.

TABLE 4.2 Groups of Low Symmetry

Group	Symmetry	Examples
C_1	No symmetry other than the identity operation	CHFClBr 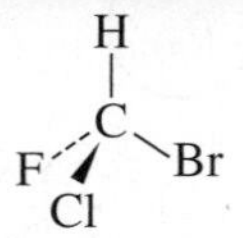
C_s	Only one mirror plane	$H_2C{=}CClBr$ 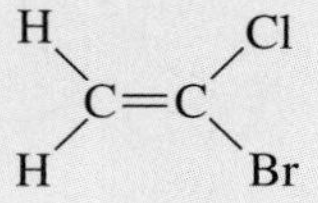
C_i	Only an inversion center; few molecular examples	HClBrC—CHClBr (staggered conformation) 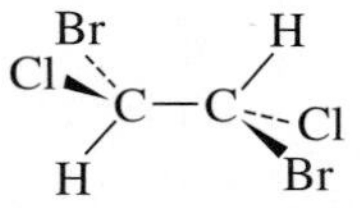

TABLE 4.3 Groups of High Symmetry

Group	Description	Examples
$C_{\infty v}$	These molecules are linear, with an infinite number of rotations and an infinite number of reflection planes containing the rotation axis. They do not have a center of inversion.	C_∞ H—Cl
$D_{\infty h}$	These molecules are linear, with an infinite number of rotations and an infinite number of reflection planes containing the rotation axis. They also have perpendicular C_2 axes, a perpendicular reflection plane, and an inversion center.	C_∞ O=C=O, C_2
T_d	Most (but not all) molecules in this point group have the familiar tetrahedral geometry. They have four C_3 axes, three C_2 axes, three S_4 axes, and six σ_d planes. They have no C_4 axes.	CH_4 (H, C, H, H, H)
O_h	These molecules include those of octahedral structure, although some other geometrical forms, such as the cube, share the same set of symmetry operations. Among their 48 symmetry operations are four C_3 rotations, three C_4 rotations, and an inversion.	SF_6 (F, F, F—S—F, F, F)
I_h	Icosahedral structures are best recognized by their six C_5 axes, as well as many other symmetry operations—120 in all.	$B_{12}H_{12}^{2-}$ with BH at each vertex of an icosahedron

In addition, there are four other groups, T, T_h, O, and I, which are rarely seen in nature. These groups are discussed at the end of this section.

Low Symmetry

CHFClBr has no symmetry other than the identity operation and has C_1 symmetry, $H_2C{=}CClBr$ has only one mirror plane and C_s symmetry, and HClBrC—CHClBr in the staggered conformation has only a center of inversion and C_i symmetry.

High Symmetry

Molecules with many symmetry operations may fit one of the high-symmetry cases of linear, tetrahedral, octahedral, or icosahedral symmetry with the characteristics described in Table 4.3. Molecules with very high symmetry are of two types, linear and polyhedral. Linear molecules having a center of inversion have $D_{\infty h}$ symmetry; those lacking an inversion center have $C_{\infty v}$ symmetry. The highly symmetric point groups T_d, O_h, and I_h are described in Table 4.3. It is helpful to note the C_n axes of these molecules. Molecules with T_d symmetry have only C_3 and C_2 axes; those with O_h symmetry have C_4 axes in addition to C_3 and C_2; and I_h molecules have C_5, C_3, and C_2 axes.

HCl has $C_{\infty v}$ symmetry, CO_2 has $D_{\infty h}$ symmetry, CH_4 has tetrahedral (T_d) symmetry, SF_6 has octahedral (O_h) symmetry, and $B_{12}H_{12}^{2-}$ has icosahedral (I_h) symmetry.

There are now seven molecules left to be assigned to point groups out of the original 15.

4.2.2 Other Groups

2. Find the rotation axis with the highest n, the highest order C_n (or principal) axis for the molecule.

The rotation axes for the examples are shown in **Figure 4.9**. Some molecules feature multiple equivalent rotation axes of highest order. In this case, any one can be chosen as the principal axis.

3. Does the molecule have any C_2 axes perpendicular to the principal C_n axis?

Examples of the C_2 axes are shown in **Figure 4.10**.

Yes: *D Groups*

PF_5, H_3CCH_3, $[Co(en)_3]^{3+}$

Molecules with C_2 axes perpendicular to the principal axis are in one of the groups designated by the letter D; there are n of these perpendicular C_2 axes.

No: *C or S Groups*

NH_3, 1,5-dibromonaphthalene, H_2O_2, 1,3,5,7-tetrafluorocyclooctatetraene

Molecules with no perpendicular C_2 axes are in one of the groups designated by the letters C or S.

While point group assignments have not yet been made, the molecules are now divided into two major categories, the D set and the C or S set.

FIGURE 4.9 Rotation Axes.

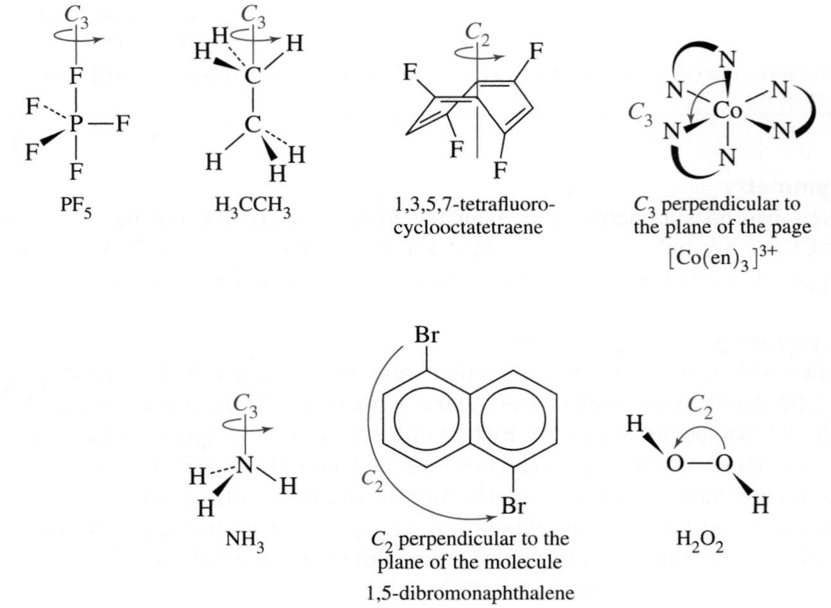

NH_3 No | 1,5-dibromonaphthalene No | H_2O_2 No | 1,3,5,7-tetrafluorocylooctatetraene No

PF_5 Yes | H_3CCH_3 Yes | $[Co(en)_3]^{3+}$ Yes

FIGURE 4.10 Perpendicular C_2 Axes. While all three perpendicular C_2 axes are shown for PF_5, only one out of three of these axes is shown for both H_3CCH_3 and $[Co(en)_3]^{3+}$.

4. Does the molecule have a mirror plane (σ_h horizontal plane) perpendicular to the principal C_n axis?

The horizontal mirror planes are shown in **Figure 4.11**.

***D* Groups**

Yes $\boxed{D_{nh}}$

PF_5 is D_{3h}

***C* and *S* Groups**

Yes $\boxed{C_{nh}}$

1,5-dibromonaphthalene is C_{2h}

Point groups are now assigned to these molecules. Both have horizontal mirror planes.

No D_n or D_{nd}

H_3CCH_3, $[Co(en)_3]^{3+}$

No C_n, C_{nv}, or S_{2n}

NH_3, H_2O_2, 1,3,5,7-tetrafluorocyclooctatetraene

None of these have horizontal mirror planes; they must be carried further in the process.

***D* Groups**

Yes $\boxed{D_{nd}}$

H_3CCH_3 (staggered) is D_{3d}

***C* and *S* Groups**

Yes $\boxed{C_{nv}}$

NH_3 is C_{3v}

D Groups

H_3CCH_3 No | $[Co(en)_3]^{3+}$ No

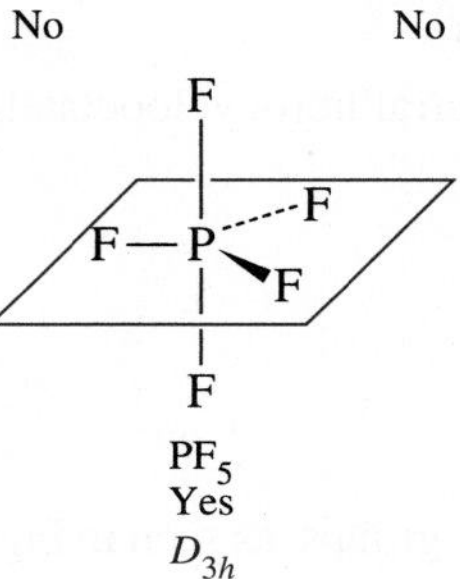

PF_5
Yes
D_{3h}

C and *S* Groups

NH_3 No | H_2O_2 No | 1,3,5,7-tetrafluoro-cyclooctatetraene No

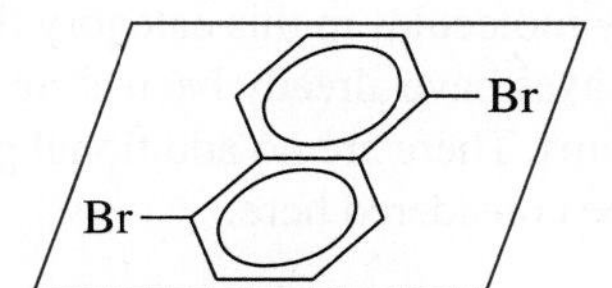

1,5-dibromonaphthalene
Yes
C_{2h}

FIGURE 4.11 Horizontal Mirror Planes.

5. Does the molecule have any mirror planes that contain the principal C_n axis?

These mirror planes are shown in **Figure 4.12**.

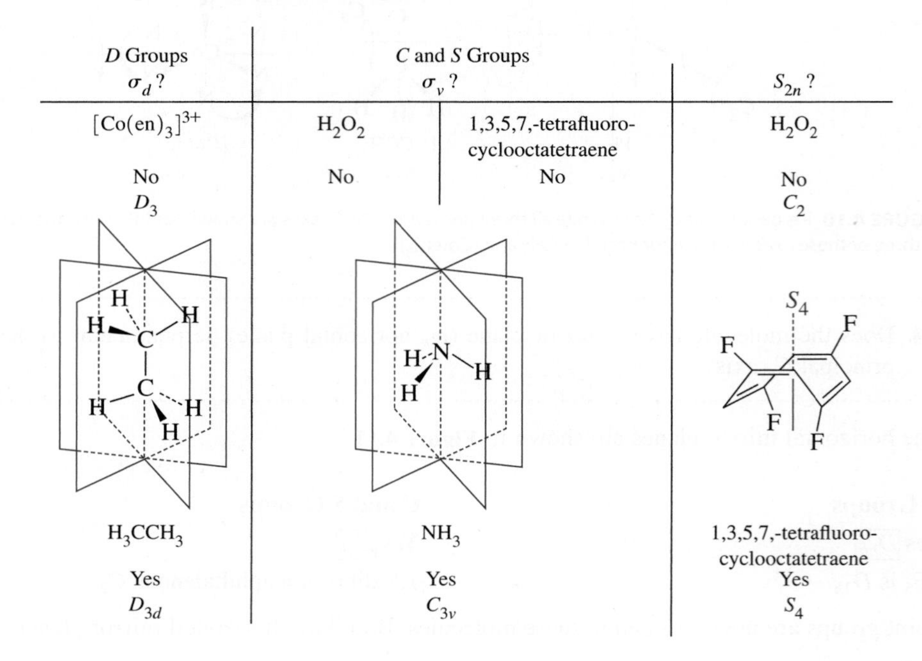

D Groups
σd ?
C and S Groups
σv ?
S2n ?
[Co(en)3]3+
No
D3
H2O2
No
1,3,5,7,-tetrafluoro-
cyclooctatetraene
No
H2O2
No
C2
S4
H3CCH3
Yes
D3d
NH3
Yes
C3v
1,3,5,7,-tetrafluoro-
cyclooctatetraene
Yes
S4

FIGURE 4.12 Vertical or Dihedral Mirror Planes or S_{2n} Axes.

These molecules have mirror planes containing the principal C_n axis, but no horizontal mirror planes, and are assigned to the corresponding point groups. There are *n* of these mirror planes.

No $\boxed{D_n}$

$[Co(en)_3]^{3+}$ is D_3

No C_n or S_{2n}

H_2O_2, 1,3,5,7-tetrafluorocyclooctatetraene

These molecules are in the simpler rotation groups D_n, C_n, and S_{2n} because they do not have any mirror planes. D_n and C_n point groups have *only* C_n axes. S_{2n} point groups have C_n and S_{2n} axes and may have an inversion center.

6. Is there an S_{2n} axis collinear with the principal C_n axis?

***D* Groups**

Any molecules in this category that have S_{2n} axes have already been assigned to groups. There are no additional groups to be considered here.

***C* and *S* Groups**

Yes $\boxed{S_{2n}}$ 1,3,5,7-tetrafluorocyclooctatetraene is S_4

No $\boxed{C_n}$

H_2O_2 is C_2

We have only one example in our list that falls into the S_{2n} groups, as seen in Figure 4.12.

A flowchart that summarizes this point group assignment method is given in Figure 4.7, and more examples are given in **Table 4.4**.

TABLE 4.4 Further Examples of *C* and *D* Point Groups

General Label	Point Group and Example		
C_{nh}	C_{2h}	difluorodiazene	F—N=N—F
	C_{3h}	$B(OH)_3$, planar	H, O, B, O, H, O, H
C_{nv}	C_{2v}	H_2O	H, O, H
	C_{3v}	PCl_3	Cl, P, Cl, Cl
	C_{4v}	BrF_5 (square pyramid)	F, F, F, Br, F, F
	$C_{\infty v}$	HF, CO, HCN	H—F C≡O H—C≡N
C_n	C_2	N_2H_4, which has a *gauche* conformation	H, N—N, H, H, H
	C_3	$P(C_6H_5)_3$, which is like a three-bladed propeller distorted out of the planar shape by a lone pair on the P	P
D_{nh}	D_{3h}	BF_3	F, B, F, F
	D_{4h}	$PtCl_4^{2-}$	Cl, Cl, Pt, Cl, Cl, 2−
	D_{5h}	$Os(C_5H_5)_2$ (eclipsed)	Os
	D_{6h}	benzene	

(*continues*)

TABLE 4.4 Further Examples of *C* and *D* Point Groups *(cont.)*

General Label	Point Group and Example		
	$D_{\infty h}$	F_2, N_2 acetylene (C_2H_2)	F—F N≡N H—C≡C—H
D_{nd}	D_{2d}	$H_2C{=}C{=}CH_2$, allene	
	D_{4d}	Ni(cyclobutadiene)$_2$ (staggered)	Ni
	D_{5d}	$Fe(C_5H_5)_2$ (staggered)	Fe
D_n	D_3	$[Ru(NH_2CH_2CH_2NH_2)_3]^{2+}$ (treating the $NH_2CH_2CH_2NH_2$ group as a planar ring)	Ru

EXAMPLE 4.2

Determine the point groups of the following molecules from Figures 3.13 and 3.16.

XeF_4

1. XeF_4 is not in a low or high symmetry group.
2. Its highest order rotation axis is C_4.
3. It has four C_2 axes perpendicular to the C_4 axis and is therefore in the D set of groups.
4. It has a horizontal plane perpendicular to the C_4 axis. Therefore its point group is D_{4h}.

SF_4

1. SF_4 is not in a high or low symmetry group.
2. Its highest order (and only) rotation axis is a C_2 axis passing through the lone pair.
3. The ion has no other C_2 axes and is therefore in the C or S set.
4. It has no mirror plane perpendicular to the C_2.
5. It has two mirror planes containing the C_2 axis. Therefore, the point group is C_{2v}.

IOF_3

1. The molecule has has only a mirror plane. Its point group is C_s.

EXERCISE 4.3 Use the procedure described previously to verify the point groups of the molecules in Table 4.4.

C Versus *D* Point Group Classifications

All molecules having these classifications must have a C_n axis. If more than one C_n axis is found, the highest order (principal) axis (largest value of n) is used as the reference axis. It is generally useful to orient this axis vertically.

	D Classifications	*C* Classifications
General Case:		
Look for C_2 axes perpendicular to the highest order C_n axis.	nC_2 axes $\perp$ C_n axis	No C_2 axes $\perp$ C_n axis
Subcategories:		
If a horizontal plane of symmetry exists:	$\mathbf{D}_{nh}$	$\mathbf{C}_{nh}$
If n vertical planes exist:	$\mathbf{D}_{nd}$	$\mathbf{C}_{nv}$
If no planes of symmetry exist:	$\mathbf{D}_n$	$\mathbf{C}_n$

NOTES:

1. Vertical planes often contain the highest order C_n axis. In the D_{nd} case, these planes are designated *dihedral* (thus, the subscript *d*) because they are between the C_2 axes.
2. The presence of a C_n axis does not guarantee that a molecule will be in a *D* or *C* category; the high-symmetry T_d, O_h, and I_h point groups and related groups have a large number of C_n axes.
3. When in doubt, check the character tables (Appendix C) for a complete list of symmetry elements for any point group.

Groups Related to I_h, O_h, and T_d Groups

The high-symmetry point groups I_h, O_h, and T_d are ubiquitous in chemistry and are represented by the classic molecules C_{60}, SF_6, and CH_4. For each of these point groups, there is a purely rotational subgroup (*I*, *O*, and *T*, respectively) in which the only symmetry operations other than the identity operation are proper axes of rotation. The symmetry operations for these point groups are in **Table 4.5**.

One more high-symmetry point group, T_h, remains. The T_h point group is derived by adding a center of inversion to the *T* point group; adding *i* generates the additional symmetry operations S_6, S_6^5, and σ_h. T_h symmetry is known for only a few molecules. The compound shown in **Figure 4.13** is an example. *I*, *O*, and *T* symmetry are rarely encountered in chemistry.

That's all there is to it! It takes a fair amount of practice, preferably using molecular models, to learn the point groups well, but once you know them, they are extremely useful. Several practical applications of point groups are discussed in this chapter, and additional applications are included in later chapters.

FIGURE 4.13 $W[N(CH_3)_2]_6$, a Molecule with T_h Symmetry.

TABLE 4.5 Symmetry Operations for High-Symmetry Point Groups and Their Rotational Subgroups

Point Group	Symmetry Operations										
I_h	E	$12C_5$		$12C_5^2$	$20C_3$	$15C_2$	i	$12S_{10}$	$12S_{10}^3$	$20S_6$	15σ
I	E	$12C_5$		$12C_5^2$	$20C_3$	$15C_2$					
O_h	E	$8C_3$		$6C_2$	$6C_4$	$3C_2(\equiv C_4^2)$	i	$6S_4$	$8S_6$	$3\sigma_h$	$6\sigma_d$
O	E	$8C_3$		$6C_2$	$6C_4$	$3C_2(\equiv C_4^2)$					
T_d	E	$8C_3$		$3C_2$				$6S_4$			$6\sigma_d$
T	E	$4C_3$	$4C_3^2$	$3C_2$							
T_h	E	$4C_3$	$4C_3^2$	$3C_2$			i	$4S_6$	$4S_6^5$	$3\sigma_h$	

4.3 Properties and Representations of Groups

All mathematical groups, including point groups, must have certain properties. These properties are listed and illustrated in **Table 4.6**, using the symmetry operations of NH_3 in **Figure 4.14** as an example.

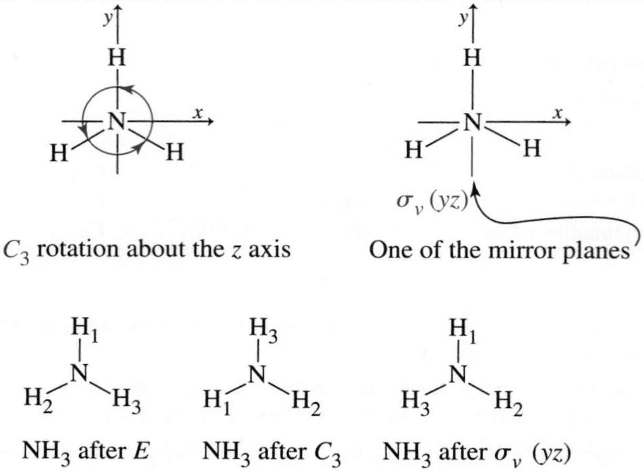

FIGURE 4.14 Symmetry Operations for Ammonia. (Top view) NH_3 is of point group C_{3v}, with the symmetry operations E, C_3, C_3^2, σ_v, σ_v', and σ_v'' usually written as E, $2C_3$, and $3\sigma_v$ (note that $C_3^3 \equiv E$).

TABLE 4.6 Properties of a Group

Property of Group	Examples from Point Group
1. Each group must contain an **identity** operation that commutes (in other words, $EA = AE$) with all other members of the group and leaves them unchanged ($EA = AE = A$).	C_{3v} molecules (and *all* molecules) contain the identity operation E.
2. Each operation must have an **inverse** that, when combined with the operation, yields the identity operation (sometimes a symmetry operation may be its own inverse). *Note*: By convention, we perform sequential symmetry operations *from right to left* as written.	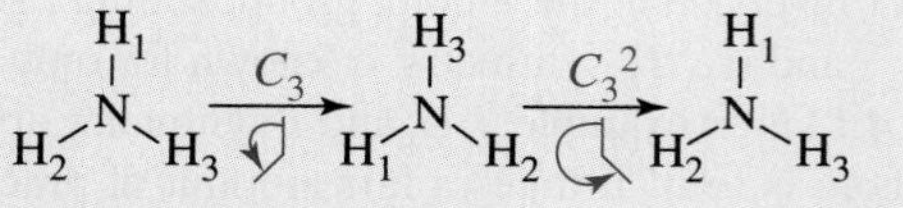 $C_3^2C_3 = E$ (C_3 and C_3^2 are inverses of each other) 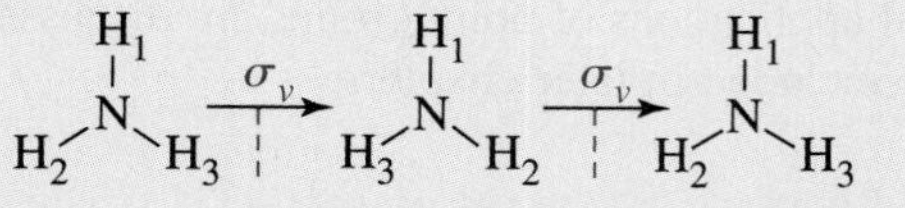 $\sigma_v\sigma_v = E$ (mirror planes are shown as dashed lines; σ_v is its own inverse)
3. The product of any two group operations must also be a member of the group. This includes the product of any operation with itself.	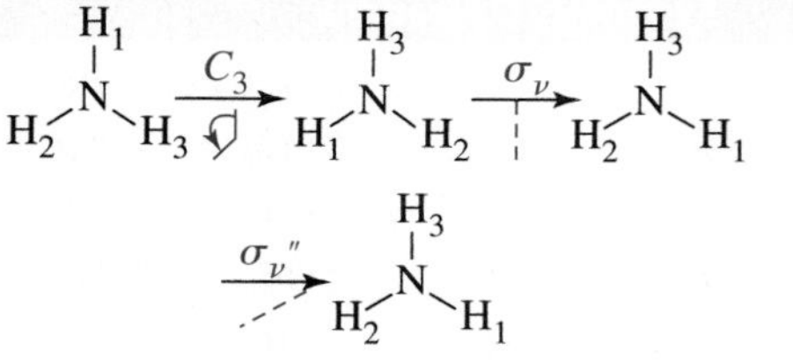 σ_vC_3 has the same overall effect as σ_v'', therefore we write $\sigma_vC_3 = \sigma_v''$. It can be shown that the products of any two operations in C_{3v} are also members of C_{3v}.
4. The associative property of combination must hold. In other words, $A(BC) = (AB)C$.	$C_3(\sigma_v\sigma_v') = (C_3\sigma_v)\sigma_v'$

4.3.1 Matrices

Important information about the symmetry aspects of point groups is summarized in character tables, described later in this chapter. To understand the construction and use of character tables, we must consider the properties of matrices, which are the basis for the tables.*

A **matrix** is an ordered array of numbers, such as

$$\begin{bmatrix} 3 & 7 \\ 2 & 1 \end{bmatrix} \text{ or } [2 \quad 0 \quad 1 \quad 3 \quad 5]$$

While the properties of the symmetry operations that comprise a group can be examined by using a molecule (as described previously), the most rigorous tests require application of the matrices that define the operations. To multiply matrices, the number of vertical columns of the first matrix must be equal to the number of horizontal rows of the second matrix. To find the product, add, term by term, the products of each *row* of the first matrix by each *column* of the second (each term in a row must be multiplied by its corresponding term in the appropriate column of the second matrix). Place the resulting sum in the product matrix with the row determined by the row of the first matrix and the column determined by the column of the second matrix:

$$C_{ij} = \Sigma A_{ik} \times B_{kj}$$

Here, C_{ij} = product matrix, with i rows and j columns
A_{ik} = initial matrix, with i rows and k columns
B_{kj} = initial matrix, with k rows and j columns

EXAMPLE 4.3

$$i\overset{k}{\begin{bmatrix} 1 & 5 \\ 2 & 6 \end{bmatrix}} \times \overset{j}{\begin{bmatrix} 7 & 3 \\ 4 & 8 \end{bmatrix}}k = \overset{j}{\begin{bmatrix} (1)(7) + (5)(4) & (1)(3) + (5)(8) \\ (2)(7) + (6)(4) & (2)(3) + (6)(8) \end{bmatrix}}i = \overset{j}{\begin{bmatrix} 27 & 43 \\ 38 & 54 \end{bmatrix}}i$$

This example has two rows and two columns in each initial matrix, so it has two rows and two columns in the product matrix; $i = j = k = 2$.

$$i\overset{k}{[1 \quad 2 \quad 3]}\overset{j}{\begin{bmatrix} 1 & 0 & 0 \\ 0 & -1 & 0 \\ 0 & 0 & 1 \end{bmatrix}}k =$$

$$\overset{j}{[(1)(1)+(2)(0)+(3)(0) \quad (1)(0)+(2)(-1)+(3)(0) \quad (1)(0)+(2)(0)+(3)(1)]}i = \overset{j}{[1 \quad -2 \quad 3]}i$$

Here, $i = 1$, $j = 3$, and $k = 3$, so the product matrix has one row (i) and three columns (j).

$$i\overset{k}{\begin{bmatrix} 1 & 0 & 0 \\ 0 & -1 & 0 \\ 0 & 0 & 1 \end{bmatrix}}\overset{j}{\begin{bmatrix} 1 \\ 2 \\ 3 \end{bmatrix}}k = \overset{j}{\begin{bmatrix} (1)(1) + (0)(2) + (0)(3) \\ (0)(1) + (-1)(2) + (0)(3) \\ (0)(1) + (0)(2) + (1)(3) \end{bmatrix}}i = \overset{j}{\begin{bmatrix} 1 \\ -2 \\ 3 \end{bmatrix}}i$$

Here $i = 3, j = 1$, and $k = 3$, so the product matrix has three rows (i) and one column (j).

*More details on matrices and their manipulation are available in Appendix 1 of F. A. Cotton, *Chemical Applications of Group Theory*, 3rd ed., John Wiley & Sons, New York, 1990, and in linear algebra and finite mathematics textbooks.

EXERCISE 4.4 Do the following multiplications:

a. $\begin{bmatrix} 5 & 1 & 3 \\ 4 & 2 & 2 \\ 1 & 2 & 3 \end{bmatrix} \times \begin{bmatrix} 2 & 1 & 1 \\ 1 & 2 & 3 \\ 5 & 4 & 3 \end{bmatrix}$

b. $\begin{bmatrix} 1 & -1 & -2 \\ 0 & 1 & -1 \\ 1 & 0 & 0 \end{bmatrix} \times \begin{bmatrix} 2 \\ 1 \\ 3 \end{bmatrix}$

c. $[1 \quad 2 \quad 3] \times \begin{bmatrix} 1 & -1 & -2 \\ 2 & 1 & -1 \\ 3 & 2 & 1 \end{bmatrix}$

4.3.2 Representations of Point Groups

Symmetry Operations: Matrix Representations

We will now consider how the C_{2v} point group symmetry operations transform a set of x, y, and z coordinates. The water molecule possesses C_{2v} symmetry. It has a C_2 axis through the oxygen and in the plane of the molecule, no perpendicular C_2 axes, and no horizontal mirror plane; but it does have two vertical mirror planes, as shown in Table 4.1 and **Figure 4.15**. The z axis is usually chosen as the axis of highest rotational symmetry; for H_2O, this is the *only* rotational axis. The other axes are arbitrary. We will use the xz plane as the plane of the molecule.* This set of axes is chosen to obey the right-hand rule (the thumb and first two fingers of the right hand, held perpendicular to each other, are labeled x, y, and z, respectively).

Each symmetry operation can be expressed as a **transformation matrix** as follows:

$$[\text{New coordinates}] = [\text{transformation matrix}]\,[\text{old coordinates}]$$

As examples, consider how transformation matrices can be used to represent the symmetry operations of the C_{2v} point group:

C_2: Rotate a point having coordinates (x, y, z) about the $C_2(z)$ axis. The new coordinates are given by

$$\begin{aligned} x' &= \text{new } x = -x \\ y' &= \text{new } y = -y \\ z' &= \text{new } z = z \end{aligned} \qquad \begin{bmatrix} -1 & 0 & 0 \\ 0 & -1 & 0 \\ 0 & 0 & 1 \end{bmatrix} \qquad \text{Transformation matrix for } C_2$$

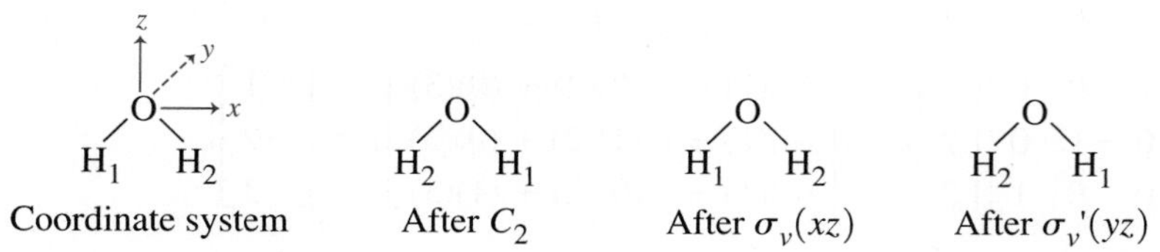

Coordinate system | After C_2 | After $\sigma_v(xz)$ | After $\sigma_v'(yz)$

FIGURE 4.15 Symmetry Operations of the Water Molecule.

*Some sources use yz as the plane of the molecule. The assignment of B_1 and B_2 in Section 4.3.3 is reversed with this choice.

In matrix notation,

$$\begin{bmatrix} x' \\ y' \\ z' \end{bmatrix} = \begin{bmatrix} -1 & 0 & 0 \\ 0 & -1 & 0 \\ 0 & 0 & 1 \end{bmatrix} \begin{bmatrix} x \\ y \\ z \end{bmatrix} = \begin{bmatrix} -x \\ -y \\ z \end{bmatrix} \quad \text{or} \quad \begin{bmatrix} x' \\ y' \\ z' \end{bmatrix} = \begin{bmatrix} -x \\ -y \\ z \end{bmatrix}$$

$$\begin{bmatrix} \text{New} \\ \text{coordinates} \end{bmatrix} = \begin{bmatrix} \text{transformation} \\ \text{matrix} \end{bmatrix} \begin{bmatrix} \text{old} \\ \text{coordinates} \end{bmatrix} = \begin{bmatrix} \text{new coordinates} \\ \text{in terms of old} \end{bmatrix}$$

$\sigma_v(xz)$: Reflect a point with coordinates (x, y, z) through the xz plane.

$$\begin{matrix} x' = \text{new } x = x \\ y' = \text{new } y = -y \\ z' = \text{new } z = z \end{matrix} \quad \begin{bmatrix} 1 & 0 & 0 \\ 0 & -1 & 0 \\ 0 & 0 & 1 \end{bmatrix} \qquad \text{Transformation matrix for } \sigma_v(xz)$$

The matrix equation is

$$\begin{bmatrix} x' \\ y' \\ z' \end{bmatrix} = \begin{bmatrix} 1 & 0 & 0 \\ 0 & -1 & 0 \\ 0 & 0 & 1 \end{bmatrix} \begin{bmatrix} x \\ y \\ z \end{bmatrix} = \begin{bmatrix} x \\ -y \\ z \end{bmatrix} \quad \text{or} \quad \begin{bmatrix} x' \\ y' \\ z' \end{bmatrix} = \begin{bmatrix} x \\ -y \\ z \end{bmatrix}$$

The transformation matrices for the four symmetry operations of the group are

$$E: \begin{bmatrix} 1 & 0 & 0 \\ 0 & 1 & 0 \\ 0 & 0 & 1 \end{bmatrix} \quad C_2: \begin{bmatrix} -1 & 0 & 0 \\ 0 & -1 & 0 \\ 0 & 0 & 1 \end{bmatrix} \quad \sigma_v(xz): \begin{bmatrix} 1 & 0 & 0 \\ 0 & -1 & 0 \\ 0 & 0 & 1 \end{bmatrix} \quad \sigma_v'(yz): \begin{bmatrix} -1 & 0 & 0 \\ 0 & 1 & 0 \\ 0 & 0 & 1 \end{bmatrix}$$

EXERCISE 4.5

Verify the transformation matrices for the E and $\sigma_v'(yz)$ operations of the C_{2v} point group.

This set of matrices satisfies the properties of a mathematical **group**. We call this a **matrix representation** of the C_{2v} point group. This representation is a set of matrices, each corresponding to an operation in the group; these matrices combine in the same way as the operations themselves. For example, multiplying two of the matrices is equivalent to carrying out the two corresponding operations and results in a matrix that directly transforms the coordinates as does the combination of symmetry operations (the operations are carried out right to left, so $C_2 \times \sigma_v$ means σ_v followed by C_2):

$$C_2 \times \sigma_v(xz) = \begin{bmatrix} -1 & 0 & 0 \\ 0 & -1 & 0 \\ 0 & 0 & 1 \end{bmatrix} \begin{bmatrix} 1 & 0 & 0 \\ 0 & -1 & 0 \\ 0 & 0 & 1 \end{bmatrix} = \begin{bmatrix} -1 & 0 & 0 \\ 0 & 1 & 0 \\ 0 & 0 & 1 \end{bmatrix} = \sigma_v'(yz)$$

The matrices of the matrix representation of the C_{2v} group also describe the operations of the group shown in Figure 4.15. The C_2 and $\sigma_v'(yz)$ operations interchange H_1 and H_2, whereas E and $\sigma_v(xz)$ leave them unchanged.

Characters

The **character**, defined only for a square matrix, is the trace of the matrix, or the sum of the numbers on the diagonal from upper left to lower right. For the C_{2v} point group, the following characters are obtained from the preceding matrices:

E	C_2	$\sigma_v(xz)$	$\sigma_v'(yz)$
3	-1	1	1

This set of characters also forms a **representation**, a shorthand version of the matrix representation. This representation is called a **reducible representation**, a combination of more fundamental **irreducible representations** as described in the next section. Reducible representations are frequently designated with a capital gamma (Γ).

Reducible and Irreducible Representations

Each transformation matrix in the C_{2v} set can be "block diagonalized"; that is, it can be broken down into smaller matrices along the diagonal, with all other matrix elements equal to zero:

$$E: \begin{bmatrix} [1] & 0 & 0 \\ 0 & [1] & 0 \\ 0 & 0 & [1] \end{bmatrix} \quad C_2: \begin{bmatrix} [-1] & 0 & 0 \\ 0 & [-1] & 0 \\ 0 & 0 & [1] \end{bmatrix} \quad \sigma_v(xz): \begin{bmatrix} [1] & 0 & 0 \\ 0 & [-1] & 0 \\ 0 & 0 & [1] \end{bmatrix} \quad \sigma_v'(yz): \begin{bmatrix} [-1] & 0 & 0 \\ 0 & [1] & 0 \\ 0 & 0 & [1] \end{bmatrix}$$

All the nonzero elements become 1×1 matrices along the principal diagonal.

When matrices are block diagonalized in this way, the x, y, and z coordinates are also block diagonalized. As a result, the x, y, and z coordinates are independent of each other. The matrix elements in the 1,1 positions (numbered as row, column) describe the results of the symmetry operations on the x coordinate, those in the 2,2 positions describe the results of the operations on the y coordinate, and those in the 3,3 positions describe the results of the operations on the z coordinate. The four matrix elements for x form a representation of the group, those for y form a second representation, and those for z form a third representation, all shown in the following table:

	E	C_2	$\sigma_v(xz)$	$\sigma_v'(yz)$	*Coordinate Used*
	1	-1	1	-1	x
	1	-1	-1	1	y
	1	1	1	1	z
Γ	3	-1	1	1	

These irreducible representations of the C_{2v} point group add to make up the reducible representation Γ.

Each row is an irreducible representation: it cannot be simplified further. The characters of these three irreducible representations added together under each operation (column) make up the characters of the reducible representation Γ, just as the combination of all the matrices for the x, y, and z coordinates makes up the matrices of the reducible representation. For example, the sum of the three characters for x, y, and z under the C_2 operation is -1, the character for Γ under this same operation.

The set of 3×3 matrices obtained for H_2O is called a reducible matrix representation because it is the sum of irreducible representations (the block-diagonalized 1×1 matrices), which cannot be reduced to smaller component parts. The set of characters of these matrices also forms the reducible representation Γ, for the same reason.

4.3.3 Character Tables

Three of the representations for C_{2v}, labeled A_1, B_1, and B_2, have now been determined. The fourth, called A_2, can be found by using the group properties described in **Table 4.7**. A complete set of irreducible representations for a point group is called the **character table** for that group. The character table for each point group is unique; character tables for common point groups are in Appendix C.

TABLE 4.7 Properties of Characters of Irreducible Representations in Point Groups

Property	Example: C_{2v}
1. The total number of symmetry operations in the group is called the **order** (h). To determine the order of a group, simply total the number of symmetry operations listed in the top row of the character table.	Order = 4 four symmetry operations: E, C_2, $\sigma_v(xz)$, and $\sigma_v'(yz)$
2. Symmetry operations are arranged in **classes**. All operations in a class have identical characters for their transformation matrices and are grouped in the same column in character tables.	Each symmetry operation is in a separate class; therefore, there are four columns in the character table.
3. The number of irreducible representations equals the number of classes. This means that character tables have the same number of rows and columns (they are square).	Because there are four classes, there must also be four irreducible representations—and there are.
4. The sum of the squares of the **dimensions** (characters under E) of each of the irreducible representations equals the order of the group. $h = \sum_i [\chi_i(E)]^2$	$1^2 + 1^2 + 1^2 + 1^2 = 4 = h$, the order of the group.
5. For any irreducible representation, the sum of the squares of the characters multiplied by the number of operations in the class (see Table 4.8 for an example), equals the order of the group. $h = \sum_R [\chi_i(R)]^2$	For A_2, $1^2 + 1^2 + (-1)^2 + (-1)^2 = 4 = h$. Each operation is its own class in this group.
6. Irreducible representations are **orthogonal** to each other. The sum of the products of the characters, multiplied together for each class, for any pair of irreducible representations is 0. $h = \sum_R \chi_i(R)\chi_j(R) = 0$ when $i \neq j$ Taking any pair of irreducible representations, multiplying together the characters for each class, multiplying by the number of operations in the class (see Table 4.8 for an example), and adding the products gives zero.	B_1 and B_2 are orthogonal: $(1)(1) + (-1)(-1) + (1)(-1) + (-1)(1) = 0$ E C_2 $\sigma_v(xz)$ $\sigma_v'(yz)$ Each operation is its own class in this group.
7. All groups include a **totally symmetric representation**, with characters of 1 for all operations.	C_{2v} has A_1, in which all characters = 1.

The complete character table for C_{2v}, with the irreducible representations in the order commonly used, is:

C_{2v}	E	C_2	$\sigma_v(xz)$	$\sigma_v'(yz)$		
A_1	1	1	1	1	z	x^2, y^2, z^2
A_2	1	1	−1	−1	R_z	xy
B_1	1	−1	1	−1	x, R_y	xz
B_2	1	−1	−1	1	y, R_x	yz

The labels used with character tables are:

x, y, z	transformations of the x, y, z coordinates or combinations thereof
R_x, R_y, R_z	rotation about the x, y, and z axes
$\boldsymbol{R}$	any symmetry operation, such as C_2 or $\sigma_v(xz)$
χ	character of an operation
i and j	designation of different representations, such as A_1 or A_2
h	order of the group (the total number of symmetry operations in the group)

The labels in the left column that designate the irreducible representations will be described later in this section. Other useful terms are defined in Table 4.7.

The A_2 representation of the C_{2v} group can now be explained. The character table has four columns; it has four classes of symmetry operations (Property 2 in Table 4.7). It must therefore have four irreducible representations (Property 3). The sum of the products of the characters of any two representations must equal zero (orthogonality, Property 6). Therefore, a product of A_1 and the unknown representation must have 1 for two of the characters and −1 for the other two. The character for the identity operation of this new representation must be 1 [$\chi(E) = 1$] to have the sum of the squares of these characters equal 4 (required by Property 4). Because no two representations can be the same, A_2 must then have $\chi(E) = \chi(C_2) = 1$, and $\chi(\sigma_{xz}) = \chi(\sigma_{yz}) = -1$. This representation is also orthogonal to B_1 and B_2, as required.

The relationships among symmetry operations, matrix representations, reducible and irreducible representations, and character tables are conveniently illustrated in a flowchart, as shown for C_{2v} symmetry in **Table 4.8**.

Prepare a representation flowchart according to the format of Table 4.8 for *trans*-N_2F_2, which has C_{2h} symmetry.

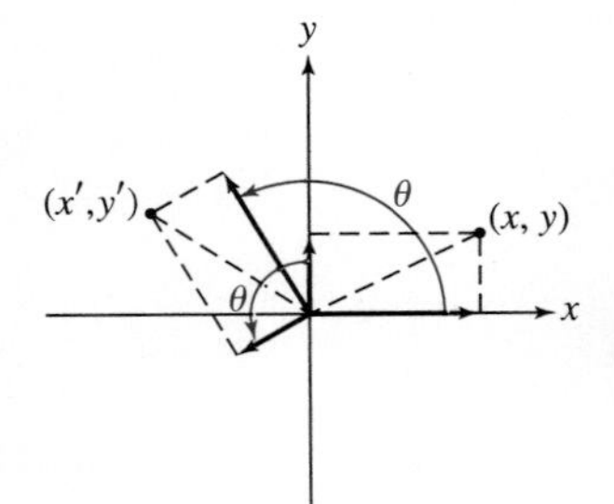

General case: $x' = x\cos\theta - y\sin\theta$
$y' = x\sin\theta + y\cos\theta$
For C_3: $\theta = 2\pi/3 = 120°$

General Transformation Matrix for rotation by $\theta°$ about z axis:

$$\begin{bmatrix} \cos\theta & -\sin\theta & 0 \\ \sin\theta & \cos\theta & 0 \\ 0 & 0 & 1 \end{bmatrix}$$

FIGURE 4.16 Effect of Rotation on Coordinates of a Point.

Another Example: C_{3v}(NH_3)

Full descriptions of the matrices for the operations in this group will not be given, but the characters can be found by using the properties of a group. Consider the C_3 rotation shown in **Figure 4.16**. Counterclockwise rotation of 120° results in a new x' and y' as shown, which can be described in terms of the vector sums of x and y by using trigonometric functions:

$$x' = x\cos\frac{2\pi}{3} - y\sin\frac{2\pi}{3} = -\frac{1}{2}x - \frac{\sqrt{3}}{2}y$$

$$y' = x\sin\frac{2\pi}{3} + y\cos\frac{2\pi}{3} = \frac{\sqrt{3}}{2}x - \frac{1}{2}y$$

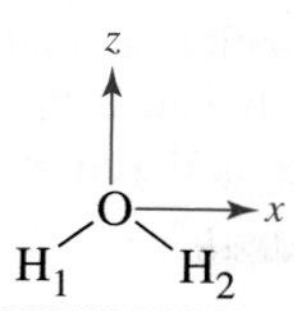

TABLE 4.8 Representation Flowchart: $H_2O(C_{2v})$

Symmetry Operations

H_1 O H_2	H_2 O H_1	H_1 O H_2	H_2 O H_1
after E	after C_2	after $\sigma_v(xz)$	after $\sigma_v{}'(yz)$

Reducible Matrix Representations

$$E: \begin{bmatrix} 1 & 0 & 0 \\ 0 & 1 & 0 \\ 0 & 0 & 1 \end{bmatrix} \quad C_2: \begin{bmatrix} -1 & 0 & 0 \\ 0 & -1 & 0 \\ 0 & 0 & 1 \end{bmatrix} \quad \sigma_v(xz): \begin{bmatrix} 1 & 0 & 0 \\ 0 & -1 & 0 \\ 0 & 0 & 1 \end{bmatrix} \quad \sigma_v'(yz): \begin{bmatrix} -1 & 0 & 0 \\ 0 & 1 & 0 \\ 0 & 0 & 1 \end{bmatrix}$$

Characters of Matrix Representations

3	–1	1	1

Block Diagonalized Matrices

$$\begin{bmatrix} [1] & 0 & 0 \\ 0 & [1] & 0 \\ 0 & 0 & [1] \end{bmatrix} \quad \begin{bmatrix} [-1] & 0 & 0 \\ 0 & [-1] & 0 \\ 0 & 0 & [1] \end{bmatrix} \quad \begin{bmatrix} [1] & 0 & 0 \\ 0 & [-1] & 0 \\ 0 & 0 & [1] \end{bmatrix} \quad \begin{bmatrix} [-1] & 0 & 0 \\ 0 & [1] & 0 \\ 0 & 0 & [1] \end{bmatrix}$$

Irreducible Representations

E	C_2	$\sigma_v(xz)$	$\sigma_v{}'(yz)$	Coordinate Used
1	–1	1	–1	x
1	–1	–1	1	y
1	1	1	1	z
3	–1	1	1	

Character Table

C_{2v}	E	C_2	$\sigma_v(xz)$	$\sigma_v{}'(yz)$	Matching Functions	
A_1	1	1	1	1	z	x^2, y^2, z^2
A_2	1	1	–1	–1	R_z	xy
B_1	1	–1	1	–1	x, R_y	xz
B_2	1	–1	–1	1	y, R_x	yz

The transformation matrices for the symmetry operations are

$$E: \begin{bmatrix} 1 & 0 & 0 \\ 0 & 1 & 0 \\ 0 & 0 & 1 \end{bmatrix} \quad C_3: \begin{bmatrix} \cos\frac{2\pi}{3} & -\sin\frac{2\pi}{3} & 0 \\ \sin\frac{2\pi}{3} & \cos\frac{2\pi}{3} & 0 \\ 0 & 0 & 1 \end{bmatrix} = \begin{bmatrix} -\frac{1}{2} & -\frac{\sqrt{3}}{2} & 0 \\ \frac{\sqrt{3}}{2} & -\frac{1}{2} & 0 \\ 0 & 0 & 1 \end{bmatrix} \quad \sigma_{v(xz)}: \begin{bmatrix} 1 & 0 & 0 \\ 0 & -1 & 0 \\ 0 & 0 & 1 \end{bmatrix}$$

In the C_{3v} point group $\chi(C_3^2) = \chi(C_3)$, which means that they are in the same class and listed as $2C_3$ in the character table. In addition, the three reflections have identical characters and are in the same class, listed as $3\sigma_v$.

The transformation matrices for C_3 and $C_3{}^2$ cannot be block diagonalized into 1×1 matrices, because the C_3 matrix has off-diagonal entries; however, the matrices can be block diagonalized into 2×2 and 1×1 matrices, with all other matrix elements equal to zero:

$$E: \begin{bmatrix} \begin{bmatrix} 1 & 0 \\ 0 & 1 \end{bmatrix} & \begin{matrix} 0 \\ 0 \end{matrix} \\ \begin{matrix} 0 & 0 \end{matrix} & [1] \end{bmatrix} \quad C_3: \begin{bmatrix} \begin{bmatrix} -\frac{1}{2} & -\frac{\sqrt{3}}{2} \\ \frac{\sqrt{3}}{2} & -\frac{1}{2} \end{bmatrix} & \begin{matrix} 0 \\ 0 \end{matrix} \\ \begin{matrix} 0 & 0 \end{matrix} & [1] \end{bmatrix} \quad \sigma_{v(xz)}: \begin{bmatrix} \begin{bmatrix} 1 & 0 \\ 0 & -1 \end{bmatrix} & \begin{matrix} 0 \\ 0 \end{matrix} \\ \begin{matrix} 0 & 0 \end{matrix} & [1] \end{bmatrix}$$

The C_3 matrix must be blocked this way because the (x, y) combination is needed for the new x' and y'; determination of *each* of the transformed coordinates x' and y' requires *both* original x and y coordinates. This is the case with most rotations, as defined by the general transformation matrix in Figure 4.16. The other C_{3v} matrices must follow the same pattern for consistency across the representation. In this case, x and y are not independent of each other.

The characters of the matrices are the sums of the numbers on the principal diagonal (from upper left to lower right). The set of 2×2 matrices has the characters corresponding to the E representation in the following character table; the set of 1×1 matrices matches the A_1 representation. The third irreducible representation, A_2, can be found by using the defining properties of a mathematical group, as in the C_{2v} previously shown. **Table 4.9** gives the properties of the characters for the C_{3v} point group.

C_{3v}	E	$2C_3$	$3\sigma_v$		
A_1	1	1	1	z	$x^2 + y^2, z^2$
A_2	1	1	-1	R_z	
E	2	-1	0	$(x, y), (R_x, R_y)$	$(x^2 - y^2, xy), (xz, yz)$

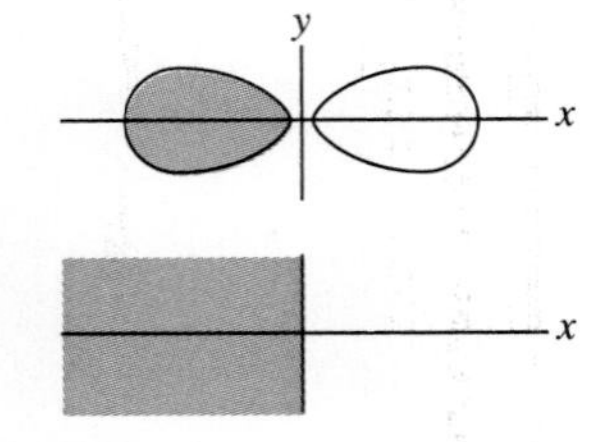

p_x orbitals have the same symmetry as x (positive in half the quadrants, negative in the other half).

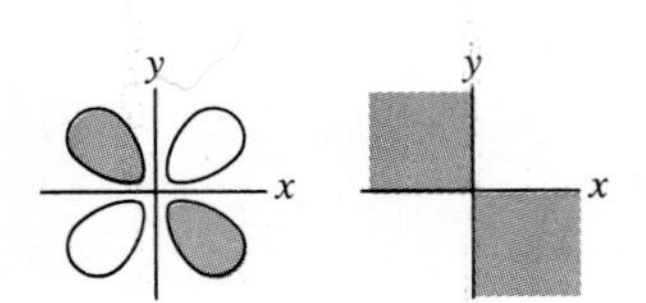

d_{xy} orbitals have the same symmetry as the function xy (sign of the function in the four quadrants).

FIGURE 4.17 Orbitals and Representations.

Additional Features of Character Tables

1. When operations such as C_3 and C_3^2 are in the same class, the listing in a character table is $2C_3$, indicating that the characters are the same, whether rotation is in a clockwise or counterclockwise direction (or, alternately, that C_3 and C_3^2 give the same characters). In either case, this is equivalent to two columns in the table being shown as one. Similar notation is used for multiple reflections.
2. When necessary, the C_2 axes perpendicular to the principal axis (in a D group) are designated with primes; a single prime indicates that the axis passes through several atoms of the molecule, whereas a double prime indicates that it passes between the atoms.
3. When the mirror plane is perpendicular to the principal axis, or horizontal, the reflection is called σ_h. Other planes are labeled σ_v or σ_d (see the character tables in Appendix C).
4. The expressions listed to the right of the characters indicate the symmetry properties of the point group for the x, y, and z axes, other mathematical functions, and rotation about these axes (R_x, R_y, R_z). These are used to find atomic orbitals with symmetries that match the representation. For example, the x axis with its positive and negative directions matches the p_x orbital (with a node defined by the yz plane). The function xy, with alternating signs in the four quadrants within the xy plane matches the lobes of the d_{xy} orbital. These useful connections are shown in **Figure 4.17**. The totally symmetric s orbital always matches the first representation in the group, one of the A set. The irreducible representations that describe rotations about the axes (R_x, R_y, R_z)

TABLE 4.9 Properties of the Characters for the C_{3v} Point Group

Property	C_{3v} Example
1. Order	6 (6 symmetry operations)
2. Classes	3 classes: E $2C_3\,(= C_3, C_3^2)$ $3\sigma_v\,(= \sigma_v, \sigma_v', \sigma_v'')$
3. Number of irreducible representations	3 (A_1, A_2, E)
4. Sum of squares of dimensions equals the order of the group	$1^2 + 1^2 + 2^2 = 6$
5. Sum of squares of characters multiplied by the number of operations in each class equals the order of the group	E $\quad 2C_3 \quad 3\sigma_v$ A_1: $1^2 + 2(1)^2 + 3(1)^2 = 6$ A_2: $1^2 + 2(1)^2 + 3(-1)^2 = 6$ E: $2^2 + 2(-1)^2 + 3(0)^2 = 6$ (Multiply the squares by the number of symmetry operations in each class.)
6. Orthogonal representations	The sum of the products of any two representations multiplied by the number of operations in each class equals 0. Example of $A_2 \times E$: $(1)(2) + 2(1)(-1) + 3(-1)(0) = 0$
7. Totally symmetric representation	A_1, with all characters $= 1$

also describe the rotational motions of molecules relative to these axes. Rotation and other motions of the water molecule are discussed in Section 4.4.2

In the C_{3v} example, the x and y coordinates appeared together in the E irreducible representation. The notation for this is to group them as (x, y) in this section of the table. This means that x and y together have the same symmetry properties as the E irreducible representation. Consequently, the p_x and p_y orbitals together have the same symmetry as the E irreducible representation in this point group.

5. Matching the symmetry operations of a molecule with those listed in the top row of the character table will confirm any point group assignment.

6. Irreducible representations are assigned labels according to the following rules, in which *symmetric* means a character of 1 and *antisymmetric* a character of -1 (see the character tables in Appendix C for examples).

a. Letters are assigned according to the dimension of the irreducible representation (the character for the identity operation).

Dimension	Symmetry Label
1	A If the representation is symmetric to the principal rotation operation ($\chi(C_n) = 1$).
	B If it is antisymmetric ($\chi(C_n) = -1$).*
2	E
3	T

*In a few cases, such as D_{nd} (n = even) and S_{2n} point groups, the highest order axis is an S_{2n}. This axis takes priority, so the classification is B if the character is -1 for the S_{2n} operation, even if the character is $+1$ for the highest order C_n axis.

b. Subscript 1 designates a representation symmetric to a C_2 rotation perpendicular to the principal axis, and subscript 2 designates a representation antisymmetric to the C_2. If there are no perpendicular C_2 axes, 1 designates a representation symmetric to a vertical plane, and 2 designates a representation antisymmetric to a vertical plane.

c. Subscript *g* (*gerade*) designates representations symmetric to inversion, and subscript *u* (*ungerade*) designates representations antisymmetric to inversion.

d. Single primes are symmetric to σ_h and double primes are antisymmetric to σ_h when a distinction between representations is needed (C_{3h}, C_{5h}, D_{3h}, D_{5h}).

4.4 Examples and Applications of Symmetry

Here we will consider two applications of symmetry and group theory, in the realms of chirality and molecular vibrations. In Chapter 5 we will also examine how symmetry can be used to understand chemical bonding, perhaps the most important application of symmetry in chemistry.

4.4.1 Chirality

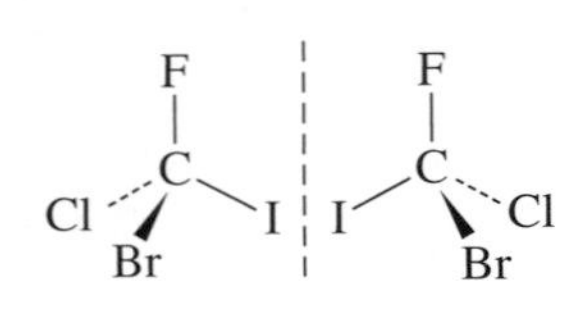

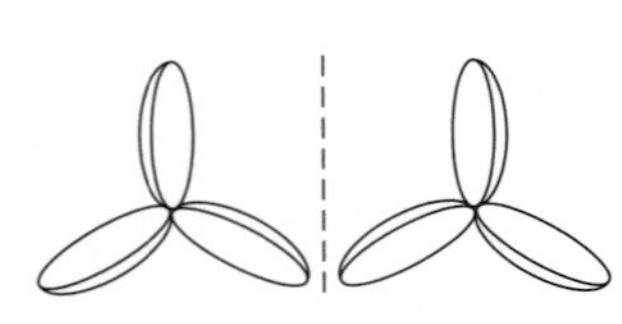

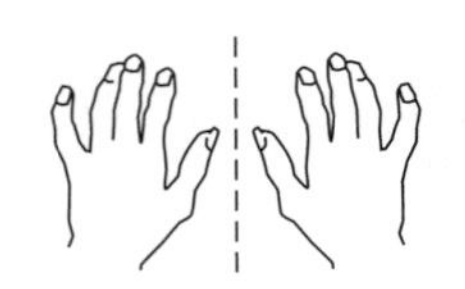

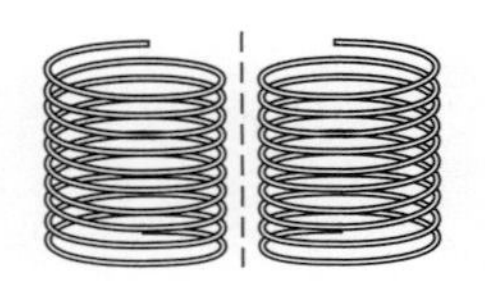

FIGURE 4.18 A Chiral Molecule and Other Chiral Objects.

Many molecules are not superimposable on their mirror image. Such molecules, labeled **chiral** or **dissymmetric**, may have important chemical properties as a consequence of this nonsuperimposability. One chiral molecule is CBrClFI, and many examples of chiral objects can also be found on the macroscopic scale, as in **Figure 4.18.**

Chiral objects are termed *dissymmetric*. This term does not imply that these objects necessarily have *no* symmetry. For example, the propellers in Figure 4.18 each have a C_3 axis, yet they are nonsuperimposable (if both were spun in a clockwise direction, they would move an airplane in opposite directions). In general, a molecule or object is chiral if it has no symmetry operations (other than E), or if it has *only proper rotation axes.*

EXERCISE 4.7

Which point groups are possible for chiral molecules? (Hint: refer to the character tables in Appendix C.)

Air blowing past the stationary propellers in Figure 4.18 will be rotated in either a clockwise or counterclockwise direction. By analogy, plane-polarized light will be rotated on passing through chiral molecules (**Figure 4.19**); clockwise rotation is designated **dextrorotatory**, and counterclockwise rotation is **levorotatory**. The ability of chiral molecules to rotate plane-polarized light is termed **optical activity**, which may be measured experimentally.

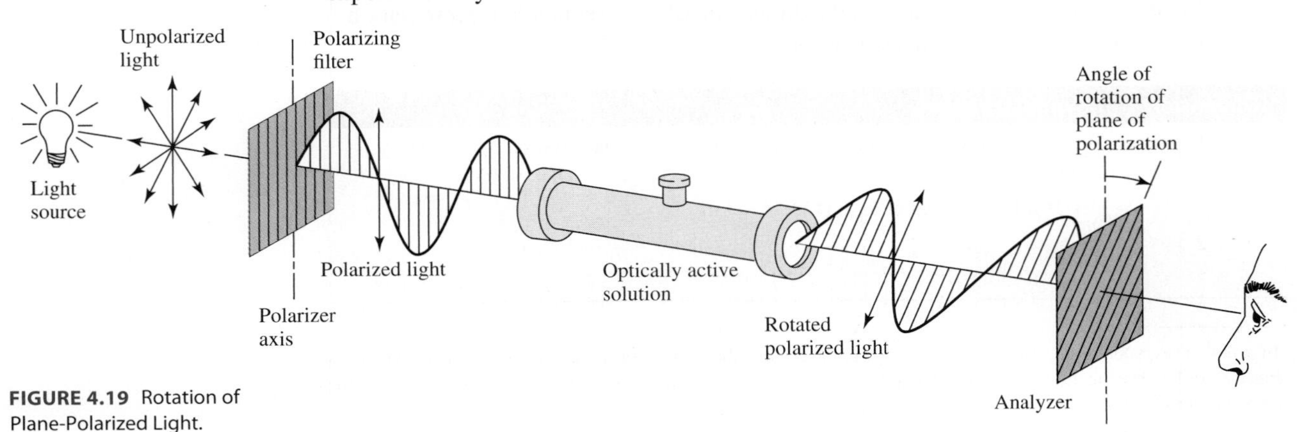

FIGURE 4.19 Rotation of Plane-Polarized Light.

Many coordination compounds are chiral and exhibit optical activity if they can be resolved into the two isomers. One of these is $[Ru(NH_2CH_2CH_2NH_2)_3]^{2+}$, with D_3 symmetry (**Figure 4.20**). Mirror images of this molecule look much like left- and right-handed three-bladed propellers. Further examples will be discussed in Chapter 9.

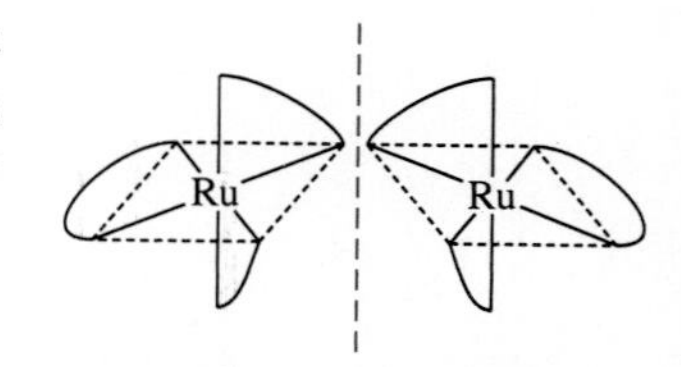

FIGURE 4.20 Chiral Isomers of $[Ru(NH_2CH_2CH_2NH_2)_3]^{2+}$.

4.4.2 Molecular Vibrations

Symmetry is helpful to determine the modes of vibration of molecules. The vibrational modes of water and the stretching modes of CO in carbonyl complexes can be treated quite simply, as described in the following pages. Other molecules can be studied using the same methods.

Water (C_{2v} Symmetry)

Because the study of vibrations is the study of motion of the atoms in a molecule, we must first attach a set of x, y, and z coordinates to each atom. For convenience, we assign the z axes parallel to the C_2 axis of the molecule, the x axes in the plane of the molecule, and the y axes perpendicular to the plane (**Figure 4.21**). Each atom can move in all three directions, so a total of nine transformations (motion of each atom in the x, y, and z directions) must be considered. For N atoms in a molecule, there are $3N$ total motions, known as **degrees of freedom**. Degrees of freedom for different geometries are summarized in **Table 4.10**. Because water has three atoms, there must be nine different motions.

We will use transformation matrices to determine the symmetry of all nine motions within the C_{2v} point group and then assign them to translation, rotation, and vibration. Fortunately, it is only necessary to determine the characters of the transformation matrices (and only one matrix for a symmetry operation of a unique class), not the individual matrix elements.

In this case, the initial axes make a column matrix with nine elements, and each transformation matrix is 9×9. A nonzero entry appears along the diagonal of the matrix only for an atom that does not change position. If the atom changes position during the symmetry operation, a zero is entered. If the atom remains in its original location and the vector direction is unchanged, a 1 is entered. If the atom remains, but the vector direction is reversed, a -1 is entered. (Because all the operations change vector directions by 0° or 180° in the C_{2v} point group, these are the only possibilities.) When all nine vectors are summed, the character of the reducible representation Γ is obtained. The full 9×9 matrix for C_2 is shown as an example; note that only the diagonal entries are used in finding the character.

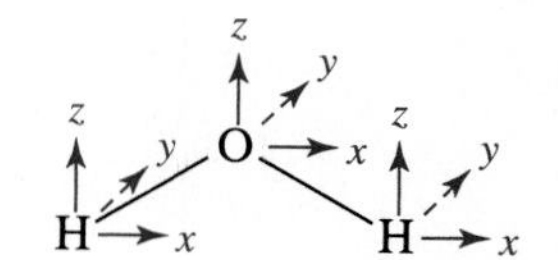

FIGURE 4.21 A Set of Axes for the Water Molecule.

$$
\begin{matrix} \text{O} \\ \\ \\ \text{H}_a \\ \\ \\ \text{H}_b \\ \\ \end{matrix}
\begin{bmatrix} x' \\ y' \\ z' \\ x' \\ y' \\ z' \\ x' \\ y' \\ z' \end{bmatrix}
=
\begin{bmatrix}
-1 & 0 & 0 & 0 & 0 & 0 & 0 & 0 & 0 \\
0 & -1 & 0 & 0 & 0 & 0 & 0 & 0 & 0 \\
0 & 0 & 1 & 0 & 0 & 0 & 0 & 0 & 0 \\
0 & 0 & 0 & 0 & 0 & 0 & -1 & 0 & 0 \\
0 & 0 & 0 & 0 & 0 & 0 & 0 & -1 & 0 \\
0 & 0 & 0 & 0 & 0 & 0 & 0 & 0 & 1 \\
0 & 0 & 0 & -1 & 0 & 0 & 0 & 0 & 0 \\
0 & 0 & 0 & 0 & -1 & 0 & 0 & 0 & 0 \\
0 & 0 & 0 & 0 & 0 & 1 & 0 & 0 & 0
\end{bmatrix}
\begin{bmatrix} x \\ y \\ z \\ x \\ y \\ z \\ x \\ y \\ z \end{bmatrix}
\begin{matrix} \\ \text{O} \\ \\ \\ \text{H}_a \\ \\ \\ \text{H}_b \\ \\ \end{matrix}
$$

TABLE 4.10 Degrees of Freedom

Number of Atoms	Total Degrees of Freedom	Translational Modes	Rotational Modes	Vibrational Modes
N(Linear)	$3N$	3	2	$3N - 5$
3 (HCN)	9	3	2	4
N(Nonlinear)	$3N$	3	3	$3N - 6$
3(H_2O)	9	3	3	3

The H_a and H_b entries are not on the principal diagonal, because H_a and H_b exchange with each other in a C_2 rotation, and $x'(H_a) = -x(H_b)$, $y'(H_a) = -y(H_b)$, and $z'(H_a) = z(H_b)$. Only the oxygen atom contributes to the character for this operation, for a total of -1.

The other entries for Γ can also be found without writing out the matrices, as follows:

E: All nine vectors are unchanged in the identity operation, so the character is 9.

C_2: The hydrogen atoms change position in a C_2 rotation, so all their vectors have zero contribution to the character. The oxygen atom vectors in the x and y directions are reversed, each contributing -1, and the vector in the z direction remains the same, contributing 1 for a total of -1. The sum of the principal diagonal $= \chi(C_2) = (-1) + (-1) + (1) = -1$.

$\sigma_v(xz)$: Reflection in the plane of the molecule changes the direction of all the y vectors and leaves the x and z vectors unchanged, for a total of $3 - 3 + 3 = 3$.

$\sigma_v'(yz)$: Finally, reflection perpendicular to the plane of the molecule changes the position of the hydrogens so their contribution is zero; the x vector on the oxygen changes direction, and the y and z vectors are unchanged for a total of 1.

EXERCISE 4.8

Write the corresponding 9×9 transformation matrices for the $\sigma(xz)$ and $\sigma(yz)$ operations in C_{2v} symmetry.

Because all nine direction vectors are included in this representation, it represents all the motions of the molecule: three translations, three rotations, and (by difference) three vibrations. The characters of the reducible representation Γ are shown as the last row below the irreducible representations in the C_{2v} character table.

C_{2v}	E	C_2	$\sigma_v(xz)$	$\sigma_v'(yz)$		
A_1	1	1	1	1	z	x^2, y^2, z^2
A_2	1	1	-1	-1	R_z	xy
B_1	1	-1	1	-1	x, R_y	xz
B_2	1	-1	-1	1	y, R_x	yz
Γ	9	-1	3	1		

Reducing Representations to Irreducible Representations

The next step is to determine how the irreducible representations sum to give the reducible representation. This requires another property of groups. The number of times any irreducible representation contributes to a reducible representation is equal to the sum of the products of the characters of the reducible and irreducible representations multiplied by the number of operations in the class, taken one operation at a time, divided by the order of the group. This may be expressed in equation form, with the sum taken over all symmetry operations of the group.*

$$\begin{pmatrix}\text{Number of irreducible}\\ \text{representations of}\\ \text{a given type}\end{pmatrix} = \frac{1}{\text{order}}\sum_{R}\left[\begin{pmatrix}\text{number}\\ \text{of operations}\\ \text{in the class}\end{pmatrix} \times \begin{pmatrix}\text{character of}\\ \text{reducible}\\ \text{representation}\end{pmatrix} \times \begin{pmatrix}\text{character of}\\ \text{irreducible}\\ \text{representation}\end{pmatrix}\right]$$

*This procedure should yield an integer for the number of irreducible representations of each type; obtaining a fraction in this step indicates a calculation error.

In the water example, the order of C_{2v} is 4, with one operation in each class (E, C_2, σ_v, σ_v'). The results are as follows:

$$n_{A_1} = \tfrac{1}{4}[(9)(1) + (-1)(1) + (3)(1) + (1)(1)] = 3$$

$$n_{A_2} = \tfrac{1}{4}[(9)(1) + (-1)(1) + (3)(-1) + (1)(-1)] = 1$$

$$n_{B_1} = \tfrac{1}{4}[(9)(1) + (-1)(-1) + (3)(1) + (1)(-1)] = 3$$

$$n_{B_2} = \tfrac{1}{4}[(9)(1) + (-1)(-1) + (3)(-1) + (1)(1)] = 2$$

The reducible representation for all motions of the water molecule is therefore reduced to $3A_1 + A_2 + 3B_1 + 2B_2$.

Examination of the columns on the far right in the character table shows that translation along the x, y, and z directions is $A_1 + B_1 + B_2$ (translation is motion along the x, y, and z directions, so it transforms similarly as the three axes). Rotation in the three directions (R_x, R_y, R_z) is $A_2 + B_1 + B_2$. Subtracting these from the total given previously leaves $2A_1 + B_1$, the three vibrational modes shown in **Table 4.11**. The number of vibrational modes equals $3N - 6$, as described earlier. Two of the modes are totally symmetric (A_1) and do not change the symmetry of the molecule, but one is antisymmetric to C_2 rotation and to reflection perpendicular to the plane of the molecule (B_1). These modes are illustrated as symmetric stretch, symmetric bend, and antisymmetric stretch in **Table 4.12**. It is noteworthy that the complex motion of a water molecule in three-dimensional space (a gas phase molecule will often be simultaneously rotating and translating while vibrating) could be described in terms of varying contributions from each of these fundamental modes.

TABLE 4.11 Symmetry of Molecular Motions of Water

All Motions	Translation (x, y, z)	Rotation (R_x, R_y, R_z)	Vibration (remaining modes)
$3A_1$	A_1		$2A_1$
A_2		A_2	
$3B_1$	B_1	B_1	B_1
$2B_2$	B_2	B_2	

TABLE 4.12 The Vibrational Modes of Water

A_1	H O H	Symmetric stretch: change in dipole moment; more distance between positive hydrogens and negative oxygen *IR active*
B_1	H O H	Antisymmetric stretch: change in dipole moment; change in distances between positive hydrogens and negative oxygen *IR active*
A_1	H O H	Symmetric bend: change in dipole moment; angle between H—O vectors changes *IR active*

EXAMPLE 4.4

Using the x, y, and z coordinates for each atom in XeF_4, determine the reducible representation for all molecular motions; reduce this representation to its irreducible components; and classify these representations into translational, rotational, and vibrational modes.

First, it is useful to assign x, y, and z coordinate axes to each atom, as shown in **Figure 4.22**.

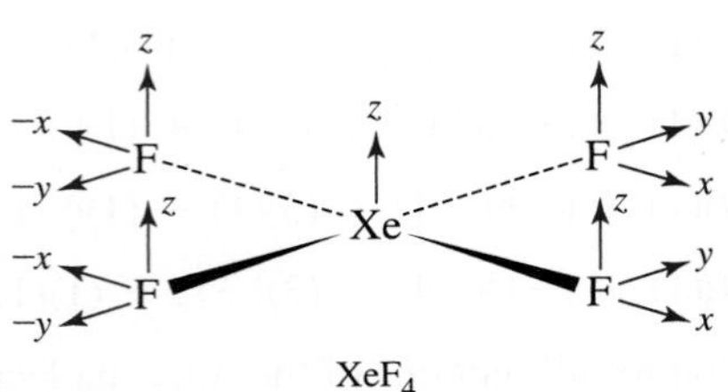

FIGURE 4.22 Coordinate Axes for XeF_4.

It is essential to recognize that *only the coordinates on atoms that do* not *move when symmetry operations are applied can give rise to nonzero elements along the diagonals of transformation matrices.* For example, if a symmetry operation applied to XeF_4 causes all F atoms to change position, these atoms cannot give rise to elements along the diagonal, so they can be ignored; only coordinates of Xe would need to be considered.

In addition: If a symmetry operation leaves the direction of a coordinate unchanged, it gives a character of 1 along the diagonal.

For example, the identity operation on XeF_4 leaves the coordinates x, y, and z unchanged; each of these has a diagonal element of 1 for each atom.

$$\underset{1}{x \longrightarrow x} \quad \underset{1}{y \longrightarrow y} \quad \underset{1}{z \longrightarrow z}$$

If a symmetry operation reverses the direction of a coordinate, this corresponds to a diagonal element of −1.

The σ_h operation on XeF_4 reverses the direction of the z axis for each atom.

$$\underset{-1}{z \longrightarrow -z}$$

If a symmetry operation transforms a coordinate into another coordinate, this gives a diagonal element of zero.

If XeF_4 is rotated about its C_4 axis, the x and y coordinates of Xe are interchanged; they contribute zero to the character.

Examining each of the D_{4h} symmetry operations in turn generates the following reducible representation for all the molecular motions of XeF_4:

D_{4h}	E	$2C_4$	C_2	$2C_2{}'$	$2C_2{}''$	i	$2S_4$	σ_h	$2\sigma_v$	$2\sigma_d$
Γ	15	1	−1	−3	−1	−3	−1	5	3	1

The character under E indicates that there are 15 possible motions to be considered. By the procedure illustrated in the preceding example, this representation reduces to

$$\Gamma = A_{1g} + A_{2g} + B_{1g} + B_{2g} + E_g + 2\,A_{2u} + B_{2u} + 3\,E_u$$

These can be classified as follows:

Translational Motion. This is motion through space with x, y, and z components. The irreducible representations matching these components have the labels x, y, and z

on the right side of the D_{4h} character table: A_{2u} (matching z) and E_u (doubly degenerate, matching x and y together). These three motions can be represented as shown in **Figure 4.23.**

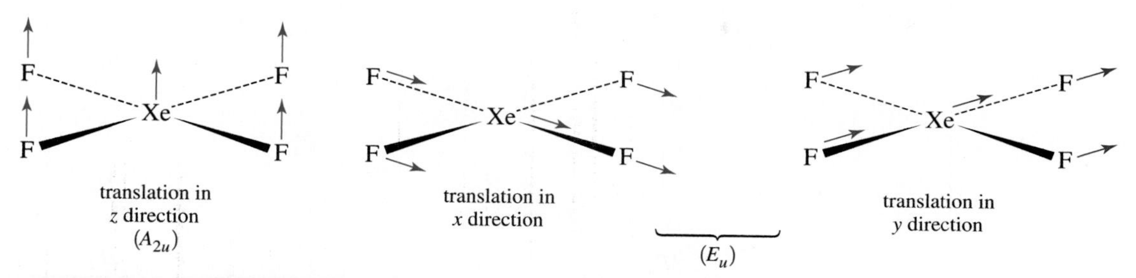

FIGURE 4.23 Translational Modes of XeF_4.

Rotational Motion. This type of motion can be factored into rotation about the mutually orthogonal x, y, and z axes. The matching expressions in the character table are R_x, R_y, and R_z, representing rotation about these three axes, respectively. The irreducible representations are A_{2g} (R_z, rotation about the z axis), and E_g ((R_x, R_y), doubly degenerate rotations about the x and y axes) as shown in **Figure 4.24**.

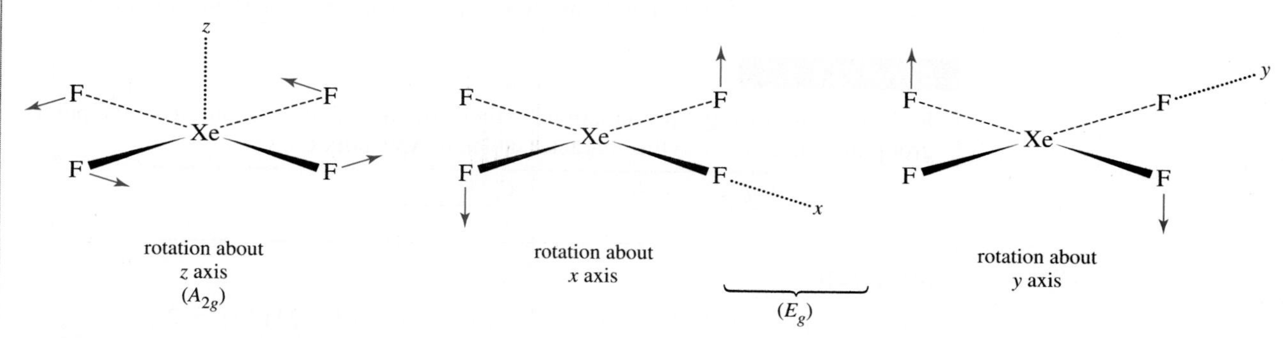

FIGURE 4.24 Rotational Modes of XeF_4.

Vibrational Motion. The remaining nine motions (15 total − 3 translations − 3 rotations) are vibrational. They involve changes in bond lengths and angles and motions both within and out of the molecular plane. For example, symmetrical stretching of all four Xe—F bonds matches the A_{1g} irreducible representation, symmetrical stretching of opposite bonds matches B_{1g}, and simultaneous opening of opposite bond angles matches B_{2g} as shown in **Figure 4.25.**

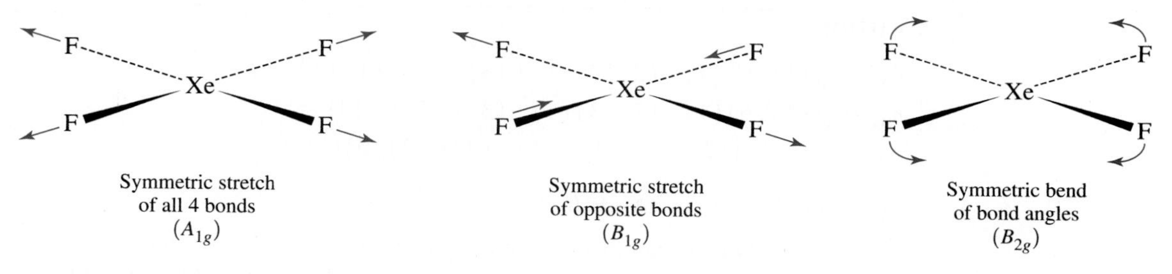

FIGURE 4.25 Selected Vibrational Modes of XeF_4.

TABLE 4.13 Symmetry of Molecular Motions of XeF_4

	Γ (all modes)	Translation	Rotation	Vibration
	A_{1g}			A_{1g}
	A_{2g}		A_{2g}	
	B_{1g}			B_{1g}
	B_{2g}			B_{2g}
	E_g		E_g	
	$2A_{2u}$	A_{2u}		A_{2u}
	B_{2u}			B_{2u}
	$3E_u$	E_u		$2E_u$
Total	15	3	3	9

Table 4.13 summarizes the classification of irreducible representations according to mode.

EXERCISE 4.9 Using the x, y, and z coordinates for each atom in N_2O_4, which is planar and has a nitrogen–nitrogen bond, determine the reducible representation for all molecular motions. Reduce this representation to its irreducible components, and classify these representations into translational, rotational, and vibrational modes.

EXAMPLE 4.5

Reduce the following representations to their irreducible representations in the point group indicated (refer to the character tables in Appendix C):

C_{2h}	E	C_2	i	σ_h
Γ	4	0	2	2

SOLUTION

$$n_{A_g} = \tfrac{1}{4}[(1)(4)(1)+(1)(0)(1)+(1)(2)(1)+(1)(2)(1)] = 2$$
$$n_{B_g} = \tfrac{1}{4}[(1)(4)(1)+(1)(0)(-1)+(1)(2)(1)+(1)(2)(-1)] = 1$$
$$n_{A_u} = \tfrac{1}{4}[(1)(4)(1)+(1)(0)(1)+(1)(2)(-1)+(1)(2)(-1)] = 0$$
$$n_{B_u} = \tfrac{1}{4}[(1)(4)(1)+(1)(0)(-1)+(1)(2)(-1)+(1)(2)(1)] = 1$$

Therefore, $\Gamma = 2A_g + B_g + B_u$.

C_{3v}	E	$2C_3$	$3\sigma_v$
Γ	6	3	−2

SOLUTION

$$n_{A_1} = \tfrac{1}{6}[(1)(6)(1)+(2)(3)(1)+(3)(-2)(1)] = 1$$
$$n_{A_2} = \tfrac{1}{6}[(1)(6)(1)+(2)(3)(1)+(3)(-2)(-1)] = 3$$
$$n_E = \tfrac{1}{6}[(1)(6)(2)+(2)(3)(-1)+(3)(-2)(0)] = 1$$

Therefore, $\Gamma = A_1 + 3A_2 + E$.

Be sure to include the number of symmetry operations in a class (column) of the character table. This means that the second term in the C_{3v} calculation must be multiplied by 2 ($2C_3$; there are two operations in this class); the third term must be multiplied by 3, as shown.

EXERCISE 4.10 Reduce the following representations to their irreducible representations in the point groups indicated:

T_d	E	$8C_3$	$3C_2$	$6S_4$	$6\sigma_d$
Γ_1	4	1	0	0	2

D_{2d}	E	$2S_4$	C_2	$2C_2'$	$2\sigma_d$
Γ_2	4	0	0	2	0

C_{4v}	E	$2C_4$	C_2	$2\sigma_v$	$2\sigma_d$
Γ_3	7	−1	−1	−1	−1

Infrared Spectra

A molecular vibration is infrared active (i.e., excitation of the vibrational mode can be measured via an IR spectrum absorption) only if it results in a change in the dipole moment of the molecule. The three vibrations of water (Table 4.12) can be analyzed in this way to determine their infrared behavior.

Group theory in principle can account for the possible infrared activity of *all* vibrational modes of a molecule. In group theory terms, a vibrational mode is active in the infrared *if it corresponds to an irreducible representation that has the same symmetry (or transforms) as the Cartesian coordinates x, y, or z,* because a vibrational motion that shifts the center of charge of the molecule in any of the *x*, *y*, or *z* directions results in a change in dipole moment. Otherwise, the vibrational mode is not infrared active.

EXERCISE 4.11

Which of the nine vibrational modes of XeF_4 (Table 4.13) are infrared active?

EXERCISE 4.12

Analysis of the *x*, *y*, and *z* coordinates of each atom in NH_3 gives the following representation:

C_{3v}	E	$2C_3$	$3\sigma_v$
Γ	12	0	2

a. Reduce Γ to its irreducible representations.
b. Classify the irreducible representations into translational, rotational, and vibrational modes.
c. Show that the total number of degrees of freedom $= 3N$.
d. Which vibrational modes are infrared active?

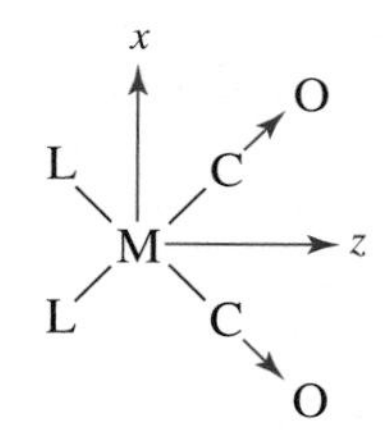

cis-Dicarbonyl complex

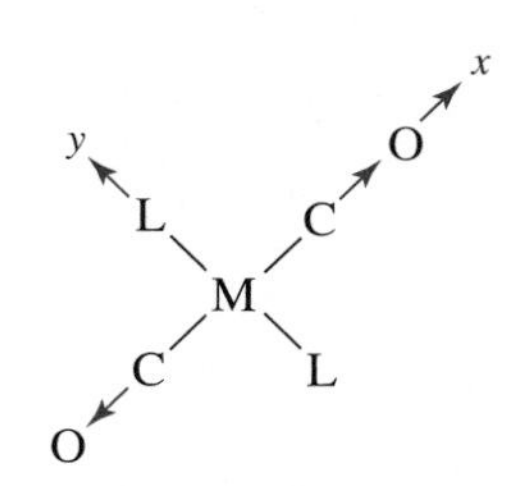

trans-Dicarbonyl complex

FIGURE 4.26 Carbonyl Stretching Vibrations of *cis*- and *trans*-Dicarbonyl Square Planar Complexes.

Selected Vibrational Modes

It is often useful to consider a particular vibrational mode for a compound. For example, useful information often can be obtained from the C—O stretching bands in infrared spectra of metal complexes containing CO (carbonyl) ligands. The following example of *cis*- and *trans*-dicarbonyl square planar complexes shows the procedure.

For these complexes,* a routine IR spectrum can distinguish whether a sample is *cis*- or *trans*-$ML_2(CO)_2$; the number of C—O stretching bands is determined by the geometry of the complex (**Figure 4.26**).

***cis*-$ML_2(CO)_2$, point group C_{2v}.** The principal axis (C_2) is the *z* axis, with the *xz* plane assigned as the plane of the molecule. Possible C—O stretching motions are shown by arrows in **Figure 4.27**. These vectors are used to create the reducible representation

*M represents any metal and L any ligand other than CO in these formulas.

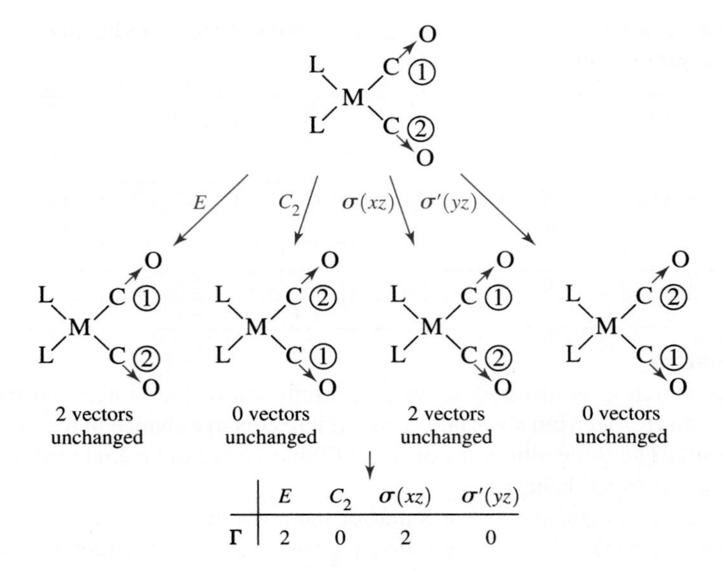

FIGURE 4.27 Symmetry Operations and Characters for *cis* – $ML_2(CO)_2$.

that follows, using the symmetry operations of the C_{2v} point group. A C—O bond will transform with a character of 1 *if it remains unchanged* by the symmetry operations, and with a character of 0 *if it is changed*. These operations and their characters are shown in Figure 4.27. Both stretches are unchanged in the identity operation and in the reflection through the plane of the molecule, so each contributes 1 to the character, for a total of 2 for each operation. Both vectors move to new locations on rotation or reflection perpendicular to the plane of the molecule, so these two characters are 0.

The reducible representation Γ reduces to $A_1 + B_1$:

C_{2v}	E	C_2	$\sigma_v(xz)$	$\sigma_v'(yz)$		
Γ	2	0	2	0		
A_1	1	1	1	1	z	x^2, y^2, z^2
B_1	1	−1	1	−1	x, R_y	xz

A_1 is an appropriate irreducible representation for an IR-active band, because it transforms as (has the symmetry of) the Cartesian coordinate z. Furthermore, the vibrational mode corresponding to B_1 should be IR active, because it transforms as the Cartesian coordinate x.

In summary, there are two vibrational modes for C—O stretching, one having A_1 symmetry and one B_1 symmetry. Both modes are IR active, and we therefore expect to see two C—O stretches in the IR. This assumes that the C—O stretches are not sufficiently similar in energy to overlap in the infrared spectrum.

***trans*-$ML_2(CO)_2$, point group D_{2h}.** The principal axis, C_2, is again chosen as the z axis, which renders the plane of the molecule the xy plane. Using the symmetry operation of the D_{2h} point group, we obtain a reducible representation for C—O stretches that reduces to $A_g + B_{3u}$:

D_{2h}	E	$C_2(z)$	$C_2(y)$	$C_2(x)$	i	$\sigma(xy)$	$\sigma(xz)$	$\sigma(yz)$	
Γ	2	0	0	2	0	2	2	0	
A_g	1	1	1	1	1	1	1	1	x^2, y^2, z^2
B_{3u}	1	−1	−1	1	−1	1	1	−1	x

The vibrational mode of A_g symmetry is not IR active, because it does not have the same symmetry as a Cartesian coordinate *x*, *y*, or *z* (this is the IR-inactive symmetric stretch). The mode of symmetry B_{3u}, on the other hand, is IR active, because it has the same symmetry as *x*.

In summary, there are two vibrational modes for C—O stretching, one having the same symmetry as A_g, and one the same symmetry as B_{3u}. The A_g mode is IR inactive (it does not have the symmetry of *x*, *y*, or *z*); the B_{3u} mode is IR active (it has the symmetry of *x*). We therefore expect to see one C—O stretch in the IR.

It is therefore possible to distinguish *cis*- and *trans*-$ML_2(CO)_2$ by taking an IR spectrum. If one C—O stretching band appears, the molecule is *trans*; if two bands appear, the molecule is *cis*. A significant distinction can be made by a very simple measurement.

EXAMPLE 4.6

Determine the number of IR-active CO stretching modes for *fac*-$Mo(CO)_3(CH_3CH_2CN)_3$, as shown in the margin.

This molecule has C_{3v} symmetry. The operations to be considered are E, C_3, and σ_v. E leaves the three bond vectors unchanged, giving a character of 3. C_3 moves all three vectors, giving a character of 0. Each σ_v plane passes through one of the CO groups, leaving it unchanged, while interchanging the other two. The resulting character is 1.

The representation to be reduced, therefore, is as follows:

E	$2C_3$	$3\sigma_v$
3	0	1

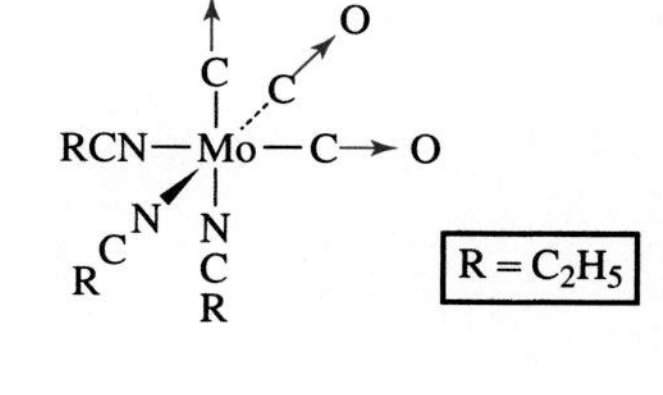

This reduces to $A_1 + E$. A_1 has the same symmetry as the Cartesian coordinate *z* and is therefore IR active. E has the same symmetry as the *x* and *y* coordinates together and is also IR active. It represents a degenerate pair of vibrations, which appear as one absorption band, as shown in **Figure 4.28**. Infrared absorptions associated with specific bonds are commonly designated as ν(XY), where XY is the bond that contributes most significantly to the vibrational modes responsible for the absorptions. In Figure 4.28, the ν(CO) spectrum features absorptions at 1920 and 1790 cm^{-1}.

EXERCISE 4.13 Determine the number of IR-active C—O stretching modes for $Mn(CO)_5Cl$.

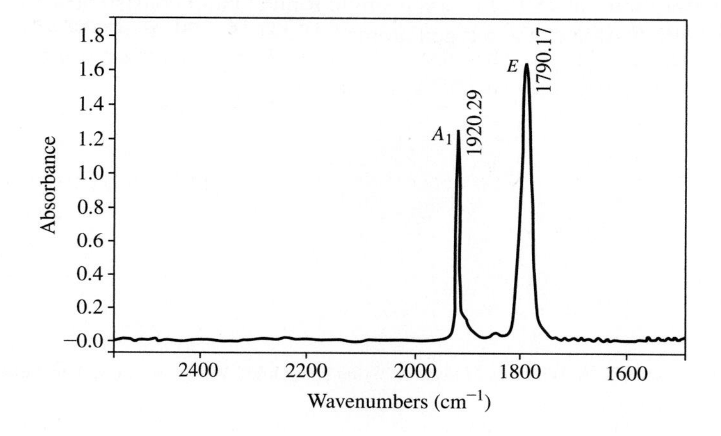

FIGURE 4.28 Infrared spectrum of *fac*-$Mo(CO)_3(CH_3CH_2CN)_3$.

Raman Spectroscopy

This spectroscopic method uses a different approach to observe molecular vibrations. Rather than directly observing absorption of infrared radiation as in IR spectroscopy, in Raman spectroscopy higher energy radiation, ordinarily from a laser, excites molecules to higher electronic states, envisioned as short-lived "virtual" states. Scattered radiation from decay of these excited states to the various vibrational states provides information about vibrational energy levels that is complementary to information gained from IR spectroscopy. In general, a vibration can give rise to a line in a Raman spectrum if it causes a change in polarizability.* From a symmetry standpoint, vibrational modes are Raman active if they match the symmetries of the functions xy, xz, yz, x^2, y^2, or z^2 or a linear combination of any of these; if vibrations match these functions, they also occur with a change in polarizability. These functions are among those commonly listed in character tables. In some cases—when molecular vibrations match both these functions and x, y, or z—molecular vibrations can be both IR and Raman active.

EXAMPLE 4.7

Vibrational spectroscopy has played a role in supporting the tetrahedral structure of the highly explosive XeO_4. Raman spectroscopy has shown two bands in the region expected for Xe—O stretching vibrations, at 776 and 878 cm^{-1}.[1] Is this consistent with the proposed T_d structure?

To address this question, we need to once again create a representation, this time using the Xe=O stretches as a basis in the T_d point group. The resulting representation is

	E	$8C_3$	$3C_2$	$6S_4$	$6\sigma_d$
Γ	4	1	0	0	2

This reduces to $A_1 + T_2$:

A_1	1	1	1	1	1		$x^2+y^2+z^2$
T_2	3	0	−1	−1	1	(x, y, z)	$(xy, xz\ yz)$

Both the A_1 and T_2 representations match functions necessary for Raman activity. The presence of these two bands is consistent with the proposed T_d symmetry.

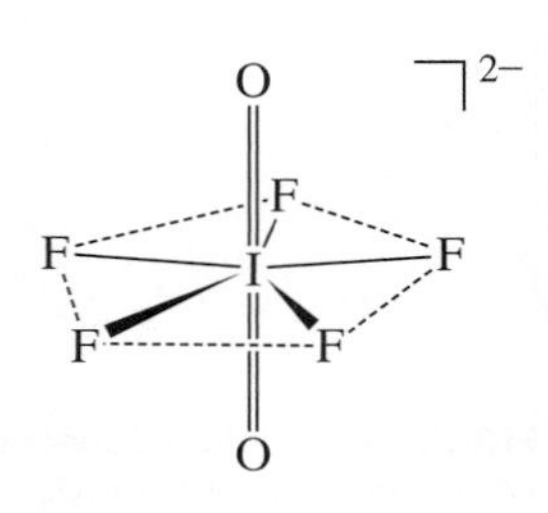

EXERCISE 4.14 Vibrational spectroscopy has played a role in supporting the pentagonal bipyramidal structure of the ion $IO_2F_5^{2-}$.[2] Raman spectroscopy of the tetramethylammonium salt of this ion shows a single absorption in the region expected for I=O stretching vibrations, at 789 cm^{-1}. Is a single Raman band consistent with the proposed *trans* orientation of the oxygen atoms?

*For more details, see D. J. Willock, *Molecular Symmetry*, John Wiley & Sons, Chichester, UK, 2009, pp. 177–184.

References

1. M. Gerken, G. J. Schrobilgen, *Inorg. Chem.*, **2002**, *41*, 198.
2. J. A. Boatz, K. O. Christe, D. A. Dixon, B. A. Fir, M. Gerken, R. Z. Gnann, H. P. A. Mercier, G. J. Schrobilgen, *Inorg. Chem.*, **2003**, *42*, 5282.

General References

There are several helpful books on molecular symmetry and its applications. Good examples are D. J. Willock, *Molecular Symmetry*, John Wiley & Sons, Chichester, UK, 2009; F. A. Cotton, *Chemical Applications of Group Theory*, 3rd ed., John Wiley & Sons, New York, 1990; S. F. A. Kettle, *Symmetry and Structure: Readable Group Theory for Chemists*, 2nd ed., John Wiley & Sons, New York, 1995; and I. Hargittai and M. Hargittai, *Symmetry Through the Eyes of a Chemist*, 2nd ed., Plenum Press, New York, 1995. The last two also provide information on space groups used in solid state symmetry, and all give relatively gentle introductions to the mathematics of the subject. For an explanation of situations involving complex numbers in character tables, see S. F. A. Kettle, *J. Chem. Educ.*, **2009**, *86*, 634.

Problems

4.1 Determine the point groups for
- **a.** Ethane (staggered conformation)
- **b.** Ethane (eclipsed conformation)
- **c.** Chloroethane (staggered conformation)
- **d.** 1,2-Dichloroethane (staggered *anti* conformation)

4.2 Determine the point groups for
- **a.** Ethylene
- **b.** Chloroethylene
- **c.** The possible isomers of dichloroethylene

4.3 Determine the point groups for
- **a.** Acetylene
- **b.** $H{-}C{\equiv}C{-}F$
- **c.** $H{-}C{\equiv}C{-}CH_3$
- **d.** $H{-}C{\equiv}C{-}CH_2Cl$
- **e.** $H{-}C{\equiv}C{-}Ph$ (Ph = phenyl)

4.4 Determine the point groups for
- **a.** Naphthalene

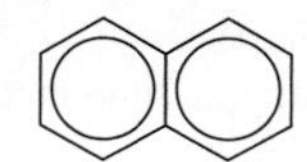

- **b.** 1,8-Dichloronaphthalene
- **c.** 1,5-Dichloronaphthalene

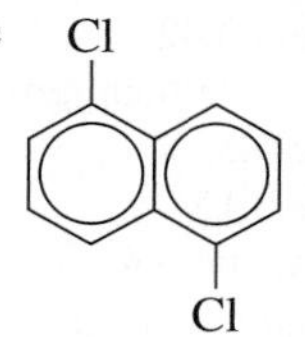

- **d.** 1,2-Dichloronaphthalene

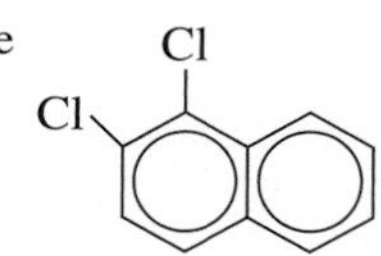

4.5 Determine the point groups for
- **a.** 1,1′ − Dichloroferrocene

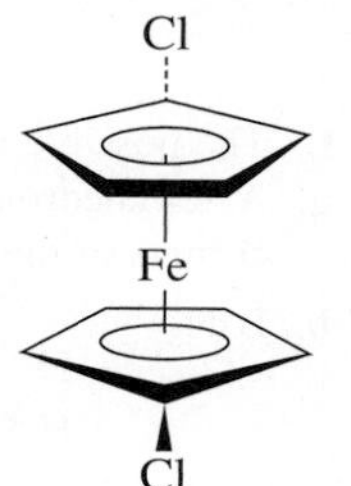

- **b.** Dibenzenechromium (eclipsed conformation)

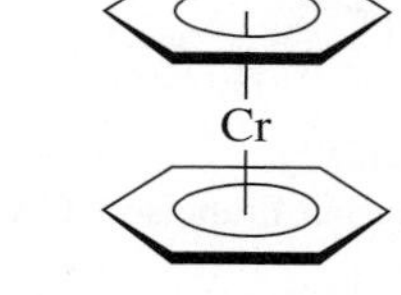

- **c.**

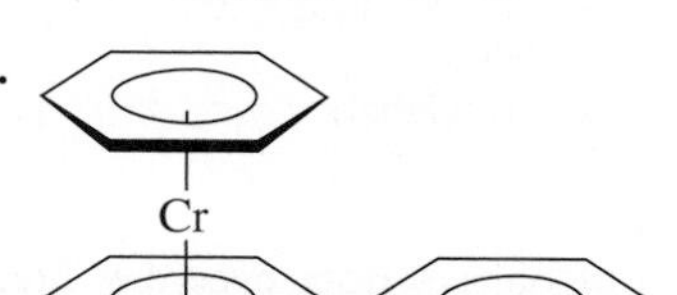

- **d.** H_3O^+
- **e.** O_2F_2
- **f.** Formaldehyde, H_2CO

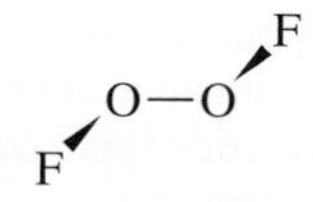

- **g.** S_8 (puckered ring)
- **h.** Borazine (planar)

i. $[Cr(C_2O_4)_3]^{3-}$

j. A tennis ball (ignoring the label, but including the pattern on the surface)

4.6 Determine the point groups for

a. Cyclohexane (chair conformation)

b. Tetrachloroallene $Cl_2C{=}C{=}CCl_2$

c. SO_4^{2-}

d. A snowflake

e. Diborane

f. The possible isomers of tribromobenzene

g. A tetrahedron inscribed in a cube in which alternate corners of the cube are also corners of the tetrahedron.

h. B_3H_8

i. A mountain swallowtail butterfly.

j. The Golden Gate Bridge, in San Francisco, CA

4.7 Determine the point groups for

a. A sheet of typing paper

b. An Erlenmeyer flask (no label)

c. A screw

d. The number 96

e. Five examples of objects from everyday life; select items from five different point groups

f. A pair of eyeglasses, assuming lenses of equal strength

g. A five-pointed star

h. A fork with no decoration

i. Wilkins Micawber, *David Copperfield* character who wore a monocle

j. A metal washer

4.8 Determine the point groups for

a. A flat oval running track

b. A jack (child's toy)

c. A person's two hands, palm to palm

d. A rectangular towel, blue on front, white on back

e. A hexagonal pencil with a round eraser

f. The recycle symbol, in three dimensions

g. The meander motif

h. An open, eight-spoked umbrella with a straight handle

i. A round toothpick

j. A tetrahedron with one green face, the others red

4.9 Determine the point groups for

a. A triangular prism

b. A plus sign

c. A t-shirt with the letter T on the front

d. Set of three wind turbine blades

e. A spade design (as on a deck of playing cards)

f. A sand dollar

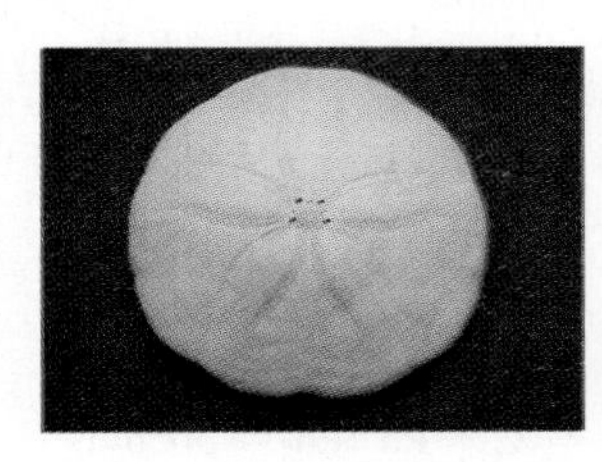

g. Flying Mercury sculpture, by Giambologna at the Louvre in Paris, France

h. An octahedron with one blue face, the others yellow

i. A hula hoop

j. A coiled spring

4.10 Determine the point groups for the examples of symmetry in Figure 4.1.

4.11 Determine the point groups of the molecules in the following end-of-chapter problems from Chapter 3:

a. Problem 3.41

b. Problem 3.42

4.12 Determine the point groups of the molecules and ions in

a. Figure 3.8

b. Figure 3.15

4.13 Determine the point groups of the following atomic orbitals, including the signs on the orbital lobes:

a. p_x

b. d_{xy}

c. $d_{x^2-y^2}$

d. d_{z^2}

e. f_{xyz}

4.14 **a.** Show that a cube has the same symmetry elements as an octahedron.
b. Suppose a cube has four dots arranged in a square on each face as shown. What is the point group?
c. Suppose that this set of dots is rotated as a set 10° clockwise on each face. Now what is the point group?

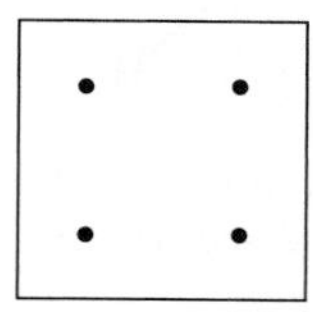

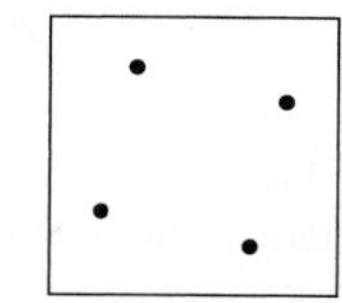

4.15 Suppose an octahedron can have either yellow or blue faces.
a. What point groups are possible if exactly two faces are blue?
b. What points are possible if exactly three faces are blue?
c. Now suppose the faces have four different colors. What is the point group if pairs of opposite faces have identical colors?

4.16 What point groups are represented by the symbols of chemical elements?

4.17 Baseball is a wonderful game, particularly for someone interested in symmetry. Where else can one watch a batter step from an on-deck circle of **a** symmetry to a rectangular batter's box of **b** symmetry, adjust a cap of **c** symmetry (it has OO on the front, for the Ozone City Oxygens), swing a bat of **d** symmetry (ignoring the label and grain of the wood) across a home plate of **e** symmetry at a baseball that has **f** symmetry (also ignoring the label) that has been thrown by a chiral pitcher having **g** symmetry, hit a towering fly ball that bounces off the fence, and race around the bases, only to be called out at home plate by an umpire who may have no appreciation for symmetry at all.

4.18 Determine the point groups for the following flags or parts of flags. You will need to look up images of flags not shown.
a. Botswana

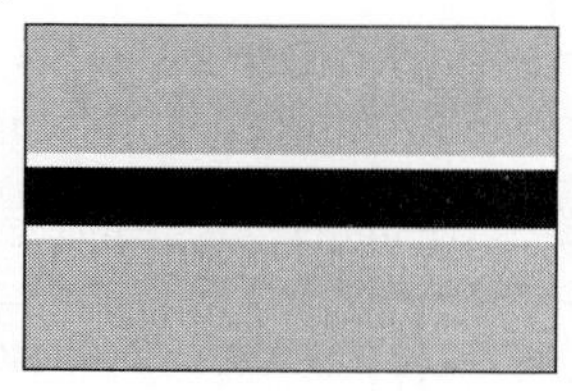

b. Finland

c. Honduras

d. Field of stars in flag of Micronesia

e. Central design on the Ethiopian flag:

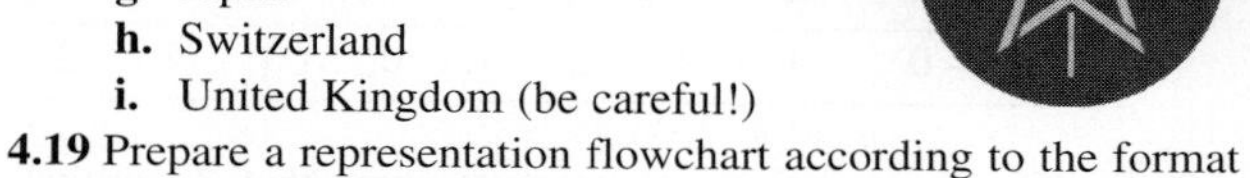

f. Turkey
g. Japan
h. Switzerland
i. United Kingdom (be careful!)

4.19 Prepare a representation flowchart according to the format of Table 4.8 for SNF_3.

4.20 For *trans*-1,2-dichloroethylene, which has C_{2h} symmetry,
a. List all the symmetry operations for this molecule.
b. Write a set of transformation matrices that describe the effect of each symmetry operation in the C_{2h} group on a set of coordinates x, y, z for a point (your answer should consist of four 3×3 transformation matrices).
c. Using the terms along the diagonal, obtain as many irreducible representations as possible from the transformation matrices. You should be able to obtain three irreducible representations in this way, but two will be duplicates. You may check your results using the C_{2h} character table.
d. Using the C_{2h} character table, verify that the irreducible representations are mutually orthogonal.

4.21 Ethylene has D_{2h} symmetry.
a. List all the symmetry operations of ethylene.
b. Write a transformation matrix for each symmetry operation that describes the effect of that operation on the coordinates of a point x, y, z.
c. Using the characters of your transformation matrices, obtain a reducible representation.
d. Using the diagonal elements of your matrices, obtain three of the D_{2h} irreducible representations.
e. Show that your irreducible representations are mutually orthogonal.

4.22 Using the D_{2d} character table,
a. Determine the order of the group.
b. Verify that the E irreducible representation is orthogonal to each of the other irreducible representations.

c. For each of the irreducible representations, verify that the sum of the squares of the characters equals the order of the group.

d. Reduce the following representations to their component irreducible representations:

D_{2d}	E	$2S_4$	C_2	$2C_2'$	$2\sigma_d$
Γ_1	6	0	2	2	2
Γ_2	6	4	6	2	0

4.23 Reduce the following representations to irreducible representations:

C_{3v}	E	$2C_3$	$3\sigma_v$
Γ_1	6	3	2
Γ_2	5	-1	-1

O_h	E	$8C_3$	$6C_2$	$6C_4$	$3C_2$	i	$6S_4$	$8S_6$	$3\sigma_h$	$6\sigma_d$
Γ	6	0	0	2	2	0	0	0	4	2

4.24 For D_{4h} symmetry use sketches to show that d_{xy} orbitals have B_{2g} symmetry and that $d_{x^2-y^2}$ orbitals have B_{1g} symmetry. (Hint: you may find it useful to select a molecule that has D_{4h} symmetry as a reference for the operations of the D_{4h} point group. Observe how the signs on the orbital lobes change as the symmetry operations are applied.)

4.25 Which items in Problems 4.5 through Problem 4.9 are chiral? List three items *not* from this chapter that are chiral.

4.26 $XeOF_4$ has one of the more interesting structures among noble gas compounds. On the basis of its symmetry,

a. Obtain a representation based on *all* the motions of the atoms in $XeOF_4$.

b. Reduce this representation to its component irreducible representations.

c. Classify these representations, indicating which are for translational, rotational, and vibrational motion.

d. Determine the irreducible representation matching the xenon–oxygen stretching vibration. Is this vibration IR active?

4.27 Repeat the procedure from the previous problem, parts a through c, for the SF_6 molecule and determine which vibrational modes are IR active.

4.28 For the following molecules, determine the number of IR-active C—O stretching vibrations:

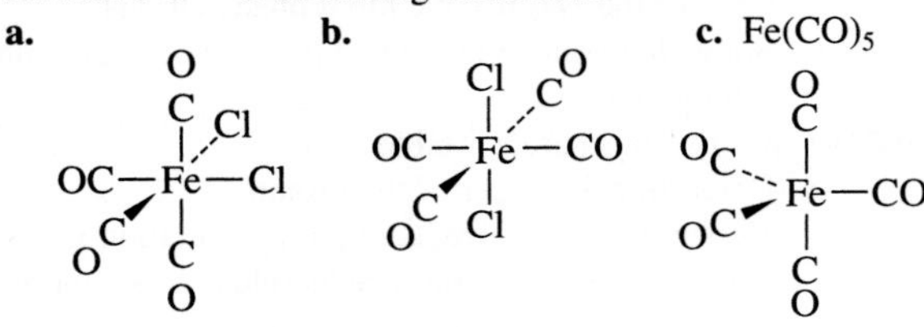

4.29 Repeat Problem 4.28 to determine the number of Raman-active C—O stretching vibrations.

4.30 The structure of 1,1,2,2-tetraiododisilane is shown here. (Reference: T. H. Johansen, K. Hassler, G. Tekautz, K. Hagen, J. *Mol. Struct.*, **2001**, *598*, 171.)

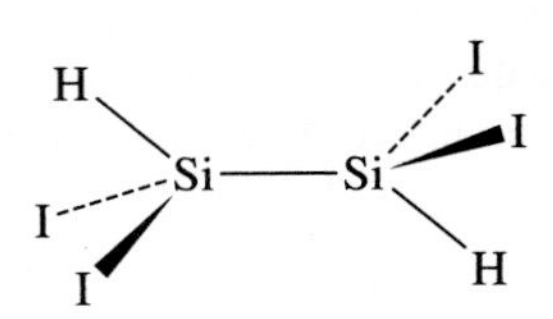

a. What is the point group of this molecule?

b. Predict the number of IR-active Si—I stretching vibrations.

c. Predict the number of Raman-active Si—I stretching vibrations.

4.31 Both *cis* and *trans* isomers of $IO_2F_4^-$ have been observed. Can IR spectra distinguish between these? Explain, supporting your answer on the basis of group theory. (Reference: K. O. Christe, R. D. Wilson, C. J. Schack, *Inorg. Chem.*, **1981**, *20*, 2104.)

4.32 White elemental phosphorus consists of tetrahedral P_4 molecules and is an important source of phosphorus for synthesis. In contrast, tetrahedral As_4 (yellow arsenic) is unstable, and decomposes to a grey As allotrope with a sheet structure. However, AsP_3, previously only observed at high temperature in the gas phase, has been isolated at ambient temperature as a white solid, where an As atom replaces one vertex of the tetrahedron.

a. The Raman spectrum of AsP_3, shown next, exhibits four absorptions. Is this consistent with the proposed structure? (Facile Synthesis of AsP_3, Brandi M. Cossairt, Mariam-Céline Diawara, Christopher C. Cummins. © 2009. The American Association for the Advancement of Science. Reprinted with permission from AAAS.)

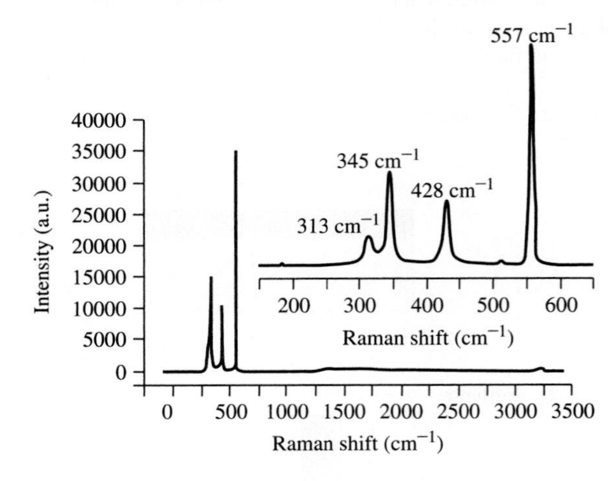

b. If As_2P_2 is ever isolated as a pure substance, how many Raman absorptions would be expected? (Reference: B. M. Cossairt, C. C. Cummins, *J. Am. Chem. Soc.* **2009**, *131*, 15501.)

c. Could a pure sample of P_4 be distinguished from pure AsP_3 simply on the basis of the number of Raman absorptions? Explain.

4.33 Complexes of the general formula $Fe(CO)_{5-x}(PR_3)_x$ are long known. The bimetallic $Fe_2(CO)_9$ reacts with triphenylphosphine in refluxing diethyl ether to afford a monosubstituted product $Fe(CO)_4(PPh_3)$ that exhibits ν (CO) absorptions at 2051, 1978, and 1945 cm^{-1} in hexane. (N. J. Farrer, R. McDonald, J. S. McIndoe, *Dalton Trans.*, **2006**, 4570.) Can these data be used to unambiguously establish whether the PPh_3 ligand is bound in either an equatorial or axial site in this trigonal bipyramidal complex? Support your decision by determining the number of IR-active CO stretching modes for these isomers.

4.34 Disubstituted $Fe(CO)_3(PPh_3)_2$ (ν(CO): 1883 cm^-; M. O. Albers, N. J. Coville, T. V. Ashworth, E. J. Singleton, *Organomet. Chem.*, **1981**, *217*, 385.) is also formed in the reaction described in Problem 4.33. Which of the following molecular geometries is supported by this spectrum? Support your decision by determining the number of IR-active CO stretching modes for these isomers. What does R. L. Keiter, E. A. Keiter, K. H. Hecker, C. A. Boecker, *Organometallics* **1988**, *7*, 2466 indicate about the infallibility of group theoretical CO stretching mode infrared spectroscopic prediction in the case of $Fe(CO)_3(PPh_3)_2$?

4.35 The reaction of $[Ti(CO)_6]^{2-}$ and chlorotriphenylmethane, Ph_3CCl, results in rapid oxidation of $[Ti(CO)_6]^{2-}$ to afford a trityltitanium tetracarbonyl complex (P. J. Fischer, K. A. Ahrendt, V. G. Young, Jr., J. E. Ellis, *Organometallics*, **1998**, *17*, 13). On the basis of the IR spectrum (ν(CO)): 1932, 1810 cm^{-1}) acquired in tetrahydrofuran solution, shown next, is this complex expected to exhibit a square planar or a square pyramidal arrangement of four CO ligands bound to titanium? Does the spectrum rule out either of these possible geometries?

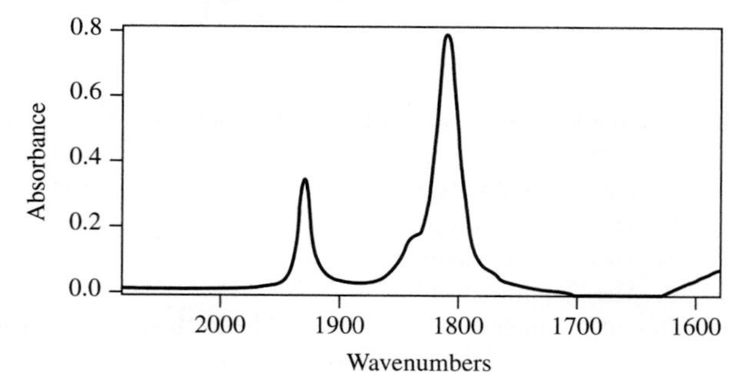

4.36 A related reaction to the one described in Problem 4.34, in which *cis* – $Mo(CO)_4(POPh_3)_2$ rearranges to *trans*-$Mo(CO)_4(POPh_3)_2$, has been probed mechanistically (D. J. Darensbourg, J. R. Andretta, S. M. Stranahan, J. H. Reibenspies, *Organometallics* **2007**, *26*, 6832.) When this reaction is conducted under an atmosphere of carbon monoxide, *cis/trans* isomerization does not occur. Instead a new complex (ν(CO) (hexane): 2085, 2000 (very weak), 1972, 1967 cm^{-1}) is formed (D. J. Darensbourg, T. L. Brown, *Inorg. Chem.* **1968**, *7*, 959.) Propose the formula of this Mo carbonyl complex consistent with the ν(CO) IR spectra data. Support your answer by determining its expected number of CO stretching modes for comparison with the published spectrum.

4.37 Three isomers of $W_2Cl_4(NHEt)_2(PMe_3)_2$ have been reported. These isomers have the core structures shown here. Determine the point group of each. (Reference: F. A. Cotton, E. V. Dikarev, W-Y. Wong, *Inorg. Chem.*, **1997**, *36*, 2670.)

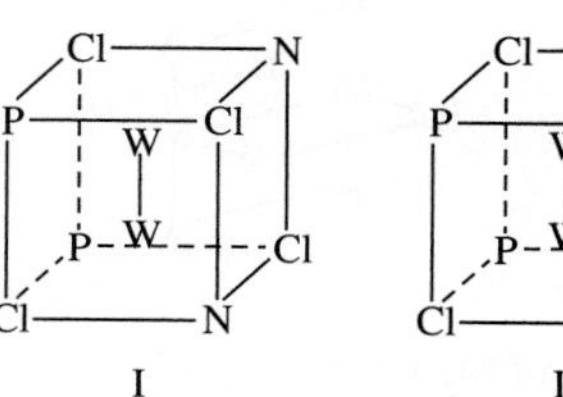

I

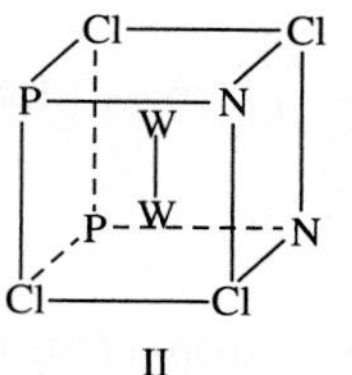

II

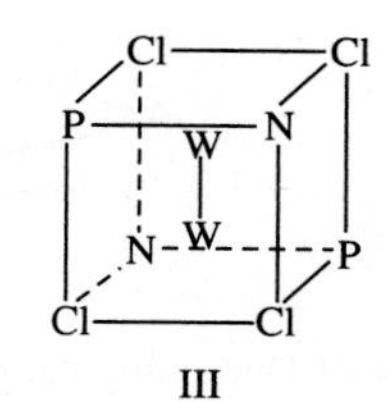

III

4.38 Derivatives of methane can be obtained by replacing one or more hydrogen atoms with other atoms, such as F, Cl, or Br. Suppose you had a supply of methane and the necessary chemicals and equipment to make derivatives of methane containing all possible combinations of the elements H, F, Cl, and Br. What would be the point groups of the molecules you could make? There are many possible molecules, and they can be arranged into five sets for assignment of point groups.

4.39 Determine the point groups of the following molecules:

a. F_3SCCF_3, with a triple S — C bond

b. $C_6H_6F_2Cl_2Br_2$, a derivative of cyclohexane, in a chair conformation

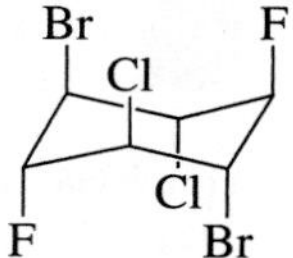

c. $M_2Cl_6Br_4$, where M is a metal atom

d. $M(NH_2C_2H_4PH_2)_3$, considering the $NH_2C_2H_4PH_2$ rings as planar

e. PCl_2F_3 (the most likely isomer)

4.40 Assign the point groups of the four possible structures for asymmetric bidentate ligands bridging two metals in a "paddlewheel" arrangement: (Reference: Y. Ke, D. J. Collins, H. Zhou, *Inorg. Chem.* **2005**, *44*, 4154.)

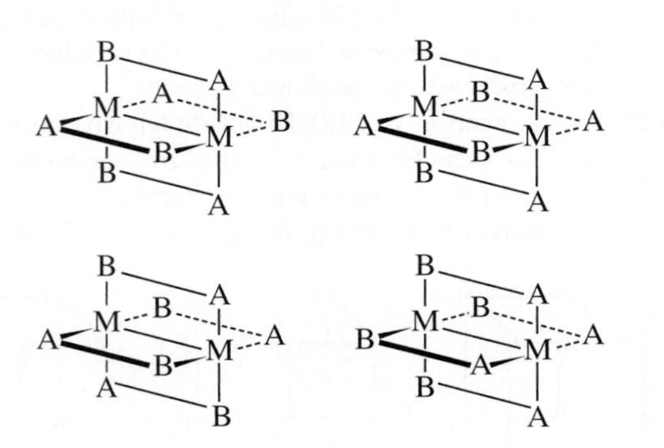

4.41 Determine the point groups of the following:

a. The cluster anion $[Re_3(\mu_3 - S)(\mu - S)_3Br_9]^{2-}$ (Reference: H. Sakamoto, Y. Watanabe, T. Sato, *Inorg. Chem.*, **2006**, *45*, 4578.)

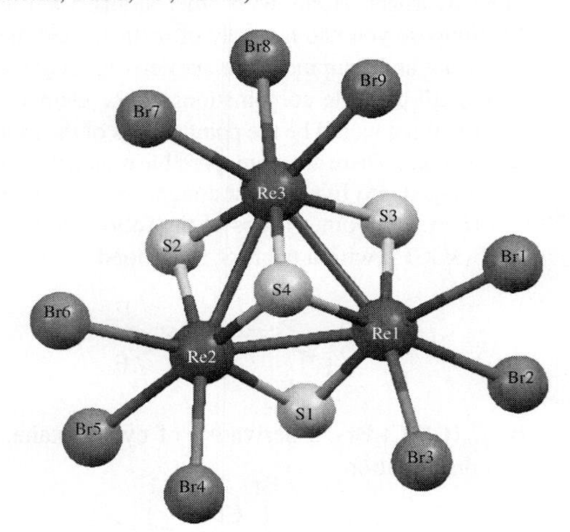

b. The cluster anion $[Fe@Ga_{10}]^{3-}$ (Reference: B. Zhou, M. S. Denning, D. L. Kays, J. M. Goicoechea, *J. Am. Chem. Soc.*, **2009**, *131*, 2802).

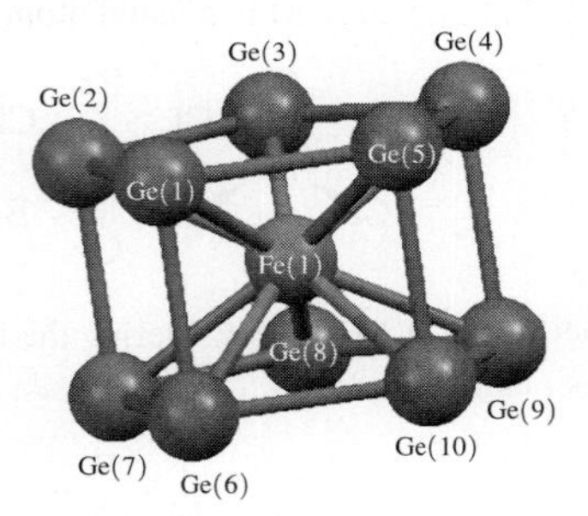

c. The "corner" and "square" structures: (Reference: W. H. Otto, M. H. Keefe, K. E. Splan, J. T. Hupp, C. K. Larive, *Inorg. Chem.* **2002**, *41*, 6172.)

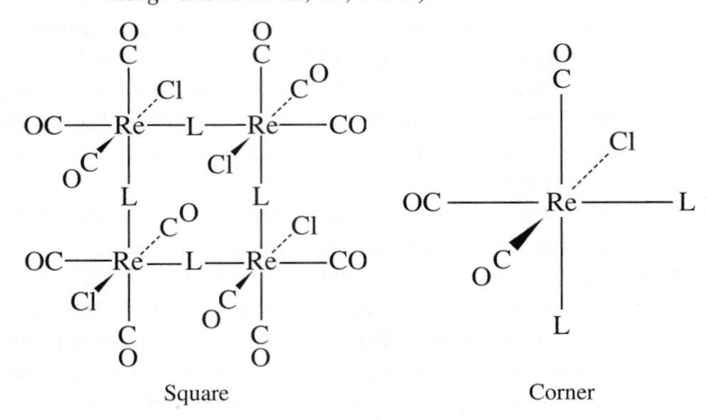

(L = N◯N)

d. The $[Bi_7I_{24}]^{3-}$ ion. (Reference: K.Y. Monakhov, C. Gourlaouen, R. Pattacini, P. Braunstein, *Inorg. Chem.*, **2012**, *51*, 1562. This reference also has alternative depictions of this structure.

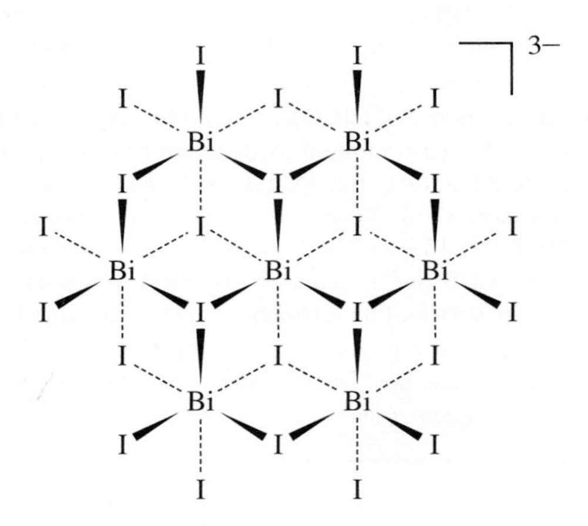

4.42 Use the Internet to search for molecules with the symmetry of

a. The S_6 point group

b. The T point group

c. The I_h point group

d. The T_h point group

Report the molecules, the URL of the Web site where you found them, and the search strategy you used.

Chapter 5

Molecular Orbitals

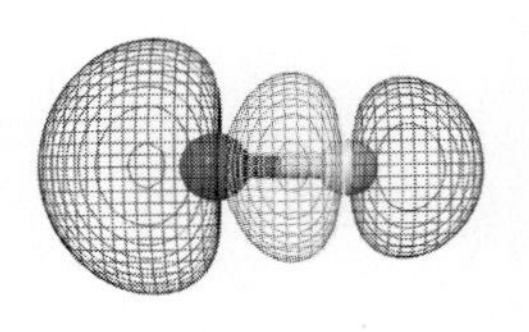

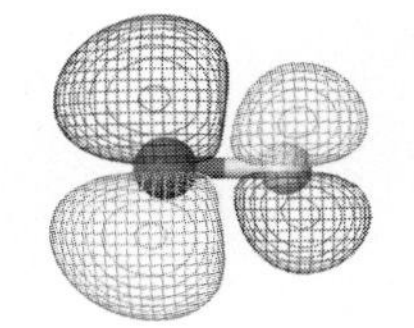

Molecular orbital theory uses group theory to describe the bonding in molecules; it complements and extends the introductory bonding models in Chapter 3. In molecular orbital theory the symmetry properties and relative energies of atomic orbitals determine how these orbitals interact to form molecular orbitals. The molecular orbitals are then occupied by the available electrons according to the same rules used for atomic orbitals as described in Sections 2.2.3 and 2.2.4. The total energy of the electrons in the molecular orbitals is compared with the initial total energy of electrons in the atomic orbitals. If the total energy of the electrons in the molecular orbitals is less than in the atomic orbitals, the molecule is stable relative to the separate atoms; if not, the molecule is unstable and predicted not to form. We will first describe the bonding, or lack of it, in the first 10 homonuclear diatomic molecules (H_2 through Ne_2) and then expand the discussion to heteronuclear diatomic molecules and molecules having more than two atoms.

A less rigorous pictorial approach is adequate to describe bonding in many small molecules and can provide clues to more complete descriptions of bonding in larger ones. A more elaborate approach, based on symmetry and employing group theory, is essential to understand orbital interactions in more complex molecular structures. In this chapter, we describe the pictorial approach and develop the symmetry methodology required for complex cases.

5.1 Formation of Molecular Orbitals from Atomic Orbitals

As with atomic orbitals, Schrödinger equations can be written for electrons in molecules. Approximate solutions to these molecular Schrödinger equations can be constructed from **linear combinations of atomic orbitals (LCAO)**, the sums and differences of the atomic wave functions. For diatomic molecules such as H_2, such wave functions have the form

$$\Psi = c_a\psi_a + c_b\psi_b$$

where Ψ is the molecular wave function, ψ_a and ψ_b are atomic wave functions for atoms a and b, and c_a and c_b are adjustable coefficients that quantify the contribution of each atomic orbital to the molecular orbital. The coefficients can be equal or unequal, positive or negative, depending on the individual orbitals and their energies. As the distance between two atoms is decreased, their orbitals overlap, with significant probability for electrons from both atoms being found in the region of overlap. As a result, **molecular orbitals** form. Electrons in bonding molecular orbitals have a high probability of occupying the space between the nuclei; the electrostatic forces between the electrons and the two positive nuclei hold the atoms together.

Three conditions are essential for overlap to lead to bonding. First, the symmetry of the orbitals must be such that regions with the same sign of ψ overlap. Second, the atomic orbital energies must be similar. When the energies differ greatly, the change in the energy of electrons upon formation of molecular orbitals is small, and the net reduction in energy of the electrons is too small for significant bonding. Third, the distance between the atoms must be short enough to provide good overlap of the orbitals, but not so short that repulsive forces of other electrons or the nuclei interfere. When these three conditions are met, the overall energy of the electrons in the occupied molecular orbitals is lower in energy than the overall energy of the electrons in the original atomic orbitals, and the resulting molecule has a lower total energy than the separated atoms.

5.1.1 Molecular Orbitals from s Orbitals

Consider the interactions between two *s* orbitals, as in H_2. For convenience, we label the atoms of a diatomic molecule *a* and *b*, so the atomic orbital wave functions are $\psi(1s_a)$ and $\psi(1s_b)$. We can visualize the two atoms approaching each other, until their electron clouds overlap and merge into larger molecular electron clouds. The resulting molecular orbitals are linear combinations of the atomic orbitals, the sum of the two orbitals and the difference between them.

In general terms — for H_2

$$\Psi(\sigma) = N[c_a\psi(1s_a) + c_b\psi(1s_b)] = \frac{1}{\sqrt{2}}[\psi(1s_a) + \psi(1s_b)]\ (\mathrm{H}_a + \mathrm{H}_b)$$

$$\text{and } \Psi(\sigma^*) = N[c_a\psi(1s_a) - c_b\psi(1s_b)] = \frac{1}{\sqrt{2}}[\psi(1s_a) - \psi(1s_b)]\ (\mathrm{H}_a - \mathrm{H}_b)$$

where $N = \text{normalizing factor, so } \int \Psi\Psi^*\, d\tau = 1$

$c_a \text{ and } c_b = \text{adjustable coefficients}$

In this case, the two atomic orbitals are identical, and the coefficients are nearly identical as well.* These orbitals are depicted in **Figure 5.1**. In this diagram, as in all the orbital diagrams in this book (such as Table 2.3 and Figure 2.6), the signs of orbital lobes are indicated by shading or color. Light and dark lobes or lobes of different color indicate opposite signs of Ψ. The choice of positive and negative for specific atomic orbitals is arbitrary; what is important is how they combine to form molecular orbitals. In the diagrams in **Figure 5.2**, the different colors show opposite signs of the wave function, both

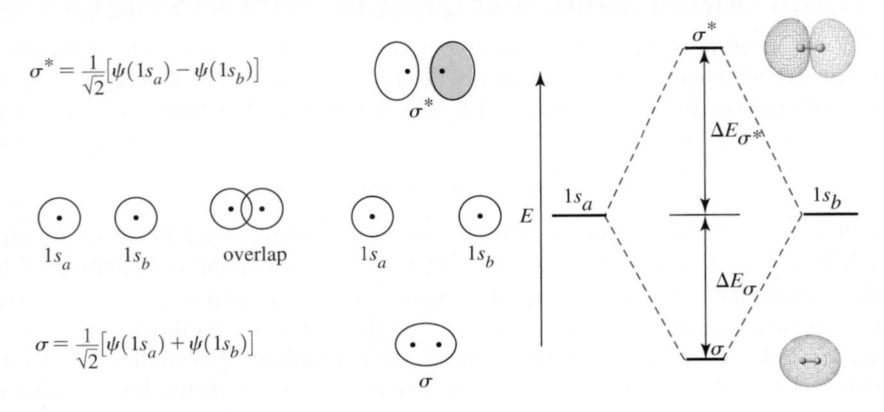

FIGURE 5.1 Molecular Orbitals from Hydrogen 1*s* Orbitals. The σ molecular orbital is a bonding molecular orbital, and has a lower energy than the original atomic orbitals, since this combination of atomic orbitals results in an increased concentration of electrons between the two nuclei. The σ^* orbital is an antibonding orbital at higher energy since this combination of atomic orbitals results in a node with zero electron density between the nuclei.

*More precise calculations show that the coefficients of the σ^* orbital are slightly larger than those for the σ orbital; but for the sake of simplicity, we will generally not focus on this. For identical atoms, we will use $c_a = c_b = 1$ and $N = \frac{1}{\sqrt{2}}$. The difference in coefficients for the σ and σ^* orbitals also results in a larger change in energy (increase) from the atomic to the σ^* molecular orbitals than for the σ orbitals (decrease). In other words, $\Delta E_{\sigma^*} > \Delta E_{\sigma}$, as shown in Figure 5.1.

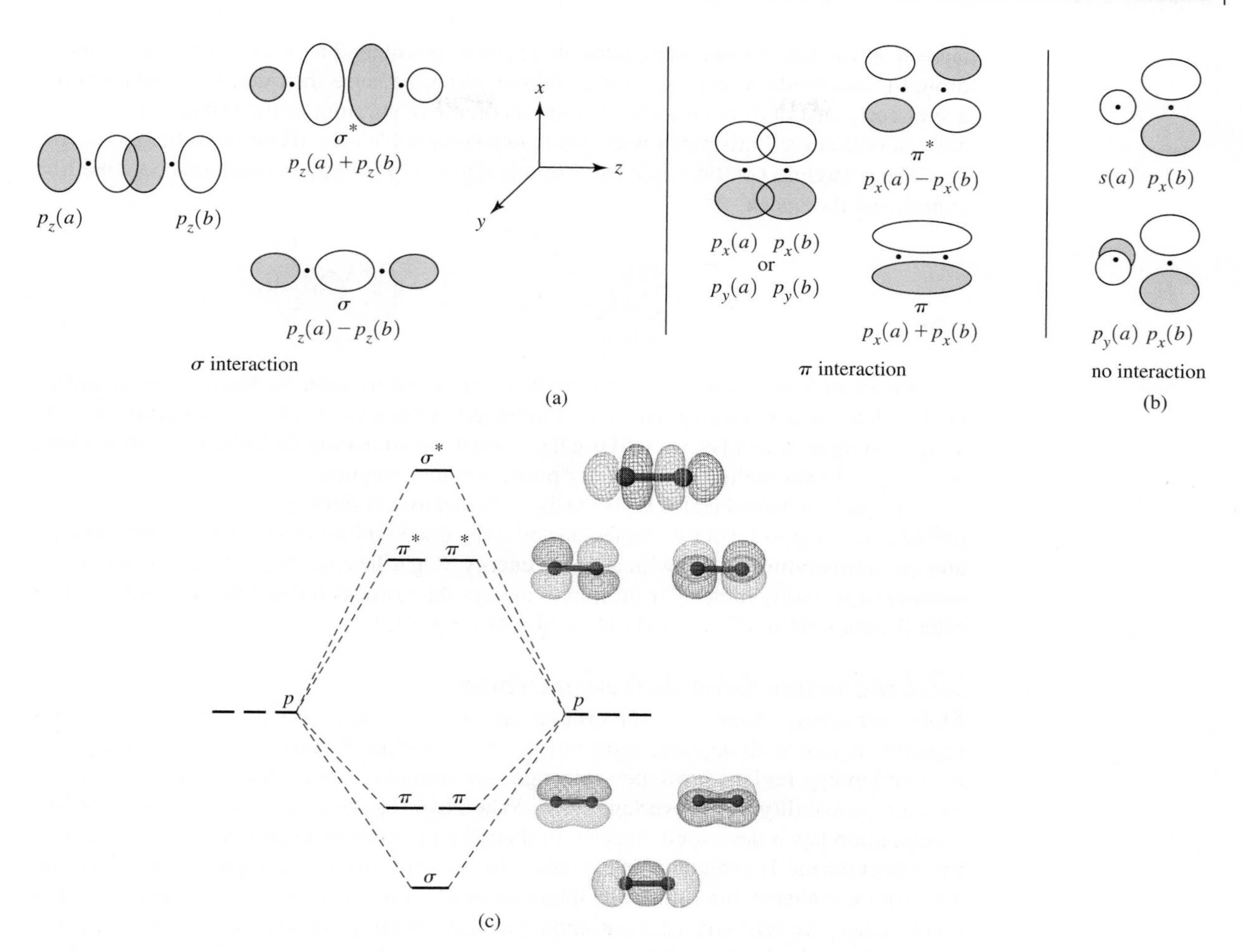

FIGURE 5.2 Interactions of *p* Orbitals. (a) Formation of molecular orbitals. (b) Orbitals that do not form molecular orbitals. (c) Energy-level diagram. (Interactions of *p* Orbitals by Kaitlin Hellie. Reprinted by permission.)

in the schematic sketches on the left of the energy level diagram and in the calculated molecular orbital images on the right.*

Because the σ molecular orbital is the sum of two atomic orbitals, $\frac{1}{\sqrt{2}}[\psi(1s_a) + \psi(1s_b)]$, and results in an increased concentration of electrons between the two nuclei, it is a **bonding molecular orbital** and has a lower energy than the original atomic orbitals. The σ^* molecular orbital is the difference of the two atomic orbitals, $\frac{1}{\sqrt{2}}[\psi(1s_a) - \psi(1s_b)]$. It has a node with zero electron density between the nuclei, due to cancellation of the two wave functions, and a higher energy; it is therefore called an **antibonding orbital**. Electrons in bonding orbitals are concentrated between the nuclei and attract the nuclei, holding them together. Antibonding orbitals have one or more nodes between the nuclei; electrons in these orbitals are destabilized relative to the parent atomic orbitals; the electrons do not have access to the region between the nuclei where they could experience the maximum

---

*Molecular orbital images in this chapter were prepared using Scigress Explorer Ultra, Version 7.7.0.47, © 2000–2007 Fujitsu Limited, © 1989–2000 Oxford Molecular Ltd.

nuclear attraction. **Nonbonding orbitals** are also possible. The energy of a nonbonding orbital is essentially that of an atomic orbital, either because the orbital on one atom has a symmetry that does not match any orbitals on the other atom or the orbital on one atom has a severe energy mismatch with symmetry-compatible orbitals on the other atom.

The σ (sigma) notation indicates orbitals that are symmetric to rotation about the line connecting the nuclei:

C_2 z C_2 z

σ^* from s orbital σ^* from p_z orbital

An asterisk is frequently used to indicate antibonding orbitals. Because the bonding, nonbonding, or antibonding nature of a molecular orbital is not always straightforward to assign in larger molecules, we will use the asterisk notation only for those molecules where bonding and antibonding orbital descriptions are unambiguous.

The pattern described for H_2 is the usual model for combining two orbitals: two atomic orbitals combine to form two molecular orbitals, one bonding orbital with a lower energy and one antibonding orbital with a higher energy. Regardless of the number of orbitals, the number of resulting molecular orbitals is always the same as the initial number of atomic orbitals; the total number of orbitals is always conserved.

5.1.2 Molecular Orbitals from *p* Orbitals

Molecular orbitals formed from p orbitals are more complex since each p orbital contains separate regions with opposite signs of the wave function. When two orbitals overlap, and the overlapping regions have the same sign, the sum of the two orbitals has an increased electron probability in the overlap region. When two regions of opposite sign overlap, the combination has a decreased electron probability in the overlap region. Figure 5.1 shows this effect for the $1s$ orbitals of H_2; similar effects result from overlapping lobes of p orbitals with their alternating signs. The interactions of p orbitals are shown in Figure 5.2. For convenience, we will choose a common z axis connecting the nuclei and assign x and y axes as shown in the figure.

When we draw the z axes for the two atoms pointing in the same direction,* the p_z orbitals subtract to form σ and add to form σ^* orbitals, both of which are symmetric to rotation about the z axis, with nodes perpendicular to the line that connects the nuclei. Interactions between p_x and p_y orbitals lead to π and π^* orbitals. The π (pi) notation indicates a change in sign of the wave function with C_2 rotation about the bond axis:

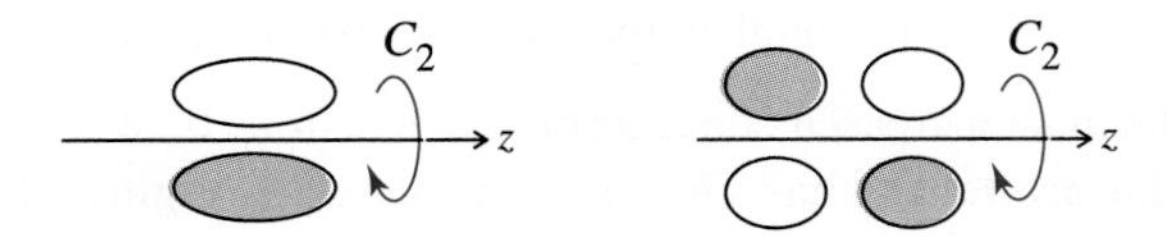

As with the s orbitals, the overlap of two regions with the same sign leads to an increased concentration of electrons, and the overlap of two regions of opposite signs leads to a node of zero electron density. In addition, the nodes of the atomic orbitals become the

*The choice of direction of the z axes is arbitrary. When both are positive in the same direction, the difference between the p_z orbitals is the bonding combination. When the positive z axes are chosen to point toward each other, the sum of the p_z orbitals is the bonding combination. We have chosen to have the p_z orbitals positive in the same direction for consistency with our treatment of triatomic and larger molecules.

nodes of the resulting molecular orbitals. In the π^* antibonding case, four lobes result that are similar in appearance to a d orbital, as in Figure 5.2(c).

The p_x, p_y, and p_z orbital pairs need to be considered separately. Because the z axis was chosen as the internuclear axis, the orbitals derived from the p_z orbitals are symmetric to rotation around the bond axis and are labeled σ and σ^* for the bonding and antibonding orbitals, respectively. Similar combinations of the p_y orbitals form orbitals whose wave functions change sign with C_2 rotation about the bond axis; they are labeled π and π^*. In the same way, the p_x orbitals also form π and π^* orbitals.

It is common for s and p atomic orbitals on different atoms to be sufficiently similar in energy for their combinations to be considered. However, if the symmetry properties of the orbitals do not match, no combination is possible. For example, when orbitals overlap equally with both the same and opposite signs, as in the $s + p_x$ example in Figure 5.2(b), the bonding and antibonding effects cancel, and no molecular orbital results. If the symmetry of an atomic orbital does not match *any* orbital of the other atom, it is called a nonbonding orbital. Homonuclear diatomic molecules have only bonding and antibonding molecular orbitals; nonbonding orbitals are described further in Sections 5.1.4, 5.2.2, and 5.4.3.

5.1.3 Molecular Orbitals from *d* Orbitals

In the heavier elements, particularly the transition metals, d orbitals can be involved in bonding. **Figure 5.3** shows the possible combinations. When the z axes are collinear, two d_{z^2} orbitals can combine end-on for σ bonding. The d_{xz} and d_{yz} orbitals form π orbitals. When atomic orbitals meet from two parallel planes and combine side to side, as do the $d_{x^2-y^2}$ and d_{xy} orbitals with collinear z axes, they form (δ) delta orbitals (Figure 1.2). (The δ notation indicates sign changes on C_4 rotation about the bond axis.) Sigma orbitals have no nodes that include the line connecting the nuclei, pi orbitals have one node that includes the line connecting the nuclei, and delta orbitals have two nodes that include the line connecting the nuclei. Again, some orbital interactions are forbidden on the basis of symmetry; for example, p_z and d_{xz} have zero net overlap if the z axis is chosen as the bond axis since the p_z would approach the d_{xz} orbital along a d_{xz} node (Example 5.1). It is noteworthy in this case that p_x and d_{xz} would be eligible to interact in a π fashion on the basis of the assigned coordinate system. This example emphasizes the importance of maintaining a consistent coordinate system when assessing orbital interactions.

EXAMPLE 5.1

Sketch the overlap regions of the following combination of orbitals, all with collinear z axes, and classify the interactions.

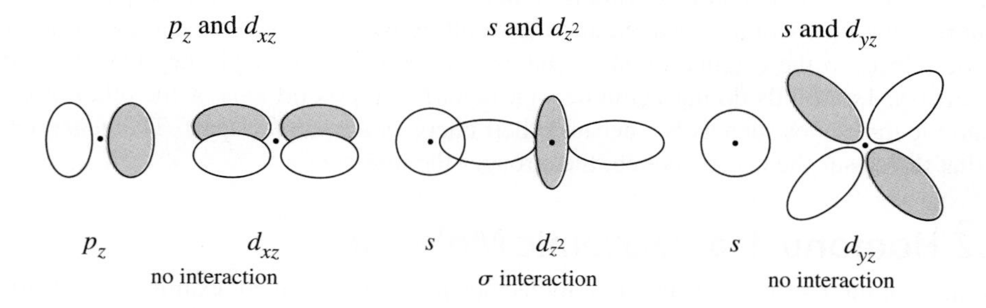

EXERCISE 5.1 Repeat the process in the preceding example for the following orbital combinations, again using collinear z axes.

p_x and d_{xz} $\quad$ p_z and d_{z^2} $\quad$ s and $d_{x^2-y^2}$

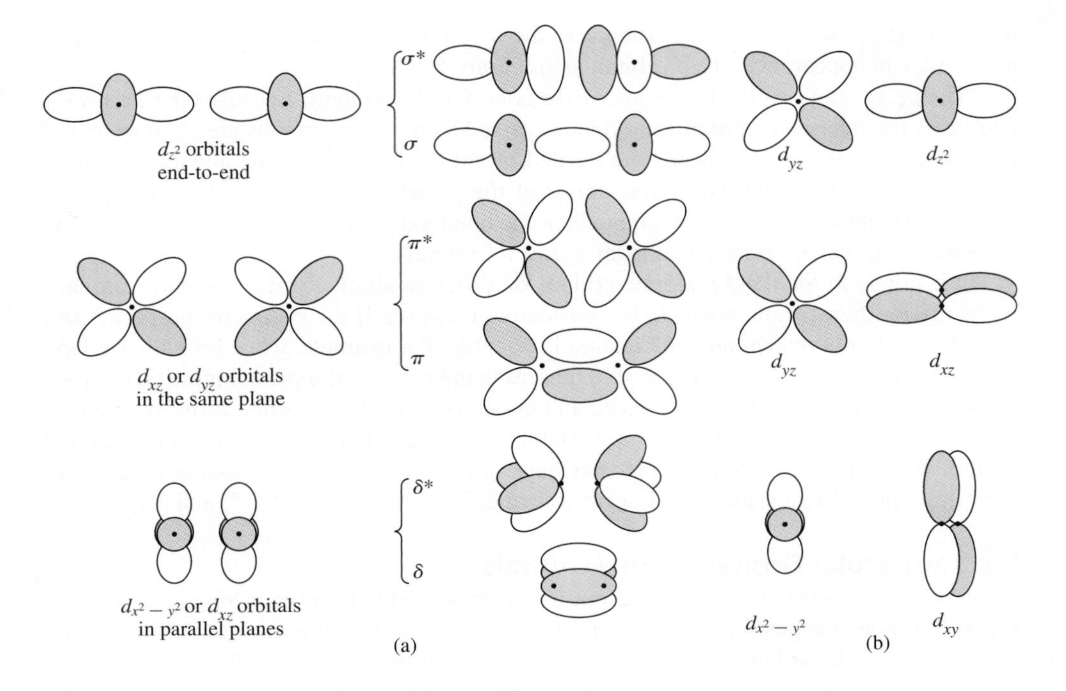

FIGURE 5.3 Interactions of *d* Orbitals. (a) Formation of molecular orbitals. (b) Atomic orbital combinations that do not form molecular orbitals.

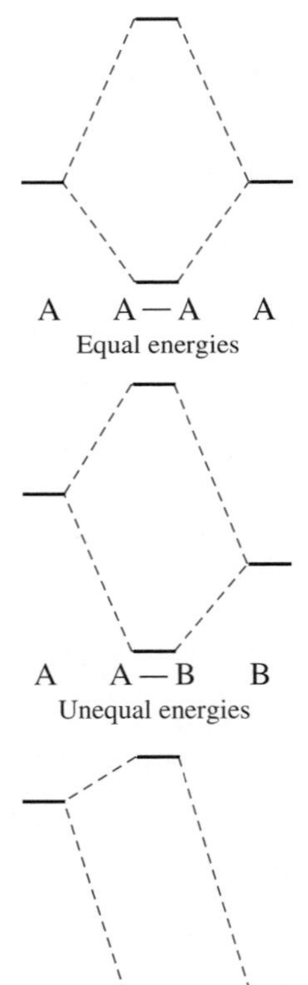

FIGURE 5.4 Energy Match and Molecular Orbital Formation.

5.1.4 Nonbonding Orbitals and Other Factors

As mentioned previously, nonbonding molecular orbitals have energies essentially equal to that of atomic orbitals. These can form in larger molecules, for example when there are three atomic orbitals of the same symmetry and similar energies, a situation that requires the formation of three molecular orbitals. Most commonly, one molecular orbital formed is a low-energy bonding orbital, one is a high-energy antibonding orbital, and one is of intermediate energy and is a nonbonding orbital. Examples will be considered in Section 5.4 and in later chapters.

In addition to symmetry, the second major factor that must be considered in forming molecular orbitals is the relative energy of the atomic orbitals. As shown in **Figure 5.4**, when the interacting atomic orbitals have the same energy, the interaction is strong, and the resulting molecular orbitals have energies well below (bonding) and above (antibonding) that of the original atomic orbitals. When the two atomic orbitals have quite different energies, the interaction is weaker, and the resulting molecular orbitals have energies and shapes closer to the original atomic orbitals. For example, although they have the same symmetry, 1*s* orbitals do not combine significantly with 2*s* orbitals of the other atom in diatomic molecules such as N_2, because their energies are too far apart. The general rule is that the closer the energy match, the stronger the interaction.

5.2 Homonuclear Diatomic Molecules

Because of their simplicity, diatomic molecules provide convenient examples to illustrate how the orbitals of individual atoms interact to form orbitals in molecules. In this section, we will consider **homonuclear** diatomic molecules such as H_2 and O_2; in Section 5.3 we will examine **heteronuclear** diatomics such as CO and HF.

5.2.1 Molecular Orbitals

Although apparently satisfactory Lewis electron-dot structures of N_2, O_2, and F_2 can be drawn, the same is not true with Li_2, Be_2, B_2, and C_2, which violate the octet rule. In addition, the Lewis structure of O_2 predicts a double-bonded, diamagnetic (all electrons paired) molecule (:Ö=Ö:), but experiment has shown O_2 to have two unpaired electrons, making it paramagnetic. As we will see, the molecular orbital description predicts this paramagnetism, and is more in agreement with experiment. **Figure 5.5** shows the full set of molecular orbitals for the homonuclear diatomic molecules of the first 10 elements, based on the energies appropriate for O_2. The diagram shows the order of energy levels for the molecular orbitals, assuming significant interactions only between atomic orbitals of identical energy. The energies of the molecular orbitals change in a periodic way with atomic number, since the energies of the interacting atomic orbitals decrease across a period (**Figure 5.7**), but the general order of the molecular orbitals remains similar (with some subtle changes, as will be described in several examples) even for heavier atoms lower in the periodic table. Electrons fill the molecular orbitals according to the same rules that govern the filling of atomic orbitals, filling from lowest to highest energy (aufbau principle), maximum spin

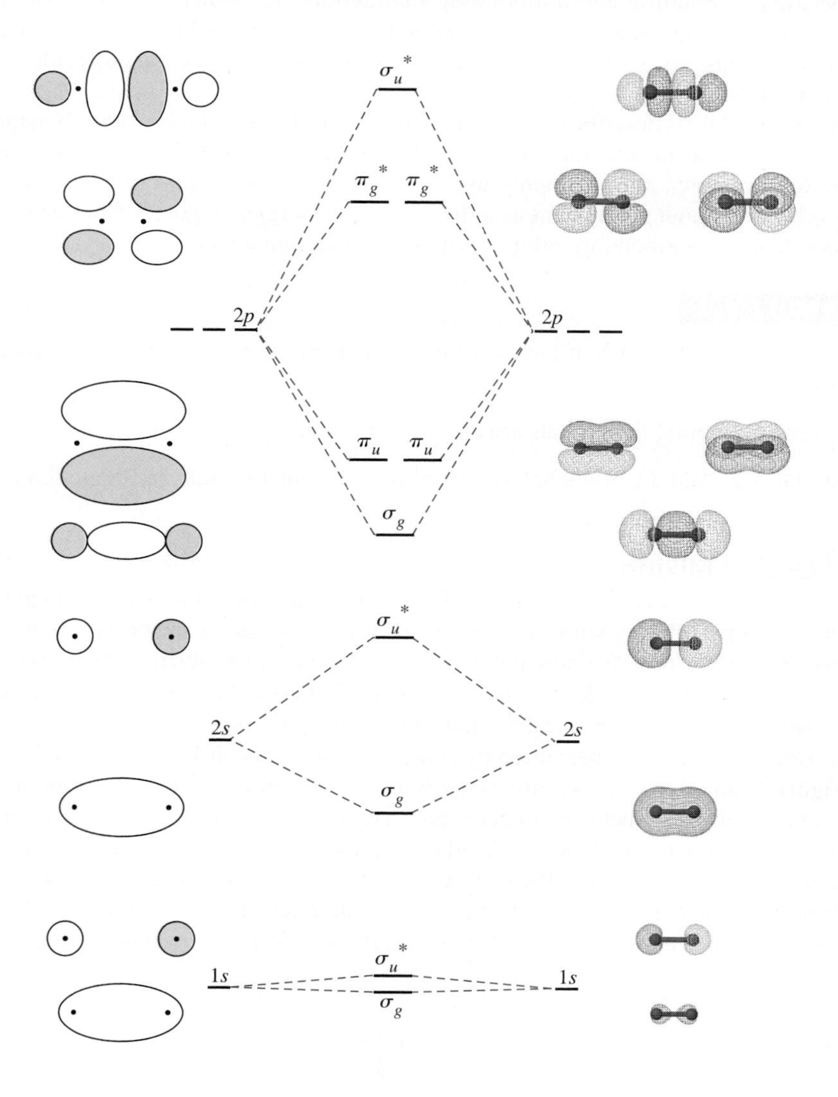

FIGURE 5.5 Molecular Orbitals for the First 10 Elements, Assuming Significant Interactions Only between the Valence Atomic Orbitals of Identical Energy. (Molecular Orbitals for the First 10 Elements by Kaitlin Hellie. Reprinted by permission.)

multiplicity consistent with the lowest net energy (Hund's rules), and no two electrons with identical quantum numbers (Pauli exclusion principle). The most stable configuration of electrons in the molecular orbitals is always the configuration with minimum energy, and the greatest net stabilization of the electrons.

The overall number of bonding and antibonding electrons determines the number of bonds (bond order):

$$\text{Bond order} = \frac{1}{2}\left[\begin{pmatrix}\text{number of electrons} \\ \text{in bonding orbitals}\end{pmatrix} - \begin{pmatrix}\text{number of electrons} \\ \text{in antibonding orbitals}\end{pmatrix}\right]$$

It is generally sufficient to consider only valence electrons. For example, O_2, with 10 electrons in bonding orbitals and 6 electrons in antibonding orbitals, has a bond order of 2, a double bond. Counting only valence electrons, 8 bonding and 4 antibonding, gives the same result. Because the molecular orbitals derived from the 1*s* orbitals have the same number of bonding and antibonding electrons, they have no net effect on the bond order. Generally electrons in atomic orbitals lower in energy than the valence orbitals are considered to reside primarily on the original atoms and to engage only weakly in bonding and antibonding interactions, as shown for the 1*s* orbitals in Figure 5.5; the difference in energy between the σ_g and σ_u^* orbitals is slight. Because such interactions are so weak, we will not include them in other molecular orbital energy level diagrams.

Additional labels describe the orbitals. The subscripts *g* for *gerade*, orbitals symmetric to inversion, and *u* for *ungerade*, orbitals antisymmetric to inversion (those whose signs change on inversion), are commonly used.* The *g* or *u* notation describes the symmetry of the orbitals without a judgment as to their relative energies. Figure 5.5 has examples of both bonding and antibonding orbitals with *g* and *u* designations.

EXAMPLE 5.2

Add a *g* or *u* label to each of the molecular orbitals in the energy-level diagram in Figure 5.2.

From top to bottom, the orbitals are σ_u^*, π_g^*, π_u, and σ_g.

EXERCISE 5.2 Add a *g* or *u* label to each of the molecular orbitals in Figure 5.3(a).

5.2.2 Orbital Mixing

In Figure 5.5, we only considered interactions between atomic orbitals of identical energy. However, atomic orbitals with similar, but unequal, energies can interact if they have appropriate symmetries. We now outline two approaches to analyzing this phenomenon, one in which we first consider the atomic orbitals that contribute most to each molecular orbital before consideration of additional interactions and one in which we consider all atomic orbital interactions permitted by symmetry simultaneously.

Figure 5.6(a) shows the familiar energy levels for a homonuclear diatomic molecule where only interactions between **degenerate** (having the same energy) atomic orbitals are considered. However, when two molecular orbitals of the same symmetry have similar energies, they interact to lower the energy of the lower orbital and raise the energy of the higher orbital. For example, in the homonuclear diatomics, the $\sigma_g(2s)$ and $\sigma_g(2p)$ orbitals both have σ_g symmetry (symmetric to infinite rotation and inversion); these orbitals

*See the end of Section 4.3.3 for more details on symmetry labels.

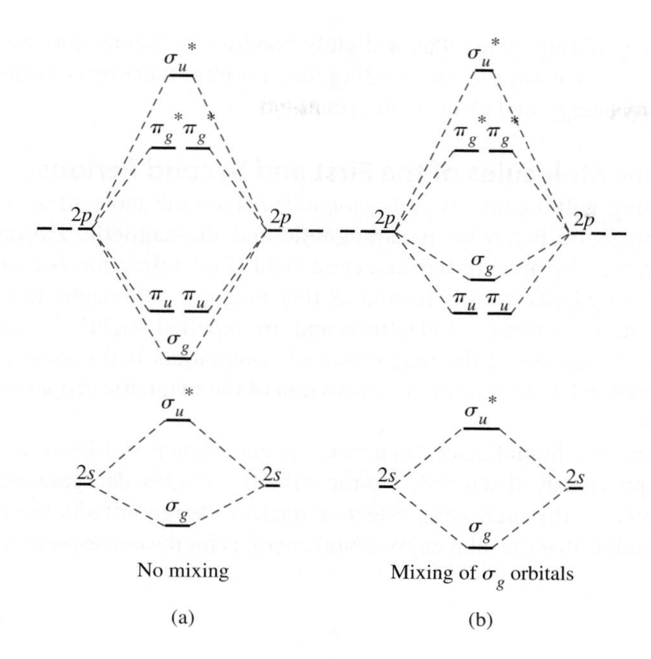

FIGURE 5.6 Interaction between Molecular Orbitals. Mixing molecular orbitals of the same symmetry results in a greater energy difference between the orbitals. The σ orbitals mix strongly; the σ^* orbitals differ more in energy and mix weakly.

interact to lower the energy of the $\sigma_g(2s)$ and to raise the energy of the $\sigma_g(2p)$ as shown in Figure 5.6(b). Similarly, the $\sigma_u{}^*(2s)$ and $\sigma_u{}^*(2p)$ orbitals interact to lower the energy of the $\sigma_u{}^*(2s)$ and to raise the energy of the $\sigma_u{}^*(2p)$. This phenomenon is called **mixing**, which takes into account that molecular orbitals with similar energies interact if they have appropriate symmetry, a factor ignored in Figure 5.5. When two molecular orbitals of the same symmetry mix, the one with higher energy moves still higher in energy, and the one with lower energy moves lower. Mixing results in additional electron stabilization, and enhances the bonding.

A perhaps more rigorous approach to explain mixing considers that the four σ molecular orbitals (MOs) result from combining the four atomic orbitals (two 2*s* and two $2p_z$) that have similar energies. The resulting molecular orbitals have the following general form, where *a* and *b* identify the two atoms, with appropriate normalization constants for each atomic orbital:

$$\Psi = c_1\psi(2s_a) \pm c_2\psi(2s_b) \pm c_3\psi(2p_a) \pm c_4\psi(2p_b)$$

For homonuclear diatomic molecules, $c_1 = c_2$ and $c_3 = c_4$ in each of the four MOs. The lowest energy MO has larger values of c_1 and c_2, the highest has larger values of c_3 and c_4, and the two intermediate MOs have intermediate values for all four coefficients. The symmetry of these four orbitals is the same as those without mixing, but their shapes are changed somewhat by having significant contributions from both the *s* and *p* atomic orbitals. In addition, the energies are shifted relative to their placement if the upper two exhibited nearly exclusive contribution from $2p_z$ while the lower two exclusive contribution from 2*s*, as shown in Figure 5.6.

It is clear that *s-p* mixing often has a detectable influence on molecular orbital energies. For example, in early second period homonuclear diatomics (Li_2 to N_2), the σ_g orbital formed from $2p_z$ orbitals is higher in energy than the π_u orbitals formed from the $2p_x$ and $2p_y$ orbitals. This is an inverted order from that expected without *s-p* mixing (Figure 5.6). For B_2 and C_2, *s-p* mixing affects their magnetic properties. Mixing also changes the bonding–antibonding nature of some orbitals. The orbitals with intermediate energies may, on the basis of *s-p* mixing, gain either a slightly bonding or slightly antibonding character and contribute in minor ways to the bonding. Each orbital must be considered separately on the basis of its energy and electron distribution.

5.2.3 Diatomic Molecules of the First and Second Periods

Before proceeding with examples of homonuclear diatomic molecules, we must define two types of magnetic behavior, **paramagnetic** and **diamagnetic**. Paramagnetic compounds are attracted by an external magnetic field. This attraction is a consequence of one or more unpaired electrons behaving as tiny magnets. Diamagnetic compounds, on the other hand, have no unpaired electrons and are repelled slightly by magnetic fields. (An experimental measure of the magnetism of compounds is the **magnetic moment**, a concept developed in Chapter 10 in the discussion of the magnetic properties of coordination compounds.)

H_2, He_2, and the homonuclear diatomic species shown in **Figure 5.7** will now be discussed. As previously discussed, atomic orbital energies decrease across a row in the Periodic Table as the increasing effective nuclear charge attracts the electrons more strongly. The result is that the molecular orbital energies for the corresponding homonuclear

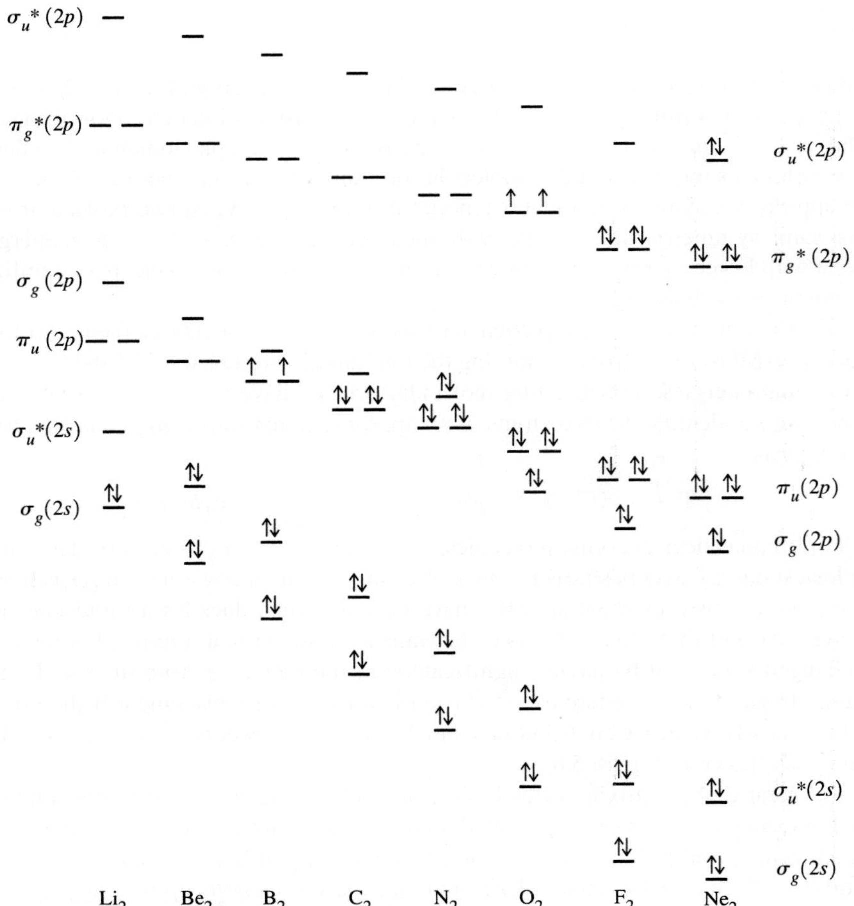

FIGURE 5.7 Energy Levels of the Homonuclear Diatomics of the Second Period.

	Li_2	Be_2	B_2	C_2	N_2	O_2	F_2	Ne_2
Bond order	1	0	1	2	3	2	1	0
Unpaired e^-	0	0	2	0	0	2	0	0

diatomics also decrease across the row. As shown in Figure 5.7, this decrease in energy is larger for σ orbitals than for π orbitals, due to the greater overlap of the atomic orbitals that participate in σ interactions.

$H_2[\sigma_g^{\,2}(1s)]$

This is the simplest diatomic molecule. The MO description (Figure 5.1) shows a single σ orbital containing one electron pair; the bond order is 1, representing a single bond. The ionic species H_2^+, with a single electron in the a σ orbital and a bond order of $\frac{1}{2}$, has been detected in low-pressure gas-discharge systems. As expected, H_2^+ has a weaker bond than H_2 and therefore a considerably longer bond distance than H_2 (105.2 pm vs. 74.1 pm).

$He_2[\sigma_g^{\,2}\sigma_u^{\,*2}(1s)]$

The molecular orbital description of He_2 predicts two electrons in a bonding orbital and two in an antibonding orbital, with a bond order of zero—in other words, no bond. This is what is observed experimentally. The noble gas He has no significant tendency to form diatomic molecules and, like the other noble gases, exists in the form of free atoms. He_2 has been detected only in very low-pressure and low-temperature molecular beams. It has an extremely low binding energy,[1] approximately 0.01 J/mol; for comparison, H_2 has a bond energy of 436 kJ/mol.

$Li_2[\sigma_g^{\,2}(2s)]$

As shown in Figure 5.7, the MO model predicts a single Li—Li bond in Li_2, in agreement with gas-phase observations of the molecule.

$Be_2[\sigma_g^{\,2}\sigma_u^{\,*2}(2p)]$

Be_2 has the same number of antibonding and bonding electrons and consequently a bond order of zero. Hence, like He_2, Be_2 is an unstable species.*

$B_2[\pi_u^{\,1}\pi_u^{\,1}(2p)]$

Here is an example in which the MO model has a distinct advantage over the Lewis dot model. B_2 is a gas-phase species; solid boron exists in several forms with complex bonding, primarily involving B_{12} icosahedra.

B_2 is paramagnetic. This behavior can be explained if its two highest energy electrons occupy separate π orbitals, as shown. The Lewis dot model cannot account for the paramagnetic behavior of this molecule.

The energy-level shift caused by *s-p* mixing is vital to understand the bonding in B_2. In the absence of mixing, the $\sigma_g(2p)$ orbital would be expected to be lower in energy than the $\pi_u(2p)$ orbitals, and the molecule would likely be diamagnetic.** However, mixing of the $\sigma_g(2s)$ orbital with the $\sigma_g(2p)$ orbital (Figure 5.6b) lowers the energy of the $\sigma_g(2s)$ orbital and increases the energy of the $\sigma_g(2p)$ orbital to a higher level than the π orbitals, giving the order of energies shown in Figure 5.7. As a result, the last two electrons are unpaired in the degenerate π orbitals, as required by Hund's rule of maximum multiplicity, and the molecule is paramagnetic. Overall, the bond order is one, even though the two π electrons are in different orbitals.

---

*Be_2 is calculated to have a very weak bond when effects of higher energy, unoccupied orbitals are taken into account. See A. Krapp, F. M. Bickelhaupt, and G. Frenking, *Chem. Eur. J.*, **2006**, *12*, 9196.

**This presumes that the energy difference between $\sigma_g(2p)$ and $\pi_u(2p)$ would be greater than Π_c (Section 2.2.3), a reliable expectation for molecular orbitals discussed in this chapter, but sometimes not true in transition metal complexes, as discussed in Chapter 10.

C—C Distance	(pm)
C=C (gas phase)	124.2
H—C≡C—H	120.5
CaC_2	119.1

$C_2[\pi_u^2\pi_u^2(2p)]$

The MO model of C_2 predicts a doubly bonded molecule, with all electrons paired, but with both **highest occupied molecular orbitals (HOMOs)** having π symmetry. C_2 is unusual because it has two π bonds and no σ bond. Although C_2 is a rarely encountered allotrope of carbon (carbon is significantly more stable as diamond, graphite, fullerenes and other polyatomic forms described in Chapter 8), the acetylide ion, C_2^{2-}, is well known, particularly in compounds with alkali metals, alkaline earths, and lanthanides. According to the molecular orbital model, C_2^{2-} should have a bond order of 3 (configuration $\pi_u^2\pi_u^2\sigma_g^2$). This is supported by the similar C—C distances in acetylene and calcium carbide (acetylide)[2,3].

$N_2[\pi_u^2\pi_u^2\sigma_g^2(2p)]$

N_2 has a triple bond according to both the Lewis and the molecular orbital models. This agrees with its very short N—N distance (109.8 pm) and extremely high bond-dissociation energy (942 kJ/mol). Atomic orbitals decrease in energy with increasing nuclear charge Z as discussed in Section 2.2.4, and further described in Section 5.3.1; as the effective nuclear charge increases, the energies of all orbitals are reduced. The varying shielding abilities of electrons in different orbitals and electron–electron interactions cause the difference between the $2s$ and $2p$ energies to increase as Z increases, from 5.7 eV for boron to 8.8 eV for carbon and 12.4 eV for nitrogen. (These energies are given in Table 5.2 in Section 5.3.1.) The radial probability functions (Figure 2.7) indicate that $2s$ electrons have a higher probability of being close to the nucleus relative to $2p$ electrons, rendering the $2s$ electrons more susceptible to the increasing nuclear charge as Z increases. As a result, the $\sigma_g(2s)$ and $\sigma_g(2p)$ levels of N_2 interact (mix) less than the corresponding B_2 and C_2 levels, and the N_2 $\sigma_g(2p)$ and $\pi_u(2p)$ are very close in energy. The order of energies of these orbitals has been controversial and will be discussed in more detail in Section 5.2.4.

$O_2[\sigma_g^2\pi_u^2\pi_u^2\pi_g^{*1}\pi_g^{*1}(2p)]$

O_2 is paramagnetic. As for B_2, this property cannot be explained by the Lewis dot structure :Ö=Ö:, but it is evident from the MO picture, which assigns two electrons to the degenerate π_g^* orbitals. The paramagnetism can be demonstrated by pouring liquid O_2 between the poles of a strong magnet; O_2 will be held between the pole faces until it evaporates. Several charged forms of diatomic oxygen are known, including O_2^+, O_2^-, and O_2^{2-}. The internuclear O—O distance can be conveniently correlated with the bond order predicted by the molecular orbital model, as shown in the following table.*

	Bond Order	Internuclear Distance (pm)
O_2^+ (dioxygenyl)	2.5	111.6
O_2 (dioxygen)	2.0	120.8
O_2^- (superoxide)	1.5	135
O_2^{2-} (peroxide)	1.0	150.4

Oxygen–oxygen distances in O_2^- and O_2^{2-} are influenced by the cation. This influence is especially strong in the case of O_2^{2-} and is one factor in its unusually long bond distance, which should be considered approximate. The disproportionation of KO_2 to O_2 and O_2^{2-} in the presence of hexacarboximide cryptand (similar molecules will be discussed in Chapter 8) results in encapsulation of O_2^{2-} in the cryptand via hydrogen-bonding interactions. This O–O peroxide distance was determined as 150.4(2) pm.[4]

*See Table 5.1 for references.

The extent of mixing is not sufficient in O_2 to push the $\sigma_g(2p)$ orbital to higher energy than the $\pi_u(2p)$ orbitals. The order of molecular orbitals shown is consistent with the photoelectron spectrum, discussed in Section 5.2.4.

$F_2[\sigma_g^2 \pi_u^2 \pi_u^2 \pi_g^{*2} \pi_g^{*2}(2p)]$

The MO model of F_2 shows a diamagnetic molecule having a single fluorine–fluorine bond, in agreement with experimental data.

The bond order in N_2, O_2, and F_2 is the same whether or not mixing is taken into account, but the order of the $\sigma_g(2p)$ and $\pi_u(2p)$ orbitals is different in N_2 than in O_2 and F_2. As stated previously and further described in Section 5.3.1, the energy difference between the 2*s* and 2*p* orbitals of the second row main group elements increases with increasing *Z*, from 5.7 eV in boron to 21.5 eV in fluorine. As this difference increases, the *s*-*p* interaction (mixing) decreases, and the "normal" order of molecular orbitals returns in O_2 and F_2. The higher $\sigma_g(2p)$ orbital (relative to $\pi_u(2p)$) occurs in many heteronuclear diatomic molecules, such as CO, described in Section 5.3.1.

Ne_2

All the molecular orbitals are filled, there are equal numbers of bonding and antibonding electrons, and the bond order is therefore zero. The Ne_2 molecule is a transient species, if it exists at all.

One triumph of molecular orbital theory is its prediction of two unpaired electrons for O_2. Oxygen had long been known to be paramagnetic, but early explanations for this phenomenon were unsatisfactory. For example, a special "three-electron bond"[5] was proposed. The molecular orbital description directly explains why two unpaired electrons are required. In other cases, experimental observations (paramagnetic B_2, diamagnetic C_2) require a shift of orbital energies, raising σ_g above π_u, but they do not require major modifications of the model.

Bond Lengths in Homonuclear Diatomic Molecules

Figure 5.8 shows the variation of bond distance with the number of valence electrons in second-period *p*-block homonuclear diatomic molecules having 6 to 14 valence electrons. Beginning at the left, as the number of electrons increases the number in bonding orbitals also increases; the bond strength becomes greater, and the bond length becomes shorter.

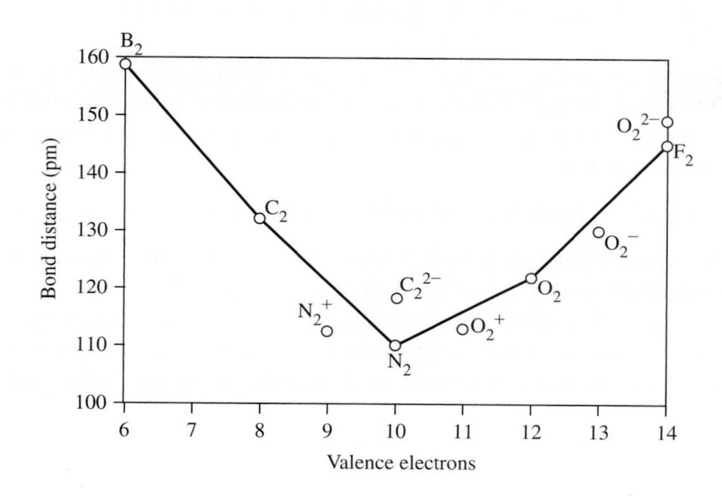

FIGURE 5.8 Bond Distances of Homonuclear Diatomic Molecules and Ions.

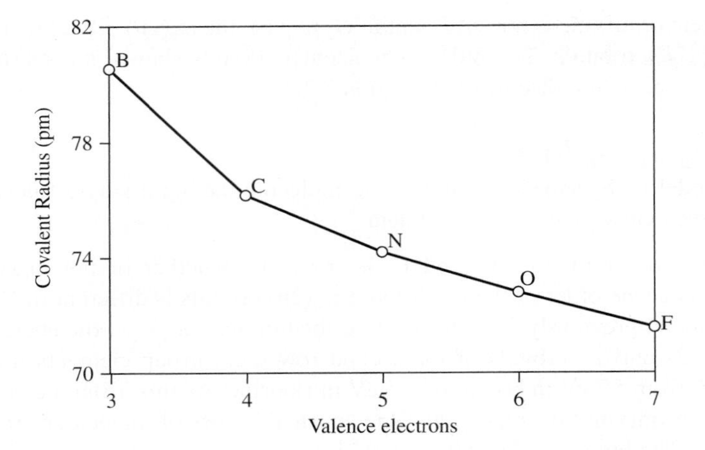

FIGURE 5.9 Covalent Radii of Second-Period Atoms.

This continues up to 10 valence electrons in N_2, where the trend reverses, because the additional electrons occupy antibonding orbitals. The ions N_2^+, O_2^+, O_2^-, and O_2^{2-} are also shown in the figure and follow a similar trend.

The minimum in Figure 5.8 occurs even though the radii of the free atoms decrease steadily from B to F. **Figure 5.9** shows the change in covalent radius for these atoms (defined for single bonds), decreasing as the number of valence electrons increases, primarily because the increasing nuclear charge pulls the electrons closer to the nucleus. For the elements boron through nitrogen, the trends shown in Figures 5.8 and 5.9 are similar: as the covalent radius of the atom decreases, the bond distance of the matching diatomic molecule also decreases. However, beyond nitrogen these trends diverge. Even though the covalent radii of the free atoms continue to decrease ($N > O > F$), the bond distances in their diatomic molecules increase ($N_2 < O_2 < F_2$) with the increasing population of antibonding orbitals. In general the bond order is the more important factor, overriding the covalent radii of the component atoms. Bond lengths of homonuclear and heteronuclear diatomic species are given in **Table 5.1**.

5.2.4 Photoelectron Spectroscopy

In addition to data on bond distances and energies, specific information about the energies of electrons in orbitals can be determined from photoelectron spectroscopy.[6] In this technique, ultraviolet (UV) light or X-rays eject electrons from molecules:

$$O_2 + h\nu(\text{photons}) \rightarrow O_2^+ + e^-$$

The kinetic energy of the expelled electrons can be measured; the difference between the energy of the incident photons and this kinetic energy equals the ionization energy (binding energy) of the electron:

$$\text{Ionization energy} = h\nu(\text{energy of photons}) - \text{kinetic energy of the expelled electron}$$

UV light removes outer electrons; X-rays are more energetic and can remove inner electrons. **Figures 5.10** and **5.11** show photoelectron spectra for N_2 and O_2, respectively, and the relative energies of the highest occupied orbitals of the ions. The lower energy peaks (at the top in the figure) are for the higher energy orbitals (less energy required to remove electrons). If the energy levels of the ionized molecule are assumed to be essentially the same as those of

TABLE 5.1 Bond Distances in Diatomic Species[a]

Formula	Valence Electrons	Internuclear Distance (pm)
H_2^+	1	105.2
H_2	2	74.1
B_2	6	159.0
C_2	8	124.2
C_2^{2-}	10	119.1[b]
N_2^+	9	111.6
N_2	10	109.8
O_2^+	11	111.6
O_2	12	120.8
O_2^-	13	135
O_2^{2-}	14	150.4[c]
F_2	14	141.2
CN	9	117.2
CN^-	10	115.9[d]
CO	10	112.8
NO^+	10	106.3
NO	11	115.1
NO^-	12	126.7

[a]Except as noted in footnotes, data are from K. P. Huber and G. Herzberg, *Molecular Spectra and Molecular Structure. IV. Constants of Diatomic Molecules*, Van Nostrand Reinhold Company, New York, 1979. Additional data on diatomic species can be found in R. Janoscheck, *Pure Appl. Chem.*, **2001**, *73*, 1521.
[b]Distance in CaC_2 in M. J. Atoji, *J. Chem. Phys.*, **1961**, *35*, 1950.
[c]Reference 4.
[d]Distance in low-temperature orthorhombic phase of NaCN in T. Schräder, A. Loidl, T. Vogt, *Phys. Rev. B*, **1989**, *39*, 6186.

the uncharged molecule,* the observed energies can be directly correlated with the molecular orbital energies. The levels in the N_2 spectrum are more closely spaced than those in the O_2 spectrum, and some theoretical calculations have disagreed about the order of the highest occupied orbitals in N_2. Stowasser and Hoffmann[7] have compared different calculation methods and showed that the different order of energy levels was simply a function of the calculation method; the methods favored by the authors agree with the experimental results, with σ_g above π_u.

The photoelectron spectrum shows the π_u lower than σ_g in N_2 (Figure 5.10). In addition to the ionization energies of the orbitals, the spectrum provides evidence of the quantized electronic and vibrational energy levels of the molecule. Because vibrational energy levels

*This perspective on photoelectron spectroscopy is oversimplified; a rigorous treatment of this technique is beyond the scope of this text. The interpretation of photoelectron specta is challenging since these spectra provide differences between energy levels of the ground electronic state in the neutral molecule and energy levels in the ground and excited electronic states of the ionized molecule. Rigorous interpretation of photoelectron spectra requires consideration of how the energy levels and orbital shapes vary between the neutral and ionized species.

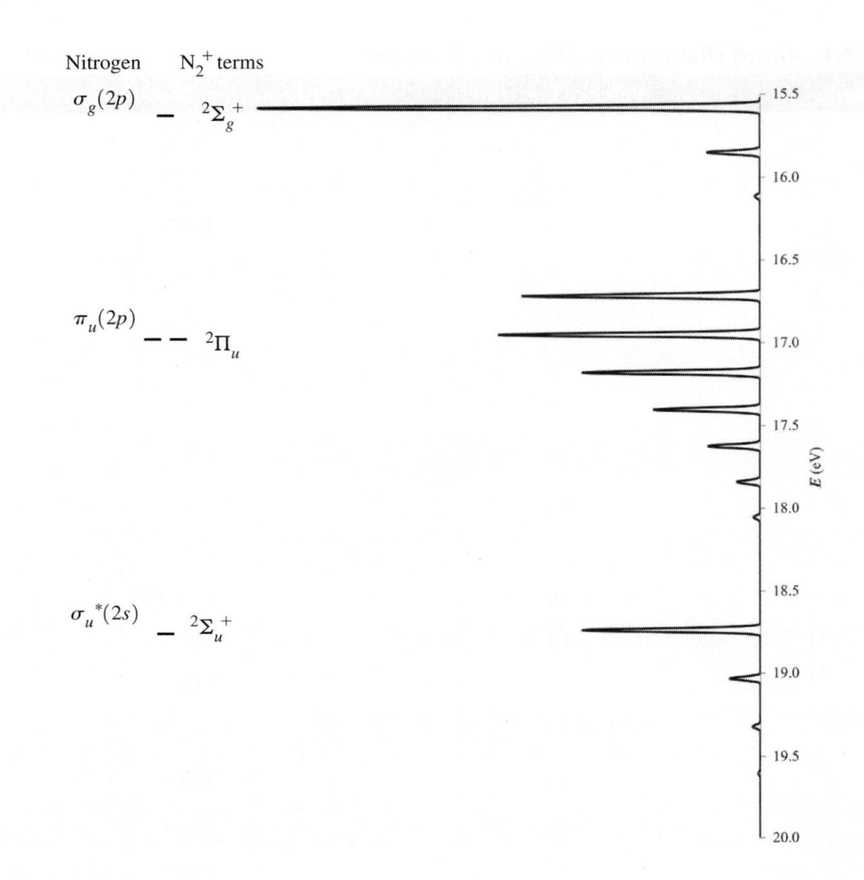

FIGURE 5.10 Photoelectron Spectrum and Molecular Orbital Energy Levels of N_2. Spectrum simulated by Susan Green using FCF program available at R. L. Lord, L. Davis, E. L. Millam, E. Brown, C. Offerman, P. Wray, S. M. E. Green, reprinted with permission from *J. Chem. Educ.*, **2008**, *85*, 1672 © 2008, American Institute of Physics and data from "Constants of Diatomic Molecules" by K.P. Huber and G. Herzberg (data prepared by J.W. Gallagher and R.D. Johnson, III) in **NIST Chemistry WebBook, NIST Standard Reference Database Number 69,** Eds. P.J. Linstrom and W.G. Mallard, National Institute of Standards and Technology, Gaithersburg MD, 20899, http://webbook.nist.gov, (retrieved July 22, 2012). (Reprinted with permission from *J. Chem Phys.* 62 (4) 1447 (1975), Copyright 1975, American Institute of Physics.)

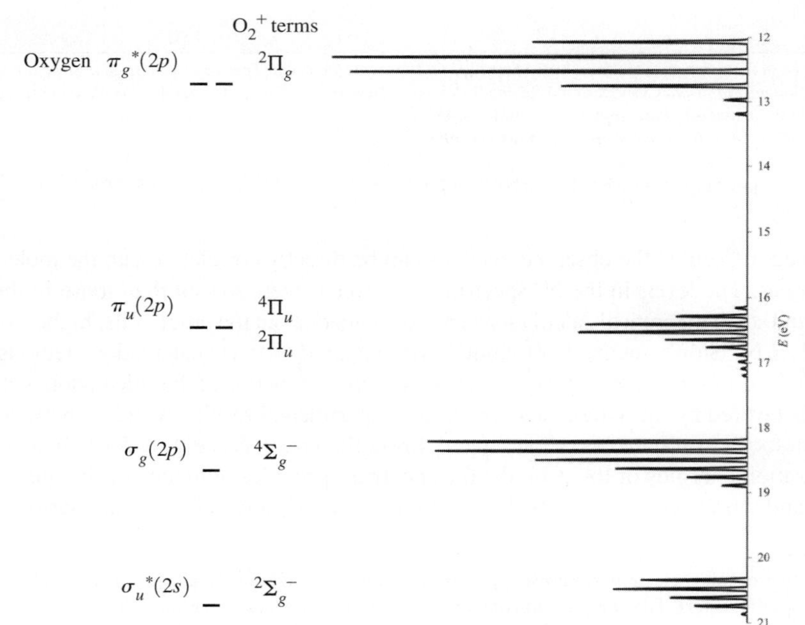

FIGURE 5.11 Photoelectron Spectrum and Molecular Orbital Energy Levels of O_2. Spectrum simulated by Susan Green using FCF program available at R. L. Lord, L. Davis, E. L. Millam, E. Brown, C. Offerman, P. Wray, S. M. E. Green, *J. Chem. Educ.*, **2008**, *85*, 1672 and data from "Constants of Diatomic Molecules" by K.P. Huber and G. Herzberg (data prepared by J.W. Gallagher and R.D. Johnson, III) in **NIST Chemistry WebBook, NIST Standard Reference Database Number 69**, Eds. P.J. Linstrom and W.G. Mallard, National Institute of Standards and Technology, Gaithersburg MD, 20899, http://webbook.nist.gov, (retrieved July 22, 2012).

are much more closely spaced than electronic levels, any collection of molecules will include molecules with different vibrational energies even when the molecules are in the ground electronic state. Therefore, transitions from electronic levels can originate from different vibrational levels, resulting in multiple peaks for a single electronic transition. Orbitals that are strongly involved in bonding have vibrational fine structure (multiple peaks); orbitals that are less involved in bonding have only a few peaks at each energy level.[8] The N_2 spectrum indicates that the π_u orbitals are more involved in the bonding than either of the σ orbitals. The CO photoelectron spectrum (**Figure 5.13**) has a similar pattern. The O_2 photoelectron spectrum (Figure 5.11) has much more vibrational fine structure for all the energy levels, with the π_u levels again more involved in bonding than the other orbitals. The photoelectron spectra of O_2 and of CO show the expected order of energy levels for these molecules.[8]

5.3 Heteronuclear Diatomic Molecules

The homonuclear diatomic molecules discussed Section 5.2 are nonpolar molecules. The electron density within the occupied molecular orbitals is evenly distributed over each atom. A discussion of heteronuclear diatomic molecules provides an introduction into how molecular orbital theory treats molecules that are polar, with an unequal distribution of the electron density in the occupied orbitals.

5.3.1 Polar Bonds

The application of molecular orbital theory to heteronuclear diatomic molecules is similar to its application to homonuclear diatomics, but the different nuclear charges of the atoms require that interactions occur between orbitals of unequal energies and shifts the resulting molecular orbital energies. In dealing with these heteronuclear molecules, it is necessary to estimate the energies of the atomic orbitals that may interact. For this purpose, the orbital potential energies, given in **Table 5.2** and **Figure 5.12**, are useful.* These potential energies are negative, because they represent attraction between valence electrons and atomic nuclei. The values are the average energies for all electrons in the same level (for example, all $3p$ electrons), and they are weighted averages of all the energy states that arise due to electron–electron interactions discussed in Chapter 11. For this reason, the values do not

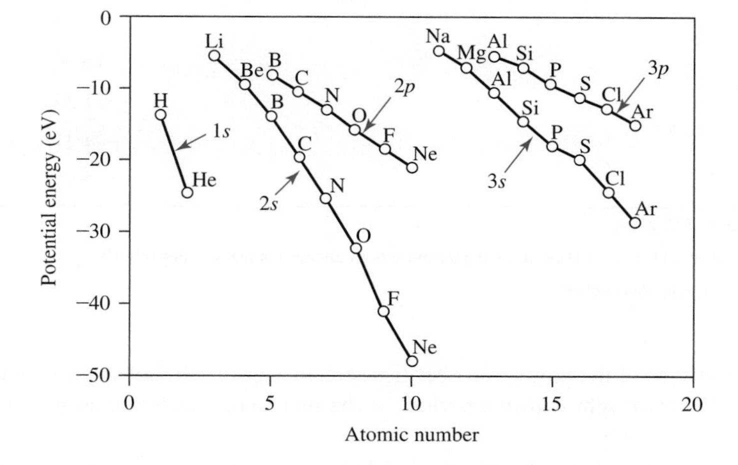

FIGURE 5.12 Orbital Potential Energies.

*A more complete listing of orbital potential energies is in Appendix B-9, available online at pearsonhighered.com/advchemistry

TABLE 5.2 Orbital Potential Energies

		Orbital Potential Energy (eV)						
Atomic Number	Element	1s	2s	2p	3s	3p	4s	4p
1	H	−13.61						
2	He	−24.59						
3	Li		−5.39					
4	Be		−9.32					
5	B		−14.05	−8.30				
6	C		−19.43	−10.66				
7	N		−25.56	−13.18				
8	O		−32.38	−15.85				
9	F		−40.17	−18.65				
10	Ne		−48.47	−21.59				
11	Na				−5.14			
12	Mg				−7.65			
13	Al				−11.32	−5.98		
14	Si				−15.89	−7.78		
15	P				−18.84	−9.65		
16	S				−22.71	−11.62		
17	Cl				−25.23	−13.67		
18	Ar				−29.24	−15.82		
19	K						−4.34	
20	Ca						−6.11	
30	Zn						−9.39	
31	Ga						−12.61	−5.93
32	Ge						−16.05	−7.54
33	As						−18.94	−9.17
34	Se						−21.37	−10.82
35	Br						−24.37	−12.49
36	Kr						−27.51	−14.22

J. B. Mann, T. L. Meek, L. C. Allen, *J. Am. Chem. Soc.*, 2000, *122*, 2780.

All energies are negative, representing average attractive potentials between the electrons and the nucleus for all terms of the specified orbitals.

Additional orbital potential energy values are available in the online Appendix B-9.

show the variations of the ionization energies seen in Figure 2.10 but steadily become more negative from left to right within a period, as the increasing nuclear charge attracts all the electrons more strongly.

The atomic orbitals of the atoms that form homonuclear diatomic molecules have identical energies, and both atoms contribute equally to a given MO. Therefore, in the molecular orbital equations, the coefficients associated with the same atomic orbitals of

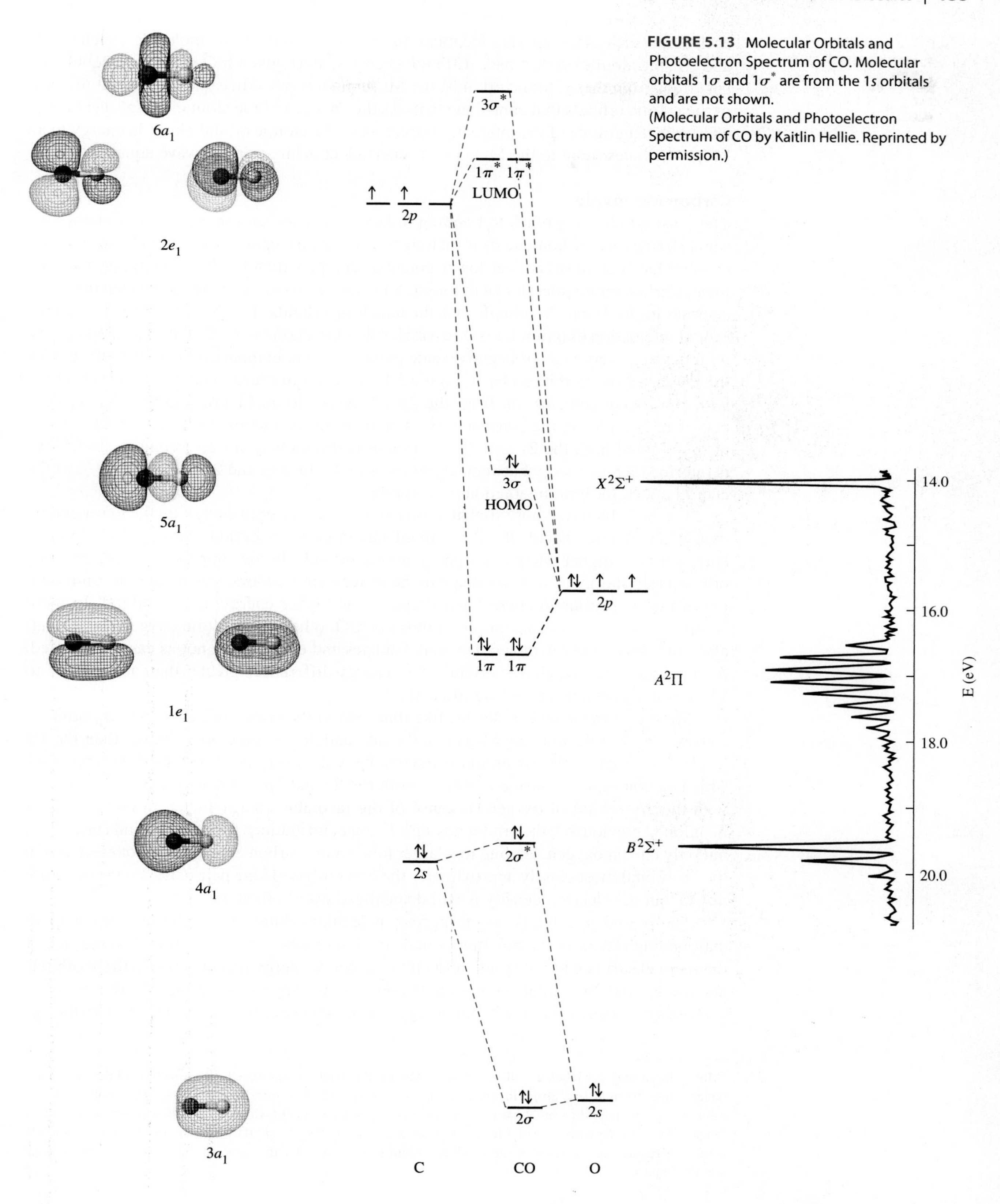

FIGURE 5.13 Molecular Orbitals and Photoelectron Spectrum of CO. Molecular orbitals 1σ and $1\sigma^*$ are from the 1s orbitals and are not shown.
(Molecular Orbitals and Photoelectron Spectrum of CO by Kaitlin Hellie. Reprinted by permission.)

each atom (such as the $2p_z$) are identical. In heteronuclear diatomic molecules, such as CO and HF, the atomic orbitals have different energies, and a given MO receives unequal contributions from these atomic orbitals; the MO equation has a different coefficient for each of the atomic orbitals that contribute to it. As the energies of the atomic orbitals get farther apart, the magnitude of the interaction decreases. The atomic orbital closer in energy to an MO contributes more to the MO, and its coefficient is larger in the wave equation.

Carbon Monoxide

The most efficient approach to bonding in heteronuclear diatomic molecules employs the same strategy as for homonuclear diatomics with one exception: the more electronegative element has atomic orbitals at lower potential energies than the less electronegative element. Carbon monoxide, shown in Figure 5.13, shows this effect, with oxygen having lower energies for its $2s$ and $2p$ orbitals than the matching orbitals of carbon. The result is that the orbital interaction diagram for CO resembles that for a homonuclear diatomic (Figure 5.5), with the right (more electronegative) side pulled down in comparison with the left. In CO, the lowest set of π orbitals (1π in Figure 5.13) is lower in energy than the lowest σ orbital with significant contribution from the $2p$ subshells (3σ in Figure 5.13); the same order occurs in N_2. This is the consequence of significant interactions between the $2p_z$ orbital of oxygen and both the $2s$ and $2p_z$ orbitals of carbon. Oxygen's $2p_z$ orbital (-15.85 eV) is intermediate in energy between carbon's $2s$ (-19.43 eV) and $2p_z$ (-10.66 eV), so the energy match for both interactions is favorable.

The 2σ orbital has more contribution from (and is closer in energy to) the lower energy oxygen $2s$ atomic orbital; the $2\sigma^*$ orbital has more contribution from (and is closer in energy to) the higher energy carbon $2s$ atomic orbital.* In the simplest case, the bonding orbital is similar in energy and shape to the lower energy atomic orbital, and the antibonding orbital is similar in energy and shape to the higher energy atomic orbital. In more complicated cases, such as the $2\sigma^*$ orbital of CO, other orbitals (the oxygen $2p_z$ orbital) also contribute, and the molecular orbital shapes and energies are not as easily predicted. As a practical matter, atomic orbitals with energy differences greater than about 10 eV to 14 eV usually do not interact significantly.

Mixing of the σ and σ^* levels, like that seen in the homonuclear σ_g and σ_u orbitals, causes a larger split in energy between the $2\sigma^*$ and 3σ, and the 3σ is higher than the 1π levels. The shape of the 3σ orbital is interesting, with a very large lobe on the carbon end. This is a consequence of the ability of both the $2s$ and $2p_z$ orbitals of carbon to interact with the $2p_z$ orbital of oxygen (because of the favorable energy match in both cases, as mentioned previously); the orbital has significant contributions from two orbitals on carbon but only one on oxygen, leading to a larger lobe on the carbon end. The pair of electrons in the 3σ orbital most closely approximates the carbon-based lone pair in the Lewis structure of CO, but the electron density is still delocalized over both atoms.

The p_x and p_y orbitals also form four molecular orbitals, two bonding (1π) and two antibonding ($1\pi^*$). In the bonding orbitals the larger lobes are concentrated on the side of the more electronegative oxygen, reflecting the better energy match between these orbitals and the $2p_x$ and $2p_y$ orbitals of oxygen. In contrast, the larger lobes of the π^* orbitals are on carbon, a consequence of the better energy match of these antibonding orbitals with the $2p_x$

*Molecular orbitals are labeled in different ways. Most in this book are numbered within each set of the same symmetry ($1\sigma_g$, $2\sigma_g$ and $1\sigma_u$, $2\sigma_u$). In some figures of homonuclear diatomics, $1\sigma_g$ and $1\sigma_u$ MOs from $1s$ atomic orbitals are understood to be at lower energies than the MOs from the valence orbitals and are omitted. It is noteworthy that interactions involving core orbitals are typically very weak; these interactions feature sufficiently poor overlap that the energies of the resulting orbitals are essentially the same as the energies of the original atomic orbitals.

and $2p_y$ orbitals of carbon. The distribution of electron density in the 3σ and $1\pi^*$ orbitals is vital to understand how CO binds to transition metals, a topic to be discussed further in this section. When the electrons are filled in, as in Figure 5.13, the valence orbitals form four bonding pairs and one antibonding pair for a net bond order of 3.*

EXAMPLE 5.3

Molecular orbitals for HF can be found by using the approach used for CO. The $2s$ orbital of the fluorine atom is more than 26 eV lower than that of the hydrogen $1s$, so there is very little interaction between these orbitals. The fluorine $2p_z$ orbital (-18.65 eV) and the hydrogen $1s$ (-13.61 eV), on the other hand, have similar energies, allowing them to combine into bonding and antibonding σ^* orbitals. The fluorine $2p_x$ and $2p_y$ orbitals remain nonbonding, each with a pair of electrons. Overall, there is one bonding pair and three lone pairs; however, the lone pairs are not equivalent, in contrast to the Lewis dot approach. The occupied molecular orbitals of HF predict a polar bond since all of these orbitals are biased toward the fluorine atom. The electron density in HF is collectively distributed with more on the fluorine atom relative to the hydrogen atom, and fluorine is unsurprisingly the negative end of the molecule.

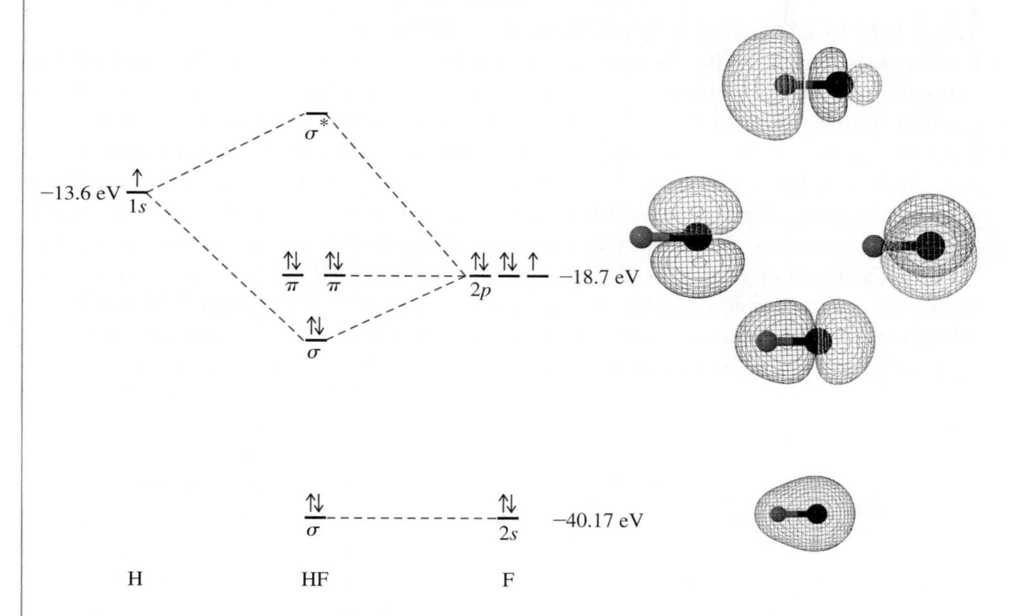

EXERCISE 5.3 Use a similar approach to the discussion of HF to explain the bonding in the hydroxide ion OH^-. (Exercise 5.3 by Kaitlin Hellie. Reprinted by permission.)

The molecular orbitals that are typically of greatest interest for reactions are the **highest occupied molecular orbital (HOMO)** and the **lowest unoccupied molecular orbital (LUMO)**, collectively known as **frontier orbitals** because they lie at the occupied–unoccupied frontier. The MO diagram of CO helps explain its reaction chemistry with transition metals, which is different than that predicted by electronegativity considerations that would suggest more electron density on the oxygen. On the sole basis of the carbon–oxygen electronegativity

*The classification of the filled σ orbitals as "bonding" and "antibonding" in CO is not as straightforward as in, for example, H_2, since the 2σ and $2\sigma^*$ orbitals are only changed modestly in energy relative to the $2s$ orbitals of oxygen and carbon, respectively. However, these orbital classifications are consistent with a threefold bond order for CO.

difference (and without considering the MO diagram), compounds in which CO is bonded to metals, called *carbonyl complexes*, would be expected to bond as M—O—C with the more electronegative oxygen attached to the electropositive metal. One impact of this electronegativity difference within the MO model is that the 2σ and 1π molecular orbitals in CO feature greater electron density on the more electronegative oxygen (Figure 5.13). However, the structure of the vast majority of metal carbonyl complexes, such as $Ni(CO)_4$ and $Mo(CO)_6$, has atoms in the order M—C—O. The HOMO of CO is 3σ, with a larger lobe, and therefore higher electron density, on the carbon (as explained above on the basis of *s*-*p* mixing). The electron pair in this orbital is more concentrated on the carbon atom, and can form a bond with a vacant orbital on the metal. The electrons of the HOMO are of highest energy (and least stabilized) in the molecule; these are generally the most energetically accessible for reactions with empty orbitals of other reactants. The LUMOs are the $1\pi^*$ orbitals; like the HOMO, these are concentrated on the less electronegative carbon, a feature that also predisposes CO to coordinate to metals via the carbon atom. Indeed, the frontier orbitals can both donate electrons (HOMO) and accept electrons (LUMO) in reactions. These tremendously important effects in organometallic chemistry are discussed in more detail in Chapters 10 and 13.

5.3.2 Ionic Compounds and Molecular Orbitals

Ionic compounds can be considered the limiting form of polarity in heteronuclear diatomic molecules. As mentioned previously, as the atoms forming bonds differ more in electronegativity, the energy gap between the interacting atomic orbitals also increases, and the concentration of electron density is increasingly biased toward the more electronegative atom in the bonding molecular orbitals. At the limit, the electron is transferred completely to the more electronegative atom to form a negative ion, leaving a positive ion with a high-energy vacant orbital. When two elements with a large difference in their electronegativities (such as Li and F) combine, the result is an ionic bond. However, in molecular orbital terms, we can treat an ion pair like we do a covalent compound. In **Figure 5.14**, the atomic orbitals and an approximate indication of molecular orbitals for such a diatomic molecule, LiF, are given. On formation of the diatomic molecule LiF, the electron from the Li 2*s*

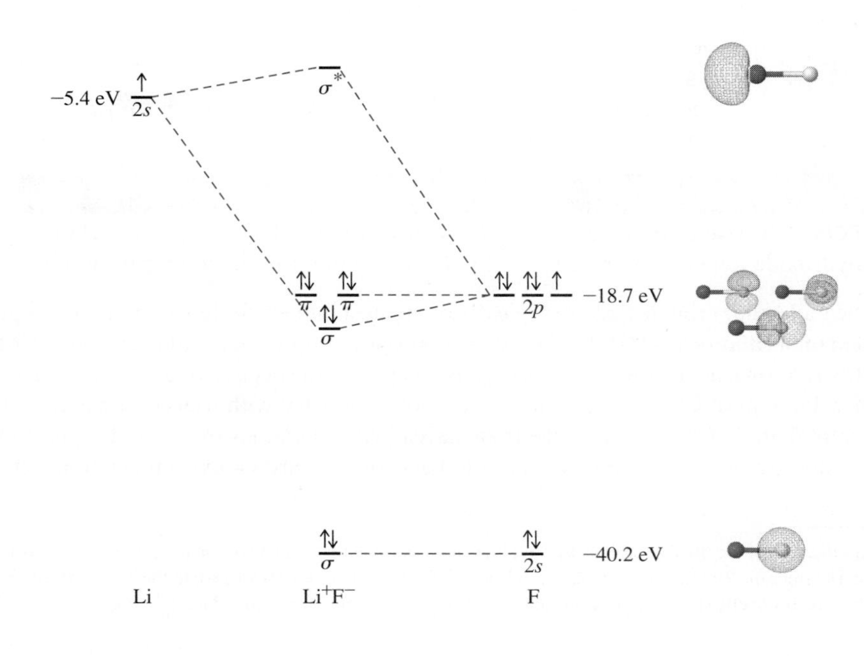

FIGURE 5.14 Approximate LiF Molecular Orbitals. (Approximate LiF Molecular Orbitals by Kaitlin Hellie. Reprinted by permission.)

orbital is transferred to the bonding orbital formed from interaction between the Li $2s$ orbital and the F $2p_z$ orbital. Indeed, the σ orbital deriving from the $2s/2p_z$ interaction has a significantly higher contribution from $2p_z$ relative to $2s$ on the basis of the large energy gap. Both electrons, the one originating from Li and the one originating from the F $2p_z$ orbital, are stabilized. Note that the level of theory used to calculate the orbital surfaces in Figure 5.14 suggests essentially no covalency in diatomic LiF.

Lithium fluoride exists as a crystalline solid; this form of LiF has significantly lower energy than diatomic LiF. In a three-dimensional crystalline lattice containing many formula units of a salt, the ions are held together by a combination of electrostatic (ionic) attraction and covalent bonding. There is a small covalent bonding contribution in all ionic compounds, but salts do not exhibit directional bonds, in contrast to molecules with highly covalent bonds that adopt geometries predicted by the VSEPR model. In the highly ionic LiF, each Li^+ ion is surrounded by six F^- ions, each of which in turn is surrounded by six Li^+ ions. The orbitals in the crystalline lattice form energy bands, described in Chapter 7.

Addition of these elementary steps, beginning with solid Li and gaseous F_2, results in formation of the corresponding gas-phase ions, and provides the net enthalpy change for this chemical change:

Elementary Step	Chemical/Physical Change	$\Delta H°$(kJ/mol)*
$Li(s) \longrightarrow Li(g)$	Sublimation	161
$Li(g) \longrightarrow Li^+(g) + e^-$	Ionization	520
$\frac{1}{2}F_2(g) \longrightarrow F(g)$	Bond dissociation	79
$F(g) + e^- \longrightarrow F^-(g)$	–Electron affinity	–328
$Li(s) + \frac{1}{2}F_2(g) \longrightarrow Li^+(g) + F^-(g)$	Formation of gas-phase ions from the elements in their standard states	432

The free energy change $(\Delta G = \Delta H - T\Delta S)$ must be negative for a reaction to proceed spontaneously. The very large and positive ΔH (432 kJ/mol) associated with the formation of these gas-phase ions renders the ΔG positive for this change despite its positive ΔS. However, if we combine these isolated ions, the large coulombic attraction between them results in a dramatic decrease in electronic energy, releasing 755 kJ/mol on formation of gaseous Li^+F^- ion pairs and 1050 kJ/mol on formation of a LiF crystal containing 1 mol of each ion.

Elementary Step	Chemical Change	$\Delta H°$(kJ/mol)
$Li^+(g) + F^-(g) \longrightarrow LiF(g)$	Formation of gaseous ion pairs	–755
$Li^+(g) + F^-(g) \longrightarrow LiF(s)$	Formation of crystalline solid	–1050

The **lattice enthalpy** for crystal formation is sufficiently large and negative to render ΔG for $Li(s) + \frac{1}{2}F_2(g) \longrightarrow LiF(s)$ negative. Consequently, the reaction is spontaneous,

*While ionization energy and electron affinity are formally internal energy changes (ΔU), these are equivalent to enthalpy changes since $\Delta H = \Delta U + P\Delta V$ and $\Delta V = 0$ for the processes that define ionization energy and electron affinity.

despite the net endothermic contribution for generating gaseous ions from the parent elements and the negative entropy change associated with gas-phase ions coalescing into a crystalline solid.

5.4 Molecular Orbitals for Larger Molecules

The methods described previously for diatomic molecules can be extended to molecules consisting of three or more atoms, but this approach becomes more challenging as molecules become more complex. We will first consider several examples of linear molecules to illustrate the concept of group orbitals and then proceed to molecules that benefit from the application of formal group theory methods.

5.4.1 FHF^-

The linear ion FHF^-, an example of very strong hydrogen bonding that can be described as a covalent interaction,[9] provides a convenient introduction to the concept of **group orbitals**, collections of matching orbitals on outer atoms. To generate a set of group orbitals, we will use the valence orbitals of the fluorine atoms, as shown in **Figure 5.15**. We will then examine which central-atom orbitals have the proper symmetry to interact with the group orbitals.

The lowest energy group orbitals are composed of the $2s$ orbitals of the fluorine atoms. These orbitals either have matching signs of their wave functions (group orbital 1) or opposite signs (group orbital 2). These group orbitals should be viewed as sets of orbitals that potentially could interact with central atom orbitals. Group orbitals are the same combinations that formed bonding and antibonding orbitals in diatomic molecules (e.g., $p_{xa} + p_{xb}$, $p_{xa} - p_{xb}$), but now are separated by the central hydrogen atom. Group orbitals 3 and 4 are derived from the fluorine $2p_z$ orbitals, in one case having lobes with matching signs pointing toward the center (orbital 3), and in the other case having opposite signs pointing toward the center (orbital 4). Group orbitals 5 through 8 are derived from the $2p_x$ and $2p_y$ fluorine orbitals, which are mutually parallel and can be paired according to matching (orbitals 5 and 7) or opposite (orbitals 6 and 8) signs of their wave functions.

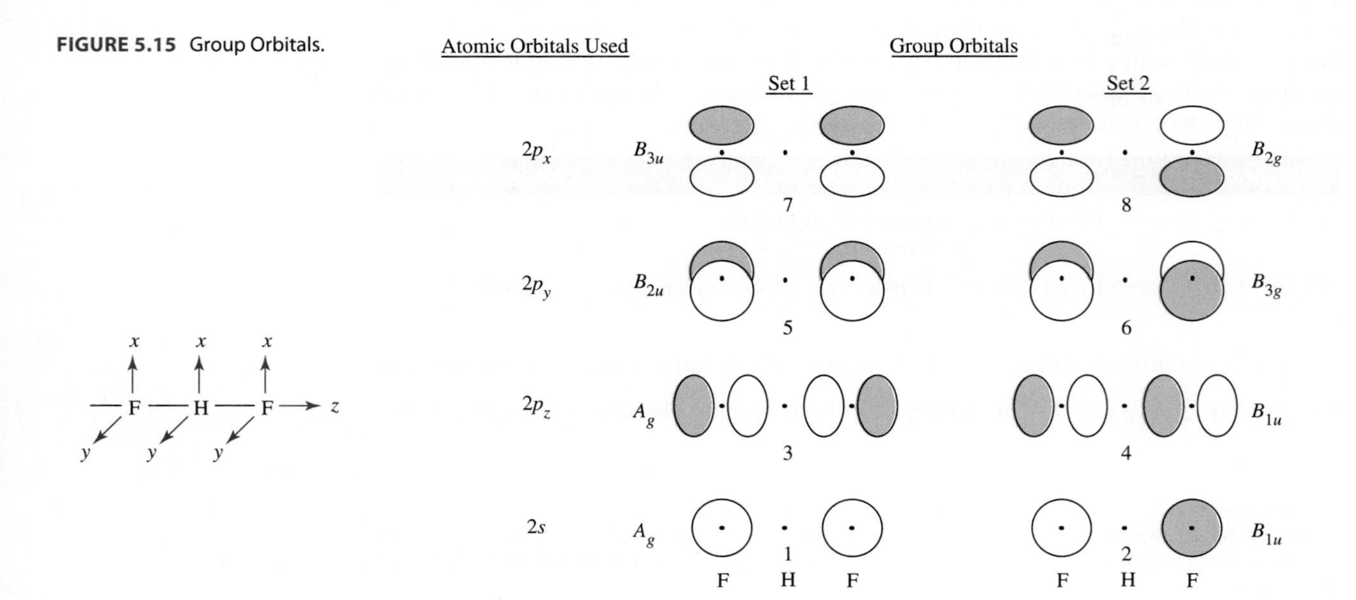

FIGURE 5.15 Group Orbitals.

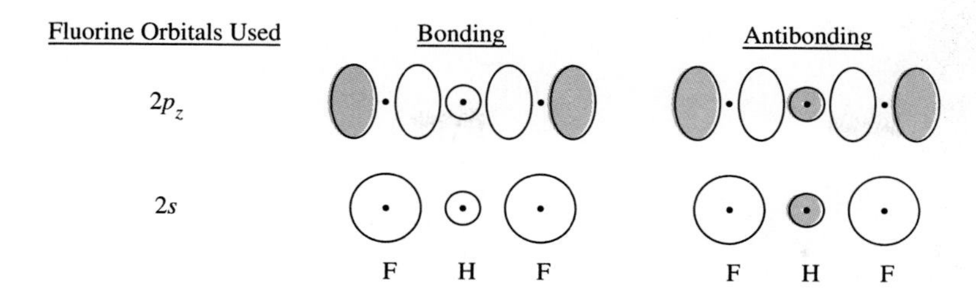

FIGURE 5.16 Interaction of Fluorine Group Orbitals with the Hydrogen 1*s* Orbital.

The central hydrogen atom in FHF^-, with only its 1*s* orbital available for bonding, is only eligible on the basis of its symmetry to interact with group orbitals 1 and 3; the 1*s* orbital is nonbonding with respect to the other group orbitals. These bonding and antibonding combinations are shown in **Figure 5.16**.

Which interaction, the hydrogen 1*s* orbital with group orbital 1 or 3, respectively, is likely to be stronger? The potential energy of the 1*s* orbital of hydrogen (−13.61 eV) is a much better match for the fluorines' $2p_z$ orbitals (−18.65 eV) than their 2*s* orbitals (−40.17 eV). Consequently, we expect the interaction with the $2p_z$ orbitals (group orbital 3) to be stronger than with the 2*s* orbitals (group orbital 1). The 1*s* orbital of hydrogen cannot interact with group orbitals 5 through 8; these orbitals are nonbonding.

The molecular orbitals for FHF^- are shown in **Figure 5.17**. In sketching molecular orbital energy diagrams of polyatomic species, we will show the orbitals of the central atom on the far left, the group orbitals of the surrounding atoms on the far right, and the resulting molecular orbitals in the middle.

Five of the six group orbitals derived from the fluorine 2*p* orbitals do not interact with the central atom; these orbitals remain essentially nonbonding and contain lone pairs of electrons.

It is important to recognize that these "lone pairs" are each delocalized over two fluorine atoms, a different perspective from that afforded by the Lewis dot model where these pairs are associated with single atoms. There is a slight interaction between orbitals on the non-neighboring fluorine atoms, but not enough to change their energies significantly because the fluorine atoms are so far apart (229 pm, more than twice the 91.7 pm distance between hydrogen and fluorine in HF). As already described, the sixth 2*p* group orbital, the remaining $2p_z$ (number 3), interacts with the 1*s* orbital of hydrogen to give two molecular orbitals, one bonding and one antibonding. An electron pair occupies the bonding orbital. The group orbitals from the 2*s* orbitals of the fluorine atoms are much lower in energy than the 1*s* orbital of the hydrogen atom and are essentially nonbonding. Group orbitals 2 and 4, while both essentially nonbonding, are slightly stabilized, and destabilized, respectively, due to a mixing phenomenon analogous to *s-p* mixing. These group orbitals are eligible to interact since they possess the same symmetry; this general issue will be discussed later in this chapter.

The Lewis approach to bonding requires two electrons to represent a single bond between two atoms and would result in four electrons around the hydrogen atom of FHF^-. The molecular orbital model, on the other hand, suggests a 2-electron bond delocalized over *three* atoms (a 3-center, 2-electron bond). The bonding MO in Figure 5.17 formed from group orbital 3 shows how the molecular orbital approach represents such a bond: two electrons occupy a low-energy orbital formed by the interaction of all three atoms (a central atom and a two-atom group orbital). The remaining electrons are in the group orbitals derived from the 2*s*, p_x, and p_y orbitals of the fluorine at essentially the same energy as that of the atomic orbitals.

In general, bonding molecular orbitals derived from three or more atoms, like the one in Figure 5.17, are stabilized more relative to their parent orbitals than bonding molecular

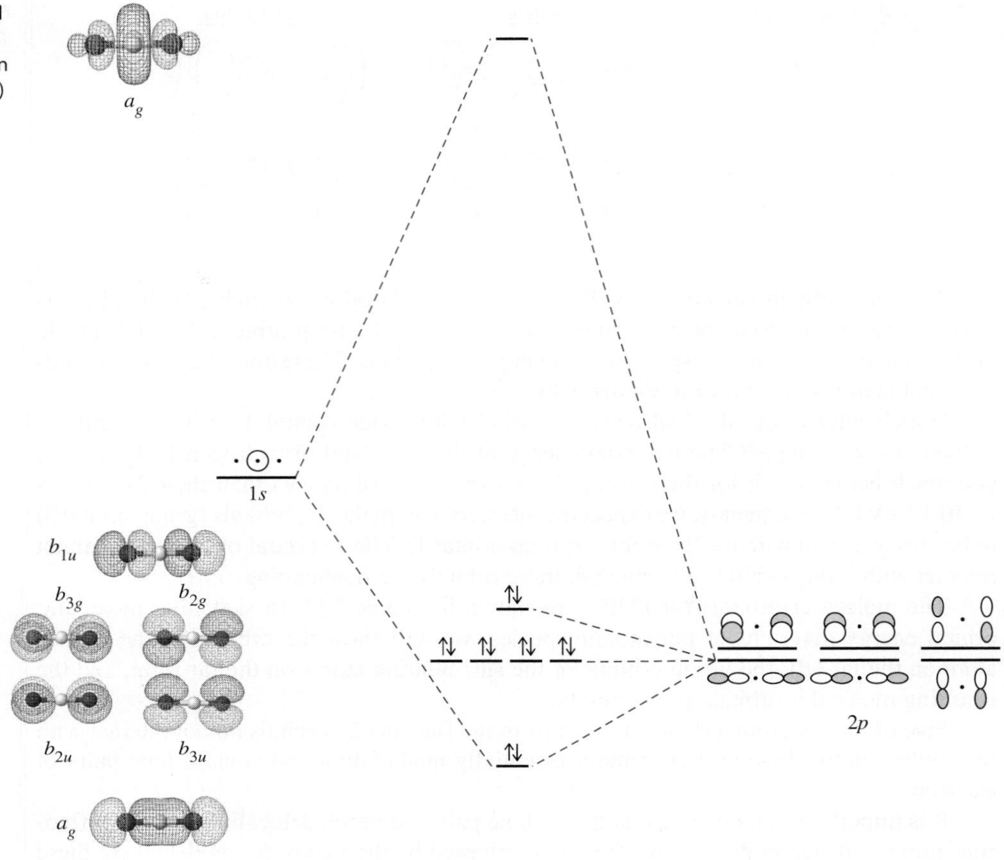

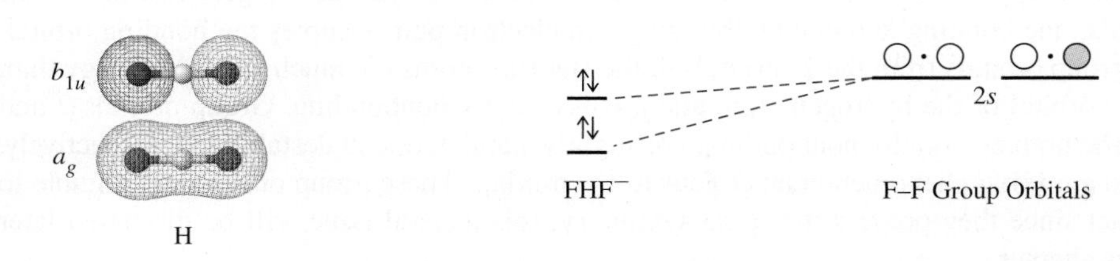

FIGURE 5.17 Molecular Orbital Diagram of FHF^-. (Molecular Orbital Diagram of FHF by Kaitlin Hellie. Reprinted by permission.)

orbitals that arise from orbitals on only two atoms. Electrons in bonding molecular orbitals consisting of more than two atoms experience attraction from multiple nuclei, and are delocalized over more space relative to electrons in bonding MOs composed of two atoms. Both of these features lead to additional stabilization in larger systems. However, the total energy of a molecule is the sum of the energies of *all* of the electrons in *all* the occupied orbitals. FHF^- has a bond energy of 212 kJ/mol and F—H distances of 114.5 pm. HF has a bond energy of 574 kJ/mol and an F—H bond distance of 91.7 pm.[10]

EXERCISE 5.4

Sketch the energy levels and the molecular orbitals for the linear H_3^+ ion.

5.4.2 CO_2

The approach used so far can be applied to other linear species—such as CO_2, N_3^-, and BeH_2—to consider how molecular orbitals can be constructed on the basis of interactions of group orbitals with central atom orbitals. However, we also need a method to understand the bonding in more complex molecules. We will first illustrate this approach using carbon dioxide, another linear molecule with a more complicated molecular orbital description than FHF^-. The following stepwise approach permits examination of more complex molecules:

1. Determine the point group of the molecule. If it is linear, substituting a simpler point group that retains the symmetry of the orbitals (ignoring the wave function signs) makes the process easier. It is useful to substitute D_{2h} for $D_{\infty h}$ and C_{2v} for $C_{\infty v}$. This substitution retains the symmetry of the orbitals without the need to use infinite-fold rotation axes.*
2. Assign x, y, and z coordinates to the atoms, chosen for convenience. Experience is the best guide here. A general rule is that *the highest order rotation axis of the molecule is assigned as the z axis of the central atom.* In nonlinear molecules, the y axes of the outer atoms are chosen to point toward the central atom.
3. Construct a (reducible) representation for the combination of the valence s orbitals on the outer atoms. If the outer atom is not hydrogen, repeat the process, finding the representations for each of the other sets of outer atom orbitals (for example, p_x, p_y, and p_z). As in the case of the vectors described in Chapter 4, any orbital that changes position during a symmetry operation contributes zero to the character of the resulting representation; any orbital that remains in its original position—such as a p orbital that maintains its position and direction (signs of its orbital lobes)—contributes 1; and any orbital that remains in the original position, with the signs of its lobes reversed, contributes -1.
4. Reduce each representation from Step 3 to its irreducible representations. This is equivalent to finding the symmetry of the **group orbitals** or the **symmetry-adapted linear combinations (SALCs)** of the orbitals. The group orbitals are then the combinations of atomic orbitals that match the symmetry of the irreducible representations.
5. Identify the atomic orbitals of the central atom with the same symmetries (irreducible representations) as those found in Step 4.
6. Combine the atomic orbitals of the central atom and those of the group orbitals with matching symmetry and similar energy to form molecular orbitals. The total number of molecular orbitals formed must equal the number of atomic orbitals used from all the atoms.**

In summary, the process used in creating molecular orbitals is to match the symmetries of the group orbitals, using their irreducible representations, with the symmetries of the central atom orbitals. If the symmetries match and the energies are similar, there is an interaction—both bonding and antibonding—if not, there is no interaction.

In CO_2 the group orbitals for the oxygen atoms are identical to the group orbitals for the fluorine atoms in FHF^- (Figure 5.15), but the central carbon atom in CO_2 has both *s and p* orbitals capable of interacting with the oxygen atom group orbitals. As in the discussion of FHF^-, the group orbital–atomic orbital interactions of CO_2 will be the focus.

1. **Point Group:** CO_2 has $D_{\infty h}$ symmetry so the D_{2h} point group will be used.
2. **Coordinate System:** The z axis is chosen as the C_∞ axis, and the y and z coordinates are chosen similarly to the FHF^- example (**Figure 5.18**).

---

*This approach is sometimes referred to as a "Descent in Symmetry."

**We use lowercase labels on the molecular orbitals, with uppercase for the atomic orbitals and for representations in general. This practice is common but not universal.

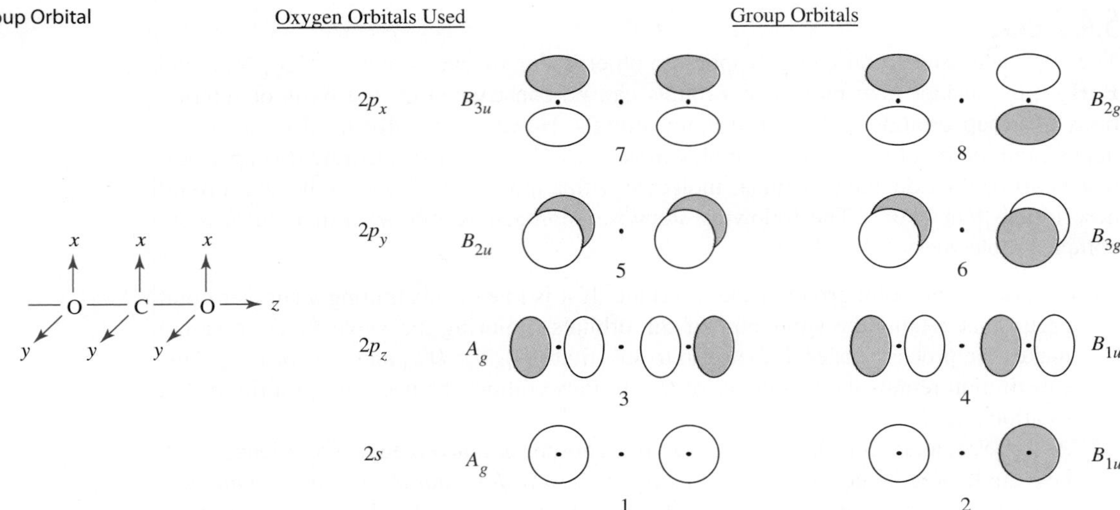

FIGURE 5.18 Group Orbital Symmetry in CO_2.

3. **Reducible Representations for Outer Atom Orbitals:** In CO_2 these are the $2s$ and $2p$ oxygen orbitals. These can be grouped into four sets (Figure 5.18). For example, the pair of $2s$ orbitals on the oxygen atoms has the following representation:

D_{2h}	E	$C_2(z)$	$C_2(y)$	$C_2(x)$	i	$\sigma(xy)$	$\sigma(xz)$	$\sigma(yz)$
$\Gamma(2s)$	2	2	0	0	0	0	2	2

The other group orbitals have the following representations:

D_{2h}	E	$C_2(z)$	$C_2(y)$	$C_2(x)$	i	$\sigma(xy)$	$\sigma(xz)$	$\sigma(yz)$
$\Gamma(2p_z)$	2	2	0	0	0	0	2	2
$\Gamma(2p_x)$	2	−2	0	0	0	0	2	−2
$\Gamma(2p_y)$	2	−2	0	0	0	0	−2	2

4. **Group Orbitals from Reducible Representations:** Each of the representations from Step 3 can be reduced by the procedure described in Section 4.4.2. For example, the representation $\Gamma(2s)$ reduces to $A_g + B_{1u}$:

D_{2h}	E	$C_2(z)$	$C_2(y)$	$C_2(x)$	i	$\sigma(xy)$	$\sigma(xz)$	$\sigma(yz)$
A_g	1	1	1	1	1	1	1	1
B_{1u}	1	1	−1	−1	−1	−1	1	1

When this procedure is conducted for each representation, the representations describe the symmetry properties of the oxygen atom group orbitals for CO_2, shown with the appropriate D_{2h} labels in Figure 5.18. Note that these group orbitals are the same as those deduced by inspection for the fluorine atoms in FHF^-.

EXERCISE 5.5

Using the D_{2h} character table shown, verify that the group orbitals in Figure 5.18 match their irreducible representations.

D_{2h}	E	$C_2(z)$	$C_2(y)$	$C_2(x)$	i	$\sigma(xy)$	$\sigma(xz)$	$\sigma(yz)$		
A_g	1	1	1	1	1	1	1	1		x^2, y^2, z^2
B_{1g}	1	1	−1	−1	1	1	−1	−1	R_z	xy
B_{2g}	1	−1	1	−1	1	−1	1	−1	R_y	xz
B_{3g}	1	−1	−1	1	1	−1	−1	1	R_x	yz
A_u	1	1	1	1	−1	−1	−1	−1		
B_{1u}	1	1	−1	−1	−1	−1	1	1	z	
B_{2u}	1	−1	1	−1	−1	1	−1	1	y	
B_{3u}	1	−1	−1	1	−1	1	1	−1	x	

5. **Matching Orbitals on the Central Atom:** To determine which atomic orbitals of carbon are of correct symmetry to interact with the group orbitals, we will consider each group orbital separately. The carbon atomic orbitals are shown in **Figure 5.19** with their symmetry labels within the D_{2h} point group.

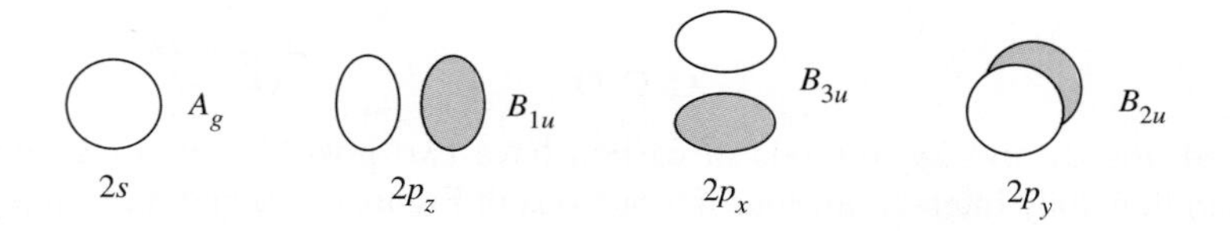

FIGURE 5.19 Symmetry of the Carbon Atomic Orbitals in the D_{2h} Point Group.

The D_{2h} character table shows the symmetry of these orbitals. For example, B_{1u} has the symmetry of the z axis and of the p_z oxygen orbitals; they are unchanged by the E, $C_2(z)$, $\sigma(xz)$, and $\sigma(yz)$ operations, and the $C_2(y)$, $C_2(x)$, i, and $\sigma(xy)$ operations reverse their signs.

6. **Formation of Molecular Orbitals:** Group orbitals 1 and 2 in **Figure 5.20**, formed by adding and subtracting the oxygen $2s$ orbitals, have A_g and B_{1u} symmetry, respectively. Group orbital 1 is of appropriate symmetry to interact with the $2s$ orbital of carbon (both have A_g symmetry), and group orbital 2 is of appropriate symmetry to interact with the $2p_z$ orbital of carbon (both have B_{1u} symmetry).

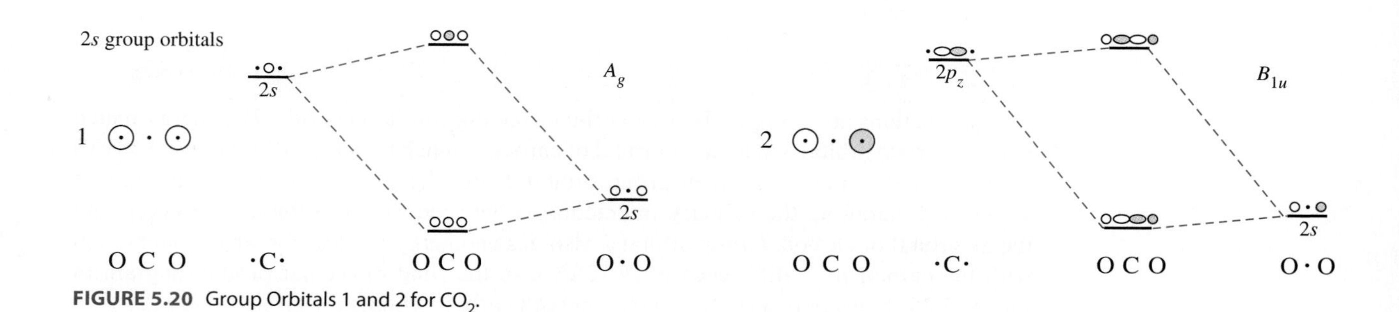

FIGURE 5.20 Group Orbitals 1 and 2 for CO_2.

Group orbitals 3 and 4 in **Figure 5.21**, formed by adding and subtracting the oxygen $2p_z$ orbitals, also have A_g and B_{1u} symmetries. Therefore, group orbital 3 can also interact with the $2s$ of carbon, and group orbital 4 can also interact with the carbon $2p_z$.

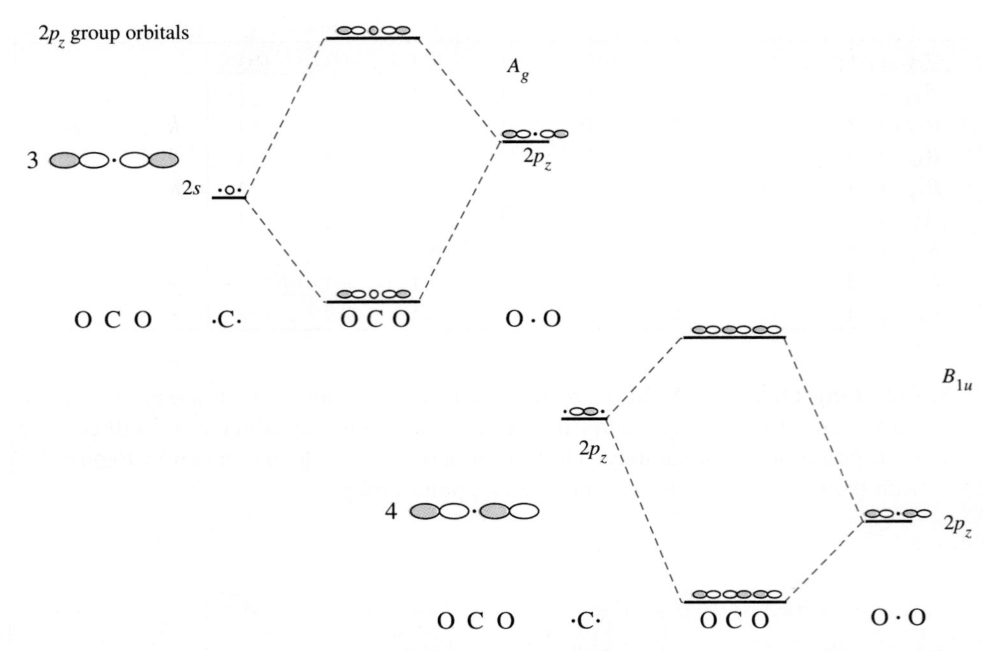

FIGURE 5.21 Group Orbitals 3 and 4 for CO_2.

Indeed, the $2s$ and $2p_z$ orbitals of carbon have two possible sets of group orbitals with which they may interact; all four interactions in Figures 5.20 and 5.21 are symmetry allowed. It is necessary to estimate which interactions are expected to be the strongest from the potential energies of the $2s$ and $2p$ orbitals of carbon and oxygen given in **Figure 5.22**.

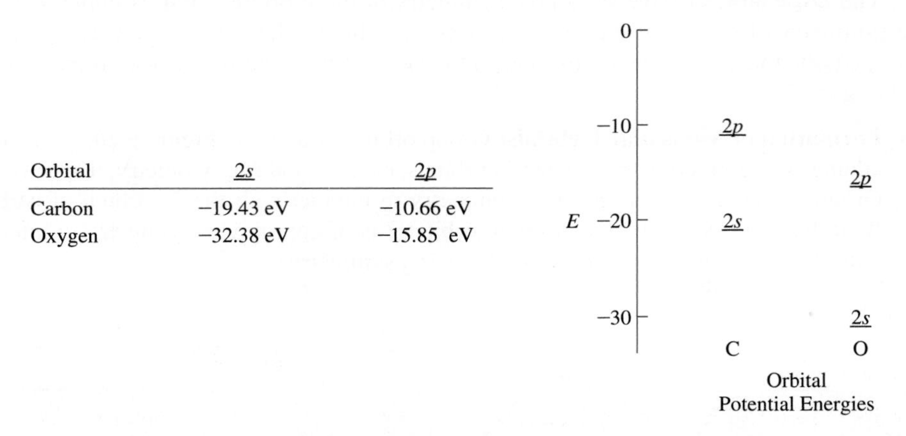

Orbital	2s	2p
Carbon	−19.43 eV	−10.66 eV
Oxygen	−32.38 eV	−15.85 eV

FIGURE 5.22 Orbital Potential Energies of Carbon and Oxygen.

Interactions are strongest between orbitals having similar energies. The energy match between group orbital 3 and the $2s$ orbital of carbon is much better (a difference of 3.58 eV) than the energy match between group orbital 1 and the $2s$ of carbon (a difference of 12.95 eV); therefore, the primary interaction is between the $2p_z$ orbitals of oxygen and the $2s$ orbital of carbon. Group orbital 2 also has an energy too low for strong interaction with the carbon p_z (a difference of 21.72 eV), so the final molecular orbital diagram in Figure 5.25 shows only a slight interaction with carbon orbitals for group orbital 2.

EXERCISE 5.6

Using orbital potential energies, show that group orbital 4 is more likely than group orbital 2 to interact strongly with the $2p_z$ orbital of carbon.

The $2p_y$ orbital of carbon has B_{2u} symmetry and interacts with group orbital 5 (**Figure 5.23**). The result is the formation of two π molecular orbitals, one bonding and one antibonding. However, there is no carbon orbital with B_{3g} symmetry to interact with group orbital 6, also formed by combining $2p_y$ orbitals of oxygen. Therefore, group orbital 6 is nonbonding.

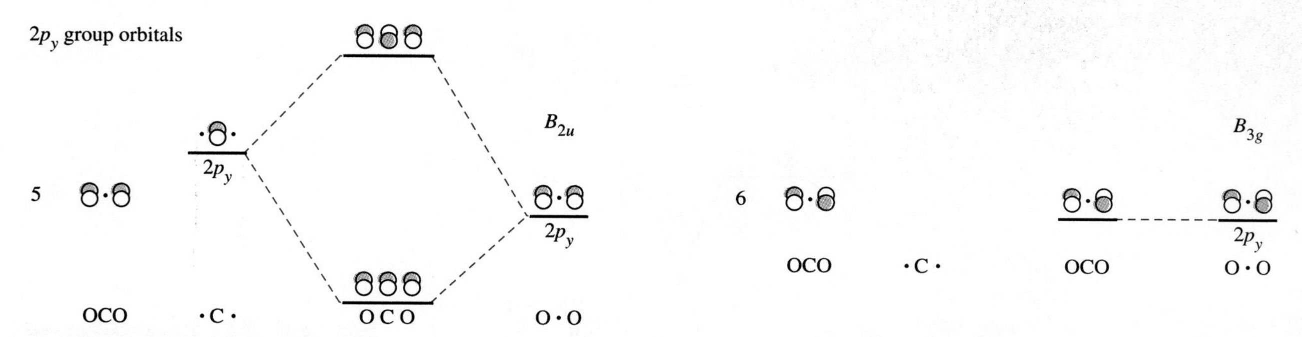

FIGURE 5.23 Group Orbitals 5 and 6 for CO_2.

Interactions of the $2p_x$ orbitals are similar to those of the $2p_y$ orbitals. Group orbital 7, with B_{3u} symmetry, interacts with the $2p_x$ orbital of carbon to form π bonding and antibonding orbitals, whereas group orbital 8 is nonbonding (**Figure 5.24**).

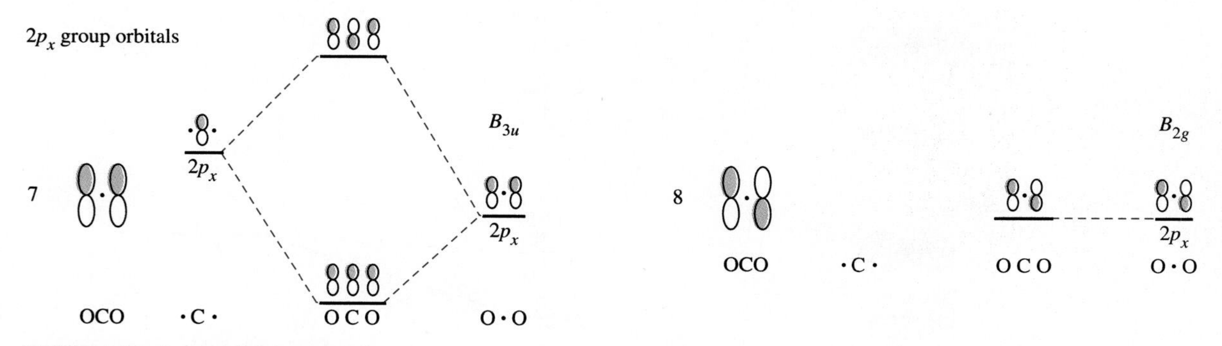

FIGURE 5.24 Group Orbitals 7 and 8 for CO_2.

The molecular orbital diagram of CO_2 is shown in **Figure 5.25**. The 16 valence electrons occupy, from the bottom, two essentially nonbonding σ orbitals, two bonding σ orbitals, two bonding π orbitals, and two nonbonding π orbitals. In other words, two of the bonding electron pairs are in σ orbitals, two are in π orbitals, and there are four bonds in the molecule, as expected. As in FHF^-, all the occupied molecular orbitals are 3-center, 2-electron orbitals.

The molecular orbitals of related linear triatomic species—such as N_3^-, CS_2, and OCN^-—can be determined similarly. The molecular orbitals of longer polyatomic species can be determined by a similar method. Examples of bonding in linear and cyclic π systems are considered in Chapter 13.

EXERCISE 5.7

Prepare a molecular orbital diagram for the azide ion, N_3^-.

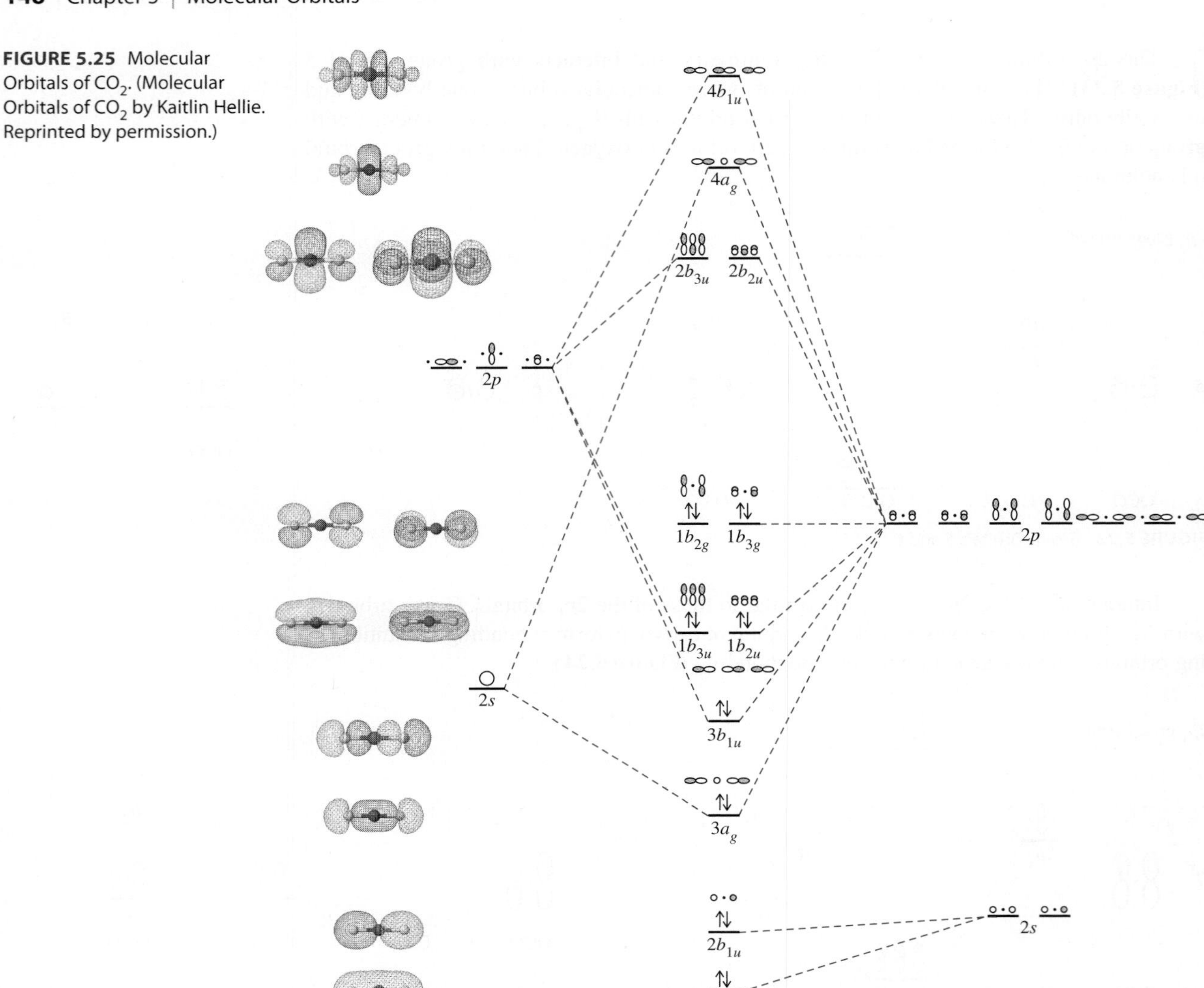

FIGURE 5.25 Molecular Orbitals of CO_2. (Molecular Orbitals of CO_2 by Kaitlin Hellie. Reprinted by permission.)

EXERCISE 5.8

Prepare a molecular orbital diagram for the BeH_2 molecule. (Assume an orbital potential energy of –6.0 eV for 2*p* orbitals of Be. This orbital set should be taken into account, even though it is unoccupied in a free Be atom.)

This process can be extended to obtain numerical values of the coefficients of the atomic orbitals used in the molecular orbitals.[11,12] The coefficients may be small or large, positive or negative, similar or quite different, depending on the characteristics of the orbital under consideration. Computer software packages are available to calculate these coefficients and generate the pictorial diagrams that describe the molecular orbitals. Examples of problems that use molecular modeling software to generate molecular orbitals of a variety of molecules are included in the problems at the end of this chapter and in later chapters. Discussion of these computational methods is beyond the scope of this text.

5.4.3 H_2O

Molecular orbitals of nonlinear molecules can be determined by similar procedures. Water is a useful example:

1. Water is a bent molecule with a C_2 axis through the oxygen and two mirror planes that intersect along this axis (**Figure 5.26**). The point group is C_{2v}.
2. The C_2 axis is chosen as the z axis and the xz plane as the plane of the molecule.* Because the hydrogen 1s orbitals have no directionality, it is not necessary to assign axes to the hydrogen atoms.
3. The hydrogen atoms determine the symmetry of the molecule, and their 1s orbitals form the basis of a reducible representation. The characters for each operation for the hydrogen atom 1s orbitals are readily obtained, Γ. The sum of the contributions to the character (1, 0, or -1, as described previously) for each symmetry operation is the character for that operation in the representation. The complete list for all operations of the group is the reducible representation for the atomic orbitals:

   The E operation leaves both hydrogen orbitals unchanged, for a character of 2.

   C_2 rotation interchanges the orbitals, so each contributes 0, for a total character of 0.

   Reflection in the plane of the molecule (σ_v) leaves both hydrogens unchanged, for a character of 2.

   Reflection perpendicular to the plane (σ_v') switches the two orbitals, for a character of 0.

   Step 3 is summarized in **Table 5.3**.

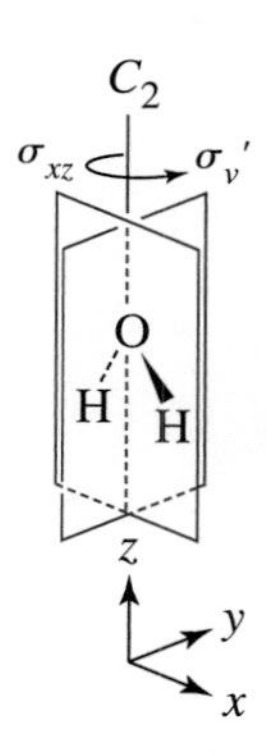

FIGURE 5.26 Symmetry of the Water Molecule.

TABLE 5.3 Representations for C_{2v} Symmetry Operations for Hydrogen Atoms in Water

C_{2v} Character Table

C_{2v}	E	C_2	$\sigma_v(xz)$	$\sigma_v'(yz)$		
A_1	1	1	1	1	z	x^2, y^2, z^2
A_2	1	1	-1	-1	R_z	xy
B_1	1	-1	1	-1	x, R_y	xz
B_2	1	-1	-1	1	y, R_x	yz

$$\begin{bmatrix} H_a' \\ H_b' \end{bmatrix} = \begin{bmatrix} 1 & 0 \\ 0 & 1 \end{bmatrix} \begin{bmatrix} H_a \\ H_b \end{bmatrix} \text{ for the identity operation}$$

$$\begin{bmatrix} H_a' \\ H_b' \end{bmatrix} = \begin{bmatrix} 0 & 1 \\ 1 & 0 \end{bmatrix} \begin{bmatrix} H_a \\ H_b \end{bmatrix} \text{ for the } C_2 \text{ operation}$$

$$\begin{bmatrix} H_a' \\ H_b' \end{bmatrix} = \begin{bmatrix} 1 & 0 \\ 0 & 1 \end{bmatrix} \begin{bmatrix} H_a \\ H_b \end{bmatrix} \text{ for the } \sigma_v \text{ reflection } (xz) \text{ plane}$$

$$\begin{bmatrix} H_a' \\ H_b' \end{bmatrix} = \begin{bmatrix} 0 & 1 \\ 1 & 0 \end{bmatrix} \begin{bmatrix} H_a \\ H_b \end{bmatrix} \text{ for the } \sigma_v' \text{ } (yz \text{ plane})$$

The reducible representation $\Gamma = A_1 + B_1$:

C_{2v}	E	C_2	$\sigma_v(xz)$	$\sigma_v'(yz)$	
Γ	2	0	2	0	
A_1	1	1	1	1	z
B_1	1	-1	1	-1	x

*One can also select the yz plane as the plane of the molecule. This results in $\Gamma = A_1 + B_2$ and switches the b_1 and b_2 labels of the molecular orbitals.

4. The representation Γ can be reduced to the irreducible representations $A_1 + B_1$, representing the symmetries of the group orbitals. In Step 5 these group orbitals will be matched with oxygen orbitals of matching symmetries.

5. The first step in deducing the molecular orbitals is to combine the two hydrogen 1*s* orbitals to afford group orbitals. Their sum, $\frac{1}{\sqrt{2}}[\Psi(H_a) + \Psi(H_b)]$, has A_1 symmetry (this is the group orbital in which the 1*s* wave functions have matching signs); their difference, $\frac{1}{\sqrt{2}}[(\Psi(H_a) - \Psi(H_b)]$, has symmetry B_1 (the group orbital in which the 1*s* wave functions have opposite signs) as shown in **Figure 5.27**. These equations define the group orbitals, or symmetry-adapted linear combinations. In this case, the combining 1*s* atomic orbitals are identical and contribute equally to each group orbital. This means that the coefficients for each unique 1*s* atomic orbital in the group orbital equations have the same magnitude. These coefficients also reflect the normalization requirement discussed in Section 2.2; the sum of squares of the coefficients for each atomic orbital must equal 1 when all group orbitals including a given atomic orbital are considered. In this case, the normalizing factor is $\frac{1}{\sqrt{2}}$. In general, the normalizing factor (N) for a group orbital is

$$N = \frac{1}{\sqrt{\sum c_i^2}}$$

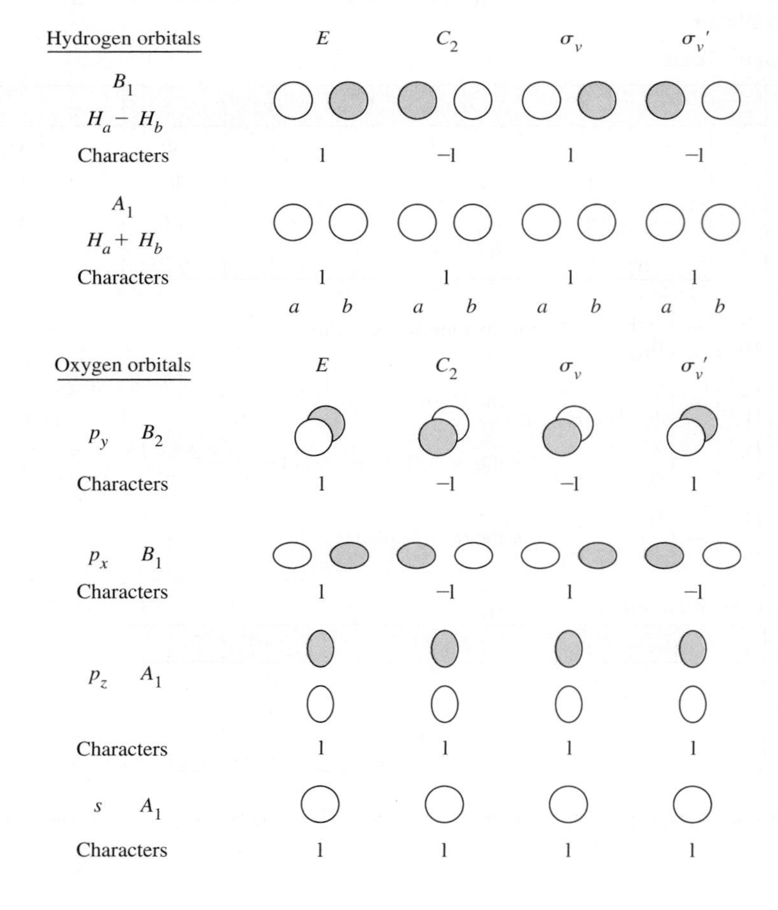

FIGURE 5.27 Symmetry of Atomic and Group Orbitals in the Water Molecule.

TABLE 5.4 Molecular Orbitals for Water

Symmetry	Molecular Orbitals		Oxygen Atomic Orbitals		Group Orbitals from Hydrogen Atoms	Description
B_1	ψ_6	=	$c_9\psi(p_x)$	+	$c_{10}[\psi(H_a) - \psi(H_b)]$	Antibonding (c_{10} is negative)
A_1	ψ_5	=	$c_7\psi(s)$	+	$c_8[\psi(H_a) + \psi(H_b)]$	Antibonding (c_8 is negative)
B_2	ψ_4	=	$\psi(p_y)$			Nonbonding
A_1	ψ_3	=	$c_5\psi(p_z)$	+	$c_6[\psi(H_a) + \psi(H_b)]$	Slightly bonding (c_6 is small)
B_1	ψ_2	=	$c_3\psi(p_x)$	+	$c_4[\psi(H_a) - \psi(H_b)]$	Bonding (c_4 is positive)
A_1	ψ_1	=	$c_1\psi(s)$	+	$c_2[\psi(H_a) + \psi(H_b)]$	Bonding (c_2 is positive)

where $c_i =$ the coefficients on the atomic orbitals. Each group orbital is treated as a single orbital for combination with the oxygen atomic orbitals.

The symmetries of oxygen's 2*s* and 2*p* atomic orbitals can be assigned and confirmed using the C_{2v} character table. The *x*, *y*, and *z* axes and the more complex functions assist in assigning representations to the atomic orbitals. In this case:

The *s* orbital is unchanged by all the operations, so it has A_1 symmetry; an *s* orbital is totally symmetric.

The p_x orbital has the B_1 symmetry of the *x* axis.

The p_y orbital has the B_2 symmetry of the *y* axis.

The p_z orbital has the A_1 symmetry of the *z* axis.

6. The atomic and group orbitals with the same symmetry are combined into molecular orbitals, as listed in **Table 5.4** and shown in **Figure 5.28**. They are numbered Ψ_1 through Ψ_6 in order of their energy, with 1 the lowest and 6 the highest.

The A_1 group orbital combines with the *s* and p_z orbitals of the oxygen to form three molecular orbitals (three atomic or group orbitals forming three molecular orbitals), Ψ_1, Ψ_3, and Ψ_5. The energy of molecular orbital Ψ_1 is only slightly lower in energy relative the oxygen 2*s* orbital; Ψ_1 can be regarded as relatively nonbonding.[13] The electrons that occupy Ψ_1 represent one of the lone pairs in the Lewis electron-dot structure of water, a pair with high probability of being found on the oxygen. As expected on the basis of the relatively close energy match between the oxygen 2*p* subshell and the hydrogen group orbitals, Ψ_3 has significant contribution from oxygen p_z; Ψ_3 is a bonding orbital. Ψ_5 is antibonding and has significantly more contribution from the oxygen p_z relative to the oxygen 2*s*.

The hydrogen B_1 group orbital combines with the oxygen p_x orbital to form two MOs, one bonding and one antibonding (Ψ_2 and Ψ_6). The oxygen p_y (Ψ_4, with B_2 symmetry) does not match the symmetry of the hydrogen 1*s* group orbitals and is therefore nonbonding. This pair of electrons represents the second lone pair in the Lewis structure of water. It is noteworthy that the nonbonding pairs afforded by the MO model of water are not equivalent as in the Lewis model. Overall, there are two bonding orbitals (Ψ_2, Ψ_3), one nonbonding (Ψ_4), one essentially nonbonding orbital (Ψ_1), and two antibonding orbitals (Ψ_5, Ψ_6). The oxygen 2*s* orbital (-32.38 eV) is nearly 20 eV below the hydrogen orbitals in energy (-13.61 eV), so it has very little interaction with the group orbitals. The oxygen 2*p* orbitals (-15.85 eV) are a good match for the hydrogen 1*s* energy, allowing formation of the bonding b_1 and a_1 molecular orbitals. When the eight valence electrons are added, two pairs occupy bonding orbitals, and two pairs occupy nonbonding orbitals; this complements the two bonds and two lone pairs of the Lewis electron-dot structure.

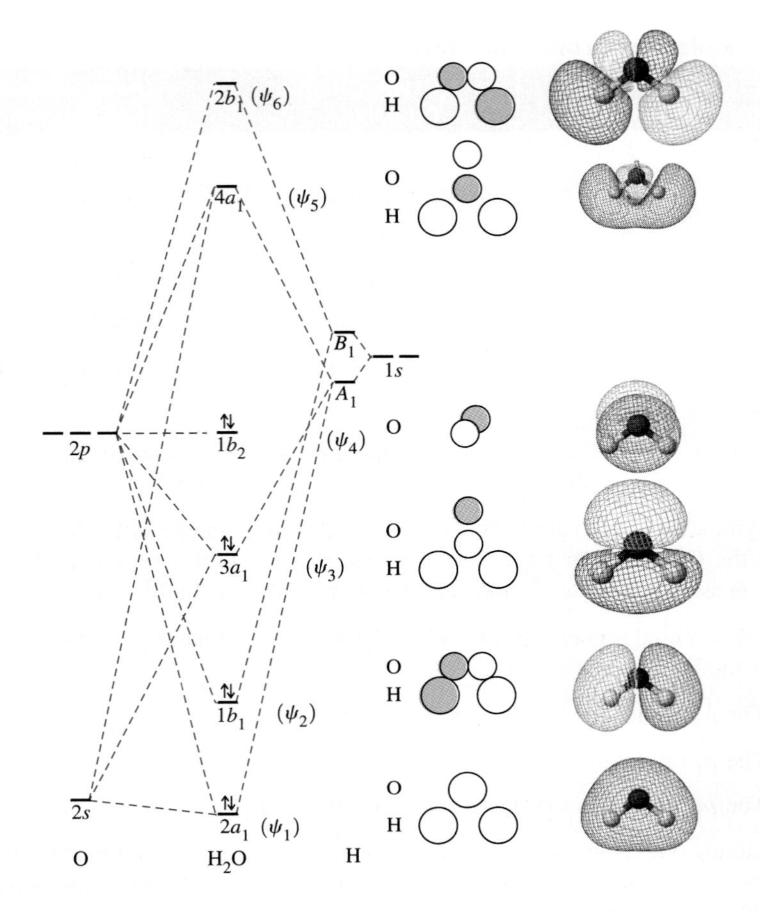

FIGURE 5.28 Molecular Orbitals of H_2O. (Molecular Orbitals of H_2O by Kaitlin Hellie. Reprinted by permission.)

As mentioned previously, the molecular orbital perspective differs from the common conception of the water molecule as having two equivalent lone electron pairs and two equivalent O—H bonds. In the MO model, the highest energy electron pair, designated b_2, is truly nonbonding, occupying the oxygen $2p_y$ orbital with its axis perpendicular to the plane of the molecule. The two pairs next highest in energy are bonding pairs, resulting from overlap of the $2p_z$ and $2p_x$ orbital with the $1s$ orbitals of the hydrogen atoms. The lowest energy pair is concentrated on the $2s$ orbital of oxygen. All four occupied molecular orbitals are different.

5.4.4 NH_3

The VSEPR approach describes ammonia as a pyramidal molecule with a lone pair of electrons and C_{3v} symmetry. To obtain a molecular orbital description of NH_3, it is convenient to view this molecule down the C_3, or z, axis and with the yz plane passing through one of the hydrogen atoms, as shown in **Figure 5.29**. The reducible representation for the three hydrogen atom $1s$ orbitals is given in **Table 5.5**. It can be reduced to the A_1 and E irreducible representations. Because three hydrogen $1s$ orbitals are considered, there must be three group orbitals formed from them, one with A_1 symmetry and two with E symmetry.

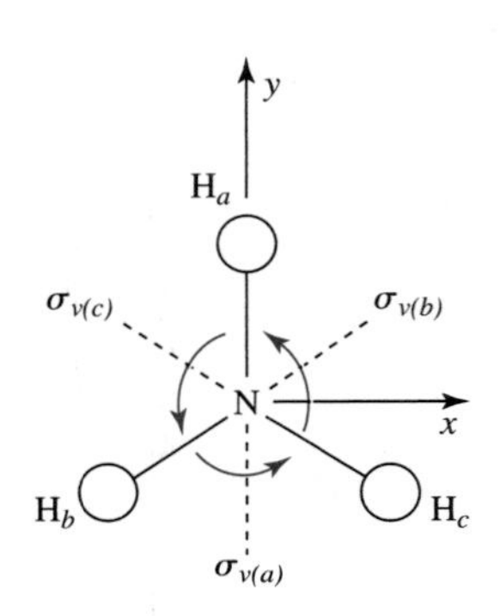

FIGURE 5.29 Coordinate System for NH_3.

Deducing group orbitals thus far has been relatively straightforward; each polyatomic example considered (FHF^-, CO_2, H_2O) has two atoms attached to the central atom, and the group orbitals could be obtained by combining identical atomic orbitals on the terminal atoms in both a bonding and antibonding sense. This is no longer possible with NH_3. To address situations such as NH_3, the **projection operator method**, a systematic approach for deduction of group orbitals, is the preferred strategy.

TABLE 5.5 Representations for Atomic Orbitals in Ammonia

C_{3v} Character Table

C_{3v}	E	$2C_3$	$3\sigma_v$		
A_1	1	1	1	z	$x^2 + y^2, z^2$
A_2	1	1	-1		
E	2	-1	0	$(x, y), (R_x, R_y)$	$(x^2 - y^2, xy), (xz, yz)$

The reducible representation $\Gamma = A_1 + E$:

C_{3v}	E	$2C_3$	$3\sigma_v$		
Γ	3	0	1		
A_1	1	1	1	z	$x^2 + y^2, z^2$
E	2	-1	0	$(x, y), (R_x, R_y)$	$(x^2 - y^2, xy), (xz, yz)$

The projection operator method permits elucidation of how atomic orbitals should be combined to afford the symmetry-adapted linear combinations (SALCs) that define the group orbitals. This method requires determining the impact of each point group symmetry operation on one atomic orbital (e.g., the hydrogen 1*s* orbital of H_a) within a set of identical atomic orbitals (e.g., the set of three hydrogen 1*s* orbitals in Figure 5.29). For example, the E operation leaves hydrogen 1*s* orbital H_a unchanged while C_3 transforms H_a to H_b. These outcomes are best tabulated; note that each unique symmetry operation is considered without their usual groupings into classes as in Table 5.5.

Original Orbital	E	C_3	C_3^2	$\sigma_{v(a)}$	$\sigma_{v(b)}$	$\sigma_{v(c)}$
H_a becomes...	H_a	H_b	H_c	H_a	H_c	H_b

Linear combinations of these hydrogen 1*s* atomic orbitals that match the symmetries of the A_1, A_2, and E irreducible representations can be obtained via

(1) multiplication of each outcome by the characters associated with each operation for these irreducible representations, followed by

(2) addition of the results. This approach affords the following:

	$E \quad C_3 \quad C_3^2 \quad \sigma_{v(a)} \quad \sigma_{v(b)} \quad \sigma_{v(c)}$
A_1	$H_a + H_a + H_c + H_a + H_c + H_b = 2H_a + 2H_b + 2H_c$
A_2	$H_a + H_b + H_c - H_a - H_c - H_b = 0$
E	$2H_a - H_b - H_c + 0 + 0 + 0 = 2H_a - H_b - H_c$

EXERCISE 5.9

The same general SALCs are obtained regardless of the initial atomic orbital examined. Show that if hydrogen 1*s* orbital H_b is chosen as the basis (instead of H_a), the resulting A_1 and A_2 linear combinations would be identical to those shown previously, and the E linear combination would feature the same relative contributions and signs of the three wave functions as in SALC(E) generated via H_a.

The zero sum associated with A_2 confirms that a group orbital of this symmetry cannot be obtained using these three hydrogen 1*s* orbitals, consistent with the $A_1 + E$ reducible representation for the group orbitals. The symmetry-adapted linear combination for A_1

(SALC(A_1)) indicates that each $1s$ orbital contributes equally to this group orbital, as one would visualize for SALC(A_1), a situation not the case with E.

Recall that the sum of the squares of the coefficients of each unique atomic orbital in the SALCs must equal 1. To meet this requirement, it is necessary that each group orbital equation be normalized, so that the sum of the squares of the coefficients within each equation affords 1. This requires that the wave functions for the $1s$ orbitals of H_a, H_b, and H_c each have a contribution of $\frac{1}{\sqrt{3}}$ in the normalized A_1 group orbital.

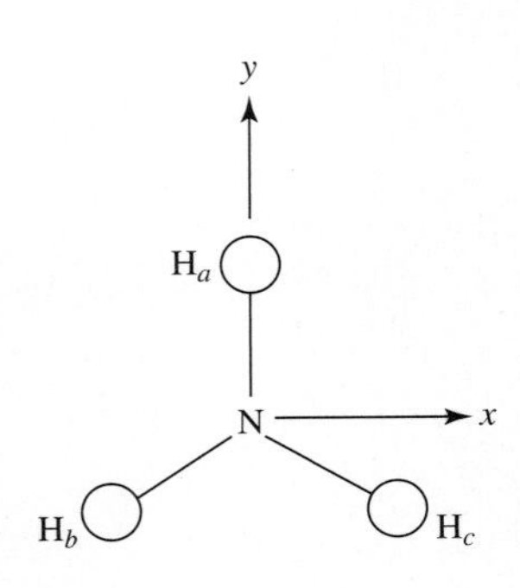

A_1 group orbital

$$\frac{1}{\sqrt{3}}[\Psi(H_a) + \Psi(H_b) + \Psi(H_c)]$$

As described in Section 5.4.3, the normalization factor is formally calculated via $N = \left(\sqrt{(c_a^2 + c_b^2 + c_c^2)}\right)^{-1} = \left(\sqrt{(1^2 + 1^2 + 1^2)}\right)^{-1} = \frac{1}{\sqrt{3}}$, where c_a, c_b, and c_c are the lowest common integer coefficients for the hydrogen $1s$ atomic orbital wave functions within SALC(A_1). This uniform contribution for each atomic orbital is expected for a totally symmetric group orbital. The SALC(A_1) equation $\frac{1}{\sqrt{3}}[\Psi(H_a) + \Psi(H_b) + \Psi(H_c)]$ also indicates that each $1s$ atomic orbital in the A_1 group orbital will exhibit the same wave function sign, since the signs of all three terms are positive.

Normalization of SALC(E) derived from the projection operator method must account for the doubled contribution of H_a relative to H_b and H_c, while maintaining the opposite wave function signs for H_a relative to H_b and H_c.

$$\frac{1}{\sqrt{6}}[2\Psi(H_a) - \Psi(H_b) - \Psi(H_c)]$$

$$N = \left(\sqrt{(c_a^2 + c_b^2 + c_c^2)}\right)^{-1} = \left(\sqrt{(2^2 + (-1)^2 + (-1)^2)}\right)^{-1} = \frac{1}{\sqrt{6}}$$

$E(y)$ group orbital

The second E group orbital can be motivated by remembering that the symmetry of central atom orbitals (in this case the N atom) must match the symmetry of the group orbitals with which they are combined to form molecular orbitals. The C_{3v} character table indicates that E describes the symmetries of the pair of atomic orbitals p_x and p_y, so the E group orbitals must be compatible with these orbitals. On the basis of the coordinate system defined in Figure 5.29, the E group orbital shown previously has the same symmetry as p_y (with the xz plane defining a node); it will interact with the N p_y orbital to create molecular orbitals. In this way, we see that hydrogen $1s$ orbital H_a was the most convenient basis for deduction of the SALCs; since hydrogen $1s$ orbital H_a lies on the y axis, the first E group orbital is compatible with this axis. The other E group orbital must match the symmetry of x, requiring zero contribution from H_a due to the orthogonal node defined by the yz plane; this means that only H_b and H_c can contribute to the second E group orbital. The H_b and H_c wave function contributions can be deduced by cataloging the squares of the coefficients for the normalized equations (**Table 5.6**). The coefficients for H_b and H_c must be $\frac{1}{\sqrt{2}}$ and $-\frac{1}{\sqrt{2}}$, respectively, to satisfy the normalization requirement while leading to identical total contributions from all three $1s$ wave functions across the three group orbitals. The positive and negative coefficients are necessary to match the symmetry of the p_x atomic orbital.

TABLE 5.6 SALC Coefficients and Evidence of Normalization

	Coefficients in Normalized SALCs			Squares of SALC Coefficients			Sum of the Squares = 1 for Normalization Requirement
	c_a	c_b	c_c	c_a^2	c_b^2	c_c^2	
A_1	$\frac{1}{\sqrt{3}}$	$\frac{1}{\sqrt{3}}$	$\frac{1}{\sqrt{3}}$	$\frac{1}{3}$	$\frac{1}{3}$	$\frac{1}{3}$	1
$E(y)$	$\frac{2}{\sqrt{6}}$	$-\frac{1}{\sqrt{6}}$	$-\frac{1}{\sqrt{6}}$	$\frac{2}{3}$	$\frac{1}{6}$	$\frac{1}{6}$	1
$E(x)$	0	$\frac{1}{\sqrt{2}}$	$-\frac{1}{\sqrt{2}}$	0	$\frac{1}{2}$	$\frac{1}{2}$	1
Sum of the squares for each 1s wave function must total 1 for an identical contribution of each atomic orbital to the group orbitals				1	1	1	

$$\frac{1}{\sqrt{2}}[\Psi(H_b) - \Psi(H_c)]$$

$$N = \left(\sqrt{(c_a^2 + c_b^2 + c_c^2)}\right)^{-1} = \left(\sqrt{0^2 + 1^2 + (-1)^2}\right)^{-1} = \frac{1}{\sqrt{2}}$$

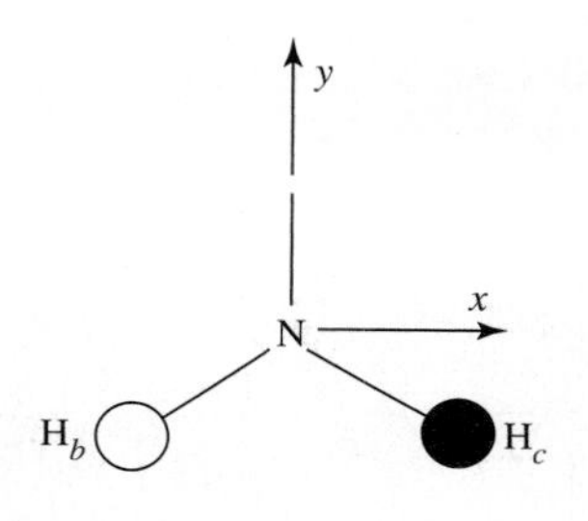

$E(x)$ group orbital

The s and p_z orbitals of nitrogen both have A_1 symmetry, and the pair p_x, p_y has E symmetry, exactly the same as the representations for the hydrogen 1s group orbitals; there are symmetry matches for both A_1 and E. As with the previous examples, each group orbital is treated as a single orbital in combining with the nitrogen orbitals (**Figure 5.30**). The nitrogen s and p_z orbitals combine with the hydrogen A_1 group orbital to give three a_1 orbitals, one bonding, one nonbonding, and one antibonding. The nonbonding orbital is almost entirely nitrogen p_z, with the nitrogen s orbital combining with the hydrogen group orbital for the bonding and antibonding orbitals. The nitrogen p_x and p_y orbitals combine with the E group orbitals to form four e orbitals, two bonding and two antibonding (e has a dimension of 2, which requires a pair of degenerate orbitals).

When eight electrons are put into the lowest energy levels, three bonds and one essentially nonbonding pair are indicated. The 1s orbital energies (-13.61 eV) of the hydrogen atoms match well with the energies of the nitrogen 2p orbitals (-13.18 eV), resulting in large differences between the bonding and antibonding orbital energies. The nitrogen 2s has such a sufficiently low energy (-25.56 eV) that its interaction with the hydrogen orbitals is quite small, and the $2a_1$ molecular orbital has nearly the same energy as the nitrogen 2s orbital.

The HOMO of NH_3 is slightly bonding, because it contains an electron pair in an orbital resulting from interaction of the $2p_z$ orbital of nitrogen with the 1s orbitals of the hydrogens (from the zero-node A_1 group orbital). The $2p_z$ orbital participates in weak overlap with the A_1 group orbital. One half of the $2p_z$ orbital points away from the hydrogen atoms, while the other half points at the center of the triangle formed by the three hydrogen nuclei. The HOMO is the lone pair of the electron-dot and VSEPR models. It is also the pair donated by ammonia when it functions as a Lewis base (discussed in Chapter 6).

5.4.5 CO_2 Revisited with Projection Operators

Section 5.4.2 outlines the process for determining group orbitals in the linear case where the outer atoms employ both s and p orbitals; Figure 5.18 illustrates the group orbitals comprised of $2p_x$, $2p_y$, $2p_z$, and 2s orbitals, respectively. While these orbitals can be deduced by

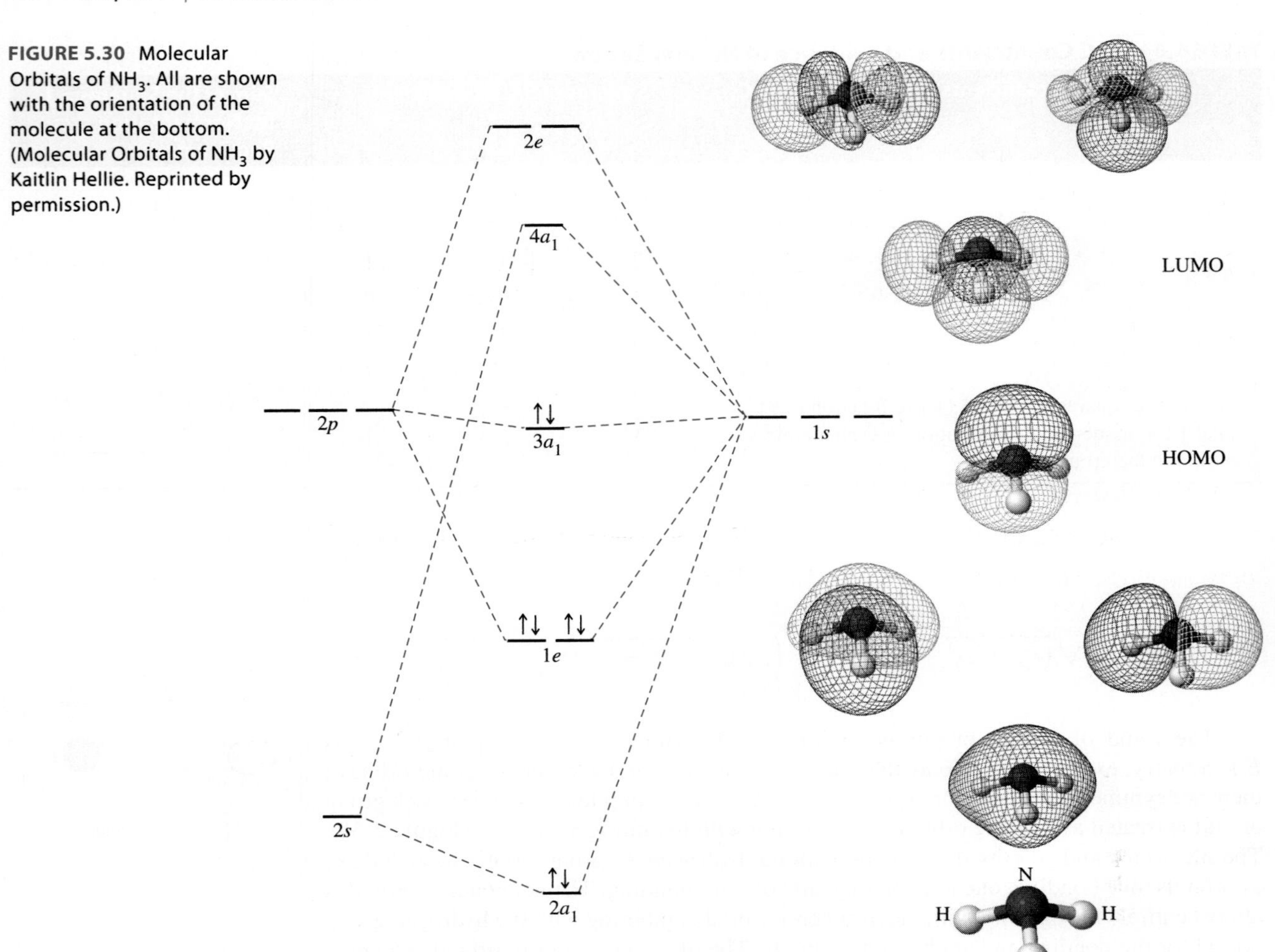

FIGURE 5.30 Molecular Orbitals of NH_3. All are shown with the orientation of the molecule at the bottom. (Molecular Orbitals of NH_3 by Kaitlin Hellie. Reprinted by permission.)

matching their symmetries to the corresponding irreducible representations, a complementary strategy is to employ projection operators. For example, the group orbitals composed of oxygen atom 2*s* atomic orbitals have A_g and B_{1u} symmetry in the D_{2h} point group.* As in the NH_3 example, consider the impact of each D_{2h} point group symmetry operation on one atomic orbital (in this case the oxygen 2*s* orbital of O_A) within a set of two 2*s* atomic orbitals. With linear molecules, the group orbital with matching orbital lobe signs toward the center is always chosen as the basis, in this case the A_g orbital. This general strategy was also employed in Section 5.4.4, where the A_1 group orbital (Figure 5.29) was used as the basis.

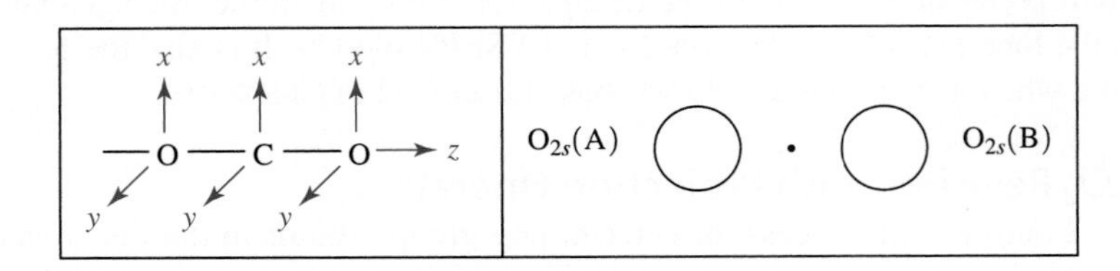

*We "descend in symmetry" from $D_{\infty h}$ to D_{2h} for convenience.

Original Orbital	E	$C_2(z)$	$C_2(y)$	$C_2(x)$	i	$\sigma(xy)$	$\sigma(xz)$	$\sigma(yz)$
$O_{2s(A)}$ becomes…	$O_{2s(A)}$	$O_{2s(A)}$	$O_{2s(B)}$	$O_{2s(B)}$	$O_{2s(B)}$	$O_{2s(B)}$	$O_{2s(A)}$	$O_{2s(A)}$

In this D_{2h} case, unlike in the C_{3v} point group, each operation is in its own class, and the number of columns above is identical to that in the D_{2h} character table. SALC(A_g) and SALC(B_{1u}) of these oxygen 2s wave functions can be obtained by multiplication of each outcome by the characters associated with each operation of these irreducible representations, followed by addition of the results:

	E	$C_2(z)$	$C_2(y)$	$C_2(x)$	i	$\sigma(xy)$	$\sigma(xz)$	$\sigma(yz)$	
A_g	$O_{2s(A)}$	$+\ O_{2s(A)}$	$+\ O_{2s(B)}$	$+\ O_{2s(B)}$	$+\ O_{2s(B)}$	$+\ O_{2s(B)}$	$+\ O_{2s(A)}$	$+\ O_{2s(A)}$	$= 4(O_{2s(A)}) + 4(O_{2s(B)})$
B_{1u}	$O_{2s(A)}$	$+\ O_{2s(A)}$	$-\ O_{2s(B)}$	$-\ O_{2s(B)}$	$-\ O_{2s(B)}$	$-\ O_{2s(B)}$	$+\ O_{2s(A)}$	$+\ O_{2s(A)}$	$= 4(O_{2s(A)}) - 4(O_{2s(A)})$

Normalization results in the expected group orbitals. In all molecules with two identical outer atoms, normalization always leads to $\pm\frac{1}{\sqrt{2}}$ coefficients as two atomic orbitals equally contribute toward two different SALCs.

A_g	$O_{2s(A)}$		$O_{2s(B)}$	$\frac{1}{\sqrt{2}}[\psi(O_{2s(A)}) + \psi(O_{2s(B)})]$
B_{1u}	$O_{2s(A)}$		$O_{2s(B)}$	$\frac{1}{\sqrt{2}}[\psi(O_{2s(A)}) - \psi(O_{2s(B)})]$

The $2p_z$ group orbitals also possess A_g and B_{1u} symmetry. The basis for the projection operator method is again the A_g group orbital, with the same signs of the orbital lobes pointing toward the center.

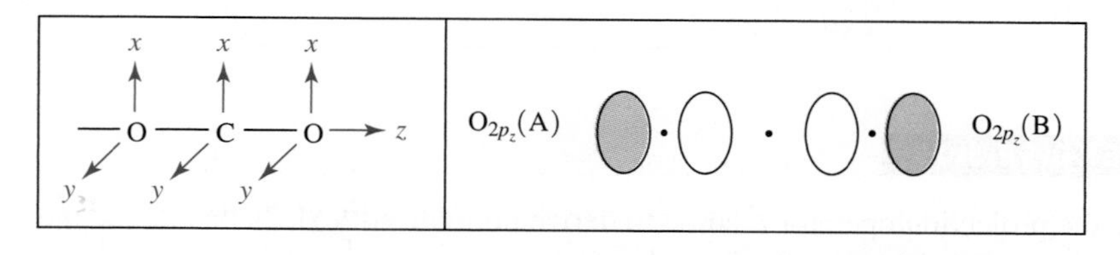

Original Orbital	E	$C_2(z)$	$C_2(y)$	$C_2(x)$	i	$\sigma(xy)$	$\sigma(xz)$	$\sigma(yz)$
$O_{2p_z(A)}$ becomes…	$O_{2p_z(A)}$	$O_{2p_z(A)}$	$O_{2p_z(B)}$	$O_{2p_z(B)}$	$O_{2p_z(B)}$	$O_{2p_z(B)}$	$O_{2p_z(A)}$	$O_{2p_z(A)}$

Extension to the SALCs affords the anticipated wave function equations and group orbitals; the signs are defined on the basis of the orientation of the orbitals relative to the central atom.

A_g	$O_{2p_z(A)}$		$O_{2p_z(B)}$	$\frac{1}{\sqrt{2}}[\psi(O_{2p_z(A)}) + \psi(O_{2p_z(B)})]$
B_{1u}	$O_{2p_z(A)}$		$O_{2p_z(B)}$	$\frac{1}{\sqrt{2}}[\psi(O_{2p_z(A)}) - \psi(O_{2p_z(B)})]$

The SALCs for the $2p_x$ orbitals exhibit B_{3u} and B_{2g} symmetries. With the B_{3u} orbital as the basis, we encounter situations, common with group orbitals designated for π bonding, where the original orbital (in this case $O_{2p_z(A)}$) becomes its own inverse or the inverse of another orbital upon transformation by some symmetry operations.

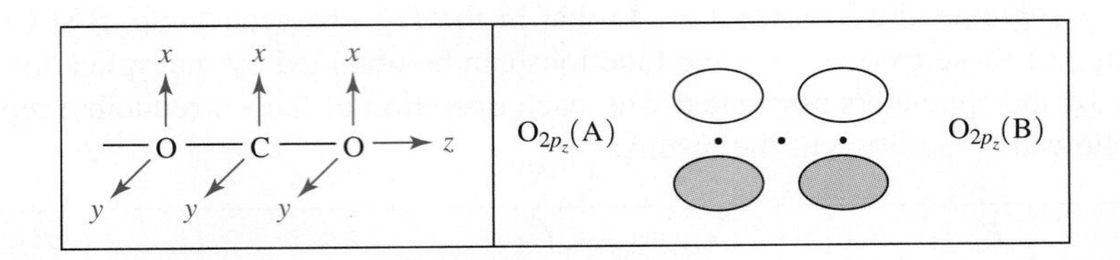

Original Orbital	E	$C_2(z)$	$C_2(y)$	$C_2(x)$	i	$\sigma(xy)$	$\sigma(xz)$	$\sigma(yz)$
$O_{2p_x(A)}$ becomes...	$O_{2p_x(A)}$	$-O_{2p_x(A)}$	$-O_{2p_x(B)}$	$O_{2p_x(B)}$	$-O_{2p_x(B)}$	$O_{2p_x(B)}$	$O_{2p_x(A)}$	$-O_{2p_x(A)}$
B_{3u}	$O_{2p_x(A)} + O_{2p_x(A)} + O_{2p_x(B)} + O_{2p_x(B)} + O_{2p_x(B)} + O_{2p_x(B)} + O_{2p_x(A)} + O_{2p_x(A)} = 4(O_{2p_x(A)}) + 4(O_{2p_x(B)})$							
B_{2g}	$O_{2p_x(A)} + O_{2p_x(A)} - O_{2p_x(B)} - O_{2p_x(B)} - O_{2p_x(B)} - O_{2p_x(B)} + O_{2p_x(A)} + O_{2p_x(A)} = 4(O_{2p_x(A)}) - 4(O_{2p_x(B)})$							

B_{3u}	$O_{2p_x(A)}$	$O_{2p_x(B)}$	$\frac{1}{\sqrt{2}}[\psi(O_{2p_x(A)}) + \psi(O_{2p_x(B)})]$
B_{2g}	$O_{2p_x(A)}$	$O_{2p_x(B)}$	$\frac{1}{\sqrt{2}}[\psi(O_{2p_x(A)}) - \psi(O_{2p_x(B)})]$

EXERCISE 5.10

Use the projection operator method to derive normalized SALCs that define the group orbitals for CO_2 based on the $2p_y$ orbitals.

5.4.6 BF_3

Boron trifluoride is a Lewis acid, an electron-pair acceptor. Molecular orbital theory of BF_3 must provide an orbital capable of acting as such an acceptor to be consistent with this chemical property. The VSEPR-predicted shape is trigonal, consistent with experimental observations.

The procedure for describing molecular orbitals of BF_3 differs from NH_3, because the fluorine atoms surrounding the central boron atom have $2p$ as well as $2s$ electrons to be considered. We assign the highest order rotation axis, the C_3, to be the z axis. The fluorine p_y axes are chosen to point toward the boron atom; the p_x axes are in the molecular plane. The group orbitals and their symmetry in the D_{3h} point group are shown in **Figure 5.31**. The molecular orbitals are shown in **Figure 5.32**.

As discussed in Chapter 3, consideration of all resonance structures for BF_3 suggests that this molecule possesses some double-bond character in the B—F bonds. The molecular orbital view of BF_3 has an electron pair in a bonding π orbital with a_2'' symmetry delocalized over all four atoms; this is the orbital slightly lower in energy than the five nonbonding electron pairs. Overall, BF_3 has three bonding σ orbitals (a_1' and e') and one slightly bonding π orbital (a_2'') occupied by electron pairs, together with eight nonbonding pairs on the fluorine atoms. The greater than 10 eV difference between the boron and fluorine p orbital energies renders this π orbital weakly bonding, but not insignificant.

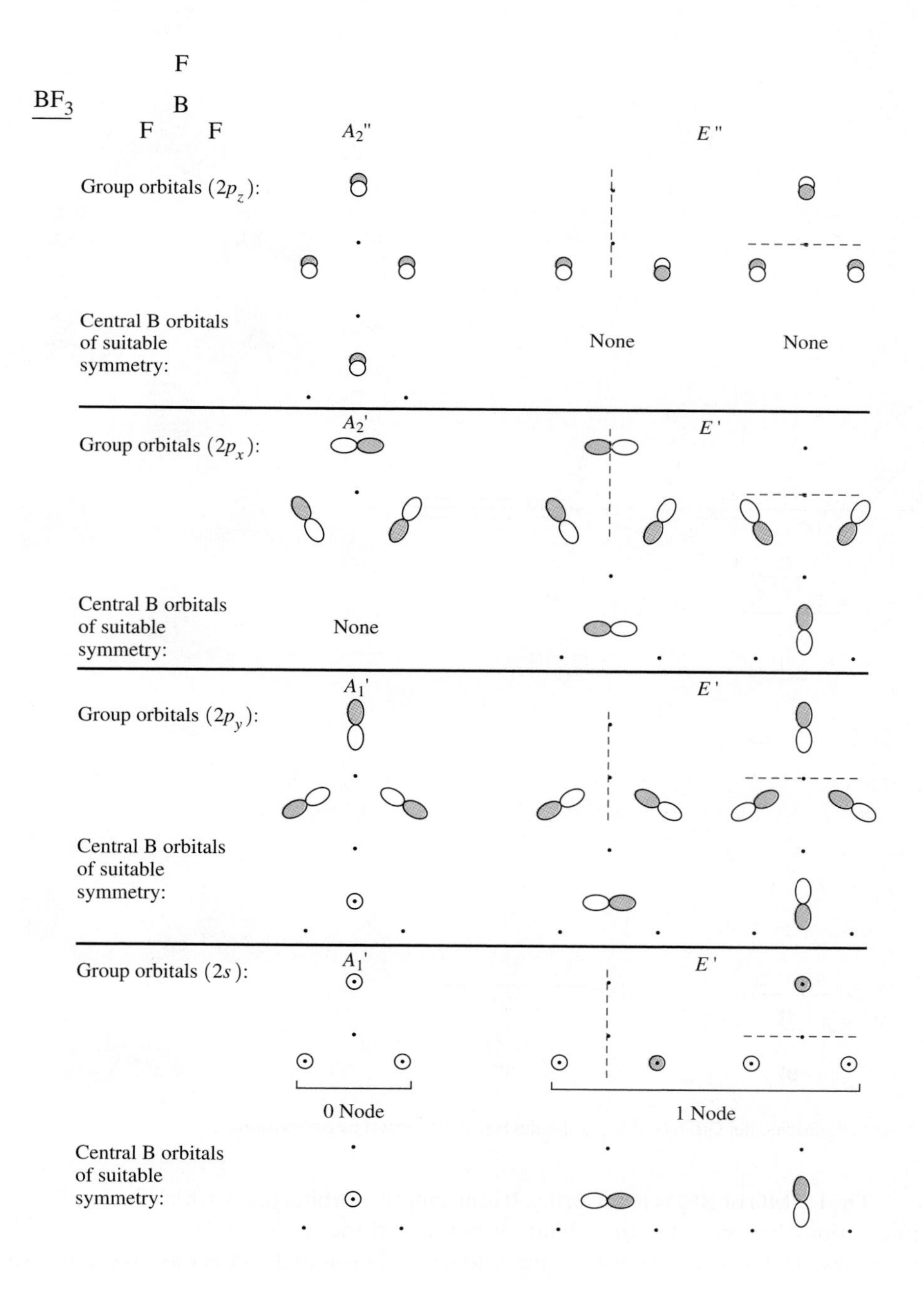

FIGURE 5.31 Group Orbitals for BF_3.

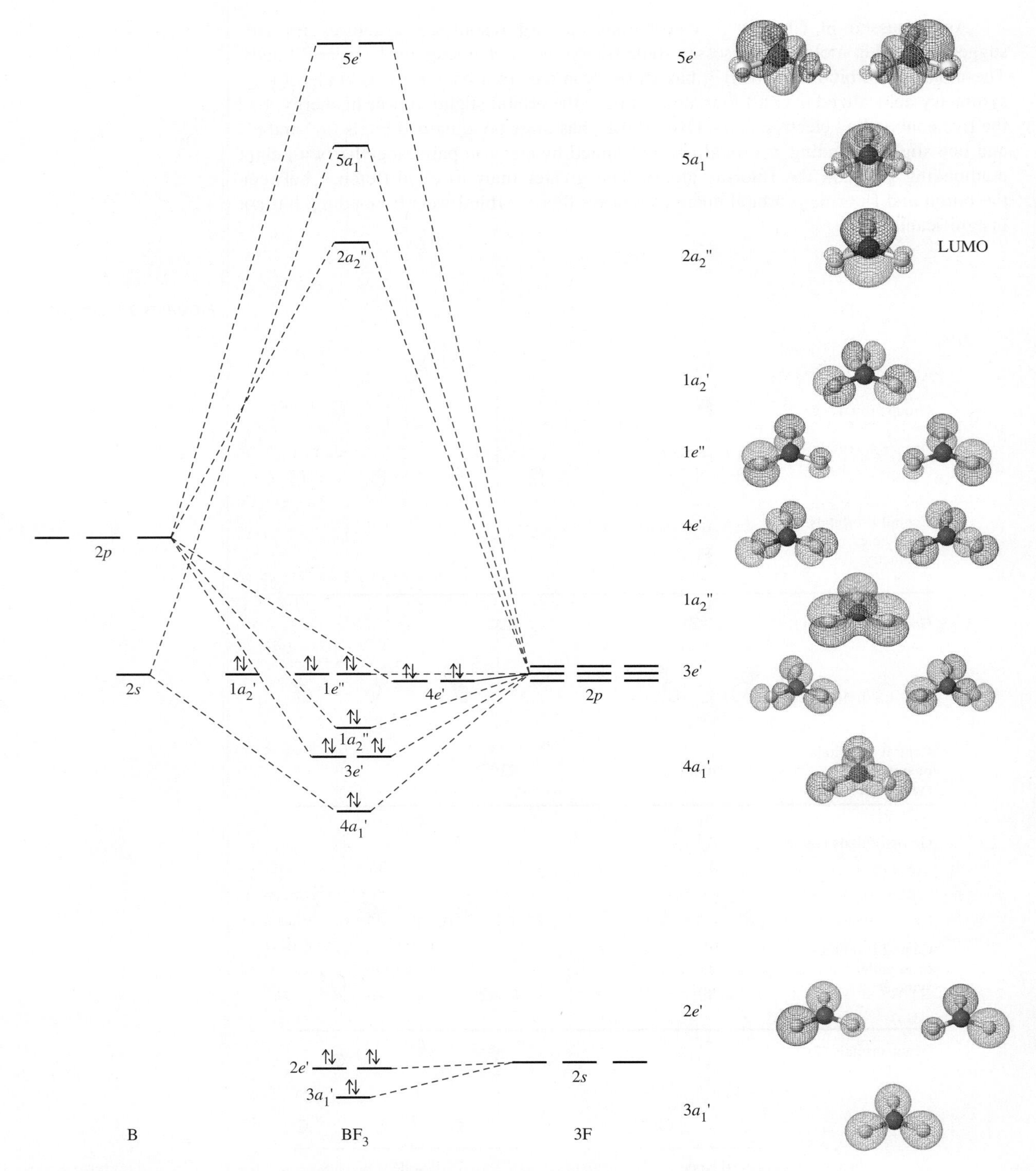

FIGURE 5.32 Molecular Orbitals of BF_3. (Molecular Orbitals of BF_3 by Kaitlin Hellie. Reprinted by permission.)

The LUMO of BF_3 is noteworthy. It is an empty π orbital (a_2''), which has antibonding interactions between the $2p_z$ orbital on boron and the $2p_z$ orbitals of the surrounding fluorines. This orbital has large, empty lobes on boron and can act as an electron-pair acceptor (for example, from the HOMO of NH_3) using these lobes in Lewis acid–base interactions.

Other trigonal species can be treated using this molecular orbital theory approach. The trigonal planar SO_3, NO_3^-, and CO_3^{2-} are isoelectronic with BF_3, with three electron pairs in σ-bonding orbitals and one pair in a π-bonding orbital with contributions from all four atoms. The resonance forms of these oxygen-containing species all predict delocalized π-electron density as an important aspect of their electronic ground states.

Because the extent of orbital overlap in π interactions is generally less than that in most σ interactions, a double bond composed of one filled σ orbital and one filled π orbital is not twice as strong as a single bond. Single bonds between the same atoms, but within different chemical environments in different molecules, can have widely different energies. An "average" C—C bond is usually described as having an approximate energy of 345 kJ/mol; a large number of molecules containing C—C bonds in different environments contribute to this average energy. These individual values vary tremendously; some are as low as 69 and as high as 649 kJ/mol.[14] The low value is for hexaphenyl ethane, $(C_6H_5)_3C—C(C_6H_5)_3$, and the high is for diacetylene, H—C≡C—C≡C—H, examples of extremes in steric crowding and bonding, respectively, adjacent to the C—C bond.

The group orbital approach described in this chapter, despite its modest use of group theory, conveniently provides a qualitatively useful description of bonding in simple molecules. Computational chemistry methods are necessary for more complex molecules and to obtain wave equations for the molecular orbitals. These advanced methods also apply molecular symmetry and group theory concepts.

While a qualitative group orbital approach does not allow the determination of the precise energies of molecular orbitals, we can generally place the MOs in approximate order on the basis of their shapes and expected orbital overlaps. Relatively nonbonding energy levels at intermediate energies can be particularly difficult to rank in energy. Proficiency at estimating orbital energies is only gained by experience, and by always attempting to correlate molecular orbitals and their energies with experimental properties of the molecules under consideration. Mastery of the interactions that define the basic molecular shapes described in this chapter provides a foundation that can be extended to other geometries.

5.4.7 Hybrid Orbitals

An oversimplified bonding model, and one engrained in the chemistry vernacular, that also employs molecular symmetry and group theory involves **hybrid orbitals**. The merits of hybrid atomic orbital theory have been recently debated.[15] While the modern inorganic literature almost exclusively employs molecular orbital theory to gain insights regarding structure and bonding, chemists evoke the hydridization model when it is convenient to help visualize an interaction. In the hybrid concept, the orbitals of the central atom are combined into sets of equivalent hybrids. These hybrid orbitals form bonds with orbitals of other atoms. The hybrid model is especially useful in organic chemistry where, for example, it predicts four equivalent C—H bonds in methane. A traditional criticism of the hybrid orbital description of methane was its alleged inconsistency with photoelectron spectroscopic data. However, the assumptions associated with these conclusions have been criticized, and the utility of the hybrid orbital model in the interpretation of methane's photoelectron spectrum has been validated.[16] Like all bonding models, hybrids are useful so long as their limits are recognized.

Hybrid orbitals are localized in space and are directional, pointing in a specific direction. In general, these hybrids point from a central atom toward surrounding atoms or lone pairs. Therefore, the symmetry properties of a set of hybrid orbitals will be identical to the properties of a set of vectors with origins at the nucleus of the central atom and pointing toward the surrounding atoms.

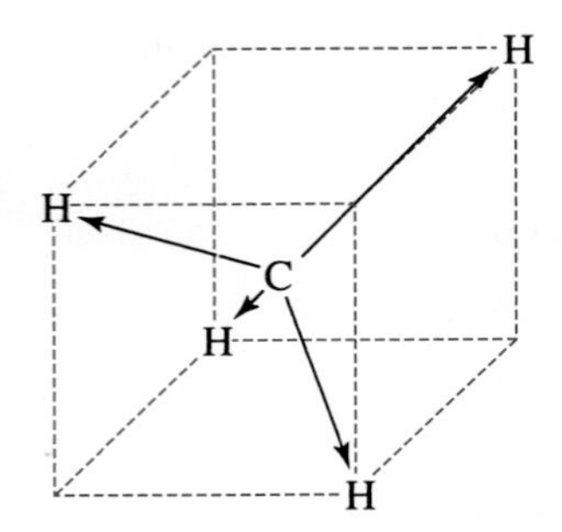

FIGURE 5.33 Bond Vectors in Methane.

For example, in methane, the vectors point at the corners of a tetrahedron or at alternate corners of a cube (**Figure 5.33**).

Using the T_d point group, these four vectors form the basis of a reducible representation. As usual, the character for each vector is 1 if it remains unchanged by the symmetry operation, and 0 if it changes position (reversing direction is not an option for hybrids). The reducible representation for these four vectors is $\Gamma = A_1 + T_2$:

T_d	E	$8C_3$	$3C_2$	$6S_4$	$6\sigma_d$		
Γ	4	1	0	0	2		
A_1	1	1	1	1	1		$x^2 + y^2 + z^2$
T_2	3	0	-1	-1	1	(x, y, z)	(xy, xz, yz)

In terms of hybrids, this means that the atomic orbitals of carbon used in the hybrids *must* have symmetry matching $A_1 + T_2$; more specifically, one orbital must match A_1, and a set of three (degenerate) orbitals must match T_2.

A_1, the totally symmetric representation, has the same symmetry as the $2s$ orbital of carbon; T_2 has the same symmetry as the three $2p$ orbitals taken together (x, y, z) or the d_{xy}, d_{xz}, and d_{yz} orbitals taken together. Because the $3d$ orbitals of carbon are at much higher energy, and are therefore a poor match for the energies of the $1s$ orbitals of the hydrogens, the hybridization for methane must be sp^3, combining four atomic orbitals—one $2s$ and three $2p$—into four equivalent hybrid orbitals, one directed toward each hydrogen atom.

Ammonia fits the same pattern. Bonding in NH_3 uses all the nitrogen valence orbitals, so the hybrids are sp^3, incorporating one s orbital and all three p orbitals, with overall tetrahedral symmetry. The predicted HNH angle is 109.5°, reduced to the actual 106.6° by repulsion from the lone pair, which is also viewed to occupy an sp^3 orbital.

There are two hybridization descriptions for the water molecule. A commonly taught idea in general chemistry is that the electron pairs around the oxygen atom in water can be considered as adopting nearly tetrahedral symmetry (counting the two lone pairs and the two bonds equally). All four valence orbitals of oxygen are used, and the hybrid orbitals are sp^3. The predicted bond angle is then the tetrahedral angle of 109.5°, compared with the experimental value of 104.5°. Repulsion by the lone pairs explains this smaller angle.

In another approach, which complements the molecular orbital description of Section 5.4.3, the bent planar shape indicates that the oxygen orbitals used in molecular orbital bonding in water are the $2s$, $2p_x$, and $2p_z$ (in the plane of the molecule). As a result, the hybrids could be described as sp^2, a combination of one s orbital and two p orbitals. Three sp^2 orbitals have trigonal symmetry and a predicted H—O—H angle of 120°, considerably larger than the experimental value. Repulsion by the lone pairs on the oxygen—one in an sp^2 orbital, one in the remaining p_y orbital—forces the angle to be smaller. Note that the $1b_2$ orbital in the molecular orbital picture of H_2O (Figure 5.28) is a filled nonbonding $2p_y$ orbital.

Similarly, CO_2 uses sp hybrids, and SO_3 uses sp^2 hybrids. Only the σ bonding is considered when determining the orbitals used in hybridization; p orbitals not used in the hybrids are available for π interactions. The number of atomic orbitals used in the hybrids is frequently the same as the steric number in the VSEPR method. The common hybrids are summarized in **Figure 5.34**. The group theory approach to hybridization is described in the following example.

Geometry	Atomic orbitals used	Hybrid orbitals
Linear	*s*, *p*	Two *sp* hybrid orbitals
Trigonal	*s*, *p*, *p*	Three sp^2 hybrid orbitals
Tetrahedral	*s*, *p*, *p*, *p*	Four sp^3 hybrid orbitals
Trigonal bipyramidal	*s*, *p*, *p*, *p*, *d*	Five dsp^3 hybrid orbitals
Octahedral	*s*, *p*, *p*, *p*, *d*, *d*	Six d^2sp^3 hybrid orbitals

FIGURE 5.34 Hybrid Orbitals. Each single hybrid has the general shape . The figures here show all the resulting hybrids combined, omitting the smaller lobe in the sp^3 and higher orbitals.

EXAMPLE 5.11

Determine the hybrid orbitals for boron in BF_3.

For a trigonal planar molecule such as BF_3, the orbitals likely to be involved in bonding are the $2s$, $2p_x$, and $2p_y$. This can be confirmed by finding the reducible representation in the D_{3h} point group of vectors pointing at the three fluorines and reducing it to its irreducible representations. The procedure is as follows.

Step 1 Determine the shape of the molecule (VSEPR), considering each sigma bond and lone pair on the central atom to be a vector pointing out from the center.

Step 2 Determine the reducible representation for the vectors, and deduce the irreducible representations that combine to form the reducible representation.

Step 3 The atomic orbitals that match the irreducible representations are those used in the hybrid orbitals.

Using the symmetry operations of the D_{3h} group, we find that the reducible representation $\Gamma = A_1' + E'$.

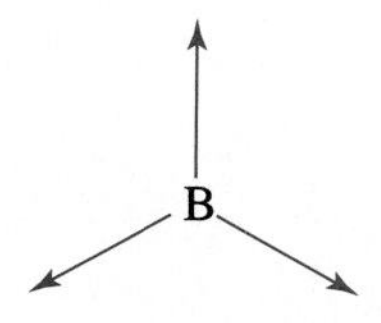

D_{3h}	E	$2C_3$	$3C_2$	σ_h	$2S_3$	$3\sigma_v$		
Γ	3	0	1	3	0	1		
A_1'	1	1	1	1	1	1		$x^2 + y^2, z^2$
E'	2	−1	0	2	−1	0	(x, y)	$(x^2 - y^2, xy)$

The atomic orbitals that combine to afford the hybrids must have the same symmetry as A_1' and E'. One orbital must have A_1' symmetry and two orbitals must have E' symmetry. We must therefore select one orbital with A_1' symmetry and one *pair* of orbitals that collectively have E' symmetry. Examining the functions in the right-hand column of the character table, we see that the *s* orbital (not listed, but understood to be present for the totally symmetric representation) and the d_{z^2} orbital possess A_1' symmetry. However, the 3*d* orbitals, the lowest energy *d* orbitals, are too high in energy for bonding in BF_3 compared with the 2*s*. Therefore, the 2*s* orbital is the A_1' contributor.

The functions listed for E' symmetry match the p_x, p_y set and the $d_{x^2-y^2}$, d_{xy} set. The *d* orbitals are energetically inaccessible; the $2p_x$ and $2p_y$ orbitals are used by the central atom.*

Overall, the orbitals used in the hybridization are the 2*s*, $2p_x$, and $2p_y$ orbitals of boron, comprising the familiar sp^2 hybrids. The difference between this approach and the molecular orbital approach is that these orbitals are combined to form the hybrids before considering their interactions with the fluorine orbitals. Because the overall symmetry is trigonal planar, the resulting hybrids must have that same symmetry, so the three sp^2 orbitals point at the three corners of a triangle, and each interacts with a fluorine *p* orbital to form the three σ bonds. The $2p_z$ orbital is not involved in the bonding and, according to the hybrid approach, is empty; this orbital serves as an acceptor in acid–base reactions.

EXERCISE 5.11 Determine the types of hybrid orbitals that are consistent with the symmetry of the central atom in

a. PF_5
b. $[PtCl_4]^{2-}$, a square planar ion

The procedure for determining hybrids is in some respects similar to that used in the molecular orbital approach. Hybridization uses vectors pointing toward the outlying atoms and usually deals only with σ bonding. Once the σ hybrids are known, π bonding is added, using orbitals that do not participate in the hybridization. It is also possible to use hybridization techniques for π bonding.[17] As an approximate approach, hybridization may be quicker than the molecular orbital approach, because the molecular orbital approach uses all the atomic orbitals of the atoms and includes both σ and π bonding directly. Molecular orbital theory has gained prominence on the basis of its ability to predict the relative energies of electrons in molecules much more successfully than the hybrid orbital approach.[17]

EXERCISE 5.12

Determine the reducible representation for all the σ bonds, reduce it to its irreducible representations, and determine the sulfur orbitals used in bonding for $SOCl_2$.

---

*A combination of one *p* orbital and one *d* orbital cannot be chosen, because orbitals in parentheses must always be taken together.

References

1. F. Luo, G. C. McBane, G. Kim, C. F. Giese, W. R. Gentry, *J. Chem. Phys.*, **1993**, *98*, 3564; see also L. L. Lohr, S. M. Blinder, *J. Chem. Educ.*, **2007**, *84*, 860, and references cited therein.
2. M. Atoji, *J. Chem. Phys.*, **1961**, *35*, 1950.
3. J. Overend, H. W. Thompson, *Proc. R. Soc. London*, **1954**, *A234*, 306.
4. N. Lopez, D. J. Graham, R. McGuire, Jr., G. E. Alliger, Y. Shao-Horn, C. C. Cummins, D. G. Nocera, *Science*, **2012**, *335*, 450.
5. L. Pauling, *The Nature of the Chemical Bond*, 3rd ed., Cornell University Press, Ithaca, NY, 1960, pp. 340–354.
6. E. A. V. Ebsworth, D. W. H. Rankin, and S. Cradock, *Structural Methods in Inorganic Chemistry*, 2nd ed., CRC Press, Boca Raton, FL, 1991, pp. 255–279. Pages 274 and 275 discuss the spectra of N_2 and O_2.
7. R. Stowasser, R. Hoffmann, *J. Am. Chem. Soc.*, **1999**, *121*, 3414.
8. R. S. Drago, *Physical Methods in Chemistry*, 2nd ed., Saunders College Publishing, Philadelphia, 1992, pp. 671–677.
9. J. H. Clark, J. Emsley, D. J. Jones, R. E. Overill, *J. Chem. Soc.*, **1981**, 1219; J. Emsley, N. M. Reza, H. M. Dawes, M. B. Hursthouse, *J. Chem. Soc. Dalton Trans.*, **1986**, 313; N. Elghobashi, L. González, *J. Chem. Phys.*, **2006**, *124*, 174308, and references cited therein.
10. M. Mautner, *J. Am. Chem. Soc.*, **1984**, *106*, 1257.
11. F. A. Cotton, *Chemical Applications of Group Theory*, 3rd ed., John Wiley & Sons, New York, 1990, pp. 133–188.
12. D. J. Willock, *Molecular Symmetry*, John Wiley & Sons, Chichester, UK, 2009, pp. 195–212.
13. R. Stowasser, R. Hoffmann, *J. Am. Chem. Soc.*, **1999**, *121*, 3414.
14. A. A. Zavitsas, *J. Phys. Chem. A*, **2003**, *107*, 897.
15. (a) A. Grushow, *J. Chem. Educ.*, **2011**, *88*, 860. (b) R. L. DeKock, J.R. Strikwerda, *J. Chem. Educ.*, **2012**, 10.1021/ed200472t. (c) D. G. Truhlar, *J. Chem. Educ.*, **2012**, 10.1021/ed200565h.
16. C. R. Landis, F. Weinhold, *J. Chem. Educ.*, **2012**, 10.1021/ed200491q.
17. F. A. Cotton, *Chemical Applications of Group Theory*, 3rd ed., John Wiley & Sons, New York, 1990, pp. 227–230.

General References

There are many books describing bonding and molecular orbitals, with levels ranging from those even more descriptive and qualitative than the treatment in this chapter to those designed for the theoretician interested in the latest methods. A classic that starts at the level of this chapter and includes many more details is R. McWeeny's revision of *Coulson's Valence*, 3rd ed., Oxford University Press, Oxford, 1979. A different approach that uses the concept of generator orbitals is that of J. G. Verkade in *A Pictorial Approach to Molecular Bonding and Vibrations*, 2nd ed., Springer-Verlag, New York, 1997. The group theory approach in this chapter is similar to that of F. A. Cotton in *Chemical Applications of Group Theory*, 3rd ed., John Wiley & Sons, New York, 1990. A more recent book that extends the description is Y. Jean and F. Volatron, *An Introduction to Molecular Orbitals*, translated and edited by J. K. Burdett, Oxford University Press, Oxford, 1993. J. K. Burdett's *Molecular Shapes*, John Wiley & Sons, New York, 1980, and B. M. Gimarc's, *Molecular Structure and Bonding*, Academic Press, New York, 1979, are both good introductions to the qualitative molecular orbital description of bonding.

This is the first chapter that includes problems involving molecular modeling software. Discussion of such software is beyond the scope of this text. E. G. Lewars, *Computational Chemistry*, 2nd ed., Springer, New York, 2011 provides a readable introduction to the theory and applications of molecular modeling. In addition, examples of uses of molecular modeling in chemistry curricula are discussed in L. E. Johnson, T. Engel, *J. Chem. Educ.*, **2011**, *88*, 569 and references cited therein.

Problems

5.1 Expand the list of orbitals considered in Figures 5.2 and 5.3 by using all three *p* orbitals of atom A and all five *d* orbitals of atom B. Which of these have the necessary match of symmetry for bonding and antibonding orbitals? These combinations are rarely seen in simple molecules but can be important in transition metal complexes.

5.2 On the basis of molecular orbitals, predict the shortest bond, and provide a brief explanation.

a. Li_2^+ Li_2

b. F_2^+ F_2

c. He_2^+ HHe^+ H_2^+

5.3 On the basis of molecular orbitals, predict the weakest bond, and provide a brief explanation.

a. P_2 S_2 Cl_2

b. S_2^+ S_2 S_2^-

c. NO^- NO NO^+

5.4 Compare the bonding in O_2^{2-}, O_2^-, and O_2. Include Lewis structures, molecular orbital structures, bond lengths, and bond strengths in your discussion.

5.5 Although the peroxide ion, O_2^{2-}, and the acetylide ion, C_2^{2-}, have long been known, the diazenide ion N_2^{2-} has only been prepared much more recently. By comparison with the other diatomic species, predict the bond order, bond distance, and number of unpaired electrons for N_2^{2-}. (See G. Auffermann, Y. Prots, R. Kniep, *Angew. Chem., Int. Ed.*, **2001**, *40*, 547.)

5.6 High-resolution photoelectron spectroscopy has provided information on the energy levels and bond distance in the ion Ar_2^+. Prepare a molecular orbital energy-level diagram for this ion. How would you expect the bond distance in Ar_2^+ to compare with 198.8 pm, the bond distance in Cl_2? (See A. Wüst, F. Merkt, *J. Chem. Phys.*, **2004**, *120*, 638.)

5.7 **a.** Prepare a molecular orbital energy-level diagram for NO, showing clearly how the atomic orbitals interact to form MOs.
b. How does your diagram illustrate the difference in electronegativity between N and O?
c. Predict the bond order and the number of unpaired electrons.
d. NO^+ and NO^- are also known. Compare the bond orders of these ions with the bond order of NO. Which of the three would you predict to have the shortest bond? Why?

5.8 **a.** Prepare a molecular orbital energy-level diagram for the cyanide ion. Use sketches to show clearly how the atomic orbitals interact to form MOs.
b. What is the bond order for cyanide, and how many unpaired electrons does cyanide have?
c. Which molecular orbital of CN^- would you predict to interact most strongly with a hydrogen 1*s* orbital to form an H—C bond in the reaction $CN^- + H^+ \rightarrow HCN$? Explain.

5.9 NF is a known molecule!
a. Construct a molecular orbital energy-level diagram for NF, being sure to include sketches that show how the valence orbitals of N and F interact to form molecular orbitals.
b. What is the most likely bond order for NF?
c. What are the point groups of the molecular orbitals of this molecule?
(See D. J. Grant, T-H. Wang, M. Vasiliu, D. A. Dixon, K. O. Christe, *Inorg. Chem.* **2011**, *50*, 1914 for references and theoretical calculations of numerous small molecules and ions having formula N_xF_y.)

5.10 The hypofluorite ion, OF^-, can be observed only with difficulty.
a. Prepare a molecular orbital energy level diagram for this ion.
b. What is the bond order, and how many unpaired electrons are in this ion?
c. What is the most likely position for adding H^+ to the OF^- ion? Explain your choice.

5.11 Reaction of KrF_2 with AsF_5 at temperatures between −78 and −53 °C yields $[KrF][AsF_6]$, a compound in which KrF^+ interacts strongly with AsF_6^- through a fluorine bridge, as shown. Would you predict the Kr—F bond to be shorter in KrF^+ or in KrF_2? Provide a brief explanation. (See J. F. Lehmann, D. A. Dixon, G. J. Schrobilgen, *Inorg. Chem.*, **2001**, *40*, 3002.)

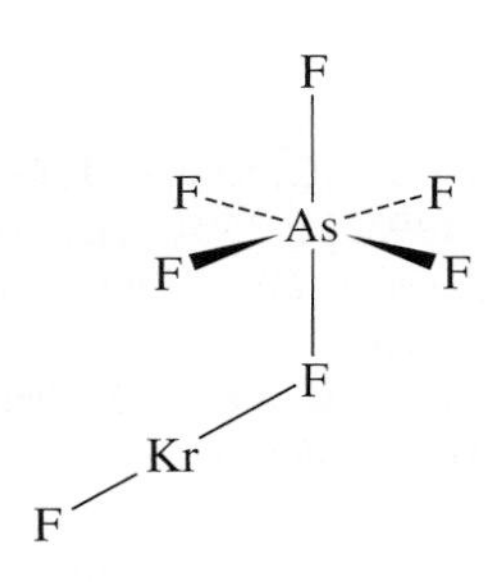

5.12 Although KrF^+ and XeF^+ have been studied, $KrBr^+$ has not yet been prepared. For $KrBr^+$:
a. Propose a molecular orbital diagram, showing the interactions of the valence shell *s* and *p* orbitals to form molecular orbitals.
b. Toward which atom would the HOMO be polarized? Why?
c. Predict the bond order.
d. Which is more electronegative, Kr or Br? Explain your reasoning.

5.13 Prepare a molecular orbital energy level diagram for SH^-, including sketches of the orbital shapes and the number of electrons in each of the orbitals. If a program for calculating molecular orbitals is available, use it to confirm your predictions or to explain why they differ.

5.14 Methylene, CH_2, plays an important role in many reactions. One possible structure of methylene is linear.
a. Construct a molecular orbital energy-level diagram for this species. Include sketches of the group orbitals, and indicate how they interact with the appropriate orbitals of carbon.
b. Would you expect linear methylene to be diamagnetic or paramagnetic?

5.15 Beryllium hydride, BeH_2, is linear in the gas phase.
a. Construct a molecular orbital energy level diagram for BeH_2. Include sketches of the group orbitals, and indicate how they interact with the appropriate orbitals of Be.
b. If you have worked Problem 5.14, compare the results of these two problems.

5.16 In the gas phase, BeF_2 forms linear monomeric molecules. Prepare a molecular orbital energy-level diagram for BeF_2, showing clearly which atomic orbitals are involved in bonding and which are nonbonding.

5.17 For the compound XeF_2 do the following:
a. Sketch the valence shell group orbitals for the fluorine atoms (with the *z* axes collinear with the molecular axis).
b. For each of the group orbitals, determine which outermost *s*, *p*, and *d* orbitals of xenon are of suitable symmetry for interaction and bonding.

5.18 TaH_5 has been predicted to have C_{4v} symmetry, with a calculated axial H—Ta—H angle of approximately 117.5°. Using the six-step approach described in Section 5.4.2, describe the bonding in TaH_5 on the basis of matching group orbitals and central atom orbitals according to their symmetry. (See C. A. Bayse, M. B. Hall, *J. Am. Chem. Soc.*, **1999**, *121*, 1348.)

5.19 Describe the bonding in ozone, O_3, on the basis of matching group orbitals and central-atom orbitals according to their symmetry. Include both σ and π interactions, and try to put the resulting orbitals in approximate order of energy.

5.20 Describe the bonding in SO_3 by using group theory to find the molecular orbitals. Include both σ and π interactions, and try to put the resulting orbitals in approximate order of energy. (The actual results are more complex because of mixing of orbitals, but a simple description can be found by the methods given in this chapter.)

5.21 The ion H_3^+ has been observed, but its structure has been the subject of some controversy. Prepare a molecular orbital energy level diagram for H_3^+, assuming a cyclic structure. (The same problem for a linear structure is given in Exercise 5.4 in Section 5.4.1.)

5.22 Use molecular orbital arguments to explain the structures of SCN^-, OCN^-, and CNO^-, and compare the results with the electron-dot pictures of Chapter 3.

5.23 Thiocyanate and cyanate ions both bond to H^+ through the nitrogen atoms (HNCS and HNCO), whereas SCN^- forms bonds with metal ions through either nitrogen or sulfur, depending on the rest of the molecule. What does this suggest about the relative importance of S and N orbitals in the MOs of SCN^-? (Hint: See the discussion of CO_2 bonding in Section 5.4.2.)

5.24 The thiocyanate ion, SCN^-, can form bonds to metals through either S or N (see Problem 5.23). What is the likelihood of cyanide, CN^-, forming bonds to metals through N as well as C?

5.25 The isomeric ions NSO^- (thiazate) and SNO^- (thionitrite) ions have been reported. (S. P. So, *Inorg. Chem.*, **1989**, *28*, 2888).

a. On the basis of the resonance structures of these ions, predict which would be more stable.

b. Sketch the approximate shapes of the π and π^* orbitals of these ions.

c. Predict which ion would have the shorter N—S bond and which would have the higher energy N—S stretching vibration? (Stronger bonds have higher energy vibrations.)

5.26 Apply the projection operator method to derive the group orbital SALCs for H_2O given in Section 5.4.3. Confirm using the squares of the coefficients that the group orbital wave function equations are normalized and that each 1*s* orbital contributes equally to the two group orbitals.

5.27 Apply the projection operator method to derive the group orbital SALCs for BF_3 on the basis of the irreducible representations given in Figure 5.31 for sets of $2s$, $2p_x$, $2p_y$, and $2p_z$ orbitals, respectively. Employ a set of three identical orbitals where all have the same bias (i.e., the group orbitals with A_1', A_2', and A_2'' symmetry in Figure 5.31) as a starting point. For each determination, provide a table like that in Section 5.4.4 to tabulate wave function coefficients, their squares, and how these values simultaneously satisfy the normalization requirement and confirm that each atomic orbital contributes equally to each set of group orbitals.

5.28 A set of four group orbitals derived from four 3*s* atomic orbitals is necessary to examine the bonding in $[PtCl_4]^-$, a square planar complex. Deduce the wave function equations for these four SALCs using the 3*s* labeling scheme specified, starting with the irreducible representations for these group orbitals. Using sketches of the deduced orbitals, symmetry characteristics of the representations, and a coefficient table like that in Section 5.4.4, deduce the SALCs not derived initially from the character table analysis. Provide normalized equations and a sketch for each group orbital.

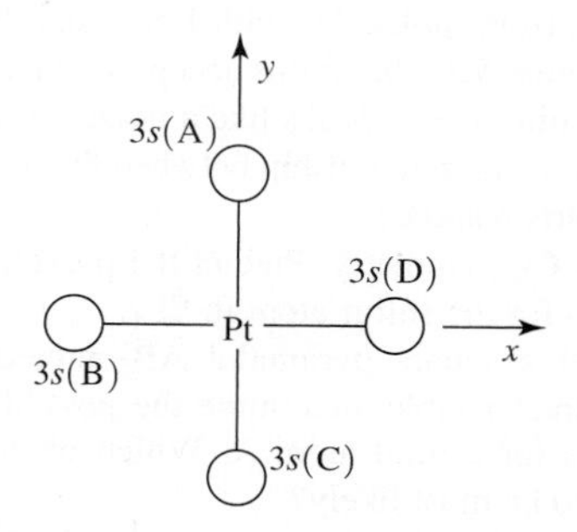

5.29 The projection operator method has applications beyond the deduction of group orbital SALCs. Deduce the wave function equations for the six π molecular orbitals of benzene, using the labels specified for each $2p_z$ orbital. First, derive initial SALCs using *each* representation of the D_{6h} point group; some combinations will afford zero. Using sketches of the deduced orbitals, symmetry characteristics of the representations, and a coefficient table like that in Section 5.4.4, deduce the SALCs not derived initially from the character table analysis. Provide normalized equations and a sketch for each π molecular orbital.

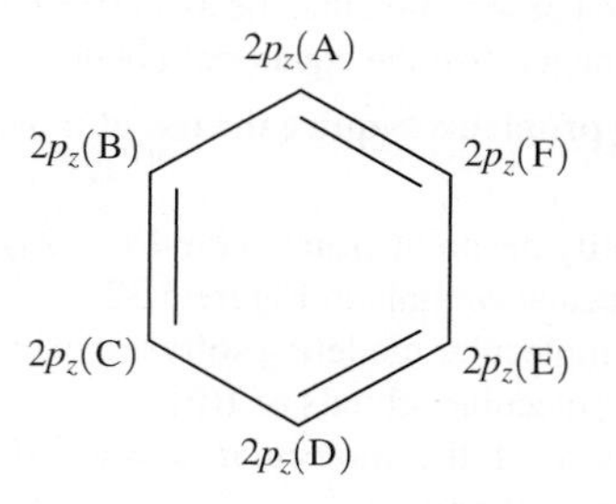

5.30 Although the Cl_2^+ ion has not been isolated, it has been detected in the gas phase by UV spectroscopy. An attempt to prepare this ion by reaction of Cl_2 with IrF_6 yielded not Cl_2^+, but the rectangular ion Cl_4^+. (See S. Seidel, K. Seppelt, *Angew. Chem., Int. Ed.*, **2000**, *39*, 3923.)

a. Compare the bond distance and bond energy of Cl_2^+ with Cl_2.

b. Account for the bonding in Cl_4^+. This ion contains two short Cl—Cl bonds and two much longer ones. Would you expect the shorter Cl—Cl distances in Cl_4^+ to be longer or shorter than the Cl—Cl distance in Cl_2? Explain.

5.31 BF_3 is often described as a molecule in which boron is electron deficient, with an electron count of six. However, resonance structures can be drawn in which boron has an octet, with delocalized π electrons.

a. Draw these structures.

b. Find the molecular orbital in Figure 5.32 that shows this delocalization and explain your choice.

c. BF_3 is *the* classic Lewis acid, accepting a pair of electrons from molecules with lone pairs. Find the orbital in Figure 5.32 that is this acceptor; explain your choice, including why it looks like a good electron acceptor.

d. What is the relationship between the orbitals identified in Parts b and c?

5.32 SF_4 has C_{2v} symmetry. Predict the possible hybridization schemes for the sulfur atom in SF_4.

5.33 Consider a square pyramidal AB_5 molecule. Using the C_{4v} character table, determine the possible hybridization schemes for central atom A. Which of these would you expect to be most likely?

5.34 In coordination chemistry, many square-planar species are known (for example, $[PtCl_4]^{2-}$). For a square planar molecule, use the appropriate character table to determine the types of hybridization possible for a metal surrounded in a square-planar fashion by four ligands; consider hybrids used in σ bonding only.

5.35 For the molecule PCl_5:

a. Using the character table for the point group of PCl_5, determine the possible type(s) of hybrid orbitals that can be used by P in forming σ bonds to the five Cl atoms.

b. What type(s) of hybrids can be used in bonding to the axial chlorine atoms? To the equatorial chlorine atoms?

c. Considering your answer to part b, explain the experimental observation that the axial P—Cl bonds (219 pm) are longer than the equatorial bonds (204 pm).

The following problems require the use of molecular modeling software.

5.36 **a.** Identify the point group of the $1a_2''$, $2a_2''$, $1a_2'$, and $1e''$ molecular orbitals in Figure 5.32.

b. Use molecular modeling software to calculate and view the molecular orbitals of BF_3.

c. Do any of the molecular orbitals show interactions between B and F?

d. Print out the contributions of the atomic orbitals to the $3a_1'$, $4a_1'$, $1a_2''$, $1a_2'$, and $2a_2''$ molecular orbitals, confirming (if you can) the atomic orbital combinations shown in Figure 5.32.

5.37 The ions and molecules NO^+, CN^-, CO, and N_2 form an isoelectronic series. The changing nuclear charges will also change the molecular energy levels of the orbitals formed from the $2p$ atomic orbitals (1π, 3σ, and $1\pi^*$). Use molecular modeling software for the following:

a. Calculate and display the shapes of these three molecular orbitals for each species (CO and N_2 are included in this chapter).

b. Compare the shapes of each of the orbitals for each of the species (for example, the shapes of the 1π orbitals for each). What trends do you observe?

c. Compare the energies of each of the orbitals. For which do you see evidence of mixing?

5.38 Molecular modeling software is typically capable of calculations on molecules that are hypothetical, even seemingly bizarre, in their structures. Beginning with N_2, calculate and display molecular orbitals of the isoelectronic CO, BF, and BeNe (which is truly hypothetical!). Compare the shapes of the matching molecular orbital in this series What trends do you observe?

5.39 Calculate and display the orbitals for the linear molecule BeH_2. Describe how they illustrate the interaction of the outer group orbitals with the orbitals on the central atom. Compare your results with the answer to Problem 5.15.

5.40 Calculate and display the orbitals for the linear molecule BeF_2. Compare the orbitals and their interactions with those of BeH_2 from Problem 5.39. In particular, indicate the outer group orbitals that do not interact with orbitals on the central atom.

5.41 The azide ion, N_3^-, is another linear triatomic species. Calculate and display the orbitals for this ion, and compare the three highest energy occupied orbitals with those of BeF_2. How do the outer atom group orbitals differ in their interactions with the central atom orbitals? How do the orbitals compare with the CO_2 orbitals discussed in Section 5.4.2?

5.42 Calculate and display the molecular orbitals of the ozone molecule, O_3. Which orbitals show π interactions? Compare your results with your answer to Problem 5.19.

5.43 **a.** Calculate and display the molecular orbitals for linear and cyclic H_3^+.

b. Which species is more likely to exist (i.e., which is more stable)?

5.44 Diborane, B_2H_6, has the structure shown.

a. Using the point group of the molecule, create a representation using the $1s$ orbitals on the hydrogens as a basis. Reduce this representation, and sketch group orbitals matching each of the irreducible representations. (Suggestion: Treat the bridging and terminal hydrogens separately.)

b. Calculate and display the molecular orbitals. Compare the software-generated images with the group orbital sketches from part a, and explain how hydrogen can form "bridges" between two B atoms. (This type of bonding is discussed in Chapter 8.)

Chapter 6

Acid–Base and Donor–Acceptor Chemistry

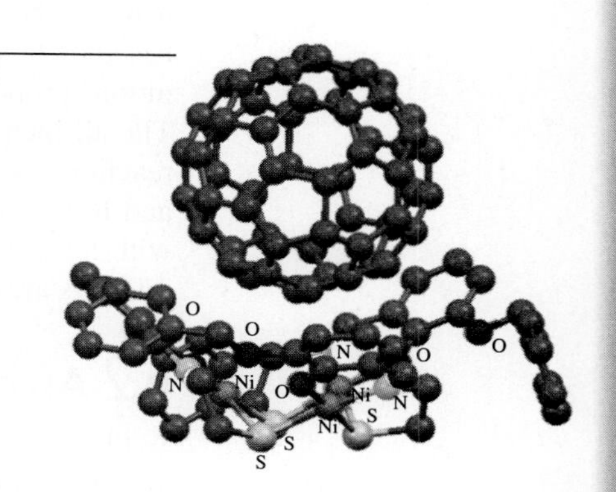

6.1 Acid–Base Models as Organizing Concepts

A long-standing chemical objective is to organize reactions by using models to account for trends and gain insight into what properties of reactants are prerequisites for chemical change. Analyzing trends among similar reactions permits discovery of structure–function relationships (for example, how do molecular geometry and electronic structure influence reactivity?) and guides the design of molecules for practical use.

Classifying substances as acids and bases has been important since ancient times; alchemists used neutralization—the ubiquitous reaction of an acid and base to form salt and water—to compile observations about different substances that engaged in similar reactions. Without modern structural analysis tools, such as X-ray crystallography and NMR spectroscopy, alchemists used their senses: they observed the tastes of acids (sour) and bases (bitter) and color changes of indicators. Many acid–base definitions have been devised, but only a few have been widely adopted.

This chapter discusses the major acid–base models and their application in inorganic chemistry. After a historical introduction (**Section 6.1.1**), the models are presented in the rough order of their development. Among these are the ones attributed to Arrhenius (**Section 6.2**), Brønsted–Lowry (**Section 6.3**), and Lewis (**Section 6.4**). These sections emphasize the challenges associated with quantifying acidity and basicity, and relationships between acid/base strength and molecular structure. The 1960s application of molecular orbitals (i.e., HOMO/LUMO interactions) to frame Lewis acid–base reactions (**Section 6.4.1**) permeates inorganic chemistry and dramatically expands the perspective on what constitutes an acid–base reaction. Extension of HOMO/LUMO interactions to intermolecular forces is covered in **Section 6.5**. For example, acid–base molecular orbital concepts permit rationalization of host–guest interactions (**Section 6.5.2**) involving C_{60} (figure above*). Finally, the concept of "hard" and "soft" acids and bases is discussed in **Section 6.6**.

6.1.1 History of Acid–Base Models

The history of chemistry is marked with many acid–base models. A limitation of most early models is that they are applicable to only specific classes of compounds or a narrow set of conditions. One such limiting idea in the eighteenth century was that all acids contained oxygen; oxides of nitrogen, phosphorus, sulfur, and the halogens all form aqueous acids. However, by the early nineteenth century, this definition was regarded as too narrow; many compounds had been discovered that did not contain oxygen but showed behavior associated

---

*Molecular structure drawing created with CIF data from E. C. Constable, G. Zhang, D. Häussinger, C. E. Housecroft, J. A. Zampese, *J. Am. Chem. Soc.*, **2011**, *133*, 10776, with hydrogen atoms omitted for clarity.

with acids. By 1838, Liebig broadened the definition of acids to "compounds containing hydrogen, in which the hydrogen can be replaced by a metal."[1] The early twentieth century featured the introduction of some models rarely evoked today. The Lux–Flood definition[2] is based on oxide (O^{2-}) as the unit transferred between acids and bases. The Usanovich definition[3] proposes that classification of an acid–base reaction only requires salt formation. The all-inclusive aspect of the Usanovich definition, which included oxidation–reduction reactions, was criticized as too broad. The electrophile–nucleophile approach of Ingold[4] and Robinson,[5] part of the organic chemistry vernacular, is essentially the Lewis theory with terminology related to reactivity: electrophilic reagents are acids and nucleophilic reagents are bases. **Table 6.1** summarizes the history of acid–base definitions.

6.2 Arrhenius Concept

Acid–base chemistry was first satisfactorily explained in molecular terms after Ostwald and Arrhenius established the existence of ions in aqueous solution in the late nineteenth century (Arrhenius received the 1903 Nobel Prize in Chemistry). **Arrhenius acids** yield hydrogen ions* in aqueous solution; **Arrhenius bases** yield hydroxide ions in aqueous solution. The neutralization of hydrogen and hydroxide ions to form water, the net ionic equation of

$$HCl(aq) + NaOH(aq) \xrightarrow{H_2O(l)} NaCl(aq) + H_2O(l)$$

is a classic Arrhenius acid–base reaction, with a salt (in this case NaCl) and water as products. The Arrhenius concept is useful in aqueous solutions, but does not apply to the many reactions that occur in other inorganic solvents, organic solvents, the gas phase, or the solid state.

TABLE 6.1 Acid–Base Definition History

		Definitions		Examples	
Description	Date	Acid	Base	Acid	Base
Liebig	~1776	Oxide of N, P, S	Reacts with acid	SO_3	NaOH
	1838	H replaceable by metal	Reacts with acid	HNO_3	NaOH
Arrhenius	1894	Forms hydronium ion	Forms hydroxide ion	HCl	NaOH
Brønsted–Lowry	1923	Hydrogen ion donor	Hydrogen ion acceptor	H_3O^+	H_2O
				H_2O	OH^-
				NH_4^+	NH_3
Lewis	1923	Electron-pair acceptor	Electron-pair donor	Ag^+	NH_3
Ingold–Robinson	1932	Electrophile (electron-pair acceptor)	Nucleophile (electron-pair donor)	BF_3	NH_3
Lux–Flood	1939	Oxide ion acceptor	Oxide ion donor	SiO_2	CaO
Usanovich	1939	Electron acceptor	Electron donor	Cl_2	Na
Solvent system	1950s	Solvent cation	Solvent anion	BrF_2^+	BrF_4^-
Frontier orbitals	1960s	LUMO of acceptor	HOMO of donor	BF_3	NH_3

*The original Arrhenius concept did not include solvation. In modern practice, H_3O^+, *the hydronium* ion, is commonly used as an abbreviation of $H^+(aq)$, and this is the practice in this book. The International Union of Pure and Applied Chemistry (IUPAC) recommends *oxonium* for H_3O^+. Frequently the shorthand H^+ notation is used, for which the IUPAC recommends the term *hydrogen ion*, rather than *proton*.

6.3 Brønsted–Lowry Concept

Brønsted[6] and Lowry[7] defined an **acid** as *a species with a tendency to lose a hydrogen ion* and a **base** as *a species with a tendency to gain a hydrogen ion*. For example, the Brønsted–Lowry reaction of a strong acid with the weak base nitrite in aqueous solution would be:

$$\underbrace{H_3O^+(aq) + NO_2^-(aq) \rightleftharpoons H_2O(l)}_{\text{conjugate pair}} + HNO_2(aq)$$

(conjugate pairs: H_3O^+/H_2O and NO_2^-/HNO_2)

The strong acid hydronium loses (donates) H^+ to the base NO_2^- to form H_2O (the conjugate base of H_3O^+) and HNO_2 (the conjugate acid of NO_2^-).

In principle, the Brønsted–Lowry acidity of a general acid HB could be measured in the gas phase by determining the equilibrium constant of

$$HB(g) \rightleftharpoons H^+(g) + B^-(g)$$

However, typically large and positive ΔG° values associated with gaseous ionization render this acidity measurement problematic. Examination of the explicit HB ionization above is impossible in solution; independent H^+ ions do not exist in solution because of their strong interaction with solvent molecules.

The Brønsted–Lowry model explicitly dictates that acids generate different conjugate acids depending on the solvent used. For example, H_2SO_4 yields H_3O^+ on ionization in water but $H_3SO_4^+$ on ionization in sulfuric acid. Acid strength in solution is inherently tied to the solvent. The acidity or basicity ranking of a series of solutes determined in one solvent may therefore be different in another solvent. The Brønsted–Lowry strategy is to compare *conjugate acids and bases*, species differing only in the presence or absence of a proton, and describe reactions as occurring between a stronger acid and a stronger base to form a weaker acid and a weaker base, as in the example of H_3O^+ reacting with NO_2^-. A stronger acid has a greater tendency to transfer a hydrogen ion than a weaker acid, and a stronger base has a greater tendency to accept a proton than a weaker base. The equilibrium always favors the formation of *weaker* acids and bases. In the example above, H_3O^+ is a stronger acid than HNO_2, and NO_2^- is a stronger base than H_2O; the equilibrium lies to the right.

Brønsted–Lowry examples can be shown in **amphoteric** solvents that can function as an acid or a base and whose conjugates play vital roles. Examples of amphoteric solvents are in **Table 6.2**.

TABLE 6.2 Properties of Amphoteric Solvents

Solvent	Acid Cation	Base Anion	pK_{ion}(25° C)	Boiling Point (°C)
Sulfuric acid, H_2SO_4	$H_3SO_4^+$	HSO_4^-	3.4 (10°)	330
Hydrogen fluoride, HF	H_2F^+	HF_2^-	~12 (0°)	19.5
Water, H_2O	H_3O^+	OH^-	14.0	100
Acetic acid, CH_3COOH	$CH_3COOH_2^+$	CH_3COO^-	14.45	118.2
Methanol, CH_3OH	$CH_3OH_2^+$	CH_3O^-	16.6	64.7
Ammonia, NH_3	NH_4^+	NH_2^-	27	−33.4
Acetonitrile, CH_3CN	CH_3CNH^+	CH_2CN^-	34.4	81

Data from W. L. Jolly, *The Synthesis and Characterization of Inorganic Compounds*, Prentice Hall, Englewood Cliffs, NJ, 1970, pp. 99–101 and M. Rosés, *Anal. Chim. Acta*, **1993**, *276*, 223 (pK_{ion} for acetonitrile).

EXERCISE 6.1

Calculate the concentration of CH_3CNH^+ in CH_3CN at 25° C.

Consider the net ionic equation for the reaction of a strong acid (HCl) and strong base (NaOH) in aqueous solution. In water, the acid H_3O^+ from the ionization of HCl and base OH^- from the dissociation of NaOH engage in Brønsted–Lowry proton transfer to form H_2O. The net ionic equation features the conjugate acid and base of water as reactants:

$$H_3O^+(aq) + OH^-(aq) \rightleftharpoons H_2O(l) + H_2O(l)$$

This class of Brønsted–Lowry reaction can be carried out in any amphoteric solvent. For example, liquid ammonia, NH_3 (the conjugate acid of NH_2^- and conjugate base of NH_4^+), is a useful solvent for reactions impossible to carry out in the stronger oxidant water. While NH_2^- reacts vigorously with H_2O to yield its conjugate acid NH_3, and OH^- (the conjugate base of H_2O), NH_4Cl and $NaNH_2$ react in liquid ammonia via the net ionic Brønsted–Lowry equation

$$NH_4^+ + NH_2^- \xrightarrow{NH_3(l)} 2\ NH_3(l)$$

Because NH_4^+ is a stronger acid than its conjugate (NH_3), and NH_2^- is a stronger base than its conjugate (also NH_3), the products are favored.

The Brønsted–Lowry concept can be applied in any solvent, regardless whether the solvent possesses hydrogen atoms that can participate. For example, cyclopentadienide, $[C_5H_5]^-$, a common anion in organometallic chemistry, can be prepared in tetrahydrofuran (THF, C_4H_8O) via reaction of sodium hydride and cyclopentadiene (C_5H_6). The Brønsted–Lowry acidity of C_5H_6 is enhanced because the resulting negative charge is delocalized within aromatic $[C_5H_5]^-$; hydride is an extremely strong base that reacts vigorously with water, and this reaction must be carried out under anhydrous conditions.

$$NaH + \text{cyclopentadiene} \xrightarrow{THF} Na^+\ \text{cyclopentadienide} + H_2(g)$$

cyclopentadiene cyclopentadienide

The classification of H_2 as a conjugate acid (of the strong base hydride) is noteworthy. It highlights that any molecule containing hydrogen can in principle function as a Brønsted–Lowry acid, even if some (for example, aliphatic hydrocarbons, H_2) only function as acids under extraordinary conditions.*

EXERCISE 6.2

Organolithium reagents are sources of strong Brønsted–Lowry bases in nonaqueous media. The equilibria between hydrocarbons and organolithium reagents can be predicted via the Brønsted–Lowry concept. Which side of the following Li/H exchange equilibrium should be favored if C_6H_6 is a stronger acid than n-butane?**

$$C_6H_6 + n\text{-}C_4H_9Li \rightleftharpoons C_6H_5Li + n\text{-}C_4H_{10}$$

6.3.1 Nonaqueous Solvents and Acid–Base Strength

The $H_3O^+/H_2O/OH^-$ reference for quantifying relative acid and base strengths is only useful when the examined acid is inherently weaker than H_3O^+ or the examined base is weaker than OH^-. Hydronium ion and hydroxide ion are the strongest acid and base,

*The solvent tetrahydrofuran is another molecule with exceedingly low Brønsted–Lowry acidity that resists deprotonation by sodium hydride.

**The Brønsted–Lowry preference is achieved very slowly in this case, but rapid Li/H exchange occurs upon addition of tetramethylethylenediamine (TMEDA) that enhances n-C_4H_9Li reactivity.

respectively, than can exist in water. Acids inherently stronger than H_3O^+ cannot be differentiated by their aqueous ionization; this is called the **leveling effect**. Due to leveling, nitric, sulfuric, perchloric, and hydrochloric acids are all equally strong acids in dilute aqueous solutions, essentially ionizing quantitatively to H_3O^+ and the corresponding conjugate base. In these cases, more strongly acidic solvents are necessary to differentiate acidity. For example, acetic acid, like water, is amphoteric, and can accept protons from acids classified as strong in water, resulting in *partial* ionization. The $CH_3COOH_2^+$ ion is the strongest acid that can exist in glacial acetic acid (100% acetic acid). The solvent sets a limit on the effective strength of the dissolved acid.

$$H_2SO_4 + CH_3COOH \xrightleftharpoons{CH_3COOH} CH_3COOH_2^+ + HSO_4^-$$

In glacial acetic acid, the relative acid strength of $HClO_4 > HCl > H_2SO_4 > HNO_3$ can be determined. In the same way, basic solvents permit the differentiation of bases that could not be accomplished in a less basic solvent.

A key perspective on leveling is that the classification of a substance as "weak" or "strong" is stringently tied to the solvent. A weak aqueous base may be strong in a more acidic solvent. Weak aqueous acids appear strong in basic solvents. For example, the equilibrium position of

$$NH_3 + CH_3COOH(l) \xrightleftharpoons{CH_3COOH} NH_4^+ + C_2H_3O_2^-$$

lies much further to the right in glacial acetic acid than the ammonia ionization equilibrium in water.

Nonamphoteric solvents, with neither Brønsted–Lowry acidic nor basic properties, do not limit solute acidity or basicity because the solute does not react with the solvent. In these solvents, the inherent solute acid or base strength determines the reactivity, without a leveling effect. For example, hydride sources (e.g., $LiAlH_4$, NaH) are commonly employed as Brønsted–Lowry bases or reducing agents in organic solvents (for example, Et_2O, hydrocarbons) where no acid–base reaction with the solvent is possible. In these cases, reaction conditions are often heterogeneous; the hydride source remains insoluble due to its lack of significant interaction with the solvent. The acid–base effects of the solvent, and the compatibility of reactants with intended solvents must be always considered when planning reactions.

6.3.2 Brønsted–Lowry Superacids

From a Bronsted–Lowry perspective, designing molecules with exceedingly weak conjugate bases results in acids that could potentially transfer protons to species that could not be appreciably protonated by leveled acids in either water (H_3O^+) or sulfuric acid ($H_3SO_4^+$). George Olah won the Nobel Prize in Chemistry in 1994 for the discovery and application of **superacids**, acid solutions more acidic than sulfuric acid. Olah proposed using superacids to protonate monocationic species (for example, nitronium ion, below) to produce useful concentrations of dicationic (charge = 2+) ions with increased reactivity. He coined the term *superelectrophilic activation* to describe the result of generating small organic ions bearing a large amount of positive charge.[8]

$$\left[O{=}N{=}O\right]^+ \xrightleftharpoons{\textit{super}\text{-HA}} \left[O{=}N{=}OH\right]^{2+} A^-$$

A variety of dicationic superelectrophiles have been synthesized,[9] and tricationic species formed via protonation of triaryl methanols were reported recently.[10] These species exhibit new reactions resulting from the effects of closely oriented positive charges.

Superacid acidity is measured by the Hammett acidity function:[11]

$$H_0 = pK_{BH^+} - \log\frac{[BH^+]}{[B]}$$

where B and BH^+ are a nitroaniline indicator and its conjugate acid, respectively. The stronger the acid, the more negative its H_0 value. On this scale, the H_0 of 100% sulfuric acid is –11.9 (Table 6.3).

The term *Magic Acid* was coined upon observation that a mixture of antimony pentafluoride and fluorosulfonic acid dissolved a candle, demonstrating the ability of the acid to protonate hydrocarbons. The ability of superacids to activate hydrocarbons, and particularly methane, via protonation, is an area of intense interest, since methane is the primary component of natural gas. The abundance of methane makes it attractive as a feedstock for synthesis of more complex molecules. Computational studies have proposed structures with 2-center 3-electron bonds for CH_5^+, CH_6^{2+}, and even CH_7^{3+}, but these species have not been isolated.*

Methanesulfonyl chloride, CH_3SO_2Cl, can be prepared from CH_4 and SO_2Cl_2 in sulfuric acid or triflic acid in the presence of a radical initiator.[12] Sulfonation of CH_4 to methanesulfonic acid (CH_3SO_3H) has been achieved with SO_3 in H_2SO_4 in the presence of a radical initiator.[13] Dissolving SO_3 in sulfuric acid results in "fuming sulfuric acid." This superacidic solution contains $H_2S_2O_7$ and higher polysulfuric acids, all of them stronger than H_2SO_4.

Complex fluorinated anions are formed in solutions of Magic Acid and HF-AF_5 (A = As, Sb) that serve as counterions to superelectrophilic cations.

$$2\,HF + 2\,SbF_5 \rightleftharpoons H_2F^+ + Sb_2F_{11}^-$$

$$2\,HSO_3F + 2\,SbF_5 \rightleftharpoons H_2SO_3F^+ + Sb_2F_{10}(SO_3F)^-$$

Superacid media from AsF_5 and SbF_5 in HF can protonate H_2S, H_2Se, AsH_3, SbH_3, and H_2O_2.[14] An example of the utility of one such reaction has been protonation of H_2S in superacid media to yield $[H_3S][SbF_6]$, useful as a reagent in the synthesis of $[(CH_3S)_3S][SbF_6]$, the first trithiosulfonium salt with three RS substituents.[15]

Water is a strong base in superacid media; the presence of H_2O in HF-AsF_5 or HF-SbF_5 results quantitatively in the hydronium salts $[H_3O][AsF_6]$ and $[H_3O][SbF_6]$.[16] This ability of superacids to generate an anhydrous environment (but not aprotic!) by converting trace amounts of water to hydronium has been exploited to prepare solutions of low oxidation state metal cations from metal(II) oxides. These oxides immediately react to form water, which is protonated to form hydronium. Divalent metal ions have been isolated as $[H_3O][M][AsF_6]_3$ (M = Mn, Co, Ni) by conversion of the corresponding metal oxides in HF-AsF_5 media.[17] Related reactions have been achieved with Ln_2O_3 (Ln = lanthanide element)[18] and CdO;[19] some of the resulting salts incorporate fluoride, $[SbF_6]^-$, and $[Sb_2F_{11}]^-$.

TABLE 6.3 Common Superacids and Their Acidities

Acid		H_0
Hydrofluoric acid[a]	HF	−11.0
Sulfuric acid	H_2SO_4	−11.9
Perchloric acid	$HClO_4$	−13.0
Trifluoromethanesulfonic acid (triflic acid)	HSO_3CF_3	−14.6
Fluorosulfonic acid	HSO_3F	−15.6
Magic Acid[b]	$HSO_3F—SbF_5$	−21 to −25[c]
Fluoroantimonic acid	$HF—SbF_5$	−21 to −28[c]

[a] HF is not a superacid but is included for comparison. It is noteworthy that while HF is a weak acid in dilute aqueous solution, concentrated HF is significantly stronger.
[b] *Magic Acid* is a registered trademark of Cationics, Inc., Columbia, SC.
[c] Depending on concentration (how much SbF_5 has been added)

*G. Rasul, G. A. Olah, G. K. Surya Prakash, *J. Phys. Chem.* A., **2012**, *116*, 756. Protonated methane, CH_5^+, has been reported in superacid solutions.

6.3.3 Thermodynamic Measurements in Solution

Various thermodynamic approaches have been used to probe acidity and basicity in solutions. The impact of solvation effects must always be considered when designing these experiments.

Comparing Acidity

A defining property for any acid is its strength. One way to assess the strengths of aqueous acids is to quantify the enthalpy change of

$$HA(aq) + H_2O(l) \longrightarrow H_3O^+(aq) + A^-(aq)$$

Direct measurement of this enthalpy change is complicated since weak acids do not ionize completely (that is, the above reaction is generally an equilibrium with a relatively large concentration of unionized HA). A traditional strategy is to apply Hess's law using thermodynamic data from reactions that essentially go to completion. For example, the enthalpy of ionization of weak acid HA can be determined by measuring (1) the enthalpy change for the reaction of HA with NaOH, and (2) the enthalpy change for the reaction of H_3O^+ and NaOH:

$$(1)\ HA(aq) + OH^-(aq) \longrightarrow A^-(aq) + H_2O(l) \qquad \Delta H_1$$

$$(2)\ H_3O^+(aq) + OH^-(aq) \longrightarrow 2\,H_2O(l) \qquad \Delta H_2$$

$$(3)\ HA(aq) + H_2O(l) \longrightarrow H_3O^+(aq) + A^-(aq) \qquad \Delta H_3 = \Delta H_1 - \Delta H_2$$

This strategy is not straightforward since HA is partly ionized before OH^- is added, complicating ΔH_1 determination, but this approach is an acceptable starting point. It is also possible to measure K_a (via titration curves) at different temperatures and use the van't Hoff equation

$$\ln K_a = \frac{-\Delta H_3}{RT} + \frac{\Delta S_3}{R}$$

to simultaneously determine ΔH_3 and ΔS_3. The slope of a plot of $\ln K_a$ versus $\frac{1}{T}$ is $\frac{-\Delta H_3}{R}$ and the intercept is $\frac{\Delta S_3}{R}$. The accuracy of this method requires that the acid ionization ΔH_3 and ΔS_3 do not change appreciably over the temperature range used. Data for the $\Delta H°$, $\Delta S°$, and K_a for acetic acid are given in **Table 6.4**.

EXERCISE 6.3

Use the data in Table 6.4 to calculate the enthalpy and entropy changes associated with aqueous acetic acid ionization (the third equation in the table), and examine the temperature dependence of K_a by graphing $\ln K_a$ versus $\frac{1}{T}$. How do the $\Delta H°$ values obtained via these two approaches compare?

TABLE 6.4 Thermodynamics of Acetic Acid Ionization

	$\Delta H°$(kJ/mol)	$\Delta S°$(J/mol·K)
$H_3O^+(aq) + OH^-(aq) \longrightarrow 2\,H_2O(l)$	−55.9	80.4
$CH_3COOH(aq) + OH^-(aq) \longrightarrow H_2O(l) + C_2H_3O_2^-(aq)$	−56.3	−12.0

$CH_3COOH(aq) + H_2O(l) \rightleftharpoons H_3O^+(aq) + C_2H_3O_2^-(aq)$

T (K)	303	308	313	318	323
$K_a(\times 10^{-5})$	1.750	1.728	1.703	1.670	1.633

NOTE: $\Delta H°$ and $\Delta S°$ for these reactions change rapidly with temperature. Calculations based on these data are valid only over the limited temperature range given above.

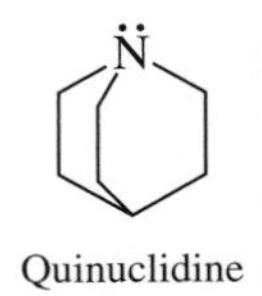

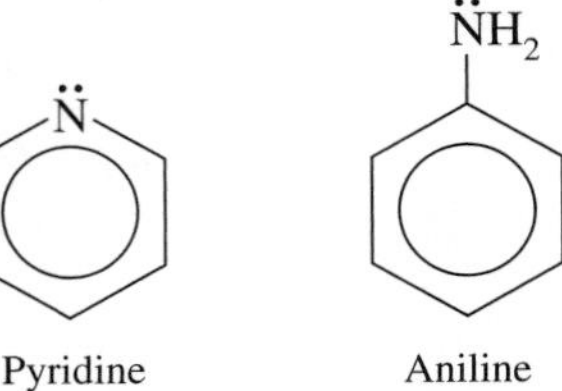

Comparing Basicity

Basicity has also been probed by measuring the enthalpies of proton transfer reactions between weak bases and strong acids. A Brønsted basicity scale has been established by measuring the enthalpy changes associated with the protonation of weak bases in fluorosulfonic acid (HSO_3F, a superacid [**Section 6.3.2**]). Enthalpies of protonation for a series of nitrogen bases (**Table 6.5**; increasingly negative ΔH values correspond to stronger basicity towards HSO_3F) rank in a way that compares favorably with the ranking of the corresponding conjugate acid pK_{BH^+} values (for aqueous solution).[20] The more positive the pK_{BH^+}, the weaker the conjugate acid and the stronger the conjugate nitrogen base.* These data imply that various molecular properties, for example inductive and steric effects (**Section 6.3.6**), are significant in influencing acid/base behavior.

6.3.4 Brønsted–Lowry Gas-Phase Acidity and Basicity

The purest measures of acid–base strength are gas-phase acidity and basicity parameters, where solvent effects are not applicable:

$$\mathrm{HA}(g) \longrightarrow \mathrm{A}^-(g) + \mathrm{H}^+(g) \qquad \begin{array}{l} \Delta G = \text{Gas-Phase Acidity (GA)} \\ \Delta H = \text{Proton Affinity (PA)} \end{array}$$

$$\mathrm{BH}^+(g) \longrightarrow \mathrm{B}(g) + \mathrm{H}^+(g) \qquad \begin{array}{l} \Delta G = \text{Gas-Phase Basicity (GB)} \\ \Delta H = \text{Proton Affinity (PA)} \end{array}$$

TABLE 6.5 Basicity of Selected Nitrogen Bases with Water and Fluorosulfonic Acid

Base	pK_{BH^+} (H_2O)	$-\Delta H$ (HSO_3F, kJ/mol)
Di-*n*-butylamine	11.25	194.1
Quinuclidine	11.15	191.6
Diethylamine	11.02	199.5
Dimethylamine	10.78	197.4
Triethylamine	10.72	205.7
Ethylamine	10.68	195.9
Methylamine	10.65	193.9
Tri-*n*-butylamine	9.93	189.2
Trimethylamine	9.80	196.8
2,4,6-Trimethylpyridine	7.43	178.5
2,6-Dimethylpyridine	6.72	170.3
4-Methylpyridine	6.03	163.4
Pyridine	5.20	161.3
Aniline	4.60	142.3
3-Bromopyridine	2.85	144.9
2-Bromopyridine	0.90	126.2
2-Chloropyridine	0.72	132.5
3,5-Dichloropyridine	0.67	128.4

Data from C. Laurence and J.-F. Gal, *Lewis Basicity and Affinity Scales Data and Measurement*, John Wiley and Sons, United Kingdom, 2010, p. 5. The pK_{BH^+} of ammonia is 9.25.

*Note that these bases are ranked on their basicity towards water; the ranking on the basis of measurements in HSO_3F is different. This is one example of the challenges associated with basicity determinations; basicity rankings of the same bases typically vary with the solvent used.

Proton affinities and gas-phase basicities have been determined for thousands of neutral organic bases; the literature on these parameters is significantly more extensive than on gas-phase acidities. For the majority of bases, the thermodynamic parameters PA and GB are large and positive; the reaction is essentially bond breaking without the benefit of any solvation of the products. Increasing proton affinity and gas-phase basicity magnitudes indicate increasing difficulty to remove the hydrogen; the more positive these values, the stronger B as a base, and the weaker the acid BH^+ in the gas phase.[21] Laurence and Gal have criticized the terminology associated with proton affinity and gas-phase basicity since an affinity is formally a chemical potential while proton affinity is defined as an enthalpy.*

While direct measurement of PA and GB via the reactions shown above is practically impossible, these values have been estimated, with the accuracy improving with advances in chemical instrumentation. In the early twentieth century, Born–Haber thermodynamic cycles were employed to estimate proton affinities. As with all approaches based on thermodynamic cycles (**Section 5.3.2**), uncertainties in the data used to construct a cycle propagate to the calculated values (in this case the proton affinity).

Modern mass spectrometry, photoionization techniques, and ion cyclotron resonance spectroscopy[22] revolutionized gas-phase basicity determination. These techniques have permitted extremely accurate absolute gas-phase basicities to be obtained for a few molecules via thermodynamic cycles using electron affinity and ionization energy data.[23, **] Absolute gas-phase basicities for even a few molecules provide valuable references for determination of GB values for bases for which direct GB determination is problematic. The mathematical approach is conceptually straightforward. Consider the general GB equations for B_1 and B_2:

$$B_1H^+ \longrightarrow B_1 + H^+ \qquad \Delta G_1 = \text{Gas-Phase Basicity of } B_1$$

$$B_2H^+ \longrightarrow B_2 + H^+ \qquad \Delta G_2 = \text{Gas-Phase Basicity of } B_2$$

Subtraction of the second reaction from the first leads to

$$B_1H^+ + B_2 \longrightarrow B_1 + B_2H^+ \quad \Delta G = \Delta G_1 - \Delta G_2$$

The ΔG for this reaction can be calculated from the equilibrium constant

$$B_1H^+ + B_2 \rightleftharpoons B_1 + B_2H^+$$

$$\Delta G = \Delta G_1 - \Delta G_2 = -RT\ln K_{eq}$$

Ion-trapping and flow reactor mass spectrometry permit gas-phase ions and neutral molecules to be confined and attain equilibrium after a sufficient number of collisions. The equilibrium constant can then be deduced by measurement of gas partial pressures (for B_1 and B_2) and mass spectrometric ion intensities (for gaseous B_1H^+ and B_2H^+). The resulting ΔG provides the *difference* in gas-phase basicity between B_1 and B_2. If the absolute gas-phase basicity of either base is known, then the GB value of the other can be determined. Proton affinities are subsequently determined via $\Delta G = \Delta H - T\Delta S$, with the "entropy of basicity" approximated via quantum chemical approaches. **Table 6.6** lists proton affinity and gas-phase basicities for nitrogen bases, with the bases ranked in order of descending gas-phase basicity. The majority of organic bases exhibit GB values between 700 and 1000 kJ/mol. Compilations of PA and GB values are available.[24]

---

*Chemical potential is defined as $\left(\frac{\partial G}{\partial n}\right)_{P,T}$.

**The details of these measurements are beyond the scope of this text. Because they require an ionizing electron beam, measured proton affinities and gas-phase basicities for many species have large uncertainties, because the molecules involved frequently are in excited states (with excess energy above their ground states), and some species do not yield the necessary acid as a gaseous fragment. Relatively few molecules are ideally suited for this analysis.

TABLE 6.6 Gas-Phase Basicity and Proton Affinities for Nitrogen Bases

Base	GB (kJ/mol)	PA (kJ/mol)
Tri-*n*-butylamine	967.6	998.5
Quinuclidine	952.5	983.3
Triethylamine	951.0	981.8
Di-*n*-butylamine	935.3	968.5
2,6-Dimethylpyridine	931.1	963.0
Diethylamine	919.4	952.4
Trimethylamine	918.1	948.9
4-Methylpyridine	915.3	947.2
Pyridine	898.1	930.0
Dimethylamine	896.5	929.5
3-Bromopyridine	878.2	910.0
Ethylamine	878.0	912.0
2-Bromopyridine	873.0	904.8
2-Chloropyridine	869.0	900.9
Methylamine	864.5	899.0
Aniline	850.6	882.5
Ammonia	819.0	853.6

Data from C. Laurence and J.-F. Gal, *Lewis Basicity and Affinity Scales Data and Measurement*, John Wiley and Sons, United Kingdom, 2010, p. 5.

6.3.5 Brønsted–Lowry Superbases

Thermodynamic data for quinuclidine (PA = 983.3 kJ/mol) and its conjugate acid (pK_a = 11.15 in water) underscore its relatively high basicity. What is the upper limit on Brønsted–Lowry basicity that is still compatible with high selectivity for deprotonation? While leveling limits the operative strength of all Brønsted–Lowry bases inherently stronger than hydroxide in water, extremely basic carbanions (for example, Grignard and organolithium reagents) are ubiquitous in organic synthesis. The high nucleophilicity of these carbanions lowers their tolerance towards many functional groups, and motivates the synthesis of strong Brønsted–Lowry bases that exhibit broader tolerance and extremely high selectivity for deprotonation reactions.[25] The aforementioned carbanions are *both* strongly Brønsted–Lowry basic and nucleophilic; one virtue of uncharged organic bases is their lowered nucleophilicity. There is also motivation to conduct deprotonations in syntheses without inorganic hydroxides.[26]

Superbases have been classified as those with gas-phase proton affinities > 1000 kJ/mol,[27] greater than the bases listed in Table 6.6. Examples of organic superbases are shown in **Figure 6.1**. These are weak bases in water but exhibit superbasic characteristics in organic solvents.

The organic superbase, 1,8-diazabicyclo[5.4.0]undec-7-ene (DBU), is a workhorse for organic synthesis (PA = 1048 kJ/mol). 1,8-Bis(dimethylamino)naphthalene (PA = 1028 kJ/mol) is sometimes called a "proton sponge." Its strong Brønsted–Lowry basicity is believed to arise from two effects: (1) the relief of steric hindrance of two dimethylamino substituents in close proximity (if the nonbonding pairs adopt opposite positions to reduce *lp–lp* repulsion, the methyl groups are brought into close proximity), and (2) the formation of a strong intramolecular hydrogen bond upon protonation.[23] It is instructive that the considerably more flexible 1,3-bis(dimethylamino) propane is also a superbase (PA = 1035 kJ/mol) even though this molecule is not susceptible to the sterically enforced *lp–lp* repulsion present

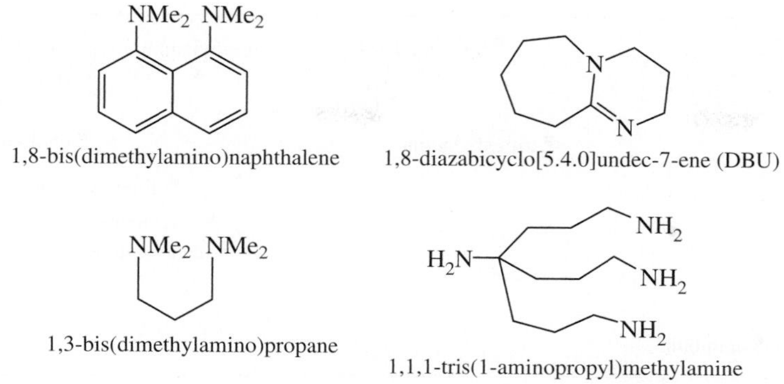

NMe_2 NMe_2

1,8-bis(dimethylamino)naphthalene

1,8-diazabicyclo[5.4.0]undec-7-ene (DBU)

NMe_2 NMe_2

1,3-bis(dimethylamino)propane

H_2N NH_2 NH_2 NH_2

1,1,1-tris(1-aminopropyl)methylamine

FIGURE 6.1 Organic Superbases.

in 1,8-bis(dimethylamino)naphthalene. The inductive effect of alkyl vs. aryl substitution appears more important than the relief of *lp–lp* repulsion in affording superbasic characteristics when these two molecules are compared. The addition of propylamine substituents to methylamine affords $NH_2C(CH_2CH_2CH_2NH_2)_3$ (PA = 1072 kJ/mol).[28] Upon protonation, the propylamine arms are postulated to wrap around, bringing more nitrogen atoms into contact with the proton, stabilizing the conjugate acid.

The possibility of using 2,6-disubstituted pyridines and 2,6,7-trisubstituted quinuclidines, where the substituents feature remote atoms with lone pairs to stabilize the hydrogen upon protonation, are proposed superbases that have been explored by computational approaches.[29] There is interest in synthesizing macrocyclic proton chelaters as catalytically active organic superbases,[30] and a new structural motif for superbases featuring caged secondary amines has been reported.[31] The alkali metal hydroxides, of equal basicity in aqueous solution, have proton affinities* in the order LiOH (1000 kJ/mol) < NaOH < KOH < CsOH (1118 kJ/mol). This order matches the increasing ionic character of the alkali metal-hydroxide bonds.

6.3.6 Trends in Brønsted–Lowry Basicity

Correlations between gas-phase and aqueous basicity data provide a starting point to consider the importance of electronic, steric, and solvation effects on proton transfer reactions. **Figures 6.2** and **6.3** provide such correlations, by plotting gas-phase basicity versus aqueous basicity (on the basis of conjugate acid pK_a) for nitrogen bases listed in Tables 6.5 and 6.6. Higher placement on the *y*-axis indicates higher basicity in the gas phase, while increasing conjugate acid pK_a indicates higher basicity in water. Initial inspection of these graphs reveals that higher gas-phase basicity does not necessarily translate to higher aqueous basicity (for example, tri-*n*-butylamine relative to di-*n*-butylamine), just as higher aqueous basicity does not always correlate well with higher gas-phase basicity (for example, ammonia relative to 2,6-dimethylpyridine). Exploring these data uncovers trends in Brønsted–Lowry basicity and highlights the essential role of the solvent in influencing basicity.

Inductive effects are useful to rationalize trends in Figures 6.2 and 6.3. For example, both gas-phase and aqueous basicity increase as:

$$NH_3 < NH_2Me < NH_2Et < NHMe_2 < NHEt_2 < NHBu_2$$

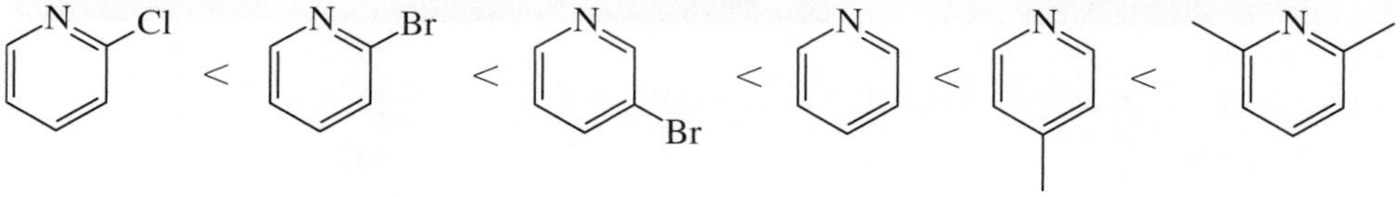

N Cl < N Br < N Br < N < N < N

*Proton affinities of inorganic hydroxides cannot be obtained from direct proton transfer measurements, but rather from Born–Haber cycles using other thermodynamic data.

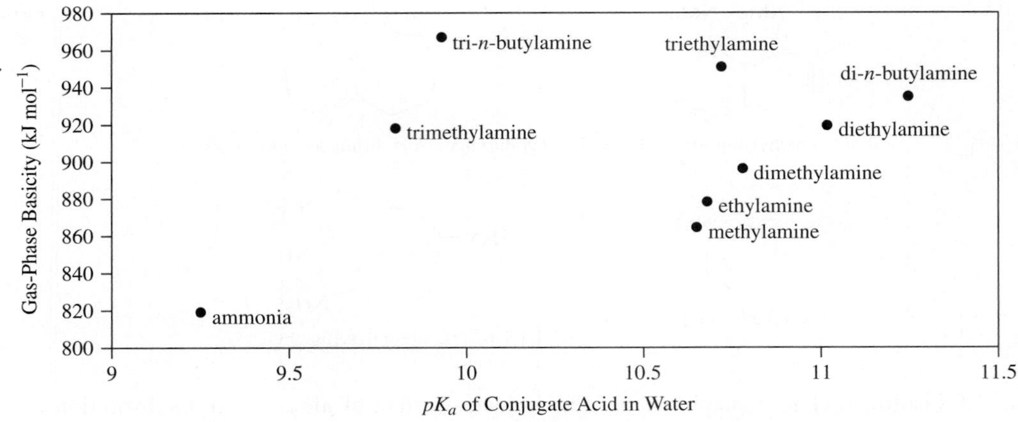

FIGURE 6.2 Gas-Phase Basicity vs. pK_a for Ammonia and Alkyl-substituted Amines.

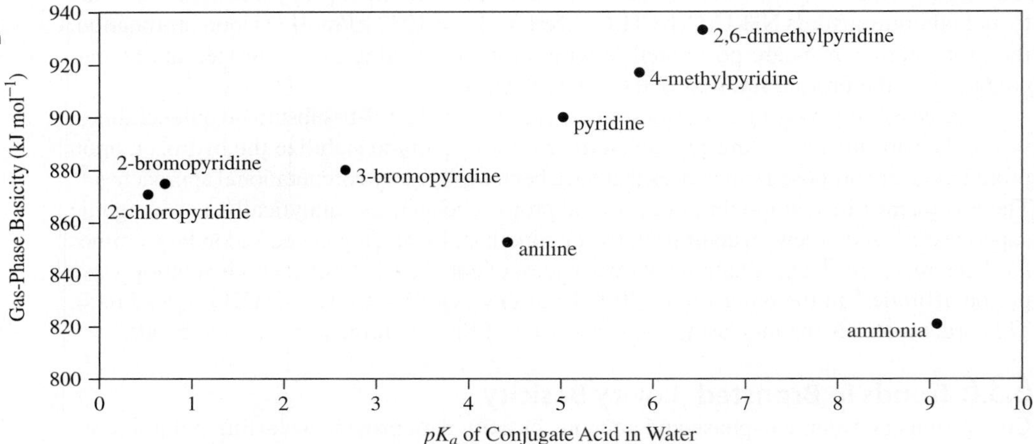

FIGURE 6.3 Gas-Phase Basicity vs. pK_a for Ammonia and Aromatic Amines.

The substitution of alkyl groups for hydrogen within the series of ammonia to primary amines to secondary amines results in progressively more electron-rich nitrogen centers and stronger Brønsted–Lowry bases. Within this series, a longer alkyl chain enhances the effect. Similarly, methylpyridines are stronger Brønsted–Lowry bases than pyridine. The substitution of highly electronegative atoms or groups (for example, fluorine, chlorine, CF_3 or CF_3SO_2) results in weaker bases by drawing electron density away from the Brønsted basic atom. The halopyridines are dramatically weaker bases than pyridine. The gas-phase acidities in **Table 6.7** illustrate the impact of increasing CF_3SO_2 substitution.

TABLE 6.7 Impact of CF_3SO_2 Substitution on Gas-Phase Acidity

$HA(g) \longrightarrow A^-(g) + H^+(g)$ ΔG = Gas-Phase Acidity (GA)

Acid	GA (kJ/mol)
$CF_3SO_2CH_3$	1422
$CF_3SO_2NH_2$	1344
$(CF_3SO_2)_2NH$	1221
$(CF_3SO_2)_2CH_2$	1209

Data from J.-F. Gal, P.-C. Maria, E. D. Raczyńska, *J. Mass Spectrom.*, **2001**, *36*, 699.

Inductive effects provide a reasonable way to rationalize this gas-phase basicity ranking:

$$NMe_3 < NHEt_2 < NHBu_2 < NEt_3 < NBu_3$$

Since tri-*n*-butylamine is more basic than triethylamine in the gas phase, we postulate that trimethylamine is less basic than the secondary amines since two longer alkyl groups and hydrogen enrich the electron density at nitrogen more than three methyl groups. Perhaps more interesting is the aqueous basicity ranking of the amines below, which appears to contradict the inductive rationalization of their gas-phase values; the tertiary amines are weaker than expected.

$$NMe_3 < NBu_3 < NEt_3 < NHEt_2 < NHBu_2$$

In addition, in aqueous solution, the methyl-substituted amines have basicities in the order $NHMe_2 > NH_2Me > NMe_3 > NH_3$ as given in Table 6.5, and shown in Figure 6.2. The ethyl-substituted amines are in the order $NHEt_2 > NEt_3 > NH_2Et > NH_3$. In these series, the tertiary amines are weaker than expected because of the reduced solvation of their protonated cations. Solvation enthalpy magnitudes for the general reaction

$$NH_{4-n}R_n^+(g) \xrightarrow{H_2O} NH_{4-n}R_n^+(aq)$$

are in the order $NH_3R^+ > NH_2R_2^+ > NHR_3^+$.* Solvation is dependent on the number of hydrogen atoms available to form O···H—N hydrogen bonds with water. With fewer hydrogen atoms available for hydrogen bonding, the more highly substituted molecules are rendered less basic. Competition between these induction and solvation effects gives the scrambled order of solution basicity. The maximal opportunity for hydrogen bonding with aqueous NH_4^+ plays an important role in the solution basicity of NH_3 being stronger than all the bases in Figure 6.3 even though NH_3 exhibits the lowest gas-phase basicity among these bases. Figure 6.3 also shows that pyridine and aniline have higher gas-phase basicities than ammonia, but are weaker bases than NH_3 in aqueous solution. The higher basicity of NH_3 in water is also attributed to enhanced hydrogen bonding with NH_4^+ relative to the pyridinium or anilinium ions.[32]

Steric effects are less obvious from these correlations. For example, 2,6-dimethylpyridine, despite the steric bulk adjacent the nitrogen atom, is more basic than 4-methylpyridine in both solution and the gas phase. However, consider the following basicity ranking in aqueous solution:

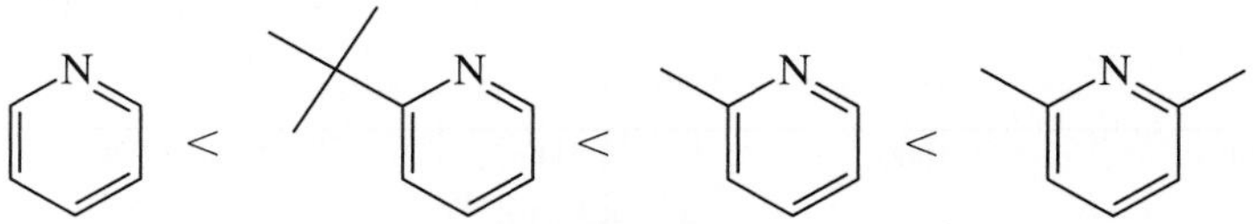

N < N < N < N

One might expect 2-*t*-butylpyridine to be more basic than 2-methylpyridine on the basis of inductive effects, but the tertiary butyl steric bulk attenuates the basicity on steric grounds by making the nitrogen less accessible and more difficult to solvate upon protonation. Another steric effect deals with a geometry change that accompanies protonation; the potential of steric hindrance in the conjugate acid may attenuate basicity. While steric hindrance often plays major roles in understanding Lewis acidity/basicity (**Section 6.4.7**), the small proton size renders these steric effects less important in gauging the Brønsted–Lowry basicity of amines. For example, the gas-phase basicities of quinuclidine (1-azabicyclo[2.2.2]octane, 952.5 kJ/mol) and triethylamine (951.0 kJ/mol) are nearly identical, even though the geometric parameters associated with the cyclic quinuclidine system remains essentially unchanged upon protonation, while a greater structural change accompanies conversion of triethylamine to triethylammonium ion.

---

*E. M. Arnett, *J. Chem. Ed.*, **1985**, *62*, 385 reviews the effects of solvation, with many references.

6.3.7 Brønsted–Lowry Acid Strength of Binary Hydrogen Compounds

The binary hydrogen compounds (those containing only hydrogen and one other element) range from the strong acids HCl, HBr, and HI to the weak base NH_3. Others, such as CH_4, show almost no acid–base properties. Some of these molecules—in order of increasing gas-phase acidities, from left to right—are shown in **Figure 6.4**.

Two seemingly contradictory trends are seen in these data. Acidity increases with increasing numbers of electrons in the central atom, either going across the table or down; but the electronegativity effects are opposite for the two directions (**Figure 6.5**). Within each column of the periodic table, acidity increases going down the series, as in $H_2Se > H_2S > H_2O$. The strongest acid is the largest, heaviest member, low in the periodic table, containing the nonmetal of lowest electronegativity of the group. An explanation of this is that the conjugate bases (SeH^-, SH^-, and OH^-) with the larger main group atoms have lower charge density and therefore a smaller attraction for hydrogen ions (the H—O bond is stronger than the H—S bond, which in turn is stronger than the H—Se bond). As a result, the larger molecules are stronger acids, and their conjugate bases are weaker.

On the other hand, within a period, acidity is greatest for the compounds of elements toward the right, with greater electronegativity. The electronegativity argument used above is not applicable here, because in this series, the more electronegative elements form the stronger acids. The order of acid strength follows this trend: $NH_3 < H_2O < HF$.

The same general acidity trends are observed in aqueous solution. The three heaviest hydrohalic acids—HCl, HBr, and HI—are equally strong in water because of the leveling effect. All the other binary hydrogen compounds are weaker acids, their strength decreasing toward the left in the periodic table. Methane and ammonia exhibit no acidic behavior in aqueous solution, nor do silane (SiH_4) and phosphine (PH_3).

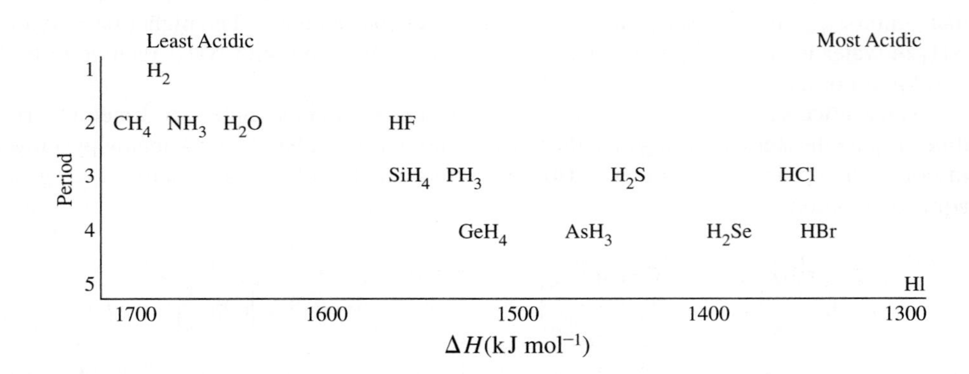

FIGURE 6.4 Acidity of Binary Hydrogen Compounds. Enthalpy of ionization in kJ/mol for the reaction $HA(g) \rightarrow A^-(g) + H^+(g)$, (the same as the proton affinity, Section 6.3.4).

(Data from J. E. Bartmess, J. A. Scott, and R. T. McIver, Jr., *J. Am. Chem. Soc.*, **1979**, *101*, 6046; AsH_3 value from J. E. Bartmess and R. T. McIver, Jr., *Gas Phase Ion Chemistry*, M. T. Bowers, ed., Academic Press, New York, 1979, p. 87.)

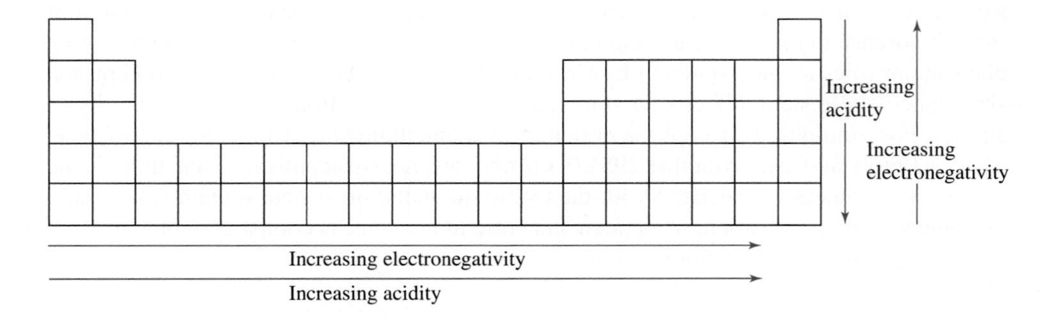

FIGURE 6.5 Trends in Acidity and Electronegativity of Binary Hydrides.

6.3.8 Brønsted–Lowry Strength of Oxyacids

The acid strength of the oxyacids of chlorine in aqueous solution rank as

$$HClO_4 > HClO_3 > HClO_2 > HOCl$$

The pK_a values of these acids are below.

$$\text{H—O—Cl(=O)(—O)—O} \quad \text{H—O—Cl(=O)—O} \quad \text{H—O—Cl(=O)} \quad \text{H—O—Cl}$$

Acid	Strongest $HClO_4$	$HClO_3$	$HClO_2$	Weakest $HOCl$
pK_a (298 K)	(−10)	−1	2	7.2

For oxyacids with multiple ionizable hydrogens, the pK_a values increase by about five units with each successive proton removal:

	H_3PO_4	$H_2PO_4^-$	HPO_4^{2-}	H_2SO_4	HSO_4^-
pK_a (298 K)	2.15	7.20	12.37	<0	2

The trends in these pK_a values are rationalized on the basis of electronegativity and resonance arguments. Oxygen atoms have a high electronegativity, and influence the distribution of electron density in molecules. In the case of oxyacids, the electronegativity of a terminal oxygen atom is greater than the group electronegativity of OH (**Section 3.2.3**). The net result is that the electron density supporting the O—H bond decreases (along with the bond strength) as the number of oxygen atoms increases. This renders the O—H bond more susceptible to the heterolytic cleavage associated with Brønsted–Lowry proton transfer. As the number of oxygen atoms increases, oxyacid acid strength increases.

The negative charge of oxyacid conjugate bases is stabilized by delocalization, represented by resonance forms where each terminal oxygen atom is progressively assigned a negative charge. The conjugate base is stabilized to a greater extent as the number of oxygen atoms increases for this negative charge delocalization. The more effectively the negative charge is delocalized, the weaker the conjugate base, and the stronger the acid.

6.3.9 Brønsted–Lowry Acidity of Aqueous Cations

Transition metal cations exhibit acidic behavior in solution; the impact of a positively charged metal ion on its bound water molecules is related to the inductive effect in oxyacids. The O—H bonds of water bound to transition metal ions are weakened since bonding electron density is drawn towards the metal. For example, aqueous Fe^{3+} is acidic, with yellow or brown iron species formed by reactions where proton transfer occurs from solvated water molecules resulting in bound hydroxide

$$[Fe(H_2O)_6]^{3+}(aq) + H_2O(l) \longrightarrow [Fe(H_2O)_5(OH)]^{2+}(aq) + H_3O^+(aq)$$

$$[Fe(H_2O)_5(OH)]^{2+}(aq) + H_2O(l) \longrightarrow [Fe(H_2O)_4(OH)_2]^+(aq) + H_3O^+(aq)$$

In more basic solutions, hydroxide or oxide bridges form between metal atoms, resulting in cations with rather high positive charge. The higher positive charge further enhances the acidity of bound water molecules, and eventually metal hydroxide precipitates. A possible first step in this process is

$$2\,[Fe(H_2O)_5(OH)]^{2+} \rightleftharpoons [(H_2O)_4Fe(\mu\text{-OH})_2Fe(H_2O)_4]^{4+} + 2\,H_2O$$

TABLE 6.8 Hydrated Metal Ion Brønsted–Lowry Acidities (298 K)

Metal Ion	K_a	Metal Ion	K_a
Fe^{3+}	6.7×10^{-3}	Fe^{2+}	5×10^{-9}
Cr^{3+}	1.6×10^{-4}	Cu^{2+}	5×10^{-10}
Al^{3+}	1.1×10^{-5}	Ni^{2+}	5×10^{-10}
Sc^{3+}	1.1×10^{-5}	Zn^{2+}	2.5×10^{-10}

NOTE : These are equilibrium constants for $[M(H_2O)_m]^{n+} + H_2O \rightleftharpoons [M(H_2O)_{m-1}(OH)]^{(n-1)+} + H_3O^+$.

Metal ions with larger charges and smaller radii are stronger acids. The alkali metal cations show essentially no acidity, the alkaline earth metal cations show it only slightly, 2+ transition-metal ions are weakly acidic, 3+ transition-metal ions are moderately acidic, and ions that would have charges of 4+ or higher as monatomic ions are such strong acids in aqueous solutions that they exist only as oxygenated ions. At this highly charged extreme, the free metal cation is no longer a detectable species. Instead, ions such as permanganate (MnO_4^-), chromate (CrO_4^{2-}), uranyl (UO_2^+), dioxovanadium (VO_2^+), and vanadyl (VO^{2+}) are formed, with oxidation numbers of 7, 6, 5, 5, and 4 for the metals, respectively. Acid-dissociation constants for transition metal ions are given in **Table 6.8**.

6.4 Lewis Acid–Base Concept and Frontier Orbitals

Lewis[33] defined a base as an **electron-pair donor** and an acid as an **electron-pair acceptor.*** Modern inorganic chemistry extensively uses the Lewis definition, which encompasses the Brønsted–Lowry definition, since H^+ accepts an electron pair from a Brønsted base during protonation. The Lewis definition dramatically expands the acid list to include metal ions and main group compounds, and provides a framework for nonaqueous reactions. The Lewis definition includes reactions such as

$$Ag^+ + 2\,{:}NH_3 \rightarrow [H_3N{:}Ag{:}NH_3]^+$$

with the silver ion as an acid and ammonia as a base. In this class of reaction, the Lewis acid and base combine to provide an **adduct**. The bond that links the Lewis acid and base is called a *coordinate covalent* or *dative bond*; this bond features a shared pair of electrons that originated from the Lewis base.** The boron trifluoride–ammonia adduct, $BF_3 \cdot NH_3$, is a classical Lewis acid–base complex. The BF_3 molecule described in Sections 3.1.4 and 5.4.6 is trigonal planar. The B—F bonds are highly polarized by virtue of the large difference in electronegativity between fluorine and boron; the boron is frequently described as electron deficient. The electrons housed in the HOMO of the ammonia molecule interact with the empty LUMO of the BF_3—which has a large contribution from the boron $2p_z$ orbital (Figure 5.32)—to form the adduct. The molecular orbitals involved are depicted in **Figure 6.6**, and their energy levels are shown in **Figure 6.7**. The driving force for adduct formation is stabilization of the electrons in the donor HOMO.

The B—F bonds in $BF_3 \cdot NH_3$ are bent away from the ammonia into a nearly tetrahedral geometry. The related boron trifluoride–diethyl ether adduct, $BF_3 \cdot O(C_2H_5)_2$, is used in synthesis. The HOMO of diethyl ether features significant electron density at the oxygen, as reflected in the Lewis structure via two nonbonding pairs at the oxygen. These electrons are relatively high in energy and can be stabilized via interaction with a suitable LUMO. In this case, the HOMO electrons attack the boron-centered LUMO, changing the

*A Lewis base is also called a **nucleophile**, and a Lewis acid is also called an **electrophile**.

**In a standard covalent bond, like that in H_2, each atom formally provides one electron to each bonding pair.

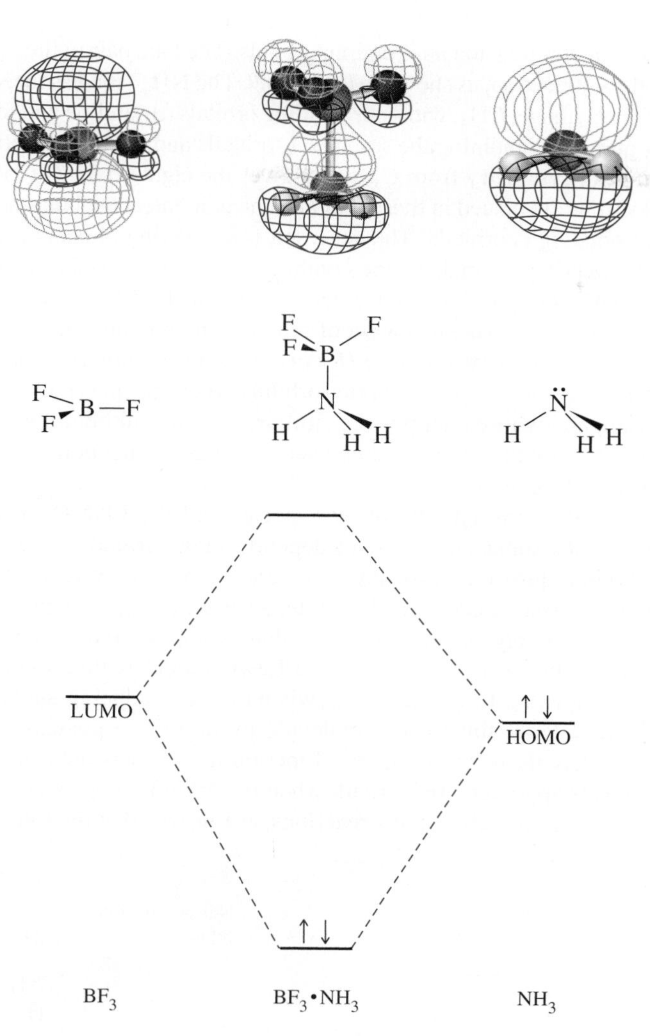

FIGURE 6.6 Donor–Acceptor Bonding in $BF_3 \cdot NH_3$.

FIGURE 6.7 Simplified Energy Level Diagram for the Donor–Acceptor Bonding in $BF_3 \cdot NH_3$.

geometry around B from planar to nearly tetrahedral, as shown in **Figure 6.8**. As a result, BF_3, with a boiling point of −99.9° C, and diethyl ether, with a boiling point of 34.5° C, form an adduct with a boiling point of roughly 125°. At this temperature the dative bond dissociates to give BF_3 and $O(CH_2CH_3)_2$. The chemical and physical properties of adducts are often dramatically different than those of the component Lewis acid and base.

Lewis acid–base adducts involving metal ions are called **coordination compounds**; their chemistry will be discussed in Chapters 9 through 14.

6.4.1 Frontier Orbitals and Acid–Base Reactions[34]

The molecular orbital description of acid–base reactions in Section 6.4 uses **frontier molecular orbitals**, those at the occupied–unoccupied frontier, which can be further illustrated by $NH_3 + H^+ \rightarrow NH_4^+$. In this reaction, the a_1 orbital containing the lone-pair electrons of the ammonia molecule (Figure 5.30) combines with the empty 1*s* orbital of the

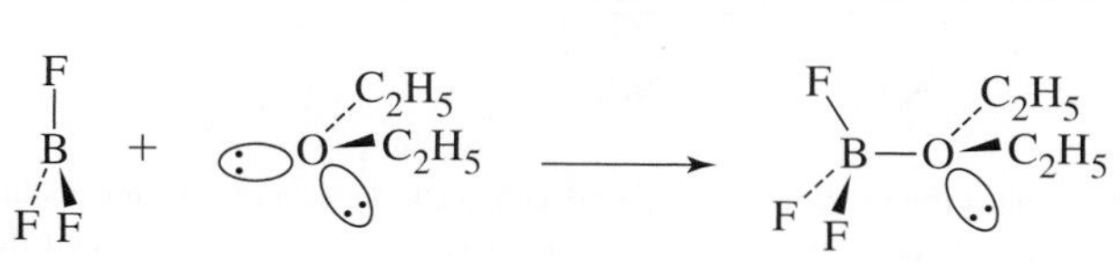

$$BF_3 + O(C_2H_5)_2 \longrightarrow F_3B{-}O(C_2H_5)_2$$

FIGURE 6.8 Boron Trifluoride–Ether Adduct Formation from a Lewis Perspective.

hydrogen ion to form bonding and antibonding orbitals. The lone pair in the a_1 orbital of NH_3 is stabilized by this interaction, as shown in **Figure 6.9**. The NH_4^+ ion has the same molecular orbital structure as methane, CH_4, with four bonding orbitals (a_1 and t_2) and four antibonding orbitals (also a_1 and t_2). Combining the seven NH_3 orbitals and the one H^+ orbital, accompanied by the change in symmetry from C_{3v} to T_d, gives the eight orbitals of NH_4^+. When the eight valence electrons are placed in these orbitals, one pair enters the bonding a_1 orbital, and three pairs enter bonding t_2 orbitals. The net result is a lowering of energy as the nonbonding a_1 becomes a bonding t_2, making the combined NH_4^+ more stable than the separated $NH_3 + H^+$. The HOMO of the base NH_3 interacts with the LUMO of the acid H^+ resulting in a change in symmetry to make a new set of orbitals, one bonding and one antibonding.

In most Lewis acid–base reactions, *a HOMO–LUMO combination forms new HOMO and LUMO orbitals of the product*. Frontier orbitals whose shapes and symmetries allow significant overlap, and whose energies are similar, form useful bonding and antibonding orbitals. If the orbital combinations have no useful overlap, no net bonding is possible, and they cannot form acid–base products.*

When the shapes of the HOMO of one species and the LUMO of another species match, whether or not a stable adduct forms depends on the orbital energies. Formation of a robust dative bond requires a reasonably close energy match between these orbitals. As the energies of these orbitals get more disparate, electron transfer from the HOMO to the LUMO becomes more likely, resulting in a possible oxidation–reduction reaction (without adduct formation). A fascinating aspect of the Lewis model is that a single species can act as an oxidizing agent, a Lewis acid, a Lewis base, or a reducing agent, depending on the other reactant. Indeed, since every molecule by definition possesses a HOMO and a LUMO, every molecule in principle can function as a Lewis acid or base. Although predictions using this approach are difficult when the orbital energies are not known, this perspective is useful to rationalize many reactions, as illustrated in the following examples.

FIGURE 6.9 $NH_3 + H^+ \longrightarrow NH_4^+$ Molecular Energy Levels.

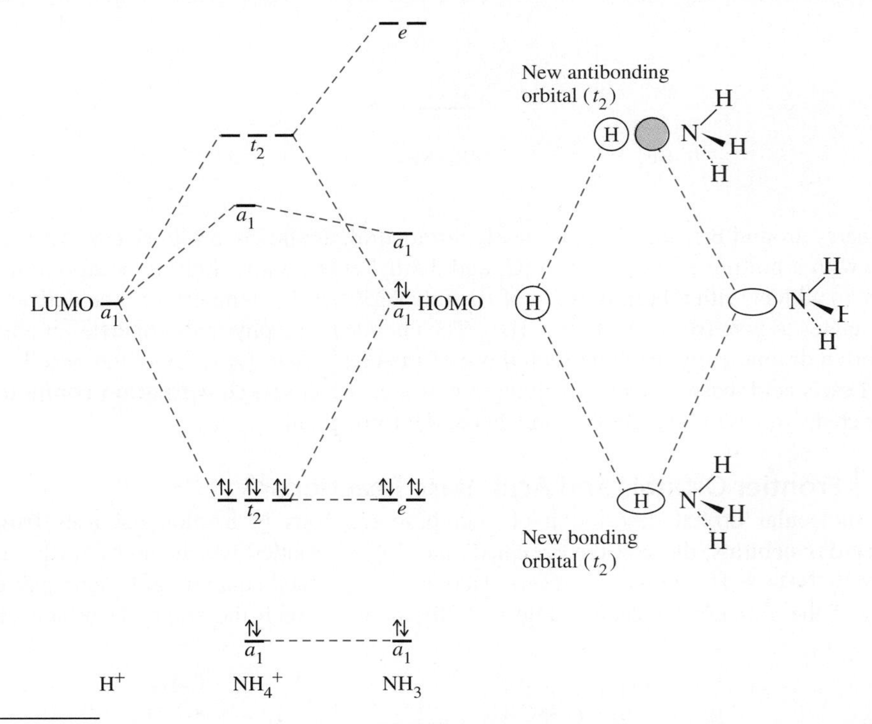

*In less common cases, the orbitals with the required geometry and energy do not include the HOMO; this possibility should be kept in mind. When this happens, the HOMO is usually a lone pair that does not have the geometry needed for bonding with the acid.

EXAMPLE 6.1

Water plays different roles that can be rationalized from the perspective of frontier orbital interactions.

Water as oxidizing agent

An example would be the reaction of water with calcium. In this situation, the water frontier orbitals are significantly lower in energy than the frontier orbitals of calcium (the alkali metals react similarly but have only one electron in their highest *s* orbital).* The energies are sufficiently different that no adduct can form, but electron transfer occurs from the Lewis base to the Lewis acid. One would never classify Ca as a Lewis base in introductory chemistry but it is one within this model!

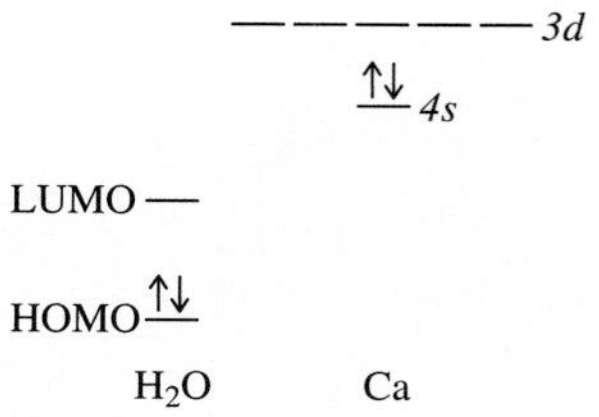

From simple electron transfer from calcium to water, we might expect formation of H_2O^-, but electron transfer into the antibonding H_2O LUMO results in O—H bond weakening, leading to formation of hydrogen gas; H_2O is reduced to H_2 and OH^-, and Ca is oxidized to Ca^{2+}:

$$2\,H_2O(l) + Ca(s) \longrightarrow Ca^{2+}(aq) + 2\,OH^-(aq) + H_2(g) \text{ (water as oxidant)}$$

While the relatively wide energy gap between the participating Ca and H_2O orbitals plays an important role in the driving force of this reaction, the thermodynamics associated with ion solvation and gas evolution are vital as well.

3d
↑↓ 4s
LUMO
HOMO ↑↓
H_2O Ca

Solvation of an anion

If orbitals with matching shapes have similar energies, the resulting adduct bonding orbitals will have lower energy than the Lewis base HOMO, and a net decrease in energy (stabilization of electrons in the new HOMO) drives the formation of an adduct. Adduct stability depends on the difference between the total energy of the product and the total energy of the reactants.

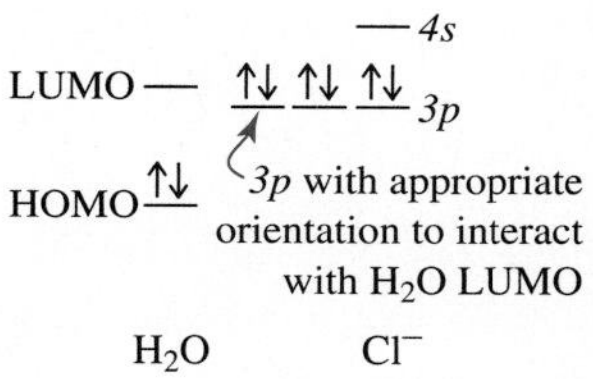

An example with water as acceptor (with lower energy frontier orbitals) is its interaction with the chloride ion:

$$n\,H_2O(l) + Cl^- \longrightarrow [Cl(H_2O)_n]^- \text{ (water as Lewis acid)}$$

The product is solvated chloride. In this case, water is the acceptor, using as LUMO an antibonding orbital centered primarily on the hydrogen atoms (Figure 5.28). The chloride HOMO is a 3*p* orbital occupied by an electron pair. This frontier orbital approach can be applied to many ion–dipole interactions.

4s
LUMO ↑↓ ↑↓ ↑↓ 3p
HOMO ↑↓ 3p with appropriate orientation to interact with H_2O LUMO
H_2O Cl^-

Solvation of cation

A reactant with frontier orbitals lower in energy than those of water (for example, Mg^{2+}) allows water to act as a donor. In this example, the resulting adduct is a solvated metal cation:

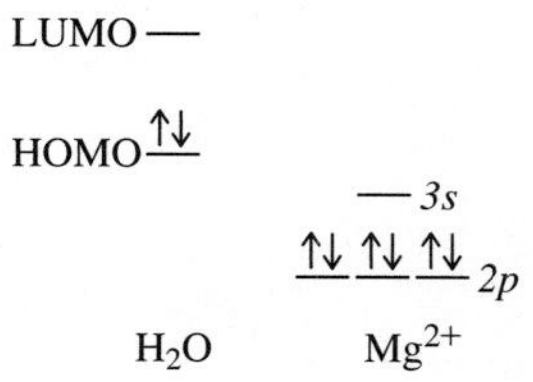

$$6\,H_2O(l) + Mg^{2+} \longrightarrow [Mg(H_2O)_6]^{2+}(aq) \text{ (water as Lewis base)}$$

Water plays its traditional role as a Lewis base, contributing a lone pair primarily from the HOMO, which is the oxygen atom $2p_y$ orbital (Figure 5.28). The magnesium ion LUMO (Mg^{2+} is the Lewis acid) is the vacant 3*s* orbital. This model provides an introductory perspective of the driving force for the formation of hydrated metal cations. More details are provided in Chapter 10.

LUMO
HOMO ↑↓
3s
↑↓ ↑↓ ↑↓ 2p
H_2O Mg^{2+}

Water as reducing agent

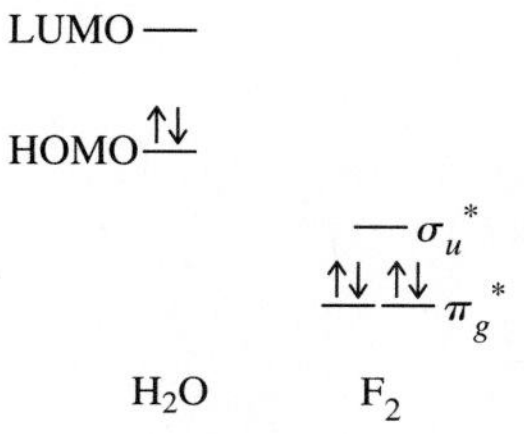

Finally, if the reactant has frontier orbitals much lower than the water orbitals (F_2, for example), water acts as a reductant and transfers electrons to the other reactant.

LUMO
HOMO ↑↓
σ_u^*
↑↓ ↑↓ π_g^*
H_2O F_2

*Frontier orbitals can be *atomic* as well as molecular orbitals.

The fate of the H_2O is not the instantaneous result of electron transfer (H_2O^+) but the formation of molecular oxygen and hydrogen ions:

$$2\ H_2O(l) + 2\ F_2(g) \longrightarrow 4\ F^-(aq) + 4\ H^+(aq) + O_2(g) \text{ (water as reductant)}$$

We can now express the Lewis definition of acids and bases in terms of frontier orbitals:

A base has an electron pair in a HOMO of suitable symmetry to interact with the LUMO of the acid.

An excellent energy match between the base's HOMO and the acid's LUMO leads to adduct formation with a coordinate covalent bond. More disparate energy gaps between the frontier orbitals can result in oxidation–reduction reactions initiated by electron transfer from the base to the acid. While this model must be considered in concert with other considerations (most notably thermodynamics) to predict the fate of potential reactants, the frontier orbital perspective provides a conceptual framework for analyzing reactions.

6.4.2 Spectroscopic Support for Frontier Orbital Interactions

Reactions of I_2 as a Lewis acid with Lewis basic solvents dramatically show the effect of adduct formation. The spectral changes caused by the changes in energy of the participating electronic energy levels (**Figures 6.10** and **6.11**) are striking. The upper I_2 energy levels are shown on the left in Figure 6.10, with a bond order of 1 due to the filled $9\sigma_g$ and $4\pi_u$ bonding orbitals and $4\pi_g^*$ antibonding orbitals. Gaseous I_2 is violet, absorbing light near 500 nm to affect electronic excitation from the $4\pi_g^*$ level to the $9\sigma_u^*$ level. This absorption, broadened due to excitation from gaseous I_2 in ground and excited vibrational and rotational states, removes photons from the yellow, green, and blue parts of the visible spectrum, transmitting red and violet that combine to afford the observed violet color.

In solvents such as hexane, with frontier orbitals neither amenable to robust adduct formation nor electron transfer with I_2, the electronic structure of iodine is essentially unchanged, and the color remains essentially the same violet; the absorption spectra of gaseous I_2 and solutions of I_2 in hexane are nearly identical in the visible range (Figure 6.11). However, in benzene and other π-electron solvents, the color becomes more reddish; and in good donors—such as ethers, alcohols, and amines—the color becomes distinctly brown. The solubility of I_2 also increases as the ability of the solvent to interact as a donor towards I_2 is enhanced. Interaction of a solvent donor orbital with the $9\sigma_u^*$ I_2 LUMO results in a lower occupied bonding orbital and a higher unoccupied antibonding orbital. As a result, the $\pi_g^* \rightarrow \sigma_u^*$ transition for the I_2 + donor adduct is shifted higher in energy, and the absorbance peak is blue-shifted. The transmitted color shifts toward brown (combined red, yellow, and green), as more of the yellow and green light passes through. Water is a poor donor towards the I_2 LUMO; I_2 is very slightly soluble in water. In contrast, I^-, an excellent donor towards the I_2 LUMO; I^- reacts with I_2 to form I_3^-, which is very soluble in water giving a brown solution. When the interaction between the donor and I_2 is strong, the adduct LUMO is shifted to higher energy, resulting in the donor–acceptor transition ($\pi_g^* \rightarrow \sigma_u^*$) increasing in energy. The coordinate covalent bonds formed in these adducts are called **halogen bonds** (**Section 6.4.5**).

In addition to the donor–acceptor absorption, a new ultraviolet band (230 to 400 nm, marked CT in Figure 6.11) appears upon adduct formation. This absorption is associated with the transition $\sigma \rightarrow \sigma^*$ between the two orbitals formed by the interaction between the frontier orbitals. Because the donor orbital (in this case, from the solvent or I^-) contributes the most to lower σ adduct orbital, and the I_2 LUMO contributes the most to the σ^* adduct orbital, the CT transition transfers an electron from an orbital that is primarily of donor composition to one that is primarily of acceptor composition; hence, the name **charge transfer (CT)** for this transition. The energy of this transition is less predictable, because it depends on the energy of the donor orbital. These transitions result in electron density being shifted

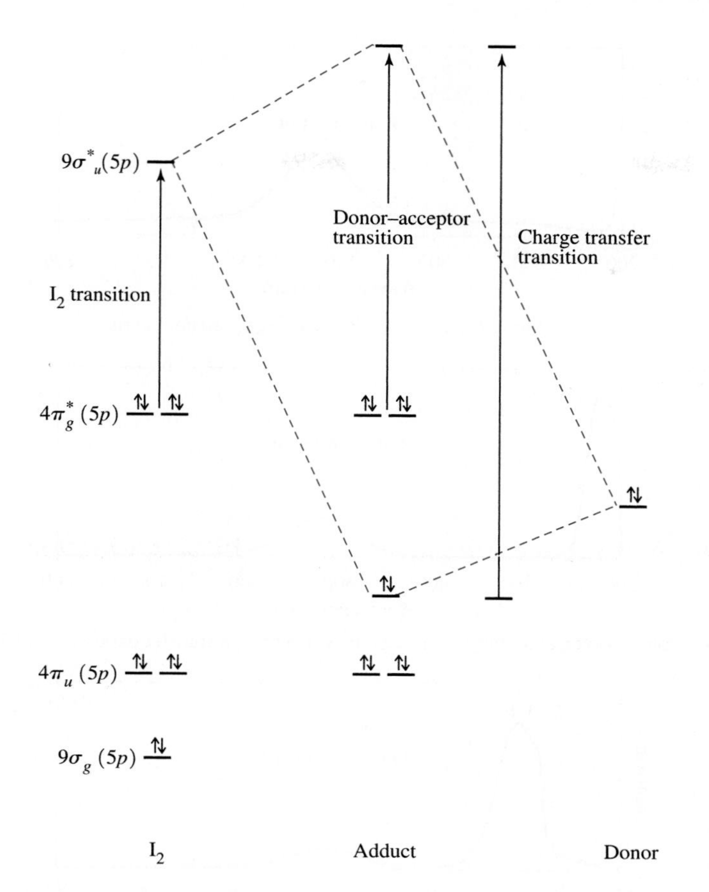

FIGURE 6.10 Electronic Transitions in I_2 Adducts.

from one adduct region to another upon excitation. Charge-transfer phenomena are evident in many adducts, and provide further experimental evidence for the utility of the frontier orbital reactivity model. Charge-transfer in transition-metal complexes is discussed in Chapter 11.

6.4.3 Quantification of Lewis Basicity

Significant effort has been devoted to quantifying Lewis basicity. From this standpoint, **Lewis basicity** is defined as *the thermodynamic tendency of a substance to act as a Lewis base.* Comparative measures of this property are provided by the equilibrium constants for adduct formation of Lewis bases with a common reference acid.[35] A major challenge is identifying a reference acid that is suitable to assess a variety of bases. Because Lewis basicity is a complicated phenomenon that is modified subtly by electronic and steric effects, the basicity ranking for a set of Lewis bases can vary depending on the reference acid used. As previously discussed for proton affinity (**Section 6.3.4**), gas-phase measurements are ideal to measure Lewis basicities without the complication of solvation effects. However, in practice, most thermodynamic data for Lewis bases have been obtained in solution, with important attention paid to solvent selection. An ideal solvent for these studies would dissolve a variety of bases yet not itself react significantly towards these solutes as a Lewis acid.* In addition, a Lewis base may be rendered less basic towards a given Lewis acid in one solvent relative to another solvent. Selecting a single solvent that meets these criteria for many Lewis bases is a tall order.

*In essence, a solvent is desired that will primarily interact with solutes via dispersion forces. Solvents in this category (for example, hydrocarbons) typically are limited to relatively nonpolar solutes. Sometimes more polar solvents must be employed, rendering the quantification of Lewis basicity more complicated as the enthalpy associated with solvation plays an increasing role.

FIGURE 6.11 Spectra of I_2 with Different Bases.

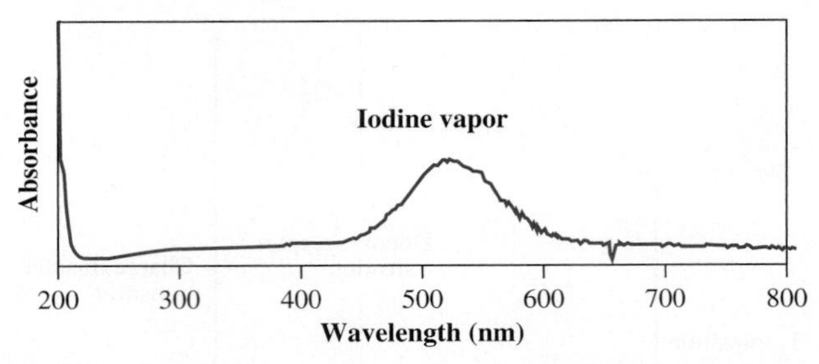

I_2 vapor is purple or violet, absorbing near 520 nm, with no charge-transfer bands.

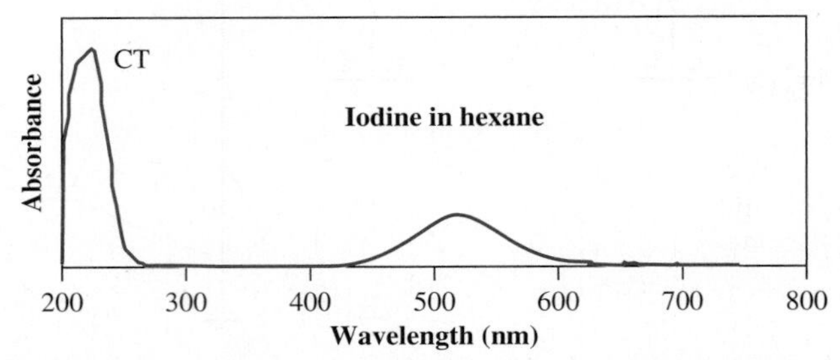

I_2 in hexane is purple or violet, absorbing near 520 nm, with a charge-transfer band at about 225 nm.

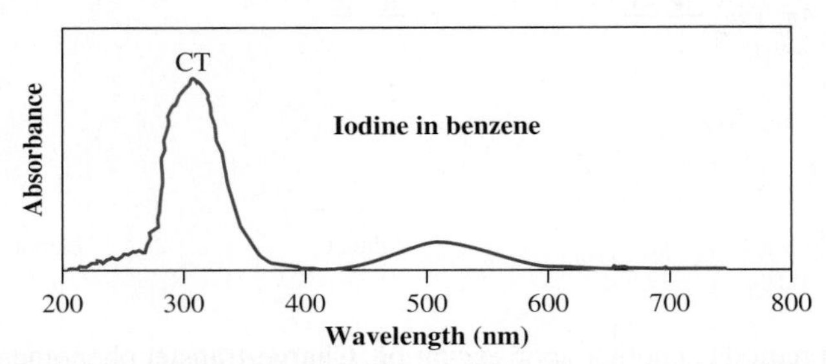

I_2 in benzene is red violet, absorbing near 500 nm, with a charge-transfer band at about 300 nm.

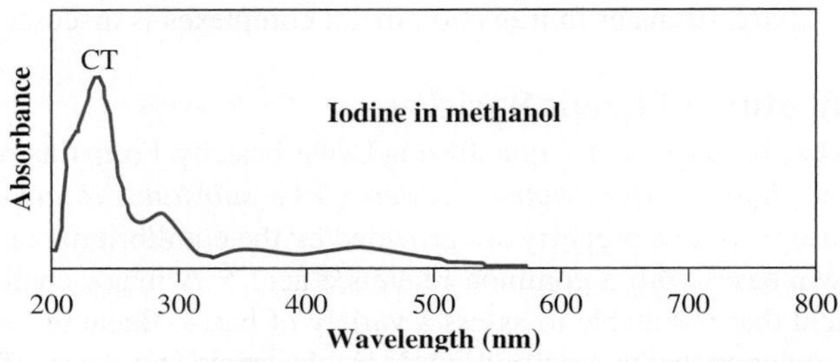

I_2 in methanol is yellow brown, absorbing near 450 nm, with a charge-transfer band near 240 nm and a shoulder at 290 nm.

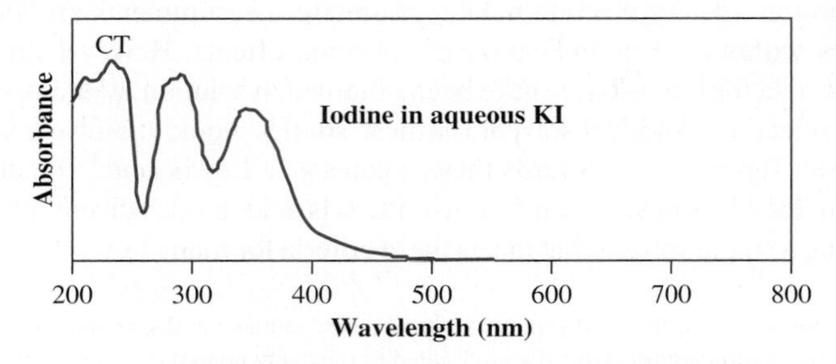

I_2 in aqueous KI is brown, absorbing near 360 nm, with charge-transfer bands at higher energy.

With this experimental design challenge in mind, quantifying Lewis basicity is simple in theory. The equilibrium constants for adduct formation (K_{BA}, more commonly expressed as $\log K_{BA}$) can be ranked via increasing K_{BA} or $\log K_{BA}$ to communicate increasing Lewis basicity of the base employed for adduct formation.

$$\underset{\text{Base}}{\text{B}} + \underset{\text{Acid}}{\text{A}} \rightleftharpoons \underset{\text{Lewis Acid–Base Adduct}}{\text{BA}}$$

$$K_{BA} = \frac{[BA]}{[B][A]}$$

The solubilities of I_2 in different solvents provide a qualitative way to assess the Lewis basicity of these solvents towards I_2. **Table 6.9** provides $\log K_{BA}$ values for the formation of adducts of the Lewis acid I_2 with Lewis bases in the solvents CCl_4 and $CHCl_3$. It is interesting that these five bases exhibit the same Lewis basicity *ranking* towards I_2 in these solvents. However, the absolute basicities are quite different; for example, *N,N*-dimethylformamide is roughly five times more basic towards I_2 in CCl_4 than in $CHCl_3$. And on the basis of these data, $(C_6H_5)_3P{=}Se$ is slightly more basic towards I_2 in $CHCl_3$ than in CCl_4, while all the other bases are more basic towards this acid in CCl_4.

The spectroscopic evidence for I_2 adduct formation (**Section 6.4.2**) suggests that spectroscopic measurement of Lewis basicity is possible. The requirement is that the reference Lewis acid must exhibit a spectroscopic change upon adduct formation (for example, an NMR chemical shift, change in UV-Vis or IR spectrum) that can be attributed primarily to the strength of the coordinate covalent bond within the adduct. While these spectral measurements are generally routine (Figure 6.11), their reliability in accurately assessing Lewis basicity must be confirmed via correlating these data (for example, how much did the chemical shift or visible absorption change?) to K_{BA} or $\Delta H°$ values for the complexation reactions.

6.4.4 The BF_3 Affinity Scale for Lewis Basicity

The Lewis acid BF_3 is the most commonly employed reference to probe Lewis basicity. The affinity of BF_3 towards many bases has been measured in dichloromethane solution where the affinity is defined as the magnitude of the enthalpy change of adduct formation:

$$BF_3 + \text{Lewis Base} \xrightarrow{CH_2Cl_2} \text{Lewis Base—}BF_3 \quad -\Delta H° = BF_3 \text{ Affinity}$$

These enthalpies must be corrected for the $\Delta H°$ for BF_3 dissolving in the solvent. Upon this adjustment, increasing BF_3 affinities indicate stronger coordinate covalent bonding, consequently increasing Lewis basicity of the base towards BF_3. Some BF_3 affinities are listed in **Table 6.10**. Steric and electronic effects suggested by these data will be discussed in **Sections 6.4.6** and **6.4.7**.

TABLE 6.9 $\log K_{BA}$ for I_2• Lewis Base Adducts in Different Solvents (298 K)

Lewis Base	$\log K_{BA}$ in CCl_4	$\log K_{BA}$ in $CHCl_3$
Tetrahydrofuran	0.12	–0.44
N,N-dimethylformamide	0.46	–0.22
$(C_6H_5)_3P{=}O$	1.38	0.89
$(C_6H_5)_3P{=}S$	2.26	2.13
$(C_6H_5)_3P{=}Se$	3.48	3.65

Data from *Lewis Basicity and Affinity Scales Data and Measurement*, John Wiley and Sons, p. 33, 91–101, 295–302.

TABLE 6.10 BF_3 Affinities for Lewis Bases in CH_2Cl_2 (298 K)

Lewis Base	BF_3 Affinity (kJ/mol)
4-dimethylaminopyridine	151.55
trimethylamine	139.53
3-methylpyridine	130.93
4-phenylpyridine	129.50
pyridine	128.08
2-methylpyridine	123.44
2-phenylpyridine	103.34
trimethylphosphine	97.43
tetrahydrofuran	90.40
2-trifluoromethylpyridine	82.46
2-*tert*-butylpyridine	80.10
tetrahydrothiophene	51.62

Data from *Lewis Basicity and Affinity Scales Data and Measurement*, John Wiley and Sons, p. 33, 91–101, 295–302.

Of the Lewis bases in Table 6.10, 4-dimethylaminopyridine has the highest BF_3 affinity and is the strongest Lewis base. What are the extremes in BF_3 affinities? This question is interesting; it communicates the wide range of molecules that can function as Lewis bases to BF_3. The super (Brønsted) base 1,8-diazabicyclo[5.4.0]undec-7-ene (DBU, Figure 6.1) also has a very large BF_3 affinity (159.36 kJ/mol). Very low BF_3 affinities have been reported for ethene (5.4 kJ/mol) and propene (6.9 kJ/mol) in liquid nitrogen. These data cannot be directly compared to the values determined in dichloromethane (Table 6.10), however. In these adducts, the unsaturated hydrocarbons presumably use their π bonding HOMOs as the donor orbitals.[36] The role of olefin–borane complexes as intermediates in chemical synthesis has been proposed.[37]

6.4.5 Halogen Bonds

The reactions of I_2 as a Lewis acid with donor solvents and Lewis bases have been discussed in **Sections 6.4.2** and **6.4.3**. The coordinate covalent bonds formed by the halogens (X_2) and interhalogens (XY, for example, ICl, discussed in **Section 8.9.1**) to Lewis bases are called **halogen bonds**.[38] These long known donor–acceptor interactions, featuring many similarities to hydrogen bonding (**Section 6.5.1**), have been "rediscovered" and show potential in drug design and material science.[39] Halogen bonds typically exhibit approximate 180° angles between the donor atom and the halogen acceptor, consistent with acceptance by the halogen σ^* LUMO, which lies along the halogen bond axis. As an example, the gas-phase structure of the ClF adduct with formaldehyde was determined by rotational spectroscopy, and an O · · · Cl—F angle of 176.8° was found, with the coordinate covalent bond in the expected location based on the formaldehyde HOMO and ClF LUMO.[40] In the gas phase acetylene—Br_2 adduct, the halogen LUMO interacts with the acetylene π bonding HOMO.[40]

F–Cl–O angle = 176.8°

As described in Section 6.4.3, the Lewis acid I_2 is used to catalog Lewis basicity via determination of the equilibrium constants (K_c) associated with adduct formation.

The experimental challenge is determining the I_2 and I_2—Lewis base adduct concentrations; these are often obtained via UV-Visible spectroscopy by examining charge transfer and donor–acceptor transitions (**Section 6.4.2**).

$$I_2 + \text{Lewis base} \underset{}{\overset{\text{alkanes}}{\rightleftharpoons}} I_2\text{—Lewis base}$$

$$K_c = \frac{[I_2\text{—Lewis base}]}{[I_2][\text{Lewis base}]}$$

An extensive compilation of K_c values is available for a variety of main group Lewis bases, with most determined in heptane.[41] There is interest in developing an I_2 affinity scale defined similarly as the BF_3 affinity scale (based on ΔH° of adduct formation). Many ΔH° values have been determined for reactions of I_2 with Lewis bases, but reconciliation of data obtained via different experimental conditions for reliable I_2 affinity comparisons remains a challenge.

How do we expect the bond within a halogen to change upon complexation with a Lewis base? Since adduct formation results in donation into the halogen σ^* LUMO, this bond should weaken and lengthen. The subsequent reduction in the halogen–halogen bond's force constant is shown by a decrease in the bond's stretching frequency, suggesting the possibility of assessing Lewis basicity by determining how much the stretching frequency decreases from that in free halogen. This initially seems a very attractive strategy; a change in spectroscopic property of only one bond could be correlated to Lewis basicity. Unfortunately, vibrational modes that feature considerable halogen or interhalogen stretching generally also include motion of the Lewis basic atom as well. Nevertheless, stretching frequency changes induced in I_2, ICN, and ICl upon complexation have been tabulated for many complexes and correlated to Lewis basicity with reasonable effectiveness.* The infrared stretching frequency of the I—C bond in ICN (485 cm^{-1} uncomplexed) experiences red shifts (to lower energy) upon complexation ranging from 5 cm^{-1} (in benzene—ICN) to 107 cm^{-1} (in quinuclidine—ICN). The range in red shifts observed in I_2 complexes is not as dramatic; a maximum red shift of only 39.5 cm^{-1} was observed in piperidine—I_2.** These red shifts correlate fairly well with $\log K_c$ values, especially when similar Lewis bases are compared.

Section 6.4.2 discussed that the donor–acceptor transition in I_2—Lewis base complexes is modified depending on the extent of the donor interaction with the I_2 LUMO. The blue shift (to higher energy) in the $4\pi_g^* \rightarrow 9\sigma_u^*$ transition upon I_2 complexation has also been correlated to Lewis base strength. As shown in Figure 6.10, this transition increases in energy as the base strength increases. **Table 6.11** lists blue shifts induced by selected bases, used to assess Lewis basicity.

6.4.6 Inductive Effects on Lewis Acidity and Basicity

Rationalization of the Lewis basicity ranking in Table 6.10 requires that inductive effects be considered. Substitution of electronegative atoms or groups, such as fluorine or chlorine, in place of hydrogen on ammonia or phosphine results in weaker bases. The electronegative atom draws electrons toward itself, and as a result, the nitrogen or phosphorus atom has less negative charge, and its lone pair is less readily donated to an acid. For example, PF_3 is a much weaker Lewis base than PH_3. A similar effect in the reverse direction results from substituting alkyl groups for hydrogen. For example, in amines, the alkyl groups contribute electrons to the nitrogen, increasing its negative character and making it a stronger base.

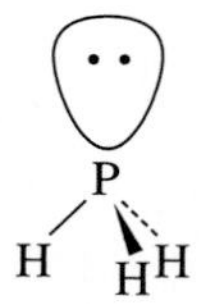

*The I_2 stretching mode is IR inactive, but vibrational modes with significant I—I contribution become active upon complexation. The free I_2 stretching band is 210 cm^{-1} on the basis on Raman spectroscopy.

**For tabulated infrared shifts of ICN, I_2, and ICl as well as $4\pi_g^* \rightarrow 9\sigma_u^*$ blue shifts: C. Laurence and J.-F. Gal, *Lewis Basicity and Affinity Scale Data and Measurement*, John Wiley and Sons, United Kingdom, 2010, pp. 286–306.

TABLE 6.11 Blue Shifts in the $4\pi_g^* \rightarrow 9\sigma_u^*$ Transition in I_2 Complexes in Heptane at 15° C

Lewis Base	Blue Shift (cm^{-1})
Pyridine	4560
Dimethylsulfide	3570
THF	2280
Diethyl ether	1950
Acetonitrile	1610
Toluene	580
Benzene	450

Data from *Lewis Basicity and Affinity Scales Data and Measurement*, John Wiley and Sons, p. 33, 91–101, 295–302.

Additional substitution increases the effect, with the following resulting order of Lewis base strength in the gas phase:

$$NMe_3 > NHMe_2 > NH_2Me > NH_3$$

These **inductive effects** are similar to the effects seen in organic molecules containing electron-contributing or electron-withdrawing groups. Caution is required in applying this idea to other compounds. The boron halides do not follow this argument, because BF_3 and BCl_3 have significant π bonding that increases the electron density on the boron atom. Exclusive consideration of halogen electronegativity would lead to a prediction that BF_3 would be the strongest Lewis acid of the boron halides; the high electronegativity of fluorine atoms would be expected to draw electron density away from the boron most strongly. However, the varying boron–halogen bond lengths also play an important role. The relatively short B—F bonds in BF_3 permit the previously mentioned π bonding to partially quench some Lewis acidity at boron. Developing strategies to gauge the Lewis acidities of boron halides remains a contemporary pursuit.[42,43] Calculations applying coupled cluster theory[41] and the determination of boron's valence deficiency[42] are consistent with the Lewis acidity increasing as $BF_3 \ll BCl_3 < BBr_3 < BI_3$. As the halogen atom size increases, the B—X bond lengthens, and the π interaction that attenuates boron Lewis acidity decreases in importance.

6.4.7 Steric Effects on Lewis Acidity and Basicity

Steric effects also influence acid–base behavior. When bulky groups are forced together by adduct formation, their mutual repulsion makes the reaction less favorable. Brown, a major contributor to these studies,[44] described molecules as having front (F) strain or back (B) strain, depending on whether the bulky groups interfere directly with the approach of an acid and a base to each other, or whether the bulky groups interfere with each other when VSEPR effects force them to bend away from the other molecule forming the adduct. He also called effects from electronic differences within similar molecules internal (I) strain. Reactions involving substituted amines and pyridines were used to sort out these effects.

EXAMPLE 6.2

As discussed in **Section 6.3.6**, reactions of a series of substituted pyridines with hydrogen ions show this order of Brønsted–Lowry base strengths:

2,6-dimethylpyridine > 2-methylpyridine > 2-*t*-butylpyridine > pyridine

This matches the expected order for electron donation by alkyl groups; the *t*-butyl group has counterbalancing inductive and steric effects. However, reaction with larger Lewis acids, such as BF_3 or BMe_3, shows the following order of Lewis basicity:

pyridine > 2-methylpyridine > 2,6-dimethylpyridine > 2-*t*-butylpyridine

Explain the difference between these two series.

The larger fluorine atoms or methyl groups attached to the boron and the groups on the *ortho* position of the substituted pyridines interfere with each other when the molecules approach each other, so reaction with the substituted pyridines is less favorable. Interference is greater with the 2,6-substituted pyridine and greater still for the *t*-butyl substituted pyridine. This is an example of F strain.

EXERCISE 6.4 Based on inductive arguments, would you expect boron trifluoride or trimethylboron to be the stronger acid in reaction with NH_3? Which of these acids would you expect to be stronger towards 2,6-dimethylpyridine or 2-*t*-butylpyridine?

Proton affinities (Table 6.6) show the gas-phase Brønsted–Lowry base strength ranking $Me_3N > Me_2NH > MeNH_2 > NH_3$, as predicted on the basis of electron donation by the methyl groups, resulting in increased electron density at nitrogen.[45] When larger Lewis acids than H^+ are used, the order changes, as shown in **Table 6.12**. With both BF_3 and BMe_3, Me_3N is a much weaker base, with very nearly the same ΔH° for adduct formation as with $MeNH_2$. With the more bulky acid tri(*t*-butyl)boron, the order is nearly reversed from the proton affinity order, although ammonia is still weaker than methylamine. Brown argued that these effects are from crowding of the methyl groups at the back of the nitrogen as the adduct is formed (B strain). It can be argued that some F strain is also present. When triethylamine is used as the base, it does not form a trimethylboron adduct, although the enthalpy change for such a reaction is weakly exothermic.* Initially, this seems to be another example of B strain, but examination of molecular models shows that one ethyl group is normally twisted out to the front of the molecule, where it interferes with adduct formation. When the alkyl chains are linked into rings, as in quinuclidine, adduct formation is more favorable, because the potentially interfering chains are pinned back and do not change on adduct formation.

The proton affinities of quinuclidine and triethylamine are nearly identical, 983.3 and 981.8 kJ/mol (Table 6.6). When mixed with trimethylboron, whose methyl groups are large enough to interfere with the ethyl groups of triethylamine, the quinuclidine reaction is twice as exothermic as that of triethylamine (−84 vs. −42 kJ/mol). Whether the

TABLE 6.12 Rankings of the Enthalpies of Formation for Adduct Formation for Amines and Lewis Acids

Amine	Proton Affinity (Order)	BF_3 (Order)	ΔH° for amine—BMe_3 adduct formation (*kJ/mol*)	BMe_3 (Order)	$B(t-Bu)_3$ (Order)
NH_3	4	4	−57.53	4	2
CH_3NH_2	3	2	−73.81	2	1
$(CH_3)_2NH$	2	1	−80.58	1	3
$(CH_3)_3N$	1	3	−73.72	3	4
$(C_2H_5)_3N$			~−42		
Quinuclidine			−84		
Pyridine			−74.9		

ΔH° Data from H. C. Brown, *J. Chem. Soc.*, **1956**, 1248.

*Keep in mind that all of these adduct formation reactions will have $\Delta S^\circ < 0$, so ΔH° must be sufficiently negative to render $\Delta G^\circ < 0$ for adduct formation.

triethylamine effect is due to interference at the front or the back of the amine is a subtle question, because the interference at the front is indirectly caused by other steric interference at the back between the ethyl groups.

6.4.8 Frustrated Lewis Pairs

Does the presence of excessive steric bulk when a Lewis acid and a Lewis base attempt to form an adduct automatically render these species inert towards each other? The unique behavior of sterically **frustrated Lewis pairs (FLPs)**, pioneered by Stephan, is a vigorous research area with applications for small molecule activation* and catalysis.[46] The highly Lewis acidic and sterically bulky tris(pentafluorophenyl)borane[47] plays a role in many FLP reactions. The great promise of FLP chemistry was revealed by reactions between tris(pentafluorophenyl)borane and tertiary and secondary phosphines, where sterics preclude formation of classic adducts. A seminal example of a frustrated Lewis pair is that of the secondary phosphine di(2,4,6-trimethylphenyl)phosphine that is precluded from forming a classic adduct with tris(pentafluorophenyl)borane. The phosphine Lewis pair is "frustrated" since it cannot interact with boron to form the adduct.

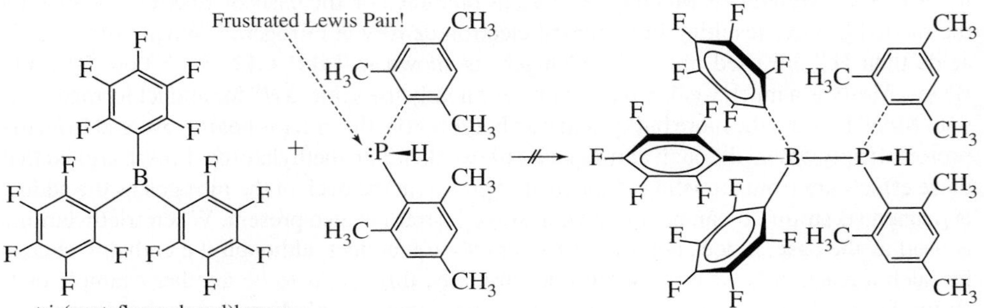

Frustrated Lewis Pair!

tris(pentafluorophenyl)borane

Remarkably, the frustrated Lewis pair engages in nucleophilic attack at the *para* carbon of the borane, affording a **zwitterionic** species** after fluoride migration. This fluoride can be substituted for a hydrogen atom to give a zwitterion that *releases hydrogen gas upon heating*.[48] The product phosphino-borane reacts with hydrogen gas at ambient temperature to reform the zwitterion and is the first non-transition-metal molecule that can reversibly activate the H—H bond and release H_2.

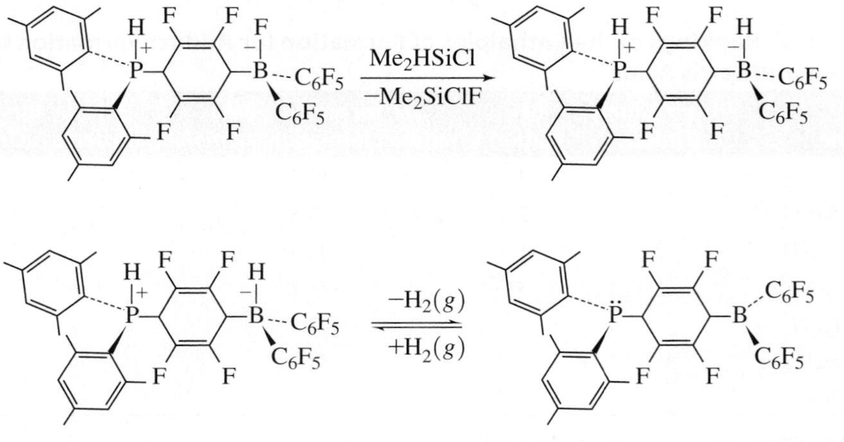

Me_2HSiCl / $-Me_2SiClF$

$-H_2(g)$ / $+H_2(g)$

*"Activation" in this sense refers to making an otherwise inert small molecule chemically reactive.

**A zwitterion is a species that contains at least one formal positive and negative charge.

Spectacular small molecule activation has been achieved with FLPs, including with H_2,[49] CO_2,[50] and N_2O.[51] The chemistry enabled by frustrated Lewis pairs is being exploited to develop non-transition-metal catalysts that promise to substitute for catalysts that contain toxic and expensive heavy metals.[52]

6.5 Intermolecular Forces

The frontier molecular orbital model for donor–acceptor complexes provides a convenient framework for a discussion of hydrogen bonding and receptor–guest interactions.

6.5.1 Hydrogen Bonding

Hydrogen bonding (introduced in Section 3.4) is relevant to most scientific disciplines. The definition of a hydrogen bond in terms of its covalent and ionic contributions continues to be debated. Its traditional definition, a force that arises between molecules that have an H atom bonded to a small, highly electronegative atom with lone pairs, usually N, O, or F, has been considerably broadened. The IUPAC Physical and Biophysical Chemistry Division has recommended a new definition[53] that includes the following themes.

A hydrogen bond X—H···B is formed from an attraction between an X—H unit (where the electronegativity of X is greater than of H), and a donor atom (B). The X—H and B components may be incorporated into larger molecular fragments. The interaction between these components can be either intermolecular or intramolecular. Hydrogen bonds can be described on the basis of varying relative contributions from three components:

- The polarity of X—H leads to an electrostatic contribution.
- The donor-acceptor nature of the interaction results in partial covalent character and charge transfer from B to X—H.
- Dispersion forces also contribute to hydrogen bonds.

The H···B strength increases as the electronegativity of X within the polar covalent X—H bond increases. For example, in the series N—H, O—H, and F—H, the H···B strength is greatest for interaction with F—H.

The IUPAC recommendations emphasize that experimental evidence is necessary to support the existence of a hydrogen bond.[54] Such evidence may be provided in a variety of ways:

- An X—H···B angle of 180° is indicative of a relatively strong hydrogen bond, with a short H···B distance. Increased deviation from X—H···B linearity indicates weaker hydrogen bonds, with longer H···B distances.
- The X—H infrared stretching frequency is red-shifted upon formation of X—H···B, consistent with X—H weakening and lengthening. As the strength of the H···B bond increases, the X—H bond strength decreases. Hydrogen bond formation results in new vibrational modes with H···B contributions.
- The NMR chemical shift of the proton linking X and B is a sensitive function of the hydrogen bond strength. Typically these protons in X—H···B are deshielded relative to X—H. The extent of deshielding is correlated to the hydrogen bond strength.
- Experimental detection of hydrogen bonding requires that the magnitude of ΔG for X—H···B formation be larger than the thermal energy of the system.

The electrostatic component, the attraction of a bonded hydrogen atom to a region of relatively high electron density of the base (B), is considered the *dominant* contribution to most hydrogen bonds. Frontier molecular orbitals and the donor–acceptor model can be applied to understand the covalent contribution.

Very strong hydrogen bonding in the symmetric FHF^- ion was described via molecular orbital theory in **Section 5.4.1**. The key interactions responsible for the covalent contribution to hydrogen bonding can be generated by combining HF molecular orbitals (see the energy-

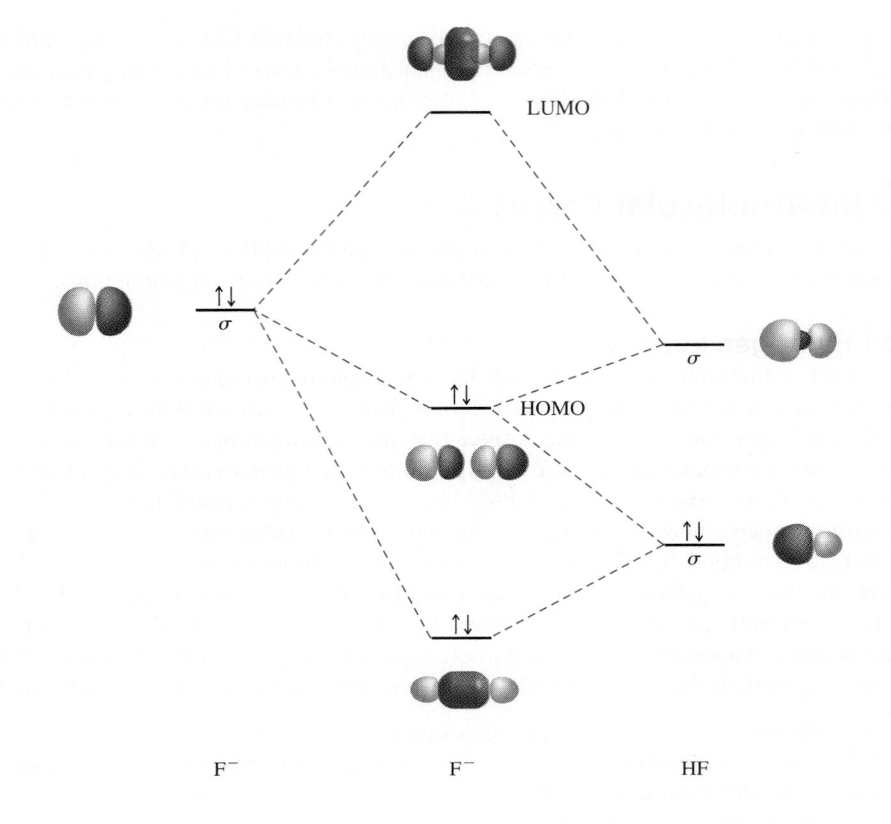

FIGURE 6.12 Orbital Interactions Responsible for the Covalent Contribution to the Hydrogen Bonding in FHF^-. Figure 5.16 shows the full set of molecular orbitals. The orbitals shown here are labeled a_g, b_{1g}, and a_g in Figure 5.16.

level diagram preceding Exercise 5.3) with an atomic orbital of F^-, as shown in **Figure 6.12**. The nonbonding p_x and p_y orbitals on the fluorines of both F^- and HF can be ignored, because there are no matching orbitals on the H atom. The shapes of the other orbitals are appropriate for bonding; overlap of the fluoride $2p_z$ with the σ and σ^* HF orbitals forms three product orbitals. In this case, these three orbitals are all symmetric about the central H nucleus. The lowest orbital is distinctly bonding, the middle (HOMO) orbital is essentially nonbonding, and the highest energy orbital (LUMO) is antibonding with respect to the hydrogen bond. From the donor–acceptor perspective, the fluoride $2p_z$ is the HOMO, and the HF σ^* orbital the LUMO. The "charge-transfer" aspect of hydrogen bonding formally refers to the creation of the HOMO in Figure 6.12, where electron density originally localized on the base (in this case fluoride) gains access to a delocalized orbital that includes both F atoms in FHF^-.

The linear arrangement of X—H· · ·B hydrogen bonds supports this covalent contribution to the bonding since maximum overlap of the donor orbital with the σ^* LUMO of X—H requires approach along the X—H bond axis.

One IUPAC criterion for hydrogen bonding concerns application of infrared spectroscopy to measure the decrease in X—H stretching frequency (from weakening of the X—H bond) that accompanies hydrogen bond formation. Indeed, red shifts in the X—H stretching frequency upon X—H· · ·B formation are used to assess hydrogen bond strengths. A scale based on the change in the $\nu(OH)$ of methanol upon CH_3OH · · ·B formation in CCl_4 has been employed. Approximately 800 $\Delta\nu(OH)$ values, ranging from ~3 cm^{-1} with chloroform as the Lewis base to 488 cm^{-1} with tri(*n*-octyl)amine *N*-oxide as the Lewis base, have been reported.*

*For tabulated $\Delta\nu(OH)$ values see C. Laurence and J.-F. Gal, *Lewis Basicity and Affinity Scales Data and Measurement*, John Wiley and Sons, United Kingdom, 2010, pp. 190–206.

Another IUPAC criterion for hydrogen bonding indicates that the hydrogen in X—H···B is deshielded (shifted downfield) in the ^{1}H NMR spectrum. ^{1}H NMR spectroscopy has emerged as an extremely sensitive probe for hydrogen bonding in solution. The NMR spectra of N—H···B and O—H···B show differences in the ^{1}H chemical shifts for the bridging hydrogen of the hydrogen-bonded complex relative to the chemical shift in free X—H. These shifts range from 8–10 ppm for weak to up to 22 ppm for very strong hydrogen bonds[55] and facilitate the assignment of these resonances. The high resolution of modern high-field NMR spectrometers allows resolution of small shifts due to very weak hydrogen bonds, and NMR methods continue to be developed to study hydrogen bonding.

Predicting the strength of hydrogen bonds solely on the basis of the structures of the participating molecules is a challenge that has been called the *H-bond puzzle*.[56] The crux of this puzzle is that seemingly very similar hydrogen bonds often exhibit dramatically different strengths. The classic example is a comparison of the O—H···O hydrogen bond between two water molecules, and the one between a hydronium ion and a water molecule. The latter $[OH_3\cdots OH_2]^+$ hydrogen bond is roughly six times stronger than the $OH_2\cdots OH_2$ hydrogen bond (approximately 125 kJ/mol vs. 20 kJ/mol) even though both are fundamentally O—H···O interactions![56]

Gilli[55,56] has contributed significantly to developing a *pK_a equalization* approach for predicting the strength of hydrogen bonds in aqueous solution. This concept envisions the hydrogen bond X—H···B as associated with the proton transfer equilibrium between X and B. Three limiting states contribute to varying extents to every hydrogen bond, with increasing contribution from limiting state X···H···B enhancing its strength.

$$\text{X—H}\cdots\text{B} \rightleftharpoons \text{X}\cdots\text{H}\cdots\text{B} \rightleftharpoons \overset{-}{\text{X}}\cdots\text{H—}\overset{+}{\text{B}}$$

1: Hydrogen bond between HX and B   **2**   **3**: Hydrogen bond between X^- and BH^+

It is noteworthy that the frontier molecule orbital approach applied to FHF^- results in a symmetrical hydrogen bond with an extremely high contribution from limiting state **2** ($[F—H—F]^-$); this can be viewed as a three-center-four-electron covalent hydrogen bond. The Brønsted–Lowry acidity of HX, relative to the acidity of the conjugate acid of the Lewis base (in this case, BH^+), assessed by the ΔpK_a of these acids, forecasts the X—H···B hydrogen bond strength:

$$\Delta pK_a(\text{X—H}\cdots\text{B}) = pK_a(\text{HX}) - pK_a(\text{BH}^+)$$

The better matched these pK_a values (the closer ΔpK_a is to 0), the stronger the predicted hydrogen bond in aqueous solution. Gilli has developed ΔpK_a ranges associated with different hydrogen bond strengths, and has supported the *pK_a equalization* approach by an extensive analysis of gas phase and crystallographic data.[56] These ranges permitted the development of a *pK_a Slide Rule* that allows convenient prediction of hydrogen bond strengths in aqueous solution. Correlations between ΔpK_a and hydrogen bond strength according to this approach are listed below.[55]

ΔpKa	Predicted H-Bond Strength
$-3 \le 0 \le 3$	Strong
−11 to −3, 3 to 11	Medium Strong
−15 to −11, 11 to 21	Medium
21 to 31	Medium Weak
> 31	Weak

EXAMPLE 6.3

The pK_a *Slide Rule* approach helps explain the *H-bond puzzle* posed earlier.

Consider the O—H···O hydrogen bond between hydronium ion and water in $[OH_3 \cdots OH_2]^+$. In this case, the hydrogen bond donor (H_3O^+) is identical to the conjugate acid of the hydrogen bond acceptor (H_3O^+ is the conjugate acid of H_2O). The ΔpK_a between these species equals 0; the hydrogen bond is "strong."

Now consider the O—H···O hydrogen bond between two water molecules in $OH_2 \cdots OH_2$. The hydrogen bond donor is now H_2O ($pK_a = 15.7$) and the conjugate acid of the hydrogen bond acceptor is H_3O^+ ($pK_a = -1.7$). The ΔpK_a equals 17.4, and this hydrogen bond falls into the "medium" category.

6.5.2 Receptor–Guest Interactions*

Another important interaction can occur between molecules having extended pi systems, when their pi systems interact with each other to hold molecules or portions of molecules together. Such interactions are important on a large scale—for example, as a component of protein folding and other biochemical processes—and on a truly small scale, as in the function of molecular electronic devices. An interesting recent area of study in the realm of π–π interactions has been the design of concave receptors that can wrap around and attach to fullerenes such as C_{60} in what has been described as a "ball-and-socket" structure. Several such receptors, sometimes called *molecular tweezers* or *clips*, have been designed, involving porphyrin rings, corannulene and its derivatives, and other pi systems.[57] The first crystal structure of such a receptor–guest complex involving C_{60}, dubbed by the authors a "double-concave hydrocarbon buckycatcher," is shown in **Figure 6.13**; it was synthesized simply by mixing approximately equimolar quantities of corannulene derivative $C_{60}H_{24}$ and C_{60} in toluene solution.[58]

The product, called an **inclusion complex**, has two concave corannulene "hands" wrapped around the buckminsterfullerene; the distance between the carbons in the corannulene rings and the matching carbons on C_{60} is consistent with the distance expected for π–π interactions between the subunits of the structure (approximately 350 pm), with a shortest C···C distance of 312.8 pm. With no direct C—C covalent bonding, the binding between the fullerene and the corannulene units is attributed to pure π–π interactions. Subsequently, corannulene itself has been found to engage in host–guest complexing with C_{60}, although the shortest distance between the corannulene and fullerene is slightly longer than in the complex shown in Figure 6.13.[59] The electronic structure of the $C_{60}H_{24}$ complex

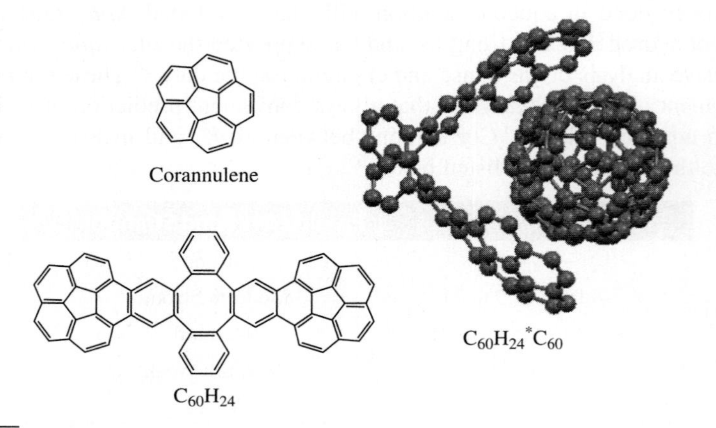

FIGURE 6.13 Receptor–Guest Complex $C_{60}H_{24}$*C_{60}. Molecular structure drawing created using CIF data.

*Also called *receptor–substrate* or *host–guest interactions*.

has been studied in connection with the potential use of such complexes as building units in molecular electronics.[60] Similar receptor–guest complexes involving other molecular tweezers and C_{60} have shown evidence of electron transfer from receptor to the fullerene on absorption of light (another example of charge-transfer transitions), and may be useful in the construction of photovoltaic devices.[61]

6.6 Hard and Soft Acids and Bases

Consider these experimental observations:

1. ***Relative solubilities of halides.*** The solubilities of silver halides in water decrease, going down the column of halogens in the periodic table:

$$AgF(s) \xrightarrow{H_2O(l)} Ag^+(aq) + F^-(aq) \qquad K_{sp} = 205$$

$$AgCl(s) \xrightarrow{H_2O(l)} Ag^+(aq) + Cl^-(aq) \qquad K_{sp} = 1.8 \times 10^{-10}$$

$$AgBr(s) \xrightarrow{H_2O(l)} Ag^+(aq) + Br^-(aq) \qquad K_{sp} = 5.2 \times 10^{-13}$$

$$AgI(s) \xrightarrow{H_2O(l)} Ag^+(aq) + I^-(aq) \qquad K_{sp} = 8.3 \times 10^{-17}$$

   Mercury(I) halides have a similar trend, with Hg_2F_2 the most soluble and Hg_2I_2 the least soluble. However, LiF is by far the *least* soluble of the lithium halides; its K_{sp} is 1.8×10^{-3}, but the other lithium halides are highly soluble in water. Similarly, MgF_2 and AlF_3 are less soluble than the corresponding chlorides, bromides, and iodides. How can one account for these divergent trends?
2. ***Coordination of thiocyanate to metals.*** As we will formally consider in Chapter 9, numerous ions and other groups can act as **ligands**, forming bonds to metal ions. Thiocyanate, SCN^-, has the capacity to bond through either its sulfur or nitrogen. When it bonds to a large, highly polarizable metal ion such as Hg^{2+}, it attaches through sulfur ($[Hg(SCN)_4]^{2-}$); but when it bonds to smaller, less polarizable metals such as Zn^{2+}, it attaches through nitrogen ($[Zn(NCS)_4]^{2-}$). How can this be explained?
3. ***Equilibrium constants of exchange reactions.*** When the ion $[CH_3Hg(H_2O)]^+$—with CH_3 and H_2O attached as ligands to Hg^{2+}—is reacted with other potential ligands, sometimes the reaction is favorable, sometimes not. For example, the reaction with HCl goes nearly to completion:

$$[CH_3Hg(H_2O)]^+ + HCl \rightleftharpoons CH_3HgCl + H_3O^+ \quad K = 1.8 \times 10^{12}$$

   But the reaction with HF does not:

$$[CH_3Hg(H_2O)]^+ + HF \rightleftharpoons CH_3HgF + H_3O^+ \quad K = 4.5 \times 10^{-2}$$

   Is it possible to predict the relative magnitudes of such equilibrium constants?

To rationalize observations such as these, Pearson presented the concept of **hard and soft acids and bases (HSABs)**, designating polarizable acids and bases as **soft** and nonpolarizable acids and bases as **hard**.[62] Much of the hard–soft distinction depends on polarizability, the degree to which a molecule or ion is distorted by interaction with other molecules or ions. Electrons in polarizable molecules can be attracted or repelled by charges on other molecules, forming slightly polar species that can then interact with the other molecules. The HSAB concept is a useful guide to explain acid–base chemistry and other chemical phenomena.* Pearson stated, "Hard acids prefer to bind to hard bases, and

*For early discussions of the principles and theories of the HSAB concept, see R. G. Pearson, *J. Chem. Educ.*, **1968**, *45*, 581, and 643.

soft acids prefer to bind to soft bases." Interactions between two hard or two soft species are stronger than those between one hard and one soft species. The three examples above can be interpreted in such terms, with reactions tending to favor hard–hard and soft–soft combinations.

Relative Solubilities

In these examples, the metal cation is the Lewis acid and the halide is the Lewis base. In the series of silver ion–halide reactions, the iodide ion is much softer (more polarizable) than the others and interacts more strongly with the silver ion, a soft cation. The result is a greater covalent contribution to the bonding in AgI relative to that with the other halides.

The lithium halides have solubilities roughly in the reverse order: LiBr > LiCl > LiI > LiF. The strong hard–hard interaction in LiF overcomes the tendency of LiF to be solvated by water. The weaker hard–soft interactions between Li^+ and the other halides are not strong enough to prevent solvation, and these halides are more soluble than LiF. LiI is out of order, probably because of the poor solvation of the very large iodide ion, but LiI is still much more soluble than LiF. Ahrland, Chatt, and Davies[63] organized these and other phenomena by dividing metal ions into two classes:

Class (a) ions	Class (b) ions
Most metals	Cu^{2+}, Pd^{2+}, Ag^+, Pt^{2+}, Au^+, $Hg_2{}^{2+}$, Hg^{2+}, Tl^+, Tl^{3+}, Pb^{2+}, and heavier transition-metal ions

Class (b) members are located primarily in a small region in the periodic table at the lower right-hand side of the transition metals. **Figure 6.14** identifies the elements that are always in class (b) and those commonly in class (b) when they have low oxidation states. The transition metals exhibit class (b) character in compounds in which their oxidation state is zero (primarily organometallic compounds). The class (b) ions form halides whose solubility is generally in the order $F^- > Cl^- > Br^- > I^-$; the solubility of class (a) halides in water is typically in the reverse order. The class (b) metal ions also have a larger enthalpy of reaction with phosphorus donors than with nitrogen donors, again the opposite of the reactions of class (a) metal ions.

Ahrland, Chatt, and Davies explained the class (b) metals as having *d* electrons available for π bonding.* Elements farther left in the table have more class (b) character in low

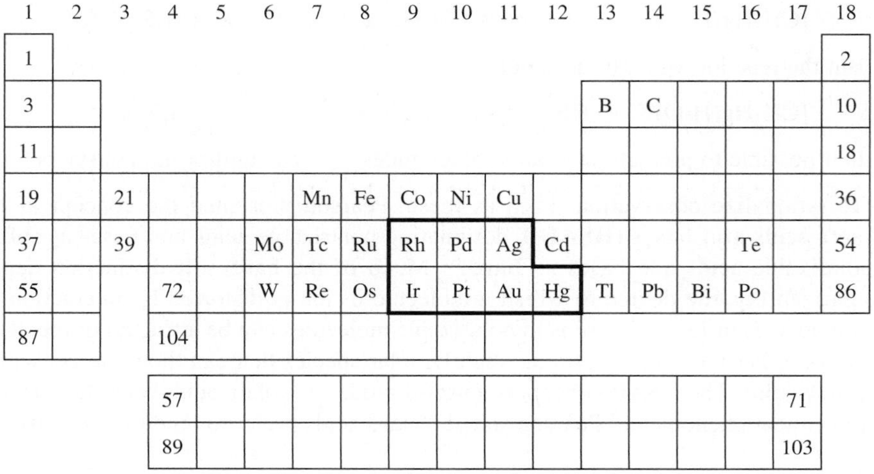

1	2	3	4	5	6	7	8	9	10	11	12	13	14	15	16	17	18
1																	2
3												B	C				10
11																	18
19		21				Mn	Fe	Co	Ni	Cu							36
37		39			Mo	Tc	Ru	Rh	Pd	Ag	Cd				Te		54
55			72		W	Re	Os	Ir	Pt	Au	Hg	Tl	Pb	Bi	Po		86
87			104														
			57													71	
			89													103	

FIGURE 6.14 Location of Class (b) Metals in the Periodic Table. Those in the outlined region are always class (b) acceptors. Others indicated by their symbols are borderline elements, whose behavior depends on their oxidation state and the donor. The remainder (blank) are class (a) acceptors.

(Adapted from *Quarterly Reviews, Chemical Society,* issue 3, 12, 265–276 with permission from The Royal Society of Chemistry.)

*Metal–ligand bonding is discussed in Chapters 10 and 13.

or zero oxidation states when more *d* electrons are present. Donor molecules or ions that have the most favorable enthalpies of reaction with class (b) metals are those that are readily polarizable and may have vacant *d* or π^* orbitals available for π bonding.

Coordination of Thiocyanate to Metals

We can now account for the two SCN^- bonding modes. The Hg^{2+} ion is much larger and more polarizable (softer) than the smaller, harder Zn^{2+}. The softer end of the thiocyanate ion is the sulfur. Consequently, the mercury ion $[Hg(SCN)_4]^{2-}$ is a soft–soft combination, and the zinc ion $[Zn(NCS)_4]^{2-}$ is a hard–hard combination, consistent with the HSAB prediction. Other soft cations such as Pd^{2+} and Pt^{2+} form thiocyanate complexes attached through the softer sulfur; harder cations such as Ni^{2+} and Cu^{2+} form N-bonded thiocyanates. Intermediate transition-metal ions can in some cases bond to either end of thiocyanate. For example, both $[Co(NH_3)_5(SCN)]^{2+}$ and $[Co(NH_3)_5(NCS)]^{2+}$ are known, forming an example of "linkage" isomers (Chapter 9).

Equilibrium Constants of Exchange Reactions

Consider the data in **Table 6.13** for reactions of aqueous methylmercury(II) cations

$$[CH_3Hg(H_2O)]^+ + BH^+ \rightleftharpoons [CH_3HgB]^+ + H_3O^+$$

These may be considered *exchange reactions* in that they involve an exchange of water and base B on the mercury. They may also be considered examples of competition between H_2O and B for coordination at mercury—and also between H_2O and B for binding to H^+. In reactions 1 through 4 the trend is clear: as the halide becomes larger and more polarizable ($F^- \rightarrow I^-$), the tendency for attachment to mercury(II) grows stronger—a soft–soft interaction between mercury(II) and the increasingly soft halide ion. Among the other examples, the relatively soft sulfur atoms in reactions 6 and 7 also can be viewed as leading to a soft–soft interaction with Hg^{2+} and a large equilibrium constant. In reactions 1 and 5, on the other hand, the harder F and O atoms from HF and H_2O are less able to compete for attachment to the soft mercury(II), and the equilibrium constants are small.

6.6.1 Theory of Hard and Soft Acids and Bases

Pearson[64] designated the class (a) metal ions of Ahrland, Chatt, and Davies as *hard acids* and the class (b) ions as *soft acids*. Bases are also classified as hard or soft on the basis of polarizability: the halide ions range from F^-, a very hard base, through less hard Cl^- and Br^- to I^-, a soft base. Reactions are more favorable for hard–hard and soft–soft interactions than for a mix of hard and soft reactants. Hard acids and bases are relatively small, compact, and nonpolarizable; soft acids and bases are larger and more polarizable. The hard acids include cations with a large positive charge (3+ or larger) or those whose

TABLE 6.13 Equilibrium Constants for Exchange Reactions of Mercury Complexes[65]

Reaction	K (298 K)
1. $[CH_3Hg(H_2O)]^+ + HF \rightleftharpoons CH_3HgF + H_3O^+$	4.5×10^{-2}
2. $[CH_3Hg(H_2O)]^+ + HCl \rightleftharpoons CH_3HgCl + H_3O^+$	1.8×10^{12}
3. $[CH_3Hg(H_2O)]^+ + HBr \rightleftharpoons CH_3HgBr + H_3O^+$	4.2×10^{15}
4. $[CH_3Hg(H_2O)]^+ + HI \rightleftharpoons CH_3HgI + H_3O^+$	1×10^{18}
5. $[CH_3Hg(H_2O)]^+ + H_2O \rightleftharpoons CH_3HgOH + H_3O^+$	5×10^{-7}
6. $[CH_3Hg(H_2O)]^+ + SH^- \rightleftharpoons [CH_3HgS]^- + H_3O^+$	1×10^{7}
7. $[CH_3Hg(H_2O)]^+ + HSCN \rightleftharpoons CH_3HgSCN + H_3O^+$	5×10^{6}

d electrons are relatively unavailable for π bonding (e.g., alkaline earth ions, Al^{3+}). Hard acid cations that do not fit this description include Cr^{3+}, Mn^{2+}, Fe^{3+}, and Co^{3+}. Soft acids are those whose *d* electrons or orbitals are readily available for π bonding (neutral and 1+ cations, heavier 2+ cations).

In addition, the larger and more massive the atom, the softer it is likely to be, because the large numbers of inner electrons shield the outer ones and make the atom more polarizable. This description fits the class (b) ions well, because they are primarily 1+ and 2+ ions with filled or nearly filled *d* orbitals, and most are in the second and third rows of the transition elements, with 45 or more electrons. **Tables 6.14** and **6.15** list bases and acids in terms of their hardness or softness.

The trends in bases are easier to see: fluoride is hard and iodide is soft. Again, more electrons and larger sizes lead to softer behavior. S^{2-} is softer than O^{2-} because it has more electrons spread over a larger volume, making S^{2-} more polarizable. Within a group, such comparisons are easy; as the electronic structure and size change, comparisons become more difficult but are still possible. Thus, S^{2-} is softer than Cl^-, which has the same electronic structure, because S^{2-} has a smaller nuclear charge and is larger. Soft acids tend to react with soft bases, and hard acids with hard bases, so the reactions produce hard–hard and soft–soft combinations. Quantitative measures of hard–soft parameters are described in **Section 6.6.2**.

TABLE 6.14 Hard and Soft Bases

Hard Bases	Borderline Bases	Soft Bases
		H^-
F^-, Cl^-	Br^-	I^-
H_2O, OH^-, O^{2-}		H_2S, SH^-, S^{2-}
ROH, RO^-, R_2O, CH_3COO^-		RSH, RS^-, R_2S
NO_3^-, ClO_4^-	NO_2^-, N_3^-	SCN^-, CN^-, RNC, CO
CO_3^{2-}, SO_4^{2-}, PO_4^{3-}	SO_3^{2-}	$S_2O_3^{2-}$
NH_3, RNH_2, N_2H_4	$C_6H_5NH_2$, C_5H_5N, N_2	PR_3, $P(OR)_3$, AsR_3, C_2H_4, C_6H_6

Classifications from R. G. Pearson, *J. Chem. Educ*, **1968**, *45*, 581.

TABLE 6.15 Hard and Soft Acids

Hard Acids	Borderline Acids	Soft Acids
H^+, Li^+, Na^+, K^+		
Be^{2+}, Mg^{2+}, Ca^{2+}, Sr^{2+}		
BF_3, BCl_3, $B(OR)_3$	$B(CH_3)_3$	BH_3, Tl^+, $Tl(CH_3)_3$
Al^{3+}, $Al(CH_3)_3$, $AlCl_3$, AlH_3		
Cr^{3+}, Mn^{2+}, Fe^{3+}, Co^{3+}	Fe^{2+}, Co^{2+}, Ni^{2+}, Cu^{2+}, Zn^{2+}, Rh^{3+}, Ir^{3+} Ru^{3+}, Os^{2+}	Cu^+, Ag^+, Au^+, Cd^{2+}, Hg_2^{2+}, Hg^{2+}, CH_3Hg^+, $[CO(CN)_5]^{2-}$, Pd^{2+}, Pt^{2+}, Pt^{4+}, Br_2, I_2
Ions with formal oxidation states of 4 or higher		Metals with zero oxidation state
HX (hydrogen-bonding molecules)		π acceptors: e.g., trinitrobenzene, quinines, tetracyanoethylene

Classifications from R. G. Pearson, *J. Chem. Educ*, **1968**, *45*, 581.

EXAMPLE 6.4

Is OH^- or S^{2-} more likely to form insoluble salts with 3+ transition-metal ions? Which is more likely to form insoluble salts with 2+ transition-metal ions?

Because OH^- is hard and S^{2-} is soft, OH^- is more likely to form insoluble salts with 3+ transition-metal ions (hard), and S^{2-} is more likely to form insoluble salts with 2+ transition-metal ions (borderline or soft).

EXERCISE 6.5 Some of the products of the following reactions are insoluble, and some form soluble adducts. Consider only the HSAB characteristics in your answers.

a. Will Cu^{2+} react more favorably with OH^- or NH_3? With O^{2-} or S^{2-}?

b. Will Fe^{3+} react more favorably with OH^- or NH_3? With O^{2-} or S^{2-}?

c. Will Ag^+ react more favorably with NH_3 or PH_3?

d. Will Fe, Fe^{2+}, or Fe^{3+} react more favorably with CO?

An acid or a base can be described as hard or soft and at the same time as strong or weak. Both characteristics must be considered to rationalize reactivity. For example, if the reactivity of two bases with similar softness is being compared, consideration of which base is stronger (from a Brønsted–Lowry perspective) can be helpful to assess which side of an equilibrium will be favored. For example, consider this classic organic reaction for the synthesis of phenyllithium.

n-BuLi + C₆H₅Br ⟶ *n*-BuBr + C₆H₅Li

soft–hard (under *n*-BuLi); *soft–hard* (under phenyllithium)

Carbon-based nucleophiles are generally regarded as soft bases, and Li^+ is a hard acid. From this perspective alone, it would be difficult to rationalize that this reaction heavily favors the products, since the softness of the bases *n*-butyl and phenyl carbanions is similar. The significantly higher Brønsted–Lowry basicity of the *n*-butyl carbanion (pK_a of *n*-butane is ~50) relative to the phenyl carbanion (pK_a of benzene is ~43) plays a vital role in rationalizing the formation of phenyllithium.

An oversimplified perspective considers the hard–hard interactions as primarily electrostatic (with a high ionic contribution), with the LUMO of the acid far above the HOMO of the base and relatively little change in orbital energies on adduct formation.[66] A soft–soft interaction involves HOMO and LUMO energies that are much closer and give a large change in orbital energies on adduct formation. Diagrams of such interactions, as shown in **Figure 6.15**, need to be used with caution. The small drop in energy in the hard–hard case only suggests a small covalent contribution to the bonding. Hard–hard interactions feature a relatively large ionic contribution that is not effectively represented in Figure 6.15. In this way, the small amount of stabilization suggested for hard–hard combinations often underestimates the bond strength. In many of the reactions described by the HSAB concept, the hard–hard interactions form stronger bonds relative to the soft–soft interactions.

6.6.2 HSAB Quantitative Measures

There are two major HSAB quantitative measures. One, developed by Pearson,[67] uses the hard–soft terminology and defines the **absolute hardness**, η, as half the difference between the ionization energy and the electron affinity (both in eV):

$$\eta = \frac{I - A}{2}$$

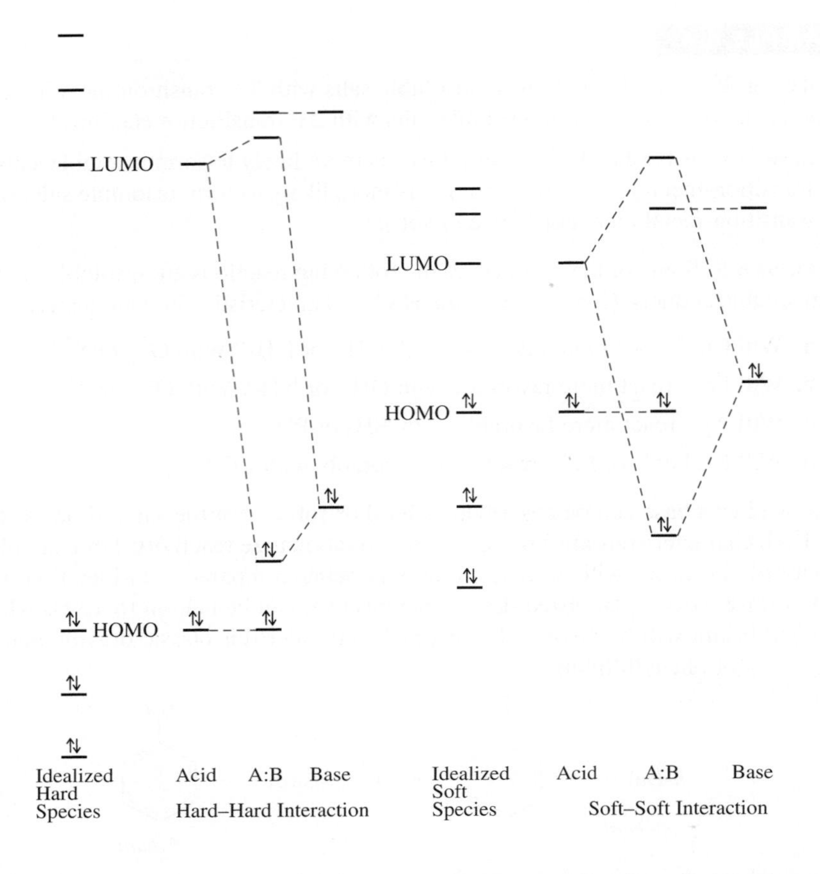

FIGURE 6.15 HOMO–LUMO Diagrams for Hard–Hard and Soft–Soft Interactions.

This definition of hardness is related to Mulliken's definition of electronegativity, called *absolute electronegativity* by Pearson:

$$\chi = \frac{I + A}{2}$$

This approach describes a hard acid or base as a species that has a large difference between its ionization energy and its electron affinity. Ionization energy is assumed to measure the energy of the HOMO, and electron affinity is assumed to measure the LUMO for a given molecule:

$$E_{HOMO} = -I \qquad E_{LUMO} = -A$$

Softness is defined as the inverse of hardness, $\sigma = \frac{1}{\eta}$. Because there are no electron affinities for anions, the values for the atoms and corresponding neutral molecules are used as approximate equivalents.

The halogens offer good examples of the use of these arguments. For the halogens, the trend in hardness (η) parallels the change in HOMO energies, because the LUMO energies are nearly the same, as shown in **Figure 6.16**. Fluorine is the most electronegative halogen. It is also the smallest and least polarizable halogen and is therefore the hardest. In orbital terms, the LUMOs of the halogen molecules have very similar energies, and the HOMOs increase in energy from F_2 to I_2. The absolute electronegativities decrease in the order $F_2 > Cl_2 > Br_2 > I_2$ as the HOMO energies increase. The hardness also decreases in the same order in which the difference between the HOMO and LUMO decreases. Data for a number of other species are given in **Table 6.16** and Appendix B-5.

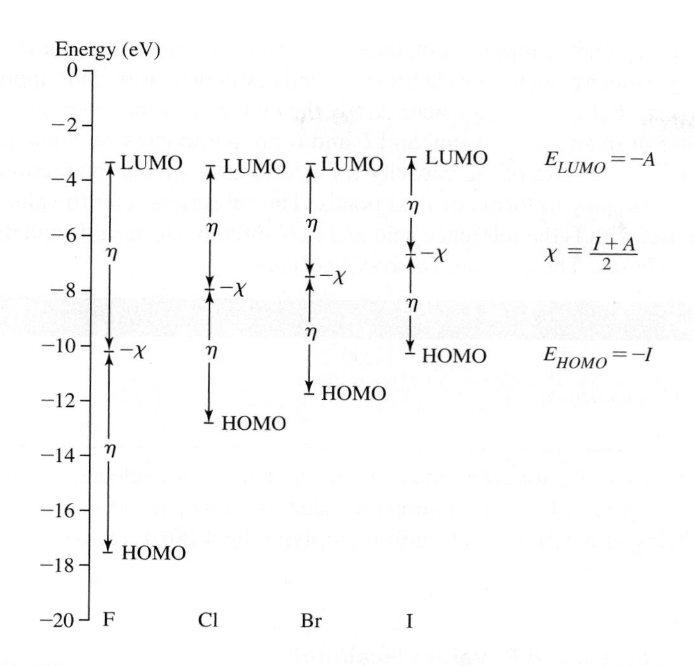

FIGURE 6.16 Energy Levels for Halogens. Relationships between absolute electronegativity (χ), absolute hardness (η), and HOMO and LUMO energies for the halogens.

EXERCISE 6.6

Confirm the absolute electronegativity and absolute hardness values for the following species, using data from Table 6.16 and Appendix B-5:

a. Al^{3+}, Fe^{3+}, Co^{3+}

b. OH^-, Cl^-, NO_2^-

c. H_2O, NH_3, PH_3

TABLE 6.16 Hardness Parameters (eV)

Ion	I	A	χ	η
Al^{3+}	119.99	28.45	74.22	45.77
Li^+	75.64	5.39	40.52	35.12
Na^+	47.29	5.14	26.21	21.08
K^+	31.63	4.34	17.99	13.64
Au^+	20.5	9.23	14.90	5.6
BF_3	15.81	−3.5	6.2	9.7
H_2O	12.6	−6.4	3.1	9.5
NH_3	10.7	−5.6	2.6	8.2
PF_3	12.3	−1.0	5.7	6.7
PH_3	10.0	−1.9	4.1	6.0
F^-	17.42	3.40	10.41	7.01
Cl^-	13.01	3.62	8.31	4.70
Br^-	11.84	3.36	7.60	4.24
I^-	10.45	3.06	6.76	3.70

NOTE: The anion values are approximated as the same as the parameters for the corresponding neutral radicals or atoms.
Data from R. G. Pearson, *Inorg. Chem.*, **1988**, *27*, 734.

Drago and Wayland[68] proposed a quantitative system of acid–base parameters to account for reactivity by explicitly including electrostatic and covalent factors. This approach uses the equation $-\Delta H = E_A E_B + C_A C_B$ where ΔH is the enthalpy of the reaction $A + B \longrightarrow AB$ in the gas phase or in an inert solvent, and E and C are parameters calculated from experimental data. E is a measure of the capacity for electrostatic (ionic) interactions, and C is a measure of the tendency to form covalent bonds. The subscripts refer to values assigned to the acid and base, with I_2 the reference acid and *N,N*-dimethylacetamide and diethyl sulfide as the reference bases. These values (in *kcal/mol*) are

	C_A	E_A	C_B	E_B
I_2	1.00	1.00		
N,N-dimethylacetamide				1.32
Diethyl sulfide			7.40	

Values of E_A and C_A for selected acids and E_B and C_B for selected bases are given in **Table 6.17** and Appendix B-6. Combining the values of these parameters for acid–base pairs gives the enthalpy of reaction (kcal/mol; multiplying by 4.184 J/cal converts to joules).

TABLE 6.17 C_A, E_A, C_B, and E_B Values (kcal/mol)

Acid	C_A	E_A
Trimethylboron, $B(CH_3)_3$	1.70	6.14
Boron trifluoride (gas), BF_3	1.62	9.88
Trimethylaluminum, $Al(CH_3)_3$	1.43	16.9
Iodine, I_2	1.00*	1.00*
Trimethylgallium, $Ga(CH_3)_3$	0.881	13.3
Iodine monochloride, ICl	0.830	5.10
Sulfur dioxide, SO_2	0.808	0.920
Phenol, C_6H_5OH	0.442	4.33
tert-butyl alcohol, C_4H_9OH	0.300	2.04
Pyrrole, C_4H_4NH	0.295	2.54
Chloroform, $CHCl_3$	0.159	3.02
Base	C_B	E_B
Quinuclidine, $HC(C_2H_4)_3N$	13.2	0.704
Trimethylamine, $(CH_3)_3N$	11.54	0.808
Triethylamine, $(C_2H_5)_3N$	11.09	0.991
Dimethylamine, $(CH_3)_2NH$	8.73	1.09
Diethyl sulfide, $(C_2H_5)_2S$	7.40*	0.399
Pyridine, C_5H_5N	6.40	1.17
Methylamine, CH_3NH_2	5.88	1.30
Ammonia, NH_3	3.46	1.36
Diethyl ether, $(C_2H_5)_2O$	3.25	0.963
N,N-dimethylacetamide, $(CH_3)_2NCOCH_3$	2.58	1.32*
Benzene, C_6H_6	0.681	0.525

*Reference values.

Data from R. S. Drago, *J. Chem. Educ.*, **1974**, *51*, 300.

Most acids have lower C_A values and higher E_A values than I_2. Because I_2 has no permanent dipole, it has little electrostatic attraction for bases and a low E_A. On the other hand, its high polarizability equips I_2 reasonably well for covalent interactions, indicated by a relatively large C_A. Because 1.00 was chosen as the reference value for both parameters for I_2, most C_A values are below 1, and most E_A values are above 1.

The example of iodine and benzene shows how these parameters can be used:

$$\underset{\text{acid}}{I_2} + \underset{\text{base}}{C_6H_6} \longrightarrow I_2 \cdot C_6H_6$$

$$-\Delta H = E_A E_B + C_A C_B \quad \text{or} \quad \Delta H = -(E_A E_B + C_A C_B)$$

$$\Delta H = -([1.00 \times 0.681] + [1.00 \times 0.525]) = -1.206 \text{ kcal/mol, or } -5.046 \text{ kJ/mol}$$

The experimental ΔH is -1.3 kcal/mol, or -5.5 kJ/mol, 9% larger than the predicted value.[69] In many cases, the agreement between calculated and experimental enthalpies is within 5%. The agreement between the calculated and experimental values tends to improve with the strength of the adduct bond; the bonding in $I_2 \cdot C_6H_6$ is rather weak.

EXAMPLE 6.5

Calculate the enthalpy of adduct formation predicted by Drago's E and C equation for the reactions of I_2 with diethyl ether and diethyl sulfide.

	E_A E_B C_A C_A ΔH(kcal/mol)	Experimental ΔH
Diethyl ether	$-([1.00 \times 0.963] + [1.00 \times 3.25]) = -4.21$	-4.2
Diethyl sulfide	$-([1.00 \times 0.339] + [1.00 \times 7.40]) = -7.74$	-7.8

Agreement is very good, with the product $C_A \times C_B$ by far the dominant factor. The softer sulfur reacts more strongly with the soft I_2.

EXERCISE 6.7 Calculate the enthalpy of adduct formation predicted by Drago's E and C equation for the following combinations, and explain the trends in terms of the electrostatic and covalent contributions:

a. BF_3 reacting with ammonia, methylamine, dimethylamine, and trimethylamine

b. Pyridine reacting with trimethylboron, trimethylaluminum, and trimethylgallium

Drago's system emphasizes the two primary electronic factors involved in acid–base strength (the capability of a species to engage in electrostatic and covalent interactions) in his two-term enthalpy of reaction equation. Pearson put greater emphasis on the covalent factor. Pearson[70] proposed the equation $\log K = S_A S_B + \sigma_A \sigma_B$, with the inherent strength S modified by a softness factor σ. Larger values of strength and softness lead to larger equilibrium constants. Although Pearson attached no numbers to this equation, it does show the need to consider more than just hardness or softness in acid–base reactions. Absolute hardness is based on orbital energies, returns to a single parameter, and considers only gas-phase reactions. When E and C parameters are available for the reactions in question, quantitative comparisons regarding adduct bond strengths can be made. The qualitative HSAB approach, while sometimes challenging to apply due to the need to make judgments, provides a guide for predicting reactions. The terms *hard* and *soft* are entrenched in the chemical vernacular.

Solvation contributes to the driving force of many reactions, but neither the Drago nor the Pearson quantitative models take this factor into account. It is very difficult to develop broadly applicable models of reactivity. However, chemists continue to examine these important questions that probe the most fundamental contributions towards bond formation.

References

1. R. P. Bell, *The Proton in Chemistry*, 2nd ed., Cornell University Press, Ithaca, NY, 1973, p. 9.
2. H. Lux, *Z. Electrochem.*, **1939**, *45*, 303; H. Flood and T. Förland, *Acta Chem. Scand.*, **1947**, *1*, 592, 718.
3. M. Usanovich, *Zh. Obshch. Khim.*, **1939**, *9*, 182; H. Gehlen, *Z. Phys. Chem.*, **1954**, *203*, 125; H. L. Finston and A. C. Rychtman, *A New View of Current Acid–Base Theories*, John Wiley & Sons, New York, 1982, pp. 140–146.
4. C. K. Ingold, *J. Chem. Soc.*, **1933**, 1120; *Chem. Rev.*, **1934**, *15*, 225; *Structure and Mechanism in Organic Chemistry*, Cornell University Press, Ithaca, NY, 1953, Chapter V; W. B. Jensen, *The Lewis Acid–Base Concepts*, Wiley Interscience, New York, 1980, pp. 58–59.
5. R. Robinson, *Outline of an Electrochemical (Electronic) Theory of the Course of Organic Reactions*, Institute of Chemistry, London, 1932, pp. 12–15.
6. J. N. Brønsted, *Rec. Trav. Chem.*, **1923**, *42*, 718; G. N. Lewis, *Valence and the Structure of Atoms and Molecules*, Chemical Catalog, New York, 1923, pp. 141–142; *J. Franklin Inst.*, **1938**, *226*, 293.
7. T. M. Lowry, *Chem. Ind. (London)*, **1923**, *42*, 43.
8. G. A. Olah, A. Germain, H. C. Lin, D. Forsyth, *J. Am. Chem. Soc.* **1975**, *97*, 2928.
9. G. A. Olah, D. A. Klumpp, *Superelectrophiles and Their Chemistry*, Wiley: New York, **2008**.
10. R. R. Naredla, C. Zheng, S. O. Nilsson Lill, D. A. Klumpp, *J. Am. Chem. Soc.* **2011**, *133*, 13169.
11. L. P. Hammett, A. J. Deyrup, *J. Am. Chem. Soc.*, **1932**, *54*, 2721.
12. S. Mukhopadhyay, M. Zerella, A. T. Bell, R. Vijay Srinivas, G. A. Smith, *Chem. Commun.* **2004**, 472.
13. S. Mukhopadhyay, A. T. Bell, *Angew. Chem., Int. Ed.* **2003**, *42*, 2990.
14. R. Minkwitz, A. Kormath, W. Sawodny, J. Hahn, *Inorg. Chem.*, **1996**, *35*, 3622 and references therein.
15. R. Minkwitz, A. Kornath, W. Sawodny, V. Typke, J. A. Boatz, *J. Am. Chem. Soc.* **2000**, *122*, 1073.
16. K. O. Christe, C. J. Schack, R. D. Wilson, *Inorg. Chem.*, **1975**, *14*, 2224; K. O. Christe, P. Charpin, E. Soulie, R. Bougon, J. Fawcett, D. R. Russell, *Inorg. Chem.*, **1984**, 3756; D. Zhang, S. J. Rettig, J. Trotter, F. Aubke, *Inorg. Chem.*, **1996**, *35*, 6113; R. Minkwitz, S. Schneider, A. Kornath, *Inorg. Chem.*, **1998**, *37*, 4662.
17. K. Lutar, B. Zemva, H. Bormann, *Eur. J. Solid-State Inorg. Chem.*, **1996**, *33*, 957; Z. Mazej, P. Benkic, K. Lutar, B. Zema, *J. Fluorine Chem.*, **2002**, *130*, 399.
18. Z. Mazej, E. Goreshnik, *J. Fluorine Chem.*, **2009**, *130*, 399.
19. G. Tavcar, Z. Mazej, *Inorg. Chim. Acta.*, **2011**, *377*, 69.
20. E. M. Arnett, E. J. Mitchell, T. S. S. R. Murty, *J. Am. Chem. Soc.*, **1974**, *96*, 3874.
21. C. Laurence and J. -F. Gal, *Lewis Basicity and Affinity Scales Data and Measurement*, John Wiley and Sons, United Kingdom, 2010.
22. R. S. Drago, *Physical Methods in Chemistry*, W. B. Saunders, Philadelphia, 1977, pp. 552–565.
23. J. -F. Gal, P. -C. Maria, E. D. Raczynska, *J. Mass Spectrom.* **2001**, *36*, 699 and references therein.
24. E. P. L. Hunter, S. G. Lias., Evaluated gas phase basicities and proton affinities of molecules: An Update. *J. Phys. Chem. Ref. Data*, **1998**, *27*, 413; E. P. L. Hunter, S. G. Lias., Proton Affinity and Basicity Data, in NIST Chemistry Web Book, Standard Reference Database, **2005**, No. 69 (eds W. G. Mallard and P. J. Linstrom), National Institutes of Standards and Technology, Gaithersburg, MD (http://webbook.nist.gov/chemistry, accessed May 2012).
25. H. A. Staab, T. Saupe, *Angew. Chem. Int. Ed. Engl.* **1998**, *27*, 865.; Alder, R. W. *Chem. Rev.* **1989**, *89*, 1251.
26. J. G. Verkade, P. B. Kisanga, *Tetrahedron*, **2003**, *59*, 7819.
27. M. Decouzon, J.-F. Gal, P.-C. Maria, E. D. Raczyńska, *Rapid Commun. Mass Spectrom*, **1993**, *7*, 599; E. D. Raczyńska, P.-C. Maria, J.-F. Gal, M. Decouzon, *J. Phys. Org. Chem.*, **1994**, *7*, 725. The definition of a superbase as one more basic than 1,8-bis(dimethylamino)naphthalene has been proposed in reference 23.
28. Z. Tian, A. Fattachi, L. Lis, S. R. Kass, *Croatica Chemica Acta.*, **2009**, *82*, 41.
29. S. M. Bachrach, C. C. Wilbanks, *J. Org. Chem.*, **2010**, *75*, 2651.
30. N. Uchida, A. Taketoshi, J. Kuwabara, T. Yamamoto, Y. Inoue, Y. Watanabe, T. Kanbara, *Org. Lett.*, **2010**, *12*, 5242.
31. J. Galeta, M. Potáček, *J. Org. Chem.*, **2012**, *77*, 1010.
32. H. L. Finston, A. C. Rychtman, *A New View of Current Acid–Base Theories*, John Wiley & Sons, New York, 1982, pp. 59–60.
33. G. N. Lewis, *Valence and the Structure of Atoms and Molecules*, Chemical Catalog, New York, 1923, pp. 141–142; *J. Franklin Inst.*, **1938**, *226*, 293.
34. W. B. Jensen, *The Lewis Acid–Base Concepts*, Wiley Interscience, New York, 1980, pp. 112–155.
35. V. Gold, *Pure. Appl. Chem.*, **1983**, *55*, 1281.
36. W. A. Herrebout, B. J. van der Veken, *J. Am. Chem. Soc.*, **1997**, *119*, 10446.
37. X. Zhao, D. W. Stephan, *J. Am. Chem. Soc.*, **2011**, *133*, 12448.
38. (a) P. Metrangolo, H. Neukirch, T. Pilati, G. Resnati, *Acc. Chem. Res.*, **2005**, *38*, 386; (b) P. Metrangolo, G. Resnati, *Chem. Eur. J.*, **2001**, *7*, 2511.
39. A.-C. C. Carlsson, J. Gräfenstein, A. Budnjo, J. L. Laurila, J. Bergquist, A. Karim, R. Kleinmaier, U. Brath, M. Erdélyi, *J. Am. Chem. Soc.*, **2012**, *134*, 5706.
40. A. C. Legon, *Angew. Chem. Int. Ed.*, **1999**, *38*, 2687.
41. C. Laurence and J.-F. Gal, *Lewis Basicity and Affinity Scales Data and Measurement*, John Wiley and Sons, United Kingdom, 2010, pp. 241–282.
42. D. J. Grant, D. A. Dixon, D. Camaioni, R. G. Potter, K. O. Christe, *Inorg. Chem.*, **2009**, *48*, 8811.
43. J. A. Plumley, J. D. Evanseck, *J. Phys. Chem. A*, **2009**, *113*, 5985.

44. H. C. Brown, *J. Chem. Soc.*, **1956**, 1248.
45. M. S. B. Munson, *J. Am. Chem. Soc.*, **1965**, *87*, 2332; J. I. Brauman and L. K. Blair, *J. Am. Chem. Soc.*, **1968**, *90*, 6561; J. I. Brauman, J. M. Riveros, and L. K. Blair, *J. Am. Chem. Soc.*, **1971**, *93*, 3914.
46. D. Stephan, *Dalton Trans.*, **2009**, 3129.
47. G. Erker, *Dalton Trans.*, **2005**, 1883.
48. G. C. Welch, R. S. S. Juan, J. D. Masuda, D. W. Stephan, *Science*, **2006**, *314*, 1124.
49. D. W. Stephan, G. Erker, *Angew. Chem., Int. Ed.*, **2010**, *49*, 46.
50. C. M. Moemming, E. Otten, G. Kehr, R. Froehlich, S, Grimme, D. W. Stephan, G. Erker, *Angew. Chem., Int. Ed.*, **2009**, *48*, 6643.
51. E. Otten, R. C. Neu, D. W. Stephan, *J. Am. Chem. Soc.*, **2009**, *131*, 8396.
52. T. Mahdi, Z. M. Heiden, S. Grimes, D. W. Stephan, *J. Am. Chem. Soc.*, **2012**, *134*, 4088; C. B. Caputo; D. W. Stephan, *Organometallics*, **2012**, *31*, 27.
53. E. Arunan, G. R. Desiraju, R. A. Klein, J. Sadlej, S. Scheiner, I. Alkorta, D. C. Cleary, R. H. Crabtree, J. J. Dannenberg, P. Hobza, H. G. Kjaergaard, A. C. Legon, B. Mennucci, D. J. Nesbit, *Pure Appl. Chem.*, **2011**, *83*, 1619.
54. E. Arunan, G. R. Desiraju, R. A. Klein., J. Sadlej, S. Scheiner, I. Alkorta, D. C. Clary, R. H. Crabtree, J. J. Dannenberg, P. Hobza, H. G. Kjaergaard, A. C. Legon, B. Mennucci, D. J. Nesbit, *Pure Appl. Chem.*, **2011**, *83*, 1637.
55. G. Gilli, P. Gilli, *The Nature of the Hydrogen Bond*, Oxford University Press, New York, 2009, p. 16.
56. P. Gilli, L. Pretto, V. Bertolasi, G. Gilli, *Acc. Chem. Res.*, **2009**, *42*, 33.
57. S. S. Gayathri, M. Wielopolski, E. M. Pérez, G. Fernández, L. Sánchez, R. Viruela, E. Ortí, D. M. Guldi, and N. Martín, *Angew. Chem., Int. Ed.,* **2009**, *48*, 815; P. A. Denis, *Chem. Phys. Lett.*, **2011**, *516*, 82.
58. A. Sygula, F. R. Fronczek, R. Sygula, P. W. Rabideau, M. M. Olmstead, *J. Am. Chem. Soc.*, **2007**, *129*, 3842.
59. L. N. Dawe, T. A. AlHujran, H.-A. Tran, J. I. Mercer, E. A. Jackson, L. T. Scott, P. E. Georghiou, *Chem. Commun.*, **2012**, *48*, 5563.
60. A. A. Voityuk, M. Duran, *J. Phys. Chem. C*, **2008**, *112*, 1672.
61. A. Molina-Ontoria, G. Fernández, M. Wielopolski, C. Atienza, L. Sánchez, A. Gouloumis, T. Clark, N. Martín, D. M. Guldi, *J. Am. Chem. Soc.,* **2009**, *131*, 12218.
62. R. G. Pearson, *J. Am. Chem. Soc.*, **1963**, *85*, 3533.
63. S. Ahrland, J. Chatt, and N. R. Davies, *Quart. Rev. Chem. Soc.*, **1958**, *12*, 265.
64. R. G. Pearson, *J. Am. Chem. Soc.*, **1963**, *85*, 3533; *Chem. Br.*, **1967**, *3*, 103; R. G. Pearson, ed., *Hard and Soft Acids and Bases*, Dowden, Hutchinson & Ross, Stroudsburg, PA, 1973. The terms *hard* and *soft* are attributed to D. H. Busch in the first paper of this footnote.
65. G. Schwarzenbach, M. Schellenberg, *Helv. Chim Acta*, **1965**, *48*, 28.
66. Jensen, pp. 262–265; C. K. Jørgensen, *Struct. Bonding (Berlin)*, **1966**, *1*, 234.
67. R. G. Pearson, *Inorg. Chem.*, **1988**, *27*, 734.
68. R. S. Drago, B. B. Wayland, *J. Am. Chem. Soc.*, **1965**, *87*, 3571; R. S. Drago, G. C. Vogel, T. E. Needham, *J. Am. Chem. Soc.*, **1971**, *93*, 6014; R. S. Drago, *Struct. Bonding (Berlin)*, **1973**, *15*, 73; R. S. Drago, L. B. Parr, C. S. Chamberlain, *J. Am. Chem. Soc.*, **1977**, *99*, 3203.
69. R. M. Keefer, L. J. Andrews, *J. Am. Chem. Soc.*, **1955**, *77*, 2164.
70. R. G. Pearson, *J. Chem. Educ.*, **1968**, *45*, 581.

General References

W. B. Jensen, *The Lewis Acid–Base Concepts: An Overview*, Wiley Interscience, New York, 1980, and H. L. Finston and Allen C. Rychtman, *A New View of Current Acid–Base Theories*, John Wiley & Sons, New York, 1982, provide good overviews of the history of acid–base theories and critical discussions of the different theories. R. G. Pearson's *Hard and Soft Acids and Bases*, Dowden, Hutchinson, & Ross, Stroudsburg, PA, 1973, is a review by one of the leading proponents of HSAB. For other viewpoints, the references provided in this chapter should be consulted. In particular, C. Laurence and J.-F. Gal, *Lewis Basicity and Affinity Scales Data and Measurement*, John Wiley & Sons, New York, 2010, is an outstanding reference with respect to both tabulated data and discussion of methodology. G. Gilli and P. Gilli, *The Nature of the Hydrogen Bond*, Oxford University Press, New York, 2009, provides an in-depth discussion of this topic. For an excellent review of superbases: *Superbases for Organic Synthesis: Guanidines, Amidines, and Phosphazenes and Related Organocatalysts*, Ishikawa, T., eds., Wiley, New York, 2009.

Problems

Additional acid–base problems are at the end of Chapter 8.

6.1 For each of the following reactions, identify the acid and the base. Also indicate which acid–base definition (Lewis, Brønsted–Lowry) applies. In some cases, more than one definition may apply.

a. $AlBr_3 + Br^- \rightarrow AlBr_4^-$
b. $HClO_4 + CH_3CN \rightarrow CH_3CNH^+ + ClO_4^-$
c. $Ni^{2+} + 6\,NH_3 \rightarrow [Ni(NH_3)_6]^{2+}$
d. $NH_3 + ClF \rightarrow H_3N\cdots ClF$
e. $2\,ClO_3^- + SO_2 \rightarrow 2\,ClO_2 + SO_4^{2-}$
f. $C_3H_7COOH + 2\,HF \rightarrow [C_3H_7C(OH)_2]^+ + HF_2^-$

6.2 For each of the following reactions, identify the acid and the base. Also indicate which acid–base definition (Lewis, Brønsted–Lowry) applies. In some cases, more than one definition may apply.

a. $XeO_3 + OH^- \rightarrow [HXeO_4]^-$
b. $Pt + XeF_4 \rightarrow PtF_4 + Xe$
c. $C_2H_5OH + H_2SeO_4 \rightarrow C_2H_5OH_2^+ + HSeO_4^-$

d. $[CH_3Hg(H_2O)]^+ + SH^- \rightleftharpoons [CH_3HgS]^- + H_3O^+$
e. $(benzyl)_3N + CH_3COOH \rightarrow (benzyl)_3NH^+ + CH_3COO^-$
f. $SO_2 + HCl \rightarrow OSO\cdots HCl$

6.3 Baking powder is a mixture of aluminum sulfate and sodium hydrogencarbonate, which generates a gas and makes bubbles in biscuit dough. Explain what the reactions are.

6.4 The conductivity of BrF_3 is increased by adding KF. Explain this increase, using appropriate chemical equations.

6.5 The following reaction can be conducted as a titration in liquid BrF_5:

$$2\,Cs[\quad]^- + [\quad]^+[Sb_2F_{11}]^- \rightarrow 3\,BrF_5 + 2\,CsSbF_6$$

a. The ions in brackets contain both bromine and fluorine. Fill in the most likely formulas of these ions.
b. What are the point groups of the cation and anion identified in part a?
c. Is the cation in part a serving as an acid or base?

6.6 Anhydrous H_2SO_4 and anhydrous H_3PO_4 both have high electrical conductivities. Explain.

6.7 The gas-phase basicities for the nitrogen bases listed in Table 6.6 are uniformly less positive than the corresponding proton affinities. Explain.

6.8 The proton affinities of acetone, diethylketone, and benzophenone are 812.0, 836.0, and 882.3 kJ/mol, respectively. Rationalize the ranking of these values. (Data from C. Laurence and J.-F. Gal, *Lewis Basicity and Affinity Scales Data and Measurement*, John Wiley and Sons, United Kingdom, 2010, p. 5.)

6.9 The gas-phase basicity of triphenylamine (876.4 kJ/mol) is less than that of triphenylphosphine (940.4 kJ/mol). Do you expect electronic or steric factors to contribute more to this difference? Explain. (Data from C. Laurence and J.-F. Gal, *Lewis Basicity and Affinity Scales Data and Measurement*, John Wiley and Sons, United Kingdom, 2010, p. 5.)

6.10 Correlation of gas-phase and aqueous-solution basicity data is instructive. Construct a graph of gas-phase basicity vs. pK_a of conjugate acids in water (like Figures 6.2 and 6.3) for the following Brønsted–Lowry bases using these data from C. Laurence and J.-F. Gal, *Lewis Basicity and Affinity Scales Data and Measurement*, John Wiley and Sons, United Kingdom, 2010, p. 5. Label each point clearly with the identity of the base.

	pK_a of Conjugate Acid	Gas-Phase Basicity (kJ/mol)
Methanol	−2.05	724.5
Ethanol	−1.94	746.0
Water	−1.74	660.0
Dimethylether	−2.48	764.5
Diethylether	−2.39	801.0

a. Qualitatively, how well do these gas phase and solution data correlate? Explain.
b. Rationalize the positions of the ethers on your graph relative to the alcohols and water.
c. Qualitatively, how well do the gas phase and solution data correlate for the two ethers and the two alcohols? Are these trends the result of inductive or steric effects? Explain.
d. Rationalize the seemingly paradoxical location of water in your graph relative to the other bases.

6.11 Consider these BF_3 affinities for various sulfoxides from C. Laurence and J.-F. Gal, *Lewis Basicity and Affinity Scales Data and Measurement*, John Wiley and Sons, United Kingdom, 2010, p. 99. Rationalize this BF_3 affinity trend with inductive and resonance arguments.

Lewis base	Formula	BF_3 Affinity (kJ/mol)
Diphenylsulfoxide	Ph_2SO	90.34
Methyl phenyl sulfoxide	PhSOMe	97.37
Dimethyl sulfoxide	Me_2SO	105.34
Di-*n*-butyl sulfoxide	$(n\text{-}Bu)_2SO$	107.60
Tetramethylene sulfoxide	*cyclo*-$(CH_2)_4SO$	108.10

6.12 The development of new Lewis basicity scales is of ongoing interest. Maccarrone and Di Bella recently reported a scale of Lewis basicity that employs a zinc(II) complex as a reference Lewis acid (I. P. Oliveri, G. Maccarrone, S. Di Bella, *J. Org. Chem.*, **2011**, *76*, 8879).
a. What are the ideal characteristics of a reference acid discussed by these authors?
b. Compare the relative Lewis basicities of quinuclidine and pyridine found by these authors to the BF_3 affinity for pyridine (Table 6.10) and the value for quinuclidine (150.01 kJ/mol). To what feature do these authors attribute the high basicities of these two nitrogen bases to their zinc(II) reference?
c. What general trends do these authors report for alicyclic (that is, aliphatic and cyclic) and acyclic amines? How are these trends rationalized?

6.13 If an equimolar mixture of $P(t\text{-}C_4H_9)_3$ and $B(C_6F_5)_3$ is mixed with 1 bar of the gas N_2O in bromobenzene solution, a white product is formed in good yield. A variety of NMR evidence has been gathered on the product: there is a single ^{31}P NMR resonance; ^{11}B and ^{19}F NMR are consistent with a 4-coordinate boron atom; and ^{15}N NMR indicates two nonequivalent nitrogen atoms. In addition, no gas is released in the reaction.
a. Suggest the role of N_2O in this reaction.
b. Propose a structure of the product. (See E. Otten, R. C. Neu, D. W. Stephan, *J. Am. Chem. Soc.*, **2009**, *131*, 9918.)

6.14 FLP chemistry continues to afford remarkable reactions that proceed without transition metals. Metal-free aromatic hydrogenation of *N*-bound phenyl rings can be achieved in the presence of H_2 and $B(C_6F_5)_3$ to form *N*-cyclohexylammonium hydridoborate salts (T. Mahdi, Z. M. Heiden, S. Grimme, D. W. Stephan, *J. Am. Chem. Soc.*, **2012**, *134*, 4088).

a. Sketch the reaction coordinate diagram in Figure 2 of this report, including the structures of intermediates and transition states.

b. Discuss the hypothesized steps for initial addition of H^+ to the aromatic ring of *t*-BuNHPh.

c. How does this diagram support the outcome of the reaction of *t*-BuNHPh, $B(C_6F_5)_3$, and H_2 in pentane (at 298 K) versus in refluxing toluene (383 K)?

d. What happens if the utilized amine becomes *too* basic?

6.15 The ability of frustrated Lewis pairs to capture NO (nitric oxide) to afford aminoxyl radicals is a recent triumph of FLP chemistry (M. Sajid, A. Stute, A. J. P. Cardenas, B. J. Culotta, J. A. M. Hepperle, T. H. Warren, B. Schirmer, S. Grimme, A. Studer, C. G. Daniliuc, R. Fröhlich, J. L. Peterson, G. Kehr, G. Erker, *J. Am. Chem. Soc.*, **2012**, *134*, 10156.) Use a molecular orbital argument to hypothesize why the N—O bond lengthens in complex **2b** relative to nitric oxide (HINT: Which orbital is likely the acceptor?).

a. Use the arrow-pushing formalism to propose mechanisms for the reaction of **2b** with 1,4-cyclohexadiene, and of **2b** with toluene.

b. Explain why C—O bond formation in the toluene reaction occurs exclusively at the primary carbon and not at a carbon atom within the aromatic ring.

2b H H $(Mes)_2\overset{+}{P}$ $\overset{-}{B}(C_6F_2)_2$ N O Mes =

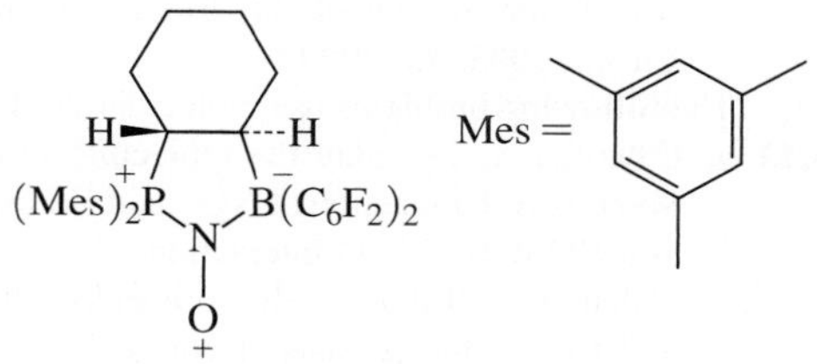

6.16 Use the pK_a *Slide Rule* in P. Gilli, L. Pretto, V. Bertolasi, G. Gilli, *Acc. Chem. Res.*, **2009**, *42*, 33 to answer these questions:

a. Which forms a stronger hydrogen bond with water, HCN or HSCN?

b. Identify two inorganic acids that are predicted to form strong hydrogen bonds with organic nitriles.

c. Which organic acid in the *Slide Rule* is predicted to form the strongest hydrogen bond to organic sulfides?

d. Water is a prototypical hydrogen bond donor. Classify the strengths of $A\cdots H_2O$ hydrogen bonds (for example, as strong, medium strong, medium, medium weak, or weak) with the following classes of acceptors: amines, triphosphines, sulfoxides, ketones, and nitro compounds.

6.17 The X-ray structure of $Br_3As \cdot C_6Et_6 \cdot AsBr_3$ (Et = ethyl) has been reported (H. Schmidbaur, W. Bublak, B. Huber, G. Müller, *Angew. Chem., Int. Ed.*, **1987**, *26*, 234).

a. What is the point group of this structure?

b. Propose an explanation of how the frontier orbitals of $AsBr_3$ and C_6Et_6 can interact to form chemical bonds that stabilize this structure.

6.18 When $AlCl_3$ and $OPCl_3$ are mixed, the product, $Cl_3Al—O—PCl_3$ has a nearly linear Al—O—P arrangement (bond angle 176°).

a. Suggest an explanation for this unusually large angle.

b. The O—P distance in $Cl_3Al—O—PCl_3$ is only slightly longer than the comparable distance in $OPCl_3$ even though the latter has a formal double bond. Suggest why there is so little difference in these bond distances. (See N. Burford, A. D. Phillips, R. W. Schurko, R. E. Wasylishen, J. F. Richardson, *Chem. Commun.*, **1997**, 2363.)

6.19 Of the donor–acceptor complexes $(CH_3)_3N—SO_3$ and $H_3N—SO_3$ in the gas phase,

a. Which has the longer N—S bond?

b. Which has the larger N—S—O angle?

Explain your answers briefly. (See D. L. Fiacco, A. Toro, K. R. Leopold, *Inorg. Chem.*, **2000**, *39*, 37.)

6.20 Xenon difluoride, XeF_2, can act as a Lewis base toward metal cations such as Ag^+ and Cd^{2+}.

a. In these cases, do you expect the XeF_2 to exert its basicity through the lone pairs on Xe or those on F?

b. $[Ag(XeF_2)_2]AsF_6$ and $[Cd(XeF_2)_2](BF_4)_2$ have both been synthesized. In which case, AsF_6^- or BF_4^-, do you expect the fluorines to act as stronger Lewis bases? Explain briefly. (See G. Tavcar, B. Zemva, *Inorg. Chem.*, **2005**, *44*, 1525.)

6.21 The ion NO^- can react with H^+ to form a chemical bond. Which structure is more likely, HON or HNO? Explain your reasoning.

6.22 The absorption spectra of solutions containing Br_2 are solvent dependent. When elemental bromine is dissolved in nonpolar solvents such as hexane, a single absorption band in the visible spectrum is observed near 500 nm. When Br_2 is dissolved in methanol, however, this absorption band shifts and a new band is formed.

a. Account for the appearance of the new band.

b. Is the 500 nm band likely to shift to a longer or shorter wavelength in methanol? Why?

In your answers, you should show clearly how appropriate orbitals of Br_2 and methanol interact.

6.23 AlF_3 is insoluble in liquid HF but dissolves if NaF is present. When BF_3 is added to the solution, AlF_3 precipitates. Explain.

6.24 Why were most of the metals used in antiquity class (b) (*soft*, in HSAB terminology) metals?

6.25 The most common source of mercury is cinnabar (HgS), whereas Zn and Cd in the same group occur as sulfide, carbonate, silicate, and oxide. Why?

6.26 The difference between melting point and boiling point (in °C) is given below for each of the Group IIB halides.

	F^-	Cl^-	Br^-	I^-
Zn^{2+}	630	405	355	285
Cd^{2+}	640	390	300	405
Hg^{2+}	5	25	80	100

What deductions can you draw?

6.27 **a.** Use Drago's *E* and *C* parameters to calculate Δ*H* for the reactions of pyridine and BF_3 and of pyridine and $B(CH_3)_3$. Compare your results with the reported experimental values of −71.1 and −64 kJ/mol for pyridine—$B(CH_3)_3$ and −105 kJ/mol for pyridine—BF_3.

b. Explain the differences found in part a in terms of the structures of BF_3 and $B(CH_3)_3$.

c. Explain the differences in terms of HSAB theory.

6.28 Repeat the calculations of the preceding problem using NH_3 as the base, and put the four reactions in order of the magnitudes of their ΔH values.

6.29 Compare the results of Problems 6.20 and 6.21 with the absolute hardness parameters of Appendix B-5 for BF_3, NH_3, and pyridine (C_5H_5N). What value of η would you predict for $B(CH_3)_3$? Compare NH_3 and $N(CH_3)_3$ as a guide.

6.30 CsI is much less soluble in water than CsF, and LiF is much less soluble than LiI. Why?

6.31 Rationalize the following data in HSAB terms:

	ΔH(kcal)
$CH_3CH_3 + H_2O \rightarrow CH_3OH + CH_4$	12
$CH_3COCH_3 + H_2O \rightarrow CH_3COOH + CH_4$	−13

6.32 Predict the order of solubility in water of each of the following series, and explain the factors involved.

a. $MgSO_4$ $CaSO_4$ $SrSO_4$ $BaSO_4$

b. $PbCl_2$ $PbBr_2$ PbI_2 PbS

6.33 In some cases CO can act as a bridging ligand between main-group and transition-metal atoms. When it forms a bridge between Al and W in the compound having the formula $(C_6H_5)_3$Al—[bridging CO]—$W(CO)_2(C_5H_5)$, is the order of atoms in the bridge Al—CO—W or Al—OC—W? Briefly explain your choice.

6.34 Choose and explain:

a. Strongest Brønsted acid: SnH_4 SbH_3 TeH_2

b. Strongest Brønsted base: NH_3 PH_3 SbH_3

c. Strongest base to H^+ (gas phase):
NH_3 CH_3NH_2 $(CH_3)_2NH$ $(CH_3)_3N$

d. Strongest base to BMe_3: pyridine 2-methylpyridine 4-methylpyridine

6.35 B_2O_3 is acidic, Al_2O_3 is amphoteric, and Sc_2O_3 is basic. Why?

6.36 Predict the reactions of the following hydrogen compounds with water, and explain your reasoning.

a. CaH_2

b. HBr

c. H_2S

d. CH_4

6.37 List the following acids in order of their acid strength when reacting with NH_3.

BF_3 $B(CH_3)_3$ $B(C_2H_5)_3$ $B[C_6H_2(CH_3)_3]_3$

$[C_6H_2(CH_3)_3]$ = 2,4,6-trimethylphenyl

6.38 Choose the stronger acid or base in the following pairs, and explain your choice.

a. CH_3NH_2 or NH_3 in reaction with H^+

b. Pyridine or 2-methylpyridine in reaction with trimethylboron

c. Triphenylboron or trimethylboron in reaction with ammonia

6.39 List the following acids in order of acid strength in aqueous solution:

a. $HMnO_4$ H_3AsO_4 H_2SO_3 H_2SO_4

b. HClO $HClO_4$ $HClO_2$ $HClO_3$

6.40 Solvents can change the acid–base behavior of solutes. Compare the acid–base properties of dimethylamine in water, acetic acid, and 2-butanone.

6.41 HF has $H_0 = -11.0$. Addition of 4% SbF_5 lowers H_0 to −21.0. Explain why SbF_5 should have such a strong effect and why the resulting solution is so strongly acidic that it can protonate alkenes.

$$(CH_3)_2C{=}CH_2 + H^+ \rightarrow (CH_3)_3C^+$$

6.42 The reasons behind the relative Lewis acidities of the boron halides BF_3, BCl_3, and BBr_3 with respect to NH_3 have been controversial. Although BF_3 might be expected to be the strongest Lewis acid on the basis of electronegativity, the Lewis acidity order is $BBr_3 > BCl_3 > BF_3$. Consult the references listed below to address the following questions. (See also J. A. Plumley, J. D. Evanseck, *J. Phys. Chem. A*, **2009**, *113*, 5985.)

a. How does the LCP approach account for a Lewis acidity order of $BBr_3 > BCl_3 > BF_3$? (See B. D. Rowsell, R. J. Gillespie, G. L. Heard, *Inorg. Chem.*, **1999**, *38*, 4659.)

b. What explanation has been offered on the basis of the calculations presented in F. Bessac, G. Frenking, *Inorg. Chem.*, **2003**, *42*, 7990?

The following problems use molecular modeling software.

6.43 **a.** Calculate and display the molecular orbitals of NO^-. Show how the reaction of NO^- and H^+ can be described as a HOMO–LUMO interaction.

b. Calculate and display the molecular orbitals of HNO and HON. On the basis of your calculations, and your answer to part a, which structure is favored?

6.44 Calculate and display the frontier orbitals of Br_2, methanol, and the Br_2—methanol adduct to show how the orbitals of the reactants interact.

6.45 **a.** Calculate and display the molecular orbitals of BF_3, NH_3, and the F_3B—NH_3 Lewis acid–base adduct.

b. Examine the bonding and antibonding orbitals involved in the B—N bond in F_3B—NH_3. Is the bonding orbital polarized toward the B or the N? The antibonding orbital? Explain briefly.

6.46 Section 6.4.5 includes a diagram of a halogen bond between Br_2 and acetylene.

a. Use sketches to show how a π orbital of acetylene interacts with the LUMO of Br_2 to form the adduct.

b. Calculate and display the molecular orbitals of the acetylene—Br_2 adduct. Describe the interactions that you observe between the π orbitals of acetylene and orbitals of Br_2.

Chapter 7

The Crystalline Solid State

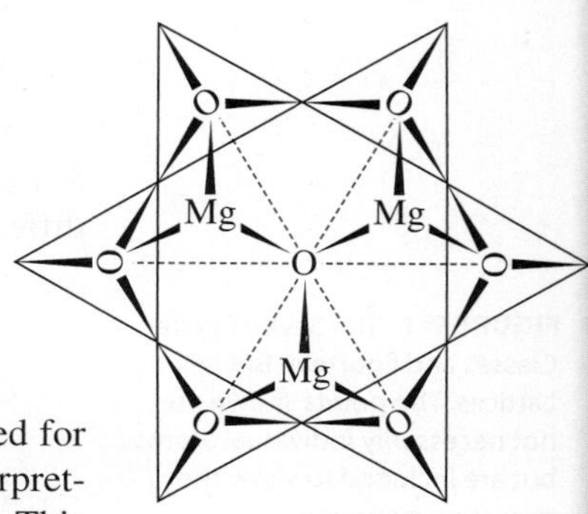

Solid-state chemistry uses the same general principles for bonding as those employed for molecules. The physical and chemical properties of a macroscopic crystal can be interpreted by envisioning the crystal as consisting of molecular orbitals extending throughout. This bonding model requires an alternate perspective on molecular orbitals than that in Chapter 5, but one that is useful in explaining the unique properties of bulk materials compared to small molecules. There are two major classifications of solids: crystals and amorphous materials. The focus of this chapter is on crystalline solids composed of atoms or ions.

We will first describe the common structures of crystals and then consider their bonding according to the molecular orbital approach. Finally, we will describe some of the thermodynamic and electronic properties of these materials and their uses.

7.1 Formulas and Structures

Crystalline solids have atoms, ions, or molecules packed in regular geometric arrays, with the simplest repeating structural unit called the **unit cell**. Some of the common crystal geometries are described in this section. In addition, we will consider the role of the relative sizes of the components in determining structures. Use of a model kit, such as the one available from ICE,* facilitates the study of these structures.

7.1.1 Simple Structures

The crystal structures of metals are relatively simple. Those of some minerals can be complex, but minerals usually have simpler structures that can be recognized within the more complex structure. The unit cell is a structural component that, when repeated in all directions, results in a macroscopic crystal. Structures of the 14 possible crystal structures (Bravais lattices) are shown in **Figure 7.1**. Several different unit cells are possible for some structures; the one used may be chosen for convenience, depending on the particular application. The atoms (or ions) on the corners, edges, or faces of the unit cell are shared with other unit cells as follows:

> Atoms at the corners of rectangular unit cells are shared equally by eight unit cells and contribute $\frac{1}{8}$ to each ($\frac{1}{8}$ of the atom is counted as part of each cell). The total for a single unit cell is $8 \times \frac{1}{8} = 1$ atom for all of the corners.

*Institute for Chemical Education, Department of Chemistry, University of Wisconsin–Madison, 1101 University Ave., Madison, WI 53706. Sources for other model kits are given in A. B. Ellis, M. J. Geselbracht, B. J. Johnson, G. C. Lisensky, and W. R. Robinson, *Teaching General Chemistry: A Materials Science Companion*, American Chemical Society, Washington, DC, 1993.

Atoms at the corners of nonrectangular unit cells also contribute one atom total to the unit cell; small fractions on one corner are matched by larger fractions on another.

Atoms on edges of unit cells are shared by four unit cells—two in one layer, two in the adjacent layer—and contribute $\frac{1}{4}$ to each.

Atoms on faces of unit cells are shared between two unit cells and contribute $\frac{1}{2}$ to each.

Figure 7.1 demonstrates that unit cells need not have equal dimensions or angles. For example, triclinic crystals are defined by three different angles and may have three different distances for the unit cell dimensions.

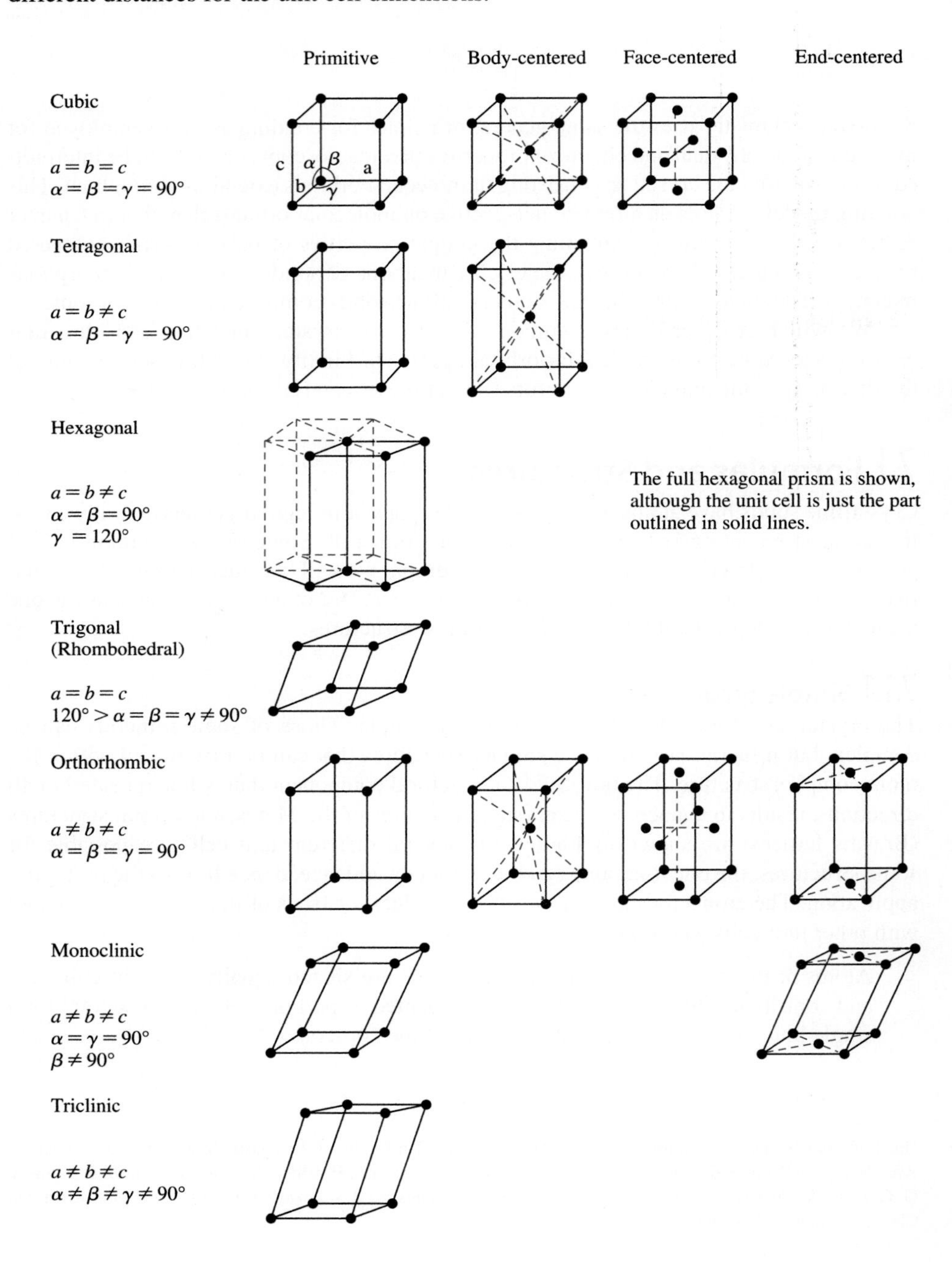

FIGURE 7.1 The Seven Crystal Classes and Fourteen Bravais Lattices. The points shown are not necessarily individual atoms but are included to show the necessary symmetry.

EXAMPLE 7.1

The diagram below shows a space-filling diagram of a face-centered cubic unit cell cut to show only the part of each atom that is inside the unit cell boundaries. The corner atoms are each shared among eight unit cells, so $\frac{1}{8}$ of the atom is in the unit cell shown. The face-centered atoms are shared between two unit cells, so $\frac{1}{2}$ of the atom is in the unit cell shown. The eight corners of the unit cell then total $8 \times \frac{1}{8} = 1$ atom, and the six faces total $6 \times \frac{1}{2} = 3$ atoms; a total of four atoms are in the unit cell.

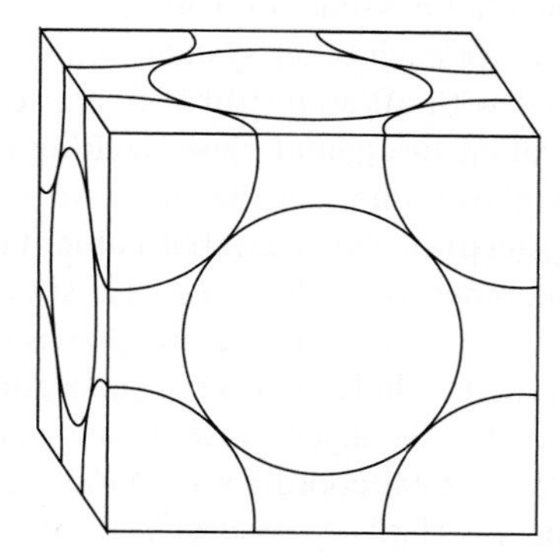

EXERCISE 7.1 Calculate the number of atoms in each unit cell (Figure 7.1) of

a. A body-centered cubic structure
b. A hexagonal structure

The positions of atoms are frequently described in **lattice points**, expressed as fractions of the unit cell dimensions. For example, the body-centered cube has atoms at the origin $[$the corner at which $x = 0, y = 0, z = 0$, or $(0, 0, 0)]$ and at the center of the cube $[x = \frac{1}{2}, y = \frac{1}{2}, z = \frac{1}{2}, (\frac{1}{2}, \frac{1}{2}, \frac{1}{2})]$. The other atoms can be generated by moving these two atoms in each direction in increments of one cell length.

Cubic

The most basic crystal class is the simple cube, called the **primitive cubic** structure, with atoms at the eight corners. It can be described by specifying the length of one side, the angle 90°, and the single lattice point (0, 0, 0). Because each of the atoms is shared between eight cubes, four in one layer and four in the layer above or below, the total number of atoms in the unit cell is $8 \times \frac{1}{8} = 1$, the number of lattice points required to define the primitive cube. Six atoms surround each identical lattice point atom; the **coordination number** (CN) of each atom is 6. This structure is not efficiently packed, because the spheres occupy only 52.4% of the total volume. In the center of the cube is a vacant space that has eight nearest neighbors. If an atom were present in this void, its coordination number would be 8. Calculation shows that a sphere with a radius $0.73r$, where r is the radius of the corner spheres, would fit in the center of this cube if the corner spheres were in contact with each other.

Body-Centered Cubic

If another sphere is added in the center of the simple cubic structure, **body-centered cubic (bcc)** is the result. If the added sphere has the same radius as the others, the size of the unit cell expands (relative to primitive cubic) because the radius of the central atom is larger than $0.73r$ so that the diagonal distance through the cube is $4r$, where r is the radius of the spheres. The corner atoms are no longer in contact with each other. The new unit cell is $2.31r$ on each side and contains two atoms, because the body-centered atom is completely within the unit cell. This cell has two lattice points, at the origin (0, 0, 0) and at the center of the cell $(\frac{1}{2}, \frac{1}{2}, \frac{1}{2})$.

EXERCISE 7.2

Show that the side of the unit cell for a body-centered cubic crystal is 2.31 times the radius of the atoms in the crystal.

Close-Packed Structures

When marbles or ball bearings are poured into a flat box, they tend to form a close-packed layer, in which each sphere is surrounded by six others in the same plane. This arrangement provides the most efficient packing possible for a single layer. When three or more close-packed layers are placed on top of each other systematically, two structures are possible. When the third layer is placed with all atoms directly above those of the first layer, the result is an ABA structure called **hexagonal close packing (hcp)**. When the third layer is displaced, so each atom is above a hole in the first layer, the resulting ABC structure is called **cubic close packing (ccp)** or **face-centered cubic (fcc)**. In both **hcp** and **ccp/fcc** structures, the coordination number for each atom is 12, six in its own layer, three in the layer above, and three in the layer below. These are shown in **Figure 7.2**. In both these structures, there are two tetrahedral holes per atom (coordination number 4, formed by three atoms in one layer and one in the layer above or below) and one octahedral hole per atom (three atoms in each layer, a total coordination number of 6).

Hexagonal close packing is relatively easy to see, with hexagonal prisms sharing vertical faces in the larger crystal (**Figure 7.3**). The minimal unit cell is smaller than the hexagonal prism (Figure 7.1); taking any four atoms that all touch each other in one layer and extending lines up to the third layer will generate a unit cell with a parallelogram as the base. As shown in Figure 7.3, this cell contains half an atom in the first layer (four

FIGURE 7.2 Close-Packed Structures.

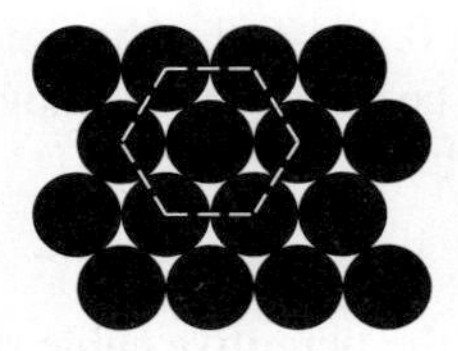

A single close-packed layer, A, with the hexagonal packing outlined.

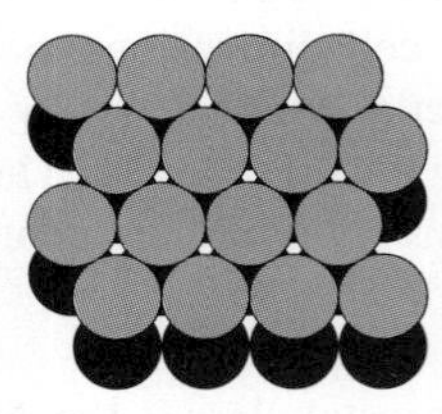

Two close-packed layers, A and B. Octahedral holes can be seen extending through both layers surrounded by three atoms in each layer. Tetrahedral holes are under each atom of the second layer and over each atom of the bottom layer. Each is made up of three atoms from one layer and one from the other.

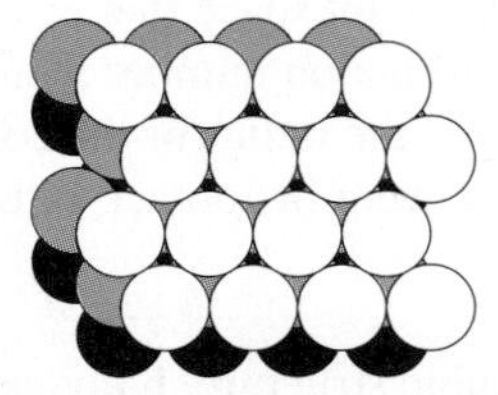

Cubic close-packed layers, in an ABC pattern. Octahedral holes are offset, so no hole extends through all three layers.

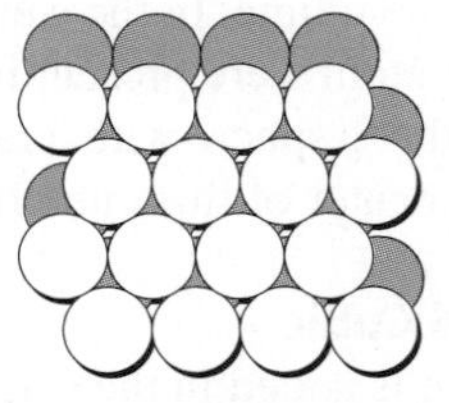

Hexagonal close-packed layers. The third layer is exactly over the first layer in this ABA pattern. Octahedral holes are aligned exactly over each other, one set between the first two layers A and B, the other between the second and third layers, B and A.

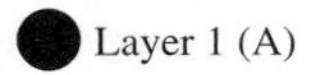

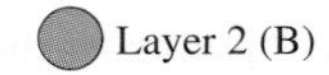

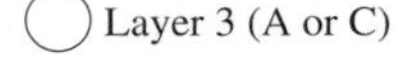

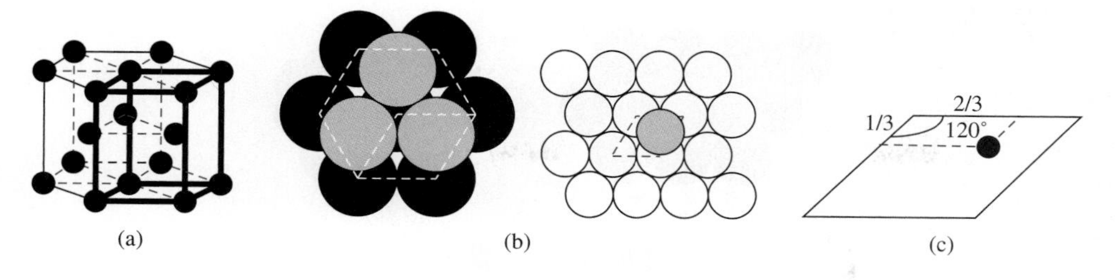

FIGURE 7.3 Hexagonal Close Packing. (a) The hexagonal prism with the unit cell outlined in bold. (b) Two layers of an hcp unit cell. The parallelogram is the base of the unit cell. The third layer is identical to the first. (c) Location of the atom in the second layer.

atoms averaging $\frac{1}{8}$ each), four similar atoms in the third layer, and one atom from the second layer—whose center is within the unit cell—for a total of two atoms in the unit cell. The unit cell has dimensions of $2r$, $2r$, and $2.83r$, an angle of 120° between the first two axes in the basal plane, and 90° between each of these axes and the third, vertical axis. The atoms are at the lattice points (0, 0, 0) and $\left(\frac{1}{3}, \frac{2}{3}, \frac{1}{2}\right)$.

The cube in cubic close packing is more difficult to see when each of the layers is close-packed. The unit cell cube rests on one corner, with four close-packed layers required to complete the cube. The first layer has only one sphere, and the second has six in a triangle, as shown in **Figure 7.4(a)**. The third layer has another six-membered triangle, with the vertices rotated 60° from the one in the second layer, and the fourth layer again has one sphere. The cubic shape of the cell is easier to see if the faces are placed in the conventional horizontal and vertical directions, as in **Figure 7.4(b)**.

The unit cell of the cubic close-packed structure is a face-centered cube, with spheres at the corners and in the middle of each of the six faces. The lattice points are at (0, 0, 0), $\left(\frac{1}{2}, \frac{1}{2}, 0\right)$, $\left(\frac{1}{2}, 0, \frac{1}{2}\right)$, and $\left(0, \frac{1}{2}, \frac{1}{2}\right)$, for a total of four atoms in the unit cell, $\frac{1}{8} \times 8$ at the corners and $\frac{1}{2} \times 6$ in the faces. In both close-packed structures, the spheres occupy 74.0% of the total volume.

Ionic crystals can also be described in terms of the interstices, or holes, in the structures. **Figure 7.5** shows the location of tetrahedral and octahedral holes in close-packed structures. Whenever an atom is placed in a new layer over a close-packed layer, it creates a tetrahedral hole surrounded by three atoms in the first layer and one in the second (CN = 4). When additional atoms are added to the second layer, they also create tetrahedral holes surrounded by one atom in the one layer and three in the other. In addition, there are octahedral holes (CN = 6) surrounded by three atoms in each layer. Overall, close-packed structures have two tetrahedral holes and one octahedral hole per atom. These holes can be filled by smaller ions. For example, the tetrahedral holes created by larger ions of radius r can be filled with smaller ions of radius $0.225r$ (for ions in contact). Octahedral holes can similarly be filled by ions with radius $0.414r$. These reduced ionic radii require the smaller ions in their holes to be in contact with their nearest four and six neighbors,

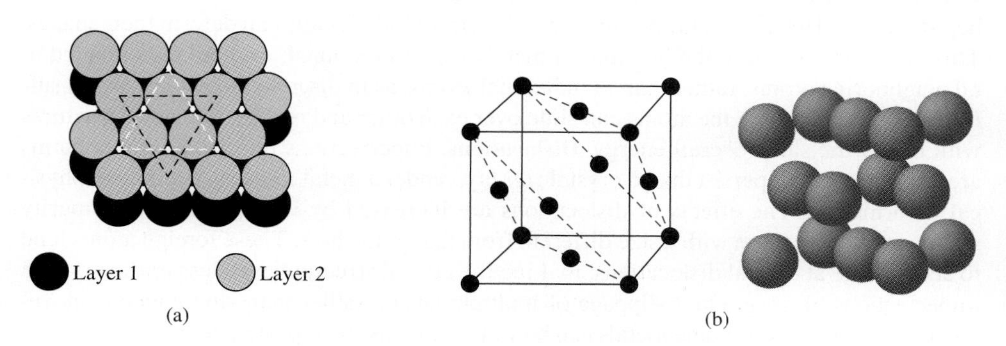

FIGURE 7.4 Cubic Close Packing. (a) Two layers of a ccp (or fcc) cell. The atom in the center of the triangle in the first layer and the six atoms connected by the triangle form half the unit cell. The other half, in the third and fourth layers, is identical, but with the direction of the triangle reversed. (b) Two views of the unit cell, with the close-packed layers marked in the first.

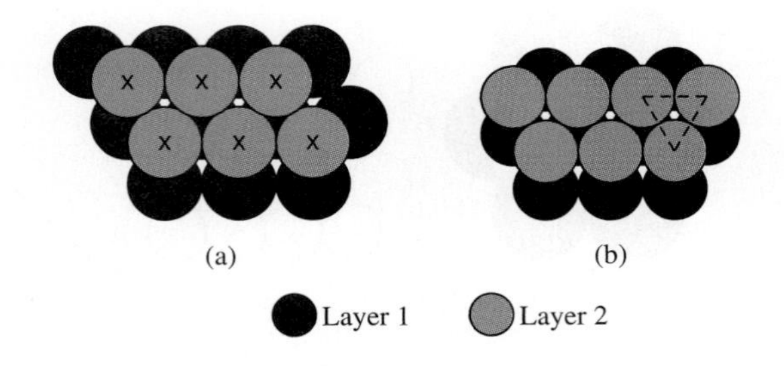

FIGURE 7.5 Tetrahedral and Octahedral Holes in Close-Packed Layers. (a) Tetrahedral holes are under each x and at each point where an atom of the first layer appears in the triangle between three atoms in the second layer. (b) An octahedral hole is outlined, surrounded by three atoms in each layer.

respectively. In more complex crystals, even if the ions are not in contact with each other, the geometry is described in the same terminology. For example, NaCl has chloride ions in a cubic close-packed array, with sodium ions, also in a ccp array, in the octahedral holes. The sodium ions have a radius 0.695 times the radius of the chloride ion ($r_+ = 0.695r_-$) large enough to force the chloride ions apart, but not large enough to allow a coordination number larger than 6.

Metallic Crystals

Except for the actinides, most metals crystallize in body-centered cubic, cubic close-packed, and hexagonal close-packed structures, with approximately equal numbers of each type. Interestingly, changes in pressure or temperature can change the structure of many metallic crystals. This variability and dynamic behavior emphasize that it is inappropriate to think of these metal atoms as hard spheres that pack together in crystals independent of their electronic structure and external conditions. Instead, the packing of atoms into crystals for metal atoms is challenging to predict, and different arrangements are sometimes separated by relatively small differences in energy. Atoms attract each other at moderate distances and repel each other when they are close enough that their electron clouds overlap too much. The balance between these forces, modulated by the specific electronic configuration of the atoms, determines the most stable structure. Simple geometric considerations are not sufficient to predict the crystal structures of metals.

Properties of Metals

Conductivity is the property that best distinguishes metals from nonmetals. Metals have high conductivity (low resistance) to the passage of electricity and heat; nonmetals have low conductivity (high resistance). A notable exception is the nonmetal diamond, which has low electrical conductivity and high heat conductivity. Conductivity, from the perspective of the electronic structure of metals and semiconductors, is discussed in Section 7.3.

Aside from conductivity, metals have remarkably varied properties. Some are soft and easily deformed by pressure or impact (i.e., malleable), such as Cu (an fcc structure), whereas others are hard and brittle, more likely to break rather than bend, such as Zn (an hcp structure). However, hammering or bending most bulk metals can deform their shapes. This is possible because the bonding in metals is nondirectional; each atom is bonded to all neighboring atoms, rather than to individual atoms, as in discrete molecules. When sufficient force is applied, the atoms can slide over each other and realign into new structures with nearly the same overall energy. **Dislocations**, imperfections in crystals where atoms are out of place, but persist due to crystal rigidity, render a metal more susceptible to physical deformation. The effects of dislocations are increased by incorporation of impurity atoms, especially those with a size different from that of the host. These foreign atoms tend to accumulate at crystal dislocations, making the crystal structure even less uniform. Such imperfections allow gradual slippage of multiple layers, rather than simultaneous movement of an entire layer. Some metals can be work-hardened by repeated deformation. When the metal is hammered, the defects tend to group together, rendering uniform the majority of the metal structure, resulting in increased resistance to deformation. Heating can restore flexibility by redistributing the dislocations and reducing their numbers. Indeed, the impact of dislocations on a metal's physical properties is difficult to predict. For different metals or alloys (mixtures of metals) heat treatment and slow or fast cooling result in significantly different outcomes. Some metals can be tempered to be harder and hold a sharp edge better, others can be heat-treated to be more resilient and able to flex without being permanently bent. Still others can be treated to have "shape memory." These alloys can be bent but return to their initial shape on moderate heating.

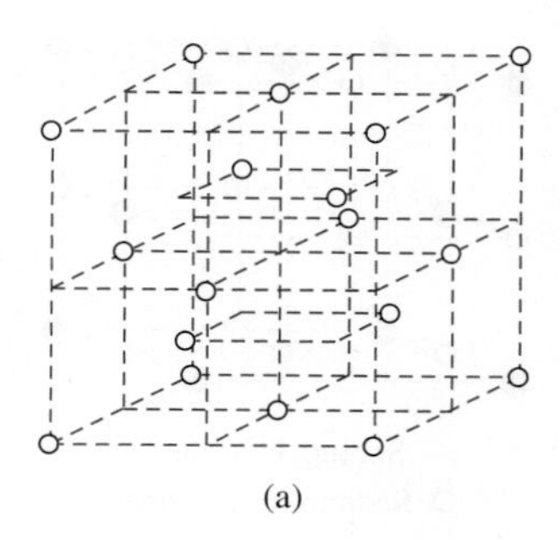

(a)

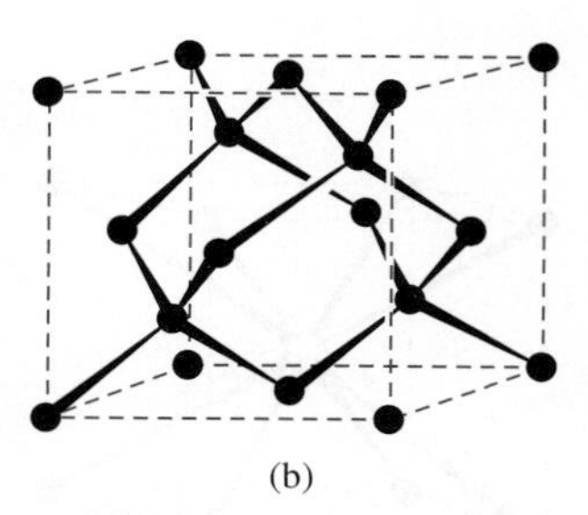

(b)

FIGURE 7.6 The Structure of Diamond. (a) Subdivision of the unit cell, with atoms in alternating smaller cubes. (b) The tetrahedral coordination of carbon is shown for the four interior atoms.

Diamond

The final simple structure we will consider is that of diamond (**Figure 7.6**), which has the same overall geometry as zinc blende (described later) but with all atoms identical. If a face-centered cubic crystal is divided into eight smaller cubes by planes cutting through the middle, and additional atoms are added in the centers of four of the smaller cubes, none of them adjacent, the diamond structure is the result. Each carbon atom is bonded tetrahedrally to its four nearest neighbors, and the bonding between them is similar to ordinary carbon–carbon single bonds. The strength of the crystal is a consequence of the covalent nature of the bonding; each carbon has its complete set of four bonds. Although there are potential cleavage planes in diamond, the structure has essentially the same strength in all directions. In addition to carbon, three other elements in the same group—silicon, germanium, and α-tin—have the same structure. Ice also has the same crystal symmetry (see Figure 3.27), with O—H—O hydrogen bonds between all the oxygens. The ice structure is more open because of the greater distance between oxygen atoms, and ice is less dense than liquid water.

7.1.2 Structures of Binary Compounds

Binary compounds, those consisting of two elements, may have very simple crystal structures and can be described in several different ways. Two simple structures are shown in **Figure 7.7**. As described in Section 7.1.1, there are two tetrahedral holes and one octahedral hole per atom in close-packed structures. If the larger ions (usually the anions) are in close-packed structures, ions of the opposite charge occupy these holes, depending primarily on two factors:

1. *The relative sizes of the atoms or ions.* The **radius ratio** (usually r_+/r_- but sometimes r_-/r_+, where r_+ is the radius of the cation and r_- is the radius of the anion) is generally used to assess the relative sizes of ions. Small cations often fit in the tetrahedral or octahedral holes of a close-packed anion lattice. Somewhat larger cations can fit in the octahedral holes, but not in tetrahedral holes, of the same lattice. Still larger cations force a change in structure (as described in Section 7.1.4).
2. *The relative numbers of cations and anions.* For example, a formula of M_2X will not allow a close-packed anion lattice with cations exclusively in each octahedral hole because the number of cations is twice the number of octahedral holes! The structure must have the cations in tetrahedral holes (this is theoretically possible because the ratio of tetrahedral holes to each anion is 2:1) or have an anion lattice that is not close-packed.

The structures described in this section are generic, named for the most common compound with the structure. The focus in this section is on structures where the constituent ions exhibit a high degree of ionic bonding. The effect of a relatively high covalent component on the structure of crystals is not considered here. The electronic structure of the ions, and the resulting ionic and covalent contributions to their bonds, must be considered to fully rationalize crystal structures.

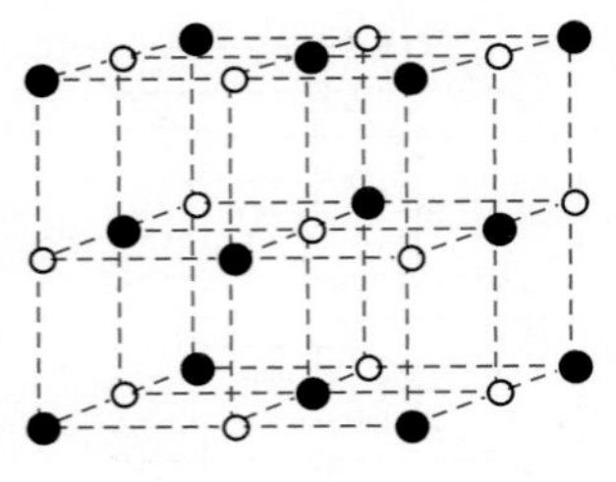

(a)

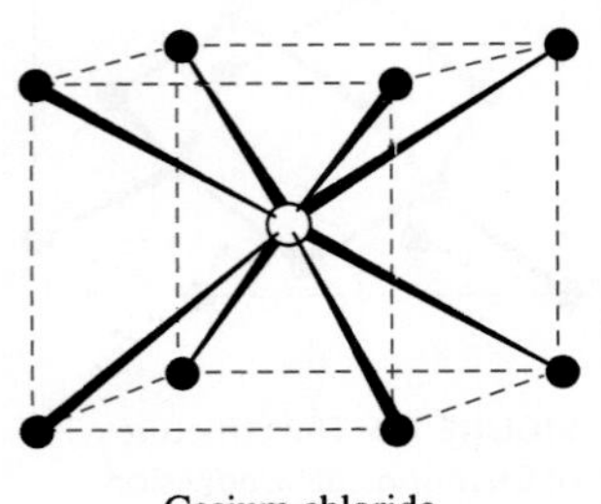

(b)

FIGURE 7.7 Sodium Chloride and Cesium Chloride Unit Cells.

Sodium Chloride, NaCl

NaCl is made up of face-centered cubes of sodium ions and face-centered cubes of chloride ions, offset by half a unit cell length in one direction, so that the sodium ions are centered in the edges of the chloride lattice and vice versa (Figure 7.7(a)). If all the ions were identical, the NaCl unit cell would be made up of eight cubic unit cells. Many alkali halides share this structure. For these crystals, the ions tend to have quite different sizes, usually with the anions larger than the cations. Each sodium ion is surrounded by six nearest-neighbor chloride ions, and each chloride ion is surrounded by six nearest-neighbor sodium ions.

Cesium Chloride, CsCl

As mentioned previously, a sphere of radius $0.73r$ will fit exactly in the center of a cubic structure. Although the fit is not perfect, this is what happens in CsCl, Figure 7.7(b), where the chloride ions form simple cubes with cesium ions in the centers.* In the same way, the cesium ions form simple cubes with chloride ions in the centers. The average chloride ion radius is 0.83 times as large as the cesium ion (167 pm and 202 pm, respectively), but the interionic distance in CsCl is 356 pm, about 3.5% smaller than the sum of the average ionic radii. Only CsCl, CsBr, CsI, TlCl, TlBr, TlI, and CsSH have this structure at ordinary temperatures and pressures, although some other alkali halides have this structure at high pressure and high temperature. The cesium salts can be crystallized in the NaCl lattice on NaCl or KBr substrates, and CsCl converts to the NaCl lattice at about 469 °C.

Zinc Blende, ZnS

ZnS has two common crystalline forms, both with coordination number 4. Zinc blende is the most common zinc ore and has essentially the same geometry as diamond, with alternating layers of zinc and sulfide (**Figure 7.8(a)**). It can also be described as having zinc ions and sulfide ions, each in face-centered lattices, so that each ion is in a tetrahedral hole of the other lattice. The stoichiometry requires half of these tetrahedral holes to be occupied, with alternating occupied and vacant sites.

Wurtzite, ZnS

The wurtzite form of ZnS is much rarer than zinc blende and is formed at higher temperatures. Its zinc and sulfide ions each occupy tetrahedral holes of the other ion's hexagonal close-packed lattice (**Figure 7.8(b,c)**). As in zinc blende, half of the tetrahedral holes in each lattice are occupied.

FIGURE 7.8 ZnS Crystal Structures. (a) Zinc blende. (b, c) Wurtzite.

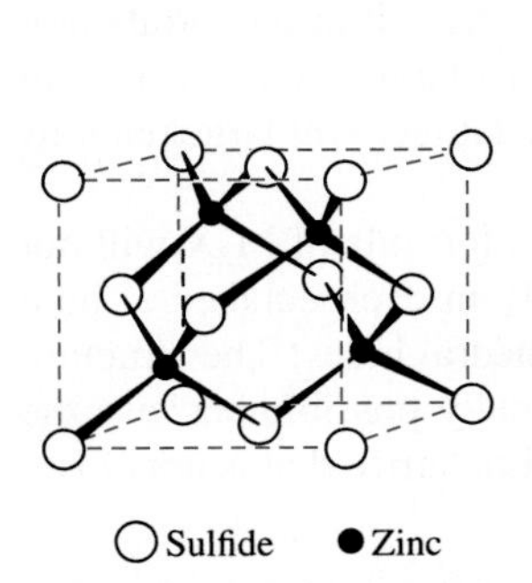

(a) Zinc blende (the same structure results if the Zn and S positions are reversed)

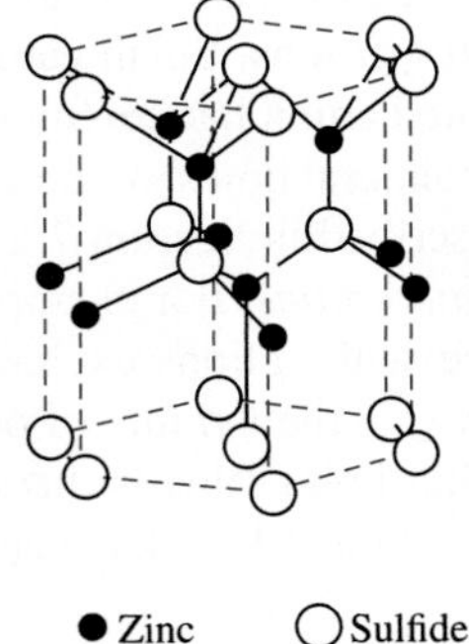

(b) Wurtzite

(c) One sulfide layer and one zinc layer of wurtzite. The third layer contains sulfide ions, directly above the zinc ions. The fourth layer is zinc ions, directly above the sulfides of the first layer.

*CsCl is not body-centered cubic because the ions in the center and on the corners of the unit cell are different.

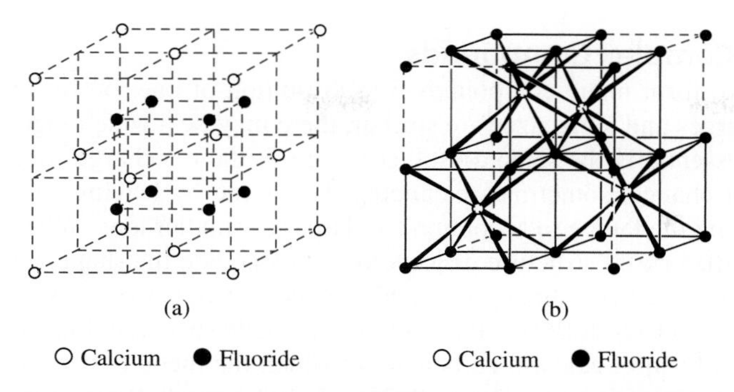

FIGURE 7.9 Fluorite and Antifluorite Crystal Structures. (a) Fluorite shown as Ca^{2+} in a cubic close-packed lattice, each surrounded by eight F^- in the tetrahedral holes. (b) Fluorite shown as F^- in a simple cubic array, with Ca^{2+} in alternate body centers. Solid lines enclose the cubes containing Ca^{2+} ions. If the positive and negative ion positions are reversed, as in Li_2O, the structure is known as antifluorite.

Fluorite, CaF_2

The fluorite structure in **Figure 7.9(a)** can be described as having the calcium ions in a cubic close-packed lattice, with eight fluoride ions surrounding each calcium ion and occupying all of the tetrahedral holes. An alternative description of the same structure, shown in **Figure 7.9(b)**, has the fluoride ions in a simple cubic array, with calcium ions in alternate body centers. The calcium and fluoride ionic radii are nearly ideal for this adopted geometry. In an *antifluorite* structure the cation–anion stoichiometry is reversed. This structure is found in all the oxides and sulfides of Li, Na, K, and Rb and in Li_2Te and Be_2C. In the antifluorite structure, every tetrahedral hole in the anion lattice is occupied by a cation, in contrast to the ZnS structures, in which half the tetrahedral holes of the sulfide ion lattice are occupied by zinc ions.

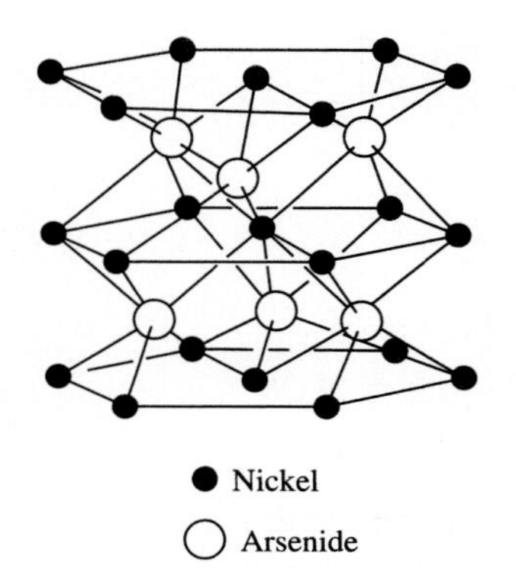

FIGURE 7.10 NiAs Crystal Structure.

Nickel Arsenide, NiAs

The nickel arsenide structure (**Figure 7.10**) has arsenic atoms in identical close-packed layers stacked directly over each other with nickel atoms in all the octahedral holes. The larger arsenic atoms are in the center of trigonal prisms of nickel atoms. The nickel and arsenic atoms both have coordination numbers of 6, with layers of nickel atoms close enough that each nickel can also be considered as bonded to two others. An alternate description is that the nickel atoms occupy all the octahedral holes of a hexagonal close-packed arsenic lattice. This structure is also adopted by many MX compounds, where M is a transition metal and X is from Groups 14, 15, or 16 (Sn, As, Sb, Bi, S, Se, or Te). This structure is easily changed to allow for larger amounts of the nonmetal to be incorporated into nonstoichiometric materials.

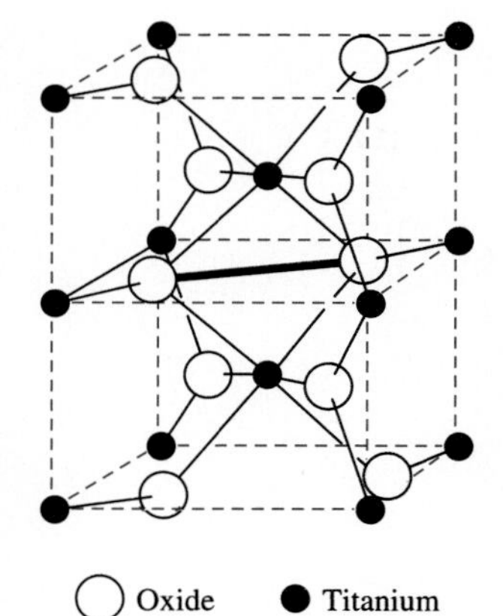

FIGURE 7.11 Rutile (TiO_2) Crystal Structure. The figure shows two unit cells of rutile. The heavy line across the middle shows the edge shared between two (TiO_6) octahedra.

Rutile, TiO_2

TiO_2 in the rutile structure (**Figure 7.11**) has distorted TiO_6 octahedra that form columns by sharing edges, resulting in coordination numbers of 6 and 3 for titanium and oxygen, respectively. The oxide ions have three nearest-neighbor titanium ions in a planar configuration, one at a slightly greater distance than the other two. The unit cell has titanium ions at the corners and in the body center, two oxygens in opposite quadrants of the bottom face, two oxygens directly above the first two in the top face, and two oxygens in the plane with the body-centered titanium forming the final two positions of the oxide octahedron. The salts MgF_2, ZnF_2, and some transition-metal fluorides exhibit the rutile structure. Compounds such as CeO_2 and UO_2 that contain larger metal ions adopt the fluorite structure with coordination numbers of 8 and 4.

7.1.3 More Complex Compounds

It is possible to form many compounds by substitution of one ion for another in a lattice. If the charges and ionic sizes are similar, there may be a wide range of possibilities that result in essentially the same crystal structure. If the ion charges or sizes differ, the structure must change, sometimes balancing the charge by leaving vacancies and frequently adjusting the lattice to accommodate larger or smaller ions. When the anions are nonspherical, the crystal structure distorts to accommodate the shape, and large cations may require increased coordination numbers. A large number of salts ($LiNO_3$, $NaNO_3$, $MgCO_3$, $CaCO_3$, $FeCO_3$, $InBO_3$, YBO_3) adopt the calcite structure, **Figure 7.12(a)**, named for a hexagonal form of calcium carbonate, in which the metal has six nearest-neighbor oxygens. A smaller number (KNO_3, $SrCO_3$, $LaBO_3$), with larger cations, adopt the aragonite structure shown in **Figure 7.12(b)**, an orthorhombic form of $CaCO_3$ that has 9-coordinate metal ions.

7.1.4 Radius Ratio

Coordination numbers in different crystals depend on the sizes and shapes of the ions or atoms, their electronic structures, and, in some cases, on the temperature and pressure under which they were formed. An oversimplified and approximate approach to predicting coordination numbers uses the radius ratio, r_+/r_-. Simple calculation from tables of ionic radii allows prediction of possible structures by modeling the ions as hard spheres. For hard spheres, the ideal size for a smaller cation in an octahedral hole of an anion lattice is a radius of $0.414r_-$. Calculations for other geometries result in the radius ratios and coordination number predictions shown in **Table 7.1**.

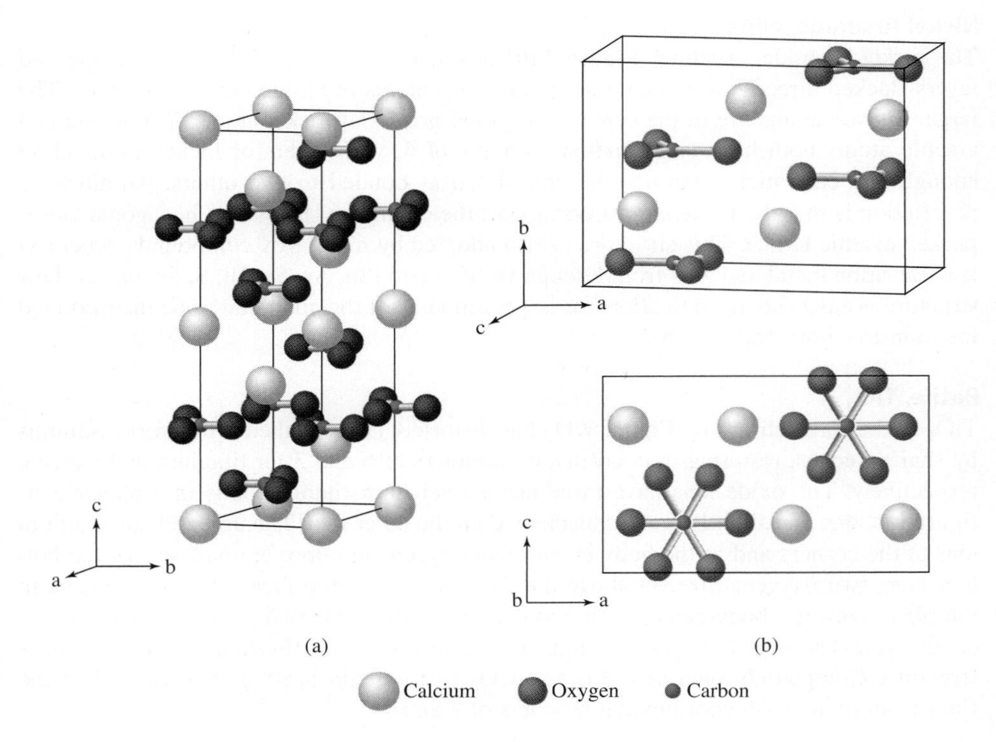

FIGURE 7.12 Structures of Calcium Carbonate, $CaCO_3$. (a) Calcite. (b) Two views of aragonite.

TABLE 7.1 Radius Ratios (r_+/r_-) and Predicted Coordination Numbers

Radius Ratio Limiting Values	Predicted Coordination Number	Geometry	Examples
	4	Tetrahedral	ZnS
0.414			
	4	Square planar	None
	6	Octahedral	NaCl, TiO_2 (rutile)
0.732			
	8	Cubic	CsCl, CaF_2 (fluorite)
1.00			
	12	Cubooctahedron	No ionic examples, but many metals are 12-coordinate.

EXAMPLE 7.3

NaCl

Using the radius of the Na^+ cation (Appendix B-1) for either CN = 4 or CN = 6, $r_+/r_- = 113/167 = 0.677$ or $116/167 = 0.695$, both of which predict CN = 6. The Na^+ cation fits easily into the octahedral holes of the Cl^- lattice, which is ccp.

ZnS

The zinc ion radius varies more with coordination number. The radius ratios are $r_+/r_- = 74/170 = 0.435$ for the CN = 4 and $r_+/r_- = 88/170 = 0.518$ for the CN = 6 radius. Both predict CN = 6, but the smaller one is close to the tetrahedral limit of 0.414. Experimentally, the Zn^{2+} cation fits into the tetrahedral holes of the S^{2-} lattice, which is either ccp (zinc blende) or hcp (wurtzite).

EXERCISE 7.3 Fluorite (CaF_2) has fluoride ions in a simple cubic array and calcium ions in alternate body centers, with $r_+/r_- = 0.97$. What coordination numbers of the two ions are predicted by the radius ratio? What coordination numbers are observed? Predict the coordination number of Ca^{2+} in $CaCl_2$ and $CaBr_2$.

The predictions of the example and exercise match reasonably well with the facts for these two compounds, even though ZnS is largely covalent rather than ionic. However, all radius ratio predictions should be used with caution, because ions are not hard spheres, and there are many cases in which the radius ratio predictions are incorrect. One study[1] reported that the actual structure matches the predicted structure in about two thirds of the cases, with a higher fraction correct at CN = 8 and a lower fraction correct at CN = 4.

There are also compounds in which the cations are larger than the anions; in these cases, the appropriate radius ratio is r_-/r_+, which determines the CN of the anions in the holes of a cation lattice. Cesium fluoride is an example, with $r_-/r_+ = 119/181 = 0.657$, which places it in the six-coordinate range, consistent with the NaCl structure observed for this compound.

When the ionic radii for the cation and anion are nearly equal, a cubic arrangement of anions with the cation in the body center results, as in cesium chloride with CN = 8. Although a close-packed structure (ignoring the difference between cations and anions) would seem to give greater attractive forces, the CsCl structure separates ions of the same charge, reducing the repulsive forces between them.

Some salts without a 1:1 ratio of formula units, such as CaF_2 and Na_2S, may either have different coordination numbers for the cations and anions or structures in which only a fraction of the possible sites are occupied. Details of such structures are available in Wells[2] and other references.

7.2 Thermodynamics of Ionic Crystal Formation

Formation of ionic compounds from the elements can be written as a series of steps that sum to afford the overall reaction; this procedure can provide additional insight into the driving force for salt formation. The Born–Haber cycle considers the series of component reactions that can be imagined as the individual steps for compound formation. In the example of lithium fluoride, the first five reactions added together result in the sixth overall reaction.*

Reaction				
$Li(s) \longrightarrow Li(g)$	$\Delta H_{sub} =$	161 kJ/mol	Sublimation	(1)
$\frac{1}{2}F_2(g) \longrightarrow F(g)$	$\Delta H_{dis} =$	79 kJ/mol	Dissociation	(2)
$Li(g) \longrightarrow Li^+(g) + e^-$	$\Delta H_{ion} =$	520 kJ/mol	Ionization energy	(3)
$F(g) + e^- \longrightarrow F^-(g)$	$\Delta H_{ion} =$	–328 kJ/mol	–Electron affinity	(4)
$Li^+(g) + F^-(g) \longrightarrow LiF(s)$	$\Delta H_{xtal} =$	–1050 kJ/mol	Lattice enthalpy	(5)
$Li(s) + \frac{1}{2}F_2(g) \longrightarrow LiF(s)$	$\Delta H_{form} =$	–618 kJ/mol	Formation	(6)

Such calculations were used historically to determine electron affinities after the enthalpies for the other reactions had been either measured or calculated. The improved experimental accuracy of modern electron affinity determinations permits these cycles to provide even more accurate lattice enthalpies. Despite the simplicity of this approach, it can be powerful in calculating thermodynamic properties for reactions that are difficult to measure directly.

7.2.1 Lattice Energy and the Madelung Constant

At first glance, calculation of the crystal lattice energy may seem straightforward: just take every pair of ions and calculate the sum of the electrostatic energy between each pair, using the following equation.

$$\Delta U = \frac{Z_i Z_j}{r_0}\left(\frac{e^2}{4\pi\,\varepsilon_0}\right)$$

where
Z_i, Z_j = ionic charges in electron units
r_0 = distance between ion centers
e = electronic charge = 1.602×10^{-19} C
$4\pi\,\varepsilon_0$ = permittivity of a vacuum = 1.11×10^{-10} C^{2}J^{-1}m^{-1}
$\dfrac{e^2}{4\pi\,\varepsilon_0} = 2.307 \times 10^{-28}$ J m

However, this approach is problematic because it assumes that each cation only experiences attraction from one anion within the crystalline lattice. Indeed, even summing the energy for interactions between *nearest neighbors* is insufficient, because significant energy is involved in longer-range interactions between the ions. For NaCl, the closest neighbors to a sodium ion are six chloride ions at half the unit cell distance, but the set of next-nearest neighbors is a set of 12 sodium ions at 0.707 times the unit cell distance, and the numbers rise rapidly from there. Consideration of all these geometric factors carried out at increasing distances between ions until the interactions become infinitesimal results in a correction factor called the **Madelung constant**. It is used below to determine

*Although ionization energy and electron affinity are formally internal energy changes (ΔU), these are equivalent to enthalpy changes because $\Delta H = \Delta U + P\Delta V$, and $\Delta V = 0$ for the processes that define ionization energy and electron affinity.

the lattice energy, the net stabilization due to attraction of the ions within the lattice of one mole of a salt,

$$\Delta U = \frac{NMZ_+Z_-}{r_0}\left(\frac{e^2}{4\pi\,\varepsilon_0}\right)$$

where N is Avogadro's number and M is the Madelung constant. Repulsion between close neighbors is a more complex function, frequently involving an inverse sixth- to twelfth-power dependence on the distance. The Born–Mayer equation corrects for this using only the distance and a constant, ρ:

$$\Delta U = \frac{NMZ_+Z_-}{r_0}\left(\frac{e^2}{4\pi\,\varepsilon_0}\right)\left(1 - \frac{\rho}{r_0}\right)$$

For simple compounds, $\rho = 30$ pm works well when r_0 is also in pm. Lattice energies are twice as large when charges of 2 and 1 are present, and four times as large when both ions are doubly charged. Madelung constants for some crystal structures are given in **Table 7.2**.

Although the preceding equations provide the change in internal energy associated with lattice formation from gas phase ions, the more commonly used **lattice enthalpy** is $\Delta H_{xtal} = \Delta U + \Delta(PV) = \Delta U + \Delta nRT$, where Δn is the change in the number of moles of gas phase ions upon formation of the crystal (e.g., –2 for AB compounds, –3 for AB_2 compounds). The value of ΔnRT generally is relatively small at 298 K (–4.95 kJ/mol for AB, –7.43 kJ/mol for AB_2) and $\Delta H_{xtal} \approx \Delta U$.

EXERCISE 7.4

Calculate the lattice energy for NaCl, using the ionic radii from Appendix B-1.

TABLE 7.2 Madelung Constants

Crystal Structure	Madelung Constant, M
NaCl	1.74756
CsCl	1.76267
ZnS (zinc blende)	1.63805
ZnS (wurtzite)	1.64132
CaF_2	2.51939
TiO_2 (rutile)	2.3850
Al_2O_3 (corundum)	4.040

Source: D. Quane, *J. Chem. Educ.*, **1970**, *47*, 396, has described this definition and several others, which include all or part of the charge (Z) in the constant. Caution is needed when using M because of the different possible definitions.

7.2.2 Solubility, Ion Size, and HSAB

The effects of solvation and solubility can also be explored using thermodynamic calculations. For the overall reaction $RbCl(s) \xrightarrow{H_2O} Rb^+(aq) + Cl^-(aq)$ the following reactions are used:*

$RbCl(s) \longrightarrow Rb^+(g) + Cl^-(g)$	$\Delta H =$ 689 kJ/mol	–Lattice enthalpy
$Rb^+(g) \xrightarrow{H_2O} Rb^+(aq)$	$\Delta H =$ –358 kJ/mol	Solvation
$Cl^-(g) \xrightarrow{H_2O} Cl^-(aq)$	$\Delta H =$ –316 kJ/mol	Solvation
$RbCl(s) \xrightarrow{H_2O} Rb^+(aq) + Cl^-(aq)$	$\Delta H =$ 15 kJ/mol	Dissolution

*Solvation data from J. V. Coe, *Chem. Phys. Lett.*, **1994**, *229*, 161.

If the enthalpy changes of any three of the four reactions can be determined, the fourth enthalpy change can be found by completing the cycle.* The solvation effects of many ions have been estimated by comparing similar measurements on different compounds. The entropy of solvation needs to be included to calculate the free energy change associated with dissolution.

Many factors are involved in the thermodynamics of solubility, including ionic size and charge, the hardness or softness of the ions (HSAB), the crystal structure of the solid, and the electronic structure of each ion. In general, small ions have a strong electrostatic attraction for each other and for water molecules. Large ions have a weaker attraction for each other and for water molecules, but can accommodate more water molecules around each ion.[3] The balance between these factors is often evoked to rationalize why compounds formed of two large ions (soft) or of two small ions (hard) are generally less soluble than compounds containing one large ion and one small ion, particularly when they have the same charge magnitude. In the examples given by Basolo, LiF, with two small ions, and CsI, with two large ions, are less soluble than LiI and CsF, which have one large and one small ion. For salts consisting of small ions, the exceptionally large lattice energies apparently cannot be compensated by the relatively large hydration enthalpies, rendering these salts less soluble. For salts of large ions, the lower solubility is rationalized by relatively small hydration enthalpies that cannot compensate for the lattice energy. In the case of large ions, even though the lattice energy is relatively low in magnitude, the smaller hydration enthalpies still allow the lattice energy to dominate.

Cation	Hydration Enthalpy (kJ/mol)	Anion	Hydration Enthalpy (kJ/mol)	Lattice Energy (kJ/mol)	Net Enthalpy of Solution (kJ/mol)
Li^+	–519	F^-	–506	–1025	0
Li^+	–519	I^-	–293	–745	–67
Cs^+	–276	F^-	–506	–724	–58
Cs^+	–276	I^-	–293	–590	+21

In this same set of four compounds, the reaction $LiI(s) + CsF(s) \longrightarrow CsI(s) + LiF(s)$ is exothermic ($\Delta H = -146$ kJ/mol) because of the large lattice enthalpy of LiF. This is contrary to the simple electronegativity notion that the most electropositive and the most electronegative elements form the most stable compounds. However, this result is consistent with the hard–soft model, with LiF, the hard–hard combination, and CsI, the soft–soft combination, the least soluble salts (Section 6.6).

One caution associated with these helpful generalizations is that they exclusively consider the enthalpy terms associated with dissolution, and ignore the entropy contribution. Although the destruction of the crystalline lattice upon dissociation of its ions contributes positively to the dissolution entropy, the resulting change in hydrogen bonding, a large contributor to the entropy of dissolution, is a function of ions involved. Pure water features an extensive network of hydrogen bonding (Chapter 6) that is perturbed when salts dissolve. Small cations with high charge densities (for example, Li^+) have been classified as "structure-making" in that they remove some water molecules from this network, but form an even stronger network around the cations (these are called hydration shells). In contrast, "structure-breaking" ions (for example, Cs^+) have lower charge densities and do

*For RbCl, an enthalpy of solution for infinite aqueous dilution of approximately 16.7 kJ/mol has been reported (A. Sanahuja; J. L. Gómez-Estévez, *Thermochimica Acta*, **1989**, *156*, 85) which supports application of the cycle to approximate this value.

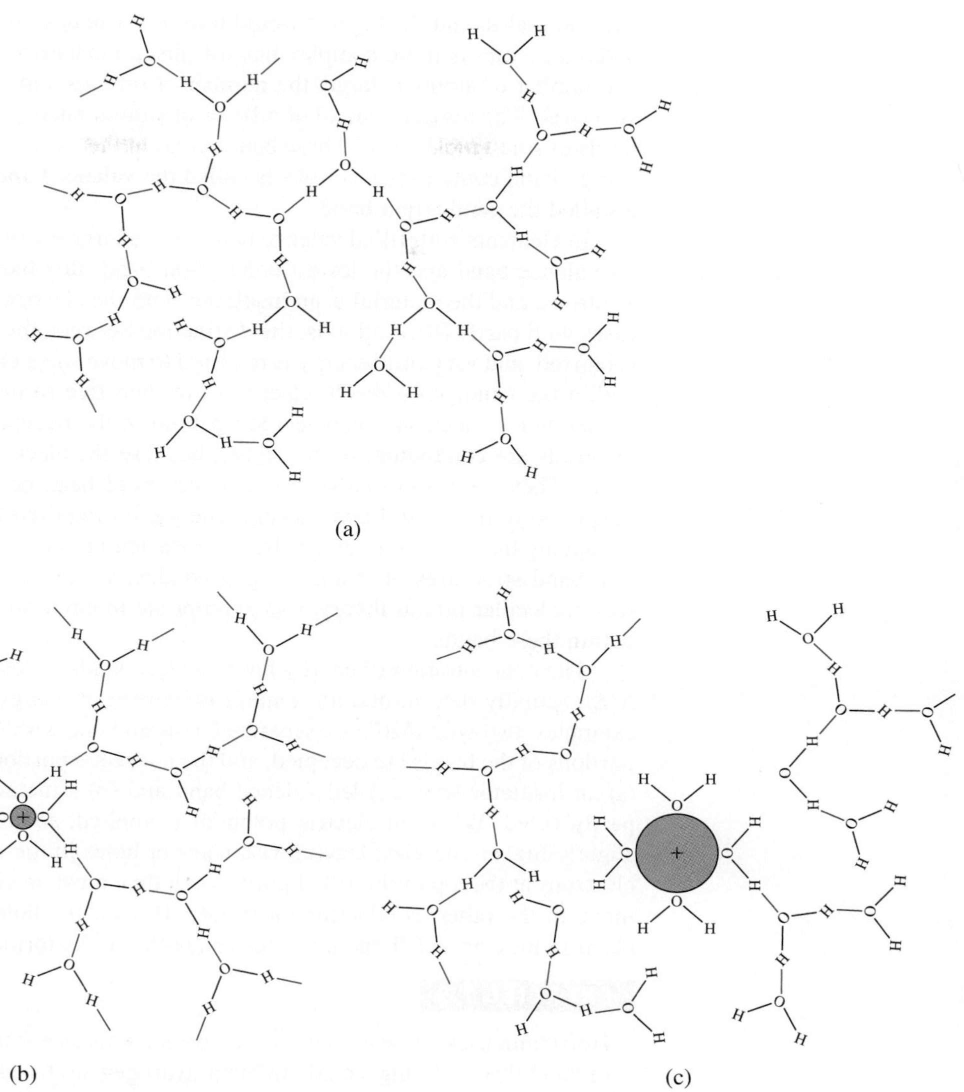

FIGURE 7.13 Effect of Cations on Hydrogen Bonding Networks. (a) The hydrogen-bonding network in pure water. (b) A "structure-making" cation is encapsulated by an even stronger structural network. (c) A "structure-breaking" cation has less charge density and creates a less extended hydration shell. A "structure-making" cation renders the entropy of dissolution relatively less positive. (Reproduced with permission from J. Mähler, I. Persson, *Inorg. Chem.*, *51*, 425. Copyright © 2012 American Chemical Society.)

not perturb the initial hydrogen-bonding network as much.[3] Cations with higher charge densities often lead to less positive solution entropies (the ΔS is still positive due to the dominant contribution from the breakdown of the crystalline lattice) relative to those with lower charge densities. **Figure 7.13** provides two-dimensional representations of these solvation perspectives. An example of the significance of entropy is that a saturated CsI solution is more than 60 times as concentrated as a solution of LiF (in molarity) in spite of the less favorable enthalpy change for the former.

7.3 Molecular Orbitals and Band Structure

The concept of orbital conservation (Chapter 5) requires that when molecular orbitals are formed from two atoms, each interacting pair of atomic orbitals (such as 2*s*) gives rise to two molecular orbitals (σ_{2s} and σ_{2s}^*). When *n* atoms are used, the same approach results in *n* molecular orbitals. In the case of solids, *n* is very large—Avogadro's number for a mole of atoms. If the atoms were all in a one-dimensional row, the lowest energy orbital would have no nodes, and the highest would have $n - 1$ nodes; in a three-dimensional solid, the nodal structure is more complex but still just an extension of this linear model. Because the number of atoms is large, the number of orbitals and closely spaced energy levels is also large. The result is a **band** of orbitals of similar energy, rather than the discrete energy levels of small molecules.[4] These bands contain the electrons from the atoms. The highest energy band containing electrons is called the **valence band**; the next higher, empty band is called the **conduction band**.

In elements with filled valence bands and a large energy difference between the highest valence band and the lowest conduction band, this **band gap** prevents motion of the electrons, and the material is an **insulator**, with the electrons restricted in their motion. In cases with partly filled orbitals, the distinction between the valence and conduction bands is blurred, and very little energy is required to move some electrons to higher energy levels within the band. As a result, electrons are then free to move throughout the crystal, as are the **holes** (electron vacancies) left behind in the occupied portion of the band. These materials are **conductors** of electricity, because the electrons and holes are both free to move. They are also usually good conductors of heat, because the electrons are free to move within the crystal and transmit energy. As required by the usual rules of electrons occupying the lowest energy levels, the holes tend to be in the upper levels within a band. The band structures of insulators and conductors are shown in **Figure 7.14**. Consistent with molecular orbital theory, it is appropriate to envision electrons as being delocalized within these bands.

The concentration of energy levels within bands is described as the **density of states**, *N(E)*, actually determined for a small increment of energy, *dE*. **Figure 7.15** shows three examples, two with distinctly separate bands and one with overlapping bands. The shaded portions of the bands are occupied, and the unshaded portions are empty. The figure shows (a) an insulator with a filled valence band and (b) a metal in which the valence band is partly filled. When an electric potential is applied, some of the electrons can move to slightly higher energies, leaving vacancies or holes in the lower part of the band (c). The electrons at the top of the filled portion can then move in one direction, and the holes can move in the other, conducting electricity. In fact, the holes appear to move because an electron moving to fill one hole creates another in its former location.

EXERCISE 7.5

Hoffmann uses a linear chain of hydrogen atoms as a starting model for his explanation of band theory. Using a chain of eight hydrogen atoms, sketch the phase relationships (positive and negative signs) of all the molecular orbitals that can be formed. These orbitals, bonding at the bottom and antibonding at the top, form a band.

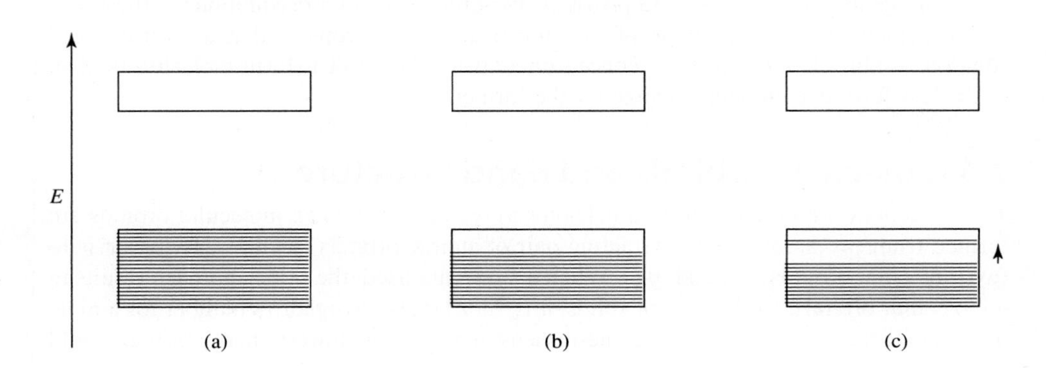

FIGURE 7.14 Band Structure of Insulators and Conductors. (a) Insulator. (b) Metal with no voltage applied. (c) Metal with electrons excited by applied voltage.

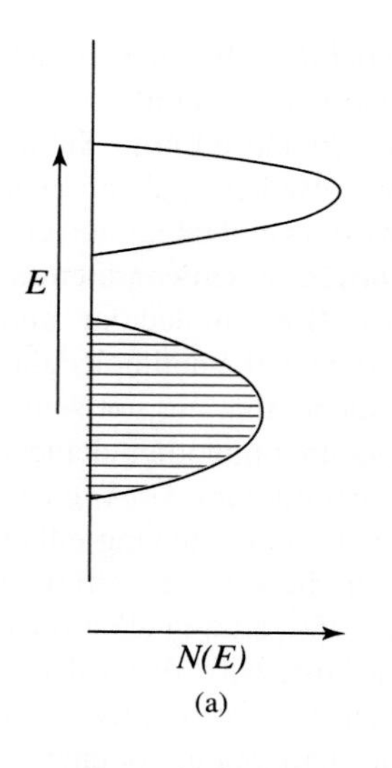

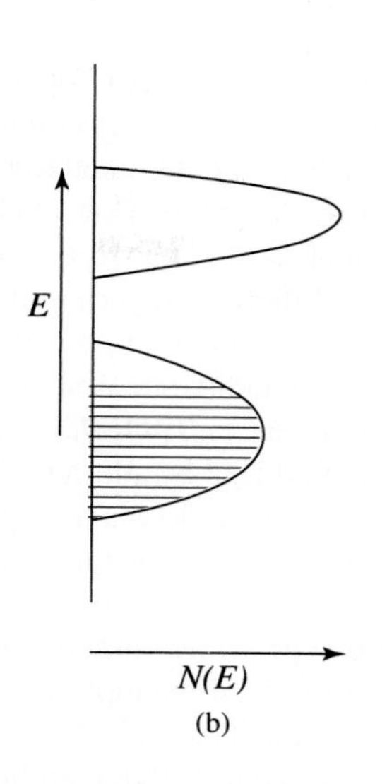

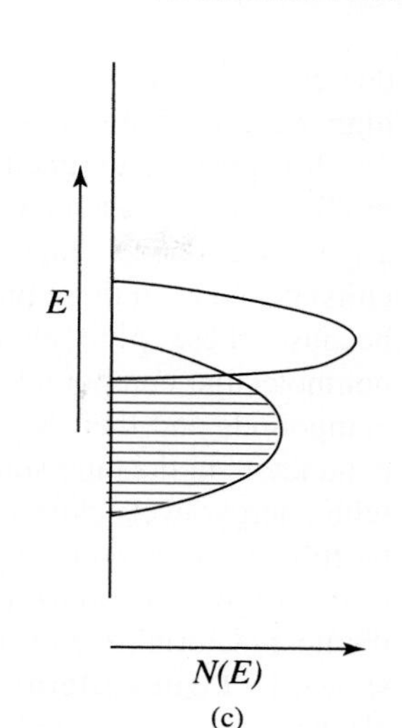

FIGURE 7.15 Energy Bands and Density of States. (a) An insulator with a filled valence band. (b) A metal with a partly filled valence band and a separate empty band. (c) A metal with overlapping bands caused by similar energies of the initial atomic orbitals.

The conductance of metals decreases with increasing temperature, because the increasing vibrational motion of the atoms interferes with the motion of the electrons and increases the resistance to electron flow. High conductance (low resistance) in general, and decreasing conductance with increasing temperature, are characteristics of metals. Some elements have bands that are either completely filled or completely empty, but they differ from insulators by having the bands very close in energy (approximately 2 eV or less). Silicon and germanium are examples: their diamond structure crystals have bonds that are more nearly like ordinary covalent bonds, with four bonds to each atom. At very low temperatures, they are insulators, but the conduction band is very near the valence band in energy. At higher temperatures, when a potential is placed across the crystal, a few electrons can jump into the higher (vacant) conduction band, as in **Figure 7.16(a)**. These electrons are then free to move through the crystal. The vacancies, or holes, left in the lower energy band can also appear to move as electrons move into them. In this way, a small amount of current can flow. When the temperature is raised, more electrons are excited into the upper band, more holes are created in the lower band, and conductance *increases* (resistance

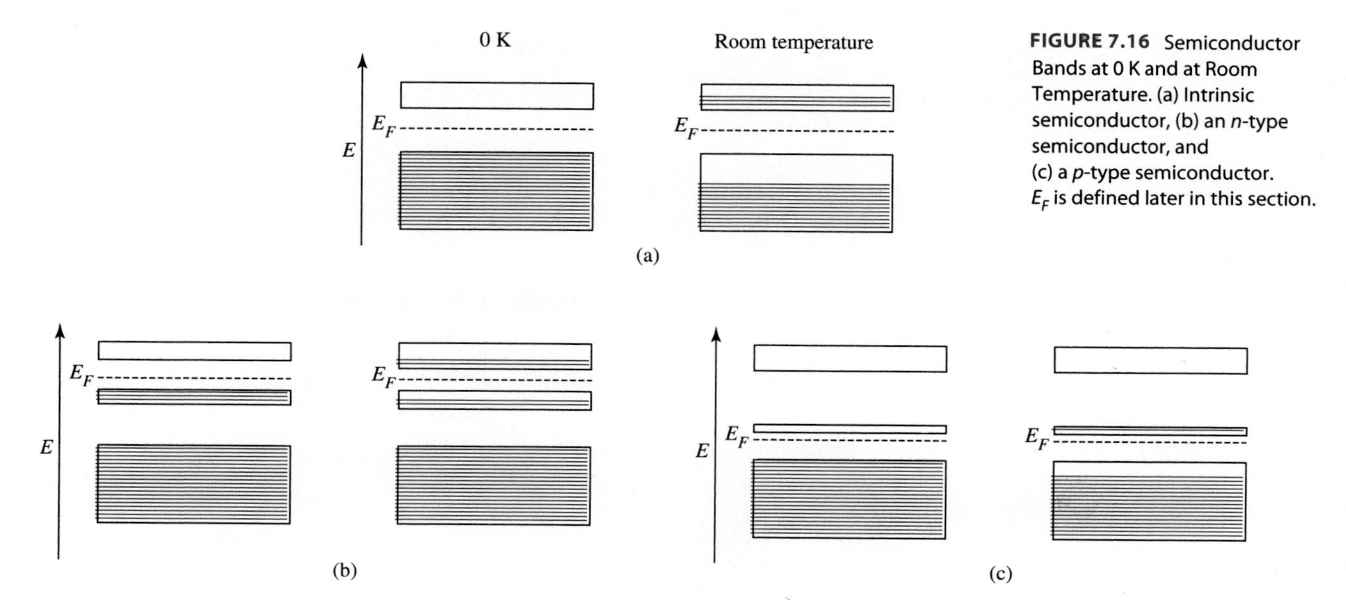

FIGURE 7.16 Semiconductor Bands at 0 K and at Room Temperature. (a) Intrinsic semiconductor, (b) an *n*-type semiconductor, and (c) a *p*-type semiconductor. E_F is defined later in this section.

decreases). This is the distinguishing characteristic of **semiconductors**. They have much higher conductivity than insulators and much lower conductivity than conductors.

It is possible to tune the properties of semiconductors within very close limits. Subtle modifications to a semiconductor can change the flow of electrons caused by application of a specific applied voltage. The field of solid-state electronics (transistors and integrated circuits) depends on these phenomena. Silicon and germanium are **intrinsic semiconductors**, because these pure elements have semiconductive properties. Both molecular and nonmolecular compounds can be semiconductors. A short list of some of the nonmolecular compounds and their band gaps is given in **Table 7.3**. Some elements that are not semiconductors in the pure state can be modified by adding a small amount of another element with energy levels close to those of the host, to make **doped semiconductors**. *Doping* can be thought of as replacing a few atoms of the original element with atoms having either more or fewer electrons. If the added material has more electrons in the valence shell than the host material, the result is an ***n*-type semiconductor** (*n* for *negative*, adding electrons), shown in **Figure 7.16(b)**. Phosphorus is an example in a silicon host, with five valence electrons, compared with four in silicon. These electrons have energies just slightly lower than that of the conduction band of silicon. With the addition of a small amount of energy, electrons from this added energy level can jump up into the empty band of the host material, resulting in higher conductance.

If the added material has fewer valence electrons than the host, it adds positive holes and the result is a ***p*-type semiconductor**, shown in **Figure 7.16(c)**. Aluminum is a *p*-type dopant in a silicon host, with three electrons instead of four in a band very close in energy to that of the silicon valence band. Addition of a small amount of energy boosts electrons from the host valence band into this new level and generates more holes in the valence band of the host, thus increasing the conductance. With careful doping, the conductance

TABLE 7.3 Semiconductors

Material	Minimum Room Temperature Band Gap (eV)
	Elemental
Si	1.110
Ge	0.664
α-Sn	0.08
	Group 13–15 Compounds
GaP	2.272
GaAs	1.441
GaSb	0.70
InP	1.34
InAs	0.356
InSb	0.180
	Group 12–16 Compounds
ZnS	3.80
ZnSe	2.713
ZnTe	2.26
CdS	2.485
CsSe	1.751
CdTe	1.43

Data from: W. M. Haynes, editor-in-chief, *CRC Handbook of Chemistry and Physics*, 92nd ed., electronic version, CRC Press, **2012**, p. 12–90.

can be carefully tailored to the need. Layers of intrinsic, *n*-type, and *p*-type semiconductors, together with insulating materials, are used to create the integrated circuits that are essential to the electronics industry. Controlling the voltage applied to the junctions between the different layers controls conductance through the device.

The number of electrons that are able to jump between the valence and the conduction band depends on the temperature and on the energy gap between the two bands. In an intrinsic semiconductor, the **Fermi level** (E_F, Figure 7.16), the energy at which an electron is equally likely to be in each of the two levels, is near the middle of the band gap. Addition of an *n*-type dopant raises the Fermi level to near the middle of the band gap between the new band and the conduction band of the host. Addition of a *p*-type dopant lowers the Fermi level to a point near the middle of the band gap between the new conduction band and the valence band of the host.

7.3.1 Diodes, the Photovoltaic Effect, and Light-Emitting Diodes

Putting layers of *p*-type and *n*-type semiconductors together creates a *p–n* junction. A few of the electrons in the conduction band of the *n*-type material can migrate to the valence band of the *p*-type material, leaving the *n* type positively charged and the *p* type negatively charged. An equilibrium is quickly established, because the electrostatic forces are too large to allow much charge to accumulate. The separation of charges then prevents transfer of any more electrons. At this point, the Fermi levels are at the same energy, as shown in **Figure 7.17**. The band gap remains the same in both layers, with the energy levels of the *n*-type layer lowered by the buildup of positive charge. If a negative potential is applied to the *n*-type side of the junction and a positive potential is applied to the *p*-type side, it is called a *forward bias*. The excess electrons raise the level of the *n*-type conduction band and then have enough energy to move into the *p*-type side. Holes move toward the junction from the left, and electrons move toward the junction from the right, canceling each other at the junction, and current can flow readily. If the potential is reversed (reverse bias), the energy of the *n*-type levels is lowered compared with the *p*-type levels, the holes and electrons both move away from the junction, and very little current flows. This is the description of a **diode**, which allows current to flow readily in one direction but has a high resistance to current flow in the opposite direction, as in **Figure 7.18**.

A junction of this sort can be used as a light-sensitive switch. With a reverse bias applied (extra electrons supplied to the *p* side), no current would flow, as described for diodes. However, if the difference in energy between the valence band and the conduction band of a semiconductor is small enough, light of visible wavelengths is energetic enough

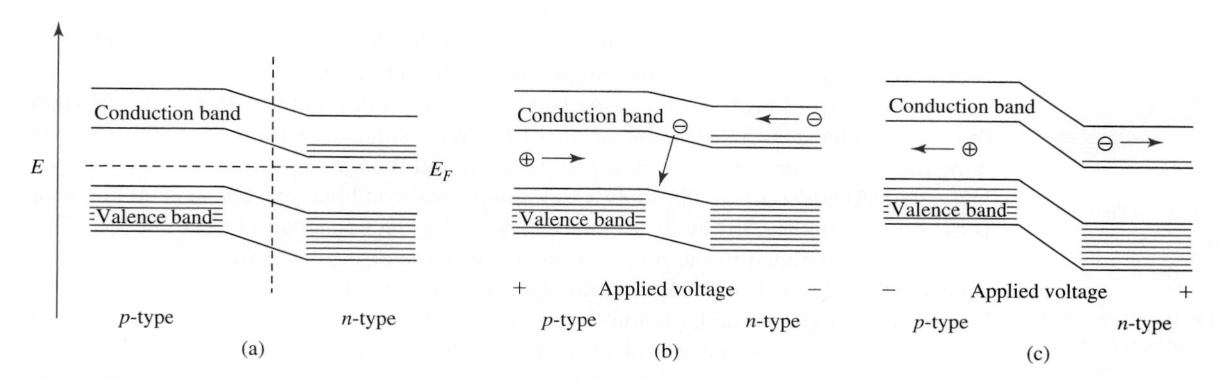

FIGURE 7.17 Band-Energy Diagram of a *p–n* Junction. (a) At equilibrium, the two Fermi levels are at the same energy, changing from the pure *n*-type or *p*-type Fermi levels, because a few electrons can move across the boundary (vertical dashed line). (b) With forward bias, current flows readily. (c) With reverse bias, very little current flows.

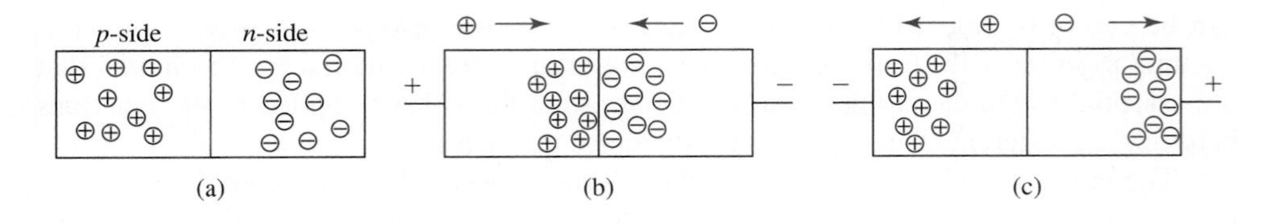

FIGURE 7.18 Diode Behavior. (a) With no applied voltage, no current flows, and few charges are neutralized near the junction by transfer of electrons. (b) Forward bias: current flows readily, with holes and electrons combining at the junction. (c) Reverse bias: very little current can flow, because the holes and electrons move away from each other.

to lift electrons from the valence band into the conduction band, as shown in **Figure 7.19**. Light falling on the junction increases the number of electrons in the conduction band and the number of holes in the valence band, allowing current to flow in spite of the reverse bias. Such a junction then acts as a photoelectric switch, passing current when light strikes it.

If no external voltage is applied, and if the gap has the appropriate energy, light falling on the junction can increase the transfer of electrons from the *p*-type material into the conduction band of the *n*-type material. If external connections are made to the two layers, current can flow through this external circuit. **Photovoltaic cells** of this sort are commonly used in calculators and other "solar" devices—for example, emergency telephones and low-energy lighting—and are increasingly being used to generate electricity for home and commercial use.

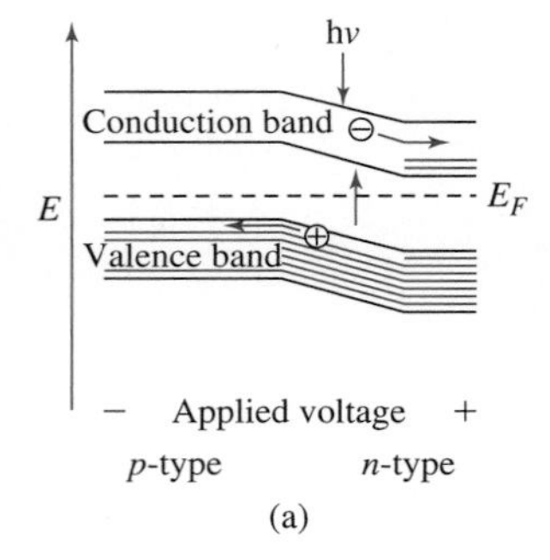

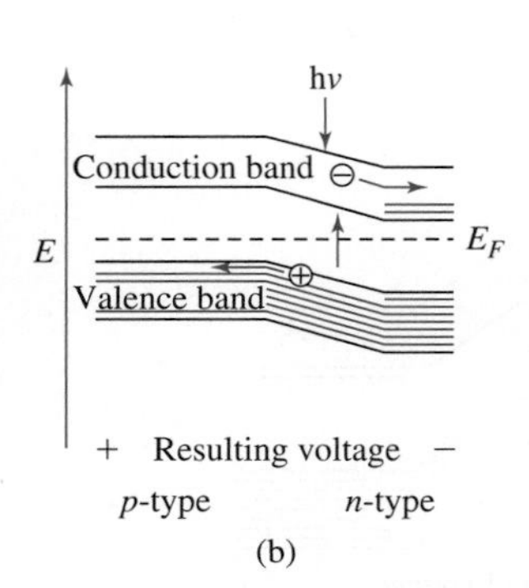

FIGURE 7.19 The Photovoltaic Effect. (a) As a light-activated switch. (b) Generating electricity. Light promotes electrons into the conduction band in the junction.

A forward-biased junction can reverse this process and emit light as a **light-emitting diode (LED)**. The current is carried by holes on the *p*-type side and by electrons on the *n*-type side. When electrons move from the *n*-type layer to the *p*-type layer, they recombine with the holes. If the resulting energy change is of the right magnitude, it can be released as visible light (luminescence), and an LED results. In practice, $GaP_x As_{1-x}$ with $x = 0.40$ to 1.00 can be used for LEDs that emit red light (band gap of 1.88 eV) to green light (2.27 eV). The energy of the light emitted can be changed by adjusting the composition of the material. GaAs has a band gap of 1.441 eV; GaP has a band gap of 2.272 eV. The band gap increases steadily as the fraction of phosphorus is increased, with an abrupt change in slope at $x = 0.45$, where there is a change from a direct band gap to an indirect band gap.[5] In arsenic-rich materials, the electrons drop directly across the energy gap into holes in the lower level (a direct band gap), and the light is emitted with high efficiency. In phosphorus-rich materials, this process must be accompanied by a change in vibrational energy of the crystal (an indirect band gap). This indirect process is less efficient and requires addition of a dopant to give efficient emission by relaxing these rules. These materials also have more complex emission and absorption spectra because of the addition of the dopant, in contrast to the arsenic-rich materials, which have spectra with one simple band. The efficiency of emission is also improved for both types at lower temperatures, at which the intensity of vibrations is reduced. Similar behavior is observed in $Al_x Ga_{1-x} As$ LEDs; emission bands (from 840 nm for $x = 0.05$ to 625 nm for $x = 0.35$) shift to shorter wavelengths, and much greater intensity, on cooling to the temperature of liquid nitrogen (77 K).

Adding a third layer with a larger band gap and making the device exactly the right dimensions changes the behavior of an LED into a solid-state laser. Gallium arsenide doped to provide an *n*-type layer, a *p*-type layer, and then a larger band gap in a *p*-type layer, with Al added to the GaAs, is a commonly used combination. The general behavior is the same as in the LED, with a forward bias on the junction creating luminescence. The larger band gap added to the *p*-type layer prevents the electrons from moving out of the middle *p*-type layer. If the length of the device is exactly a half-integral number of wavelengths of the light emitted, photons released by recombination of electrons and holes are reflected by the edges and stimulate the release of more photons in phase with the first ones. The net result is a large increase in the number of photons and a specific direction for their release in a laser beam. The commonly seen red laser light used for pointers and in supermarket scanners uses this phenomenon.

7.3.2 Quantum Dots

If samples of semiconductors are prepared in smaller and smaller sizes, at some point the bulk properties of the sample will no longer show a continuum of states, as described in previous sections, but will begin to exhibit quantized energy states; the limiting case would be a single molecule with its molecular orbitals. This is called the *quantum confinement* effect and results in very small particles showing a discrete energy level structure rather than a continuum. Nanoparticles showing this effect that have diameters smaller than approximately 10 nm are often called **quantum dots**; because of their size, they behave differently than bulk solids.

The energy-level spacings of quantum dots are related to their size; experiments have shown that the difference in energy between the valence band and conduction band increases as the particle size gets smaller, and the bulk semiconductor becomes more like a single molecule. Consequently, for smaller particles, more energy is needed for excitation and, similarly, more energy is emitted as electrons return to the valence band. More specifically, when an electron is excited, it leaves a hole in the valence band. This excited electron-hole combination is called an **exciton**, and it has an energy slightly lower than the lowest energy of the conduction band. Decay of the exciton to the valence band causes photoemission of a specific energy. Because the energy level spacing is a function of nanoparticle size, it is possible to take advantage of this effect to prepare particles that have the capacity to emit electromagnetic radiation of specific (quantized) energies, for example, where light of a particular color is needed.

The relationship between the size of the quantum dots and their electronic emission spectra has been elegantly shown with zinc selenide,[6] cadmium sulfide,[7] indium phosphide,[8] and indium arsenide[7] nanocrystals. The emission maxima of CdS, InP, and InAs quantum dots (**Table 7.4**) provide an excellent example of how these maxima are dependent on both

TABLE 7.4 Quantum Dot Emission Wavelength as a Function of Nanocrystal Composition and Size

Diameter of Quantum Dot (nm)	Wavelength of Maximum Emission (nm)
CdS	
2.1	484
2.4	516
3.1	550
3.6	576
4.6	606
InP	
3.0	660
3.5	672
4.6	731
InAs	
2.8	905
3.6	1004
4.6	1132
6.0	1333

CdS and InAs Data from X. Peng, J. Wickham, A. P. Alivisatos, *J. Am. Chem. Soc.*, **1998**, *120*, 5343. InP Data from A. A. Guzelian, J. E. B. Katari, A. V. Kadavanich, U. Banin, K. Hamad, E. Juban, A. P. Alivisatos, R. H. Wolters, C. C. Arnold, J. R. Heath, *J. Phys. Chem.*, **1996**, *100*, 7212.

nanocrystal composition and size. For each quantum dot formulation, as the nanocrystal size increases, the band gap grows smaller, and the emission maximum moves to higher wavelength (lower energy), consistent with closer energy-level spacing as the particle size increases. These quantum dots emit within unique regions across the entire visible spectrum, and into the infrared.

Considerable effort has been devoted to developing processes for preparing quantum dots of consistent and reproducible sizes and shapes and using different materials for optimum optical properties.[9] For example, ZnSe quantum dots emit in the violet and UV, PbS in the near-IR and visible, and CdSe throughout the range of visible light. A challenge associated with quantum dots is their tendency to "blink" by not continuously emitting light despite being continuously excited. This "blinking" phenomenon reduces the utility of these semiconductor nanocrystals, and research is underway to determine the origin of "blinking" with the hope of rationally preparing quantum dots that do not exhibit spontaneous emission fluctuations. A spectroelectrochemical technique revealed two mechanisms responsible for "blinking" in CdSe quantum dots.[10] In one mechanism, the energy emitted by an excited quantum dot as a photon sometimes can be transferred to an electron that is subsequently emitted (auger electron emission, a nonradiative exiton decay pathway). The emitted electron renders the quantum dot as "dark", causing the apparent "blinking" even though the quantum dot is still relaxing. The second, and more prevalent, mechanism is due to *surface electron traps,* essentially "dangling" orbitals of surface atoms with reduced coordination number, that can intercept electrons and cause nonradiative decay. Mixed CdSe/ZnSe semiconductor nanocrystals with a composition gradient from the core to the external surface do not blink but emit photons at multiple wavelengths.[11] Exploring the correlation between quantum dot composition and spectral behavior is an active research area.

Numerous applications of quantum dots have been proposed, for example the conversion of solar energy to electricity, in data processing and recording, and in a variety of uses as biosensors. Among medical applications has been the tracking of the uptake of nanoparticles of different sizes by tumors, to investigate whether there is an optimum size for drug delivery,[12] and labeling cell surface proteins to track their motion within cell membranes.[13] Because many nanoparticles are potentially toxic, effort has also been directed to coating them to reduce potential medical and environmental effects.[14] One of the broadest potential impacts of quantum dot technology may be in the realm of highly efficient lighting. Currently LEDs have a wide variety of uses—automobiles, video displays, sensors, traffic signals—but they emit light of too narrow a spectrum to be attractive as high-efficiency general replacements for incandescent and fluorescent lighting; in addition, LEDs remain expensive. Coupling LEDs with quantum dots, perhaps in coatings that use a variety of particle sizes for a range of emission colors, may provide a pathway to more efficient and whiter solid-state lighting for general use.[15] In addition, methods of preparing multicolor inorganic quantum dot LEDs for digital displays in electronic devices have shown promise in the development of displays that may be more robust than organic-based systems.[16]

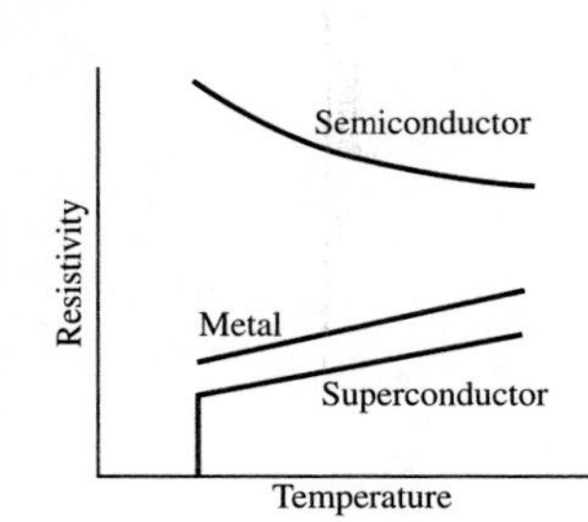

FIGURE 7.20 Temperature Dependence of Resistivity in Semiconductors, Metals, and Superconductors.

7.4 Superconductivity

The conductivity of some metals changes abruptly near liquid helium temperatures (frequently below 10 K), as in **Figure 7.20**, and they become **superconductors**, an effect discovered by Kammerling Onnes in 1911[17] while studying mercury at liquid helium temperature. In this state, these metals offer no resistance to the flow of electrons, and currents started in a loop will continue to flow indefinitely—several decades at least—without significant change. The elements lead, niobium, and tin are other metals that exhibit superconductivity at temperatures near absolute zero. For chemists, one of the most common uses of this effect is in superconducting magnets used in nuclear magnetic resonance instruments, in which it allows generation of much larger magnetic fields than can be obtained with ordinary electromagnets.

7.4.1 Low-Temperature Superconducting Alloys

Some of the most common superconducting materials are alloys of niobium, particularly Nb–Ti alloys, which can be formed into wire and handled with relative ease. These Type I superconductors have the additional property of expelling all magnetic flux when cooled below the **critical temperature**, T_c. This abrupt change is called the *Meissner effect*. It prevails until the magnetic field reaches a critical value, H_c, at which point the applied field destroys the superconductivity. This temperature-dependent change, like that between superconducting and normal conduction, is also sudden rather than gradual. The highest T_c found for niobium alloys is 23.3 K for Nb_3Ge.[18]

Type II superconductors have a more complicated field dependence. Below a given critical temperature they exclude the magnetic field completely. Between this first critical temperature and a second critical temperature, they allow partial penetration by the field; above this second critical temperature, they lose their superconductivity and display normal conductance behavior. In the intermediate temperature region, these materials seem to have a mixture of superconducting and normal regions.

The Meissner effect is being explored for practical use in many areas, including magnetic levitation of trains, although other electromagnetic effects are presently being used for this. A common demonstration is to cool a small piece of superconducting material below its critical temperature and then place a small magnet above it. The magnet is suspended above the superconductor, because the superconductor repels the magnetic flux of the magnet. As long as the superconductor remains below its critical temperature, it expels the magnetic flux from its interior and holds the magnet at a distance.

The levitation demonstration works only with Type II superconductors, because the magnetic field lines that do enter the superconductor resist sideways motion and allow the balance of magnetic repulsion and gravitation to "float" the magnet above the superconductor. With Type I superconductors, the magnetic field lines cannot enter the superconductor at all and, because there is no resistance to sideways motion, the magnet will not remain stationary over the superconductor.

The materials used in the coils of superconducting magnets are frequently Nb–Ti–Cu or Nb_3Sn–Cu mixtures, providing a balance between T_c, which is about 10 K for these materials, and ductility for easier formation into wire.

Superconducting magnets allow very high currents to flow with no change indefinitely as long as the magnet is kept cold enough. In practice, an outer Dewar flask containing liquid nitrogen (boiling point 77.3 K) reduces boil-off of liquid helium (boiling point 4.23 K) from the inner Dewar flask surrounding the magnet coils. A power supply is attached to the magnet, and electrical current is supplied to bring it to the proper field. When the power supply is removed, current flows continuously, maintaining the magnetic field.

A major goal of superconductor research is finding a material that is superconducting at higher temperatures, to remove the need for liquid helium and liquid nitrogen for cooling.

7.4.2 The Theory of Superconductivity (Cooper Pairs)

In the late 1950s, more than 40 years after its discovery, Bardeen, Cooper, and Schrieffer[19] (BCS) provided a theory to explain superconductivity. Their BCS theory postulated that electrons travel through a material in pairs, in spite of their mutual electrostatic repulsion, as long as the two have opposite spins. The formation of these *Cooper pairs* is assisted by small vibrations of the atoms in the lattice; as one electron moves past, the nearest positively charged atoms are drawn very slightly toward it. This increases the positive charge density, which attracts the second electron. This effect then continues through the crystal, in a manner somewhat analogous to a sports crowd doing the wave. The attraction between the two electrons is small, and they change partners frequently, but the overall effect is that the lattice helps them on their way rather than interfering, as is the case with metallic conductivity. If the temperature rises above T_c, the thermal motion of the atoms is sufficient to overcome the slight attraction between the electrons, and the superconductivity ceases.

7.4.3 High-Temperature Superconductors: $YBa_2Cu_3O_7$ and Related Compounds

In 1986, Bednorz and Müller discovered that the ceramic oxide La_2CuO_4 was superconducting above 30 K when doped with Ba, Sr, or Ca to form compounds such as $(La_{2-x}Sr_x)CuO_4$.[20] The seminal realization that cuprates could exhibit superconductivity stimulated intense research in this area. In 1987, $YBa_2Cu_3O_7$ was discovered to have an even higher T_c, 93 K, which was particularly groundbreaking because relatively inexpensive liquid nitrogen (which boils at 77 K) can be used to achieve this temperature.[21] This material, called *1-2-3* for the stoichiometry of the metals in it, is a Type II superconductor, which expels magnetic flux at low fields, but allows some magnetic field lines to enter at higher fields, and consequently ceases to be superconducting at high fields. A number of other similar compounds have since been prepared and found to be superconducting at these or even higher temperatures. These high-temperature superconductors are of great practical interest, because they would allow cooling with liquid nitrogen, rather than liquid helium, a much more expensive coolant. However, the ceramic nature of these materials makes them more difficult to work with than metals. They are brittle and cannot be drawn into wire, making fabrication a problem. Researchers are working to overcome these problems by modifying the formulas or by depositing the materials on a flexible substrate. The present record is a critical temperature of 164 K for $HgBa_2Ca_2Cu_3O_{8-\delta}$ under pressure.[22]

The structures of all the high-temperature cuprate superconductors are related, most with copper oxide planes and chains, as shown in **Figure 7.21**. In $YBa_2Cu_3O_7$, these are stacked with square-pyramidal, square-planar, and inverted square-pyramidal units. In a related tetragonal structure, the oxygen atoms of the top and bottom planes in Figure 7.21 are randomly dispersed in the four equivalent edges of the plane; the resulting material is not superconducting. Oxygen-deficient structures are also possible and are superconducting until about $\delta = 0.65$; materials closer to the formula $YBa_2Cu_3O_6$ are not superconducting.

The discovery of non-copper oxide superconducting materials is an intense research area. The remarkably simple MgB_2 was discovered in 2001 to be a superconductor with a T_c of 39 K,[23] but this appears to be an anomaly; superconductors with related compositions have not been reported. Although magnetic iron was not anticipated to permit superconductivity, the rare earth iron arsenide LaOFeAs, doped with fluoride ions, contradicted this expectation, with a T_c of 26 K.[24] The strategic doping of fluoride for oxide is believed to facilitate electron transfer into the iron–arsenic layer to permit high conductivity. This serendipitous discovery has prompted exploration of so-called *unconventional* superconductors based on a solid-state structural motif similar to that of LaOFeAs, with transition metals arranged in a square lattice with each metal tetrahedrally coordinated.[25]

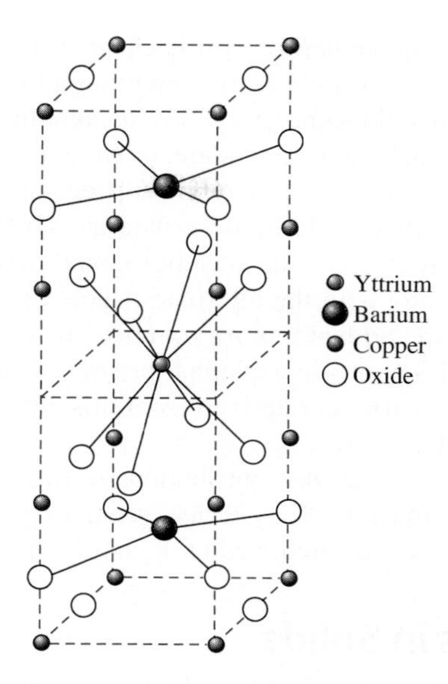

FIGURE 7.21 Unit Cell of Orthorhombic $YBa_2Cu_3O_7$. The yttrium atom in the middle is in a reflection plane.

The understanding of superconductivity in high-temperature superconductors is incomplete, but at this point, an extension of the BCS theory seems to fit much of the known data. The mechanism of electron pairing and the details of the behavior of the electron pairs are less clear. It is hoped that the discovery of new families of superconducting materials (such as the rare-earth iron arsenides) will lead to greater conceptual understanding of this phenomenon, and allow more rational design of these materials. In this regard, a useful correlation between the *a*-axis lattice constant and T_c in FeAs superconductors has been reported.*

7.5 Bonding in Ionic Crystals

The simplest bonding model in ionic crystals is of hard-sphere ions held together by purely electrostatic forces. This model is oversimplified, even for compounds such as NaCl that are strongly ionic in character. The deviation from this simple model makes questions about ion sizes difficult. For example, the Pauling radius of Li^+ is 60 pm, and the crystal radius given by Shannon (Appendix B-1) for Li^+ in a six-coordinate structure is 90 pm. The latter value is much closer to the position of minimum electron density between ions, determined by X-ray crystallography. Four-coordinate Li^+ has a radius of 73 pm, and estimates by Goldschmidt and Ladd are between 73 and 90 pm.[26] The sharing of electrons or the transfer of charge from the anion to the cation varies from a few percent in NaCl to as much as 0.33 electrons per atom in LiI. Each set of radii is self-consistent, but mixing some radii from one set with some from another does not work.

*Figure 3 in Shirage, P. M.; Kikou, K.; Lee, C.-H.; Kito, H.; Eisaki, H.; Iyo, A., *J. Am. Chem. Soc.*, **2011**, *133*, 9630 provides this plot and defines the *a*-axis lattice constant for these materials.

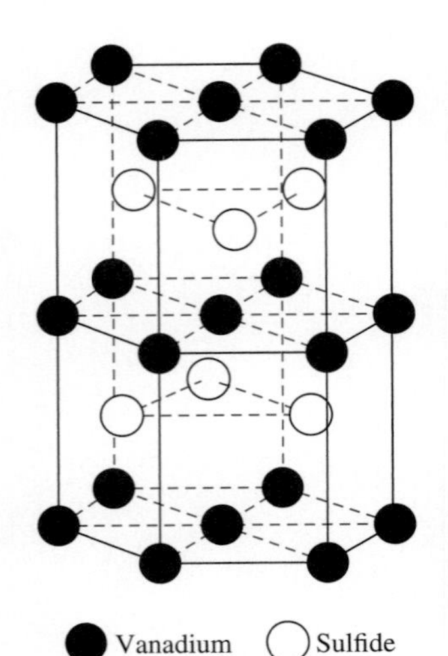

FIGURE 7.22 Structure of Vanadium Sulfide.

Some of the structures shown earlier in this chapter (Figures 7.7 through 7.11) are given as if the components were simple ions, even though the bonding is strongly covalent. In any of these structures, this ambiguity must be kept in mind. The band structures described previously are much more complete in their descriptions of the bonding. Hoffmann[27] has described the bands in vanadium sulfide, an example of the NiAs structure. The crystal has layers that could be described as ABACA in the hexagonal unit cell, with the identical A layers made up of a hexagonal array of V atoms, and the B and C layers made up of S atoms in the alternate trigonal prisms formed by the metal (**Figure 7.22**). In this structure, both atoms are six-coordinate, with V atoms octahedrally coordinated to S atoms and S atoms in a trigonal prism of V atoms. Hoffmann has analyzed the very complex band structure derived from this arrangement of atoms in terms of smaller components of the crystal.

Hoffmann has also shown that the contributions to the density of states of specific orbitals can be calculated.[28] In rutile, TiO_2, a clear separation of the d orbital contribution into t_{2g} and e_g parts can be seen, as predicted by ligand field theory (Chapter 10).

7.6 Imperfections in Solids

In practice, all crystals have imperfections. If a substance crystallizes rapidly, it is likely to have many more imperfections, because crystal growth starts at many sites almost simultaneously. Each small crystallite grows until it runs into its neighbors; the boundaries between these small crystallites are called *grain boundaries*, which can be seen on microscopic examination of a polished surface. Slow crystal growth reduces the number of grain boundaries, because crystal growth starts from a smaller number of sites. However, even if a crystal appears to be perfect, it will likely have imperfections on an atomic level caused by impurities in the material or by dislocations within the lattice.

Vacancies and Self-Interstitials

Vacancies are missing atoms and are the simplest defects. Because higher temperatures increase vibrational motion and expand a crystal, more vacancies are formed at higher temperatures. However, even near the melting point, the number of vacancies is small relative to the total number of atoms—on the order of 1 in 10,000. The effect of a vacancy on the rest of the lattice is small, because it is a localized defect, and the rest of the lattice remains unaffected. Self-interstitials are atoms displaced from their normal location that appear in one of the interstices* in the lattice. Here, the distortion spreads at least a few layers in the crystal, because the atoms are much larger than the available space. In most cases, the number of these defects is much smaller than the number of vacancies.

Substitutions

Substitution of one atom for another is a common phenomenon. Such mixtures are also called *solid solutions*. For example, nickel and copper atoms have similar sizes and electronegativities and form the same FCC crystal structures. Mixtures of the two are stable in any proportion, with random arrangement of the atoms in the alloys. Other combinations that can work well have a very small atom in a lattice of larger atoms. In this case, the small atom occupies one of the interstices in the larger lattice, with small effects on the rest of the lattice but potentially large effects on the behavior of the mixture.

*Interstices refers to the spaces between adjacent atoms in a crystalline lattice.

If the impurity atoms are larger than the holes, lattice strains result, and a new solid phase may be formed.

Dislocations

Edge dislocations result when atoms in one layer do not match up precisely with those of the next. As a result, the distances between the dislocated atoms and atoms in adjacent rows are larger than usual, and the angles between atoms are distorted for a number of rows on either side of the dislocation. A screw dislocation is one that has part of one layer shifted a fraction of a cell dimension. This kind of dislocation frequently causes a rapidly growing site during crystal growth, and it forms a helical path, hence the name. Because they provide sites that allow atoms from the solution or melt to fit into a corner where attractions from three directions can hold them in place, screw dislocations are frequently growth sites for crystals.

In general, dislocations are undesirable in crystals. Mechanically, they can lead to weakness that can cause fracture. Electrically, they interfere with conduction of electrons and reduce reliability, reproducibility, and efficiency in semiconductor devices. For example, one of the challenges of photocell manufacture is to raise the efficiency of cells made of polycrystalline silicon to levels that are reached by single crystals.

7.7 Silicates

Oxygen, silicon, and aluminum are the most abundant elements in the surface of the earth, and more than 80% of the atoms in the solid crust are oxygen or silicon, mostly in the form of silicates. The number of compounds and minerals that contain these elements is very large, and their importance in industrial uses matches their number. We will focus on a few of the silicates.

Silica, SiO_2, has three crystalline forms: quartz at temperatures below 870 °C, tridymite from 870° to 1470 °C, and cristobalite from 1470° to 1710 °C, at which temperature it melts. The high viscosity of molten silica makes crystallization slow; instead of crystallizing, it frequently forms a glass, which softens near 1500 °C. Conversion from one crystalline form to another is difficult and slow, even at high temperatures, because it requires breaking Si–O bonds. All forms contain SiO_4 tetrahedra sharing oxygen atoms, with Si—O—Si angles of 143.6°.

Quartz is the most common form of silica and contains helical chains of SiO_4 tetrahedra, which are chiral with clockwise or counterclockwise twists. Each full turn of the helix contains three Si atoms and three O atoms, and six of these helices combine to form the overall hexagonal shape (**Figure 7.23**).*

Four-coordinate silicon is also present in the silicates, forming chains, double chains, rings, sheets, and three-dimensional arrays. Al^{3+} can substitute for Si^{4+} but requires the addition of another cation to maintain charge balance. Aluminum, magnesium, iron, and titanium are common cations that occupy octahedral holes in the aluminosilicate structure, although any metal cation can be present. Some of the simpler examples of silicate structures are shown in **Figure 7.24**. These subunits pack together to form octahedral holes to accommodate the cations required to balance the charge. As mentioned previously, aluminum can substitute for silicon. A series of minerals is known with similar structures but different ratios of silicon to aluminum.

*Part (b) of this figure was prepared with the assistance of Robert M. Hanson's Origami program and the Chime plug-in (MDL) to Netscape.

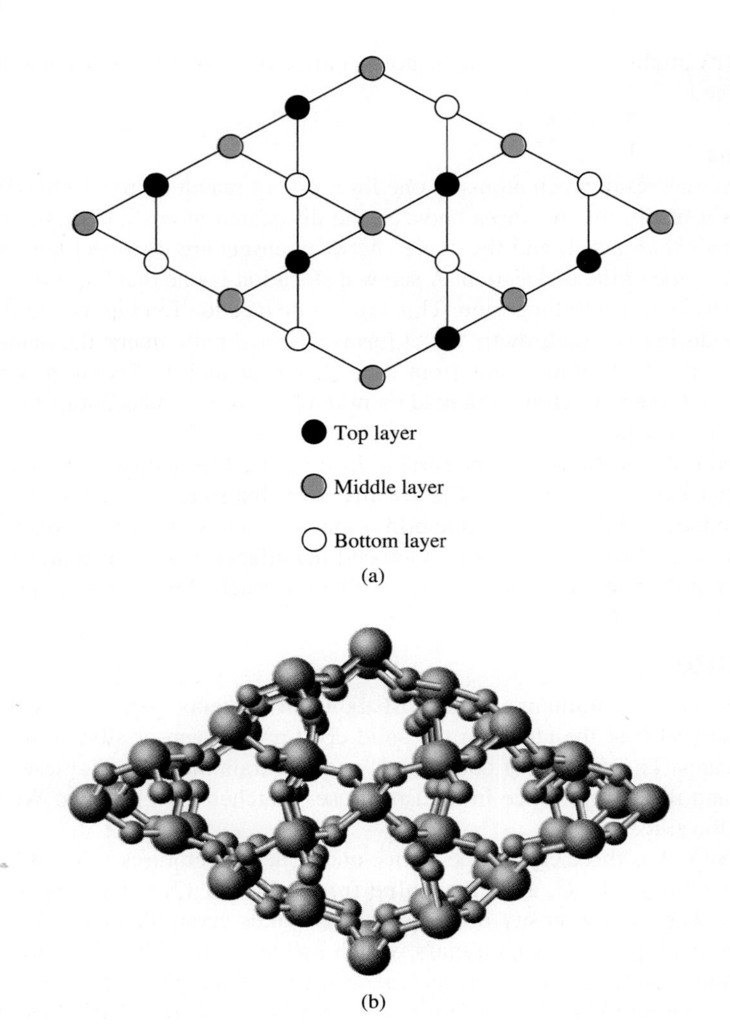

FIGURE 7.23 Crystal Structure of *β*-Quartz. (a) Overall structure, showing silicon atoms only. (b) Three-dimensional representation with both silicon (larger) and oxygen atoms. There are six triangular units surrounding and forming each hexagonal unit. Each triangular unit is helical, with a counterclockwise twist, and three silicon atoms and three oxygen atoms per turn; *α*-Quartz has a similar, but less regular, structure.

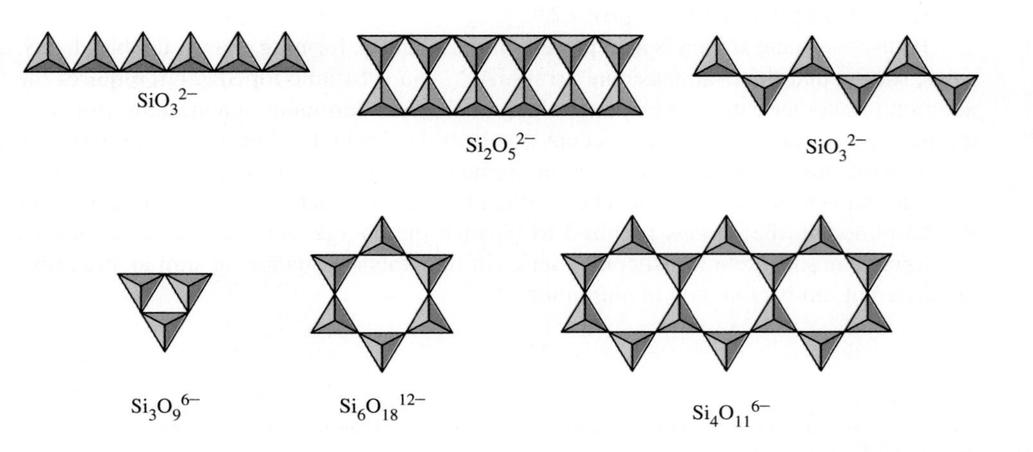

FIGURE 7.24 Common Silicate Structures. (Data from N. N. Greenwood and A. Earnshaw, *Chemistry of the Elements*, Pergamon Press, Elmsford, NY, 1984, pp. 403, 405 and from A. F. Wells, *Structural Inorganic Chemistry*, 5th ed., Oxford University Press, New York, 1984, pp. 1022.)

EXAMPLE 7.6

Relate the formulas of SiO_3^{2-} and $Si_2O_5^{2-}$ to the number of corners shared in the structures shown in Figure 7.24.

Consider the first tetrahedron in the chains of SiO_3^{2-} to have four oxygen atoms, or SiO_4. Extending the chain by adding SiO_3 units, with the fourth position sharing an oxygen atom of the previous tetrahedron, results in an infinite chain with the formula SiO_3. The charge can be calculated based on Si^{4+} and O^{2-}.

$Si_2O_5^{2-}$ can be described similarly. Beginning with one Si_2O_7 unit can start the chain. Adding Si_2O_5 units—two tetrahedra sharing one corner, and each with a vacant corner for sharing with the previous unit—can continue the chain indefinitely. Again, the charge can be calculated from the formula, Si_2O_5 based on Si^{4+} and O^{2-}.

EXERCISE 7.6 Describe the structure of $Si_3O_9^{6-}$ in a similar fashion.

One common family has units of two layers of silicates in the $Si_4O_{11}^{6-}$ geometry bound together by Mg^{2+}, Al^{3+}, or other metal ions, and hydroxide ions to form $Mg_3(OH)_4Si_2O_5$ or $Al_4(OH)_8Si_4O_{10}$ (kaolinite). Kaolinite is a china-clay mineral that forms very small hexagonal plates. If three magnesium ions substitute for two aluminum ions (for charge balance), the result is talc, $Mg_3(OH)_2Si_4O_{10}$. In either case, the oxygen atoms of the silicate units that are not shared between silicon atoms are in a hexagonal array that fits with the positions of hydroxide ions around the cation. The result is hydroxide ion bridging between Al or Mg and Si, as shown in **Figure 7.25(a)**. The layers in talc are (1) all oxygen, the three shared by silicate tetrahedra; (2) all silicon; (3) oxygen and hydroxide in a 2:1 ratio, shared by silicon and magnesium; (4) magnesium; and (5) hydroxide shared between magnesium ions. If another silicate layer—made up of individual layers 3, 2, and 1—is on top of these layers, as in kaolinite, the mineral is called *pyrophyllite*. In both pyrophyllite and talc, the outer surfaces of these layered structures are the oxygen atoms of silicate tetrahedra, resulting in weak attractive forces and very soft materials. Soapstone and the talc used in cosmetics, paints, and ceramics are commercial products with these structures.

Hydrated montmorillonite has water between the silicate–aluminate–silicate layers. The micas (e.g., muscovite) have potassium ions in comparable positions and also have aluminum substituting for silicon in about 25% of the silicate sites. Changes in the proportions of aluminum and silicon in either of these allow the introduction of other cations and the formation of a large number of minerals. The layered structures of some micas are pronounced, allowing them to be cleaved into sheets used for high-temperature applications in which a transparent window is needed. They also have valuable insulating properties and are used in electrical devices.

If the *octahedral* Al^{3+} is partially replaced by Mg^{2+}, additional cations with charges of 1+ or 2+ are also added to the structures, and montmorillonites are the result. These clays swell on the absorption of water, act as cation exchangers, and have **thixotropic** properties; they are gels when undisturbed but become liquid when stirred, making them useful as oil field "muds" and in paints. Their formulas are variable, with $Na_{0.33}[Mg_{0.33}Al_{1.67}(OH)_2(Si_4O_{10})] \cdot n\,H_2O$ as an example. The cations can include Mg, Al, and Fe in the framework and H, Na, K, Mg, or Ca in the exchangeable positions.

The term *asbestos* is usually applied to a fibrous group of minerals that includes the amphiboles, such as tremolite, $Ca_2(OH)_2Mg_5(Si_4O_{11})_2$, with double-chain structures, and chrysotile, $Mg_3(OH)_4Si_2O_5$. In chrysotile, the dimensions of the silicate and magnesium layers are different, resulting in a curling that forms the characteristic cylindrical fibers.

(a)

FIGURE 7.25 Layer Structure of $Mg(OH)_2$-Si_2O_5 Minerals. (a) Side view of the separate layers. (b) Separate views of the layers. (c) The two layers superimposed, showing the sharing of O and OH between them.

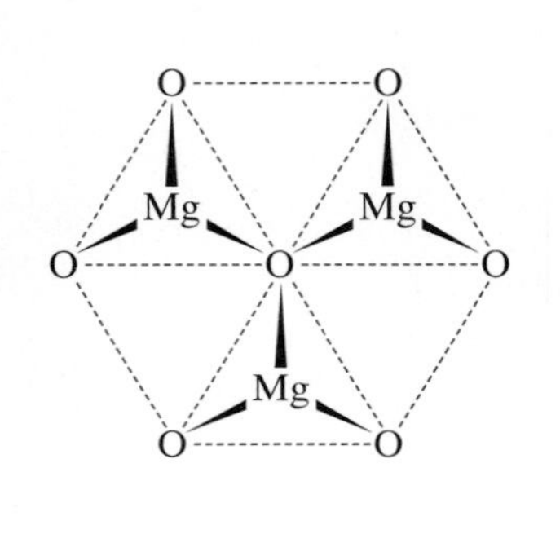

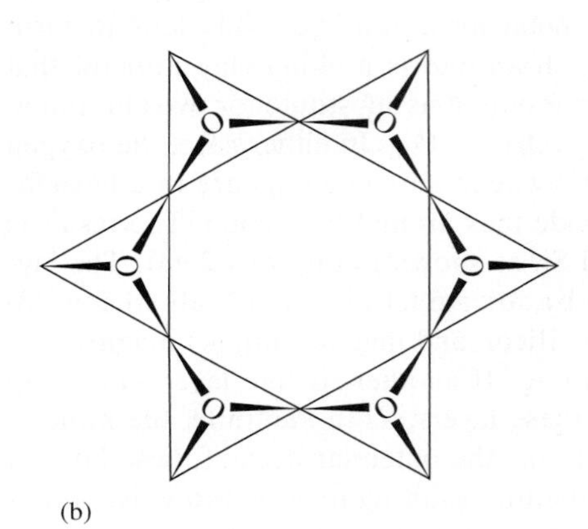

(b)

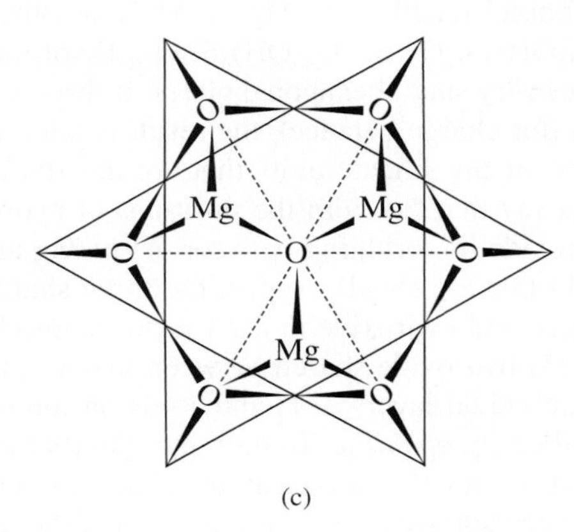

(c)

The final group we will consider are the *zeolites*, mixed aluminosilicates containing $(Si, Al)_nO_{2n}$ frameworks with cations added to maintain charge balance. These minerals contain cavities that are large enough for other molecules to enter. Synthetic zeolites can be made with cavities tailored for specific purposes. The holes that provide entrances to the cavities can have from 4 to 12 silicon atoms around them. A common feature of many of these is a cubo-octahedral cavity formed from 24 silicate tetrahedra, each sharing oxygens on three corners. These units can then be linked to form cubic or tetrahedral units with still larger cavities by sharing of the external oxygen atoms. These minerals exhibit ion-exchange properties in which alkali and alkaline earth metal cations can exchange, depending on concentration. They were used in water softeners to remove excess Ca^{2+} and Mg^{2+} before the development of polystyrene ion-exchange resins. They can also be used to absorb water, oil, and other molecules and are known in the laboratory as "molecular sieves." A larger commercial market is as cat litter and oil absorbent, and they are also used in the petroleum industry as catalysts and as supports for other surface catalysts. Application of zeolites for the conversion of methanol to hydrocarbons is potentially important for the realization of biomass as an energy source. Hydrocarbon product selectivity in these conversions depends on zeolite cavity and pore size.[29] Many zeolites have been described and illustrated in the *Atlas of Zeolite Structure Types.*[30] The references by Wells, Greenwood, and Earnshaw cited previously also provide more information about these essential materials.

Figure 7.26 shows an example of a zeolite structure. Others have larger or smaller pores and larger or smaller entries into the pores.

The extreme range of sizes for the pores (260 to 1120 pm) makes it possible to control entry to and escape from the pores based on the size and branching geometry of the added material. In addition, the surfaces of the pores can be prepared with reactive metal atoms, providing opportunities for surface-catalyzed reactions. Although much of the design of these catalytic zeolites is of the "try it and see what happens" variety, patterns are emerging from the extensive base of data, and planned synthesis of catalysts is possible in some cases.

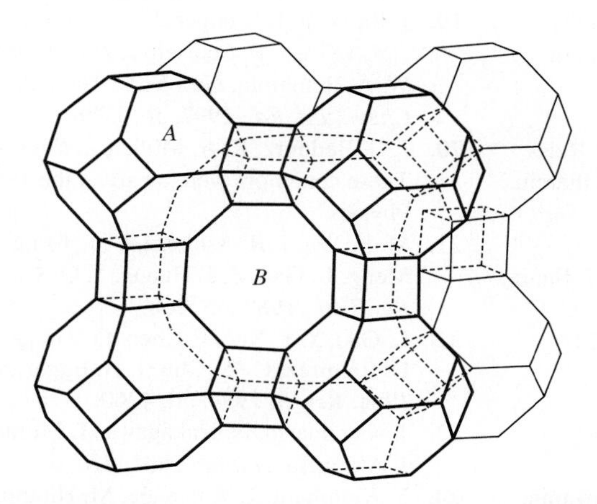

FIGURE 7.26 An Example of an Aluminosilicate Structure. Illustrated is the space-filling arrangement of truncated octahedra, cubes, and truncated cubooctahedra. (Adapted from A. F. Wells, *Structural Inorganic Chemistry*, 5th ed., p. 1039, 1975, by permission of Oxford University Press.)

References

1. L. C. Nathan, *J. Chem. Educ.*, **1985**, *62*, 215.
2. A. F. Wells, *Structural Inorganic Chemistry*, 5th ed., Oxford University Press, New York, 1988.
3. J. Mähler, I. Persson, *Inorg. Chem.*, **2012**, *51*, 425.
4. R. Hoffmann, *Solids and Surfaces: A Chemist's View of Bonding in Extended Structures*, VCH Publishers, New York, 1988, pp. 1–7.
5. A. G. Thompson, M. Cardona, K. L. Shaklee, J. C. Wooley, *Phys. Rev.*, **1966**, *146*, 601; H. Mathieu, P. Merle, and E. L. Ameziane, *Phys. Rev.*, **1977**, *B15*, 2048; M. E. Staumanis, J. P. Krumme, M. Rubenstein, *J. Electrochem. Soc.*, **1977**, *146*, 640.
6. V. V. Nikesh, A. D. Lad, S. Kimura, S. Nozaki, S. Mahamuni, *J. Appl. Phys.*, **2006**, *100*, 113520; P. Reiss, *New. J. Chem.*, **2007**, *31*, 1843 and references cited therein.
7. X. Peng, J. Wickham, A. P. Alivisatos, *J. Am. Chem. Soc.*, **1998**, *120*, 5343.
8. A. A. Guzelian, J. E. B. Katari, A. V. Kadavanich, U. Banin, K. Hamad, E. Juban, A. P. Alivisatos, R. H. Wolters, C. C. Arnold, J. R. Heath, *J. Phys. Chem.*, **1996**, *100*, 7212.
9. G. D. Scholes, *Adv. Funct. Mater.*, **2008**, *18*, 1157.
10. C. Galland, Y. Ghosh, A. Steinbrück, M. Sykora, J. A. Hollingsworth, V. I. Klimov, H. Htoon, *Nature*, **2011**, *479*, 203.
11. X. Wang, X. Ren, K. Kahen, M. A. Hahn, M. Rajeswaran, S. Maccagnano-Zacher, J. Silcox, G. E. Cragg, A. L. Efros, T. D. Krauss, *Nature*, **2009**, *459*, 686.
12. M. Stroh, J. P. Zimmer, D. G. Duda, T. S. Levchenko, K. S. Cohen, E. B. Brown, D. T. Scadden, V. P. Torchilin, M. G. Bawendi, D. Fukumora, R. K. Jain, *Nature Medicine*, **2005**, *11*, 678.
13. M. Howarth, K. Takao, Y. Hayashi, A. Y. Ting, *Proc. Natl. Acad. Sci. U.S.A.*, **2005**, *102*, 7583.
14. J. Drbohlavova, V. Adam, R. Kizek, J. Hubalek, *Int. J. Mol. Sci.*, **2009**, *10*, 656.
15. (a) N. Khan, N. Abas, *Renew. Sustain. Energy Rev.* **2011**, *15*, 1, 296. (b) S. Pimputkar; J. S. Speck, S. P. Denbaars, S. Nakamura, *Nat. Photonics*, **2009**, *3*, 180. (c) M. Molaei, M. Marandi, E. Saievar-Iranizad, N. Taghavinia, B. Liu, H. D. Sun, X. W. Sun, *J. Luminescence*, **2012**, *132*, 467.
16. S. Pickering, A. Kshirsagar, J. Ruzyllo, J. Xu, *Opto-Electron. Rev.*, **2012**, *20*, 148.
17. H. Kammerlingh Onnes, *Akad. Van Wetenschappen (Amsterdam)*, **1911**, *14*, 113, and *Leiden Comm.*, **1911**, *122b*, 124c. Onnes was awarded the 1913 Nobel Prize in Physics for his contributions to the field of superconductivity.
18. C. P. Poole, Jr., H. A. Farach, and R. J. Creswick, *Superconductivity*, Academic Press, San Diego, 1995, p. 22.
19. J. Bardeen, L. Cooper, J. R. Schrieffer, *Phys. Rev.*, **1957**, *108*, 1175; J. R. Schrieffer, *Theory of Superconductivity*, W. A. Benjamin, New York, 1964; A. Simon, *Angew. Chem., Int. Ed.*, **1997**, *36*, 1788.
20. J. G. Bednorz, K. A. Müller, *Z. Phys. B*, **1986**, *64*, 189. These coauthors were awarded the 1987 Nobel Prize in Physics.
21. M. K. Wu, J. R. Ashburn, C. J. Torng, P. H. Hor, R. L. Meng, L. Gao, Z. J. Huang, Y. Q. Wang, C. W. Chu, *Phys. Rev. Lett.*, **1987**, *58*, 908.
22. L. Gao, Y. Y. Xue, F. Chen, Q. Ziong, R. L. Meng, D. Ramirez, C. W. Chu, J. H. Eggert, H. K. Mao, *Phys. Rev. B*, **1994**, *50*, 4260.
23. J. Nagamatsu, N. Nakagawa, T. Muranaka, Y. Zenitani, J. Akimitsu, *Nature,* **2001**, *410*, 63.
24. Y. Kamihara, T. Watanabe, M. Hirano, H. Hosono, *J. Am. Chem. Soc.,* **2008**, *130*, 3296.
25. N. Imamura, H. Mizoguchi, H. Hosono, *J. Am. Chem. Soc.,* **2012**, *134*, 2516.
26. N. N. Greenwood and A. Earnshaw, *Chemistry of the Elements*, 2nd ed., Butterworth–Heinemann, Oxford, 1997, p. 81.
27. R. Hoffmann, *Solids and Surfaces: A Chemist's View of Bonding in Extended Structures*, VCH Publishers, New York, 1988, pp. 102–107.
28. R. Hoffmann, *Solids and Surfaces*, p. 34.
29. U. Olsbye, S. Svelle, M. Bjørgen, P. Beato; T. V. W. Janssens, F. Joensen; S. Bordiga, K. P. Lillerud, *Angew. Chem. Int. Ed.*, **2012**, *51*, 5810.
30. W. M. Meier and D. H. Olson, *Atlas of Zeolite Structure Types*, 2nd ed., Structure Commission of the International Zeolite Commission, Butterworths, London, 1988.

General References

Good introductions to most of the topics in this chapter are in A. B. Ellis et al., *Teaching General Chemistry: A Materials Science Companion*, American Chemical Society, Washington, DC, 1993; P. A. Cox, *Electronic Structure and Chemistry of Solids*, Oxford University Press, Oxford, 1987; and L. Smart and E. Moore, *Solid State Chemistry*, Chapman & Hall, London, 1992. Cox presents more of the theory, and Smart and Moore present more description of structures and their properties. Superconductivity is described by C. P. Poole, Jr., H. A. Farach, and R. J. Creswick in *Superconductivity*, Academic Press, San Diego, 1995; and G. Burns' *High-Temperature Superconductivity*, Academic Press, San Diego, 1992. A. F. Wells' *Structural Inorganic Chemistry*, 5th ed., Clarendon Press, Oxford, 1984; and N. N. Greenwood and A. Earnshaw's *Chemistry of the Elements*, 2nd ed., Butterworth–Heinemann, Oxford, 1997 describe the structures of a very large number of solids and discuss the bonding in them. A very good Web site on superconductors is superconductors.org.

Problems

7.1 Determine the point groups of the following unit cells:
a. Face-centered cubic
b. Body-centered tetragonal
c. CsCl (Figure 7.7)
d. Diamond (Figure 7.6)
e. Nickel arsenide (Figure 7.10)

7.2 Show that atoms occupy only 52.4% of the total volume in a primitive cubic structure in which all the atoms are identical.

7.3 Show that a sphere of radius $0.73r$, where r is the radius of the corner atoms, will fit in the center of a primitive cubic structure.

7.4 **a.** Show that spheres occupy 74.0% of the total volume in a face-centered cubic structure in which all atoms are identical.
b. What percent of the total volume is occupied by spheres in a body-centered cube in which all atoms are identical?

7.5 Using the diagrams of unit cells shown below, count the number of atoms at each type of position (corner, edge, face, internal) and each atom's fraction in the unit cell to determine the formulas (M_mX_n) of the compounds represented. Open circles represent cations, and closed circles represent anions.

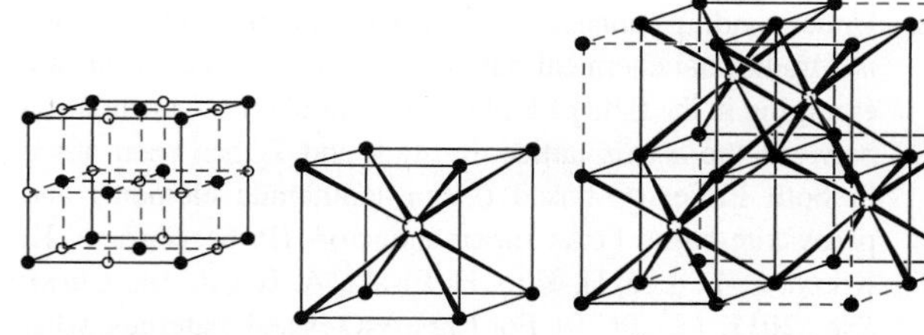

7.6 LiBr has a density of 3.464 g/cm^3 and the NaCl crystal structure. Calculate the interionic distance, and compare your answer with the value from the sum of the ionic radii found in Appendix B.1.

7.7 Compare the CsCl and CaF_2 lattices, particularly their coordination numbers.

7.8 Show that the zinc blende structure can be described as having zinc and sulfide ions each in face-centered lattices, merged so that each ion is in a tetrahedral hole of the other lattice.

7.9 Graphite has a layered structure, with each layer made up of six-membered rings of carbon fused with other similar rings on all sides. The Lewis structure shows alternating single and double bonds. Diamond is an insulator, and graphite is a moderately good conductor. Explain these facts in terms of the bonding in each. (Conductance of graphite is significantly lower than metals but is higher than most nonmetals.) What behavior would you predict for carbon nanotubes, the cylindrical form of fullerenes?

7.10 What experimental evidence is there for the model of alkali halides as consisting of positive and negative ions?

7.11 Mercury(I) chloride and all other Hg(I) salts are diamagnetic. Explain how this can be true. You may want to check the molecular formulas of these compounds.

7.12 **a.** Formation of anions from neutral atoms results in an increase in size, but formation of cations from neutral atoms results in a decrease in size. What causes these changes?
b. The oxide ion and the fluoride ion both have the same electronic structure, but the oxide ion is larger. Why?

7.13 Calculate the radius ratios for the alkali halides. Which fit the radius ratio rules, and which violate them? (See L. C. Nathan, *J. Chem. Educ.*, **1985**, *62*, 215.)

7.14 Comment on the trends in the following values for interionic distances (pm):

LiF	201	NaF	231	AgF	246
LiCl	257	NaCl	281	AgCl	277
LiBr	275	NaBr	298	AgBr	288

7.15 Calculate the electron affinity of Cl from the following data for NaCl, and compare your result with the value in Appendix B.3: Cl_2 bond energy = 239 kJ/mol; ΔH_f(NaCl) = −413 kJ/mol; ΔH_{sub}(Na) = 109 kJ/mol; IE(Na) = 5.14 eV; and $r_+ + r_- = 274$ pm.

7.16 CaO is harder and has a higher melting point than KF, and MgO is harder and has a higher melting point than CaF_2. CaO, KF, and MgO have the NaCl structure. Explain these differences.

7.17 Calculate the lattice energies of the hypothetical compounds $NaCl_2$ and MgCl, assuming that the Mg^+ and Na^+ ions and the Na^{2+} and Mg^{2+} ions have the same radii. How do these results explain the compounds that are found experimentally? Use the following data in the calculation: Second ionization energies ($M^+ \longrightarrow M^{2+} + e^-$): Na, 4562 kJ/mol; Mg, 1451 kJ/mol; enthalpy of formation: NaCl, –411 kJ/mol; $MgCl_2$, –642 kJ/mol.

7.18 Use the Born–Haber cycle to calculate the enthalpy of formation of KBr, which crystallizes in the NaCl lattice. Use these data in the calculation: $\Delta H_{vap}(Br_2)$ = 29.8 kJ/mol; Br_2 bond energy = 190.2 kJ/mol; and ΔH_{sub}(K) = 79 kJ/mol.

7.19 Use the Born–Haber cycle to calculate the enthalpy of formation of MgO, which crystallizes in the NaCl lattice. Use these data in the calculation: O_2 bond energy = 494 kJ/mol; ΔH_{sub}(Mg) = 37 kJ/mol. Second ionization energy of Mg = 1451 kJ/mol; second electron affinity of O = −744 kJ/mol.

7.20 Using crystal radii from Appendix B-1, calculate the lattice energy for PbS, which crystallizes in the NaCl structure. Compare the results with the Born–Haber cycle values obtained using the ionization energies and the following data for enthalpies of formation. Remember that enthalpies of formation are calculated beginning with the elements in their standard states. Use these data: ΔH_f: $S^{2-}(g)$, 535 kJ/mol;

Pb(*g*), 196 kJ/mol; PbS, –98 kJ/mol. The second ionization energy of Pb = 15.03 eV.

7.21 In addition to the doping described in this chapter, *n*-type semiconductors can be formed by increasing the amount of metal in ZnO or TiO_2, and *p*-type semiconductors can be formed by increasing the amount of nonmetal in Cu_2S, CuI, or ZnO. Explain how this is possible.

7.22 Explain how Cooper pairs can exist in superconducting materials, even though electrons repel each other.

7.23 Referring to other references if necessary, explain how zeolites that contain sodium ions can be used to soften water.

7.24 CaC_2 is an insulating ionic crystal. However, $Y_2Br_2C_2$, which can be described as containing C_2^{4-} ions, is metallic in two dimensions and becomes superconducting at 5 K. Describe the possible electronic structure of C_2^{4-}. In the crystal, monoclinic crystal symmetry leads to distortion of the Y_6 surrounding structure. How might this change the electronic structure of the ion? (See A. Simon, *Angew. Chem., Int. Ed.*, **1997**, *36*, 1788.)

7.25 Gallium arsenide is used in LEDs that emit red light. Would gallium nitride be expected to emit higher or lower energy light than gallium arsenide? How might such a process (emission of light from gallium nitride) be useful?

7.26 A series of ZnSe quantum dots was prepared of a range of sizes, with diameters from approximately 1.5 to 4.5 nm, and the photoluminescence emission spectra were recorded. Were the lowest energy emission bands produced by the largest or smallest quantum dots? Explain. (See V. V. Nikesh, A. D. Lad, S. Kimura, S. Nozaki, S. Mahamuni, *J. Appl. Phys.*, **2006**, *100*, 113520.)

7.27 Medical studies on applications of quantum dots are progressing rapidly. Using appropriate search tools, such as Web of Science and SciFinder, find and briefly describe medical applications of quantum dots other than those mentioned in this chapter. Cite the references you consult.

7.28 The correlation between the size of quantum dots (QDs) and their emission spectra motivates the development of methodology to prepare QDs with specific sizes. The size of CdS QDs can be controlled by employing the reaction of *L*-cysteine and CS_2 as a source of S^{2-} (in the form of H_2S) for combination with Cd^{2+}. The *R*-2-thiotetrahydrothiazole-4-carboxylic acid formed in these reactions further modulates QD size by binding to the QD surface, effectively stabilizing it towards further growth. Which *L*-cysteine/Cd^{2+} molar ratio affords the smallest CdS quantum dots? Which *L*-cysteine/Cd^{2+} molar ratio affords CdS quantum dots that absorb UV-vis radiation of highest energy? (Y.-M. Mo, Y. Tang, F. Gao, J. Yang, Y.-M. Zhang, *Ind. Eng. Chem. Res.*, **2012**, 2012, 51, 5995.)

7.29 A variety of sulfide precursors can be used to prepare CdS QDs, including thiourea (A. Aboulaich, D. Billaud, M. Abyan, L. Balan, J.-J. Gaumet, G. Medjadhi, J. Ghanbaja, R. Schneider, *ACS Appl. Mater. Interfaces*, **2012,** *4*, 2561). Provide a reaction scheme and describe the reaction conditions employed to prepare 3-mercaptopropionic acid-capped CdS QDs. What reaction time is optimum to achieve the highest energy emission obtained when these QDs are excited with UV radiation? What is the primary influence of the reaction time on the size of these QDs?

7.30 "Digestive ripening" is a technique used for narrowing quantum dot size distributions. The properties of water-soluble CdSe and CdTe QDs change during the ripening process (M. Kalita, S. Cingarapu, S. Roy, S. C. Park, D. Higgins, R. Janlowiak, V. Chikan, K. J. Klabunde, S. H. Bossmann, *Inorg. Chem.*, **2012**, *51*, 4521). What two methods did these chemists employ to assess whether their CdSe or CdTe QDs were larger after digestive ripening? Which quantum dots were larger, and how did the data support this claim?

7.31 The superconductivity of a binary barium germanide, $BaGe_3$, has been established (H. Fukuoka, Y. Tomomitsu, K. Inumaru, *Inorg. Chem.*, **2011**, *50*, 6372). What data were used to determine that $BaGe_3$ was metallic, and what observation prompted the assignment of the 4.0 K critical temperature? Density of states (DOS) calculations help assess orbital contributions to superconductor band structure. Which orbitals contribute the most to the conduction band of $BaGe_3$? Which orbitals contribute most near the $BaGe_3$ Fermi level?

7.32 Understanding connections between "structure and function" is a traditional chemical pursuit, and similar correlations are emerging in the field of FeAs superconductors. A correlation between the *a*-axis lattice constant and T_c has been noted in both LnFeAsO-based (Ln = lanthanide element) and perovskite-based FeAs superconductors (P. M. Shirage, K. Kikou, C.-H. Lee, H. Kito, H. Eisaki, A. Iyo, *J. Am. Chem. Soc.*, **2011**, *133*, 9630). For LnFeAsO-based materials, what is the optimum *a*-axis lattice constant that results in the highest T_c? How do the T_c values vary on the basis of the identity of Ln? Do these T_c values exhibit a periodic trend? Explain. For perovskite-based (As) materials, what is the predicted *a*-axis lattice constant upper limit above which superconductivity would not be expected at any temperature?

7.33 Determine the formulas of the following silicates (*c* is a chain, extending vertically).

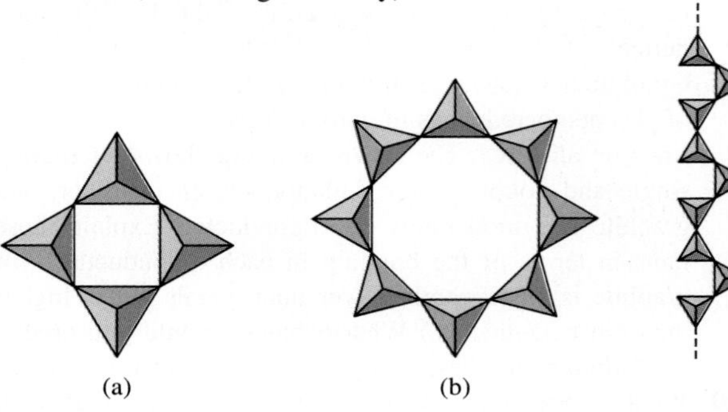

Chapter 8

Chemistry of the Main Group Elements

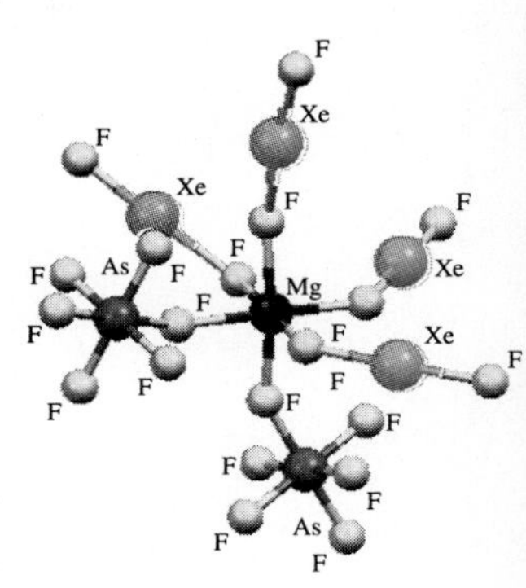

The 20 chemicals produced in the greatest amounts by the United States chemical industry are main group elements or compounds (Table 8.1), and eight of the top ten are "inorganic"; main group compounds are of great commercial importance. These compounds also exhibit fascinating and sometimes unexpected properties and reactivity. Chapter 8 presents descriptive chemistry for each of the main groups of elements, treating hydrogen first and continuing in sequence from Groups 1, 2, and 13 through 18.

A discussion of main group chemistry provides the opportunity to introduce topics that are particularly characteristic of main group chemistry but may also be applicable to transition metals. For example, many examples of atoms that form bridges between other atoms are known. Three main group examples are shown in the margin. In this chapter, we discuss one type of bridge in depth: the hydrogens that bridge boron atoms in boranes. A similar approach can be used to describe bridges formed by other atoms and by groups such as CO (CO bridges between transition-metal atoms are discussed in Chapter 13).

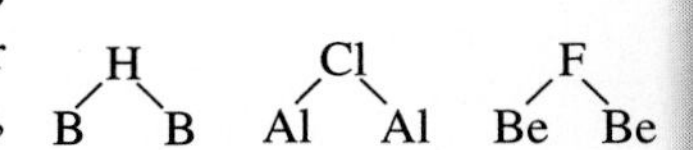

Chapter 8 provides examples in which chemistry has developed in ways surprisingly different from previously held ideas. These groundbreaking examples have included compounds in which carbon is bonded to more than four atoms, alkali metal anions, and the chemistry of noble gases. The fascinating development of carbon's chemistry in the last three decades, including fullerenes ("buckyballs"), nanotubes, graphene, and other previously unknown forms, holds the tantalizing promise to transform a wide variety of aspects of electronics, medicine, and other realms. Chapter 8 introduces intriguing aspects of main group chemistry and catalogs classic compounds for useful reference. An understanding of the bonding and structures of main group compounds (Chapters 3 and 5), and acid–base reactions involving these compounds (Chapter 6) provides a conceptual basis for this chapter.

8.1 General Trends in Main Group Chemistry

Section 8.1 broadly discusses properties of the main group elements that influence their physical and chemical behavior.

8.1.1 Physical Properties

The main group elements fill their *s* and *p* valence orbitals to complete their electron configurations. The last digit in the group numbers recommended by the IUPAC (Groups 1, 2, and 13 through 18) conveniently provides this number of electrons.[1] These elements range from the most metallic to the most nonmetallic, with elements of intermediate properties in between. On the far left, the alkali metals and alkaline earths exhibit the metallic characteristics of luster, high ability to conduct heat and electricity, and malleability. The distinction between metals and nonmetals is illustrated by their difference in conductance.

TABLE 8.1 Top 20 Industrial Chemicals Produced in the United States, 2010*

Rank	Chemical	Production ($\times 10^9$ kg)
1	Sodium chloride, $NaCl$	45.0[a]
2	Sulfuric acid, H_2SO_4	32.5
3	Phosphate rock, MPO_4	26.1
4	Ethylene, $H_2C{=}CH_2$	24.0
5	Lime, CaO	18.0[a]
6	Propylene, $H_2C{=}CH{-}CH_3$	14.1
7	Sodium carbonate, Na_2CO_3	10.0[a]
8	Chlorine, Cl_2	9.7
9	Phosphoric acid, H_3PO_4	9.4
10	Sulfur, S_8	9.1[a]
11	Dichloroethane, $ClH_2C{-}CH_2Cl$	8.8
12	Ammonia, NH_3	8.6[a]
13	Sodium hydroxide, $NaOH$	7.5
14	Ammonium hydrogen phosphate, $(NH_4)_2HPO_4$	7.4
15	Ammonium nitrate, NH_4NO_3	6.9
16	Nitric acid, HNO_3	6.3[a]
17	Ammonium dihydrogen phosphate, $NH_4(H_2PO_4)$	4.3
18	Ethylbenzene, $C_2H_5C_6H_5$	4.2
19	Styrene, $C_6H_5CH{=}CH_2$	4.1
20	Hydrochloric acid, HCl	3.5

Sources: Data from *Chem. Eng. News*, July 4, 2011, pp. 55–63; U. S. Department of the Interior, U.S. Geological Survey, *Mineral Commodity Summaries 2011*.

[a]Estimated value.

In **Figure 8.1**, electrical resistivities (inversely proportional to conductivities) of the solid main group elements are plotted.** At the far left are the alkali metals, having low resistivities (high conductances); at the far right are the nonmetals. Metals contain valence electrons that are relatively free to move and thereby conduct current. In most cases, nonmetals contain much more localized electrons and covalently bonded pairs that are less mobile. Graphite (Section 8.6.1), an allotrope (elemental form) of the nonmetal carbon, is an exception that has a much greater ability to conduct than most nonmetals because of delocalized electrons.

Elements along a rough diagonal from boron to polonium are intermediate in behavior, in some cases having both metallic and nonmetallic allotropes; these elements are designated as **metalloids** or **semimetals**. Some elements, such as silicon and germanium, are capable of having their conductivity finely tuned by the addition of small amounts of impurities and are consequently of enormous importance in the manufacture of semiconductors (Chapter 7) in the electronics industry.

*Additional production data for select high-volume inorganic and organic chemicals from other countries in 2011 are available in *Chem. Eng. News*, **2012**, *90*(27), 59.

**The electrical resistivity shown for carbon is for the diamond allotrope. Graphite, another allotrope of carbon, has a resistivity between that of metals and semiconductors.

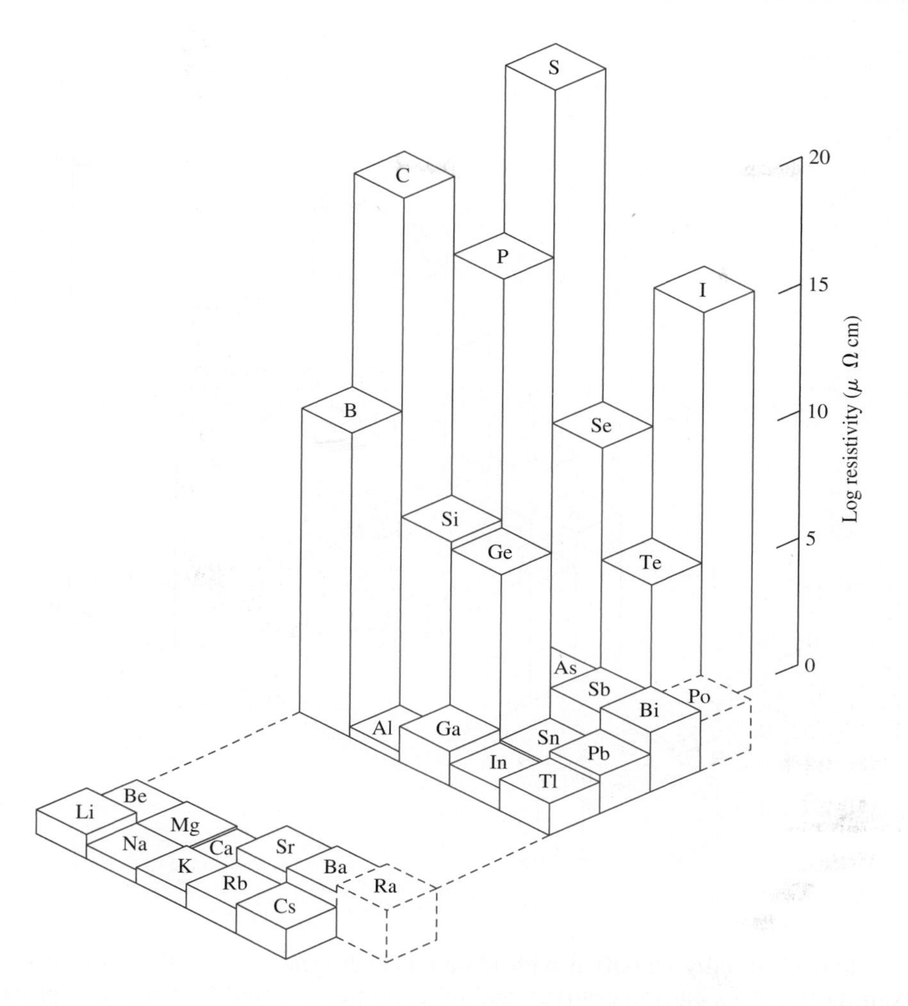

FIGURE 8.1 Electrical Resistivities of the Main Group Elements. Dashed lines indicate estimated values.

(Data from J. Emsley, *The Elements*, Oxford University Press, 1989.)

Some of the columns of main group elements have long been designated by common names; names for others have been suggested, and some have been used more frequently in recent years:

Group	Common Name	Group	Common Name
1(I)	Alkali metals	15(V)	Pnicogens, pnictogens
2(II)	Alkaline earths	16(VI)	Chalcogens
13(III)	Triel elements	17(VII)	Halogens
14(IV)	Tetrel elements	18(VIII)	Noble gases

8.1.2 Electronegativity

Electronegativity provides a guide to the chemical behavior of the main group elements (**Figure 8.2**). The extremely high electronegativity of fluorine and the noble gases helium and neon are evident, with a steady decline in electronegativity toward the left and the bottom of the periodic table. The semimetals form a diagonal of intermediate electronegativity. Definitions of electronegativity are given in Section 3.2.3 and tabulated values for the elements are given in **Table 3.3** and Appendix B.4.

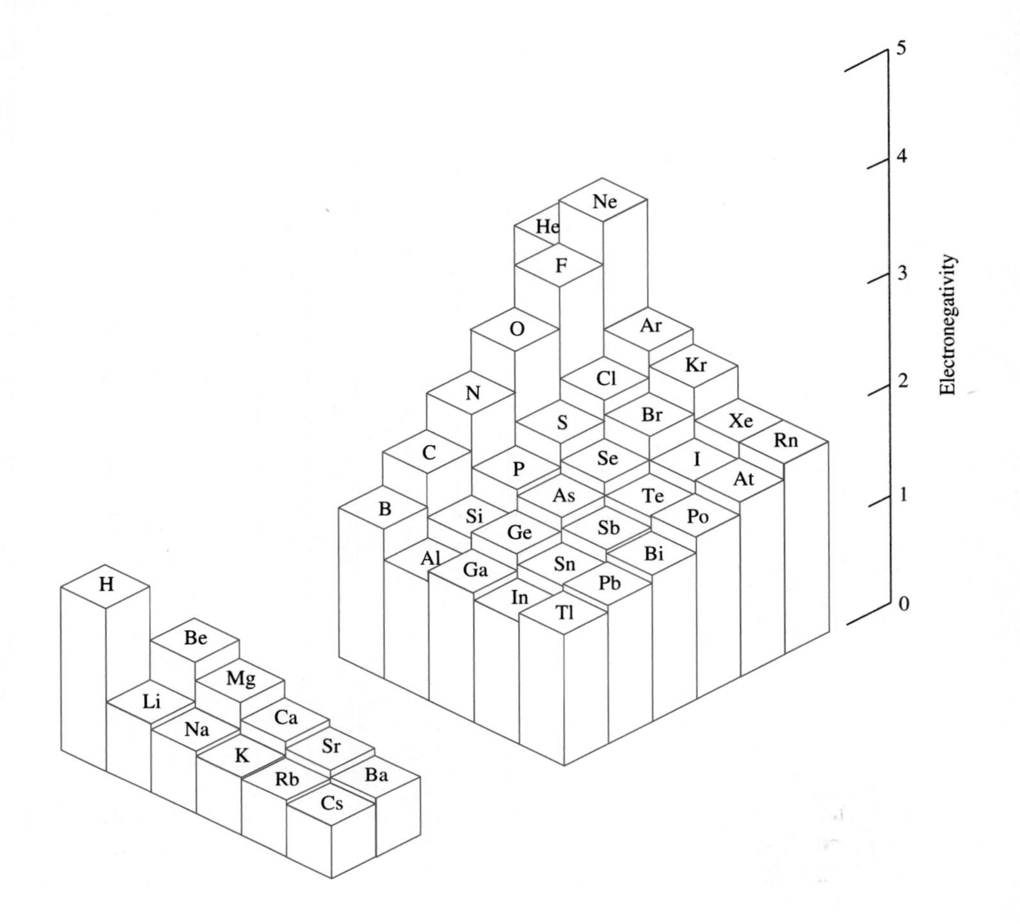

FIGURE 8.2 Electronegativities of the Main Group Elements. (Data from J. B. Mann, T. L. Meek, L. C. Allen. *J. Am. Chem. Soc., 122*, 2780, 2000.)

Although usually classified with Group 1, hydrogen is quite dissimilar from the alkali metals in its electronegativity and other properties, both chemical and physical. Hydrogen's chemistry is distinctive from all the groups.

The noble gases have higher ionization energies than the halogens, and calculations have suggested that the electronegativities of the noble gases match or even exceed those of the halogens.[2] The noble gas atoms are somewhat smaller than the neighboring halogen atoms (e.g., Ne is smaller than F) as a consequence of a greater effective nuclear charge. This charge, which attracts noble gas electrons strongly toward the nucleus, is likely to exert a strong attraction on electrons of neighboring atoms; the high electronegativities predicted for the noble gases are reasonable. Estimated values of these electronegativities are in Figure 8.2 and Appendix B.4.

8.1.3 Ionization Energy

Ionization energies of the main group elements (Section 2.3.1 and **Figure 8.3**) exhibit trends similar to those of electronegativity, with some subtle differences.

Although a general increase in ionization energy occurs toward the upper right-hand corner of the periodic table, three of the Group 13 elements have lower ionization energies than the preceding Group 2 elements, and three Group 16 elements have lower ionization energies than the preceding Group 15 elements. For example, the ionization energy of boron is lower than that of beryllium, and the ionization energy of oxygen is

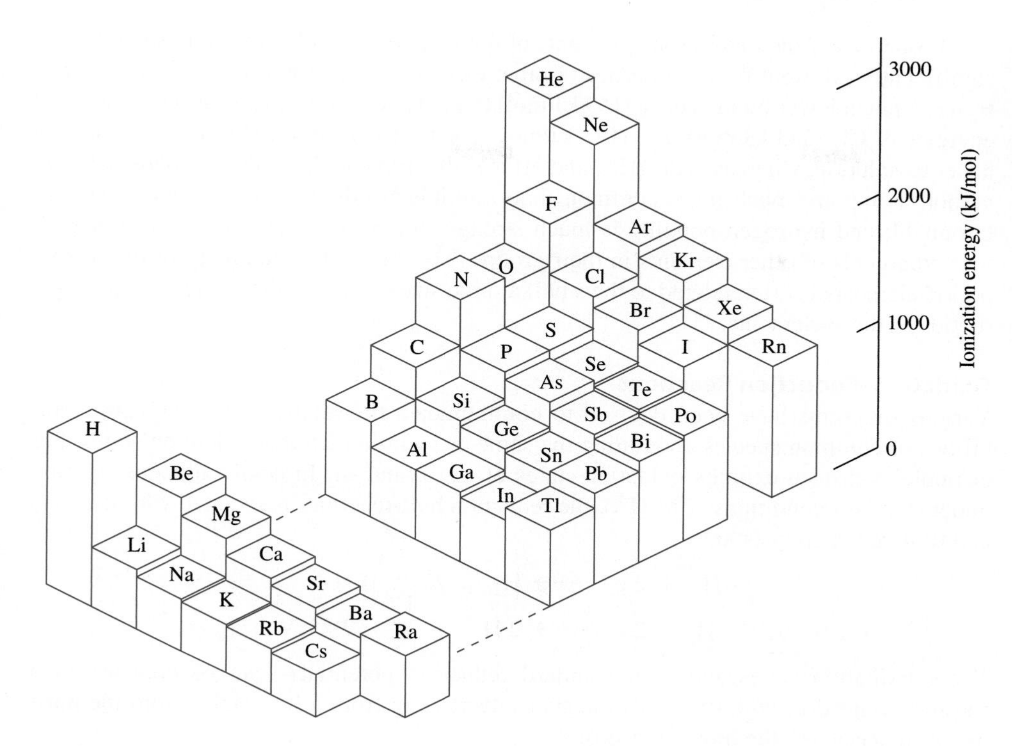

FIGURE 8.3 Ionization Energies of the Main Group Elements. (Data from C. E. Moore, *Potentials and Ionization Limits Derived from the Analyses of Optical Spectra*, National Standard Reference Data Series, U. S. National Bureau of Standards, NSRDS-NBS 34, Washington, DC, 1970.)

lower than that of nitrogen. Be and N have electron subshells that are completely filled ($2s^2$ for Be) or half-filled ($2p^3$ for N). The next atoms (B and O) have an additional electron that can be removed with comparative ease. In boron, the outermost electron, a $2p$, has significantly higher energy (higher quantum number l) than the filled $1s$ and $2s$ orbitals and is thus more easily removed than a $2s$ electron of Be. In oxygen, the fourth $2p$ electron must pair with another $2p$ electron; occupation of this orbital by two electrons is accompanied by an increase in electron–electron repulsion (the Π_c term described in Section 2.2.3) that facilitates loss of an electron. Figures 8.3 and 2.13 provide other examples of this phenomenon, and tabulated values of ionization energies are in Appendix B.2.

8.1.4 Chemical Properties

Efforts to find similarities in the chemistry of the main group elements began before the formulation of the modern periodic table. The strongest parallels are within each group. For example, the alkali metals and the halogens, respectively, exhibit many similar properties within their groups. Similarities have also been recognized between some elements along periodic table diagonals (upper left to lower right). Figure 8.2 shows that electronegativities along diagonals are similar; for example, values along the diagonal from B to Te range from 1.9 to 2.2. Other "diagonal" similarities include the unusually low solubilities of LiF and MgF_2 (a consequence of the small sizes of Li^+ and Mg^{2+}, which lead to high lattice energies in these ionic compounds), similarities in solubilities of carbonates and hydroxides of Be and Al, and the formation of complex three-dimensional structures based on SiO_4 and BO_4 tetrahedra. These parallels can often be explained on the basis of similarities in sizes and electronic structures of the compounds in question.

Properties of the main group elements of the second period (Li through Ne) are often significantly different from properties of other elements in the same group. For example, F_2 has a much lower bond energy (159 kJ/mol) than expected by extrapolation of the bond energies of Cl_2 (243 kJ/mol), Br_2 (193 kJ/mol) and I_2 (151 kJ/mol); HF is a weak acid in aqueous solution, whereas HCl, HBr, and HI are all strong acids; multiple bonds between carbon atoms are much more common than multiple bonds between other elements in Group 14; and hydrogen bonding is much stronger for compounds of F, O, and N than for compounds of other elements in their groups. The distinctive chemistry of the second period elements is often related to the small atomic sizes and the related high electronegativities of these elements.

Oxidation–Reduction Reactions

Various diagrams have been devised to communicate the relative thermodynamic stabilities of common species containing the same element in different oxidation states. For example, hydrogen exhibits oxidation states of –1, 0, and +1. In acidic aqueous solution under standard conditions (1 M H^+), the reduction half-reactions associated with the +1/0 and 0/–1 redox couples are

$$2\,H^+ + 2\,e^- \longrightarrow H_2 \qquad E^\circ = 0\text{ V}$$

$$H_2 + 2\,e^- \longrightarrow 2\,H^- \qquad E^\circ = -2.25\text{ V}$$

These oxidation states and their standard reduction potentials* can be depicted in a **Latimer diagram**,[3] with the oxidation states decreasing from left to right (from the most oxidized species to the most reduced):

$$\overset{+1}{H^+} \xrightarrow{0} \overset{0}{H_2} \xrightarrow{-2.25} \overset{-1}{H^-} \quad \longleftarrow \text{Oxidation states}$$

The half-reactions associated with these hydrogen redox couples in basic solution (1 M OH^-) are

$$H_2O + e^- \longrightarrow OH^- + \tfrac{1}{2}H_2 \qquad E^\circ = -0.828\text{ V}$$

$$H_2 + 2e^- \longrightarrow 2\,H^- \qquad E^\circ = -2.25\text{ V}$$

and the matching Latimer diagram is

$$\overset{+1}{H_2O} \xrightarrow{-0.828} \overset{0}{H_2} \xrightarrow{-2.25} \overset{-1}{H^-}$$

The Latimer diagrams for oxygen are

$$\overset{0}{O_2} \xrightarrow{0.695} \overset{-1}{H_2O_2} \xrightarrow{1.763} \overset{-2}{H_2O} \quad \text{in acid}$$

$$\overset{0}{O_2} \xrightarrow{-0.0649} \overset{-1}{HO_2^-} \xrightarrow{0.867} \overset{-2}{OH^-} \quad \text{in base}$$

Potentials for nonadjacent pairs of substances (those that share a common species in their reduction half-reactions) can be conveniently derived, as the following example describes.

---

*The half-reaction $2\,H^+ + 2\,e^- \longrightarrow H_2$ is used as the standard reference for electrode potentials in acid solutions; its voltage under standard conditions is zero.

EXAMPLE 8.1

Determine the standard reduction potential for the O_2/H_2O redox couple in acid:

$$O_2 + 4\,H^+ + 4e^- \longrightarrow 2\,H_2O \quad E_3^o = ?$$

These species share H_2O_2 in their corresponding reduction half-reactions; addition of these half-reactions provides the O_2/H_2O half-reaction:

$$\begin{array}{llll} (1) & O_2 + 2H^+ + 2e^- \longrightarrow & H_2O_2 & E_1^o = 0.695\ V \\ (2) & H_2O_2 + 2H^+ + 2e^- \longrightarrow & 2H_2O & E_2^o = 1.763\ V \\ \hline (3) & O_2 + 4H^+ + 4e^- \longrightarrow & 2H_2O & E_3^o = ? \end{array}$$

Because voltage is not a state function, we obtain E_3^o by adding the ΔG^o values, noting that $\Delta G^o = -nFE^o$:

$$\Delta G_3^o = \Delta G_1^o + \Delta G_2^o$$

$$-nFE_3^o = (-nFE_1^o) + (-nFE_2^o) \qquad (F = 96485\ C/mol)$$

$$(-4\ mol)F(E_3^o) = (-2\ mol)F(0.695\ \text{V}) + (-2\ mol)F(1.763\ \text{V})$$

$$E_3^o = 1.23\ \text{V}$$

(E_3^o is the average of E_1^o and E_2^o in this case only because the summed half-reactions each require the same number of moles of electrons.)

EXERCISE 8.1 Determine the standard reduction potential for $H_2O + 2e^- \longrightarrow OH^- + H^-$ in basic solution.

The Latimer diagram for oxygen in acidic solution shows that a lower voltage is associated with the 0/–1 couple than for the –1/–2 couple. This indicates that H_2O_2 is susceptible to **disproportionation** in acid, where H_2O_2 decomposes via $2\,H_2O_2 \longrightarrow 2\,H_2O + O_2$. The spontaneity (negative ΔG^o) associated with disproportionation can be shown by determining ΔG^o using the standard voltages (Exercise 8.2).

EXERCISE 8.2

Use the standard voltages in the Latimer diagram for oxygen to show that $\Delta G^o < 0$ for the disproportionation of H_2O_2 in acid solution.

Latimer diagrams (Appendix B.7) summarize many half-reactions. They also permit construction of **Frost diagrams** to illustrate the relative potency of species as oxidizing and reducing agents. For example, consider the Latimer diagram for nitrogen in acidic solution:

$$\overset{+5}{NO_3^-} \xrightarrow{0.803} \overset{+4}{N_2O_4} \xrightarrow{1.07} \overset{+3}{HNO_2} \xrightarrow{0.996} \overset{+2}{NO} \xrightarrow{1.59} \overset{+1}{N_2O} \xrightarrow{1.77} \overset{0}{N_2} \xrightarrow{-1.87} \overset{-1}{NH_3OH^+} \xrightarrow{1.41} \overset{-2}{N_2H_5^+} \xrightarrow{1.275} \overset{-3}{NH_4^+}$$

The **Frost diagram** in **Figure 8.4** is constructed from this Latimer diagram. A Frost diagram is a plot of nE^o versus *oxidation state*, where nE^o is proportional to ΔG^o of the half-reactions (because $\Delta G^o = -nFE^o$). When plotting points on a Frost diagram, n is equal to the *change in oxidation state* of the element in question, with nE^o expressed in volts. For creation of the Frost diagram, each reduction is formally considered as a half-reaction with *one mole of atoms* of the element undergoing the oxidation state change; the moles of electrons involved is equal in magnitude to the oxidation state change of the element being examined. This is illustrated by describing how to construct the Frost diagram for nitrogen (Figure 8.4).

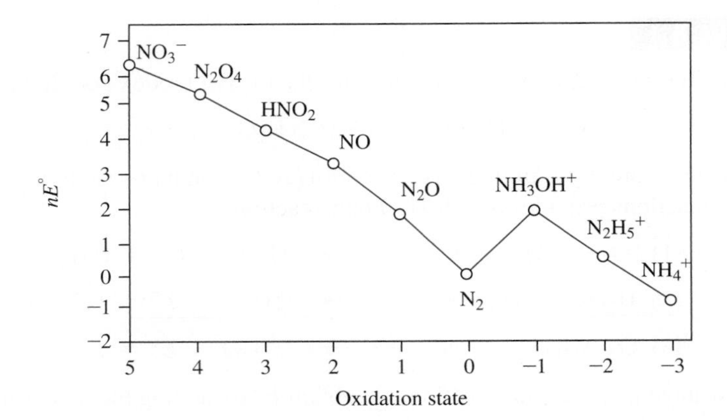

FIGURE 8.4 Frost Diagram for Nitrogen Compounds in Acid.

We begin by placing the substance with the element in the zero oxidation state (in this case, N_2) at (0,0) as a reference. The other points are plotted sequentially from this reference point using the reduction half-reactions from the Latimer diagram. For example, the Latimer diagram indicates that E° for $N_2O \longrightarrow N_2$ equals 1.77 V. The balanced half-reaction is

$$N_2O + 2H^+ + 2e^- \rightarrow N_2 + H_2O \qquad E^\circ = 1.77 \text{ V}$$

as N is reduced from an oxidation state of +1 to 0. For the Frost diagram, we need to scale this to one mole of nitrogen atoms undergoing a change in oxidation state:

$$\frac{1}{2}N_2O + 1H^+ + 1e^- \rightarrow \frac{1}{2}N_2 + \frac{1}{2}H_2O \qquad E^\circ = 1.77 \text{ V}$$

The number of moles of electrons equals the magnitude of the oxidation state change for nitrogen. E° is unchanged because voltage is a ratio (Joules/coulomb).

The N oxidation state change is –1 (decrease from 1 to 0), so $nE^\circ = (-1)(1.77 \text{ V}) = -1.77 \text{ V}$, and the point for N_2O in the Frost diagram is placed at (1,1.77), where 1 is the oxidation state of N in N_2O. The reduction $N_2O \longrightarrow N_2$ is accompanied by a *decrease* in nE° of 1.77 V; N_2O is 1.77 V higher relative to N_2 on the nE° axis.

Next, consider the location of NH_3OH^+ on the Frost diagram. Because $N_2 \longrightarrow NH_3OH^+$ features an N oxidation state change of –1 (0 to –1), and $E^\circ = -1.87$ V, $nE^\circ = (-1)(-1.87 \text{ V}) = 1.87$ V; the point for NH_3OH^+ is placed at (–1,1.87), 1.87 V higher on the nE° axis relative to N_2.

The reduction $NH_3OH^+ \longrightarrow N_2H_5^+$ features an N oxidation state change of –1 (–1 to –2), $E^\circ = 1.41$ V, and $nE^\circ = (-1)(1.41 \text{ V}) = -1.41$ V. The point for $N_2H_5^+$ is placed at (–2,0.46), where 0.46 V is 1.41 V lower than 1.87 V (the nE° value for NH_3OH^+). (Recall that –2 is the oxidation state of nitrogen in $N_2H_5^+$.)

The other points are deduced similarly by considering each successive reduction in the Latimer diagram. Connecting the points completes the Frost diagram.

EXERCISE 8.3

Deduce the positions of the points for NO_3^-, N_2O_4, HNO_2, NO, and NH_4^+ in Figure 8.4.

How is a Frost diagram useful? The slopes of the lines connecting two species are proportional to ΔG° and $-E^\circ$ for the half-reaction involving these species. One application is the determination of reduction potentials not directly provided in the Latimer diagram. For example, the reduction potential for $N_2O \longrightarrow NH_3OH^+$ can be deduced. Because the slope of the line connecting these points (not shown in Figure 8.4) is positive ($\Delta(nE^\circ) > 0$), and

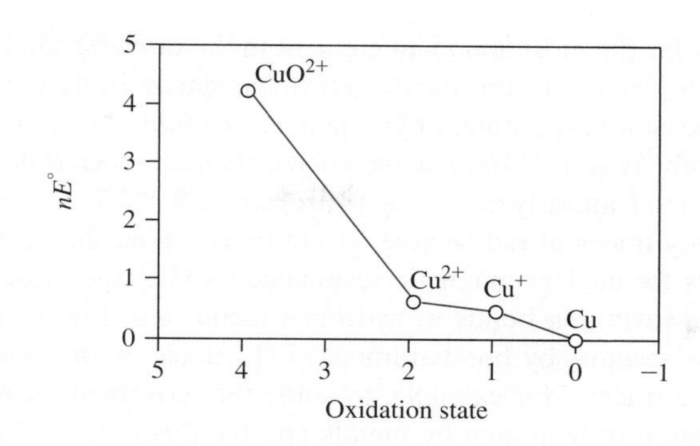

FIGURE 8.5 Frost Diagram for Copper in Acid.

the change in oxidation state is negative (+1 to –1, n = –2) we can immediately assert that the potential for $N_2O \longrightarrow NH_3OH^+$ will be negative. The coordinates for N_2O (1,1.77) and NH_3OH^+ (–1,1.87) indicate that the difference in $nE°$ between these species is 0.10. If $\Delta(nE°) = 0.10$, and $n = -2$, then $E° = \frac{0.10\text{ V}}{-2} = -0.05$ V and $N_2O \xrightarrow{-0.05} NH_3OH^+$.

Frost diagrams are more commonly used to make descriptive statements about reactions. For example, consider adding Cu to 1 M HNO_3. The Frost diagram for Cu (**Figure 8.5**) indicates that the slope associated with Cu^+/Cu is less steep than that of NO_3^-/N_2O_4 in Figure 8.4; this means that NO_3^- will spontaneously oxidize Cu to Cu^+ in acidic solution. The subsequent Cu^{2+}/Cu^+ slope is even lower in magnitude, so immediate oxidation to Cu^{2+} occurs; a significant Cu^+ concentration never accumulates.* The CuO^{2+}/Cu^{2+} slope is steeper than that connecting any potential oxidant in Figure 8.4, so Cu^{2+} cannot be oxidized further, and +2 is the most stable copper oxidation state in this reaction. If Cu is present in excess, Figure 8.4 indicates that N_2 is the expected thermodynamic nitrogen-containing product because all the redox couples from NO_3^-/N_2O_4 to N_2O/N_2 possess steeper slopes than does the Cu^{2+}/Cu couple. However, NO evolution and the slow rate of N_2 formation kinetically prevent N_2 from being a product when excess Cu and 1 M HNO_3 are mixed. The rich chemistry of nitrogen is discussed in Section 8.7.

EXERCISE 8.4

Use the standard voltages in the Latimer diagrams for hydrogen in acidic and basic solutions to construct Frost diagrams.

8.2 Hydrogen

The position of hydrogen in the periodic table has been disputed. Its electron configuration, $1s^1$, is similar to the valence electron configurations of the alkali metals (ns^1); hence, hydrogen is most commonly listed in the periodic table at the top of Group 1. However, hydrogen has little chemical similarity to the alkali metals. Hydrogen is also one electron short of a noble gas configuration and could be classified with the halogens. However, hydrogen has only limited similarities with the halogens—for example, in forming a diatomic molecule and an ion of 1– charge. A third possibility is to place hydrogen in Group 14 above carbon: both elements have half-filled valence electron shells, are of similar electronegativity, and usually form covalent rather than ionic bonds. We prefer not to fit hydrogen into any particular group; hydrogen is sufficiently unique to deserve separate consideration.

---

*In this case, the slope associated with the Cu^{2+}/Cu couple is lower in magnitude than that of the NO_3^-/N_2O_4 couple, so essentially immediate oxidation of Cu to Cu^{2+} is expected.

$$^3_1H \longrightarrow ^3_2He + ^{0}_{-1}e$$

$$^6_3Li + ^1_0n \longrightarrow ^4_2He + ^3_1H$$

Hydrogen is by far the most abundant element in the universe (and the sun) and is the third most abundant element in the Earth's crust (primarily in its compounds). The element occurs as three isotopes: ordinary hydrogen, or *protium*, ^1H; *deuterium*, ^2H or D; and *tritium*, ^3H or T. Both ^1H and ^2H have stable nuclei; ^3H undergoes β decay (left) and has a half-life of 12.35 years. Naturally occurring hydrogen is 99.985% ^1H and essentially all the remainder is ^2H; only traces of radioactive ^3H are found on earth. Deuterium compounds are used as solvents for nuclear magnetic resonance (NMR) spectroscopy and in kinetic studies on reactions involving bonds to hydrogen (deuterium isotope effects). Tritium is produced in nuclear reactors by bombardment of ^6Li nuclei with neutrons (left). Tritium has applications as a tracer—for example, to study the movement of ground water—and to study the *ab*sorption of hydrogen by metals and the *ad*sorption of hydrogen on metal surfaces. Tritium also has been used as an energy source for "tritium watches" and other timing devices. Some of the important physical properties of the isotopes of hydrogen are listed in Table 8.2.

8.2.1 Chemical Properties

Hydrogen can gain an electron to achieve a noble gas configuration in forming hydride, H^-. The alkali and alkaline earth metals form hydrides that are essentially ionic and contain discrete H^- ions. Hydride ion is a powerful reducing agent; it reacts with water and other protic solvents to generate H_2:

$$H^- + H_2O \longrightarrow H_2 + OH^-$$

Hydride ions also act as ligands in bonding to metals, with up to nine hydrogens on a single metal, as in ReH_9^{2-}. Main group hydrides, such as BH_4^- and AlH_4^-, are important reducing agents for chemical synthesis. Although such complexes are named "hydrides," their bonding is essentially covalent, without discrete H^- ions.

Reference to the "hydrogen ion," H^+, is common. However, in solution, the extremely small size of the proton (radius approximately 1.5×10^{-3} pm) leads to an extremely high charge density and requires that H^+ be associated with solvent or solute molecules via hydrogen bonds. In aqueous solution, a more correct description is $H_3O^+(aq)$, although larger species such as $H_9O_4^+$ have been identified.

The combustibility of hydrogen, together with the lack of polluting by-products, renders hydrogen attractive as a fuel. As a fuel for automobiles and buses, H_2 provides more energy per unit mass than gasoline without producing environmentally damaging by-products such as carbon monoxide, sulfur dioxide, and unburned hydrocarbons.

TABLE 8.2 Properties of Hydrogen, Deuterium, and Tritium

			Properties of Molecules, X_2			
Isotope	Abundance (%)	Atomic Mass	Melting Point (K)	Boiling Point (K)	Critical Temperature (K)[a]	Enthalpy of Dissociation (kJ mol^{-1} at 25 °C)
Protium (^1H), H	99.985	1.007825	13.957	20.30	33.19	435.88
Deuterium (^2H), D	0.015	2.014102	18.73	23.67	38.35	443.35
Tritium (^3H), T	$\sim 10^{-16}$	3.016049	20.62	25.04	40.6 (calc)	446.9

Sources: Data from Abundance and atomic mass data from I. Mills, T. Cuitoš, K. Homann, N. Kallay, and K. Kuchitsu, eds., *Quantities, Units, and Symbols in Physical Chemistry*, International Union of Pure and Applied Chemistry, Blackwell Scientific Publications, Oxford UK, 1988. Other data are from N. N. Greenwood and A. Earnshaw *Chemistry of the Elements*, Pergamon Press, Elmsford, NY, 1984.

[a]The highest temperature at which a gas can be condensed to a liquid

A challenge is to develop practical thermal or photochemical processes for generating hydrogen from its most abundant source—water.[4]

H_2 is prepared industrially by "cracking" petroleum hydrocarbons with solid catalysts, also forming alkenes

$$C_2H_6 \longrightarrow C_2H_4 + H_2$$

or by steam re-forming of natural gas, typically using a nickel catalyst

$$CH_4 + H_2O \longrightarrow CO + 3\,H_2$$

Molecular hydrogen is an important reagent, especially in the industrial hydrogenation of unsaturated organic molecules. Transition–metal catalyzed hydrogenations are discussed in Chapter 14.

8.3 Group 1: The Alkali Metals

Alkali metal salts, in particular sodium chloride, have been used since antiquity. In early times, salt was used in the preservation and flavoring of food. However, because of the difficulty of reducing alkali metal ions, the elements were not isolated until comparatively recently. Two of the alkali metals, sodium and potassium, are essential for human life; their careful regulation is important in treating a variety of medical conditions.

8.3.1 The Elements

Potassium and sodium were first isolated in 1807 by Davy as products of the electrolysis of molten KOH and NaOH. In 1817, Arfvedson recognized similarities between the solubilities of compounds of lithium and those of sodium and potassium. The following year, Davy also isolated lithium by electrolysis of molten Li_2O. Cesium and rubidium were discovered with the help of the spectroscope in 1860 and 1861, respectively; they were named after the colors of the most prominent emission lines (Latin, *caesius*, "sky blue," *rubidus*, "deep red"). Francium was not identified until 1939 as a short-lived radioactive isotope from the nuclear decay of actinium.

The alkali metals are silvery—except for cesium, which has a golden appearance—highly reactive solids having low melting points. They are ordinarily stored under nonreactive oil to prevent oxidation and are soft enough to be cut with a knife or spatula. Their melting points decrease with increasing atomic number because metallic bonding between the atoms becomes weaker with increasing atomic size. Physical properties of the alkali metals are summarized in **Table 8.3**.

8.3.2 Chemical Properties

The alkali metals are very similar in their chemical properties, which are governed by the ease with which they can lose one electron (the alkali metals have the lowest ionization energies of all the elements) and achieve a noble gas configuration. All are excellent reducing agents. The metals react vigorously with water to form hydrogen; for example,

$$2\,Na + 2\,H_2O \longrightarrow 2\,NaOH + H_2$$

This reaction is highly exothermic, and the hydrogen formed may ignite in air, sometimes explosively. The vigor of this reaction increases going down the alkali metal group. Special precautions must be taken to prevent these metals from coming into contact with water.

TABLE 8.3 Properties of the Group 1 Elements: The Alkali Metals

Element	Ionization Energy (kJ mol^{-1})	Electron Affinity (kJ mol^{-1})	Melting Point (°C)	Boiling Point (°C)	Electro-negativity	$E°$ ($M^+ \longrightarrow M$) (V)[a]
Li	520	60	180.5	1347	0.912	−3.04
Na	496	53	97.8	881	0.869	−2.71
K	419	48	63.2	766	0.734	−2.92
Rb	403	47	39.0	688	0.706	−2.92
Cs	376	46	28.5	705	0.659	−2.92
Fr	400[b, c]	60[b, d]	27		0.7[b]	−2.9[d]

Sources: Data from Ionization energies cited in this chapter are from C. E. Moore, *Ionization Potentials and Ionization Limits Derived from the Analyses of Optical Spectra*, National Standard Reference Data Series, U.S. National Bureau of Standards, NSRDS-NBS 34, Washington, DC, 1970, unless noted otherwise. Electron affinity values listed in this chapter are from H. Hotop and W. C. Lineberger, *J. Phys. Chem. Ref. Data,* **1985,** *14,* 731. Standard electrode potentials listed in this chapter are from A. J. Bard, R. Parsons, and J. Jordan, eds., *Standard Potentials in Aqueous Solutions,* Marcel Dekker (for IUPAC), New York, 1985. Electronegativities cited in this chapter are from J. B. Mann, T. L. Meek, and L. C. Allen, *J. Am. Chem. Soc.*, **2000,** *122,* 2780, Table 2. Other data are from N. N. Greenwood and A. Earnshaw, *Chemistry of the Elements,* Pergamon Press, Elmsford, NY, 1984, except where noted. J. Emsley, The Elements, Oxford University Press, New York, 1989. S. G. Bratsch, *J. Chem. Educ.*, **1988,** *65,* 34.

[a]Aqueous solution, 25 °C; [b]Approximate value; [c]J. Emsley, *The Elements*, Oxford University Press, New York, 1989; [d]S. G. Bratsch, *J. Chem. Educ.*, **1988**, *65*, 34.

Alkali metals react with oxygen to form oxides, peroxides, and superoxides, depending on the metal. Combustion in air yields the following products:*

Alkali Metal	Principal Combustion Product (Minor Product) Oxide	Peroxide	Superoxide
Li	Li_2O	(Li_2O_2)	
Na	(Na_2O)	Na_2O_2	
K			KO_2
Rb			RbO_2
Cs			CsO_2

Alkali metals dissolve in liquid ammonia and other donor solvents, such as aliphatic amines (NR_3, in which R = alkyl) and $OP(NMe_2)_3$, hexamethylphosphoramide, to give blue solutions believed to contain solvated electrons:

$$Na + x\,NH_3 \longrightarrow Na^+ + e(NH_3)_x^-$$

Because of these solvated electrons, dilute solutions of alkali metals in ammonia conduct electricity better than dissociated ionic compounds in aqueous solutions. As the concentration of the alkali metals is increased in liquid ammonia, the conductivity first declines, then increases. At sufficiently high concentration, the solution acquires a bronze metallic luster and a conductivity comparable to the molten metal. Dilute solutions are paramagnetic, with approximately one unpaired electron per metal atom, corresponding to one solvated electron per metal atom; this paramagnetism decreases at higher concentrations. These solutions are less dense than liquid ammonia itself. The solvated electrons may be viewed as creating cavities for themselves (estimated radius of approximately 300 pm) in the solvent, thus increasing the volume significantly. The blue color, corresponding to a broad absorption band near 1500 nm that extends into the visible range, is attributed to the solvated electron (alkali metal ions are colorless). At higher concentrations the bronze-colored solutions contain alkali metal anions, M^-.

*Additional information on peroxide, superoxide, and other oxygen-containing ions is provided in Table 8.12.

Solutions of alkali metals in liquid ammonia are excellent reducing agents. Examples of reductions that can be effected by these solutions include:

$$RC\equiv CH + e^- \longrightarrow RC\equiv C^- + \tfrac{1}{2}H_2$$

$$NH_4^+ + e^- \longrightarrow NH_3 + \tfrac{1}{2}H_2$$

$$S_8 + 2\,e^- \longrightarrow S_8^{2-}$$

$$Fe(CO)_5 + 2\,e^- \longrightarrow [Fe(CO)_4]^{2-} + CO$$

The solutions of alkali metals are unstable and undergo slow decomposition to form amides:

$$M + NH_3 \longrightarrow MNH_2 + \tfrac{1}{2}H_2$$

The alkaline earths Ca, Sr, and Ba and the lanthanides Eu and Yb (both of which form 2+ ions) also dissolve in liquid ammonia to give solvated electrons. The alkali metals undergo this reaction more efficiently and are used more extensively for synthetic purposes.

Alkali metal cations can form complexes with a variety of Lewis bases. Of particular interest are cyclic Lewis bases that have several donor atoms that can complex, or trap, cations. Examples of such molecules are shown in **Figure 8.6**. The first of these is one of a large group of cyclic ethers, known as "crown" ethers, which donate electron density to metals through their oxygen atoms. The second, one of a family of cryptands (or cryptates), can be even more effective as a cage with eight donor atoms surrounding a central metal. Metallacrowns, which incorporate metals into the crown structure, have also been developed.[5] An example of the framework structure of an iron-containing metallacrown is in Figure 8.6. The importance of these structures was recognized when Cram, Pedersen, and Lehn won the Nobel Prize in Chemistry in 1987.*

The ability of a cryptand to trap an alkali metal cation depends on the sizes of both the cage and the metal ion: the better the match between these sizes, the more effectively the ion can be trapped. This effect is shown for the alkali metal ions in **Figure 8.7**.

The largest of the alkali metal cations, Cs^+, is trapped most effectively by the largest cryptand ([3.2.2]), and the smallest, Li^+, by the smallest cryptand ([2.1.1]).** Related correlations are in Figure 8.7. Cryptands play a vital role in the study of alkali metal anions (alkalides).

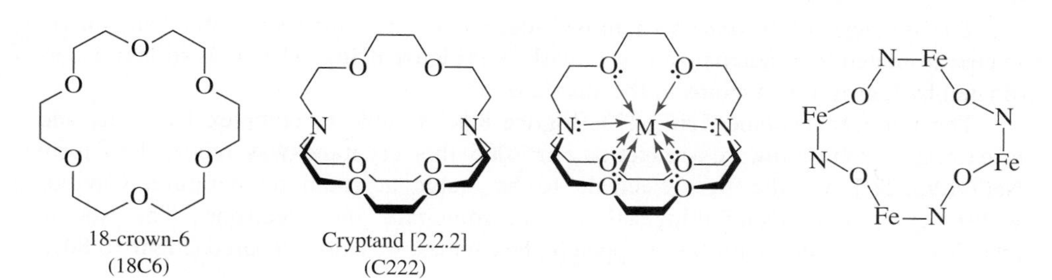

FIGURE 8.6 A Crown Ether, a Cryptand, a Metal Ion Encased in a Cryptand, and a Metallacrown.

*Their Nobel Prize lectures: D. J. Cram, *Angew. Chem.*, **1988**, *100*, 1041; C. J. Pedersen, *Angew. Chem.*, **1988**, *100*, 1053; J.-M. Lehn, *Angew. Chem.*, **1988**, *100*, 91.

**The numbers indicate the number of oxygen atoms in each bridge between the nitrogens. Thus, cryptand [3.2.2] has one bridge with three oxygens and two bridges with two oxygens, as shown in Figure 8.7.

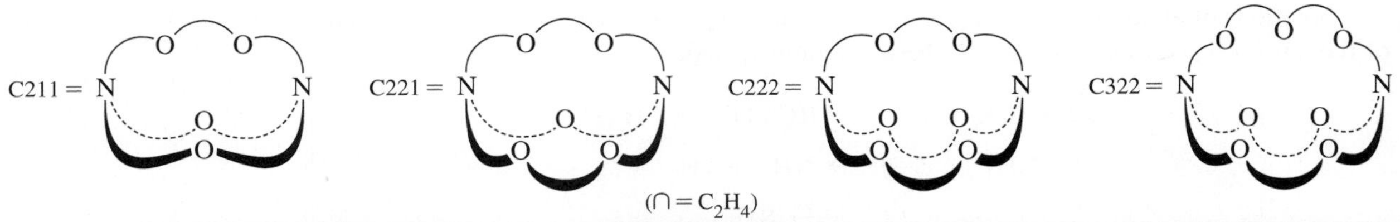

C211 = C221 = C222 = C322 =

(∩ = C_2H_4)

	Alkali Metal Ion (Radius, pm)				
Cryptand (Cavity Radius)	**Li^+ (79)**	**Na^+ (107)**	**K^+ (138)**	**Rb^+ (164)**	**Cs^+ (173)**
C211 (80)	**7.58**	6.08	2.26	<2.0	<2.0
C221 (110)	4.18	**8.84**	7.45	5.80	3.90[a]
C222 (140)	1.8	7.21	**9.75**	8.40	3.54
C322 (180)	<2.0	4.57	7.0	**7.30**	7.0

NOTE: These formation constants were obtained in 95:5 methanol:water

[a]2:1 cryptand:cation complexes may also be present in this mixture.

FIGURE 8.7 Alkali Metal Cryptands and their Formation Constants (reported as log K_c for the complexation reactions). The highest formation constant for each cryptand is indicated in boldface. Data from J. L. Dye, *Progr. Inorg. Chem.*, **1984,** *32,* 337.

Numerous alkalides have been reported. The first of these was the sodide ion, Na^-, formed from the reaction of sodium with the cryptand $N\{(C_2H_4O)_2C_2H_4\}_3N$ in the presence of ethylamine:

$$2\,Na + \underset{\text{cryptand[2.2.2]}}{N\{(C_2H_4O)_2C_2H_4\}_3N} \longrightarrow \underset{[\text{Na(cryptand[2.2.2])}]^+}{[Na\,N\{(C_2H_4O)_2C_2H_4\}_3N]^+} + Na^-$$

In this complex, Na^- occupies a site sufficiently remote from the coordinating N and O atoms of the cryptand that it can be viewed as a separate entity; it is formed as the result of disproportionation of Na into Na^+ (surrounded by the cryptand) plus Na^-. Alkalide ions are known for the other members of Group 1 and for other metals, especially those for which a 1– charge gives rise to an s^2d^{10} electron configuration. Alkalide ions are powerful reducing agents. The complexing agent must be highly resistant to reduction to avoid being reduced by the alkalide. Even with robust complexing agents, most alkalides are unstable and subject to decomposition.

The crystal structure of the crown ether sandwich electride $Cs^+(15C5)_2e^-$ in **Figure 8.8a** shows both the coordination of two 15C5 rings to each Cs^+ ion and the cavity occupied by the electron.[6]

The synthesis of "inverse sodium hydride," which contains sodide, Na^-, and an H^+ ion encapsulated in 3^6 adamanzane is remarkable (**Figure 8.8b**).[7] The H^+ is strongly coordinated by four nitrogen atoms in the adamanzane.

The per-aza cryptand TriPip222 (**Figure 8.8c**) is able to complex Na^+ ions and is extremely robust towards reduction. In 2005 this cryptand was reported to form $Na(TriPip222)]^+e^-$, the first electride to be stable at room temperature. Cavities within crystalline $[Na(TriPip222)]^+e^-$ accommodate the electrons. The sodide $[Na(TriPip222)]^+Na^-$, with Na^- trapped in the cavities, was also prepared (**Figure 8.8d**).[8]

8.4 Group 2: The Alkaline Earths

8.4.1 The Elements

Magnesium and calcium compounds have been used since antiquity. The ancient Romans used mortars containing lime (CaO) mixed with sand, and ancient Egyptians used gypsum

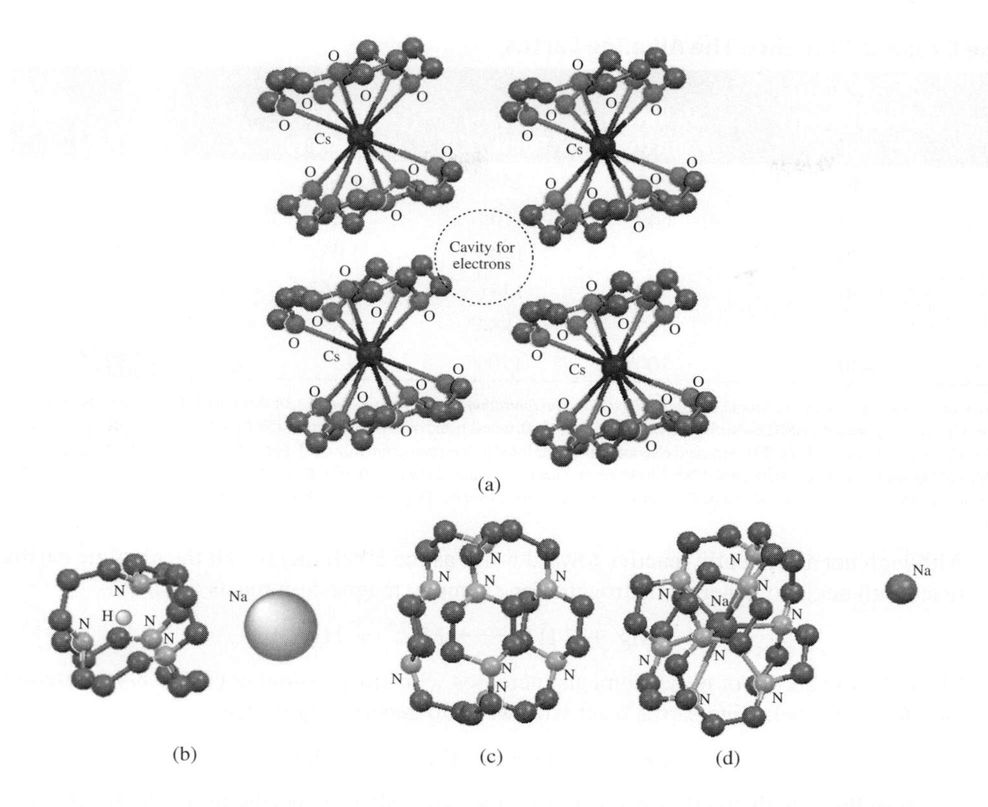

FIGURE 8.8 (a) $Cs^+(15C5)_2e^-$, a crown ether electride. The cesium cation is bound to the oxygen atoms of the 15C5 molecules. (b) Na^-H^+ 3^6 adamanzane complex. (c) TriPip222 cryptand. (d) $[Na(TriPip222)]^+Na^-$. (Molecular structure drawings created with CIF data. The structure of (b) is an approximation because only CIF data for protonated 3^6 adamanzane as a glycolate ($[HOCH_2CO_2]^-$) salt have been reported. Hydrogen atoms are omitted for clarity.)

($CaSO_4 \cdot 2\,H_2O$) in plasters to decorate their tombs. These two alkaline earths are among the most abundant elements in the Earth's crust (calcium is fifth and magnesium sixth, by mass), and they occur in a wide variety of minerals. Strontium and barium are less abundant; but like magnesium and calcium, they commonly occur as sulfates and carbonates in their mineral deposits. Beryllium is fifth in abundance of the alkaline earths and is obtained primarily from the mineral beryl, $Be_3Al_2(SiO_3)_6$. All radium isotopes are radioactive (the longest lived isotope is ^{226}Ra, with a half-life of 1600 years). Pierre and Marie Curie first isolated radium from the uranium ore pitchblende in 1898. Physical properties of the alkaline earths are given in **Table 8.4**.

Atoms of the Group 2 elements are smaller than the neighboring Group 1 elements as a consequence of the greater nuclear charge of the Group 2 elements. As a result, the Group 2 elements are more dense and have higher ionization energies than the Group 1 elements. They also have higher melting and boiling points and higher enthalpies of fusion and vaporization (Tables 8.3 and 8.4). Beryllium is used in alloys with copper, nickel, and other metals. When added in small amounts to copper, beryllium increases the strength of the metal dramatically, improves the corrosion resistance, and preserves high conductivity. Emeralds and aquamarine are obtained from two types of beryl; the green and blue colors of these stones are the result of small amounts of chromium and other impurities. Magnesium-containing alloys are used as strong, but relatively light, construction materials. Radium was historically used in the treatment of cancerous tumors.

8.4.2 Chemical Properties

The Group 2 elements, with the exception of beryllium, have very similar chemical properties; much of their chemistry is governed by their tendency to lose two electrons to achieve a noble gas electron configuration. The Group 2 elements are good reducing agents.

TABLE 8.4 Properties of the Group 2 Elements: The Alkaline Earths

Element	Ionization Energy (kJ mol^{-1})	Electron Affinity (kJ mol^{-1})^b	Melting Point (°C)	Boiling Point (°C)	Electro-negativity	$E°$ ($M^{2+} + 2e^- \longrightarrow M$) (V)^a
Be	899	−50	1287	2500^b	1.576	−1.97
Mg	738	−40	649	1105	1.293	−2.36
Ca	590	−30	839	1494	1.034	−2.84
Sr	549	−30	768	1381	0.963	−2.89
Ba	503	−30	727	1850^b	0.881	−2.92
Ra	509	−30	700^b	1700^b	0.9^b	−2.92

Source: Data from: Ionization energies cited in this chapter are from C. E. Moore, *Ionization Potentials and Ionization Limits Derived from the Analyses of Optical Spectra,* National Standard Reference Data Series, U.S. National Bureau of Standards, NSRDS-NBS 34, Washington, DC, 1970, unless noted otherwise. Electron affinity values listed in this chapter are from H. Hotop and W. C. Lineberger, *J. Phys. Chem. Ref. Data*, **1985**, *14*, 731. Standard electrode potentials listed in this chapter are from A. J. Bard, R. Parsons, and J. Jordan, eds., *Standard Potentials in Aqueous Solutions,* Marcel Dekker (for IUPAC), New York, 1985. Electronegativities cited in this chapter are from J. B. Mann, T. L. Meek, and L. C. Allen, *J. Am. Chem. Soc.*, **2000**, *122*, 2780, Table 2. Other data are from N. N. Greenwood and A. Earnshaw, *Chemistry of the Elements*, Pergamon Press, Elmsford, NY, 1984, except where noted.
^{a}Aqueous solution, 25 °C; ^{b}Approximate values

Although not as violently reactive toward water as the alkali metals, all the alkaline earths react with acids to generate hydrogen. For example, magnesium reacts as:

$$Mg + 2\,H^+ \longrightarrow Mg^{2+} + H_2$$

The reducing ability of these elements increases with atomic number (Table 8.4). Calcium and the heavier alkaline earths react with water to generate hydrogen:

$$Ca + 2\,H_2O \longrightarrow Ca(OH)_2 + H_2$$

Beryllium is distinctly different from the other alkaline earths in its chemical properties. The smallest of the alkaline earths, it participates primarily in covalent bonding. Although $[Be(H_2O)_4]^{2+}$ is known, Be^{2+} ions are rarely encountered. Beryllium and its compounds are extremely toxic. Although beryllium halides (BeX_2) may be monomeric and linear in the gas phase (Section 3.1.4), in the solid phase the molecules polymerize to form halogen-bridged chains with tetrahedral coordination around beryllium (**Figure 8.9**). Beryllium hydride (BeH_2) is also polymeric in the solid with bridging hydrogens. The three-center bonding involved in bridging by halogens, hydrogen, and other atoms and groups is also commonly encountered in the chemistry of the Group 13 elements (Section 8.5).

Among the most chemically useful magnesium compounds are the Grignard reagents (RMgX (R = alkyl or aryl, X = halide)). These reagents are complex in their structure and function, consisting of a variety of species in solution related by equilibria (**Figure 8.10**). The relative positions of these equilibria, and the concentrations of the various species, are affected by the nature of the R group and the halogen, solvent, and temperature. Grignard reagents can be used to synthesize a vast range of organic compounds, including alcohols, aldehydes, ketones, carboxylic acids, esters, thiols, and amines.*

Chlorophylls contain magnesium and are essential in photosynthesis. The abundance, worldwide accessibility, low cost, and environmental compatibility of calcium motivate its application in catalysts.[9]

FIGURE 8.9 Structure of $BeCl_2$.

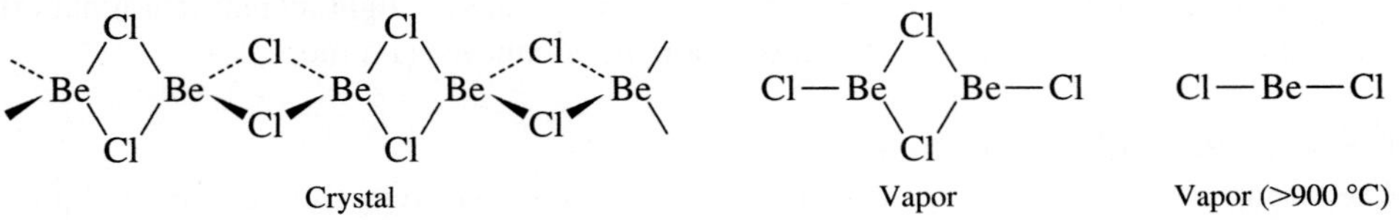

Crystal

Cl—Be(μ-Cl)₂Be—Cl

Vapor

Cl—Be—Cl

Vapor (>900 °C)

*The development of these reagents since their original discovery by Victor Grignard in 1900 has been reviewed. See *Bull. Soc. Chim. France*, **1972**, 2127–2186. See also D. Seyferth, *Organometallics*, **2009**, *28*, 1598 for a discussion of highlights of the history of Grignard reagents.

$$2\,RMg^+ + 2\,X^-$$

$$\rightleftharpoons$$

$$R{-}Mg(\mu{-}X)_2Mg{-}R \rightleftharpoons 2\,RMgX \rightleftharpoons MgR_2 + MgX_2$$

$$\rightleftharpoons \qquad\qquad\qquad \rightleftharpoons$$

$$RMg^+ + RMgX_2^- \rightleftharpoons Mg(\mu{-}X)_2MgR_2$$

FIGURE 8.10 Grignard Reagent Equilibria.

Portland cement—a mixture of calcium silicates, aluminates, and ferrates—is one of the world's most important construction materials, with annual worldwide production in excess of 10^{12} kg. When mixed with water and sand, it changes by slow hydration to concrete. Water and hydroxide link the other components into very strong crystals.

8.5 Group 13

8.5.1 The Elements

Elements in this group include one nonmetal, boron, and four elements that are primarily metallic. Physical properties of these elements are shown in **Table 8.5**.

Boron

Boron's chemistry is so different from that of the other elements in this group that it deserves separate discussion. Chemically, boron is a nonmetal; in its tendency to form covalent bonds, it shares more similarities with carbon and silicon than with aluminum and the other Group 13 elements. Like carbon, boron forms many hydrides; like silicon, it forms oxygen-containing minerals with complex structures (borates). Compounds of boron have been used since ancient times in the preparation of glazes and borosilicate glasses, but the element itself has proven extremely difficult to purify. The pure element has a wide diversity of allotropes, many of which are based on the icosahedral B_{12} unit.

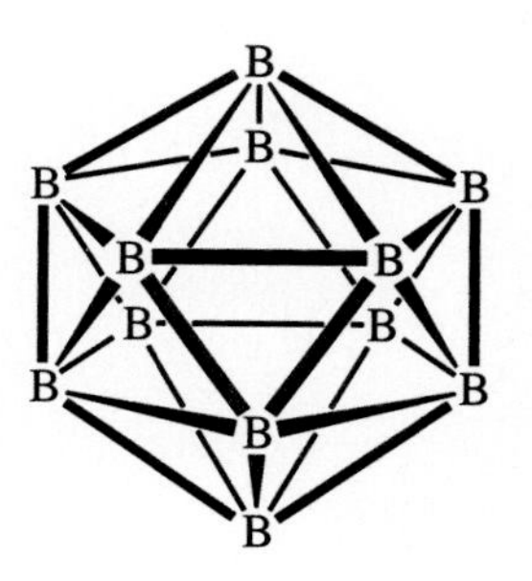

In boron hydrides, called *boranes*, hydrogen often serves as a bridge between boron atoms, a function rarely performed by hydrogen in carbon chemistry. How is it possible for

TABLE 8.5 Properties of the Group 13 Elements

Element	Ionization Energy (kJ mol^{-1})	Electron Affinity (kJ mol^{-1})	Melting Point (°C)	Boiling Point (°C)	Electro-negativity
B	801	27	2180	3650[a]	2.051
Al	578	43	660	2467	1.613
Ga	579	30[a]	29.8	2403	1.756
In	558	30[a]	157	2080	1.656
Tl	589	20[a]	304	1457	1.789

Source: Data from: Ionization energies cited in this chapter are from C. E. Moore, *Ionization Potentials and Ionization Limits Derived from the Analyses of Optical Spectra,* National Standard Reference Data Series, U.S. National Bureau of Standards, NSRDS-NBS 34, Washington, DC, 1970, unless noted otherwise. Electron affinity values listed in this chapter are from H. Hotop and W. C. Lineberger, *J. Phys. Chem. Ref. Data,* **1985,** *14,* 731. Standard electrode potentials listed in this chapter are from A. J. Bard, R. Parsons, and J. Jordan, eds., *Standard Potentials in Aqueous Solutions,* Marcel Dekker (for IUPAC), New York, 1985. Electronegativities cited in this chapter are from J. B. Mann, T. L. Meek, and L. C. Allen, *J. Am. Chem. Soc.,* **2000,** *122,* 2780, Table 2. Other data are from N. N. Greenwood and A. Earnshaw, *Chemistry of the Elements,* Pergamon Press, Elmsford, NY, 1984, except where noted.
[a]Approximate values

Diborane

hydrogen to serve as a bridge? One way to address this question is to consider the bonding in diborane (B_2H_6, structure in margin). Diborane has 12 valence electrons. By the Lewis approach, eight of these electrons are involved in bonding to the terminal hydrogens. Thus, four electrons remain to account for bonding in the bridges. This type of bonding, involving three atoms and two bonding electrons per bridge, is described as *three-center, two-electron bonding.*[10] To understand this type of bonding, we need to consider the orbital interactions in diborane.

We will use the group orbital approach described in Section 5.4, focusing on the boron atoms and the bridging hydrogens. Group orbitals for these atoms and their matching irreducible representations in diborane's D_{2h} symmetry are shown in **Figure 8.11**. The possible interactions between the boron group orbitals and the group orbitals of the bridging hydrogens can be determined by matching the labels of the irreducible representations. For example, one group orbital in each set has B_{3u} symmetry. This involves the hydrogen group orbital with lobes of opposite sign and one of the boron group orbitals derived from p_x atomic orbitals. The results (**Figure 8.12**) are two molecular orbitals of b_{3u} symmetry, one bonding and one antibonding. The bonding orbital, with lobes spanning both B—H—B bridges, is chiefly responsible for the stability of the bridges.

The other hydrogen group orbital has A_g symmetry. Two boron group orbitals have A_g symmetry: one is derived from p_z orbitals, and one is derived from s orbitals. These three group orbitals have similar energy. The result of the A_g interactions is the formation of three

FIGURE 8.11 Group Orbitals of Diborane.

Reducible representation for p orbitals involved in bonding with bridging hydrogens:

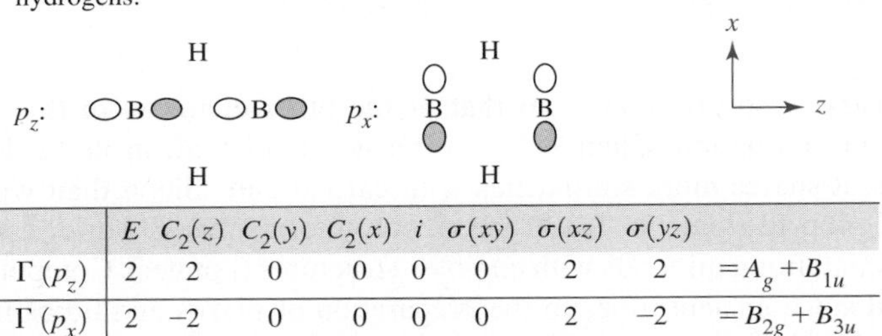

	E	$C_2(z)$	$C_2(y)$	$C_2(x)$	i	$\sigma(xy)$	$\sigma(xz)$	$\sigma(yz)$	
$\Gamma(p_z)$	2	2	0	0	0	0	2	2	$= A_g + B_{1u}$
$\Gamma(p_x)$	2	−2	0	0	0	0	2	−2	$= B_{2g} + B_{3u}$

The irreducible representations have the following symmetries:

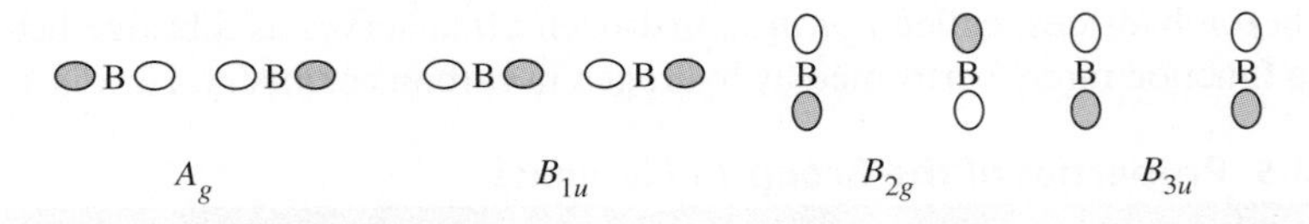

Reducible representation for 1s orbitals of bridging hydrogens:

	E	$C_2(z)$	$C_2(y)$	$C_2(x)$	i	$\sigma(xy)$	$\sigma(xz)$	$\sigma(yz)$
$\Gamma(1s)$	2	0	0	2	0	2	2	0

This reduces to $A_g + B_{3u}$, which have the following symmetries:

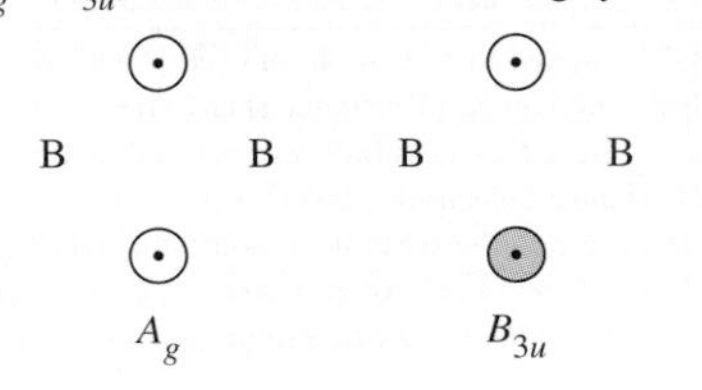

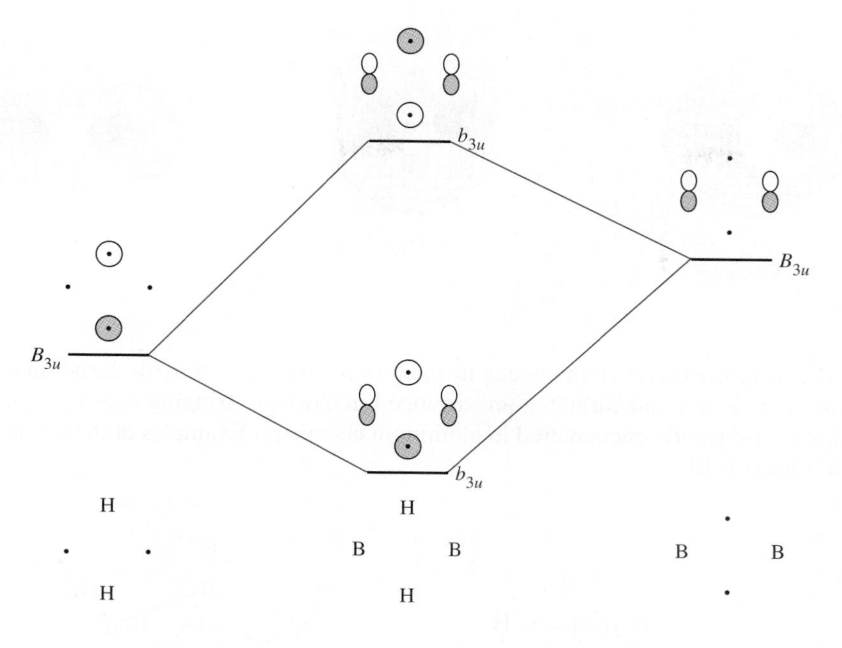

FIGURE 8.12 B_{3u} Orbital Interactions in Diborane.

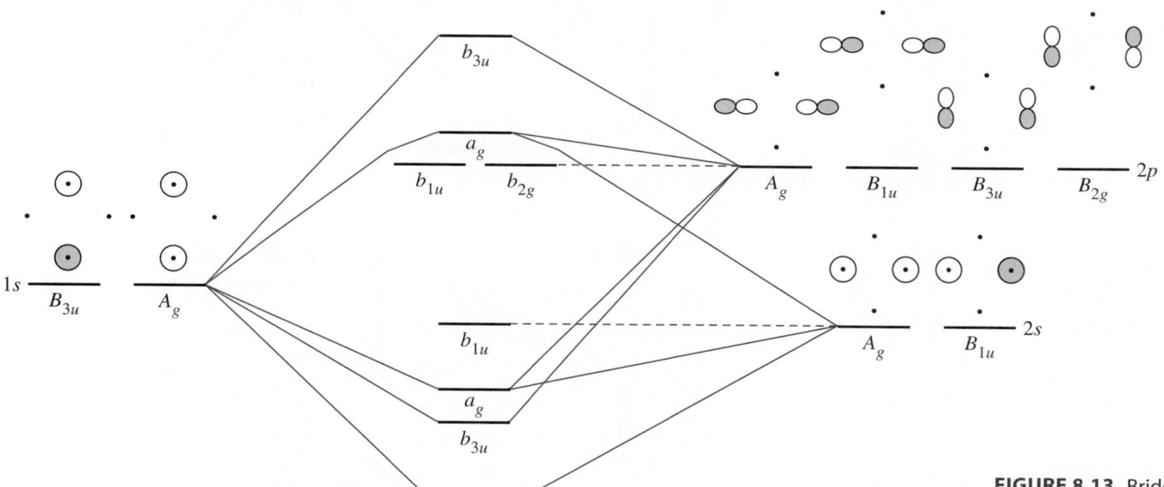

FIGURE 8.13 Bridging Orbital Interactions in Diborane.

molecular orbitals: one strongly bonding, one weakly bonding, and one antibonding.* The other boron group orbitals, with B_{1u} and B_{2g} symmetry, do not participate in interactions with the bridging hydrogens. These interactions are summarized in **Figure 8.13**.** Three bonding orbitals (**Figure 8.14**) play a significant role in joining the boron atoms through the hydride bridges, two of a_g symmetry and one of b_{3u} symmetry.

---

*One of the group orbitals on the terminal hydrogens also has A_g symmetry. The interaction of this group orbital with the other orbitals of A_g symmetry influences the energy and shape of the a_g molecular orbitals shown in Figure 8.13, and generates a fourth, antibonding a_g molecular orbital (not shown).

**Figure 8.13 does not show interactions with terminal hydrogens. One terminal atom group orbital has B_{1u} symmetry and therefore interacts with the B_{1u} group orbitals of boron, resulting in molecular orbitals that are no longer nonbonding.

FIGURE 8.14 Bonding Orbitals Involved in Hydrogen Bridges in Diborane. The a_g orbitals also include contributions from the terminal H atoms.

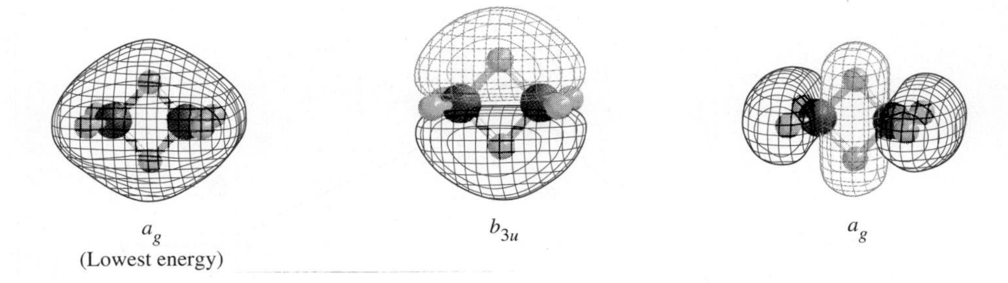

Bridging hydrogen atoms occur in many boranes, as well as in carboranes, which contain both boron and carbon atoms arranged in clusters. Bridging hydrogens and alkyl groups are frequently encountered in aluminum chemistry. Examples of these compounds are in **Figure 8.15**.

FIGURE 8.15 Boranes, Carboranes, and Bridged Aluminum Compounds.

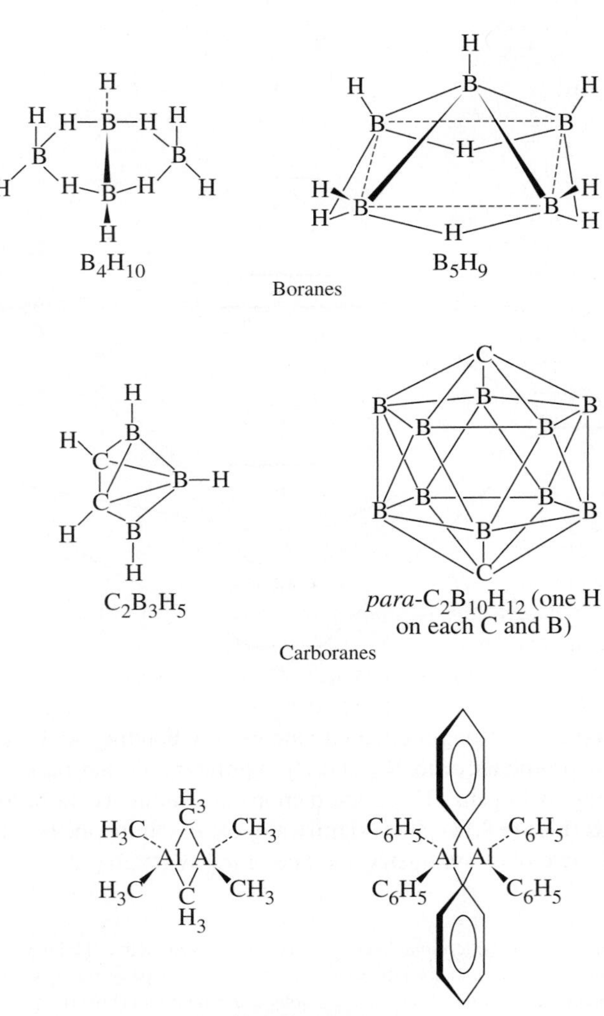

The boranes, carboranes, and related compounds are of interest in the field of cluster chemistry. The bonding in cluster compounds is discussed in Chapter 15.

Boron has two stable isotopes, ^{11}B (80.4% abundance) and ^{10}B (19.6%). ^{10}B is a good absorber of neutrons. This property has been developed for use in the treatment of cancerous tumors in a process called *boron neutron capture therapy* (BNCT).[11] Boron-containing compounds having a strong preference for attraction to tumor sites, rather than healthy sites, can be irradiated with beams of neutrons. The subsequent nuclear decay emits high-energy particles, ${}^{7}_{3}Li$ and ${}^{4}_{2}He$ (alpha particles), which can kill the adjacent cancerous tissue. The development of boron-containing reagents that can be selectively concentrated in cancerous tissue while avoiding healthy tissue is a challenge. Various strategies have been attempted.[12]

$$ {}^{10}_{5}B + {}^{1}_{0}n \longrightarrow {}^{11}_{5}B $$

$$ {}^{11}_{5}B \longrightarrow {}^{7}_{3}Li + {}^{4}_{2}He $$

8.5.2 Other Chemistry of the Group 13 Elements

The boron trihalides, BX_3, are Lewis acids (Chapter 6). These compounds are monomeric and planar—unlike diborane, B_2H_6, and the aluminum halides, Al_2X_6 (Section 3.1.4). As Lewis acids, boron trihalides can accept an electron pair from a halide to form tetrahaloborate ions, BX_4^-. Boron halide catalysts act as halide ion acceptors, as in the Friedel–Crafts alkylation of aromatic hydrocarbons (in margin).

$$ \begin{array}{lll} BF_3 + RX & \longrightarrow & R^+ + BF_3X^- \\ R^+ + PhH & \longrightarrow & H^+ + RPh \\ H^+ + BF_3X^- & \longrightarrow & HX + BF_3 \\ \hline RX + PhH & \longrightarrow & RPh + HX \end{array} $$

The metallic nature of the elements in Group 13 increases down the group. Aluminum, gallium, indium, and thallium form 3+ ions by loss of their valence *p* electron and both valence *s* electrons. Thallium also forms a 1+ ion by losing its *p* electron and retaining its two *s* electrons. This is the **inert pair effect**, in which a metal has an oxidation state that is two less than expected on the basis of the electronic configuration of the atom. For example, Pb forms both 2+ and 4+ ions. This effect is commonly ascribed to the stability of an electron configuration with entirely filled subshells: in the inert pair effect, a metal loses all the *p* electrons in its outermost subshell, leaving a filled s^2 subshell; the pair of *s* electrons seems relatively "inert" and is less easily removed. The actual reasons for this effect are considerably more complex.*

Parallels between main group and organic chemistry can be instructive. One of these parallels is between the organic molecule benzene and the isoelectronic borazine ("inorganic benzene"), $B_3N_3H_6$. Some of the striking similarities in physical properties between these two are in **Table 8.6**.

Despite these parallels, the chemistry of these two compounds is quite different. In borazine, the difference in electronegativity between boron (2.051) and nitrogen (3.066) adds considerable polarity to the B—N bonds and makes the molecule much more susceptible to attack by nucleophiles (at the more positive boron) and electrophiles (at the more negative nitrogen) than benzene.

Parallels between benzene and isoelectronic inorganic rings remain of interest. Some examples include boraphosphabenzenes (containing B_3P_3 rings)[13] and $[(CH_3)AlN(2,6\text{-diisopropylphenyl})]_3$ containing an Al_3N_3 ring.[14] Calculations on borazine, $B_3P_3H_6$, and a variety of other candidate "inorganic benzenes" have indicated that borazine is not aromatic, but $B_3P_3H_6$, Si_6H_6, N_6, and P_6 have some aromatic character.[15]

Another interesting parallel between boron–nitrogen and carbon chemistry is offered by boron nitride, BN. Like carbon (Section 8.6), boron nitride exists in a diamond-like form and in a form similar to graphite. In the diamond-like (cubic) form, each nitrogen is coordinated tetrahedrally by four borons, and each boron by four nitrogens. As in diamond, such coordination gives high rigidity to the structure and makes BN comparable to diamond

*See, for example, N. N. Greenwood and A. Earnshaw, *Chemistry of the Elements*, 2nd ed., Butterworth Heinemann, Oxford, 1997, pp. 226–227.

TABLE 8.6 Benzene and Borazine

Property	Benzene	Borazine
Melting point (°C)	6	−57
Boiling point (°C)	80	55
Density (g cm^{-3})[a]	0.81	0.81
Surface tension (N m^{-1})[a]	0.0310	0.0311
Dipole moment	0	0
Internuclear distance in ring (pm)	142	144
Internuclear distance, bonds to H (pm)	C—H: 108	B—H: 120 N—H: 102

Source: Data from N. N. Greenwood and A. Earnshaw, *Chemistry of the Elements*, Pergamon Press, Elmsford, NY, 1984, p. 238.
[a]At the melting point

in hardness. In the graphite-like hexagonal form, BN also occurs in extended fused-ring systems. However, there is much less delocalization of pi electrons in this form and, unlike graphite, hexagonal BN is a poor conductor. As in the case of diamond, the harder and more dense form (cubic) can be formed from the less dense form (hexagonal) under high pressure.

Are multiple bonds between Group 13 elements possible? For example, can the simple neutral *diborene*, HB═BH, be synthesized? Diborene is predicted to be extremely reactive because it exhibits two one-electron π bonds (similar to B_2 as described in Section 5.2.3). However, a suitable Lewis base can be envisioned to potentially stabilize this species (in margin). A sterically demanding and strongly donating N-heterocyclic carbene (**Figure 8.16a**) permitted isolation of the first neutral diborene (**Figure 8.16b**) in 2007.[16] Structural data and computational evidence support the existence of a B═B double bond in this complex.

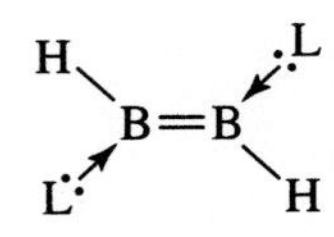

The first *triple bond* between main group metals caused considerable controversy upon its report in 1997.[17] A gallium dichloride with a substituted phenyl group (**Figure 8.17a**) was reduced with sodium to form a complex containing an alleged gallium—gallium triple bond (a *gallyne*, **Figure 8.17b**). The sodium ions reside on each side of the bond, forming a nearly planar Ga_2Na_2 four-membered ring.[18] Reconciliation of a bond order of three despite the non-linear (*trans*-bent) C—Ga—Ga—C core of the complex sparked an intellectual debate about the fundamentals of chemical bonding. Eight years after the report of this gallyne, further support for a gallium—gallium bond order of three was presented.[19]

(a)

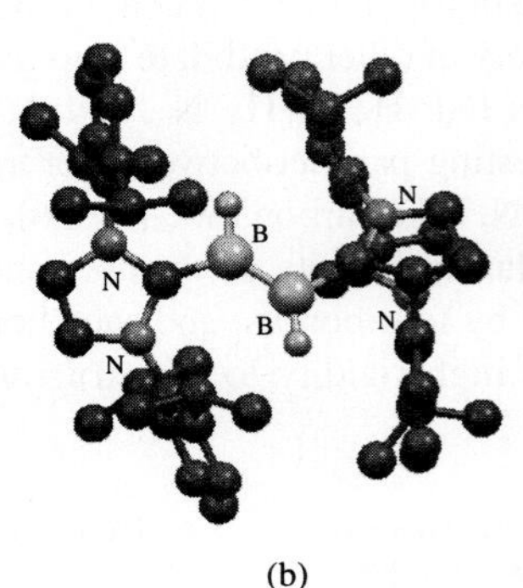

(b)

FIGURE 8.16 (a) N-heterocyclic carbene. (b) The first neutral diborene containing a B═B double bond. Molecular structure created with CIF data, with hydrogen atoms omitted for clarity.

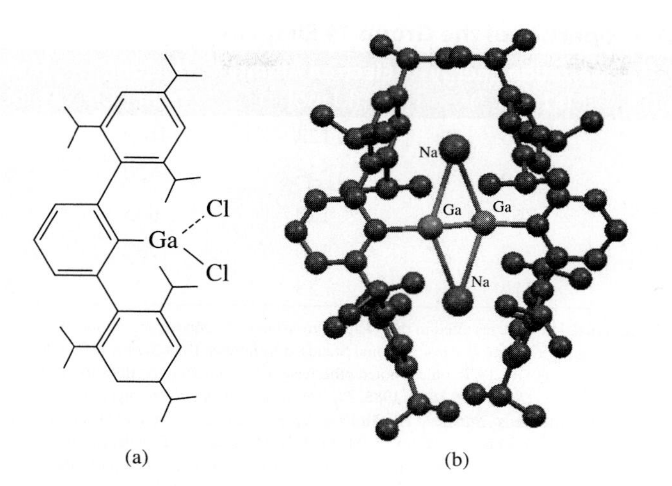

FIGURE 8.17 The Ga(III) precursor (a) used to prepare the first gallyne (b). Molecular structure created with CIF data, with hydrogen atoms omitted for clarity.

Multiple bonding between main group metals is a vigorous research area, with many seminal synthetic achievements published since the turn of the century. The diborene and gallyne complexes are provided here as a brief introduction to this exciting area; helpful reviews are available for more details.[20]

8.6 Group 14

8.6.1 The Elements

Group 14 elements range from a nonmetal, carbon, to the metals tin and lead, with the intervening elements showing semimetallic behavior. Carbon has been known from prehistory as the charcoal resulting from partial combustion of organic matter. Diamonds have been prized as precious gems for thousands of years. Neither form of carbon, however, was recognized as a chemical element until late in the eighteenth century. Tools made of flint, primarily SiO_2, were used in the Stone Age. However, free silicon was not isolated until 1823, when Berzelius obtained it by reducing K_2SiF_6 with potassium. Tin and lead have also been known since ancient times. A major early use of tin was in combination with copper in the alloy bronze; weapons and tools containing bronze date back more than 5000 years. Lead was used by the ancient Egyptians in pottery glazes and by the Romans for plumbing and other purposes. The toxicity of lead and lead compounds has gained increasing attention and has led to restrictions on the use of lead compounds, for example in paint pigments and in gasoline additives, primarily tetraethyllead, $(C_2H_5)_4Pb$. Germanium was a "missing" element for a number of years. Mendeleev predicted the properties of this then-unknown element in 1871 ("eka-silicon"), but it was not discovered until 1886. Properties of the Group 14 elements are summarized in **Table 8.7**.

Although carbon occurs primarily as the isotope ^{12}C, whose atomic mass serves as the basis the atomic mass unit (amu) and therefore atomic mass in the periodic table, two other isotopes, ^{13}C and ^{14}C, are important as well. ^{13}C, which has a natural abundance of 1.11 percent, has a nuclear spin of $\frac{1}{2}$ in contrast to ^{12}C, which has zero nuclear spin. Although ^{13}C comprises only about 1 part in 90 of naturally occurring carbon, it is used as the basis of NMR observations for the characterization of carbon-containing compounds. ^{13}C NMR spectrometry is a valuable tool in both organic and inorganic chemistry. Uses of ^{13}C NMR in organometallic chemistry are described in Chapter 13.

TABLE 8.7 Properties of the Group 14 Elements

Element	Ionization Energy (kJ mol^{-1})	Electron Affinity (kJ mol^{-1})	Melting Point (°C)	Boiling Point (°C)	Electro-negativity
C	1086	122	4100	[a]	2.544
Si	786	134	1420	3280[b]	1.916
Ge	762	120	945	2850	1.994
Sn	709	120	232	2623	1.824
Pb	716	35	327	1751	1.854

Source: Data from: Ionization energies cited in this chapter are from C. E. Moore, *Ionization Potentials and Ionization Limits Derived from the Analyses of Optical Spectra,* National Standard Reference Data Series, U.S. National Bureau of Standards, NSRDS-NBS 34, Washington, DC, 1970, unless noted otherwise. Electron affinity values listed in this chapter are from H. Hotop and W. C. Lineberger, *J. Phys. Chem. Ref. Data,* **1985,** *14,* 731. Standard electrode potentials listed in this chapter are from A. J. Bard, R. Parsons, and J. Jordan, eds., *Standard Potentials in Aqueous Solutions,* Marcel Dekker (for IUPAC), New York, 1985. Electronegativities cited in this chapter are from J. B. Mann, T. L. Meek, and L. C. Allen, *J. Am. Chem. Soc.,* **2000,** *122,* 2780, Table 2. Other data are from N. N. Greenwood and A. Earnshaw, *Chemistry of the Elements,* Pergamon Press, Elmsford, NY, 1984, except where noted.

[a]Sublimes [b]Approximate value

$${}^{14}_{7}N + {}^{1}_{0}n \longrightarrow {}^{14}_{6}C + {}^{1}_{1}H$$

$${}^{14}_{6}C \longrightarrow {}^{14}_{7}N + {}^{0}_{-1}e$$

^{14}C is formed in the atmosphere from nitrogen by neutrons from cosmic rays. Although ^{14}C is formed in comparatively small amounts (approximately 1.2×10^{-10} percent of atmospheric carbon), it is incorporated into plant and animal tissues by biological processes. When a plant or animal dies, the process of exchange of its carbon with the environment by respiration and other biological processes ceases, and the ^{14}C in its system is effectively trapped. However, ^{14}C decays by beta emission, with a half-life of 5730 years. Therefore, by measuring the remaining amount of ^{14}C, one can determine to what extent this isotope has decayed and, in turn, the time elapsed since death. Often called *radiocarbon dating*, this procedure has been used to estimate the ages of many archeological samples, including Egyptian remains, charcoal from early campfires, and the Shroud of Turin.

EXAMPLE 8.2

What fraction of ^{14}C remains in a sample that is 50,000 years old?

This is 50,000/5,730 = 8.73 half-lives. For first-order reactions, such as radioactive decay, the initial amount decreases by $\frac{1}{2}$ during each half-life, so the fraction remaining is $(\frac{1}{2})^{8.73} = 2.36 \times 10^{-3}$.

EXERCISE 8.5 A sample of charcoal from an archeological site has a remaining fraction of ^{14}C of 3.5×10^{-2}. What is its age?

Diamond and Graphite

Carbon was encountered primarily in two allotropes, diamond and graphite, until 1985. The diamond structure is rigid, with each atom surrounded tetrahedrally by four other atoms in a structure that has a cubic unit cell. As a result, diamond is extremely hard, the hardest of all naturally occurring substances. Graphite, on the other hand, consists of layers of fused, six-membered rings of carbon atoms. The carbon atoms in these layers may be viewed as being sp^2 hybridized. The remaining, unhybridized p orbitals are perpendicular to the layers and participate in extensive pi bonding, with pi electron density delocalized over the layers. Because of the relatively weak interactions between the layers, the layers are free to slip with respect to each other, and pi electrons are free to move within each layer, making graphite a good lubricant and electrical conductor, respectively. Potassium graphite (KC_8), formed in an inert atmosphere by mixing molten potassium metal and graphite, is a common reducing agent in synthesis. The KC_8 structure features potassium atoms between the graphite layers; these atoms are *intercalated* within the graphite. The structures of diamond and graphite are shown in **Figure 8.18**, and their important physical properties are in **Table 8.8**.

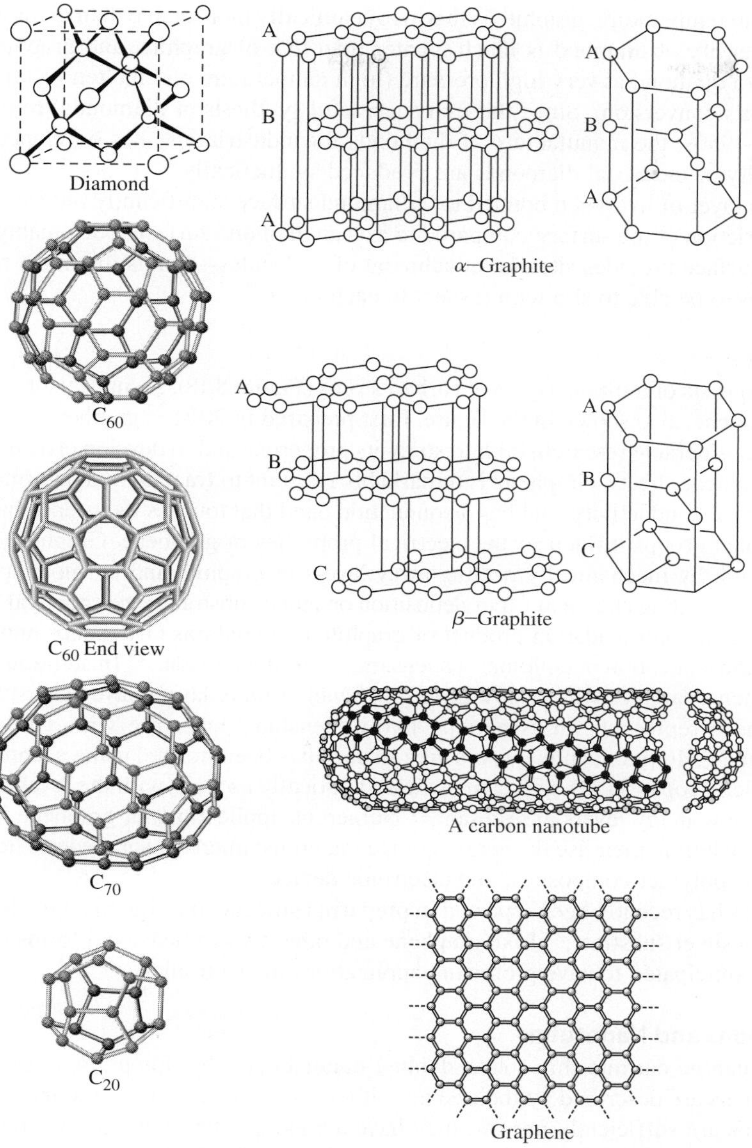

FIGURE 8.18 Diamond, Graphite, Fullerenes, and Graphene.

TABLE 8.8 **Physical Properties of Diamond and Graphite**

Property	Diamond	Graphite
Density (g cm^{-3})	3.513	2.260
Electrical resistivity (Ω m)	10^{11}	1.375×10^{-5}
Standard molar entropy (J mol^{-1} K^{-1})	2.377	5.740
C_p at 25 °C (J mol^{-1} K^{-1})	6.113	8.527
C—C distance (pm)	154.4	141.5 (within layer) 335.4 (between layers)

Data from J. Elmsley, *The Elements*, Oxford University Press, New York, 1989, p. 44.

At room temperature, graphite is thermodynamically more stable than diamond. However, the density of diamond is much greater than that of graphite, and graphite can be converted to diamond at very high pressure (high temperature and molten metal catalysts facilitate this conversion). Since the first successful synthesis of diamonds from graphite in the mid-1950s, the manufacture of diamonds for industrial use has developed rapidly. The majority of industrial diamonds are produced synthetically.

A thin layer of hydrogen bonded to a diamond surface significantly reduces the coefficient of friction of the surface compared to a clean diamond surface, presumably because the clean surface provides sites for attachment of molecules—bonds that must be broken for surfaces to be able to slip with respect to each other.[21]

Graphene

Graphite consists of multiple layers of carbon atoms (Figure 8.18). A single, isolated layer is called **graphene**, also shown in the figure. First prepared in 2004,[22] graphene has been the center of considerable research, both to study its properties and to develop efficient ways to prepare graphene sheets. Graphene is remarkably resistant to fracture and deformation, has a high thermal conductivity, and has a conduction band that touches its valence band. Adding functional groups influences the electrical properties of graphene. Graphene has been prepared both by mechanically peeling away layers of graphite and by more specialized techniques, including chemical vapor deposition on metal substrates, the chemical reduction of graphite oxide (an oxidation product of graphite that contains OH groups and bridging oxygens), and sonication of colloidal suspensions of graphite oxide.[*,23] In addition to single-layer graphene, samples with two (bilayer graphene) or more layers have been studied.

Graphene represents an essentially two-dimensional structure with a thickness of approximately 340 pm. Its honeycomb-like surface has been imaged using scanning transmission microscopy (STM), and can be viewed optically as well, with the level of contrast indicating how many layers are present.[24] Numerous applications of graphene have been proposed, including their use in energy-storage materials, microsensor devices, liquid crystal displays, polymer composites, and electronic devices.

Success has recently been reported in preparing silicene, the silicon equivalent of graphene, on a silver substrate.[25] Like graphene and other two-dimensional forms of carbon, silicene is anticipated to have promising applications in electronics.

Nanoribbons and Nanotubes

Graphene can be cut into thin strips, dubbed **nanoribbons**, by lithographic techniques.[26] These ribbons are described by their edges, either *zigzag* or *armchair* (**Figure 8.19**). If the nanoribbons are sufficiently narrow, they have a band gap between their conduction and valence bands. As in the case of quantum dots (Section 7.3.2), the quantum confinement effect becomes more pronounced as the dimension—in this case, the width of the ribbon—becomes smaller, and the band gap in these semiconductors becomes wider.[27] In orbital terms, the smaller the dimension, the more the energy levels show distinct, quantized energies, rather than a continuum of energy levels, as in a larger structure.

One can envision joining the edges of nanoribbons together to form tubes. Such **nanotubes** can be formed in several ways: If the edges that are joined are zigzag, the ends of the tube have carbons in a chair arrangement; if armchair edges are joined, the ends of the tube have a zigzag arrangement (Figures 8.19 and **8.20**). If the edges that are brought

---

[*]See C. N. R. Rao, K. Biswas, K. S. Subrahmanyam, A. Govindaraj, *J. Mater. Chem.*, **2009**, *19*, 2457 for a useful review of graphene's synthesis and properties, including electronic properties that are beyond the scope of this text. This review includes images of graphene in single and multiple layers.

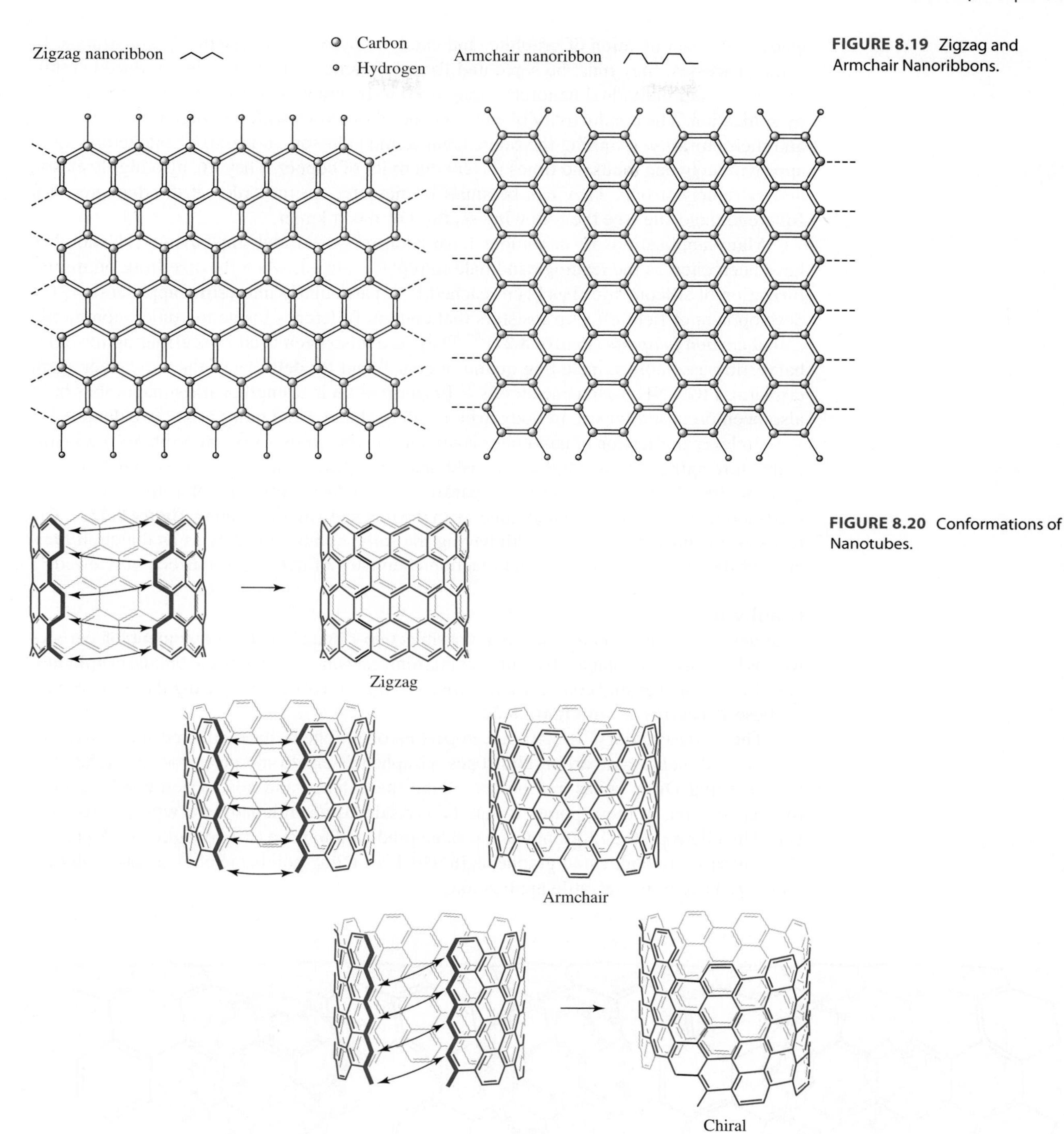

FIGURE 8.19 Zigzag and Armchair Nanoribbons.

FIGURE 8.20 Conformations of Nanotubes.

together are from different rows, the nanotube that is formed will be helical, and therefore chiral (Figure 8.20).[28]

Both single-walled nanotubes (Figure 8.18) and multiple-walled nanotubes with multiple, concentric layers of carbons have been studied extensively. Commercially they are prepared by a variety of methods, including application of electric arcs to graphite electrodes, laser ablation of graphite, and catalytic pyrolysis of gaseous hydrocarbons. In some processes, they must be separated from fullerenes, which may be produced simultaneously; and individual nanotubes may need to be unbundled from others, for example by sonication. The conductivity of carbon nanotubes varies with the diameter of the tubes and their chirality, spanning the range from semiconductors to metallic conductors, up to approximately one thousand times the conductivity of copper. They are also highly elastic and extremely strong. They can be small in diameter, on the order of the diameter of a fullerene molecule (see the following section) or much larger.*

Many applications of nanotubes have been proposed and implemented. Nanotubes have been cited as the leading candidate to replace silicon, when the size limit on miniaturization of silicon chips has been reached.[29] A particularly interesting application is the development of field-effect transistors that contain fullerenes inside the tubes, commonly called *carbon peapods* (**Figure 8.21**).[30] They have also been used as scaffolding for light-harvesting assemblies, including quantum dots,[31] and for delivery of the anticancer agent cisplatin, $PtCl_2(NH_3)_2$, to cancer cells.[32] Because of their strength, carbon nanotubes have also been proposed for use in body armor, safety shields, and other protective devices.

A clever application of nanotubes is their use in the synthesis of nanoribbons by chemically "unzipping" the nanotubes into ribbons. This process involves the strong oxidizing agent $KMnO_4/H_2SO_4$ to form rows of parallel carbonyls on both sides of a slit, as it is being made along the length of the nanotube, with the mechanism proposed in **Figure 8.22**.[33] The result is a nanoribbon with a width having the same number of carbons as the circumference of the original nanotube. The reaction chemistry of nanotubes has been reviewed.[34]

Graphyne

Graphene is by no means the only possible two-dimensional arrangement of carbon. Recently a variety of planar structures containing carbon—carbon triple bonds, **graphynes**, have been proposed and considerable effort has been devoted to preparing them. Examples of these structures are in **Figure 8.23**.

The "extraordinary" electronic properties of graphene have spurred the search for other two-dimensional carbon allotropes. Graphene's electronic properties are related to its exhibiting Dirac cones and points, where the valence and conduction bands meet at the Fermi level; at these points it may be considered a semiconductor with a zero band gap. The allotrope 6,6,12-graphyne has been predicted to have two nonequivalent types of Dirac points—in contrast to graphene, in which all Dirac points are equivalent—and may therefore have more versatile applications.[35]

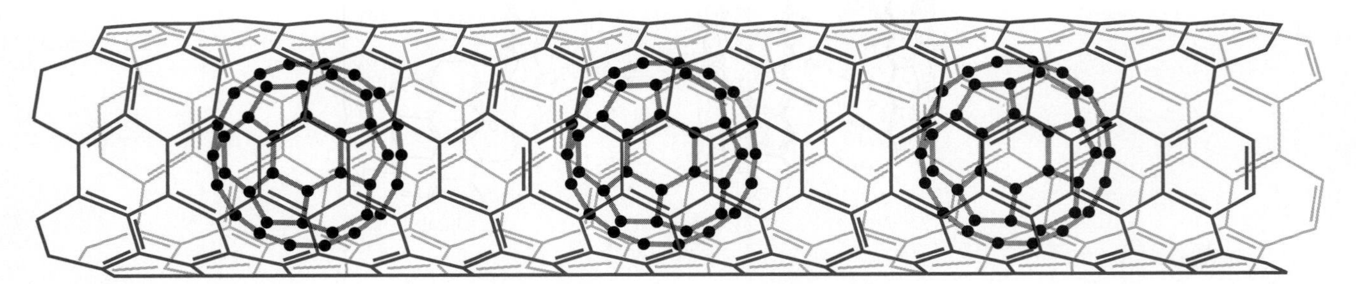

FIGURE 8.21 A Carbon Peapod.

*See D. R. Mitchell, R. M. Brown, Jr., T. L. Spires, D. K. Romanovicz, R. J. Lagow, *Inorg. Chem.*, **2001**, *40*, 2751 for a variety of electronic microscopic images of larger diameter nanotubes.

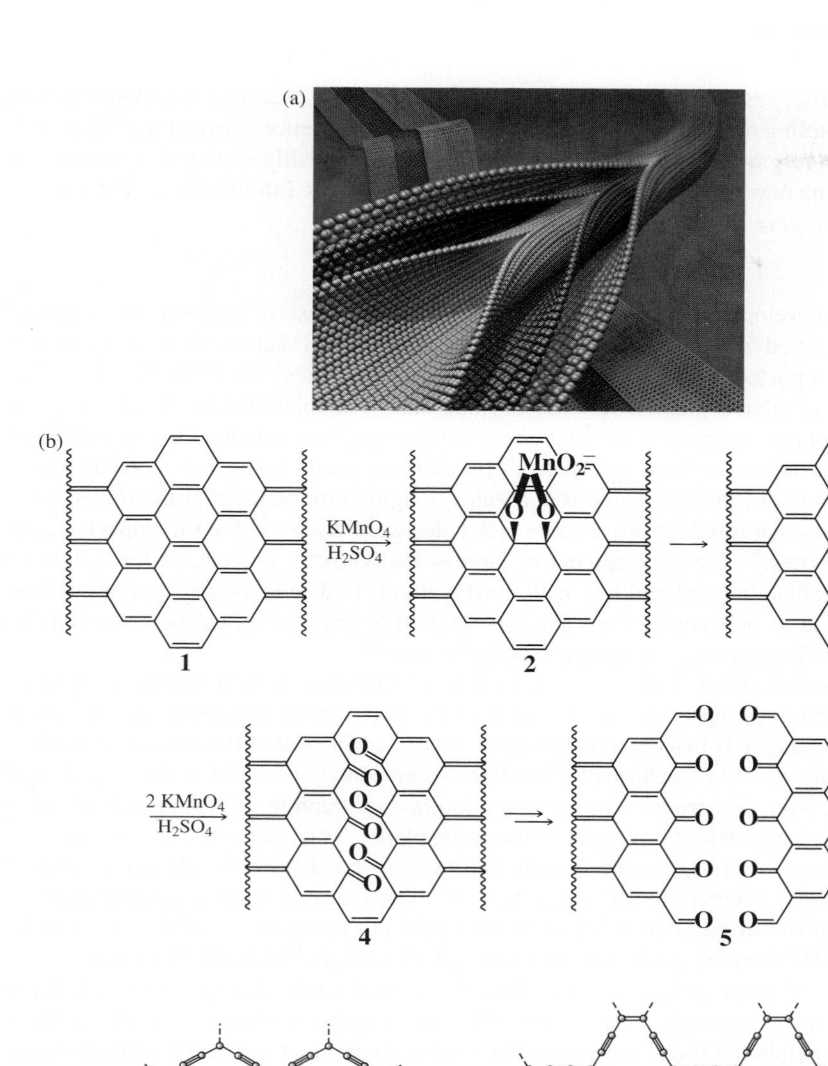

FIGURE 8.22 (a) Representation of nanotube unzipping. (b) Proposed mechanism of unzipping of a carbon nanotube. Illustration by Paul Quade.

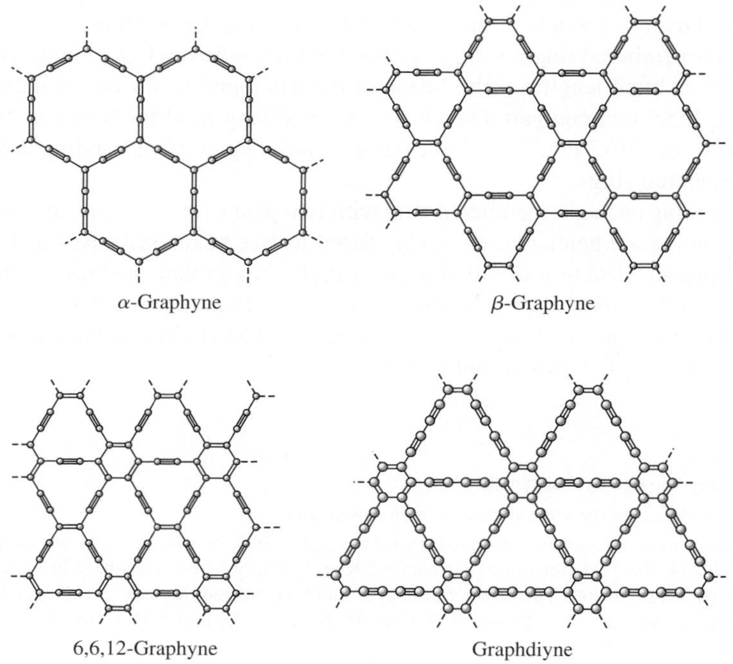

FIGURE 8.23 Graphyne Structures.

α-Graphyne

β-Graphyne

6,6,12-Graphyne

Graphdiyne

Films of the graphdiyne allotrope (Figure 8.23) have been prepared on copper surfaces by a cross-coupling reaction of hexaethynylbenzene in the presence of pyridine.[36] The binding of lithium to graphyne and graphdiyne has been theoretically explored in connection with developing new anode materials for lithium batteries. The lithium sites in these materials may also serve as potential sites for hydrogen storage.[37]

Fullerenes

A fascinating development in modern chemistry was the synthesis of buckminsterfullerene,* C_{60}, and the related *fullerenes*, molecules having near-spherical shapes resembling geodesic domes. First reported by Kroto, Curl, Smalley, and colleagues[38] in 1985, C_{60}, C_{70}, C_{80}, and a variety of related species were soon synthesized; examples of their structures are in Figure 8.18. Many compounds of fullerenes containing groups attached to the outside of these large clusters have been synthesized. In addition, small atoms and molecules have been trapped inside fullerene cages. Remarkably, roughly nine years after the first synthesis of fullerenes, natural deposits of these molecules were discovered at the impact sites of ancient meteorites.[39] The development of large-scale synthetic procedures for fullerenes has been a challenging undertaking, with most methods to date involving condensation of carbon, from laser or other high-energy vaporization of graphite, in an inert atmosphere, or from controlled pyrolysis of aromatic hydrocarbons.[40]

The prototypical fullerene, C_{60}, consists of fused five- and six-membered carbon rings. Each six-membered ring is surrounded alternately by hexagons and pentagons of carbon atoms; each pentagon is fused to five hexagons. The consequence of this structural motif is that each hexagon is like the base of a bowl: the three pentagons fused to this ring, linked by hexagons, force the structure to curve (in contrast to graphite, in which each hexagon is fused to six surrounding hexagons in the same plane). This phenomenon, best seen by assembling a model of C_{60}, results in a domelike structure that eventually curves around on itself to give a structure resembling a sphere.** The shape resembles a soccer ball (the most common soccer ball has an identical arrangement of pentagons and hexagons on its surface); all 60 atoms are equivalent and give rise to a single ^{13}C NMR resonance.

Although all atoms in C_{60} are equivalent, the bonds are not. Two types of bonds occur (best viewed using a model), at the fusion of two six-membered rings and at the fusion of five- and six-membered rings. X-ray crystallographic studies on C_{60} complexes have shown that the C—C bond lengths at the fusion of two six-membered rings in these complexes are shorter, 135.5 pm, compared to the C—C bond lengths at the fusion of five- and six-membered rings, 146.7 pm.† This indicates a greater degree of pi bonding at the fusion of the six-membered rings.

Surrounding each six-membered ring with two pentagons (on opposite sides) and four hexagons (with each pentagon, as in C_{60}, fused to five hexagons) gives a slightly larger, somewhat prolate structure (the ball used in rugby is a prolate spheroid) with 70 carbon atoms. C_{70} is often obtained as a by-product of the synthesis of C_{60} and is among the most stable of the fullerenes. Unlike C_{60}, five different carbon environments are present in C_{70}, giving rise to five ^{13}C NMR resonances.[41]

---

*More familiarly known as a "buckyball."

**The structure of C_{60} has the same symmetry as an icosahedron.

†These distances were obtained for a twinned crystal of C_{60} at 110 K. (S. Liu, Y. Lu, M. M. Kappes, J. A. Ibers, *Science*, **1991**, *254*, 408.) Neutron diffraction data at 5 K give slightly different results: 139.1 pm at the fusion of the 6-membered rings and 145.5 pm at the fusion of 5- and 6-membered rings (W. I. F. David, R. M. Ibberson, J. C. Matthew, K. Pressides, T. J. Dannis, J. P. Hare, H. W. Kroto, R. Taylor, D. C. M. Walton, *Nature*, **1991**, *353*, 147).

Structural variations on fullerenes have evolved beyond the individual clusters themselves. The following are a few examples.

Polymers

The rhombohedral polymer of C_{60} acts as a ferromagnet at room temperature.[42] Linear chain polymers have also been reported.[43] Reduced and oxidized fullerenes (fullerides and fullereniums, respectively) form solid-state polymers (**Figures 8.24a** and **8.24b**).

Nano "onions"

These are spherical particles based on multiple carbon layers surrounding a C_{60} or other fullerene core. One proposed use is in lubricants.[44]

Other linked structures

These include fullerene rings,[45] dimers (**Figure 8.24c** and **8.24d**),[46] and chiral fullerenes.[47]

Endofullerenes[48]

The diameter of C_{60} is 3.7 Å. Endofullerenes feature atoms or molecules trapped (or encapsulated) inside, including metal ions, noble gases, nitrogen atoms, and hydrogen gas. A remarkable synthetic achievement was the synthesis of $H_2O@C_{60}$,* where a single water molecule is trapped inside C_{60} (**Figure 8.24e**).[49] The synthesis required the creation in

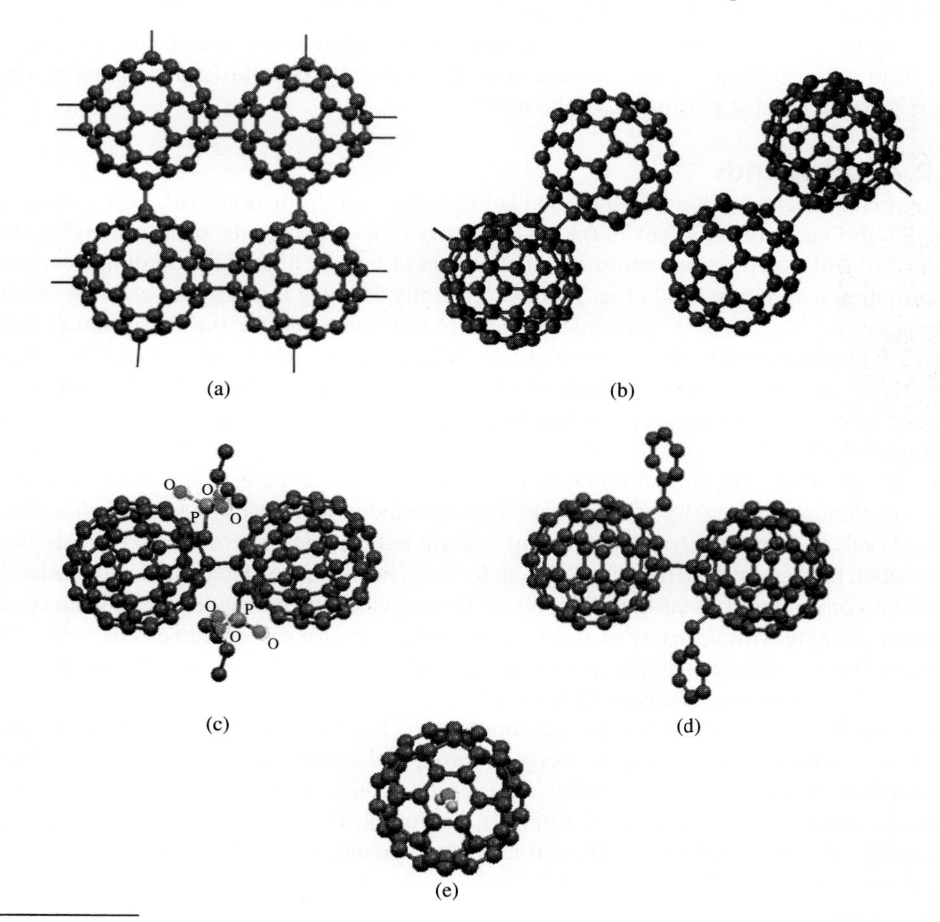

FIGURE 8.24 (a) Portion of the crystal structure of polymerized Li_4C_{60} with Li ions omitted.[54] (b) Portion of the "zigzag" polymeric structure of $C_{60}(AsF_6)_2$ (with $AsF_6{}^{2-}$ omitted) that features chains of $[C_{60}]^{2+}$ connected alternatively by single C—C bonds and four-membered carbon rings.[55] (c) Singly bonded C_{60} dimer with diethylphosphorylmethyl groups ($CH_2P(O)(OEt)_2$).[56] (d) Singly bonded $PhCH_2C_{60}—C_{60}CH_2Ph$ dimer.[57] (e) $H_2O@C_{60}$.[49] ((a) Data from S. Margadonna, D. Pontiroli, M. Belli, T. Shiroka, M. Riccò, M. Brunelli, *J. Am. Chem. Soc.*, **2004,** *126,* 15032. (b) Data from M. Riccò, D. Pontiroli, M. Mazzani, F. Gianferrari, M. Pagliari, A. Goffredi, M. Brunelli, G. Zandomeneghi, B. H. Meier, T. Shiroka, *J. Am. Chem. Soc.*, **2010,** *132,* 2064. (c-e) Data from F. Cheng, Y. Murata, K. Komatsu, *Org. Lett.,* **2002,** *4,* 2541. Molecular structure drawings (c-e) were generated from CIF data, with hydrogen atoms omitted for clarity.)

*The @ symbol designates encapsulation within the cage structure.

C_{60} of a sufficiently large hole that was adorned with oxygen atoms to facilitate hydrogen bonding (to render the C_{60} somewhat hydrophilic), followed by treatment at elevated pressure and temperature to coax a water molecule into the cavity, and subsequent synthetic steps to close the hole. NMR studies indicate that the water molecule is free to move inside $H_2O@C_{60}$.

The smallest known fullerene is C_{20} (Figure 8.18), synthesized by replacing the hydrogen atoms of dodecahedrane, $C_{20}H_{20}$, with bromines, followed by debromination.[50] The chemistry of small fullerenes ($< C_{60}$) has been reviewed.[51]

Since 1985, fullerene chemistry has developed from its infancy into a broadly studied realm of science that is increasingly focused on practical applications. Fullerenes also form a variety of chemical compounds via reactions on their surfaces and can trap atoms and small molecules inside; these features of fullerene chemistry are considered in Chapter 13. It was discovered in 2006 that a water-soluble C_{60} derivative could be attached to a melanoma antibody; loading this derivative with anticancer drug molecules can deliver the drug directly into the melanoma.[52] The use of fullerenes, nanotubes, quantum dots, and other nanoparticles for drug delivery has been reviewed.[53]

Silicon and germanium crystallize in the diamond structure. They have somewhat weaker covalent bonds than carbon as a consequence of less efficient orbital overlap. These weaker bonds result in lower melting points for silicon—1420 °C for Si and 945 °C for Ge, compared with 4100 °C for diamond—and greater chemical reactivity. Silicon and germanium are semiconductors (Chapter 7).

Tin has two allotropes: a diamond (α) form more stable below 13.2 °C and a metallic (β) form more stable at higher temperatures.* Lead is entirely metallic and is among the most dense, and most poisonous, of the metals.

8.6.2 Compounds

A common misconception is that carbon is limited to 4-coordination. Although carbon is bonded to four or fewer atoms in the vast majority of its compounds, many examples are known in which carbon has coordination numbers of 5, 6, or higher. 5-coordinate carbon is common, with methyl and other groups frequently forming bridges between two metal atoms, as in $Al_2(CH_3)_6$ (Figure 8.15). There is also evidence for the 5-coordinate ion CH_5^+.[58] Organometallic cluster compounds contain carbon atoms surrounded by polyhedra of metal atoms; these *carbide clusters* are discussed in Chapter 15. Examples of carbon atoms having coordination numbers of 5, 6, 7, and 8 in such clusters are shown in **Figure 8.25**.

CO and CO_2, the most familiar oxides of carbon, are colorless and odorless gases. Carbon monoxide is a rarity of sorts, a stable compound in which carbon formally has only three bonds. It is extremely toxic, forming a bright red complex with the iron in hemoglobin, which has a greater affinity for CO than for O_2. The highest occupied molecular orbital of CO is concentrated on carbon (Chapter 5); this provides the molecule an opportunity to interact strongly with a variety of metal atoms, which in turn can donate electron density through their *d* orbitals to empty π^* orbitals (LUMOs) on CO. The details of such interactions are described more fully in Chapter 13.

Carbon dioxide is familiar as a component of Earth's atmosphere—although only fifth in abundance, after nitrogen, oxygen, argon, and water vapor—and as the product of respiration, combustion, and other natural and industrial processes. It was the first gaseous component to be isolated from air, by Joseph Black in 1752. CO_2 has gained international attention because of its role in the "greenhouse" effect and the potential

*These forms are *not* similar to the α and β forms of graphite (Figure 8.18).

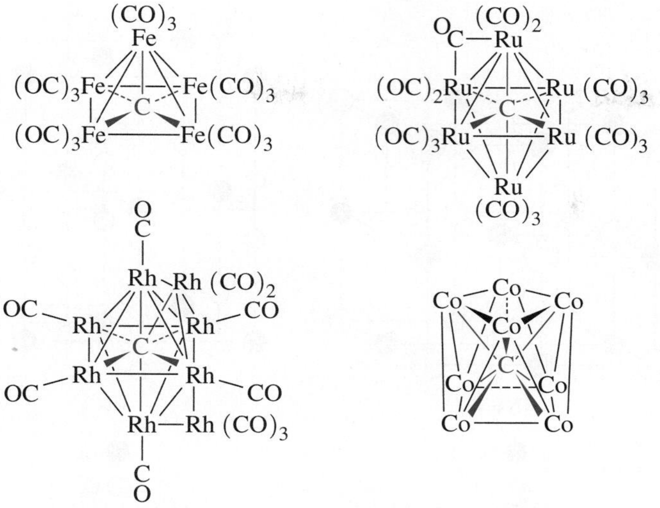

FIGURE 8.25 High Coordination Numbers of Carbon.

$Rh_8C(CO)_{19}$
Not shown: eight COs bridging edges of polyhedron

$[Co_8C(CO)_{18}]^{2-}$
COs not shown: one on each cobalt; ten bridging edges of polyhedron

atmospheric warming and other climatic consequences of an increase in CO_2 abundance. Because of the energies of carbon dioxide's vibrational levels, it absorbs a significant amount of thermal energy and, hence, acts as a sort of atmospheric blanket. Since the beginning of the Industrial Revolution, the carbon dioxide concentration in the atmosphere has increased substantially, an increase that will continue indefinitely unless major policy changes are made by the industrialized nations. The consequences of a continuing increase in atmospheric CO_2 are difficult to forecast; the dynamics of the atmosphere are extremely complex, and the interplay between atmospheric composition, human activity, the oceans, solar cycles, and other factors is not yet well understood.

Carbon forms several anions, especially in combination with the most electropositive metals. In these *carbide* compounds, there is considerable covalent as well as ionic bonding, with the proportion of each depending on the metal. The best characterized carbide ions are:

Ion	Common Name	Systematic Name	Example	Major Hydrolysis Product
C^{4-}	Carbide or methanide	Carbide	Al_4C_3	CH_4
C_2^{2-}	Acetylide	Dicarbide (2–)	CaC_2	$H—C{\equiv}C—H$
C_3^{4-}		Tricarbide (4–)	Mg_2C_3[a]	$H_3C—C{\equiv}C—H$

[a]This is the only known compound containing the C_3^{4-} ion.

Carbides can liberate organic molecules in reactions with water. For example:

$$Al_4C_3 + 12\ H_2O \longrightarrow 4\ Al(OH)_3 + 3\ CH_4$$

$$CaC_2 + 2\ H_2O \longrightarrow Ca(OH)_2 + HC{\equiv}CH$$

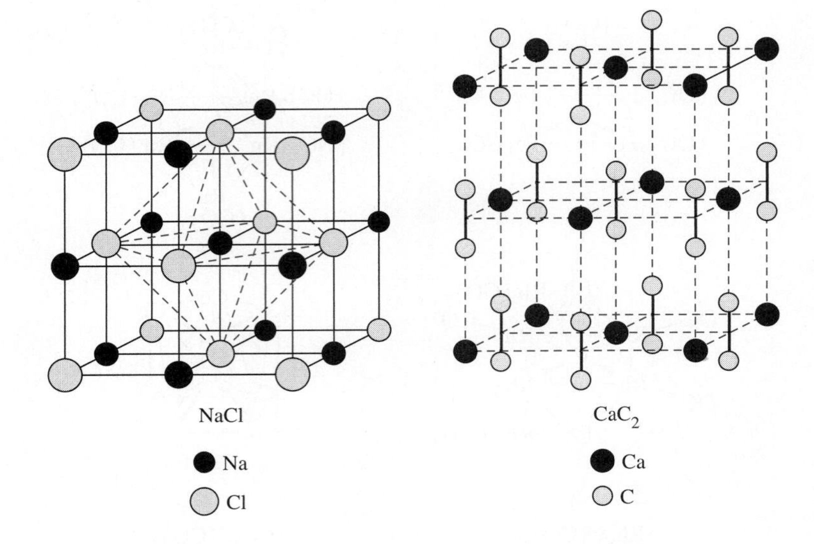

FIGURE 8.26 Crystal Structures of NaCl and CaC_2.

Calcium carbide, CaC_2, is the most important of the metal carbides. Its crystal structure resembles that of NaCl, with parallel C_2^{2-} units (**Figure 8.26**). Before compressed gases were readily available, calcium carbide was used as a source of acetylene for lighting and welding; early automobiles had carbide headlights.

It may seem surprising that carbon, with its vast range of literally millions of compounds, is not the most abundant element in this group. By far the most abundant Group 14 element on Earth is silicon, which comprises 27% of Earth's crust by mass and is second in abundance after oxygen; carbon is only seventeenth in abundance. Silicon, with its semimetallic properties, is of enormous importance in the semiconductor industry.

In nature, silicon occurs almost exclusively in combination with oxygen, with many minerals containing tetrahedral SiO_4 structural units. Silicon dioxide, SiO_2, occurs in a variety of forms in nature, the most common of which is α-quartz, a major constituent of sandstone and granite. SiO_2 is industrially important as the major component of glass; in finely divided form as a chromatographic support (silica gel) and catalyst substrate; as a filtration aid (as diatomaceous earth, the remains of diatoms, tiny unicellular algae); and many other applications.

The SiO_4 structural units occur in nature in silicates, compounds in which these units may be fused by sharing corners, edges, or faces in diverse ways. Examples of silicate structures are shown in **Figure 8.27**. Extensive discussions of these structures are in the chemical literature.[59]

With carbon forming the basis for the colossal number of organic compounds, it is interesting to consider whether silicon or other members of this group can form an equally vast array of compounds. This does not seem the case; the ability to catenate (form bonds with other atoms of the same element) is much lower for the other Group 14 elements than for carbon, and the hydrides of these elements are also much less stable.

Silane, SiH_4, is stable and, like methane, tetrahedral. However, although silanes (of formula Si_nH_{n+2}) up to eight silicon atoms in length have been synthesized, their stability decreases markedly with chain length; Si_2H_6, disilane, undergoes very slow decomposition, but Si_8H_{18} decomposes rapidly. In recent years, a few compounds containing

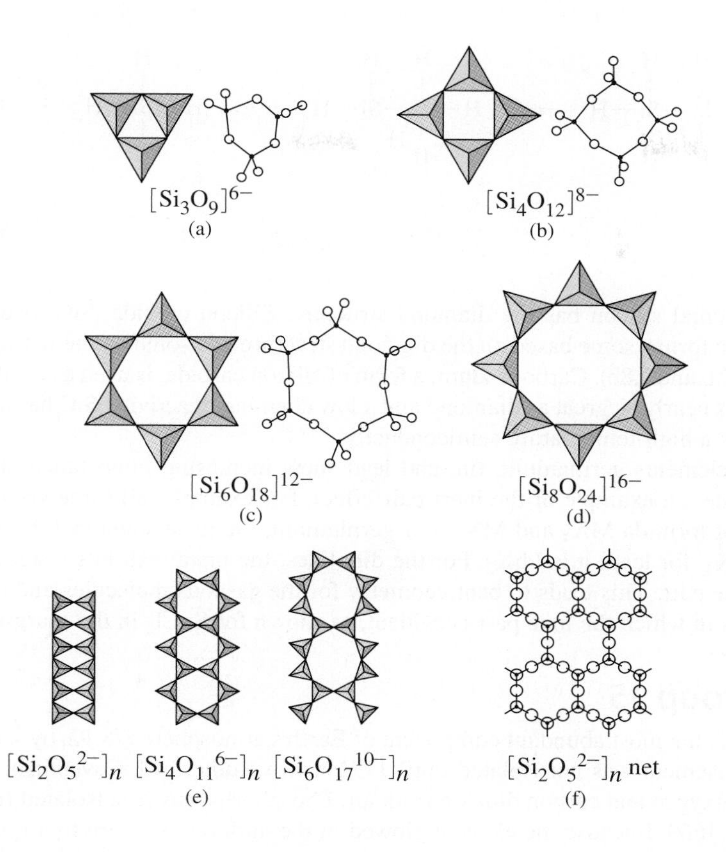

FIGURE 8.27 Silicate Structures. (Published in "Chemistry of the Elements", Greenwood et al., pp. 403, 405, figures 9.7, 9.9, and 9.10, Copyright Elsevier (1984).)

Si═Si bonds have been synthesized, but there is no promise of a chemistry of multiply bonded Si species comparable at all in diversity with the chemistry of unsaturated organic compounds. Germanes of formulas GeH_4 to Ge_5H_{12} have been made, as have SnH_4 (stannane), Sn_2H_6, and small amounts of the highly reactive PbH_4 (plumbane) and other group 14 hydrides,[60] but the chemistry in these cases is even more limited than of the silanes.

Why are the silanes and other analogous compounds less stable (more reactive) than the corresponding hydrocarbons? First, the Si—Si bond is weaker than the C—C bond (approximate bond energies, 340 and 368 kJ mol^{-1}, respectively), and Si—H bonds are weaker than C—H bonds (393 versus 435 kJ mol^{-1}). Silicon is less electronegative (1.92) than hydrogen (2.30) and is, therefore, more susceptible to nucleophilic attack; this is in contrast to carbon, which is more electronegative (2.54) than hydrogen. Silicon atoms are also larger and therefore provide greater surface area for attack by nucleophiles. In addition, silicon atoms have low-lying *d* orbitals that can act as acceptors of electron pairs from nucleophiles. Similar arguments can be used to describe the high reactivity of germanes, stannanes, and plumbanes. Silanes are believed to decompose by elimination of $:SiH_2$ via a transition state having a bridging hydrogen (**Figure 8.28**). This reaction can be used to prepare silicon of extremely high purity.

FIGURE 8.28 Decomposition of Silanes.

$$H_3Si{-}SiH_3 \longrightarrow H_3Si{\cdots}(H){\cdots}SiH_2 \longrightarrow SiH_4 + :SiH_2 \qquad :SiH_2 \longrightarrow Si + H_2$$

Elemental silicon has the diamond structure. Silicon carbide, SiC, occurs in many crystalline forms, some based on the diamond structure and some on the wurtzite structure (Figures 7.6 and 7.8b). Carborundum, a form of silicon carbide, is used as an abrasive, with a hardness nearly as great as diamond and a low chemical reactivity. SiC has also garnered interest as a high-temperature semiconductor.

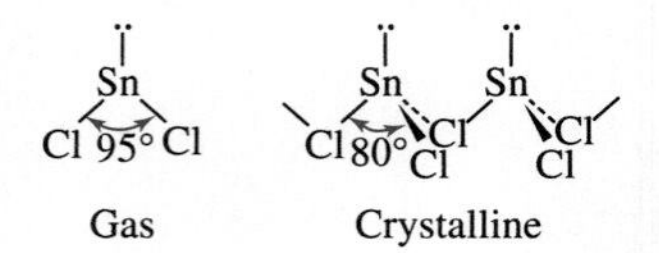

The elements germanium, tin, and lead show increasing importance of the 2+ oxidation state, an example of the inert pair effect. For example, all three show two sets of halides, of formula MX_4 and MX_2. For germanium, the most stable halides have the formula GeX_4; for lead, it is PbX_2. For the dihalides, the metal exhibits a stereochemically active lone pair. This leads to bent geometry for the gaseous molecules and to crystalline structures in which the lone pair is evident, as shown for $SnCl_2$ in the margin.

8.7 Group 15

Nitrogen is the most abundant component of Earth's atmosphere (78.1% by volume). However, the element was not isolated until 1772, when Rutherford, Cavendish, and Scheele removed oxygen and carbon dioxide from air. Phosphorus was first isolated from urine by Brandt in 1669. Because the element glowed in the dark on exposure to air, it was named after the Greek *phos*, "light," and *phoros*, "bringing." The three heaviest Group 15 elements* were isolated before nitrogen and phosphorus. The dates of discovery of arsenic, antimony, and bismuth are unknown, but alchemists studied these elements before the fifteenth century. The Group 15 elements span the range from nonmetallic (nitrogen and phosphorus) to metallic (bismuth) behavior, with the elements in between (arsenic and antimony) having intermediate properties. Selected physical properties are in **Table 8.9**.

TABLE 8.9 Properties of the Group 15 Elements

Element	Ionization Energy ($kJ\ mol^{-1}$)	Electron Affinity ($kJ\ mol^{-1}$)	Melting Point (°C)	Boiling Point (°C)	Electro-negativity
N	1402	−7	−210	−195.8	3.066
P	1012	72	44[a]	280.5	2.053
As	947	78	[b]	[b]	2.211
Sb	834	103	631	1587	1.984
Bi	703	91	271	1564	2.01[c]

Sources: Data from: Ionization energies cited in this chapter are from C. E. Moore, *Ionization Potentials and Ionization Limits Derived from the Analyses of Optical Spectra,* National Standard Reference Data Series, U.S. National Bureau of Standards, NSRDS-NBS 34, Washington, DC, 1970, unless noted otherwise. Electron affinity values listed in this chapter are from H. Hotop and W. C. Lineberger, *J. Phys. Chem. Ref. Data,* **1985,** *14,* 731. Standard electrode potentials listed in this chapter are from A. J. Bard, R. Parsons, and J. Jordan, eds., *Standard Potentials in Aqueous Solutions,* Marcel Dekker (for IUPAC), New York, 1985. Electronegativities cited in this chapter are from J. B. Mann, T. L. Meek, and L. C. Allen, *J. Am. Chem. Soc.,* **2000,** *122,* 2780, Table 2. Other data are from N. N. Greenwood and A. Earnshaw, *Chemistry of the Elements,* Pergamon Press, Elmsford, NY, 1984, except where noted.

[a]α-P_4; [b]Sublimes at 615° C; [c]Approximate value

*Elements in this group are sometimes called the *pnicogens* or *pnictogens*.

8.7.1 The Elements

Nitrogen, a colorless diatomic gas, features a very short (109.8 pm) nitrogen–nitrogen triple bond. The unusual stability of this bond is responsible for the low reactivity of N_2. Nitrogen is suitable as an inert atmosphere for reactions that are oxygen or moisture sensitive. Liquid nitrogen, at 77 K, is an inexpensive coolant for studying reactions, trapping solvent vapors, and cooling superconducting magnets (for preserving the liquid helium coolant, which boils at 4 K).

Phosphorus has many allotropes, with white phosphorus, which exists in two modifications, α-P_4 (cubic), and β-P_4 (hexagonal), the most common. Condensation of gaseous or liquid phosphorus, both of which contain tetrahedral P_4 molecules, gives primarily the α form, which slowly converts to the β form above –76.9° C. During oxidation in air, α-P_4 emits a yellow-green light, an example of phosphorescence known since antiquity; white phosphorus is commonly stored underwater to slow its oxidation.

Heating white phosphorus in the absence of air gives red phosphorus, an amorphous, polymeric material. Black phosphorus is the most thermodynamically stable form, obtained by heating white phosphorus at very high pressures. Black phosphorus converts to other forms at still higher pressures. Examples of these structures are shown in **Figure 8.29**. Detailed information on allotropes of phosphorus is available.[61]

Phosphorus exists as tetrahedral P_4 molecules in the liquid and gas phases. At very high temperatures, P_4 dissociates into P_2. At approximately 1800 °C, this dissociation reaches 50 percent.

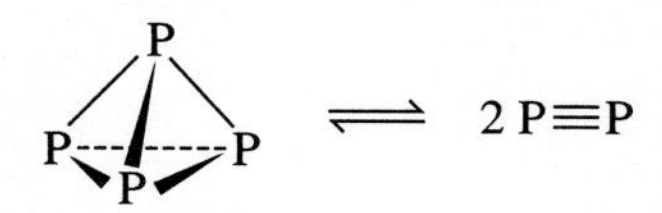

White phosphorus (P_4) is readily accessible industrially from phosphate minerals, hence the high demand for phosphate rock (Table 8.1). White phosphorus is chlorinated or oxygenated to afford P(III) and P(V) molecules (for example, PCl_3, PCl_5, $POCl_3$). These are starting materials for high-demand phosphorus compounds (for example, phosphoric acid (Section 8.7.2) and phosphines). A vigorous research area motivated by sustainability concerns is to directly incorporate phosphorus atoms from white phosphorus into these desired compounds without the need to synthesize chlorinated or oxygenated intermediates.[62] In this regard, "P_4 activation" is being actively pursued.[63]

Arsenic, antimony, and bismuth also exhibit allotropes. The most stable allotrope of arsenic is the gray (α) form, which is similar to the rhombohedral form of phosphorus.

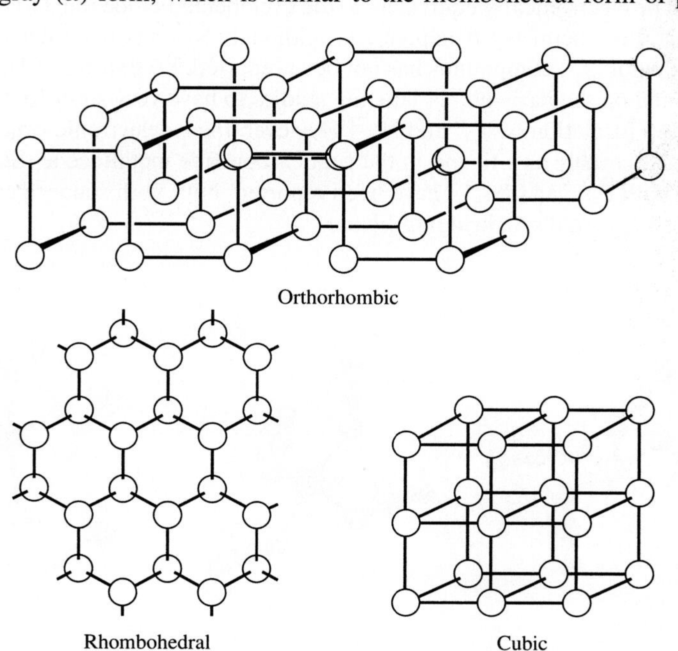

FIGURE 8.29 Allotropes of Phosphorus. (Published in "Chemistry of the Elements", Greenwood et al., Greenwood et al, p. 558, Copyright Elsevier (1984).)

In the vapor phase, arsenic, like phosphorus, exists as tetrahedral As_4. Antimony and bismuth also have similar α forms. These three elements have a somewhat metallic appearance but are only moderately good conductors. Arsenic is the best conductor in this group but has an electrical resistivity nearly 20 times as great as copper. Bismuth is the heaviest element to have a nonradioactive nucleus; polonium and all heavier elements are radioactive.

Solution phase syntheses and isolation of AsP_3 and SbP_3, tetrahedra similar to P_4, were reported in 2009.[64] These interpnictides are potentially useful reagents in materials science as stoichiometric 1:3 sources of Group 15 elements.

Ions

For more than a century, the only isolable species containing exclusively nitrogen were N_2, nitride (N^{3-}) and azide (N_3^-). Nitrides of primarily ionic character are formed by lithium and the Group 2 elements. The synthesis of *sodium* nitride, Na_3N was a longstanding challenge. This explosive compound was not successfully prepared until 2002, from atomic beams of sodium and nitrogen at liquid nitrogen temperatures.[65] Nitride is a strong pi-donor ligand toward transition metals (Chapter 10).

Ionic Group 1 and 2 metal compounds of linear N_3^- are known. Many covalent azide compounds are also known. For example, the heavier Group 15 elements form structurally characterized anions $[E(N_3)_4]^-$ (E = As, Sb) and $[Bi(N_3)_5(DMSO)]^-$ (**Figure 8.30**).[66]

Although C_2^{2-} and O_2^{2-} were long known, N_2^{2-} (diazenide) was not characterized until 2001.[67] In SrN_2, the N—N bond distance is 122.4 pm, comparable to 120.8 pm in the isoelectronic O_2 molecule and much longer than the 109.8 pm in N_2. The first alkali metal diazenide, Li_2N_2, was isolated upon treatment of LiN_3 at 750 K and approximately 89,000 atm![68]

A new species, N_5^+, was reported in 1999:

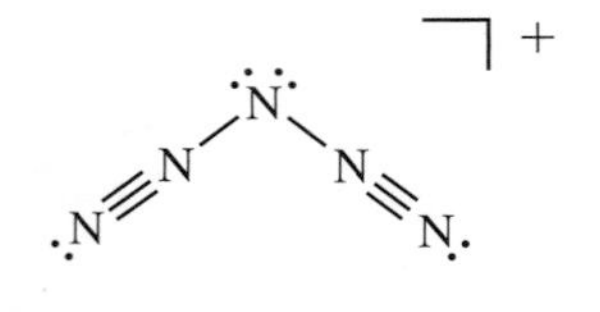

$$N_2F^+[AsF_6]^- + HN_3 \longrightarrow N_5^+[AsF_6]^- + HF$$

$N_5^+[AsF_6]^-$ is not stable at room temperature but can be preserved for weeks at –78 °C. The N_5^+ ion has a V-shaped structure, bent at the central nitrogen and linear at the neighboring atoms.[69]

There has been considerable interest in synthesizing the pentazolate ion, *cyclo*-N_5^-, which would be electronically equivalent to the cyclopentadienide ion, $C_5H_5^-$, a key ligand in organometallic chemistry (Chapter 13). Although N_5^- has been detected in the gas phase, isolation of N_5^- compounds has not been reported.[70] Mixtures of HN_5/N_5^- in solution were found on the basis of ^{15}N labeling studies to have very short lifetimes at –40 °C with respect to formation of N_3^- and N_2.[71] However, the isoelectronic *cyclo*-P_5^- has been prepared, most notably as a ligand in the first carbon-free metallocene (Section 15.2.2).[72] In addition, both P_4^{2-} and As_4^{2-} have been reported, both with square geometries,[73] and N_4^{4-} has been reported as a bridging ligand.[74]

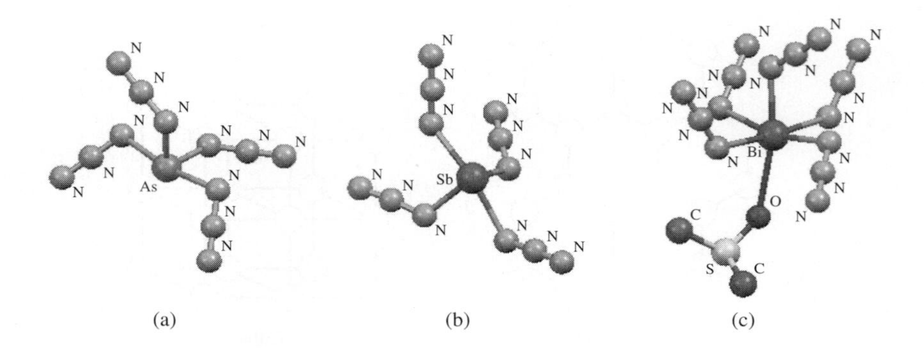

FIGURE 8.30 Molecular structures of (a) $[As(N_3)_4]^-$, (b) $[Sb(N_3)_4]^-$, and (c) $[Bi(N_3)_5(DMSO)]^-$ (DMSO = Dimethylsulfoxide). (Molecular structures created from CIF data with hydrogen atoms omitted for clarity.)

Although phosphides, arsenides, and other Group 15 compounds are known with formulas that suggest that they are ionic (e.g., Na_3P, Ca_3As_2), such compounds are generally lustrous and have good thermal and electrical conductivity, properties more consistent with metallic bonding than with ionic bonding.

8.7.2 Compounds

Hydrides

In addition to ammonia, nitrogen forms the hydrides N_2H_4 (hydrazine), N_2H_2 (diazene or diimide), and HN_3 (hydrazoic acid) (**Figure 8.31**).

Ammonia is of immense industrial importance. More than 80 percent of the ammonia produced is used in fertilizers, with additional uses that include the synthesis of explosives, synthetic fibers—such as rayon, nylon, and polyurethanes—and a wide variety of compounds. Liquid ammonia is used as a nonaqueous ionizing solvent.

In nature, ammonia is produced by the action of nitrogen-fixing bacteria on atmospheric N_2 under very mild conditions (room temperature and 0.8 atm N_2 pressure). These bacteria contain nitrogenases, iron- and molybdenum-containing enzymes that catalyze the formation of NH_3. Industrially, ammonia is synthesized from its elements by the Haber–Bosch process, which typically uses finely divided iron as catalyst, enhanced in activity by addition of potassium salts. Even with a catalyst, this process is far more difficult than the nitrogenase-catalyzed route in bacteria; typically, temperatures above 380 °C and pressures of approximately 200 atm are necessary. Fritz Haber won the 1918 Nobel Prize for this discovery and is credited both with making commercial fertilizers possible and for helping Germany in World War I to replace imported nitrates used in explosives. The N_2 for this process is obtained from fractional distillation of liquid air and H_2 is obtained from hydrocarbons (Section 8.2.1).

$$N_2 + 3\,H_2 \longrightarrow 2\,NH_3$$

On the basis of the high global energy consumption required to produce NH_3, there is great interest of synthesizing NH_3 from its parent elements under milder conditions. The production of N_2-derived ammonia via an iron complex, a potassium reductant, and H_2, reported by Holland in 2011, is a seminal achievement.[75]

Oxidation of hydrazine is highly exothermic:

$$N_2H_4 + O_2 \longrightarrow N_2 + 2\,H_2O \qquad \Delta H^\circ = -622\ \text{kJ mol}^{-1}$$

The major use of hydrazine and its methyl derivatives is in rocket fuels. Hydrazine is a versatile reducing agent in acidic (as the protonated hydrazonium ion, $N_2H_5^+$, see Latimer and Frost diagrams in Section 8.1.4) and basic solutions.

The *cis* and *trans* isomers of diazene are stable only at very low temperatures. The fluoro derivatives, N_2F_2, are more stable. Both isomers of N_2F_2 show N—N distances

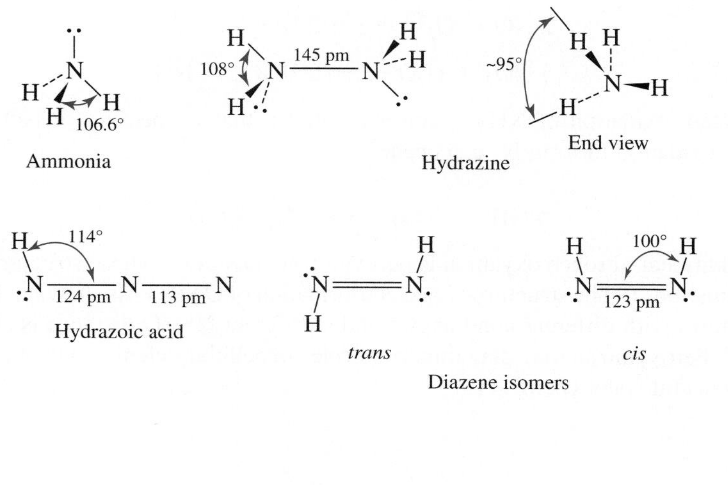

106.6°
Ammonia
108°
145 pm
~95°
End view
Hydrazine
114°
124 pm
113 pm
Hydrazoic acid
100°
123 pm
trans
cis
Diazene isomers

FIGURE 8.31 Nitrogen Hydrides. Multiple bonds in hydrazoic acid are not shown. (Data from Bond angles and distances are from A. F. Wells, *Structural Inorganic Chemistry*, 5th ed., Oxford University Press, New York, 1984.)

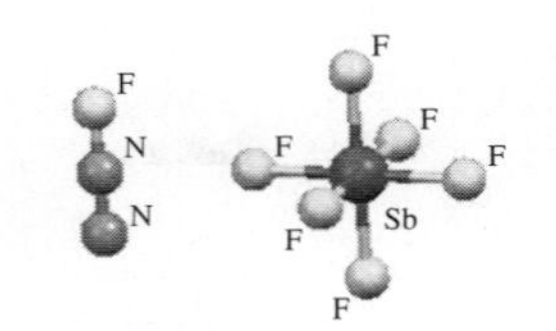

FIGURE 8.32 Molecular structure of $[N_2F][SbF_6]$ generated from CIF data.

consistent with double bonds (*cis*, 120.9 pm; *trans*, 122.4 pm). Improved syntheses of *cis*-N_2F_2 and *trans*-N_2F_2, and a rigorous examination of the *cis–trans* isomerization mechanism, were reported in 2010.[76] In a technically challenging reaction, *cis*-N_2F_2 reacts with SbF_5 in anhydrous HF to afford $[N_2F][SbF_6]$ (**Figure 8.32**). The average N—N distance of 110 pm in N_2F^+ compounds is nearly identical to the triple bond distance in N_2, 109.8 pm.

Phosphine, PH_3, is a highly poisonous gas. It has significantly weaker intermolecular attractions than NH_3 in the solid state; its melting and boiling points are much lower than those of ammonia (–133.5 °C and –87.5 °C for PH_3 versus –77.8 °C and –34.5 °C for NH_3). Phosphines (PR_3 (R = H, alkyl, or aryl)), phosphites ($P(OR)_3$), arsines AsR_3, and stibines, SbR_3, are important ligands in coordination and organometallic chemistry.

Nitrogen Oxides and Oxyions

Nitrogen oxides and ions containing nitrogen and oxygen are summarized in **Table 8.10**. Nitrous oxide, N_2O, is used as a dental anesthetic and an aerosol propellant. On atmospheric decomposition, N_2O yields its innocuous parent gases and is an environmentally acceptable substitute for chlorofluorocarbons. On the other hand, N_2O contributes to the greenhouse effect and its atmospheric concentration is increasing. The behavior of nitrous oxide as a ligand has been reviewed.[77] Nitric oxide, NO, is also a ligand (Chapter 13) and has many biological functions.

$$N_2O_4(g) \rightleftharpoons 2\,NO_2(g)$$

$$\Delta H^\circ = 57.20 \text{ kJ mol}^{-1}$$

The gases N_2O_4 and NO_2 form an interesting pair. At ordinary temperatures and pressures, both exist in significant amounts in equilibrium. Colorless, diamagnetic N_2O_4 has a weak N—N bond that can readily dissociate to give the brown, paramagnetic NO_2.

Nitric oxide is formed in the combustion of fossil fuels and is present in automobile and power plant exhausts; it can also be formed from the action of lightning on atmospheric N_2 and O_2. In the atmosphere, NO is oxidized to NO_2. These gases, often collectively designated NO_x, contribute to acid rain, primarily because NO_2 reacts with atmospheric water to form nitric acid. Nitrogen oxides are also believed to be instrumental in the destruction of the Earth's ozone layer (Section 8.8.1).

$$3\,NO_2 + H_2O \longrightarrow 2\,HNO_3 + NO$$

Nitric acid is of immense industrial importance, especially in the synthesis of ammonium nitrate, a salt used primarily as a fertilizer. NH_4NO_3 is thermally unstable and undergoes violently exothermic decomposition at elevated temperature. Its use in commercial explosives is now second in importance to its use as a fertilizer.

$$2\,NH_4NO_3 \longrightarrow 2\,N_2 + O_2 + 4\,H_2O$$

Nitric acid is synthesized commercially via two nitrogen oxides. First, ammonia is reacted with oxygen using a platinum-rhodium gauze catalyst to form nitric oxide, NO:

$$4\,NH_3 + 5\,O_2 \longrightarrow 4\,NO + 6\,H_2O$$

The nitric oxide is then oxidized by air and water:

$$2\,NO + O_2 \longrightarrow 2\,NO_2$$

$$3\,NO_2 + H_2O \longrightarrow 2\,HNO_3 + NO$$

The first step, oxidation of NH_3, requires a catalyst that is specific for NO generation; otherwise, oxidation to form N_2 can occur:

$$4\,NH_3 + 3\,O_2 \longrightarrow 2\,N_2 + 6\,H_2O$$

An additional nitrogen oxyanion is peroxynitrite, $ONOO^-$, whose structure has different conformations.[78] The structure of one conformation of $ONOO^-$ is shown in **Figure 8.33**; a twisted form with different bond angles and a different N—O distance is also found in the crystal. Peroxynitrite may play important roles in cellular defense against infection and in environmental water chemistry.[79]

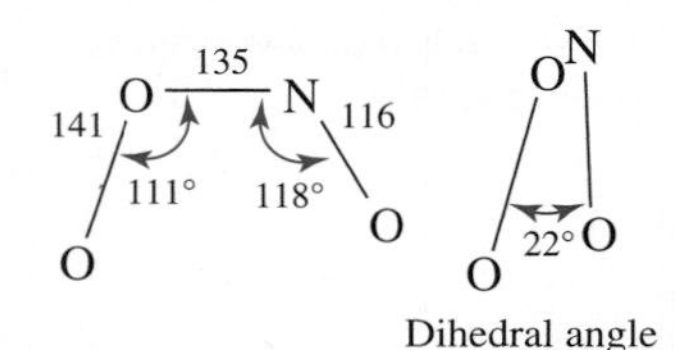

FIGURE 8.33 Peroxynitrite Structure.

TABLE 8.10 Compounds and Ions Containing Nitrogen and Oxygen

Formula	Name	Structure[a]	Notes
N_2O	Nitrous oxide	N≡N≡O (113, 118)	mp = −90.9 °C; bp = −88.5 °C
NO	Nitric oxide	N≡O (115)	mp = −163.6 °C; bp = −151.8 °C; bond order approximately 2.5; paramagnetic
NO_2	Nitrogen dioxide	O–N–O (119; 134°)	Brown, paramagnetic gas; exists in equilibrium with N_2O_4; $2\ NO_2 \rightleftharpoons N_2O_4$
N_2O_3	Dinitrogen trioxide	O=N–N(=O)O (114, 186, 120, 122; 105°, 130°, 117°)	mp = −100.1 °C; dissociates above melting point: $N_2O_3 \rightleftharpoons NO + NO_2$
N_2O_4	Dinitrogen tetroxide	O₂N–NO₂ (175, 121; 135°)	mp = −11.2 °C; bp = −21.15 °C; dissociates into 2 NO_2 [ΔH(dissociation) = 57 kJ/mol]
N_2O_5	Dinitrogen pentoxide	O₂N–O–NO₂	N—O—N bond may be bent; consists of $NO_2^+NO_3^-$ in the solid
NO^+	Nitrosonium or nitrosyl	N≡O (106)	Isoelectronic with CO
NO_2^+	Nitronium or nitryl	O=N=O (115)	Isoelectronic with CO_2
NO_2^-	Nitrite	O–N–O	N—O distance varies from 113 to 123 pm, and bond angle varies from 116° to 132° depending on cation; versatile ligand (see Chapter 9)
NO_3^-	Nitrate	N(O)₃ (122; 120°)	Forms compounds with nearly all metals; as ligand, has a variety of coordination modes
$N_2O_2^{2-}$	Hyponitrite	O–N=N–O	Reducing agent
NO_4^{3-}	Orthonitrate	N(O)₄ (139)	Na and K salts known; decomposes in presence of H_2O and CO_2
HNO_2	Nitrous acid	H–O–N=O (143, 118; 102°, 111°)	Weak acid (pK_a = 3.3 at 25 °C); disproportionates: $3\ HNO_2 \rightleftharpoons H_3O^+ + 2\ NO + NO_3^-$ in aqueous solution
HNO_3	Nitric acid	H–O–NO₂ (141, 121; 102°, 130°, 114°)	Strong acid in aqueous solution; concentrated aqueous solutions are strong oxidizing agents

Source: Data from N. N. Greenwood and A. Earnshaw, *Chemistry of the Elements*, Pergamon Press, Elmsford, NY, 1984, pp. 508–545.

[a]Distances in pm

Nitrogen exhibits rich redox chemistry in aqueous solution, as indicated by the Latimer and Frost diagrams in Section 8.1.4.

EXERCISE 8.6

Use the Latimer and Frost diagrams for nitrogen in acidic solution (Section 8.1.4) to find the potential for $HNO_2 \longrightarrow N_2H_5^+$.

EXERCISE 8.7

Show whether the decomposition of NH_4NO_3 ($2\ NH_4NO_3 \longrightarrow 2\ N_2 + O_2 + 4\ H_2O$) can be a spontaneous reaction, based on the potentials given in Appendix B.7, by determining both E^o and ΔG^o.

$P_4 + 5\ O_2 \longrightarrow P_4O_{10}$

$P_4O_{10} + 6\ H_2O \longrightarrow 4\ H_3PO_4$

Among acids, phosphoric acid, H_3PO_4, is second only to sulfuric acid in industrial production. A common preparation method involves the combustion of molten phosphorus, sprayed into a mixture of air and steam in a stainless steel chamber. The P_4O_{10} intermediate is converted into H_3PO_4. Phosphoric acid is also synthesized by treating phosphate minerals with sulfuric acid:

$$Ca_3(PO_4)_2 + 3\ H_2SO_4 \longrightarrow 2\ H_3PO_4 + 3\ CaSO_4$$

8.8 Group 16

The group 16 elements, or *chalcogens*, span the range from nonmetallic (oxygen, sulfur, and selenium) to metallic (polonium) behavior, with tellurium having intermediate properties. Selected physical properties are in **Table 8.11**.

8.8.1 The Elements

The first two group 16 elements are familiar as O_2, the colorless gas that comprises about 21 percent of Earth's atmosphere, and sulfur, a nonmetallic yellow solid. The third element, selenium, is important in the xerography process. A brilliant red formed by a combination of CdS and CdSe is used in colored glasses. Although elemental selenium is highly poisonous, trace amounts are essential for life. Tellurium is used in small amounts in metal alloys, tinting of glass, and catalysts in the rubber industry. All isotopes of polonium are radioactive. Its highly exothermic radioactive decay has made it a power source for satellites.

TABLE 8.11 Properties of the Group 16 Elements

Element	Ionization Energy (kJ mol^{-1})	Electron Affinity (kJ mol^{-1})	Melting Point (°C)	Boiling Point (°)	Electronegativity
O	1314	141	−218.8	−183.0	3.610
S	1000	200	112.8	444.7	2.589
Se	941	195	217	685	2.424
Te	869	190	452	990	2.158
Po	812	180[a]	250[a]	962	2.19[a]

Source: Data from: Ionization energies cited in this chapter are from C. E. Moore, *Ionization Potentials and Ionization Limits Derived from the Analyses of Optical Spectra,* National Standard Reference Data Series, U.S. National Bureau of Standards, NSRDS-NBS 34, Washington, DC, 1970, unless noted otherwise. Electron affinity values listed in this chapter are from H. Hotop and W. C. Lineberger, *J. Phys. Chem. Ref. Data,* **1985,** *14,* 731. Standard electrode potentials listed in this chapter are from A. J. Bard, R. Parsons, and J. Jordan, eds., *Standard Potentials in Aqueous Solutions,* Marcel Dekker (for IUPAC), New York, 1985. Electronegativities cited in this chapter are from J. B. Mann, T. L. Meek, and L. C. Allen, *J. Am. Chem. Soc.,* **2000,** *122,* 2780, Table 2. Other data are from N. N. Greenwood and A. Earnshaw, *Chemistry of the Elements,* Pergamon Press, Elmsford, NY, 1984, except where noted.

[a]Approximate value

Sulfur, which occurs as the free element in numerous natural deposits, has been known since prehistoric times; it is the "brimstone" of the Bible. Sulfur was of considerable interest to alchemists and, following the development of gunpowder (a mixture of sulfur, KNO_3, and powdered charcoal) in the thirteenth century, to military leaders as well. Although oxygen is widespread in Earth's atmosphere and, combined with other elements, in the Earth's crust (which contains 46 percent oxygen by mass) and in bodies of water, pure O_2 was not isolated and characterized until the 1770s by Scheele and Priestley. Priestley's classic synthesis of oxygen by heating HgO with sunlight focused by a magnifying glass is a landmark in experimental chemistry. Selenium (1817) and tellurium (1782) were soon discovered and, because of their chemical similarities, were named after the moon (Greek, *selene*) and Earth (Latin, *tellus*). Marie Curie discovered polonium in 1898; like radium, it was isolated in trace amounts from tons of uranium ore. Important physical properties of these elements are summarized in Table 8.11.

Oxygen

Oxygen exists primarily in the diatomic form O_2, but traces of ozone, O_3, are found in the upper atmosphere and in the vicinity of electrical discharges. O_2 is paramagnetic and O_3 is diamagnetic. The paramagnetism of O_2 is the consequence of two electrons with parallel spin occupying $\pi^*(2p)$ orbitals (Section 5.2.3). In addition, the two known excited states of O_2 have π^* electrons of opposite spin and are higher in energy as a consequence of the effects of pairing energy and exchange energy (see Section 2.2.3):

		Relative Energy (kJ mol^{-1})
Excited states:	$\uparrow\downarrow$ __ __	157.85
	$\uparrow$ __ $\downarrow$ __	94.72
Ground state:	$\uparrow$ __ $\uparrow$ __	0

These electronic excited states of O_2 can be achieved when photons are absorbed in the liquid phase during molecular collisions; under these conditions, a single photon can simultaneously excite two colliding molecules. These absorptions occur in the visible region of the spectrum, at 631 and 474 nm, and gives rise to the blue color of the liquid.[80] These excited states are important in many oxidation processes. Of course, O_2 is essential for respiration; hemoglobin plays a vital role in facilitating oxygen transport to cells.

Although molecular oxygen phases have been studied for many years, the molecular structure of a high-pressure phase, ε-oxygen, which consists of tetrameric $(O_2)_4$, or O_8 molecules, was not determined until 2006.[81] A dark red allotrope observed at pressures between 10 and 96 GPa, O_8 consists of collections of parallel O_2 units (**Figure 8.34a**). The O_2 units have bond distances of 120(3) pm, comparable to the oxygen–oxygen distance in free O_2, 120.8 pm; the bond order within each O_2 molecule is not changed significantly upon formation of this phase. The oxygen–oxygen distances between O_2 units are much longer (218(1) pm).

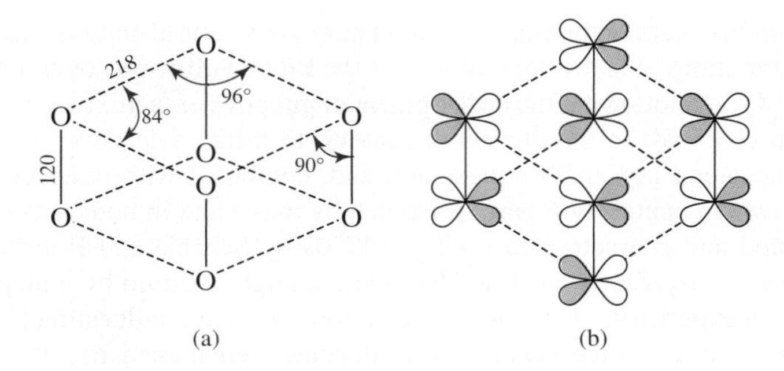

FIGURE 8.34 (a) Dimensions of an O_8 unit in ε-O_2. (b) Hypothesized orbital interactions involving π^* orbitals of O_2 in O_8.

The bonding between O_2 units in ε-oxygen can be described as involving the π^* orbitals of O_2. These orbitals in free O_2 are singly occupied, rendering the molecules paramagnetic. When four O_2 molecules combine to form a unit of O_8, two singly occupied π^* orbitals from each O_2 overlap, generating four bonding and four antibonding orbitals. **Figure 8.34b** shows one of these orbitals. The eight available electrons occupy the bonding orbitals, pairing all these electrons, consistent with the diamagnetism observed for O_8.[82]

Ozone absorbs ultraviolet radiation below 320 nm, forming a shield in the upper atmosphere that protects Earth's surface from hazardous effects of high-energy electromagnetic radiation. There is increasing concern because atmospheric pollutants are depleting the ozone layer, with the most serious depletion over Antarctica as a result of seasonal variations in high-altitude air circulation. In the upper atmosphere, ozone is formed from O_2:

$$O_2 \xrightarrow{h\nu} 2\,O \qquad \lambda \le 242 \text{ nm}$$

$$O + O_2 \longrightarrow O_3$$

Absorption of ultraviolet radiation by O_3 causes it to decompose to O_2. Therefore, a steady-state concentration of ozone is achieved in the upper atmosphere, a concentration ordinarily sufficient to provide significant ultraviolet protection of Earth's surface. However, some pollutants catalyze ozone decomposition. Examples include nitrogen oxides from high-flying aircraft (nitrogen oxides are also produced in trace amounts naturally) and chlorine atoms from photolytic decomposition of chlorofluorocarbons from aerosols and refrigerants. The processes governing the concentration of ozone in the atmosphere are extremely complex. The following reactions are examples of the processes believed to be involved in the atmosphere. Although these reactions can be carried out in the laboratory, computational chemistry is often used to model reactions in the atmosphere and propose alternate pathways for ozone depletion.

$$NO_2 + O_3 \longrightarrow NO_3 + O_2$$

$$NO_3 \longrightarrow NO + O_2$$

$$NO + O_3 \longrightarrow NO_2 + O_2$$

$$\text{Net:} \quad 2\,O_3 \longrightarrow 3\,O_2$$

$$Cl + O_3 \longrightarrow ClO + O_2 \qquad \leftarrow \text{Cl formed from photodecomposition of chlorofluorocarbons}$$

$$ClO + O \longrightarrow Cl + O_2$$

$$\text{Net:} \quad O_3 + O \longrightarrow 2\,O_2$$

Ozone is a more potent oxidizing agent than O_2; in acidic solution, ozone is exceeded only by fluorine among the elements as an oxidizing agent. Several diatomic and triatomic oxygen ions are known (**Table 8.12**).

TABLE 8.12 Neutral and Ionic O_2 and O_3 Species

Formula	Name	O—O Distance (pm)	Notes
O_2^+	Dioxygenyl	111.6	Bond order 2.5
O_2	Dioxygen	120.8	Coordinates to transition metals; singlet O_2 (excited state) important in photochemical reactions; oxidizing agent
O_2^-	Superoxide	135	Moderate oxidizing agent; most stable compounds are KO_2, RbO_2, CsO_2
O_2^{2-}	Peroxide	149	Forms ionic compounds with alkali metals, Ca, Sr, Ba; strong oxidizing agent
O_3	Ozone	127.8	Bond angle 116.8°; strong oxidizing agent; absorbs in UV (below 320 nm)
O_3^-	Ozonide	134	Formed from reaction of O_3 with dry alkali metal hydroxides, decomposes to O_2^-

Source: Data from N. N. Greenwood and A. Earnshaw, *Chemistry of the Elements*, Pergamon Press, Elmsford, NY, 1984; K. P. Huber and G. Herzberg, *Molecular Spectra and Molecular Structure. IV. Constants of Diatomic Molecules*, Nostrand Reinhold Company, New York, 1979.

Sulfur

More allotropes are known for sulfur than for any other element, with the most stable form at room temperature (orthorhombic, α-S_8) having eight sulfur atoms arranged in a puckered ring. **Figure 8.35** shows three of the most common sulfur allotropes.[83]

Heating sulfur results in interesting viscosity changes. At approximately 119 °C, sulfur melts to give a yellow liquid, whose viscosity gradually decreases until approximately 155 °C (**Figure 8.36**). Further heating causes the viscosity to increase dramatically. Above 159 °C the liquid pours very sluggishly. Above about 200 °C, the viscosity slowly decreases, with the liquid eventually acquiring a reddish hue at higher temperatures.[84]

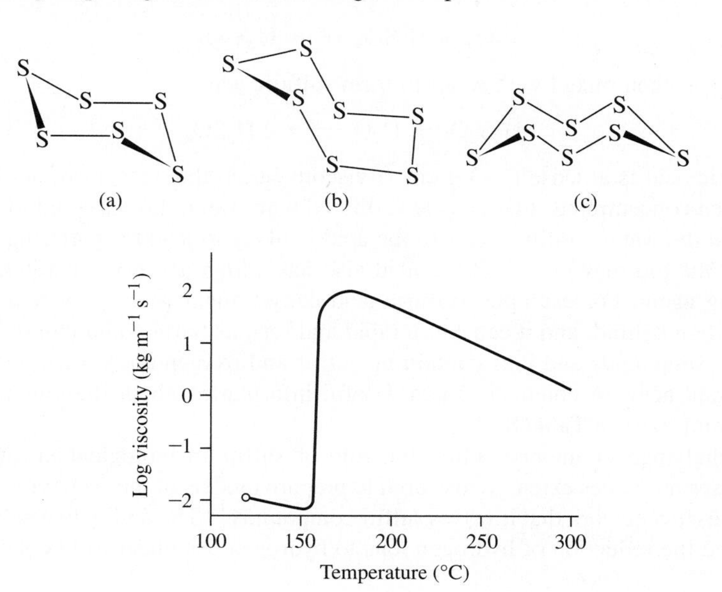

FIGURE 8.35 Structures of Common Allotropes of Sulfur. (a) S_6 (b) S_7 (c) α-S_8

FIGURE 8.36 The Viscosity of Sulfur.

The explanation of these viscosity changes involves the tendency of S—S bonds to break and re-form at high temperatures. Above 159 °C, the S_8 rings begin to open; the resulting S_8 chains can react with other S_8 rings to open them and form S_{16} chains, S_{24} chains, and so on:

$$S_8 \text{ (ring)} \longrightarrow \text{S–S–S–S–S–S–S–S (chain)} \longrightarrow S_{16} \longrightarrow S_{24} \longrightarrow \cdots$$

The longer the chains, the more the chains can intertwine with each other, and the greater the viscosity. Large rings can also form by the linking of ends of chains. Chains exceeding 200,000 sulfur atoms are formed at the temperature of maximum viscosity, near 180 °C. At higher temperatures, thermal breaking of sulfur chains occurs more rapidly than propagation of chains, and the average chain length decreases, accompanied by a decrease in viscosity. At very high temperatures, brightly colored species such as S_3 increase in abundance, and the liquid takes on a reddish coloration. When molten sulfur is poured into cold water, it forms a rubbery solid that can be molded readily. However, this form eventually converts to the yellow crystalline α form, the most thermodynamically stable allotrope, which consists again of the S_8 rings.

Sulfuric acid has been manufactured for approximately 400 years. The modern process for producing H_2SO_4 begins with the synthesis of SO_2, either by combustion of sulfur or by roasting (heating in the presence of oxygen) of sulfide minerals:

$$S + O_2 \longrightarrow SO_2 \qquad \text{Combustion of sulfur}$$

$$M_xS_y + O_2 \longrightarrow y\,SO_2 + M_xO_y \qquad \text{Roasting of sulfide ore}$$

SO_2 is then converted to SO_3 by the exothermic reaction

$$2\,SO_2 + O_2 \longrightarrow 2\,SO_3$$

using V_2O_5 or another suitable catalyst in a multiple-stage catalytic converter (multiple stages are necessary to achieve high yields of SO_3). The SO_3 then reacts with water to form sulfuric acid:

$$SO_3 + H_2O \longrightarrow H_2SO_4$$

If SO_3 is passed directly into water, a fine aerosol of H_2SO_4 droplets is formed. To avoid this, the SO_3 is absorbed into 98 percent H_2SO_4 solution to form disulfuric acid, $H_2S_2O_7$ (oleum):

$$SO_3 + H_2SO_4 \longrightarrow H_2S_2O_7$$

The $H_2S_2O_7$ is then mixed with water to form sulfuric acid:

$$H_2S_2O_7 + H_2O \longrightarrow 2\,H_2SO_4$$

Sulfuric acid is a dense (1.83 g cm^{-3}) viscous liquid that reacts exothermically with water. When concentrated sulfuric acid is diluted with water, it is essential to add the acid carefully to the water; adding water to the acid is likely to lead to spattering because the solution at the top may boil. Sulfuric acid also has a high affinity for water, it acts as a dehydrating agent. For example, sulfuric acid causes sugar to char by removing water, leaving carbon behind, and it can cause rapid and serious burns to human tissue.

Many compounds and ions containing sulfur and oxygen are known; many of these are important acids or conjugate bases. Useful information about these compounds and ions is summarized in **Table 8.13**.

The challenge of understanding the role of sulfur in biological nitrogenases and hydrogenases motivates extensive research to prepare models of these enzymes, and drives much of the current work that involves sulfur compounds.[85] The ability to catalyze nitrogen fixation and the reduction of hydrogen ions to hydrogen gas under mild conditions on an

TABLE 8.13 Molecules and Ions Containing Sulfur and Oxygen

Formula	Name	Structure[a]	Notes
SO_2	Sulfur dioxide	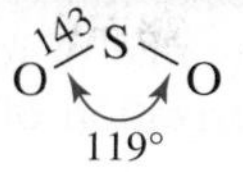 O–S–O; 143; 119°	mp = −75.5 °C; bp = −10.0 °C; colorless, choking gas; product of combustion of elemental sulfur
SO_3	Sulfur trioxide	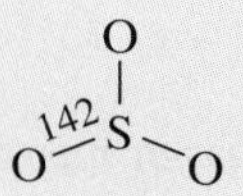 SO_3 (planar); 142	mp = −16.9 °C; bp = 44.6 °C; formed from oxidation of SO_2: $SO_2 + \frac{1}{2} O_2 \longrightarrow SO_3$; in equilibrium with trimer S_3O_9 in liquid and gas phases; reacts with water to form sulfuric acid
	Trimer	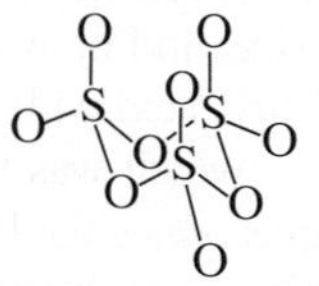 S_3O_9 (ring)	
SO_3^{2-}	Sulfite	 151; 106°	Conjugate base of HSO_3^-, formed when SO_2 dissolves in water
SO_4^{2-}	Sulfate	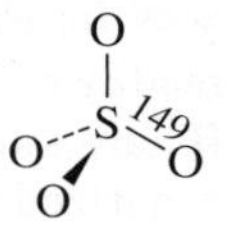 149	T_d symmetry, extremely common ion, used in gravimetric analysis
$S_2O_3^{2-}$	Thiosulfate	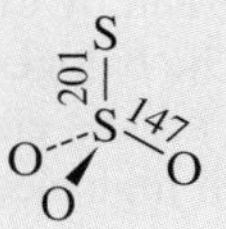 S–S 201; S–O 147	Moderate reducing agent, used in analytical determination of I_2: $I_2 + 2\ S_2O_3^{2-} \longrightarrow 2\ I^- + S_4O_6^{2-}$
$S_2O_4^{2-}$	Dithionite	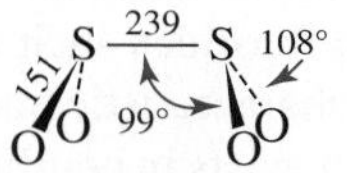 S–S 239; S–O 151; 108°; 99°	Very long S—S bond; dissociates into SO_2^-: $S_2O_4^{2-} \rightleftharpoons 2\ SO_2^-$; Zn and Na salts used as reducing agents
$S_2O_8^{2-}$	Peroxodisulfate	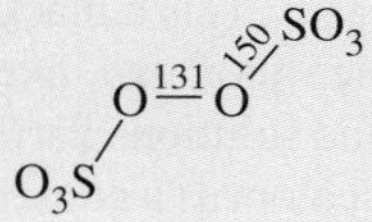 O_3S–O–O–SO_3; O–O 131; 150	Useful oxidizing agent, readily reduced to sulfate: $S_2O_8^{2-} + 2\ e^- \rightleftharpoons 2\ SO_4^{2-}$, $E° = 2.01$ V
H_2SO_4	Sulfuric acid	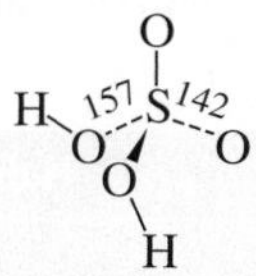 157; 142	C_2 symmetry; mp = 10.4 °C; bp = ~300 °C (dec); strong acid in aqueous solution; undergoes autoionization: $2\ H_2SO_4 \rightleftharpoons H_3SO_4^+ + HSO_4^-$, $pK = 3.57$ at 25 °C

Source: Data from N. N. Greenwood and A. Earnshaw, *Chemistry of the Elements*, Pergamon Press, Elmsford, NY, 1984, pp. 821–854.
[a]Distances in pm

industrial scale would save significant amounts of energy; these challenges are also at the forefront of modern inorganic chemistry research.

Other Elements

Selenium, a poisonous element, and tellurium also exist in a variety of allotropic forms, whereas polonium, a radioactive element, exists in two metallic allotropes. Selenium is a photoconductor—a poor conductor ordinarily, but a good conductor in the presence of light. It is used extensively in xerography, photoelectric cells, and semiconductor devices.

8.9 Group 17: The Halogens

Compounds containing the halogens (Greek, *halos* + *gen*, "salt former") have been used since antiquity, with the first use probably that of rock or sea salt (primarily NaCl) as a food preservative. Isolation and characterization of the neutral elements, however, has occurred comparatively recently.[86]

8.9.1 The Elements

Van Helmont first recognized chlorine as a gas in approximately 1630. Scheele conducted careful studies on chlorine in the 1770s (hydrochloric acid, which was used in these syntheses, had been prepared by alchemists around 900 CE). Iodine was obtained by Courtois in 1811 by subliming the product of the reaction of sulfuric acid with seaweed ash. Balard obtained bromine in 1826 by reacting chlorine with $MgBr_2$, which was present in saltwater marshes. Although hydrofluoric acid had been used to etch glass since the late seventeenth century, elemental fluorine was not isolated until 1886, when Moissan obtained this reactive gas by electrolysis of KHF_2 in anhydrous HF. Astatine, one of the last of the nontransuranium elements to be produced, was first synthesized in 1940 by Corson, Mackenzie, and Segre by bombardment of ^{209}Bi with alpha particles. All isotopes of astatine are radioactive (the longest lived isotope has a half-life of 8.1 hours) and, consequently, the chemistry of this element has been studied with difficulty.

All neutral halogens are diatomic and readily reduced to halide ions. All combine with hydrogen to form gases that, except for HF, are strong acids in aqueous solution. Some physical properties of the halogens are summarized in Table 8.14.

The chemistry of the halogens is governed by their tendency to acquire an electron to attain a noble gas electron configuration. The halogens are excellent oxidizing agents, with F_2 being the strongest oxidizing agent of all the elements. The tendency of the halogen atoms to attract electrons is reflected in their high electron affinities and electronegativities.

F_2 is extremely reactive and must be handled with special techniques; its standard synthesis is by electrolysis of molten fluorides such as KF. Gaseous Cl_2 is yellow and has an odor that is recognizable as the characteristic scent of "chlorine" bleach (an alkaline solution of hypochlorite, ClO^-, which exists in equilibrium with small amounts of Cl_2). Br_2 is a dark red liquid that evaporates easily. I_2 is a black, lustrous solid, readily sublimable at room temperature to produce a purple vapor, and, like the other halogens, highly soluble in nonpolar solvents. The color of iodine solutions varies significantly with the donor ability of the solvent, typically giving vivid colors as a consequence of charge-transfer interactions (Chapter 6). Iodine is a moderately good oxidizing agent but the weakest of the halogens.

TABLE 8.14 Properties of the Group 17 Elements: The Halogens[a]

				Halogen Molecules, X_2			
Element	Ionization Energy (kJ mol^{-1})	Electron Affinity (kJ mol^{-1})	Electro-negativity	Melting Point (°C)	Boiling Point (°C)	X—X Distance (pm)	ΔH of Dissociation (kJ mol^{-1})
F	1681	328	4.193	−218.6	−188.1	143	158.8
Cl	1251	349	2.869	−101.0	−34.0	199	242.6
Br	1140	325	2.685	−7.25	59.5	228	192.8
I	1008	295	2.359	113.6[a]	185.2	266	151.1
At	930[b]	270[b]	2.39[b]	302[b]			

Source: Data from: Ionization energies cited in this chapter are from C. E. Moore, *Ionization Potentials and Ionization Limits Derived from the Analyses of Optical Spectra,* National Standard Reference Data Series, U.S. National Bureau of Standards, NSRDS-NBS 34, Washington, DC, 1970, unless noted otherwise. Electron affinity values listed in this chapter are from H. Hotop and W. C. Lineberger, *J. Phys. Chem. Ref. Data,* **1985,** *14,* 731. Standard electrode potentials listed in this chapter are from A. J. Bard, R. Parsons, and J. Jordan, eds., *Standard Potentials in Aqueous Solutions,* Marcel Dekker (for IUPAC), New York, 1985. Electronegativities cited in this chapter are from J. B. Mann, T. L. Meek, and L. C. Allen, *J. Am. Chem. Soc.,* **2000,** *122,* 2780, Table 2. Other data are from N. N. Greenwood and A. Earnshaw, *Chemistry of the Elements,* Pergamon Press, Elmsford, NY, 1984, except where noted.
[a]Sublimes readily [b]Approximate value

Solutions of iodine in alcohol ("tincture of iodine") are household antiseptics. Because of its radioactivity, astatine has not been studied extensively; it would be interesting to be able to compare its properties and reactions with those of the other halogens.

Several trends in physical properties of the halogens are immediately apparent (Table 8.14). As the atomic number increases, the ability of the nucleus to attract outermost electrons decreases; consequently, fluorine is the most electronegative and has the highest ionization energy, and astatine is lowest in both properties. With increasing size and number of electrons of the diatomic molecules in going down the periodic table, the London interactions between the molecules increase: F_2 and Cl_2 are gases, Br_2 is a liquid, and I_2 is a solid as a consequence of these interactions. The trends are not entirely predictable, because fluorine and its compounds exhibit some behavior that is substantially different than would be predicted by extrapolation of the characteristics of the other members of the group.

A striking property of F_2 is its remarkably low bond dissociation enthalpy, an important factor in the high reactivity of this molecule. Extrapolation from the bond dissociation enthalpies of the other halogens would yield a value of approximately 290 kJ mol^{-1}, nearly double the actual value. It is likely that the weakness of the F—F bond is largely a consequence of repulsions between the nonbonding electron pairs.[87] The small size of the fluorine atom brings these pairs into close proximity when F—F bonds are formed. Electrostatic repulsions between these pairs on neighboring atoms result in weaker bonding and an equilibrium bond distance significantly greater than would be expected in the absence of such repulsions.

For example, the covalent radius obtained for other compounds of fluorine is 64 pm, and an F—F distance of 128 pm would be expected in F_2. However, the actual distance is 143 pm. It is significant that oxygen and nitrogen share similar anomalies with fluorine; the O—O bonds in peroxides and the N—N bonds in hydrazines are longer than the sums of their covalent radii, and these bonds are weaker than the corresponding S—S and P—P bonds in the respective groups of these elements. In the case of oxygen and nitrogen, it is likely that the repulsion of electron pairs on neighboring atoms also plays a major role in the weakness of these bonds.* The very high reactivity of F_2 is attributed to both the weakness of the fluorine–fluorine bond and the small size and high electronegativity of fluorine.

Of the hydrohalic acids, HF is the weakest in aqueous solution (pK_a = 3.2 at 25 °C); HCl, HBr, and HI are all strong acids. Although HF reacts with water, strong hydrogen bonding occurs between F^- and the hydronium ion to form the ion pair $H_3O^+F^-$, reducing the activity coefficient of H_3O^+. As the concentration of HF increases, however, its tendency to form H_3O^+ increases as a result of further reaction of this ion pair with HF.

$$H_3O^+F^- + HF \rightleftharpoons H_3O^+ + HF_2^-$$

This view is supported by X-ray crystallographic studies of the ion pairs $H_3O^+F^-$ and $H_3O^+F_2^-$.[88]

Chlorine and chlorine compounds are used as bleaching and disinfecting agents. Perhaps the most commonly known of these compounds is hypochlorite, OCl^-, a household bleach prepared by dissolving chlorine gas in sodium or calcium hydroxide.

$$Cl_2 + 2\,OH^- \longrightarrow Cl^- + ClO^- + H_2O$$

The redox potentials associated with this reaction are provided in the Latimer diagram for chlorine in basic solution (Appendix B.7); the Frost diagram for chlorine is in **Figure 8.37**. The disproportionation of Cl_2 to Cl^- and OCl^- in basic solution is predicted by Figure 8.37, because Cl_2 is above the line between Cl^- and OCl^-. The free-energy

---

*Anomalous properties of fluorine, oxygen, and nitrogen have been discussed by P. Politzer in *J. Am. Chem. Soc.*, **1969**, *91*, 6235, and *Inorg. Chem.*, **1977**, *16*, 3350.

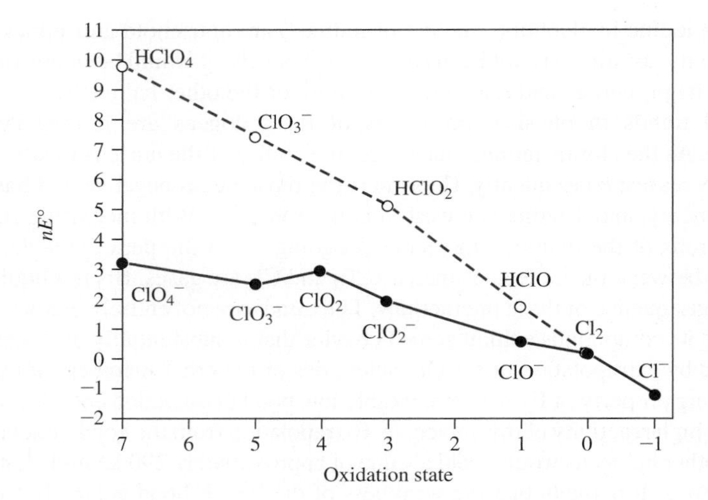

FIGURE 8.37 Frost Diagram for Chlorine Species. The dashed line is for acidic solutions, and the solid line is for basic solutions.

change from Cl_2 to OCl^- is positive (higher on the nE° scale), but the free-energy change from Cl_2 to Cl^- is negative and larger in magnitude, resulting in a net negative free-energy change and a spontaneous disproportionation reaction. Perchlorate is an extremely strong oxidizing agent, and ammonium perchlorate is used as a rocket fuel. In the fall of 2001, chlorine dioxide, ClO_2, was used to disinfect U.S. mail and at least one congressional office that may have been infected with anthrax. This gas is also used as an alternative to Cl_2 for purifying drinking water and as a bleaching agent in the paper industry.

EXERCISE 8.8

Write balanced half reactions for the following redox couples in acidic solution: ClO_4^-/ClO_3^-, $ClO_3^-/HClO_2$, $HClO_2/HClO$, $HClO/Cl_2$, Cl_2/Cl^-. (The only other species required for balancing are H_2O and H^+.)

Polyatomic Ions

In addition to the common monatomic halide ions, numerous polyatomic species, both cationic and anionic, have been prepared. The brown triiodide ion, I_3^-, is formed from I_2 and I^-:

$$I_2 + I^- \rightleftharpoons I_3^- \qquad K \approx 698 \text{ at } 25\ ^\circ\text{C in aqueous solution}$$

Other polyiodide ions have been characterized; in general, these may be viewed as aggregates of I_2 and I^- (sometimes I_3^-). Examples are shown in **Figure 8.38**.

The halogens Cl_2, Br_2, and I_2 can also be oxidized to cationic species. Examples include the diatomic ions Br_2^+ and I_2^+ (Cl_2^+ has been characterized in low-pressure discharge tubes but is much less stable), I_3^+ and I_5^+. I_2^+ dimerizes into I_4^{2+}.

Interhalogens

Halogens form many compounds containing two or more different halogens. Like the halogens themselves, these may be diatomic, such as ClF, or polyatomic, such as ClF_3, BrF_5, or IF_7. Many polyatomic ions containing two or more halogens have been synthesized. **Table 8.15** lists neutral (boxed) and ionic interhalogen species. The effect of the size of the central atom can readily be seen, with iodine the only element able to have up to seven fluorine atoms in a neutral molecule, whereas chlorine and bromine have a maximum of five fluorines. The effect of size is also evident in the ions, with iodine being the only halogen large enough to exhibit ions of formula XF_6^+ and XF_8^-.

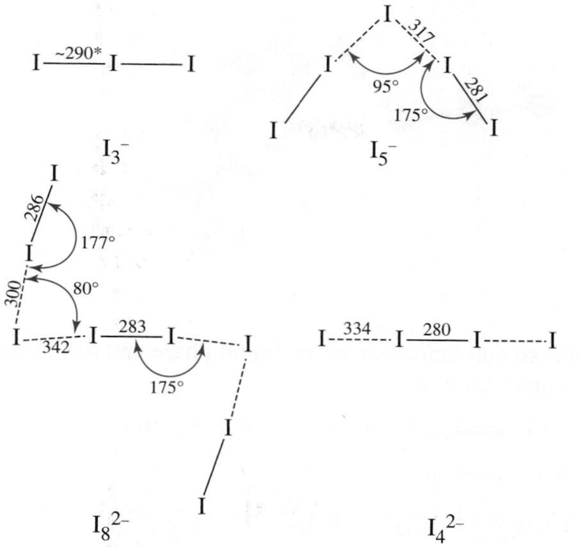

FIGURE 8.38 Polyiodide Ions. (Data from N. N. Greenwood and A. Earnshaw, *Chemistry of the Elements*, Pergamon Press, Elmsford, NY, 1984, pp. 821–854.)
*Distances in triiodide vary depending on the cation. In some cases both I—I distances are identical, but in the majority of cases they are different. Differences in I—I distances as great as 33 pm have been reported.

Neutral interhalogens can be prepared by direct reaction of the elements (the favored product often depending on the molar ratio of halogens used) and reaction of halogens with metal halides or other halogenating agents. Examples include:

$$Cl_2 + F_2 \longrightarrow 2\,ClF \qquad T = 225\ ^\circ C$$

$$I_2 + 5\,F_2 \longrightarrow 2IF_5 \qquad \text{Room temperature}$$

$$I_2 + 3\,XeF_2 \longrightarrow 2\,IF_3 + 3\,Xe \qquad T < -30\ ^\circ C$$

$$I_2 + AgF \longrightarrow IF + AgI \qquad 0\ ^\circ C$$

Interhalogens also serve as intermediates in the synthesis of other interhalogens:

$$ClF + F_2 \longrightarrow ClF_3 \qquad T = 200\ ^\circ C \text{ to } 300\ ^\circ C$$

$$ClF_3 + F_2 \longrightarrow ClF_5 \qquad h\nu, \text{ room temperature}$$

TABLE 8.15 Interhalogen Species

Formal Oxidation State of Central Atom	Number of Lone Pairs on Central Atom	Compounds and Ions						
+7	0			IF_7				
				IF_6^+				
				IF_8^-				
+5	1	ClF_5	BrF_5	IF_5				
		ClF_4^+	BrF_4^+	IF_4^+				
			BrF_6^-	IF_6^-				
+3	2	ClF_3	BrF_3	IF_3		I_2Cl_6		
		ClF_2^+	BrF_2^+	IF_2^+		ICl_2^+	IBr_2^+	$IBrCl^+$
		ClF_4^-	BrF_4^-	IF_4^-		ICl_4^-		
+1	3	ClF	BrF	IF	$BrCl$	ICl	IBr	
		ClF_2^-	BrF_2^-	IF_2^-	$BrCl_2^-$	ICl_2^-	IBr_2^-	
					Br_2Cl^-	I_2Cl^-	I_2Br^-	$IBrCl^-$

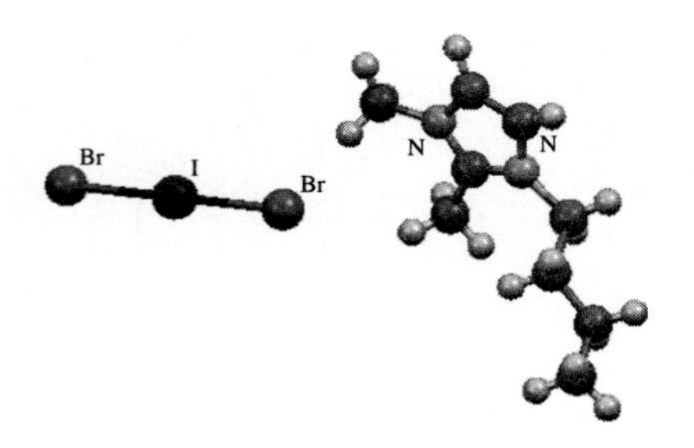

FIGURE 8.39 Molecular structure of 1,2-dimethyl-3-butylimidazolium iododibromide, generated from CIF data.

Several interhalogens undergo autoionization in the liquid phase and have been studied as nonaqueous solvents. Examples include:

$$3\ IX \rightleftharpoons I_2X^+ + IX_2^- \quad (X = Cl, Br)$$

$$2\ BrF_3 \rightleftharpoons BrF_2^+ + BrF_4^-$$

$$I_2Cl_6 \rightleftharpoons ICl_2^+ + ICl_4^-$$

$$IF_5 \rightleftharpoons IF_4^+ + IF_6^-$$

The use of trihalide salts, including salts of interhalogen anions, as electrolytes in solar cells motivates current research in this area.[89] In the pursuit of this objective, 1,2-dimethyl-3-butylimidazolium iododibromide has been structurally characterized (Figure 8.39).[90]

Pseudohalogens

Parallels have been observed between the chemistry of the halogens and other dimeric species. Dimeric molecules showing considerable similarity to the halogens are often called **pseudohalogens.** Chapter 15 considers parallels between the halogens and pseudohalogens, including examples from organonmetallic chemistry. However, the concept of pseudohalogens is also instructive in main group chemistry. For example, there are many similarities between the halogens and cyanogen, NCCN. The monoanion, CN^-, is well known; it combines with hydrogen to form the weak acid HCN and, with Ag^+ and Pb^{2+}, to form precipitates of low solubility in water. Interhalogen compounds such as FCN, ClCN, BrCN, and ICN are known. Cyanogen, like the halogens, can add across double or triple carbon–carbon bonds. The pseudohalogen idea is a useful classification tool that will be explored further in Chapter 15.*

8.10 Group 18: The Noble Gases

The elements in Group 18, traditionally designated the "inert" or "rare" gases, no longer satisfy these early labels. These elements have an interesting chemistry and are rather abundant. Helium, for example, is the second most abundant element in the universe, and argon is the third most abundant component of dry air, approximately 24 times as abundant by volume as carbon dioxide.

8.10.1 The Elements

Cavendish obtained the first experimental evidence for the noble gases in 1766. In a series of experiments on air, he was able to sequentially remove nitrogen (then known as "phlogisticated air"), oxygen ("dephlogisticated air"), and carbon dioxide ("fixed air") from air

*For additional examples of pseudohalogens, see J. Ellis, *J. Chem. Educ.*, **1976**, *53*, 2.

by chemical means; but a small residue, no more than one part in 120, resisted all attempts at reaction.[91] The nature of this unreactive fraction of air remained a mystery for more than a century; it was eventually shown to be a mixture of argon and other noble gases.*

During a solar eclipse in 1868, a new emission line, matching no known element, was found in the spectrum of the solar corona. Locklear and Frankland proposed the existence of a new element, helium (Greek, *helios*, sun). The same spectral line was subsequently observed in the gases of Mount Vesuvius.

In the early 1890s, Lord Rayleigh and William Ramsay observed a discrepancy in the apparent density of nitrogen isolated from air and from ammonia. The two researchers independently performed painstaking experiments to isolate and characterize what seemed to be either a new form of nitrogen (the formula N_3 was one suggestion) or a new element. Eventually the two worked cooperatively, with Ramsay apparently the first to suggest that the unknown gas might fit into the periodic table after the element chlorine. In 1895, they reported the details of their experiments and evidence for the element they had isolated: argon (Greek, *argos*, no work, lazy).[92]

Within three years, Ramsay and Travers had isolated three additional elements by low-temperature distillation of liquid air: neon (Greek, *neos*, new), krypton (Greek, *kryptos*, concealed), and xenon (Greek, *xenos*, strange). The last of the noble gases, radon, was isolated as a nuclear decay product in 1902.

Helium is fairly rare on Earth, but it is the second most abundant element in the universe (76% H, 23% He) and is a major component of stars. Crude natural gas often contains helium. Commercially available helium is obtained via its separation from these gaseous mixtures by fractional distillation. The other noble gases, with the exception of radon, are present in small amounts in air (Table 8.16) and are commonly obtained by fractional distillation of liquid air. Helium is used as an inert atmosphere for arc welding, in weather balloons, and in gas mixtures used in deep-sea diving, but it is less soluble in blood than nitrogen. Liquid helium (with a boiling point of 4.2 K) is used as a coolant for superconducting magnets in NMR and MRI instruments. Argon, the least expensive noble gas, is used as an inert atmosphere

TABLE 8.16 Properties of the Group 18 Elements: The Noble Gases

Element	Ionization Energy ($kJ\ mol^{-1}$)	Melting Point (°C)	Boiling Point (°C)	Enthalpy of Vaporization ($kJ\ mol^{-1}$)	Electronegativity	Abundance in Dry Air (% by Volume)
He[a]	2372	–	−268.93	0.08	4.160	0.000524
Ne	2081	−248.61	−246.06	1.74	4.787	0.001818
Ar	1521	−189.37	−185.86	6.52	3.242	0.934
Kr	1351	−157.20	−153.35	9.05	2.966	0.000114
Xe	1170	−111.80	−108.13	12.65	2.582	0.0000087
Rn	1037	−71	−62	18.1	2.60[b]	Trace

Source: Data from: Ionization energies cited in this chapter are from C. E. Moore, *Ionization Potentials and Ionization Limits Derived from the Analyses of Optical Spectra,* National Standard Reference Data Series, U.S. National Bureau of Standards, NSRDS-NBS 34, Washington, DC, 1970, unless noted otherwise. Electron affinity values listed in this chapter are from H. Hotop and W. C. Lineberger, *J. Phys. Chem. Ref. Data,* **1985,** *14,* 731. Standard electrode potentials listed in this chapter are from A. J. Bard, R. Parsons, and J. Jordan, eds., *Standard Potentials in Aqueous Solutions,* Marcel Dekker (for IUPAC), New York, 1985. Electronegativities cited in this chapter are from J. B. Mann, T. L. Meek, and L. C. Allen, *J. Am. Chem. Soc.,* **2000,** *122,* 2780, Table 2. Other data are from N. N. Greenwood and A. Earnshaw, *Chemistry of the Elements,* Pergamon Press, Elmsford, NY, 1984, except where noted.3

[a]Helium cannot be frozen at 1 atm pressure [b]Approximate value

*Cavendish's experiments and other early developments in noble gas chemistry have been described in E. N. Hiebert, "Historical Remarks on the Discovery of Argon: The First Noble Gas", in H. H. Hyman, ed., *Noble Gas Compounds*, University of Chicago Press, Chicago, 1963, pp. 3–20.

for studying chemical reactions, for high-temperature metallurgical processes, and for filling incandescent bulbs. The noble gases emit light of vivid colors when an electrical discharge is passed through them; the color depends on the gases used. Neon's emission spectrum is responsible for the bright orange-red of neon signs. All isotopes of radon are radioactive; the longest lived isotope, ^{222}Rn, has a half-life of only 3.825 days. There is concern regarding the level of radon in many homes. A potential cause of lung cancer, radon is formed from the decay of trace amounts of uranium in certain rock formations and itself undergoes alpha decay, generating radioactive daughter isotopes. Radon can enter homes through basement walls and floors. Important properties of the noble gases are summarized in Table 8.16.

8.10.2 Chemistry of Group 18 Elements

The group 18 elements were once believed to be totally unreactive as a consequence of the very stable "octet" valence electron configurations of their atoms. Their chemistry was simple: they had none!

The first chemical compounds containing noble gases to be discovered were **clathrates**, "cage" compounds in which noble gas atoms could be trapped. Experiments in the late 1940s showed that when water or solutions containing hydroquinone (*p*-dihydroxybenzene, $HO{-}C_6H_4{-}OH$) were crystallized under high pressures of certain gases, hydrogen-bonded lattices having large cavities could be formed, with gas molecules trapped in the cavities. Clathrates containing the noble gases argon, krypton, and xenon—as well as those containing small molecules such as SO_2, CH_4, and O_2—have been prepared. No clathrates have been found for helium and neon; these atoms are too small to be trapped.

Even though clathrates of three noble gases had been prepared by the beginning of the 1960s, no compounds containing covalently bonded noble gas atoms had been synthesized. Attempts had been made to react xenon with elemental fluorine, but without success. This situation changed dramatically in 1962. Neil Bartlett had observed that PtF_6 changed color on exposure to air. He demonstrated that PtF_6 was serving as an oxidizing agent in this reaction and that the color change was due to the formation of O_2^+ $[PtF_6]^-$.[93] Bartlett noted the similarity of the ionization energies of xenon (1169 kJ mol^{-1}) and O_2 (1175 kJ mol^{-1}) and reacted Xe with PtF_6. He observed a color change from the deep red of PtF_6 to orange yellow and reported the product as Xe^+ $[PtF_6]^-$.[94] Although this formulation was ultimately proved incorrect (the reaction actually provided a complex mixture of xenon compounds), these compounds were the first covalently bonded noble gas compounds to be synthesized. The compounds XeF_2 and XeF_4 were soon characterized, and other noble gas compounds followed.*

Noble gas compounds of xenon are the most diverse;[95] most of the other chemistry of this group is of compounds of krypton.[96] There is evidence for RnF_2, but the study of radon chemistry is hampered by the element's high radioactivity. The first "stable" compound of argon, HArF, was reported in 2000.[97] This compound was synthesized by condensing a mixture of argon and an HF–pyridine polymer onto a CsI substrate at 7.5 K. Although stable at a low temperature, HArF decomposes at room temperature. Transient species containing helium and neon have been observed using mass spectrometry. However, most known noble gas compounds are those of xenon with the elements F, O, and Cl; a few compounds have also been reported with Xe—N, Xe—C, and even Xe—transition metal bonds. Examples of noble gas compounds and ions are in **Table 8.17**.

Structures of xenon compounds provide tests for models of bonding. For example, XeF_2 and XeF_4 have structures entirely in accord with their VSEPR descriptions: XeF_2 is linear, with three lone pairs on Xe, and XeF_4 is planar, with two lone pairs (**Figure 8.40**).

The application of VSEPR to rationalize the structures of XeF_6 and XeF_8^{2-} is more challenging. Each Lewis structure has a lone pair on the central xenon. The VSEPR model would

*For a discussion of the development of the chemistry of xenon compounds, see P. Laszlo, G. J. Schrobilgen, *Angew. Chem., Int. Ed.*, **1988**, *27*, 479.

TABLE 8.17 Example of Noble Gas Compounds and Ions

Formal Oxidation State of Noble Gas	Number of Lone Pairs on Central Atom	Compounds and Ions
+2	3	KrF^+ XeF^+ $Xe_3OF_3^+$
		KrF_2 XeF_2
+4	2	XeF_3^+
		XeF_4 $XeOF_2$ XeO_2
		XeF_5^-
+6	1	XeF_5^+ $XeOF_4$ XeO_3
		XeF_6 XeO_2F_2
		XeF_7^- XeO_2F^+
		XeF_8^{2-} XeO_3F^-
		$XeOF_5^-$
+8	0	XeO_3F_2 XeO_4
		XeO_6^{4-}

XeF_2 XeF_4 XeF_6 XeF_8^{2-}

FIGURE 8.40 Structures of Xenon Fluorides.

predict this lone pair to occupy a definite position on the xenon. However, the central lone pair of XeF_6 or XeF_8^{2-} seems to occupy no definite location.* One explanation of these phenomena is that on the basis of the large number of fluorines attached to the Xe, repulsions between the electrons in the xenon–fluorine bonds are extensive, and perhaps too strong to enable a lone pair to occupy a well-defined position. This is contrary to the VSEPR model prediction that lone-pair–bonding-pair repulsion is more important than bonding-pair–bonding-pair repulsion in influencing molecular geometry. The central lone pair does play a role, however. In XeF_6, the structure is not octahedral but is somewhat distorted. Spectroscopic evidence of gas phase XeF_6 indicates that its lowest energy structure has C_{3v} symmetry. The molecule is fluxional; it undergoes rapid rearrangement from one C_{3v} structure to another—the lone pair appears to move from the center of one face to another.[98] Theoretical studies support a very low barrier between C_{3v} and O_h structures.[99] Solid XeF_6 exists in at least six different modifications, with different forms favored at different temperatures.[100] The modification present at room temperature features a trimeric unit of three XeF_6 interacting with a fourth XeF_6. The structure contains square-pyramidal XeF_5^+ ions bridged by fluoride ions (**Figure 8.41**).

The structure of XeF_8^{2-} is slightly distorted; it is an approximate square antiprism (D_{4d} symmetry), but with one face slightly larger than the opposite face, resulting in approximate C_{4v} symmetry (Figure 8.40).[101] Although this distortion may be a consequence of the way these ions pack in the crystal, a lone pair exerting some influence on the size of the larger face may cause the distortion.**

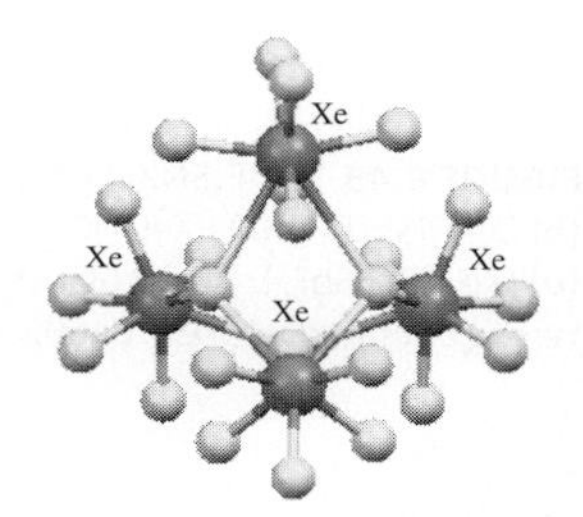

FIGURE 8.41 Xenon Hexafluoride modification most stable at room temperature. (Molecular structure generated from CIF data.) (CSD 416317 obtained from the Crystal Structure Deposition at Fachinformationszentrum Karlsruhe, www.fiz-karlsruhe.de/crystal_structure_dep.html)

*A lone pair that appears to have no impact on the molecular structure is classified as *stereochemically inactive*.

The effect of lone pairs can be difficult to predict. For examples of stereochemically active and inactive lone pairs in ions of formula AX_6^{n-}, see K. O. Christe, W. Wilson, *Inorg. Chem.*, **1989, *28*, 3275 and references therein.

The newest member of Table 8.17 is XeO_2. A longstanding mystery has been the very low abundance of xenon in the atmospheres of Earth and Mars in comparison with the lighter noble gases, in contrast to the higher abundance of xenon in chondritic meteorites. The quest for the "missing xenon" has led to the proposal that xenon may have been incorporated into olivines (silicates containing magnesium, iron, and sometimes other metals) and quartz (SiO_2) under high pressures and temperatures in Earth's mantle.[102] Treatment of olivines with high pressures of xenon results in xenon uptake, leading to the hypothesized existence of XeO_2, formed by proposed substitution of silicon by xenon in SiO_2. Although XeO_4 and XeO_3 had been prepared by 1964, the structure of XeO_2 was not reported until 2011. In contrast to the other oxides, XeO_2 was found by Raman spectroscopy to have an extended structure with square planar XeO_4 units and oxygen bridges, as shown in **Figure 8.42**.[103] Reinterpretation of earlier spectroscopic data is consistent with the possibility that polymeric XeO_2 and other forms of xenon may occur in deeply buried silicate minerals and account for the "missing" xenon, consistent with the proposal that xenon depletion from the Earth's atmosphere may occur by xenon insertion at high temperatures and pressures. The situation on Mars awaits future exploration, possibly by you or another reader of this text.

Positive ions of xenon are known. Bartlett's original reaction of Xe with PtF_6 was likely:

$$Xe + 2\,PtF_6 \longrightarrow [XeF]^+[PtF_6]^- + PtF_5 \longrightarrow [XeF]^+[Pt_2F_{11}]^-$$

XeF^+ does not occur independently but attaches covalently to other species in its salts. For example, this cation coordinates to a nitrogen atom of $F_3S{\equiv}N$ in $[F_3SNXeF][AsF_6]$ (**Figure 8.43a**), a rare example of a Xe—N bond;[104] to a fluorine atom of $[Sb_2F_{11}]^-$ in $[XeF][Sb_2F_{11}]$[105] (**Figure 8.43b**); and to an oxygen atom of nitrate in $FXeONO_2$ **Figure 8.43c**.[106] The fascinating Z-shaped xenon(II) oxide fluoride in $[Xe_3OF_3][SbF_6]$ (**Figure 8.43d**) was obtained by reaction of $[H_3O][SbF_6]$ and XeF_2 in HF.[107] Compression of XeF_2 to 200 GPa results in dissociation into an ionic solid with formula [XeF][F].[108]

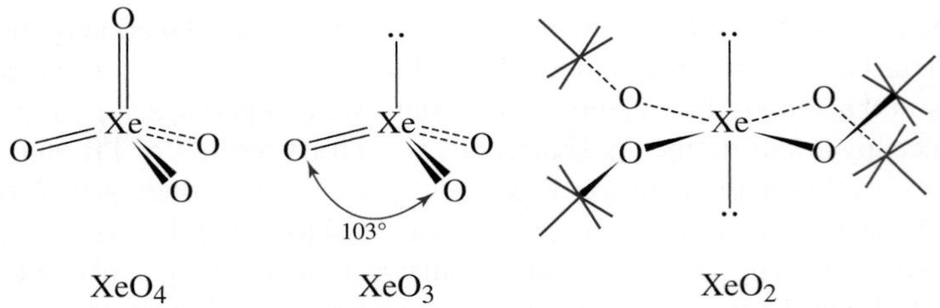

FIGURE 8.42 Xenon Oxides.

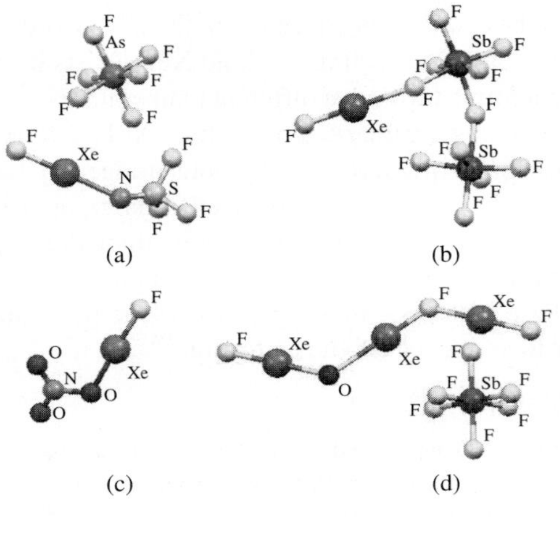

FIGURE 8.43 **(a)** $[F_3SNXeF][AsF_6]$. **(b)** $[XeF][Sb_2F_{11}]$. **(c)** $FXeONO_2$. **(d)** $[Xe_3OF_3][SbF_6]$. (Molecular structures generated with CIF data).

Xenon acts as a ligand toward Au^{2+}. Square planar $AuXe_4{}^{2+}$ (**Figure 8.44a**) was synthesized as an $[Sb_2F_{11}]^-$ salt in the superacid HF/SbF_5, in which Xe displaces HF from $[Au(HF)_n]^{2+}$ complexes. The linear cation $[AuXe_2]^{2+}$ was isolated in *trans*-$AuXe_2(SbF_6)_2$, where the $[SbF_6]^-$ coordinates to Au^{2+} via fluorine atoms (**Figure 8.44b**).[109] Notable xenon compounds include salts of the perfluorophenylxenon ion $[C_6F_5Xe]^+$ (**Figure 8.44c**).[110] Using these ions in synthesis is of current interest.[111] Linear HXeCCH was formed by insertion of xenon into a C—H bond of acetylene at extremely low temperatures (approximately 40 K),[112] and a similar strategy was used to prepare halogenated xenon cyanides (for example, ClXeCN).[113] Many compounds have been synthesized that feature XeF_2 as a ligand, bound via a fluorine atom to a metal. Examples include complexes of Group 1 and 2 metals and transition metals.[114] The structure of one of these, $[Mg(XeF_2)_4](AsF_6)_2$, is in **Figure 8.44d.**[115] The dimeric $[Mg(XeF_2)(XeF_4)(AsF_6)]_2$ (**Figure 8.44e**) features both XeF_2 and XeF_4 as ligands.[115] The crystal structure of $[H_3O][AsF_6] \cdot 2\ XeF_2$ contains XeF_2 molecules interacting with hydronium ion.

Krypton forms several species with fluorine, including the ions KrF^+ (for example, in $[KrF][AsF_6]$, **Figure 8.45a**) and $Kr_2F_3{}^+$ (for example, in $[Kr_2F_3][SbF_6] \cdot KrF_2$, **Figure 8.45b**, the crystal structure of this salt also contains linear KrF_2).[116] Pure KrF_2 has been structurally characterized.[116] A rare example of KrF_2 serving a ligand was observed in $[BrOF_2][AsF_6] \cdot 2\ KrF_2$ (**Figure 8.45c**).[117] Examples of bonding to elements other than fluorine include $[F—Kr—N{\equiv}CH]^+AsF_6{}^-$,[118] $Kr(OTeF_5)_2$,[119] and HKrCCH.[120]

Significant progress has been made in the development of noble gas hydrides, with 23 neutral species reported by 2009.[121] Typically prepared by UV photolysis of precursors in frozen noble gas matrices, hydrides are known for the elements argon (HArF, mentioned previously), krypton, and xenon and include both a dihydride (HXeH) and compounds with bonds between noble gases and F, Cl, Br, I, C, N, O, and S.

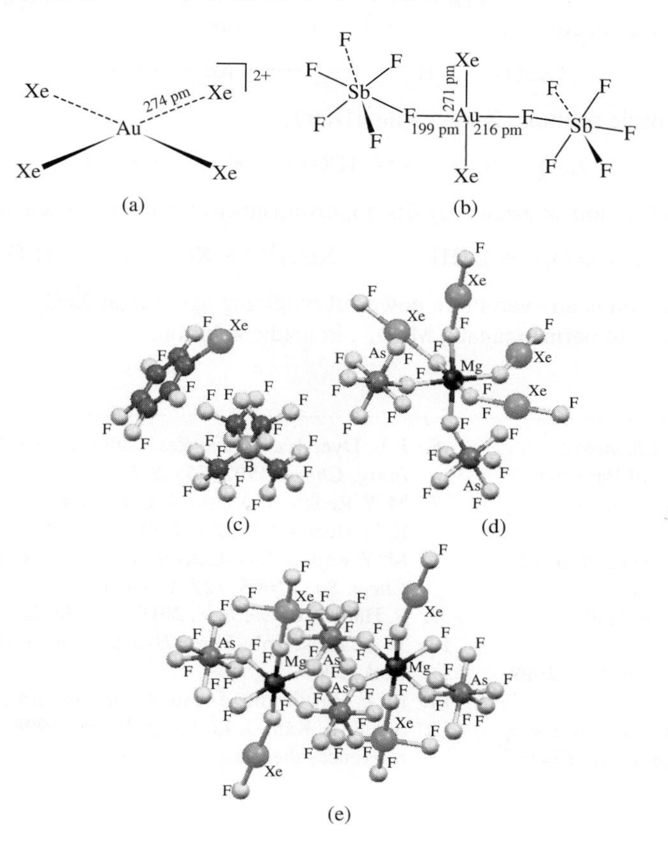

FIGURE 8.44 (a) $AuXe_4{}^{2+}$. (b) trans-$AuXe_2(SbF_6)_2$. (c) $[C_6F_5Xe][B(CF_3)_4]$. (d) $[Mg(XeF_2)_4](AsF_6)_2$. (e) $[Mg(XeF_2)(XeF_4)(AsF_6)]_2$. (Molecular structures created using CIF data.)

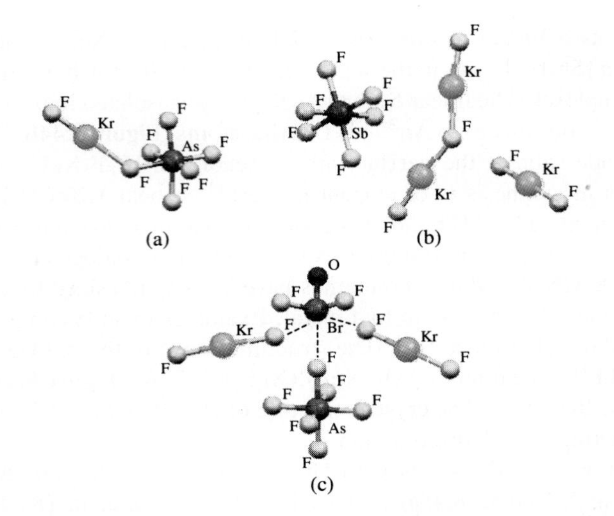

FIGURE 8.45 (a) $[KrF][AsF_6]$. (b) $[Kr_2F_3][SbF_6] \cdot KrF_2$. (c) $[BrOF_2][AsF_6] \cdot 2\ KrF_2$. (Molecular structures created using CIF data).

The radioactivity of radon renders its study difficult; RnF_2 and a few other compounds have been observed through tracer studies.

Interest in using noble gas compounds as reagents in organic and inorganic synthesis is of interest because the by-product of such reactions is often the noble gas itself. The xenon fluorides XeF_2, XeF_4, and XeF_6 have been used as fluorinating agents. For example,

$$2\ SF_4 + XeF_4 \longrightarrow 2\ SF_6 + Xe \qquad C_6H_5I + XeF_2 \longrightarrow C_6H_5IF_2 + Xe$$

XeF_4 can selectively fluorinate aromatic positions in arenes such as toluene.

The oxides XeO_3 and XeO_4 are explosive. XeO_3 is a powerful oxidizing agent in aqueous solution. The electrode potential of the half-reaction

$$XeO_3 + 6\ H^+ + 6\ e^- \longrightarrow Xe + 3\ H_2O$$

is 2.10 V. In basic solution, XeO_3 forms $HXeO_4^-$:

$$XeO_3 + OH^- \rightleftharpoons HXeO_4^- \quad K = 1.5 \times 10^{-3}$$

The $HXeO_4^-$ ion subsequently disproportionates to form the perxenate ion, XeO_6^{4-}:

$$2\ HXeO_4^- + 2\ OH^- \longrightarrow XeO_6^{4-} + Xe + O_2 + 2\ H_2O$$

The perxenate ion is an even more powerful oxidizing agent than XeO_3 and is capable of oxidizing Mn^{2+} to permanganate, MnO_4^-, in acidic solution.

References

1. G. J. Leigh, ed., *Nomenclature of Inorganic Chemistry, Recommendations 1990*, International Union of Pure and Applied Chemistry, Blackwell Scientific Publications, Oxford UK, pp. 41–43.
2. L. C. Allen, J. E. Huheey, *J. Inorg. Nucl. Chem.*, **1980**, *42*, 1523; T. L. Meek, *J. Chem. Educ.*, **1995**, *72*, 17.
3. W. M. Latimer, *Oxidation Potentials*, Prentice Hall, Englewood Cliffs, NJ, 1952.
4. D. Gust, T. A. Moore, A. L. Moore, *Acc. Chem. Res.*, **2009**, *42*, 1890.
5. V. L. Pecoraro, A. J. Stemmler, B. R. Gibney, J. J. Bodwin, H. Wang, J. W. Kampf, A. Barwinski, *Progr. Inorg. Chem.*, **1997**, *45*, 83.
6. J. L. Dye, *Acc. Chem. Res.*, **2009**, *42*, 1564. J. L. Dye, *Inorg. Chem.*, **1997**, *36*, 3816.
7. M. Y. Redko, M. Vlassa, J. E. Jackson, A. W. Misiolek, R. H. Huang, J. L. Dye, *J. Am. Chem. Soc.*, **2002**, *124*, 5928.
8. M. Y. Redko, J. E. Jackson, R. H. Huang, J. L. Dye, *J. Am. Chem. Soc.*, **2005**, *127*, 12416.
9. S. Harder, *Chem. Rev.*, **2010**, *110*, 3852.
10. W. N. Lipscomb, *Boron Hydrides*, W. A. Benjamin, New York, 1963.
11. M. F. Hawthorne, *Angew. Chem., Int. Ed.*, **1993**, *32*, 950.
12. See S. B. Kahl, J. Li, *Inorg. Chem.*, **1996**, *35*, 3878, and references therein.

13. H. V. R. Dias, P. P. Power, *Angew. Chem., Int. Ed.*, **1987**, *26*, 1270; *J. Am. Chem. Soc.*, **1989**, *111*, 144.

14. K. M. Waggoner, H. Hope, P. P. Power, *Angew. Chem., Int. Ed.*, **1988**, *27*, 1699.

15. J. J. Engelberts, R. W. A. Havenith, J. H. van Lenthe, L. W. Jenneskens, P. W. Fowler, *Inorg. Chem.*, **2005**, *44*, 5266.

16. Y. Wang, B. Quillian, P. Wei, C. S. Wannere, Y. Xie, R. B. King, H. F. Schaefer, III, P. v. R. Schleyer, G. H. Robinson, *J. Am. Chem. Soc.*, **2007**, *129*, 12412.

17. R. Dagani, *Chem. Eng. News*, **1997**, *75*(16), 9. R. Dagani, *Chem. Eng. News*, **1998**, *76*(16), 31.

18. J. Su, X.-W. Li, R. C. Crittendon, G. H. Robinson, *J. Am. Chem. Soc.*, **1997**, *119*, 5471.

19. P. Pyykkö, S. Riedel, M. Patzschke, *Chem. Eur. J.*, **2005**, *11*, 3511.

20. Y. Wang, G.H. Robinson, *Dalton Trans.*, **2012**, *41*, 337. Y. Wang, G. H. Robinson, *Inorg. Chem.*, **2011**, *50*, 12326. R. C. Fischer, P. P. Power, *Chem. Rev.*, **2010**, *110*, 3877. Y. Wang, G. H. Robinson, *Chem. Commun.*, **2009**, 5201.

21. R. J. A. van den Oetelaar, C. F. J. Flipse, *Surf. Sci.*, **1997**, *384*, L828.

22. A. K. Geim, et al., *Science*, **2004**, *306*, 666.

23. S. Park, R. S. Ruoff, *Nature Nanotech.*, **2009**, *4*, 217.

24. Z. H. Ni, H. M. Wang, J. Kasim, H. M. Fan, T. Yu, Y. H. Wu, Y. P. Feng, Z. X. Shen, *Nano Lett.*, **2007**, *7*, 2758.

25. P. Vogt, P. De Padova, C. Quaresima, J. Avila, E. Frantzeskakis, M. C. Asensio, A. Resta, B. Ealet, G. Le Lay, *Phys. Rev. Lett.*, **2012**, *108*, 155501.

26. W. A. de Heer, *Science*, **2006**, *312*, 1191.

27. Y.-W. Son, M. L. Cohen, S. G. Louie, *Phys. Rev. Lett.*, **2006**, *97*, 216803.

28. M. S. Dresselhaus, G. Dresselhaus, R. Saito, *Carbon*, **1995**, *33*, 883.

29. V. Derycke, R. Martel, J. Appenzeller, P. Avouris, *Nano Lett.*, **2001**, *1*, 453.

30. L. Ge, J. H. Jefferson, B. Montanari, N. M. Harrison, D. G. Pettifor, G. A. D. Briggs, *ACS Nano*, **2009**, *3*, 1069 and references cited therein.

31. A. Kongkanand, P. V. Kamat, *ACS Nano*, **2007**, *1*, 13.

32. A. A. Bhirde, V. Patel, J. Gavard, G. Zhang, A. A. Sousa, A. Masedunskas, R. D. Leapman, R. Weigert, J. S. Gutkind, J. F. Rusling, *ACS Nano*, **2009**, *3*, 307.

33. D. V. Kosynkin, A. L. Higginbotham, A. Sinitskii, J. R. Lomeda, A. Dimiev, B. K. Price, J. M. Tour, *Nature*, **2009**, *458*, 872.

34. D. Tasis, N. Tgmatarchis, A. Bianco, M. Prato, *Chem. Rev.*, **2006**, *106*, 1105.

35. D. Malko, C. Neiss, F. Viñes, A. Görling, *Phys. Rev. Lett.*, **2012**, *108*, 086804.

36. G. Li, Y. Li, H. Liu, Y. Guo, Y. Li, D. Zhu, *Chem. Commun.*, **2010**, *46*, 3256.

37. K. Srinivasu, S. K. Ghosh, *J. Phys. Chem. C*, **2012**, *116*, 5951.

38. H. W. Kroto, J. R. Heath, S. C. O'Brien, R. F. Curl, R. E. Smalley, *Nature (London)*, **1985**, *318*, 162.

39. L. Becker, J. L. Bada, R. E. Winans, J. E. Hunt, T. E. Bunch, B. M. French, *Science*, **1994**, *265*, 642; D. Heymann, L. P. F. Chibante, R. R. Brooks, W. S. Wolbach, R. E. Smalley, *Science*, **1994**, *265*, 645.

40. J. R. Bowser, *Adv. Inorg. Chem.*, **1994**, *36*, 61–62 and references therein.

41. R. Taylor, J. P. Hare, A. K. Abdul-Sada, H. W. Kroto, *Chem. Commun.*, **1990**, 1423.

42. T. L. Makarova, B. Sundqvist, R. Höhne, P. Esquinazi, Y. Kopelevich, P. Scharff, V. A. Davydov, L. S. Kashevarova, A. V. Rakhmanina, *Nature (London)*, **2001**, *413*, 716; *Chem. Eng. News*, **2001**, *79*, 10. M. Núñez-Regueiro, L. Marques, J.-L. Hodeau, O. Béthoux, M. Perroux, *Phys. Rev., Lett.*, **1995**, *74*, 278.

43. F. Giacaloine, N. Martín, *Chem. Rev.*, **2006**, *106*, 5136. H. Brumm, E. Peters, M. Jansen, *Angew. Chem., Int. Ed.*, **2001**, *40*, 2069.

44. N. Sano, H. Wang, M. Chhowalla, I. Alexandrou, G. A. J. Amaratunga, *Nature (London)*, **2001**, *414*, 506.

45. Y. Li, Y. Huang, S. Du, R. Liu, *Chem. Phys. Lett.*, **2001**, *335*, 524.

46. A. A. Shvartsburg, R. R. Hudgins, R. Gutierrez, G. Jungnickel, T. Frauenheim, K. A. Jackson, M. F. Jarrold, *J. Phys. Chem. A*, **1999**, *103*, 5275.

47. C. Thilgen, F. Diederich, *Chem. Rev.*, **2006**, *106*, 5049.

48. T. Akasaka, S. Nagase, Eds., *Endofullerenes: A New Family of Carbon Clusters*, Kluwer Academic, Netherlands, 2002.

49. K. Kurotobi, Y. Murata, *Science*, **2011**, *333*, 613.

50. H. Prinzbach, A. Weller, P. Landenberger, F. Wahl, J. Wörth, L. T. Scott, M. Gelmont, D. Olevano, B. Issendorff, *Nature (London)*, **2000**, *407*, 60.

51. X. Lu, Z. Chen, *Chem. Rev.*, **2005**, *105*, 3643.

52. J. M. Ashcroft, D. A. Tsyboulski, K. B. Hartman, T. Y. Zakharian, J. W. Marks, R. B. Weisman, M. G. Rosenblum, L. J. Wilson, *Chem. Commun.*, **2006**, 3004.

53. R. Singh, J. W. Lillard, Jr., *Exp. Mol. Pathology*, **2009**, *86*, 215.

54. S. Margadonna, D. Pontiroli, M. Belli, T. Shiroka, M. Riccò, M. Brunelli, *J. Am. Chem. Soc.*, **2004**, *126*, 15032.

55. M. Riccò, D. Pontiroli, M. Mazzani, F. Gianferrari, M. Pagliari, A. Goffredi, M. Brunelli, G. Zandomeneghi, B. H. Meier, T. Shiroka, *J. Am. Chem. Soc.*, **2010**, *132*, 2064.

56. F. Cheng, Y. Murata, K. Komatsu, *Org. Lett.*, **2002**, *4*, 2541.

57. W.-W Yang, Z.-J. Li, X. Gao, *J. Org. Chem.*, **2011**, *76*, 6067.

58. G. A. Olah, G. Rasul, *Acc. Chem. Res.*, **1997**, *30*, 245.

59. A. F. Wells, *Structural Inorganic Chemistry*, 5th ed., Clarendon Press, Oxford, 1984, pp. 1009–1043.

60. X. Wang, L. Andrews, *J. Am. Chem. Soc.*, **2003**, *125*, 6581.

61. A. Pfitzer, *Angew. Chem., Int. Ed.*, **2006**, *45*, 699. A. F. Wells, *Structural Inorganic Chemistry*, 5th ed., Clarendon Press, Oxford, 1984, pp. 838–840 and references therein.

62. M. Sheer, G. Balázs, A. Seitz, *Chem. Rev.*, **2010**, *110*, 4236.

63. B. M. Cossairt, N. A. Piro, C. C. Cummins, *Chem. Rev.*, **2010**, *110*, 4164. M. Caporali, L. Gonsalvi, A. Possin, M. Peruzzini, *Chem. Rev.*, **2010**, *110*, 4178.

64. B. M. Cossairt, M.-C. Diawara, C. C. Cummins, *Science*, **2009**, *323*, 602. B. M. Cossairt, C. C. Cummins, *J. Am. Chem. Soc.*, **2009**, *131*, 15501.

65. D. Fischer, M. Jansen, *Angew. Chem. Int. Ed.*, **2002**, *41*, 1755. G. V. Vajenine, *Inorg. Chem.*, **2007**, *46*, 5146

66. A. Schulz, A. Villinger, *Chem. Eur. J.*, **2012**, *18*, 2902.

67. G. Auffermann, Y. Prots, R. Kniep, *Angew. Chem., Int. Ed.*, **2001**, *40*, 547.

68. S. B. Schneider, R. Frankovsky, W. Schnick, *Angew. Chem., Int. Ed.*, **2012**, *51*, 1873.

69. K. O. Christe, W. W. Wilson, J. A. Sheehy, J. A. Boatz, *Angew. Chem., Int. Ed.*, **1999**, *38*, 2004; for more information on nitrogen-containing species, see T. M. Klapötke, *Angew. Chem., Int. Ed.*, **1999,** *38*, 2536.

70. I. Kobrsi, W. Zheng, J. E. Knox, M. J. Heeg, H. B. Schlegel, C. H. Winter, *Inorg. Chem.*, **2006**, *45*, 8700. T. Schroer, R. Haiges, S. Schneider, K. O. Christe, *Chem. Commun.*, **2005**, 1607.

71. R. N. Butler, J. M. Hanniffy, J. C. Stephens, L. A. Burke, *J. Org. Chem.*, **2008**, *73*, 1354.

72. E. Urnezius, W. W. Brennessel, C. J. Cramer, J. E. Ellis, P. v. R. Schleyer, *Science*, **2002**, *295*, 832.

73. F. Kraus, T. Hanauer, N. Korber, *Inorg. Chem.*, **2006**, *45*, 1117.

74. W. Massa, R. Kujanek, G. Baum, K. Dehnicke, *Angew. Chem., Int. Ed.*, **1984**, *23*, 149.

75. M. M. Rodriguez, E. Bill, W. W. Brennessel, P. L. Holland, *Science*, **2011**, *334*, 780.

76. K. O. Christe, D. A. Dixon, D. J. Grant, R. Haiges, F. S. Tham, A. Vij, V. Vij, T.-H. Wang, W. W. Wilson, *Inorg. Chem.*, **2010**, *49*, 6823.

77. W. B. Tolman, *Angew. Chem., Int. Ed.*, **2010**, *49*, 1018.

78. M. Wörle, P. Latal, R. Kissner, R. Nesper, W. H. Koppenol, *Chem. Res. Toxicol.*, **1999**, *12*, 305.

79. O. V. Gerasimov, S. V. Lymar, *Inorg. Chem.*, **1999**, *38*, 4317; *Chem. Res. Toxicol.*, **1998**, *11*, 709.

80. E. A. Ogryzlo, *J. Chem. Educ.*, **1965**, *42*, 647.

81. L. F. Lundegaard, G. Weck, M. I. McMahon, S. Desgreniers, P. Loubeyre, *Nature*, **2006**, *443*, 201.

82. R. Steudel, M. W. Wong, *Angew. Chem, Int. Ed.*, **2007**, *46*, 1768.

83. B. Meyer, *Chem. Rev.*, **1976**, *76*, 367.

84. W. N. Tuller, ed., *The Sulphur Data Book*, McGraw-Hill, New York, 1954.

85. K. Grubel, P. L. Holland, *Angew. Chem., Int. Ed.*, **2012**, *51*, 3308. C. Tard, C. J. Pickett, *Chem. Rev.*, **2009**, *109*, 2245.

86. M. E. Weeks, "The Halogen Family," in *Discovery of the Elements*, 7th ed, revised by H. M. Leicester, Journal of Chemical Education, Easton, PA, 1968, pp. 701–749.

87. J. Berkowitz, A. C. Wahl, *Adv. Fluorine Chem.*, **1973**, *7*, 147. See also R. Ponec, D. L. Cooper, *J. Phys. Chem. A*, **2007**, *111*, 11294 and references cited therein.

88. D. Mootz, *Angew. Chem., Int. Ed.*, **1981**, *20*, 791.

89. A. Hagfeldt, G. Boschloo, L. Sun, L. Kloo, H. Pettersson, *Chem. Rev.*, **2010**, *110*, 6595.

90. M. Gorlov, H. Pettersson, A. Hagfeldt, L. Kloo, *Inorg. Chem.*, **2007**, *46*, 3566.

91. H. Cavendish, *Philos. Trans.*, **1785**, *75*, 372.

92. Lord Rayleigh, W. Ramsay, *Philos. Trans. A*, **1895**, *186*, 187.

93. N. Bartlett, D. H. Lohmann, *Proc. Chem. Soc.*, **1962**, 115.

94. N. Bartlett, *Proc. Chem. Soc.*, **1962**, 218.

95. W. Grochala, *Chem. Soc. Rev.*, **2007**, *36*, 1632.

96. J. F. Lehmann, H. P. A. Mercier, G. J. Schrobilgen, *Coord. Chem. Rev.*, **2002**, *233*, 1.

97. L. Khriachtchev, M. Pettersson, N. Runeberg, J. Lundell, M. Räsänen, *Nature (London)*, **2000**, *406*, 874.

98. K. Seppelt, D. Lentz, *Progr. Inorg. Chem.*, **1982**, *29*, 172–180; E. A. V. Ebsworth, D. W. H. Rankin, and S. Craddock, *Structural Methods in Inorganic Chemistry*, Blackwell Scientific Publications, Oxford, 1987, pp. 397–398.

99. D. A. Dixon, W. A. de Jong, K. A. Peterson, K. O. Christe, G. J. Schrobilgen, *J. Am. Chem. Soc.*, **2005**, *127*, 8627.

100. S. Hoyer, T. Emmler, K. Sepplet, *J. Fluor. Chem.*, **2006**, *127*, 1415. CSD 416317 obtained from the Crystal Structure Deposition at Fachinformationszentrum Karlsruhe, www.fiz-karlsruhe.de/crystal_structure_dep.html

101. S. W. Peterson, J. H. Holloway, B. A. Coyle, J. M. Williams, *Science*, **1971**, *173*, 1238.

102. C. Sanloup, B. C. Schmidt, G. Gudfinnsson, A. Dewaele, M. Mezouar, *Geochem. Cosmochim. Acta*, **2011**, *75*, 6271 and references cited therein.

103. D. S. Brock, G. J. Schrobilgen, *J. Am. Chem. Soc.*, **2011**, *133*, 6265.

104. G. L. Smith, H. P. A. Mercier, G. J Schrobilgen, *Inorg. Chem.*, **2007**, *46*, 1369.

105. H. St. A. Elliott, J. F. Lehmann, H. P. A. Mercier, H. D. B. Jenkins, G. J. Schrobilgen, *Inorg. Chem.*, **2010**, *49*, 8504.

106. M. D. Moran, D. S. Brock, H. P. A. Mercier, G. J. Schrobilgen, *J. Am. Chem. Soc.*, **2010**, *132*, 13823.

107. M. Gerken, M. D. Moran, H. P. A. Mercier, B. E. Pointner, G. J. Schrobilgen, B. Hoge, K. O. Christe, J. A. Boatz, *J. Am. Chem. Soc.*, **2009**, *131*, 13474.

108. D. Kurzydiowski, P. Zaleski-Ejgierd, W. Grochala, R. Hoffmann, *Inorg. Chem.*, **2011**, *50*, 3832.

109. S. Seidel, K. Seppelt, *Science*, **2000**, *290*, 117; T. Drews, S. Seidel, K. Seppelt, *Angew. Chem., Int. Ed.*, **2002**, *41*, 454.

110. K. Koppe, V. Bilir, H.-J. Frohn, H. P. A. Mercier, G. J. Schrobilgen, *Inorg. Chem.*, **2007**, *46*, 9425.

111. H.-J. Frohn, V. V. Bardin, *Inorg. Chem.*, **2012**, *51*, 2616.

112. L. Khriachtchev, H. Tanskanen, J. Lundell, M. Pettersson, H. Kiljunen, M. Räsänen, *J. Am. Chem. Soc.*, **2003**, *125*, 4696.

113. T. Arppe, L. Khriachtchev, A. Lignell, A. V. Domanskaya, M. Räsänen, *Inorg. Chem.*, **2012**, *51*, 4398.

114. G. Tavcar, B. Žemva, *Angew. Chem. Int. Ed.*, **2009**, *48*, 1432. M. Tramšek, B. Žemva, *J. Fluor. Chem.*, **2006**, *127*, 1275. G. Tavčar, M. Tramšek, T. Bunič, P. Benkič, B. Žemva, *J. Fluor. Chem.*, **2004**, *125*, 1579.

115. M. Tramšek, P. Benkič, B. Žemva, *Inorg. Chem.*, **2004**, *43*, 699.

116. J. F. Lehmann, D. A. Dixon, G. J. Schrobilgen, *Inorg. Chem.*, **2001**, *40*, 3002.

117. D. S. Brock, J. J. Casalis de Pury, H. P. A. Mercier, G. J. Schrobilgen, B. Silvi, *J. Am. Chem. Soc.*, **2010**, *132*, 3533.

118. P. J. MacDougall, G. J. Schrobilgen, R. F. W. Bader, *Inorg. Chem.*, **1989**, *28*, 763.

119. J. C. P. Saunders, G. J. Schrobilgen, *Chem. Commun.*, **1989**, 1576.

120. L. Khriachtchev, H. Tanskanen, A. Cohen, R. B. Gerber, J. Lundell, M. Pettersson, H. Kilijunen, M. Räsänen, *J. Am. Chem. Soc.*, **2003**, *125*, 6876.

121. L. Khriachtchev, M. Räsänen, R. B. Gerber, *Acc. Chem. Res.*, **2009**, *42*, 183.

General References

More detailed descriptions of the chemistry of the main group elements can be found in N. N. Greenwood and A. Earnshaw, *Chemistry of the Elements,* 2nd ed., Butterworth-Heinemann, London, 1997, and in F. A. Cotton, G. Wilkinson, C. A. Murillo, and M. Bochman, *Advanced Inorganic Chemistry,* 6th ed., Wiley InterScience, New York, 1999. A handy reference on the properties of the elements themselves, including many physical properties, is J. Emsley, *The Elements*, 3rd ed., Oxford University Press, 1998. For extensive structural information on inorganic compounds, see A. F. Wells, *Structural Inorganic Chemistry,* 5th ed., Clarendon Press, Oxford, 1984. Three useful references on the chemistry of nonmetals are R. B. King, *Inorganic Chemistry of Main Group Element*, VCH Publishers, New York, 1995; P. Powell and P. Timms, *The Chemistry of the Nonmetals*, Chapman and Hall, London, 1974; and R. Steudel, *Chemistry of the Non-Metals*, Walter de Gruyter, Berlin, 1976, English edition by F. C. Nachod and J. J. Zuckerman. The most complete reference on chemistry of the main group compounds through the early 1970s is the five-volume set by J. C. Bailar, Jr., H. C. Emeléus, R. Nyholm, and A. F. Trotman-Dickinson, editors, *Comprehensive Inorganic Chemistry*, Pergamon Press, Oxford, 1973. Two recent references on fullerene and related chemistry are A. Hirsch and M. Brettreich, *Fullerenes*, Wiley-VCH, Weinheim, Germany, 2005, and F. Langa and J.-F. Nierengarten, editors, *Fullerenes, Principles, and Applications*, RSC Publishing, Cambridge, UK, 2007. For entertaining and informative commentary and numerous photographs of all the elements that lend themselves to photos, see T. Gray, *The Elements*, Black Dog & Leventhal, New York, 2009.

Problems

8.1 The ions H_2^+ and H_3^+ have been observed in gas discharges.

a. H_2^+ has been reported to have a bond distance of 106 pm and a bond dissociation enthalpy of 255 kJ mol^{-1}. Comparable values for the neutral molecule are 74.2 pm and 436 kJ mol^{-1}. Are these values for H_2^+ in agreement with the molecular orbital picture of this ion? Explain.

b. Assuming H_3^+ to be triangular (the probable geometry), describe the molecular orbitals of this ion and determine the expected H—H bond order.

8.2 The species He_2^+ and HeH^+ have been observed spectroscopically. Prepare molecular orbital diagrams for these two ions. What would you predict for the bond order of each?

8.3 The chemical species IF_4^- and XeF_4 are known. Is the isoelectronic ion CsF_4^+ plausible? If it could be made, predict and sketch its most likely shape. Comment on whether such an ion might or might not be possible.

8.4 The equilibrium constant for the formation of the cryptand $[Sr\{cryptand(2.2.1)\}]^{2+}$ is larger than the equilibrium constants for the analogous calcium and barium cryptands. Suggest an explanation. (See E. Kauffmann, J-M. Lehn, J-P. Sauvage, *Helv. Chim. Acta*, **1976**, *59*, 1099.)

8.5 Gas phase BeF_2 is monomeric and linear. Prepare a molecular orbital description of the bonding in BeF_2.

8.6 In the gas phase, $BeCl_2$ forms a dimer of the following structure:

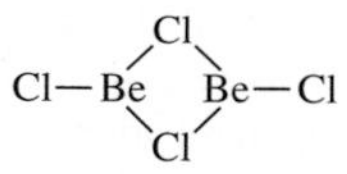

Cl—Be(μ-Cl)₂Be—Cl

Describe the bonding of the chlorine bridges in this dimer in molecular orbital terms.

8.7 BF can be obtained by reaction of BF_3 with boron at 1850 °C and low pressure; BF is highly reactive but can be preserved at liquid nitrogen temperature (77 K). Prepare a molecular orbital diagram of BF. How would the molecular orbitals of BF differ from CO, with which BF is isoelectronic?

8.8 N-heterocyclic carbenes, such as the example shown here, have become increasingly important in both main group and transition-metal chemistry. For example, the first report of a stable neutral molecule having a boron–boron double bond used this N-heterocyclic carbene to bond to each of the borons in a product of the reaction of $\mathbf{R}BBr_3$ with the powerful reducing agent KC_8 (potassium graphite) in diethylether solvent. (See Y. Wang, B. Quillian, P. Wei, C. S. Wannere, Y. Xie, R. B. King, H. F. Schaefer, III, P. v. R. Schleyer, G. H. Robinson, *J. Am. Chem. Soc.*, **2007**, *129*, 12412.)

a. What was the structure of this first molecule with a B$=$B bond? What evidence was cited for a double bond?

b. If $\mathbf{R}SiCl_4$ was used in place of $\mathbf{R}BBr_3$, an equally notable silicon compound was formed. What was this compound, and what parallels were noted with the reaction in Part **a**? (See Y. Wang, Y. Xie, P. Wei, R. B. King, H. F. Schaefer III, P. v. R. Schleyer, G. H. Robinson, *Science*, **2008**, *321*, 1069.)

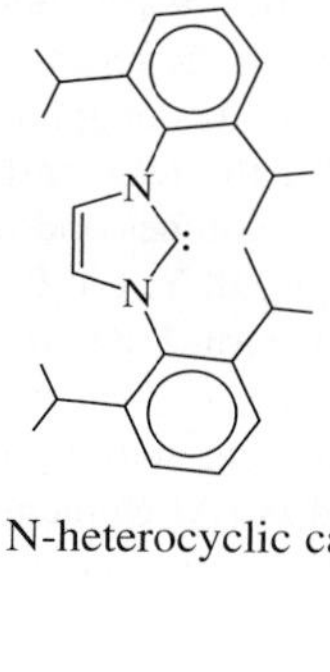

R, an N-heterocyclic carbene

8.9 $Al_2(CH_3)_6$ is isostructural with diborane, B_2H_6. On the basis of the orbitals involved, describe the Al—C—Al bonding for the bridging methyl groups in $Al_2(CH_3)_6$.

8.10 Referring to the description of bonding in diborane in Figure 8.12:

a. Show that the representation $\Gamma(p_z)$ reduces to $A_g + B_{1u}$.

b. Show that the representation $\Gamma(p_x)$ reduces to $B_{2g} + B_{3u}$.

c. Show that the representation $\Gamma(1s)$ reduces to $A_g + B_{3u}$.

d. Using the D_{2h} character table, verify that the sketches for the group orbitals match their respective symmetry designations (A_g, B_{2g}, B_{1u}, B_{3u}).

8.11 The compound $C(PPh_3)_2$ is bent at carbon; the P—C—P angle in one form of this compound has been reported as 130.1°. Account for the nonlinearity at carbon.

8.12 The C—C distances in carbides of formula MC_2 are in the range 119 to 124 pm if M is a Group 2 (IIA) metal or other metal commonly forming a 2+ ion, but in the approximate range 128 to 130 pm for Group 3 (IIIB) metals, including the lanthanides. Why is the C—C distance greater for the carbides of the Group 3 metals?

8.13 The half-life of ^{14}C is 5730 years. A sample taken for radiocarbon dating was found to contain 56 percent of its original ^{14}C. What was the age of the sample? (Radioactive decay of ^{14}C follows first-order kinetics.)

8.14 Prepare a model of buckminsterfullerene, C_{60}. By referring to the character table, verify that this molecule has I_h symmetry.

8.15 Determine the point groups of the following:

a. The unit cell of diamond

b. C_{20}

c. C_{70}

d. The nanoribbons shown in Figure 8.19

8.16 What is graph*ane*? How has it been synthesized, and what might be some of its potential uses? (See D. C. Elias, et al., *Science*, **2009**, *323*, 610.)

8.17 Prepare a sheet showing an extended graphene structure, approximately 12 by 15 fused carbon rings or larger. Use this sheet to show how the graphene structure could be rolled up to form (**a**) a zigzag nanotube, (**b**) an armchair nanotube, and (**c**) a chiral nanotube. Is more than one chiral structure possible? (See M. S. Dresselhaus, G. Dresselhaus, and R. Saito, *Carbon*, **1995**, *33*, 883.)

8.18 To what extent has using carbon nanotubes for delivery of the anticancer agent cisplatin shown promise in the killing of cancer cells? (See J. F. Rusling, J. S. Gutkind, A. A. Bhirde, et al., *ACS Nano*, **2009**, *3*, 307.) The reader is encouraged to search the recent literature to find more up to date references in this ongoing area of research.

8.19 What accounts for the relationship between the color of armchair carbon nanotubes and their diameter? See E. H. Hároz, J. G. Duque, B. Y. Lu, P. Nikolaev, S. Arepalli, R. H. Hauge, S. K. Doorn, J. Kono, *J. Am. Chem. Soc.*, **2012**, *134*, 4461.

8.20 Explain the increasing stability of the 2+ oxidation state for the Group 14 (IVA) elements with increasing atomic number.

8.21 1,2-Diiododisilane has been observed in both *anti* and *gauche* conformations. (See K. Hassler, W. Koell, K. Schenzel, *J. Mol. Struct.*, **1995**, *348*, 353.) For the *anti* conformation, shown here,

a. What is the point group?

b. Predict the number of infrared-active silicon–hydrogen stretching vibrations.

8.22 The reaction $P_4(g) \rightleftharpoons 2\ P_2(g)$ has $\Delta H = 217\ \text{kJ mol}^{-1}$. If the bond energy of a single phosphorus–phosphorus bond is 200 kJ mol^{-1}, calculate the bond energy of the P≡P bond. Compare the value you obtain with the bond energy in N_2(946 kJ mol^{-1}), and suggest an explanation for the difference in bond energies in P_2 and N_2.

8.23 The azide ion, N_3^-, is linear with equal N—N bond distances.

a. Describe the pi molecular orbitals of azide.

b. Describe in HOMO-LUMO terms the reaction between azide and H^+ to form hydrazoic acid, HN_3.

c. The N—N bond distances in HN_3 are given in Figure 8.30. Explain why the terminal N—N distance is shorter than the central N—N distance in this molecule.

8.24 In aqueous solution, hydrazine is a weaker base than ammonia. Why? (pK_b values at 25 °C: NH_3, 4.74; N_2H_4, 6.07)

8.25 The bond angles for the hydrides of the Group 15 (VA) elements are as follows: NH_3, 106.6°; PH_3, 93.2°; AsH_3, 92.1°; and SbH_3, 91.6°. Account for this trend.

8.26 Gas-phase measurements show that the nitric acid molecule is planar. Account for the planarity of this molecule.

8.27 With the exception of NO_4^{3-}, all the molecules and ions in Table 8.10 are planar. Assign their point groups.

8.28 The uncatalyzed isomerization for *cis*-N_2F_2 and *trans*-N_2F_2 has been examined via computational techniques (K. O. Christe, D. A. Dixon, D. J. Grant, R. Haiges, F. S. Tham, A. Vij, V. Vij, T.-H. Wang, W. W. Wilson, *Inorg. Chem.*, **2010**, *49*, 6823).

a. Which isomer has the lower electronic ground state at 298 K (and by how much) on the basis of the computational work?

b. Describe the proposed isomerization mechanisms. Which has a lower activation barrier (and by how much)?

c. Using SbF_5 as the Lewis acid, draw a Lewis structure of the species proposed to lower the isomerization barrier for *trans*-N_2F_2. Which mechanism is presumed to be operative in this case? Describe the expected geometric changes in the *trans*-N_2F_2/SbF_5 adduct, and how these facilitate the isomerization.

8.29 Use the Latimer diagrams for phosphorus in Appendix B.7 to construct Frost diagrams for acidic and basic conditions. Provide balanced reduction half-reactions for all adjacent couples in these Latimer diagrams.

8.30 What type of interaction holds the O_2 units together in the O_8 structure? How does this interaction stabilize the larger molecule? (Hint: consider the molecular orbitals of O_2. See R. Steudel, M. W. Wong, *Angew. Chem., Int. Ed.*, **2007**, *46*, 1768.)

8.31 The sulfur–sulfur distance in S_2, the major component of sulfur vapor above ~720° C, is 189 pm, significantly shorter than the sulfur–sulfur distance of 206 pm in S_8. Suggest an explanation for the shorter distance in S_2. (See C. L. Liao, C. Y. Ng, *J. Chem. Phys.*, **1986**, *84*, 778.)

8.32 Because of its high reactivity with most chemical reagents, F_2 is ordinarily synthesized electrochemically. However, the chemical synthesis of F_2 has been recorded via the reaction

$$2\ K_2MnF_6 + 4\ SbF_5 \longrightarrow 4\ KSbF_6 + 2\ MnF_3 + F_2$$

This reaction can be viewed as a Lewis acid–base reaction. Explain. (See K. O. Christe, *Inorg. Chem.*, **1986**, *25*, 3722.)

8.33 **a.** Chlorine forms a variety of oxides, among them Cl_2O and Cl_2O_7. Cl_2O has a central oxygen; Cl_2O_7 also has a central oxygen atom, bridging two ClO_3 groups. Of these two compounds, which would you predict to have the smaller Cl—O—Cl angle? Explain briefly.

b. The dichromate ion, $Cr_2O_7^{2-}$, has the same structure as Cl_2O_7, with oxygen now bridging two CrO_3 groups. Which of these two oxygen-bridged species do you expect to have the smaller outer atom–O–outer atom angle? Explain briefly.

8.34 The triiodide ion I_3^- is linear, but I_3^+ is bent. Explain.

8.35 Although B_2H_6 has D_{2h} symmetry, I_2Cl_6 is planar. Account for the difference in the structures of these two molecules.

8.36 BrF_3 undergoes autodissociation according to the equilibrium

$$2\ BrF_3 \rightleftharpoons BrF_2^+ + BrF_4^-$$

Ionic fluorides such as KF behave as bases in BrF_3, whereas some covalent fluorides such as SbF_5 behave as acids. On the basis of the solvent system concept, write balanced chemical equations for these acid–base reactions of fluorides with BrF_3.

8.37 The diatomic cations Br_2^+ and I_2^+ are both known.

a. On the basis of the molecular orbital model, what would you predict for the bond orders of these ions? Would you predict these cations to have longer or shorter bonds than the corresponding neutral diatomic molecules?

b. Br_2^+ is red, and I_2^+ is bright blue. What electronic transition is most likely responsible for absorption in these ions? Which ion has the more closely spaced HOMO and LUMO?

c. I_2 is violet, and I_2^+ is blue. On the basis of frontier orbitals (identify them), account for the difference in their colors.

8.38 I_2^+ exists in equilibrium with its dimer I_4^{2+} in solution. I_2^+ is paramagnetic and the dimer is diamagnetic. Crystal structures of compounds containing I_4^{2+} have shown this ion to be planar and rectangular, with two short I—I distances (258 pm) and two longer distances (326 pm).

a. Using molecular orbitals, propose an explanation for the interaction between two I_2^+ units to form I_4^{2+}.

b. Which form is favored at high temperature, I_2^+ or I_4^{2+}? Why?

8.39 How many possible isomers are there of the ion $IO_2F_3^{2-}$? Sketch these, and indicate the point group for each. The observed ion has IR-active iodine–oxygen stretches at 802 and 834 cm^{-1}. What does this indicate about the most likely structure? (See J. P. Mack, J. A. Boatz, M. Gerken, *Inorg. Chem.*, **2008**, *47*, 3243.

8.40 What is a **superhalogen**? Does it contain a halogen? By searching the literature, find the earliest reference to this term and the rationale for the name "superhalogen." Also provide examples of modern chemical applications of superhalogens.

8.41 Bartlett's original reaction of xenon with PtF_6 apparently yielded products other than the expected $Xe^+PtF_6^-$. However, when xenon and PtF_6 are reacted in the presence of a large excess of sulfur hexafluoride, $Xe^+PtF_6^-$ is apparently formed. Suggest the function of SF_6 in this reaction. (See: K. Seppelt, D. Lentz, *Progr. Inorg. Chem.*, **1982**, *29*, 170.)

8.42 On the basis of VSEPR, predict the structures of $XeOF_2$, $XeOF_4$, XeO_2F_2, and XeO_3F_2. Assign the point group of each.

8.43 The sigma bonding in the linear molecule XeF_2 may be described as a three-center, four-electron bond. If the z axis is assigned as the internuclear axis, use the p_z orbitals on each of the atoms to prepare a molecular orbital description of the sigma bonding in XeF_2.

8.44 The $OTeF_5$ group can stabilize compounds of xenon in formal oxidation states IV and VI. On the basis of VSEPR, predict the structures of $Xe(OTeF_5)_4$ and $O{=}Xe(OTeF_5)_4$.

8.45 Write a balanced equation for the oxidation of Mn^{2+} to MnO_4^- by the perxenate ion in acidic solution; assume that neutral Xe is formed.

8.46 The ion XeF_5^- is a rare example of pentagonal planar geometry. On the basis of the symmetry of this ion, predict the number of IR-active Xe–F stretching bands.

8.47 $XeOF_4$ has one of the more interesting structures among noble gas compounds. On the basis of its symmetry,

a. Obtain a representation based on *all* the motions of the atoms in $XeOF_4$.

b. Reduce this representation to its component irreducible representations.

c. Classify these representations, indicating which are for translational, rotational, and vibrational motion.

8.48 The ion XeF_2^{2-} has not been reported. Would you expect this ion to be bent or linear? Suggest why it remains elusive.

8.49 Photolysis of a solid mixture of H_2O, N_2O, and Xe at 9 K, followed by annealing, gave a product identified as HXeOXeH (L. Khriachtchev, K. Isokoski, A. Cohen, M. Räsänen, R. B. Gerber, *J. Am. Chem. Soc.*, **2008**, *130*, 6114). Supporting evidence included infrared data, with a band at 1379.7 cm^{-1}, assigned to Xe—H stretching. When the experiment was repeated using deuterated water containing both D_2O and H_2O, bands at 1432.7 cm^{-1}, 1034.7 cm^{-1}, and 1003.3 cm^{-1}, were observed, together with the band at 1379.7 cm^{-1}. Account for the infrared data from the experiment conducted using deuterated water.

8.50 Use the Latimer diagram for xenon in basic solution in Appendix B.7 to:

a. Construct balanced half-reactions for the $HXeO_6^{3-}/HXeO_4^-$ and $HXeO_4^-/Xe$ redox couples.

b. Use these half-reactions to show that the disproportionation of $HXeO_4^-$ in basic solution is spontaneous by calculating E^o and ΔG^o for the reaction.

8.51 Although atmospheric ozone depletion is a widely known phenomenon, the estimate that as much as 90% of the xenon originally in Earth's atmosphere is now absent is largely unknown. The synthesis and study of XeO_2 (D. S. Brock, G. J. Schrobilgen, *J. Am. Chem. Soc.*, **2011**, *133*, 6265) supports the hypothesis that the reaction of SiO_2 from the Earth's core with Xe may be responsible for the depletion of this noble gas.

a. Describe how the synthesis of XeO_2 is carried out, including the reactants, conditions, temperature sensitivity, and hazards associated with the synthesis.

b. Explain why monomeric XeO_2 would likely be soluble in water.

c. Raman spectroscopic studies were vital to characterize XeO_2. What experimental observation suggested that XeO_2 only contains Xe—O bonding? What experimental observation suggested that XeO_2 did not contain any hydrogen atoms?

8.52 Liquid NSF_3 reacts with $[XeF][AsF_6]$ to form $[F_3SNXeF][AsF_6]$, **1**. On mild warming in the solid state, compound **1** rearranges to $[F_4SNXe][AsF_6]$, **2**. Reaction of compound **2** with HF yields $[F_4SNH_2][AsF_6]$, **3**, $[F_5SN(H)Xe][AsF_6]$, **4**, and XeF_2. (See G. L. Smith, H. P. A. Mercier, G. J. Schrobilgen, *Inorg. Chem.*, **2007**, *46*, 1369; *Inorg. Chem.*, **2008**, *47*, 4173; and *J. Am. Chem. Soc.*, **2009**, *131*, 4173.)

a. What is the bond order in $[XeF]^+$?

b. Of compounds **1** through **4**, which has the longest S—N distance? The shortest?

c. Which of the compounds **1** through **4** is most likely to have linear bonding around the nitrogen atom?

d. By the VSEPR approach, is the bonding around Xe expected to be linear or bent in compound **1**?

e. In compounds **2** and **3**, are the NXe and NH_2 groups likely to occupy axial or equatorial sites on sulfur?

f. In compounds **2** and **3**, which bonds are more likely to be longer, the $S—F_{axial}$ bonds or $S—F_{equatorial}$ bonds?

8.53 Determine the point groups:

a. O_8 (Figure 8.34)

b. S_8 (Figure 8.35)

c. A chiral nanotube

d. The section of β-graphyne shown in Figure 8.23

e. $Xe_3OF_3^+$

f. The proposed very high pressure form of oxygen, spiral-chain O_4 (L. Zhu, Z. Wang, Y. Wang, G. Zou, H. Mao, Y. Ma, *Proc. Nat. Acad. Sci.*, **2012**, *109*, 751).

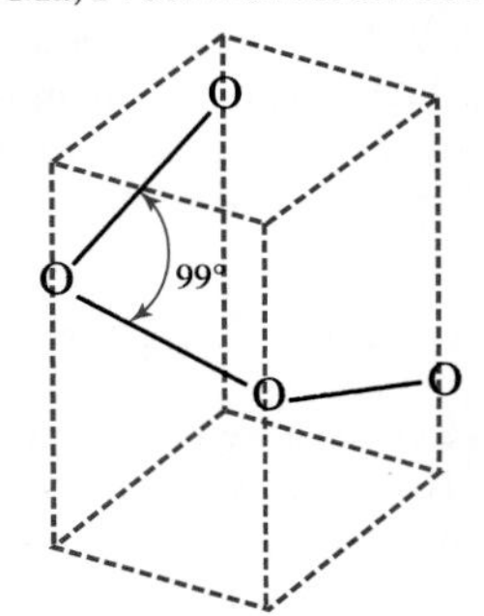

The following problems require the use of molecular modeling software.

8.54 It has been proposed that salts containing the cation $[FBeNg]^+$, where Ng = He, Ne, or Ar, may be stable. Use molecular modeling software to calculate and display the molecular orbitals of $[FBeNe]^+$. Which molecular orbitals would be the primary ones engaged in bonding in this ion? (See M. Aschi, F. Grandinetti, *Angew. Chem., Int. Ed.*, **2000**, *39*, 1690.)

8.55 The dixenon cation Xe_2^+ has been characterized structurally. Using molecular modeling software, calculate and observe the energies and shapes of the molecular orbitals of this ion. Classify each of the seven highest-energy occupied orbitals as σ, π, or δ, and as bonding or antibonding. The bond in this compound was reported as the longest main group–main group bond observed to date. Account for this very long bond. (See T. Drews, K. Seppelt, *Angew. Chem., Int. Ed.*, **1997**, *36*, 273.)

8.56 To further investigate the bonding in the O_8 structure, construct an O_8 unit, and calculate and display its molecular orbitals. Which orbitals are involved in holding the O_2 units together? In each of these orbitals, identify the atomic orbitals that are primarily involved. (See Figure 8.34; additional structural details can be found in L. F. Lundegaard, G. Weck, M. I. McMahon, S. I. Desgreniers, P. Loubeyre, *Science*, **2006**, *443*, 201 and R. Steudel, M. W. Wong, *Angew. Chem., Int. Ed.*, **2007**, *46*, 1768.)

Chapter 9

Coordination Chemistry I: Structures and Isomers

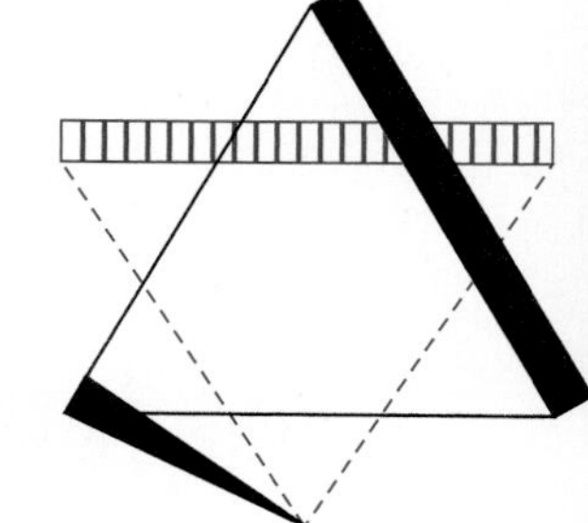

Coordination compounds are composed of a metal atom or ion and one or more **ligands** (atoms, ions, or molecules) that donate electrons to the metal. This definition includes compounds with metal–carbon bonds, or **organometallic compounds**, described in Chapters 13 to 15.

Coordination compound comes from the coordinate covalent bond, which historically was considered to form by donation of a pair of electrons from one atom to another. In coordination compounds the donors are usually the ligands, and the acceptors are the metals. Coordination compounds are examples of acid–base adducts (Chapter 6), frequently called **complexes** or, if charged, **complex ions**.

9.1 History

Although the formal study of coordination compounds really begins with Alfred Werner (1866–1919), coordination compounds have been used as pigments and dyes since antiquity. Examples include Prussian blue ($KFe[Fe(CN)_6]$), aureolin ($K_3[Co(NO_2)_6] \cdot 6H_2O$, yellow), and alizarin red dye (the calcium aluminum salt of 1,2-dihydroxy-9,10-anthraquinone). The tetraamminecopper(II) ion—actually $[Cu(NH_3)_4(H_2O)_2]^{2+}$ in solution, which has a striking royal blue color—was known in prehistoric times. The formulas of these compounds were deduced in the late nineteenth century, providing background for the development of bonding theories.

Inorganic chemists tried to use existing theories applied to organic molecules and salts to explain bonding in coordination compounds, but these theories were found inadequate. For example, in hexaamminecobalt(III) chloride, $[Co(NH_3)_6]Cl_3$, early bonding theories allowed only three other atoms to be attached to the cobalt because of its "valence" of 3. By analogy with salts, such as $FeCl_3$, the chlorides were assigned this role. It was necessary to develop new ideas to explain the bonding involving the ammonia. Blomstrand[1] (1826–1894) and Jørgensen[2] (1837–1914) proposed that the nitrogens could form chains (**Table 9.1**) with those atoms having a valence of 5. According to this theory, chloride ions attached directly to cobalt were bonded more strongly than chloride bonded to nitrogen. Werner[3] proposed that all six ammonias could bond directly to the cobalt ion. Werner allowed for a looser bonding of the chloride ions; we now consider them independent ions.

Table 9.1 illustrates how chain theory and Werner's coordination theory predict the number of ions afforded by dissociation by various cobalt complexes. Blomstrand's theory allowed dissociation of chlorides attached to ammonia but not of chlorides attached to cobalt. Werner's theory also included two kinds of chlorides. The first kind were attached to the cobalt (these metal-bound chlorides were believed not to dissociate); these plus the number of ammonia molecules totaled six. The other chlorides were considered less firmly bound, permitting their dissociation.

TABLE 9.1 Comparison of Blomstrand's Chain Theory and Werner's Coordination Theory

Werner Formula (Modern Form)	Number of Ions Predicted	Blomstrand Chain Formula	Number of Ions Predicted
$[Co(NH_3)_6]\mathit{Cl}_3$	4	$NH_3-\mathit{Cl}$ $Co-NH_3-NH_3-NH_3-NH_3-\mathit{Cl}$ $NH_3-\mathit{Cl}$	4
$[Co(NH_3)_5Cl]\mathit{Cl}_2$	3	$NH_3-\mathit{Cl}$ $Co-NH_3-NH_3-NH_3-NH_3-\mathit{Cl}$ Cl	3
$[Co(NH_3)_4Cl_2]\mathit{Cl}$	2	Cl $Co-NH_3-NH_3-NH_3-NH_3-\mathit{Cl}$ Cl	2
$[Co(NH_3)_3Cl_3]$	0	Cl $Co-NH_3-NH_3-NH_3-\mathit{Cl}$ Cl	2

The italicized chlorides dissociate in solution, according to the two theories.

Except for the last compound, the predicted number of ions upon dissociation match. Even with the last compound, experimental challenges left some ambiguity. The debate between Jørgensen and Werner continued for years. This case illustrates good features of scientific controversy. Werner was forced to develop his theory further, and synthesize new compounds to test his ideas, because Jørgensen vigorously defended his chain theory. Werner proposed an octahedral structure for compounds such as those in Table 9.1. He prepared and characterized many isomers, including both green and violet forms of $[Co(H_2NC_2H_4NH_2)_2Cl_2]^+$. He claimed that these compounds had the chlorides arranged ***trans*** (opposite each other) and ***cis*** (adjacent to each other) respectively, in an overall octahedral geometry, as in **Figure 9.1**. Jørgensen offered alternative isomeric structures but accepted Werner's model in 1907, when Werner synthesized the green *trans* and the violet *cis* isomers of $[Co(NH_3)_4Cl_2]^+$. Chain theory could not account for two *different* structures with the same formula for this complex ion.

Werner's syntheses of $[Co(NH_3)_4Cl_2]^+$ and discovery of optically active, carbon-free, coordination compounds did not convince all chemists, even when chain theory could not

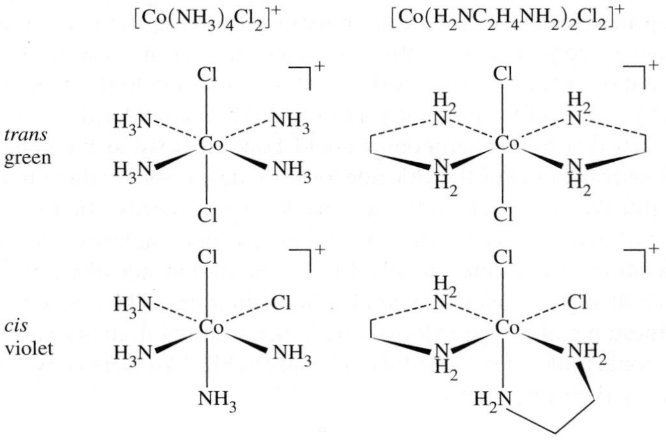

FIGURE 9.1 *cis* and *trans* Isomers.

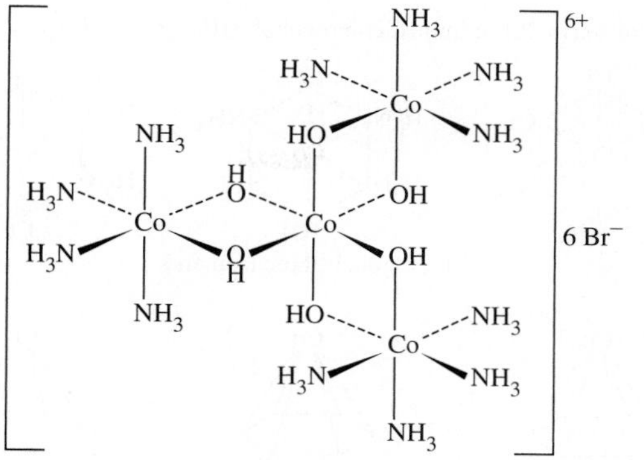

FIGURE 9.2 Werner's Carbon-Free Optically Active Compound, $[Co\{Co(NH_3)_4(OH)_2\}_3]Br_6$.

be applied. Some argued that Werner was mistaken that his optically active compounds were carbon-free; these chemists speculated that the chirality of Werner's isomers was due to undetected carbon atoms. Werner validated his hypothesis by resolving a racemic mixture of Jørgensen's $[Co\{Co(NH_3)_4(OH)_2\}_3]Br_6$ (**Figure 9.2**) into its two optically active forms, using *d*- and *l*-α-bromocamphor-π-sulfonate as resolving agents. With definitive proof of optical activity without carbon, Werner's theory was accepted. Pauling[4] extended the theory in terms of hybrid orbitals. Later theories[5] adapted arguments used for electronic structures of ions in crystals to coordination compounds.

Werner studied compounds that are relatively slow to react in solution to develop his theories. He synthesized compounds of Co(III), Rh(III), Cr(III), Pt(II), and Pt(IV), which are kinetically inert.[*] Subsequent examination of more reactive compounds confirmed his theories.

Werner's theory required so-called primary bonding, in which the positive charge of the metal ion is balanced by negative ions, and secondary bonding, in which molecules or ions (**ligands**) are attached directly to the metal ion. The secondary bonded unit is called the **complex ion** or the **coordination sphere**; modern formulas are written with this part in brackets. The words *primary* and *secondary* no longer bear the same significance. In the Table 9.1 examples, the coordination sphere acts as a unit; the ions outside the brackets balance the charge and dissociate in solution. Depending on the metal and the ligands, the metal can have from one up to at least 16 atoms attached to it, with four and six the most common.[**] Chapter 9 concentrates on the coordination sphere. The ions outside the coordination sphere, sometimes called **counterions**, can often be exchanged for others without changing the bonding or ligands within the complex ion coordination sphere.

Werner developed his theories using compounds with four or six ligands. The shapes of the coordination compounds were established by the synthesis of isomers. For example, he was able to synthesize only two isomers of $[Co(NH_3)_4Cl_2]^+$. Possible structures with six ligands are hexagonal, hexagonal pyramidal, trigonal prismatic, trigonal antiprismatic, and octahedral. Because there are two possible isomers for the octahedral shape and three for each of the others (**Figure 9.3**), Werner claimed the structure was octahedral. Such an argument is not irrefutable, because additional isomers may be difficult to synthesize or isolate. However, later experiments confirmed the octahedral shape, with *cis* and *trans* isomers as shown.

Werner's synthesis and separation of optical isomers (Figure 9.2) proved the octahedral shape conclusively; none of the other 6-coordinate geometries could have similar optical activity.

---

*Kinetically inert coordination compounds are discussed in Chapter 12.

**N. N. Greenwood and A. Earnshaw, *Chemistry of the Elements*, 2nd ed., Butterworth–Heinemann, Oxford, UK, 1997, p. 912. The larger numbers depend on how the number of donors in organometallic compounds are counted; some would assign smaller coordination numbers because of the special nature of the organic ligands.

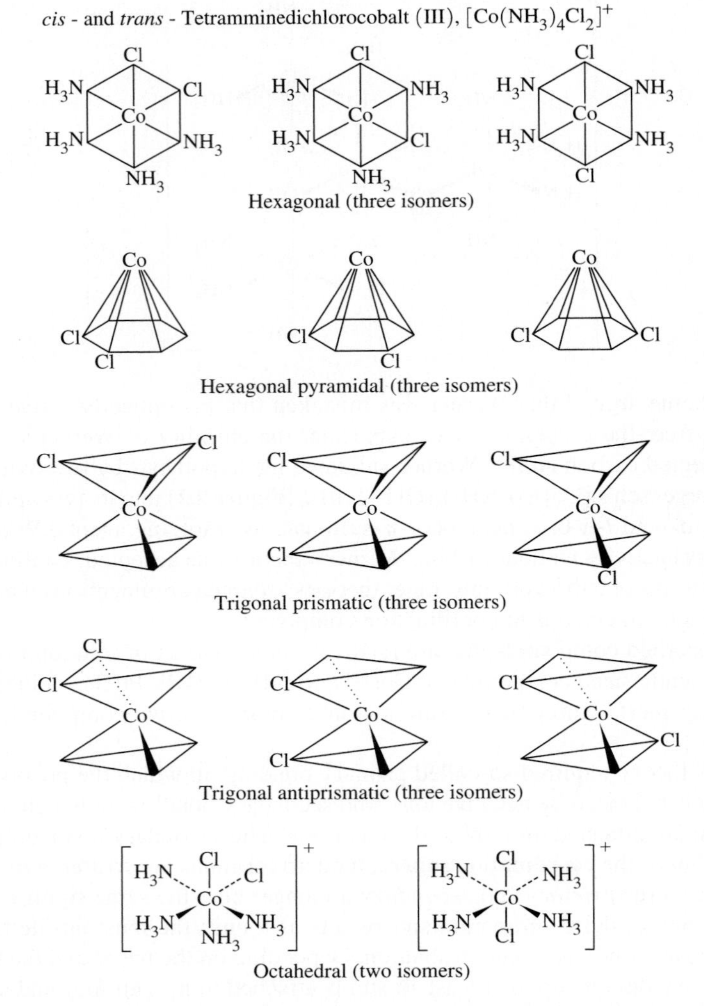

FIGURE 9.3 Possible Hexacoordinate Isomers for $[Co(NH_3)_4Cl_2]^+$ considered by Werner. Only the octahedral structure allows for only two isomers.

Other experiments were consistent with square-planar Pt(II) compounds, with the four ligands at the corners of a square. Werner found only two isomers for $[Pt(NH_3)_2Cl_2]$. These isomers conceivably could have different shapes (tetrahedral and square-planar are just two examples (**Figure 9.4**)), but Werner assumed they had the same shape. Because only one tetrahedral structure is possible for $[Pt(NH_3)_2Cl_2]$, he argued that the two isomers had square-planar shapes with *cis* and *trans* geometries. His theory was correct, although his evidence could not be conclusive.

Werner's evidence for these structures required a theory to rationalize these metal-ligand bonds, and how more than four atoms could bond to a single metal center. Transition-metal

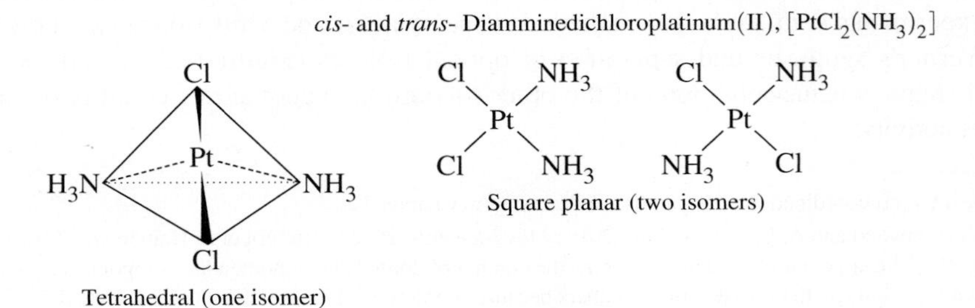

FIGURE 9.4 Possible Structures for $[Pt(NH_3)_2Cl_2]$ considered by Werner.

compounds with six ligands cannot fit the Lewis theory with eight electrons around each atom, and even expanding the shell to 10 or 12 electrons does not work in cases such as $[Fe(CN)_6]^{4-}$, with a total of 18 electrons to accommodate. The **18-electron rule** simply accounts for the bonding in many coordination compounds; the total number of valence electrons around the central atom is counted, with 18 a common result. (This approach is applied to organometallic compounds in Chapter 13.)

Pauling[6] used his **valence bond** approach to explain differences in magnetic behavior among coordination compounds by use of either metal ion 3*d* or 4*d* orbitals. Griffith and Orgel[7] developed **ligand field theory**, derived from the **crystal field theory** of Bethe[8] and Van Vleck[9] on the behavior of metal ions in crystals and from the molecular orbital treatment of Van Vleck.[10] Chapter 10 discusses these theories.

This chapter describes the different shapes of coordination compounds. It can be difficult to confidently predict shapes with only knowledge of complex formulas; subtle electronic and steric factors often govern these structures. The differences in energy between the observed and unobserved structures of complexes are often small. It is useful to correlate structures with the factors that dictate their shapes. This chapter also describes isomeric possibilities for coordination compounds and experimental methods used to study them. The structures of organometallic complexes (Chapters 13 through 15) are also challenging to predict.

9.2 Nomenclature

The nomenclature of coordination chemistry has changed over time. The older literature features multiple nomenclature styles. Contemporary rules used for naming coordination compounds are discussed in this chapter. More complete sources are available to explore classic nomenclature approaches necessary to examine older literature and additional nomenclature schemes not covered in this introductory section.[11]

Ligands are frequently named using older trivial names rather than the International Union of Pure and Applied Chemistry (IUPAC) names. **Tables 9.2**, **9.3**, and **9.4** list common ligands. Those with two or more points of attachment to metal atoms are called **chelating ligands**, and their compounds are called **chelates** (pronounced key′-lates), a name derived from the Greek *khele*, the claw of a crab. Ligands such as ammonia are **monodentate**, with one point of attachment (literally, "one tooth"). Ligands are

TABLE 9.2 Classic Monodentate Ligands

Common Name	IUPAC Name	Formula
hydrido	hydrido	H^-
fluoro	fluoro	F^-
chloro	chloro	Cl^-
bromo	bromo	Br^-
iodo	iodo	I^-
nitrido	nitrido	N^{3-}
azido	azido	N_3^-
oxo	oxido	O^{2-}
cyano	cyano	CN^-
thiocyano	thiocyanato-*S* (S-bonded)	SCN^-
isothiocyano	thiocyanato-*N* (N-bonded)	NCS^-

(continues)

TABLE 9.2 Classic Monodentate Ligands *(cont.)*

Common Name	IUPAC Name	Formula
hydroxo	hydroxo	OH^-
aqua	aqua	H_2O
carbonyl	carbonyl	CO
thiocarbonyl	thiocarbonyl	CS
nitrosyl	nitrosyl	NO^+
nitro	nitrito-*N* (N-bonded)	NO_2^-
nitrito	nitrito-*O* (O-bonded)	ONO^-
methyl isocyanide	methylisocyanide	CH_3NC
phosphine	phosphane	PR_3
pyridine	pyridine (abbrev. py)	C_5H_5N
ammine	ammine	NH_3
methylamine	methylamine	$MeNH_2$
amido	azanido	NH_2^-
imido	azanediido	NH^{2-}

TABLE 9.3 Chelating Amines

Chelating Points	Common Name	IUPAC Name	Abbrev.	Formula
bidentate	ethylenediamine	1,2-ethanediamine	en	$NH_2CH_2CH_2NH_2$
tridentate	diethylenetriamine	1,4,7-triazaheptane	dien	$NH_2CH_2CH_2NHCH_2CH_2NH_2$
		1,3,7-triazacyclononane	tacn	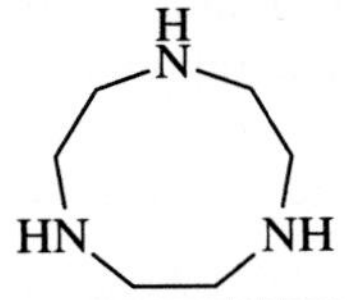 H N HN NH
tetradentate	triethylenetetraamine	1,4,7,10-tetraazadecane	trien	$NH_2CH_2CH_2NHCH_2CH_2NHCH_2CH_2NH_2$
	β, β', β''-triaminotriethylamine	β, β', β''-tris(2-aminoethyl)amine	tren	$NH_2CH_2CH_2NCH_2CH_2NH_2$ | $CH_2CH_2NH_2$
	tetramethylcyclam	1,4,8,11-tetramethyl-1,4,8,11-tetraazacyclotetradecane	TMC	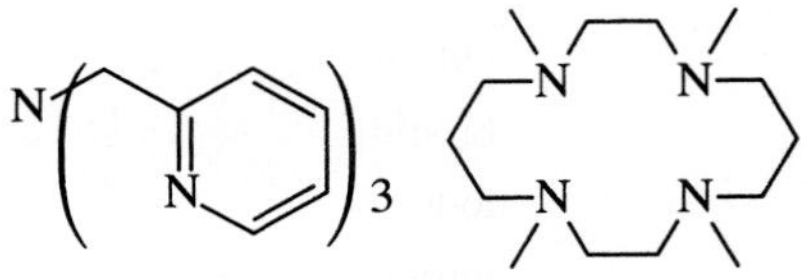 TMC
	tris(2-pyridylmethyl) amine	tris(2-pyridylmethyl) amine	TPA	TPA
pentadentate	tetraethylene-pentamine	1,4,7,10,13-pentaazatridecane		$NH_2CH_2CH_2NHCH_2CH_2NHCH_2CH_2NHCH_2CH_2NH_2$
hexadentate	ethylenediamine-tetraacetate	1,2-ethanediyl (dinitrilo) tetraacetate	EDTA	$^-OOCCH_2$ CH_2COO^- NCH_2CH_2N $^-OOCCH_2$ CH_2COO^-

TABLE 9.4 Multidentate (Chelating) Ligands

Common Name	IUPAC Name	Abbreviation	Formula and Structure
acetylacetonato	2,4-pentanediono	acac	$CH_3COCHCOCH_3^-$ 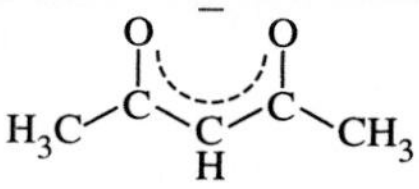
2,2′-bipyridine	2,2′-bipyridyl	bipy	$C_{10}H_8N_2$ 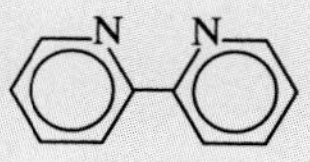
nacnac	*N,N′*-diphenyl-2,4-pentanediiminato	nacnac	$C_{17}H_{17}N_2^-$ 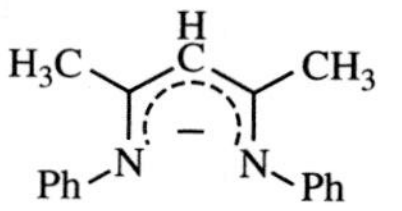
1,10-phenanthroline, *o*-phenanthroline	1,10-diaminophenanthrene	phen, *o*-phen	$C_{12}H_8N_2$ 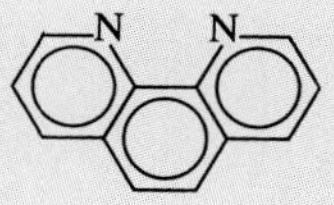
oxalato	oxalato	ox	$C_2O_4^{2-}$ 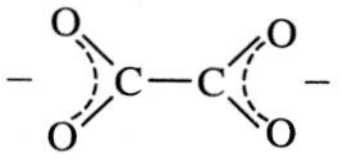
dialkyldithiocarbamato	dialkylcarbamodithioato	dtc	$S_2CNR_2^-$ 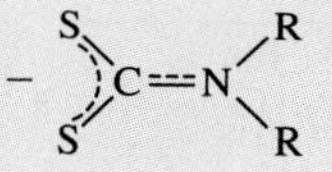
ethylenedithiolate	1,2-ethenedithiolate	dithiolene	$S_2C_2H_2^{2-}$ 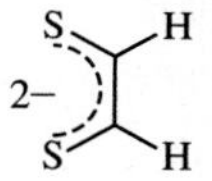
1,2-bis (diphenylphosphino) ethane	1,2-ethanediylbis-(diphenylphosphane)	dppe	$Ph_2PC_2H_4PPh_2$ 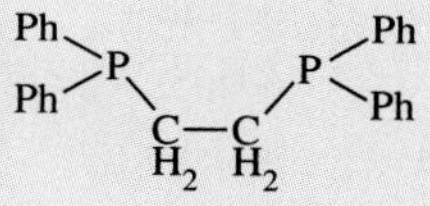
BINAP	2,2′-bis (diphenylphopshino) -1,1′-binapthyl	BINAP	$Ph_2P(C_{10}H_6)_2PPh_2$ 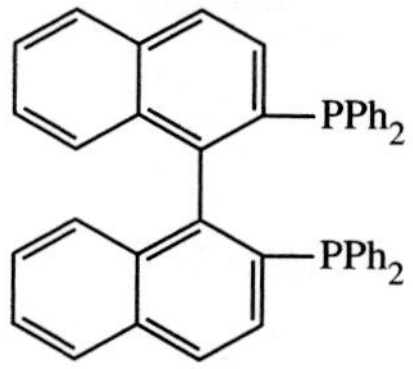
dimethylglyoximato	butanediene dioxime	DMG	$HONCC(CH_3)C(CH_3)NO^-$ 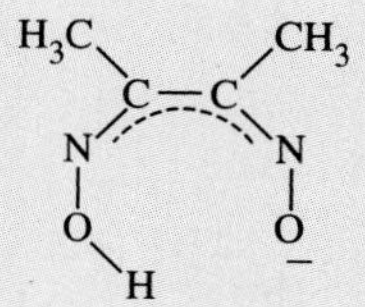
pyrazolylborato (scorpionate)	hydrotris-(pyrazo-1-yl)borato	Tp	$[HB(C_3H_3N_2)_3]^-$ 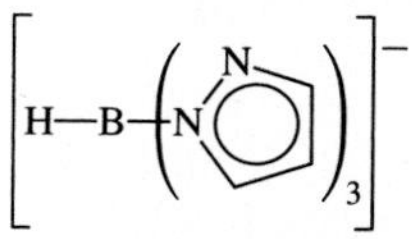
salen	2,2′-Ethylenebis-(nitrilomethylidene)-diphenoxide	salen	$^-OPh(CHNCH_2CH_2NCH)PhO^-$ 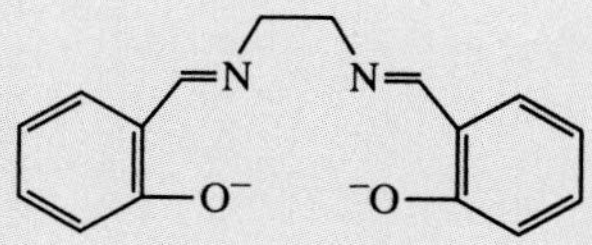

described as **bidentate** if they have two points of attachment, as in ethylenediamine ($NH_2CH_2CH_2NH_2$), which can bond to metals through the two nitrogen atoms. The prefixes tri-, tetra-, penta-, and hexa- are used for three through six bonding positions (Table 9.3). **Chelate rings** may have any number of atoms; the most common contain five or six atoms, including the metal. Smaller rings have angles and distances that lead to strain; larger rings frequently result in crowding, both within the ring and between adjoining ligands. Some ligands form more than one ring; ethylenediaminetetraacetate (EDTA) can form five via its carboxylate groups and two amine nitrogen atoms.

Nomenclature Rules

1. The cation comes first, followed by the anion.

   Examples: diamminesilver(I) chloride, $[Ag(NH_3)_2]Cl$

   potassium hexacyanoferrate(III), $K_3[Fe(CN)_6]$

2. The inner coordination sphere is enclosed in square brackets. Although the metal is provided first within the brackets, the ligands within the coordination sphere are written before the metal in the formula name.

   Examples: tetraamminecopper(II) sulfate, $[Cu(NH_3)_4]SO_4$

   hexaamminecobalt(III) chloride, $[Co(NH_3)_6]Cl_3$

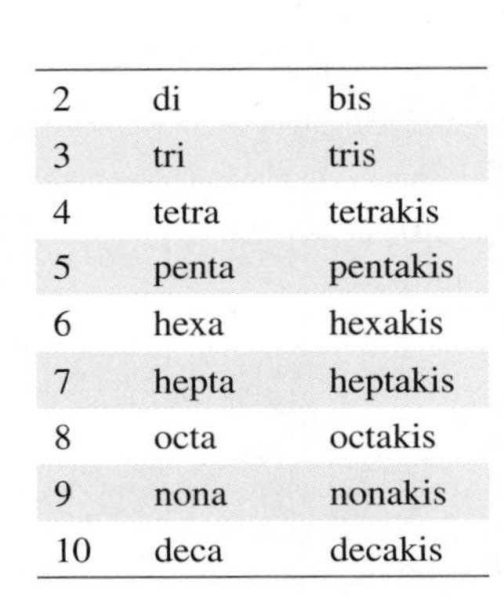

2	di	bis
3	tri	tris
4	tetra	tetrakis
5	penta	pentakis
6	hexa	hexakis
7	hepta	heptakis
8	octa	octakis
9	nona	nonakis
10	deca	decakis

3. The number of ligands of each kind is indicated by prefixes (in margin). In simple cases, the prefixes in the second column are used. If the ligand name already includes these prefixes or is complicated, it is set off in parentheses, and prefixes in the third column (ending in *–is*) are used.

   Examples: dichlorobis(ethylenediamine)cobalt(III), $[Co(NH_2CH_2CH_2NH_2)_2Cl_2]^+$

   tris(2,2′-bipyridine)iron(II), $[Fe(C_{10}H_8N_2)_3]^{2+}$

4. Ligands are generally written in alphabetical order—according to the ligand name, not the prefix.

   Examples: tetraamminedichlorocobalt(III), $[Co(NH_3)_4Cl_2]^+$

   (tetraammine is alphabetized by *a* and dichloro by *c*, not by the prefixes)

   amminebromochloromethylamineplatinum(II), $Pt(NH_3)BrCl(CH_3NH_2)$

5. Anionic ligands are given an *o* suffix. Neutral ligands retain their usual name. Coordinated water is called *aqua* and coordinated ammonia is called *ammine*. Examples are in Table 9.2.

6. Two systems exist for designating charge or oxidation number:

   a. The *Stock system* puts the calculated oxidation number of the metal as a Roman numeral in parentheses after the metal name. Although this is the most commonly employed method, its drawback is that the oxidation state of a metal within a complex can be ambiguous, and difficult to specify.

   b. The *Ewing-Bassett system* puts the charge on the coordination sphere in parentheses after the name of the metal. This convention offers an unambiguous identification of the species.

   In either case, if the charge is negative, the suffix *-ate* is added to the name.

Examples: tetraammineplatinum(II) or tetraammineplatinum(2+), $[Pt(NH_3)_4]^{2+}$

tetrachloroplatinate(II) or tetrachloroplatinate(2–), $[PtCl_4]^{2-}$

hexachloroplatinate(IV) or hexachloroplatinate(2–), $[PtCl_6]^{2-}$

7. Prefixes designate adjacent (*cis*-) and opposite (*trans*-) geometric locations (Figures 9.1 and **9.5**). Other prefixes will be introduced as needed.

Examples: *cis*- and *trans*-diamminedichloroplatinum(II), $[PtCl_2(NH_3)_2]$

cis- and *trans*-tetraamminedichlorocobalt(III), $[CoCl_2(NH_3)_4]^+$

8. Bridging ligands between two metal ions (Figures 9.2 and **9.6**) have the prefix μ-.

Examples: tris(tetraammine-μ-dihydroxocobalt)cobalt(6+), $[Co(Co(NH_3)_4(OH)_2)_3]^{6+}$

μ-amido-μ-hydroxobis(tetramminecobalt)(4+), $[(NH_3)_4Co(OH)(NH_2)Co(NH_3)_4]^{4+}$

9. When the complex is negatively charged, the names for these metals are derived from the sources of their symbols:

iron (Fe)	ferrate	lead (Pb)	plumbate
silver (Ag)	argentate	tin(Sn)	stannate
		gold (Au)	aurate

Examples: tetrachloroferrate(III) or tetrachloroferrate(1–), $[FeCl_4]^-$

dicyanoaurate(I) or dicyanoaurate(1–), $[Au(CN)_2]^-$

FIGURE 9.5 *cis* and *trans* Isomers of Diamminedichloroplatinum(II), $[PtCl_2(NH_3)_2]$. The *cis* isomer, also known as *cisplatin*, is used in cancer treatment.

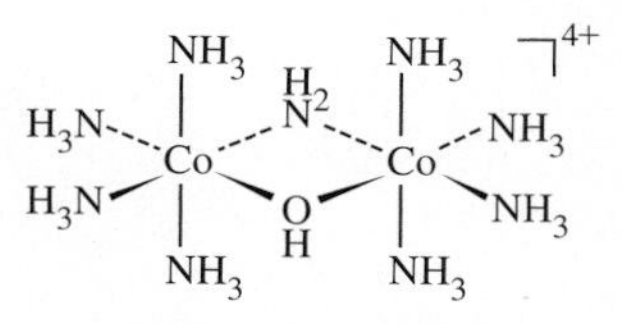

FIGURE 9.6 Bridging Amide and Hydroxide Ligands in μ-amido-μ-hydroxobis (tetraamminecobalt) (4+), $[(NH_3)_4Co(OH)(NH_2)Co(NH_3)_4]^{4+}$.

EXERCISE 9.1

Name these coordination complexes:

a. $Cr(NH_3)_3Cl_3$

b. $Pt(en)Cl_2$

c. $[Pt(ox)_2]^{2-}$

d. $[Cr(H_2O)_5Br]^{2+}$

e. $[Cu(NH_2CH_2CH_2NH_2)Cl_4]^{2-}$

f. $[Fe(OH)_4]^-$

EXERCISE 9.2

Give the structures of these coordination complexes:

a. Tris(acetylacetonato)iron(III)

b. Hexabromoplatinate(2–)

c. Potassium diamminetetrabromocobaltate(III)

d. Tris(ethylenediamine)copper(II) sulfate

e. Hexacarbonylmanganese(I) perchlorate

f. Ammonium tetrachlororuthenate(1–)

9.3 Isomerism

The variety of coordination numbers in these complexes provides a large number of **isomers**. As the coordination number increases so does the number of possible isomers. We will focus on the common coordination numbers, primarily 4 and 6. We will not discuss isomerism where the ligands *themselves* are isomers. For example, coordination compounds of the ligands 1-aminopropane and 2-aminopropane are isomers, but we will not include them in our discussion.

Hydrate or **solvent isomers**, **ionization isomers**, and **coordination isomers** have the same overall formula but have different ligands attached to the central atom or ion. The terms **linkage** or **ambidentate isomerism** are used for cases of bonding through different atoms of the same ligand. **Stereoisomers** have the same ligands, but differ in their geometric arrangement. **Figure 9.7** provides a flowchart that describes the most fundamental ways in which these isomers are distinguished from each other.

9.3.1 Stereoisomers

Stereoisomers include *cis* and *trans* isomers, chiral isomers, compounds with different conformations of chelate rings, and other isomers that differ only in the geometry of attachment to the metal. The study of stereoisomers provided much of the experimental evidence used to develop and defend the Werner coordination theory. X-ray crystallography allows facile elucidation of isomeric structures as long as suitable crystals can be obtained.

9.3.2 4-Coordinate Complexes

Cis and *trans* isomers of square-planar complexes are common; many platinum(II) examples are known. The isomers of $[Pt(NH_3)_2Cl_2]$ are shown in Figure 9.5. The *cis* isomer is used in medicine as the antitumor agent cisplatin. Chelate rings can enforce a *cis* structure if the chelating ligand is too small to span the *trans* positions. The distance across the two

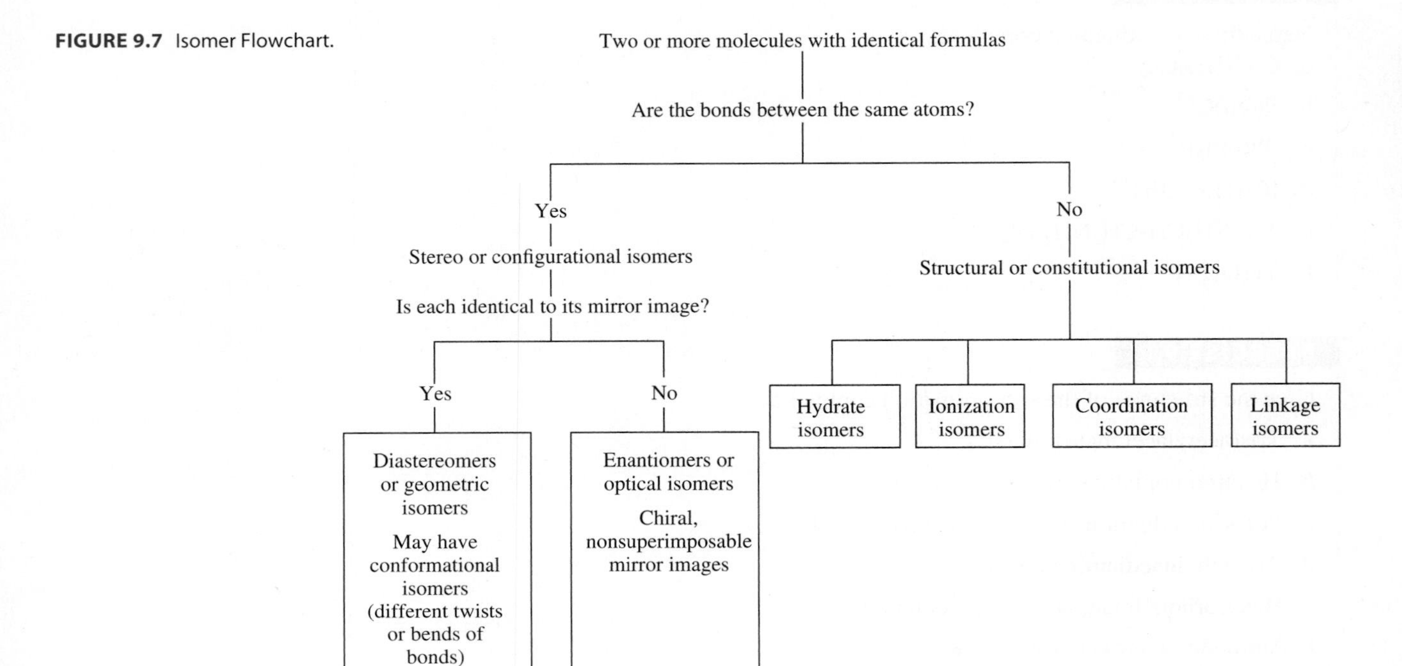

FIGURE 9.7 Isomer Flowchart.

trans positions is too large for most ligands, and the longer the span between donor sites within a ligand, the greater the possibility of these sites binding to different metals rather than chelating the same metal.

No chiral isomers are possible when the molecule has a mirror plane. When determining whether a square-planar molecule has a mirror plane, we usually ignore minor changes in the ligand, such as rotation of substituent groups, conformational changes in ligand rings, and bending of bonds. Examples of chiral square-planar complexes are the platinum(II) and palladium(II) isomers in **Figure 9.8**, where the ligand geometry rules out mirror planes. If the complexes were tetrahedral, only one structure would be possible, with a mirror plane bisecting the two ligands between the two phenyl groups and between the two methyl groups.

9.3.3 Chirality

Chiral molecules have nonsuperimposable mirror images, a condition that can be expressed in terms of symmetry elements. A molecule is chiral only if it has no rotation-reflection (S_n) axes.* This means that chiral molecules (Section 4.4.1) either have no symmetry elements (except identity, C_1) or have only axes of proper rotation (C_n). Tetrahedral molecules with four different ligands or with unsymmetrical chelating ligands are chiral. All isomers of tetrahedral complexes are chiral. Octahedral molecules with bidentate or higher chelating ligands, or with [$Ma_2b_2c_2$], [$Mabc_2d_2$], [$Mabcd_3$], [$Mabcde_2$], or [Mabcdef] structures, where M = metal and a, b, c, d, e, f are monodentate ligands, can be chiral. Not all isomers of these molecules with coordination number of 6 are chiral, but the possibility must be considered.

9.3.4 6-Coordinate Complexes

$ML_3L'_3$ complexes where L and L′ are monodentate ligands, have two isomers called *fac*- (facial) and *mer*- (meridional). *Fac* isomers have three identical ligands on one triangular face; *mer* isomers have three identical ligands in a plane bisecting the molecule. Similar isomers are possible with chelating ligands; examples with monodentate and tridentate ligands are shown in **Figure 9.9**.

Special nomenclature has been proposed for related isomers. For example, triethylenetetramine compounds have three forms: α, with all three chelate rings in different

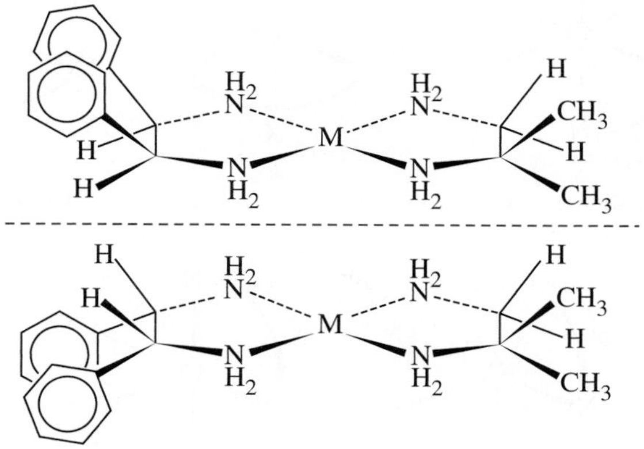

FIGURE 9.8 Chiral Isomers of Square-Planar Complexes. (*Meso*-stilbenediamine)(*iso*-butylenediamine) platinum(II) and palladium(II).

(Data from W. H. Mills, T. H. H. Quibell, *J. Chem. Soc.*, **1935**, 839; A. G. Lidstone, W. H. Mills, *J. Chem. Soc.*, **1939**, 1754.)

*Because $S_1 \equiv \sigma$ and $S_2 \equiv i$, locating a mirror plane or inversion center in a structure indicates that it is not chiral. A structure may be achiral by virtue of an S_n axis where $n > 2$ even without the presence of a mirror plane or inversion center as symmetry elements.

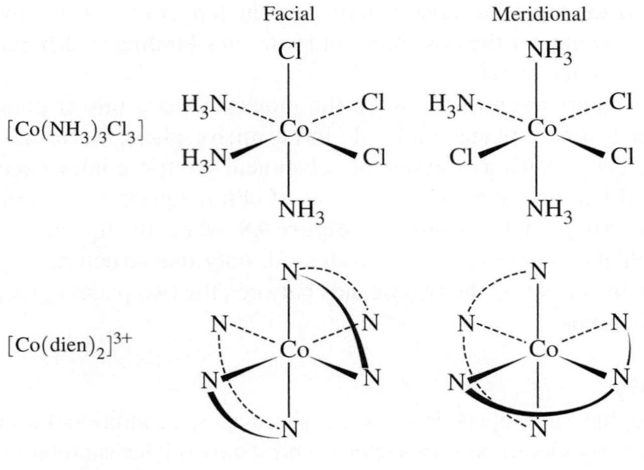

FIGURE 9.9 Facial and Meridional Isomers of $[Co(NH_3)_3Cl_3]$ and $[Co(dien)_2]^{3+}$.

planes; β, with two of the rings coplanar, and *trans*, with all three rings coplanar (**Figure 9.10**). Additional isomers are possible that will be discussed later (both α and β are chiral, and all three have additional isomers that depend on the chelate ring conformations). Even when a multidentate ligand exhibits the same binding mode, the incorporation of other ligands can result in isomers. For example, in **Figure 9.11**, the β, β', β''-triaminotriethylamine (tren) ligand bonds to four adjacent sites, but an asymmetric ligand such as salicylate can then bond in the two ways, with the carboxylate either *cis* or *trans* to the tertiary nitrogen.

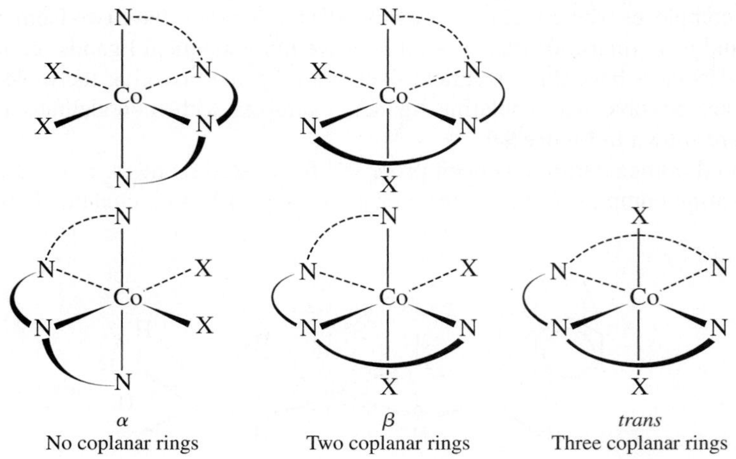

FIGURE 9.10 Isomers of Triethylenetetramine (trien) Complexes.

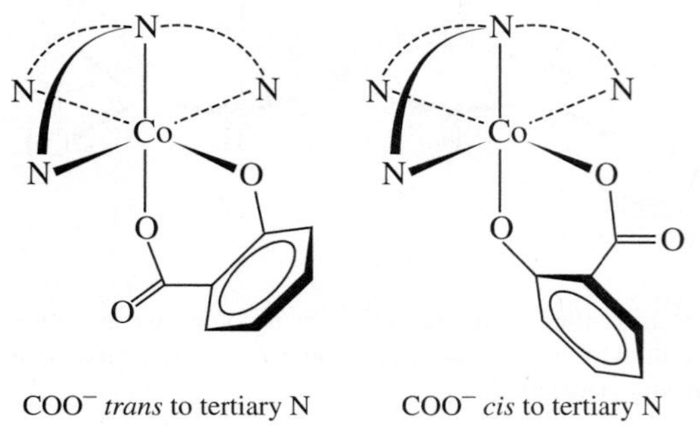

FIGURE 9.11 Isomers of $[Co(tren)(sal)]^+$.

The number of possible isomers generally increases with the number of different ligands. Strategies have been developed for calculating the maximum number of isomers on the basis of an initial structure,[12] but complete isomer lists were difficult to obtain until computer programs were used. Pólya used group theory to calculate the number of isomers.[13]

One approach to tabulating isomers is shown in **Figure 9.12** and **Table 9.5**. The notation <ab> indicates that a and b are *trans* to each other; M is the metal; and a, b, c, d, e, and f are monodentate ligands. The six octahedral positions are commonly numbered as in Figure 9.12, with positions 1 and 6 in axial positions and 2 through 5 in counterclockwise order as viewed from the 1 position.

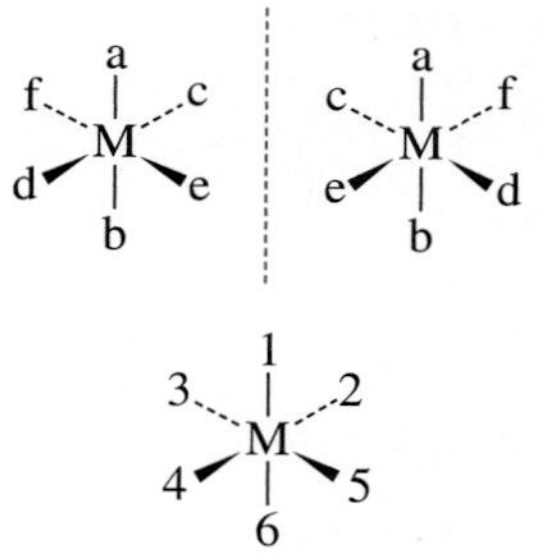

If the [Mabcdef] ligands are completely scrambled, there are 15 different diastereoisomers—structures that are not mirror images of each other—each of which has an enantiomer, or nonsuperimposable mirror image. This means that a complex with six different ligands in an octahedral shape has 30 different isomers! The isomers of [Mabcdef] are in Table 9.5. Each of the 15 entries represents an enantiomeric pair, for a total of 30 isomers. Note that each unique set of *trans* ligands in this [Mabcdef] case generates three diastereomers, where each diastereomer is chiral.

Identifying all isomers of a given complex involves systematically listing the possible structures, then checking for duplicates and chirality. Bailar suggested a systematic method, where one *trans* pair, such as <ab>, is held constant; the second pair has one component constant, and the other is systematically changed; and the third pair is whatever is left over. Then, the second component of the first pair is changed, and the process is continued. This procedure generates the Table 9.5 results. The pair of enantiomers indicated in Table 9.5 box A1 is shown in Figure 9.12.

The same approach can be used for chelating ligands, with limits on the chelate ring location. For example, a normal bidentate chelate ring cannot connect *trans* positions. After listing all the isomers without this restriction, those that are sterically impossible can be eliminated and the others checked for duplicates and enantiomers. **Table 9.6** lists the number of isomers and enantiomers for many general formulas.[14]

FIGURE 9.12 [M<ab><cd><ef>] Enantiomers and the Octahedral Numbering System.

TABLE 9.5 [Mabcdef] Isomers[a]

	A	B	C
1	ab cd ef	ab ce df	ab cf de
2	ac bd ef	ac be df	ac bf de
3	ad bc ef	ad be cf	ad bf ce
4	ae bc df	ae bf cd	ae bd cf
5	af bc de	af bd ce	af be cd

[a] Each 1 × 3 box is a set of three *trans* pairs of ligands. For example, box C3 represents the two enantiomers of [M < ad > < bf > < ce >].

TABLE 9.6 Number of Possible Isomers for Specific Complexes

Formula	Number of Stereoisomers	Pairs of Enantiomers
Ma_6	1	0
Ma_5b	1	0
Ma_4b_2	2	0
Ma_3b_3	2	0
Ma_4bc	2	0
Ma_3bcd	5	1
Ma_2bcde	15	6
Mabcdef	30	15
$Ma_2b_2c_2$	6	1
Ma_2b_2cd	8	2
Ma_3b_2c	3	0
M(AA)(BC)de	10	5
M(AB)(AB)cd	11	5
M(AB)(CD)ef	20	10
$M(AB)_3$	4	2
M(ABA)cde	9	3
$M(ABC)_2$	11	5
M(ABBA)cd	7	3
M(ABCBA)d	7	3

Uppercase letters represent chelating ligands, and lowercase letters represent monodentate ligands.

EXAMPLE 9.1

The isomers of $Ma_2b_2c_2$ can be found by Bailar's method. In each row below, the first pair of ligands is held constant: <aa>, <ab>, and <ac> in rows 1, 2, and 3, respectively. In column B, one component of the second pair is traded for a component of the third pair (for example, in row 2, <ab> and <cc> become <ac> and <bc>).

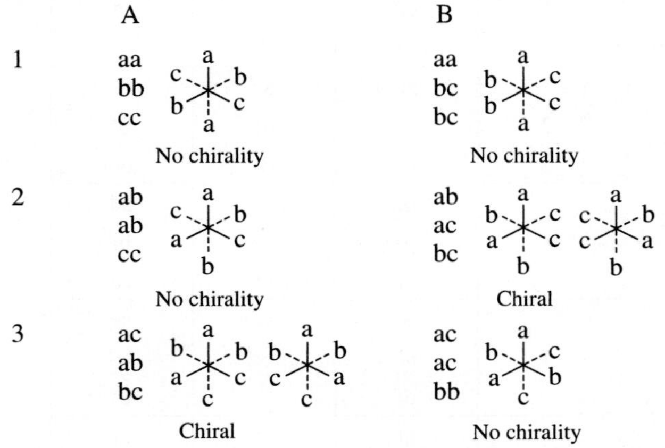

	A	B
1	aa bb cc No chirality	aa bc bc No chirality
2	ab ab cc No chirality	ab ac bc Chiral
3	ac ab bc Chiral	ac ac bb No chirality

Once all the *trans* arrangements are listed and drawn, we check for chirality. Entries A1, B1, A2, and B3 possess mirror plane symmetry; they are achiral. Entries A3 and B2 do not have mirror plane symmetry; these are chiral and have nonsuperimposable mirror images. However, we must check for duplicates that can arise via this systematic method. In this case, A3 and B2 are identical as each set has *trans* ac, ab, and bc ligands. Overall, there are four nonchiral isomers and one chiral pair, for a total of six isomers.

EXERCISE 9.3 Find the number and identity of the isomers of $[Ma_2b_2cd]$.

EXAMPLE 9.2

A methodical approach is important in finding isomers. Consider M(AA)(BB)cd. AA and BB must be in *cis* positions, because they are linked in the chelate ring. For M(AA)(BB)cd, we first try c and d in *cis* positions. One A and one B must be *trans* to each other:

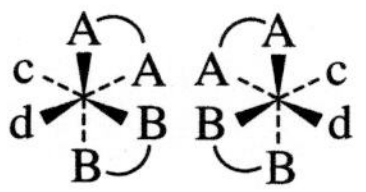

c opposite B
d opposite A

The mirror image is different, so there is a chiral pair. These mirror images have no improper axes of rotation, including neither an inversion center nor mirror planes.

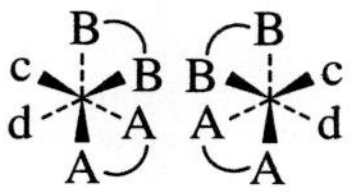

c opposite A
d opposite B

The mirror image is different, so there is a chiral pair. These mirror images have no improper axes of rotation, including neither an inversion center nor mirror planes.

Then, trying c and d in *trans* positions, where AA and BB are in the horizontal plane:

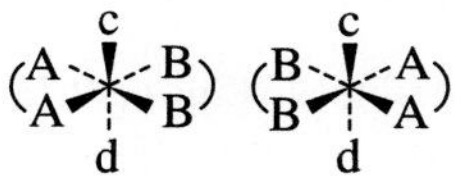

The mirror images are identical, and the diastereomer used to generate the mirror image has a mirror plane, so there is only one isomer. There are two chiral pairs and one achiral diastereomer, for a total of five isomers.

EXERCISE 9.4 Find the number and identity of all isomers of [M(AA)bcde], where AA is a bidentate ligand with identical coordinating groups.

9.3.5 Combinations of Chelate Rings

Before discussing nomenclature rules for ring geometry, we need to establish the handedness of propellers and helices. Consider the propellers in **Figure 9.13**. The first is a left-handed propeller; rotating it *counterclockwise* in air or water would move it away from the observer. The second, a right-handed propeller, moves away on *clockwise* rotation. The tips of the propeller blades describe left- and right-handed helices, respectively. With rare exceptions, the threads on screws and bolts are right-handed helices; a clockwise twist with a screwdriver or wrench drives them into a nut or piece of wood. The same clockwise motion drives a nut onto a stationary bolt. Another example of a helix is a coil spring, which can usually have either handedness without affecting its operation.

Complexes with three rings formed via chelating ligands, such as $[Co(en)_3]^{3+}$, can be treated like three-bladed propellers by looking at the molecule down a threefold axis. **Figure 9.14** shows a number of different, but equivalent, ways to draw these structures. The procedure for assigning the counterclockwise (Λ) or clockwise (Δ) notation is described in the next paragraph.

Complexes with two or more nonadjacent chelate rings (not sharing a common atom bonded to the metal) may be chiral. Any two non-coplanar and nonadjacent chelate rings

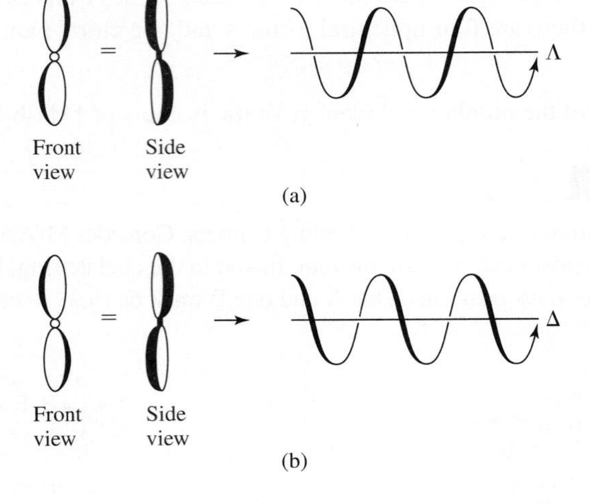

FIGURE 9.13 Right- and Left-Handed Propellers. (a) Left-handed propeller and helix traced by the tips of the blades. (b) Right-handed propeller and helix traced by the tips of the blades.

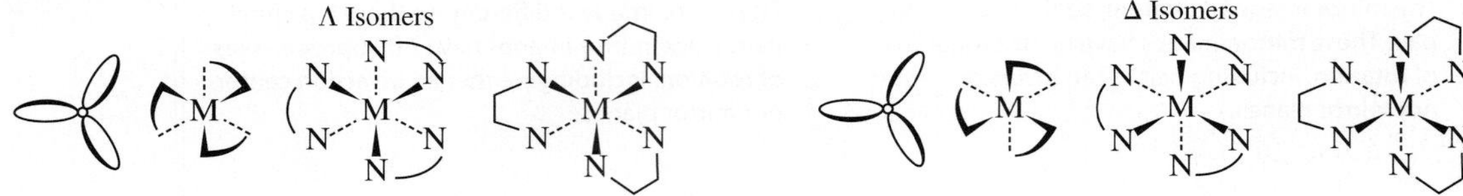

FIGURE 9.14 Left- and Right-Handed Chelates.

can be used to determine the handedness. **Figure 9.15** illustrates the process, which can be summarized as follows:

1. Rotate the figure to place one ring horizontally across the back, at the top of one of the triangular faces.
2. Imagine the ring in the front triangular face as having originally been parallel to the ring at the back. Determine what rotation of the front face is required to obtain the actual configuration.
3. If the rotation from Step 2 is counterclockwise, the structure is designated lambda (Λ). If the rotation is clockwise, the designation is delta (Δ).

A molecule with more than one pair of rings may require more than one label. The handedness of each pair of skew rings is determined; the final description includes all the designations. For example, an EDTA complex wherein the ligand is fully bound has six

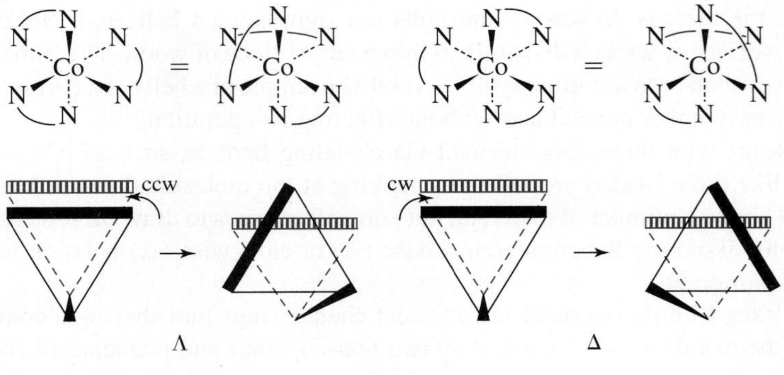

FIGURE 9.15 Procedure for Determining Handedness.

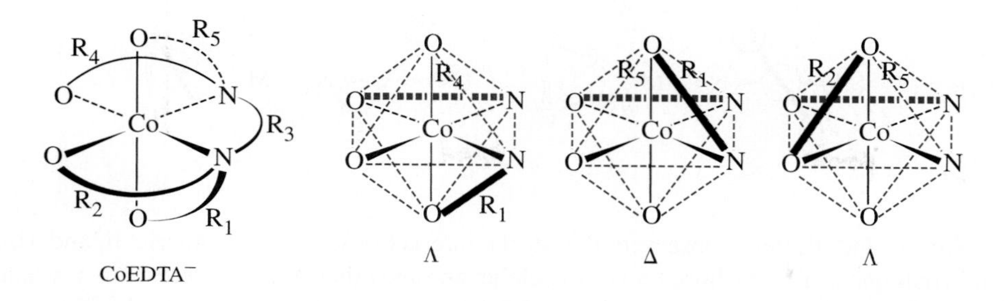

FIGURE 9.16 Labeling of Chiral Rings. The rings are numbered arbitrarily R_1 through R_5. The combination R_1-R_4 is Λ, R_1-R_5 is Δ, and R_2-R_5 is Λ. The notation for this structure is $\Lambda\Delta\Lambda$-(ethylenediaminetetraacetato) cobaltate(III).

points of attachment and five rings. One isomer is shown in **Figure 9.16**, where the rings are numbered arbitrarily R_1 through R_5. All ring pairs that are not coplanar and are not connected at the same atom are used in the description. The N—N ring (R_3) is omitted, because it is connected at the same atom with each of the other rings. Considering only the four O—N rings, there are three useful pairs, R_1-R_4, R_1-R_5, and R_2-R_5. The fourth pair, R_2-R_4, is not used because the two rings are coplanar. The method described above gives Λ for R_1-R_4, Δ for R_1-R_5, and Λ for R_2-R_5. The notation for the compound given is then $\Lambda\Delta\Lambda$-(ethylenediaminetetraacetato)cobaltate(III). The order of the designations is arbitrary and could also be $\Lambda\Lambda\Delta$ or $\Delta\Lambda\Lambda$.

EXAMPLE 9.3

Determine the chirality label(s) for:

Rotating the figure 180° about the vertical axis puts one ring across the back and the other connecting the top and the front right positions. If this front ring were originally parallel to the back one, a clockwise rotation would put it into the correct position. Therefore, the structure is Δ-*cis*-dichlorobis(ethylenediamine)cobalt(III).

EXERCISE 9.5 Determine the chirality label(s) for:

9.3.6 Ligand Ring Conformation

Because many chelate rings are not planar, they can have different conformations in different molecules, even in otherwise identical molecules. In some cases, these different conformations are also chiral. The notation used in these situations requires using two lines to establish the handedness and the labels λ and δ. The first line connects the atoms bonded to the metal. In the case of ethylenediamine, this line connects the two nitrogen atoms. The second line connects the two carbon atoms of the ethylenediamine, and the handedness of the two rings is found by the method described in Section 9.3.5 for separate rings. A counterclockwise rotation of the second line is called λ, and a clockwise rotation is called δ, as shown in **Figure 9.17**. Complete description of a complex requires identification of the overall chirality and the chirality of each ring.

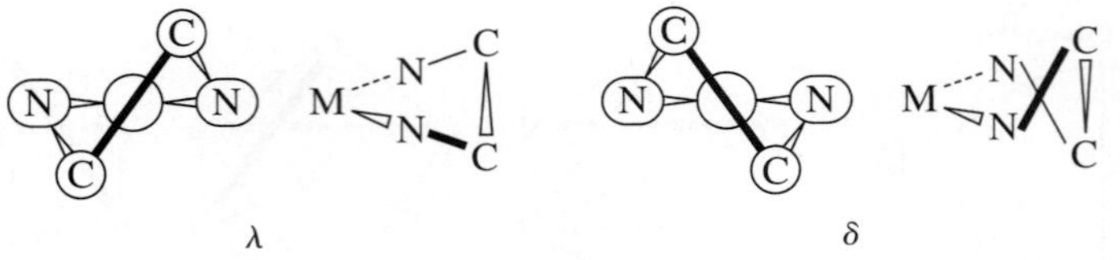

FIGURE 9.17 Chelate Ring Conformations.

λ δ

Corey and Bailar[15] observed the steric interactions due to isomeric ligand ring conformations similar to those found in cyclohexane and other ring structures. For example, $\Delta\lambda\lambda\lambda$-$[Co(en)_3]^{3+}$ was calculated to be 7.5 kJ/mol more stable than the $\Delta\delta\delta\delta$ isomer because of interactions between the NH_2 groups of different ethylenediamine ligands. For the Λ isomers, the $\delta\delta\delta$ ring conformation is more stable. Experimental results have confirmed these calculations. The small difference in energy leads to an equilibrium between the λ and δ ligand conformations in solution, and the most abundant configuration for the Λ isomer is $\delta\delta\lambda$.[16]

Determining the relative energies of diastereomers arising from ring conformations formed by multidentate ligands bound to lanthanides is important in the development of MRI (magnetic resonance imaging) contrast agents.[17] The subtle steric changes imparted by different chelate ring conformations can modify the aqua ligand substitution exchange rate in these complexes; this water exchange rate influences the performance of contrast agents. Chelate ring conformations also dictate the fate of insertion reactions (Chapter 14)[18] used for asymmetric syntheses (syntheses designed to introduce specific chirality in the products).

An additional isomeric possibility arises because the ligand symmetry can be changed by coordination. An example is a secondary amine in diethylenetriamine (dien) or triethylenetetraamine (trien). Inversion at the nitrogen has a very low energy barrier in the free ligands; only one isomer of each molecule exists. Upon coordination, the nitrogen becomes 4-coordinate, and there may be chiral isomers. If there are chiral centers on the ligands, either inherent in their structure or created by coordination, their structure is described by the R and S notation from organic chemistry.[19] Some trien complex structures are in **Figures 9.18** and **9.19**; the *trans* isomers are described in the following example. The α, β, and *trans* structures of the Figures 9.18 and 9.19 complexes appear in Figure 9.10 without consideration of ring conformations.

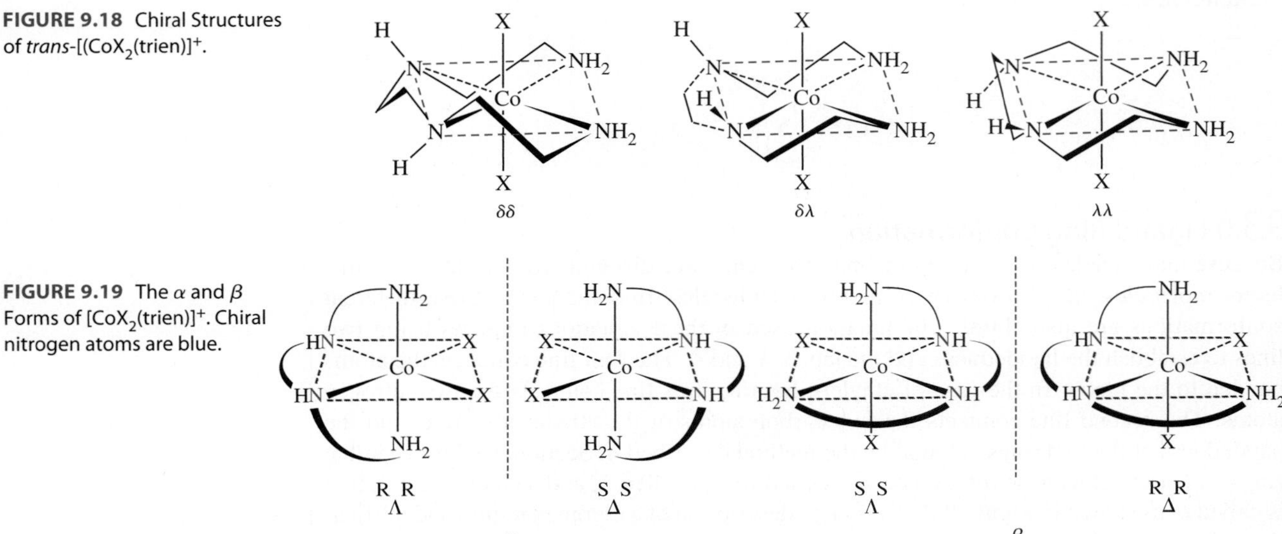

FIGURE 9.18 Chiral Structures of *trans*-$[(CoX_2(trien)]^+$.

FIGURE 9.19 The α and β Forms of $[CoX_2(trien)]^+$. Chiral nitrogen atoms are blue.

EXAMPLE 9.4

Confirm the chirality on the basis of ring confirmations in the Figure 9.18 *trans*-$[CoX_2trien]^+$ structures.

Take the ring on the front edge of the first structure, with an imaginary line connecting the two nitrogen atoms as a reference. If the line connecting the two carbons was originally parallel to this N—N line, a clockwise rotation is required to reach the actual conformation, so the conformation is δ. Viewing the ring shown on the back of the molecule *from the outside, looking toward the metal* is the same so this ring also is δ. The tetrahedral geometry of the ligand N forces the hydrogens on the two secondary nitrogens into the positions shown. Viewing this middle ring from the outside shows that the conformation is opposite that of the front and back rings, so the classification is λ. Because there is no other possibility, inclusion of a λ label would be redundant. The label for this isomer is therefore $\delta\delta$.

The same procedure on the other two structures results in labels of $\delta\lambda$ and $\lambda\lambda$, respectively. Again, the middle ring conformation is dictated by the other two, so it need not be labeled. It is noteworthy that these *trans* isomers are not chiral on the basis of the co-planar arrangement of the three chelate rings (Section 9.3.5). (In all of these cases, use of molecular models is strongly encouraged!)

EXERCISE 9.6 $[Co(dien)_2]^{3+}$ can have several forms, two of which are shown below. Identify the Δ or Λ chirality of the rings, using all unconnected pairs. Each complex may have three labels.

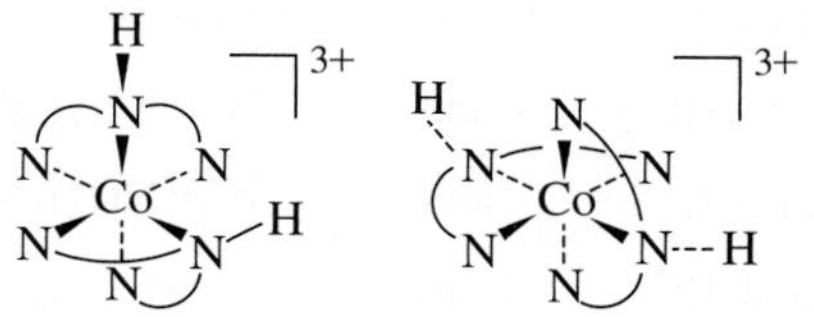

9.3.7 Constitutional Isomers

Hydrate Isomerism

Hydrate isomerism requires water to play two roles, as (1) a ligand and as (2) an additional occupant (or solvate) within the crystal structure.* *Solvent isomerism* broadens the definition to allow for the possibility of ammonia or other ligands participating as solvates.

$CrCl_3 \cdot 6\,H_2O$ is a classic example. Three different crystalline compounds that each feature 6-coordinate Cr(III) have this empirical formula: $[Cr(H_2O)_6]Cl_3$ (violet), $[CrCl(H_2O)_5]Cl_2 \cdot H_2O$ (blue-green), and $[CrCl_2(H_2O)_4]Cl \cdot 2\,H_2O$ (dark green). These three hydrate isomers can be separated from commercial $CrCl_3 \cdot 6\,H_2O$, with *trans*-$[CrCl_2(H_2O)_4]Cl \cdot 2\,H_2O$ the major component.** Other examples of hydrate isomers are:

$[Co(NH_3)_4(H_2O)Cl]Cl_2$ and $[Co(NH_3)_4Cl_2]Cl \cdot H_2O$

$[Co(NH_3)_5(H_2O)](NO_3)_3$ and $[Co(NH_3)_5(NO_3)](NO_3)_2 \cdot H_2O$

---

*For example, hydrates of sodium sulfate ($Na_2SO_4 \cdot 7\,H_2O$ and $Na_2SO_4 \cdot 10\,H_2O$) feature varying numbers of water molecules within their crystal structures. However, these salts are not hydrate isomers because their empirical formulas are different. The ability of anhydrous sodium sulfate and magnesium sulfate to accommodate water molecules within their crystalline lattices permit application of these salts as drying agents in organic synthesis.

The related neutral $[CrCl_3(H_2O)_3]$ (yellow-green) can be generated in high concentrations of HCl. See S. Diaz-Moreno, A. Muñoz-Paez, J. M. Martinez, R. R. Pappalardo, E. S. Marcos, *J. Am. Chem. Soc.*, **1996, *118*, 12654.

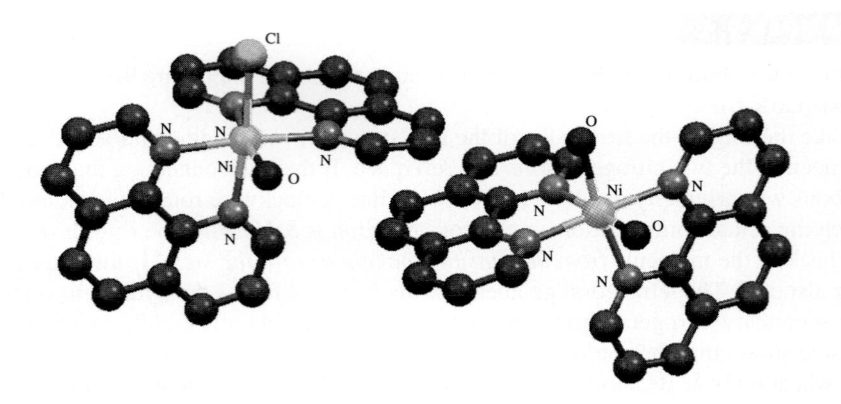

FIGURE 9.20 The cations $[cis\text{-Ni(phen)}_2\text{Cl(H}_2\text{O)}]^+$ and $[cis\text{-Ni(phen)}_2\text{(H}_2\text{O)}_2]^{2+}$ co-crystallize with three PF_6^- counterions and H_2O solvate molecules (not shown). If the counterions were Cl^-, these would be hydrate isomers. (Molecular structure drawing created from CIF data, with hydrogen atoms omitted for clarity.)

Discovering hydrate isomers is often serendipitous. The crystallization of $\{[cis\text{-M(phen)}_2\text{Cl(H}_2\text{O)}][cis\text{-M(phen)}_2\text{(H}_2\text{O)}_2]\}(\text{PF}_6)_3$ (M = Co, Ni) (**Figure 9.20**) suggests that $[cis\text{-M(phen)}_2\text{Cl(H}_2\text{O)}]\text{Cl} \cdot \text{H}_2\text{O}$ and $[cis\text{-M(phen)}_2\text{(H}_2\text{O)}_2]\text{Cl}_2$ are viable hydrate isomer targets for synthesis.[20]

Ionization Isomerism

Compounds with the same formula, but which give different ions upon dissociation, exhibit ionization isomerization. The difference is in which ion is included as a ligand and which is present to balance the overall charge. Some examples are also hydrate isomers:

$[\text{Co(NH}_3)_4(\text{H}_2\text{O})\text{Cl}]\text{Br}_2$	and	$[\text{Co(NH}_3)_4\text{Br}_2]\text{Cl} \cdot \text{H}_2\text{O}$
$[\text{Co(NH}_3)_5\text{SO}_4]\text{NO}_3$	and	$[\text{Co(NH}_3)_5\text{NO}_3]\text{SO}_4$
$[\text{Co(NH}_3)_4(\text{NO}_2)\text{Cl}]\text{Cl}$	and	$[\text{Co(NH}_3)_4\text{Cl}_2]\text{NO}_2$

Coordination Isomerism

The definition of coordination isomerism depends on the context. Historically, a complete series of coordination isomers required at least two metals. The ligand:metal ratio remains the same, but the ligands attached to a specific metal ion change. For the empirical formula $\text{Pt(NH}_3)_2\text{Cl}_2$, there are three coordination isomer possibilities that contain Pt(II).

$[\text{Pt(NH}_3)_2\text{Cl}_2]$	
$[\text{Pt(NH}_3)_3\text{Cl}][\text{Pt(NH}_3)\text{Cl}_3]$	(This compound apparently has not been reported, but the individual ions are known.)
$[\text{Pt(NH}_3)_4][\text{PtCl}_4]$	(Magnus's green salt, the first platinum ammine, was discovered in 1828.)

Coordination isomers can also be composed of different metal ions, or the same metal in different oxidation states:

$[\text{Co(en)}_3][\text{Cr(CN)}_6]$	and	$[\text{Cr(en)}_3][\text{Co(CN)}_6]$
$[\text{Pt(NH}_3)_4][\text{PtCl}_6]$	and	$[\text{Pt(NH}_3)_4\text{Cl}_2][\text{PtCl}_4]$
Pt(II) Pt(IV)		Pt(IV) Pt(II)

The design of multidentate ligands that can bind to metals in different ways is of major contemporary interest. A fundamental goal of these ligands is to create alternate electronic and steric environments at metals to facilitate reactions. The flexibility of a

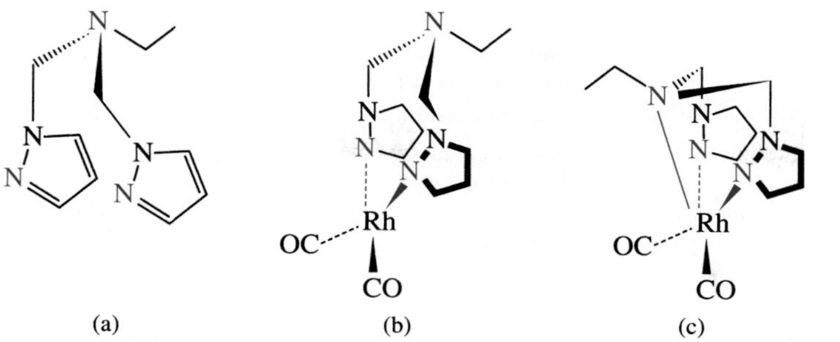

(a) (b) (c)

FIGURE 9.21 (a) The bis(1-pyrazolylmethyl)ethylamine ligand, with the nitrogen atoms eligible to bind to metals in blue. (b) κ^2-$[Rh(N\text{-ligand})(CO)_2]^+$ (multiple bonds not shown for clarity), (c) κ^3-$[Rh(N\text{-ligand})(CO)_2]^+$. These coordination isomers, as BF_4^- salts, exist in a 1:1.2 ratio in CH_2Cl_2, where the κ^3 complex is the major isomer.

ligand towards alternate binding modes results in another definition of "coordination isomer." For example, bis(1-pyrazolylmethyl)ethylamine (**Figure 9.21a**), related to the pyrazolylborato ligand (Table 9.4), participates in coordination isomerism. Two coordination isomers of $[Rh(N\text{-ligand})(CO)_2]^+$ exist in solution, with the ligand bound via either the two pyrazolyl ring nitrogens (κ^2, Figure 9.21b) or via these nitrogen atoms and the tertiary amine (κ^3, Figure 9.21c).[21] The challenges associated with isolating a desired coordination isomer can be tackled by creative synthetic approaches and separation techniques.[22]

Linkage (Ambidentate) Isomerism

Ligands such as thiocyanate, SCN^-, and nitrite, NO_2^-, can bond to the metal through different atoms. Class (a) metal ions (hard acids) tend to bond to the thiocyanate nitrogen and class (b) metal ions (soft acids) bond through the thiocyanate sulfur. Solvent can also influence the point of attachment. Compounds of rhodium and iridium with the general formula $[M(PPh_3)_2(CO)(NCS)_2]$ form M—S bonds in solvents of high dipole moment (for example, acetone and acetonitrile) and M—N bonds in solvents of low dipole moment (for example, benzene and CCl_4).[23] A related example with Pd is in **Figure 9.22a**.

The proposed application of ambidentate thiocyanate for solar energy applications has prompted detailed examination of ruthenium(II) polypyridyl thiocyanate complexes, which possess useful charge transfer prospects (Section 11.3.8).[24,25] The linkage isomers $[Ru(terpy)(tbbpy)SCN]^+$ (Figure 9.22b) and $[Ru(terpy)(tbbpy)NCS]^+$ (Figure 9.22c) exist in equilibrium in solution (ligand structures are shown in Figures 9.22d and 9.22e), with the N-bound isomer more thermodynamically stable.[25] As shown in Figures 9.22b and 9.22c, M–NCS combinations are always linear, and M–SCN combinations are always bent at the S atom. As evident in these figures, S-bound thiocyanate has greater effective steric bulk than N-bound thiocyanate because of the larger region swept out when the S-bound ligand rotates about the M—S bond.

Jørgensen and Werner studied the classic nitrite isomers of $[Co(NH_3)_5NO_2]^{2+}$. They observed ambidentate isomers of different colors (Figure 9.22f). A red form of low stability converted readily to a yellow form. The red form was hypothesized as the M—ONO nitrito isomer and the yellow form the M—NO_2 nitro isomer. This conclusion was later confirmed, and kinetic[26] and ^{18}O labeling[27] experiments showed that this isomerization is strictly intramolecular, not a result of dissociation of the NO_2^- ion followed by reattachment.

EXERCISE 9.7

Use the HSAB concept to account for the tendency of M–SCN complexes to be favored in solvents having high dipole moments and M–NCS complexes to be favored in solvents having low dipole moments.

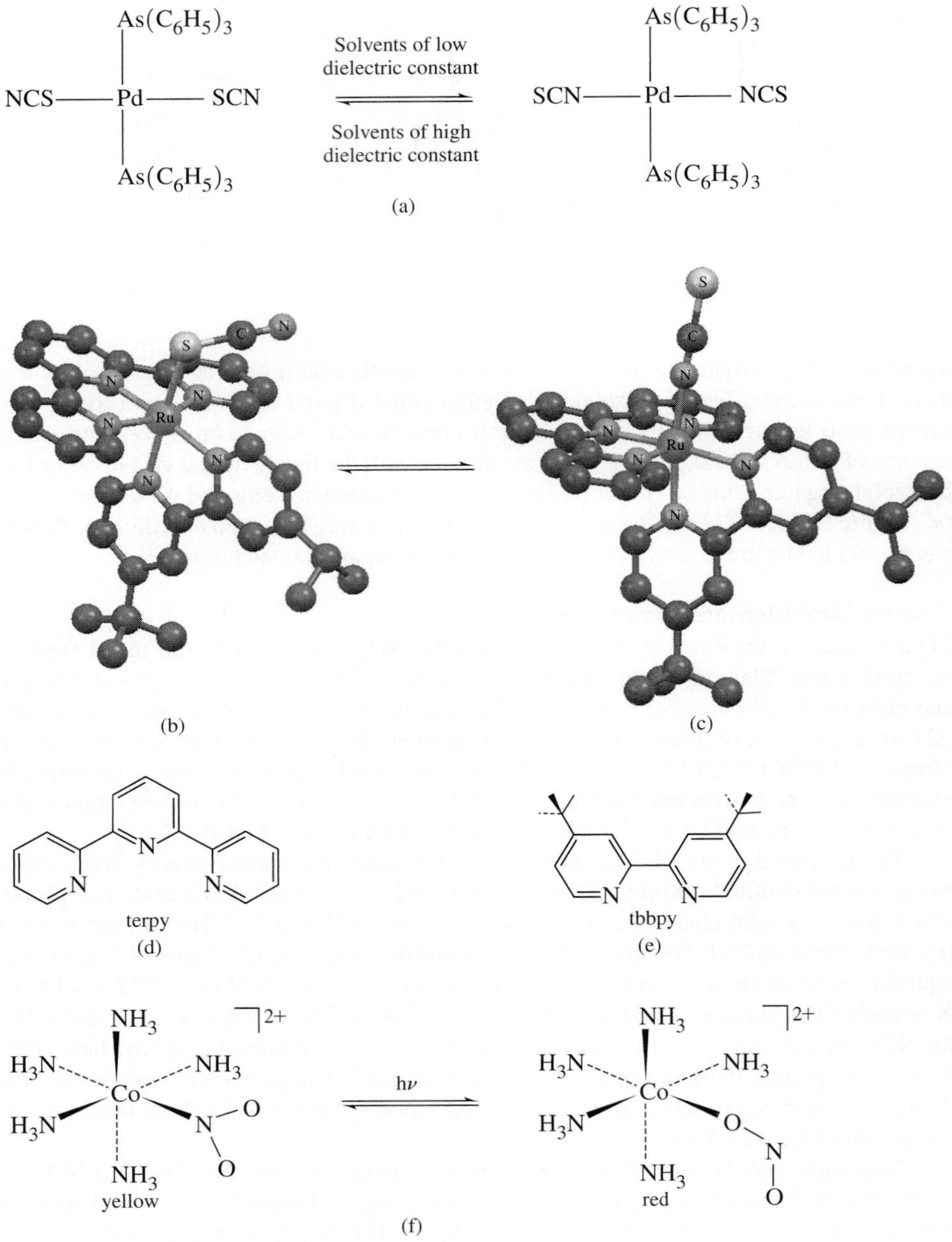

FIGURE 9.22 Linkage (Ambidentate) Isomers. (d) terpy ligand employed in isomers (b) and (c). (e) tbbpy ligand employed in isomers (b) and (c). (Molecular structure drawings (b) and (c) were generated using CIF data, with hydrogen atoms omitted for clarity.)

9.3.8 Separation and Identification of Isomers

Fractional crystallization can separate geometric isomers. This strategy assumes that the isomers will exhibit appreciably different solubilities in a specific solvent mixture, and that the isomers will not co-crystallize. For complex cations and anions, alternate counterions can be introduced (via a process called *metathesis*) to fine tune the solubilities of the resulting isomeric cation/anion combinations. One factor that dictates the solubility of an ionic complex is how effectively the ions pack into their crystals. Because geometric isomers have different shapes, the packing of isomeric ions into their respective crystals should be different. A useful guideline[32] is that ionic compounds are least soluble (have the greatest tendency to crystallize) when the positive and negative ions have the same size and magnitude of charge. For example, large cations of charge 2+ are best crystallized with large anions of charge 2−. This method suggests potentially useful cation/anions combinations that can be used to adjust complex solubility.

Separating chiral isomers requires chiral counterions. Cations are frequently resolved by using anions *d*-tartrate, antimony *d*-tartrate, and α-bromocamphor-π -sulfonate; anionic complexes are resolved by the bases brucine or strychnine or by resolved cationic complexes such as $[Rh(en)_3]^{3+}$.[33] A novel strategy using *l*- and *d*-phenylalanine to resolve Ni(II) complexes that form chiral helical chains has been reported.[34] For compounds that racemize at appreciable rates, adding a chiral counterion may shift the equilibrium, even if it does not precipitate one form; interactions between the ions in solution may be sufficient to stabilize one form over the other.[35]

The pursuit of chiral magnets (magnetism is discussed in Section 10.1.2) has exploited chiral templating to obtain enantiopure coordination complexes. For example, use of resolved $[\Delta\text{-Ru(bpy)}_3]^{2+}$ and $[\Lambda\text{-Ru(bpy)}_3]^{2+}$ results in three-dimensional optically active oxalate bridged networks of anionic $[Cu_{2x}Ni_{2(1-x)}(C_2O_4)_3]^{2-}$, where the chirality of the anion matches that of the cation.[36] Resolved chiral quaternary ammonium cations impose specific absolute configurations about the metals in a two-dimensional network of $[MnCr(ox)_3]^-$ units that exhibits ferromagnetism.[37]

X-ray crystallography is a state-of-the-art method for identifying isomers in the solid state. This method provides the coordinates for all of the atoms, allowing rapid determination of the absolute configuration. Although traditionally applied to metal complexes with relatively heavy atoms, X-ray crystallography is now often the method of choice for determining the absolute configuration of organic isomers as well.

Measurement of optical activity via polarimetry is a classic method for assigning absolute configuration to resolved chiral isomers, and one still used.[38] It is typical to examine the rotation as a function of wavelength to determine the isomer present. Optical rotation changes markedly with the wavelength of the light, and it changes sign near absorption peaks. Many organic compounds have their largest rotation in the ultraviolet, even though the sodium D wavelength (589.29 nm)* is traditionally used. Coordination compounds frequently have their major absorption (and therefore rotation) bands in the visible part of the spectrum.

Polarized light can be either circularly polarized or plane polarized. When circularly polarized, the electric or magnetic vector rotates (right-handed if clockwise rotation when viewed facing the source, left-handed if counterclockwise) with a frequency related to the frequency of the light. Plane-polarized light is made up of both right- and left-handed components; when combined, the vectors reinforce each other at 0° and 180° and cancel at 90° and 270°, leaving a planar motion of the vector. When plane-polarized light passes through a chiral substance, the plane of polarization is rotated. This **optical rotatory dispersion (ORD)**, or optical rotation, is caused by a difference in the refractive indices of the right and left circularly polarized light, according to the equation

$$\alpha = \frac{\eta_l - \eta_r}{\lambda}$$

where η_l and η_r are the refractive indices for left and right circularly polarized light, and λ is the wavelength of the light. ORD is measured by passing light through a polarizing medium, then through the substance to be measured, and then through an analyzing polarizer. The polarizer is rotated until the angle at which the maximum amount of light passing through the substance is found, and the measurement is repeated at different wavelengths. ORD frequently shows a positive value on one side of an absorption maximum and a

*Actually a doublet with emission at 588.99 and 589.59 nm.

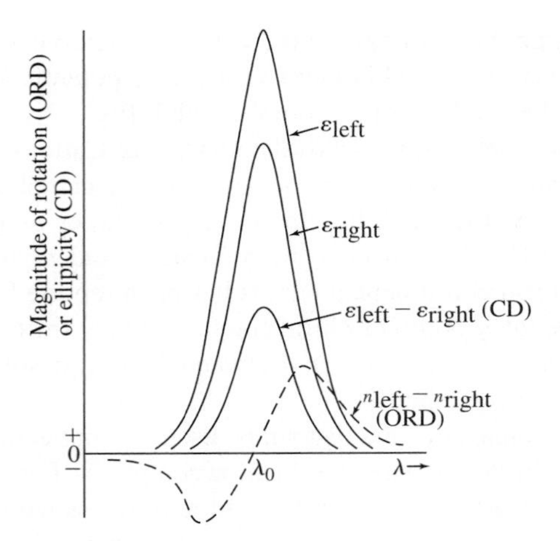

FIGURE 9.23 The Cotton Effect in ORD and CD. Idealized optical rotatory dispersion (ORD) and circular dichroism (CD) curves at an absorption peak, with a positive Cotton effect.

negative value on the other, passing through zero at or near the absorption maximum; it also frequently shows a long tail extending far from the absorption wavelength. When optical rotation of colorless compounds is measured using visible light, it is this tail that is measured, far from the ultraviolet absorption band. The variance with wavelength is known as the **Cotton effect**, positive when the rotation is positive (right-handed) at low energy and negative when it is positive at high energy.

Circular dichroism (CD), is caused by a difference in the absorption of right- and left-circularly polarized light, defined by the equation

$$\text{Circular dichroism} = \varepsilon_l - \varepsilon_r$$

where ε_l and ε_r are the molar absorption coefficients for left- and right-circularly polarized light. CD spectrometers have an optical system much like UV-visible spectrophotometers with the addition of a crystal of ammonium phosphate mounted to allow imposition of a large electrostatic field on it. When the field is imposed, the crystal allows only circularly polarized light to pass through; changing the direction of the field rapidly provides alternating left- and right-circularly polarized light. The light received by the detector is presented as the difference between the absorbances.

Circular dichroism is usually observed in the vicinity of an absorption band: a positive Cotton effect shows a positive peak at the absorption maximum and a negative effect shows a negative peak. This simple spectrum makes CD more selective and easier to interpret than ORD; CD has become the method of choice for studying chiral complexes. ORD and CD spectra are shown in **Figure 9.23**.

CD spectra are not always easily interpreted, because there may be overlapping bands of different signs. Interpretation requires determination of the overall symmetry around the metal ion and assignment of absorption spectra to specific transitions between energy levels (discussed in Chapter 11) in order to assign specific CD peaks to the appropriate transitions.

9.4 Coordination Numbers and Structures

The isomers described to this point have had octahedral or square-planar geometry. In this section, we describe other geometries. Explanations for some of the shapes are consistent with VSEPR predictions (Chapter 3), with the general assumption that the metal *d* electrons are stereochemically inactive. In these cases, 3-coordinate complexes have a trigonal-planar shape, 4-coordinate complexes are tetrahedral, and so forth, assuming that each ligand–metal bond results from a two-electron donor atom interacting with the metal. Some coordination complexes, however, exhibit geometries that VSEPR cannot explain.

The structure of a coordination compound is the result of several interacting factors, and structure predictions must always be supported with experimental data because the relative importance of these factors for a given complex can be challenging to deduce. These factors include:

1. **VSEPR considerations.**
2. **Occupancy of *d* orbitals.** Examples of how the number of *d* electrons may affect the geometry (e.g., square-planar versus tetrahedral) are discussed in Chapter 10.
3. **Steric interference,** by large ligands crowding each other around the central metal.
4. **Crystal packing effects.** This solid-state effect results from the sizes of ions and the shapes of coordination complexes. The regular shape may be distorted when it is packed into a crystalline lattice, and it may be difficult to determine whether deviations from regular geometry are caused by effects within a given unit or by packing into a crystal.

9.4.1 Coordination Numbers 1, 2, and 3

Coordination number 1 is rare for complexes in condensed phases (solids and liquids). Attempts to prepare species in solution with only one ligand are generally futile; solvent molecules often coordinate, resulting in higher coordination numbers. Two organometallic compounds with coordination number 1 are the Tl(I) (**Figure 9.24a**) and In(I) complexes of 2,6-$Trip_2C_6H_3$.[36] Although this very bulky ligand prevents dimerization (coupling) of these complexes to form metal-metal bonds (Chapter 15), the indium complex forms a complex with the less bulky $Mn(\eta^5\text{-}C_5H_5)(CO)_2$,* in which indium is 2-coordinate, with an In—Mn bond. The bulky fluorinated triazapentadienyl ligand in Figure 9.24c forms

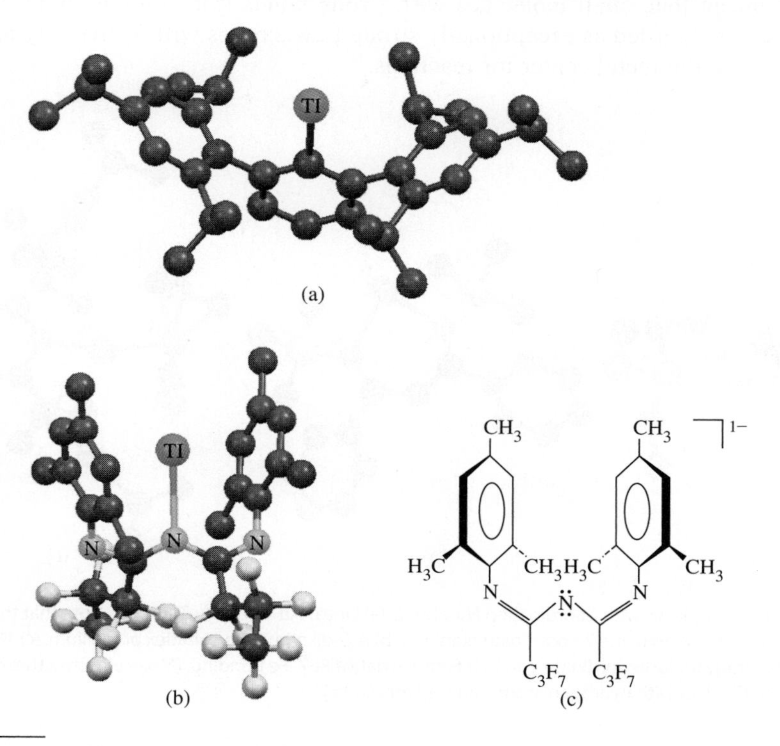

FIGURE 9.24 Coordination Number 1. (a) 2,6-$Trip_2C_6H_3Tl$ (Trip = 2,4,6-i-$Pr_3C_6H_2$). (b) A Tl(I) Triazapentadienyl complex. (c) One resonance form of the triazapentadienyl ligand. (Molecular structure drawings generated from CIF data with hydrogen atoms omitted for clarity.)

* The η^5-C_5H_5 designation is described in Section 13.2.

a 1-coordinate Tl(I) complex (Figure 9.24b).[37] Structural and computational data support only very weak interactions between the Tl(I) center and the aromatic π systems in this complex. Gas phase $Ga[C(SiMe_3)_3]$ is a monomeric, singly coordinated organometallic.[38] A transient species that seems to be singly coordinated is VO^{2+}.

$[H_3N—Ag—NH_3]^+$

$[Cl—Cu—Cl]^-$

$NC—Hg—CN$

$[CN—Au—CN]^-$

A long-known 2-coordinate complex is $[Ag(NH_3)_2]^+$. Ag^+ is d^{10} (a filled, spherical subshell); the only electrons considered in the VSEPR approach are those forming the bonds to the NH_3 ligands, and the structure is linear as predicted. Other linear d^{10} complexes include $[CuCl_2]^-$, $Hg(CN)_2$, $[Au(CN)_2]^-$, and $[Et_3PAuTi(CO)_6]^-$ (**Figure 9.25a**).[39] Initial success in preparing 2-coordinate complexes of metals with other electronic configurations employed extremely bulky ligands (for example, sterically encumbered alkyl, aryl, amido, alkoxo, and thiolato) to limit the coordination number, and shield the metal from attack.[40] For example, the silylamide $N(SiMePh_2)_2$ enforces a linear or near-linear arrangement. 2-coordinate $M[N(SiMePh_2)_2]_2$, where M = iron and manganese, are classic examples of the effectiveness of this strategy.[41]

Many interesting examples with coordination number 2 have been reported since the turn of the century.[42] The **coordinative unsaturation** of these complexes provides opportunities for small molecule activation* and magnetic properties afforded by unique atomic orbital interactions. Coordinative unsaturation indicates that the metal has open coordination sites that are kinetically blocked to varying extents, depending on the nature of the other ligands. Coordinative unsaturation can arise from both steric and electronic ligand effects. In the case of 2-coordinate complexes with bulky ligands, there are generally empty valence orbitals localized on the metal that are essentially nonbonding because of the geometry enforced by the ligands. These empty orbitals render the metal rather Lewis acidic, but the steric bulk of the ligands shields the metal from attack by bases of reasonable size. The utility of complexes with low coordination numbers for small molecule activation is based on the concept that small molecules with strong bonds (for example, N_2O, N_2, CH_4) and those not regarded as exceptionally strong Lewis bases will be the only molecules with access to the metal center for reactions.

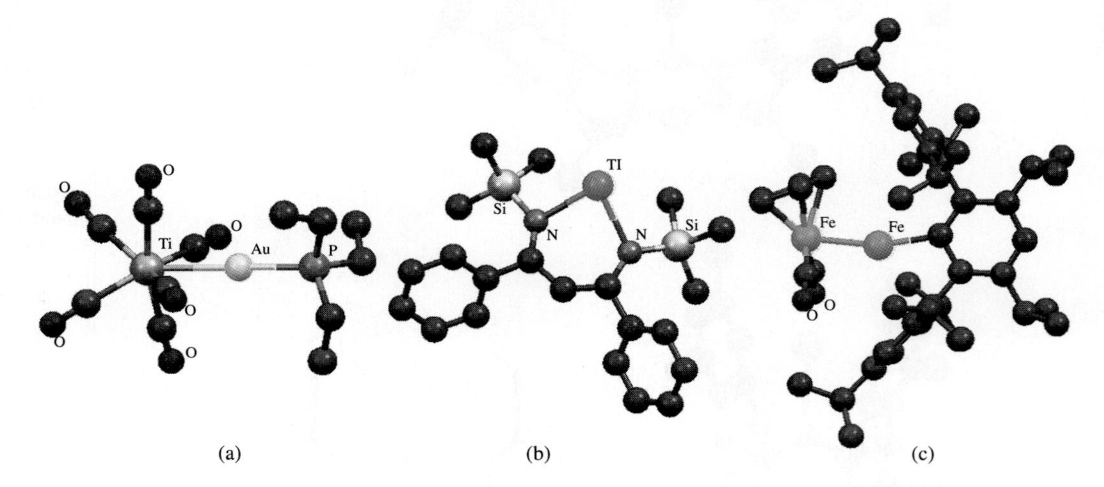

FIGURE 9.25 Complexes with Coordination Number 2. (a) Linear Au in $[Et_3PAuTi(CO)_6]^-$. Note that this structure features 2-coordinate gold and 7-coordinate titanium. (b) A β-diketiminate complex of thallium. (c) The result of a novel strategy for low coordinate iron that features dative Fe—Fe bonding. (Molecular structure drawings created from CIF data with hydrogen atoms omitted for clarity.)

Activation refers to enabling molecules to react which are otherwise unreactive under a set of conditions.

The β-diketiminate $[\{N(SiMe_3)C(Ph)\}_2CH]^-$, a nacnac ligand (Table 9.4), enforces a 2-coordinate nonlinear geometry at the main group metal Tl(I) (Figure 9.25b).[43] It is remarkable that the majority of structurally characterized 2-coordinate metal complexes are currently *nonlinear*. Well-characterized 2-coordinate d^1–d^9 transition metal complexes are known only for the first row transition metals Cr, Mn, Fe, Co, and Ni. Although the employment of sterically encumbered ligands remains a staple of these studies, the synthesis of complexes with postulated fivefold metal–metal bond orders (Chapter 15) and the discovery of dative bonding between metals has expanded the strategies available for preparing 2-coordinate transition metal complexes. For example, the diiron complex in Figure 9.25c features a dative Fe—Fe bond where $[Fe(CO)_2(\eta^5\text{-}C_5H_5)]^-$ is the donor, and the extremely bulky $[Fe\{C_6H\text{-}2,6\text{-}(C_6H_2\text{-}2,4,6\text{-}{}^iPr_3)_2\text{-}3,5\text{-}{}^iPr_2)\}]^+$ is the acceptor, resulting in 2-coordinate iron.[44]

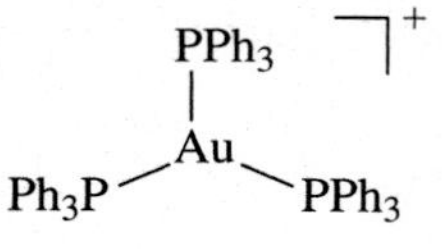

Classic 3-coordinate d^{10} Au(I) and Cu(I) complexes include $[Au(PPh_3)_3]^+$, $[Au(PPh_3)_2Cl]$, and $[Cu(SPPh_3)_3]^+$.[45,46] Bulky ligands of the same types used in 2-coordinate complexes can also favor coordination numbers of 3. All transition metals of the first transition series, as well as some heavier transition metals of groups 5, 6, 8, and 9, form 3-coordinate complexes with nearly trigonal planar metal geometries.[47] The experimental confirmation that $Fe\{N(SiMe_3)_2\}_3$ was monomeric and trigonal planar in the solid state in 1969 was a landmark achievement.[48] 3-coordinate iron complexes are of great interest as nitrogenase models.[49] In this regard, nacnac ligands with bulky substituents stabilize a 3-coordinate diiron complex (Figure 9.26a) that features a bridging dinitrogen ligand that has a substantially weakened N—N bond.[50] A similar β-diketiminate ligand enforces a trigonal planar geometry at Co(II) (Figure 9.26b),[51] and related Zn(II) and Cd(II) complexes are known.[52]

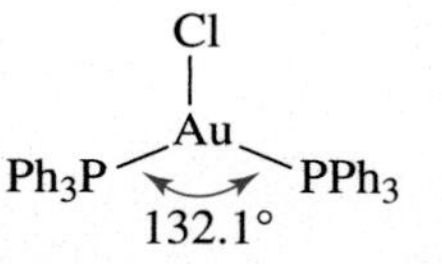

PPh3
Au
Ph3P PPh3

$[Au(PPh_3)_3]^+$

Cl
Au
Ph3P PPh3
132.1°

$Au(PPh_3)_2Cl$

9.4.2 Coordination Number 4

Tetrahedral and square-planar structures are common.[53] Another structure, with four bonds and one lone pair, appears in main group compounds—such as SF_4 and $TeCl_4$—giving a "seesaw" geometry (Chapter 3). Many d^0 and d^{10} complexes have tetrahedral

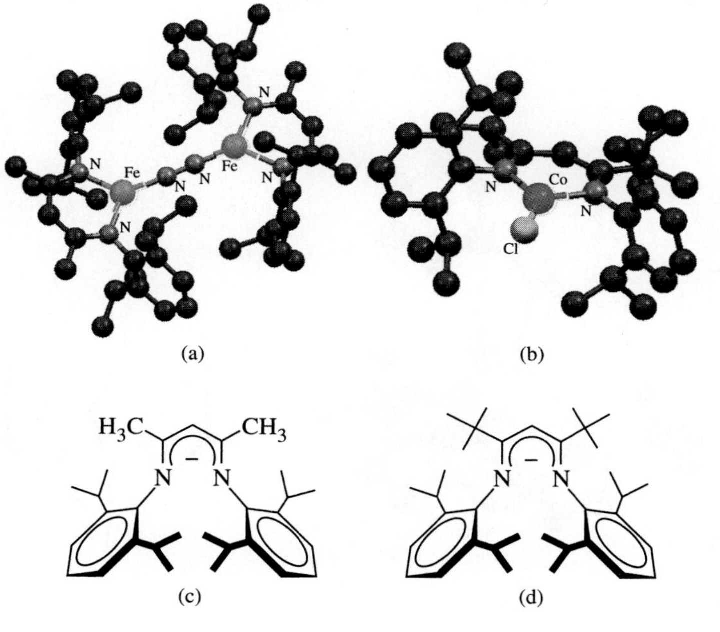

N Fe N N Fe N N N
(a)

N Co N Cl
(b)

H3C CH3 N N
(c)

N N
(d)

FIGURE 9.26 Complexes with Coordination Number 3. (a) A diiron complex with a bridging dinitrogen ligand. (b) A trigonal planar Co(II) chloride. (c) The nacnac ligand used in (a). (d) The nacnac ligand used in (b). (Molecular structure drawings generated from CIF data, with hydrogen atoms omitted for clarity.)

structures—such as MnO_4^-, CrO_4^{2-}, $TiCl_4$, $Ni(CO)_4$, and $[Cu(py)_4]^+$. A few d^5 complexes, including $MnCl_4^{2-}$, are also tetrahedral. These shapes can be explained on the basis of VSEPR, because the *d* orbital occupancy is spherically symmetrical with zero, one, or two electrons in each *d* orbital. Tetrahedral structures occur in some Co(II) d^7 and Ni(II) d^8 species, such as $CoCl_4^{2-}$, $NiCl_4^{2-}$, and $[NiCl_2(PPh_3)_2]$. Tetrahalide complexes of d^9 Cu(II), however, may be distorted from pure tetrahedral geometry. For example, in $Cs_2[CuCl_4]$ and $(NMe_4)_2[CuCl_4]$ the Jahn–Teller effect (Chapter 10) causes distortion of the $CuCl_4^{2-}$ ion, with two of the Cl—Cu—Cl bond angles near 102° and two near 125°. The bromide complexes have similar structures. Examples of tetrahedral species are in **Figure 9.27**.

Square-planar geometry is common for 4-coordinate species, with the same geometric requirements imposed by octahedral geometry; both require 90° angles between ligands. The most common square-planar complexes contain d^8 ions (for example, Ni(II), Pd(II), Pt(II), and Rh(I)). As mentioned in Section 10.5, Ni(II) and Cu(II) can have tetrahedral, square-planar, or intermediate shapes, depending on both the ligand and the counterion in the crystal. These cases indicate that the energy difference between the two structures is small; crystal packing forces can dictate the geometry adopted. Pd(II) and Pt(II) complexes are square-planar, as are the d^8 complexes $[AgF_4]^-$, $[RhCl(PPh_3)_3]$, $[Ni(CN)_4]^{2-}$, and $[NiCl_2(PMe_3)_2]$. $[NiBr_2(P(C_6H_5)_2(CH_2C_6H_5))_2]$, has both square-planar and tetrahedral isomers in the same crystal.[55] Some square-planar complexes are in **Figure 9.28**.

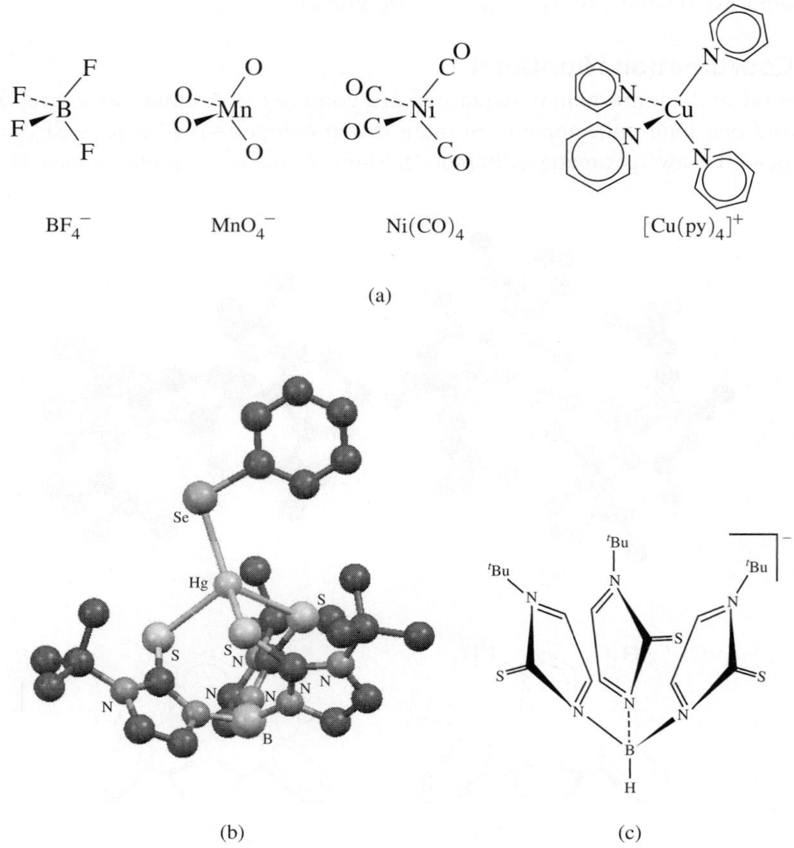

FIGURE 9.27 Complexes with Tetrahedral Geometry. (a) Classic complexes. (b) Tetrahedral Mercury(II) in tris(2-mercapto-1-*t*-butyl-imidazolyl) hydroborato ligation.[54] (c) The tris(2-mercapto-1-*t*-butyl-imidazolyl)hydroborato ligand that binds to Hg via the three sulfur atoms. This ligand is related to the pyrazolylborato ligand (Table 9.4). (Molecular structure drawing generated with CIF data. Hydrogen atoms omitted for clarity.)

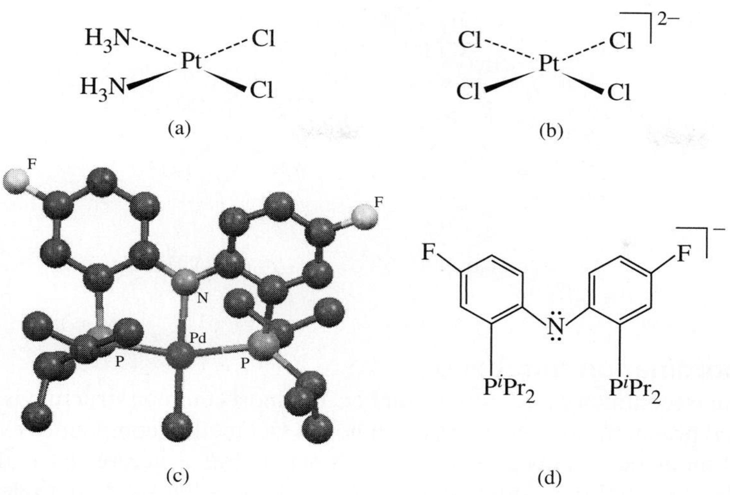

FIGURE 9.28 Complexes with Square-Planar Geometry. (a) $PtCl_2(NH_3)_2$ commonly called "cisplatin" and used as an antitumor agent. (b) $[PtCl_4]^{2-}$. (c) The $Pd(PNP)CH_3$ complex has an approximately square-planar geometry.[56] (d) The PNP ligand employed in (c). Chelating PNP "pincer" ligands based on the *bis*(*ortho*-phosphinoaryl)amine substructure are members of a very popular class of ligands.[57] (Molecular structure drawing created with CIF data with hydrogen atoms omitted for clarity.)

9.4.3 Coordination Number 5

Coordination number 5 includes trigonal bipyramidal, square pyramidal, and pentagonal planar complexes. Pentagonal planar complexes are extremely rare; only the main group ions $[XeF_5]^-$ (58) and $[IF_5]^{2-}$ (59) are known. The energy difference between the trigonal bipyramidal and square pyramidal structures is small in many 5-coordinate complexes; these molecules sometimes exhibit fluxional behavior. For example, $Fe(CO)_5$ and PF_5 have NMR spectra—using ^{13}C and ^{19}F, respectively—that show only one peak, indicating only a single carbon and a single fluorine chemical environment on the NMR time scale. Because both the trigonal bipyramidal and square pyramidal structures have ligands in two different environments, these NMR spectra suggest that the compounds isomerize from one structure to another rapidly. $Fe(CO)_5$ and PF_5 are trigonal bipyramidal in the solid state. $VO(acac)_2$ is square pyramidal, with the doubly bonded oxygen in the apical site. There is evidence that $[Cu(NH_3)_5]^{2+}$ is square pyramidal in liquid ammonia.[60] 5-coordinate complexes are known for many transition metals, including $[CuCl_5]^{3-}$ and $[FeCl(S_2C_2H_2)_2]$. 5-coordinate complexes are shown in **Figure 9.29**.

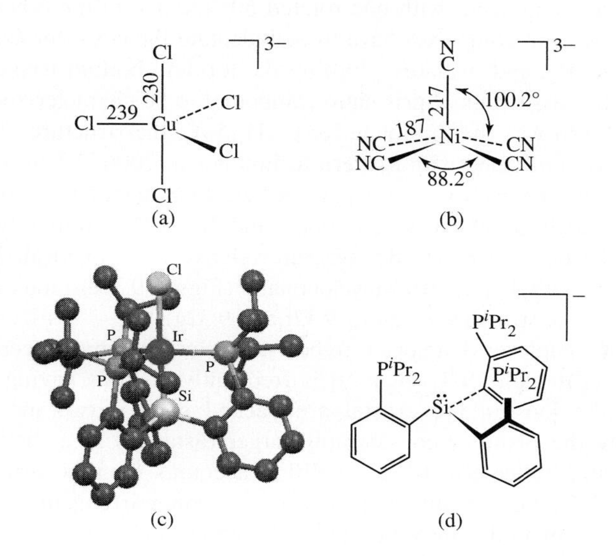

FIGURE 9.29 Complexes with Coordination Number 5. (a) $[CuCl_5]^{3-}$(61). (b) $[Ni(CN)_5]^{3-}$(62). (c) $[Si(o\text{-}C_6H_4P^iPr_2)_3]IrCl$[63]. (d) The tetradentate $[Si(o\text{-}C_6H_4P^iPr_2)_3]^-$ and related ligands create a pocket that stabilizes many 5-coordinate complexes.

$[Co(en)_3]^{3+}$ $[Co(NO_2)_6]^{3-}$

FIGURE 9.30 Complexes with Octahedral Geometry.

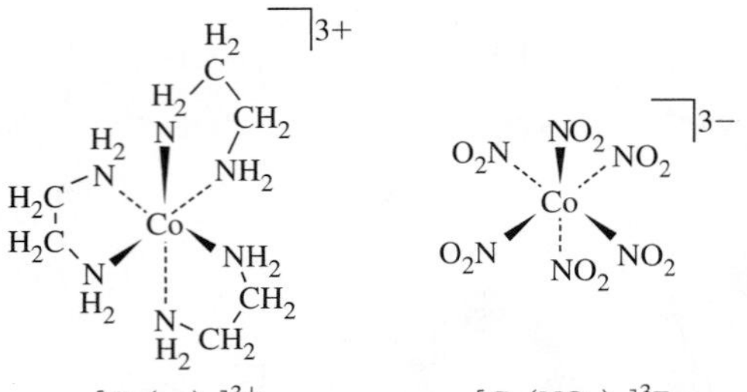

9.4.4 Coordination Number 6

Six is the most common coordination number. The most common structure is octahedral, but trigonal prismatic structures are also known. Octahedral compounds exist for d^0 to d^{10} transition metals. Many compounds with octahedral structures have already been displayed as examples in this chapter. Others include chiral tris(ethylenediamine) cobalt(III), $[Co(en)_3]^{3+}$, and hexanitritocobaltate(III), $[Co(NO_2)_6]^{3-}$, shown in **Figure 9.30**.

For complexes that are not regular octahedra, several types of distortion are possible. The first is elongation, where the bonds to two *trans* ligands elongate, but the other four metal–ligand bonds within a square plane contract in length. Alternatively, the bonds to two *trans* ligand can contract in length, with the other four bonds lengthening. These distortions, which can be driven by both steric and electronic effects, result in a tetragonal shape (**Figure 9.31**). Chromium dihalides exhibit tetragonal elongation in the solid state; crystalline CrF_2 has a distorted rutile structure, with four Cr–F distances of 200 pm and two of 243 pm, and other chromium(II) halides have similar bond distances but different crystal structures.[64] An electronic origin for tetragonal distortion is provided in Section 10.5.

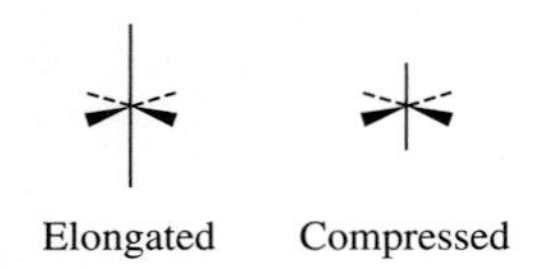

FIGURE 9.31 Tetragonal Distortions of the Octahedron.

An elongation or compression of an octahedron involving trigonal faces (recall that an octahedron has eight faces) can result in a trigonal prismatic or a trigonal antiprismatic structure, with a spectrum of possibilities in between these extremes depending on the angles between the trigonal faces. For example, a trigonal prism (**Figure 9.32a**) results when the top and bottom triangular faces are eclipsed. In a trigonal antiprism, these two triangular faces are staggered, with one rotated 60° relative to the other (Figure 9.32b). Many trigonal prismatic complexes have three bidentate ligands—for example, a variety of dithiolates, $S_2C_2R_2$, and oxalates—linking the top and bottom triangular faces. It is noteworthy that the first trigonal prismatic compound to be characterized by X-ray crystallography (in 1966) was the dithiolate $Re(C_{14}H_{10}S_2)_3$; the structure of this compound (Figure 9.32c) was confirmed with modern techniques in 2006.[65] The trigonal prismatic structures of complexes such as these may be due to π interactions between adjacent sulfur atoms in the trigonal faces. Campbell and Harris[66] summarize the arguments for stability of the trigonal prismatic structure relative to octahedral. The hexadentate 1,4,7-tris(2-mercaptoethyl)-1,4,7-triazacyclononane (Figure 9.32e) binds Fe(III) to afford a trigonal antiprimatic structure (Figures 9.32f and 9.32g).[67]

A number of complexes that appear to be 4-coordinate are more accurately described as 6-coordinate. Although $(NH_4)_2[CuCl_4]$ is frequently cited as having a square-planar $[CuCl_4]^{2-}$ ion, the ions in the crystal are packed so that two more chlorides are above and below the plane at considerably larger distances in a distorted octahedral structure. The Jahn–Teller effect (Section 10.5) accounts for this distortion. Similarly, $[Cu(NH_3)_4]SO_4 \cdot H_2O$ has the ammonias in a square-planar arrangement, but each copper is also connected to more distant water molecules above and below the plane.

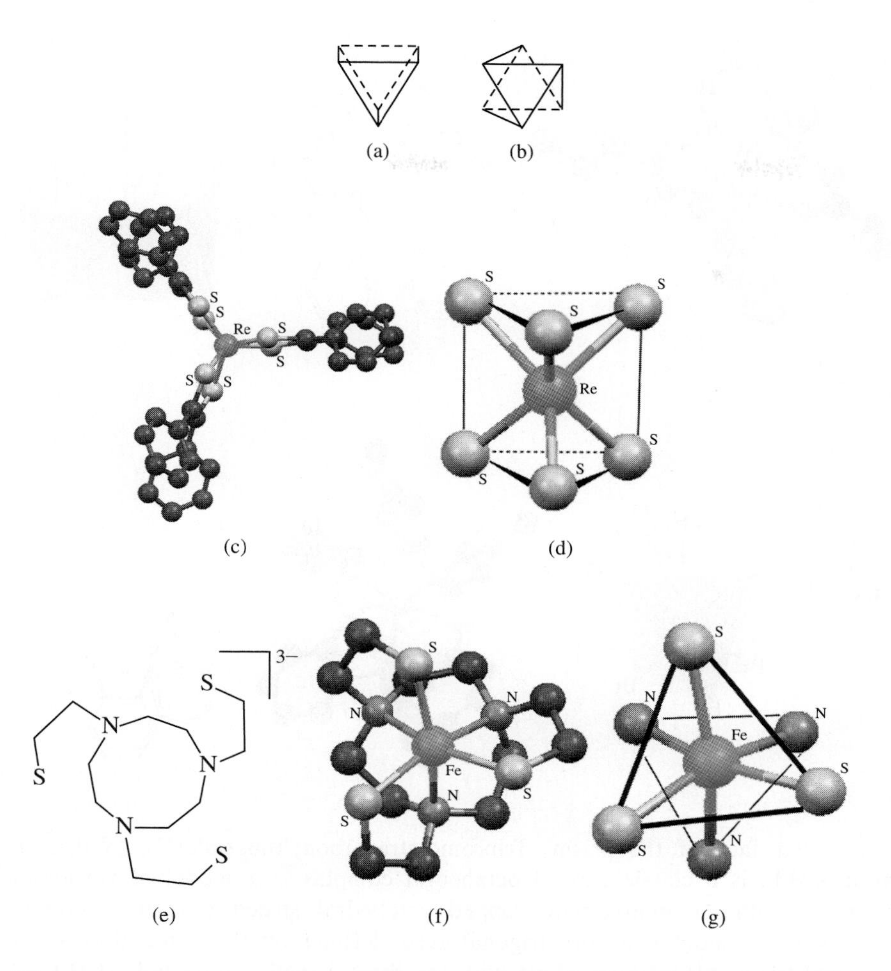

FIGURE 9.32 (a) A trigonal prism. (b) A trigonal antiprism. (c) $Re[S_2C_2(C_6H_5)_2]_3$. This structure is oriented to permit visualization of the eclipsed sulfur atoms that define the corners of the trigonal prismatic core. (d) The trigonal prismatic ReS_6 core of $Re[S_2C_2(C_6H_5)_2]_3$. (e) The trianionic 1,4,7-tris(2-mercaptoethyl)-1,4,7-triazacyclononane ligand that supports trigonal antiprismatic geometry. (f) Fe(III) complex with ligand in (e). (g) The trigonal face defined by the three sulfur atoms is staggered relative to the trigonal face defined by the three nitrogen atoms. The triangles are different sizes because these triangles are composed of different atoms. (Molecular structure diagrams generated from CIF data, with hydrogen atoms omitted for clarity.)

The structure of the 6-coordinate ion $[CuCl_6]^{4-}$ in $[\text{tris(2-aminoethyl)amine}H_4]_2[CuCl_6]Cl_4 \cdot 2H_2O$ is unique.[68] There are three different Cu—Cl bond distances in $[CuCl_6]^{4-}$, in *trans* pairs at 225.1, 236.1, and 310.5 pm, resulting in approximately D_{2h} symmetry with bond angles close to 90°. Hydrogen bonding between the chlorides and the water molecules in this crystal are partly responsible for these varying distances, and Cu(II) also exhibits a Jahn–Teller distortion.

9.4.5 Coordination Number 7

Three structures are possible for 7-coordinate complexes, the pentagonal bipyramid, capped trigonal prism, and capped octahedron.[69] In the capped shapes, the seventh ligand is simply added to a face of the core structure, with necessary adjustments in the other angles to accommodate the additional ligand. All three shapes are found experimentally. The preference for a particular 7-coordinate structure is driven primarily by steric requirements of the ligands. One example of 7-coordination was presented earlier in this chapter. The titanium–gold complex in Figure 9.25a features 2-coordinate gold and 7-coordinate titanium; the geometry at the titanium has been described as approximately capped trigonal prismatic, with the Au atom roughly capping a face defined by four carbon atoms bound to titanium.[39]

The pentagonal bipyramidal geometry is exhibited in main group and transition metal chemistry; examples include IF_7, $[UO_2F_5]^{3-}$, $[NbOF_6]^{3-}$, and the iron(II) complex[70] in **Figure 9.33a**. Examples of capped trigonal prismatic structures include $[NiF_7]^{2-}$ and $[NbF_7]^{2-}$ (Figure 9.33c) in which the seventh fluoride caps

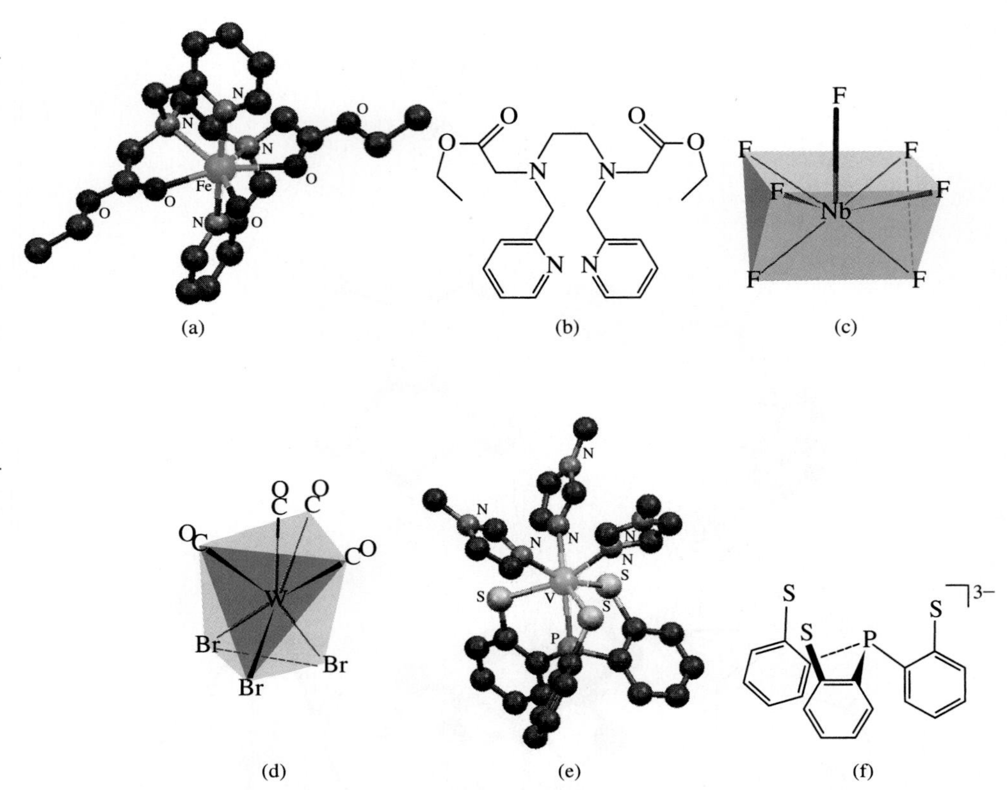

FIGURE 9.33 Coordination Number 7. (a) Approximate pentagonal bipyramidal iron(II) in aqua(*N,N'*-bis(2-pyridylmethyl)-bis(ethylacetate)-1,2-ethanediamine)iron(II). (b) Hexadentate ligand in (a) complex (the other oxygen is a water ligand). (c) Heptafluoroniobate(V), $[NbF_7]^{2-}$, a capped trigonal prism. The capping F is at the top. (d) $[W(CO)_4Br_3]^-$, a capped octahedron. (e) A V(III) complex of the trianion of tris(2-thiophenyl)phosphine and three 1-methylimidazole ligands. (f) Tetradentate ligand employed in (e). (Molecular structure drawings created with CIF data, with hydrogen atoms omitted for clarity.)

a rectangular face of the prism. Tribromotetracarbonyltungstate(II), $[W(CO)_4Br_3]^-$ (Figure 9.33d), is a classic capped octahedral complex.[71] A model of vanadium(III) nitrogenase with an approximate capped octahedral structure is in Figure 9.33e. The phosphorus atom caps the trigonal face defined by the three thiolato sulfur atoms, and these sulfur atoms along with the three 1-methyl imidazole ($CH_3C_3H_3N_2$) nitrogen donor atoms define an approximate octahedron.[72] A variety of 7-coordinate geometries have been analyzed.[73]

9.4.6 Coordination Number 8

Although a central atom or ion within a complex where each donor atom defined the corner of a cube would be 8-coordinate, this structure exists only in ionic lattices such as CsCl, and not in discrete molecular species. However, 8-coordinate square antiprismatic and dodecahedral geometries are common.[74] 8-coordination is less common with first-row transition metals because a relatively larger atomic or ionic radius can better accommodate this many ligands.

A classic example of square antiprismatic geometry is solid-state $Na_7Zr_6F_{31}$, which features square antiprismatic ZrF_8 units arranged in the crystal.[75] A noteworthy recent square antiprismatic W(V) complex is $[W(bipy)(CN)_6]^-$ (**Figure 9.34b**), used to prepare *3d–4f–5d* heterotrimetallic complexes for magnetic applications.[76] $[Yb(NH_3)_8]^{3+}$ is also square antiprismatic.[77]

The structure of $Zr(acac)_2(NO_3)_2$, 8-coordinate by virtue of bidentate nitrate ligands, is a classic dodecahedral complex.[78] Coordination of two tetradentate thiolate ligands like that in Figure 9.33e, but with trimethylsilyl substitutents *ortho* to each thiolate group, results in a dodecahedral V(V) anion (Figure 9.34e) with an interesting electronic ground state.[79]

Other unique 8-coordinate coordination geometries are possible. $[AmCl_2(H_2O)_6]^+$ exhibits a trigonal prism of water ligands with chloride caps on the trigonal faces

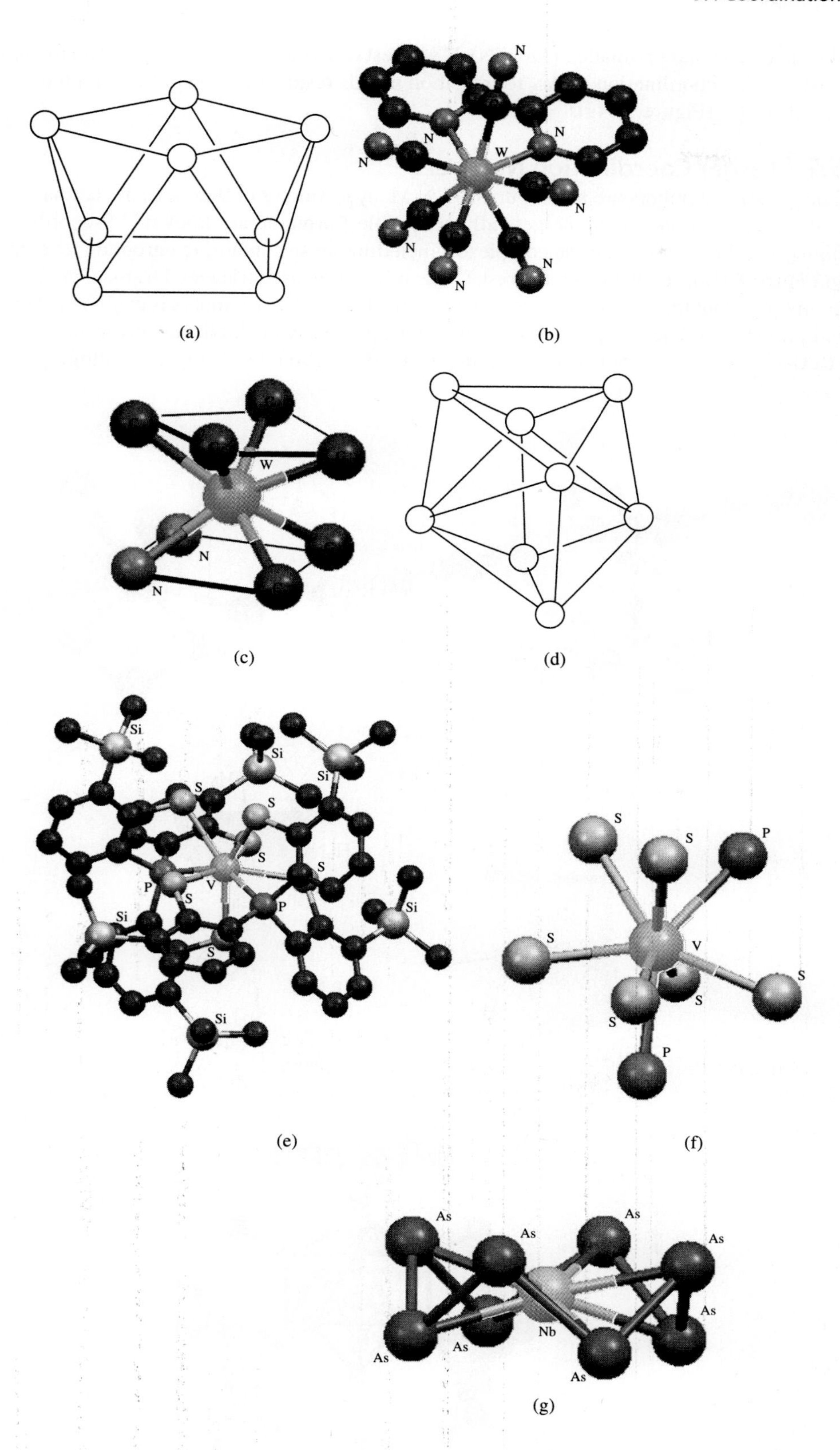

FIGURE 9.34 Coordination Number 8. (a) Square antiprism (with no central atom). (b) $[W(bipy)(CN)_6]^-$. (c) Staggered *trans* squares that define the approximate square antiprism of $[W(bipy)(CN)_6]^-$. (d) Regular trigonal dodecahedron (with no central atom). (e) A dodecahedral high-oxidation state vanadium–thiolate complex. (f) Dodecahedral core of complex in (e). (g) $[NbAs_8]^{3-}$. (Molecular structures generated with CIF data, with hydrogen atoms omitted for clarity.)

(bicapped trigonal prismatic). $[Mo(CN)_8]^{4-}$ is best described as a compressed square antiprism.[80] The coordination of As_8 to transition metals results in "crownlike" structures, as in $[NbAs_8]^{3-}$ (Figure 9.34g).[81]

9.4.7 Larger Coordination Numbers

Coordination numbers are known up to 16.[82] Many examples of 9-coordinate lanthanides and actinides, atoms with energetically accessible *f* orbitals, are known.[83] 9-coordinate highly luminescent lanthanide complexes, including those containing europium (example in **Figure 9.35a**), are of current interest.[84] The mildly distorted tricapped trigonal prismatic geometry about the 9-coordinate europium in the Figure 9.35a complex is explicitly shown in Figure 9.35b. The tricapped trigonal prismatic geometry of classic nonahydridorhenate, $[ReH_9]^{2-}$ (Figure 9.35c), originally determined in 1964 by X-ray crystallography,[85]

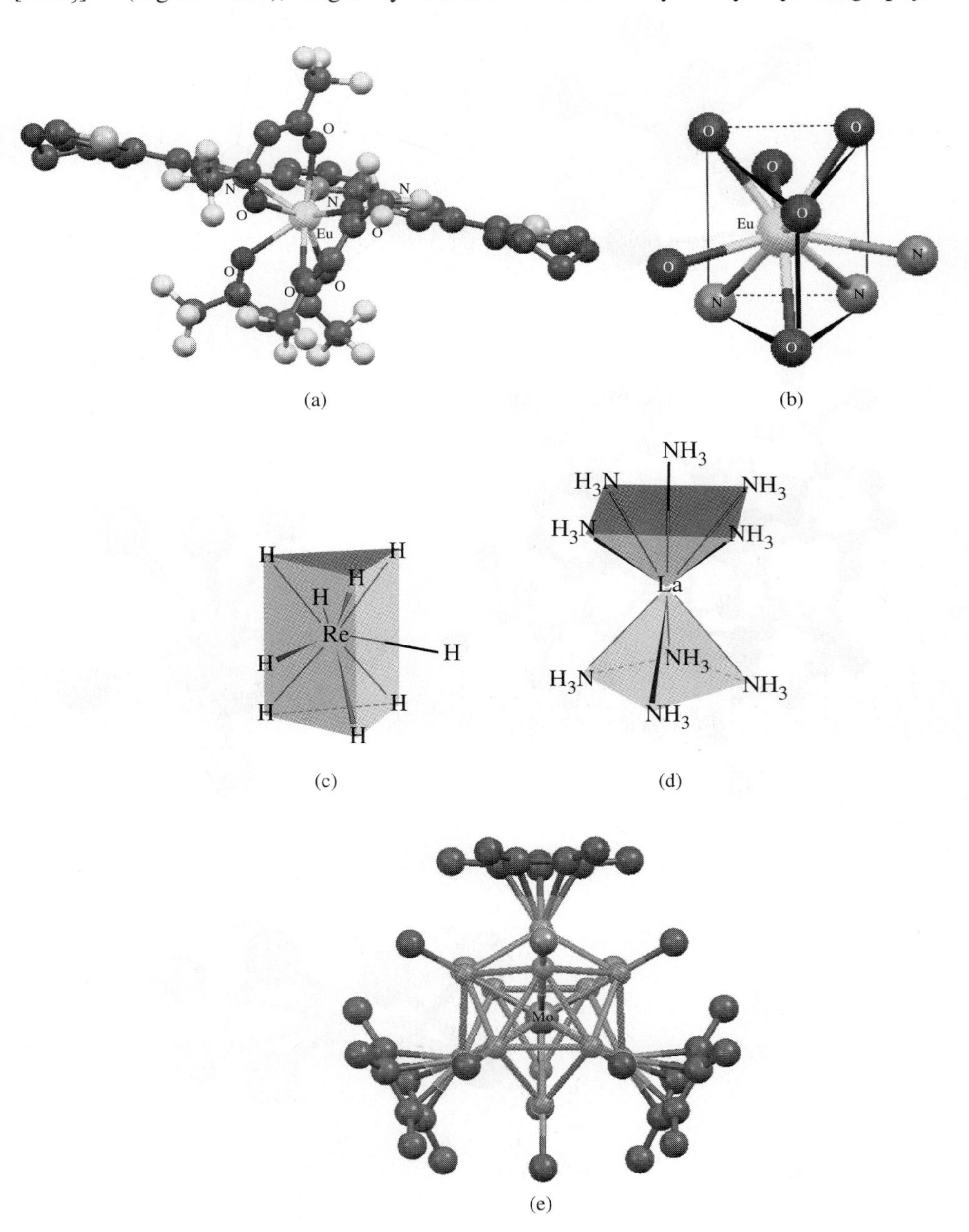

FIGURE 9.35 Complexes with Larger Coordination Numbers. (a) A 9-coordinate Eu(III) complex. (b) The Eu core of the complex in (a) is described as a distorted tricapped trigonal prism. (c) Nonahydridorhenate, $[ReH_9]^{2-}$. (d) $[La(NH_3)_9]^{3+}$, capped square antiprism. (e) Twelve ligands coordinating one Mo center in $[Mo(ZnCH_3)_9(ZnC_5Me_5)_3]$. The zinc atoms are blue. (Molecular structure drawings generated with CIF data, with hydrogen atoms omitted for clarity).

was confirmed by neutron diffraction in 1999.[86] $[La(NH_3)_9]^{3+}$ (Figure 9.35d) has a capped square antiprismatic structure.[87] The novel $[Mo(ZnCH_3)_9(ZnC_5Me_5)_3]$ (Figure 9.35e) is believed to have properties of both a coordination complex, with a 12-*coordinate* sd^5 hydridized Mo center, and a cluster (Chapter 15) with Zn—Zn bonding.[88]

9.5 Coordination Frameworks

To this point, coordination complexes that are individual entities that pack together in the solid state and exist as separated units in solution have been our focus. A burgeoning area of inorganic chemistry is the synthesis and application of substances in which ligands act as bridges to create extended structures in the solid state.

Zeolites (Chapter 7) are porous three-dimensional aluminosilicate structures used in ion exchange and catalysis. A more recent development of inorganic porous materials has been the construction of crystalline or amorphous **coordination polymers**, in which coordination complexes are linked through ligands in infinite arrays. These polymers may be "one-dimensional" chains, with linear or zigzag linkages, or they may be two- or three-dimensional; the range of possibilities is broad. We will focus on structures in which coordination complexes are linked through organic molecules and ions or through donor groups on ligands.

Metal-organic frameworks (MOFs) are three-dimensional extended structures in which metal ions or clusters are linked through organic molecules that have two or more sites through which links can be formed. Unlike coordination polymers, MOFs are exclusively crystalline; the trademarks of MOFs are their extremely high surface areas, tunable pore size, and adjustable internal surface properties.[89] Functional groups that link metals or metal clusters within MOFs are molecules or ions that have two or more Lewis basic sites (for example, carboxylates, triazolates, tetrazolates,[90] and pyrazolates[91] (**Figure 9.36**)).

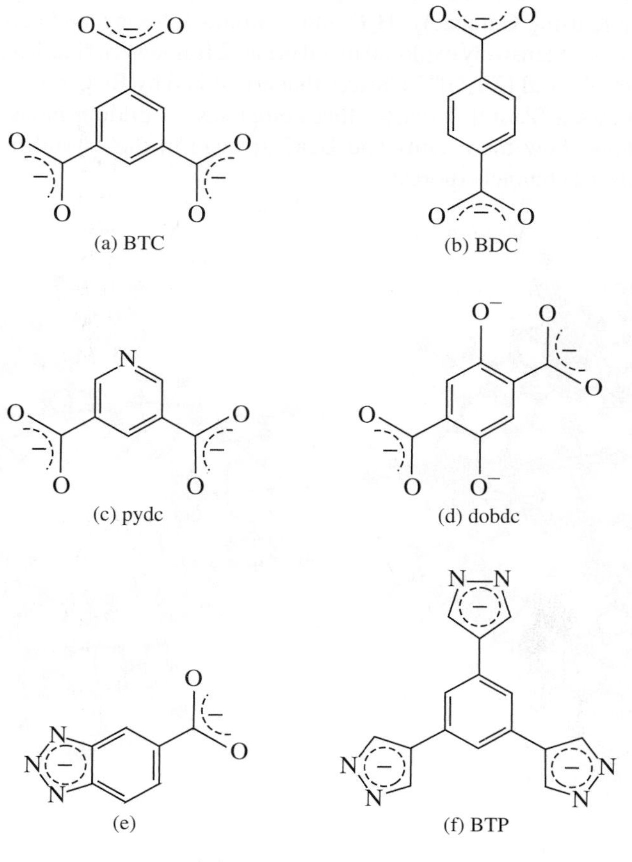

(a) BTC

(b) BDC

(c) pydc

(d) dobdc

(e)

(f) BTP

FIGURE 9.36 Examples of Linking Groups in Metal-Organic Frameworks. (a) 1,3,5-benzenetriscarboxylate. (b) 1,4-benzenedicarboxylate. (c) 3,5-pyridinedicarboxylate. (d) 1,4-dioxido-2,5-benzenedicarboxylate. (e) 5-carboxylatobenzo[1,2,3]triazolate. (f) 1,3,5-tris(pyrazol-4-yl) benzene.

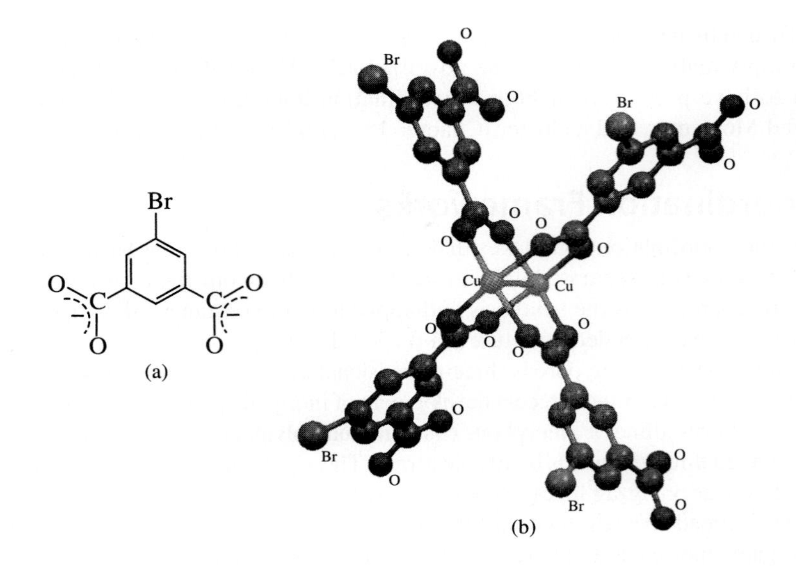

FIGURE 9.37 (a) 5-Bromo-1,3-benzenedicarboxylate. (b) Pairs of Cu^{2+} Ions Bridged by 5-Bromo-1,3-benzenedicarboxylate in a "Paddlewheel" Structure. (Molecular structure generated from CIF data, with hydrogen atoms omitted for clarity).

Construction of frameworks that exhibit desired structures and properties involves judicious assembly of such "building blocks."[92] **Figure 9.37** provides building blocks of a **metal-organic polyhedron (MOP)**, composed of 5-bromo-1,3-benzenedicarboxylate (Figure 9.37a) and copper(II). Specifically, this building block contains pairs of copper(II) ions bridged by the dicarboxylate "paddlewheel" units (Figure 9.37b). The metal-organic polyhedron structure has twelve such units and an overall structure of a great rhombicuboctahedron (**Figure 9.38a**) with a spherical void (Figure 9.38b) having an average diameter of 1.38 nm.[93] Although this MOP structure is complex, its synthesis is relatively simple using $Cu(acac)_2 \cdot H_2O$ and 5-bromo-1,3-benzenedicarboxylic acid.[93]

One of the most extensively explored metal organic frameworks has been MOF-5.[94] This MOF features tetrahedral $[Zn_4O]^{6+}$ clusters that are linked by BDC groups (Figure 9.36b). **Figure 9.39a** shows a $[Zn_4O]^{6+}$ cluster that comprises a building block of MOF-5, and Figure 9.39b shows how these units and BDC arrange in the crystalline metal-organic framework to afford channels (pores).

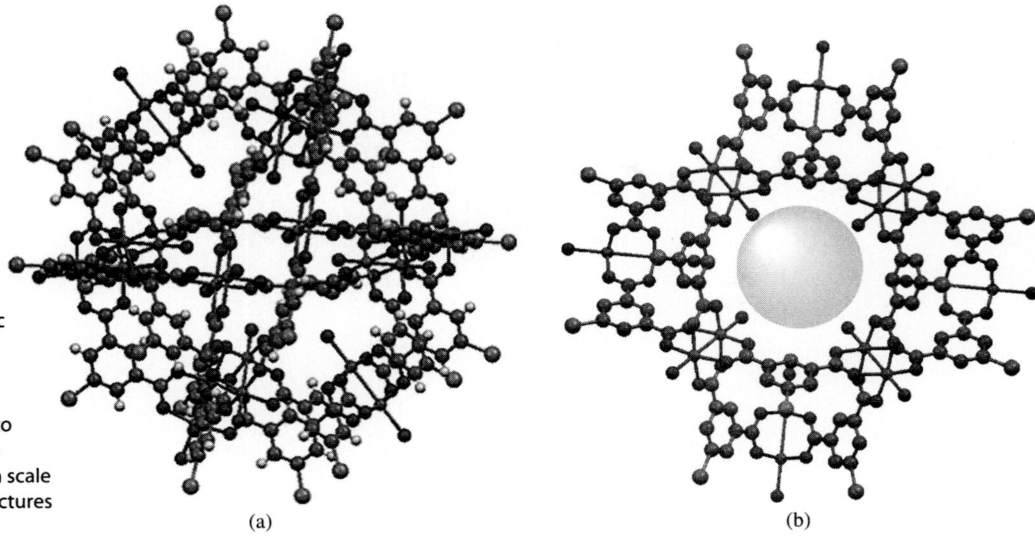

FIGURE 9.38 (a) Crystal Structure of Metal-Organic Polyhedron Composed of Cu(II) and 5-Bromo-1,3-benzenedicarboxylate. (b) Alternate perspective to show spherical cavity. The sphere shown is smaller in scale than the actual void. (Structures generated from CIF data).

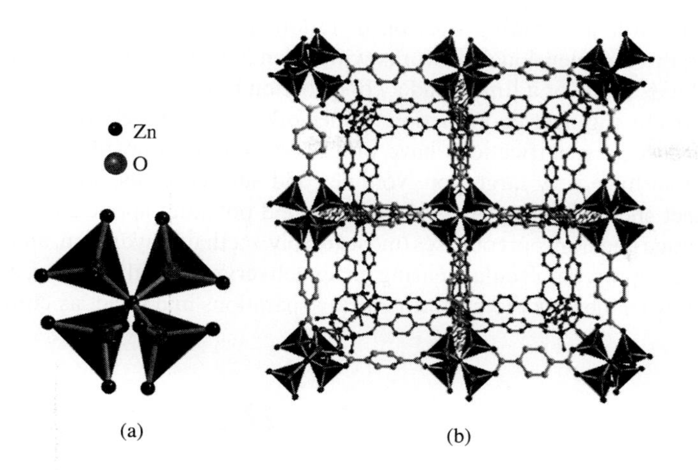

FIGURE 9.39 (a) $[Zn_4O]^{6+}$ cluster building block that comprises MOF-5.* (b) Crystal structure of MOF-5. (Structures generated from CIF data).

The dobdc ligand (Figure 9.36d) shows remarkable versatility for metal-organic frameworks. MOFs of the general formula M_2(dobdc) (M = Mg, Mn, Fe, Co, Ni, Zn) have been prepared. The chemical and physical properties of M_2(dobdc) MOFs are heavily influenced by the specific metal ion incorporated. **Figure 9.40** shows a two- and three-dimensional perspective of the pores formed by Zn-MOF-74 (Zn_2(dobdc)).[95]

A detailed classification system for networks has been devised using three-letter codes to describe topologies, for example **dia** for networks based on the diamond structure, **pcu** for primitive cubic, **bcu** for body-centered cubic, and so forth.[96]

Coordination complexes with attached donor groups on their ligands, designated **metalloligands**, can also be used as building blocks in MOFs.[97] These complexes in essence are acting as ligands themselves. For example, the coordination complex $[Cr(ox)_3]^{3-}$ (ox = oxalate, $C_2O_4^{2-}$) has electron pairs in its outer six oxygen atoms that can form bonds via donor–acceptor interactions to metal ions and serve as a metalloligand in the formation of large, two-dimensional frameworks.[98] Depending on the number and

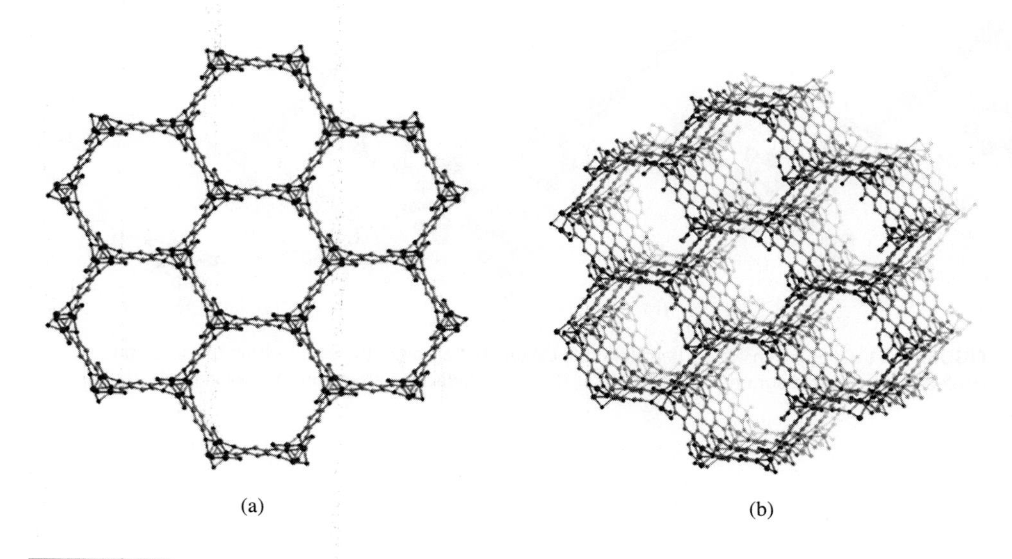

FIGURE 9.40 (a) Metal-Organic Framework forms a cavity in Zn-MOF-74. (b) Layered perspective of pores within Zn-MOF-74. (Structures generated from CIF data).

*The MOF-X identification scheme was developed by O.M. Yaghi. The magnitude of X roughly indicates the order in which these metal-organic frameworks were originally synthesized.

orientation of available bonding sites on metalloligands, one-dimensional coordination polymers and three-dimensional networks with extended channels through them, can also be formed. An example of a linker and a coordination polymer is shown in **Figure 9.41**.[99]

Procedures to engineer coordination frameworks, via a process often called modular synthesis, to desired specifications have progressed rapidly toward formation of robust frameworks that have very large pore volumes and surface areas, structures that can be crafted to meet specific needs. Areas of interest and potential applications[100] include ion exchange, storage and transport of gases (most notably, methane, hydrogen, and acetylene[101]), carbon dioxide capture, molecular sensing, drug delivery, medical imaging, and the design of chiral porous frameworks to perform chiral separations and to act as chiral catalysts.

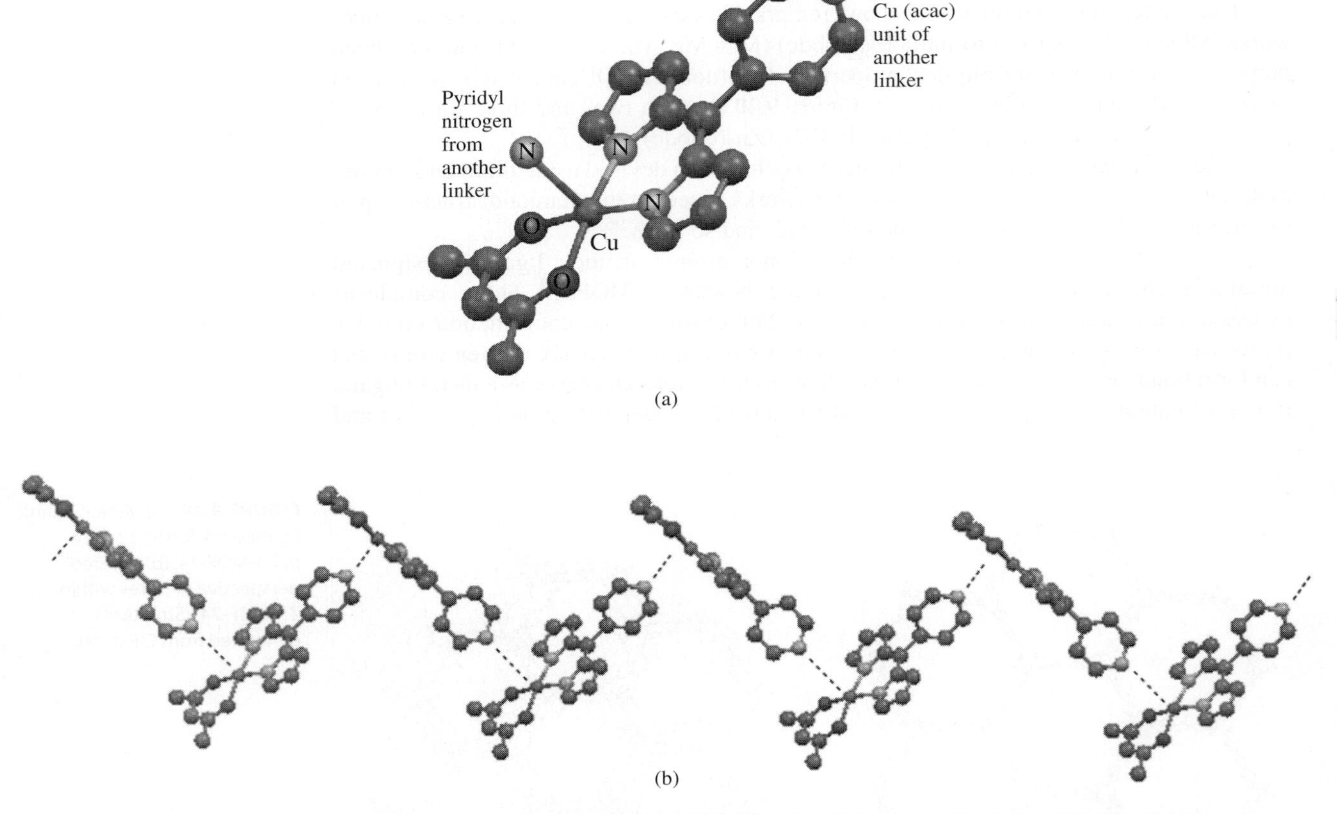

FIGURE 9.41 One-Dimensional Chain. (a) Cu(II) Linker. (b) Portion of the chain within the solid state. (Molecular structure diagrams created with CIF data, with hydrogen atoms omitted for clarity).

References

1. C. W. Blomstrand, *Berichte*, **1871**, *4*, 40; translated by G. B. Kauffman, *Classics in Coordination Chemistry, Part 2*, Dover, New York, 1976, pp. 75–93.
2. S. M. Jørgensen, *Z. Anorg. Chem.*, **1899**, *19*, 109; translated by G. B. Kauffman, *Classics in Coordination Chemistry, Part 2*, pp. 94–164.
3. A. Werner, *Z. Anorg. Chem.*, **1893**, *3*, 267; *Berichte*, **1907**, *40*, 4817; **1911**, *44*, 1887; **1914**, *47*, 3087; A. Werner, A. Miolati, *Z. Phys. Chem.*, **1893**, *12*, 35; **1894**, *14*, 506; all translated by G. B. Kauffman, *Classics in Coordination Chemistry, Part 1*, New York, 1968.
4. L. Pauling, *J. Chem. Soc.*, **1948**, 1461; *The Nature of the Chemical Bond*, 3rd ed., Cornell University Press, Ithaca, NY, 1960, pp. 145–182.
5. J. S. Griffith, L. E. Orgel, *Q. Rev. Chem. Soc.*, **1957**, *XI*, 381.
6. Pauling, *The Nature of the Chemical Bond*, pp. 145–182.
7. Griffith and Orgel, *op. cit.*; L. E. Orgel, *An Introduction to Transition-Metal Chemistry*, Methuen, London, 1960.
8. H. Bethe, *Ann. Phys.*, **1929**, *3*, 133.
9. J. H. Van Vleck, *Phys. Rev.*, **1932**, *41*, 208.
10. J. H. Van Vleck, *J. Chem. Phys.*, **1935**, *3*, 807.
11. T. E. Sloan, "Nomenclature of Coordination Compounds," in G. Wilkinson, R. D. Gillard, and J. A. McCleverty, eds., *Comprehensive Coordination Chemistry*, Pergamon Press, Oxford, 1987, Vol. 1, pp. 109–134; G. J. Leigh, ed., International Union of Pure and Applied Chemistry, *Nomenclature of Inorganic Chemistry: Recommendations 1990,* Blackwell Scientific Publications, Cambridge, MA, 1990; J. A. McCleverty and N. G. Connelly, eds., International Union of Pure and Applied Chemistry, *Nomenclature of Inorganic Chemistry II: Recommendations 2000,* Royal Society of Chemistry, Cambridge, UK, 2001.
12. J. C. Bailar, Jr., *J. Chem. Educ.*, **1957**, *34*, 334; S. A. Meyper, *J. Chem. Educ.*, **1957**, *34*, 623.
13. S. Pevac, G. Crundwell, *J. Chem. Educ.*, **2000**, *77*, 1358; I. Baraldi, D. Vanossi, *J. Chem. Inf. Comput. Sci.*, **1999**, *40*, 386.
14. W. E. Bennett, *Inorg. Chem.*, **1969**, *8*, 1325; B. A. Kennedy, D. A. MacQuarrie, C. H. Brubaker, Jr., *Inorg. Chem.*, **1964**, *3*, 265 and others have written computer programs to calculate the number of isomers for different ligand combinations.
15. E. J. Corey, J. C. Bailar, Jr., *J. Am. Chem. Soc.*, **1959**, *81*, 2620.
16. J. K. Beattie, *Acc. Chem. Res.*, **1971**, *4*, 253.
17. A. Rodríguez-Rodríguez, D. Esteban-Gómez, A. de Blas, T. Rodríguez-Blas, M. Fekete, M. Botta, R. Tripier, C. Platas-Iglesias, *Inorg. Chem.*, **2012**, *51*, 2509.
18. S. Chen, S. A. Pullarkat, Y. Li, P.-H. Leung, *Organometallics*, **2011**, *30*, 1530.
19. R. S. Cahn, C. K. Ingold, *J. Chem. Soc.*, **1951**, 612; Cahn, Ingold, V. Prelog, *Experientia*, **1956**, *12*, 81.
20. B. Brewer, N. R. Brooks, S. Abdul-Halim, A. G. Sykes, *J. Chem. Cryst.*, **2003**, *33*, 651.
21. G. Aullón, G. Esquius, A. Lledós, F. Maseras, J. Pons, J. Ros, *Organometallics*, **2004**, *23*, 5530.
22. W. R. Browne, C. M. O'Connor, H. G. Hughes, R. Hage, O. Walter, M. Doering, J. F. Gallagher, J. G. Vos, *J. Chem. Soc., Dalton Trans.*, **2002**, 4048.
23. J. L. Burmeister, R. L. Hassel, R. J. Phelen, *Inorg. Chem.*, **1971**, *10*, 2032; J. E. Huheey, S. O. Grim, *Inorg. Nucl. Chem. Lett.*, **1974**, *10*, 973.
24. S. Ghosh, G. K. Chaitanya, K. Bhanuprakash, M. K. Nazeeruddin, M. Grätzel, P. Y. Reddy, *Inorg. Chem.*, **2006**, *45*, 7600.
25. T. P. Brewster, W. Ding, N. D. Schley, N. Hazari, V. S. Batista, R. H. Crabtree, *Inorg. Chem.*, **2011**, *50*, 11938.
26. B. Adell, *Z. Anorg. Chem.*, **1944**, *252*, 277.
27. R. K. Murmann, H. Taube, *J. Am. Chem. Soc.*, **1956**, *78*, 4886.
28. F. Basolo, *Coord. Chem. Rev.*, **1968**, *3*, 213.
29. R. D. Gillard, D. J. Shepherd, D. A. Tarr, *J. Chem. Soc., Dalton Trans.*, **1976**, 594.
30. G.-C. Ou, L. Jiang, X.-L. Feng, T.-B. Lu, *Inorg. Chem.*, **2008**, *47*, 2710.
31. J. C. Bailar, ed., *Chemistry of the Coordination Compounds*, Reinhold Publishing, New York, **1956**, pp. 334–335, cites several instances, specifically $[Fe(phen)_3]^{2+}$ (Dwyer), $[Cr(C_2O_4)_3]^{3-}$ (King), and $[Co(en)_3]^{3+}$ (Jonassen, Bailar, and Huffmann).
32. F. Pointillart, C. Train, M. Gruselle, F. Villain, H. W. Schmalle, D. Talbot, P. Gredin, S. Decurtins, M. Verdaguer, *Chem. Mater.*, **2004**, *16*, 832.
33. C. Train, R. Gheorghe, V. Krstic, L.-M. Chamoreau, N. S. Ovanesyan, G. L. J. A. Rikken, M. Gruselle, M. Verdaguer, *Nature Mat.*, **2008**, *7*, 729.
34. Z.-H. Zhou, H. Zhao, K.-R. Tsai, *J. Inorg. Biochem.*, **2004**, *98*, 1787.
35. R. D. Gillard, "Optical Rotatory Dispersion and Circular Dichroism," in H. A. O. Hill and P. Day, eds., *Physical Methods in Advanced Inorganic Chemistry*, Wiley InterScience, New York, 1968, pp. 183–185; C. J. Hawkins, *Absolute Configuration of Metal Complexes*, Wiley InterScience, New York, 1971, p. 156.
36. M. Niemeyer, P. P. Power, *Angew. Chem., Int. Ed.*, **1998**, *37*, 1277; S. T. Haubrich, P. P. Power, *J. Am. Chem. Soc.*, **1998**, *120*, 2202.
37. H. V. Rasika Dias, S. Singh, T. R. Cundari, *Angew. Chem., Int. Ed.*, **2005**, *44*, 4907.
38. A. Haaland, K.-G. Martinsen, H. V. Volden, W. Kaim, E. Waldhör, W. Uhl, U. Schütz, *Organometallics*, **1996**, *15*, 1146.
39. P. J. Fischer, V. G. Young, Jr., J. E. Ellis, *Chem. Commun.*, **1997**, 1249.
40. P. P. Power, *J. Organomet. Chem.*, **2004**, *689*, 3904.
41. H. Chen, R. A. Bartlett, H. V. R. Dias, M. M. Olmstead, P. P. Power, *J. Am. Chem. Soc.*, **1989**, *111*, 4338.
42. P. P. Power, *Chem. Rev.*, **2012**, *112*, 3482. D. L. Kays, *Dalton Trans.*, **2011**, *40*, 769.

43. Y. Cheng, P. B. Hitchcock, M. F. Lappert, M. Zhou, *Chem. Commun.*, **2005**, 752.
44. H. Lei, J.-D. Guo, J. C. Fettinger, S. Nagase, P. P. Power, *J. Am. Chem. Soc.*, **2010**, *132*, 17399.
45. F. Klanberg, E. L. Muetterties, L. J. Guggenberger, *Inorg. Chem.*, **1968**, *7*, 2273.
46. N. C. Baenziger, K. M. Dittemore, J. R. Doyle, *Inorg. Chem.*, **1974**, *13*, 805.
47. C. C. Cummins, *Prog. Inorg. Chem.*, **1998**, *47*, 885.
48. D. C. Bradley, M. B. Hursthouse, P. F. Rodesiler, *Chem. Commun.*, **1969**, 14.
49. (a) P. L. Holland, *Acc. Chem. Res.*, **2008**, *41*, 905. (b) P. L. Holland, *Can. J. Chem.*, **2005**, *83*, 296.
50. J. M. Smith, A. R. Sadique, T. R. Cundari, K. R. Rodgers, G. Lukat-Rodgers, R. J. Lachicotte, C. J. Flaschenriem, J. Vela, P. L. Holland, *J. Am. Chem. Soc*., **2006**, *128*, 756.
51. P. L. Holland, T. R. Cundari, L. L. Perez, N. A. Eckert, R. J. Lachicotte, *J. Am. Chem. Soc.*, **2002**, *124*, 14416.
52. K. Pang, Y. Rong, G. Parkin, *Polyhedron*, **2010**, *29*, 1881.
53. M. C. Favas, D. L. Kepert, *Prog. Inorg. Chem.*, **1980**, *27*, 325.
54. J. G. Meinick, K. Yurkerwich, G. Parkin, *J. Am. Chem. Soc.*, **2010**, *132*, 647.
55. B. T. Kilbourn, H. M. Powell, J. A. C. Darbyshire, *Proc. Chem. Soc.*, **1963**, 207.
56. O. V. Ozerov, C. Guo, L. Fan, B. M. Foxman, *Organometallics*, **2004**, *23*, 5573.
57. D. Morales-Morales and C. M. Jensen (eds.), *The Chemistry of Pincer Compounds*, 2007, Elsevier, Amsterdam, The Netherlands.
58. K. O. Christe, E. C. Curtis, D. A. Dixon, H. P. Mercier, J. C. P. Sanders, G. J. Schrobilgen *J. Am. Chem. Soc.*, **1991**, *113*, 3351.
59. K. O. Christe, W. W. Wilson, G. W. Drake, D. A. Dixon, J. A. Boatz, R. Z. Gnann, J. Am. Chem. Soc., **1998**, *120*, 4711.
60. M. Valli, S. Matsuo, H. Wakita, Y. Yamaguchi, M. Nomura, *Inorg. Chem.*, **1996**, *35*, 5642.
61. K. N. Raymond, D. W. Meek, J. A. Ibers, *Inorg. Chem.*, **1968**, *7*, 111.
62. K. N. Raymond, P. W. R. Corfield, J. A. Ibers, *Inorg. Chem.*, **1968**, *7*, 1362.
63. M. T. Whited, N. P. Mankad, Y. Lee, P. F. Oblad, J. C. Peters, *Inorg. Chem.*, **2009**, *48*, 2507.
64. A. F. Wells, *Structural Inorganic Chemistry*, 5th ed., Oxford University Press, Oxford, 1984, p. 413.
65. R. Eisenberg, W. W. Brennessel, *Acta. Cryst.*, **2006**, *C62*, m464.
66. S. Campbell, S. Harris, *Inorg. Chem.*, **1996**, *35*, 3285.
67. J. Notni, K. Pohle, J. A. Peters, H. Görls, C. Platas-Iglesias, *Inorg. Chem.*, **2009**, *48*, 3257.
68. M. Wei, R. D. Willett, K. W. Hipps, *Inorg. Chem.*, **1996**, *35*, 5300.
69. D. L. Kepert, *Prog. Inorg. Chem.*, **1979**, *25*, 41.
70. Q. Zhang, J. D. Gorden, R. J. Beyers, C. R. Goldsmith, *Inorg. Chem.*, **2011**, *50*, 9365.
71. M. G. B. Drew, A. P. Wolters, *Chem. Commun.*, **1972**, 457.
72. S. Ye, F. Neese, A. Ozarowski, D. Smirnov, J. Krzystek, J. Telser, J.-H. Liao, C.-H. Hung, W.-C. Chu, Y.-F. Tsai, R.-C. Wang, K.-Y. Chen, H.-F. Hsu, *Inorg. Chem.*, **2010**, *49*, 977.
73. D. Casanova, P. Alemany, J. M. Bofill, S. Alvarez, *Chem. Eur. J.*, **2003**, *9*, 1281. Z. Lin, I. Bytheway, *Inorg. Chem.*, **1996**, *35*, 594.
74. D. L. Kepert, *Prog. Inorg. Chem.*, **1978**, *24*, 179.
75. J. H. Burns, R. D. Ellison, H. A. Levy, *Acta Cryst.*, **1968**, *B24*, 230.
76. M.-G. Alexandru, D. Visinescu, A. M. Madalan, F. Lloret, M. Julve, M. Andruh, *Inorg. Chem.*, **2012**, *51*, 4906.
77. D. M. Young, G. L. Schimek, J. W. Kolis, *Inorg. Chem.*, **1996**, *35*, 7620.
78. V. W. Day, R. C. Fay, *J. Am. Chem. Soc.*, **1975**, *97*, 5136.
79. Y.-H. Chang, C.-L. Su, R.-R. Wu, J.-H. Liao, Y.-H. Liu, H.-F Hsu, *J. Am. Chem. Soc.*, **2011**, *133*, 5708.
80. W. Meske, D. Babel, *Z. Naturforsch., B: Chem. Sci.*, **1999**, *54*, 117.
81. B. Kesanli, J. Fettinger, B. Scott, B. Eichhorn, *Inorg. Chem.*, **2004**, *43*, 3840.
82. A. Ruiz-Martínez, S. Alvarez, *Chem. Eur. J.*, **2009**, *15*, 7470. M. C. Favas, D. L. Kepert, *Prog. Inorg. Chem.*, **1981**, *28*, 309.
83. T. Harada, H. Tsumatori, K. Nishiyama, J. Yuasa, Y. Hasegawa, T. Kawai, *Inorg. Chem.*, **2012**, *51*, 6476. A. Figuerola, J. Ribas, M. Llunell, D. Casanova, M. Maestro, S Alvarez, C. Diaz, *Inorg. Chem.*, **2005**, *44*, 6939.
84. J. M. Stanley, X. Zhu, X. Yang, B. J. Holliday, *Inorg. Chem.*, **2010**, *49*, 2035.
85. S. C. Abrahams, A. P. Ginsberg, K. Knox, *Inorg. Chem.*, **1964**, *3*, 558.
86. W. Bronger, L. à Brassard, P. Müller, B. Lebech, Th. Schultz, *Z. Anorg. Allg. Chem.*, **1999**, *625*, 1143.
87. D. M. Young, G. L. Schimek, J. W. Kolis, *Inorg. Chem.*, **1996**, *35*, 7620.
88. T. Cadenbach, T. Bollermann, C. Gemel, I. Fernandez, M. von Hopffgarten, G. Frenking, R. A. Fischer, *Angew. Chem., Int. Ed.*, **2008**, *47*, 9150.
89. H.-C. Zhou, J. R. Long, O. M. Yaghi, *Chem. Rev.*, **2012**, *112*, 673.
90. G. Aromí, L. A. Barrios, O. Roubeau, P. Gamez, *Coord. Chem. Rev.*, **2011**, *255*, 485. A. Phan, C. J. Doonan, F. J. Uribe-Romo, C. B. Knobler, M. O'Keeffe, O. M. Yaghi, *Acc. Chem. Res.*, **2010**, *43*, 58.
91. V. Colombo, S. Galli, H. J. Choi, G. D. Han, A. Maspero, G. Palmisano, N. Masciocchi, J. R. Long, *Chem. Sci.*, **2011**, *2*, 1311.
92. O. M. Yaghi, M. O'Keeffe, N. W. Ockwig, H. K. Chae, M. Eddaoudi, J. Kim, *Nature*, **2003**, *423*, 705.
93. H. Furukawa, J. Kim, N. W. Ockwig, M. O'Keeffe, O. M. Yaghi, *J. Am. Chem. Soc.*, **2008**, *130*, 11650.
94. M. Eddaoudi, H. Li, O. M. Yaghi, *J. Am. Chem. Soc.*, **2000**, *122*, 1391. S. S. Kaye, A. Dailly, O. M. Yaghi, J. R. Long, *J. Am. Chem. Soc.*, **2007**, *129*, 14176. L. Hailian, M. Eddaoudi, M. O'Keeffe, O. M. Yaghi, *Nature*, **1999**, *402*, 276.

95. N. L. Rosi, J. Kim, M. Eddaoudi, B. Chen, M. O'Keeffe, O. M. Yaghi, *J. Am. Chem. Soc.*, **2005**, *127*, 1504.

96. N. W. Ockwig, O. Delgado-Friedrichs, M. O'Keeffe, O. M. Yaghi, *Acc. Chem. Res.*, **2005**, *38*, 176. Supplemental information in this reference has illustrations of nets based on coordination numbers 3 through 6; M. O'Keeffe, M. A. Peskov, S. J. Ramsden, O. M. Yaghi, *Acc. Chem. Res.*, **2008**, *41*, 1782.

97. S. J. Garibay, J. R. Stork, S. M. Cohen, *Prog. Inorg. Chem.*, **2009**, *56*, 335.

98. R. P. Farrell, T. W. Hambley, P. A. Lay, *Inorg. Chem.*, **1995**, *34*, 757.

99. S. R. Halper, M. R. Malachowski, H. M. Delaney, and S. M. Cohen, *Inorg. Chem.*, **2004**, *43*, 1242.

100. K. Sumida, D. L. Rogow, J. A. Mason, T. M. McDonald, E. R. Bloch, Z. R. Herm, T.-H. Bae, J. R. Long, *Chem. Rev.*, **2012**, *112*, 724. J.-R. Li, R. J. Kuppler, H.-C. Zhou, *Chem. Soc. Rev.*, **2009**, *38*, 1477. R. E. Morris, P. S. Wheatley, *Angew. Chem., Int. Ed.*, **2008**, *47*, 4966. S. Kitagawa, R. Kitaura, S.-I. Noro, *Angew. Chem., Int. Ed.*, **2004**, *43*, 2334. G. Férey, *Chem. Soc. Rev.*, **2008**, *37*, 191. J. Y. Lee, O. K. Farha, J. Roberts, K. A. Scheidt, S. T. Nguyen, J. T. Hupp, *Chem. Soc. Rev.*, **2009**, *38*, 1450. L. J. Murray, M. Dincă, J. R. Long, *Chem. Soc. Rev.*, **2009**, *38*, 1294. D. Farrusseng, S. Aguado, C. Pinel, *Angew. Chem., Int. Ed.*, **2009**, *48*, 7502. A. Corma, H. García, F. X. Llabrés i Xamena, *Chem. Rev.*, **2010**, *110*, 4606.

101. R. B. Getman, Y.-S. Bae, C. E. Wilmer, R. Q. Snurr, *Chem. Rev.*, **2012**, *112*, 703.

General References

The official documents on IUPAC nomenclature are G. J. Leigh, editor, *Nomenclature of Inorganic Chemistry*, Blackwell Scientific Publications, Oxford, England, 1990 and J. A. McCleverty and N. G. Connelly, editors, *IUPAC, Nomenclature of Inorganic Chemistry II: Recommendations 2000*, Royal Society of Chemistry, Cambridge, UK, 2001. The best single reference for isomers and geometric structures is G. Wilkinson, R. D. Gillard, and J. A. McCleverty, editors, *Comprehensive Coordination Chemistry*, Pergamon Press, Oxford, 1987. The reviews cited in the individual sections are also comprehensive. A useful recent reference on coordination polymers, metal-organic frameworks, and related topics is S. R. Batten, S. M. Neville, and D. R. Turner, *Coordination Polymers*, RSC Publishing, Cambridge, UK, 2009. *Chemical Reviews*, **2012**, volume 112, issue 2 is devoted to Metal-Organic Frameworks. *Dalton Transactions*, **2012**, issue 14 is devoted to Coordination Chemistry in the Solid State.

Problems

9.1 By examining the symmetry, determine if any of the first four proposed structures for hexacoordinate complexes in Figure 9.3 would show optical activity.

9.2 Give chemical names for the following:
- **a.** $[Fe(CN)_2(CH_3NC)_4]$
- **b.** $Rb[AgF_4]$
- **c.** $[Ir(CO)Cl(PPh_3)_2]$ (two isomers)
- **d.** $[Co(N_3)(NH_3)_5]SO_4$
- **e.** $[Ag(NH_3)_2][BF_4]$

9.3 Give chemical names for the following:
- **a.** $[V(C_2O_3)_3]^{3-}$
- **b.** $Na[AlCl_4]$
- **c.** $[Co(en)_2(CO_3)]Cl$
- **d.** $[Ni(bipy)_3](NO_3)_2$
- **e.** $Mo(CO)_6$

9.4 Give chemical names for the following:
- **a.** $[Cu(NH_3)_4]^{2+}$
- **b.** $[PtCl_4]^{2-}$
- **c.** $Fe(S_2CNMe_2)_3$
- **d.** $[Mn(CN)_6]^{4-}$
- **e.** $[ReH_9]^{2-}$

9.5 Name all the complexes in Problem 9.12, omitting isomer designations.

9.6 Name all the complexes in Problem 9.19, omitting isomer designations.

9.7 Give structures for the following:
- **a.** Bis(en)Co(III)-μ-amido-μ-hydroxobis(en)Co(III) ion
- **b.** DiaquadiiododinitritoPd(IV), all isomers
- **c.** $Fe(dtc)_3$, all isomers

9.8 Show structures for the following:
- **a.** Triammineaquadichlorocobalt(III) chloride, all isomers
- **b.** μ-oxo-bis[pentaamminechromium(III)] ion
- **c.** Potassium diaquabis(oxalato)manganate(III)

9.9 Show structures for the following:
- **a.** *cis*-Diamminebromochloroplatinum(II)
- **b.** Diaquadiiododinitritopalladium(IV), all ligands *trans*
- **c.** Tri- μ-carbonylbis(tricarbonyliron(0))

9.10 Glycine has the structure NH_2CH_2COOH. It can lose a proton from the carboxyl group and form chelate rings bonded through both the N and one of the O atoms. Draw structures for all possible isomers of tris(glycinato) cobalt(III).

9.11 Sketch structures of all isomers of $M(AB)_3$, in which AB is a bidentate unsymmetrical ligand, and label the structures *fac* or *mer*.

9.12 Sketch all isomers of the following. Indicate clearly each pair of enantiomers.

a. $[Pt(NH_3)_3Cl_3]^+$
b. $[Co(NH_3)_2(H_2O)_2Cl_2]^+$
c. $[Co(NH_3)_2(H_2O)_2BrCl]^+$
d. $[Cr(H_2O)_3BrClI]$
e. $[Pt(en)_2Cl_2]^{2+}$
f. $[Cr(o\text{-phen})(NH_3)_2Cl_2]^+$
g. $[Pt(bipy)_2BrCl]^{2+}$
h. $Re(arphos)_2Br_2$

arphos = $As(CH_3)_2$ / $P(CH_3)_2$

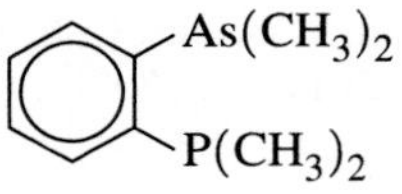

i. $Re(dien)Br_2Cl$

9.13 Determine the number of stereoisomers for the following. Sketch these isomers, and identify pairs of enantiomers. ABA, CDC, and CDE represent tridentate ligands.

a. M(ABA)(CDC)
b. M(ABA)(CDE)

9.14 In Table 9.6 the number of stereoisomers for the formula $M(ABC)_2$ where ABC is a tridentate ligand is given as 11, including five pairs of enantiomers. However, not all literature sources agree. Use sketches and models to verify (the authors hope!) the numbers cited in the table.

9.15 The (2-aminoethyl)phosphine ligand has the structure shown below; it often acts as a bidentate ligand toward transition metals. (See N. Komine, S. Tsutsuminai, M. Hirano, S. Komiya, *J. Organomet. Chem.*, **2007**, *692*, 4486.)

H_2N PH_2

a. When this ligand forms monodentate complexes with palladium, it bonds through its phosphorus atom rather than its nitrogen. Suggest an explanation.
b. How many possible isomers of dichlorobis [(2-aminoethyl)phosphine]nickel(II), an octahedral coordination complex in which (2-aminoethyl)phosphine is bidentate, are there? Sketch each isomer, and identify any pairs of enantiomers.
c. Classify the configuration of chiral isomers as Λ or Δ.

9.16 An octahedrally coordinated transition metal M has the following ligands:

Two chloro ligands
One (2-aminoethyl)phosphine ligand (see Problem 9.15)
One $[O{-}CH_2{-}CH_2{-}S]^{2-}$ ligand

a. Sketch all isomers, clearly indicating pairs of enantiomers.
b. Classify the configuration of chiral isomers as Λ or Δ.

9.17 Suppose a complex of formula $[Co(CO)_2(CN)_2Br_2]^-$ has been synthesized. In the infrared spectrum, it shows two bands attributable to C—O stretching but only one band attributable to C—N stretching. What is the most likely structure of this complex? (See Section 4.4.2.)

9.18 How many possible isomers are there of an octahedral complex having the formula $M(ABC)(NH_3)(H_2O)Br$, where ABC is the tridentate ligand $H_2N{-}C_2H_4{-}PH{-}C_2H_4{-}AsH_2$? How many of these consist of pairs of enantiomers? Sketch all isomers, showing clearly any pairs of enantiomers. The tridentate ligand may be abbreviated as N—P—As for simplicity.

9.19 Assign absolute configurations (Λ or Δ) to the following:

a. S⌒S = dimethyldithiocarbamate

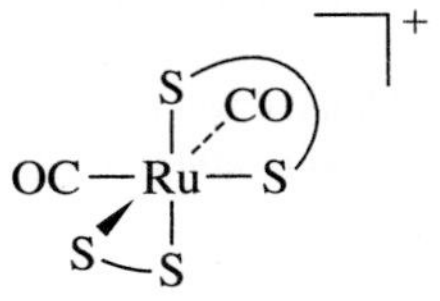

b. O⌒O = oxalate

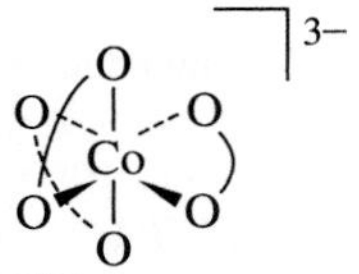

c. N⌒N = ethylenediamine

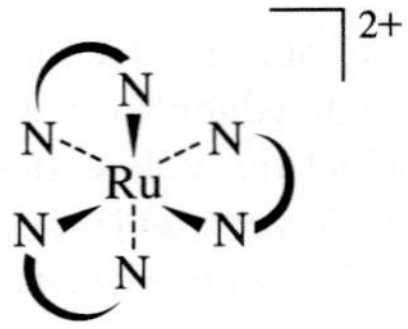

d. N⌒N = 2,2'-bipyridine

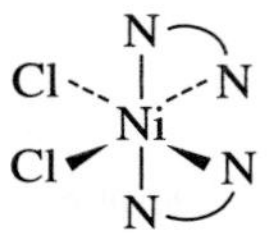

9.20 Which of the following molecules are chiral?

a. Ligand = EDTA

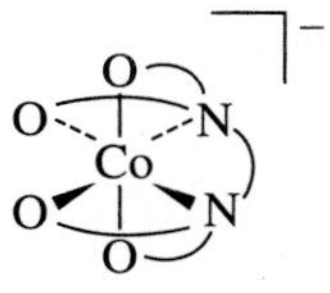

b.

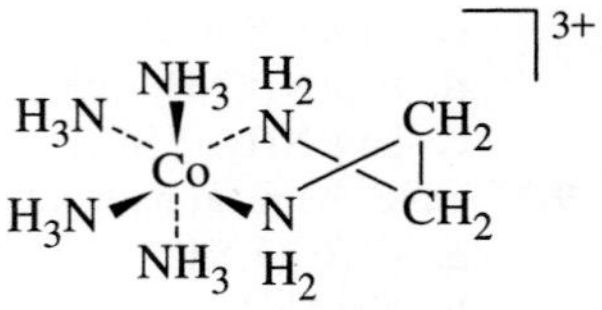

c. 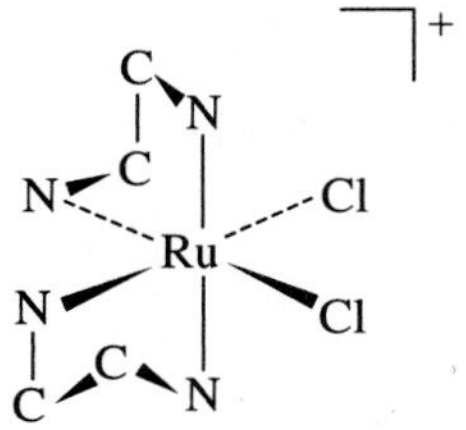 Hydrogens omitted for clarity

9.21 Give the symmetry designations (λ or δ) for the chelate rings in Problem 9.20b and 9.20c.

9.22 Numerous compounds containing central **cubane** structures, formally derived from the cubic organic molecule cubane, C_8H_8, have been prepared. The core structures typically have four metals at opposite corners of a distorted cube, with nonmetals such as O and S at the other corners, as shown (E = nonmetal).

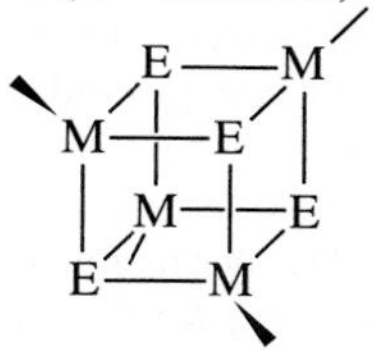

In addition to cubanes in which all metals and nonmetals are identical, they have been prepared with more than one metal and/or more nonmetal in the central 8-atom core; attached groups on the outside may also vary.

a. How many isomers are possible if the core has the following formulas:

1. Mo_3WO_3S?

2. $Mo_3WO_2S_2$?

3. $CrMo_2WO_2SSe$?

b. Assign the point groups for each isomer identified in part a.

c. Is it possible for the central 8-atom core of a cubane structure to be chiral? Explain.

9.23 When cis-OsO_2F_4 is dissolved in SbF_5, the cation $OsO_2F_3^+$ is formed. The ^{19}F NMR spectrum of this cation shows two resonances, a doublet and a triplet having relative intensities of 2:1. What is the most likely structure of this ion? What is its point group? (See W. J. Casteel, Jr., D. A. Dixon, H. P. A. Mercier, G. J. Schrobilgen, *Inorg. Chem.*, **1996**, *35*, 4310.)

9.24 When solid $Cu(CN)_2$ was ablated with 1064 nm laser pulses, various ions containing 2-coordinate Cu^{2+} bridged by cyanide ions were formed. These ions collectively have been dubbed a metal cyanide "abacus." What are the likely structures of such ions, including the most likely geometry around the copper ion? (See I. G. Dance, P. A. W. Dean, K. J. Fisher, *Inorg. Chem.*, **1994**, *33*, 6261.)

9.25 Complexes with the formula $[Au(PR_3)_2]^+$, where R is mesityl, exhibit "propeller" isomerism at low temperature as a consequence of crowding around the phosphorus. How many such isomers are possible? (See A. Bayler, G. A. Bowmaker, H. Schmidbaur, *Inorg. Chem.*, **1996**, *35*, 5959.)

9.26 One of the more striking hydride complexes is $[ReH_9]^{2-}$, which has tricapped trigonal prismatic geometry (Figure 9.35c). Construct a representation using the hydrogen orbitals as a basis. Reduce this to its component irreducible representations, and indicate which orbitals of Re are of suitable symmetry to interact with the hydrogen orbitals.

9.27 The chromium(III) complex $[Cr(bipy)(ox)_2]^-$ can act as a metalloligand to form a coordination polymer chain with Mn(II) ions, in which each manganese ion is 8-coordinate (in flattened square-antiprism geometry) and bridges four $[Cr(bipy)(ox)_2]^-$ units; the ratio of Mn to Cr is 1:2. Sketch two units of this chain. (See F. D. Rochon, R. Melanson, M. Andruh, *Inorg. Chem.*, **1996**, *35*, 6086.)

9.28 The metalloligand $Cu(acacCN)_2$ forms a two-dimensional "honeycomb" sheet with 2′,4′,6′-tri(pyridyl)triazine (tpt); each honeycomb "cell" has sixfold symmetry. Show how six metalloligands and six tpt molecules can form such a structure. (See J. Yoshida, S.-I. Nishikiori, R. Kuroda, *Chem. Lett.*, **2007**, *36*, 678.)

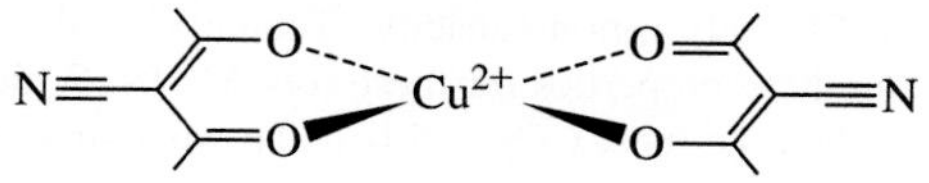

$Cu(acacCN)_2$

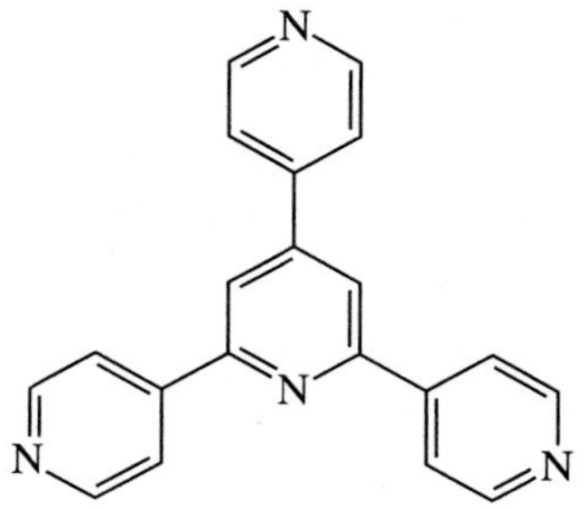

tpt

9.29 Determine the point groups:

a. $Cu(acacCN)_2$ and tpt in Problem 9.28. (Assume delocalization of electrons in the O ··· O part of the acacCN ligands and in the aromatic rings of tpt.)

b. A molecular cartwheel (note orientation of rings). (See H. P. Dijkstra, P. Steenwinkel, D. M. Grove, M. Lutz, A. L. Spek, G. van Koten, *Angew. Chem., Int. Ed.*, **1999**, *38*, 2186.)

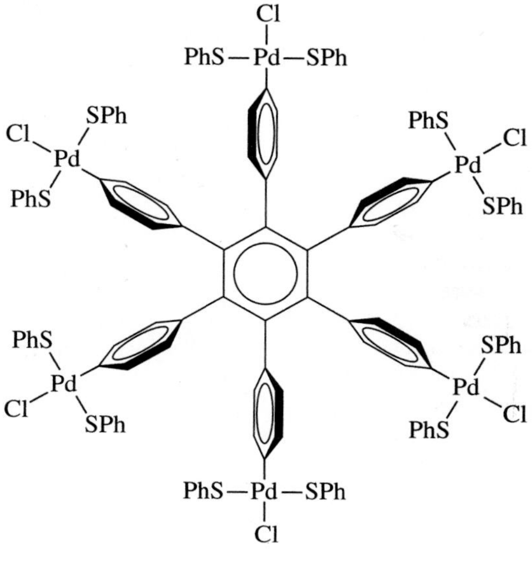

9.30 The separation of carbon dioxide from hydrogen gas is a promising industrial application of MOFs. A variety of metal-organic frameworks (MOF-177, Co(BDP), Cu-BTTri, and Mg_2(dobdc)) were screened to assess their relative abilities for adsorption of these gases at partial pressures up to 40 bar and at 313 K (Z. R. Herm, J. A. Swisher, B. Smit, R. Krishna, J. R. Long, *J. Am. Chem. Soc.*, **2011**, *133*, 5664). Which gas, CO_2 or H_2, is adsorbed more effectively by all four MOFs? Which two metal-organic framework properties most strongly correlate with CO_2 adsorption capacity? Tabulate the data quantifying these properties for these four MOFs. Which MOF adsorbed the most CO_2 at 5 bar? To what structural feature was this high adsorption at low pressures attributed? Which of these MOFs were identified as the best prospects for CO_2/H_2 separation?

9.31 The capture of CO_2 by MOFs from post-combustion gas mixtures has been proposed to reduce CO_2 emissions from coal-fired power plants. The challenge is to engineer MOFs that will selectively adsorb CO_2 from these mixtures at relatively low temperatures and pressures, and subsequently permit facile CO_2 removal to regenerate the MOF for re-use. Describe the synthetic strategy used in T. M. McDonald, W. R. Lee, J. A. Mason, B. M. Wiers, C. S. Hong, J. R. Long, *J. Am. Chem. Soc.*, **2012**, *134*, 7056 to prepare MOFs with excellent CO_2 adsorption capability. Why do these MOFs require "activation," and how is this carried out? Why were early attempts with M_2(dobdc) likely unsuccessful, and how did this inform the design of a new MOF linker?

9.32 The exceptional stability of metal-organic frameworks containing Zr(IV) render them attractive for applications. One strategy for tuning MOF properties is to incorporate additional metals into the framework. Discuss the metalation options for MOF-525 and MOF-545 attempted in W. Morris, B. Volosskiy, S. Demir, F. Gándara, P. L. McGrier, H. Furukawa, D. Cascio, J. F. Stoddart, O. M. Yaghi, *Inorg. Chem.*, **2012**, *51*, 6443. Draw the structure of the linker that permits metal incorporation. How are MOF-525 and MOF-545 similar in their ability to be metalated? How are they different?

Chapter 10

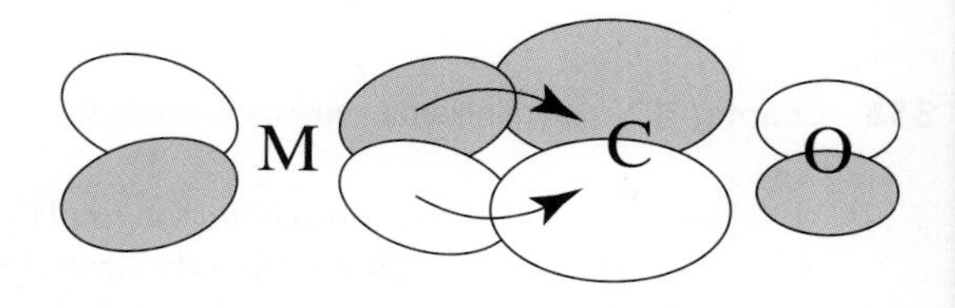

Coordination Chemistry II: Bonding

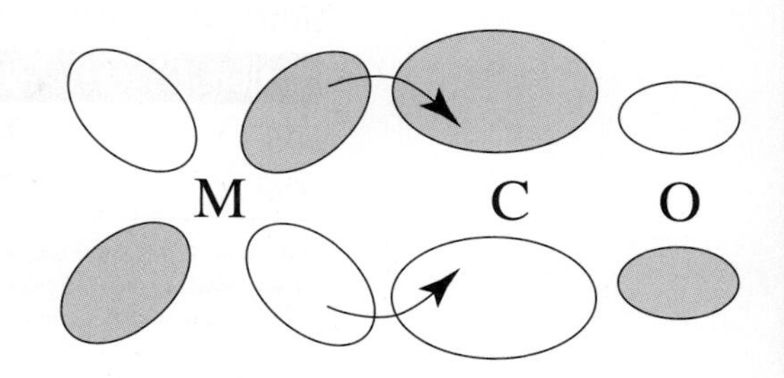

10.1 Evidence for Electronic Structures

A successful bonding theory must be consistent with experimental data. This chapter reviews experimental observations that have been made on coordination complexes, and describes electronic structure and bonding theories used to account for the properties of these complexes.

10.1.1 Thermodynamic Data

A critical objective of any bonding theory is to explain the energies of chemical compounds. Inorganic chemists frequently use **stability constants**, sometimes called **formation constants**, as indicators of bonding strength. These are equilibrium constants for reactions that form coordination complexes. Here are two examples of the formation of coordination complexes and their stability constant expressions:*

$$[Fe(H_2O)_6]^{3+}(aq) + SCN^-(aq) \rightleftharpoons [Fe(SCN)(H_2O)_5]^{2+}(aq) + H_2O\,(l) \qquad K_1 = \frac{[FeSCN^{2+}]}{[Fe^{3+}][SCN^-]} = 9 \times 10^2$$

$$[Cu(H_2O)_6]^{2+}(aq) + 4\,NH_3\,(aq) \rightleftharpoons [Cu(NH_3)_4(H_2O)_2]^{2+}(aq) + 4\,H_2O\,(l) \qquad K_4 = \frac{[Cu(NH_3)_4{}^{2+}]}{[Cu^{2+}][NH_3]^4} = 1 \times 10^{13}$$

In these reactions in aqueous solution, the large stability constants indicate that bonding of the metal ions with incoming ligands is much more favorable than bonding with water, even though water is present in large excess. In other words, the incoming ligands, SCN^- and NH_3, win the competition with H_2O to form bonds to the metal ions.

Table 10.1 provides equilibrium constants for reactions of hydrated Ag^+ and Cu^{2+} with different ligands to form coordination complexes where an incoming ligand has replaced a water molecule. The variation in these equilibrium constants involving the same ligand but different metal ion is striking. Although Ag^+ and Cu^{2+} discriminate significantly between each of the ligands relative to water molecules, the differences are dramatic if the formation constants are compared. For example, the metal ion–ammonia constants are relatively similar (K for Cu^{2+} is ~8.5 times larger than the value for Ag^+), as are the metal ion–fluoride constants (K for Cu^{2+} is ~12 times larger than the value for Ag^+), but the metal ion–chloride and metal ion–bromide constants are very different (by factors of 1,000 and more than 22,000 with Ag^+ now exhibiting a larger K than Cu^{2+}). Chloride and bromide compete much more effectively with water for bonding to Ag^+ than does fluoride, whereas fluoride competes more effectively with water bound to Cu^{2+} relative to Ag^+. This can be

*Water molecules within the formulas of the coordination complexes are omitted from the equilibrium constant expressions for simplicity.

rationalized via the HSAB concept:* silver ion is a soft cation, and copper(II) is borderline. Neither bonds strongly to the hard fluoride ion, but Ag^+ bonds much more strongly with the softer bromide ion than does Cu^{2+}. Such qualitative descriptions are useful, but it is difficult to completely understand the origin of these preferences without additional data.

TABLE 10.1 Formation Constants (K) at 25° C for $[M(H_2O)_n]^z + X^m \longrightarrow [M(H_2O)_{n-1}X]^{z+m} + H_2O\ (l)$

Cation	NH_3	F^-	Cl^-	Br^-
Ag^+	2,000	0.68	1,200	20,000
Cu^{2+}	17,000	8	1.2	0.9

Data from: R. M. Smith and A. E. Martell, *Critical Stability Constants, Vol. 4, Inorganic Complexes*, Plenum Press, New York, 1976, pp. 40–42, 96–119. Not all ionic strengths were identical for these determinations, but the trends in K values shown here are consistent with determinations at a variety of ionic strengths.

An additional consideration appears when a ligand has two donor sites, such as ethylenediamine (en), $NH_2CH_2CH_2NH_2$. After one amine nitrogen bonds with a metal ion, the proximity of the second nitrogen facilitates its simultaneous interaction with the metal. The attachment of multiple donor sites of the same ligand (chelation) generally increases formation constants relative to those for complexes of the same metal ion containing electronically similar monodentate ligands by rendering ligand dissociation more difficult; it is more difficult to separate a ligand from a metal if there are multiple sites of attachment. For example, $[Ni(en)_3]^{2+}$ is stable in dilute solution; but under similar conditions, the monodentate methylamine complex $[Ni(CH_3NH_2)_6]^{2+}$ dissociates methylamine, and nickel hydroxide precipitates:

$$[Ni(CH_3NH_2)_6]^{2+}(aq) + 6\ H_2O\ (l) \longrightarrow Ni(OH)_2(s) + 6\ CH_3NH_3^+(aq) + 4\ OH^-(aq)$$

The formation constant for $[Ni(en)_3]^{2+}$ is clearly larger in magnitude than that for $[Ni(CH_3NH_2)_6]^{2+}$, as the latter is thermodynamically unstable in water with respect to ligand dissocation. This **chelate effect** has the largest impact on formation constants when the ring size formed by ligand atoms and the metal is five or six atoms; smaller rings are strained, and for larger rings, the second complexing atom is farther away, and formation of the second bond may require the ligand to contort. A more complete understanding of this effect requires the determination of the enthalpies and entropies of these reactions.

Enthalpies of reaction can be measured by calorimetric techniques. Alternatively, the temperature dependence of equilibrium constants can be used to determine ΔH° and ΔS° for these ligand substitution reactions by plotting $\ln K$ versus $1/T$.

Thermodynamic parameters such as ΔH°, ΔS°, and the dependence of K with T are useful for comparing reactions of different metal ions reacting with the same ligand or a series of different ligands reacting with the same metal ion. When these data are available for a set of related reactions, correlations between these thermodynamic parameters and the electronic structure of the complexes can sometimes be postulated. However, exclusive knowledge of the ΔH° and ΔS° for a formation reaction is rarely sufficient to predict important characteristics of coordination complexes such as their structures or formulas.

The complexation of Cd^{2+} with methylamine and ethylenediamine are compared in **Table 10.2** for:

$$[Cd(H_2O)_6]^{2+} + 4\ CH_3NH_2 \longrightarrow [Cd(CH_3NH_2)_4(H_2O)_2]^{2+} + 4\ H_2O$$

(no change in number of molecules)

$$[Cd(H_2O)_6]^{2+} + 2\ en \longrightarrow [Cd(en)_2(H_2O)_2]^{2+} + 4\ H_2O$$

(increase of two molecules)

*The HSAB concept is discussed in Chapter 6.

TABLE 10.2 Thermodynamic Data for Monodentate vs. Bidentate Ligand Substitution Reactions at 25 °C

Reactants	Product	ΔH° (kJ/mol)	ΔS° (J/mol K)	ΔG° (kJ/mol) $\Delta H^\circ - T\Delta S^\circ$	K
$[Cd(H_2O)_6]^{2+}$					
$4\ CH_3NH_2$	$[Cd(CH_3NH_2)_4(H_2O)_2]^{2+}$	−57.3	−67.3	−37.2	3.3×10^6
2 en	$[Cd(en)_2(H_2O)_2]^{2+}$	−56.5	+14.1	−60.7	4.0×10^{10}
$[Cd(H_2O)_6]^{2+}$					
$2\ NH_3$	$[Cu(NH_3)_2(H_2O)_4]^{2+}$	−46.4	−8	−43.9	4.5×10^7
en	$[Cu(en)(H_2O)_4]^{2+}$	−54.4	+23	−61.1	4.4×10^{10}

Sources: Data from F. A. Cotton, G. Wilkinson, *Advanced Inorganic Chemistry*, 6th ed., 1999, Wiley InterScience, New York, p. 28; M. Ciampolini, P. Paoletti, L. Sacconi, J. Chem. Soc., 1960, 4553.

Because the ΔH° for these reactions are similar, the large difference in equilibrium constants (over four orders of magnitude!) is a consequence of the large difference in ΔS°: the second reaction has a positive ΔS° accompanying a net increase of two moles in the reaction, in contrast to the first reaction, in which the number of moles is unchanged. In this case, the chelation of ethylenediamine, with one ligand occupying two coordination sites that were previously occupied by two ligands, is the dominant factor in rendering the ΔS° more positive, leading to a more negative ΔG° and more positive formation constant.

Another example in Table 10.2 compares substitution of a pair of aqua ligands in $[Cu(H_2O)_6]^{2+}$ with either two NH_3 ligands or one ethylenediamine. Again, the substantial increase in entropy in the reaction with ethylenediamine plays a very important role in the greater formation constant of this reaction, this time by three orders of magnitude. This is also an example in which the chelating ligand also has a significant enthalpy effect.[1]

10.1.2 Magnetic Susceptibility

The magnetic properties of a coordination compound can provide indirect evidence of its orbital energy levels, similarly to that described for diatomic molecules in Chapter 5. Hund's rule requires the maximum number of unpaired electrons in energy levels with equal, or nearly equal, energies. Diamagnetic compounds, with all electrons paired, are slightly repelled by a magnetic field. When there are unpaired electrons, a compound is paramagnetic and is attracted into a magnetic field. The measure of this magnetism is called the **magnetic susceptibility**, χ.[2] The larger the magnetic susceptibility, the more dramatically a sample of a complex is magnetized (that is, becomes a magnet) when placed in an external magnetic field.

A defining characteristic of a paramagnetic substance is that its magnetization increases linearly with the strength of the externally applied magnetic field at a constant temperature. In contrast, the magnetization of a diamagnetic complex *decreases* linearly with increasing applied field; the induced magnet is oriented in the opposite direction relative to the applied field. Magnetic susceptibility is related to the **magnetic moment**, $\boldsymbol{\mu}$, according to the relationship

$$\mu = 2.828(\chi T)^{\frac{1}{2}}$$

where χ = magnetic susceptibility (cm^3/mol)
T = temperature (Kelvin)
The unit of magnetic moment is the Bohr magneton, μ_B
$1\ \mu_B = 9.27 \times 10^{-24}\ J\ T^{-1}$ (joules/tesla)

Paramagnetism arises because electrons, modeled as negative charges in motion, behave as tiny magnets. Although there is no direct evidence for spinning movement by electrons, a spinning charged particle would generate a **spin magnetic moment**, hence the term **electron spin**. Electrons with $m_s = -\frac{1}{2}$ are said to have a negative spin, and those with $m_s = +\frac{1}{2}$ a positive spin (Section 2.2.2). The total spin magnetic moment for a configuration of electrons is characterized by the spin quantum number S, which is equal to the maximum total spin, the sum of the m_s values.

For example, a ground state oxygen atom with electron configuration $1s^2\,2s^2\,2p^4$ has one electron in each of two $2p$ orbitals and a pair in the third. The maximum total spin is $S = +\frac{1}{2} + \frac{1}{2} + \frac{1}{2} - \frac{1}{2} = 1$. The orbital angular momentum, characterized by the quantum number L, where L is equal to the maximum possible sum of the m_l values for an electronic configuration, results in an additional orbital magnetic moment. For the oxygen atom, the maximum possible sum of the m_l values for the p^4 electrons occurs when two electrons have $m_l = +1$ and one each has $m_l = 0$ and $m_l = -1$. In this case, $L = +1 + 0 - 1 + 1 = 1$. The combination of these two contributions to the magnetic moment, added as vectors, is the total magnetic moment of the atom or molecule. Chapter 11 provides additional details on quantum numbers S and L.

EXERCISE 10.1

Calculate L and S for the nitrogen atom.

The magnetic moment in terms of S and L is

$$\mu_{S+L} = g\sqrt{[S(S+1)] + [\tfrac{1}{4}L(L+1)]}$$

where
μ = magnetic moment
g = gyromagnetic ratio (conversion to magnetic moment)
S = spin quantum number
L = orbital quantum number

Although detailed electronic structure determination requires including the orbital moment, for most complexes of the first transition series, the spin-only moment is sufficient, because orbital contribution is small. The **spin-only magnetic moment, μ_S**, is

$$\mu_S = g\sqrt{S(S+1)}$$

Fields from other atoms and ions may effectively quench the orbital moment in these complexes. For the heavier transition metals and the lanthanides, the orbital contribution is larger and must be taken into account. Because we are usually concerned primarily with the number of unpaired electrons in a compound, and the possible values of μ differ significantly for different numbers of unpaired electrons, the errors introduced by considering only the spin moment are usually not large enough to affect confident predictions of the number of unpaired electrons.

In Bohr magnetons, the gyromagnetic ratio, g, is 2.00023, frequently rounded to 2. The equation for μ_S then becomes

$$\mu_S = 2\sqrt{S(S+1)} = \sqrt{4S(S+1)}$$

Because $S = \frac{1}{2}, 1, \frac{3}{2}, \ldots$ for 1, 2, 3, . . . unpaired electrons, this equation can also be written

$$\mu_S = \sqrt{n(n+2)}$$

where n = number of unpaired electrons. This is the equation that is used most frequently. **Table 10.3** shows the change in μ_S and μ_{S+L} with n, and some experimental moments.

TABLE 10.3 Calculated and Experimental Magnetic Moments

Ion	n	S	L	μ_S	μ_{S+L}	Observed
V^{4+}	1	$\frac{1}{2}$	2	1.73	3.00	1.7–1.8
Cu^{2+}	1	$\frac{1}{2}$	2	1.73	3.00	1.7–2.2
V^{3+}	2	1	3	2.83	4.47	2.6–2.8
Ni^{2+}	2	1	3	2.83	4.47	2.8–4.0
Cr^{3+}	3	$\frac{3}{2}$	3	3.87	5.20	~3.8
Co^{2+}	3	$\frac{3}{2}$	3	3.87	5.20	4.1–5.2
Fe^{2+}	4	2	2	4.90	5.48	5.1–5.5
Co^{3+}	4	2	2	4.90	5.48	~5.4
Mn^{2+}	5	$\frac{5}{2}$	0	5.92	5.92	~5.9
Fe^{3+}	5	$\frac{5}{2}$	0	5.92	5.92	~5.9

Data from F. A. Cotton and G. Wilkinson, *Advanced Inorganic Chemistry*, 4th ed., Wiley, New York, 1980, pp. 627–628.
NOTE: All moments are given in Bohr magnetons.

EXERCISE 10.2

Show that $\sqrt{4S(S+1)}$ and $\sqrt{n(n+2)}$ are equivalent expressions.

EXERCISE 10.3

Calculate the spin-only magnetic moment for the following atoms and ions. (Remember the rules for electron configurations associated with the ionization of transition metals (Section 2.2.4)).

Fe Fe^{2+} Cr Cr^{3+} Cu Cu^{2+}

Measuring Magnetic Susceptibility

The Gouy method[3] is a traditional approach for determining magnetic susceptibility. This method, rarely used in modern laboratories, requires an analytical balance and a small magnet (**Figure 10.1**).[4] The solid sample is packed into a glass tube. A small high-field U-shaped magnet is weighed four times: (1) alone, (2) with the sample suspended between the poles of the magnet, (3) with a reference compound of known magnetic susceptibility suspended in the gap, and finally (4) with the empty tube suspended in the gap (to correct for any magnetism induced in the sample tube). With a diamagnetic sample, the sample and magnet repel each other, and the magnet appears slightly heavier. With a paramagnetic sample, the sample and magnet attract each other, and the magnet appears lighter. The measurement of the reference compound provides a standard from which the **mass susceptibility** (susceptibility per gram) of the sample can be calculated and converted to the **molar susceptibility**.*

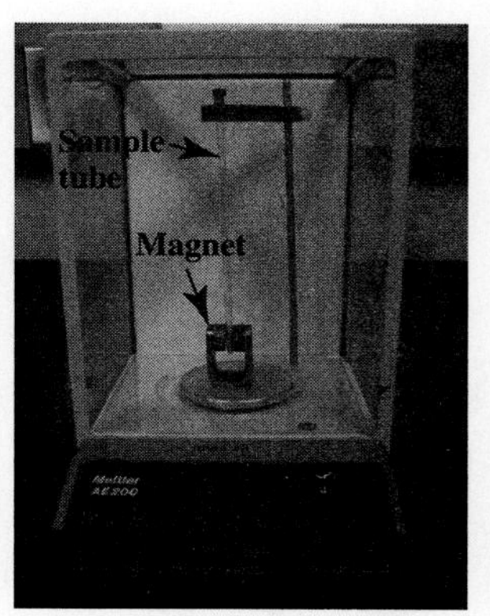

FIGURE 10.1 Modified Gouy Magnetic Susceptibility Apparatus within an Analytical Balance Chamber. (Modeled after the design in S. S. Eaton, G. R. Eaton, *J. Chem. Educ.*, **1979**, *56*, 170.) (Photo Credit: Paul Fischer)

Modern magnetic susceptibility measurements are determined via a magnetic susceptibility balance for solids and via the Evans NMR method for solutes. A magnetic susceptibility balance, like a Gouy balance, assesses the impact of a solid sample on a magnet, but without the magnet being stationary. In a magnetic susceptibility balance, a current is applied to counter (or balance) the deflection of a movable magnet induced by the suspension of the solid sample between the magnet poles. The applied current required to restore the magnet to

*Our objective is to introduce the fundamentals of magnetic susceptibility measurements. The reader is encouraged to examine the cited references for details regarding the calculations involved when applying these methods.

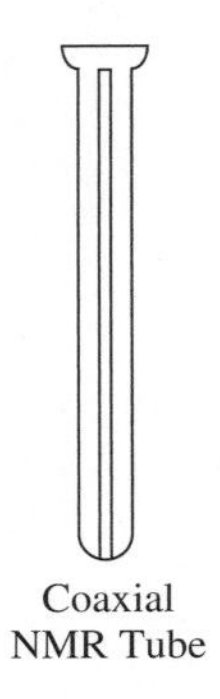
Coaxial NMR Tube

its original position when the sample is suspended is proportional to the mass susceptibility. Like the Gouy method, a magnetic susceptibility balance requires calibration with a reference compound of known susceptibility. $Hg[Co(SCN)_4]$ is a commonly employed reference.

The Evans NMR method[5] requires a coaxial NMR tube where two solutions can be physically separated.* One chamber in the tube contains a solution of a reference solute and the other contains a solution of the paramagnetic analyte and the reference solute. The reference solute must be inert toward the analyte. Because the chemical shift(s) for the resonances of the reference solute in the resulting NMR spectrum will be different for that in the solution with the paramagnetic analyte than in the solution without the analyte, resonances are observed for each chamber. The frequency shift of the selected reference resonance (measured in Hz) is proportional to the mass susceptibility of the analyte.[6] Application of high-field NMR spectrometers is ideal for these studies because rather small chemical shift differences can be resolved.

The superconducting magnets used in modern high-field NMR spectroscopy are also used in Superconducting Quantum Interference Devices (SQUID magnetometer) that measure the magnetic moment of complexes, from which magnetic susceptibility can be determined. In a SQUID, the sample magnetic moment induces an electrical current in superconducting detection coils that subsequently generate a magnetic field. The intensity of this magnetic field is correlated to the sample magnetic moment, and a SQUID instrument has extremely high sensitivity to magnetic field fluctuations.[7] SQUID permits measurement of a sample's magnetic moment over a range of temperatures. The magnetization of a sample (and hence the magnetic susceptibility) as a function of temperature is an important measurement that provides more details about the magnetic properties of the substance.**

Ferromagnetism and Antiferromagnetism

Paramagnetism and diamagnetism represent only two types of magnetism. These substances only become magnetized when placed in an external magnetic field. However, when most people think of magnets, for example those that attach themselves to iron, they are envisioning a persistent magnetic field without the requirement of an externally applied field. This is called **ferromagnetism**. In a ferromagnet, the magnetic moments for each component particle (for example, each iron atom) are aligned in the same direction as a result of the long range order in the bulk solid.† These magnetic moments couple to afford a magnetic field. Common ferromagnets include the metals iron, nickel, and cobalt, as well as alloys (solid solutions) of these metals. **Antiferromagnetism** results from an alternate long-range arrangement of these magnetic moments, where adjacent moments line up in opposite directions. Chromium metal is antiferromagnetic, but this property is most commonly observed in metal oxides (for example NiO). The interested reader is encouraged to examine other resources that treat magnetism in more depth.[8]

10.1.3 Electronic Spectra

Evidence of orbital energy levels can be obtained from electronic spectra. The energy of the photons absorbed as electrons are raised to higher levels is the difference in energy between

---

*An inexpensive approach is to place a sealed capillary tube containing the solution of the reference solute in a standard NMR tube.

**A complex with one unpaired electron exhibits ideal *Curie paramagnetism* if the inverse of the molar susceptibility (for a given applied external field) increases linearly with temperature and has a *y*-intercept of 0. It is common to use SQUID to determine how closely the complex can be described by the *Curie*, or the related *Curie–Weiss*, relationships. The temperature dependence associated with paramagetism can be nonideal and complicated, and is beyond the scope of this text.

†In a paramagnetic complex the magnetic moments of individual species do not effectively couple, but act more or less independently of each other.

the states, which depends on the orbital energy levels and their occupancy. Because of electron–electron interactions, these spectra are frequently more complex than the energy-level diagrams in this chapter would suggest. Chapter 11 describes these interactions, and therefore gives a more complete picture of electronic spectra of coordination compounds.

10.1.4 Coordination Numbers and Molecular Shapes

Although multiple factors influence the number of ligands bonded to a metal and the shapes of the resulting species, in some cases we can predict which structure is favored from electronic structure information. For example, two four-coordinate structures are possible, tetrahedral and square planar. Some metals, such as Pt(II), almost exclusively form square-planar complexes. Others, such as Ni(II) and Cu(II), exhibit both structures, and sometimes intermediate structures, depending on the ligands. Subtle differences in electronic structure help to explain these differences.

10.2 Bonding Theories

Various theoretical approaches to the electronic structure of coordination complexes have been developed. We will discuss three of these bonding models.

Crystal Field Theory

This is an electrostatic approach, used to describe the split in metal *d*-orbital energies within an octahedral environment. It provides an approximate description of the electronic energy levels often responsible for the ultraviolet and visible spectra of coordination complexes, but it does not describe metal–ligand bonding.

Ligand Field Theory

This is a description of bonding in terms of the interactions between metal and ligand frontier orbitals to form molecular orbitals. It uses some crystal field theory terminology but focuses on orbital interactions rather than attractions between ions.

Angular Overlap Method

This is a method of estimating the relative magnitudes of molecular orbital energies within coordination complexes. It explicitly takes into account the orbitals responsible for ligand binding as well as the relative orientation of the frontier orbitals.

Modern computational chemistry allows calculations to predict geometries, orbital shapes and energies, and other properties of coordination complexes. Molecular orbital calculations are typically based on the Born–Oppenheimer approximation, which considers nuclei to be in fixed positions in comparison with rapidly moving electrons. Because such calculations are "many-body" problems that cannot be solved exactly, approximate methods have been developed to simplify the calculations and require less calculation time. The simplest of these approaches, using Extended Hückel Theory, generates useful three-dimensional images of molecular orbitals. Details of molecular orbital calculations are beyond the scope of this text; however, the reader is encouraged to make use of molecular modeling software to supplement the topics and images—some of which were generated using molecular modeling software—in this text. Suggested references on this topic are provided.*

We will now briefly describe crystal field theory to provide a historical context for more recent developments. Ligand field theory and the method of angular overlap are then emphasized.

---

*A brief introduction and comparison of various computational methods is in G. O. Spessard and G. L. Miessler, *Organometallic Chemistry*, Oxford University Press, New York, 2010, pp. 42–49.

10.2.1 Crystal Field Theory

Crystal field theory[9] was originally developed to describe the electronic structure of metal ions in crystals, where they are surrounded by anions that create an electrostatic field with symmetry dependent on the crystal structure. The energies of the *d* orbitals of the metal ions are split by the electrostatic field, and approximate values for these energies can be calculated. No attempt was made to deal with covalent bonding, because covalency was assumed nonexistent in these crystals. Crystal field theory was developed in the 1930s. Shortly afterward, it was recognized that the same arrangement of electron-pair donor species around a metal ion existed in coordination complexes as well as in crystals, and a more complete molecular orbital theory was developed.[10] However, neither was widely used until the 1950s, when interest in coordination chemistry increased.

When the *d* orbitals of a metal ion are placed in an octahedral field of ligand electron pairs, any electrons in these orbitals are repelled by the field. As a result, the $d_{x^2-y^2}$ and d_{z^2} orbitals, which have e_g symmetry, are directed at the surrounding ligands and are raised in energy. The d_{xy}, d_{xz}, and d_{yz} orbitals (t_{2g} symmetry), directed between the ligands, are relatively unaffected by the field. The resulting energy difference is identified as Δ_o (*o* for octahedral; older references use $10Dq$ instead of Δ_o). This approach provides an elementary means of identifying the *d*-orbital splitting found in coordination complexes.

The average energy of the five *d* orbitals is above that of the free ion orbitals, because the electrostatic field of the ligands raises their energy. The t_{2g} orbitals are $0.4\Delta_o$ below and the e_g orbitals are $0.6\Delta_o$ above this average energy, as shown in **Figure 10.2**. The three t_{2g} orbitals then have a total energy of $-0.4\Delta_o \times 3 = -1.2\Delta_o$ and the two e_g orbitals have a total energy of $+0.6\Delta_o \times 2 = +1.2\Delta_o$ compared with the average. The energy difference between the actual distribution of electrons and that for the hypothetical configuration with all electrons in the uniform (or spherical) field level is called the **crystal field stabilization energy (CFSE)**. The CFSE quantifies the energy difference between the electronic configurations due to (1) the *d* orbitals experiencing an octahedral ligand field that discriminates among the *d* orbitals, and (2) the *d* orbitals experiencing a spherical field that would increase their energies uniformly.

This model does not rationalize the electronic stabilization that is the driving force for metal–ligand bond formation. As we have seen in all our discussions of molecular orbitals, any interaction between orbitals leads to formation of both higher and lower energy molecular orbitals, and bonds form if the electrons are stabilized in the resulting occupied molecular orbitals relative to their original atomic orbitals. On the basis of Figure 10.2, the electronic energy of the free ion configuration can *at best* be unchanged in energy upon the free ion interacting with an octahedral ligand field; the stabilization resulting from the metal ion interacting with the ligands is absent. Because this approach does not include the lower (bonding) molecular orbitals, it fails to provide a complete picture of the electronic structure.

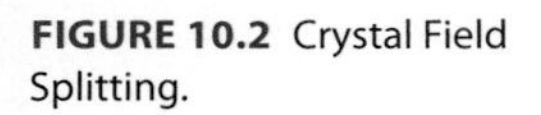
FIGURE 10.2 Crystal Field Splitting.

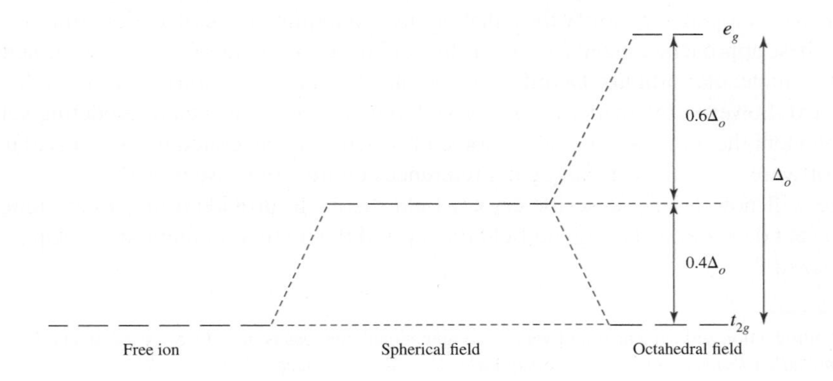

10.3 Ligand Field Theory

Crystal field theory and molecular orbital theory were combined into **ligand field theory** by Griffith and Orgel.[11] Many of the details presented here come from their work.

10.3.1 Molecular Orbitals for Octahedral Complexes

For octahedral complexes, ligands can interact with metals in a sigma fashion, donating electrons directly to metal orbitals, or in a pi fashion, with ligand–metal orbital interactions occurring in two regions off to the side. Examples of such interactions are shown in **Figure 10.3**.

As in Chapter 5, we will first consider group orbitals on ligands based on O_h symmetry, and then consider how these group orbitals can interact with orbitals of matching symmetry on the central atom, in this case a transition metal. We will consider sigma interactions first. The character table for O_h symmetry is provided in **Table 10.4**.

TABLE 10.4 Character Table for O_h

O_h	E	$8C_3$	$6C_2$	$6C_4$	$3C_2(=C_4^2)$	i	$6S_4$	$8S_6$	$3\sigma_h$	$6\sigma_d$		
A_{1g}	1	1	1	1	1	1	1	1	1	1		
A_{2g}	1	1	−1	−1	1	1	−1	1	1	−1		
E_g	2	−1	0	0	2	2	0	−1	2	0		$(2z^2 - x^2 - y^2, x^2 - y^2)$
T_{1g}	3	0	−1	1	−1	3	1	0	−1	−1	(R_x, R_y, R_z)	
T_{2g}	3	0	1	−1	−1	3	−1	0	−1	1		(xy, xz, yz)
A_{1u}	1	1	1	1	1	−1	−1	−1	−1	−1		
A_{2u}	1	1	−1	−1	1	−1	1	−1	−1	1		
E_u	2	−1	0	0	2	−2	0	1	−2	0		
T_{1u}	3	0	−1	1	−1	−3	−1	0	1	1	(x, y, z)	
T_{2u}	3	0	1	−1	−1	−3	1	0	1	−1		

Sigma Interactions

The basis for a reducible representation is a set of six donor orbitals on the ligands as, for example, σ-donor orbitals on six NH_3 ligands.* Using this set as a basis—or equivalently

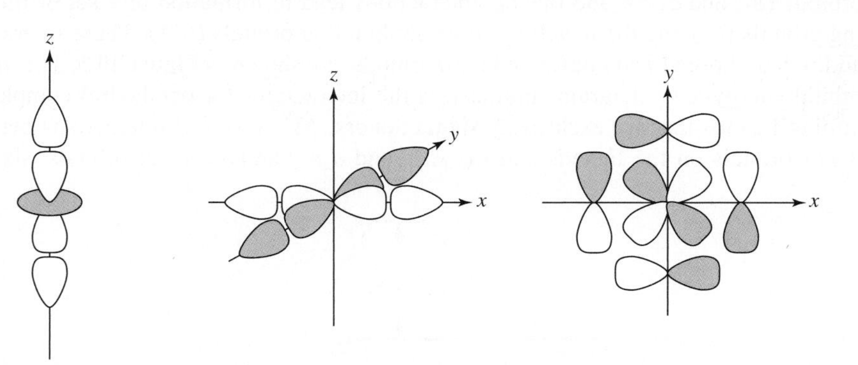

Sigma bonding interaction between two ligand orbitals and metal d_{z^2} orbital

Sigma bonding interaction between four ligand orbitals and metal $d_{x^2-y^2}$ orbital

Pi bonding interaction between four ligand orbitals and metal d_{xy} orbital

FIGURE 10.3 Orbital Interactions in Octahedral Complexes.

*In the case of molecules as ligands, the ligand HOMO often serves as the basis for these group orbitals. Ligand field theory is an extension of the frontier molecular orbital theory discussed in Chapter 6.

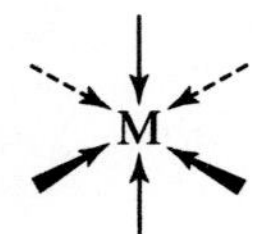

in terms of symmetry, a set of six vectors pointing toward the metal, as shown at left—the following representation can be obtained:

O_h	E	$8C_3$	$6C_2$	$6C_4$	$3C_2(=C_4^2)$	i	$6S_4$	$8S_6$	$3\sigma_h$	$6\sigma_d$
Γ_σ	6	0	0	2	2	0	0	0	4	2

This representation reduces to $A_{1g} + T_{1u} + E_g$:

O_h	E	$8C_3$	$6C_2$	$6C_4$	$3C_2(=C_4^2)$	i	$6S_4$	$8S_6$	$3\sigma_h$	$6\sigma_d$		
A_{1g}	1	1	1	1	1	1	1	1	1	1		$x^2 + y^2 + z^2$
T_{1u}	3	0	−1	1	−1	−3	−1	0	1	1	(x, y, z)	
E_g	2	−1	0	0	2	2	0	−1	2	0		$(2z^2 - x^2 - y^2, x^2 - y^2)$

EXERCISE 10.4

Verify the characters of this reducible representation Γ_σ and that it reduces to $A_{1g} + T_{1u} + E_g$.

The *d* Orbitals

The *d* orbitals play key roles in transition-metal coordination chemistry, so it is useful to examine them first. According to the O_h character table, the *d* orbitals match the irreducible representations E_g and T_{2g}. The E_g ($d_{x^2-y^2}$ and d_{z^2}) orbitals match the E_g ligand orbitals. Because the symmetries match, there is an interaction between the two sets of E_g orbitals to form a pair of bonding orbitals (e_g) and the counterpart pair of antibonding orbitals (e_g*). It is not surprising that significant interaction occurs between the $d_{x^2-y^2}$ and d_{z^2} orbitals and the sigma-donor ligands; the lobes of these *d* orbitals and the σ-donor orbitals of the ligands point toward each other. On the other hand, there are no ligand orbitals matching the T_{2g} symmetry of the d_{xy}, d_{xz}, and d_{yz} orbital—whose lobes point between the ligands—so these metal orbitals are nonbonding. The overall *d* interactions are shown in **Figure 10.4**.

The *s* and *p* Orbitals

The valence *s* and *p* orbitals of the metal have symmetry that matches the two remaining irreducible representations: *s* matches A_{1g} and the set of *p* orbitals matches T_{1u}. Because of the symmetry match, the A_{1g} interactions lead to the formation of bonding and antibonding orbitals (a_{1g} and a_{1g}*), and the T_{1u} interactions lead to formation of a set of three bonding orbitals (t_{1u}) and the matching three antibonding orbitals (t_{1u}*). These interactions, in addition to those already described for *d* orbitals, are shown in **Figure 10.5**. This molecular orbital energy-level diagram summarizes the interactions for octahedral complexes containing ligands that are exclusively sigma donors. As a result of interactions between the donor orbitals on the ligands and the *s*, *p*, and $d_{x^2-y^2}$ and d_{z^2} metal orbitals, six bonding

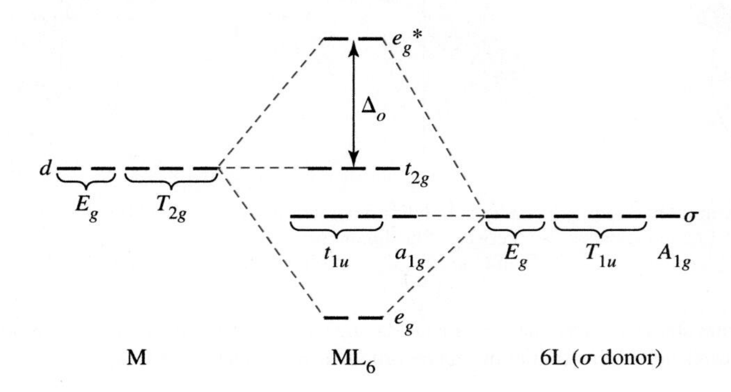

FIGURE 10.4 Sigma-Donor Interactions with Metal *d* Orbitals.

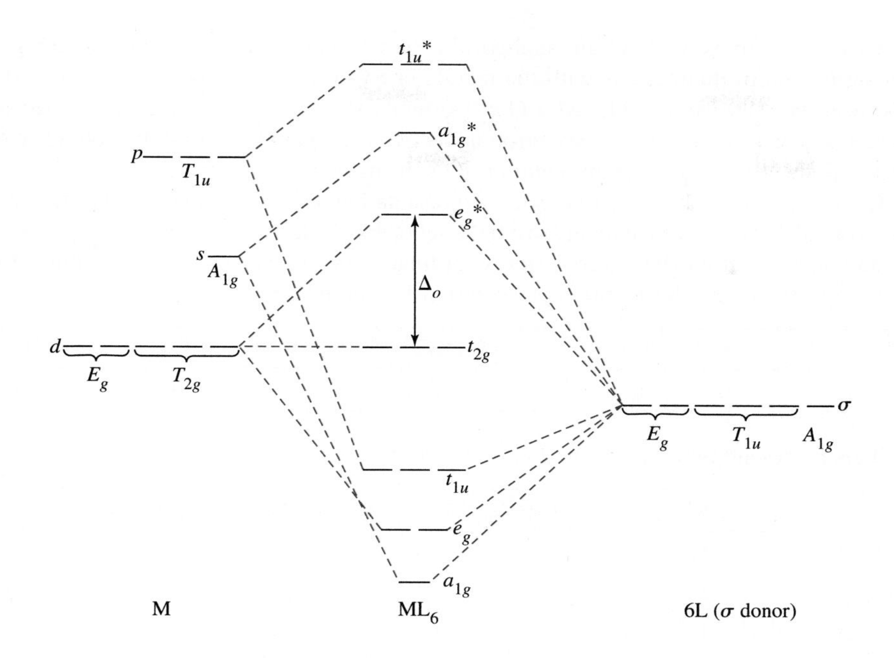

FIGURE 10.5 Sigma-Donor Interactions with Metal *s*, *p*, and *d* Orbitals. As in Chapter 5, the symmetry labels of the atomic orbitals are capitalized, and the labels of the molecular orbitals are in lowercase. The six filled ligand donor orbitals contribute 12 electrons to the lowest six molecular orbitals in this diagram. The metal valence electrons occupy the t_{2g} and, possibly, e_g^* orbitals.

orbitals are formed, occupied by the electrons donated by the ligands. These six electron pairs are stabilized in energy; they represent the sigma bonds stabilizing the complex. The stabilization of these ligand pairs contributes greatly to the driving force for coordination complex formation. This critical aspect is absent from crystal field theory.

The d_{xy}, d_{xz}, and d_{yz} orbitals are nonbonding, so their energies are unaffected by the σ donor orbitals; they are shown in the molecular orbital diagam with the symmetry label t_{2g}. At higher energy, above the t_{2g}, are the antibonding partners to the six bonding molecular orbitals.

One example of a complex that can be described by the energy-level diagram in Figure 10.5 is the green $[Ni(H_2O)_6]^{2+}$. The six bonding orbitals (a_{1g}, e_g, t_{1u}) are occupied by the six electron pairs donated by the aqua ligands. In addition, the Ni^{2+} ion has eight *d* electrons.* In the complex, six of these electrons fill the t_{2g} orbitals, and the final two electrons occupy the e_g^* (separately, with parallel spin).

The beautiful colors of many transition-metal complexes are due in part to the energy difference between the t_{2g} and e_g^* orbitals in these complexes, which is often equal to the energy of photons of visible light. In $[Ni(H_2O)_6]^{2+}$ the difference in energy between the t_{2g} and e_g^* is an approximate match for red light. Consequently, when white light passes through a solution of $[Ni(H_2O)_6]^{2+}$, red light is absorbed and excites electrons from the t_{2g} to the e_g^* orbitals; the light that passes through, now with some of its red light removed, is perceived as green, the complementary color to red. This phenomenon, which is more complicated than its oversimplified description here, will be discussed in Chapter 11.

In Figure 10.5 we again see Δ_o, a symbol introduced in crystal field theory; Δ_o is also used in ligand field theory as a measure of the magnitude of metal–ligand interactions.

Pi Interactions

Although Figure 10.5 can be used as a guide to describe energy levels in octahedral transition-metal complexes, it must be modified when ligands that can engage in pi interactions with metals are involved; pi interactions can have dramatic effects on the t_{2g} orbitals.

---

*Recall that in transition metal ions the valence electrons are all *d* electrons.

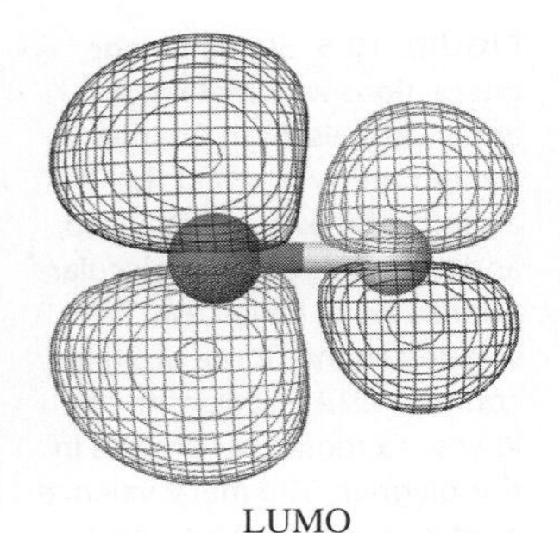

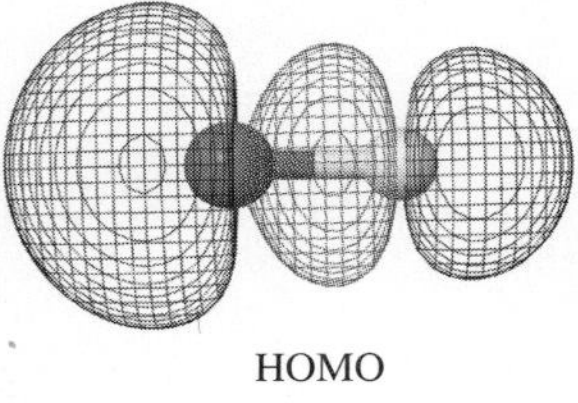

$Cr(CO)_6$ is an example of an octahedral complex that has ligands that can engage in both sigma and pi interactions with the metal. The CO ligand has large lobes on carbon in both its HOMO (3σ) and LUMO ($1\pi^*$) orbitals (Figure 5.13); it is both an effective σ donor and π acceptor. As a π acceptor, it has two orthogonal π^* orbitals, both of which can accept electron density from metal orbitals of matching symmetry.

Once again it is necessary to create a representation, this time using as basis the set of 12 π^* orbitals, two from each ligand, from the set of six CO ligands. In constructing this representation, it is useful to have a consistent coordinate scheme, such as the one in **Figure 10.6**.

Using this set as a basis, the representation Γ_π can be obtained:

O_h	E	$8C_3$	$6C_2$	$6C_4$	$3C_2(=C_4^2)$	i	$6S_4$	$8S_6$	$3\sigma_h$	$6\sigma_d$
Γ_π	12	0	0	0	−4	0	0	0	0	0

This representation reduces to $T_{1g} + T_{2g} + T_{1u} + T_{2u}$:

O_h	E	$8C_3$	$6C_2$	$6C_4$	$3C_2(=C_4^2)$	i	$6S_4$	$8S_6$	$3\sigma_h$	$6\sigma_d$	
T_{1g}	3	0	−1	1	−1	3	1	0	−1	−1	
T_{2g}	3	0	1	−1	−1	3	−1	0	−1	1	(xy, xz, yz)
T_{1u}	3	0	−1	1	−1	−3	−1	0	1	1	(x, y, z)
T_{2u}	3	0	1	−1	−1	−3	1	0	1	−1	

FIGURE 10.6 Coordinate System for Octahedral π Orbitals.

EXERCISE 10.5

Verify the characters of Γ_π and that it reduces to $T_{1g} + T_{2g} + T_{1u} + T_{2u}$.

The most important consequence of this analysis is that it generates a representation that has T_{2g} symmetry, a match for the T_{2g} set of orbitals (d_{xy}, d_{xz}, and d_{yz}) that is nonbonding for ligands that are σ donors only. If the ligand is a π acceptor such as CO, the net effect is to lower the energy of the t_{2g} orbitals, in forming bonding molecular orbitals, and to raise the energy of the (empty) t_{2g}^* orbitals, with high contribution from the ligands, in forming antibonding orbitals. The overlap of the T_{1u} group orbitals of the ligands and the set of p orbitals on the metal is relatively weak because there is also a T_{1u} sigma interaction. The T_{1g} and T_{2u} orbitals have no matching metal orbitals and are nonbonding. The overall result is shown in **Figure 10.7**.

Strong π acceptor ligands have the ability to increase the magnitude of Δ_o by lowering the energy of the t_{2g} orbitals. In the example of $Cr(CO)_6$ there are 12 electrons in the bonding a_{1g}, e_g, and t_{1u} orbitals at the bottom of the diagram; these are formally the six donor pairs from the CO ligands that are stabilized by interacting with the metal. The next six electrons, formally from Cr, fill the three t_{2g} orbitals, which are also stabilized (and bonding) by virtue of the π acceptor interactions. Because the energy difference between the t_{2g} and e_g^* is increased by virtue of the π acceptor capability of CO, it takes more energy to excite an electron between these levels in $Cr(CO)_6$ than, for example, between the t_{2g} and e_g^* levels in $[Ni(H_2O)_6]^{2+}$. Indeed, $Cr(CO)_6$ is colorless, and absorbs ultraviolet radiation because its frontier energy levels are too far apart to absorb visible light.

Electrons in the lower bonding orbitals are largely concentrated on the ligands. It is the stabilization of these ligand electrons that is primarily responsible for why the ligands bind to the metal center. Electrons in the higher levels generally are in orbitals with high metal valence orbital contribution. These electrons are affected by ligand field effects and determine structural details, magnetic properties, electronic spectrum absorptions, and coordination complex reactivity.

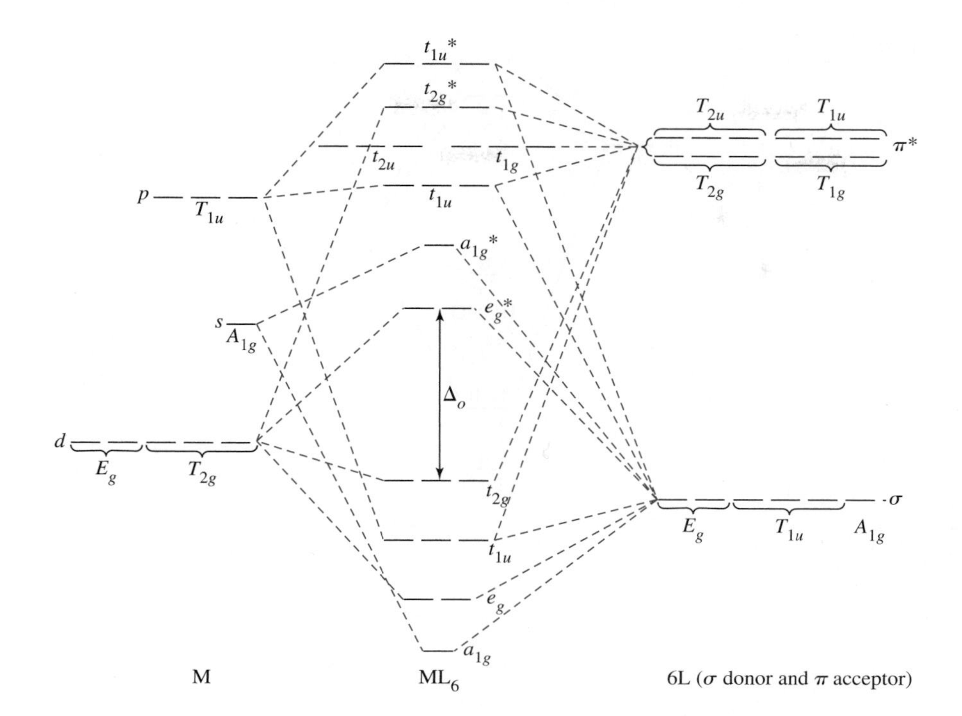

FIGURE 10.7 Sigma-Donor and π-Acceptor Interactions in an Octahedral Complex.* The six filled ligand donor orbitals contribute 12 electrons to the lowest six molecular orbitals in this diagram. The metal valence electrons occupy the t_{2g}, now π-bonding orbitals, and possibly the e_g^* orbitals.

Cyanide can also engage in sigma and pi interactions in its coordination complexes. The energy levels of CN^- (**Figure 10.8**) are intermediate between those of N_2 and CO (Chapter 5), because the energy differences between the C and N valence orbitals are less than the corresponding differences between C and O orbitals. The CN^- HOMO is a σ bonding orbital with electron density concentrated on the carbon. This is the CN^- donor orbital used to form σ orbitals in cyanide complexes. The CN^- LUMOs are two empty π^* orbitals that can be used for π bonding with the metal. A schematic comparison of the π overlap of various ligand orbitals with metal d orbitals is shown in **Figure 10.9**.

The CN^- ligand π^* orbitals have energies higher than those of the metal t_{2g} (d_{xy}, d_{xz}, d_{yz}) orbitals, with which they overlap. As a result, when they form molecular orbitals, the bonding orbitals are lower in energy than the initial metal t_{2g} orbitals. The corresponding antibonding orbitals are higher in energy than the e_g^* orbitals. Metal d electrons occupy the bonding orbitals, resulting in a larger Δ_o and increased metal–ligand bonding, as shown in **Figure 10.10(a)**. Significant electronic stabilization can result from this π bonding. This **metal-to-ligand (M $\longrightarrow$ L) π bonding** is also called **π back-bonding**. In back-bonding, electrons from d orbitals of the metal (electrons that would be localized on the metal if sigma interactions exclusively were involved) now occupy π orbitals with contribution from the ligands. Via this π interaction, the metal transfers some electron density "back" to the ligands in contrast to the sigma interactions, in which the metal is the acceptor and the ligands function as the donors. Ligands that have empty orbitals that can engage the metal in these π interactions are therefore called **π acceptors**.

*This diagram is simplified; it does not show interactions of the CO group orbitals composed of its π bonding molecular orbitals; these also have $T_{1g} + T_{2g} + T_{1u} + T_{2u}$ symmetry, and are similar in energy to the HOMO of CO. Any ligand with empty π^* orbitals also has filled π bonding orbitals that can interact with the metal. In complexes with strong π-acceptor ligands, the impact of these π bonding orbitals on the metal–ligand bonding is relatively small, and these interactions are sometimes ignored. This phenomenon, called π-donation, will be discussed later in this chapter.

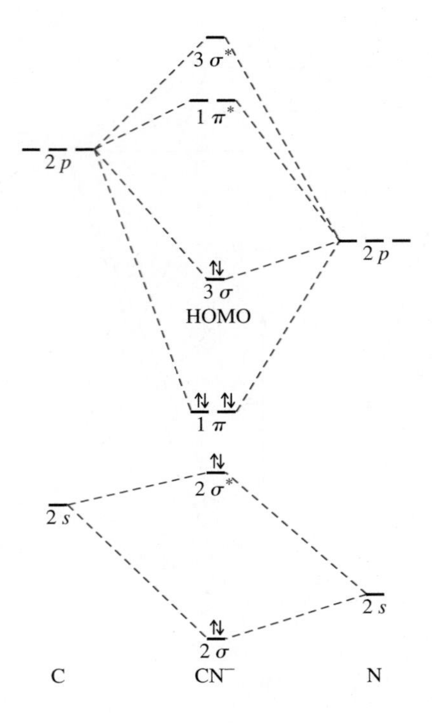

FIGURE 10.8 Cyanide Energy Levels.

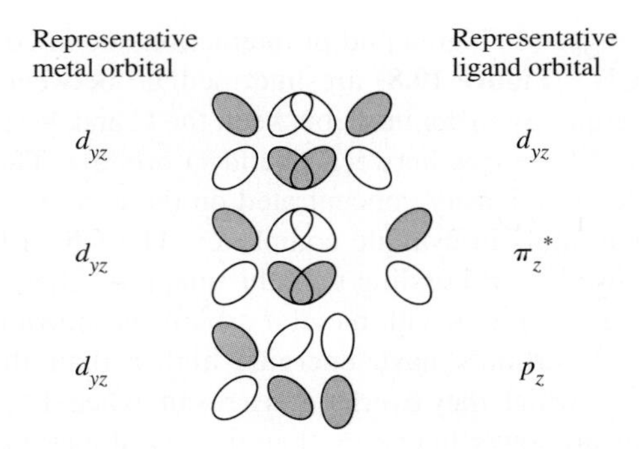

FIGURE 10.9 Overlap of d, π^*, and p Orbitals with Metal d Orbitals. Overlap is good with ligand d and π^* orbitals but poorer with ligand p orbitals.

It was mentioned previously that any ligand with π^* orbitals will also have π orbitals that can interact with the metal. Although the impact of the latter interactions is relatively modest when π acceptor ligands are employed, filled π orbitals can be a very important aspect of the electronic structure with ligands that are poor π acceptors. For example, ligands such as F^- or Cl^- have electrons in p orbitals that are not used for sigma bonding but form the basis of group orbitals with $T_{1g} + T_{2g} + T_{1u} + T_{2u}$ symmetry in octahedral complexes.* These filled T_{2g} orbitals interact with the metal T_{2g} orbitals to generate a bonding and antibonding set. These t_{2g} bonding orbitals, with high ligand orbital contribution, strengthen the ligand–metal linkage slightly, and the corresponding $t_{2g}{}^*$ levels, with high metal d-orbital contribution, are raised in energy and are antibonding. This reduces Δ_o (**Figure 10.10(b)**), and the metal ion d electrons occupy the higher $t_{2g}{}^*$ orbitals. This is

*These are electrons that would be represented as lone pairs on the halides in a Lewis structure of the coordination complex. These ligands are poor π acceptors because the necessary empty ligand orbitals are too high in energy to engage in meaningful interactions with the metal.

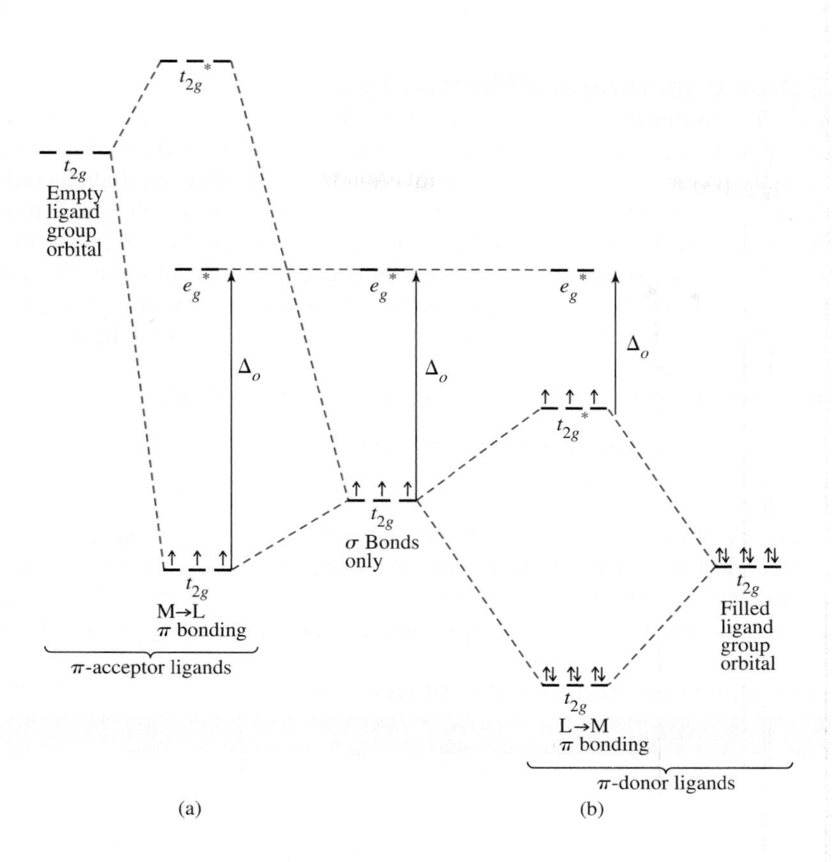

FIGURE 10.10 Effects of π Bonding on Δ_o using a d^3 Ion. Figure 10.10(a) is representative of $[Cr(CN)_6]^{3-}$ and Figure 10.10(b) is representative of $[CrF_6]^{3-}$.

described as **ligand-to-metal (L $\longrightarrow$ M) π bonding**, with the π electrons from the ligands being donated to the metal ion. Ligands participating in such interactions are called **π-donor** ligands. The decrease in the energy of the bonding orbitals is partly counterbalanced by the increase in the energy of the t_{2g}* orbitals. The combined σ and π donation from the ligands gives the metal more negative charge, which may be resisted by metals on the basis of their relatively low electronegativities. However, as with any orbital interaction, π donation will occur to the extent necessary to lower the overall electronic energy of the complex.

Overall, filled ligand p orbitals, or even filled π* orbitals, that have energies compatible with metal valence orbitals, result in L $\longrightarrow$ M π bonding and a smaller Δ_o for the complex. Empty higher-energy π or d orbitals on the ligands with comparable energies relative to the metal valence orbitals can result in M $\longrightarrow$ L π bonding and a larger Δ_o for the complex. Extensive ligand-to-metal π bonding usually favors high-spin configurations, and metal-to-ligand π bonding favors low-spin configurations, consistent with the effect on Δ_o caused by these interactions.*

Part of the stabilizing effect of π back-bonding is a result of transfer of negative charge away from the metal center. The metal, with relatively low electronegativity, accepts electrons from the ligands to form σ bonds. The metal is then left with a relatively large amount of electron density. When empty ligand π orbitals can be used to transfer some electron density back to the ligands, the net result is stronger metal–ligand bonding and increased electronic stabilization for the complex. However, because the lowered t_{2g} orbitals are largely composed of antibonding π* ligand orbitals, occupation of these backbonding orbitals results in *weakening* of the π bonding within the ligand. These π-acceptor ligands are extremely important in organometallic chemistry and are discussed further in Chapter 13.

---

*Low-spin and high-spin configurations are discussed in Section 10.3.2.

10.3.2 Orbital Splitting and Electron Spin

In octahedral coordination complexes, electrons from the ligands fill all six bonding molecular orbitals, and the metal valence electrons occupy the t_{2g} and e_g^* orbitals. Ligands whose orbitals interact strongly with the metal orbitals are called **strong-field ligands**; with these, the split between the t_{2g} and e_g^* orbitals (Δ_o) is large. Ligands with weak interactions are called **weak-field ligands**; the split between the t_{2g} and e_g orbitals (Δ_o) is smaller. For d^0 through d^3 and d^8 through d^{10} metal centers, only one electron configuration is possible. In contrast, the d^4 through d^7 metal centers exhibit **high-spin** and **low-spin** states, as shown in **Table 10.5**. Strong ligand fields lead to low-spin complexes, and weak ligand fields lead to high-spin complexes.

Terminology for these configurations is summarized as follows:

$$\text{Strong ligand field} \rightarrow \text{large } \Delta_o \rightarrow \text{low spin}$$

$$\text{Weak ligand field} \rightarrow \text{small } \Delta_o \rightarrow \text{high spin}$$

The energy of pairing two electrons depends on the Coulombic energy of repulsion between two electrons in the same region of space, Π_c, and the quantum mechanical exchange energy, Π_e (Section 2.2.3). The relationship between the t_{2g} and e_g energy level separation, the Coulombic energy, and the exchange energy—Δ_o, Π_c, and Π_e,

TABLE 10.5 Spin States and Ligand Field Strength

Complex with Weak-Field Ligands (High Spin)	d^1	d^2	d^3	d^4	d^5	d^6	d^7	d^8	d^9	d^{10}
e_g	— —	— —	— —	↑ —	↑ ↑	↑ ↑	↑ ↑	↑ ↑	↑↓ ↑	↑↓ ↑↓
Δ_o										
t_{2g}	↑ — —	↑ ↑ —	↑ ↑ ↑	↑ ↑ ↑	↑ ↑ ↑	↑↓ ↑ ↑	↑↓ ↑↓ ↑	↑↓ ↑↓ ↑↓	↑↓ ↑↓ ↑↓	↑↓ ↑↓ ↑↓

Complex with Strong-Field Ligands (Low Spin)	d^1	d^2	d^3	d^4	d^5	d^6	d^7	d^8	d^9	d^{10}
e_g	— —	— —	— —	— —	— —	— —	↑ —	↑ ↑	↑↓ ↑	↑↓ ↑↓
Δ_o										
t_{2g}	↑ — —	↑ ↑ —	↑ ↑ ↑	↑↓ ↑ ↑	↑↓ ↑↓ ↑	↑↓ ↑↓ ↑↓	↑↓ ↑↓ ↑↓	↑↓ ↑↓ ↑↓	↑↓ ↑↓ ↑↓	↑↓ ↑↓ ↑↓

respectively—determines the orbital configuration of the electrons. The configuration with the lower total energy is the ground state for the complex. Because Π_c involves electron–electron repulsions within orbitals, an increase in Π_c increases the energy of a configuration, thereby reducing its stability. An increase in Π_e corresponds to an increase in the number of exchanges of electrons with parallel spin and increases the stability of a configuration.

For example, a d^5 metal center could have five unpaired electrons, three in t_{2g} and two in e_g orbitals, as a high-spin case; or it could have only one unpaired electron, with all five electrons in the t_{2g} levels, as a low-spin case. The possibilities for all cases, d^1 through d^{10}, are given in Table 10.5.

EXAMPLE 10.1

Determine the exchange energies for high-spin and low-spin d^6 ions in an octahedral complex.

In the high-spin complex, the electron spins are as shown on the right. The five ↑ electrons have exchangeable pairs 1-2, 1-3, 2-3, and 4-5, for a total of four. The exchange energy is therefore $4\Pi_e$. Only electrons at the same energy can exchange.

↑4 ↑5

↑1↓1 ↑2 ↑3

In the low-spin complex, as shown on the right, each set of three electrons with the same spin has exchangeable pairs 1-2, 1-3, and 2-3, for a total of six, and the exchange energy is $6\Pi_e$.

The difference between the high-spin and low-spin complexes is two exchangeable pairs, and the low-spin configuration is stabilized more via its exchange contribution.

↑1↓1 ↑2↓2 ↑3↓3

EXERCISE 10.6 Determine the exchange energy for a d^5 ion, both as a high-spin and as a low-spin complex.

Relative to the total pairing energy Π, Δ_o is strongly dependent on the ligands and the metal. **Table 10.6** presents Δ_o values for aqueous ions, in which water is a relatively weak-field ligand (small Δ_o). The number of unpaired electrons in the complex depends on the balance between Δ_o and Π:

> When $\Delta_o > \Pi$, pairing electrons in the lower levels results in reduced electronic energy for the complex; the low-spin configuration is more stable.
> When $\Delta_o < \Pi$, pairing electrons in the lower levels would increase the electronic energy of the complex; the high-spin configuration is more stable.

In Table 10.6, only $[Co(H_2O)_6]^{3+}$ has Δ_o near the size of Π, and $[Co(H_2O)_6]^{3+}$ is the only low-spin aqua complex. All the other first-row transition metal ions require a stronger field ligand than water to achieve a low-spin configuration electronic ground state. The tabulated Δ_o and Π energies for $[Co(H_2O)_6]^{3+}$ indicate that the relative magnitudes of these values provide a useful conceptual framework to rationalize high- and low-spin states, but that experimental measurements, such as the determination of magnetic susceptibility, provide the most reliable data for assessing electronic configurations. Comparing Δ_o to Π is an approximate way to rationalize high spin versus low spin configurations. The references in Table 10.6 describe other important factors that determine the electronic ground state.

In general, the strength of the ligand–metal interaction is greater for metals having higher charges. This can be seen in the table: Δ_o for 3+ ions is larger than for 2+ ions. Also, values for d^5 ions are smaller than for d^4 and d^6 ions.

Another factor that influences electron configurations is the position of the metal in the periodic table. Metals from the second and third transition series form low-spin complexes more readily than metals from the first transition series. This is a consequence of two cooperating effects: one is the greater overlap between the larger 4d and 5d orbitals and the ligand orbitals, and the other is a decreased pairing energy due to the larger volume available for electrons in the 4d and 5d orbitals as compared with 3d orbitals.

TABLE 10.6 Orbital Splitting (Δ_o, cm^{-1}) and Mean Pairing Energy (Π, cm^{-1}) for Aqueous Ions

	Ion	Δ_o	Π	Ion	Δ_o	Π
d^1				Ti^{3+}	18,800	
d^2				V^{3+}	18,400	
d^3	V^{2+}	12,300		Cr^{3+}	17,400	
d^4	Cr^{2+}	9,250	23,500	Mn^{3+}	15,800	28,000
d^5	Mn^{2+}	7,850[b]	25,500	Fe^{3+}	14,000	30,000
d^6	Fe^{2+}	9,350	17,600	Co^{3+}	16,750	21,000
d^7	Co^{2+}	8,400	22,500	Ni^{3+}		27,000
d^8	Ni^{2+}	8,600				
d^9	Cu^{2+}	7,850				
d^{10}	Zn^{2+}	0				

Data Sources: From D. A. Johnson and P. G. Nelson, *Inorg. Chem.*, **1995**, 34, 5666; D. A. Johnson and P. G. Nelson, *Inorg. Chem.*, **1999**, 38, 4949; D. S. McClure, The Effects of Inner-orbitals on Thermodynamic Properties, in T. M. Dunn, D. S. McClure, and R. G. Pearson, *Some Aspects of Crystal Field Theory*, Harper & Row, New York, 1965, p. 82.
[b] Estimated value

10.3.3 Ligand Field Stabilization Energy

The difference between (1) the energy of the t_{2g}/e_g electronic configuration resulting from the ligand field splitting and (2) the hypothetical energy of the t_{2g}/e_g electronic configuration with all five orbitals degenerate and equally populated is called the **ligand field stabilization energy (LFSE)**. The LFSE is a traditional way to calculate the stabilization of the *d* electrons because of the metal–ligand environment. A common way to determine LFSE is shown for d^4 in **Figure 10.11**.

The interaction of the *d* orbitals of the metal with the ligand orbitals results in lower energy for the t_{2g} set of orbitals ($-\frac{2}{5}\Delta_o$ relative to the average energy of the five t_{2g} and e_g orbitals) and increased energy for the e_g set ($\frac{3}{5}\Delta_o$). The total LFSE of a one-electron system would then be $-\frac{2}{5}\Delta_o$, and the total LFSE of a high-spin four-electron system would be $\frac{3}{5}\Delta_o + 3(-\frac{2}{5}\Delta_o) = -\frac{3}{5}\Delta_o$. Cotton provided an alternative method of arriving at these energies.[12]

EXERCISE 10.7

Determine the LFSE for a d^6 ion for both high-spin and low-spin cases.

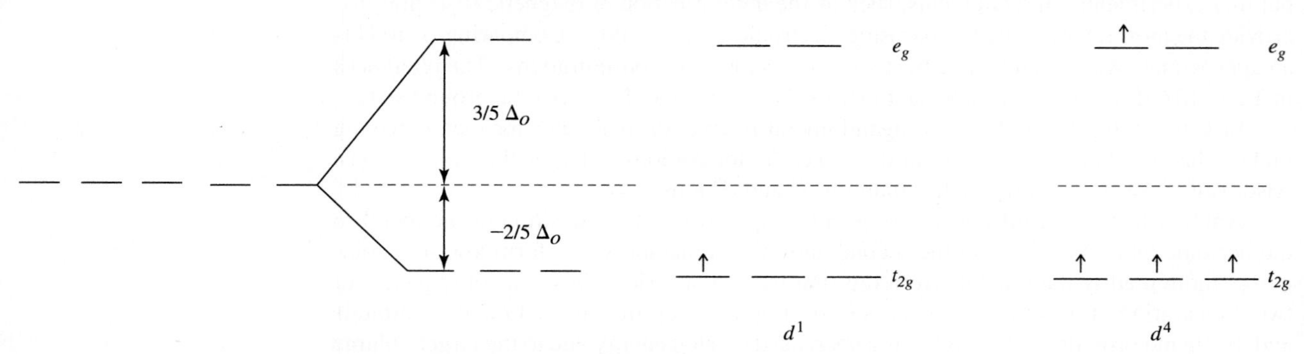

FIGURE 10.11 Splitting of Orbital Energies in a Ligand Field.

Table 10.7 lists LFSE values for *σ-bonded* octahedral complexes with 1-10 *d* electrons in both high- and low-spin arrangements. The final columns show the pairing energies and the difference in LFSE between low-spin and high-spin complexes with the same total number of *d* electrons. For one to three and eight to ten electrons, there is no difference in the number of unpaired electrons or the LFSE. For four to seven electrons, there is a significant difference in both, and high- and low-spin arrangements are possible.

A famous example of LFSE in thermodynamic data appears in the exothermic enthalpy of hydration of bivalent ions of the first transition series, assumed to have six aqua ligands:

$$M^{2+}(g) + 6\,H_2O(l) \longrightarrow [M(H_2O)_6]^{2+}\,(aq)$$

Experimental information on enthalpies of hydration has been measured for related reactions of the form:[13]

$$M^{2+}(g) + 6\,H_2O(l) + 2H^{+}(aq) + 2e^{-} \longrightarrow [M(H_2O)_6]^{2+}\,(aq) + H_2\,(g)$$

TABLE 10.7 Ligand Field Stabilization Energies

Number of *d* Electrons	Weak-Field Arrangement t_{2g}			e_g		LFSE (Δ_o)	Coulombic Energy	Exchange Energy
1	↑					$-\frac{2}{5}$		
2	↑	↑				$-\frac{4}{5}$		Π_e
3	↑	↑	↑			$-\frac{6}{5}$		$3\Pi_e$
4	↑	↑	↑	↑		$-\frac{3}{5}$		$3\Pi_e$
5	↑	↑	↑	↑	↑	0		$4\Pi_e$
6	↑↓	↑	↑	↑	↑	$-\frac{2}{5}$	Π_c	$4\Pi_e$
7	↑↓	↑↓	↑	↑	↑	$-\frac{4}{5}$	$2\Pi_c$	$5\Pi_e$
8	↑↓	↑↓	↑↓	↑	↑	$-\frac{6}{5}$	$3\Pi_c$	$7\Pi_e$
9	↑↓	↑↓	↑↓	↑↓	↑	$-\frac{3}{5}$	$4\Pi_c$	$7\Pi_e$
10	↑↓	↑↓	↑↓	↑↓	↑↓	0	$5\Pi_c$	$8\Pi_e$

Number of *d* Electrons	Strong-Field Arrangement t_{2g}			e_g		LFSE (Δ_o)	Coulombic Energy	Exchange Energy	Strong Field-Weak Field
1	↑					$-\frac{2}{5}$			0
2	↑	↑				$-\frac{4}{5}$		Π_e	0
3	↑	↑	↑			$-\frac{6}{5}$		$3\Pi_e$	0
4	↑↓	↑	↑			$-\frac{8}{5}$	Π_c	$3\Pi_e$	$-\Delta_o + \Pi_c$
5	↑↓	↑↓	↑			$-\frac{10}{5}$	$2\Pi_c$	$4\Pi_e$	$-2\Delta_o + 2\Pi_c$
6	↑↓	↑↓	↑↓			$-\frac{12}{5}$	$3\Pi_c$	$6\Pi_e$	$-2\Delta_o + 2\Pi_c + 2\Pi_e$
7	↑↓	↑↓	↑↓	↑		$-\frac{9}{5}$	$3\Pi_c$	$6\Pi_e$	$-\Delta_o + \Pi_c + \Pi_e$
8	↑↓	↑↓	↑↓	↑	↑	$-\frac{6}{5}$	$3\Pi_c$	$7\Pi_e$	0
9	↑↓	↑↓	↑↓	↑↓	↑	$-\frac{3}{5}$	$4\Pi_c$	$7\Pi_e$	0
10	↑↓	↑↓	↑↓	↑↓	↑↓	0	$5\Pi_c$	$8\Pi_e$	0

NOTE: In addition to the LFSE, each pair formed has a positive Coulombic energy, Π_c, and each set of two electrons with the same spin has a negative exchange energy, Π_e. When $\Delta_o > \Pi_c$ for d^4 or d^5, or when $\Delta_o > \Pi_c + \Pi_e$ for d^6 or d^7, the strong-field arrangement (low spin) is favored.

Transition metal ions are expected to exhibit increasingly exothermic hydration reactions (more negative ΔH) across the transition series. This prediction is based on the decreasing ionic radius with increasing nuclear charge, leading to each ion being a more concentrated source of positive charge, in turn resulting in an expected increase in electrostatic attraction for the ligands. A graph of ΔH for hydration reactions going across a row of transition metals might then be expected to show a steady decrease as the metal ion–ligand interaction becomes stronger. Instead, the enthalpies show the characteristic double-loop shape shown in **Figure 10.12**, with the d^3 and d^8 ions exhibiting significantly more negative ΔH values than expected solely on the basis of decreasing ionic radius. Table 10.7 shows that these configurations in a weak-field octahedral ligand arrangement result in the largest magnitude LFSE.

The almost linear curve of the "expected" enthalpy changes is shown by blue dashed lines in the figure for hydration reactions of M^{2+} and M^{3+} ions. The differences between this curve and the double-humped experimental values are approximately equal to the LFSE values in Table 10.7 for high-spin complexes,[14] with corrections for (1) spin-orbit coupling (0 to 16 kJ/mol)*, (2) a relaxation effect caused by contraction of the metal–ligand distance (0 to 24 kJ/mol), and (3) an interelectronic repulsion energy** that depends on the exchange interactions between electrons with the same spins (0 to –19 kJ/mol for M^{2+}, 0 to –156 kJ/mol for M^{3+}).[15] In addition, small corrections must be made for cases in which the complexes undergo Jahn–Teller distortion. These corrections affect the shape of the curve for the corrected values significantly to reflect the predicted trend on the basis of increasing ionic radius after the LFSE for each complex is taken into account; collectively they account for much of the difference between the experimental values of ΔH and the values that would be expected solely on the basis of electrostatic attractions between the metal ions and ligands.

One more consideration is necessary to understand the trends in these enthalpies. The interelectron repulsion energies for electrons in metal valence atomic orbitals are different (higher in magnitude) than for these electrons in the coordination complex orbitals.

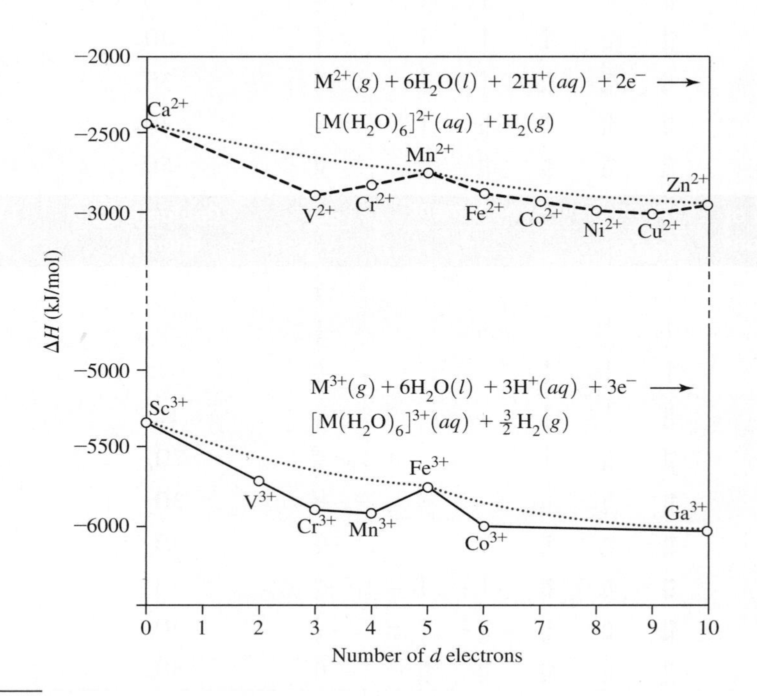

FIGURE 10.12 Enthalpies of Hydration of Transition-Metal Ions. The lower curves show experimental values for individual ions; the blue upper curves result when the LFSE, as well as contributions from spin-orbit splitting, a relaxation effect from contraction of the metal–ligand distance, and interelectronic repulsion energy are subtracted. (Data from D. A. Johnson and P. G. Nelson, *Inorg. Chem.*, **1995**, *34*, 5666 (M^{2+}data); and D. A. Johnson and P. G. Nelson, *Inorg. Chem.*, **1999**, 4949 (M^{3+} data).)

*Spin-orbit coupling is discussed in Section 11.2.1.

**This repulsion term is quantified by the Racah parameter described in Section 11.3.3.

The reduction in this repulsion term between that in the free ion and the complex is a function of both the ligands and the metal ion. The magnitude of this reduction, sometimes called the *nephelauxetic effect*, is used to assess the extent of covalency of the metal-ligand interactions. It should not be surprising that softer ligands generally result in a larger nephelauxetic effect than harder ligands. The relative decrease in the interelectron repulsion energy (the difference between these terms within the free ion and the complex) tends to be larger as the metal oxidation state increases. This decrease contributes to more negative enthalpies for complex formation with higher oxidation state metal ions. In the case of the hexaaqua complexes of the 3+ transition-metal ions, the enhanced nephelauxetic effect relative to the 2+ transition-metal ions contributes to the larger magnitude differences between the experimental and corrected values for these two series of ions in Figure 10.12.

LFSE provides a quantitative approach to assess the relative stabilities of the high- and low-spin electron configurations. It is also the basis for our discussion of the spectra of these complexes (Chapter 11). Measurements of Δ_o are commonly provided in studies of these complexes, with a goal of eventually allowing an improved understanding of metal–ligand interactions.

10.3.4 Square-Planar Complexes

Square-planar complexes are extremely important in inorganic chemistry, and we will now discuss the bonding in these complexes from the perspective of ligand field theory.

Sigma Bonding

The square-planar complex $[Ni(CN)_4]^{2-}$, with D_{4h} symmetry, provides an instructive example of how this approach can be extended to other geometries. The axes for the ligand atoms are chosen for convenience. The y axis of each ligand is directed toward the central atom, the x axis is in the plane of the molecule, and the z axis is parallel to the C_4 axis and perpendicular to the plane of the molecule, as shown in **Figure 10.13**. The p_y set of ligand orbitals is used in σ bonding. Unlike the octahedral case, there are two distinctly different sets of potential π-bonding orbitals, the parallel set ($\pi_{\|}$ or p_x, in the molecular plane) and the perpendicular set ($\pi_{\perp}$ or p_z, perpendicular to the plane). Chapter 4 techniques can be applied to find the representations that fit the symmetries of each orbital set. **Table 10.8** gives the results.

The matching metal orbitals for σ bonding in the first transition series are those with lobes in the x and y directions, $3d_{x^2-y^2}$, $4p_x$, and $4p_y$, with some contribution from the less directed $3d_{z^2}$ and $4s$. Ignoring the other orbitals for the moment, we can construct the energy-level diagram for the σ bonds, as in **Figure 10.14**. The Figure 10.14 square-planar diagram is more complex than the Figure 10.5 octahedral diagram; the lower symmetry results in orbital sets with less degeneracy than in the octahedral case. D_{4h} symmetry splits the d orbitals into three single representations (a_{1g}, b_{1g}, and b_{2g}, for d_{z^2}, $d_{x^2-y^2}$, and d_{xy}, respectively) and the degenerate e_g for the d_{xz}, d_{yz} pair. The b_{2g} and e_g levels are nonbonding (no ligand orbital matches their symmetry) and the difference between them and the antibonding a_{1g} level corresponds to Δ.

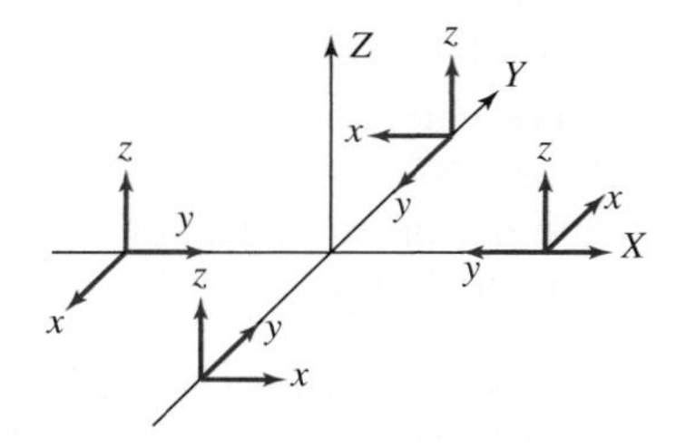

FIGURE 10.13 Coordinate System for Square-Planar Orbitals.

TABLE 10.8 Representations and Orbital Symmetry for Square-Planar Complexes

D_{4h}	E	$2C_4$	C_2	$2C_2'$	$2C_2''$	i	$2S_4$	σ_h	$2\sigma_v$	$2\sigma_d$		
A_{1g}	1	1	1	1	1	1	1	1	1	1		$x^2 + y^2, z^2$
A_{2g}	1	1	1	−1	−1	1	1	1	−1	−1	R_z	
B_{1g}	1	−1	1	1	−1	1	−1	1	1	−1		$x^2 - y^2$
B_{2g}	1	−1	1	−1	1	1	−1	1	−1	1		xy
E_g	2	0	−2	0	0	2	0	−2	0	0	(R_x, R_y)	(xz, yz)
A_{1u}	1	1	1	1	1	−1	−1	−1	−1	−1		
A_{2u}	1	1	1	−1	−1	−1	−1	−1	1	1	z	
B_{1u}	1	−1	1	1	−1	−1	1	−1	−1	1		
B_{2u}	1	−1	1	−1	1	−1	1	−1	1	−1		
E_u	2	0	−2	0	0	−2	0	2	0	0	(x, y)	

D_{4h}	E	$2C_4$	C_2	$2C_2'$	$2C_2''$	i	$2S_4$	σ_h	$2\sigma_v$	$2\sigma_d$
Γ_σ (y)	4	0	0	2	0	0	0	4	2	0
$\Gamma_\parallel$ (x)	4	0	0	−2	0	0	0	4	−2	0
$\Gamma_\perp$ (z)	4	0	0	−2	0	0	0	−4	2	0

$\Gamma_\sigma = A_{1g} + B_{1g} + E_u$ — (σ) Matching orbitals on the central atom: $s, d_{z^2}, d_{x^2-y^2}, (p_x, p_y)$

$\Gamma_\parallel = A_{2g} + B_{2g} + E_u$ — ($\parallel$) Matching orbitals on the central atom: $d_{xy}, (p_x, p_y)$

$\Gamma_\perp = A_{2u} + B_{2u} + E_g$ — ($\perp$) Matching orbitals on the central atom: $p_z, (d_{xz}, d_{yz})$

EXERCISE 10.8

Derive the reducible representations for square-planar bonding, and show that their component irreducible representations are those in Table 10.8.

Pi Bonding

The π-bonding orbitals are also shown in Table 10.8. The $d_{xy}(b_{2g})$ orbital interacts with the $p_x(\pi_\parallel)$ ligand orbitals, and the d_{xz} and $d_{yz}(e_g)$ orbitals interact with the $p_z(\pi_\perp)$ ligand orbitals, as shown in **Figure 10.15**. The b_{2g} orbital is in the plane of the molecule, and the two e_g orbitals have lobes above and below the plane. The results of these interactions are shown in **Figure 10.16**, as calculated for $[Pt(CN)_4]^{2-}$.

This diagram emphasizes how complex molecular orbitals can be!* However, key aspects of the orbitals can be discovered by examining the sets of orbitals set off by boxes:

The lowest energy set contains the σ bonding orbitals, as in Figure 10.14. Eight electrons from ligand σ-donor orbitals fill them.

The next higher set has orbitals with contributions from the eight π-donor orbitals, for example filled π orbitals on CN^- or lone pairs on a halide. Their interaction with the metal orbitals is small and has the net effect of decreasing the energy difference between the orbitals of the next higher set.

*In addition, these diagrams of *homoleptic* complexes (with all the ligands identical) are simpler than those of *heteroleptic* complexes with variation within the ligand set. The angular overlap method (Section 10.4) provides a strategy for predicting electronic structure information for heteroleptic complexes.

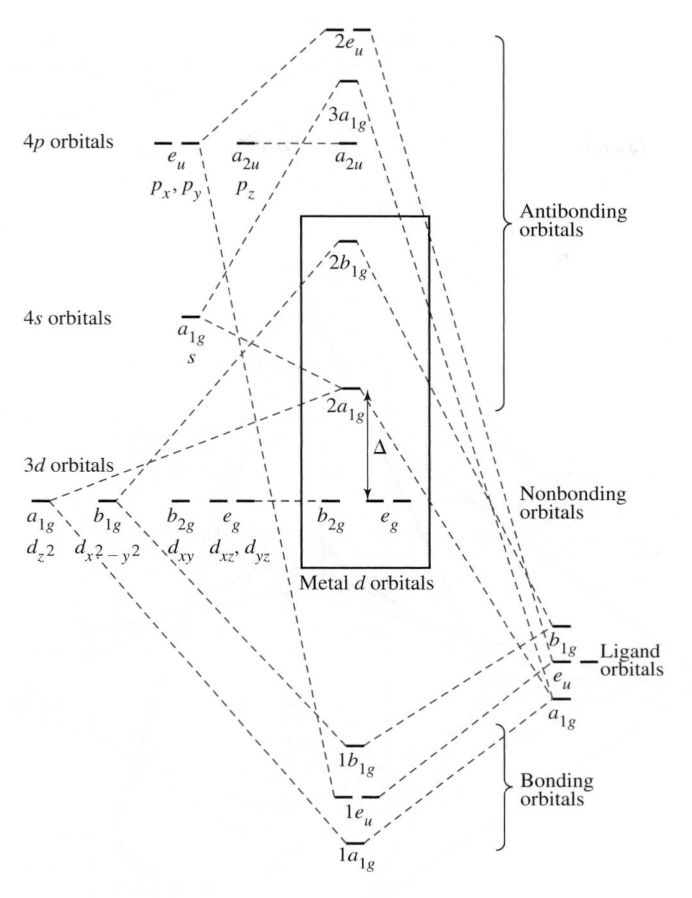

FIGURE 10.14 D_{4h} Molecular Orbitals, σ Orbitals Only. Based on *Orbital Interactions in Chemistry*, p. 296. The four pairs of electrons from the sigma orbitals occupy the four lowest molecular orbitals, and the metal valence electrons occupy the nonbonding and antibonding orbitals within the boxed region.

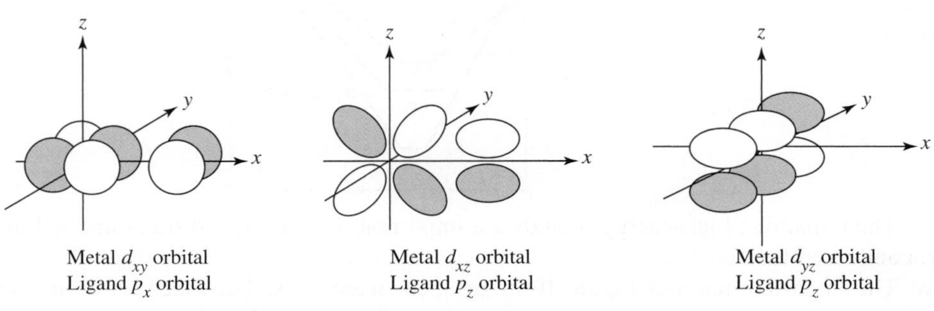

FIGURE 10.15 Pi Bonding Interactions in D_{4h} Molecules.

The third set has orbitals with high contribution from the metal and an a_{2u} orbital arising mostly from the metal p_z orbital, modified by interaction with the ligand orbitals. The energy differences between the orbitals in this set are labeled Δ_1, Δ_2, and Δ_3 from top to bottom. The order of these orbitals has been described in several ways, depending on the computational method used.[16] In all cases, there is agreement that the b_{2g}, e_g, and a_{1g} orbitals are low within this set and have small differences in energy, and the b_{1g} orbital has a much higher energy than all the others. In $[Pt(CN)_4]^{2-}$, the b_{1g} is described as being higher in energy than the a_{2u} (mostly from the metal p_z).

The relative energies of molecular orbitals derived from d orbital interactions vary with different metals and ligands. For example, the order in $[Ni(CN)_4]^{2-}$ matches that for d orbitals in Figure 10.16 ($x^2 - y^2 \gg z^2 > xz, yz > xy$), but the a_{2u}, involving a p_z interaction in $[Ni(CN)_4]^{2-}$ is calculated to be higher in energy than the $d_{x^2-y^2}(b_{1g})$.[17]

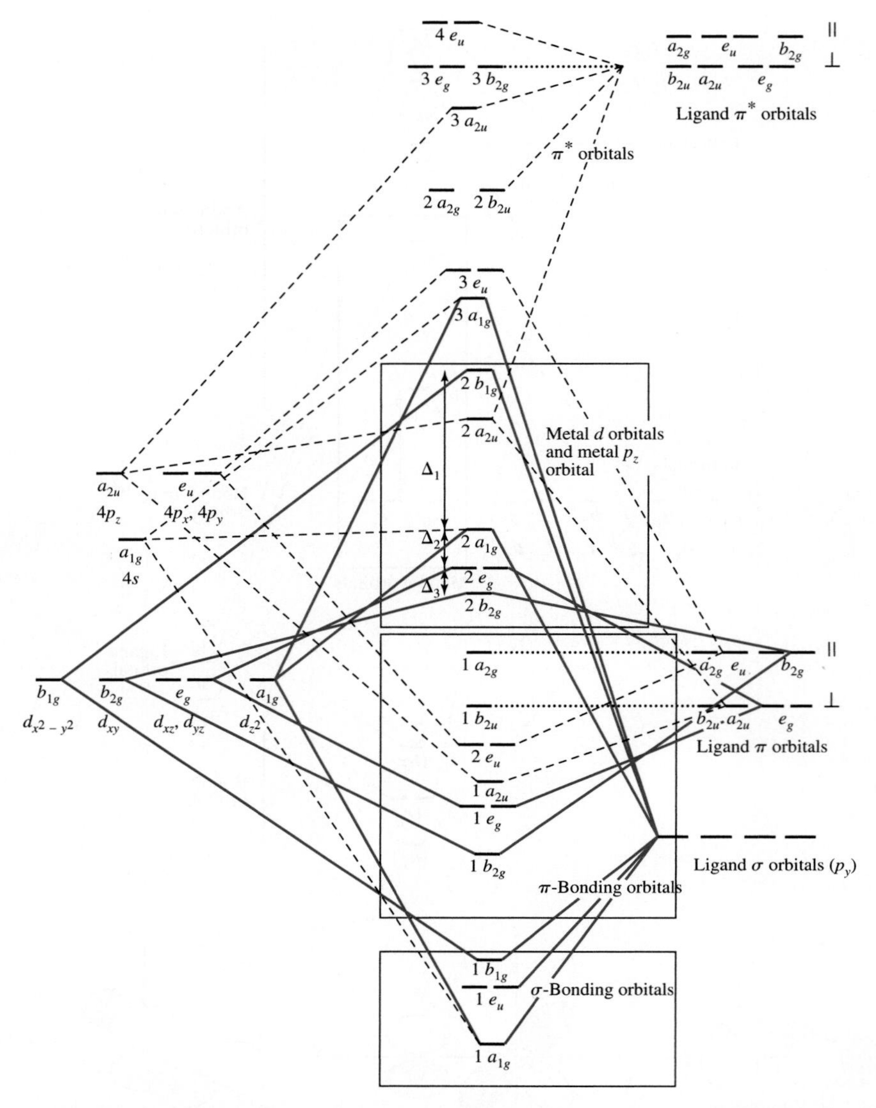

FIGURE 10.16 D_{4h} Molecular Orbitals, Including π Orbitals. Interactions with metal d orbitals are indicated by solid lines, interactions with metal s and p orbitals by dashed lines, and nonbonding orbitals by dotted lines.

The remaining high-energy orbitals are important only in excited states and will not be considered further.

The important parts of Figure 10.16 are these major sets. Two electrons from each ligand form the σ bonds, the next four electrons from each ligand can either π bond slightly or remain essentially nonbonding, and the remaining electrons from the metal occupy the third set. In the case of Ni^{2+} and Pt^{2+}, there are eight d electrons, and there is a large gap in energy between their orbitals and the LUMO (either $2a_{2u}$ or $2b_{1g}$), leading to diamagnetic complexes. The effect of the π^* orbitals of the ligands is to increase the difference in energy between these orbitals in the third set. For example, in $[PtCl_4]^{2-}$, with negligible effect from π-acceptor orbitals, the energy difference between the $2b_{2g}$ and $2b_{1g}$ orbitals is about 33,700 cm^{-1}; this corresponds to the sum of $\Delta_1 + \Delta_2 + \Delta_3$ in Figure 10.16. The $\Delta_1 + \Delta_2 + \Delta_3$ in $[Pt(CN)_4]^{2-}$, with excellent π-acceptor ligands, is more than 46,740 cm^{-1}.[18]

Because b_{2g} and e_g are π orbitals, their energies change significantly if the ligands are changed. Δ_1 is related to Δ_o, is usually much larger than Δ_2 and Δ_3, and is almost always larger than Π, the pairing energy. This means that the b_{1g} or a_{2u} level, whichever is lower, is usually empty for metal ions with fewer than nine electrons.

10.3.5 Tetrahedral Complexes

The orbital interactions associated with the tetrahedral geometry are important in both organic and inorganic chemistry.

Sigma Bonding

The σ-bonding orbitals for tetrahedral complexes are determined via symmetry analysis, using the **Figure 10.17** coordinate system to give the results (**Table 10.9**). The reducible representation includes the A_1 and T_2 irreducible representations, allowing for four bonding MOs. The energy level picture for the d orbitals, shown in **Figure 10.18**, is inverted from the octahedral levels, with e the nonbonding and t_2 the bonding and antibonding levels. In addition, the split, now called Δ_t, is smaller than for octahedral geometry; a guideline is that $\Delta_t \approx \frac{4}{9}\,\Delta_o$ when the same ligands are employed.*

Pi Bonding

The π orbitals are challenging to visualize, but if the y axis of the ligand orbitals is chosen along the bond axis, and the x and z axes are arranged to allow the C_2 operation to work properly, the results in Table 10.9 are obtained. The reducible representation includes the E, T_1, and T_2 irreducible representations. The T_1 has no matching metal atom orbitals, E matches d_{z^2} and $d_{x^2-y^2}$, and T_2 matches d_{xy}, d_{xz}, and d_{yz}. The E and T_2 interactions lower the energy of the bonding orbitals and raise the corresponding antibonding orbitals, for a net increase in Δ_t. An additional complication appears when the ligands possess bonding and antibonding π orbitals whose energies are compatible with the metal valence orbitals,

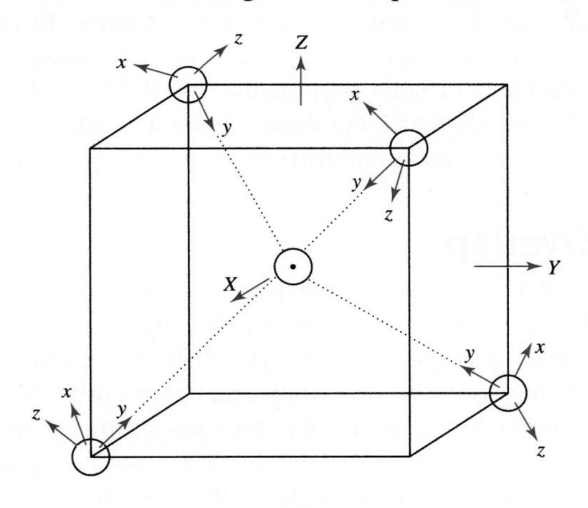

FIGURE 10.17 Coordinate System for Tetrahedral Orbitals.

TABLE 10.9 Representations of Tetrahedral Orbitals

T_d	E	$8C_3$	$3C_2$	$6S_4$	$6\sigma_d$		
A_1	1	1	1	1	1		$x^2+y^2+z^2$
A_2	1	1	1	−1	−1		
E	2	−1	2	0	0		$(2z^2-x^2-y^2,\ x^2-y^2)$
T_1	3	0	−1	1	−1	(R_x, R_y, R_z)	
T_2	3	0	−1	−1	1	(x, y, z)	(xy, yz, xz)
Γ_σ	4	1	0	0	2	A_1+T_2	
Γ_π	8	−1	0	0	0	$E+T_1+T_2$	

e_g, Δ_o, t_{2g}, O_h — t_2, $\Delta_t = \frac{4}{9}\Delta_o$, e, T_d

FIGURE 10.18 Orbital Splitting in Octahedral and Tetrahedral Geometries.

*This is the ratio predicted by the angular overlap approach, discussed in the following section.

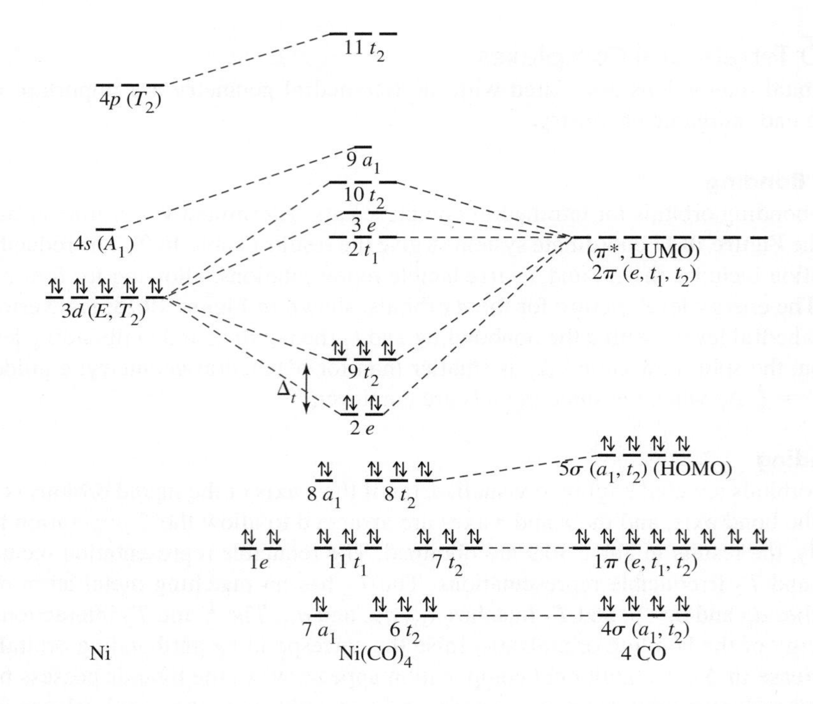

FIGURE 10.19 Molecular Orbitals for Tetrahedral $Ni(CO)_4$. C. W. Bauschlicher, Jr., P. S. Bagus, *J. Chem. Phys.*, **1984**, *81*, 5889 argues that there is almost no σ bonding from the 4*s* and 4*p* orbitals of Ni, and that the d^{10} configuration is the best starting place for the calculations, as shown here. G. Cooper, K. H. Sze, C. E. Brion, *J. Am. Chem. Soc.*, **1989**, *111*, 5051 includes the metal 4*s* as a significant part of σ bonding but with essentially the same net result in molecular orbitals.

common in tetrahedral complexes with CO and CN^-. **Figure 10.19** shows the orbitals and their relative energies for $Ni(CO)_4$, in which the interactions of the CO σ- and π-donor orbitals with the metal orbitals are probably small. Much of the bonding is from M $\longrightarrow$ L π bonding. In cases in which the *d* orbitals are not fully occupied, σ bonding is likely to be more important, with resulting shifts of the a_1 and t_2 orbitals to lower energies.

10.4 Angular Overlap

The **angular overlap** model is a useful approach for making estimates of orbital energies in coordination complexes, while having the flexibility to deal with a variety of geometries and ligands, including heteroleptic complexes, with different ligands.[19,20] This approach estimates the strength of interaction between individual ligand orbitals and metal *d* orbitals based on their mutual overlap. Both sigma and pi interactions are considered, and different coordination numbers and geometries can be treated. The term *angular overlap* is used because the amount of overlap depends strongly on the angular arrangement of the metal orbitals and the angles at which the ligands interact with metal orbitals.

In this approach, the energy of a metal *d* orbital in a coordination complex, or more specifically a molecular orbital with very high metal *d* orbital contribution, is determined by summing the effects of each ligand on the parent metal *d* orbital. Some ligands have a strong effect, some have a weaker effect, and some have no effect at all, because of their angular dependence. Both sigma and pi interactions must be taken into account to determine the final orbital energy. By systematically considering each of the five *d* orbitals, we can use this approach to determine the overall energy pattern of the five molecular orbitals that have the highest contribution from the *d* orbitals for a particular coordination geometry. This model is limited because it exclusively focuses on the metal *d* orbitals and omits the role of the metal *s* and *p* valence orbitals. However, because these molecular orbitals with high *d* orbital contribution are often the frontier orbitals of coordination complexes, the angular overlap result efficiently provides useful information for complexes that would be more difficult to treat via ligand field theory.

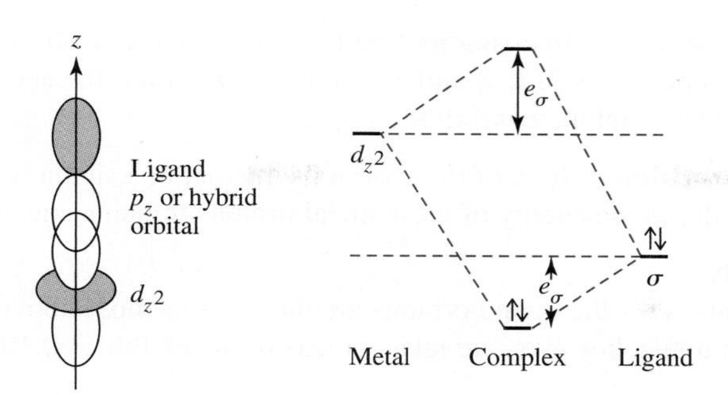

FIGURE 10.20 Sigma Interaction for Angular Overlap.

10.4.1 Sigma-Donor Interactions

In the angular overlap model the strongest sigma interaction is defined as between a metal d_{z^2} orbital and a ligand p orbital (or a hybrid ligand orbital of the same symmetry), as shown in **Figure 10.20**. The strength of all other sigma interactions is determined relative to the strength of this reference interaction. Interaction between these two orbitals results in a bonding orbital, which has a larger component of the ligand orbital, and an antibonding orbital, which is largely metal orbital in composition. Although the observed increase in energy of the antibonding orbital is greater than the decrease in energy of the bonding orbital, this model approximates the molecular orbital energies by an increase in the antibonding (mostly metal d) orbital of e_σ and a decrease in energy of the bonding (mostly ligand) orbital of e_σ.

Similar changes in orbital energy result from other interactions between metal d orbitals and ligand orbitals, with the magnitude dependent on the ligand location and the specific d orbital being considered. **Table 10.10** gives values of these energy changes (in e_σ units) for a variety of shapes. Calculation of the Table 10.10 numbers is beyond the scope of this book, but the reader should be able to justify the numbers qualitatively by comparing the amount of overlap between the orbitals being considered.

Our first example is for octahedral geometry.

EXAMPLE 10.2

$[M(NH_3)_6]^{n+}$

These are octahedral ions with only sigma interactions. The ammonia ligands have no π orbitals available for significant bonding with the metal ion. The donor orbital of NH_3 is mostly nitrogen p_z orbital in composition, and the other p orbitals are used in bonding to the hydrogens.*

In calculating the orbital energies in a complex, the value for a given d orbital is the sum of the numbers for the appropriate ligands in the vertical column for that orbital in Table 10.10. The change in energy for a specific ligand orbital is the sum of the numbers for all d orbitals in the horizontal row for the ligand position.

Metal *d* Orbitals

d_{z^2} orbital: The interaction is strongest with ligands in positions 1 and 6, along the z axis. Each interacts with the orbital to raise its energy by e_σ. The ligands in positions 2, 3, 4, and 5 interact more weakly with the d_{z^2} orbital, each raising the energy of the orbital by $\frac{1}{4}e_\sigma$. Overall, the energy of the d_{z^2} orbital is increased by the sum of all these interactions, for a total of $3e_\sigma$.

*It is common to refer to ammonia as a "σ-only ligand" despite the $1e$ orbitals (Figure 5.30) that could be used as the basis for a set of π bonding group orbitals. These $1e$ orbitals are assumed to play only a negligible role in the bonding in $[M(NH_3)_6]^{n+}$ complexes.

$d_{x^2-y^2}$ orbital: The ligands in positions 1 and 6 do not interact with this metal orbital, but the ligands in positions 2, 3, 4, and 5 each interact to raise the energy of the metal orbital by $\frac{3}{4}$ e_σ, for a total increase of $3e_\sigma$.

d_{xy}, d_{xz}, and d_{yz} orbitals: None of these orbitals interact in a sigma fashion with any of the ligand orbitals, so the energy of these metal orbitals remains unchanged.

Ligand Orbitals

The energy changes for the ligand orbitals are the same as those above for each interaction. The totals, however, are taken *across a row* of Table 10.10, including each of the *d* orbitals.

Ligands in positions 1 and 6 interact strongly with d_{z^2} and are lowered by e_σ. Ligands in these positions do not interact with the other *d* orbitals.

Ligands in positions 2, 3, 4, and 5 are lowered by $\frac{1}{4}e_\sigma$ by interaction with d_{z^2} and by $\frac{3}{4}e_\sigma$ by interaction with $d_{x^2-y^2}$, for a total stabilization of e_σ for each donor orbital.

Overall, each ligand orbital is lowered by e_σ.

The resulting energy pattern is shown in **Figure 10.21**. This is the same pattern obtained by the ligand field approach for the molecular orbitals with high *d* orbital contribution. Both the angular overlap and ligand field theory models provide similar electronic structures: two of the metal *d* orbitals increase in energy, and three remain unchanged; the six ligand orbitals and their electron pairs are stabilized in the formation of ligand–metal σ bonds.

TABLE 10.10 Angular Overlap Parameters: Sigma Interactions

Octahedral Positions	Tetrahedral Positions	Trigonal Bipyramidal Positions

Ligand Positions for Coordination Geometries

CN	Shape	Positions
2	Linear	1, 6
3	Trigonal	2, 11, 12
3	T shape	1, 3, 5
4	Tetrahedral	7, 8, 9, 10
4	Square planar	2, 3, 4, 5
5	Trigonal bipyramidal	1, 2, 6, 11, 12
5	Square pyramidal	1, 2, 3, 4, 5
6	Octahedral	1, 2, 3, 4, 5, 6

Sigma Interactions (in units of e_σ) Metal *d* Orbital

Ligand Position	z^2	x^2-y^2	xy	xz	yz
1	1	0	0	0	0
2	$\frac{1}{4}$	$\frac{3}{4}$	0	0	0
3	$\frac{1}{4}$	$\frac{3}{4}$	0	0	0
4	$\frac{1}{4}$	$\frac{3}{4}$	0	0	0
5	$\frac{1}{4}$	$\frac{3}{4}$	0	0	0
6	1	0	0	0	0
7	0	0	$\frac{1}{3}$	$\frac{1}{3}$	$\frac{1}{3}$
8	0	0	$\frac{1}{3}$	$\frac{1}{3}$	$\frac{1}{3}$
9	0	0	$\frac{1}{3}$	$\frac{1}{3}$	$\frac{1}{3}$
10	0	0	$\frac{1}{3}$	$\frac{1}{3}$	$\frac{1}{3}$
11	$\frac{1}{4}$	$\frac{3}{16}$	$\frac{9}{16}$	0	0
12	$\frac{1}{4}$	$\frac{3}{16}$	$\frac{9}{16}$	0	0

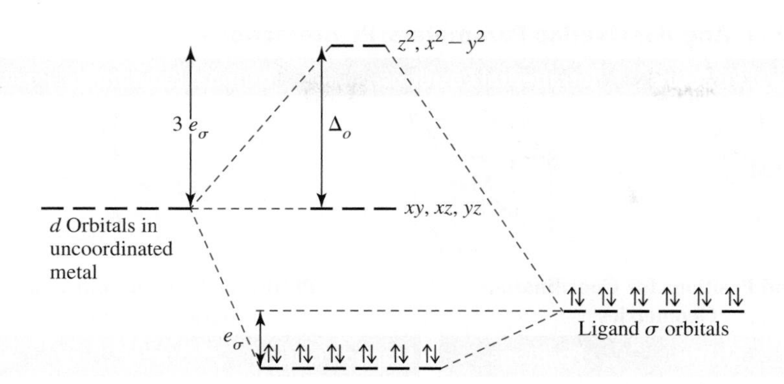

FIGURE 10.21 Energies of *d* Orbitals in Octahedral Complexes: Sigma-Donor Ligands. $\Delta_o = 3e_\sigma$. Metal *s* and *p* orbitals also contribute to the bonding molecular orbitals.

The angular overlap approach quantifies the energies of these levels: the net stabilization is $12e_\sigma$ for the bonding pairs; any *d* electrons in the upper ($e_g{}^*$) level are destabilized by $3e_\sigma$ each. A major difference between the angular overlap and ligand field model is that each of the ligand donor pairs is stabilized to the same extent in the angular overlap model instead of populating levels with three different energies in the ligand field model. The most useful feature of the angular overlap model is its reliable prediction of the *d* orbital splitting. The (more complete) ligand field theory result that includes the metal *s* and *p* orbitals in the formation of molecular orbitals is shown in Figure 10.5 for octahedral geometry.

EXERCISE 10.9

Using the angular overlap model, determine the relative energies of *d* orbitals in a metal complex of formula ML_4 having tetrahedral geometry. Assume that the ligands are capable of sigma interactions only. How does this result for Δ_t compare with the value for Δ_o?

10.4.2 Pi-Acceptor Interactions

Ligands such as CO, CN^-, and phosphines (PR_3) are π acceptors, with empty orbitals that can interact with metal *d* orbitals in a pi fashion. In the angular overlap model, the strongest pi interaction is defined as between a metal d_{xz} orbital and a ligand π^* orbital, as shown in **Figure 10.22**. The antibonding molecular orbitals with high contribution from the ligand π^* orbitals are higher in energy (by e_π) than the original π-acceptor ligand orbitals. The resulting bonding molecular orbitals (with respect to the metal–ligand bonding) are lower in energy than the metal *d* orbitals (by e_π).

Because the overlap for these orbitals is generally smaller than the sigma overlap, $e_\pi < e_\sigma$. The other pi interactions are weaker than this reference interaction, with the magnitudes depending on the degree of overlap between the orbitals. **Table 10.11** gives values for ligands at the same angles as in Table 10.10.

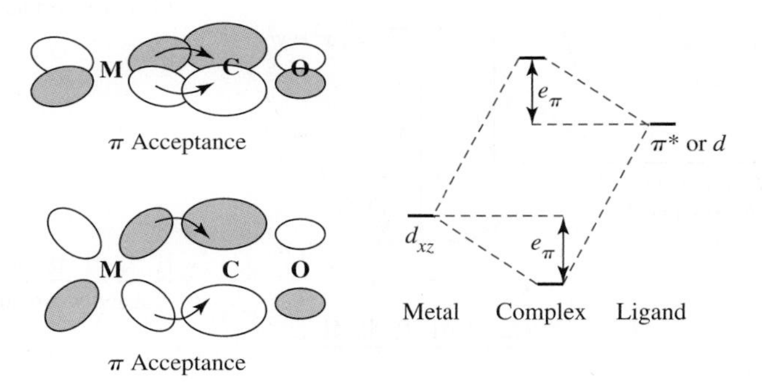

FIGURE 10.22 Pi-Acceptor Interactions.

TABLE 10.11 Angular Overlap Parameters: Pi Interactions

Octahedral Positions	Tetrahedral Positions	Trigonal Bipyramidal Positions

Ligand Positions for Coordination Geometries

CN	Shape	Positions
2	Linear	1, 6
3	Trigonal	2, 11, 12
3	T shape	1, 3, 5
4	Tetrahedral	7, 8, 9, 10
4	Square planar	2, 3, 4, 5
5	Trigonal bipyramidal	1, 2, 6, 11, 12
5	Square pyramidal	1, 2, 3, 4, 5
6	Octahedral	1, 2, 3, 4, 5, 6

Pi Interactions (in units of e_π) Metal *d* Orbital

Ligand Position	z^2	x^2-y^2	xy	xz	yz
1	0	0	0	1	1
2	0	0	1	1	0
3	0	0	1	0	1
4	0	0	1	1	0
5	0	0	1	0	1
6	0	0	0	1	1
7	$\frac{2}{3}$	$\frac{2}{3}$	$\frac{2}{9}$	$\frac{2}{9}$	$\frac{2}{9}$
8	$\frac{2}{3}$	$\frac{2}{3}$	$\frac{2}{9}$	$\frac{2}{9}$	$\frac{2}{9}$
9	$\frac{2}{3}$	$\frac{2}{3}$	$\frac{2}{9}$	$\frac{2}{9}$	$\frac{2}{9}$
10	$\frac{2}{3}$	$\frac{2}{3}$	$\frac{2}{9}$	$\frac{2}{9}$	$\frac{2}{9}$
11	0	$\frac{3}{4}$	$\frac{1}{4}$	$\frac{1}{4}$	$\frac{3}{4}$
12	0	$\frac{3}{4}$	$\frac{1}{4}$	$\frac{1}{4}$	$\frac{3}{4}$

In an octahedral complex with six π-acceptor ligands, the d_{z^2} and $d_{x^2-y^2}$ orbitals do not engage in pi interactions with the ligands in positions 1 through 6 (their parameters in the table are all zero). However, the d_{xy}, d_{xz}, and d_{yz} orbitals all have total interactions of $4e_\pi$; in the formation of molecular orbitals, these three *d* orbitals undergo stabilization by this quantity, a change of energy of $-4e_\pi$, and the ligand orbitals involved in pi interactions are raised in energy. The *d* electrons then occupy the bonding MOs, with a net energy change of $-4e_\pi$ for each electron, as in **Figure 10.23**.

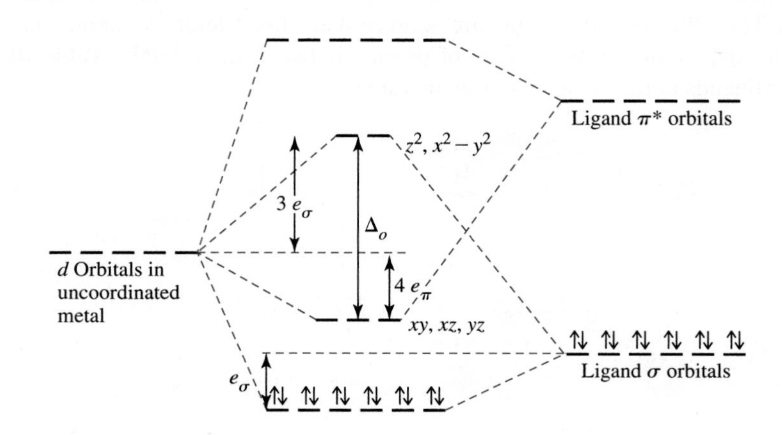

FIGURE 10.23 Energies of *d* Orbitals in Octahedral Complexes: Sigma-Donor and Pi-Acceptor Ligands. $\Delta_o = 3e_\sigma + 4e_\pi$. Metal *s* and *p* orbitals also contribute to the bonding molecular orbitals, but these contributions are omitted in the angular overlap model.

EXAMPLE 10.3

$[M(CN)_6]^{n-}$

The result of these interactions for $[M(CN)_6]^{n-}$ is shown in Figure 10.23. The d_{xy}, d_{xz}, and d_{yz} orbitals are lowered by $4e_\pi$ each, and each of the six molecular orbitals with high ligand π^* orbital contribution increases in energy by $2e_\pi$ (from summing the rows for positions 1 through 6 in Table 10.11). These π^* molecular orbitals have high energies and can be involved in charge transfer transitions (Chapter 11). The net t_{2g}/e_g split is $\Delta_o = 3e_\sigma + 4e_\pi$.

10.4.3 Pi-Donor Interactions

The interactions between occupied ligand p, d, or π orbitals and metal d orbitals are similar to those in the π-acceptor case. In other words, the angular overlap model treats π-donor ligands similarly to π-acceptor ligands except that *for π-donor ligands, the signs of the changes in energy are reversed*, as shown in **Figure 10.24**. The molecular orbitals with high d orbital contribution are raised in energy, whereas the molecular orbitals with high ligand π-donor orbital character are lowered in energy. The overall effect is shown in **Figure 10.25**.

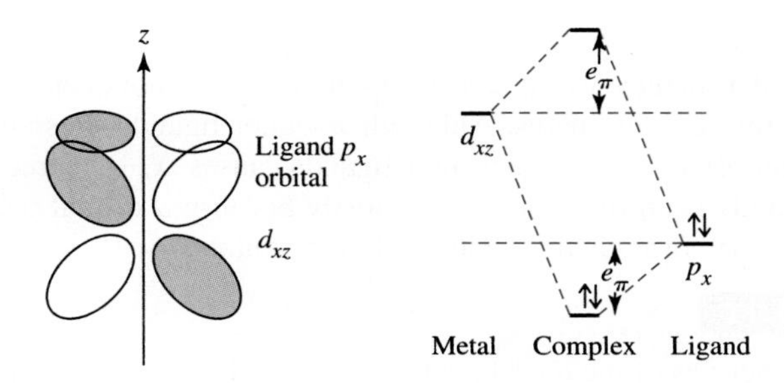

FIGURE 10.24 Pi-Donor Interactions.

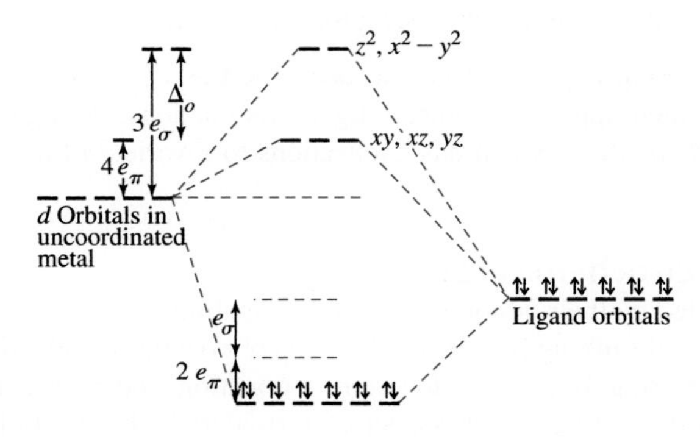

FIGURE 10.25 Energies of d Orbitals in Octahedral Complexes: Sigma-Donor and Pi-Donor Ligands. $\Delta_o = 3e_\sigma - 4e_\pi$. Metal s and p orbitals also contribute to the bonding molecular orbitals.*

*An inconsistency between the angular overlap model and ligand field theory is the treatment of the stabilization of ligand electrons due to pi donation. Note in Figure 10.25 that the *same* six pairs of electrons are stabilized via *both* sigma and pi donation in this oversimplified model, whereas separate pairs of electrons are stabilized via these interactions within ligand field theory. Again, the angular overlap model is an approximation that is most useful for determining the d orbital splittings.

EXAMPLE 10.4

$[MX_6]^{n-}$

Halide ions donate electron density to a metal via p_y orbitals, a sigma interaction; the ions also have p_x and p_z orbitals that can interact with metal orbitals and donate additional electron density by pi interactions. We will use $[MX_6]^{n-}$, where X is a halide ion or other ligand that is simultaneously a σ and a π donor.

d_{z^2} and $d_{x^2-y^2}$ orbitals: Neither of these orbitals has the correct orientation for pi interactions; therefore, the π orbitals have no effect on the energies of these d orbitals.

d_{xy}, d_{xz}, and d_{yz} orbitals: Each of these orbitals interacts in a pi fashion with four of the ligands. For example, the d_{xy} orbital interacts with ligands in positions 2, 3, 4, and 5 with a strength of $1e_\pi$, resulting in a total increase of the energy of the d_{xy} orbital of $4e_\pi$ (the interaction with ligands at positions 1 and 6 is zero). The reader should verify that the d_{xz} and d_{yz} orbitals are also raised in energy by $4e_\pi$.

EXERCISE 10.10

Using the angular overlap model, determine the splitting pattern of d orbitals for a tetrahedral complex of formula MX_4, where X is a ligand that can act as σ donor and π donor.

With ligands that behave as both π acceptors and π donors (such as CO and CN^-), the π-acceptor nature predominates. Although π-donor ligands cause the value of Δ_o to decrease, the larger effect of the π-acceptor ligands causes Δ_o to increase. The net result of pi-acceptor ligands is an increase in Δ_o, mostly because d orbital overlap is generally more effective with π^* orbitals than with π-donor orbitals.

EXERCISE 10.11

Determine the energies of the d orbitals predicted by the angular overlap model for a square-planar complex:

a. Considering σ interactions only.
b. Considering both σ-donor and π-acceptor interactions.

The angular overlap approach has also been used as a component in a more mathematically sophisticated approach to metal–ligand interactions, the ligand field molecular mechanics (LFMM) method, which has applications to a variety of concepts discussed in this chapter.[21]

10.4.4 The Spectrochemical Series

Ligands are classified by their donor and acceptor capabilities. There is a long tradition in inorganic chemistry of ranking ligands on the basis of how these collective ligand abilities result in d orbital splitting. Because σ donation, π donation, and π acceptance have unique impacts on d orbital splitting, a key aspect of formulating these rankings is to classify ligands on the basis of their general tendencies to engage in these interactions with metals. Some ligands, such as ammonia, are classified as σ donors only; these engage in negligible π interactions with metals. To a first approximation, bonding by these ligands to metals is relatively simple, using only the σ orbitals identified in Figure 10.3. The ligand field split, Δ, depends on the (1) relative energies of the metal ion and ligand orbitals and (2) on the degree of overlap. Ethylenediamine has a stronger effect than NH_3 among these ligands, generating a larger Δ. This is also the order of their proton basicity:

$$\text{en} > NH_3$$

The halide ions have ligand field strengths in the order

$$F^- > Cl^- > Br^- > I^-$$

which is also their order of proton basicity.

Ligands that have occupied *p* orbitals, such as the halides, can function as π donors. They donate these electrons to the metal while simultaneously donating their σ bonding electrons. As shown in Section 10.4.3, π donation decreases Δ, and most halide complexes have high-spin configurations. Other ligands that are π-donor candidates include H_2O, OH^-, and RCO_2^-. They fit into the series, from a tendency to create a larger Δ to a smaller Δ, in the order

$$H_2O > F^- > RCO_2^- > OH^- > Cl^- > Br^- > I^-$$

with OH^- below H_2O in the series because OH^- has more π-donating tendency.*

When ligands have vacant π^* or *d* orbitals of suitable energy, there is the possibility of π back-bonding, and the ligands may be π acceptors. This capability tends to increase Δ. Very effective π acceptors include CN^-, CO, and others with conjugated or aromatic π systems; these will be discussed within the context of organometallic chemistry in Chapter 13. A selected list of π-acceptor ligands within the coordination chemistry realm in order of tendency to increase Δ is

$$CO, CN^- > \text{phenanthroline (phen)} > NO_2^- > NCS^-$$

When these lists of ligands are combined, the result is the **spectrochemical series** (which is actually older than crystal field theory!**), which runs roughly in order from strong π-acceptor ligands to strong π-donor ligands:

$$CO, CN^- > \text{phen} > NO_2^- > \text{en} > NH_3 > NCS^- > H_2O > F^- > RCO_2^- > OH^- > Cl^- > Br^- > I^-$$

Low spin		High spin
Strong field		Weak field
Large Δ		Small Δ
π acceptors	σ donor only	π donors

Ligands high in the spectrochemical series tend to cause large splitting of *d*-orbital energies (large values of Δ) and to favor low-spin complexes; ligands low in the series are not as effective at causing *d*-orbital splitting and yield lower values of Δ.

10.4.5 Magnitudes of e_σ, e_π, and Δ

A variety of factors, involving both the metals and the ligands, can affect the degree of sigma and π interactions in coordination complexes. These factors are important in explaining the magnitude of splitting of energy levels in the complexes and in predicting the ground state electron configurations.

Charge on Metal

Because changing the ligand or the metal affects the magnitudes of e_σ and e_π, the value of Δ also changes. One consequence may be a change in the number of unpaired electrons. For example, water is a relatively weak-field ligand. When combined with Co^{2+} in an octahedral geometry, the result is high-spin $[Co(H_2O)_6]^{2+}$ with three unpaired electrons. Combined with Co^{3+}, water forms a low-spin complex with no unpaired electrons. The increase in charge on the metal changes Δ_o sufficiently to favor low spin, as shown in **Figure 10.26**.

---

*If a water molecule uses its $1b_2$ HOMO to form a σ bond to a metal (Figure 5.28), the $3a_1$ orbital is a candidate to engage in π donation to the metal. This interaction will weaken the O—H bonds in water because the electron density in the bonding $3a_1$ is delocalized over more atoms.

The spectrochemical series is attributed to Tsuchsida (R. Tsuchida, *Bull. Chem. Soc. Jpn.*, **1938, *13*, 388) and was formulated on the basis of the electronic spectroscopy of octahedral Co(III) coordination complexes.

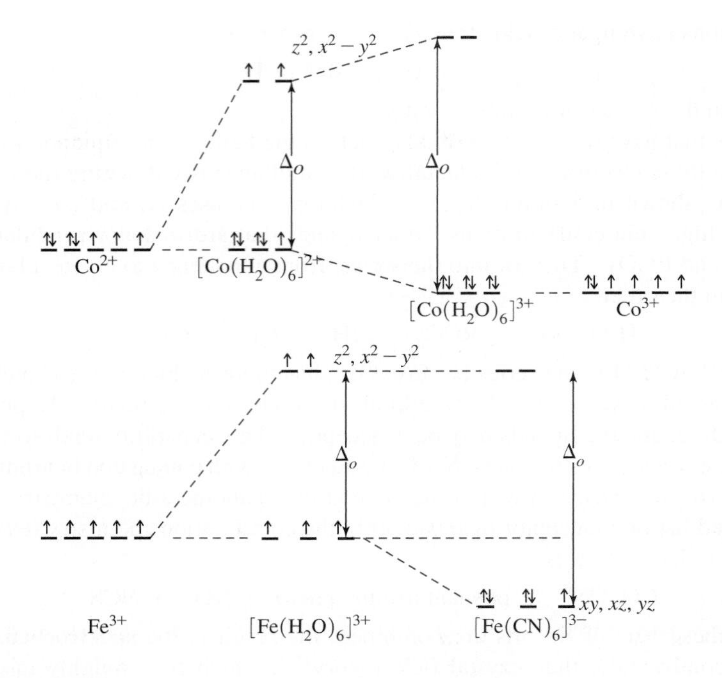

FIGURE 10.26 $[Co(H_2O)_6]^{2+}$, $[Co(H_2O)_6]^{3+}$, $[Fe(H_2O)_6]^{3+}$, $[Fe(CN)_6]^{3-}$, and Unpaired Electrons.

Different Ligands

The introduction of different ligands clearly can have a dramatic impact on the spin state of the complex. For example, $[Fe(H_2O)_6]^{3+}$ is a high-spin species, and $[Fe(CN)_6]^{3-}$ is low spin. Replacing H_2O with CN^- is enough to favor low spin; the change in Δ_o is caused solely by the ligand. As described in Section 10.3.2, the lowest energy configuration when Δ, Π_c, and Π_e are considered determines whether a complex is high or low spin.

Because Δ_t is small, low-spin tetrahedral complexes are rare. The first such complex of a first row transition metal, tetrakis(1-norbornyl)cobalt (1-norbornyl is an organic ligand, C_7H_{11}),* containing a low-spin d^5 Co(IV) center, was reported in 1986 (**Figure 10.27**).[22] The corresponding anionic d^6 Co(III) ($[Co(1\text{-nor})_4]^-$) and cationic d^4 Co(V) ($[Co(1\text{-nor})_4]^+$) low-spin tetrahedral complexes were also prepared.[23] Another organometallic complex, $Ir(Mes)_4$ (Mes = 2,4,6-trimethylphenyl), is approximately tetrahedral and features a low spin d^5 Ir(IV) center.[24]

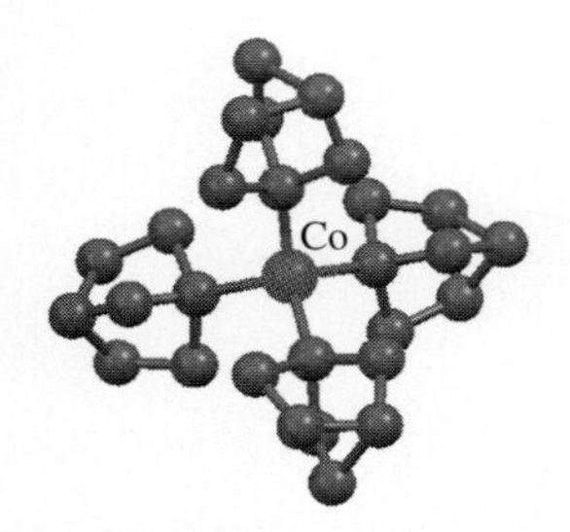

FIGURE 10.27 Structure of Tetrakis(1-norbornyl)cobalt. Tetrakis(1-norbornyl)cobalt. (Drawing generated from CIF file, reference 22. Hydrogen atoms omitted for clarity).

Tables 10.12 and **10.13** list some angular overlap parameters derived from electronic spectra, and provide some trends. First, e_σ is always larger than e_π, in some cases by a factor as large as 9, in others less than 2. This is as expected; σ interactions are more direct, with orbital overlaps directly between nuclei, in contrast to π interactions, which have smaller overlap because the interacting orbitals are not directed toward each other. In addition, the magnitudes of both the σ and π parameters decrease with increasing size and decreasing electronegativity of the halide ions. Increasing the size of the ligand and the corresponding bond length leads to a smaller overlap with the metal d orbitals. In addition, decreasing the electronegativity decreases the nuclear attraction that a ligand exerts on the metal d electrons; the two effects reinforce each other.

In Table 10.12, ligands in each group are listed in their order in the spectrochemical series. For example, for octahedral complexes of Cr^{3+}, CN^- is listed first; it causes the highest Δ_o for these Cr^{3+} complexes and is a π acceptor (e_π is negative). Ethylenediamine and NH_3 are next, listed in order of their e_σ values (which measure σ-donor ability). The halide ions are π donors as well as σ donors and as a group are at the bottom of the series.

*This is an organometallic complex. Chapter 13 describes alkyl ligands as very strong σ donors.

TABLE 10.12 Angular Overlap Parameters

Metal	X	e_σ(cm^{-1})	e_π(cm^{-1})	$\Delta_o = 3e_\sigma - 4e_\pi$
Octahedral MX_6 Complexes				
Cr^{3+}	CN^-	7530	−930	26,310
	en	7260		21,780
	NH_3	7180		21,540
	H_2O	7550	1850	15,250
	F^-	8200	2000	16,600
	Cl^-	5700	980	13,180
	Br^-	5380	950	12,340
	I^-	4100	670	9,620
Ni^{2+}	en	4000		12,000
	NH_3	3600		10,800

Data from B. N. Figgis and M. A. Hitchman, *Ligand Field Theory and Its Applications*, Wiley-VCH, New York, 2000, p. 71, and references therein.

TABLE 10.13 Angular Overlap Parameters for MA_4B_2 Complexes

		Equatorial Ligands (A)			Axial Ligands (B)		
		e_σ(cm^{-1})	e_π(cm^{-1})		e_σ(cm^{-1})	e_π(cm^{-1})	Reference
Cr^{3+}, D_{4h}	en	7233	0	F^-	7811	2016	a
		7233		F^-	8033	2000	c
		7333		Cl^-	5558	900	a
		7500		Cl^-	5857	1040	c
		7567		Br^-	5341	1000	a
		7500		Br^-	5120	750	c
		6987		I^-	4292	594	b
		6840		OH^-	8633	2151	a
		7490		H_2O	7459	1370	a
		7833		H_2O	7497	1410	c
		7534		DMSO	6769	1653	b
	H_2O	7626	1370 (assumed)	F^-	8510	2539	a
	NH_3	6967	0	F^-	7453	1751	a
Ni^{2+}, D_{4h}							
	py	4670	570	Cl^-	2980	540	c
		4500	500	Br^-	2540	340	c
	pyrazole	5480	1370	Cl^-	2540	380	c
		5440	1350	Br^-	1980	240	c
$[CuX_4]^{2-}$, D_{2d}							
	Cl^-	6764	1831				c
	Br^-	4616	821				c

Data from: M. Keeton, B. Fa-chun Chou, and A. B. P. Lever, *Can. J. Chem.* **1971**, *49*, 192; erratum, *ibid.*, **1973**, *51*, 3690; T. J. Barton and R. C. Slade, *J. Chem. Soc. Dalton Trans.*, **1975**, 650; M. Gerloch and R. C. Slade, *Ligand Field Parameters*, Cambridge University Press, London, 1973, p. 186.

Special Cases

The angular overlap model describes the electronic energy of complexes with a wide variety of shapes or with combinations of different ligands. The magnitudes of e_σ and e_π with different ligands can be estimated to predict the electronic structure of complexes such as $[Co(NH_3)_4Cl_2]^+$. This complex, like nearly all Co(III) complexes except $[CoF_6]^{3-}$ and $[Co(H_2O)_3F_3]$, is low spin, so the magnetic properties do not depend on Δ_o. However, the magnitude of Δ_o does have a significant effect on the visible spectrum (Chapter 11). Angular overlap can be used to compare the energies of different geometries—for example, to predict whether a four-coordinate complex is likely to be tetrahedral or square planar (Section 10.6). The angular overlap model can also be used to estimate the energy change for reactions in which the transition state results in either a higher or lower coordination number (Chapter 12).

10.4.6 A Magnetochemical Series

The spectrochemical series has been used for decades, but is it reliable for all metals and ligand environments? Is it most useful for octahedral complexes with d^6 metal ions, such as the Co^{3+} complexes examined by Tsuchida? Reed has exploited the magnetic properties of iron(III) porphyrin complexes to develop a ligand ranking correlated to Δ coined a **magnetochemical series.**[25] Although Reed's series is specific to approximately square pyramidal iron(III) porphyrin complexes, it suggests that similar strategies could be developed for other metal–ligand environments.

The iron(III) porphyrin complexes employed by Reed have two easily accessible electronic states when the axial ligand is rather weak that are sufficiently close in energy to mix and create a unique (or *admixed*) electronic ground state (**Figure 10.28**). Although the quantum chemical details that underlie this mixing phenomenon are beyond the scope of this text,* the magnetic properties of these admixed complexes lie along a continuum between the extremes expected for 5 and 3 unpaired electrons in the ground state, and provide a sensitive assessment of axial ligand (X) field impact.[26]

The ^1H chemical shift of the eight pyrrole hydrogens of the porphyrin ligand is extremely sensitive to the contributions of the Figure 10.28 electronic states to the admixed state, ranging from +80 ppm (downfield, for an admixed complex with very high contribution from the 5 unpaired electrons ground state) to −60 ppm (upfield, for a complex with an admixed ground state composed of roughly equal contributions from the two states in Figure 10.28).** This method is particularly useful to rank extremely weak field ligands,

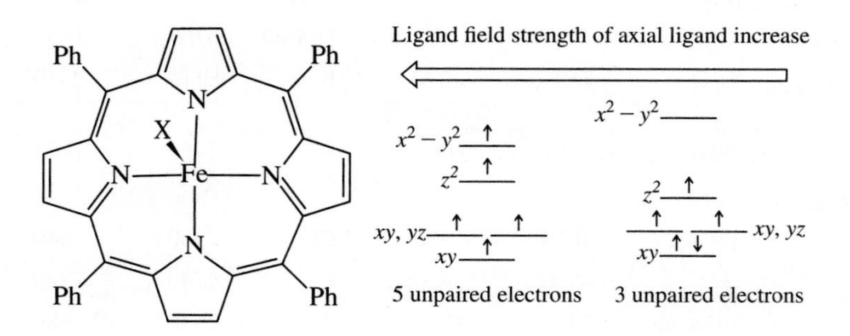

FIGURE 10.28 (Tetraphenylporphinato)iron(III) with an axially bound X ligand (X is directly above the Fe(III) center coming out of the page). The *d* orbital splitting is a very sensitive function of the ligand field strength of X. With ligands that are very weak, an admixed electronic ground state is observed with properties intermediate between the continuum of these two states.†

*The mixing of electronic states in coordination complexes is discussed in Chapter 11.

**The paramagnetism of these complexes is one factor that causes this large ^1H chemical shift range.

†A seemingly contradictory aspect of these diagrams is that introduction of a *weaker* axial ligand results in electron pairing due to *widening* of the $x^2 - y^2/z^2$ gap. This is a unique feature of the tetragonal distortion that occurs in this class of complex. For details see C. A. Reed, T. Mashiko, S. P. Bentley, M. E. Kastner, W. R. Scheidt, K. Spartalian, G. Lang, *J. Am. Chem. Soc.*, **1979**, *101*, 2948. Section 10.5 discusses another example of tetragonal distortion.

those even weaker than iodide. The magnetochemical series for some exceedingly weak ligands (X) on the basis of the pyrrole ^1H chemical shift (ppm, given below the ligand) of their Fe(III) tetraphenylporphinato complexes in C_6D_6 is[25]

$$\underset{(66.7)}{I^-} > \underset{(47.9)}{ReO_4^-} > \underset{(27.7)}{CF_3SO_3^-} > \underset{(-31.5)}{ClO_4^-} > \underset{(-58.5)}{AsF_6^-} > CB_{11}H_{12}^-$$

Details regarding the $CB_{11}H_{12}^-$ ligand, sometimes classified as a *weakly coordinating anion* on the basis of its very weak perturbation of the ligand field, are provided in Chapter 15. This series was confirmed within this Fe(III) system by magnetic susceptibility measurements, and other porphyrins were used to extend the series to ligands stronger than ReO_4^-.

The assessment of ligand field strengths is an ongoing challenge. Recent work by Gray[27] and Scheidt[28] has suggested that CN^- is weaker than CO within two different classes of coordination complexes. Computational work suggests that backbonding to CN^- is less effective relative to CO as a consequence of the CN^- negative charge.

10.5 The Jahn–Teller Effect

The Jahn–Teller theorem[29] states that degenerate orbitals (those with identical energies) cannot be unequally occupied. To avoid these unfavorable electronic configurations, molecules distort (lowering their symmetry) to render these orbitals no longer degenerate. For example, an octahedral Cu(II) complex, containing a d^9 ion, would have three electrons in the two e_g levels, as in the center of **Figure 10.29**, but an octahedral structure is not observed. Instead, the shape of the complex changes slightly, resulting in changes in the energies of the orbitals that would be degenerate within an octahedral ligand environment. The resulting distortion is usually elongation along one axis, but compression along one axis is also possible. In ideally octahedral complexes that experience Jahn–Teller distortion, the (formally) e_g^* orbitals change more in energy relative to the (formally) t_{2g} orbitals. More significant Jahn–Teller distortions occur when e_g^* orbitals would be unequally occupied within an octahedral geometry. Much more modest distortions, sometimes difficult to observe experimentally, occur to prevent unequal occupation of t_{2g} orbitals within an octahedral geometry. The general effects of elongation and compression on *d*-orbital energies are shown in Figure 10.29, and the expected degrees of Jahn–Teller distortion for different electronic configurations and spin states are summarized in the following table:

Number of Electrons	1	2	3	4	5	6	7	8	9	10
High-spin Jahn–Teller	w	w		s		w	w		s	
Low-spin Jahn–Teller	w	w		w	w		s		s	

w = weak Jahn–Teller effect expected (t_{2g} orbitals unevenly occupied); s = strong Jahn–Teller effect expected (e_g orbitals unevenly occupied); No entry = no Jahn–Teller effect expected.

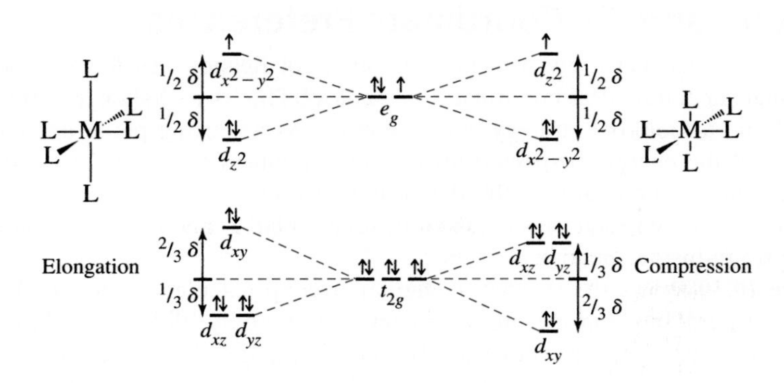

FIGURE 10.29 Jahn–Teller Effect on a d^9 Complex. Elongation along the *z* axis is coupled with a slight decrease in bond length for the other four bonding directions. Similar changes in energy result when the axial ligands have shorter bond distances. The resulting splits are larger for the e_g orbitals than for the t_{2g} orbitals. The energy differences are exaggerated in this figure.

EXERCISE 10.12

Using the d-orbital splitting diagram in Table 10.5, show that the Jahn–Teller effects in the table match the guidelines in the preceding paragraph.

Significant Jahn–Teller effects are observed in complexes of high-spin Cr(II) (d^4), high-spin Mn(III) (d^4), Cu(II) (d^9), Ni(III) (d^7), and low-spin Co(II) (d^7).

Low-spin Cr(II) complexes feature tetragonal distortion (distorted from O_h to D_{4h} symmetry). They show two absorption bands, one in the visible and one in the near-infrared region, because of this distortion. In an ideal octahedral field, there should be only one d–d transition (see Chapter 11 for more details). Cr(II) also forms dimeric complexes with Cr—Cr bonds. $Cr_2(OAc)_4$ contains acetate ions that bridge the two chromiums, with significant Cr—Cr bonding. Metal–metal bonding is discussed in Chapter 15.

$[Mn(H_2O)_6]^{3+}$ remarkably exhibits an undistorted octahedron in $CsMn(SO_4)_2 \cdot 12\,H_2O$, although other Mn(III) complexes show the expected distortion.[30]

Cu(II) complexes generally exhibit significant Jahn–Teller effects; the distortion is most often elongation of two bonds. Elongation, which results in weakening of some metal–ligand bonds, also affects equilibrium constants for complex formation. For example, [*trans*–$Cu(NH_3)_4(H_2O)_2]^{2+}$ is readily formed in aqueous solution as a distorted octahedron with two water molecules at greater distances than the ammonia ligands; liquid ammonia is the required solvent for $[Cu(NH_3)_6]^{2+}$ formation. The formation constants for these reactions show the difficulty of putting the fifth and sixth ammonias on the metal:[31]

$$[Cu(H_2O)_6]^{2+} + NH_3 \rightleftharpoons [Cu(NH_3)(H_2O)_5]^{2+} + H_2O \qquad K_1 = 20{,}000$$

$$[Cu(NH_3)(H_2O)_5]^{2+} + NH_3 \rightleftharpoons [Cu(NH_3)_2(H_2O)_4]^{2+} + H_2O \qquad K_2 = 4{,}000$$

$$[Cu(NH_3)_2(H_2O)_4]^{2+} + NH_3 \rightleftharpoons [Cu(NH_3)_3(H_2O)_3]^{2+} + H_2O \qquad K_3 = 1{,}000$$

$$[Cu(NH_3)_3(H_2O)_3]^{2+} + NH_3 \rightleftharpoons [Cu(NH_3)_4(H_2O)_2]^{2+} + H_2O \qquad K_4 = 200$$

$$[Cu(NH_3)_4(H_2O)_2]^{2+} + NH_3 \rightleftharpoons [Cu(NH_3)_5(H_2O)]^{2+} + H_2O \qquad K_5 = 0.3$$

$$[Cu(NH_3)_5(H_2O)]^{2+} + NH_3 \rightleftharpoons [Cu(NH_3)_6]^{2+} + H_2O \qquad K_6 = \text{very small}$$

Which factor is the cause and which the result is uncertain, but the bottom line is that octahedral Cu(II) complexes are difficult to synthesize with some ligand sets because the bonds to two *trans* ligands in the resulting complexes are weaker (longer) than the other bonds to the ligands. In fact, many Cu(II) complexes have square-planar or nearly square-planar geometries, with tetrahedral shapes also possible. $[CuCl_4]^{2-}$ exhibits cation-dependent structures ranging from tetrahedral through square planar to distorted octahedral.[32] The crystal structures of $(C_6N_2H_{10})CuX_4$ (X = Cl, Br) exhibit various anion geometries between the extremes of square planar and tetrahedral within the same lattice.[33]

10.6 Four- and Six-Coordinate Preferences

Can one predict whether a given metal ion and set of ligands will form an octahedral, square-planar, or tetrahedral coordination complex? This is a challenging fundamental question,[34] and a foolproof strategy does not exist. As a starting point, angular overlap calculations of the energies expected for different numbers of d electrons and different geometries indicate the relative stabilities of the resulting electronic configurations. The coordination complex geometries octahedral, square planar, and tetrahedral will be analyzed here from this perspective.

Figure 10.30 shows the results of angular overlap calculations for d^0 through d^{10} electron configurations considering sigma interactions only. Figure 10.30(a) compares octahedral and square-planar geometries. Because of the greater number of bonds formed in the octahedral complexes, they are more stable (lower energy) for all configurations

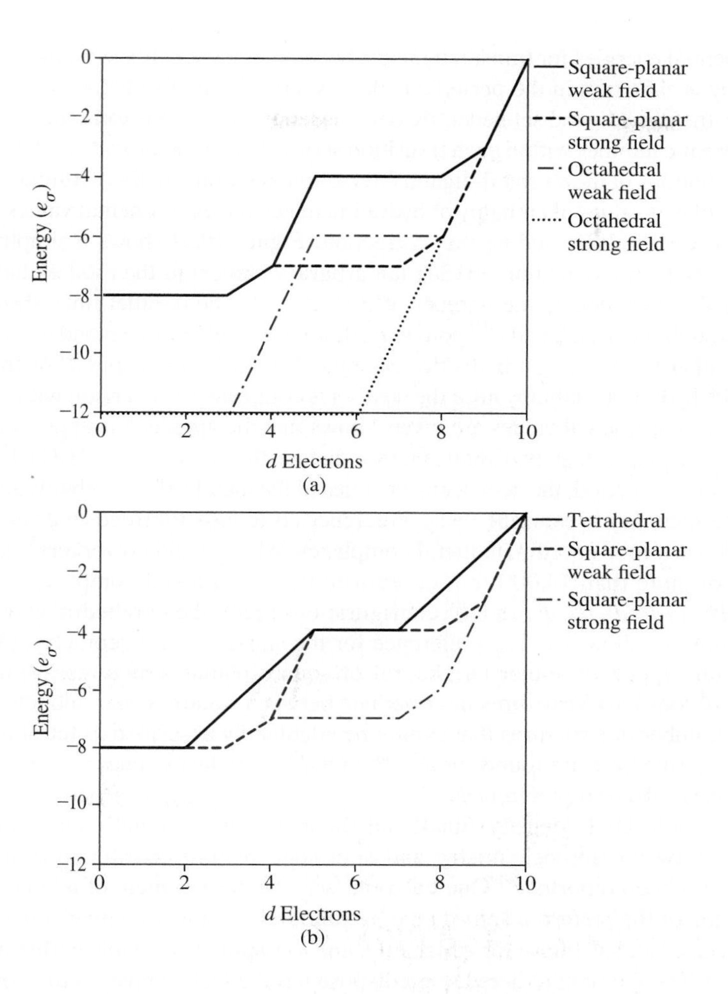

FIGURE 10.30 Angular Overlap Energies of Four- and Six-Coordinate Complexes across a Transition Series. Only sigma bonding is considered. (a) Octahedral and square-planar geometries, both strong- and weak-field cases. (b) Tetrahedral and square-planar geometries, both strong- and weak-field cases.

except d^8, d^9, and d^{10}. A low-spin square-planar geometry has the same net energy as either a high- or low-spin octahedral geometry for these three configurations. This suggests that these configurations are the most likely to have square-planar structures when sigma-only ligands are employed, although octahedral is equally probable on the basis of this approach.

Figure 10.30(b) compares square-planar and tetrahedral structures. For strong-field ligands, square planar is preferred in all cases except d^0, d^1, d^2, and d^{10}. In those cases, the angular overlap approach predicts that square-planar and tetrahedral geometries have equally stable electronic configurations. For weak-field ligands, tetrahedral and square-planar structures also have equal energies in the d^5, d^6, and d^7 cases. The success of these predictions is limited; the angular overlap model does not consider all the variables that influence geometries. For example, bond lengths (and therefore e_σ) for the same ligand–metal pair depend on the geometry of the complex. The omission of the metal s and p valence orbital interactions with ligand orbitals in the angular overlap model also reduces the utility of this approach. The bonding orbitals from these s and p orbital interactions are lower in energy than those from the d orbital interactions and are completely filled (as in Figures 10.5 and 10.14). The stabilization of these electrons also plays a role in dictating the preferred geometry for a coordination complex. The steric bulk of the ligands, as well as the possibility of chelation also play roles in governing coordination complex geometries.

The energies of all of these bonding molecular orbitals depend on the energies of the metal atomic orbitals (approximated by their orbital potential energies) and the ligand orbitals.

Orbital potential energies for transition metals become more negative with increasing atomic number across each row in the periodic table. As a result, the formation enthalpy for complexes with the same ligand set generally becomes more negative (exothermic) with increasing metal atomic number within each transition series. This trend provides a downward slope to the contributions of the *d* orbital–ligand interaction in Figure 10.30(a). Burdett[35] has shown that the calculated values of enthalpy of hydration match the experimental values for enthalpy of hydration very well by adding this correction. **Figure 10.31** shows a simplified version of this approach, simply adding $-0.3e_\sigma$ (an arbitrary choice) to the total enthalpy for each increase in Z (which equals the number of d electrons). The parallel lines show this slope running through the d^0, d^5, and d^{10} points. Addition of a d electron beyond a completed spin set within either the t_{2g} or e_g orbitals (for example, from d^3 to high-spin d^4 or from d^8 to d^9) increases the hydration enthalpy until the next set is complete. Comparison with Figure 10.12, in which the experimental values are given, shows that the approach is approximately valid. Figure 10.31 suggests that two main factors dictate the trend in $[M(H_2O)_6]^{2+}$ hydration enthalpies across a period, the decreasing energies of the metal valence orbitals and the LFSE.

An alternate way to examine these preferences is to look for trends within the vast collection of known four-coordinate metal complexes. Alvarez and coworkers[36] analyzed the structures of more than 13,000 four-coordinate transition-metal complexes and reported these trends: (1) d^0, d^1, d^2, d^5, and d^{10} configurations prefer the tetrahedral geometry, (2) d^8 and d^9 complexes show a strong preference for the square planar geometry, (3) d^3, d^4, d^6, and d^7 metals appear in either tetrahedral or square planar structures, (4) a significant fraction of d^9 ions have structures intermediate between square planar and tetrahedral, and (5) a large number of structures that cannot be adequately described as tetrahedral, square planar, or intermediate are found for d^3, d^6, and d^{10} complexes. These trends build on the angular overlap-derived preferences.

A systematic DFT (density functional theory) computational study examining the correlation between stereochemistry and spin state in four-coordinate transition-metal complexes has been reported.[34] One outcome was the development of a "magic cube" for the prediction of the preferred spin state of tetrahedral complexes with electron configurations between d^3 and d^6 (those for which high and low spin are possible within a tetrahedral ligand field). The factors predicted to predispose tetrahedral complexes to a low spin-state include (1) no π-donor ligands, (2) a metal oxidation state $\geq +4$, and (3) a metal from the second or third transition series. This DFT study[34] provides an excellent discussion of the factors that dictate the geometry of four-coordinate complexes.

As expected on the basis of the work of Alvarez,[36] Cu(II) (d^9) complexes show great variability in geometry. Overall, the two structures most commonly seen for Cu(II) (d^9) complexes are tetragonal—four ligands in a square-planar geometry, with two axial ligands at greater distances—and tetrahedral, sometimes flattened to approximately square planar. There are even trigonal-bipyramidal $[CuCl_5]^{3-}$ ions in $[Co(NH_3)_6][CuCl_5]$. By careful selection of ligands, many transition-metal ions can form complexes with geometries other than octahedral. The d^8 ions Au(III), Pt(II), Pd(II), Rh(I), and Ir(I) often form square-planar

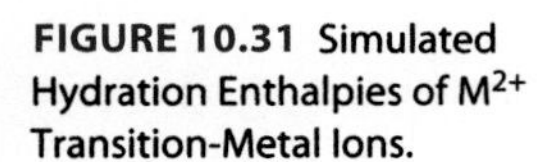
FIGURE 10.31 Simulated Hydration Enthalpies of M^{2+} Transition-Metal Ions.

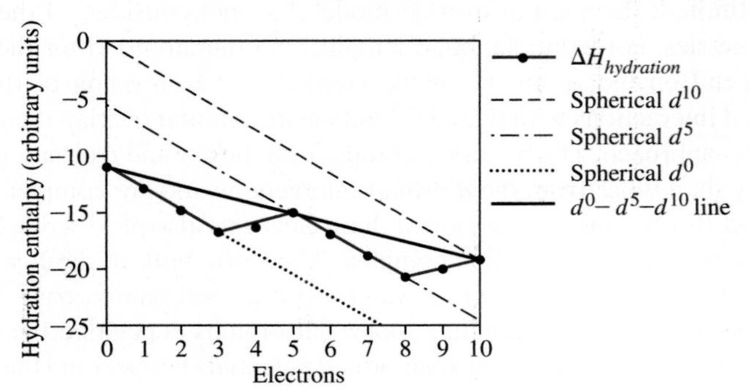

complexes. The d^8 Ni(II) forms tetrahedral $[NiCl_4]^{2-}$, octahedral $[Ni(en)_3]^{2+}$, and square-planar $[Ni(CN)_4]^{2-}$ complexes, as well the square-pyramidal $[Ni(CN)_5]^{3-}$. The d^7 Co(II) ion forms tetrahedral blue and octahedral pink complexes, $[CoCl_4]^{2-}$ and $[Co(H_2O)_6]^{2+}$, as well as square-planar complexes when the ligands have strong planar tendencies, such as [Co(salen)], where salen = bis (salicylaldehydeethylenediimine); and a few trigonal-bipyramidal structures such as $[Co(CN)_5]^{3-}$. Descriptive works[37] provide considerable information about the versatility of the same metal ion to adopt many different coordination geometries.

10.7 Other Shapes

Group theory and angular overlap can also be used to determine which *d* orbitals interact with ligand σ orbitals and to approximate the relative energies of the resulting molecular orbitals for a wide variety of geometries. As usual, the reducible representation for the ligand σ orbitals is reduced to its irreducible representation. The character table is then used to determine which *d* orbitals match the representations. A qualitative estimate of the energies can usually be determined by examination of the shapes of the orbitals and their apparent overlap, and confirmed by using angular overlap tables.

As an example, consider a trigonal-bipyramidal complex ML_5, in which L is a σ donor only. The point group is D_{3h}, and the reducible and irreducible representations are:

D_{3h}	E	$2C_3$	$3C_2$	σ_h	$2S_3$	$3\sigma_v$	Orbitals
Γ	5	2	1	3	0	3	
A_1'	1	1	1	1	1	1	s
A_1'	1	1	1	1	1	1	d_{z^2}
A_2''	1	1	−1	−1	−1	1	p_z
E'	2	−1	0	2	−1	0	$(p_x, p_y), (d_{x^2-y^2}, d_{xy})$

The d_{z^2} orbital has two ligand orbitals overlapping with it and forms the highest energy molecular orbital. The $d_{x^2-y^2}$ and d_{xy} are in the plane of the three equatorial ligands, but overlap is small because of the angles. These two orbitals in the *xy* plane form molecular orbitals relatively high in energy, but not as high as the d_{z^2}. The remaining two orbitals, d_{xz} and d_{yz}, do not have symmetry matching the ligand orbitals. This analysis is sufficient to draw the **Figure 10.32** diagram. The angular overlap method is consistent with these more qualitative results, with strong sigma interaction with d_{z^2}, somewhat weaker interaction with $d_{x^2-y^2}$ and d_{xy}, and no interaction with the d_{xz} and d_{yz} orbitals.

EXERCISE 10.13

Using the angular overlap model, predict the energies of the *d* orbitals in a trigonal bipyramidal complex in which the ligands are sigma donors only, and compare the results with Figure 10.32. Also, using the D_{3h} character table, assign symmetry labels for the irreducible representations of the *d* orbitals in Figure 10.32.

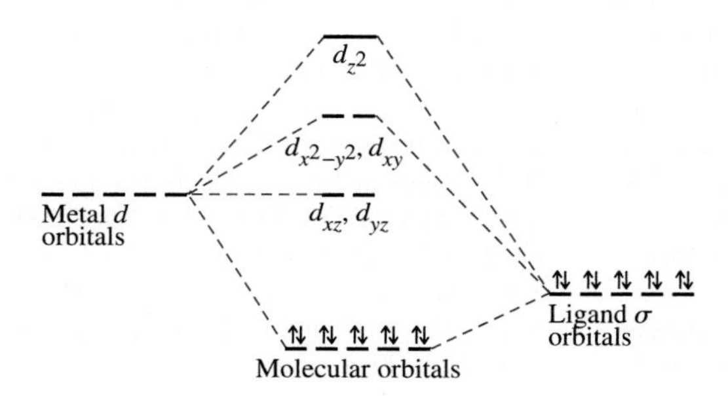

FIGURE 10.32 Trigonal-Bipyramidal Energy Levels. Metal *s* and *p* orbitals also contribute to the bonding molecular orbitals.

References

1. J. J. R. Fausto da Silva, *J. Chem. Educ.*, **1983**, *60*, 390; R. D. Hancock, *J. Chem. Educ.*, **1992**, *69*, 615.
2. D. P. Shoemaker, C. W. Garland, and J. W. Nibler, *Experiments in Physical Chemistry*, 5th ed., McGraw-Hill, New York, 1989, pp. 418–439.
3. B. Figgis and J. Lewis, in H. Jonassen and A. Weissberger, eds., *Techniques of Inorganic Chemistry*, Vol. IV, Interscience, New York, 1965, p. 137.
4. S. S. Eaton, G. R. Eaton, *J. Chem. Educ.*, **1979**, *56*, 170.
5. D. F. Evans, *J. Chem. Soc.*, **1959**, 2003.
6. E. M. Schubert, *J. Chem. Educ.*, **1992**, *69*, 62.
7. S. Tumanski, *Handbook of Magnetic Measurements (Series in Sensors)*, B. Jones and H. Huang, eds., CRC Press (Taylor and Francis Group), Boca Raton, FL (USA), 2011.
8. J. D. M. Coey, *Magnetism and Magnetic Materials*, Cambridge University Press, New York, 2009; D. Giles, *Introduction to Magnetism and Magnetic Materials*, 2nd ed., Chapman & Hall/CRC Press, Boca Raton, FL (USA), 1998.
9. H. Bethe, *Ann. Phys.*, **1929**, *3*, 133.
10. H. Van Vleck, *J. Chem. Phys.*, **1935**, *3*, 807.
11. J. S. Griffith, L. E. Orgel, *Q. Rev. Chem. Soc.*, **1957**, *XI*, 381.
12. F. A. Cotton, *J. Chem. Educ.*, **1964**, *41*, 466.
13. D. A. Johnson, P. G. Nelson, *Inorg. Chem.*, **1995**, *34*, 5666; *Inorg. Chem.*, **1999**, *38*, 4949.
14. L. E. Orgel, *J. Chem. Soc.*, **1952**, 4756; P. George, D. S. McClure, *Prog. Inorg. Chem.*, **1959**, *1*, 381.
15. D. A. Johnson, P. G. Nelson, *Inorg. Chem.*, **1995**, *34*, 3253; **1995**, *34*, 5666; **1999**, *38*, 4949.
16. T. Ziegler, J. K. Nagle, J. G. Snijders, E. J. Baerends, *J. Am. Chem. Soc.*, **1989**, *111*, 5631, and the references cited therein.
17. P. Hummel, N. W. Halpern-Manners, H. B. Gray, *Inorg. Chem.*, **2006**, *45*, 7397.
18. H. B. Gray, C. J. Ballhausen, *J. Am. Chem. Soc.*, **1963**, *85*, 260.
19. E. Larsen, G. N. La Mar, *J. Chem. Educ.*, **1974**, *51*, 633. (Note: There are misprints on pp. 635 and 636.)
20. J. K. Burdett, *Molecular Shapes*, Wiley InterScience, New York, 1980.
21. R. J. Deeth, A. Anastasi, C. Diedrich, K. Randell, *Coord. Chem. Rev.*, **2009**, *253*, 795.
22. E. K. Byrne, D. S. Richeson, K. H. Theopold, *J. Chem. Soc., Chem. Commun.*, **1986**, 1491.
23. E. K. Byrne, K. H. Theopold, *J. Am. Chem. Soc.*, **1989**, 111, 3887.
24. R. S. Hay-Motherwell, G. Wilkinson, B. Hussain-Bates, M. B. Hursthouse, *Polyhedron*, **1991**, *10*, 1457.
25. C. A. Reed, D. Guiset, *J. Am. Chem. Soc.*, **1996**, *118*, 3281.
26. M. Nakamura, *Angew. Chem. Int. Ed.*, **2009**, *48*, 2638.
27. P. M. Hummel, J. Oxagard, W. A. Goddard III, H. B. Gray, *J. Coord. Chem.*, **2005**, *58*, 41.
28. J. Li, R. L. Lord, B. C. Noll, M.-H. Baik, C. E. Schulz, W. R. Scheidt, *Angew Chem.* **2008**, *120*, 10298.
29. H. A. Jahn, E. Teller, *Proc. R. Soc. London*, **1937**, *A161*, 220.
30. A. Avdeef, J. A. Costamagna, J. P. Fackler, Jr., *Inorg. Chem.*, **1974**, *13*, 1854; J. P. Fackler, Jr., and A. Avdeef, *Inorg. Chem.*, **1974**, *13*, 1864.
31. R. M. Smith and A. E. Martell, *Critical Stability Constants, Vol. 4, Inorganic Complexes*, Plenum Press, New York, 1976, p. 41.
32. N. N. Greenwood and A. Earnshaw, *Chemistry of the Elements*, Pergamon Press, Elmsford, NY, 1984, pp. 1385–1386.
33. G. S. Long, M. Wei, R. D. Willett, *Inorg. Chem.*, **1997**, *36*, 3102.
34. J. Cirera, E. Ruiz, S. Alvarez, *Inorg. Chem.*, **2008**, *47*, 2871.
35. J. K. Burdett, *J. Chem. Soc. Dalton Trans.*, **1976**, 1725.
36. J. Cirera, P. Alemany, S. Alvarez, *Chem. Eur. J.*, **2004**, *10*, 190.
37. N. N. Greenwood and A. Earnshaw, *Chemistry of the Elements*, 2nd ed., Butterworth-Heinemann, Oxford, 1997.

General References

One of the best sources is G. Wilkinson, R. D. Gillard, and J. A. McCleverty, editors, *Comprehensive Coordination Chemistry*, Pergamon Press, Elmsford, NY, 1987; Vol. 1, *Theory and Background*, and Vol. 2, *Ligands*, are particularly useful. Others include the books cited in Chapter 4, which include chapters on coordination compounds. Some older but still useful sources are C. J. Ballhausen, *Introduction to Ligand Field Theory*, McGraw-Hill, New York, 1962; T. M. Dunn, D. S. McClure, and R. G. Pearson, *Crystal Field Theory*, Harper & Row, New York, 1965; and C. J. Ballhausen and H. B. Gray, *Molecular Orbital Theory*, W. A. Benjamin, New York, 1965.

More recent volumes include T. A. Albright, J. K. Burdett, and M. Y. Whangbo, *Orbital Interactions in Chemistry*, Wiley InterScience, New York, 1985; and the related text by T. A. Albright and J. K. Burdett, *Problems in Molecular Orbital Theory*, Oxford University Press, Oxford, 1992, which offers examples of many problems and their solutions. Discussions of angular overlap and a variety of other aspects of orbital interactions in coordination complexes are provided in J. K. Burdett, *Molecular Shapes*, John Wiley & Sons, New York, 1980, and B. N. Figgis and M. A. Hitchman, *Ligand Field Theory and Its Applications*, Wiley-VCH, New York, 2000. Discussions of computational method are provided in A. Leach, *Molecular Modeling: Principles and Applications*, 2nd ed., Prentice Hall, Upper Saddle River, NJ, 2001, and C. J. Cramer, *Essentials of Computational Chemistry: Theory and Models*, Wiley, Chichester, UK, 2002.

Problems

10.1 Predict the number of unpaired electrons for each of the following:

a. a tetrahedral d^6 ion
b. $[Co(H_2O)_6]^{2+}$
c. $[Cr(H_2O)_6]^{3+}$
d. a square-planar d^7 ion
e. a coordination compound with a magnetic moment of 5.1 Bohr magnetons

10.2 Identify the *first-row* transition metal **M** that satisfies the requirements given (more than one answer may be possible):

a. $[\mathbf{M}(H_2O)_6]^{3+}$ having one unpaired electron
b. $[\mathbf{M}Br_4]^-$ having the most unpaired electrons
c. diamagnetic $[\mathbf{M}(CN)_6]^{3-}$
d. $[\mathbf{M}(H_2O)_6]^{2+}$ having LFSE $= -\frac{3}{5}\,\Delta_o$

10.3 Identify the most likely transition metal **M**:

a. $K_3[\mathbf{M}(CN)_6]$, in which **M** is a first-row transition metal and the complex has three unpaired electrons
b. $[\mathbf{M}(H_2O)_6]^{3+}$, in which **M** is a second-row transition metal and LFSE $= -2.4\,\Delta_o$
c. tetrahedral $[\mathbf{M}Cl_4]^-$, which has five unpaired electrons and first-row transition metal **M**
d. the third row d^8 transition metal in the square-planar complex $\mathbf{M}Cl_2(NH_3)_2$, which has two **M**–Cl stretching bands in the IR

10.4 The stepwise stability constants in aqueous solution at 25°C for the formation of the ions $[M(en)(H_2O)_4]^{2+}$, $[M(en)_2(H_2O)_2]^{2+}$, and $[M(en)_3]^{2+}$ for copper and nickel are given in the table. Why is there such a difference in the third values? (Hint: Consider the special nature of d^9 complexes.)

	$[M(en)(H_2O)_4]^{2+}$	$[M(en)_2(H_2O)_2]^{2+}$	$[M(en)_3]^{2+}$
Cu	3×10^{10}	1×10^9	0.1 (estimated)
Ni	2×10^7	1×10^6	1×10^4

10.5 A first-row transition-metal complex of formula $[\mathbf{M}(H_2O)_6]^{2+}$ has a magnetic moment of 3.9 Bohr magnetons. Determine the most likely number of unpaired electrons and the identity of the metal.

10.6 Predict the magnetic moments (spin-only) of the following species:

a. $[Cr(H_2O)_6]^{2+}$
b. $[Cr(CN)_6]^{4-}$
c. $[FeCl_4]^-$
d. $[Fe(CN)_6]^{3-}$
e. $[Ni(H_2O)_6]^{2+}$
f. $[Cu(en)_2(H_2O)_2]^{2+}$

10.7 A compound with the empirical formula $Fe(H_2O)_4(CN)_2$ has a magnetic moment corresponding to $2\frac{2}{3}$ unpaired electrons per iron. How is this possible? (Hint: Two octahedral Fe(II) species are involved, each containing a single type of ligand.)

10.8 What are the possible magnetic moments of Co(II) in tetrahedral, octahedral, and square-planar complexes?

10.9 Monothiocarbamate complexes of Fe(III) have been prepared. (See K. R. Kunze, D. L. Perry, L. J. Wilson, *Inorg. Chem.*, **1977**, *16*, 594.) For the methyl and ethyl complexes, the magnetic moment, μ, is 5.7 to 5.8 μ_B at 300 K; it changes to 4.70 to 5 μ_B at 150 K, and drops still further to 3.6 to 4 μ_B at 78 K. The color changes from red to orange as the temperature is lowered. With larger R groups (propyl, piperidyl, pyrrolidyl), $\mu > 5.3\,\mu_B$ at all temperatures and greater than 6 μ_B in some. Explain these changes. Monothiocarbamate complexes have the following structure.

R₂N—C(S)(O) chelating M (structure: $R_2N{-}C$ with S and O bonded to M)

10.10 Show graphically how you would expect ΔH for the reaction $[M(H_2O)_6]^{2+} + 6\,NH_3 \longrightarrow [M(NH_3)_6]^{2+} + 6\,H_2O$ to vary for the first transition series (M = Sc through Zn).

10.11 Using the coordinate system in Figure 10.17, verify the characters of Γ_σ and Γ_π in Table 10.9 and show that these representations reduce to $A_1 + T_2$ and $E + T_1 + T_2$, respectively.

10.12 Using the angular overlap model, determine the energies of the *d* orbitals of the metal for each of the following geometries, first for ligands that act as σ donors only and second for ligands that act as both σ donors and π acceptors. Angular overlap energy-level diagrams are required for each possibility.

a. linear ML_2
b. trigonal-planar ML_3
c. square-pyramidal ML_5
d. trigonal-bipyramidal ML_5
e. cubic ML_8 (Hint: A cube is two superimposed tetrahedra.)

10.13 Use the angular overlap method to calculate the energies of both ligand and metal orbitals for *trans*-$[Cr(NH_3)_4Cl_2]^+$, taking into account that ammonia is a stronger σ-donor ligand than chloride, but chloride is a stronger π donor. Use the 1 and 6 positions for the chloride ions.

10.14 Consider a transition-metal complex of formula ML_4L'. Using the angular overlap model and assuming trigonal-bipyramidal geometry, determine the energies of the *d* orbitals

a. considering sigma interactions only (assume L and L′ are similar in donor ability).
b. considering L′ as a π acceptor as well. Consider L′ in both (1) axial and (2) equatorial positions.
c. based on the preceding answers, would you expect π-acceptor ligands to preferentially occupy axial or equatorial positions in five-coordinate complexes? What other factors should be considered in addition to angular overlap?

10.15 Which geometry, square-pyramidal or trigonal-bipyramidal, is predicted to be more likely for five-coordinate complexes by the angular overlap model? Consider both σ-donor and combined σ-donor and π-acceptor ligands.

10.16 The common structures having CN = 4 for transition-metal complexes are tetrahedral and square planar. However, these are not the only conceivable structures. Examples of main group compounds having *seesaw* structures are known, and *trigonal-pyramidal* structures may also be possible in some cases.

Seesaw **Trigonal Pyramidal**

a. For these structures, determine the relative energies of the d orbitals of a transition-metal complex of formula ML_4 in which L is a σ donor only.

b. Considering both high-spin and low-spin possibilities, calculate the energy of each configuration, d^1 to d^{10}, in terms of e_σ.

c. For which configurations is the seesaw structure favored? The trigonal-pyramidal structure? Neither?

10.17 A possible geometry for an eight-coordinate complex ML_8 might be a hexagonal bipyramid:

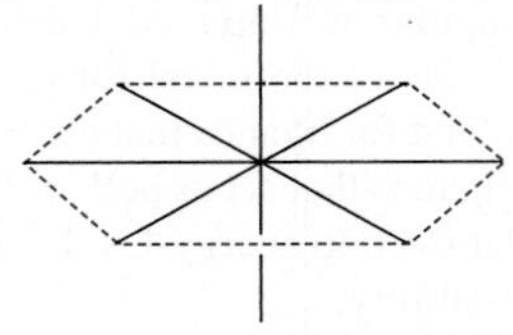

a. Predict the effect of the eight ligands on the energies of the d orbitals of a metal M, using the angular overlap model and assuming that the ligands are sigma donors only. [Note: To determine the values of e_σ, you will need to add two more positions to Table 10.10.]

b. Assign the symmetry labels of the d orbitals (labels of the irreducible representations).

c. Repeat the calculations in part **a** for a ligand that can act both as a σ donor and a π acceptor.

d. For this geometry, and assuming low spin, which d^n configurations would be expected to give rise to Jahn–Teller distortions?

10.18 $[Co(H_2O)_6]^{3+}$ is a strong oxidizing agent that will oxidize water, but $[Co(NH_3)_6]^{3+}$ is stable in aqueous solution. Rationalize this observation by comparing the difference in LFSE for each pair of oxidized and reduced complexes (that is, the differences in LFSE between $[Co(H_2O)_6]^{3+}$ and $[Co(H_2O)_6]^{2+}$, and between $[Co(NH_3)_6]^{3+}$ and $[Co(NH_3)_6]^{2+}$). Table 10.6 gives data on the aqueous complexes; Δ_o for $[Co(NH_3)_6]^{2+}$ is 10,200 cm^{-1}, and Δ_o for $[Co(NH_3)_6]^{3+}$ is about 24,000 cm^{-1}. The Co(III) complexes are low-spin complexes and the Co(II) complexes are high spin.

10.19 Explain the order of the magnitudes of the following Δ_o values for Cr(III) complexes in terms of the σ and π donor and acceptor properties of the ligands.

Ligand	F^-	Cl^-	H_2O	NH_3	en	CN^-
Δ_o (cm^{-1})	15,200	13,200	17,400	21,600	21,900	33,500

10.20 Oxygen is more electronegative than nitrogen; fluorine is more electronegative than the other halogens. Fluoride is a stronger field ligand than the other halides, but ammonia is a stronger field ligand than water. Provide a model consistent with these observations.

10.21 **a.** Explain the effect on the d-orbital energies when an octahedral complex is compressed along the z axis.

b. Explain the effect on the d-orbital energies when an octahedral complex is stretched along the z axis. In the limit, this results in a square-planar complex.

10.22 Solid CrF_3 contains a Cr(III) ion surrounded by six F^- ions in an octahedral geometry, all at distances of 190 pm. However, MnF_3 is in a distorted geometry, with Mn–F distances of 179, 191, and 209 pm (two of each). Explain.

10.23 **a.** Determine the number of unpaired electrons, magnetic moment, and ligand field stabilization energy for each of the following complexes:

$[Co(CO)_4]^-$ $[Cr(CN)_6]^{4-}$ $[Fe(H_2O)_6]^{3+}$ $[Co(NO_2)_6]^{4-}$
$[Co(NH_3)_6]^{3+}$ MnO_4^- $[Cu(H_2O)_6]^{2+}$

b. Why are two of these complexes tetrahedral and the rest octahedral?

c. Why is tetrahedral geometry more stable for Co(II) than for Ni(II)?

10.24 The 2+ ions in the first transition series generally show a preference for octahedral geometry over tetrahedral geometry. Nevertheless, the number of tetrahedral complexes formed is in the order Co > Fe > Ni.

a. Calculate the ligand field stabilization energies for tetrahedral and octahedral symmetries for these ions. Use the differences in LFSE for octahedral versus tetrahedral geometries to assess the relative stabilities of the possible configurations. Use the estimate that $\Delta_t = \frac{4}{9}\Delta_o$ to express LFSE values in terms of Δ_o. Consider both high- and low-spin cases as appropriate for the octahedral complexes. Do these numbers explain this order?

b. Does the angular overlap model offer any advantage in explaining this order? To make this assessment, determine the differences between the electronic configuration energies for octahedral and tetrahedral geometries using the angular overlap model. Consider both high- and low-spin situations as appropriate.

10.25 Except in cases in which ligand geometry requires it, square-planar geometry occurs most commonly with

d^7, d^8, and d^9 ions and with strong-field, π-acceptor ligands. Suggest why these conditions support square-planar geometry.

10.26 Use the group theory approach of Section 10.7 to prepare an energy-level diagram for
a. a square-pyramidal complex.
b. a pentagonal-bipyramidal complex.

10.27 Cobalt(I) complexes are relatively rare compared to Co(II) and Co(0), but the complexes $CoX(PPh_3)_3$ (X = Cl, Br, I) are known with approximate tetrahedral coordination geometry about the high spin d^8 metal center. The angular overlap model was used to analyze the electronic structure of $CoCl(PPh_3)_3$, where three independent molecules (with very similar yet statistically different bond lengths and angles) were observed in the unit cell (J. Krzystek, A. Ozarowski, S. A. Zvyagin, J. Tesler, *Inorg. Chem.*, **2012**, *51*, 4954). Using the angular overlap parameters for molecule 1 in Table 3 of this reference, generate an energy-level diagram for $CoCl(PPh_3)_3$. Does the electronic structure predicted by this method surprise you? Explain. On the basis of Table 3, the chloride ligands are better π donors, and the triphenylphosphine ligands better σ donors in $NiCl_2(PPh_3)_2$ relative to $CoCl(PPh_3)_3$. What is probably the most important factor that causes these differences?

10.28 Nitrogen monofluoride, NF, can serve as a ligand in transition-metal complexes.
a. Prepare a molecular orbital energy-level diagram of the NF molecule, showing clearly how the atomic orbitals interact.
b. If NF can interact with a transition-metal ion to form a chemical bond, what type(s) of ligand–metal interactions would be most important? Would you expect NF to be high or low in the spectrochemical series? Explain.

10.29 Calculations have been reported on the changes that occur when the following compounds are oxidized by one electron. (See T. Leyssens, D. Peeters, A. G. Orpen, J. N. Harvey, *New J. Chem.*, **2005**, *29*, 1424–1430.)

CO, CO, OC—Cr—PH_3, OC, CO   CO, CO, OC—Cr—NH_3, OC, CO

a. When these compounds are oxidized, what is the effect on the C–O distances? Explain.
b. When these compounds are oxidized, what is the effect on the Cr–P distance? On the Cr–N distance? Explain.

10.30 One of the more striking hydride complexes is enneahydridorhenate, $[ReH_9]^{2-}$, which has tricapped trigonal-pyramidal geometry. What is the point group of this ion? Construct a representation using the hydrogen orbitals as a basis. Reduce this to its component irreducible representations, and determine which orbitals of Re are of suitable symmetry to interact with the hydrogen group orbitals.

10.31 The linear molecule FeH_2 has been observed in the gas phase. (See H. Korsgen, W. Urban, J. M. Brown, *J. Chem. Phys.* **1999**, *110*, 3861.) Assume that the iron atom can potentially use *s*, *p*, and *d* orbitals to interact with the hydrogens. If the *z* axis is collinear with the molecular axis:
a. Sketch the group orbitals of the hydrogen atoms that potentially could interact with the iron.
b. Show how the group orbitals and the central atom would interact.
c. Which interaction would you expect to be the strongest? The weakest? Explain briefly. (Note: Orbital potential energies can be found in Appendix B.9.)

10.32 Although CrH_6 has not been synthesized, this should not prevent us from considering what its properties might be! The bonding in CrH_6 and other related small molecules of theoretical interest has been investigated. (R. J. Gillespie, S. Noury, J. Pilme, B. Silvi, *Inorg. Chem.*, **2004**, *43*, 3248). Sketch a molecular orbital diagram (energy-level diagram), showing interactions between metal *d* orbitals and relevant ligand orbitals in CrH_6, assuming octahedral geometry. Be sure to take into account orbital potential energies (Appendix B.9).

10.33 The ion $[Pt_2D_9]^{5-}$, shown below, has eclipsed geometry.
a. What is the point group of this ion?
b. Assume that the platinums can potentially use *s*, *p*, and *d* orbitals to interact with the central deuterium. If the *z* axis is chosen to be collinear with the principal axis of rotation:
1. Sketch the group orbitals of the platinum atoms that potentially could interact with the central D. Be sure to label all orbitals.
2. Show how the group orbitals and the central atom would interact.
3. Which interaction would you expect to be the strongest, and why?

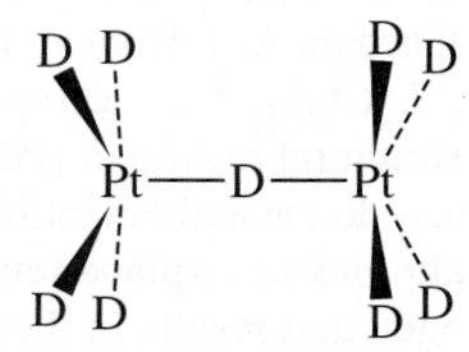

10.34 On the basis of molecular orbitals, explain why the Mn–O distance in $[MnO_4]^{2-}$ is longer (by 3.9 pm) than in $[MnO_4]^-$. (See G. J. Palenik, *Inorg. Chem.*, **1967**, *6*, 503, 507.)

10.35 The square-planar preference of Pd(II) is clearly evident in (N4)Pd(CH_3)Cl (N4 = *N,N'*-di-*tert*-butyl-2, 11-diaza[3.3](2,6)pyridinophane) (J.R. Khusnutdinova, N. P. Rath, L. M. Mirica, *J. Am. Chem. Soc.*, **2010**, *132*, 7303). Although N4 is a potentially tetradentate ligand, two of the nitrogen atoms point away from the Pd center, yielding a square-planar coordination geometry. This situation changes upon oxidation to afford a Pd(III) complex with a tetragonal geometry. Sketch the structure of the (N4)Pd(CH_3)Cl oxidation product, and label the bond lengths with their distances to provide the argument for the Jahn–Teller distortion. Which metal *d* orbital is believed to contribute the most to the HOMO of the oxidized complex? Without changing any of the donor atoms, how might the ligand be subtly modified to cause this HOMO to *increase* in energy? Sketch the modified ligand you would introduce and explain the reasoning for your selection in terms of why it might raise the HOMO energy.

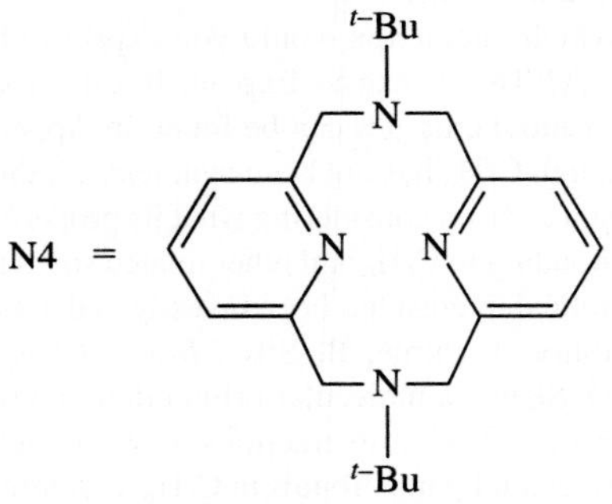

t-Bu
N
N4 = N N
N
t-Bu

10.36 Clusters containing multiple metal ions with large spin values and Jahn–Teller distortions can exhibit ferromagnetism independently, and are called *single-molecule magnets*. Reaction of {2-[(3-methylaminoethylimino)-methyl]-phenol} (HL^1) with manganese(II) acetate and sodium azide in methanol resulted in the Mn(III) dimer $[MnL^1(N_3)(OCH_3)]_2$. Sketch this complex and use bond length arguments to discuss the most obvious result of Jahn–Teller distortion (S. Naiya, S. Biswas, M. G. B. Drew, C. J. Gómez-García, A. Ghosh, *Inorg. Chem.*, **2012**, *51*, 5332).

10.37 The potentially tridentate 1,4,7-triazacyclononane (tacn) is used extensively. Tri-protonated tacn reacts with K_2PdCl_4 affording salts of $[Pd(tacn)(Htacn)]^{3+}$ with one of the unbound nitrogen atoms cationic (A. J. Blake, L. M. Gordon, A. J. Holder, T. I. Hyde, G. Reid, M. Schröder, *J. Chem. Soc., Chem. Commun.*, **1988**, 1452). What structural feature of $[Pd(tacn)(Htacn)]^{3+}$ is surprising? How do the authors rationalize this feature? Propose an alternative explanation keeping in mind that the hydrogen that results in the positively charged and non-Pd bound nitrogen could not be located in the Figure 1 drawing. Oxidation of $[Pd(tacn)_2]^{2+}$ affords $[Pd(tacn)_2]^{3+}$. Sketch the $[Pd(tacn)_2]^{3+}$ structure (incorrectly identified in Figure 3 of the reference), and label the bond lengths (with their distances) that define its Jahn–Teller distortion.

The following problems require the use of molecular modeling software.

10.38 The ion $[TiH_6]^{2-}$ has been found to have O_h symmetry. (See I. B. Bersuker, N. B. Balabanov, D. Pekker, J. E. Boggs, *J. Chem. Phys.* **2002**, *117*, 10478.)

a. Using the H orbitals of the ligands as a basis, construct a reducible representation (the symmetry equivalent of a collection of group orbitals) for this ion.

b. Reduce this representation to its irreducible components.

c. Which orbitals of Ti are suitable for interaction with each of the results from part **b**?

d. Show the interactions of *d* orbitals of Ti with the appropriate group orbital(s), labeled to show the matching irreducible representations, in an energy-level diagram. Identify Δ_o on this diagram. (Note: Useful orbital potential energies can be found in Appendix B.9.)

e. Now use molecular modeling software to calculate and display the molecular orbitals of $[TiH_6]^{2-}$. Compare the results with your work in part **d** and with Figure 10.5, and comment on the similarities and differences.

10.39 Calculate and view the molecular orbitals of the octahedral ion $[TiF_6]^{3-}$.

a. Identify the t_{2g} and e_g bonding and antibonding orbitals, and indicate which *d* orbitals of Ti are involved in each.

b. Compare your results with Figures 10.5 and 10.7. Do they indicate that fluoride is acting as a π donor as well as σ donor?

10.40 Reaction of many iron(III) compounds with hydrochloric acid yield the tetrahedral $[FeCl_4]^-$ ion. Calculate and view the molecular orbitals of this ion.

a. Identify the *e* and t_2 orbitals involved in Fe–Cl bonding (see Figures 10.18 and 10.19), and indicate which *d* orbitals of Fe are involved in each.

b. Compare your results with Figure 10.19. Comment on the similarities and differences.

10.41 Table 10.12 provides values of Δ_o for eight octahedral complexes of chromium(III). Select three of the ligands listed, draw the structures of their octahedral complexes of Cr(III), and calculate and view the molecular orbitals. Identify the t_{2g} and *eg* orbitals, record the energy of each, and determine the Δ_o values. Is your trend consistent with the values in the table? (Note: The results are likely to vary significantly with the level of sophistication of the software used. If you have several molecular modeling programs available, you may want to try different ones to compare their results.)

Chapter 11

Coordination Chemistry III: Electronic Spectra

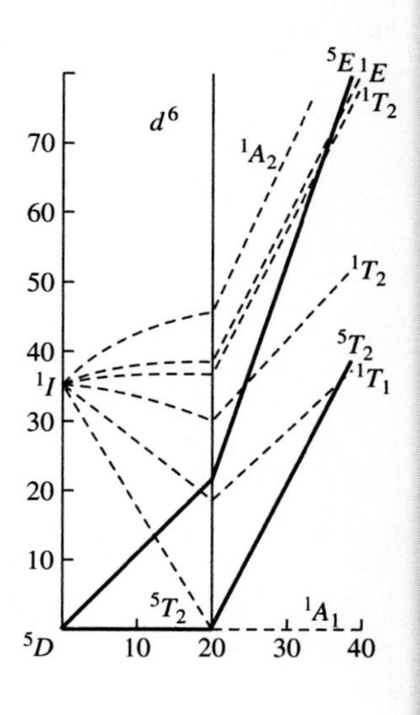

Perhaps the most striking aspect of many coordination compounds of transition metals is their vivid colors. The dye Prussian blue, for example, has been used as a pigment for more than two centuries and is still used in blueprints; it is a complicated coordination compound involving iron(II) and iron(III) coordinated octahedrally by cyanide. Many precious gems exhibit colors resulting from transition-metal ions incorporated into their crystalline lattices. For example, emeralds are green as a consequence of the incorporation of small amounts of chromium(III) into crystalline $Be_3Al_2Si_6O_{18}$; amethysts are violet as a result of the presence of small amounts of iron(II), iron(III), and titanium(IV) in an Al_2O_3 lattice; and rubies are red because of chromium(III), also in a lattice of Al_2O_3. The red heme group, a coordination compound of iron present in hemoglobin, is responsible for the color of blood. Most readers are probably familiar with blue $CuSO_4 \cdot 5\ H_2O$, a compound often used to demonstrate the growing of large, highly symmetric crystals.

Why are so many coordination compounds colored, in contrast to many organic compounds, which are transparent, or nearly so, in the visible spectrum? We will first review the concept of light absorption and how it is measured. The ultraviolet and visible spectra of coordination compounds of transition metals involve transitions between the *d* orbitals of the metals. Therefore, we will need to look closely at the energies of these orbitals (as discussed in Chapter 10) and at the possible ways that electrons can be raised from lower to higher energy levels. The energy levels of *d* electron configurations—as opposed to the energies of *individual* electrons—are somewhat more complicated than might be expected, and we need to consider how electrons in atomic orbitals can interact with each other.

For many coordination compounds, the electronic absorption spectrum provides a convenient method for determining the magnitude of the effect of ligands on the *d* orbitals of the metal. Although in principle we can study this effect for coordination compounds of any geometry, we will concentrate on the most common geometry, octahedral, and will examine how the absorption spectrum can be used to determine the magnitude of the octahedral ligand field parameter Δ_o for a variety of complexes.

11.1 Absorption of Light

In explaining the colors of coordination compounds, we are dealing with the phenomenon of *complementary colors*: if a compound absorbs light of one color, we see the complement of that color. For example, when white light (containing a broad spectrum of all visible wavelengths) passes through a substance that absorbs red light, the color observed is green. Green is the complement of red, so green predominates visually when red light is subtracted from white. Complementary colors can conveniently be remembered as the color pairs on opposite sides of the color wheel shown in the margin.

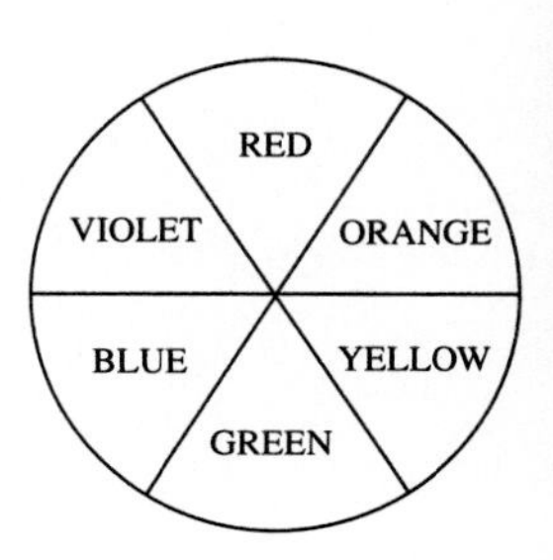

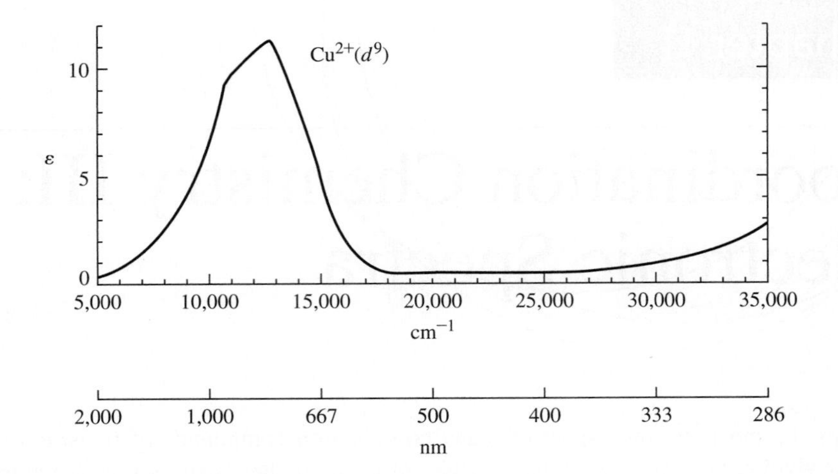

FIGURE 11.1 Absorption Spectrum of $[Cu(H_2O)_6]^{2+}$. (*Introduction to Ligand Fields*, p. 221. Used by permission from Brian Figgis)

An example from coordination chemistry is the deep blue color of aqueous solutions of copper(II) compounds, containing the ion $[Cu(H_2O)_6]^{2+}$. The blue color is a consequence of the absorption of light between approximately 600 and 1000 nm (maximum near 800 nm; **Figure 11.1**), in the yellow to infrared region of the spectrum. The color observed, blue, is the average complementary color of the light absorbed.

It is not always possible to make a simple prediction of color directly from the absorption spectrum, in large part because many coordination compounds contain two or more absorption bands of different energies and intensities. The net color observed is the color predominating after the various absorptions are removed from white light.

For reference, the approximate wavelengths and complementary colors to the principal colors of the visible spectrum are given in **Table 11.1**.

11.1.1 Beer–Lambert Absorption Law

If light of intensity I_o at a given wavelength passes through a solution containing a species that absorbs light, the light emerges with intensity I, which may be measured by a suitable detector (**Figure 11.2**).

The Beer–Lambert law may be used to describe the absorption of light (ignoring scattering and reflection of light from cell surfaces) at a given wavelength by an absorbing species in solution:

$$\log \frac{I_o}{I} = A = \varepsilon l c$$

TABLE 11.1 Visible Light and Complementary Colors

Wavelength Range (nm)	Wave Numbers (cm^{-1})	Color	Complementary Color
< 400	> 25,000	Ultraviolet	
400–450	22,000–25,000	Violet	Yellow
450–490	20,000–22,000	Blue	Orange
490–550	18,000–20,000	Green	Red
550–580	17,000–18,000	Yellow	Violet
580–650	15,000–17,000	Orange	Blue
650–700	14,000–15,000	Red	Green
> 700	< 14,000	Infrared	

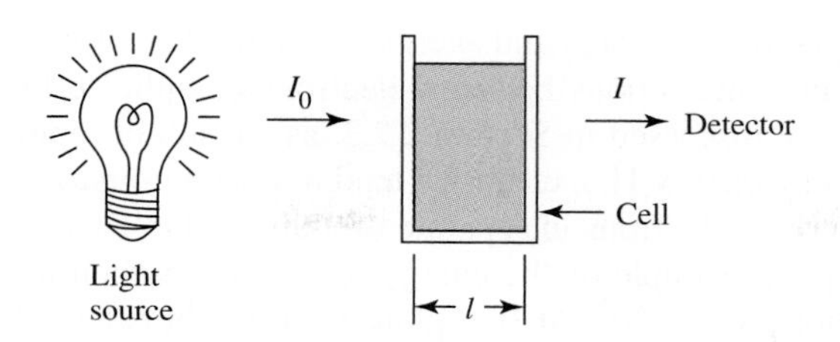

FIGURE 11.2 Absorption of Light by Solution.

where A = absorbance
ε = molar absorptivity (L mol^{-1} cm^{-1}) (also known as molar extinction coefficient)
l = length through solution (cm)
c = concentration of absorbing species (mol L^{-1})

Absorbance is a dimensionless quantity. An absorbance of 1.0 corresponds to 90% absorption at a given wavelength,* an absorbance of 2.0 corresponds to 99% absorption, and so on.

Spectrophotometers commonly obtain spectra as plots of absorbance versus wavelength. The molar absorptivity is a characteristic of the species that is absorbing the light and is highly dependent on wavelength. A plot of molar absorptivity versus wavelength gives a spectrum characteristic of the molecule or ion in question, as in Figure 11.1. As we will see, this spectrum is a consequence of transitions between states of different energies and can provide valuable information about those states and, in turn, about the structure and bonding of the molecule or ion.

Although the quantity most commonly used to describe absorbed light is the wavelength, energy and frequency are also used. In addition, the wavenumber, the number of waves per centimeter (a quantity proportional to the energy), is frequently used, especially in reference to infrared light. For reference, the relations between these quantities are given by the equations

$$E = h\nu = \frac{hc}{\lambda} = hc\left(\frac{1}{\lambda}\right) = hc\bar{\nu}$$

where E = energy

h = Planck constant = 6.626×10^{-34} J s

c = speed of light = 2.998×10^{8} m s^{-1}

ν = frequency (s^{-1})

λ = wavelength (often reported in nm)

$\frac{1}{\lambda} = \bar{\nu}$ = wavenumber (cm^{-1})

11.2 Quantum Numbers of Multielectron Atoms

Absorption of light results in the excitation of electrons from lower to higher energy states; because such states are quantized, we observe absorption in "bands" (as in Figure 11.1), with the energy of each band corresponding to the difference in energy between the initial and final states. To gain insight into these states and the energy transitions between them, we first need to consider how electrons in atoms can interact with each other.

*For absorbance = 1.0, log (I_o/I) = 1.0. Therefore, $I_o/I = 10$, and $I = 0.10\ I_o = 10\% \times I_o$; 10% of the light is transmitted, and 90% is absorbed.

Although the quantum numbers and energies of individual electrons can be described in fairly simple terms, interactions between electrons complicate this picture. Some of these interactions were discussed in Section 2.2.3: as a result of repulsions between electrons (characterized by energy Π_c), electrons tend to occupy separate orbitals; as a result of exchange energy (Π_e), electrons in separate orbitals tend to have parallel spins.

Consider again the example of the energy levels of a carbon atom. Carbon has the electron configuration $1s^2\,2s^2\,2p^2$. At first glance, we might expect the p electrons to be degenerate, and have the same energy. However, there are three major energy levels for the p^2 electrons, differing in energy by pairing and exchange energies (Π_c and Π_e). In addition, the lowest major energy level is split into three slightly different energies, for a total of five energy levels. As an alternative to the discussion presented in Section 2.2.3, each energy level can be described as a combination of the m_l and m_s values of the $2p$ electrons.

Independently, each of the $2p$ electrons could have any of six possible m_l, m_s combinations:

$n = 2,\ l = 1$	(quantum numbers defining $2p$ orbitals)
$m_l = +1, 0, \text{ or } -1$	(three possible values)
$m_s = +\frac{1}{2} \text{ or } -\frac{1}{2}$	(two possible values)

The $2p$ electrons are not independent of each other, however; the orbital angular momenta (characterized by m_l values) and the spin angular momenta (characterized by m_s values) of the $2p$ electrons interact in a manner called **Russell–Saunders coupling** or **LS coupling**.* An oversimplified view is to consider an electron as a particle; its orbital motion and spin generate magnetic fields (a charge in motion generates a magnetic field) and these fields created by multiple electrons can interact. These interactions produce atomic states called **microstates** that can be described by new quantum numbers:

$M_L = \sum m_l$	Total orbital angular momentum
$M_S = \sum m_s$	Total spin angular momentum

Because the quantum numbers m_l and m_s provide information about the magnetic fields generated by electrons due to their orbital and spin, respectively, we need to determine how many possible combinations of m_l and m_s values there are for a p^2 configuration to assess the different possible interactions between these fields.** These combinations allow determination of the corresponding values of M_L and M_S. For shorthand, we will designate the m_s value of each electron by a superscript +, representing $m_s = +\frac{1}{2}$, or –, representing $m_s = -\frac{1}{2}$. For example, an electron having $m_l = +1$ and $m_s = +\frac{1}{2}$ will be written as 1^+.

One possible set of values for the two electrons in the p^2 configuration would be

$$\left.\begin{array}{llll} \text{First electron:} & m_l = +1 & \text{and} & m_s = +\frac{1}{2} \\ \text{Second electron:} & m_l = 0 & \text{and} & m_s = -\frac{1}{2} \end{array}\right\} \text{Notation: } 1^+0^-$$

Each set of possible quantum numbers, such as 1^+0^-, which communicates each unique possible coupling of magnetic fields of the electrons, is called a microstate.

The next step is to tabulate the possible microstates. In doing this, we need to take two precautions: (1) to be sure that no two electrons in the same microstate have identical

*For a more advanced discussion of coupling and its underlying theory, see M. Gerloch, *Orbitals, Terms, and States*, Wiley InterScience, New York, 1986.

**Electrons in filled orbitals can be ignored, because their net spin and orbital angular momenta are both zero.

TABLE 11.2 Microstate Table for p^2

		M_S = −1	M_S = 0	M_S = +1
M_L	+2		1^+1^-	
	+1	1^-0^-	1^+0^- 1^-0^+	1^+0^+
	0	-1^-1^-	-1^+1^- 0^+0^- -1^-1^+	-1^+1^+
	−1	-1^-0^-	-1^+0^- -1^-0^+	-1^+0^+
	−2		-1^+-1^-	

quantum numbers (the Pauli exclusion principle applies); and (2) to count only the *unique* microstates. For example, the microstates 1^+0^- and 0^-1^+, 0^+0^- and 0^-0^+ in a p^2 configuration are duplicates and only one of each pair will be listed.

If we determine all possible microstates and tabulate them according to their M_L and M_S values, we obtain a total of 15 microstates.* These microstates can be arranged according to their M_L and M_S values and listed conveniently in a microstate table, as shown in **Table 11.2**.

EXAMPLE 11.1

Determine the possible microstates for an s^1p^1 configuration, and use them to prepare a microstate table.

The s electron can have $m_l = 0$ and $m_s = \pm\frac{1}{2}$.
The p electron can have $m_l = +1, 0, -1$ and $m_s = \pm\frac{1}{2}$.
The resulting microstate table is then

		M_S = −1	M_S = 0	M_S = +1
M_L	+1	0^-1^-	0^-1^+ 0^+1^-	0^+1^+
	0	0^-0^-	0^+0^- 0^-0^+	0^+0^+
	−1	0^--1^-	0^--1^+ 0^+-1^-	0^+-1^+

In this case, 0^+0^- and 0^-0^+ are different microstates, because the first electron is an s and the second electron is a p; both must be counted.

EXERCISE 11.1 Determine the possible microstates for a d^2 configuration and use them to prepare a microstate table. (Your table should contain 45 microstates!)

*The number of microstates $= i!/[j!(i-j)!]$, where i = number of m_l, m_s combinations (six here, because m_l can have values of 1, 0, and −1, and m_s can have values of $+\frac{1}{2}$ and $-\frac{1}{2}$) and j = number of electrons.

We have now seen how the quantum numbers m_l and m_s associated with individual electrons may be combined into quantum numbers M_L and M_S, which describe atomic microstates. M_L and M_S in turn give quantum numbers L, S, and J. These quantum numbers collectively describe the energy and symmetry of an atom or ion and determine the possible transitions between states of different energies. These transitions account for the colors observed for many coordination complexes, as will be discussed later in this chapter.

The quantum numbers that describe states of multielectron atoms are defined as follows:

L = total orbital angular momentum quantum number

S = total spin angular momentum quantum number

J = total angular momentum quantum number

These total angular momentum quantum numbers are determined by vector sums of the individual quantum numbers; determination of their values is described in this section and the next.

Quantum numbers L and S describe collections of microstates, whereas M_L and M_S describe the microstates themselves. L and S are the largest possible values of M_L and M_S. M_L is related to L much as m_l is related to l, and the values of M_S and m_s are similarly related:

Atomic States	Individual Electrons
$M_L = 0, \pm 1, \pm 2, \ldots, \pm L$	$m_l = 0, \pm 1, \pm 2, \ldots, \pm l$
$M_S = S, S-1, S-2, \ldots, -S$	$m_s = +\frac{1}{2}, -\frac{1}{2}$

Just as the quantum number m_l describes the z-component of the magnetic field due to an electron's orbital motion, the quantum number M_L describes the z-component of the magnetic field associated with a microstate. Similarly, m_s describes the magnetic field due to an electron's spin in a reference direction (usually defined as the z direction), and M_S describes the analogous component of the magnetic field produced by electron spin for a microstate.

The values of L correspond to atomic states described as S, P, D, F, and higher states in a manner similar to the designation of atomic orbitals as s, p, d, and f. The values of S (arising from M_S) are used to calculate the **spin multiplicity**, defined as $2S + 1$. For example, states having spin multiplicities of 1, 2, 3, and 4 are described as *singlet*, *doublet*, *triplet*, and *quartet* states. The spin multiplicity is designated as a left superscript. Examples of atomic states are given in **Table 11.3** and in the examples that follow.*

$L = 0$	S state
$L = 1$	P state
$L = 2$	D state
$L = 3$	F state

Atomic states characterized by S and L are often called **free-ion terms** (sometimes Russell–Saunders terms) because they describe individual atoms or ions, free of ligands. Their labels are often called **term symbols**.** Term symbols are composed of a letter relating to the value of L and a left superscript for the spin multiplicity. For example, the term symbol ^3D corresponds to a state in which $L = 2$ and the spin multiplicity $(2S + 1)$ is 3; ^5F marks a state in which $L = 3$ and $2S + 1 = 5$.

*Unfortunately, S is used in two ways: to designate the atomic spin quantum number and to designate a state having $L = 0$. Chemists are not always wise in choosing their symbols!

**Although *term* and *state* are often used interchangeably, *term* is suggested as the preferred label for the results of Russell–Saunders coupling just described, and *state* for the results of spin-orbit coupling, described in the following section, including the quantum number J. In most cases, the meaning of *term* and *state* can be deduced from the context. (See B. N. Figgis, "Ligand Field Theory," in G. Wilkinson, R. D. Gillard, and J. A. McCleverty, eds. *Comprehensive Coordination Chemistry*, Vol. 1, Pergamon Press, Elmsford, NY, 1987, p. 231.)

TABLE 11.3 Examples of Atomic States (Free-Ion Terms) and Quantum Numbers

Term	L	S
^1S	0	0
^2S	0	$\frac{1}{2}$
^3P	1	1
^4D	2	$\frac{3}{2}$
^5F	3	2

Free-ion terms are very important in the interpretation of the spectra of coordination compounds. The following examples show how to determine the values of L, M_L, S, and M_S for a given term and how to prepare microstate tables from them.

EXAMPLE 11.2

^1S (singlet S)

An S term has $L = 0$ and must therefore have $M_L = 0$. The spin multiplicity (the superscript) is $2S + 1$. Because $2S + 1 = 1$, S must equal 0 (and $M_S = 0$). There can be only one microstate having $M_L = 0$ and $M_S = 0$ for a ^1S term. For the minimum configuration of two electrons we have the following:

M_L \ M_S	0
0	0^+0^-

or

M_L \ M_S	0
0	x

Each microstate is designated by x in the second form of the table.

^2P (doublet P)

A P term has $L = 1$; therefore, M_L can have three values: +1, 0, and –1. The spin multiplicity is $2 = 2S + 1$. Therefore, $S = \frac{1}{2}$, and M_S can have two values: $+\frac{1}{2}$ and $-\frac{1}{2}$. There are six microstates in a ^2P term (3 rows $\times$ 2 columns). For the minimum case of one electron we have the following:

M_L \ M_S	$-\frac{1}{2}$	$+\frac{1}{2}$
1	1^-	1^+
0	0^-	0^+
–1	-1^-	-1^+

or

M_L \ M_S	$-\frac{1}{2}$	$+\frac{1}{2}$
1	x	x
0	x	x
–1	x	x

The spin multiplicity is equal to the number of possible values of M_S; therefore, the spin multiplicity is simply the number of columns in the microstate table.

EXERCISE 11.2 For each of the following free-ion terms, determine the values of L, M_L, S, and M_S. Diagram the microstate table as in the preceding examples: ^2D, ^1P, and ^2S.

At last, we are in a position to return to the p^2 microstate table and reduce it to its constituent atomic states (terms). To do this, it is sufficient to designate each microstate simply by x; it is important to tabulate the number of microstates, but it is not necessary to write out each microstate in full.*

To reduce this microstate table into its component free-ion terms, note that each of the terms described in the examples and in Exercise 11.2 consists of a rectangular array of microstates. To reduce the p^2 microstate table into its terms, all that is necessary is to find the rectangular arrays. This process is illustrated in **Table 11.4**. For each term, the spin multiplicity is the same as the number of columns of microstates: a singlet term (such as ^1D) has a single column, a doublet term has two columns, a triplet term (such as ^3P) has three columns, and so forth.

Therefore, the p^2 electron configuration gives rise to three free-ion terms, designated ^3P, ^1D, and ^1S. These terms have different energies; they represent three states with different degrees of electron–electron interactions. For our example of a p^2 configuration for a carbon atom, the ^3P, ^1D, and ^1S terms have three distinct energies—the three major energy levels observed experimentally.

TABLE 11.4 The Microstate Table for p^2 and Its Reduction to Free-Ion Terms

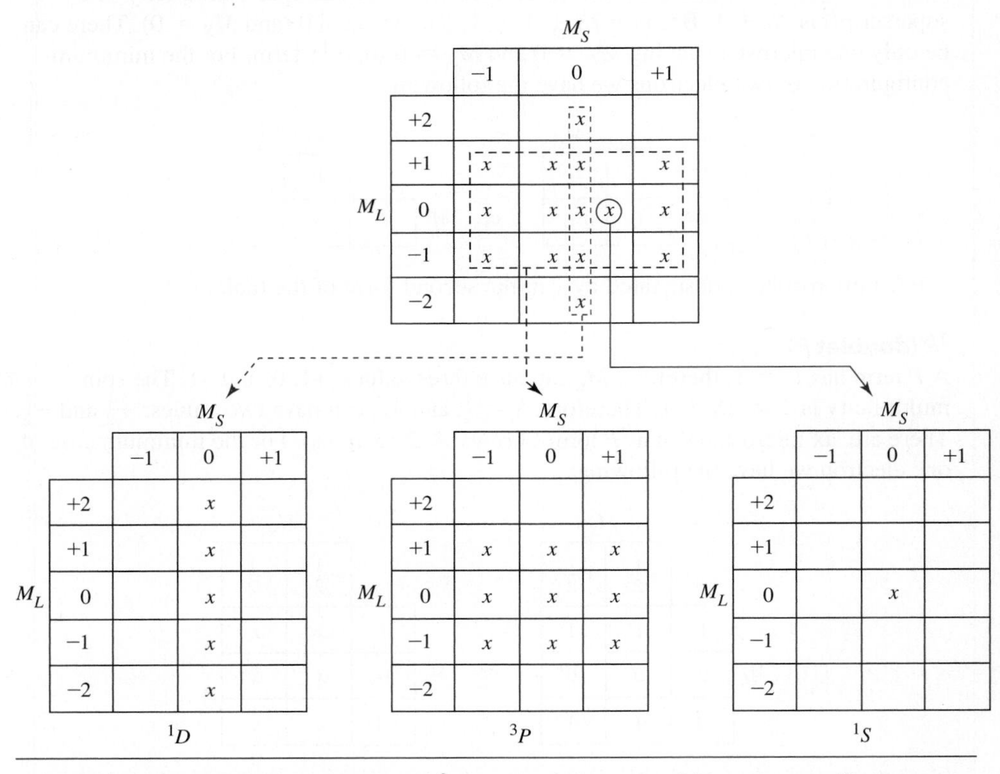

M_L \ M_S	−1	0	+1
+2		x	
+1	x	x x	x
0	x	x x (x)	x
−1	x	x x	x
−2		x	

^1D

M_L \ M_S	−1	0	+1
+2		x	
+1		x	
0		x	
−1		x	
−2		x	

^3P

M_L \ M_S	−1	0	+1
+2			
+1	x	x	x
0	x	x	x
−1	x	x	x
−2			

^1S

M_L \ M_S	−1	0	+1
+2			
+1			
0		x	
−1			
−2			

NOTE: The ^1S and ^1D terms have higher energy than the ^3P term. The relative energies of higher energy terms like these cannot be determined by simple rules.

*For an alternative approach where each microstate is generated in full, see D. A. McQuarrie, *Quantum Chemistry*, University Science Books, Mill Valley, CA, 1983, p. 313.

The final step in this procedure is to determine which term has the lowest energy by using two of **Hund's rules**:

1. The ground term (term of lowest energy) has the highest spin multiplicity.
   In our example of p^2, the ground term is the ^3P. This term can be identifed as having the configuration in the margin. This is sometimes called *Hund's rule of maximum multiplicity*, introduced in Section 2.2.3.
2. If two or more terms share the maximum spin multiplicity, the ground term is the one having the highest value of L.
   For example, if ^4P and ^4F terms are both found for an electron configuration, the ^4F has lower energy: ^4F has $L = 3$, and ^4P has $L = 1$.

2p ↑ ↑ __

2s ↑↓

1s ↑↓

EXAMPLE 11.3

Reduce the microstate table for the $s^1 p^1$ configuration to its component free-ion terms, and identify the ground-state term.

The microstate table (prepared in Example 11.1) is the sum of the microstate tables for the ^3P and ^1P terms:

^3P:

M_L \ M_S	−1	0	+1
+1	x	x	x
0	x	x	x
−1	x	x	x

^1P:

M_L \ M_S	−1	0	+1
+1		x	
0		x	
−1		x	

Hund's rule of maximum multiplicity requires ^3P as the ground state.

EXERCISE 11.3 In Exercise 11.1, you obtained a microstate table for the d^2 configuration. Reduce this to its component free-ion terms, and identify the ground-state term.

11.2.1 Spin-Orbit Coupling

Up to this point in the discussion of multielectron atoms, the spin and orbital angular momenta have been treated separately. An additional factor is important: the spin and orbital angular momenta (or the magnetic fields associated with them) couple with each other, a phenomenon known as *spin-orbit coupling*. In multielectron atoms, the S and L quantum numbers combine into the total angular momentum quantum number J. The quantum number J may have the following values:

$$J = L + S, L + S - 1, L + S - 2, \ldots, |L - S|$$

The value of J is given as a subscript.

EXAMPLE 11.4

Determine the possible values of J for the carbon terms.

For the term symbols just described for carbon, the ^1D and ^1S terms each have only one J value, whereas the ^3P term has three slightly different energies, each described by a different J. J can have only the value 0 for the ^1S term $(0 + 0)$ and only the value 2 for the ^1D term $(2 + 0)$. For the ^3P term, J can have the three values 2, 1, and 0 $(1 + 1, 1 + 1 - 1, \text{and } 1 + 1 - 2)$.

EXERCISE 11.4 Determine the possible values of J for the terms obtained from a d^2 configuration in Exercise 11.3.

Spin-orbit coupling splits free-ion terms into states of different energies. The ^3P term therefore splits into states of three different energies, and the energy level diagram for the carbon atom is as follows:

LS coupling only	Spin-orbit coupling (exaggerated scale for ^3P)	Energy (cm^{-1})
^1S	$^1S_0$	21648.8
^1D	$^1D_2$	10193.7
^3P	$^3P_2$	43.5
	$^3P_1$	16.4
	$^3P_0$	0

These are the five energy states for the carbon atom referred to at the beginning of this section. The state of lowest energy (spin-orbit coupling included) can be predicted from **Hund's third rule**:

3. For subshells (such as p^2) that are less than half filled, the state having the lowest J value has the lowest energy ($^3P_0$ for p^2); for subshells that are more than half filled, the state having the highest J value has the lowest energy. Half-filled subshells have only one possible J value.

An example of a half-filled subshell is a p^3 configuration, which has the ground term ^4S. ^4S has $L = 0$ and $S = \frac{3}{2}$. The maximum value of $J = L + S = 0 + \frac{3}{2} = \frac{3}{2}$, and the minimum value is $J = |L - S| = |0 - \frac{3}{2}| = \frac{3}{2}$. Because the maximum and minimum values of J are identical, only one value, $\frac{3}{2}$, is possible. The ground state term is therefore $^4S_{\frac{3}{2}}$.

Spin-orbit coupling can have significant effects on the electronic spectra of coordination compounds, especially those involving fairly heavy metals (atomic number > 40). For example, Pb, the most metallic of the elements in carbon's group, has the same terms and energy level pattern as shown for carbon. However, the consequences of spin-orbit coupling for Pb are much larger: the $^3P_2$ and $^3P_1$ levels are 10,650.5 and 7,819.4 cm^{-1}, respectively, above the $^3P_0$ level. These three levels in carbon are separated by only 43.5 cm^{-1}! In Pb the $^1S_0$ is 29,466.8 cm^{-1} and the $^1D_2$ level is 21,457.9 cm^{-1} above the $^3P_0$.

11.3 Electronic Spectra of Coordination Compounds

We can now make the connection between electron–electron interactions and the absorption spectra of coordination compounds. In Section 11.2, we considered a method for determining the microstates and free-ion terms for electron configurations. For example, a d^2 configuration gives rise to five free-ion ^3F, ^3P, ^1G, ^1D, and ^1S —with the ^3F term of lowest energy (Exercises 11.1 and 11.3). Absorption spectra of coordination compounds in most cases involve the d orbitals of the metal, and it is consequently important to know the free-ion terms for the possible d configurations. Determining the microstates and free-ion terms for configurations of three or more electrons can be a tedious process. These are listed for the possible d electron configurations in **Table 11.5**.

TABLE 11.5 Free-Ion Terms for d^n Configurations

Configuration	Free-Ion Terms						
d^1	^2D						
d^2		$^1S\,^1D\,^1G$	$^3P\,^3F$				
d^3	^2D		$^4P\,^4F$	$^2P\,^2D\,^2F\,^2G\,^2H$			
d^4	^5D	$^1S\,^1D\,^1G$	$^3P\,^3F$	$^3P\,^3D\,^3F\,^3G\,^3H$	$^1S\,^1D\,^1F\,^1G\,^1I$		
d^5	^2D		$^4P\,^4F$	$^2P\,^2D\,^2F\,^2G\,^2H$	$^2S\,^2D\,^2F\,^2G\,^2I$	$^4D\,^4G$	^6S
d^6	Same as d^4						
d^7	Same as d^3						
d^8	Same as d^2						
d^9	Same as d^1						
d^{10}	^1S						

NOTE: For any configuration, the free-ion terms are the sum of those listed; for example, for the d^2 configuration, the free-ion terms are $^1S + {}^1D + {}^1G + {}^3P + {}^3F$.

In the interpretation of spectra of coordination compounds, it is often important to identify the lowest-energy term. A quick and fairly simple way to do this is given here, using as an example a d^3 configuration in octahedral symmetry.*

1. Sketch the energy levels, showing the *d* electrons.

   ___ ___

   ↑ ↑ ↑

2. Spin multiplicity of lowest-energy state = number of unpaired electrons +1.**

   Spin multiplicity = 3 + 1 = 4

3. Determine the maximum possible value of M_L (sum of m_l values) for the configuration as shown. This determines the type of free-ion term (e.g., *S*, *P*, *D*).

   Maximum possible value of M_L for three electrons as shown: 2 + 1 + 0 = 3; therefore, *F* term

4. Combine results of Steps 2 and 3 to get the ground term.

   ^4F

Step 3 deserves elaboration. The maximum value of m_l for the first electron would be 2, the highest value possible for a *d* electron. Because the electron spins are parallel, the second electron cannot also have $m_l = 2$ (it would violate the exclusion principle); the highest value it can have is $m_l = 1$. Finally, the third electron cannot have $m_l = 2$ or 1, because it would then have the same quantum numbers as one of the first two electrons; the highest m_l value this electron could have would therefore be 0. Consequently, the maximum value of $M_L = 2 + 1 + 0 = 3$.

---

*Note that this approach approximates the t_{2g} and e_g molecular orbitals as metal atomic orbitals even though these molecular orbitals include some contribution from the ligands. In effect, this approach considers metal–ligand bonds to have a large degree of ionic character.

**This is equivalent to the spin multiplicity = $2S + 1$, as shown previously.

EXAMPLE 11.5

What is the ground term for d^4 (low spin)?

1.
```
 ___   ___
 ↑↓    ↑     ↑
 ___   ___   ___
```

2. Spin multiplicity $= 2 + 1 = 3$
3. Highest possible value of $M_L = 2 + 2 + 1 + 0 = 5$; therefore, H term.

   Note that here, $m_l = 2$ for the first two electrons does not violate the exclusion principle because the electrons have opposite spins.
4. Therefore, the ground term is ^3H.

EXERCISE 11.5 Determine the ground terms for high-spin and low-spin d^6 configurations in O_h symmetry.

With this review of atomic states, we may now consider the electronic states of coordination compounds and how transitions between these states give rise to the observed spectra. Before considering specific spectra, however, we must also consider which types of transitions are most probable and, therefore, give rise to the most intense absorptions.

11.3.1 Selection Rules

The relative intensities of absorption bands are governed by a series of selection rules. On the basis of the symmetry and spin multiplicity of ground and excited electronic states, two of these rules are as follows:[1,2]

1. Transitions between states of the same parity (symmetry with respect to a center of inversion) are forbidden. For example, transitions between states that arise from d orbitals are forbidden ($g \longrightarrow g$ transitions; d orbitals are symmetric to inversion), but transitions between states arising from d and p orbitals are allowed ($g \longrightarrow u$ transitions; p orbitals are antisymmetric to inversion). This is known as the **Laporte selection rule**.
2. Transitions between states of different spin multiplicities are forbidden. For example, transitions between $^4A_2$ and $^4T_1$ states are "spin-allowed," but between $^4A_2$ and $^2A_2$ are "spin-forbidden." This is called the **spin selection rule**.

These rules seem to rule out most electronic transitions for transition-metal complexes. However, many such complexes are vividly colored, a consequence of various mechanisms by which these rules can be relaxed. Some of these mechanisms are as follows:

1. The bonds in transition-metal complexes, like all chemical bonds, undergo vibrations that may temporarily change the symmetry. Octahedral complexes, for example, vibrate in ways in which the center of symmetry is temporarily lost; this phenomenon, called *vibronic coupling*, can relax the first selection rule. As a consequence, *d–d* transitions having molar absorptivities in the range of approximately $5-50$ L mol^{-1} cm^{-1} commonly occur, and they are often responsible for the bright colors of these complexes.
2. Tetrahedral complexes often absorb more strongly than octahedral complexes of the same metal in the same oxidation state. Metal–ligand σ bonding in transition-metal complexes of T_d symmetry can be described as involving a combination of sp^3 and sd^3 hybridization of the metal orbitals; both types of hybridization are consistent with the symmetry. The mixing of p-orbital character (of u symmetry) with d-orbital character provides a second way of relaxing the first selection rule.

3. Spin-orbit coupling in some cases provides a mechanism of relaxing the second selection rule, with the result that transitions may be observed from a ground state of one spin multiplicity to an excited state of different spin multiplicity. Such absorption bands for first-row transition-metal complexes are usually very weak, with typical molar absorptivities less than 1 L mol^{-1} cm^{-1}. For complexes of second- and third-row transition metals, spin-orbit coupling can be more important.

Examples of spectra illustrating the selection rules and the ways in which they may be relaxed will now be given. Our first example will be a metal complex having a d^2 configuration and octahedral geometry, $[V(H_2O)_6]^{3+}$.

In discussing spectra, it will be particularly useful to relate the electronic spectra of transition-metal complexes to the ligand field splitting, Δ_o, for octahedral complexes. To do this it will be necessary to introduce **correlation diagrams** and **Tanabe–Sugano diagrams**.

11.3.2 Correlation Diagrams

Figure 11.3 is an example of a correlation diagram for the configuration d^2. These diagrams illustrate how the energies of electronic states change between two extremes:

1. **Free ions (no ligand field)**. In Exercise 11.4, the terms ^3F, ^3P, ^1G, ^1D, and ^1S were obtained for a d^2 configuration, with the ^3F term having the lowest energy. These terms describe the energy levels of a "free" d^2 ion—in our example, a V^{3+} ion—in the absence of any interactions with ligands. In correlation diagrams, we will show these free-ion terms on the far left.
2. **Strong ligand field**. There are three possible configurations for two d electrons in an octahedral ligand field:

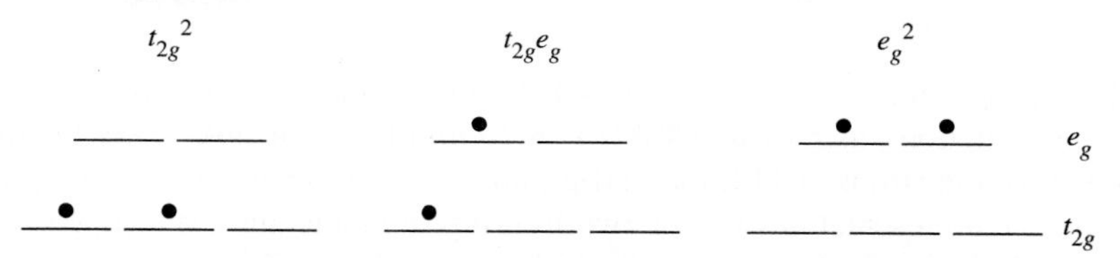

In our example, these would be the possible electron configurations of V^{3+} in an extremely strong ligand field (t_{2g}^2 would be the ground state; the others would be excited states). In correlation diagrams, we will show these states on the far right as the *strong-field limit*. Here, the effect of the ligands is so strong that it completely overrides the effects of *LS* coupling.

In actual coordination compounds, the situation is intermediate between these extremes.* At zero field, the m_l and m_s values of the individual electrons couple to form, for d^2, the five terms ^3F, ^3P, ^1G, ^1D, and ^1S, representing five atomic states with different energies. At a very high ligand field strength, the t_{2g}^2, $t_{2g}\,e_g$, and e_g^2 configurations predominate. The correlation diagram shows the full range of possibilities where the ligand field is not sufficiently strong to quench the coupling between the states generated by the metal valence electrons.

Details of the method for achieving this are beyond the scope of this text, and the interested reader should consult the literature.[3] The aspect of this problem that is important to us is that free-ion terms, shown on the far left in the correlation diagrams, have symmetry characteristics that enable them to be reduced to their constituent irreducible representations; in our example, these will be irreducible representations in the O_h point group. In an

*Note that all the energy level diagrams in Chapter 10 are based on this strong field limit. Although this perspective is useful to rationalize the bonding in transition metal complexes, understanding the spectra of these complexes requires additional considerations that result from electron–electron interactions.

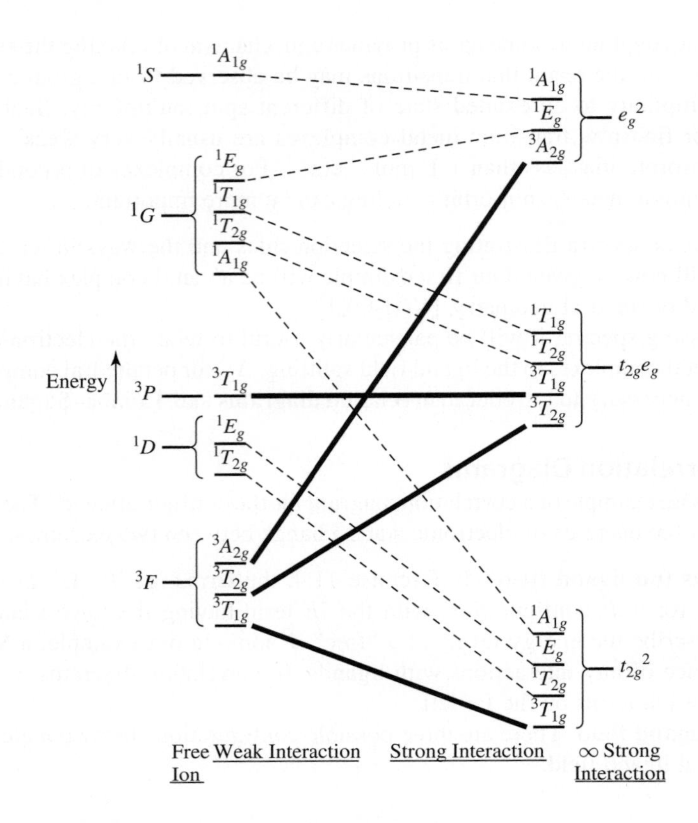

FIGURE 11.3 Correlation Diagram for d^2 in Octahedral Ligand Field.

octahedral ligand field, the free-ion terms will be split into states corresponding to the irreducible representations, as shown in **Table 11.6**. Similarly, irreducible representations may be obtained for the strong-field limit configurations (in our example, t_{2g}^2, $t_{2g}\,e_g$, and e_g^2). The irreducible representations for the two limiting situations must match; each irreducible representåtion for the free ion must match, or correlate with, a representation for the strong-field limit. This is shown in the correlation diagram for d^2 in Figure 11.3.

Note especially the following characteristics of this correlation diagram:

1. The free-ion states (terms arising from *LS* coupling) are shown on the far left.
2. The extremely strong field states are shown on the far right.

TABLE 11.6 Splitting of Free-Ion Terms in Octahedral Symmetry

Term	Irreducible Representations
S	A_{1g}
P	T_{1g}
D	$E_g + T_{2g}$
F	$A_{2g} + T_{1g} + T_{2g}$
G	$A_{1g} + E_g + T_{1g} + T_{2g}$
H	$E_g + 2T_{1g} + T_{2g}$
I	$A_{1g} + A_{2g} + E_g + T_{1g} + 2T_{2g}$

NOTE: Although representations based on atomic orbitals may have either *g* or *u* symmetry, the terms given here are for *d* orbitals and as a result have only *g* symmetry. See F. A. Cotton, *Chemical Applications of Group Theory* (3rd ed., Wiley InterScience, New York, 1990, pp. 263–264) for a discussion of these labels.

3. Both the free-ion and strong-field states can be reduced to irreducible representations. Each free-ion irreducible representation is matched with (correlates with) a strong-field irreducible representation having the same symmetry (same label). As mentioned in Section 11.3.1, transitions to excited states having the same spin multiplicity as the ground state are more likely than transitions to states of different spin multiplicity. To emphasize this, the ground state and states of the same spin multiplicity as the ground state are shown as heavy lines, and states having other spin multiplicities are shown as dashed lines.

In the correlation diagram, the states are shown in order of energy. A noncrossing rule is observed: lines connecting states of the same symmetry designation do not cross. Correlation diagrams are available for other *d*-electron configurations.[4]

11.3.3 Tanabe–Sugano Diagrams

Tanabe–Sugano diagrams are modified correlation diagrams that are useful in the interpretation of electronic spectra of coordination compounds.[5] In Tanabe–Sugano diagrams, the lowest-energy state is plotted along the horizontal axis; consequently, the vertical distance above this axis is a measure of the energy of the excited state above the ground state. For example, for the d^2 configuration, the lowest-energy state is described by the line in the correlation diagram (Figure 11.3) joining the $^3T_{1g}$ state arising from the ^3F free-ion term with the $^3T_{1g}$ state arising from the strong-field term, $t_{2g}{}^2$. In the Tanabe–Sugano diagram (**Figure 11.4**), this line is made horizontal*; it is labeled $^3T_{1g}$ (F) and is shown to arise from the ^3F term in the free-ion limit (left side of diagram).**

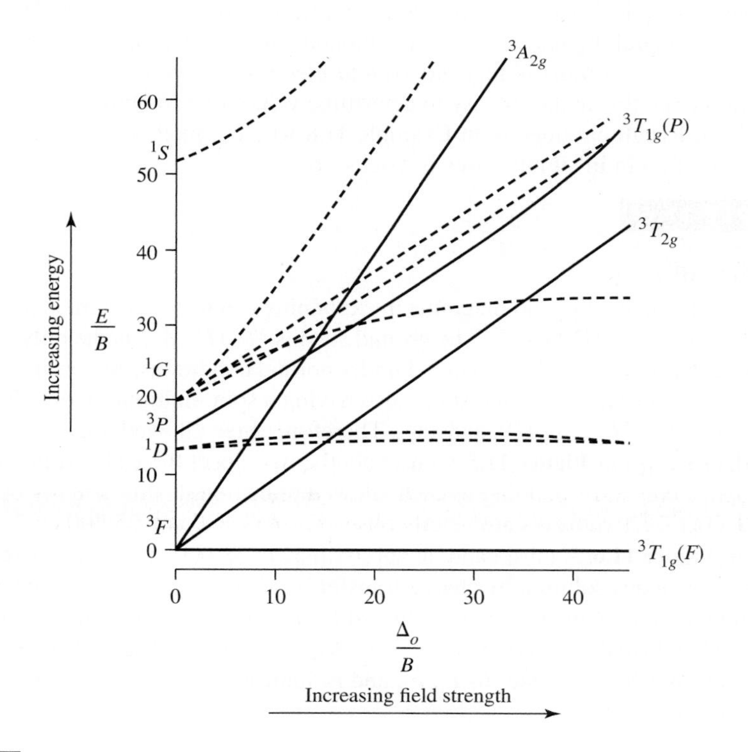

FIGURE 11.4 Tanabe–Sugano. Diagram for d^2 in Octahedral Ligand Field.

*This does not imply that there is no dependence between the absolute energy of the ground state term and the ligand field strength, but defining the horizontal axis in this way is useful from a spectroscopic perspective, where *differences* in energy between the ground and excited states are examined.

**The F in parentheses distinguishes this $^3T_{1g}$ term from the higher energy $^3T_{1g}$ term arising from the ^3P term in the free-ion limit.

The Tanabe–Sugano diagram also shows excited states. In the d^2 diagram, the excited states of the same spin multiplicity as the ground state are the ${}^3T_{2g}$, ${}^3T_{1g}(P)$, and the ${}^3A_{2g}$. The reader should verify that these are the same triplet excited states shown in the d^2 correlation diagram. Excited states of other spin multiplicities are also shown; but, as we will see, they are generally not as important for spectral interpretation.

The quantities plotted in a Tanabe–Sugano diagram are as follows:

Horizontal axis:	$\frac{\Delta_o}{B}$	where Δ_o is the octahedral ligand field splitting, described in Chapter 10.
${}^3P$ ↑ 15 B ↓ ${}^3F$	$B =$	Racah parameter, a measure of the repulsion between terms of the same multiplicity. For d^2, for example, the energy difference between ${}^3F$ and ${}^3P$ is $15B$.
Vertical axis:	$\frac{E}{B}$	where E is the energy (of excited states) above the ground state.

Racah parameters provide useful information about coordination complexes. B for a free ion is typically greater than B for the same ion in a coordination complex. This comparison is often related to how much more volume the valence electrons can access in the complex relative to the free ion. Electrons will experience less repulsion (or their magnetic fields will interact less intensely) if they occupy more volume. In this regard, the decrease in B between a free ion and a coordination complex can be used to assess the degree of covalency in the metal–ligand bonds. As mentioned, one of the most useful characteristics of Tanabe–Sugano diagrams is that *the ground electronic state is always plotted along the horizontal axis*; this makes it easy to determine values of E/B above the ground state.* We will revisit Racah parameters in Example 11.8 where a methodology that allows the determination of them in simple cases is discussed.

EXAMPLE 11.6

$[V(H_2O)_6]^{3+}(d^2)$

The utility of Tanabe–Sugano diagrams in explaining electronic spectra is provided by the d^2 complex $[V(H_2O)_6]^{3+}$. The ground state is ${}^3T_{1g}(F)$; this is the only electronic state that is appreciably occupied under normal conditions. Absorption of light should occur primarily to excited states also having a spin multiplicity of 3. There are three of these: ${}^3T_{2g}$, ${}^3T_{1g}(P)$, and ${}^3A_{2g}$. Therefore, three allowed transitions are expected, as shown in **Figure 11.5**. Consequently, we expect three absorption bands for $[V(H_2O)_6]^{3+}$, one corresponding to each allowed transition. Is this actually observed for $[V(H_2O)_6]^{3+}$? Two bands are readily observed at 17,800 and 25,700 cm^{-1}, as can be seen in **Figure 11.6**. A third band, at approximately 38,000 cm^{-1}, is apparently obscured in aqueous solution by charge-transfer bands nearby (charge-transfer bands of coordination compounds will be discussed later in this chapter).** In the solid state, however, a band attributed to the ${}^3T_{1g} \longrightarrow {}^3A_{2g}$ transition is observed at 38,000 cm^{-1}. These bands match the transitions ν_1, ν_2, and ν_3 indicated on the Tanabe–Sugano diagram (Figure 11.5).

*For a discussion of Racah parameters, see B. N. Figgis, "Ligand Field Theory," in *Comprehensive Coordination Chemistry*, Pergamon Press, Elmsford, NY, 1987, Vol. 1, p. 232.

**This third band is in the ultraviolet and is off-scale to the right in the spectrum shown; see B. N. Figgis, *Introduction to Ligand Fields*, Wiley InterScience, New York, 1966, p. 219.

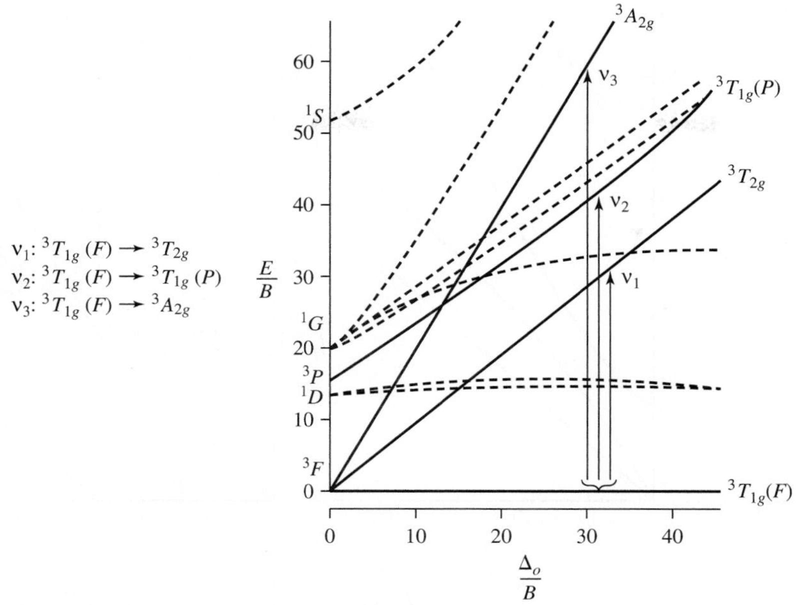

ν_1: $^3T_{1g}(F) \rightarrow {}^3T_{2g}$
ν_2: $^3T_{1g}(F) \rightarrow {}^3T_{1g}(P)$
ν_3: $^3T_{1g}(F) \rightarrow {}^3A_{2g}$

FIGURE 11.5 Spin-Allowed Transitions for d^2 Configuration.

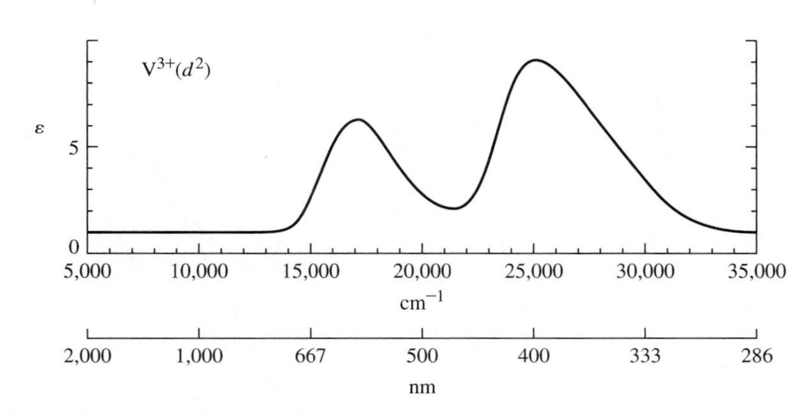

FIGURE 11.6 Absorption Spectrum of $[V(H_2O)_6]^{3+}$. (*Introduction to Ligand Fields*, p. 221. Used by permission from Brian Figgis)

Other Electron Configurations

Tanabe–Sugano diagrams for d^2 through d^8 are shown in **Figure 11.7**. The cases of d^1 and d^9 configurations will be discussed in Section 11.3.4 and illustrated in Figure 11.11. The diagrams for d^4, d^5, d^6, and d^7 have apparent discontinuities, marked by vertical lines near the center. These are configurations for which low spin and high spin are both possible. For example, consider the configuration d^4:

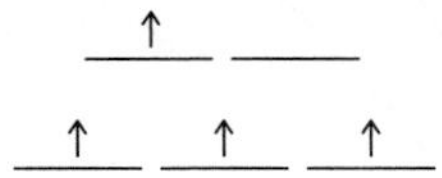

$S = 4(\frac{1}{2}) = 2$;
spin multiplicity $= 2S + 1 = 2(2) + 1 = 5$

High-spin (weak-field) d^4 has four unpaired electrons of parallel spin; such a configuration has a spin multiplicity of 5.

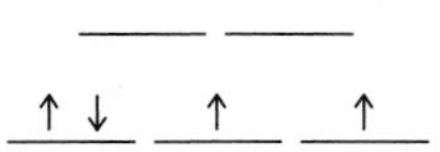

$S = 2(\frac{1}{2}) = 1$;
spin multiplicity $= 2S + 1 = 2(1) + 1 = 3$

Low-spin (strong-field) d^4 on the other hand, has only two unpaired electrons and a spin multiplicity of 3.

In the weak-field part of the d^4 Tanabe–Sugano diagram (left of $\Delta_o / B = 27$), the ground state is ^5E_g, with the expected spin multiplicity of 5. On the right (strong-field) side of the diagram, the ground state is $^3T_{1g}$ (correlating with the ^3H term in the free-ion limit),

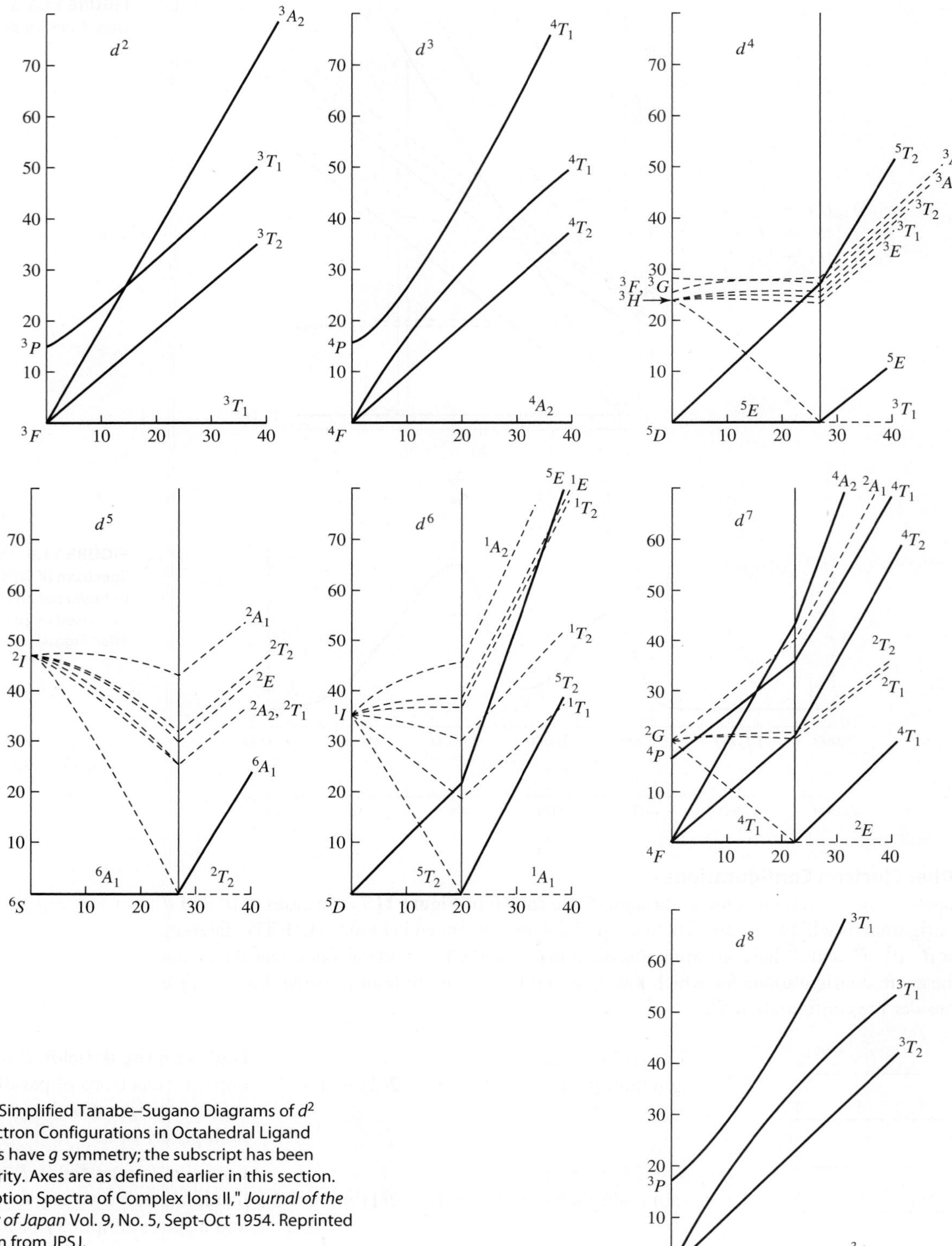

FIGURE 11.7 Simplified Tanabe–Sugano Diagrams of d^2 through d^8 Electron Configurations in Octahedral Ligand Fields. All terms have *g* symmetry; the subscript has been omitted for clarity. Axes are as defined earlier in this section. "On the Absorption Spectra of Complex Ions II," *Journal of the Physical Society of Japan* Vol. 9, No. 5, Sept-Oct 1954. Reprinted with permission from JPSJ.

with the required spin multiplicity of 3. The vertical line is thus a dividing line between weak- and strong-field cases: high-spin (weak-field) complexes are to the left of this line, and low-spin (strong-field) complexes are to the right. At the dividing line, the ground state changes from ^5E_g to $^3T_{1g}$. The spin multiplicity changes from 5 to 3 to reflect the change in the number of unpaired electrons.

Figure 11.8 shows absorption spectra of first-row transition-metal complexes of the formula $[M(H_2O)_6]^{n+}$. These are all high-spin complexes, represented by the left side of the Tanabe–Sugano diagrams, because water is a relatively weak field ligand. It is an interesting exercise to compare the number of bands in these spectra with the number of bands expected from the respective Tanabe–Sugano diagrams. Note that in some cases absorption bands are off-scale, farther into the ultraviolet than the spectral region shown.

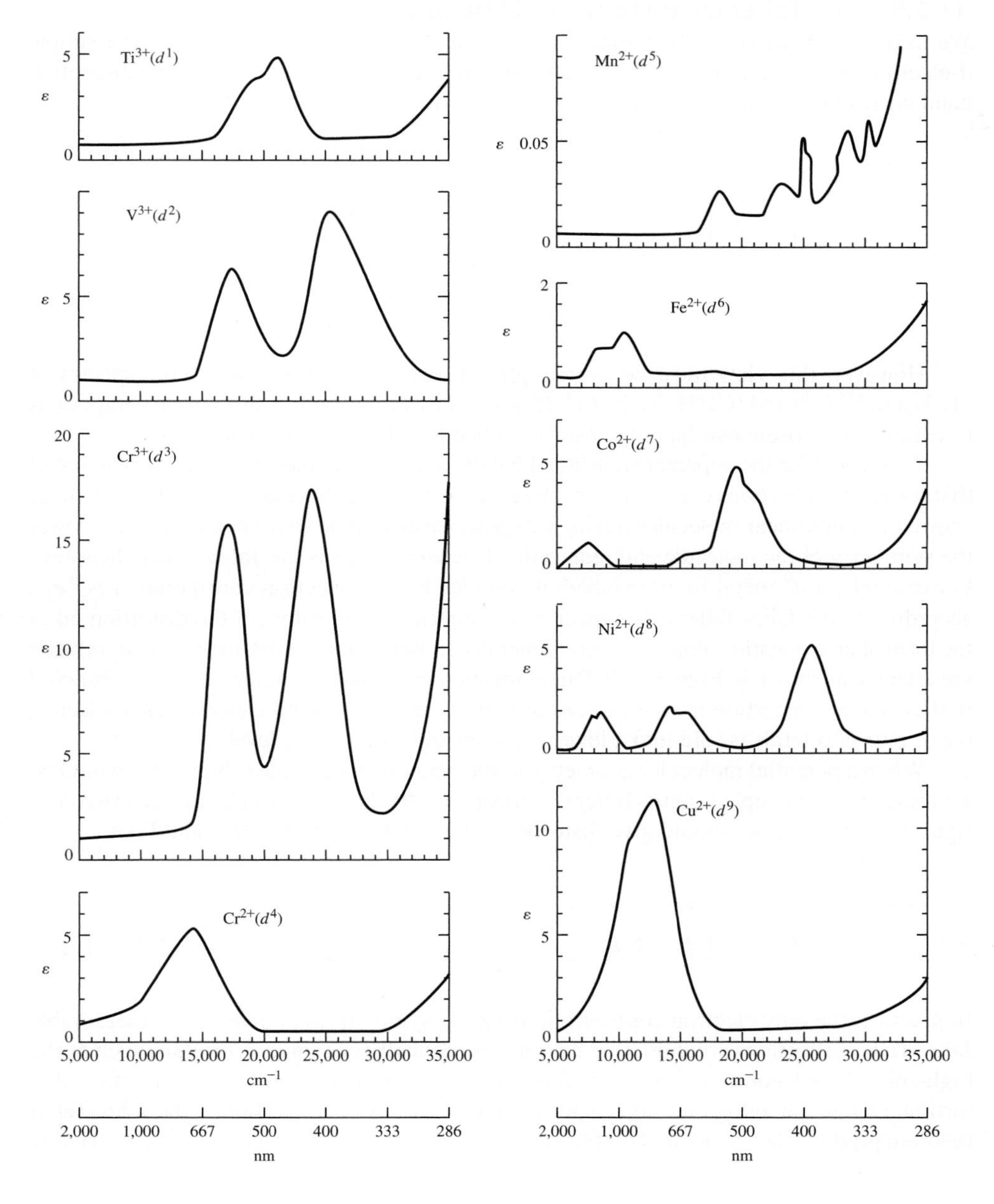

FIGURE 11.8 Electronic Spectra of First-Row Transition-Metal Complexes of Formula $[M(H_2O)_6]^{n+}$ (*Introduction to Ligand Fields*, p. 221. Used by permission from Brian Figgis)

In Figure 11.8, molar absorptivities (extinction coefficients) are shown on the vertical scale. The absorptivities for most bands are similar (1 to 20 L mol^{-1} cm^{-1}) except for the spectrum of $[Mn(H_2O)_6]^{2+}$, which has much weaker bands. Solutions of $[Mn(H_2O)_6]^{2+}$ are an extremely pale pink, much more weakly colored than solutions of the other ions shown. Why is absorption by $[Mn(H_2O)_6]^{2+}$ so weak? To answer this question, it is useful to examine the corresponding Tanabe–Sugano diagram, in this case for a d^5 configuration. We expect $[Mn(H_2O)_6]^{2+}$ to be a high-spin complex, because H_2O is a rather weak-field ligand. The ground state for weak-field d^5 is the $^6A_{1g}$. There are no excited states of the same spin multiplicity (6), and consequently there are no spin-allowed absorptions. That $[Mn(H_2O)_6]^{2+}$ is colored at all is a consequence of very weak forbidden transitions to excited states of spin multiplicity other than 6 (there are many such excited states, hence the rather complicated spectrum).

11.3.4 Jahn–Teller Distortions and Spectra

We have not yet discussed the spectra of d^1 and d^9 complexes. By virtue of the simple *d*-electron configurations for these cases, we might expect each to exhibit one absorption band corresponding to excitation of an electron from the t_{2g} to the e_g levels:

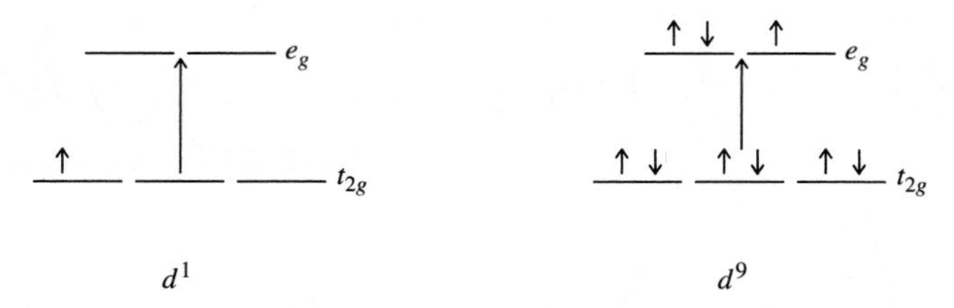

However, this view must be oversimplified, because examination of the spectra of $[Ti(H_2O)_6]^{3+}(d^1)$ and $[Cu(H_2O)_6]^{2+}(d^9)$ (Figure 11.8) shows these coordination compounds to exhibit two closely overlapping absorption bands rather than a single band.

To account for the apparent splitting of bands in these examples, it is necessary to recall that some configurations can cause complexes to be distorted (Section 10.5). Jahn and Teller showed that nonlinear molecules having a degenerate electronic state should distort to lower the symmetry of the molecule and reduce the degeneracy; this is the **Jahn–Teller theorem.**[6] For example, a d^9 metal in an octahedral complex has the electron configuration $t_{2g}{}^6 e_g{}^3$; according to the Jahn–Teller theorem, such a complex should distort. If the distortion takes the form of an elongation along one axis (generally chosen as the *z*-axis), the t_{2g} and e_g orbitals are affected as shown in **Figure 11.9**. This elongation is the most common distortion observed in these cases. Distortion from O_h to D_{4h} symmetry results in stabilization of the molecule: the e_g pair of orbitals is split into a lower a_{1g} level and a higher b_{1g} level.

When a potential molecular geometry would result in degenerate orbitals that would be asymmetrically occupied, Jahn–Teller distortions are likely. For example, the first two configurations that follow should give distortions, but the third and fourth should not:

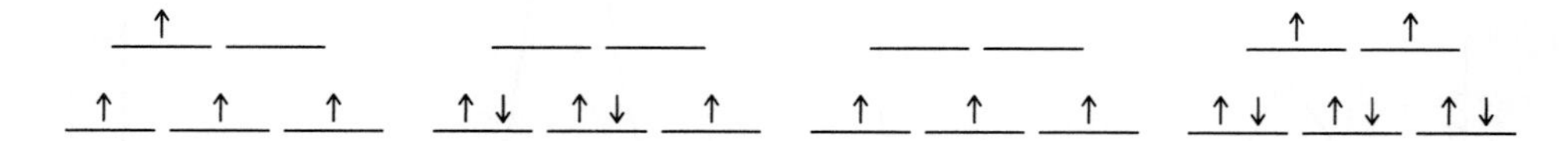

In practice, the only electron configurations for O_h symmetry that give rise to measurable Jahn–Teller distortions are those that have asymmetrically occupied e_g orbitals, such as the high-spin d^4 configuration. The Jahn–Teller theorem does not predict the nature of the distortion; elongation along the *z* axis is most commonly observed. Although the Jahn–Teller theorem predicts that structures that would result in configurations having asymmetrically

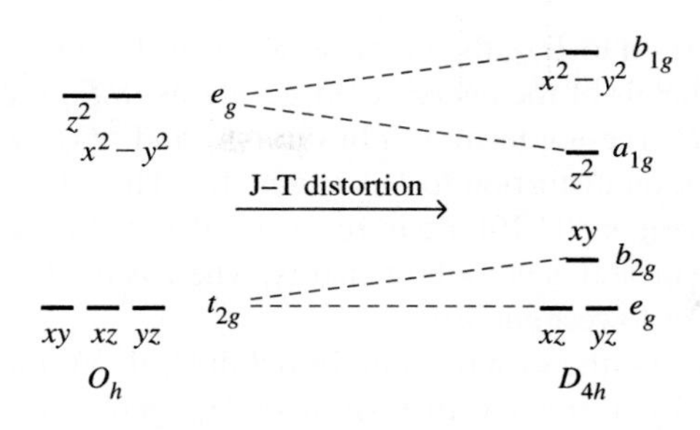

FIGURE 11.9 Effect of Jahn–Teller Distortion on *d* Orbitals of an Octahedral Complex.

occupied t_{2g} orbitals, such as the low-spin d^5 configuration, should also be distorted, such distortions are often too small to be detectable.

The Jahn–Teller effect on spectra can be seen in $[Cu(H_2O)_6]^{2+}$, a d^9 complex. Figure 11.9 shows the additional splitting of d orbitals accompanying the reduction of symmetry from O_h to D_{4h}.

Symmetry Labels for Configurations

Electron *configurations* have symmetry labels that match their degeneracies, as follows:

		Examples	
T	Designates a triply degenerate asymmetrically occupied state.	___ ___ • ___ ___	• • •• •• •
E	Designates a doubly degenerate asymmetrically occupied state.	• ___ • • •	•• • •• •• ••
A or *B*	Designate a nondegenerate state. Each set of levels in an *A* or *B* state is symmetrically occupied.	___ ___ • • •	• • • • •

EXERCISE 11.6

Identify the following configurations as *T*, *A*, or *E* states in octahedral complexes:

a. • ___ / •• •• •• **b.** • • / •• •• •• **c.** ___ ___ / • • ___

When a ^2D term for d^9 is split by an octahedral ligand field, two configurations result:

	Lower energy	Higher energy
e_g	•• •	•• ••
t_{2g}	•• •• ••	•• •• •

The lower energy configuration is doubly degenerate in the e_g orbitals (occupation of the e_g orbitals could be •• • or • ••) and has the designation ^2E_g; the higher energy configuration is triply degenerate in the t_{2g} levels (three arrangements are possible in these levels: •• • ••, •• •• •, or • •• ••) and has the designation $^2T_{2g}$. Thus, the lower energy configuration is the ^2E_g, and the higher energy configuration is the $^2T_{2g}$, as in **Figure 11.10**. This is opposite to the order of energies of the orbitals (t_{2g} lower than e_g), shown in Figure 11.9.

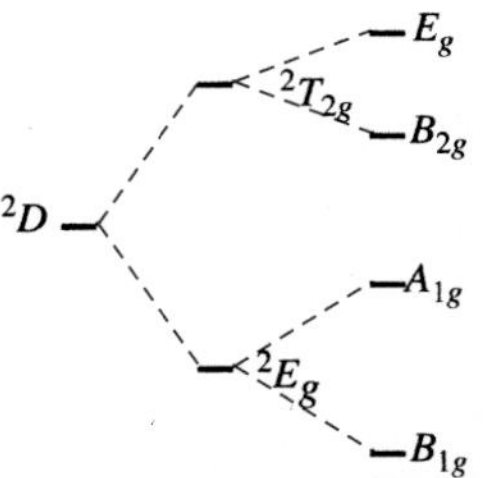

FIGURE 11.10 Splitting of Octahedral Free-Ion Terms on Jahn–Teller Distortion for d^9 Configuration.

Similarly, for distortion to D_{4h}, the order of labels of the orbitals in Figure 11.9 is the reverse of the order of labels of the energy configurations in Figure 11.10.

In summary, the ^2D free-ion term is split into ^2E_g and $^2T_{2g}$ by a field of O_h symmetry, and it is further split on distortion to D_{4h} symmetry. The labels of the states resulting from the free-ion term (Figure 11.10) are in reverse order to the labels on the orbitals; for example, the b_{1g} atomic orbital is of highest energy, whereas the B_{1g} state originating from the ^2D free-ion term is of lowest energy.[7]

For a d^9 configuration, the ground state in octahedral symmetry is a ^2E_g term, and the excited state is a $^2T_{2g}$ term. On distortion to D_{4h} geometry, these terms split, as shown in Figure 11.10. In an octahedral d^9 complex, we would expect excitation from the ^2E_g state to the $^2T_{2g}$ state and a single absorption band. Distortion of the complex to D_{4h} geometry splits the $^2T_{2g}$ level into two levels, the E_g and the B_{2g}. Excitation can now occur from the ground state (now the B_{1g} state) to the A_{1g}, the E_g, or the B_{2g} states (the splitting is exaggerated in Figure 11.10). The $B_{1g} \longrightarrow A_{1g}$ transition is too low in energy to be observed in the visible spectrum. If the distortion is strong enough, therefore, two absorption bands may be observed in the visible region, to the E_g and the B_{2g} levels (or a broadened or narrowly split peak is found, as in $[Cu(H_2O)_6]^{2+}$).

For a d^1 complex, a single absorption band, corresponding to excitation of a t_{2g} electron to an e_g orbital, might be expected:

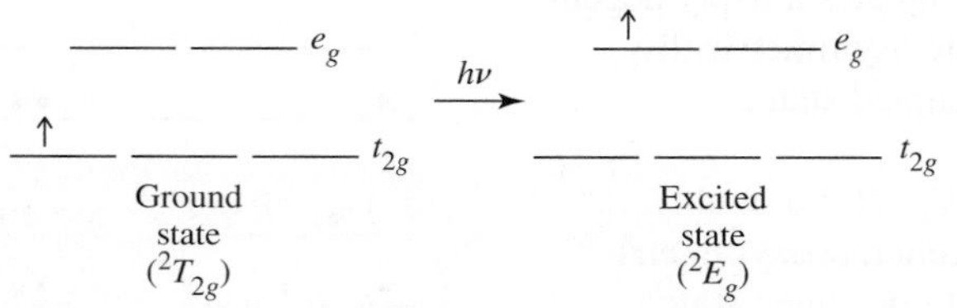

e_g e_g $h\nu$ t_{2g} t_{2g}

Ground state ($^2T_{2g}$) Excited state (^2E_g)

However, the spectrum of $[Ti(H_2O)_6]^{3+}$, an example of a d^1 complex, shows two apparently overlapping bands rather than a single band. How is this possible?

One explanation commonly used is that the excited state is Jahn–Teller distorted,[8] as in Figure 11.10. As in the examples considered previously, asymmetric occupation of the e_g orbitals can split these orbitals into two of slightly different energy (of A_{1g} and B_{1g} symmetry). Excitation can now occur from the t_{2g} level to either of these orbitals. Therefore, as in the case of the d^9 configuration, there are now two excited states of slightly different energy. The consequence may be a broadening of a spectrum into a two-humped peak, as in $[Ti(H_2O)_6]^{3+}$ or, in some cases, into two more clearly defined peaks.[9]

A crucial limitation of Tanabe–Sugano diagrams, as shown in Figure 11.7, is that they assume O_h symmetry in *excited states as well as in ground states*. The consequence is that the diagrams are useful in predicting the general properties of spectra, and many complexes do have sharply defined bands that fit the Tanabe–Sugano description well (see the d^2, d^3, and d^4 examples in Figure 11.8). However, distortions from pure octahedral symmetry are common, and the consequence can be the splitting of bands—or, in cases of severe distortion, situations in which the spectra are difficult to interpret. Additional examples of spectra showing the splitting of absorption bands can be seen in Figure 11.8.

EXERCISE 11.7

$[Fe(H_2O)_6]^{2+}$ has a two-humped absorption peak near 1000 nm. Account for (a) the most likely origin of this absorption using a Tanabe–Sugano diagram, and (b) the splitting of the absorption band.

11.3.5 Applications of Tanabe–Sugano Diagrams: Determining Δ_o from Spectra

Absorption spectra of coordination compounds can be used to determine the magnitude of the ligand field splitting, which is Δ_o for octahedral complexes. The accuracy with which Δ_o can be determined is limited by the mathematical approaches used to analyze the spectral data. Absorption spectra often have overlapping bands, and accurate determination of the positions of these bands requires special mathematical techniques to reduce the bands into their components. Such analysis is beyond the scope of this text. However, we can often obtain first approximations of Δ_o values (and sometimes values of the Racah parameter, B) by simply using the positions of the absorption maxima taken directly from the spectra.

The ease with which Δ_o can be determined depends on the d-electron configuration of the metal; in some cases, Δ_o can be deduced directly by inspection of a spectrum, but in other cases a more complicated analysis is necessary. The following discussion will proceed from the simplest cases to the most complicated.

d^1, d^4 (High Spin), d^6 (High Spin), d^9

From a simple perspective each of these cases in **Figure 11.11** corresponds to excitation of an electron from a t_{2g} to an e_g orbital, with the final (excited) electron configuration having the same spin multiplicity as the initial configuration. Our discussion in this chapter indicates that when electron–electron interactions are considered there is a single excited state of the same spin multiplicity as the ground state for each of these configurations. Consequently, there is a single spin-allowed absorption, with the energy of the absorbed light equal to Δ_o. Examples of such complexes include $[Ti(H_2O)_6]^{3+}$, $[Cr(H_2O)_6]^{2+}$, $[Fe(H_2O)_6]^{2+}$, and $[Cu(H_2O)_6]^{2+}$; note from Figure 11.8 that each of these complexes exhibits essentially a single absorption band. In some cases, splitting of bands due to Jahn–Teller distortion is observed (Section 11.3.4).

d^3, d^8

These electron configurations have a ground-state F term. In an octahedral ligand field, an F term splits into three terms: A_{2g}, T_{2g}, and T_{1g}. As shown in **Figure 11.12**, the A_{2g} is of lowest energy for d^3 or d^8. For these configurations, the difference in energy between the two lowest-energy terms, the A_{2g} and the T_{2g}, is equal to Δ_o. Therefore, to approximate

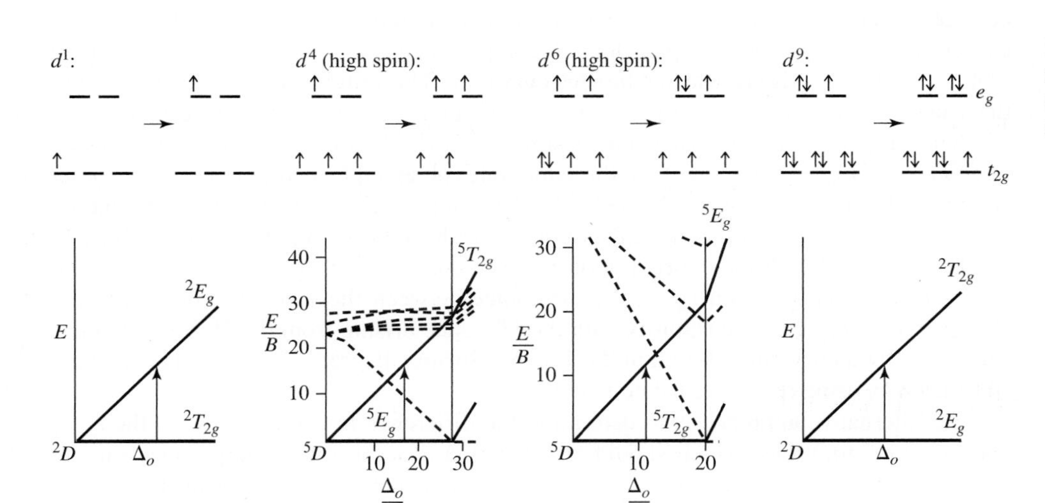

FIGURE 11.11 Determining Δ_o for d^1, d^4 (High Spin), d^6 (High Spin), and d^9 Configurations.

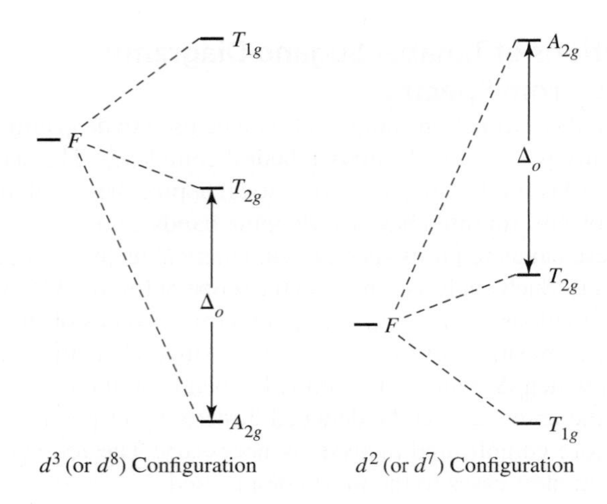

FIGURE 11.12 Splitting of *F* Terms in Octahedral Symmetry.

Δ_o, we determine the energy of the band associated with the A_{2g} to T_{2g} transition, which is typically the lowest energy absorption in the electronic spectrum. Examples include $[Cr(H_2O)_6]^{3+}$ and $[Ni(H_2O)_6]^{2+}$. In these complexes, the lowest-energy band in the spectra (Figure 11.8) is for the transition from the ${}^4A_{2g}$ ground state to the ${}^4T_{2g}$ excited state. The energies of these bands, approximately 17,500 and 8,500 cm^{-1}, respectively, are the values of Δ_o.

d^2, d^7 (High Spin)

As in the case of d^3 and d^8, the ground free-ion terms for these two configurations are *F* terms. However, the determination of Δ_o is not as simple for d^2 and d^7. It is necessary to compare the d^3 and d^2 Tanabe–Sugano diagrams to explain this complication; the d^8 and d^7 (high-spin) cases can be compared in a similar fashion (note the similarity of the d^3 and d^8 Tanabe–Sugano diagrams and of the d^2 and d^7 [high-spin region] diagrams).

In the d^3 case, the ground state is a ${}^4A_{2g}$ state. There are three excited quartet states: ${}^4T_{2g}$, ${}^4T_{1g}$ (from ${}^4F$ term), and ${}^4T_{1g}$ (from ${}^4P$ term). Note the two states of the same symmetry (${}^4T_{1g}$). States of the same symmetry can mix, just as molecular orbitals of the same symmetry can mix. The consequence of such mixing is that as the ligand field is increased, the energies of the states diverge; the lines in the Tanabe–Sugano diagram curve away from each other. This effect is visually apparent in the d^3 diagram (Figure 11.7). This mixing causes no difficulty in obtaining Δ_o for a d^3 complex, because the lowest-energy transition (${}^4A_{2g} \longrightarrow {}^4T_{2g}$) is not affected by such curvature. The Tanabe–Sugano diagram shows that the energy of the ${}^4T_{2g}$ state varies linearly with the strength of the ligand field.

The situation in the d^2 case is not as simple. For d^2, the free-ion ${}^3F$ term is also split into ${}^3T_{1g} + {}^3T_{2g} + {}^3A_{2g}$; these are the same states obtained from d^3, but in reverse order (Figure 11.12). For d^2, the ground state is ${}^3T_{1g}$. It is tempting to simply determine the energy of the ${}^3T_{1g}(F) \longrightarrow {}^3T_{2g}$ band and assign this as the value of Δ_o. After all, the ${}^3T_{1g}(F)$ can be associated with the configuration ${t_{2g}}^2$ (see correlation diagram, Figure 11.3), and ${}^3T_{2g}$ with the configuration $t_{2g}\,e_g$; the difference between these states should give Δ_o. However, the ${}^3T_{1g}(F)$ state can mix with the ${}^3T_{1g}$ state arising from the ${}^3P$ free-ion term, causing a slight curvature of both in the Tanabe–Sugano diagram. This mixing can cause some error in using the ground state to obtain Δ_o.

An alternative approach is to determine the difference in energy between the $t_{2g}\,e_g$ and ${e_g}^2$ configurations, which should also equal Δ_o (because the energy necessary to excite a single electron from a t_{2g} to an e_g orbital is equal to Δ_o). Unfortunately, direct spectroscopic measurement of a transition between *excited* electronic states is virtually

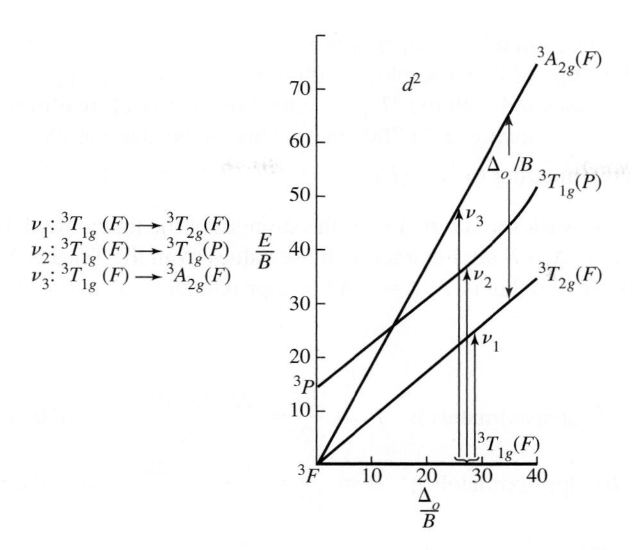

FIGURE 11.13 Spin-Allowed Transitions for d^2 Configuration.

impossible. However, we can indirectly measure the difference between ${}^3T_{2g}$ (for the $t_{2g}\,e_g$ configuration) and ${}^3A_{2g}$ (for $e_g{}^2$; see Figure 11.3):

$$\begin{array}{ll} \text{energy of transition } {}^3T_{1g} \longrightarrow {}^3A_{2g} & \\ -\text{energy of transition } {}^3T_{1g} \longrightarrow {}^3T_{2g} & \text{(see Figure 11.13)} \\ \hline \Delta_o = \text{energy difference between } {}^3A_{2g} \text{ and } {}^3T_{2g} & \end{array}$$

The difficulty with this approach is that the ${}^3A_{2g}$ and ${}^3T_{1g}\,(P)$ states cross in the Tanabe–Sugano diagram; unambiguous assignment of these bands may be challenging. Indeed, while the d^2 diagram indicates that the lowest-energy absorption band (to ${}^3T_{2g}$) is easily assigned, there are two possibilities for the next band: to ${}^3A_{2g}$ for very weak field ligands, or to ${}^3T_{1g}\,(P)$ for stronger-field ligands. Furthermore, the second and third absorption bands may overlap, making it difficult to determine their exact positions (the apparent positions of absorption maxima may be shifted if the bands overlap). In such cases a more complicated analysis—involving a calculation of the Racah parameter, B—may be necessary. This procedure is best illustrated by the following example.

EXAMPLE 11.8

$[V(H_2O)_6]^{3+}$ has absorption bands at 17,800 and 25,700 cm^{-1}. Using the Tanabe–Sugano diagram for d^2, estimate values of Δ_o and B for this complex.

From the Tanabe–Sugano diagram there are three possible spin-allowed transitions (Figure 11.13):

$$\begin{array}{llll} {}^3T_{1g}\,(F) \longrightarrow {}^3T_{2g}\,(F) & \nu_1 & & \text{(lowest energy)} \\ {}^3T_{1g}\,(F) \longrightarrow {}^3T_{1g}\,(P) & \nu_2 & \left.\right\} & \text{(one of the these must be the} \\ {}^3T_{1g}\,(F) \longrightarrow {}^3A_{2g}\,(F) & \nu_3 & & \text{higher-energy band)} \end{array}$$

It is often useful to determine the ratio of energies of the absorption bands. In this example,

$$\frac{25{,}700 \text{ cm}^{-1}}{17{,}800 \text{ cm}^{-1}} = 1.44$$

the ratio of energy of the higher-energy transition (ν_2 or ν_3) to the lowest-energy transition (ν_1) must therefore be approximately 1.44. From the Tanabe–Sugano diagram, we can see that the ratio of ν_3 to ν_1 is approximately 2, regardless of the strength of the ligand field; the slope of line associated with the $^3A_{2g}(F)$ state is approximately twice that of the line associated with the $^3T_{2g}(F)$ state. We can therefore eliminate ν_3 as the possible transition occurring at 25,700 cm^{-1}. This means that the 25,700 cm^{-1} band must be ν_2, corresponding to $^3T_{1g}\,(F) \longrightarrow {}^3T_{1g}\,(P)$, and $1.44 = \frac{\nu_2}{\nu_1}$.

The ratio $\nu_2\,/\,\nu_1$ varies as a function of the strength of the ligand field. By plotting the ratio $\nu_2\,/\,\nu_1$ versus $\Delta_o\,/\,B$ after extracting these values from the Tanabe–Sugano diagram (**Figure 11.14**), we find that $\nu_2\,/\,\nu_1 = 1.44$ at approximately $\Delta_o\,/\,B = 31$.[10,*]

At $\frac{\Delta_o}{B} = 31$,

$$\nu_2: \quad \frac{E}{B} = 42 \text{ (approximately)}; \quad B = \frac{E}{42} = \frac{25{,}700 \text{ cm}^{-1}}{42} = 610 \text{ cm}^{-1}$$

$$\nu_1: \quad \frac{E}{B} = 29 \text{ (approximately)}; \quad B = \frac{E}{29} = \frac{17{,}800 \text{ cm}^{-1}}{29} = 610 \text{ cm}^{-1}$$

Because $\frac{\Delta_o}{B} = 31$,

$$\Delta_o = 31 \times B = 31 \times 610 \text{ cm}^{-1} = 19{,}000 \text{ cm}^{-1}$$

This procedure can be followed for d^2 and d^7 complexes of octahedral geometry to estimate values for Δ_o (and B).

EXERCISE 11.8 Use the Co(II) spectrum in Figure 11.8 and the Tanabe–Sugano diagrams of Figure 11.7 to find Δ_o and B. The broad band near 20,000 cm^{-1} can be considered to have the $^4T_{1g} \longrightarrow {}^4A_{2g}$ transition in the small shoulder near 16,000 cm^{-1} and the $^4T_{1g}\,(F) \longrightarrow {}^4T_{1g}\,(P)$ transition at the peak.**

Other Configurations: d^5 (High Spin), d^4 to d^7 (Low Spin)

High-spin d^5 complexes have no excited states of the same spin multiplicity (6) as the ground state. The bands that are observed are therefore the consequence of spin-forbidden transitions and are typically very weak as, for example, in $[Mn(H_2O)_6]^{2+}$. The interested reader is referred to the literature[11] for an analysis of such spectra. In the case of low-spin

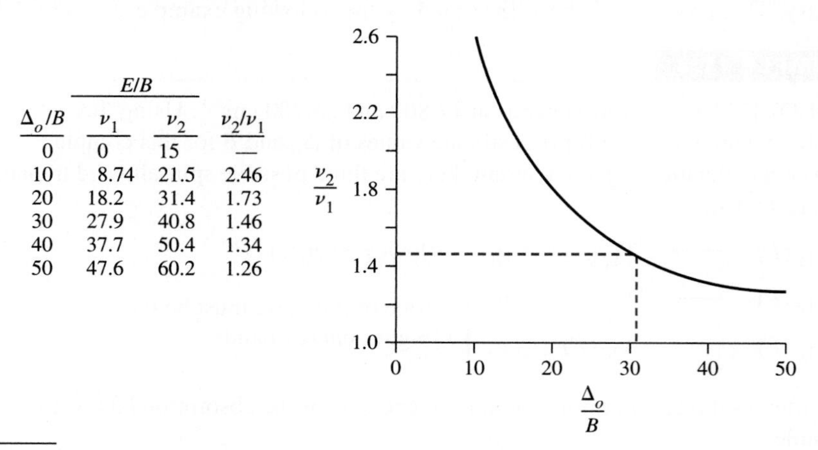

Δ_o/B	E/B ν_1	E/B ν_2	ν_2/ν_1
0	0	15	—
10	8.74	21.5	2.46
20	18.2	31.4	1.73
30	27.9	40.8	1.46
40	37.7	50.4	1.34
50	47.6	60.2	1.26

FIGURE 11.14 Value of ν_2/ν_1 Ratio for d^2 Configuration.

*Different references report slightly different positions for the absorption bands of $[V(H_2O)_6]^{3+}$ and hence slightly different values of B and Δ_o.

** The $^4T_{1g} \rightarrow {}^4A_{2g}$ transition is generally weak in octahedral complexes of Co^{2+} because such a transition corresponds to simultaneous excitation of two electrons and is less probable than the other spin-allowed transitions, which are for excitations of single electrons.

d^4 to d^7 octahedral complexes, the analysis can be difficult, because there are many excited states of the same spin multiplicity as the ground state (see right side of Tanabe–Sugano diagrams for d^4 to d^7, Figure 11.7). The chemical literature provides examples and analyses of the spectra of such compounds.[12]

An alternative approach to interpreting spectra has been presented using the effective nuclear charge of the metal instead of the ligand field parameter Δ_o. This approach, which also takes into account a relativistic spin-orbital interaction, generates diagrams comparable to Tanabe–Sugano diagrams and is proposed to be particularly useful in considering intermediate spin states for d^4–d^6 configurations.[13]

11.3.6 Tetrahedral Complexes

Tetrahedral complexes generally have more intense absorptions than octahedral complexes. This is a consequence of the first (Laporte) selection rule (Section 11.3.1): transitions between *d* orbitals in a complex having a center of symmetry are forbidden. As a result, absorption bands for octahedral complexes are weak (small molar absorptivities); that they absorb at all is partly the result of vibrational motions that spontaneously distort complexes slightly from pure O_h symmetry.

In tetrahedral complexes, the situation is different. The lack of a center of symmetry means that the Laporte selection rule does not apply. The consequence is that tetrahedral complexes often have much more intense absorption bands than octahedral complexes.*

As we have seen, the *d* orbitals for tetrahedral complexes are split in the opposite fashion to octahedral complexes:

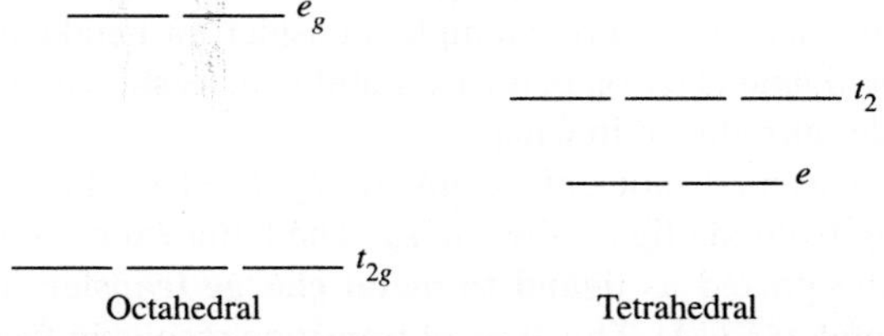

A useful comparison can be drawn between these by using the **hole formalism**. This is best illustrated by example. Consider a d^1 configuration in an octahedral complex. The one electron occupies an orbital in a triply degenerate set (t_{2g}). Now, consider a d^9 configuration in a tetrahedral complex. This configuration has a "hole" in a triply degenerate set of orbitals (t_2). It can be shown that, in terms of symmetry, the d^1 O_h configuration is analogous to the d^9 T_d configuration; the "hole" in d^9 results in the same symmetry as the single electron in d^1.

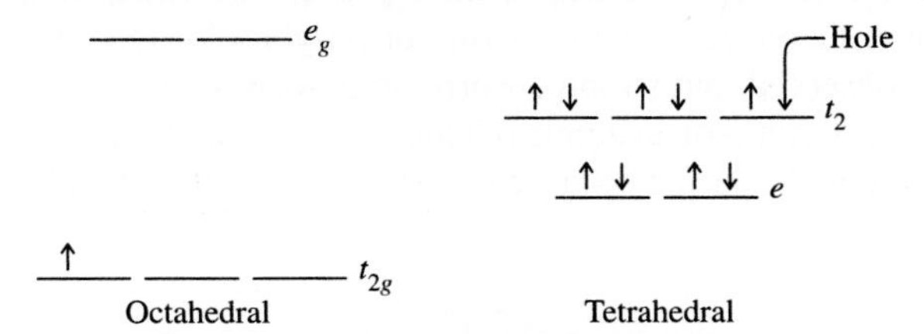

In practical terms, this means that, for tetrahedral geometry, we can use the correlation diagram for the d^{10-n} configuration in octahedral geometry to describe the d^n configuration

*Two types of hybrid orbitals are possible for a central atom of T_d symmetry: sd^3 and sp^3 (see Chapter 5). These types of hybrids may be viewed as mixing, to yield hybrid orbitals that contain some *p* character (note that *p* orbitals are not symmetric to inversion), as well as *d* character. The mixing in of *p* character can be viewed as making transitions between these orbitals more allowed. For a more thorough discussion of this phenomenon, see F. A. Cotton, *Chemical Applications of Group Theory*, 3rd ed., Wiley InterScience, New York, 1990, pp. 295–296. Pages 289–297 of this reference also give a more detailed discussion of other selection rules.

in tetrahedral geometry. Thus, for a d^2 tetrahedral case, we can use the d^8 octahedral correlation diagram; for the d^3 tetrahedral case, we can use the d^7 octahedral diagram, and so on. We can then identify the appropriate spin-allowed bands as in octahedral geometry, with allowed transitions occurring between the ground state and excited states of the same spin multiplicity.

Other geometries can also be considered according to the same principles as for octahedral and tetrahedral complexes. The interested reader is referred to the literature for a discussion of different geometries.[14]

11.3.7 Charge-Transfer Spectra

Charge-transfer absorptions in solutions of halogens are described in Chapter 6. In these cases, a strong interaction between a donor solvent and a halogen molecule, X_2, leads to the formation of a complex in which an excited state (primarily of X_2 character) can accept electrons from a HOMO (primarily of solvent character) on absorption of light of suitable energy:

$$X_2 \cdot \text{donor} \longrightarrow [\text{donor}^+][X_2^-]$$

The absorption band, known as a **charge-transfer band**, can be very intense; it is responsible for the vivid colors of some of the halogens in donor solvents.

Coordination compounds often exhibit strong charge-transfer absorptions in the ultraviolet and/or visible portions of the spectrum. These absorptions may be much more intense than *d–d* transitions (which for octahedral complexes commonly have ε values of 20 L mol^{-1} cm^{-1} or less); molar absorptivities of 50,000 L mol^{-1} cm^{-1} or greater are common for charge-transfer bands. Such absorption bands involve the transfer of electrons from molecular orbitals that are primarily ligand in character to orbitals that are primarily metal in character, or vice versa. For example, consider an octahedral d^6 complex with σ-donor ligands. The ligand electron pairs are stabilized, as shown in **Figure 11.15** via the angular overlap model introduced in Chapter 10.

Electrons can be excited, not only from the t_{2g} level to the e_g, but also from the σ orbitals originating from the ligands to the e_g. The latter excitation results in a charge-transfer transition, designated as **ligand to metal charge transfer (LMCT)** (also called charge transfer to metal, CTTM). This type of transition results in formal reduction of the metal. A LMCT excitation involving a cobalt (III) complex, for example, would exhibit an excited state having cobalt(II).

Examples of charge-transfer absorptions are numerous. The octahedral complexes $IrBr_6^{2-}$ (d^5) and $IrBr_6^{3-}$ (d^6) both show charge-transfer bands. For $IrBr_6^{2-}$, two bands appear, near 600 nm and near 270 nm; the former is attributed to transitions to the t_{2g} levels and the latter to the e_g. In $IrBr_6^{3-}$, the t_{2g} levels are filled, and the only possible LMCT absorption is therefore to the e_g. Consequently, no low-energy absorptions in the 600-nm range are observed, but strong absorption is seen near 250 nm, corresponding to charge transfer to e_g. A common example of tetrahedral geometry is the permanganate ion, MnO_4^-, which is intensely purple because of a strong absorption involving charge transfer

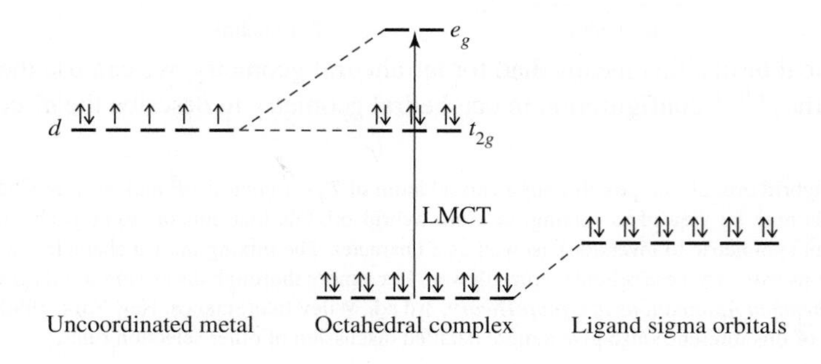

FIGURE 11.15 Ligand to Metal Charge Transfer.

from orbitals derived primarily from the filled oxygen p orbitals to empty orbitals derived primarily from the manganese(VII).

Similarly, it is possible for there to be **metal to ligand charge transfer (MLCT)** (also called charge transfer to ligand, CTTL) in coordination compounds having π-acceptor ligands. In these cases, empty π^* orbitals on the ligands become the acceptor orbitals on absorption of light. **Figure 11.16** illustrates this phenomenon for a d^5 complex, again from the oversimplified angular overlap perspective.

MLCT results in oxidation of the metal; a MLCT excitation of an iron(III) complex would give an iron(IV) excited state. MLCT most commonly occurs with ligands having empty π^* orbitals, such as CO, CN^-, SCN^-, bipyridine, and dithiocarbamate ($S_2CNR_2^-$).

In complexes such as $Cr(CO)_6$ which have both σ-donor and π-acceptor orbitals, both types of charge transfer are possible. It is not easy to determine the type of charge transfer in a given coordination compound. Many ligands give highly colored complexes that have a series of overlapping absorption bands in the ultraviolet part of the spectrum as well as the visible. In such cases, the d–d transitions may be completely overwhelmed and essentially impossible to observe.

Finally, the ligand itself may have a chromophore and an **intraligand band** may be observed. Comparing the spectra of a complex to that of the free ligand can help identify intraligand bands. However, coordination of a ligand to a metal may significantly alter the energies of the ligand orbitals, and such comparisons may be difficult, especially if charge-transfer bands overlap the intraligand bands. In addition, not all ligands exist independently; some owe their existence to the ability of metal atoms to stabilize them. Examples of these ligands will be discussed in later chapters.

EXERCISE 11.9

The isoelectronic ions VO_4^{3-}, CrO_4^{2-}, and MnO_4^- all have intense charge-transfer transitions. The wavelengths of these transitions increase in this series, with MnO_4^- having its charge-transfer absorption at the longest wavelength. Suggest a reason for this trend.

11.3.8 Charge-Transfer and Energy Applications

The pursuit of artificial photosynthesis via capture of solar radiation in solar cells[15] has prompted extensive examination of metal complexes that are photostable, feature broad absorption in the visible region, and possess significantly long-lived excited states to permit photo-promoted electron and energy transfer processes. It is not surprising that metal complexes with MLCT bands have received significant attention for these applications on the basis of their typically high MLCT molar absorptivities. Coordination complexes containing the d^6 metal ions Ru^{2+}, Os^{2+}, Re^{1+}, and Ir^{3+} with strong field N-heterocyclic organic ligands (for example, pyridine, 2,2′-bipyridine, 1,10-phenantholine) are ubiquitous in this area; these "light-harvesting" complexes offer tunable transitions, and the redox properties of these ions facilitate electron transfer steps. These low-spin complexes, such as $[Ru(bpy)_3]^{2+}$ and $[Os(phen)_3]^{2+}$, afford MLCT excited states in which the metal is formally oxidized (via excitation from metal-based t_{2g} into orbitals with extensive π^* ligand character),

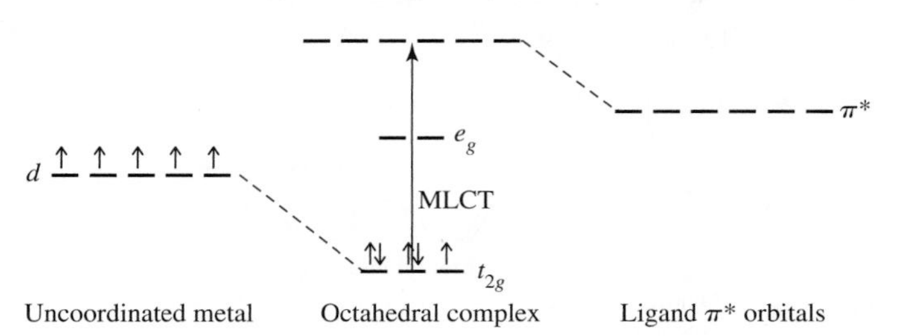

FIGURE 11.16 Metal to Ligand Charge Transfer.

d–d transitions that give rise to metal-based excited states, and intraligand bands involving $\pi \longrightarrow \pi^*$ excitation within the aromatic ligands. Varying the metal ion and the ligand field (for example, by changing ligand substituents or varying the extent of π delocalization) modulates these electronic transitions, resulting in excited states with specific energies.

These low-spin complexes feature singlet ground states, and only excitation to excited singlet states is allowed (spin selection rule, Section 11.3.1). Excitation to these singlet states generally requires higher-energy photons than would be necessary to populate spin forbidden MLCT triplet excited states. In this regard, it is fascinating that these complexes tend to *emit* from their lowest-energy triplet states. A phenomenon called **intersystem crossing**, enhanced by strong spin-orbit coupling due to the heavy metal ions, permits deactivation of the singlet excited state to a lower triplet excited state. This spin-orbit coupling also results in mixing of the excited states to relax selection rules, rendering emission possible between the MLCT triplet excited state and the singlet ground state.[16] One advantage of the intersystem process is that the excited states have sufficiently long lifetimes to allow applications of the excited complexes. It is notable that these excited states often engage in chemistry that is distinctive from the chemistry of the ground state.

Indeed, in addition to metal complex relaxation to the electronic ground state by photon emission, these excited states can be "quenched" by substrates, effecting energy transfer or electron transfer. These quenching pathways are important, because they can "sensitize" other substances to react that could not be effectively activated directly by light. In the case of simple energy transfer, the excited metal complex relaxes to its electronic ground state, and a substrate becomes electronically excited. In the electron transfer processes, the metal complex could be oxidized or reduced resulting in reduction or oxidation, respectively, of the substrate. The resulting activated substrate can then be employed to drive useful reactions, such as those necessary for sensors for glucose and other biologically relevant species,[17] and for the production of hydrogen and oxygen gas from water.[18]

The photolytic splitting of water into hydrogen and oxygen gas is an environmentally benign and cost effective potential fuel source.

$$2\,H_2O(l) \xrightarrow[\textit{catalyst}]{h\nu} 2\,H_2(g) + O_2(g)$$

The reduction of hydrogen ions to hydrogen gas, a key step in this redox process, can be promoted via a sensitization approach that begins with an MLCT event.

Figure 11.17 illustrates a representative process for the catalytic conversion of hydrogen ions to hydrogen gas. It begins (**A**, Figure 11.17) with excitation of the photosensitizer $[Ru(bpy)_3]^{2+}$ to afford $*[Ru(bpy)_3]^{2+}$, which is quenched by the dicationic electron mediator methyl viologen in step **C** to afford a monocation radical. The electron from

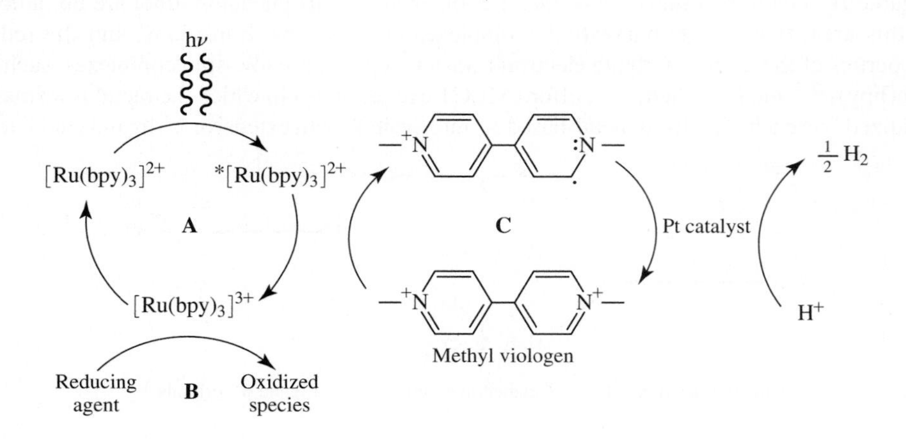

FIGURE 11.17 Photoinduced Formation of H_2 from Protons via a Sensitizer Approach.

$^*[Ru(bpy)_3]^{2+}$ enters the methyl viologen π system. The d^5 $[Ru(bpy)_3]^{3+}$ is converted to d^6 $[Ru(bpy)_3]^{2+}$ via a reducing agent in step **B** (triethylamine and EDTA have been used as reductants). Aromaticity in the monocationic methyl viologen π system is restored via electron transfer to colloidal platinum, and reduction of H^+ to H_2 occurs on the platinum surface. These cycles, initiated by MLCT in $[Ru(bpy)_3]^{2+}$, drive the conversion of hydrogen ions to hydrogen gas. Although this strategy works in solution, the incorporation of the reactants in gels and arranging the key species close together by linking them to polymer chains are approaches targeted to increase the efficiency of hydrogen generation.[19] Iron-based hydrogenase mimics* with tethered zinc-based photosensitizers (see **Figure 11.18** for an example) also facilitate hydrogen ion reduction, but with low catalytic activities. It is clear that coordination complexes with MLCT absorptions will play an important role in efforts to harness solar energy for practical applications.

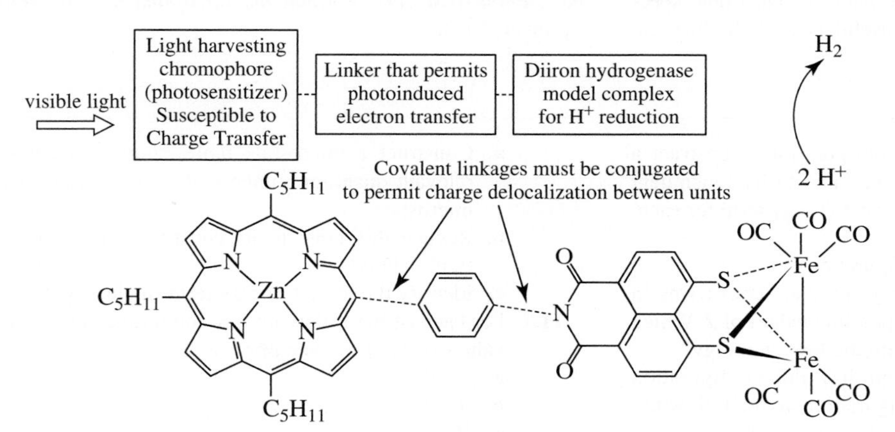

FIGURE 11.18 Schematic of the General Components of a Complex Designed to Drive a Redox Process, and Fragments That Have Been Assembled (Including a Diiron Hydrogenase Mimic and a Zinc Photosensitizer) to Investigate Catalytic Hydrogen Ion Reduction.[20] Reproduced with permission from A. P .S. Samuel, D. T. Co, C. L. Stern, M. R. Wasielewski, *J. Am. Chem. Soc.*, **2010**, *132*, 8813.

References

1. B. N. Figgis and M. A. Hitchman, *Ligand Field Theory and its Applications*, Wiley-VCH, New York, 2000, pp. 181–183.
2. B. N. Figgis, "Ligand Field Theory," in G. Wilkinson, R. D. Gillard, and J. A. McCleverty, eds., *Comprehensive Coordination Chemistry*, Vol. 1, Pergamon Press, Elmsford, NY, 1987, pp. 243–246.
3. F. A. Cotton, *Chemical Applications of Group Theory*, 3rd ed., Wiley InterScience, New York, 1990, Chapter 9, pp. 253–303.
4. B. N. Figgis and M. A. Hitchman, *Ligand Field Theory and Its Applications*, Wiley-VCH, New York, 2000, pp. 128–134.
5. Y. Tanabe, S. Sugano, *J. Phys. Soc. Japan*, **1954**, *9*, 766.
6. B. Bersucker, *Coord. Chem. Rev.*, 1975, *14*, 357.
7. B. N. Figgis, "Ligand Field Theory," in *Comprehensive Coordination Chemistry*, Pergamon Press, Elmsford, NY, 1987, Vol. 1, pp. 252–253.
8. C. J. Ballhausen, *Introduction to Ligand Field Theory*, McGraw-Hill, New York, 1962, p. 227, and references therein.
9. F. A. Cotton and G. Wilkinson, *Advanced Inorganic Chemistry*, 4th ed., Wiley InterScience, New York, 1980, pp. 680–681.
10. N. N. Greenwood and A. Earnshaw, *Chemistry of the Elements*, Pergamon Press, Elmsford, NY, 1984, p. 1161; B. N. Figgis and M. A. Hitchman, *Ligand Field Theory and Its Applications*, Wiley-VCH, New York, 2000, pp. 189–193.
11. B. N. Figgis and M. A. Hitchman, *Ligand Field Theory and Its Applications*, Wiley-VCH, New York, 2000, pp. 208–209.
12. Figgis and Hitchman, *Ligand Field Theory and Its Applications*, pp. 204–207; B. N. Figgis, in G. Wilkinson, R. D. Gillard, and J. A. McCleverty, eds., *Comprehensive Coordination Chemistry*, Vol. 1, Pergamon, Elmsford, NY, 1987, pp. 243–246.
13. K. V. Lamonova, E. S. Zhitlukhina, R. Y. Babkin, S. M. Orel, S. G. Ovchinnikov, Y. G. Pashkevich, *J. Phys. Chem. A*, **2011**, *115*, 13596.
14. Figgis and Hitchman, *Ligand Field Theory and Its Applications*, pp. 211–214; Cotton, *Chemical Applications of Group Theory*, 3rd ed., pp. 295–303.
15. (a) F. Wang, W.-G. Wang, H.-Y. Wang, G. Si, C.-H. Tung, L.-Z. Wu, *ACS Catal.*, **2012**, *2*, 407. (b) J. H. Alstrum-Acevedo, M. K. Brennaman, T. J. Meyer, *Inorg. Chem.*, **2005**, *44*, 6802. (c) T. J. Meyer, *Acc. Chem. Res.*, **1989**, *22*, 163.
16. (a) A. Juris, V. Balzani, F. Barigelletti, S. Campagna, *Coord. Chem. Rev.*, **1988**, *84*, 85. (b) B. Happ, A. Winter, M. D. Hager, U. S. Schubert, *Chem. Soc. Rev.*, **2012**, *41*, 2222.

17. (a) A. Heller, B. Feldman, *Chem. Rev.*, **2008**, *108*, 2482. (b) L. Prodi, F. Bolletta, M. Montalti, N. Zaccheroni, *Coord. Chem. Rev.*, **2000**, *205*, 59.
18. K. Kalyanasundaram, *Coord. Chem. Rev.*, **1982**, *46*, 159.
19. (a) K. Okeyoshi, R. Yoshida, *Chem. Commun.*, **2011**, *47*, 1527. (b) K. Okeyoshi, R. Yoshida, *Soft Matter*, **2009**, *5*, 4118.
20. A. P .S. Samuel, D. T. Co, C. L. Stern, M. R. Wasielewski, *J. Am. Chem. Soc.*, **2010**, *132*, 8813.

General References

B. N. Figgis and M. A. Hitchman, *Ligand Field Theory and Its Applications*, Wiley-VCH, New York, 2000; and B. N. Figgis, "Ligand Field Theory," in G. Wilkinson, R. D. Gillard, and J. A. McCleverty, eds., *Comprehensive Coordination Chemistry*, Vol. 1, Pergamon Press, Elmsford, NY, 1987, pp. 213–280, provide extensive background in the theory of electronic spectra, with numerous examples. Also useful is C. J. Ballhausen, *Introduction to Ligand Field Theory*, McGraw-Hill, New York, 1962. Important aspects of symmetry applied to this topic can be found in F. A. Cotton, *Chemical Applications of Group Theory*, 3rd ed., Wiley InterScience, New York, 1990. An excellent review of photosensitizers can be found in T. Nyokong, V. Ahsen, eds., *Photosensitizers in Medicine, Environment, and Security*, Springer, 2012.

Problems

11.1 For each of the following configurations, construct a microstate table and reduce the table to its constituent free-ion terms. Identify the lowest-energy term for each.
- **a.** p^3
- **b.** p^1d^1 (as in a $4p^1 3d^1$ configuration)

11.2 For each of the lowest-energy (ground state) terms in Problem 11.1, determine the possible values of J. Which J value describes the state with the lowest energy?

11.3 An excited state of calcium has the configuration $[Ar]4s^13d^1$. For an s^1d^1 configuration, do the following:
- **a.** Prepare a microstate table, showing each microstate.
- **b.** Reduce the table to its free ion terms.
- **c.** Determine the lowest-energy term.

11.4 The outer electron configuration of the element cerium is d^1f^1. For this configuration, do the following:
- **a.** Construct a microstate table.
- **b.** Reduce this table to its constituent free-ion terms (with labels).
- **c.** Identify the lowest-energy term (including J value).

11.5 The nitrogen atom is an example of a valence p^3 configuration. There are five energy levels associated with this configuration, with the energies shown here.

Energies (cm^{-1})
28839.31
28838.92
19233.18
19224.46
0

- **a.** Account for these five energy levels.
- **b.** Using information from Section 2.2.3, calculate Π_c and Π_e.

11.6 There is such a thing as an s^1f^1 configuration! This can occur, for example, in an excited state of a Pr^{3+} ion. For an s^1f^1 configuration, do the following:
- **a.** Construct a microstate table, showing clearly the relevant quantum numbers of each electron in each microstate.
- **b.** Reduce this table to its constituent free-ion terms (with labels).
- **c.** Identify the lowest-energy term (including J value).

11.7 For each of the following free-ion terms, determine the values of L, M_L, S, and M_S:
- **a.** $^2D(d^3)$
- **b.** $^3G(d^4)$
- **c.** $^4F(d^7)$

11.8 For each of the free-ion terms in Problem 11.7, determine the possible values of J, and decide which is the lowest in energy.

11.9 The most intense absorption band in the visible spectrum of $[Mn(H_2O)_6]^{2+}$ is at 24,900 cm^{-1} and has a molar absorptivity of 0.038 L mol^{-1} cm^{-1}. What concentration of $[Mn(H_2O)_6]^{2+}$ would be necessary to give an absorbance of 0.10 in a cell of path length 1.00 cm?

11.10
- **a.** Determine the wavelength and frequency of 24,900 cm^{-1} light.
- **b.** Determine the energy and frequency of 366 nm light.

11.11 Determine the ground terms for the following configurations:
- **a.** d^8 (O_h symmetry)
- **b.** high-spin and low-spin d^5 (O_h symmetry)
- **c.** d^4 (T_d symmetry)
- **d.** d^9 (D_{4h} symmetry, square-planar)

11.12 Identify the *first row* transition metal that satisfies the requirements given:
- **a.** $[M(H_2O)_6]^{2+}$ having two unpaired electrons (list all possibilities).
- **b.** $[M(NH_3)_6]^{3+}$ that may have splitting of spin-allowed absorption band as a consequence of distortion of excited state (provide one example).
- **c.** $[M(H_2O)_6]^{2+}$ expected to have the palest color in aqueous solution.

11.13 The spectrum of $[Ni(H_2O)_6]^{2+}$ (Figure 11.8) shows three principal absorption bands, with two of the bands showing signs of further splitting. Referring to the Tanabe–Sugano diagram, estimate the value of Δ_o. Give a likely explanation for the further splitting of the spectrum.

11.14 From the following spectral data, and using Tanabe–Sugano diagrams (Figure 11.7), calculate Δ_o for the following:

a. $[Cr(C_2O_4)_3]^{3-}$, which has absorption bands at 23,600 and 17,400 cm^{-1}. A third band occurs well into the ultraviolet.

b. $[Ti(NCS)_6]^{3-}$, which has an asymmetric, slightly split band at 18,400 cm^{-1}. (Also, suggest a reason for the splitting of this band.)

c. $[Ni(en)_3]^{2+}$, which has three absorption bands: 11,200, 18,350, and 29,000 cm^{-1}.

d. $[VF_6]^{3-}$, which has absorption bands at 14,800 and 23,250 cm^{-1}, plus a third band in the ultraviolet. Also calculate B for this ion.

e. The complex $VCl_3(CH_3CN)_3$, which has absorption bands at 694 and 467 nm. Calculate Δ_o and B for this complex.

11.15 $[Co(NH_3)_6]^{2+}$ has absorption bands at 9,000 and 21,100 cm^{-1}. Calculate Δ_o and B for this ion. (Hints: The ${}^4T_{1g} \longrightarrow {}^4A_{2g}$ transition in this complex is too weak to be observed. The graph in Figure 11.13 may be used for d^7 as well as d^2 complexes.)

11.16 Classify the following configurations as *A*, *E*, or *T* in complexes having O_h symmetry. Some of these configurations represent excited states.

a. $t_{2g}{}^4 e_g{}^2$

b. $t_{2g}{}^6$

c. $t_{2g}{}^3 e_g{}^3$

d. $t_{2g}{}^5$

e. e_g

11.17 Of the first-row transition metal complexes of formula $[M(NH_3)_6]^{3+}$, which metals are predicted by the Jahn–Teller theorem to have distorted complexes?

11.18 MnO_4^- is a stronger oxidizing agent than ReO_4^-. Both ions have charge-transfer bands; however, the charge-transfer band for ReO_4^- is in the ultraviolet, whereas the corresponding band for MnO_4^- is responsible for its intensely purple color. Are the relative positions of the charge-transfer absorptions consistent with the oxidizing abilities of these ions? Explain.

11.19 The complexes $[Co(NH_3)_5X]^{2+}$(X = Cl, Br, I) have charge transfer to metal bands. Which of these complexes would you expect to have the lowest-energy charge-transfer band? Why?

11.20 $[Fe(CN)_6]^{3-}$ exhibits two sets of charge-transfer absorptions, one of lower intensity in the visible region of the spectrum, and one of higher intensity in the ultraviolet. $[Fe(CN)_6]^{4-}$, however, shows only the high-intensity charge transfer in the ultraviolet. Explain.

11.21 The complexes $[Cr(O)Cl_5]^{2-}$ and $[Mo(O)Cl_5]^{2-}$ have C_{4v} symmetry.

a. Use the angular overlap approach (Chapter 10) to estimate the relative energies of the *d* orbitals in these complexes.

b. Using the C_{4v} character table, determine the symmetry labels (labels of irreducible representations) of these orbitals.

c. The ${}^2B_2 \longrightarrow {}^2E$ transition occurs at 12,900 cm^{-1} for $[Cr(O)Cl_5]^{2-}$ and at 14,400 cm^{-1} for $[Mo(O)Cl_5]^{2-}$. Account for the higher energy for this transition in the molybdenum complex. (See W. A. Nugent and J. M. Mayer, *Metal-Ligand Multiple Bonds*, John Wiley & Sons, New York, 1988, pp. 33–35.)

11.22 For the isoelectronic series $[V(CO)_6]^-$, $Cr(CO)_6$, and $[Mn(CO)_6]^+$, would you expect the energy of metal to ligand charge-transfer bands to increase or decrease with increasing charge on the complex? Why? (See K. Pierloot, J. Verhulst, P. Verbeke, L. G. Vanquickenborne, *Inorg. Chem.*, **1989**, *28*, 3059.)

11.23 The compound *trans*-Fe(*o*-phen)$_2$ (NCS)$_2$ has a magnetic moment of 0.65 Bohr magneton at 80 K, increasing with temperature to 5.2 Bohr magnetons at 300 K.

a. Assuming a spin-only magnetic moment, calculate the number of unpaired electrons at these two temperatures.

b. How can the increase in magnetic moment with temperature be explained? (Hint: There is also a significant change in the UV-visible spectrum with temperature.)

11.24 The absorption spectrum of the linear ion NiO_2^{2-} has bands attributed to *d–d* transitions at approximately 9,000 and 16,000 cm^{-1}.

a. Using the angular overlap model (Chapter 10), predict the expected splitting pattern of the *d* orbitals of nickel in this ion.

b. Account for the two absorption bands.

c. Calculate the approximate value of e_σ and e_π. (See M. A. Hitchman, H. Stratemeier, R. Hoppe, *Inorg. Chem.*, **1988**, *27*, 2506.)

11.25 The electronic absorption spectra of a series of complexes of formula $Re(CO)_3(L)(DBSQ)$ [DBSQ = 3,5-di*tert*-butyl-1,2-benzosemiquinone] show a single maximum in the visible spectrum. The absorption maxima for three of these complexes in benzene solution are shown; typical molar absorptivities are in the range of 5,000 to 6,000 L mol^{-1} cm^{-1}.

L	ν_{max}, cm^{-1}
$P(OPh)_3$	18,250
PPh_3	17,300
NEt_3	16,670

Are these bands more likely due to charge transfer to metal or charge transfer to ligand? Explain briefly. (See F. Hartl, A. Vlcek, Jr., *Inorg. Chem.*, **1996**, *35*, 1257.)

11.26 The d^2 ions CrO_4^{4-}, MnO_4^{3-}, FeO_4^{2-}, and RuO_4^{2-} have been reported.

a. Which of these has the largest value of Δ_t? Which has the smallest? Explain briefly.

b. Of the first three, which ion has the shortest metal–oxygen bond distance? Explain briefly.

c. The charge-transfer transitions for the first three complexes occur at 43,000, 33,000, and 21,000 cm^{-1}, respectively. Are these more likely to be ligand-to-metal or metal-to-ligand charge-transfer transitions? Explain briefly. (See T. C. Brunold, H. U. Güdel, *Inorg. Chem.*, **1997**, *36*, 2084.)

11.27 An aqueous solution of $Ni(NO_3)_2$ is green. Addition of aqueous NH_3 causes the color of the solution to change to blue. If ethylenediamine is added to the green solution, the color changes to violet. Account for the colors of these complexes. Are they consistent with the expected positions of these ligands in the spectrochemical series?

11.28 The pertechnetate ion, TcO_4^-, is often used to introduce the radioactive Tc into compounds, some of which are used as medical tracers. Unlike the isoelectronic, vividly purple permanganate ion, pertechnetate is very pale red.

a. Describe the most likely absorption that gives rise to colors in these ions. In addition to a written description, your answer should include an energy-level sketch showing how *d* orbitals on the metals interact with oxide orbitals to form molecular orbitals.

b. Suggest why TcO_4^- is red but MnO_4^- is purple.

c. The manganate ion, MnO_4^{2-}, is green. On the basis of your answers to a and b, provide an explanation for this color.

11.29 A 2.00×10^{-4} M solution of $Fe(S_2CNEt_2)_3$ ($Et = C_2H_5$) in $CHCl_3$ at 25° C has absorption bands at 350 nm ($A = 2.34$), 514 nm ($A = 0.532$), 590 nm ($A = 0.370$), and 1540 nm ($A = 0.0016$).

a. Calculate the molar absorptivity for this compound at each wavelength.

b. Are these bands more likely due to *d–d* transitions or charge-transfer transitions? Explain.

11.30 Use the following spectral data to find the term symbols for the ground and excited states of each species, and calculate Δ_o and the Racah parameter, *B*, for each.

Species	Absorption Bands (cm^{-1})		
$[Ni(H_2O)_6]^{2+}$	8,500	15,400	26,000
$[Ni(NH_3)_6]^{2+}$	10,750	17,500	28,200
$[Ni(OS(CH_3)_2)_6]^{2+}$	7,728	12,970	24,038
$[Ni(dma)_6]^{2+}$	7,576	12,738	23,809

11.31 For compounds $[Co(bipy)_3]^{2+}$ and $[Co(NH_3)_6]^{2+}$, do the following:

a. Find the ground-state term symbol.

b. Use the Tanabe–Sugano diagram to identify the predicted spectral bands.

c. Calculate the ligand field stabilization energy.

d. Do you expect broad or narrow absorption bands in the visible and UV regions?

e. Sketch an MO energy level diagram for each.

	$\nu_1(cm^{-1})$	$\nu_3(cm^{-1})$
$[Co(bipy)_3]^{2+}$	11,300	22,000
$[Co(NH_3)_6]^{2+}$	9,000	21,100

11.32 In the complexes $FeL(SC_6H_5)$ and $NiL(SC_6H_5)$, where L = hydrotris(3,5-diisopropylpyrazolylborate) $HB(3,5\text{-}i\text{-}Pr_2pz)_3^-$, strong charge-transfer bands were observed in the regions 28,000 to 32,500 and 20,100 to 30,000 cm^{-1}, respectively. Were these more likely LMCT or MLCT bands? Explain, taking into account the relative energies of the metal orbitals in these complexes. (See S. I. Gorelsky, L. Basumallick, J. Vura-Weis, R. Sarangi, K. O. Hodgson, B. Hedman, K. Fujisawa, E. I. Solomon, *Inorg. Chem.*, **2005**, *44*, 4947.)

11.33 Various models for the iron hydrogenase active site covalently linked to a ruthenium photosensitizer have been synthesized as candidates for light-driven proton reduction. A prototype for this class of complexes features a phenylacetylene linker between the iron and ruthenium portions of the molecule (S. Ott, M. Borgström, M. Kritikos, R. Lomoth, J. Bergquist, B. Åkermark, L. Hammarström, L. Sun, *Inorg. Chem.*, **2004**, *43*, 4683). Sketch the target complex. Provide three reasons for inclusion of the phenylacetylene linker, and assign the two most intense bands in its electronic absorption spectrum (solid line in spectrum). Briefly explain why this complex is unable to induce proton reduction.

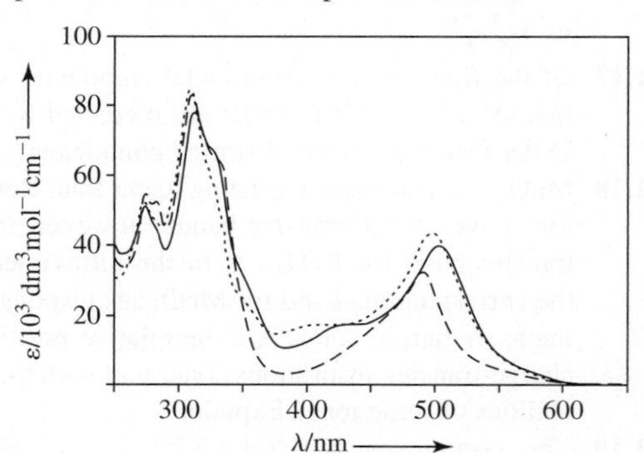

11.34 The complex in Figure 11.18 successfully induces photoinduced proton reduction, but with very low activities. Provide two reasons why using naphthalene monoimide dithiolates as the linker between the photosensitizer and the active site for proton reduction was considered desirable. What spectral argument was used to reason that the ground state zinc porphyrin moiety does not interact electronically with the diiron portion of this complex?

Chapter 12

Coordination Chemistry IV: Reactions and Mechanisms

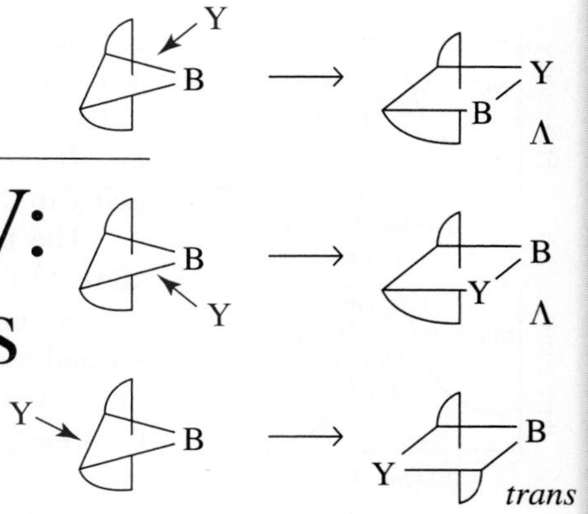

Reactions of coordination compounds are understood using the same fundamental concepts applied to all reactions. The chemistry of coordination compounds is distinctive because (1) the complexes have a relatively large diversity of geometries and more possibilities for rearrangement, and (2) metal atoms impose significant variability on the reactivity of their complexes.

Reactions of coordination complexes can be divided into (1) substitution at the metal center, (2) oxidation–reduction, and (3) reactions of the ligands that do not change the attachments to the metal center. Reactions that include more elaborate rearrangements of ligand structures are described in Chapter 14 within the context of organometallic chemistry.

12.1 Background

Although early chemists did not know the structures of the coordination compounds they worked with, they did learn how to synthesize complexes containing metals. Werner and Jørgensen prepared specific complexes to test their hypotheses about coordination geometries. These chemists initiated the contemporary strategy of designing reactions to prepare complexes with specific structures, properties, and applications. Although the ability to devise appropriate reaction conditions to obtain desired products is still an intellectual challenge, the list of known reactions is now long enough to provide considerable guidance.

The fundamental reason for our examination of reaction kinetics and mechanisms in Chapter 12 is to understand how coordination complexes can interact with other species to affect chemical change. The information gleaned from these studies allows more rational design of reactions to optimize yields and curb the formation of unwanted by-products.

We will first review the background of reaction mechanisms, then consider the major categories of such mechanisms, and finally describe the results of mechanistic studies.

Transition-state theory describes chemical reactions as moving from one energy minimum (the reactants) through higher energy structures (transition states, intermediates) to another energy minimum (the products). It is common to depict the free energy changes that accompany reactions via a reaction coordinate diagram, where free energy is plotted on the *y*-axis, and progression along the *x*-axis depicts the extent of the reaction, going in either direction. In the generic substitution $MX + Y \longrightarrow MY + X$, the diagram begins at the free energy of reactants MX and Y as separated species. As these species encounter each other and react, the free energies of the resulting structures change as the M—X distance lengthens (the bond breaks) and the M—Y distance shortens (the new bond forms). Reaction coordinate diagrams usually exhibit a saddle shape, much like a mountain pass between two valleys. Although the complexity of reaction

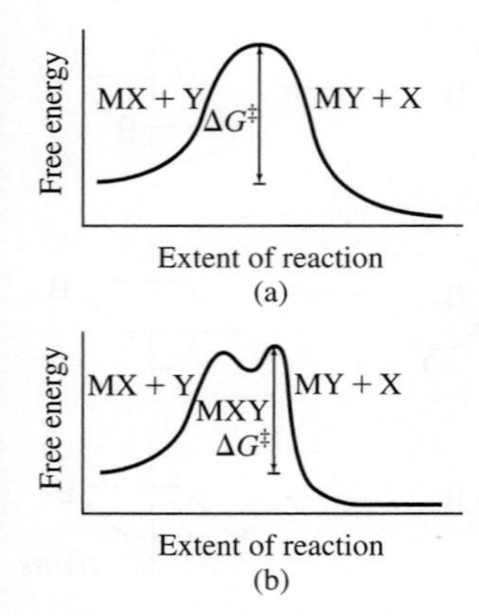

FIGURE 12.1 Energy Profiles and Intermediate Formation. (a) No intermediate. The activation energy is the energy difference between the reactants and the transition state. (b) An intermediate is present at the small minimum at the top of the curve. The activation energy is measured at the maximum point of the curve.

coordinate diagrams can vary widely, the adopted path between the reactants and the products is always the lowest energy pathway available and must be the same regardless of the direction of the reaction. This is the **principle of microscopic reversibility**; the lowest energy pathway going in one direction must also be the lowest energy pathway going in the opposite direction.

The highest energy structure along the reaction pathway is called the **transition state**. In **Figure 12.1a**, the reaction proceeds via a transition state without any structures at local energy minima. In Figure 12.1b, such a structure, called an **intermediate**, is formed along the reaction pathway. Intermediates, unlike transition states, are sometimes detectable. The presence of undetectable intemediates can allow application of the **steady-state approximation** during analysis of the reaction kinetics (Section 12.3.1), in which the concentration of the intermediate is assumed to be extremely small and essentially unchanging during much of the reaction.

Kinetics experiments are carried out to determine a number of parameters that link to the reaction mechanism. The **order** of each reactant, indicated by the power of the reactant concentration in the differential equation that describes how its concentration changes with time, indicates how the reaction rate is tied to a change in that reactant's concentration. The **rate constant**, a proportionality constant that relates the reaction rate to the concentrations of the reactants, is temperature dependent. By studying a reaction at different temperatures, the **free energy of activation** and its components, **enthalpy** and **entropy of activation**, can be found. These parameters permit further hypotheses regarding the mechanism and energy changes along the reaction pathway. Examination of pressure dependence on reaction rates provides the **volume of activation**, which offers insight into whether the transition state is larger or smaller than the reactants.

Even for thermodynamically favorable reactions ($\Delta G^\circ < 0$), a large activation energy means that the reaction will be slow. For thermodynamically unfavorable reactions ($\Delta G^\circ > 0$), even a fast reaction (with small activation energy) is unlikely to occur. The rate of reaction depends on the activation energy, as in the Arrhenius equation:

$$k = Ae^{-\frac{E_A}{RT}} \quad \text{or} \quad \ln k = \ln A - \frac{E_A}{RT}$$

Three generic reaction coordinate diagrams are in **Figure 12.2**. In (a) and (b), the reactions have large and positive equilibrium constants since $\Delta G < 0$; both reactions are spontaneous. However, in (a), E_A is large, so the reaction is slow. Reaction (b) features an intermediate at the energy minimum near the top of the curve. In (c), the reaction can occur quickly because of the low activation energy, but it has a small equilibrium constant because $\Delta G > 0$.

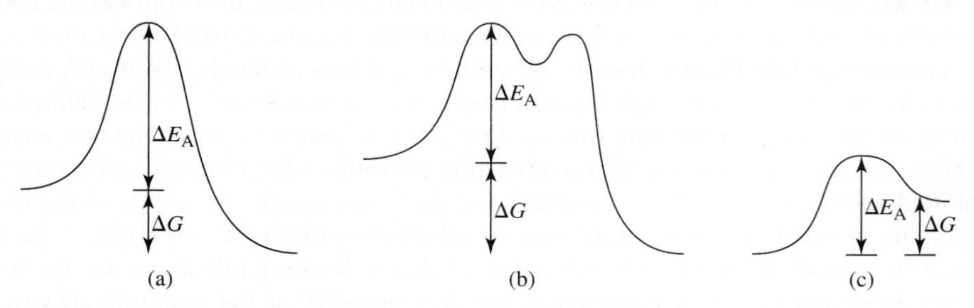

FIGURE 12.2 Reaction Coordinate Diagrams. (a), (b) Large activation energy hinders the reaction rate even though the equilibrium favors the products since $\Delta G < 0$; (c) Smaller activation energy facilitates a faster reaction, but the equilibrium favors the reactants since $\Delta G > 0$. In (b), the intermediate is potentially detectable.

12.2 Substitution Reactions

Ligand substitution is an important step in many reactions of coordination complexes. These reactions have been the subject of extensive mechanistic and kinetic studies.

12.2.1 Inert and Labile Compounds

Many reactions require substitution, replacing one ligand by another. A well-studied class of substitution reactions involves aqueous metal ions ($[M(H_2O)_m]^{n+}$) as reactants. These reactions can produce colored products used to identify metal ions:

$$\underset{\text{green}}{[Ni(H_2O)_6]^{2+}} + 6\,NH_3 \rightleftharpoons \underset{\text{blue}}{[Ni(NH_3)_6]^{2+}} + 6\,H_2O$$

$$\underset{\text{very pale violet}}{[Fe(H_2O)_6]^{3+}} + SCN^- \rightleftharpoons \underset{\text{red}}{[Fe(H_2O)_5(SCN)]^{2+}} + H_2O$$

Ligand substitution reactions involving $[M(H_2O)_m]^{n+}$ are rapid and generally form species that also undergo fast reactions. Addition of HNO_3, NaCl, H_3PO_4, KSCN, and NaF successively to a solution of $Fe(NO_3)_3 \cdot 9\,H_2O$ provides a classic example. The initial solution is yellow because of the presence of $[Fe(H_2O)_5(OH)]^{2+}$ and other Fe(III) complexes containing water and hydroxide ligands derived from the hydrolysis of $[Fe(H_2O)_6]^{3+}$. Although the exact complexes formed and their equilibrium concentrations depend on the concentrations of the ions involved, these products are representative:

$$\underset{\text{yellow}}{[Fe(H_2O)_5(OH)]^{2+}} + H^+ \longrightarrow \underset{\text{colorless (very pale violet)}}{[Fe(H_2O)_6]^{3+}}$$

$$[Fe(H_2O)_6]^{3+} + Cl^- \longrightarrow \underset{\text{yellow}}{[Fe(H_2O)_5(Cl)]^{2+}} + H_2O$$

$$[Fe(H_2O)_5(Cl)]^{2+} + PO_4^{\,3-} \longrightarrow \underset{\text{colorless}}{Fe(H_2O)_5(PO_4)} + Cl^-$$

$$Fe(H_2O)_5(PO_4) + SCN^- \longrightarrow \underset{\text{red}}{[Fe(H_2O)_5(SCN)]^{2+}} + PO_4^{\,3-}$$

$$[Fe(H_2O)_5(SCN)]^{2+} + F^- \longrightarrow \underset{\text{colorless}}{[Fe(H_2O)_5(F)]^{2+}} + SCN^-$$

While the fate of these reactions is partially governed by the relative bond strengths between Fe(III) and the incoming and departing ligands, the examination of water exchange reactions,

$$[M(H_2O)_m]^{n+} + H_2{}^{18}O \rightleftharpoons [M(H_2O)_{m-1}(H_2{}^{18}O)]^{n+} + H_2O$$

is insightful since the bonds broken and made during the substitution have essentially identical strengths.* Water exchange rate constants (**Table 12.1**) vary widely as a function of the metal ion.

Rate constants like those in Table 12.1 are known for many substitution reactions, and general trends in the speeds of these reactions have been correlated to the electronic configuration of the starting complex. The rate constants for water exchange differ by more than 13 orders of magnitude for $[Cr(H_2O)_6]^{3+}$ and $[Cr(H_2O)_6]^{2+}$! It is also intriguing that $[V(H_2O)_6]^{3+}$ undergoes water exchange roughly 6 times faster than does $[V(H_2O)_6]^{2+}$,

---

*$H_2{}^{17}O$ can also be used.

TABLE 12.1 Rate Constants for Water Exchange in $[M(H_2O)_6]^{n+}$

Complex	$k(s^{-1})$ (298 K)	Electronic Configuration*
$[Ti(H_2O)_6]^{3+}$	1.8×10^5	$t_{2g}{}^1$
$[V(H_2O)_6]^{3+}$	5.0×10^2	$t_{2g}{}^2$
$[V(H_2O)_6]^{2+}$	8.7×10^1	$t_{2g}{}^3$
$[Cr(H_2O)_6]^{3+}$	2.4×10^{-6}	$t_{2g}{}^3$
$[Cr(H_2O)_6]^{2+}$	$> 10^8$	$t_{2g}{}^3e_g{}^1$
$[Fe(H_2O)_6]^{3+}$	1.6×10^2	$t_{2g}{}^3e_g{}^2$
$[Fe(H_2O)_6]^{2+}$	4.4×10^6	$t_{2g}{}^4e_g{}^2$
$[Co(H_2O)_6]^{2+}$	3.2×10^6	$t_{2g}{}^5e_g{}^2$
$[Ni(H_2O)_6]^{2+}$	3.2×10^4	$t_{2g}{}^6e_g{}^2$
$[Cu(H_2O)_6]^{2+}$	4.4×10^9	$t_{2g}{}^6e_g{}^3$
$[Zn(H_2O)_6]^{2+}$	$> 10^7$	$t_{2g}{}^6e_g{}^4$

*These configurations assume octahedral geometry, even in cases where Jahn–Teller distortion is anticipated.
Data from R. B. Jordan, *Reaction Mechanisms of Inorganic and Organometallic Systems*, 3rd ed., Oxford (New York), 2007, p. 84.

even though ligand loss from a V(III) ion might be expected to be more difficult than from a V(II) ion on the basis of an electrostatic argument. Complexes such as $[Cr(H_2O)_6]^{2+}$, which react rapidly, essentially exchanging one ligand for another within the time of mixing the reactants, are classified as **labile**. A labile complex has a very low activation energy for ligand substitution. Taube[1] suggested a reaction half-life (the time it takes for the concentration of the initial compound to decrease by one half) of one minute or less as the criterion for lability. Compounds that react more slowly are called **inert**. Within this context, an inert compound does not resist ligand substitution; it is simply slower to react, with a higher activation energy for ligand substitution. These kinetic terms must be distinguished from the thermodynamic descriptions **stable** and **unstable**. A species such as $[Fe(H_2O)_5(F)]^{2+}$ is very stable—it has a large equilibrium constant for formation—but it is also labile. On the other hand, hexaaminecobalt (3+) is thermodynamically unstable in acid:

$$[Co(NH_3)_6]^{3+} + 6\,H_3O^+ \longrightarrow [Co(H_2O)_6]^{3+} + 6\,NH_4{}^+ \qquad (\Delta G^\circ < 0)$$

But $[Co(NH_3)_6]^{3+}$ reacts very slowly and is therefore *inert* (the activation energy for the above reaction is high). Both thermodynamic and kinetic parameters are necessary to understand chemical behavior.* For clarity, complexes can be described as **kinetically labile** or **inert**.

Werner studied Co(III), Cr(III), Pt(II), and Pt(IV) compounds because they are inert and more readily characterized than labile compounds. Much of this chapter discusses inert compounds. Labile compounds have been studied extensively, but their study requires techniques capable of dealing with complexes with very short lifetimes.

As suggested by Table 12.1, general rules can be given for inert and labile electronic structures. Inert octahedral complexes are generally those with high ligand field stabilization energies (Section 10.3.3), specifically those with d^3 or low-spin d^4 through d^6

*A classic example of this vital distinction involves the graphite and diamond allotropes of carbon. Under ambient conditions, graphite is more thermodynamically stable than diamond ($\Delta G^\circ < 0$ for diamond $\longrightarrow$ graphite), yet the rate constant for this conversion is extremely small, hence the (kinetically, inspired) adage "diamonds are forever."

TABLE 12.2 Classification of Substitution Mechanisms for Octahedral Complexes

	Stoichiometric Mechanism		
Intimate Mechanism	Dissociative 5-Coordinate Intermediate for Octahedral Reactant		Associative 7-Coordinate Intermediate for Octahedral Reactant
Dissociative activation	D	I_d	
Associative activation		I_a	A

electronic structures. These configurations have no electrons in the e_g orbitals, and at least one electron in each t_{2g} orbital. The d^3 inert classification is exhibited in Table 12.1; $[Cr(H_2O)_6]^{3+}$ undergoes water exchange exceedingly slowly relative to the high-spin d^4 $[Cr(H_2O)_6]^{2+}$, and $[V(H_2O)_6]^{2+}$ reacts slower than $[V(H_2O)_6]^{3+}$. With strong-field ligands, d^8 atoms often form inert square-planar complexes. Compounds with other d configurations tend to be labile, with a wide range of substitution reaction rate constants.

Slow Reactions (Inert)	Moderate Rate	Fast Reactions (Labile)
d^3, low-spin d^4, d^5, and d^6		d^1, d^2, high-spin d^4, d^5, and d^6
Strong-field d^8 (square planar)	Weak-field d^8	d^7, d^9, d^{10}

12.2.2 Mechanisms of Substitution

Langford and Gray[2] described a range of possibilities for substitution reactions (**Table 12.2**). At one extreme, the departing ligand leaves, and an intermediate with a lower coordination number is formed, a mechanism labeled ***D*** for **dissociation**. At the other extreme, the incoming ligand adds to the complex, and an intermediate with an increased coordination number is formed in a mechanism labeled ***A*** for **association**. Between the two extremes is **interchange, *I***, in which the incoming ligand assists in the reaction, but no detectable intermediates appear. When the degree of assistance is small and the reaction is primarily dissociative, it is called **dissociative interchange, I_d**. When the incoming ligand begins forming a bond to the central atom before the departing ligand bond is weakened appreciably, it is called **associative interchange, I_a**. Many reactions are described by I_a or I_d mechanisms, rather than by A or D, when the kinetic evidence points to association or dissociation, but detection of intermediates is not possible. The categories D, A, and I are called the **stoichiometric mechanisms**; the distinction between activation processes that are associative and dissociative is called the **intimate mechanism**. The similarities in the energy profiles for associative and dissociative reactions (**Figure 12.3**) show that unambiguously distinguishing between these mechanisms can be challenging.

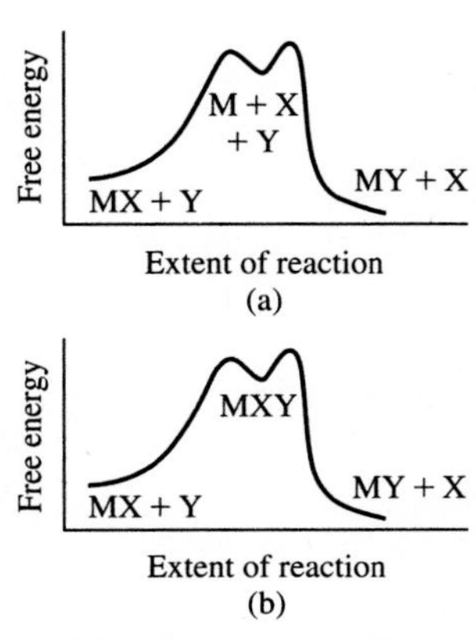

FIGURE 12.3 Energy Profiles for Dissociative and Associative Reactions. (a) Dissociative mechanism. The intermediate has a lower coordination number than the reactant. (b) Associative mechanism. The intermediate has a higher coordination number than the reactant.

12.3 Kinetic Consequences of Reaction Pathways

This chapter describes examples in which the rate law is used to propose reaction mechanisms. We provide two types of information: (1) the information used to propose mechanisms and (2) specific reactions for which mechanisms are known with fairly high levels of confidence. The first is necessary to critically examine data for other reactions. The second is helpful since it forms a knowledge base to shed light on new reactions. Each substitution mechanism, D, I, and A, will be described with its rate law.*

*In the reactions of this chapter, X will indicate the ligand that is leaving a complex, Y the ligand that is entering, and L any ligands that are unchanged during the reaction. In cases of solvent exchange, the X, Y, and L may be the same species. Charges will be omitted when using X, Y, and L, but these species may be ions. The general examples will usually be 6-coordinate; other coordination numbers can be treated similarly.

12.3.1 Dissociation (*D*)

In a dissociative (*D*) reaction, the first step is loss of a ligand to form an intermediate with a lower coordination number. Subsequent additions of either a new ligand (Y) or the leaving group (X) are two possible reaction pathways for this intermediate:

$$ML_5X \underset{k_{-1}}{\overset{k_1}{\rightleftharpoons}} ML_5 + X$$

$$ML_5 + Y \xrightarrow{k_2} ML_5Y$$

Examination of the possibility of a *D* mechanism typically requires the *steady-state approximation.* This approximation assumes that a vanishingly small (and constant) concentration of the intermediate, ML_5, is present during the reaction by assuming that the rates of formation and consumption of the intermediate are equal. If these rates are the same, the change in the $[ML_5]$ must equal zero (and this species cannot accumulate) during the reaction. Expressed as a rate equation,

$$\frac{d[ML_5]}{dt} = k_1[ML_5X] - k_{-1}[ML_5][X] - k_2[ML_5][Y] = 0$$

Solving for $[ML_5]$,

$$[ML_5] = \frac{k_1[ML_5X]}{k_{-1}[X] + k_2[Y]}$$

and substituting into the rate law for formation of the product,

$$\frac{d[ML_5Y]}{dt} = k_2[ML_5][Y]$$

leads to the rate law:

$$\frac{d[ML_5Y]}{dt} = \frac{k_2k_1[ML_5X][Y]}{k_{-1}[X] + k_2[Y]}$$

Evidence that the conditions associated with the steady-state approximation are present, and that a *D* mechanism may be operative, includes an inverse dependence between the rate of formation of $[ML_5Y]$ and the concentration of X as indicated in the rate law above. This derivation predicts that the presence of X should reduce the rate of formation of $[ML_5Y]$, as long as the intermediate ML_5 reacts with X at a rate similar to or greater than Y reacts with ML_5.

A *D* mechanism on the basis of the derived rate law will exhibit a complicated dependence upon [X] and [Y] which has two limiting cases described below. Studies that systematically vary the concentrations of both X and Y provide the best evidence for a dissociative mechanism. At high [Y], the system will show saturation kinetics in which the reaction rate depends only upon $[ML_5X]$.

$$\frac{d[ML_5Y]}{dt} = \frac{k_2k_1[ML_5X][Y]}{k_{-1}[X]} \quad and \quad \frac{d[ML_5Y]}{dt} = k_1[ML_5X]$$

If $[X] \gg [Y]$ so that $k_{-1}[X] \gg k_2[Y]$ If $[Y] \gg [X]$ so that $k_2[Y] \gg k_{-1}[X]$

While experimental data that exhibit these relationships support both the validity of the steady-state approximation and a *D* substitution mechanism, the exceedingly low $[ML_5]$ concentration that forms on the basis of this approximation renders the detection of this intermediate impossible in many cases.

12.3.2 Interchange (*I*)

An interchange (*I*) reaction in its simplest form is a direct replacement of the leaving group with the incoming group that does not proceed via an intermediate, but rather a single transition state leading to the conversion of reactants to products.

$$ML_5X + Y \xrightarrow{k_1} ML_5Y + X$$

If the substitution reaction is irreversible, an interchange mechanism will exhibit second order kinetics, being first order in $[ML_5X]$ and first order in $[Y]$.

$$\frac{d[ML_5Y]}{dt} = k_1[ML_5X][Y]$$

If the interchange reaction is reversible, a more complicated treatment is required to deal with the approach towards the resulting equilibrium. A common experimental condition is to employ high $[Y]$ and $[X]$ to approximate the reaction as a pair of opposing pseudo-first order reactions.

$$ML_5X + Y \underset{k_{-1}}{\overset{k_1}{\rightleftharpoons}} ML_5Y + X$$

$$-\frac{d[ML_5X]}{dt} = \frac{d[ML_5Y]}{dt} = k_1[ML_5X] - k_{-1}[ML_5Y]$$

If $[X]$ and $[Y]$ are large

Espenson has provided further details regarding this kinetics treatment.* The *I* and *D* mechanisms can be differentiated in principle since increasing $[Y]$ will not lead to saturation kinetics with *I* as predicted when a *D* mechanism is operative.

Two variations on the interchange mechanism are I_d, *dissociative interchange*, and I_a, *associative interchange*. The difference is the relative strengths of the M—X and M—Y bonds in the transition state. If bonding between the incoming ligand and the metal is more important in the transition state, it is an I_a mechanism. If breaking the bond between the departing ligand and the metal is more important in the transition state, it is an I_d mechanism.

12.3.3 Association (*A*)

In an associative reaction, forming an intermediate with an increased coordination number is the rate-determining step. This first step is followed by a faster reaction in which the exiting ligand is lost:

$$ML_5X + Y \underset{k_{-1}}{\overset{k_1}{\rightleftharpoons}} ML_5XY$$

$$ML_5XY \xrightarrow{k_2} ML_5Y + X$$

The steady-state approximation, which assumes $[ML_5XY]$ to be exceedingly small, results in second-order kinetics regardless of the concentration of Y:

$$\frac{d[ML_5Y]}{dt} = \frac{k_1k_2[ML_5X][Y]}{k_{-1} + k_2} = k[ML_5X][Y]$$

EXERCISE 12.1

Show that the preceding equation is the result of the steady-state approximation for an associative reaction.

*The determination of the equilibrium constant ($K = \frac{k_1}{k_{-1}}$) aids these studies as described in J. H. Espenson, *Chemical Kinetics and Reaction Mechanisms,* McGraw-Hill, New York, 1981, pp. 45–48.

Because second order kinetics are also consistent with an irreversible interchange reaction, unambiguous assignment of an A mechanism requires detection of the $[ML_5XY]$ intermediate. This detection is generally extremely challenging on the basis of the anticipated low concentration of $[ML_5XY]$ during the reaction.

12.3.4 Preassociation Complexes

As mentioned above, an aspect that complicates mechanistic studies is when multiple pathways result in similar reaction kinetics, rendering pathways difficult to distinguish from each other. One such complication is when a rapid equilibrium occurs between the incoming ligand and the 6-coordinate reactant to form an ion pair or preassociation complex.* This species then reacts to form product and release the initial ligand.

$$ML_5X + Y \underset{k_{-1}}{\overset{k_1}{\rightleftharpoons}} ML_5X \cdot Y$$

$$ML_5X \cdot Y \xrightarrow{k_2} ML_5Y + X$$

The relative magnitudes of these rate constants lead to different scenarios. If k_1 and k_{-1} are both large with respect to k_2, the first step can be considered an equilibrium (with $K_1 = k_1/k_{-1}$) that can be treated independently from the second step. The intermediate preassociation complex may be detectable on the basis of the relative magnitudes of k_1 and k_{-1}. Detection of this intermediate provides supporting evidence for this substitution mechanism, and may allow calculation of K_1.

In the event that the intermediate is undetectable (k_2 is comparable in magnitude to k_1 and k_{-1}), but the rapid preequilibrium still holds, applying the steady-state approximation to $[ML_5X \cdot Y]$ affords the relationship

$$\frac{d[ML_5X \cdot Y]}{dt} = k_1[ML_5X][Y] - k_{-1}[ML_5X \cdot Y] - k_2[ML_5X \cdot Y] = 0$$

If $[Y]_0 \cong [Y]$ during the reaction (a very high [Y] relative to $[ML_5X]$ is employed), the formation of the preassociation complex may be large enough to significantly change $[ML_5X]$ but not [Y]. In this case, it is useful to express $[ML_5X]$ in terms of the total initial metal reactant concentration $[M]_0$:

$$[M]_0 = [ML_5X] + [ML_5X \cdot Y]$$

Substitution into the steady-state approximation equation above provides

$$k_1([M]_0 - [ML_5X \cdot Y])[Y]_0 - k_{-1}[ML_5X \cdot Y] - k_2[ML_5X \cdot Y] = 0$$

The final rate equation after rearrangement and substitution of $K_1 = \frac{k_1}{k_{-1}}$ becomes

$$\frac{d[ML_5Y]}{dt} = k_2[ML_5X \cdot Y] = \frac{k_2\, K_1[M]_0[Y]_0}{1 + K_1[Y]_0 + (k_2 / k_{-1})} \cong \frac{k_2\, K_1[M]_0[Y]_0}{1 + K_1[Y]_0}$$

since k_2 / k_{-1} is often small relative to the other terms in the denominator.

K_1 can be estimated theoretically if the proposed $[ML_5X \cdot Y]$ is undetectable. While these calculations are beyond the scope of this text, the two terms in the denominator of this simplified rate law, 1 and $K_1[Y]_0$, can have the same magnitude. At high [Y], the above rate law simplifies to

$$\frac{d[ML_5Y]}{dt} \cong k_2[M]_0$$

*While an ion pair is relatively straightforward to imagine between oppositely charged species, the driving force for the formation of a preassociation complex involving neutral species can be dipole-dipole interactions.

which exhibits the same relationship as a D mechanism when [Y] is very large, rendering mechanisms involving preassociation complexes difficult to distinguish from D mechanisms. In this case, it is important to be able to control the concentration of the leaving group X to ascertain its impact on the rate.

12.4 Experimental Evidence in Octahedral Substitution

We will now consider additional evidence used to support mechanistic arguments.

12.4.1 Dissociation

In the event of a dissociative mechanism, an octahedral complex loses one ligand (X) to yield a 5-coordinate transition state, and the incoming ligand ultimately fills the vacant site to form the octahedral product. The inert and labile classifications (Section 12.2.1) were rationalized partially from ligand field theory, based on calculation of the change in LFSE between the octahedral reactant and the presumed 5-coordinate (square-pyramidal or trigonal-bipyramidal) transition state. The challenge is to assess how the molecular orbitals with high d orbital character will split upon a change in geometry from O_h to C_{4v}. The splitting of these levels is a function of the ligands and the extent with which the four ligands that comprise the square of the pyramid bend away from the axial ligand. **Figure 12.4** provides a commonly adopted possibility.[3]

Table 12.3 gives the **ligand field activation energy** (LFAE), defined as the difference between the LFSE of the square-pyramidal transition state and the LFSE of the octahedral reactant. LFAEs for trigonal-bipyramidal transition states are generally the same or larger than those for square-pyramidal transition states. These calculations give insight in regard to which electronic configurations in octahedral complexes result in lower barriers for ligand dissociation. While other factors are required to understand the kinetics of dissociative substitution reactions, the LFAE values correlate reasonably well with the Section 12.1.1 classifications. The highest (most positive) LFAE values are associated with d^3 and low-spin d^4, d^5, and d^6; all of these ions are classified as inert, supported by the predicted relatively high activation barriers for ligand dissociation from octahedral complexes of these ions. The negative LFAE parameters predict low activation barriers for dissociation, and are associated with ions classified as labile. This approach fares poorly with d^8 ions, but these ions generally do not form octahedral complexes.

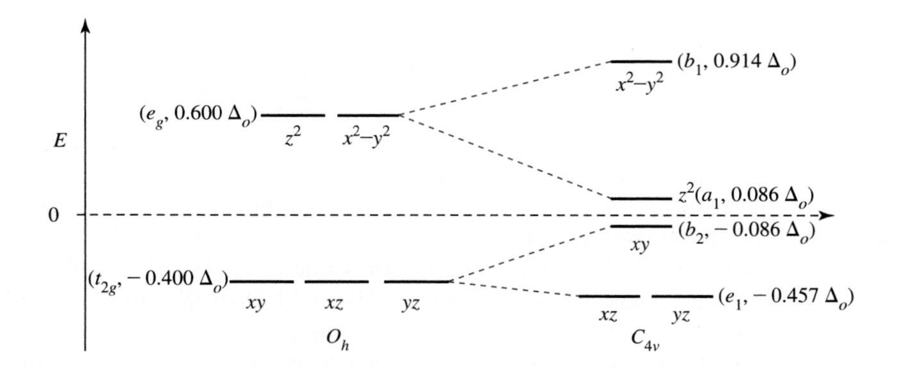

FIGURE 12.4 Change in energy levels (in terms of Δ_o) upon ligand dissociation from octahedral ML_6 to square-pyramidal ML_5.*

*Although quantifying these changes in orbital energies is beyond the scope of this text, it is reasonable that the d_{z^2} orbital should become less antibonding upon removal of a ligand along the z-axis. This orbital changes from e_g to a_1 symmetry upon ligand dissociation, and subsequently engages in mixing with the p_z and s orbitals as well, further decreasing its energy. In strong field C_{4v} complexes, electrons occupy the lower four orbitals singly for d^1–d^4 configurations, then doubly, with the b_1 orbital unoccupied until a d^9 configuration is reached.

TABLE 12.3 Ligand Field Activation Energies

	Strong Fields (units of Δ_o)			Weak Fields (units of Δ_o)		
System	Octahedral LFSE	Square-Pyramidal LFSE	LFAE	Octahedral LFSE	Square-Pyramidal LFSE	LFAE
d^0	0	0	0	0	0	0
d^1	−0.400	−0.457	−0.057	−0.400	−0.457	−0.057
d^2	−0.800	−0.914	−0.114	−0.800	−0.914	−0.114
d^3	−1.200	−1.000	0.200	−1.200	−1.000	0.200
d^4	−1.600	−0.914	0.686	−0.600	−0.914	−0.314
d^5	−2.000	−1.371	0.629	0	0	0
d^6	−2.400	−1.828	0.572	−0.400	−0.457	−0.057
d^7	−1.800	−1.914	−0.114	−0.800	−0.914	−0.114
d^8	−1.200	−1.828	−0.628	−1.200	−1.000	0.200
d^9	−0.600	−0.914	−0.314	−0.600	−0.914	−0.314
d^{10}	0	0	0	0	0	0

For a square-pyramidal transition state, LFAE = square pyramid LFSE – Octahedral LFSE, for σ donor only.

Electronic and steric factors also influence substitution reaction rates of octahedral complexes. The inequalities below indicate relative rates for ligand exchange via presumed dissociative mechanisms.

1. ***Oxidation state of the central ion.*** Central atoms with higher oxidation states have slower ligand exchange rates.

$$\underset{3+}{[AlF_6]^{3-}} > \underset{4+}{[SiF_6]^{2-}} > \underset{5+}{[PF_6]^{-}} > \underset{6+}{SF_6}$$

$$\underset{1+}{[Na(H_2O)_n]^{+}} > \underset{2+}{[Mg(H_2O)_n]^{2+}} > \underset{3+}{[Al(H_2O)_6]^{3+}}$$

2. ***Ionic radius.*** Smaller ions have slower exchange rates.

$$\underset{112\text{ pm}}{[Sr(H_2O)_6]^{2+}} > \underset{99\text{ pm}}{[Ca(H_2O)_6]^{2+}} > \underset{66\text{ pm}}{[Mg(H_2O)_6]^{2+}}$$

These two trends are attributed to a higher electrostatic attraction between the central atom and the ligands. A strong mutual attraction will slow a reaction occurring via a dissociative mechanism, because bond breaking between the metal and the departing ligand is required in the rate determining step. Despite the relatively small sizes of the first-row transition-metal 2+ ions, all of these form labile $[M(H_2O)_6]^{2+}$ complexes that rapidly exchange aqua ligands with water from the solvent. The half-lives for $[M(H_2O)_6]^{2+}$ in aqueous solution (the time required for half the complexes to undergo exchange) are shorter than 1 second; measurement of such fast reactions is done by relaxation methods.[4] The aqua complexes of the alkali metal cations have very short half-lives (10^{-9} second or less); of the first row 2+ transition-metal ions, only Be^{2+} and V^{2+} have half-lives as long as 0.01 second. For comparison, Al^{3+} has a half-life approaching 1 second, and inert Cr^{3+} has a half-life of 80 hours.

Evidence that supports a dissociative mechanism includes:[5,6,7,8]

1. The rate of reaction changes only slightly with changes in the incoming ligand. In many cases, **aquation** (substitution by water) and **anation** (substitution by an anion) rates are comparable. If ligand dissociation is the rate-determining step, the entering group should have no effect on the reaction rate. How much can a rate constant change before it is no longer considered comparable? Changes in rate constants of less than a factor of 10 are generally considered sufficiently similar when making these judgments.
2. Making the charge of the reactant complex more positive decreases the rate of substitution. The electrostatic attraction between the metal ion and the donor electrons of the ligands increases as the charge of the complex becomes more positive, decreasing the rate of ligand dissociation.
3. Steric crowding on the reactant complex increases the rate of ligand dissociation. When ligands on the reactant are crowded, loss of one of the ligands is made easier. On the other hand, if the reaction has an A or I_a mechanism, steric crowding interferes with the incoming ligand and slows the reaction.
4. The rate of reaction correlates with the metal–ligand bond strength of the leaving group, in a linear free-energy relationship (LFER, Section 12.4.2).
5. The volume of activation, ΔV_{act}, the change in volume on forming the activated complex, is measured by determining how much the rate constant changes as the pressure changes. Dissociative mechanisms generally result in positive values for ΔV_{act} because one species splits into two in the rate determining step, and associative mechanisms result in negative values because two species combine into one in this step, with a presumed transition state volume smaller than the total for the reactants. Caution is needed in interpreting volume effects to account for solvation effects, particularly for highly charged ions.

12.4.2 Linear Free-Energy Relationships

Kinetic effects are related to thermodynamic effects by **linear free-energy relationships (LFER)**.[9] A LFER can be observed when the bond strength of a metal-ligand bond (correlated to a thermodynamic parameter) plays a major role in determining the dissociation rate of a ligand (correlated to a kinetic parameter). When this is true, a plot of the logarithm of the rate constants for $[ML_5X]^{n+} + Y$ substitution reactions, where X is varied but Y is not, versus the logarithm of the equilibrium constants for $[ML_5X]^{n+} + Y \rightleftharpoons [ML_5Y]^{m+} + X$ is linear.* The Arrhenius equation for temperature dependence of rate constants and the equation for temperature dependence of equilibrium constants justify this correlation.

$$\ln k = \ln A - \frac{E_A}{RT} \quad \text{and} \quad \ln K = \frac{-\Delta H^\circ}{RT} + \frac{\Delta S^\circ}{R}$$

kinetic thermodynamic

If the preexponential factor, A, and the entropy change, ΔS°, are very similar for the series of substitution reactions considered, and the activation energy, E_A, depends on the enthalpy of reaction, ΔH°, there will be a linear correlation between ln k and ln K. A straight line on such a log–log plot is indirect evidence for a strong influence of the thermodynamic parameter, ΔH°, on the activation energy of the reaction. A stronger bond between a metal and a leaving group results in a larger activation energy, a logical connection for a dissociative mechanism. **Figure 12.5** shows an example from the hydrolysis of $[Co(NH_3)_5X]^{2+}$:

$$[Co(NH_3)_5X]^{2+} + H_2O \longrightarrow [Co(NH_3)_5(H_2O)]^{3+} + X^-$$

*The charges of these coordination complexes will be identical if Y and X have the same charge.

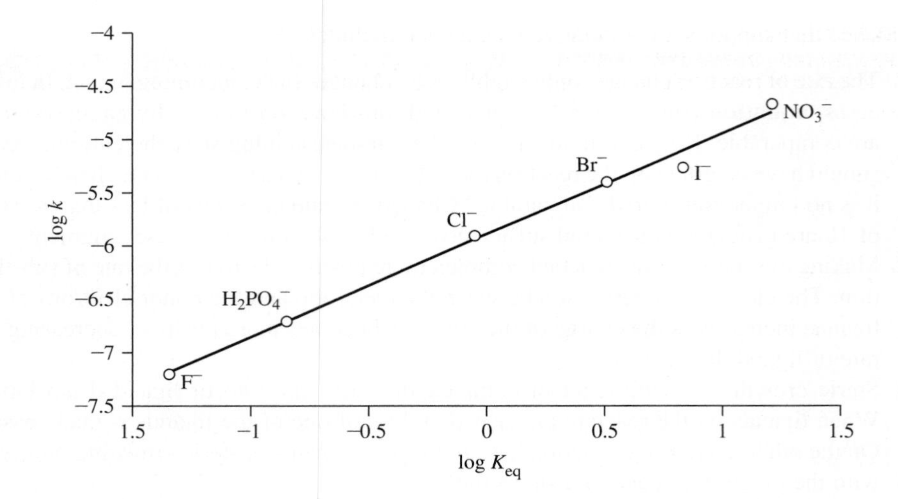

FIGURE 12.5 Linear Free Energy Relationship for $[Co(NH_3)_5X]^{2+}$ Hydrolysis at 25.0 °C (X indicated at each point). The log of the rate constant is plotted against the log of the equilibrium constant for the acid hydrolysis reaction of $[Co(NH_3)_5X]^{2+}$ ions. Data for F^- from S. C. Chan, *J. Chem. Soc.*, **1964**, 2375, and for I^- from R. G. Yalman, *Inorg. Chem.*, **1962**, *1*, 16. All other data from A. Haim, H. Taube, *Inorg. Chem.*, **1964**, *2*, 1199.

Langford[10] argued that X^- is dissociated and acts as a solvated anion in the transition state for $[Co(NH_3)_5X]^{2+}$ hydrolysis, and that water is, at most, weakly bound in the transition state.

Examples of the varying impacts of incoming ligands are given in **Tables 12.4** and **12.5**. The Table 12.4 data were obtained under *pseudo-first order conditions* (large [Y]).* The k_1 column gives the rate constants for anion exchange; the $k_1/k_1(H_2O)$ shows the ratio of k_1 to the rate for water exchange. The rate constants do not vary significantly with the substituting anion, as would be expected for a dissociative mechanism.

Table 12.5 gives data for the second-order region for anation of $[Ni(H_2O)_6]^{2+}$, which is hypothesized to occur via a preassociation mechanism.** The second-order rate constant, k_2K_1, is the product of the ion-pair equilibrium constant, K_1, and the rate constant, k_2 (Section 12.3.4):

$$[Ni(H_2O)_6]^{2+} + L^n \rightleftharpoons [Ni(H_2O)_6 \cdot L]^{2+n} \qquad K_1$$

$$[Ni(H_2O)_6 \cdot L]^{2+n} \longrightarrow [Ni(H_2O)_5L]^{2+n} + H_2O \qquad k_2$$

K_1 is calculated from an electrostatic model. The rate constants, k_2, vary by a factor of 5 or less and are close to the rate constant for water exchange. The close agreement

TABLE 12.4 Limiting Rate Constants for Anation of $[Co(NH_3)_5(H_2O)]^{3+}$ at 45 °C

$[Co(NH_3)_5(H_2O)]^{3+} + Y^{m-} \longrightarrow [Co(NH_3)_5Y]^{(3-m)+} + H_2O$

Y^{m-}	$k_1 (10^{-6} s^{-1})$	$k_1 / k_1 (H_2O)$	Reference
N_3^-	100	1.0	a
SO_4^{2-}	24	0.24	b
Cl^-	21	0.21	c
NCS^-	16	0.16	c

Data from: [a]H. R. Hunt, H. Taube, *J. Am. Chem. Soc.*, **1958**, *75*, 1463. [b]T. W. Swaddle, G. Guastalla, *Inorg. Chem.*, **1969**, *8*, 1604. [c]C. H. Langford, W. R. Muir, *J. Am. Chem. Soc.*, **1967**, *89*, 3141.

*If [Y] is large relative to [X], and also sufficiently large to be considered constant during the kinetic experiment, then the dissociative rate law $\frac{d[ML_5Y]}{dt} = \frac{k_2k_1[ML_5X][Y]}{k_{-1}[X] + k_2[Y]}$ in Section 12.3.1 simplifies to $\frac{d[ML_5Y]}{dt} = k_1[ML_5X]$. The reaction exhibits first-order kinetics.

**For a preassociation reaction, second order kinetics are observed when [Y] is low (Section 12.3.4).

TABLE 12.5 Rate Constants for $[Ni(H_2O)_6]^{2+}$ Substitution Reactions

Y	$k_2K_1(10^3\ M^{-1}\ s^{-1})$	$K_1(M^{-1})$	$k_2(10^4\ s^{-1})$
$CH_3PO_4^{2-}$	290	40	0.7
CH_3COO^-	100	3	3
NCS^-	6	1	0.6
F^-	8	1	0.8
HF	3	0.15	2
H_2O			3
NH_3	5	0.15	3
C_5H_5N, pyridine	~4	0.15	~3
$C_4H_4N_2$, pyrazine	2.8	0.15	2
$NH_2(CH_2)_2NMe_3^+$	0.4	0.02	2

Data from R. G. Wilkins, *Acc. Chem. Res.*, **1970,** *3,* 408; $C_4H_4N_2$ data are from J. M. Malin, R. E. Shepherd, *J. Inorg. Nucl. Chem.*, **1972,** *34*, 3203.

for different ligands Y shows that the effect of the incoming ligand is minor, although the difference in ion-pair formation is significant. These data are most consistent with a preassociation mechanism.

12.4.3 Associative Mechanisms

Associative reactions are less common with octahedral complexes.[11] **Table 12.6** gives data for both dissociative and associative interchanges for similar reactants. In the case of water substitution by several different anions in $[Cr(NH_3)_5(H_2O)]^{3+}$, the rate constants are similar (within a factor of 6), indicative of an I_d mechanism. Conversely, the same ligands reacting with $[Cr(H_2O)_6]^{3+}$ show a large variation in rates (more than a 2000-fold difference), indicative of an I_a mechanism. The $[Cr(NH_3)_5(H_2O)]^{3+}$ complex with the more electron rich Cr(III) center appears less reactive towards initial binding of an incoming nucleophile relative to $[Cr(H_2O)_6]^{3+}$. The very large absolute differences in these rate constants are intriguing. The varying amount of electron density at the Cr(III) centers of these complexes due to the increased donation of ammonia relative to water plays a significant role in determining both the substitution mechanism and the reaction rate.

In some cases, the typical substitution mechanism for a complex varies with the metal oxidation state. For example, reactions of Ru(III) compounds frequently have associative interchange mechanisms, and those of Ru(II) compounds generally have

TABLE 12.6 Effects of Entering Group on Rates

	Rate Constants for Anation	
Entering Ligand	$[Cr(NH_3)_5(H_2O)]^{3+}$ $k(10^{-4}\ M^{-1}\ s^{-1})$	$[Cr(H_2O)_6]^{3+}$ $k(10^{-8}\ M^{-1}\ s^{-1})$
NCS^-	4.2	180
NO_3^-	—	73
Cl^-	0.7	2.9
Br^-	3.7	0.9
I^-	—	0.08
CF_3COO^-	1.4	—

Data from D. Thusius, *Inorg. Chem.*, **1971,** *10,* 1106; T. Ramasami, A. G. Sykes, *Chem. Commun.*, **1978,** 378.

dissociative interchange mechanisms. The entropies of activation for substitution reactions of $[Ru(III)(EDTA)(H_2O)]^-$ are negative, suggesting association as part of the transition state. They also show a large range of rate constants depending on the incoming ligand (**Table 12.7**), as required for an I_a mechanism; but those of Ru(II) (**Table 12.8**) are nearly the same for different ligands, as required for an I_d mechanism. The reasons for this apparent difference in substitution mechanisms are unclear. Both complexes have a free carboxylate (the EDTA is pentadentate, with the sixth position occupied by a water molecule). Hydrogen bonding between this free carboxylate and the bound water may distort the shape sufficiently in the Ru(III) complex to open a place for entry by the incoming ligand. Although similar hydrogen bonding may be possible for the Ru(II) complex, the increased negative charge may reduce the $Ru—H_2O$ bond strength enough to promote an I_d mechanism.

12.4.4 The Conjugate Base Mechanism

Some cases in which second-order kinetics suggest an associative mechanism are believed to proceed via a **conjugate base mechanism**,[12] called S_N1CB for substitution, nucleophilic, unimolecular, conjugate base.[13] These reactions depend on amine, ammine (NH_3), or aqua ligands that can be deprotonated to form amido or hydroxo species that subsequently release a ligand dissociatively. It is often the ligand *trans* to the amido or hydroxo group that dissociates. Octahedral Co(III) complexes seem particularly predisposed to this mechanism. The small radius of low-spin Co(III) may render it sufficiently Lewis acidic to stabilize the five coordinate intermediate via π interactions.

$$[Co(NH_3)_5X]^{2+} + OH^- \rightleftharpoons [Co(NH_3)_4(NH_2)X]^+ + H_2O \quad \text{(equilibrium)} \quad (1)$$

$$[Co(NH_3)_4(NH_2)X]^+ \longrightarrow [Co(NH_3)_4(NH_2)]^{2+} + X^- \quad \text{(slow)} \quad (2)$$

$$[Co(NH_3)_4(NH_2)]^{2+} + H_2O \longrightarrow [Co(NH_3)_5(OH)]^{2+} \quad \text{(fast)} \quad (3)$$

Overall,

$$[Co(NH_3)_5X]^{2+} + OH^- \longrightarrow [Co(NH_3)_5(OH)]^{2+} + X^-$$

TABLE 12.7 Rate Constants for $[Ru(III)(EDTA)(H_2O)]^-$ Substitution

Ligand	k_1 ($M^{-1}\,s^{-1}$)	$\Delta H^\ddagger$ ($kJ\,mol^{-1}$)	$\Delta S^\ddagger$ ($J\,mol^{-1}\,K^{-1}$)
Pyrazine	20,000 ± 1,000	5.7 ± 0.5	−20 ± 3
Isonicotinamide	8,300 ± 600	6.6 ± 0.5	−19 ± 3
Pyridine	6,300 ± 500		
Imidazole	1,860 ± 100		
SCN^-	270 ± 20	8.9 ± 0.5	−18 ± 3
CH_3CN	30 ± 7	8.3 ± 0.5	−24 ± 4

Data from T. Matsubara, C. Creutz, *Inorg. Chem.*, **1979**, *18*, 1956.

TABLE 12.8 Rate Constants for $[Ru(II)(EDTA)(H_2O)]^{2-}$ Substitution

Ligand	k_1 ($M^{-1}\,s^{-1}$)
Isonicotinamide	30 ± 15
CH_3CN	13 ± 1
SCN^-	2.7 ± 0.2

Data from T. Matsubara, C. Creutz, *Inorg. Chem.*, **1979**, *18*, 1956.

In the third step, addition of Brønsted–Lowry acids other than water is possible. In basic solution, the rate constant is often designated as k_{OH}, and the equilibrium constant for the overall reaction is K_{OH}.

Other evidence for the conjugate base mechanism includes:

1. Base-catalyzed exchange of hydrogen from the amine groups validates the initial equilibrium (step 1).
2. The oxygen isotope ratio ($^{18}O/^{16}O$) in the product when a reaction is conducted in ^{18}O-enriched water is identical to the initial isotope ratio in the solvent ($H_2{}^{18}O/H_2{}^{16}O$), regardless of the leaving group ($X^- = Cl^-, Br^-, NO_3{}^-$). If an incoming water molecule had a large influence on the rate determining step (an associative mechanism), a higher proportion of ^{18}O should be in the product than in the solvent, because the equilibrium constant (K) for the reaction $H_2{}^{16}O + {}^{18}OH^- \rightleftharpoons H_2{}^{18}O + {}^{16}OH^-$ is 1.040.
3. RNH_2 complexes (R = alkyl) react faster than NH_3 complexes, possibly because steric crowding favors the 5-coordinate intermediate formed in Step 2. If the reaction had an associative mechanism, more steric bulk around the metal would be expected to lower the reaction rate.
4. The rate constants and dissociation constants for these complexes with different leaving groups form a linear free-energy relationship (LFER), in which a plot of $\ln k_{OH}$ versus $\ln K_{OH}$ is linear. This is consistent with ligand dissociation as the rate determining step.

Reactions with $[Co(tren)(NH_3)Cl]^{2+}$ isomers show that the nitrogen atom *trans* to the leaving group is most likely to be deprotonated in the conjugate base mechanism.[14] The reaction in **Figure 12.6a** is 10^4 times faster than that in Figure 12.6b. The dominant kinetic product in both reactions is hypothesized to proceed by a trigonal-bipyramidal intermediate or transition state with the amido ligand in the trigonal plane that is ultimately *trans* to the hydroxo ligand. The reaction in Figure 12.6a has a much lower barrier to obtain a trigonal-bipyramidal geometry (with only slightly widening of two N—Co—N angles than does the Figure 12.6b reaction, which requires more dramatic rearrangement of a square-pyramidal structure. The activation energy associated with hydrolysis of the square pyramidal intermediate in Figure 12.6b, where the hydroxo ligand would engage an orbital orthogonal to the amido nitrogen, is sufficiently high that the (slow) rearrangement to a trigonal-bipyramidal structure, followed by hydrolysis is still the faster pathway. Both of these routes are much slower than that in Figure 12.6a.

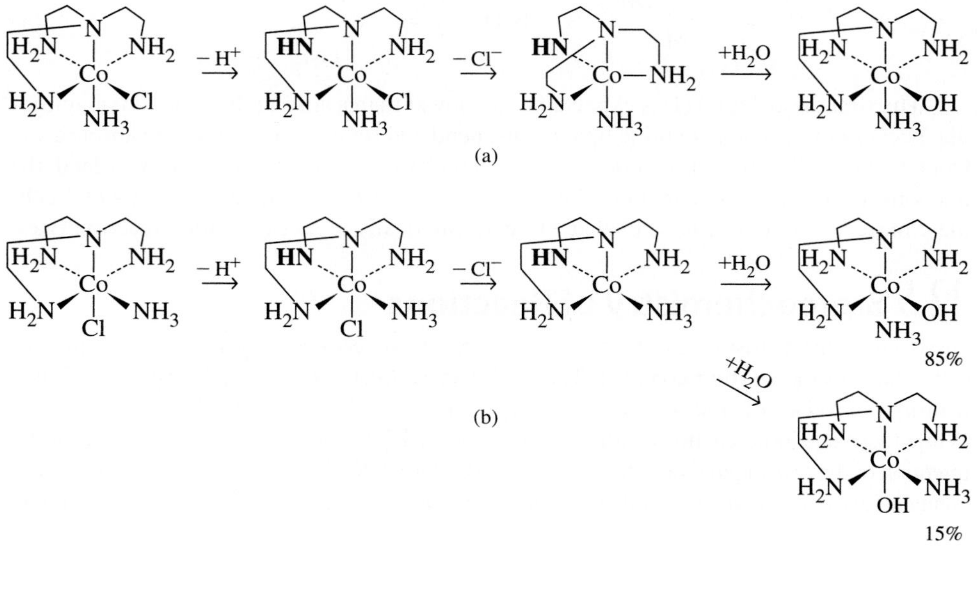

(a)

(b)

85%

15%

FIGURE 12.6 Base Hydrolysis of $[Co(tren)(NH_3)Cl]^{2+}$ Isomers. (a) Leaving group (Cl^-) *trans* to deprotonated nitrogen. (b) Leaving group (Cl^-) *cis* to deprotonated nitrogen.

(Data from D. A. Buckingham, P. J. Creswell, A. M. Sargeson, *Inorg. Chem.*, **1975**, *14*, 1485.)

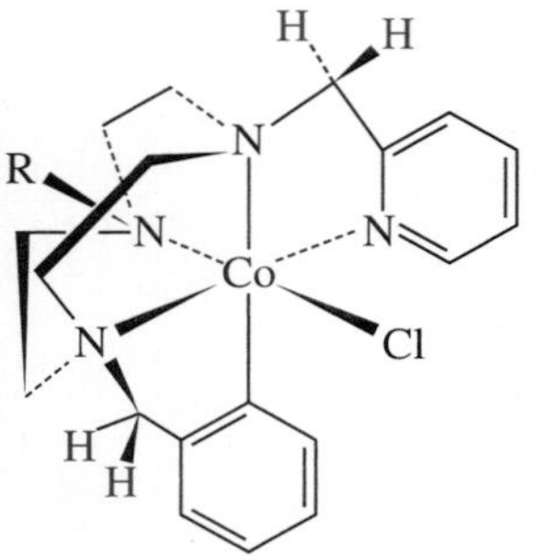

FIGURE 12.7 Base catalyzed hydrolysis of this class of pentaaminecobalt(III) complex is believed to proceed via a pseudo-aminate mechanism via deprotonation of the blue hydrogen atoms.

Why does positioning the amido ligand in the same plane to be occupied by the *trans* hydroxo ligand facilitate this reaction? Most arguments speculate that amido ligands, as strong σ and π donors, can best stabilize the Co(III) transition state for hydrolysis from this coordination site.[15]

The availability of ionizable amine hydrogens was long considered a vital prerequisite for the conjugate base mechanism to be operative. In 2003, a unique mechanism for base catalyzed hydrolysis of pentaaminecobalt(III) complexes having all tertiary amines and pyridine donors was reported.[16] This *pseudo-aminate* mechanism involves deprotonation of methylene groups linking the α carbon of the pyridine and the tertiary amine nitrogen (**Figure 12.7**).

12.4.5 The Kinetic Chelate Effect

The chelate effect (Section 10.1.1) causes polydentate complexes to be thermodynamically more stable than their monodentate counterparts.[17] Substitution for a chelated ligand is generally a slower reaction than that for a similar monodentate ligand. Explanations for this effect center on two factors. First, the ΔH associated with removal of the first bound atom is larger than for a related monodentate ligand. If this atom does separate from the metal center, its kinetic barrier for subsequent reattachment is lower than for a related monodentate ligand since the former remains in close proximity to the metal center.[18] Consider the general scheme below:

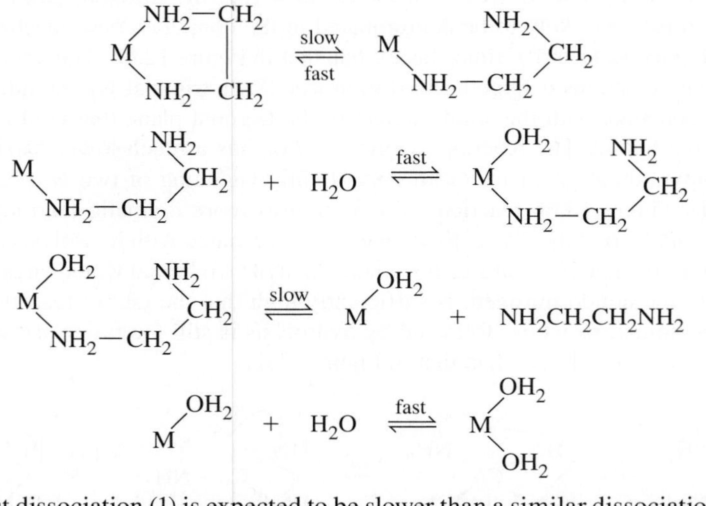

$$M(NH_2CH_2CH_2NH_2) \underset{\text{fast}}{\overset{\text{slow}}{\rightleftharpoons}} M\text{—}NH_2CH_2CH_2NH_2 \qquad (1)$$

$$M\text{—}NH_2CH_2CH_2NH_2 + H_2O \overset{\text{fast}}{\rightleftharpoons} M(OH_2)\text{—}NH_2CH_2CH_2NH_2 \qquad (2)$$

$$M(OH_2)\text{—}NH_2CH_2CH_2NH_2 \overset{\text{slow}}{\rightleftharpoons} M\text{—}OH_2 + NH_2CH_2CH_2NH_2 \qquad (3)$$

$$M\text{—}OH_2 + H_2O \overset{\text{fast}}{\rightleftharpoons} M(OH_2)_2 \qquad (4)$$

The first dissociation (1) is expected to be slower than a similar dissociation of ammonia, because the ethylenediamine ligand must bend and rotate to move the free amine away from the metal. The reverse reaction associated with the first dissocation is fast. Indeed, the uncoordinated nitrogen is held near the metal by the rest of the ligand, making reattachment more likely. This kinetic chelate effect dramatically reduces aquation reaction rates.

12.5 Stereochemistry of Reactions

Dissociative mechanisms lead to products where the stereochemistry may be the same or different than the starting complex. **Table 12.9** shows that *cis*-$[Co(en)_2L(H_2O)]^{(1+n)+}$ is a hydrolysis product of *both cis*-$[Co(en)_2LX]^{n+}$ and *trans*-$[Co(en)_2LX]^{n+}$ in acid solution. While these aquation reactions with pure *cis*-$[Co(en)_2LX]^{n+}$ lead exclusively to *cis* products, retention of the *trans* ligand orientation in *trans*-$[Co(en)_2LX]^{n+}$ depends on both L and X. The conjugate base mechanism is unlikely in these reactions; they are carried out in acidic solution.

TABLE 12.9 Stereochemistry of $[Co(en)_2LX]^{n+}$ Acid Hydrolysis

$[Co(en)_2LX]^{n+} + H_2O \longrightarrow [Co(en)_2L(H_2O)]^{(1+n)+} + X^-$					
cis-L	X	% *cis* Product	*trans*-L	X	% *cis* Product
OH^-	Cl^-	100	OH^-	Cl^-	75
OH^-	Br^-	100	OH^-	Br^-	73
Br^-	Cl^-	100	Br^-	Cl^-	50
Cl^-	Cl^-	100	Br^-	Br^-	30
Cl^-	Br^-	100	Cl^-	Cl^-	35
N_3^-	Cl^-	100	Cl^-	Br^-	20
NCS^-	Cl^-	100	NCS^-	Cl^-	50–70

Data from F. Basolo and R. G. Pearson, *Mechanisms of Inorganic Reactions*, 2nd ed., J. Wiley & Sons, New York, 1967, p. 257.

TABLE 12.10 Stereochemistry of Base Substitution

$[Co(en)_2LX]^{n+} + OH^- \longrightarrow [Co(en)_2LOH]^{n+} + X^-$						
		% *cis* Product				% *cis* Product
Δ-*cis*-L	X	Δ	Λ	*trans*-L	X	
OH^-	Cl^-	61	36	OH^-	Cl^-	94
NCS^-	Cl^-	56	24	NCS^-	Cl^-	76
NH_3	Br^-	59	26	NCS^-	Br^-	81
NH_3	Cl^-	60	24	NH_3	Cl^-	76
NO_2^-	Cl^-	46	20	NO_2^-	Cl^-	6

Data from F. Basolo and R. G. Pearson, *Mechanisms of Inorganic Reactions*, 2nd ed., J. Wiley & Sons, New York, 1967, p. 262.

In Table 12.9, the starting *cis*-$[Co(en)_2LX]^{n+}$ are racemic mixtures of Λ-*cis*-$[Co(en)_2LX]^{n+}$ and Δ-*cis*-$[Co(en)_2LX]^{n+}$, and the product mixtures, while exclusively *cis*, are also racemic. A related study (**Table 12.10**) employed optically pure Δ-*cis*-$[Co(en)_2LX]^{n+}$ for reactions with hydroxide, where a conjugate base mechanism is possible. This reaction resulted in *cis* product with partial racemization of the stereochemistry, as well as some *trans* product (% *trans* = 100% − % *cis*). When the optically inactive *trans*-$[Co(en)_2LX]^{n+}$ is used, varying yields of *cis* isomers are formed as racemic mixtures.

The extent of the inversion of configuration can sometimes be tuned by changing the reaction conditions, as in the reaction of Λ-*cis*-$[Co(en)_2Cl_2]^+$ with hydroxide. In dilute (< 0.01 M) hydroxide, the major product is Λ-*cis*-$[Co(en)_2(OH)_2]^+$, but in > 0.25 M hydroxide, the major product is Δ-*cis*-$[Co(en)_2(OH)_2]^+$.[19] The conjugate base mechanism is proposed (**Figure 12.8**), with hydroxide removing a proton from an ethylenediamine nitrogen, followed by chloride loss *trans* to the amido ligand. In more concentrated base, ion pairs ($[Co(en)_2Cl_2]^+ \cdot OH^-$) are assumed to result in a water molecule (from the OH^- and the H^+ removed from ethylenediamine), positioned for facile addition for inversion of the chiral center.[20]

12.5.1 Substitution in *trans* Complexes

Beyond the possibility of the conjugate base mechanism, substitution of Y for X in *trans*-$[M(LL)_2BX]$ (LL = a bidentate ligand) can proceed by three dissociative pathways. If dissociation of X from the reactant leaves a square-pyramidal intermediate that adds the new ligand to the vacant site, the result is retention of configuration, and the product, like the

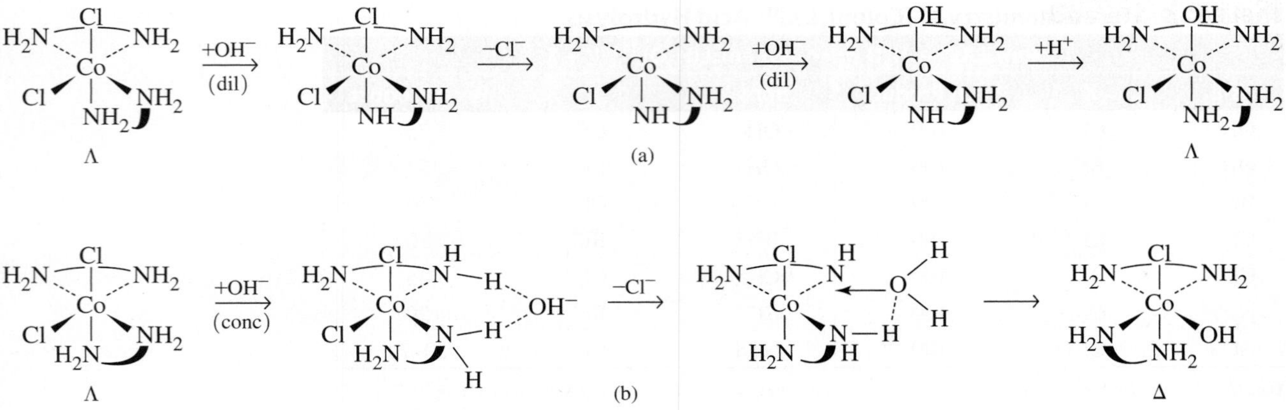

FIGURE 12.8 Mechanisms of Base Hydrolysis of Λ-*cis*-$[Co(en)_2Cl_2]^+$. (a) Retention of configuration in dilute hydroxide. (b) Inversion of configuration in concentrated hydroxide.

reactant, is *trans* (**Figure 12.9a**). A trigonal-bipyramidal intermediate with B in the trigonal plane leads to a mixture of *trans* and *cis* (Figure 12.9b). The incoming ligand can enter along any of the three sides of the triangle, resulting in two *cis* possibilities and one *trans* possibility. While only formation of Λ-*cis* isomers is shown in Figure 12.9, formation of Δ-*cis* isomers is equally likely since Λ and Δ trigonal-bipyramidal intermediates are equally likely in principle upon dissociation of X. From a purely statistical perspective, a racemic mixture of *cis* isomers should result. Dissociation to form a trigonal pyramid with B in an axial position (Figure 12.9c) allows two positions for attack by Y, both of which give *cis* products (the third side of the triangle is blocked by an LL ring). An intermediate with an axial B is less likely than one with an equatorial B, because an axial B requires more rearrangement of the ligands (a 90° change by one nitrogen and 30° changes by two others, in contrast to two 30° changes for the equatorial B) as well as a larger stretch for the LL ring in the equatorial plane.

FIGURE 12.9 Dissociation Mechanism and Stereochemical Changes for *trans*-$[M(LL)_2BX]$. (a) Square-pyramidal intermediate (retention of configuration). (b) Trigonal-bipyramidal intermediate (while only formation of the Λ isomers is shown, a racemic mixture of *cis* isomers is expected since Λ and Δ trigonal-bipyramidal intermediate are equally likely upon X dissociation). (c) Less likely trigonal-bipyramidal intermediate (two possible products).

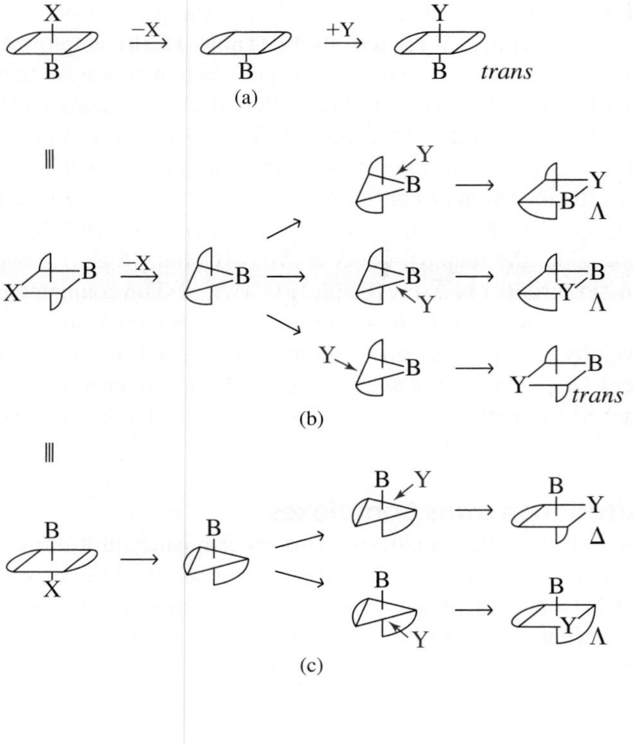

EXERCISE 12.2

Follow the example of Figure 12.9b with the structure at the right and show that the first two products could be Δ rather than Λ.

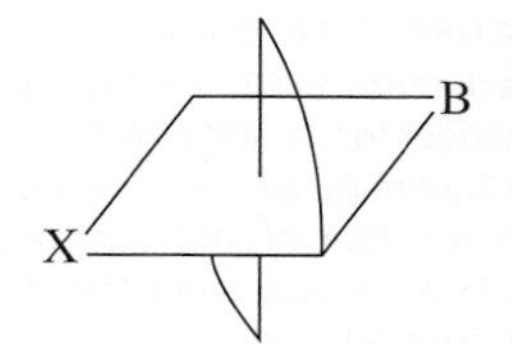

If the Figure 12.9b mechanism is operative, the statistical probability of a change from *trans* to *cis* with a *trans*-$[M(LL)_2BX]$ reactant is expected to be two thirds. This outcome is rarely achieved. With *trans*-$[Co(en)_2LX]^{n+}$, both acid aquation (Table 12.9, substitution of H_2O for X) and base substitution reactions (Table 12.10, substitution of OH^- for X) result in different fractions of *cis* and *trans* products depending on the retained ligand (B). The fraction of *cis* hydroxide substitution product from *trans*-$[Co(en)_2LX]^{n+}$ in Table 12.10 ranges from 6 to 94%. As noted above, the *cis* isomers produced from an achiral *trans* reactant are expected to be a racemic mixture of Λ and Δ.

Another factor that must be considered to rationalize the product stereochemistry is the isomerization rates of the starting complexes (as well as those of the products). **Table 12.11** lists data for $[Co(en)_2(H_2O)X]^{n+}$ isomerization and water exchange reactions. For X = Cl^-, SCN^-, and H_2O, racemization and *cis* ⟶ *trans* isomerization are nearly equal in rate.* Water exchange is faster than the other reactions for all but the hydroxide complex; simple exchange with the solvent requires no rearrangement of the ligands, and the high concentration of water makes this is a pseudo-first order process, where the rate of reaction essentially depends only on the $[Co(en)_2(H_2O)X]^{n+}$ concentration, as with the isomerization reactions of $[Co(en)_2(H_2O)X]^{n+}$. For X = NH_3, both racemization and *trans* ⟶ *cis* isomerization are mysteriously faster than *cis* ⟶ *trans* isomerization.

12.5.2 Substitution in *cis* Complexes

Substitution in *cis* complexes can also proceed by three intermediates (**Figure 12.10**). A square-pyramidal intermediate results in retention of configuration. If dissociation of X forms a trigonal bipyramid with B in the trigonal plane, there are three possible locations for the addition of Y, all within this trigonal plane. Two of these result in *cis* products, with retention of configuration, and one in a *trans* product. The less likely trigonal bipyramid with an axial B, whether derived from a *cis* or a *trans* reactant (note the identical intermediates in Figures 12.9c and 12.10c), produces a racemic mixure of *cis* products.

An optically active *cis* complex can yield products that retain the same configuration, convert to *trans* geometry, or create a racemic mixture. Statistically, the product of substitution of a *cis*-$[M(LL)_2BX]$ complex through a trigonal-bipyramidal intermediate should be one fifth *trans* if both intermediates were equally likely and one third *trans* if the axial B form is not formed at all. Experimentally, aquation of *cis*-$[M(LL)_2BX]$ in acid results in

TABLE 12.11 Rate Constants for Reactions of $[Co(en)_2(H_2O)X]^{n+}$ at 25 °C, $k(10^{-5}\ s^{-1})$

X	*cis* ⟶ *trans*	*trans* ⟶ *cis*	Racemization	H_2O Exchange
OH^-	200	300	—	160
Cl^-	2.4	7.2	2.4	—
NCS^-	0.0014	0.071	0.022	0.13
H_2O	0.012	0.68	~0.015	1.0
NH_3	<0.0001	0.002	0.003	0.10

Source: Data from M. L. Tobe, in J. H. Ridd, ed., *Studies in Structure and Reactivity*, Methuen, London, 1966, and M. N. Hughes, *J. Chem. Soc., A*, **1967**, 1284.

*It is noteworthy in Table 12.11 that the ratio of the rate constants for the *cis* ⟶ *trans* and *trans* ⟶ *cis* reactions provide the equilibrium constants for this isomerization at 25 °C. In all of these examples, the *cis* isomer is more stable than the *trans* isomer, with $K_{eq} < 1$ for *cis* ⇌ *trans*.

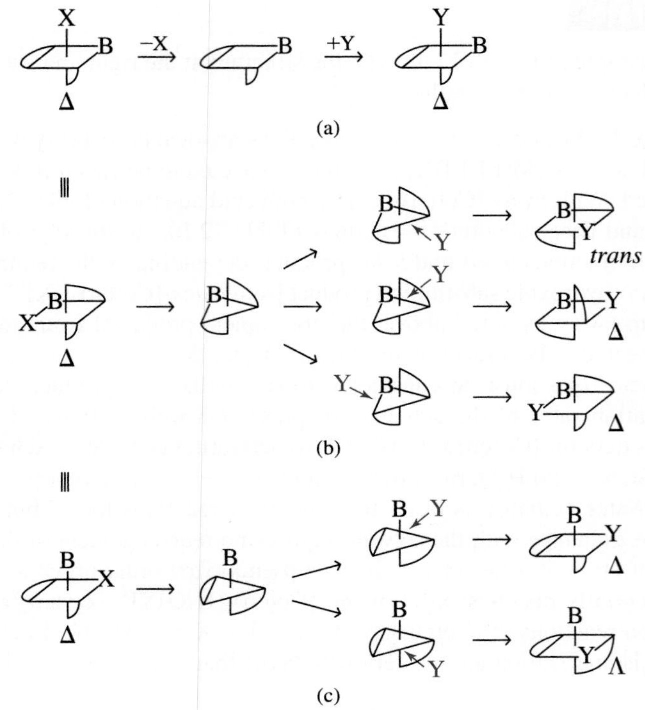

FIGURE 12.10 Dissociation Mechanism and Stereochemical Changes for *cis*-$[M(LL)_2BX]$. (a) Square-pyramidal intermediate (retention of configuration). (b) Trigonal-bipyramidal intermediate. (c) Unlikely trigonal bipyramidal intermediate.

100% *cis* isomer (Table 12.9), indicating a square-pyramidal transition state. Substitution of optically active *cis* complexes in base gives products ranging from 97% to 66% *cis*, with about 2:1 retention of the Δ configuration (Table 12.9). Among the compounds that retain their optical activity and *cis*/*trans* isomerism on hydrolysis are $[M(en)_2Cl_2]^+$ with M = Co, Rh, and Ru.[21] As a general rule, *cis* reactants give a relatively higher percentage of substitution products that retain their *cis* configuration; *trans* reactants often afford a more balanced mixture of *cis* and *trans* substitution products. The variety of factors that interplay to govern product mixtures require a significant amount of data to formulate mechanistic hypotheses.

12.5.3 Isomerization of Chelate Rings

Isomerization reactions of complexes with three bidentate ligands (for example, from Δ to Λ) can occur via initial dissociation of one point of attachment. After one end of the chelate ring dissociates, the resulting 5-coordinate intermediate can rearrange before reattachment of the loose end. This mechanism is similar to the reactions described in Sections 12.5.1 and 12.5.2; the ligand that dissociates in the first step is the same one that adds in the final step, after rearrangement.

Pseudorotation

Isomerization mechanisms involving compounds containing chelating ligands can also involve twists. The trigonal, or Bailar, twist, requires twisting the two opposite trigonal faces through a trigonal prismatic transition state to the new structure (**Figure 12.11a**). In tetragonal twists (Figures 12.11b and 12.11c), one chelate ring is held stationary, while the other two are twisted to the new structure. The tetragonal twist in Figure 12.11b has a transition state with the stationary ring perpendicular to those being twisted. The second tetragonal twist (Figure 12.11c) requires twisting the two rings through a transition state with all three rings parallel. Attempts to elucidate which specific twist is operative for a complex

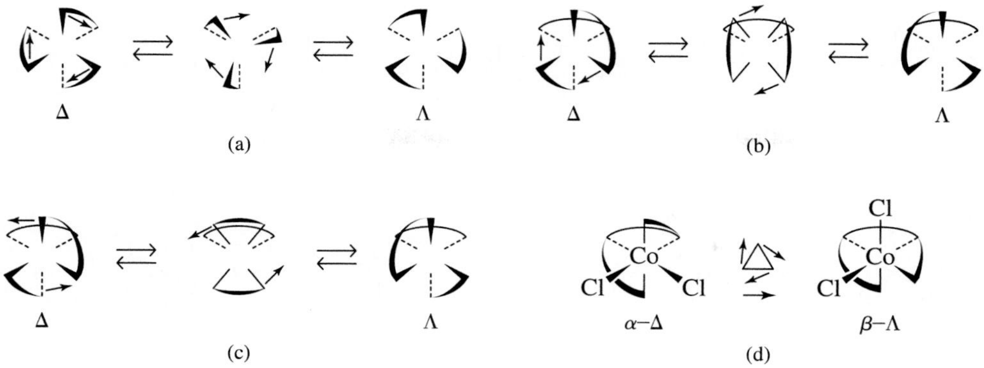

FIGURE 12.11 Twist Mechanisms for Isomerization of $M(LL)_3$ and $[Co(trien)Cl_2]^+$ Complexes. (a) Trigonal twist: the front triangular face rotates with respect to the back triangular face. (b) Tetragonal twist with perpendicular rings: the back ring remains stationary as the front two rings rotate clockwise. (c) Tetragonal twist with parallel rings: the back ring remains stationary as the front two rings rotate counterclockwise. (d) $[Co(trien)Cl_2]^+$ α–β isomerization: the connected rings limit this isomerization to a clockwise trigonal twist of the front triangular face.

is an experimental challenge. NMR study of tris(trifluoroacetylacetonato) metal(III) chelates shows that a trigonal twist mechanism is not possible for M = Al, Ga, In, and Cr but leaves it a possibility for Co.[22] The multiple-ring structure of *cis*-α-$[Co(trien)Cl_2]^+$ allows only a trigonal twist in its conversion to the β isomer (Figure 12.11d).

12.6 Substitution Reactions of Square-Planar Complexes

The products of substitution reactions of square-planar complexes have the same configuration as the reactants, with replacement of the departing ligand by the new ligand. The rates vary enormously, and different compounds can be formed, depending on the entering and the departing ligands.

12.6.1 Kinetics and Stereochemistry of Square-Planar Substitutions

Because many reactions of platinum compounds have been studied, we will use as our initial example the generic reaction

$$\text{T—Pt—X} + \text{Y} \longrightarrow \text{T—Pt—Y} + \text{X}$$

where T is the ligand *trans* to the departing ligand X, and Y is the incoming ligand. We will designate the plane of the molecule the xy plane and the Pt axis through T—Pt—X the x axis (**Figure 12.12**). The other two ligands, L, will be ignored for the moment.

It is generally accepted that reactions of square-planar complexes have significant associative character; they are classified as I_a. Two such mechanisms are in Figure 12.12. In mechanism (a) the incoming ligand approaches along the z axis. As it bonds to the Pt, the complex rearranges to approximate a trigonal bipyramid with Pt, T, X, and Y in the trigonal plane. As X leaves, Y moves into the plane of T, Pt, and the two L ligands. This same general description will fit whether the incoming ligand bonds strongly to Pt before the departing ligand bond is weakened appreciably (I_a), or the departing ligand bond is weakened considerably before the incoming ligand forms its bond (I_d). The solvent-assisted mechanism (b) follows the same pattern but requires two associative steps.

Square-planar substitution reactions are frequently described by two-term rate laws

$$\text{Rate} = k_1\,[\text{Cplx}] + k_2\,[\text{Cplx}][\text{Y}]$$

where [Cplx] = concentration of the reactant complex and [Y] = concentration of the incoming ligand. Each term in the rate law is considered to derive from an associative pathway, despite the difference in order. The k_2 term fits a standard associative mechanism (a) in which the incoming ligand Y and the reacting complex form a 5-coordinate transition state. The accepted explanation for the k_1 term is a solvent-assisted reaction (b), with solvent

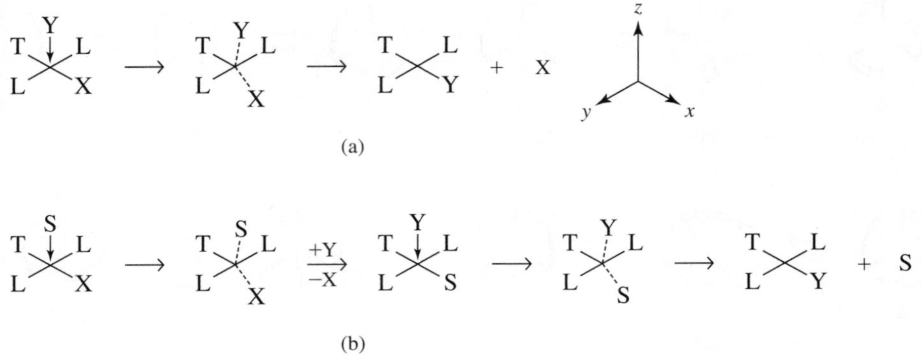

FIGURE 12.12 The Interchange Mechanism in Square-Planar Reactions. (a) Direct substitution by Y. (b) Solvent-assisted substitution.

replacing X on the complex through a similar 5-coordinate transition state, and then itself being replaced by Y. The second step of this mechanism is presumed to be faster than the first, and the concentration of solvent is large and unchanging (leading to pseudo first order conditions), so the overall rate law for this path is approximated as first order in complex.

12.6.2 Evidence for Associative Reactions

Evidence for a 5-coordinate intermediate is strong, and the transition state sometimes may even be 6-coordinate, with assistance from solvent.[23] The highest energy transition state may be either during the formation of the intermediate or as the exiting ligand dissociates from the intermediate.

This mechanism reveals an effect of the incoming ligand. Pt(II) is a soft acid, so soft ligands react more readily with it. The order of ligand reactivity depends somewhat on the other ligands on the Pt, but the rate constants for

$$\textit{trans}\text{-}PtL_2Cl_2 + Y \longrightarrow \textit{trans}\text{-}PtL_2ClY + Cl^-$$

for different Y in methanol rank as follows (**Table 12.12**):

$$PR_3 > CN^- > SCN^- > I^- > Br^- > N_3^- > NO_2^- > py > NH_3 \sim Cl^- > CH_3OH$$

Similar rankings are found for reactants with T ligands other than chloride. These rate constants vary over many orders of magnitude, with $k(PPh_3)/k(CH_3OH) = 9 \times 10^8$.

TABLE 12.12 Rate Constants and Nucleophilic Reactivity Parameters for Entering Groups

	$\textit{trans}\text{-}PtL_2Cl_2 + Y \longrightarrow \textit{trans}\text{-}PtL_2ClY + Cl^-$ k (10^{-3} M^{-1} s^{-1})		
Y	L = py ($s = 1$)	L = PEt_3 ($s = 1.43$)	η_{Pt}
PPh_3	249,000		8.93
SCN^-	180	371	5.75
I^-	107	236	5.46
Br^-	3.7	0.93	4.18
N_3^-	1.55	0.2	3.58
NO_2^-	0.68	0.027	3.22
NH_3	0.47		3.07
Cl^-	0.45	0.029	3.04

Data from U. Belluco, L. Cattalini, F. Basolo, R. G. Pearson, A. Turco, *J. Am. Chem. Soc.*, **1965**, *87*, 241; PPh_3 and η_{Pt} data from R. G. Pearson, H. Sobel, J. Songstad, *J. Am. Chem. Soc.*, **1968**, *90*, 319.

NOTE: s and η_{Pt} are nucleophilic reaction parameters explained in the text.

TABLE 12.13 Rate Constants for Leaving Groups

$[Pt(dien)X]^+ + py \longrightarrow [Pt(dien)py]^{2+} + X^-$ (Rate = $(k_1 + k_2[py])[Pt(dien)X^+]$)	
X^-	k_2 ($M^{-1}\,s^{-1}$)
NO_3^-	very fast
Cl^-	5.3×10^{-3}
Br^-	3.5×10^{-3}
I^-	1.5×10^{-3}
N_3^-	1.3×10^{-4}
SCN^-	4.8×10^{-5}
NO_2^-	3.8×10^{-6}
CN^-	2.8×10^{-6}

Rate constants calculated from data in F. Basolo, H. B. Gray, R. G. Pearson, *J. Am. Chem. Soc.*, **1960**, *82*, 4200.

Because T and Y have similar positions in the transition state, these ligands have similar effects on the rate. This ***trans*** **effect** is discussed in Section 12.7.

The leaving group X, the other species in the trigonal plane of the 5-coordinate intermediate also has a significant influence on the rate (**Table 12.13**).[24] The order of X ligands is nearly the reverse of that above, with hard ligands such as Cl^-, NH_3, and NO_3^- leaving relatively quickly. Soft ligands with considerable metal-to-ligand π bonding, such as CN^- and NO_2^-, leave relatively slowly. For example, in the reaction

$$[Pt(dien)X]^+ + py \longrightarrow [Pt(dien)(py)]^{2+} + X^-$$

the rate increases by a factor of 10^5 with X = H_2O as compared with $X^- = CN^-$ or NO_2^- as the leaving group. Metal-to-ligand π bonding to X reduces the reactivity of square-planar platinum complexes significantly towards substitution of these ligands; the M—X bonds are strengthened as a result of this π interaction. In addition π-backbonding to X uses the same orbitals as those bonding to the entering group in the trigonal plane. The extension of the HOMO from the Pt(II) center is reduced, decreasing the accessibility of this orbital towards incoming nucleophiles. These two effects result in the slow displacement of ligands that participate in metal-to-ligand π-bonding compared to ligands with only σ bonding or ligand-to-metal π bonding capability.

Good leaving groups show little discrimination between entering groups. The ease of breaking the Pt—X bond takes precedence over the formation of the Pt—Y bond. For complexes with poorer leaving groups, the other ligands bound to the Pt(II) have a significant role in dictating the rate of substitution reactions; softer PEt_3 and $AsEt_3$ ligands in the coordination sphere result in a greater range of substitution reaction rates with different entering groups Y when compared to harder dien or en ligands. The equation[25] used to assess how effectively a Pt(II) complex discriminates between a series of incoming nucleophiles (Y) is

$$\log k_Y = s\,\eta_{Pt} + \log k_S$$

where

k_Y = rate constant for reaction with Y
k_S = rate constant for reaction with solvent
s = **nucleophilic discrimination factor** (for the complex)
η_{Pt} = **nucleophilic reactivity constant** (for the entering ligand)

The nucleophilic discrimination factor s is defined as 1.00 for *trans*-$[Pt(py)_2Cl_2]$; this nucleophilicity scale has as a reference substitution reactions such as those described in the first column in Table 12.12, carried out in methanol solvent at 30 °C. Values of η_{Pt} are subsequently found with kinetic data from these reactions via

$$\eta_{Pt} = \log\left(\frac{k_Y}{k_{CH_3OH}}\right)$$

The s factors for the hard $[Pt(dien)H_2O]^{2+}$ and the soft *trans*-$[Pt(PEt_3)_2Cl_2]$, found by plotting $\log k_Y$ versus η_{Pt}, equal 0.44 and 1.43, respectively. This means that substitution reaction rates vary more for $[Pt(PEt_3)_2Cl_2]$ as the nucleophilicity of Y is changed relative to *trans*-$[Pt(py)_2Cl_2]$. This is shown in Table 12.12; the rate constants for SCN^- and Cl^- substitution vary by factors of 400 and 12,800, with *trans*-$[Pt(py)_2Cl_2]$ and $[Pt(PEt_3)_2Cl_2]$, respectively. The substitution reaction rates for $[Pt(dien)H_2O]^{2+}$ ($s = 0.44$) vary less as the nucleophilicity of Y changes relative to both *trans*-$[Pt(py)_2Cl_2]$ and $[Pt(PEt_3)_2Cl_2]$. The increase in rate constants parallels the increase in η_{Pt}. While this approach works well for Pt(II) complexes (where it predicts, for example, that chloride and ammonia, with nearly identical η_{Pt} values, should engage in substitution reactions with similar rates with a given square-planar Pt(II) complex), it must be used with caution when applied to other metals.

12.7 The *trans* Effect

Chernyaev[26] introduced the ***trans* effect** in platinum chemistry. In reactions of square-planar Pt(II) compounds, ligands *trans* to chloride are more easily replaced than those *trans* to ammonia; chloride has a stronger *trans* effect than ammonia. The *trans* effect allows the formation of isomeric Pt compounds (**Figure 12.13**).

In reaction (a), after the first ammonia is replaced, the second replacement is *trans* to the first Cl^-. In reaction (b), the second replacement is *trans* to Cl^- (replacement of ammonia by chloride is also possible, resulting in formation of the reactant $[PtCl_4]^{2-}$). The first steps in reactions (c) through (f) are the possible replacements, with nearly equal probabilities for replacement of ammonia or pyridine. The second steps of (c) through (f) depend on the *trans* effect of Cl^-. Both steps of (g) and (h) depend on the greater lability of chloride. By using reactions such as these, specific isomers can be prepared. Chernyaev prepared a wide variety of compounds and established the order of *trans*-effect ligands:

$$CN^- \sim CO \sim C_2H_4 > PH_3 \sim SH_2 > NO_2^- > I^- > Br^- > Cl^- > NH_3 \sim py > OH^- > H_2O$$

EXERCISE 12.3

Predict the products of these reactions (there may be more than one product when there are conflicting preferences).

$[PtCl_4]^{2-} + NO_2^- \longrightarrow$ (a) $\qquad$ (a) $+ NH_3 \longrightarrow$ (b)

$[PtCl_3NH_3]^- + NO_2^- \longrightarrow$ (c) $\qquad$ (c) $+ NO_2^- \longrightarrow$ (d)

$[PtCl(NH_3)_3]^+ + NO_2^- \longrightarrow$ (e) $\qquad$ (e) $+ NO_2^- \longrightarrow$ (f)

$[PtCl_4]^{2-} + I^- \longrightarrow$ (g) $\qquad$ (g) $+ I^- \longrightarrow$ (h)

$[PtI_4]^- + Cl^- \longrightarrow$ (i) $\qquad$ (i) $+ Cl^- \longrightarrow$ (j)

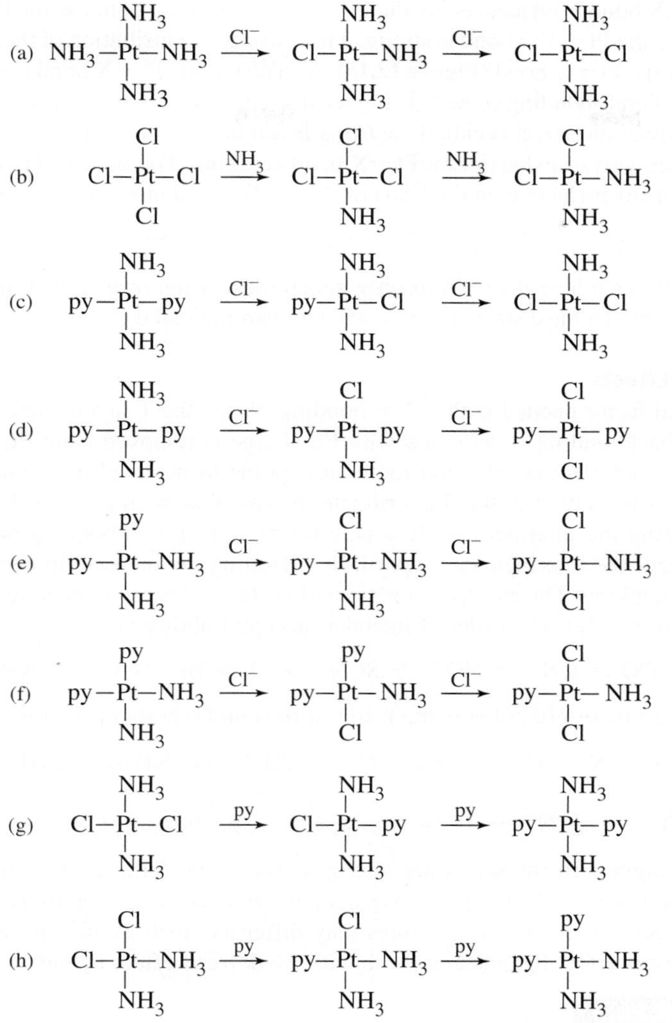

$$\text{(a)}\quad \begin{array}{c}\mathrm{NH_3}\\ | \\ \mathrm{NH_3{-}Pt{-}NH_3}\\ | \\ \mathrm{NH_3}\end{array} \xrightarrow{\mathrm{Cl^-}} \begin{array}{c}\mathrm{NH_3}\\ | \\ \mathrm{Cl{-}Pt{-}NH_3}\\ | \\ \mathrm{NH_3}\end{array} \xrightarrow{\mathrm{Cl^-}} \begin{array}{c}\mathrm{NH_3}\\ | \\ \mathrm{Cl{-}Pt{-}Cl}\\ | \\ \mathrm{NH_3}\end{array}$$

$$\text{(b)}\quad \begin{array}{c}\mathrm{Cl}\\ | \\ \mathrm{Cl{-}Pt{-}Cl}\\ | \\ \mathrm{Cl}\end{array} \xrightarrow{\mathrm{NH_3}} \begin{array}{c}\mathrm{Cl}\\ | \\ \mathrm{Cl{-}Pt{-}Cl}\\ | \\ \mathrm{NH_3}\end{array} \xrightarrow{\mathrm{NH_3}} \begin{array}{c}\mathrm{Cl}\\ | \\ \mathrm{Cl{-}Pt{-}NH_3}\\ | \\ \mathrm{NH_3}\end{array}$$

$$\text{(c)}\quad \begin{array}{c}\mathrm{NH_3}\\ | \\ \mathrm{py{-}Pt{-}py}\\ | \\ \mathrm{NH_3}\end{array} \xrightarrow{\mathrm{Cl^-}} \begin{array}{c}\mathrm{NH_3}\\ | \\ \mathrm{py{-}Pt{-}Cl}\\ | \\ \mathrm{NH_3}\end{array} \xrightarrow{\mathrm{Cl^-}} \begin{array}{c}\mathrm{NH_3}\\ | \\ \mathrm{Cl{-}Pt{-}Cl}\\ | \\ \mathrm{NH_3}\end{array}$$

$$\text{(d)}\quad \begin{array}{c}\mathrm{NH_3}\\ | \\ \mathrm{py{-}Pt{-}py}\\ | \\ \mathrm{NH_3}\end{array} \xrightarrow{\mathrm{Cl^-}} \begin{array}{c}\mathrm{Cl}\\ | \\ \mathrm{py{-}Pt{-}py}\\ | \\ \mathrm{NH_3}\end{array} \xrightarrow{\mathrm{Cl^-}} \begin{array}{c}\mathrm{Cl}\\ | \\ \mathrm{py{-}Pt{-}py}\\ | \\ \mathrm{Cl}\end{array}$$

$$\text{(e)}\quad \begin{array}{c}\mathrm{py}\\ | \\ \mathrm{py{-}Pt{-}NH_3}\\ | \\ \mathrm{NH_3}\end{array} \xrightarrow{\mathrm{Cl^-}} \begin{array}{c}\mathrm{Cl}\\ | \\ \mathrm{py{-}Pt{-}NH_3}\\ | \\ \mathrm{NH_3}\end{array} \xrightarrow{\mathrm{Cl^-}} \begin{array}{c}\mathrm{Cl}\\ | \\ \mathrm{py{-}Pt{-}NH_3}\\ | \\ \mathrm{Cl}\end{array}$$

$$\text{(f)}\quad \begin{array}{c}\mathrm{py}\\ | \\ \mathrm{py{-}Pt{-}NH_3}\\ | \\ \mathrm{NH_3}\end{array} \xrightarrow{\mathrm{Cl^-}} \begin{array}{c}\mathrm{py}\\ | \\ \mathrm{py{-}Pt{-}NH_3}\\ | \\ \mathrm{Cl}\end{array} \xrightarrow{\mathrm{Cl^-}} \begin{array}{c}\mathrm{Cl}\\ | \\ \mathrm{py{-}Pt{-}NH_3}\\ | \\ \mathrm{Cl}\end{array}$$

$$\text{(g)}\quad \begin{array}{c}\mathrm{NH_3}\\ | \\ \mathrm{Cl{-}Pt{-}Cl}\\ | \\ \mathrm{NH_3}\end{array} \xrightarrow{\mathrm{py}} \begin{array}{c}\mathrm{NH_3}\\ | \\ \mathrm{Cl{-}Pt{-}py}\\ | \\ \mathrm{NH_3}\end{array} \xrightarrow{\mathrm{py}} \begin{array}{c}\mathrm{NH_3}\\ | \\ \mathrm{py{-}Pt{-}py}\\ | \\ \mathrm{NH_3}\end{array}$$

$$\text{(h)}\quad \begin{array}{c}\mathrm{Cl}\\ | \\ \mathrm{Cl{-}Pt{-}NH_3}\\ | \\ \mathrm{NH_3}\end{array} \xrightarrow{\mathrm{py}} \begin{array}{c}\mathrm{Cl}\\ | \\ \mathrm{py{-}Pt{-}NH_3}\\ | \\ \mathrm{NH_3}\end{array} \xrightarrow{\mathrm{py}} \begin{array}{c}\mathrm{py}\\ | \\ \mathrm{py{-}Pt{-}NH_3}\\ | \\ \mathrm{NH_3}\end{array}$$

FIGURE 12.13 Stereochemistry and the *trans* Effect in Pt(II) Reactions. Charges have been omitted for clarity. In (a) through (f), the first substitution can be at any position, with the second controlled by the *trans* effect. In (g) and (h), both substitutions are controlled by the lability of chloride.

12.7.1 Explanations of the *trans* Effect[27]

Sigma-Bonding Effects

The *trans* effect is rationalized by two factors, weakening of the Pt—X bond and stabilization of the presumed 5-coordinate transition state. Pertinent energy coordinate diagrams are shown in **Figure 12.14**.

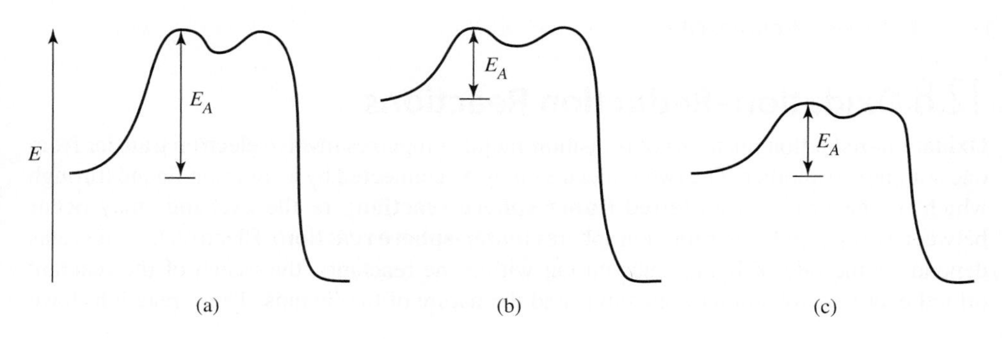

FIGURE 12.14 Activation Energy and the *trans* Effect. The depth of the energy curve for the intermediate and the relative heights of the two maxima will vary with the specific reaction. (a) Poor *trans* effect: low ground state, high transition state. (b) σ-Bonding effect: higher ground state (*trans* influence). (c) π-Bonding effect: lower transition state, (*trans* effect).

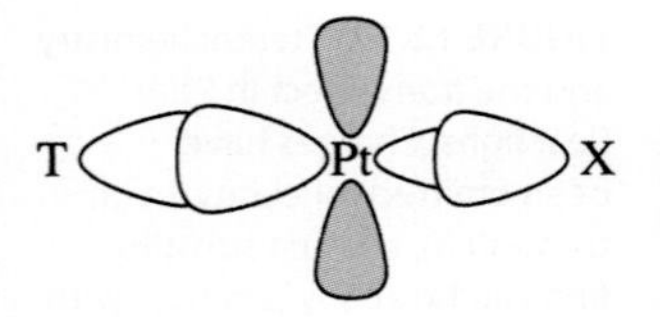

FIGURE 12.15 Sigma-Bonding Effect. A strong σ bond between Pt and T weakens the Pt—X bond.

The Pt—X bond is influenced by the Pt—T bond, because both use the Pt p_x and $d_{x^2-y^2}$ orbitals. When the Pt—T σ bond is strong, it uses a larger contribution of these orbitals and leaves less for the Pt—X bond (**Figure 12.15**). As a result, the Pt—X bond is weaker, and its ground state (sigma-bonding orbital) is higher in energy, as in Figure 12.14b. This ground-state, thermodynamic effect is called the ***trans* influence**. It contributes to the reaction rate by lowering the activation barrier for Pt—X bond breaking. The ranking below predicts the order for the *trans* influence on the basis of the relative σ-donor properties of the ligands:

$$H^- > PR_3 > SCN^- > I^- \sim CH_3^- \sim CO \sim CN^- > Br^- > Cl^- > NH_3 > OH^-$$

The order given here does not exactly match that for the *trans* effect, particularly for CO and CN^-, which have stronger *trans* effects than indicated.

Pi-Bonding Effects

The additional factor needed is Pt—T π bonding. When the T ligand engages in a strong π-acceptor (backbonding) interaction with Pt, charge is removed from Pt, rendering the metal center more electrophilic and more susceptible to nucleophilic attack. This is the prerequisite for formation of the 5-coordinate intermediate with a relatively strong Pt—Y bond, stabilizing the intermediate. It is noteworthy that π-backbonding between M and T also stabilizes the intermediate by partially offsetting the increase in energy due to the M—X bond breaking. The energy of the transition state is lowered, reducing the activation energy (Figure 12.14c). The order of ligand π-acceptor ability is

$$C_2H_4 \sim CO > CN^- > NO_2^- > SCN^- > I^- > Br^- > Cl^- > NH_3 > OH^-$$

The overall *trans* effect list is the result of the combination of the two effects:

$$CO \sim CN^- \sim C_2H_4 > PR_3 \sim H^- > CH_3^- \sim SC(NH_2)_2 > C_6H_5^- >$$

$$NO_2^- \sim SCN^- \sim I^- > Br^- > Cl^- > py, NH_3 \sim OH^- \sim H_2O$$

Ligands highest in the series are strong π acceptors, followed by strong σ donors. Ligands at the low end of the series have neither strong σ-donor nor π-acceptor abilities. The *trans* effect can be very large; rates may differ as much as 10^6 between complexes with strong *trans* effect ligands and those with weak *trans* effect ligands.

EXERCISE 12.4

It is possible to prepare isomers of Pt(II) complexes with four different ligands. Predict the products expected if 1 mole of $[PtCl_4]^{2-}$ is reacted successively with the following reagents (e.g., the product of reaction **a** is used in reaction **b**):

a. 2 moles of ammonia

b. 2 moles of pyridine (see Reactions **g** and **h** in Figure 12.13)

c. 2 moles of chloride

d. 1 mole of nitrite, NO_2^-

12.8 Oxidation–Reduction Reactions

Oxidation–reduction reactions of transition-metal complexes involve electron transfer from one complex to another. The two molecules may be connected by a common ligand through which the electron is transferred (**inner-sphere reaction**), or the exchange may occur between two separate coordination spheres (**outer-sphere reaction**). Electron transfer rates depend on the rate of ligand substitution within the reactants, the match of the reactant orbital energies, solvation of reactants, and the nature of the ligands. These reactions have been studied by chemical analysis of the products, stopped-flow spectrophotometry, and the use of radioactive and stable isotope tracers.[28]

12.8.1 Inner-Sphere and Outer-Sphere Reactions

When the ligands of both reactants are tightly held, reaction proceeds by outer-sphere electron transfer with no change in the coordination spheres. Classic examples are in **Table 12.14.**

The rate constants show large differences since they depend on the ability of the electrons to tunnel through the ligands. This is a quantum mechanical property whereby electrons can pass through potential barriers that are too high to permit ordinary transfer. Ligands with π or p orbitals that can be used in bonding provide good pathways for tunneling. Ligands like NH_3, with neither extra nonbonding pairs nor low-lying antibonding orbitals, do not provide effective tunneling pathways.

In outer-sphere reactions, the primary change on electron transfer is the metal-ligand bond distance. A higher metal oxidation state leads to shorter σ bonds, with the extent of change depending on the electronic structure. The changes in bond distance are larger when e_g electrons are involved, as in the change from high-spin Co(II) ($t_{2g}^5\, e_g^2$) to low-spin Co(III) (t_{2g}^6). Removal of electrons from these e_g antibonding orbitals results in shorter metal-ligand bond distances. Another perspective on coordination complexes functioning as outer-sphere oxidizing agents is that Δ_o decreases upon reduction of the metal oxidation state. This can be accompanied by a change from a high-spin to a low-spin electronic ground state. One way to assess the difference in energy between an outer-sphere oxidizing agent and its reduced product is by a comparison of LFSE values. The difference in LFSE between low-spin $[Co(NH_3)_6]^{2+}$ and low-spin $[Co(NH_3)_6]^{3+}$ is 39,240 cm^{-1} while the corresponding difference between high-spin $[Co(H_2O)_6]^{2+}$ and low-spin $[Co(H_2O)_6]^{3+}$ is 33,480 cm^{-1} (See Problem 10.18). The stronger field ammonia ligands render the loss in LFSE associated with reduction of $[Co(NH_3)_6]^{3+}$ greater than that for $[Co(H_2O)_6]^{3+}$. This is one factor that results in $[Co(H_2O)_6]^{3+}$ being a stronger oxidizing agent than $[Co(NH_3)_6]^{3+}$.

$$[Co(H_2O)_6]^{3+} + e^- \longrightarrow [Co(H_2O)_6]^{2+} \quad E^\circ = +1.808\text{ V}$$

$$[Co(NH_3)_6]^{3+} + e^- \longrightarrow [Co(NH_3)_6]^{2+} \quad E^\circ = +0.108\text{ V}$$

Inner-sphere reactions use the tunneling phenomenon with a ligand as the conduit. These reactions proceed in three steps: (1) a substitution reaction that leaves the oxidant

TABLE 12.14 Rate Constants for Outer-Sphere Electron Transfer Reactions[a]

	Reductants	
Oxidant	$[Cr(bipy)_3]^{2+}$	$[Ru(NH_3)_6]^{2+}$
$[Co(NH_3)_5(NH_3)]^{3+}$	6.9×10^2	1.1×10^{-2}
$[Co(NH_3)_5(F)]^{2+}$	1.8×10^3	
$[Co(NH_3)_5(OH)]^{2+}$	3×10^4	4×10^{-2}
$[Co(NH_3)_5(NO_3)]^{2+}$		3.4×10^1
$[Co(NH_3)_5(H_2O)]^{3+}$	5×10^4	3.0
$[Co(NH_3)_5(Cl)]^{2+}$	8×10^5	2.6×10^2
$[Co(NH_3)_5(Br)]^{2+}$	5×10^6	1.6×10^3
$[Co(NH_3)_5(I)]^{2+}$		6.7×10^3

$[Cr(bipy)_3]^{2+}$ data from J. P. Candlin, J. Halpern, D. L. Trimm, *J. Am. Chem. Soc.*, **1964**, *86*, 1019. $[Ru(NH_3)_6]^{2+}$ data from J. F. Endicott, H. Taube, *J. Am. Chem. Soc.*, **1964**, *86*, 1686.

NOTE: [a]Second-order rate constants in M^{-1} s^{-1} at 25 °C.

and reductant linked by a bridging ligand; (2) the electron transfer, frequently accompanied by transfer of the ligand; and (3) separation of the products:[29]

$$\underset{\text{Co(III) oxidant}}{[Co(NH_3)_5(Cl)]^{2+}} + \underset{\text{Cr(II) reductant}}{[Cr(H_2O)_6]^{2+}} \longrightarrow \underset{\text{Co(III)} \quad \text{Cr(II)}}{[(NH_3)_5Co(Cl)Cr(H_2O)_5]^{4+}} + H_2O \qquad (1)$$

$$\underset{\text{Co(III)} \quad \text{Cr(II)}}{[(NH_3)_5Co(Cl)Cr(H_2O)_5]^{4+}} \longrightarrow \underset{\text{Co(II)} \quad \text{Cr(III)}}{[(NH_3)_5Co(Cl)Cr(H_2O)_5]^{4+}} \qquad (2)$$

$$[(NH_3)_5Co(Cl)Cr(H_2O)_5]^{4+} + H_2O \longrightarrow [(NH_3)_5Co(H_2O)]^{2+} + [(Cl)Cr(H_2O)_5]^{2+} \qquad (3)$$

These are followed by formation of $[Co(H_2O)_6]^{2+}$ because Co(II) is labile:

$$[(NH_3)_5Co(H_2O)]^{2+} + 5\,H_2O \longrightarrow [Co(H_2O)_6]^{2+} + 5\,NH_3$$

This transfer of chloride to the chromium has been established experimentally. The products can be separated by ion exchange techniques, and Cr(III) complexes are inert. The isolated Cr(III) appears exclusively as $[(Cl)Cr(H_2O)_5]^{2+}$. To show that chloride transfer from a coordination complex to $[Cr(H_2O)_6]^{2+}$ is feasible to promote inner-sphere electron transfer, the $[Cr(H_2O)_6]^{2+}/[Cr(H_2O)_5Cl]^{2+}$ Cr(II)/Cr(III) exchange reaction, which results in no net change, has been studied using radioactive ^{51}Cr.[30] All the transferred chloride in this reaction comes from $[Cr(H_2O)_5Cl]^{2+}$, with none entering the Cr(II) coordination sphere even when excess Cl^- is present in the solution.

The choice between inner- and outer-sphere mechanisms is difficult to ascertain. In Table 12.14, the outer-sphere mechanism is imposed by the reducing agent. $[Ru(NH_3)_6]^{2+}$ is an inert species and does not allow formation of bridging species. Although $[Cr(bipy)_3]^{2+}$ is formally labile, the chelate effect may predispose this complex to an outer-sphere mechanism. The delocalized π systems of the bipy ligands of $[Cr(bipy)_3]^{2+}$ may lower the barrier for outer-sphere electron transfer relative to $[Ru(NH_3)_6]^{2+}$; MLCT in $[Cr(bipy)_3]^{2+}$ may facilitate the electron transfer.

The oxidant can also dictate an outer-sphere mechanism. In **Table 12.15**, $[Co(NH_3)_6]^{3+}$ and $[Co(en)_3]^{3+}$ participate in outer-sphere electron transfer; their ligands have no accessible lone pairs with which to form bonds to bridge with the reductant. The electron transfer mechanisms for the other reactions are less certain, although labile $Cr^{2+}(aq)$ is assumed to react by inner-sphere mechanisms when bridging is possible.

The range of rate constants with V^{2+} is much smaller with these oxidants than for Cr^{2+}. Since the ligands of the oxidant have a smaller impact on the electron transfer rate

TABLE 12.15 Rate Constants for Aquated Reductants[a]

	Cr^{2+}	Eu^{2+}	V^{2+}
$[Co(en)_3]^{3+}$	$\sim 2 \times 10^{-5}$	$\sim 5 \times 10^{-3}$	$\sim 2 \times 10^{-4}$
$[Co(NH_3)_6]^{3+}$	8.9×10^{-5}	2×10^{-2}	3.7×10^{-2}
$[Co(NH_3)_5(H_2O)]^{3+}$	5×10^{-1}	1.5×10^{-1}	$\sim 5 \times 10^{-1}$
$[Co(NH_3)_5(NO_3)]^{2+}$	$\sim 9 \times 10^{-1}$	$\sim 1 \times 10^{2}$	
$[Co(NH_3)_5(Cl)]^{2+}$	6×10^{5}	3.9×10^{2}	~ 5
$[Co(NH_3)_5(Br)]^{2+}$	1.4×10^{6}	2.5×10^{2}	2.5×10^{1}
$[Co(NH_3)_5(I)]^{2+}$	3×10^{6}	1.2×10^{2}	1.2×10^{2}

Data from J. P. Candlin, J. Halpern, D. L. Trimm, *J. Am. Chem. Soc.*, **1964**, *86*, 1019; data for Cr^{2+} reactions with halide complexes from J. P. Candlin, J. Halpern, *Inorg. Chem.*, **1965**, *4*, 756; data for $[Co(NH_3)_6]^{3+}$ reactions with Cr^{2+} and V^{2+} from A. Zwickel, H. Taube, *J. Am. Chem. Soc.*, **1961**, *83*, 793.

NOTE: [a]Rate constants in $M^{-1}\,s^{-1}$.

with V^{2+} in comparison with Cr^{2+}, an outer-sphere mechanism is more likely with V^{2+}. It is interesting that the rate constants for reduction of $[Co(NH_3)_5(Br)]^{2+}$ by $[Cr(bipy)_3]^{2+}$ (Table 12.14) and $[Cr(H_2O)_6]^{2+}$ (Table 12.15), respectively, are very similar; these electron transfer reactions may proceed by outer-sphere mechanisms. The rate constants with $[Co(NH_3)_5(H_2O)]^{3+}$ and these Cr(II) reductants vary by a factor of 100,000.

Eu^{2+} *(aq)* is an unusual case because the electron transfer rate *decreases* from $[Co(NH_3)_5(Cl)]^{2+}$ to $[Co(NH_3)_5(I)]^{2+}$. It has been hypothesized that the thermodynamic stability of the EuX^+ species increases as the halide size decreases, with slower rates and lower stabilities as we go down the series. Because of the smaller range of rate constants, Eu^{2+} reactions are usually classified as outer-sphere.

When $[Co(CN)_5]^{3-}$ reacts with Co(III) oxidants ($[Co(NH_3)_5X]^{2+}$) that have halide ligands as bridging candidates, the product is $[Co(CN)_5X]^{3-}$, evidence for an inner-sphere mechanism. Rate constants for these reactions are given in **Table 12.16**. Surprisingly, the reaction with $[Co(NH_3)_6]^{3+}$ has a similar rate constant even though this electron transfer likely proceeds by an outer-sphere mechanism because ammonia cannot bridge. Reactions with thiocyanate or nitrite as bridging groups show interesting behavior. The N-bonded $[(NH_3)_5Co(NCS)]^{2+}$ reacts by bridging via the free S end of the ligand. In a similar fashion, a transient O-bonded intermediate is detected[31] in reactions of $[(NH_3)_5Co(NO_2)]^{2+}$ with $[Co(CN)_5]^{3-}$.

Ligands that are easier to reduce can result in complexes that are more quickly reduced.[32] A useful comparison is between the oxidizing agents $[(NH_3)_5CoL]^{2+}$, where L is benzoate (difficult to reduce) and 4-carboxy-N-methylpyridine (easier to reduce). The rate constants for the reduction of these two complexes by $[Cr(H_2O)_6]^{2+}$ differ by a factor of 10, although both have similar structures and transition states (**Table 12.17**). Both reactions exhibit inner-sphere mechanisms, with formation of a bridging carboxylate, and Cr(II) bridging via the L carbonyl oxygen. The difference in rates reflects the electron transfer rates. Table 12.17 extends the data to the L ligands glyoxylate and glycolate, which are more easily reduced than 4-carboxy-N-methylpyridine. Electron transfer through bridging ligands is faster when those ligands are more easily reduced.

Some ligands are functionalized to permit reductants to bind to a remote site subsequent to electron transfer. For example, isonicotinamide (4-pyridine carboxylic acid amide) bonded through the pyridine nitrogen can react with Cr^{2+} through the carbonyl oxygen of the ligand. A bridging ligand results and an electron moves from the Cr(II) center to the other metal. **Table 12.18** provides the rate constants for oxidants with different metals. The rate constants for the cobalt pentaammine (17.6 $M^{-1}\,s^{-1}$) and the chromium pentaaqua

O NH2
N
isonicotinamide

TABLE 12.16 Rate Constants for Reactions with $[Co(CN)_5]^{3-}$

Oxidant	k ($M^{-1}\,s^{-1}$)
$[Co(NH_3)_5(F)]^{2+}$	1.8×10^3
$[Co(NH_3)_5(OH)]^{2+}$	9.3×10^4
$[Co(NH_3)_5(NH_3)]^{3+}$	8×10^4 [a]
$[Co(NH_3)_5(NCS)]^{2+}$	1.1×10^6
$[Co(NH_3)_5(N_3)]^{2+}$	1.6×10^6
$[Co(NH_3)_5(Cl)]^{2+}$	$\sim 5 \times 10^7$

Data from J. P. Candlin, J. Halpern, S. Nakamura, *J. Am. Chem. Soc.*, **1963**, *85*, 2517.
[a]Outer-sphere mechanism caused by the oxidant. Complexes with other potential bridging groups (PO_4^{3-}, SO_4^{2-}, CO_3^{2-}, and several carboxylic acids) also react by an outer-sphere mechanism, with constants ranging from 5×10^2 to 4×10^4.

TABLE 12.17 Ligand Reducibility and Electron Transfer

Rate Constants for the Reaction $[(NH_3)_5CoL]^{2+} + [Cr(H_2O)_6]^{2+} \longrightarrow Co^{2+} + 5\,NH_3 + [Cr(H_2O)_5L]^{2+} + H_2O$		
L	k_2 ($M^{-1}\,s^{-1}$)	Comments
$C_6H_5C(=O)-O^-$	0.15	Benzoate is difficult to reduce
$CH_3C(=O)-O^-$	0.34	Acetate is difficult to reduce
$CH_3NC_5H_4C(=O)-O^-$	1.3	4-carboxy-N-methylpyridine is more reducible
$O{=}CHC(=O)-O^-$	3.1	Glyoxylate is easy to reduce
$HOCH_2C(=O)-O^-$	7×10^3	Glycolate is very easy to reduce

Data from H. Taube, *Electron Transfer Reactions of Complex Ions in Solution*, Academic Press, New York, 1970, pp. 64–66.

TABLE 12.18 Rate Constants for Reduction of Isonicotinamide (4-Pyridine Carboxylic Acid Amide) Complexes by $[Cr(H_2O)_6]^{2+}$

Oxidant	k_2 ($M^{-1}\,s^{-1}$)
$[(NH_2C(=O)-C_5H_4N)Cr(H_2O)_5]^{3+}$	1.8
$[(NH_2C(=O)-C_5H_4N)Co(NH_3)_5]^{3+}$	17.6
$[(NH_2C(=O)-C_5H_4N)Ru(NH_3)_5]^{3+}$	5×10^5

Data from H. Taube, *Electron Transfer Reactions of Complex Ions in Solution*, Academic Press, New York, 1970, pp. 66–68.

(1.8 $M^{-1}\,s^{-1}$) complexes are much closer than usual. The rate for Co compounds with other bridging ligands is frequently as much as 10^5 faster than the rate for corresponding Cr compounds, because of the greater oxidizing power of Co(III).* With these isonicotinamide compounds, the overall rate seems to depend more on the electron transfer rate from Cr^{2+} through the bridging ligand, and the readily reducible isonicotinamide render the two reactions more nearly equal in rate. The much faster rate found for the ruthenium pentaammine has been explained as the result of the transfer of an electron through the π system of the ligand into the t_{2g} levels of Ru(III); low-spin Ru(III) has a vacancy in the t_{2g} level. A similar electron transfer to Co(III) or Cr(III) places the incoming electron in the e_g levels, which have σ symmetry.[33]

*As a familiar example, the standard reduction potential (E°) for $Co^{3+}(aq) + e^- \longrightarrow Co^{2+}(aq)$ equals 1.808 volts, while that of $Cr^{3+}(aq) + e^- \longrightarrow Cr^{2+}(aq)$ is only −0.41 volts.

12.8.2 Conditions for High and Low Oxidation Numbers

The overall stability of complexes with different charges on the metal ion depends on the LFSE, metal-ligand bonding, and redox properties of the ligands. The hard and soft character of the ligands also has an effect. All the very high oxidation numbers for the transition metals are found in combination with hard ligands, such as fluoride and oxide. Examples include MnO_4^-, CrO_4^{2-}, and FeO_4^{2-} with oxide, and RuF_5, PtF_6, and OsF_6 with fluoride. The lowest oxidation states are found with soft ligands. Extremely soft zerovalent metals are often found in complexes with carbonyl ligands; $V(CO)_6$, $Cr(CO)_6$, $Fe(CO)_5$, $Co_2(CO)_8$, and $Ni(CO)_4$ are examples (Chapter 13).

Reactions of copper complexes show these ligand effects. **Table 12.19** lists some of these reactions and their electrode potentials. If the reactions of the aquated Cu(II) and Cu(I) are taken as the basis for comparison, it can be seen that complexing Cu(II) with the hard ligand ammonia reduces the potential, stabilizing the higher Cu(II) oxidation state as compared to either Cu(I) or Cu(0). On the other hand, the soft ligand cyanide stabilizes its Cu(I) complex and renders the reduction of Cu(II) more facile. The halide cases are complicated by precipitation but still show the effects; the soft iodide ligand renders Cu(I) more stable than does the harder chloride. Table 12.19 dramatically illustrates how much the oxidizing ability of a metal ion can be tuned by coordination of ligands.

Sometimes a wide variety of ligands can kinetically stabilize ions that would normally be potent oxidizing agents. A noteworthy example is the Co(III)–Co(II) couple. The hydrated ion $[Co(H_2O)_6]^{3+}$ is a very strong oxidizing agent, reacting readily with water to form oxygen and Co(II). However, when coordinated with any ligand other than water or fluoride, Co(III) is kinetically stable, and significantly more stable thermodynamically. The oxidation of a Co(II) complex to a Co(III) complex generally results in a change from the high-spin Co(II) configuration $t_{2g}^5\, e_g^2$ to the low-spin Co(III) configuration t_{2g}^6, accompanied by a significant increase in LFSE with almost any ligand field ($-\frac{8}{5}\,\Delta_o$). This contributes to the stability of many Co(III) complexes, rendering many of them weaker oxidizing agents than expected. The reduction potentials (Table 12.19) for Co(III)–Co(II) with different ligands are in the order $H_2O > NH_3 > CN^-$; as Δ_o increases, the

TABLE 12.19 Electrode Potentials of Cobalt and Copper Species in Aqueous Solution

Cu(II)–Cu(I) Reactions	$E°$ (V)
$Cu^{2+} + 2\,CN^- + e^- \rightleftharpoons [Cu(CN)_2]^-$	+1.103
$Cu^{2+} + I^- + e^- \rightleftharpoons CuI(s)$	+0.86
$Cu^{2+} + Cl^- + e^- \rightleftharpoons CuCl(s)$	+0.538
$Cu^{2+} + e^- \rightleftharpoons Cu^+$	+0.153
$[Cu(NH_3)_4]^{2+} + e^- \rightleftharpoons [Cu(NH_3)_2]^+ + 2\,NH_3$	−0.01
Cu(II)–Cu(0) Reactions	**$E°$ (V)**
$Cu^{2+} + 2\,e^- \rightleftharpoons Cu(s)$	+0.337
$[Cu(NH_3)_4]^{2+} + 2\,e^- \rightleftharpoons Cu(s) + 4\,NH_3$	−0.05
Co(III)–Co(II) Reactions	**$E°$ (V)**
$[Co(H_2O)_6]^{3+} + e^- \rightleftharpoons [Co(H_2O)_6]^{2+}$	+1.808
$[Co(NH_3)_6]^{3+} + e^- \rightleftharpoons [Co(NH_3)_6]^{2+}$	+0.108
$[Co(CN)_6]^{3-} + e^- \rightleftharpoons [Co(CN)_6]^{4-}$	−0.83

Data from T. Moeller, *Inorganic Chemistry*, Wiley InterScience, New York, 1982, p. 742.

oxidizing strength of the Co(III) complex decreases. In this case, the LFSE effect is sufficiently important to overcome the effect of softer ligands stabilizing lower oxidation states demonstrated for the Cu(II)/Cu(I) reductions. Since CN^- is the softest ligand in this series, it would be predicted to best stabilize Co(II); $[Co(CN)_6]^{3-}$ would be expected to be the stronger oxidant solely on the basis of hard and soft arguments.

12.9 Reactions of Coordinated Ligands

Coordination to the metal changes ligand properties sufficiently to make possible reactions at the ligands that either (a) could not happen with the unbound ligand or (b) could occur without the metal but much more slowly. Reactions at coordinated ligands are a vital aspect of organometallic chemistry (Chapter 14). We will describe a few examples of these reactions within coordination chemistry.

Organic chemists have long used inorganic compounds as reagents. Lewis acids such as $AlCl_3$, $FeCl_3$, $SnCl_4$, $ZnCl_2$, and $SbCl_5$ are used in Friedel–Crafts electrophilic substitutions. The labile complexes formed by acyl or alkyl halides and these Lewis acids create positively charged carbon atoms that react readily with aromatic compounds. The reactions are generally the same as without the metal salts, but these Lewis acids speed the reactions.

12.9.1 Hydrolysis of Esters, Amides, and Peptides

Amino acid esters, amides, and peptides can be hydrolyzed in basic solution, and metal ions (Cu(II), Co(II), Ni(II), Mn(II), Ca(II), and Mg(II), and others) speed these reactions. The uncertain mechanism is either through bidentate coordination of the α-amino group and the carbonyl, or only through the amine. The rates of these reactions often exhibit complicated temperature dependence and deduction of the mechanism is difficult.[34]

Co(III) complexes promote similar reactions. When four of the six octahedral positions are occupied by amine ligands, and two *cis* positions are available for ligand substitution, these hydrolysis reactions can be examined in detail. These compounds generally catalyze the hydrolysis of N-terminal amino acids from peptides; the amino acid that is removed remains bound to the metal. The reactions apparently proceed by coordination of the free amine to cobalt, followed either by coordination of the carbonyl to cobalt and subsequent reaction with OH^- or H_2O (path 1 in **Figure 12.16**) or reaction of the carbonyl

FIGURE 12.16 Peptide Hydrolysis by $[Co(trien)(H_2O)(OH)]^{2+}$.

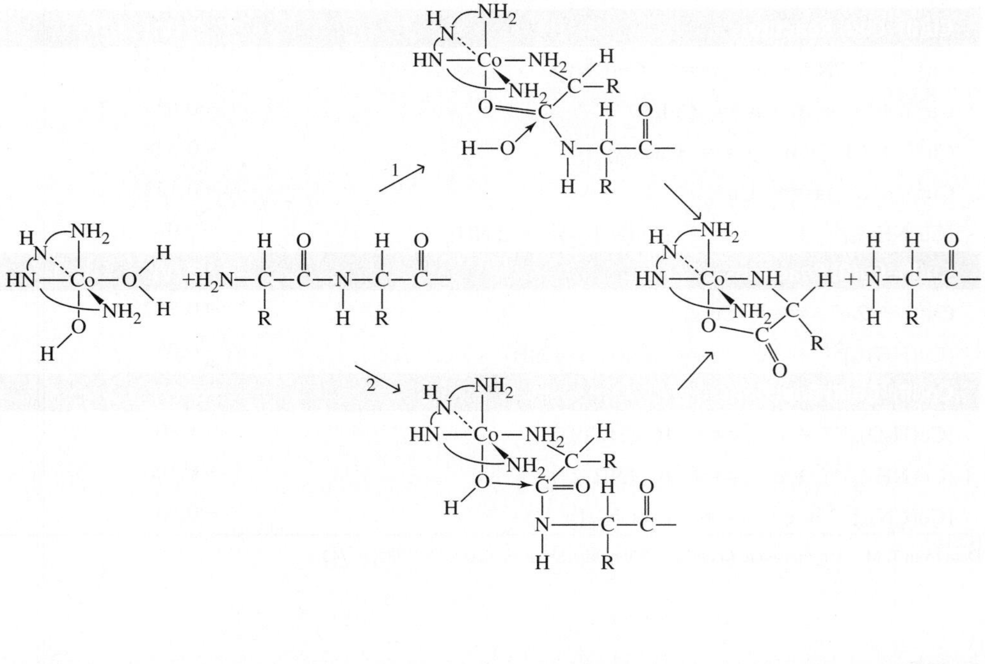

carbon with coordinated hydroxide (path 2).[35] As a result, the N-terminal amino acid is removed from the peptide and left as part of the cobalt complex in which the α-amino nitrogen and the carbonyl oxygen are bonded to the cobalt. Esters and amides are hydrolyzed by the same mechanism, with the relative importance of the two pathways dependent on the specific compounds used. Other compounds such as phosphate esters, pyrophosphates, and amides of phosphoric acid are hydrolyzed in similar reactions.

12.9.2 Template Reactions

Template reactions are those in which formation of a complex places the ligands in the correct geometry for reaction. One of the earliest was for the formation of phthalocyanines (**Figure 12.17**). The study of this chemistry began in 1928, after discovery of a blue impurity in phthalimide prepared by reaction of phthalic anhydride with ammonia in an enameled vessel. This impurity was later discovered to be an iron phthalocyanine complex, created from iron released into the mixture via a scratch in the enamel surface. A similar reaction takes place with copper; intermediates isolated from this reaction are shown in Figure 12.17. Phthalic acid and ammonia first form phthalimide, then 1-keto-3-iminoisoindoline, and then 1-amino-3-iminoisoindolenine. The cyclization reaction then occurs, probably with the assistance of the metal ion, which holds the chelated reactants in position. This is confirmed by the lack of cyclization in the absence of the metals.[36] The essential feature of these reactions is the formation of the cyclic compound by coordination to a metal ion.

Similar reactions have been used extensively in the formation of macrocyclic compounds. Imine or Schiff base complexes ($R_1N{=}CHR_2$) have been studied. In this case, the compounds can be formed without complexation, but the reaction is much faster in the presence of metal ions (**Figure 12.18**). In the absence of the Ni(II) complex, benzothiazoline is favored in the final step, rather than the imine; very little Schiff base is present at equilibrium.

A major feature of template reactions is that formation of the complex brings the reactants into close proximity with the proper orientation for reaction. Complexation also changes the electronic structure to promote the reaction. Both features are important in all coordinated ligand reactions, but the orientation factor is more obvious because the final product has a structure determined by the coordination geometry. Template reactions have been reviewed.[37]

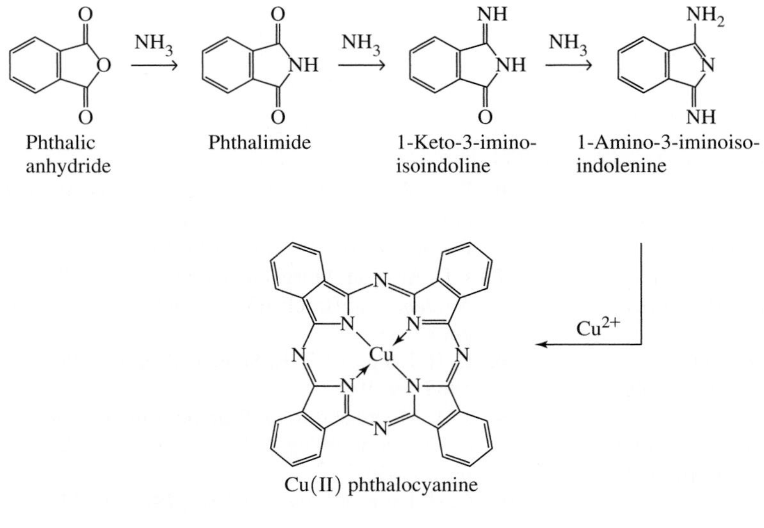

FIGURE 12.17 Phthalocyanine Synthesis.

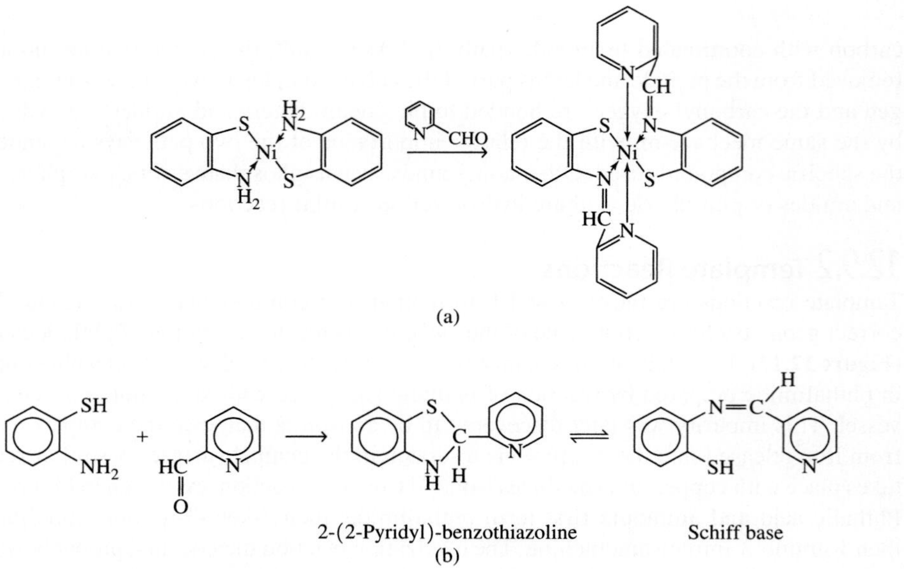

FIGURE 12.18 Schiff Base Template Reaction. (a) The Ni(II)-*o*-aminothiophenol complex reacts with pyridine-1-carboxaldehyde to form the Schiff base complex. (b) In the absence of the metal ion, the product is benzothiazoline; very little of the Schiff base is formed. Data from L.F. Lindoy, S.E. Livingstone, *Inorg. Chem.*, **1968,** *7,* 1149.[38]

(a)

2-(2-Pyridyl)-benzothiazoline Schiff base
(b)

12.9.3 Electrophilic Substitution

Acetylacetone complexes are known to undergo a wide variety of reactions that are similar to aromatic electrophilic substitutions. Bromination, nitration, and similar reactions have been studied.[39] In all cases, coordination forces the ligand into an enol form and promotes reaction at the center carbon by preventing reaction at the oxygens and concentrating negative charge on the central carbon of the acetylacetonato ligand. **Figure 12.19** shows the reactions and a possible mechanism.

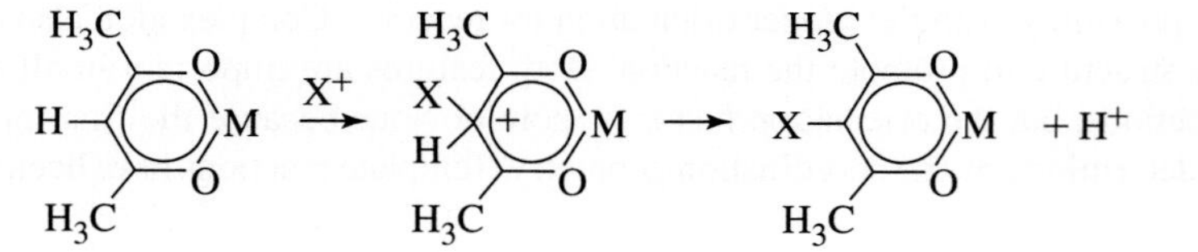

FIGURE 12.19 Electrophilic Substitution on Acetylacetone Complexes. X = Cl, Br, SCN, SAr, SCl, NO_2, CH_2Cl, $CH_2N(CH_3)_2$, COR, CHO.

References

1. H. Taube, *Chem. Rev.*, **1952**, *50,* 69.
2. C. H. Langford and H. B. Gray, *Ligand Substitution Processes*, W. A. Benjamin, New York, 1966.
3. R. B. Jordan, *Reaction Mechanisms of Inorganic and Organometallic Systems*, 3rd ed., Oxford (New York), 2007, p. 86.
4. F. Wilkinson, *Chemical Kinetics and Reactions Mechanisms*, Van Nostrand-Reinhold, New York, 1980, pp. 83–91.
5. F. Basolo and R. G. Pearson, *Mechanisms of Inorganic Reactions*, 2nd ed., John Wiley & Sons, New York, 1967, pp. 158–170.
6. R. G. Wilkins, *The Study of Kinetics and Mechanism of Reactions of Transition Metal Complexes*, Allyn and Bacon, Boston, 1974, pp. 193–196.
7. J. D. Atwood, *Inorganic and Organometallic Reaction Mechanisms*, Brooks/Cole, Monterey, CA, 1985, pp. 82–83.
8. C. H. Langford, T. R. Stengle, *Ann. Rev. Phys. Chem.*, **1968**, *19*, 193.
9. J. W. Moore and R. G. Pearson, *Kinetics and Mechanism*, 3rd ed., John Wiley & Sons, New York, 1981, pp. 357–363.
10. C. H. Langford, *Inorg. Chem.*, **1965**, *4*, 265.

11. J. D. Atwood, *Inorganic and Organometallic Reaction Mechanisms*, Brooks/Cole, Monterey, CA, 1985, p. 85.

12. Wilkins, *The Study of Kinetics and Mechanism of Reactions of Transition Metal Complexes*, pp. 207–210; Basolo and Pearson, *Mechanisms of Inorganic Reactions*, pp. 177–193.

13. C. K. Ingold, *Structure and Mechanism in Organic Chemistry*, Cornell University Press, Ithaca, NY, 1953, Chapters 5 and 7.

14. D. A. Buckingham, P. J. Cressell, A. M. Sargeson, *Inorg. Chem.*, **1975**, *14*, 1485.

15. D. A. Buckingham, P. A. Marzilli, A. M. Sargeson, *Inorg. Chem.*, **1969**, *8*, 1595.

16. A. J. Dickie, D. C. R. Hockless, A. C. Willis, J. A. McKeon, W. G. Jackson, *Inorg. Chem.*, **2003**, *42*, 3822. W. G. Jackson, A. J. Dickie, J. A. McKeon, L. Spiccia, S. J. Brudenell, D. C. R. Hockless, A. C. Willis, *Inorg. Chem.*, **2005**, *44*, 401.

17. Basolo and Pearson, *Mechanisms of Inorganic Reactions*, pp. 27, 223; G. Schwarzenbach, *Helv. Chim. Acta*, **1952**, *35*, 2344.

18. D. W. Margerum, G. R. Cayley, D. C. Weatherburn, and G. K. Pagenkopf, "Kinetics of Complex Formation and Ligand Exchange," in A. E. Martell, ed., *Coordination Chemistry*, Vol. 2, American Chemical Society Monograph 174, Washington, DC, 1978, pp. 1–220.

19. L. J. Boucher, E. Kyuno, J. C. Bailar, Jr., *J. Am. Chem. Soc.*, **1964**, *86*, 3656.

20. Basolo and Pearson, *Mechanisms of Inorganic Reactions*, p. 272.

21. S. A. Johnson, F. Basolo, R. G. Pearson, *J. Am. Chem. Soc.*, **1963**, *85*, 1741; J. A. Broomhead, L. Kane-Maguire, *Inorg. Chem.*, **1969**, *8*, 2124.

22. R. C. Fay, T. S. Piper, *Inorg. Chem.*, **1964**, *3*, 348.

23. Basolo and Pearson, *Mechanisms of Inorganic Reactions*, pp. 377–379, 395.

24. Wilkins, *The Study of Kinetics and Mechanism of Reactions of Transition Metal Complexes*, p. 231.

25. J. D. Atwood, *Inorganic and Organometallic Reaction Mechanisms*, Brooks/Cole, Monterey, CA, 1985, pp. 60–63.

26. I. I. Chernyaev, *Ann. Inst. Platine USSR.*, **1926**, *4*, 261.

27. Atwood, *Inorganic and Organometallic Reactions Mechanisms*, p. 54; Basolo and Pearson, *Mechanisms of Inorganic Reactions*, p. 355.

28. T. J. Meyer and H. Taube, "Electron Transfer Reactions," in G. Wilkinson, R. D. Gillard, and J. A. McCleverty, eds., Pergamon, *Comprehensive Coordination Chemistry*, Vol. 1, London, 1987, pp. 331–384; H. Taube, *Electron Transfer Reactions of Complex Ions in Solution*, Academic Press, New York, 1970; *Chem. Rev.*, **1952**, *50*, 69; *J. Chem. Educ.*, **1968**, *45*, 452.

29. J. P. Candlin, J. Halpern, *Inorg. Chem.*, **1965**, *4*, 766.

30. D. L. Ball, E. L. King, *J. Am. Chem. Soc.*, **1958**, *80*, 1091.

31. J. Halpern, S. Nakamura, *J. Am. Chem. Soc.*, **1965**, *87*, 3002, J. L. Burmeister, *Inorg. Chem.*, **1964**, *3*, 919.

32. Taube, *Electron Transfer Reactions of Complex Ions in Solution*, pp. 64–66; E. S. Gould, H. Taube, *J. Am. Chem. Soc.*, **1964**, *86*, 1318.

33. H. Taube, E. S. Gould, *Acc. Chem. Res.*, **1969**, *2*, 321.

34. M. M. Jones, *Ligand Reactivity and Catalysis*, Academic Press, New York, 1968. Chapter III summarizes the arguments and mechanisms.

35. J. P. Collman, D. A. Buckingham, *J. Am. Chem. Soc.*, **1963**, *85*, 3039; D. A. Buckingham, J. P. Collman, D. A. R. Hopper, L. G. Marzelli, *J. Am. Chem. Soc.*, **1967**, *89*, 1082.

36. R. Price, "Dyes and Pigments," in G. Wilkinson, R. D. Gillard, and J. A. McCleverty, eds., *Comprehensive Coordination Chemistry*, Vol. 6, Pergamon Press, Oxford, 1987, pp. 88–89.

37. D. St. C. Black, "Stoichiometric Reactions of Coordinated Ligands," in Wilkinson, Gillard, and McCleverty, *Comprehensive Coordination Chemistry*, op. cit., pp. 155–226.

38. L. F. Lindoy, S. E. Livingstone, *Inorg. Chem.*, **1968**, *7*, 1149.

39. J. P. Collman, *Angew. Chem., Int. Ed.*, **1965**, *4*, 132.

General References

The general principles of kinetics and mechanisms have been described by R. B. Jordan, *Reaction Mechanisms of Inorganic and Organometallic Systems*, 3rd ed., Oxford University Press, New York, 2007, J. W. Moore and R. G. Pearson, *Kinetics and Mechanism*, 3rd ed., Wiley InterScience, New York, 1981, and in F. Wilkinson, *Chemical Kinetics and Reaction Mechanisms*, Van Nostrand-Reinhold, New York, 1980). The classic for coordination compounds is F. Basolo and R. G. Pearson, *Mechanisms of Inorganic Reactions*, 2nd ed., John Wiley & Sons, New York, 1967. More recent books are by J. D. Atwood, *Inorganic and Organometallic Reaction Mechanisms*, Brooks/Cole, Monterey, CA, 1985, and D. Katakis and G. Gordon, *Mechanisms of Inorganic Reactions*, Wiley InterScience, New York, 1987. The reviews in G. Wilkinson, R. D. Gillard, and J. A. McCleverty, editors, *Comprehensive Coordination Chemistry*, Pergamon Press, Elmsford, NY, 1987, provide a more comprehensive collection and discussion of the data. Volume 1, *Theory and Background*, covers substitution and redox reactions, and Volume 6, *Applications*, is particularly rich in data on ligand reactions.

Problems

12.1 The high-spin d^4 complex $[Cr(H_2O)_6]^{2+}$ is *labile*, but the low-spin d^4 complex ion $[Cr(CN)_6]^{4-}$ is *inert*. Explain.

12.2 Why is the existence of a series of entering groups with different rate constants evidence for an associative mechanism (A or I_a)?

12.3 Predict whether these complexes would be labile or inert and explain your choices. The magnetic moment is given in Bohr magnetons (μ_B) after each complex.

Ammonium oxopentachlorochromate(V)	1.82
Potassium hexaiodomanganate(IV)	3.82
Potassium hexacyanoferrate(III)	2.40
Hexaammineiron(II) chloride	5.45

12.4 The yellow "prussiate of soda" $Na_4[Fe(CN)_6]$ has been added to table salt as an anticaking agent. Why have there been no apparent toxic effects, even though this compound contains cyano ligands?

12.5 Consider the half-lives of substitution reactions of the pairs of complexes:

Half-Lives Shorter than 1 Minute	Half-Lives Longer than 1 Day
$[Cr(CN)_6]^{4-}$	$[Cr(CN)_6]^{3-}$
$[Fe(H_2O)_6]^{3+}$	$[Fe(CN)_6]^{4-}$
$[Co(H_2O)_6]^{2+}$	$[Co(NH_3)_5(H_2O)]^{3+}$(H_2O exchange)

Interpret the differences in half-lives in terms of the electronic structures of each pair.

12.6 The general rate law for substitution in square-planar Pt(II) complexes is valid for the reaction

$$[Pt(NH_3)_4]^{2+} + Cl^- \longrightarrow [Pt(NH_3)_3Cl]^+ + NH_3$$

Design the experiments needed to verify this and to determine the rate constants. What experimental data are needed, and how are the data to be treated?

12.7 The graph shows plots of k_{obs} versus $[X^-]$ for the anations $[Co(en)_2(NO_2)(DMSO)]^{2+} + X^- \longrightarrow [Co(en)_2(NO_2)X]^+ + DMSO$. The rate for DMSO exchange was determined as 30×10^{-5} sec^{-1}. (Graph created using data from W. R. Muir, C. H. Langford, *Inorg. Chem.*, **1968**, *7*, 1032.)

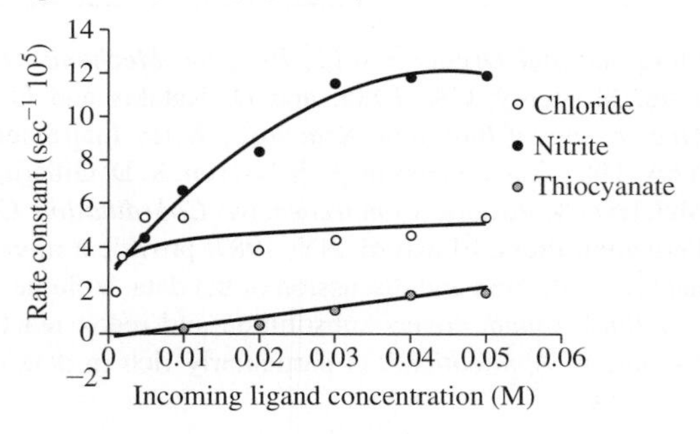

All three reactions are presumed to have the same mechanism.

a. Why is the DMSO exchange so much faster than the other reactions?

b. Why are the curves shaped as they are?

c. Explain what the limiting rate constants (at high concentration) are in terms of the rate laws for D and I_d mechanisms.

d. The limiting rate constants are $5.0 \times 10^{-5}\ s^{-1}$ and $12.0 \times 10^{-5}\ s^{-1}$ for Cl^- and NO_2^- respectively. For SCN^-, the limiting rate constant can be estimated as $1 \times 10^{-5}\ s^{-1}$. Do these values constitute evidence for an I_d mechanism?

12.8 **a.** The CO exchange reaction $Cr(^{12}CO)_6 + {}^{13}CO \longrightarrow Cr(^{12}CO)_5(^{13}CO) + {}^{12}CO$ has a rate that is first order in the concentration of $Cr(^{12}CO)_6$ but independent of the concentration of ^{13}CO. What does this imply about the mechanism of this reaction?

b. The reaction $Cr(CO)_6 + PR_3 \longrightarrow Cr(CO)_5PR_3 + CO$ [$R = P(n\text{-}C_4H_9)_3$] has the rate law of rate $= k_1[Cr(CO)_6] + k_2[Cr(CO)_6][PR_3]$. Why does this rate law have two terms?

c. For the general reaction in part **b**, will bulkier ligands tend to favor the first order or second order pathway? Explain briefly.

12.9 Account for the observation that two separate water exchange rates are found for $[Cu(H_2O)_6]^{2+}$ in aqueous solution.

12.10 The ligand exchange mechanism of the solvated lithium cations $[Li(H_2O)_4]^+$ and $[Li(NH_3)_4]^+$ has been studied using DFT calculations (R. Puchta, M. Galle, N. van E. Hommes, E. Pasgreta, R. van Eldik, *Inorg. Chem.*, **2004**, *43*, 8227.). Describe how water exchange is postulated to occur with the inclusion of the pertinent structures. What parameter most strongly supports a limiting associative substitution mechanism? How is the proposed mechanism for ligand exchange in $[Li(NH_3)_4]^+$ different? Why is the $[Li(NH_3)_4]^+$ mechanism postulated as associative interchange?

12.11 Data for the reaction

$$Co(NO)(CO)_3 + As(C_6H_5)_3 \longrightarrow Co(NO)(CO)_2[As(C_6H_5)_3] + CO$$

in toluene at 45 °C are given in the table. In all cases, the reaction is pseudo–first order in $Co(NO)(CO)_3$. Determine the rate constant(s) and discuss their probable significance (See E. M. Thorsteinson, F. Basolo, *J. Am. Chem. Soc.*, **1966**, *88*, 3929).

$[As(C_6H_5)_3]$ (M)	k ($10^{-5}\ s^{-1}$)
0.014	2.3
0.098	3.9
0.525	12
1.02	23

12.12 A plot of the log of the rate constant for substitution of CO on $Co(NO)(CO)_3$ by phosphorus and nitrogen ligands in nitromethane at 298 K, versus the half-neutralization potential of the ligands, a measure of ligand basicity, is below. Explain the linearity of such a plot and why there are two different lines. Incoming nucleophilic ligands: (1) $P(C_2H_5)_3$, (2) $P(n\text{-}C_4H_9)_3$, (3) $P(C_6H_6)(C_2H_5)_2$, (4) $P(C_6H_5)_2(C_2H_5)$, (5) $P(C_6H_5)_2(n\text{-}C_4H_9)$, (6) $P(p\text{-}CH_3OC_6H_4)_3$, (7) $P(O\text{-}n\text{-}C_4H_9)_3$, (8) $P(OCH_3)_3$, (9) $P(C_6H_5)_3$, (10) $P(OCH_2)_3CCH_3$, (11) $P(OC_6H_5)_3$, (12) 4-picoline, (13) pyridine, (14) 3-chloropyridine. (Graph created with data from E. M. Thorsteinson, F. Basolo, *J. Am. Chem. Soc.*, **1966**, *88*, 3929.)

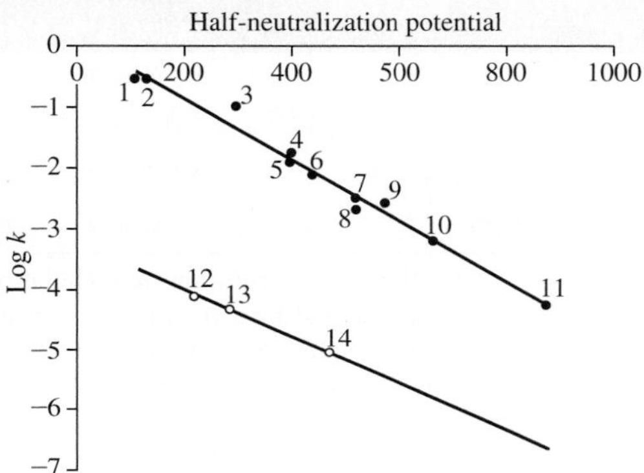

12.13 *cis*-$PtCl_2(PEt_3)_2$ is stable in benzene solution. However, small amounts of free triethylphosphine catalyze establishment of an equilibrium with the *trans* isomer:

$$cis\text{-}PtCl_2(PEt_3)_2 \rightleftharpoons trans\text{-}PtCl_2(PEt_3)_2$$

For the conversion of *cis* to *trans* in benzene at 25 °C, $\Delta H° = 10.3 \text{ kJ mol}^{-1}$ and $\Delta S° = 55.6 \text{ J mol}^{-1} \text{ K}^{-1}$.

a. Calculate the free energy change, $\Delta G°$, and the equilibrium constant for this isomerization.

b. Which isomer has the higher bond energy? Is this answer consistent with what you would expect on the basis of π bonding in the two isomers? Explain briefly.

c. Why is free triethylphosphine necessary to catalyze the isomerization?

12.14 This table shows the effect of changing ligands on the dissociation rates of CO *cis* to those ligands. Explain the effect of these ligands on the rates of dissociation. Include the effect of these ligands on Cr—CO bonding and on the transition state, presumed to be square-pyramidal. (See J. D. Atwood, T. L. Brown, *J. Am. Chem. Soc.*, **1976**, *98*, 3160).

Compound	k (s^{-1}) for CO Dissociation
$Cr(CO)_6$	1×10^{-12}
$Cr(CO)_5(PPh_3)$	3.0×10^{-10}
$[Cr(CO)_5I]^-$	$< 10^{-5}$
$[Cr(CO)_5Br]^-$	2×10^{-5}
$[Cr(CO)_5Cl]^-$	1.5×10^{-4}

12.15 When the two isomers of $Pt(NH_3)_2Cl_2$ react with thiourea [tu = $S{=}C(NH_2)_2$], one product is $[Pt(tu)_4]^{2+}$ and the other is $[Pt(NH_3)_2(tu)_2]^{2+}$. Identify the initial isomers and explain the results.

12.16 Predict the products (equimolar mixtures of reactants):

a. $[Pt(CO)Cl_3]^- + NH_3 \longrightarrow$

b. $[Pt(NH_3)Br_3]^- + NH_3 \longrightarrow$

c. $[(C_2H_4)PtCl_3]^- + NH_3 \longrightarrow$

12.17 **a.** Design a sequence of reactions, beginning with $[PtCl_4]^{2-}$, that will result in Pt(II) complexes with four different ligands—py, NH_3, NO_2^-, and CH_3NH_2—with two different sets of *trans* ligands. (CH_3NH_2 is similar to NH_3 in its *trans* effect.)

b. Pt(II) can be oxidized to Pt(IV) by Cl_2 with no change in configuration (chloride ions are added above and below the plane of the Pt(II) complex). Predict the results if the two compounds from (a) are reacted with Cl_2 and then with 1 mole of Br^- for each mole of the Pt compound.

12.18 The rates of exchange of CO on *cis* square-planar Ir complexes have been observed for the following reaction at 298 K:

$$[IrX_2(CO)_2]^- + 2\,{}^*CO \rightleftharpoons [IrX_2({}^*CO)_2]^- + 2\,CO \quad ({}^*C = {}^{13}C)$$

The observed rate constants were:

X	k (L/mol s)
Cl	1,080
Br	12,700
I	98,900

All three reactions have large negative values of entropy of activation, $\Delta S^{\ddagger}$.

a. Is this reaction associative or dissociative?

b. On the basis of the data, which halide ligand exerts the strongest *trans* effect? (See R. Churlaud, U. Frey, F. Metz, A. E. Merbach, *Inorg. Chem.*, **2000**, *39*, 304.)

12.19 The rate constant for electron exchange between $V^{2+}(aq)$ and $V^{3+}(aq)$ is observed to depend on the hydrogen ion concentration:

$$k = a + b/[H^+]$$

Propose a mechanism, and express *a* and *b* in terms of the rate constants of the mechanism. (Hint: $V^{3+}(aq)$ hydrolyzes more easily than $V^{2+}(aq)$.)

12.20 Is the reaction $[Co(NH_3)_6]^{3+} + [Cr(H_2O)_6]^{2+}$ likely to proceed by an inner-sphere or outer-sphere mechanism? Explain your answer.

12.21 The rate constants for the exchange reaction

$$CrX^{2+} + {}^*Cr^{2+} \longrightarrow {}^*CrX^{2+} + Cr^{2+}$$

(${}^*Cr$ is radioactive ${}^{51}Cr$)

are given in the table for reactions at 0 °C and 1 M $HClO_4$. Explain the differences in the rate constants in terms of the probable mechanism of the reaction.

X^-	$k\ (M^{-1}\ s^{-1})$
F^-	1.2×10^{-3}
Cl^-	11
Br^-	60
NCS^-	1.2×10^{-4} (at 24 °C)
N_3^-	> 1.2

12.22 The first complex of the ligand NSe (selenonitrosyl), $TpOs(NSe)Cl_2$ [Tp = hydrotris (1-pyrazolyl)borate], is shown. The osmium–nitrogen distances are:
Os–N(1): 210.1(7) pm
Os–N(3): 206.6(8) pm
Os–N(5): 206.9(7) pm

a. Which ligand, Cl or NSe, has the larger *trans* influence? Explain briefly.

b. The nitrogen–selenium distance in this compound is among the shortest N—Se distances known. Why is this distance so short? (See T. J. Crevier, S. Lovell, J. M. Mayer, A. L. Rheingold, I. A. Guzei, *J. Am. Chem. Soc.*, **1998**, *120*, 6607.)

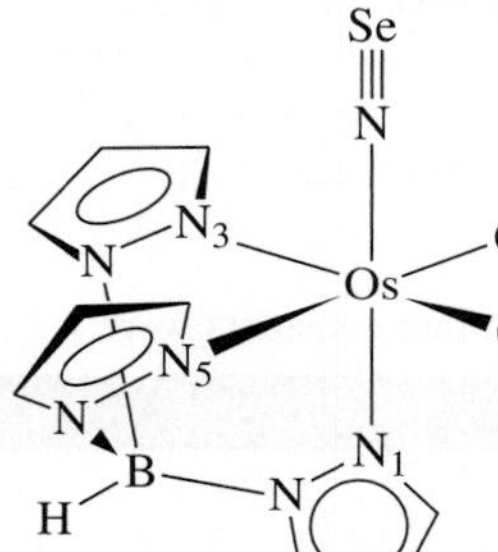

12.23 Exchange of an H_2O ligand on $[(CO)_3Mn(H_2O)_3]^+$ is much more rapid than on the analogous $[(CO)_3Re(H_2O)_3]^+$. The activation volume (change in volume on formation of the activated complex) is $-4.5 \pm 0.4\ cm^3mol^{-1}$. The Mn complex has infrared bands at 2051 and 1944 cm^{-1} that can be attributed to C—O stretching vibrations. (See U. Prinz, A. E. Merbach, O. Maas, K. Hegetschweiler, *Inorg. Chem.*, **2004**, *43*, 2387.)

a. Suggest why the Mn complex reacts more rapidly than the analogous Re complex.

b. Is the activation volume more consistent with an *A* (or I_a) or a *D* (or I_d) mechanism? Explain.

c. On the basis of the IR spectrum, is the reactant more likely a *fac* or *mer* isomer?

12.24 The mechanism of substitution reactions of *trans*-$[MnN(H_2O)(CN)_4]^{2-}$ with pyridine dicarboxylate ligands has been explored (H. J. van der Westhuizen, R. Meijboom, M. Schutte, A. Roodt, *Inorg. Chem.*, **2010**, *49*, 9599). The rate determining step is the second one, resulting in chelation and loss of cyanide. Attachment of the carboxylate oxygen to the Mn(V) center of *trans*-$[MnN(H_2O)(CN)_4]^{2-}$ is extremely fast.

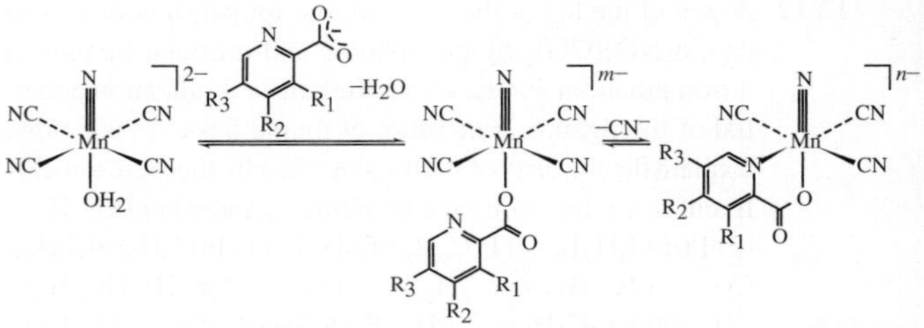

The following kinetics data were obtained at roughly 298 K, where k_1 is the rate constant associated with the overall forward reaction.

	m	*n*	$k_1\ (M^{-1}\ s^{-1})$	$\Delta S_{activation}$ ($J\ mol^{-1}K^{-1}$)
R_1 = H, R_2 = H, R_3 = H	3	2	$1.15(4) \times 10^{-3}$	43(3)
R_1 = CO_2^-, R_2 = H, R_3 = H	4	3	$1.11(1) \times 10^{-3}$	20(4)
R_1 = H, R_2 = CO_2^-, R_3 = H	4	3	$8.5(5) \times 10^{-4}$	115(14)
R_1 = H, R_2 = H, R_3 = CO_2^-	4	3	$1.08(4) \times 10^{-3}$	60(2)

Do you expect the rate determining step (chelation of the pyridine carboxylate and loss of cyanide) to proceed by dissociative or associative activation? Explain.

12.25 The two-electron oxidation of U^{4+} by $[PtCl_6]^{2-}$ in aqueous solution is postulated to proceed via an inner-sphere mechanism (R. M. Hassan, *J. Phys. Chem. A.*, **2011**, *115*, 13338).

a. What are the orders with respect to $[U^{4+}]$ and $[PtCl_6{}^{2-}]$? What data were used to determine these orders?

b. The initial combination of $[PtCl_6]^{2-}$ and U^{4+} is postulated to release two protons prior to the electron transfer step. How is this ionization proposed to increase the electron transfer rate? What is the most likely structure of the bridging species that facilitates the electron transfer from U^{4+} to Pt^{4+}?

12.26 The reduction of $[Co(edta)]^-$ by $[Fe(CN)_6]^{4-}$ is facile in water, but is remarkably "switched off" when conducted in reverse micelle microemulsions (M. D. Johnson, B. B. Lorenz, P. C. Wilkins, B. G. Lemons, B. Baruah, N. Lamborn, M. Stahla, P. B. Chatterjee, D. T. Richens, D. C. Crans, *Inorg. Chem.*, **2012**, *51*, 2757.).

a. Is the electron transfer mechanism for these two reactants believed to be inner-sphere or outer-sphere?

b. These reactions were carried out under pseudo-first order conditions. Why do plots of k_{obs} versus $[Fe(CN)_6]^{4-}$ from these experiments exhibit linear relationships? What does the *y*-intercept of 0 communicate about these electron transfer reactions? How is k_{obs} related to the bimolecular rate constant k_1 for the reduction of $[Co(edta)]^-$?

c. What is the hypothesis for the suppressed reduction of $[Co(edta)]^-$ by $[Fe(CN)_6]^{4-}$ when attempted in reverse micelle microemulsions?

Chapter 13

Organometallic Chemistry

Organometallic chemistry, the chemistry of compounds that contain metal–carbon bonds, has developed enormously as a field of study since the middle of the twentieth century. It encompasses a wide variety of compounds and their reactions, including numerous ligands that interact in sigma and pi fashions with metal atoms and ions; many cluster compounds, containing one or more metal–metal bonds; and molecules of structural types unusual or unknown in organic chemistry. Some reactions of organometallic compounds are similar to organic reactions, but in other cases, they are dramatically different. In addition to their intrinsically interesting nature, organometallic compounds form useful catalysts and are of industrial interest. In this chapter, we describe a variety of organometallic compounds, focusing on the ligands and how they interact with metal atoms. Chapter 14 describes major types of reactions of organometallic compounds and how these reactions are important in catalytic cycles. Chapter 15 focuses on parallels between organometallic chemistry and main group chemistry.

Some organometallic compounds bear similarities to coordination compounds. $Cr(CO)_6$ and $[Ni(H_2O)_6]^{2+}$, for example, are both octahedral. Both CO and H_2O are σ-donor ligands; in addition, CO is a strong π acceptor. Other ligands that exhibit both σ-donor and π-acceptor capabilities include CN^-, PPh_3, SCN^-, and many organic ligands. The metal–ligand bonding and electronic spectra of compounds containing these ligands can be described using concepts discussed in Chapters 10 and 11. However, many organometallic molecules are strikingly different from any we have considered previously. For example, cyclic organic ligands containing delocalized pi systems can team up with metal atoms to form **sandwich compounds** (**Figure 13.1**).

A characteristic of metal atoms bonded to organic ligands, especially CO, is that they often exhibit the capability to form covalent bonds to other metal atoms to form **cluster compounds**.* These clusters may contain only two or three metal atoms or many dozens. They may contain single, double, triple, or quadruple bonds—even molecules that may have quintuple bonds have been reported. In some cases cluster compounds have ligands that bridge two or more metal atoms. Examples of metal cluster compounds containing organic ligands are in **Figure 13.2**; clusters will be discussed further in Chapter 15.

Certain metal clusters encapsulate carbon atoms; the resulting carbon-centered clusters, frequently called **carbide clusters**, may contain carbon bonded to five, six, or more surrounding metals. The traditional notion of carbon forming bonds to at most four additional atoms must be reconsidered.** Two examples of carbide clusters are in Figure 13.2.

---

*Some cluster compounds contain no organic ligands.

A few examples of carbon bonded to more than four atoms are also known in organic chemistry. See, for example, G. A. Olah G. Rasul, *Acc. Chem. Res.*, **1997, *30*, 245.

FIGURE 13.1 Examples of Sandwich Compounds.

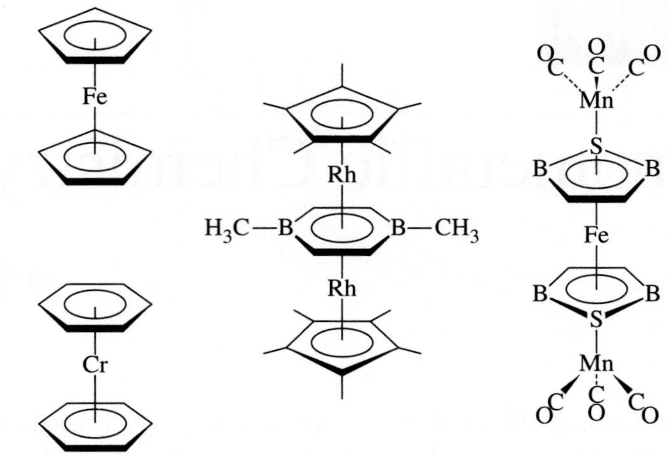

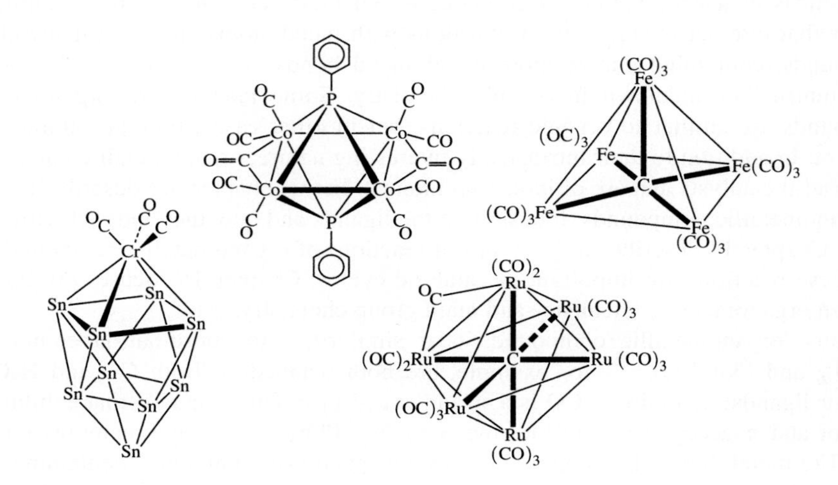

FIGURE 13.2 Examples of Cluster Compounds.

Many other types of organometallic compounds have interesting structures and chemical properties. **Figure 13.3** shows additional examples of the variety of structures that occur in this field.

Strictly speaking, the only compounds classified as organometallic should be ones that have metal–carbon bonds. In practice, however, complexes containing several other ligands similar to CO in their bonding, such as NO and N_2, are often included. (Cyanide also forms complexes in a manner similar to CO but is usually considered a nonorganic ligand.) Other π-acceptor ligands, such as phosphines, often occur in organometallic complexes. Even dihydrogen, H_2, now occurs in a new guise, as a donor–acceptor ligand, and it plays an important role in organometallic chemistry, for example in catalytic processes. We will include examples of these and other nonorganic ligands as appropriate.

13.1 Historical Background

The first organometallic compound to be reported was synthesized in 1827 by Zeise, who obtained yellow needle-like crystals after refluxing a mixture of $PtCl_4$ and $PtCl_2$ in ethanol, followed by addition of KCl solution.[1] Zeise correctly asserted that this yellow product, subsequently dubbed *Zeise's salt*, contained an ethylene group. This assertion was questioned by other chemists, most notably Liebig, and it was not verified conclusively

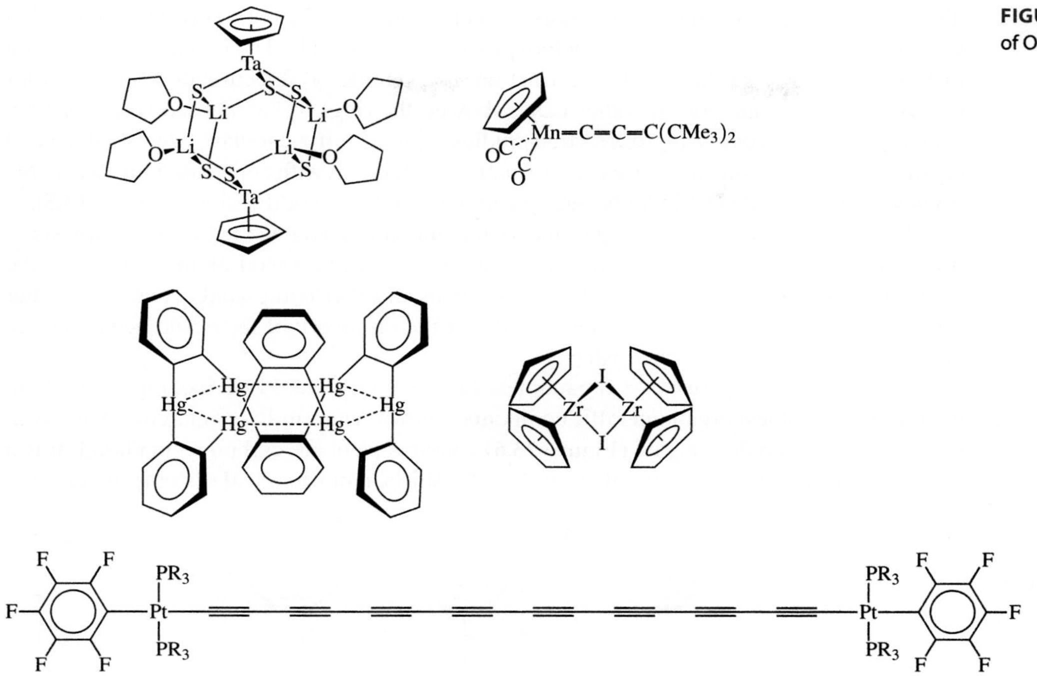

FIGURE 13.3 More Examples of Organometallic Compounds.

until experiments performed by Birnbaum in 1868. The structure of the compound was not determined until more than 100 years later![2] Zeise's salt was the first compound identified as containing an organic molecule attached to a metal using the pi electrons of the organic molecule. It is an ionic compound of formula $K[Pt(C_2H_4)Cl_3] \cdot H_2O$; the structure of the anion (**Figure 13.4**) is based on a square plane, with three chloro ligands occupying corners of the square and the ethylene occupying the fourth corner, but perpendicular to the plane.

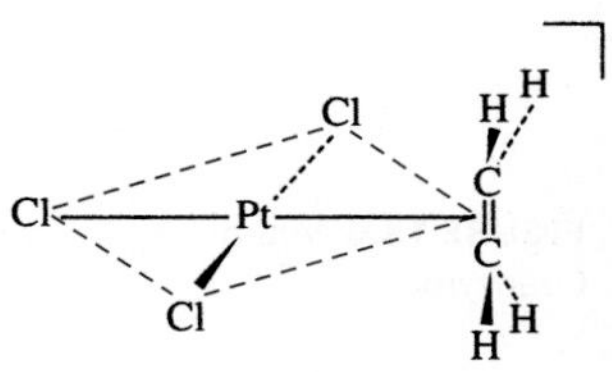

FIGURE 13.4 Anion of Zeise's Salt.

The first compound to be synthesized that contained carbon monoxide as a ligand was another platinum chloride complex, reported in 1867. In 1890, Mond reported the preparation of $Ni(CO)_4$, a compound that became commercially useful for the purification of nickel.[3] Other metal CO (carbonyl) complexes were soon obtained.

Reactions between magnesium and alkyl halides, performed by Barbier in 1898 and subsequently by Grignard,[4] led to the synthesis of alkyl magnesium complexes now known as *Grignard reagents*. These complexes often have complicated structures and contain magnesium–carbon sigma bonds. Their synthetic utility was recognized early; by 1905, more than 200 research papers had appeared on the topic. Grignard reagents and other reagents containing metal–alkyl σ bonds, such as organozinc and organolithium reagents, have been of immense importance in the development of synthetic organic chemistry.

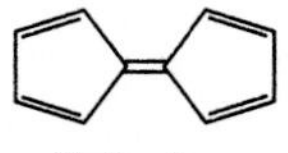

Organometallic chemistry developed slowly from the discovery of Zeise's salt in 1827 until around 1950. Some organometallic compounds, such as Grignard reagents, found utility in organic synthesis, but there was little systematic study of compounds containing metal–carbon bonds. In 1951, in an attempt to synthesize fulvalene from cyclopentadienyl bromide, Kealy and Pauson reacted the Grignard reagent *cyclo*-C_5H_5MgBr with $FeCl_3$.[5] This reaction did not yield fulvalene but an orange solid having the formula $(C_5H_5)_2Fe$, ferrocene:

Fulvalene

$$cyclo\text{-}C_5H_5MgBr + FeCl_3 \longrightarrow \underset{\text{ferrocene}}{(C_5H_5)_2Fe}$$

The product was surprisingly stable; it could be sublimed in air without decomposing and was resistant to catalytic hydrogenation and Diels–Alder reactions. In 1956, X-ray diffraction showed the structure to consist of an iron atom sandwiched between two parallel C_5H_5 rings,[6] but the details of the structure proved controversial.* The initial study indicated that the rings were in a staggered conformation (D_{5d} symmetry). Electron diffraction studies of gas-phase ferrocene, on the other hand, showed the rings to be eclipsed (D_{5h}), or very nearly so. More recent X-ray diffraction studies of solid ferrocene have identified several crystalline phases, with an eclipsed conformation at 98 K and with conformations having the rings slightly twisted (D_5) in higher-temperature crystalline modifications (**Figure 13.5**).[7]

The discovery of the prototype sandwich compound ferrocene led to the synthesis of other sandwich compounds, of other compounds containing metal atoms bonded to the C_5H_5 ring in a similar fashion, and to a vast array of other compounds containing other organic ligands. Therefore, it is often stated that the discovery of ferrocene began the era of modern organometallic chemistry.**

An introductory history of organometallic chemistry would be incomplete without mentioning the oldest organometallic compound known, vitamin B_{12} coenzyme. This naturally occurring cobalt complex (**Figure 13.6**) contains a cobalt–carbon sigma bond. It is a cofactor in a number of enzymes that catalyze 1,2 shifts in biochemical systems (in margin).

FIGURE 13.5 Conformations of Ferrocene.

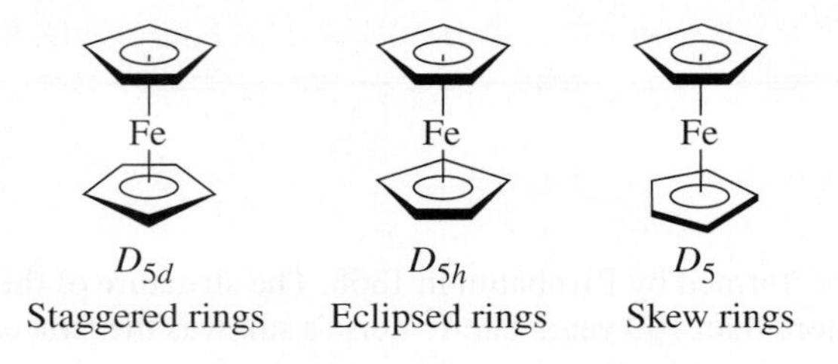

FIGURE 13.6 Vitamin B_{12} Coenzyme.

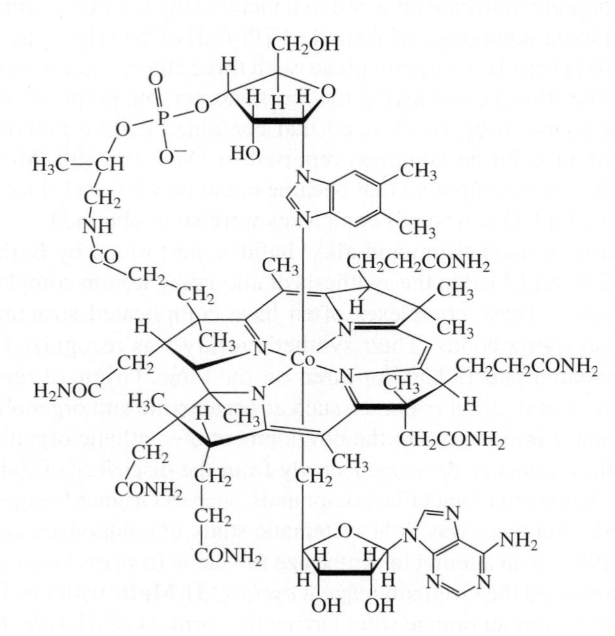

*For an interesting article on the discovery of ferrocene's structure, see P. Laszlo, R. Hoffmann, *Angew. Chem., Int. Ed.*, **2000**, *39*, 123.

A special issue of the *Journal of Organometallic Chemistry* (2002**, *637*, 1) has been devoted to ferrocene, including recollections of some involved in its discovery; a brief summary of some of these recollections appeared in *Chem. Eng. News*, **2001**, *79(49)*, 37.

13.2 Organic Ligands and Nomenclature

Some common organic ligands are shown in **Figure 13.7**. Special nomenclature has been devised to designate the manner in which some of these ligands bond to metal atoms; several of the ligands in Figure 13.7 may bond through different numbers of atoms. The number of atoms through which a ligand bonds is indicated by the Greek letter η (eta) followed by a superscript indicating the number of ligand atoms attached to the metal. For example, because the cyclopentadienyl ligands in ferrocene bond through all five atoms, they are designated η^5-C_5H_5. The formula of ferrocene may therefore be written $(\eta^5\text{-}C_5H_5)_2Fe$. The η^5-C_5H_5 ligand is designated the pentahaptocyclopentadienyl ligand. *Hapto* comes from the Greek word for fasten; therefore, *pentahapto* means "fastened in five places." C_5H_5, probably the second most frequently encountered ligand in organometallic chemistry (after CO), most commonly bonds to metals through five positions, but under certain circumstances, it may bond through only one or three positions. As a ligand, C_5H_5 is commonly abbreviated Cp.

The corresponding formulas and names are designated as follows:*

Number of Bonding Positions	Formula	Name	
1	η^1-C_5H_5	Monohaptocyclopentadienyl	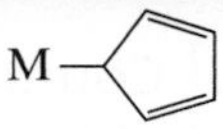 M—
3	η^3-C_5H_5	Trihaptocyclopentadienyl	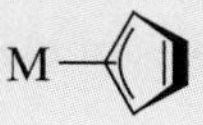 M—
5	η^5-C_5H_5	Pentahaptocyclopentadienyl	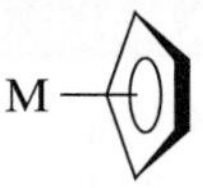 M—

Ligand	Name	Ligand	Name
CO	Carbonyl	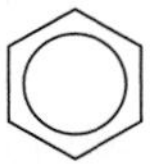	Benzene
=C<	Carbene (alkylidene)	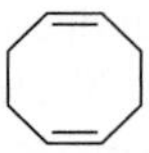	1,5-cyclooctadiene (1,5-COD) (1,3-cyclooctadiene complexes are also known)
≡C—	Carbyne (alkylidyne)	$H_2C{=}CH_2$	Ethylene
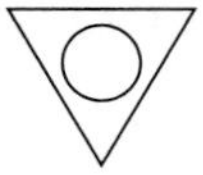	Cyclopropenyl (*cyclo*-C_3H_3)	HC≡CH	Acetylene
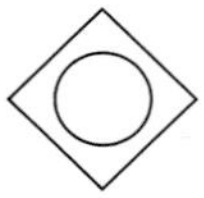	Cyclobutadiene (*cyclo*-C_4H_4)		π-Allyl (C_3H_5)
	Cyclopentadienyl (*cyclo*-C_5H_5) (Cp)	$-CR_3$	Alkyl
		—C(=O)R	Acyl

FIGURE 13.7 Classic Organic Ligands.

*For ligands having all carbons bonded to a metal, sometimes the superscript is omitted. Ferrocene may therefore be written $(\eta\text{-}C_5H_5)_2Fe$ and dibenzenechromium $(\eta\text{-}C_6H_6)_2Cr$. Similarly, π with no superscript may occasionally be used to designate that all atoms in the pi system are bonded to the metal; (for example, $(\pi\text{-}C_5H_5)_2Fe$).

As in the case of other coordination compounds, bridging ligands are designated by the prefix μ, followed by a subscript indicating the number of metal atoms bridged. Bridging carbonyl ligands, for example, are designated as follows:

Number of Atoms Bridged	Formula
None (terminal)	CO
2	μ_2-CO
3	μ_3-CO

13.3 The 18-Electron Rule

In main group chemistry, we have encountered the octet rule, in which electronic structures can be rationalized on the basis of a valence shell requirement of 8 electrons. Similarly, in organometallic chemistry, the electronic structures of many compounds are based on a total valence electron count of 18 on the central metal atom. As with the octet rule, there are many exceptions to the 18-electron rule, but the rule nevertheless provides useful guidelines to the chemistry of many organometallic complexes, especially those containing strong π-acceptor ligands.

13.3.1 Counting Electrons

Several schemes exist for counting electrons in organometallic compounds. Here are two examples of electron counting in 18-electron species:

EXAMPLE 13.1

$Cr(CO)_6$

A Cr atom has 6 electrons outside its noble gas core. Each CO is considered to act as a donor of 2 electrons. The total electron count is therefore:

Cr			6 electrons
6(CO)	6 × 2 electrons	=	12 electrons
	Total	=	18 electrons

$Cr(CO)_6$ is therefore considered an 18-electron complex. It is thermally stable and can be sublimed without decomposition. On the other hand, $Cr(CO)_5$, a 16-electron species, and $Cr(CO)_7$, a 20-electron species, are much less stable and are known only as transient species. Likewise, the 17-electron $[Cr(CO)_6]^+$ and 19-electron $[Cr(CO)_6]^-$ are far less stable than the 18-electron $Cr(CO)_6$.

The bonding in $Cr(CO)_6$, which provides a rationale for the special stability of many 18-electron systems, is discussed in Section 13.3.2.

$(\eta^5\text{-}C_5H_5)Fe(CO)_2Cl$

Electrons in this complex may be counted in two ways:

Method A: Donor Pair Method

This method considers ligands to donate electron pairs to the metal. To determine the total electron count, we must take into account the charge on each ligand and determine the formal oxidation state of the metal.

Pentahapto-C_5H_5 is considered by this method as $C_5H_5^-$, a donor of 3 electron pairs; it is a 6-electron donor. As in the first example, CO is counted as a 2-electron donor.

Chloride is considered Cl^-, a donor of two electrons. Therefore, $(\eta^5\text{-}C_5H_5)Fe(CO)_2Cl$ is formally an iron(II) complex. Iron(II) has 6 electrons beyond its noble gas core. Therefore, the electron count is

Fe(II)	6 electrons
$\eta^5\text{-}C_5H_5^-$	6 electrons
2(CO)	4 electrons
Cl^-	2 electrons
Total =	18 electrons

Method B: Neutral-Ligand Method

This method uses the number of electrons that would be donated by ligands *if they were neutral*. For simple inorganic ligands, this usually means that ligands are considered to donate the number of electrons equal to their negative charge as free ions. For example,

Cl is a 1-electron donor (charge on free ion $= -1$)

O is a 2-electron donor (charge on free ion $= -2$)

N is a 3-electron donor (charge on free ion $= -3$)

The oxidation state of the metal does not need to be specified to determine the total electron count by this method.

For $(\eta^5\text{-}C_5H_5)Fe(CO)_2Cl$, an iron *atom* has 8 electrons beyond its noble gas core, and $\eta^5\text{-}C_5H_5$ is now considered a neutral ligand (a 5-electron pi system), in which case it would contribute 5 electrons. CO is a 2-electron donor, and Cl (counted as if it were a neutral species) is a 1-electron donor. The electron count is

Fe atom	8 electrons
$\eta^5\text{-}C_5H_5$	5 electrons
2(CO)	4 electrons
Cl	1 electron
Total =	18 electrons

The results of the two methods are equivalent; both give 18 electrons.

Many organometallic complexes are charged species, and this charge must be included in determining the total electron count. In the neutral-ligand method, for anions the charge of the complex is added as a number of electrons to the total, and for cations the magnitude of the charge of the complex is subtracted as a number of electrons from the total. The reader may wish to verify, by both electron counting methods, that $[(\eta^5\text{-}C_5H_5)Fe(CO)_2]^-$ and $[Mn(CO)_6]^+$ are both 18-electron ions.

A metal–metal single bond counts as one electron per metal, a double bond counts as two electrons per metal, and so forth. For example, in dimeric $(CO)_5Mn—Mn(CO)_5$, the electron count per manganese atom is, by either method:

Mn	7 electrons
5 (CO)	10 electrons
Mn—Mn bond	1 electron
Total =	18 electrons

Electron counts for common ligands according to both schemes are in **Table 13.1**.

TABLE 13.1 Electron Counting Schemes for Common Ligands

Ligand	Method A	Method B
H	2 (H^-)	1
Cl, Br, I	2 (X^-)	1
OH, OR	2 (OH^-, OR^-)	1
CN	2 (CN^-)	1
CH_3, CR_3	2 (CH_3^-, CR_3^-)	1
NO (bent M—N—O)	2 (NO^-)	1
NO (linear M—N—O)	2 (NO^+)	3
CO, PR_3	2	2
NH_3, H_2O	2	2
=CRR′ (Carbene)	2	2
$H_2C{=}CH_2$ (Ethylene)	2	2
CNR	2	2
=O, =S	4 (O^{2-}, S^{2-})	2
η^3-C_3H_5 (π-allyl)	4 ($C_3H_5^-$)	3
≡CR (Carbyne)	3	3
≡N	6 (N^{3-})	3
Ethylenediamine (en)	4 (2 per nitrogen)	4
Bipyridine (bipy)	4 (2 per nitrogen)	4
Butadiene	4	4
η^5-C_5H_5 (Cyclopentadienyl)	6 ($C_5H_5^-$)	5
η^6-C_6H_6 (Benzene)	6	6
η^7-C_7H_7 (Cycloheptatrienyl)	6 ($C_7H_7^+$)	7

EXAMPLE 13.2

Both methods of electron counting are illustrated for the following complexes.

	Method A		Method B	
$ClMn(CO)_5$	Mn(I)	6 e⁻	Mn	7 e⁻
	Cl^-	2 e⁻	Cl	1 e⁻
	5 CO	10 e⁻	5 CO	10 e⁻
		18 e⁻		18 e⁻
$(\eta^5$-$C_5H_5)_2Fe$	Fe(II)	6 e⁻	Fe	8 e⁻
(Ferrocene)	2 η^5-$C_5H_5^-$	12 e⁻	2 η^5-C_5H_5	10 e⁻
		18 e⁻		18 e⁻
$[Re(CO)_5(PF_3)]^+$	Re(I)	6 e⁻	Re	7 e⁻
	5 CO	10 e⁻	5 CO	10 e⁻
	PF_3	2 e⁻	PF_3	2 e⁻
	+ charge	*	+ charge	−1 e⁻
		18 e⁻		18 e⁻

(continues)

(contd.)

	Method A		Method B	
$[Ti(CO)_6]^{2-}$	Ti(2–)	6 e^-	Ti	4 e^-
	6 CO	12 e^-	6 CO	12 e^-
	2– charge	*	2– charge	2 e^-
		18 e^-		18 e^-

* Charge on ion is accounted for in assignment of oxidation state to metal.

The electron-counting method of choice is a matter of individual preference. Method A includes the formal oxidation state of the metal; Method B does not. Method B may be simpler to use for ligands having extended pi systems; for example, η^5 ligands have an electron count of 5, η^3 ligands an electron count of 3, and so on. Because neither description describes the bonding in any real sense, these methods should, like the Lewis electron-dot approach in main group chemistry, be considered primarily electron bookkeeping tools. Physical measurements are necessary to provide evidence about the actual electron distribution in molecules. It is best to select one method and use it consistently.

In ligands such as CO that can interact with metal atoms in several ways, the number of electrons counted is usually based on σ donation. For example, although CO is a π acceptor and (weak) π donor, its electron-donating count of 2 is based on its σ donor ability alone. However, the π-acceptor and π-donor abilities of ligands have significant effects on the degree to which the 18-electron rule is likely to be obeyed. Linear and cyclic organic pi systems interact with metals in more complicated ways, discussed later in this chapter.

EXERCISE 13.1

Determine the valence electron counts for the transition metals in the following complexes:

a. $[Fe(CO)_4]^{2-}$

b. $[(\eta^5\text{-}C_5H_5)_2Co]^+$

c. $(\eta^3\text{-}C_5H_5)(\eta^5\text{-}C_5H_5)Fe(CO)$

d. $Co_2(CO)_8$ (has a single Co—Co bond)

EXERCISE 13.2

Identify the first-row transition metal for the following 18-electron species:

a. $[M(CO)_3(PPh_3)]^-$

b. $HM(CO)_5$

c. $(\eta^4\text{-}C_8H_8)M(CO)_3$

d. $[(\eta^5\text{-}C_5H_5)M(CO)_3]_2$ (assume single M—M bond)

13.3.2 Why 18 Electrons?

An oversimplified rationale for the special significance of 18 electrons can be made by analogy with the octet rule in main group chemistry. If the octet represents a complete valence electron shell configuration (s^2p^6), then the number 18 represents a filled valence shell for a transition metal ($s^2p^6d^{10}$). Although perhaps a useful way to relate electron configurations to the idea of valence shells of electrons for atoms, this analogy does not provide an explanation for why so many complexes violate the 18-electron rule. In particular, the valence-shell rationale does not distinguish between types of ligands (e.g., σ donors, π acceptors); this distinction is important in determining which complexes obey and which violate the rule.

A good example of a complex that adheres to the 18-electron rule is $Cr(CO)_6$. The molecular orbitals of interest in this molecule are those that result primarily from interactions between the *d* orbitals of Cr and the σ-donor (HOMO) and π-acceptor orbitals (LUMO) of the six CO ligands. The relative energies of molecular orbitals resulting from these interactions are shown in **Figure 13.8**; a more detailed diagram based on a symmetry analysis of the interacting orbitals is in Figure 10.7.

FIGURE 13.8 Molecular Orbital Energy Levels of $Cr(CO)_6$.

Molecular Orbital Energy Levels of $Cr(CO)_6$ by Gary O. Spessard and Gary L. Miessler. Reprinted by permission.

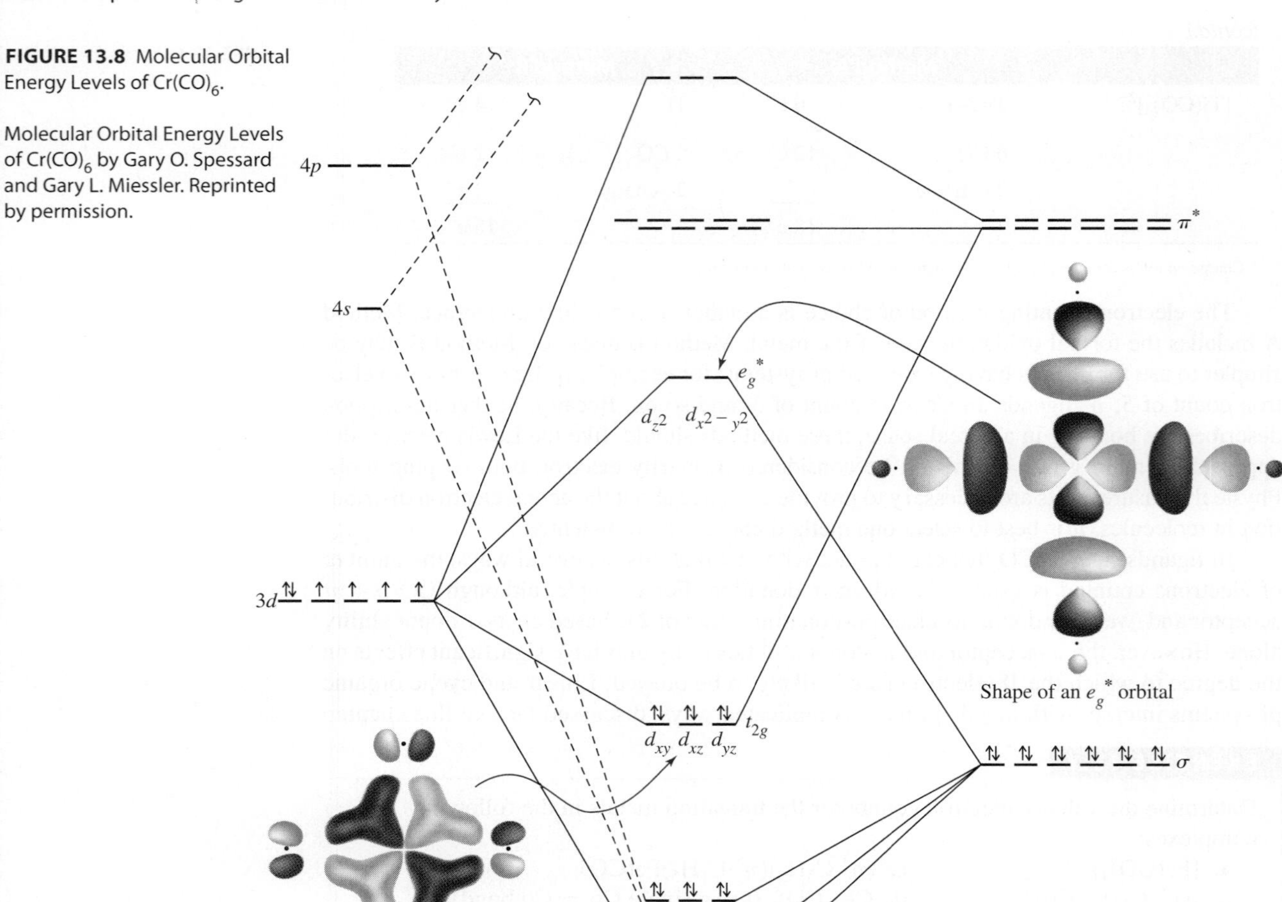

Chromium(0) has 6 electrons outside its noble gas core. Each CO contributes a pair of electrons to give a total electron count of 18. In the molecular orbital diagram, these 18 electrons appear as the 12 σ electrons—the σ electrons of the CO ligands, stabilized by their interaction with the metal orbitals—and the 6 t_{2g} electrons. Addition of one or more electrons to $Cr(CO)_6$ would populate the e_g orbitals, which are antibonding; the consequence would be destabilization of the molecule. Removal of electrons from $Cr(CO)_6$ would depopulate the t_{2g} orbitals, which are bonding as a consequence of the strong π-acceptor ability of the CO ligands; a decrease in electron density in these orbitals would also tend to destabilize the complex. The result is that the 18-electron configuration for this molecule is the most stable.

By considering 6-coordinate molecules of octahedral geometry, we can gain some insight as to when the 18-electron rule can be expected to be most valid. $Cr(CO)_6$ obeys the rule because of two factors: the strong σ-donor ability of CO raises the e_g orbitals in energy, making them considerably antibonding (and raising the energy of electrons in excess of 18); and the strong π-acceptor ability of CO lowers the t_{2g} orbitals in energy, making them bonding (and lowering the energies of electrons 13 through 18). Ligands that are both strong σ donors and π acceptors should therefore be the most effective at forcing adherence to the 18-electron rule. Other ligands, including some organic ligands, do not have these features; their compounds may or may not adhere to the rule.

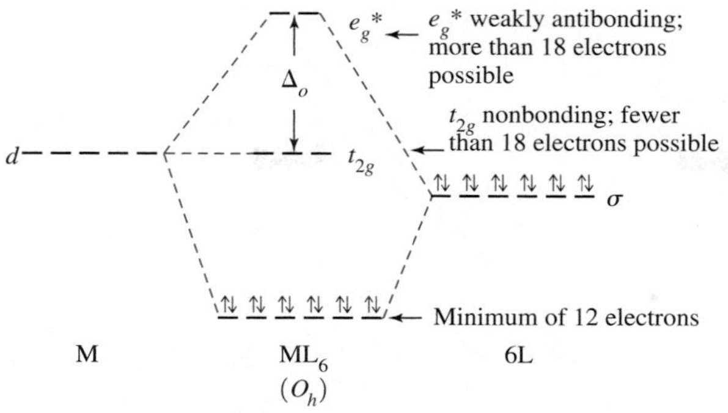

FIGURE 13.9 Exceptions to the 18-Electron Rule.

Examples of exceptions may be noted. $[Zn(en)_3]^{2+}$ is a 22-electron species; it has both the t_{2g} and e_g^* orbitals filled. Although en (ethylenediamine) is a good σ donor, it is not as strong a donor as CO. As a result, electrons in the e_g orbitals are not sufficiently antibonding to cause significant destabilization of the complex, and the 22-electron species, with 4 electrons in e_g orbitals, is stable. An example of a 12-electron species is TiF_6^{2-}. In this case, the fluoride ligand is a π donor as well as a σ donor. The π-donor ability of F^- destabilizes the t_{2g} orbitals of the complex, making them slightly antibonding. The species TiF_6^{2-} has 12 electrons in the bonding σ orbitals and no electrons in the antibonding t_{2g} or e_g^* orbitals. These exceptions to the 18-electron rule are shown schematically in **Figure 13.9**.[8]

The same type of argument can be made for complexes of other geometries; in most cases, there is an 18-electron configuration of special stability for complexes of strongly π-accepting ligands. Examples include trigonal-bipyramidal geometry, such as $Fe(CO)_5$ and tetrahedral geometry, like $Ni(CO)_4$. The most common exception is square-planar geometry, in which a 16-electron configuration may be the most stable, especially for complexes of d^8 metals.

13.3.3 Square-Planar Complexes

Examples of square-planar complexes include the d^8 16-electron complexes shown in **Figure 13.10**. To understand why 16-electron square-planar complexes might be especially stable, it is necessary to examine the molecular orbitals of such a complex. An energy diagram for the molecular orbitals of a square-planar molecule of formula ML_4, where L is a ligand that can function as both σ donor and π acceptor, is shown in **Figure 13.11**. A detailed diagram showing the symmetry analysis of the interacting orbitals is in Figure 10.16.

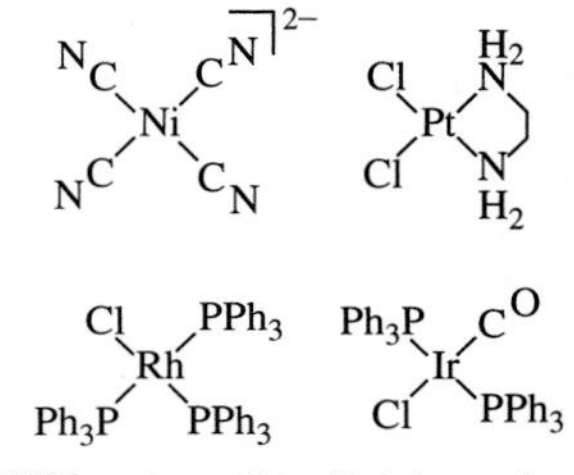

The four lowest energy molecular orbitals in this diagram result from bonding interactions between the σ-donor orbitals of the ligands and the $d_{x^2-y^2}$, d_{z^2}, p_x, and p_y metal orbitals. These molecular orbitals are filled by 8 electrons from the ligands. The next four orbitals are either slightly bonding, nonbonding, or slightly antibonding, derived primarily from the d_{xz}, d_{yz}, d_{xy}, and d_{z^2} metal orbitals.* These orbitals are occupied by a maximum of 8 electrons from the metal.** Additional electrons would occupy an orbital derived from the antibonding interaction of a metal $d_{x^2-y^2}$ orbital with the σ-donor orbitals of the ligands (the $d_{x^2-y^2}$ orbital points directly toward the ligands; its antibonding interaction is the strongest). Consequently, for square-planar complexes of ligands having both σ-donor and

Wilkinson's complex Vaska's complex

FIGURE 13.10 Examples of Square-Planar d^8 Complexes.

*The d_{z^2} orbital has A_{1g} symmetry and interacts with an A_{1g} group orbital. If this were the only metal orbital of this symmetry, the molecular orbital labeled d_{z^2} in Figure 13.11 would be antibonding. However, the next higher energy s orbital of the metal also has A_{1g} symmetry; the greater the degree to which this orbital is involved, the lower the energy of the molecular orbital.

**The relative energies of all four of these orbitals depend on the nature of the specific ligands and metal involved; in some cases in Figure 10.16, the ability of ligands to π donate can cause the order of energy levels to be different than those shown in Figure 13.11.

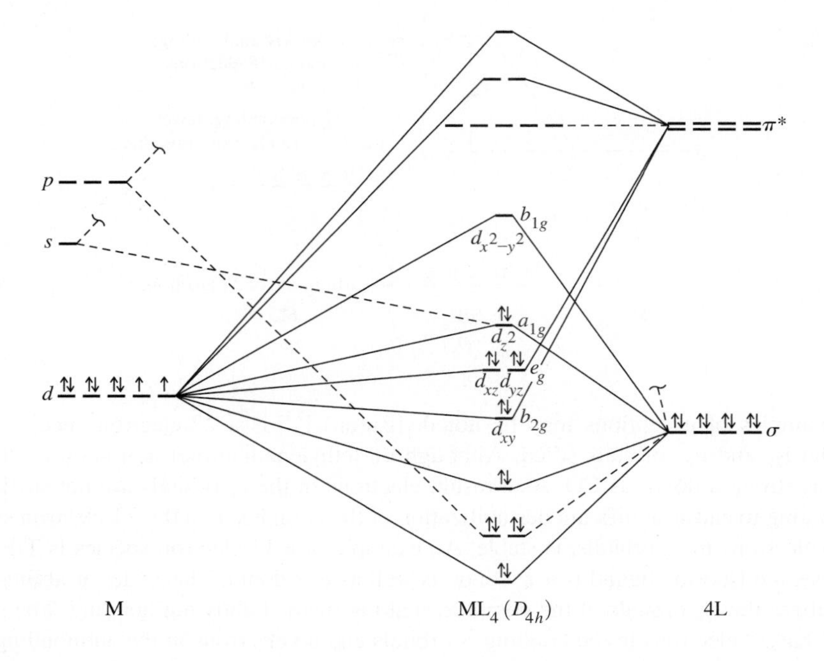

FIGURE 13.11 Molecular Orbital Energy Levels for a Square-Planar Complex.

π-acceptor characteristics, a 16-electron configuration is more stable than an 18-electron configuration. Sixteen-electron square-planar complexes may be able to accept one or two ligands at the vacant coordination sites (along the z axis) to achieve an 18-electron configuration. This is a common reaction of 16-electron square-planar complexes (Chapter 14).

EXERCISE 13.3

Verify that the complexes in Figure 13.10 are 16-electron species.

Sixteen-electron square-planar species are most commonly encountered for d^8 metals, particularly for metals having formal oxidation states of 2+ (Ni^{2+}, Pd^{2+}, and Pt^{2+}) and 1+ (Rh^+, Ir^+). Square-planar geometry is more common for second- and third-row transition-metal complexes than for first-row complexes. Two examples of square-planar d^8 complexes that are used as catalysts are Wilkinson's complex and Vaska's complex (Figure 13.10).

13.4 Ligands in Organometallic Chemistry

Hundreds of ligands are known to bond to metal atoms through carbon. Carbon monoxide forms a very large number of metal complexes and deserves special mention, along with several similar diatomic ligands. Many organic molecules containing linear or cyclic pi systems also form numerous organometallic complexes. Complexes containing such ligands will be discussed next, following a brief review of the pi systems in the ligands themselves. Finally, special attention will be paid to two types of organometallic compounds that are especially important: carbene complexes, containing metal–carbon double bonds, and carbyne complexes, containing metal–carbon triple bonds.

13.4.1 Carbonyl (CO) Complexes

Carbon monoxide is the most common ligand in organometallic chemistry. It serves as the only ligand in binary carbonyls such as $Ni(CO)_4$, $W(CO)_6$, and $Fe_2(CO)_9$ or, more commonly, in combination with other ligands, both organic and inorganic. CO may bond to a single metal, or it may serve as a bridge between two or more metals. We will consider

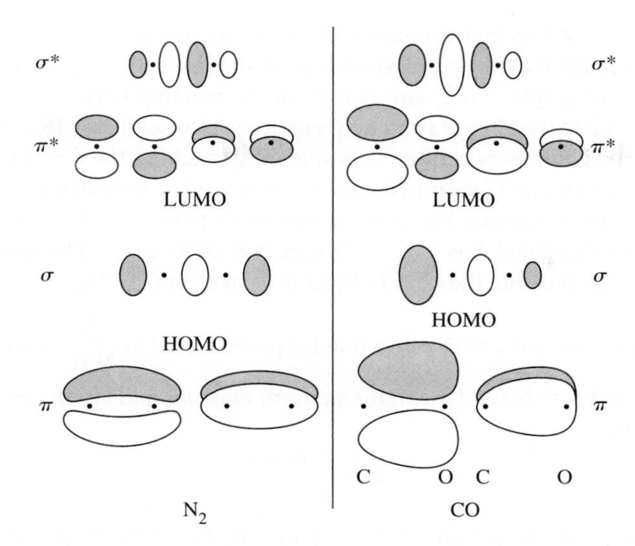

FIGURE 13.12 Selected Molecular Orbitals of N_2 and CO.

the bonding between metals and CO, the synthesis and reactions of CO complexes, and examples of various types of CO complexes.

Bonding

It is useful to review the bonding in CO. The molecular orbital picture of CO shown in Figure 5.13 is similar to that of N_2. Sketches of the molecular orbitals derived primarily from the $2p$ atomic orbitals of these molecules are shown in **Figure 13.12**.

Two features of the molecular orbitals of CO deserve attention. First, the highest-energy occupied orbital (the HOMO) has its largest lobe on carbon. It is through this orbital, occupied by an electron pair, that CO exerts its σ-donor function, donating electron density directly toward an appropriate metal orbital, such as an unfilled d or hybrid orbital. Carbon monoxide also has two empty π^* orbitals (the lowest unoccupied, or LUMO); these also have larger lobes on carbon than on oxygen. A metal atom having electrons in a d orbital of suitable symmetry can donate electron density to these π^* orbitals. These σ-donor and π-acceptor interactions are illustrated in **Figure 13.13**.

The overall effect is synergistic. CO can donate electron density via a σ orbital to a metal atom; the greater the electron density on the metal, the more effectively it can return electron density to the π^* orbitals of CO. The net effect can be strong bonding between the metal and CO; however, the strength of this bonding depends on several factors, including the charge on the complex and the ligand environment of the metal.

EXERCISE 13.4

N_2 has molecular orbitals rather similar to those of CO, as shown in Figure 13.12. Would you expect N_2 to be a stronger or weaker π acceptor than CO?

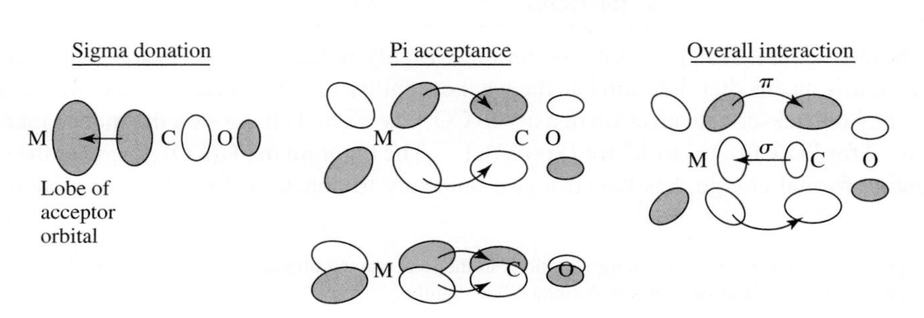

FIGURE 13.13 Sigma and Pi Interactions between CO and a Metal Atom.

If this picture of bonding between CO and metal atoms is correct, it should be supported by experimental evidence. Two sources of such evidence are infrared spectroscopy and X-ray crystallography. First, any change in the bonding between carbon and oxygen should be reflected in the C—O stretching vibration as observed by IR. The C—O stretch in organometallic complexes is often very intense (stretching the C—O bond results in a substantial change in dipole moment), and its energy often provides valuable information about the molecular structure. Free carbon monoxide has a C—O stretch at 2143 cm^{-1}. $Cr(CO)_6$, on the other hand, has its C—O stretch at 2000 cm^{-1}. The lower energy for the stretching mode means that the C—O bond is weaker in $Cr(CO)_6$.

> The energy necessary to stretch a bond is proportional to $\sqrt{\frac{k}{\mu}}$, where k = force constant and μ = reduced mass; for atoms of mass m_1 and m_2, the reduced mass is given by
>
> $$\mu = \frac{m_1 m_2}{m_1 + m_2}$$
>
> The stronger the bond between two atoms, the larger the force constant; consequently, the greater the energy necessary to stretch the bond and the higher the energy of the corresponding band (the higher the wavenumber, in cm^{-1}) in the infrared spectrum. Similarly, the more massive the atoms involved in the bond, as reflected in a higher reduced mass, the less energy necessary to stretch the bond, and the lower the energy of the absorption in the infrared spectrum.

Both σ donation (which donates electron density from a bonding orbital on CO) and π acceptance (which places electron density in C—O antibonding orbitals) would be expected to weaken the C—O bond and to decrease the energy necessary to stretch that bond.

Additional evidence is provided by X-ray crystallography. In carbon monoxide, the C—O distance has been measured at 112.8 pm. Weakening of the C—O bond would be expected to cause this distance to increase. Such an increase in bond length is found in complexes containing CO, with C—O distances approximately 115 pm for many carbonyls. Although such measurements provide definitive measures of bond distances, in practice it is far more convenient to use infrared spectra to obtain data on the strength of C—O bonds.

The charge on a carbonyl complex is also reflected in its infrared spectrum. Five isoelectronic hexacarbonyls have the following C—O stretching bands (compare with $\nu(CO) = 2143\ cm^{-1}$ for free CO):*

Complex	$\nu(CO)$, cm^{-1}
$[Ti(CO)_6]^{2-}$	1748
$[V(CO)_6]^{-}$	1859
$Cr(CO)_6$	2000
$[Mn(CO)_6]^{+}$	2100
$[Fe(CO)_6]^{2+}$	2204

Of these five, $[Ti(CO)_6]^{2-}$ contains the most highly reduced metal, formally containing Ti(2−); this means that titanium has the weakest ability to attract electrons and the greatest tendency to back-donate electron density to CO. The formal charges on the metals increase from −2 for $[Ti(CO)_6]^{2-}$ to +2 for $[Fe(CO)_6]^{2+}$. The titanium in $[Ti(CO)_6]^{2-}$, with the most negative formal charge, has the strongest tendency to donate to CO. The consequence is

*The positions of the C—O stretching vibrations in the ions may be affected by interactions with solvents or counterions, and solid and solution spectra may differ slightly.

strong population of the π^* orbitals of CO in $[Ti(CO)_6]^{2-}$ and reduction of the strength of the C—O bond. In general, the more negative the charge on the organometallic species, the greater the tendency of the metal to donate electrons to the π^* orbitals of CO, and the lower the energy of the C—O stretching vibrations.*

EXERCISE 13.5

Predict which of the complexes $[V(CO)_6]^-$, $Cr(CO)_6$, or $[Mn(CO)_6]^+$ has the shortest C—O bond.

How is it possible for cationic carbonyl complexes such as $[Fe(CO)_6]^{2+}$ to have C—O stretching bands even higher in energy than those in free CO? It seems clear that the CO ligand does not engage in significant π-acceptor activity in these complexes, so weakening of the C—O bond via this interaction should be minimal, but how does the strength of the bond *increase*? Calculations have demonstrated that a polarization effect caused by the metal cation plays a major role in these carbonyl cations.[9]

In free CO, the electrons are polarized toward the more electronegative oxygen. For example, the electrons in the π orbitals are concentrated nearer to the oxygen atom than to the carbon. The presence of a transition metal cation reduces the polarization in the C—O bond by attracting the bonding electrons:

$$\overset{+\!\!\longrightarrow}{C{\equiv}O} \qquad M{-}\overset{+\!\to}{C{\equiv}O}$$

The consequence is that the electrons in the positively charged complex are more equally shared by the carbon and the oxygen, giving rise to a stronger bond and a higher-energy C—O stretch.

Bridging Modes of CO

Many cases are known in which CO forms bridges between two or more metals. Many bridging modes are known (**Table 13.2**).

TABLE 13.2 Bridging Modes of CO

Type of CO		Approximate Range for ν(CO) in Neutral Complexes (cm^{-1})
Free CO		2143
Terminal M—CO		1850–2120
Symmetric[a] μ_2—CO	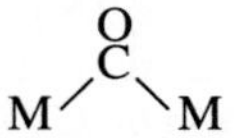 O / C / M M	1700–1860
Symmetric[a] μ_3—CO	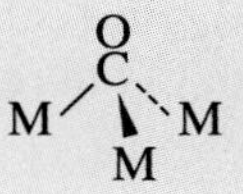 O / C / M M M	1600–1700
μ_4—CO	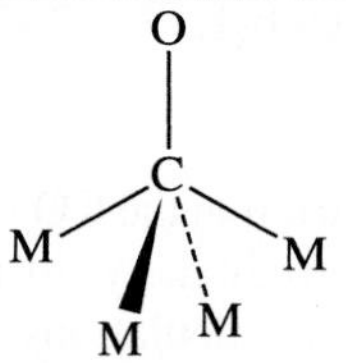 O / C / M M M M	< 1700 (few examples)

*Asymmetrically bridging μ_2- and μ_3-CO are also known.

*For reviews of metal carbonyl anions and other complexes containing metals in negative oxidation states, see J. E. Ellis, *Organometallics*, **2003**, *22*, 3322 and *Inorg. Chem.*, **2006**, *45*, 3167.

σ Donor

π Acceptor

$\nu_{CO} = 2082, 2019\ cm^{-1}$

$\nu_{CO} = 1829\ cm^{-1}$

FIGURE 13.14 Bridging CO.

The bridging mode is strongly correlated with the position of the C—O stretching band. In cases in which CO bridges two metal atoms, both metals can contribute electron density into π^* orbitals of CO to weaken the C—O bond and lower the energy of the stretch. Consequently, the C—O stretch for doubly bridging CO is at a much lower energy than for terminal COs. An example is shown in **Figure 13.14**. Interaction of three metal atoms with a triply bridging CO further weakens the C—O bond; the infrared band for the C—O stretch is still lower than in the doubly bridging case. (For comparison, carbonyl stretches in organic molecules are typically in the range 1700 to 1850 cm^{-1}, with many alkyl ketones near 1700 cm^{-1}.)

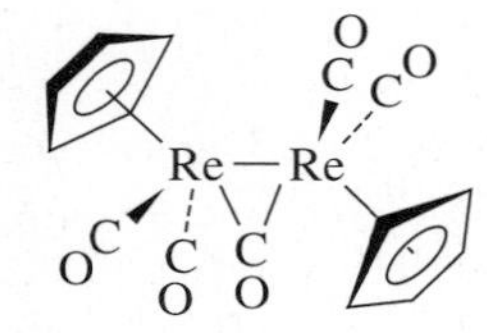

Ordinarily, terminal and bridging carbonyl ligands can be considered 2-electron donors, with the donated electrons shared by the metal atoms in the bridging cases. For example, in the complex to the left, the bridging CO is a 2-electron donor overall, with a single electron donated to each metal. The electron count for each Re atom according to method B is

Re		7 e^-
η^5-C_5H_5		5 e^-
2 (CO) (terminal)		4 e^-
$\frac{1}{2}(\mu_2$-CO)		1 e^-
M—M bond		1 e^-
	Total =	18 e^-

Nearly linear bridging carbonyls, such as those in $[(\eta^5\text{-}C_5H_5)Mo(CO)_2]_2$ are particularly interesting. When $[(\eta^5\text{-}C_5H_5)Mo(CO)_3]_2$ is heated, some carbon monoxide is released; the product, $[(\eta^5\text{-}C_5H_5)Mo(CO)_2]_2$, reacts readily with CO to reverse this reaction:[10]

$$[(\eta^5\text{-}C_5H_5)Mo(CO)_3]_2 \underset{}{\overset{\Delta}{\rightleftharpoons}} [(\eta^5\text{-}C_5H_5)Mo(CO)_2]_2 + 2\ CO$$

1960, 1915 cm^{-1} — 1889, 1859 cm^{-1}

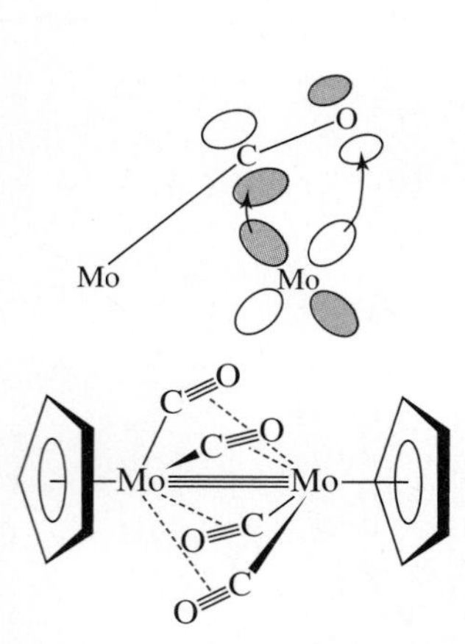

FIGURE 13.15 Bridging CO in $[(\eta^5\text{-}C_5H_5)Mo(CO)_2]_2$.

This reaction is accompanied by changes in the infrared spectrum in the CO region, as listed above. The Mo—Mo bond distance also shortens by approximately 79 pm, consistent with an increase in the metal–metal bond order from 1 to 3. Calculations have indicated that an important interaction is donation from a metal *d* orbital to the π^* orbital of CO (**Figure 13.15**).[11] Such donation weakens the carbon–oxygen bond in the ligand and results in the observed shift of the C—O stretching bands to lower energies.

Infrared spectra of carbonyl complexes are discussed further in Section 13.8.

Binary Carbonyl Complexes

Binary carbonyls, containing only metal atoms and CO, are numerous. Representative binary carbonyl complexes are in **Figure 13.16**. Most of these complexes obey the 18-electron rule. The cluster compounds $Co_6(CO)_{16}$ and $Rh_6(CO)_{16}$ do not obey the rule, however. More detailed analysis of the bonding in cluster compounds (Chapter 15) is necessary to satisfactorily account for the electron counting in these and other cluster compounds.

One other binary carbonyl does not obey the rule, the 17-electron $V(CO)_6$. This complex is one of a few cases in which strong π-acceptor ligands do not afford an 18-electron configuration. In $V(CO)_6$, the vanadium is apparently too small to permit a seventh coordination site; hence, no metal–metal bonded dimer, which would give an 18-electron configuration, is possible. However, $V(CO)_6$ is easily reduced to $[V(CO)_6]^-$, an 18-electron complex.

EXERCISE 13.6

Verify the 18-electron rule for five of the binary carbonyls—other than $V(CO)_6$, $Co_6(CO)_{16}$, and $Rh_6(CO)_{16}$—shown in Figure 13.16.

An interesting feature of the structures of binary carbonyl complexes is that the tendency of CO to bridge transition metals decreases going down the periodic table. For example, in $Fe_2(CO)_9$ there are three bridging carbonyls; but in $Ru_2(CO)_9$ and $Os_2(CO)_9$, there is a single bridging CO. A possible explanation is that the orbitals of bridging CO are less able to interact effectively with transition-metal atoms as the size of the metals increases, along with the metal–metal bond lengths.

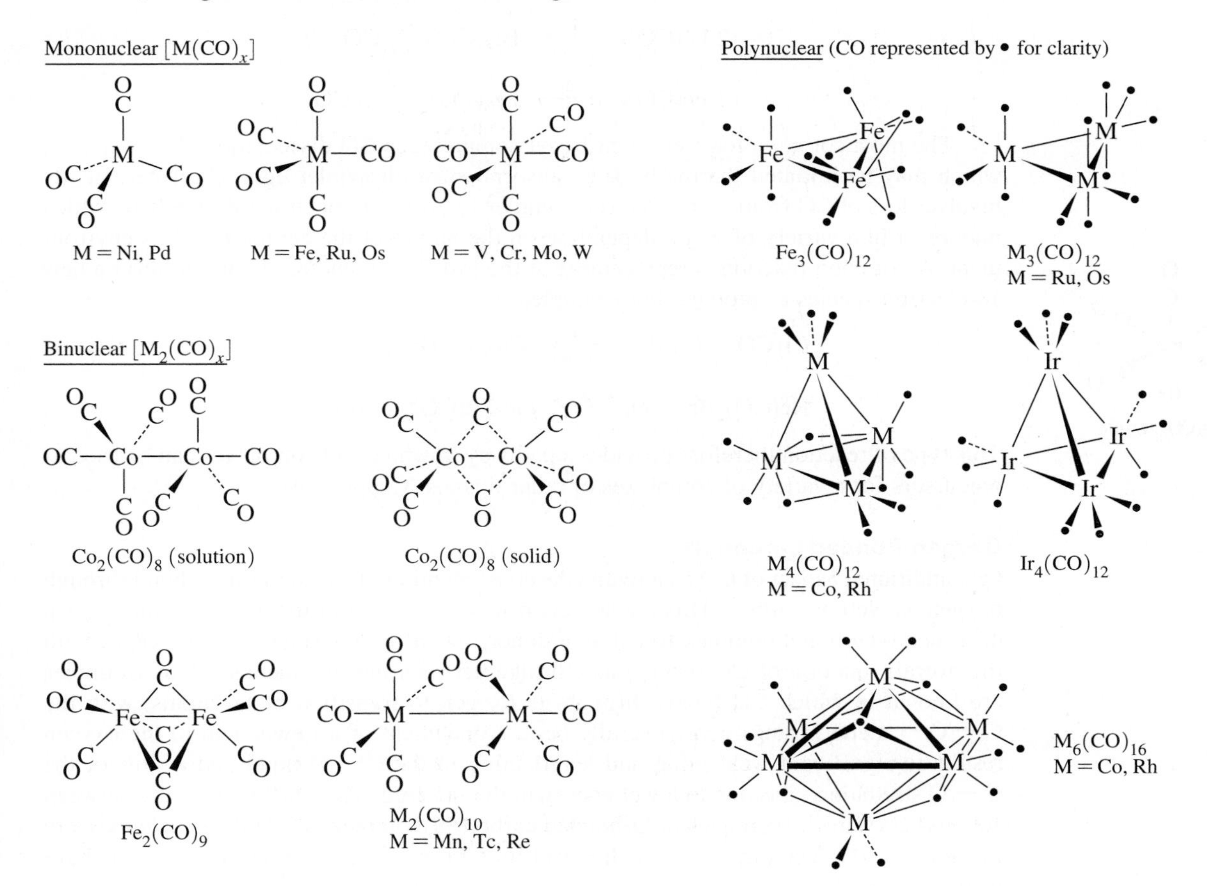

FIGURE 13.16 Binary Carbonyl Complexes.

Binary carbonyl complexes can be synthesized in many ways. Several of the most common methods are as follows:

1. *Direct reaction of a transition metal with CO.* The most facile of these reactions involves nickel, which reacts with CO at ambient temperature and 1 atm:

$$Ni + 4\ CO \longrightarrow Ni(CO)_4$$

   $Ni(CO)_4$ is a volatile, extremely toxic liquid that must be handled with great caution. It was first observed in Mond's study of the reaction of CO with nickel valves.[12] Because the reaction can be reversed at high temperature, coupling of the forward and reverse reactions has been used commercially in the Mond process for obtaining purified nickel from ores. Other binary carbonyls can be obtained from direct reaction of metal powders with CO, but elevated temperatures and pressures are needed.
2. *Reductive carbonylation:* reduction of a metal compound in the presence of CO and an appropriate reducing agent. Examples are

$$CrCl_3 + 6\ CO + Al \longrightarrow Cr(CO)_6 + AlCl_3$$

$$Re_2O_7 + 17\ CO \longrightarrow Re_2(CO)_{10} + 7\ CO_2$$

   (CO acts as a reducing agent in the second reaction; high temperature and pressure are required.)
3. *Thermal or photochemical reaction of other binary carbonyls.* Examples are

$$2\ Fe(CO)_5 \xrightarrow{h\nu} Fe_2(CO)_9 + CO$$

$$3\ Fe(CO)_5 \xrightarrow{\Delta} Fe_3(CO)_{12} + 3\ CO$$

The most common reaction of carbonyl complexes is CO dissociation. This reaction, which may be initiated thermally or by absorption of ultraviolet light, characteristically involves loss of CO from an 18-electron complex to give a 16-electron intermediate, which may react in a variety of ways, depending on the nature of the complex and its environment. A common reaction is replacement of the lost CO by another ligand to form a new 18-electron species as product. For example,

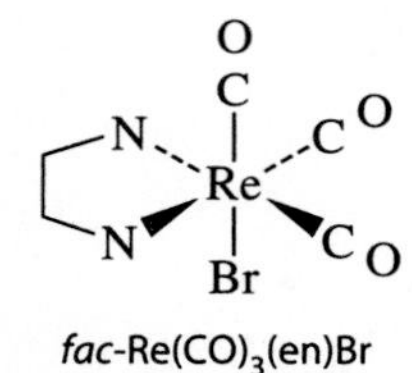

fac-$Re(CO)_3(en)Br$

$$Cr(CO)_6 + PPh_3 \xrightarrow[or\ h\nu]{\Delta} Cr(CO)_5(PPh_3) + CO$$

$$Re(CO)_5Br + en \xrightarrow{\Delta} fac\text{-}Re(CO)_3(en)Br + 2\ CO$$

This type of reaction therefore provides a pathway in which CO complexes can be used as precursors for a variety of complexes of other ligands (Chapter 14).

Oxygen-Bonded Carbonyls

One additional aspect of CO as a ligand deserves mention: it can sometimes bond through oxygen as well as carbon. This phenomenon was first noted in the ability of the oxygen of a metal–carbonyl complex to act as a donor toward Lewis acids such as $AlCl_3$, with the overall function of CO serving as a bridge between the two metals. Many examples are known in which CO bonds through its oxygen to transition metal atoms, with the C—O—metal arrangement generally bent. Attachment of a Lewis acid to the oxygen results in significant weakening and lengthening of the C—O bond and a shift of the C—O stretching vibration to lower energy in the infrared. This shift is typically between 100 and 200 cm^{-1}. Examples of O-bonded carbonyls, sometimes called *isocarbonyls*, are in Figure 13.17. The physical and chemical properties of oxygen-bonded carbonyls have been reviewed.[13]

(a) (b)

FIGURE 13.17 Oxygen-Bonded Carbonyls.

13.4.2 Ligands Similar to CO

Several diatomic ligands similar to CO are worth mention. Three—CS (thiocarbonyl), CSe (selenocarbonyl), and CTe (tellurocarbonyl)—are of interest in part for purposes of comparison with CO. The realm of CS complexes has been explored extensively since the first thiocarbonyl complex was reported in 1966,[14] but the chemistry of complexes containing CSe and CTe ligands has been more challenging.[15] This is in part because there are no sources as convenient and accessible for these two ligands as for CS complexes (CS_2 and Cl_2CS) and CO complexes. Therefore, the comparatively small number of CSe and CTe complexes should not be viewed as an indication of their relative stability. Thiocarbonyl complexes are also of interest as possible intermediates in sulfur-transfer reactions involved in removing sulfur from natural fuels. In recent years, the chemistry of complexes containing these ligands has developed more rapidly as avenues for their synthesis have been devised.

CS, CSe, and CTe are similar to CO in their bonding modes in that they behave as both σ donors and π acceptors and can bond to metals in terminal or bridging modes. The first three ligands typically function as stronger σ donors and π acceptors than CO.[16] In several cases, isostructural complexes of the ligands CO through CTe have been prepared, providing opportunities for structural and spectroscopic comparisons. Data from a set of ruthenium complexes (structure in margin) are provided in **Table 13.3**. The decrease in ruthenium—carbon distance in going down this series is consistent with increasing π-acceptor activity of the ligands, populating orbitals that are bonding with respect to the Ru—C bond. Although increasing π-acceptor ability of the ligands accounts for some of the decrease in C—E stretching frequency, the major contributor to this phenomenon is the increasing mass of the heteroatom E.

Other ligands are isoelectronic with CO and, not surprisingly, exhibit structural and chemical parallels with CO. Two examples are CN^- and N_2. Complexes of CN^- have been known longer than carbonyl complexes. Blue complexes (Prussian blue and Turnbull's blue) containing the ion $[Fe(CN)_6]^{3-}$ have been used as pigments in paints and inks for approximately three centuries. Cyanide is a stronger σ donor and a substantially weaker π acceptor than CO; overall, it is close to CO in the spectrochemical series.* Unlike most organic ligands, which bond to metals in low formal oxidation states, cyanide bonds readily

TABLE 13.3 Complexes of CO, CS, CSe, and CTe

CE	ν(C—E), cm^{-1}	Ru—C Distance, nm
CO	1934	1.829
CS	1238	1.793
CSe	1129	1.766
CTe	1024	1.748

Data from Y. Mutoh, N. Kozono, M. Araki, N. Tsuchida, K. Takano, Y. Ishii, *Organometallics*, **2010**, *29*, 519.

*A comparison of these ligands in mixed ligand complexes of Fe is in C. Loschen, G. Frenking, *Inorg. Chem.*, **2004**, *43*, 778.

to metals having higher oxidation states. As a good σ donor, CN^- interacts strongly with positively charged metal ions; as a weaker π acceptor than CO, partly a consequence of the negative charge of CN^- and the high energy of its π^* orbitals, cyanide is not as able to stabilize metals in low oxidation states. Its compounds are often studied in the context of classic coordination chemistry rather than organometallic chemistry.

The discovery that hydrogenase enzymes contain both CO and CN^- bound to iron has stimulated interest in complexes containing both ligands. Remarkably, only two iron complexes containing both CO and CN^- and a single iron atom, $[Fe(CO)(CN)_5]^{3-}$ (reported in 1887) and $[Fe(CO)_4(CN)]^-$ (reported in 1974), were known before 2001. Both the *cis* and *trans* isomers of $[Fe(CO)_2(CN)_4]^{2-}$ and *fac*-$[Fe(CO)_3(CN)_3]^-$ have been prepared. Two of the mixed ligand complexes can be made using $Fe(CO)_4I_2$:[17]

$$Fe(CO)_4I_2 \xrightarrow{3\,CN^-} fac\text{-}[Fe(CO)_3(CN)_3]^- \xrightarrow{CN^-} cis\text{-}[Fe(CO)_2(CN)_4]^{2-}$$

The complex *trans*-$[Fe(CO)_2(CN)_4]^{2-}$ can be made by the addition of cyanide to a solution of $FeCl_2$ under an atmosphere of CO:[18]

$$FeCl_2(aq \text{ or } CH_3CN) + 4\,CN^- \xrightarrow{CO} trans\text{-}[Fe(CO)_2(CN)_4]^{2-}$$

Dinitrogen is a weaker donor and acceptor than CO. However, N_2 complexes are of great interest, especially as possible intermediates in reactions that may simulate natural processes of nitrogen fixation.

NO Complexes

The NO (nitrosyl) ligand shares many similarities with CO. Like CO, it is a σ donor and π acceptor and can serve as a terminal or bridging ligand; useful information can be obtained about its compounds by analysis of its infrared spectra. Unlike CO, however, terminal NO has two common coordination modes, linear (like CO) and bent. Examples of NO complexes are in **Figure 13.18**.

A formal analogy is often drawn between the linear bonding modes of both ligands. NO^+ is isoelectronic with CO; therefore, in its bonding to metals, linear NO is considered by electron-counting scheme A as NO^+, a 2-electron donor. By the neutral ligand method (B), linear NO is counted as a 3-electron donor (it has one more electron than the 2-electron donor CO).

The bent coordination mode of NO can be considered to arise formally from NO^-, with the bent geometry suggesting sp^2 hybridization at the nitrogen. By electron-counting scheme A, therefore, bent NO is considered the 2-electron donor NO^-; by the neutral ligand model, it is considered a 1-electron donor.

Although these electron-counting methods in NO complexes are useful, they do not describe how NO actually bonds to metals. The use of NO^+, NO, or NO^- does not necessarily imply degrees of ionic or covalent character in coordinated NO; these labels are simply convenient means of counting electrons.

Useful information about the linear and bent bonding modes of NO is summarized in **Figure 13.19**. Many complexes containing each mode are known, and examples are also

FIGURE 13.18 Examples of NO and NS Complexes.

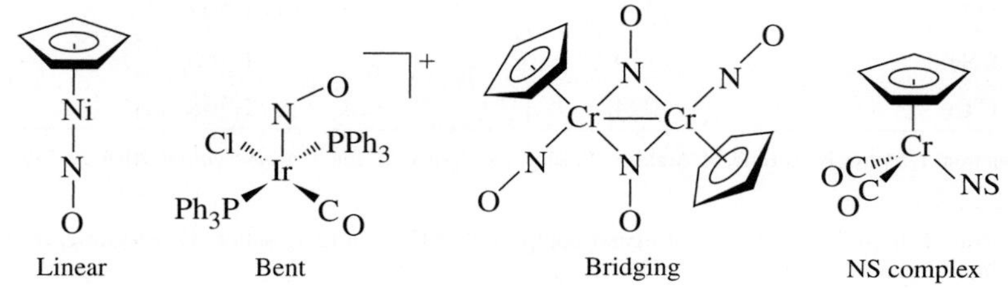

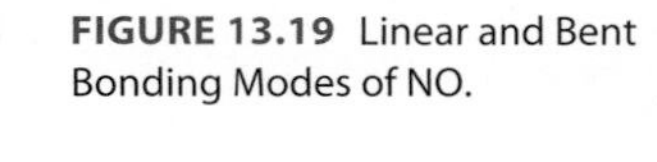

FIGURE 13.19 Linear and Bent Bonding Modes of NO.

	Linear	Bent
M—N—O angle	165°–180°	119°–140°
ν (N-O) in neutral molecules	1610–1830 cm^{-1}	1520–1720 cm^{-1}
Electron donor count	2 (as NO^+) 3 (as neutral NO)	2 (as NO^-) 1 (as neutral NO)

known in which both linear and bent NO occur in the same complex. Although linear coordination usually gives rise to N — O stretching vibrations at a higher energy than the bent mode, there is enough overlap in the ranges of these bands that infrared spectra alone may not be sufficient to distinguish between the two. Furthermore, the manner of packing in crystals may bend the metal — N — O bond considerably from 180° in the linear coordination mode.

One compound containing only NO ligands is known, $Cr(NO)_4$, a tetrahedral molecule that is isoelectronic with $Ni(CO)_4$.* Complexes containing bridging nitrosyl ligands are also known, with the neutral bridging ligand formally considered a 3-electron donor. One NO complex, the nitroprusside ion, $[Fe(CN)_5(NO)]^{2-}$, has been used as a vasodilator in the treatment of high blood pressure. Its therapeutic effect is a consequence of its ability to release its NO ligand; the NO acts as the vasodilating agent.

In recent years, several dozen compounds containing the isoelectronic NS (thionitrosyl) ligand have been synthesized; one of these is in Figure 13.18. Like NO, NS can function in linear, bent, and bridging modes. In general, NS has been reported to act as a stronger σ-donor but weaker π-acceptor ligand than NO, a consequence of the greater concentration of negative charge on the nitrogen atom in NS. The difference in polarity of the NO and NS ligands also leads to significant differences in the electronic spectra of their complexes.[19] The realm of NSe (selenonitrosyl) complex chemistry is limited; only a single complex of this ligand has been reported.[20]

13.4.3 Hydride and Dihydrogen Complexes

The simplest of all possible ligands is the hydrogen atom; similarly, the simplest possible diatomic ligand is H_2. These ligands have gained attention, by virtue of their apparent simplicity, as models for bonding schemes in coordination compounds. Moreover, both ligands have played important roles in the development of applications of organometallic chemistry to organic synthesis and catalytic processes.

Hydride Complexes

Although hydrogen atoms form bonds with nearly every element, we will specifically consider coordination compounds containing H bonded to transition metals.[21] Because the hydrogen atom only has a 1*s* orbital of suitable energy for bonding, the bond between H and a transition metal must be a σ interaction, involving metal *s*, *p*, and/or *d* orbitals. As a ligand, H may be considered a 2-electron donor as hydride ($:H^-$, method A) or a 1-electron neutral donor (H atom, method B).

Although some transition-metal complexes containing only the hydride ligand are known—an example is the 9-coordinate $[ReH_9]^{2-}$ ion (Figure 9.35), the classic example of a tricapped trigonal prism[22]—we are principally concerned with complexes containing

*Compounds containing only a single ligand, such as NO in $Cr(NO)_4$ and CO in $Mo(CO)_6$, are called *homoleptic* compounds.

H in combination with other ligands. Such complexes may be made in a variety of ways. Probably the most common synthesis is by reaction of a transition metal complex with H_2. For example,

$$Co_2(CO)_8 + H_2 \longrightarrow 2\ HCo(CO)_4$$

$$trans\text{-}Ir(CO)Cl(PEt_3)_2 + H_2 \longrightarrow Ir(CO)Cl(H)_2(PEt_3)_2$$

Carbonyl hydride complexes can also be formed by the reduction of carbonyl complexes, followed by the addition of acid. For example,

$$Co_2(CO)_8 + 2\ Na \longrightarrow 2\ Na^+[Co(CO)_4]^-$$

$$[Co(CO)_4]^- + H^+ \longrightarrow HCo(CO)_4$$

One of the most interesting aspects of transition-metal hydride chemistry is the relationship between this ligand and the chemistry of the dihydrogen ligand, H_2.

Dihydrogen Complexes

Although complexes containing H_2 molecules coordinated to transition metals had been proposed for many years, the first structural characterization of a dihydrogen complex did not occur until 1984, when Kubas synthesized $M(CO)_3(PR_3)_2(H_2)$, where M = Mo or W and R = cyclohexyl or isopropyl.[23] Subsequently, many H_2 complexes have been identified, and the chemistry of this ligand has developed rapidly.[24]

The bonding between dihydrogen and a transition metal can be described as shown in **Figure 13.20**. The σ electrons in H_2 can be donated to a suitable empty orbital on the metal (such as a *d* orbital or hybrid orbital), and the empty σ^* orbital of the ligand can accept electron density from an occupied *d* orbital of the metal. The result is an overall weakening and lengthening of the H—H bond in comparison with free H_2. Typical H—H distances in complexes containing coordinated dihydrogen are in the range of 82 to 90 pm, in comparison with 74.14 pm in free H_2.

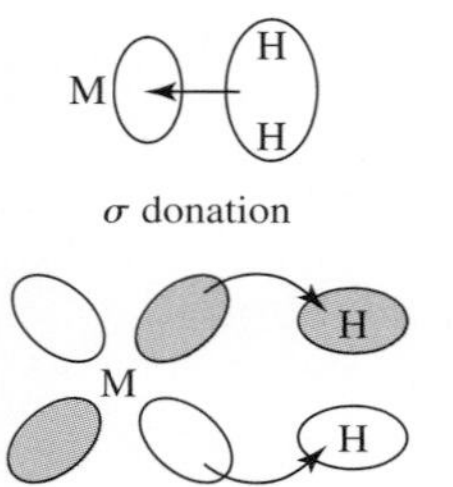

FIGURE 13.20 Bonding in Dihydrogen Complexes.

This bonding scheme leads to interesting ramifications that are distinctive from other donor–acceptor ligands such as CO. If the metal is electron rich and donates strongly to the σ^* of H_2 the H—H bond in the ligand can rupture, giving separate H atoms. Consequently, the search for stable H_2 complexes has centered on metals likely to be relatively poor donors, such as those in high oxidation states or surrounded by ligands that function as strong electron acceptors. In particular, good π acceptors, such as CO and NO, can be effective at stabilizing the dihydrogen ligand.

Explain why $Mo(PMe_3)_5H_2$ is a dihydride (contains two separate H ligands), but $Mo(CO)_3(PR_3)_2(H_2)$ contains the dihydrogen ligand (Me = methyl, R = isopropyl).

Dihydrogen complexes are frequently suggested as intermediates in a variety of reactions of hydrogen at metal centers. Some of these reactions are steps in catalytic processes of commercial interest. As this ligand becomes more completely understood, the applications of its chemistry are likely to become extremely important.

13.4.4 Ligands Having Extended Pi Systems

Although it is relatively simple to describe pictorially how ligands such as CO and PPh_3 bond to metals, explaining bonding between metals and organic ligands having extended π systems can be more complex. For example, how are the C_5H_5 rings attached to Fe in ferrocene, and how can 1,3-butadiene bond to metals? To understand the bonding between metals and π systems, we must consider the π bonding within the ligands themselves.

We will first describe linear and then cyclic π systems, after which we will consider how molecules containing such systems can bond to metals.

Linear Pi Systems*

The simplest case of an organic molecule having a linear π system is ethylene, which has a single π bond resulting from the interactions of two $2p$ orbitals on its carbon atoms. Interactions of these p orbitals result in one bonding and one antibonding π orbital, as shown:

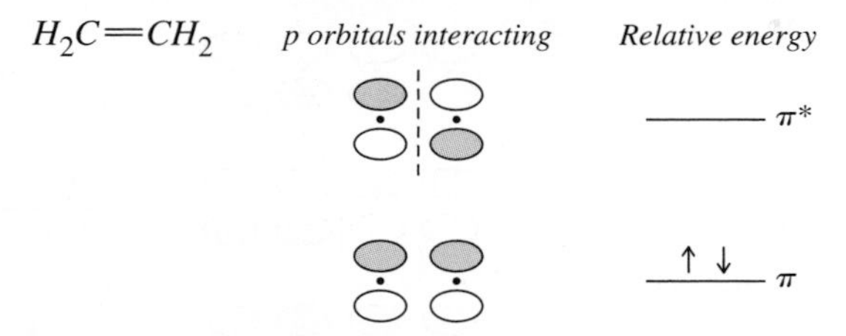

The antibonding interaction has a nodal plane perpendicular to the internuclear axis, but the bonding interaction has no such nodal plane.

Next is the three-atom π system, the π-allyl radical, C_3H_5. In this case, there are three $2p$ orbitals to be considered, one from each of the carbon atoms participating in the π system. The possible interactions are as follows:

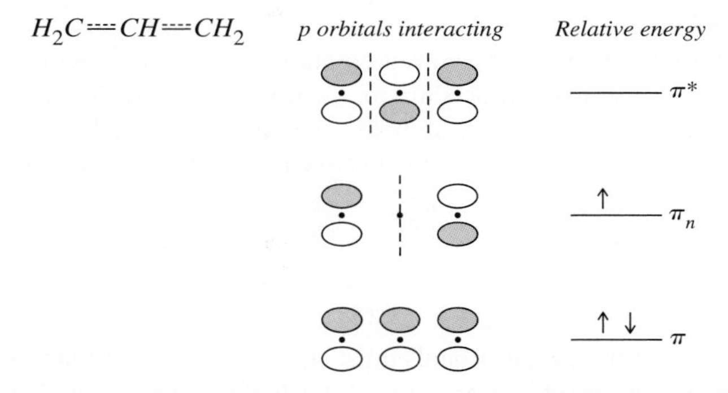

The lowest energy π molecular orbital for this system has all three p orbitals interacting constructively, to give a bonding molecular orbital. Higher in energy is the nonbonding orbital (π_n), in which a nodal plane bisects the molecule, cutting through the central carbon atom. In this case, the p orbital on the central carbon does not participate in the molecular orbital; a nodal plane passes through the center of this π orbital and thereby cancels it from participation. Highest in energy is the antibonding π^* orbital, in which there is an antibonding interaction between each neighboring pair of carbon p orbitals.

The number of nodes perpendicular to the carbon chain increases in going from lower-energy to higher-energy orbitals; for example, in the π-allyl system, the number of nodes increases from zero to one to two from the lowest to the highest energy orbital. This is a trend that will also appear in the following examples.

1,3-Butadiene may exist in *cis* or *trans* forms. For our purposes, we will treat both as linear systems; the nodal behavior of the molecular orbitals is the same in each case as in a linear π system of four atoms. The $2p$ orbitals of the carbon atoms in the chain may

*In this section the term "linear" is used broadly to include not only ligands that have carbons in a straight line but acyclic ligands that are bent at inner sp^2 carbons. For simplicity, the accompanying diagrams are also shown as straight line arrangements. In addition, the p orbitals in this section are not drawn to scale. The drawings show bonding and antibonding interactions between adjacent orbitals, but do not use different sizes to indicate the relative contributions of individual orbitals to molecular orbitals.

interact in four ways, with the lowest energy π molecular orbital having all constructive interactions between neighboring p orbitals, and the energy of the other π orbitals increasing with the number of nodes between the atoms.

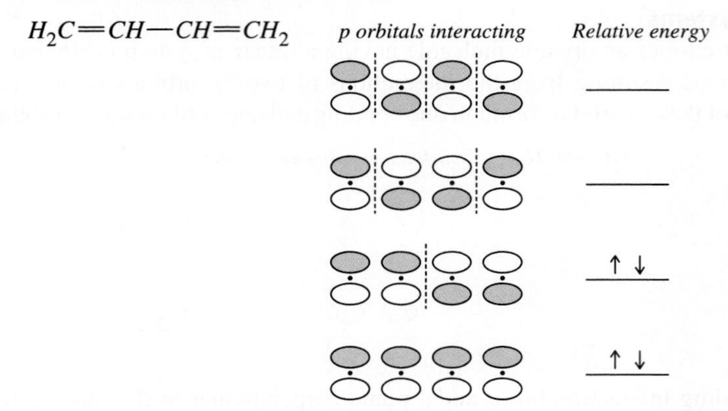

Similar patterns can be obtained for longer π systems; two more examples are included in **Figure 13.21**. The number of π molecular orbitals is equal to the number of carbons in the π system.

Cyclic Pi Systems

The procedure for obtaining a pictorial representation of the orbitals of cyclic π systems of hydrocarbons is similar to the procedure for the linear systems. The smallest such cyclic hydrocarbon is *cyclo*-C_3H_3. The lowest energy π molecular orbital for this system is the one resulting from constructive interaction between each of the $2p$ orbitals in the ring:

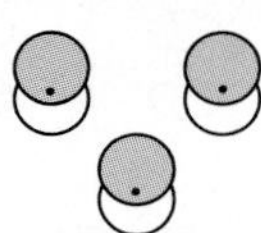

Because the number of molecular orbitals must equal the number of atomic orbitals used, two additional π molecular orbitals are needed. Each of these has a single nodal plane that

FIGURE 13.21 π Orbitals for Linear Systems.

	p orbitals interacting	Relative energy		p orbitals interacting	Relative energy
C_5H_7		↑ (3rd level); ↑↓ (4th, 5th levels)	C_6H_8		↑↓ (4th, 5th, 6th levels)

is perpendicular to the plane of the molecule and bisects the molecule; the nodes for these two molecular orbitals are perpendicular to each other:

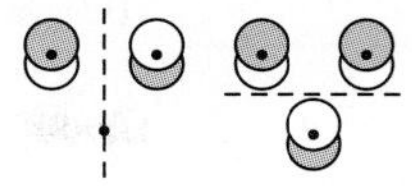

These molecular orbitals have the same energy; π molecular orbitals having the same number of nodes in cyclic π systems of hydrocarbons are degenerate (have the same energy). The total π molecular orbital diagram for *cyclo*-C_3H_3 can therefore be summarized as follows:

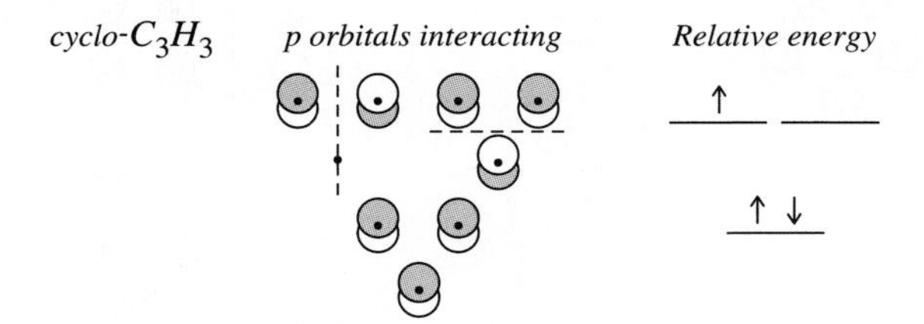

A simple way to determine the *p* orbital interactions and the relative energies of the cyclic π systems that are regular polygons is to draw the polygon with one vertex pointed down. Each vertex then corresponds to the relative energy of a molecular orbital. Furthermore, the number of nodal planes perpendicular to the plane of the molecule increases as one goes to higher energy, with the bottom orbital having zero nodes, the next pair of orbitals a single node, and so on. For example, this scheme predicts that the next cyclic π system, *cyclo*-C_4H_4 (cyclobutadiene), would have molecular orbitals as follows:*

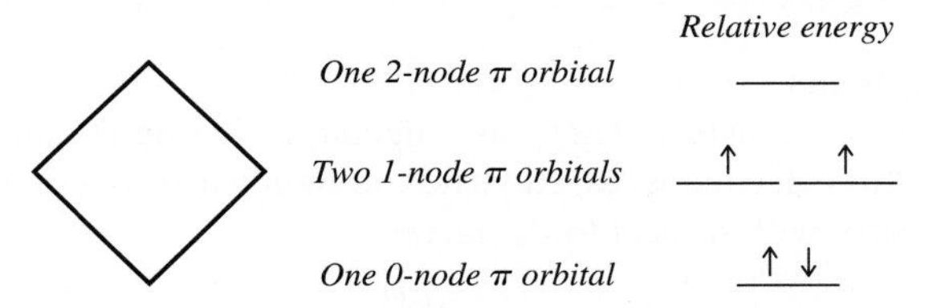

Similar results are obtained for other cyclic π systems; two of these are in **Figure 13.22**. In these diagrams, nodal planes are disposed symmetrically. For example, in *cyclo*-C_4H_4, the single-node molecular orbitals bisect the molecule through opposite sides; the nodal planes are oriented perpendicularly to each other. The 2-node orbital for this molecule also has perpendicular nodal planes.

This method may seem oversimplified, but the nodal behavior and relative energies are the same as those obtained from molecular orbital calculations. The method for obtaining equations for the molecular orbitals of cyclic hydrocarbons of formula C_nH_n ($n = 3$ to 8) is given by Cotton.[25] Throughout this discussion, we have shown not the actual shapes of the π molecular orbitals, but rather the *p* orbitals used. The nodal behavior of both sets (the π orbitals and the *p* orbitals used) is identical and therefore sufficient for the discussion of bonding with metals that follows.**

---

*This approach predicts a diradical for cyclobutadiene (one electron in each 1-node orbital). Although cyclobutadiene itself is very reactive (P. Reeves, T. Devon, R. Pettit, *J. Am. Chem. Soc.*, **1969**, *91*, 5890), complexes containing derivatives of cyclobutadiene are known. At 8 K, cyclobutadiene itself has been isolated in an argon matrix (O. L. Chapman, C. L. McIntosh, J. Pacansky, *J. Am. Chem. Soc.*, **1973**, *95*, 614; A. Krantz, C. Y. Lin, M. D. Newton, *J. Am. Chem. Soc.*, **1973**, *95*, 2746).

**Diagrams of many molecular orbitals for linear and cyclic π systems can be found in W. L. Jorgenson and L. Salem, *The Organic Chemist's Book of Orbitals*, Academic Press, New York, 1973.

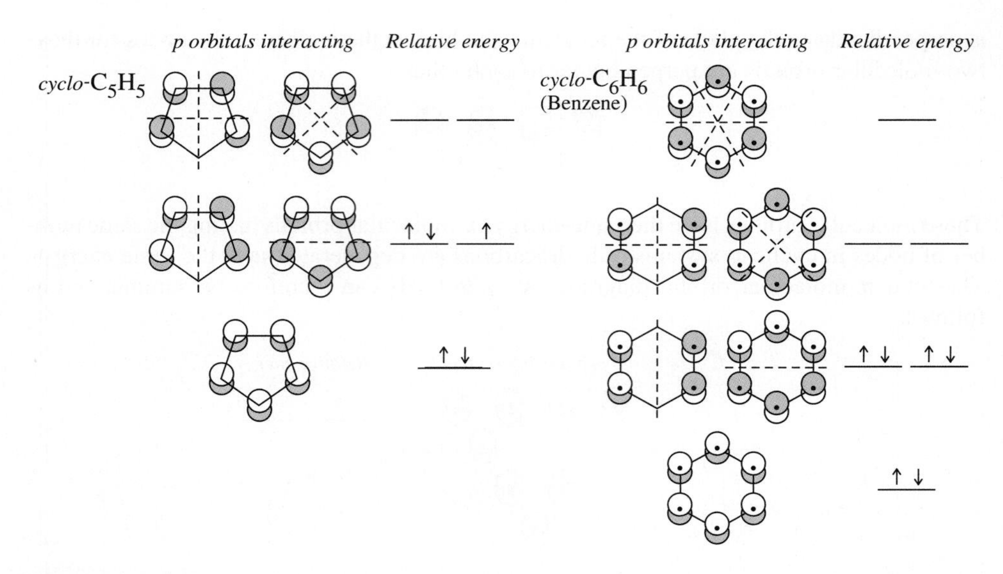

FIGURE 13.22 Molecular Orbitals for Cyclic Pi Systems.

13.5 Bonding between Metal Atoms and Organic Pi Systems

We are now ready to consider metal–ligand interactions involving such systems. We will begin with the simplest of the linear systems, ethylene, and conclude with ferrocene.

13.5.1 Linear Pi Systems

Pi–Ethylene Complexes

Many complexes involve ethylene, C_2H_4, as a ligand, including the anion of Zeise's salt, $[Pt(\eta^2\text{-}C_2H_4)Cl_3]^-$. In such complexes, ethylene commonly acts as a sidebound ligand with the following geometry with respect to the metal:

H H
C
Pt ‖
C
H H

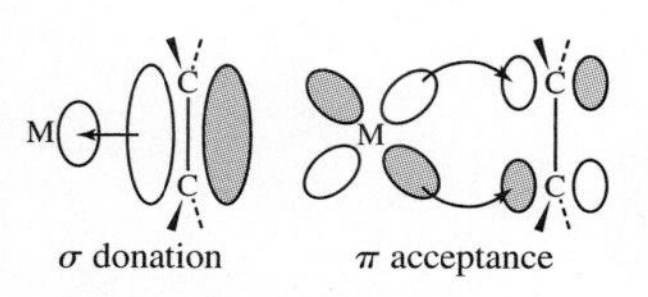

FIGURE 13.23 Bonding in Ethylene Complexes.

The hydrogens in ethylene complexes are typically bent back away from the metal, as shown. Ethylene donates electron density to the metal in a sigma fashion, using its π-bonding electron pair, as shown in **Figure 13.23**. At the same time, electron density can be donated back to the ligand in a pi fashion from a metal *d* orbital to the empty π^* orbital of the ligand. This is another example of the synergistic effect of σ donation and π acceptance encountered earlier with the CO ligand.

If this picture of bonding in ethylene complexes is correct, it should be in agreement with the measured C—C distance. The C—C distance in Zeise's salt is 137.5 pm in comparison with 133.7 pm in free ethylene. The lengthening of this bond can be explained by a combination of the two factors involved in the synergistic σ-donor, π-acceptor nature of the ligand: donation of electron density to the metal in a sigma fashion reduces the π-bonding electron density within the ligand, weakening the C—C bond. Furthermore, the back-donation of electron density from the metal to the π^* orbital of the ligand also reduces the C—C bond strength by populating the antibonding orbital. The net effect weakens and lengthens the C—C bond in the C_2H_4 ligand. In addition, vibrational frequencies of coordinated ethylene are at lower energy than in free ethylene; for example, the C=C stretch in the anion of Zeise's salt is at 1516 cm^{-1}, compared to 1623 cm^{-1} in free ethylene.

Pi–Allyl Complexes

The allyl group most commonly functions as a trihapto ligand, using delocalized π orbitals as described previously, or as a monohapto ligand, primarily σ bonded to a metal. Examples of these types of coordination are in **Figure 13.24**.

Bonding between η^3-C_3H_5 and a metal atom is shown schematically in **Figure 13.25**. The lowest energy π orbital can donate electron density in a sigma fashion to a suitable orbital on the metal. The next orbital, nonbonding in free allyl, can act as a donor or acceptor, depending on the electron distribution between the metal and the ligand. The highest energy π orbital acts as an acceptor; thus, there can be synergistic sigma and pi interactions between allyl and the metal. The C—C—C angle within the ligand is generally near 120°, consistent with sp^2 hybridization.

Allyl complexes (or complexes of substituted allyls) are intermediates in many reactions, some of which take advantage of the capability of this ligand to function in both a

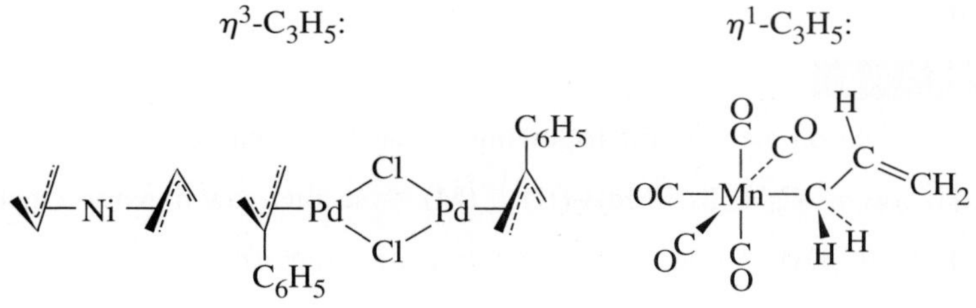

FIGURE 13.24 Examples of Allyl Complexes.

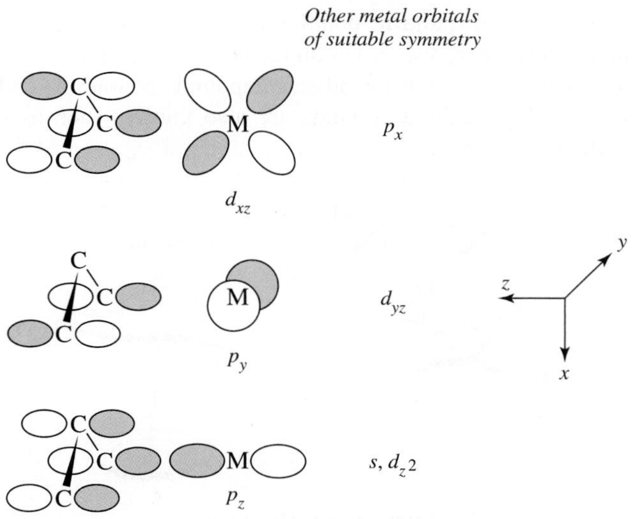

FIGURE 13.25 Bonding in η^3-Allyl Complexes.

η^3 and η^1 fashion. Loss of CO from carbonyl complexes containing η^1-allyl ligands often results in conversion of η^1- to η^3-allyl. For example,

$$[Mn(CO)_5]^- + C_3H_5Cl \longrightarrow (\eta^1\text{-}C_3H_5)Mn(CO)_5 + Cl^- \xrightarrow{\Delta \text{ or } h\nu} (\eta^3\text{-}C_3H_5)Mn(CO)_4 + CO$$

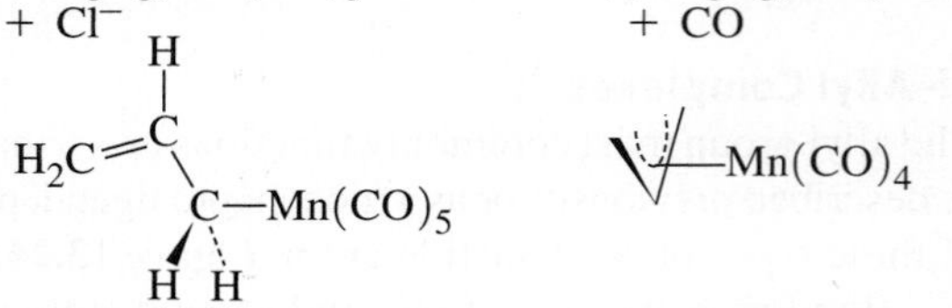

The $[Mn(CO)_5]^-$ ion displaces Cl^- from allyl chloride to give an 18-electron product containing η^1-C_3H_5. The allyl ligand switches to trihapto when a CO is lost, preserving the 18-electron count.

Other Linear Pi Systems

Many other such systems are known; several examples of organic ligands having longer π systems are in **Figure 13.26**. Butadiene and longer conjugated π systems have the possibility of isomeric ligand forms (*cis* and *trans* for butadiene). Larger cyclic ligands may have a π system extending through part of the ring. An example is cyclooctadiene (COD); the 1,3-isomer has a 4-atom π system comparable to butadiene; 1,5-cyclooctadiene has two isolated double bonds, one or both of which may interact with a metal in a manner similar to ethylene.

EXERCISE 13.8

Identify the transition metal in the following 18-electron complexes:

a. $(\eta^5\text{-}C_5H_5)(cis\text{-}\eta^4\text{-}C_4H_6)M(PMe_3)_2(H)$ (M) = second-row transition metal)

b. $(\eta^5\text{-}C_5H_5)M(C_2H_4)_2$ (M = first-row transition metal)

13.5.2 Cyclic Pi Systems

Cyclopentadienyl (Cp) Complexes

The cyclopentadienyl group, C_5H_5, may bond to metals in a variety of ways, with many examples known of the η^1-, η^3-, and η^5-bonding modes. The discovery of the first cyclopentadienyl complex, ferrocene, was a landmark in the development of organometallic chemistry and stimulated the search for other compounds containing π-bonded organic ligands. Substituted cyclopentadienyl ligands are also known, such as $C_5(CH_3)_5$, often abbreviated Cp*, and $C_5(\text{benzyl})_5$.

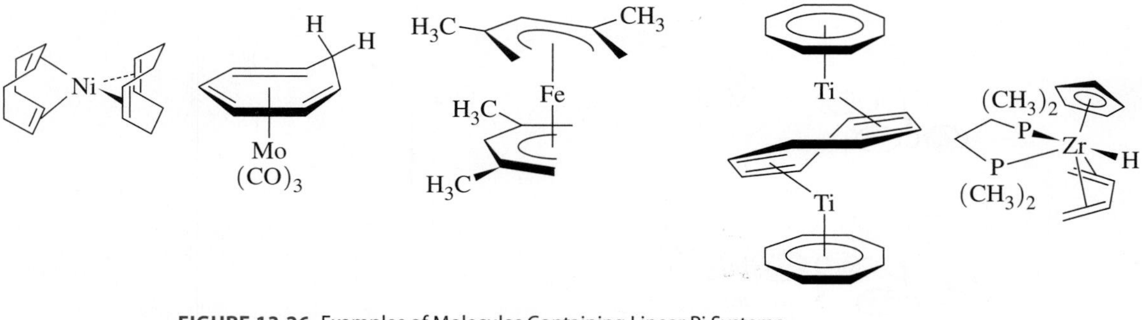

FIGURE 13.26 Examples of Molecules Containing Linear Pi Systems.

Ferrocene and other cyclopentadienyl complexes can be prepared by reacting metal salts with $C_5H_5^-$.*

$$FeCl_2 + 2\ NaC_5H_5 \longrightarrow (\eta^5\text{-}C_5H_5)_2Fe + 2\ NaCl$$

Ferrocene, $(\eta^5\text{-}C_5H_5)_2Fe$

Ferrocene is the prototype of a series of sandwich compounds, the metallocenes, with the formula $(C_5H_5)_2M$. Electron counting in ferrocene can be viewed in two ways. One possibility is to consider it an iron(II) complex with two 6-electron cyclopentadienide ($C_5H_5^-$) ions, another to view it as iron(0) coordinated by two neutral, 5-electron C_5H_5 ligands. The actual bonding situation in ferrocene is more complicated and requires an analysis of the various metal–ligand interactions. As usual, we expect orbitals on the central Fe and on the two C_5H_5 rings to interact if they have appropriate symmetry; furthermore, we expect interactions to be strongest if they are between orbitals of similar energy.

For the purposes of our analysis, it will be useful to refer to Figure 13.22 for diagrams of the π molecular orbitals of a C_5H_5 ring. Two of these rings are arranged in a parallel fashion in ferrocene to "sandwich in" the metal atom. Our discussion will be based on the eclipsed D_{5h} conformation of ferrocene, the conformation consistent with gas-phase and low-temperature data on this molecule.[26,27] The same approach using the staggered conformation would yield a similar molecular orbital picture. Descriptions of the bonding in ferrocene based on D_{5d} symmetry are common in the chemical literature because this was once believed to be the molecule's most stable conformation.**

In developing the group orbitals for a pair of C_5H_5 rings, we pair up molecular orbitals of the same energy and same number of nodes; for example, we pair the zero-node orbital of one ring with the zero-node orbital of the other.† We also must pair up the molecular orbitals in such a way that the nodal planes are coincident. Furthermore, in each pairing there are two possible orientations of the ring molecular orbitals: one in which lobes of like sign are pointed toward each other, and one in which lobes of opposite sign are pointed toward each other. For example, the zero-node orbitals of the C_5H_5 rings may be paired in the following two ways:

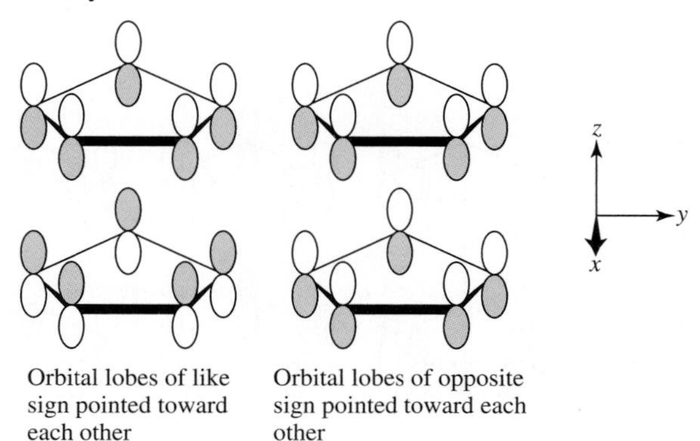

Orbital lobes of like sign pointed toward each other

Orbital lobes of opposite sign pointed toward each other

---

*Solutions of NaC_5H_5 in tetrahydrofuran are available commercially. Alternatively, NaC_5H_5 can be prepared by cracking of dicyclopentadiene, followed by reduction:

$$C_{10}H_{12}\ \text{(dicyclopentadiene)} \longrightarrow 2\ C_5H_6\ \text{(cyclopentadiene)}$$
$$2\ Na + 2\ C_5H_6 \longrightarrow 2\ NaC_5H_5 + H_2$$

The $C_5(CH_3)_5$ and C_5(benzyl)$_5$ analogs of ferrocene have staggered D_{5d} symmetry, as do several other metallocenes. See M.D. Rausch, W-M. Tsai, J. W. Chambers, R. D. Rogers, H. G. Alt, *Organometallics*, **1989, *8*, 816. For calculations that compare the energies of the conformations of ferrocene, see S. Coriani, A. Haaland, T. Helgaker, P. Jørgensen, *Chem Phys Chem*, **2007**, *7*, 245.

†Not counting the nodal planes that are coplanar with the C_5H_5 rings.

The ten group orbitals arising from the C_5H_5 ligands are shown in **Figure 13.27**.

The process of developing the molecular orbital picture of ferrocene now becomes one of matching the group orbitals with the *s*, *p*, and *d* orbitals of appropriate symmetry on Fe.

We will illustrate one of these interactions, between the d_{yz} orbital of Fe and its appropriate group orbital (the 1-node group orbitals are shown in Figure 13.27). This interaction results in a bonding and an antibonding orbital:

Selected Atomic Orbitals by Gary O. Spessard and Gary L. Miessler. Reprinted by permission.

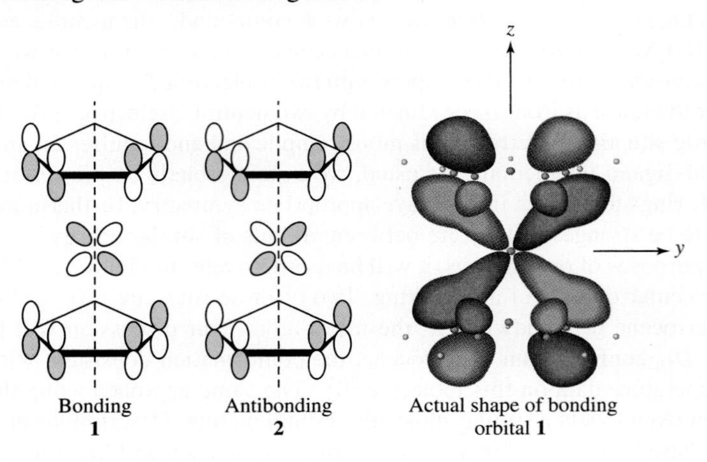

FIGURE 13.27 Group Orbitals for C_5H_5 Ligands of Ferrocene by Gary O. Spessard and Gary L. Miessler. Reprinted by permission.

2-Node group orbitals

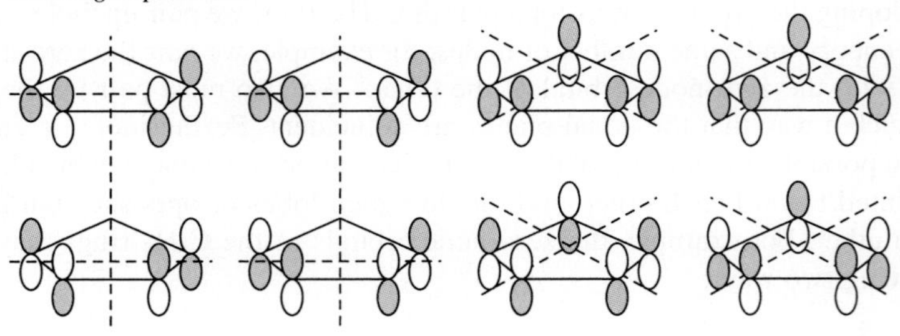

1-Node group orbitals

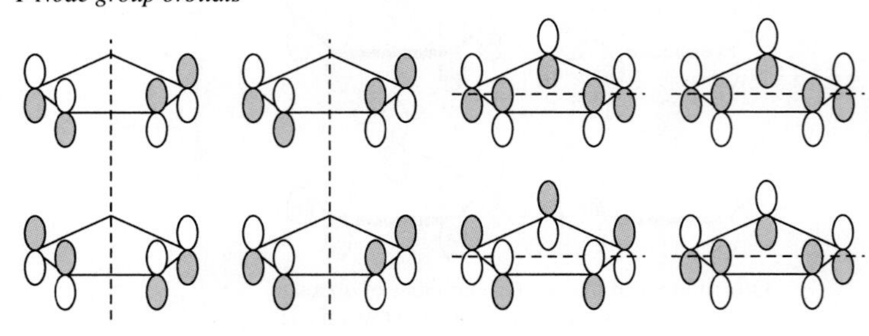

0-Node group orbitals

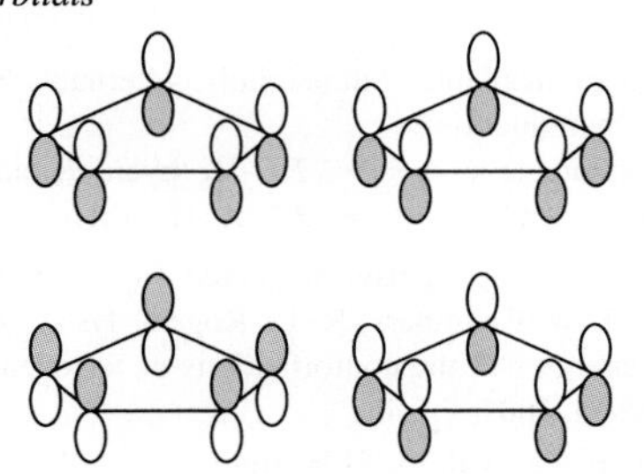

EXERCISE 13.9

Determine which orbitals on Fe are appropriate for interaction with each of the remaining group orbitals in Figure 13.27.

The complete energy-level diagram for the molecular orbitals of ferrocene is shown in **Figure 13.28**. The molecular orbital resulting from the d_{yz} bonding interaction, labeled **1** in the MO diagram, contains a pair of electrons. Its antibonding counterpart, **2**, is empty. It is a useful exercise to match the other group orbitals from Figure 13.27 with the molecular orbitals in Figure 13.28 to verify the metal–ligand interactions.

The orbitals of ferrocene having the greatest *d*-orbital character are also the highest occupied and lowest unoccupied orbitals (HOMO and LUMO). These orbitals are

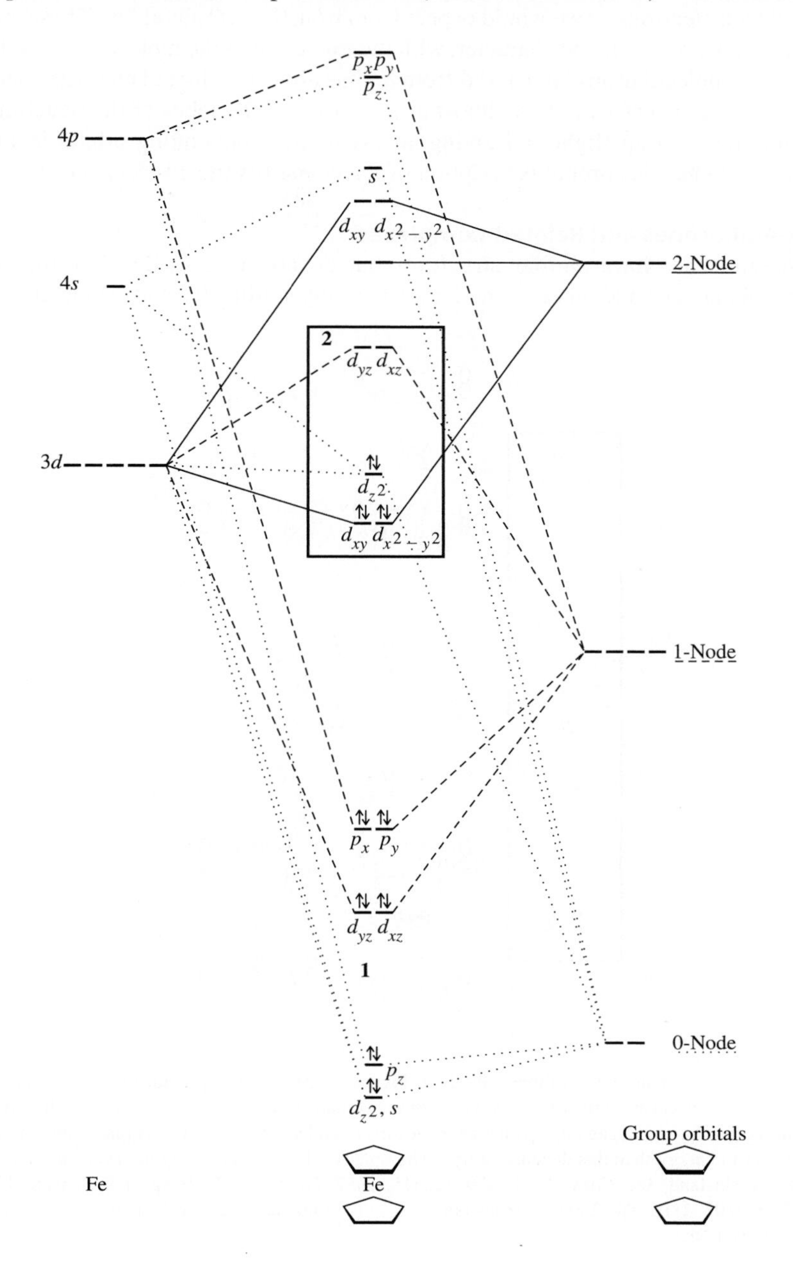

FIGURE 13.28 Molecular Orbital Energy Levels of Ferrocene.

highlighted in the box in Figure 13.28. Two orbitals, the degenerate pair having largely d_{xy} and $d_{x^2-y^2}$ character, are weakly bonding and are occupied by electron pairs; one, having largely d_{z^2} character, is essentially nonbonding and is also occupied by an electron pair; and two, having primarily d_{xy} and d_{yz} character, are empty. The relative energies of these orbitals and their *d* orbital–group orbital interactions are shown in **Figure 13.29.**[*,28]

The overall bonding in ferrocene can now be summarized. The occupied orbitals of the η^5-C_5H_5 ligands are stabilized by their interactions with iron. Note the stabilization of 0-node and 1-node group orbitals that have bonding interactions with the metal, forming molecular orbitals that are primarily ligand in nature (these are the orbitals, labeled from lowest to highest energy (d_{z^2}, s), p_z, (d_{yz}, d_{xz}), and (p_x, p_y)).

The orbitals next highest in energy are largely derived from iron *d* orbitals; they are populated by 6 electrons as we would expect from iron(II), a d^6 metal ion. These molecular orbitals also have some ligand character, with the exception of the molecular orbital derived from d_{z^2}. The molecular orbital derived from d_{z^2} has almost no ligand character because its cone-shaped nodal surface points almost directly toward the lobes of the matching group orbital, making overlap slight and giving an essentially nonbonding orbital localized on the iron. The molecular orbital description of ferrocene fits the 18-electron rule.

Other Metallocenes and Related Complexes

Other metallocenes have similar structures but do not necessarily obey the rule. For example, cobaltocene and nickelocene are structurally similar 19- and 20-electron species.

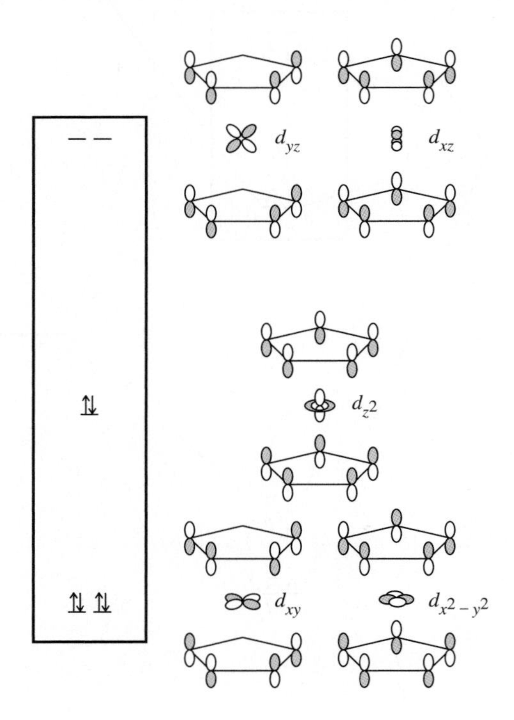

FIGURE 13.29 Molecular Orbitals of Ferrocene Having Greatest *d* Character.

*The relative energies of the lowest three orbitals in Figure 13.29 have been controversial. UV photoelectron spectroscopy is consistent with the order shown, with the orbital having largely d_{z^2} character slightly higher in energy than the pair having d_{xy} and $d_{x^2-y^2}$ character. However, a relatively recent report places the orbital with d_{z^2} character lower in energy than this degenerate pair. The order of these orbitals may be reversed for some metallocenes. See A. Haaland, *Acc. Chem. Res.*, **1979**, *12*, 415, and Z. Xu, Y. Xie, W. Feng, H. F. Schaefer III, *J. Phys. Chem. A*, **2003**, *107*, 2716. The 2003 paper also discusses the orbital energies for the metallocenes $(\eta^5\text{-}C_5H_5)_2V$ through $(\eta^5\text{-}C_5H_5)_2Ni$.

TABLE 13.4 Comparative Data for Selected Metallocenes

Complex	Electron Count	M—C Distance (pm)	ΔH for M^{2+}-$C_5H_5^-$ Dissociation (kJ/mol)
$(\eta^5\text{-}C_5H_5)_2Fe$	18	206.4	1470
$(\eta^5\text{-}C_5H_5)_2Co$	19	211.9	1400
$(\eta^5\text{-}C_5H_5)_2Ni$	20	219.6	1320

The extra electrons have chemical and physical consequences, as can be seen from comparative data in **Table 13.4**.

Electrons 19 and 20 of the metallocenes occupy slightly antibonding orbitals (largely d_{yz} and d_{xz} in character); as a consequence, the metal–ligand distance increases, and ΔH for metal–ligand dissociation decreases. Ferrocene shows much more chemical stability than cobaltocene and nickelocene; many of the chemical reactions of the latter are characterized by a tendency to yield 18-electron products. For example, ferrocene is unreactive toward iodine and rarely participates in reactions in which other ligands substitute for the cyclopentadienyl ligand. However, cobaltocene and nickelocene undergo reactions to give 18-electron products:

$$\underset{19\ e^-}{2(\eta^5\text{-}C_5H_5)_2Co} + I_2 \longrightarrow \underset{\substack{18\ e^- \\ \text{cobalticinium ion}}}{2[(\eta^5\text{-}C_5H_5)_2Co]^+} + 2\ I^-$$

$$\underset{20\ e^-}{(\eta^5\text{-}C_5H_5)_2Ni} + 4PF_3 \longrightarrow \underset{18\ e^-}{Ni(PF_3)_4} + \text{organic products}$$

Cobalticinium reacts with hydride to give a neutral, 18-electron sandwich compound in which one cyclopentadienyl ligand has been modified into η^4-C_5H_6 (**Figure 13.30**).

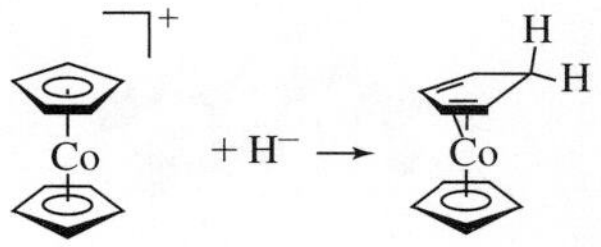

Ferrocene, however, is by no means chemically inert. It undergoes a variety of reactions, including many on the cyclopentadienyl rings. A good example is that of electrophilic acyl substitution (**Figure 13.31**), a reaction paralleling that of benzene and its derivatives. In general, electrophilic aromatic substitution reactions are much more rapid for ferrocene than for benzene, an indication of greater concentration of electron density in the rings of the sandwich compound.

FIGURE 13.30 Reaction of Cobalticinium with Hydride.

Among the most interesting ferrocene-containing compounds is a molecule that was sought for many years before its synthesis in 2006, hexaferrocenylbenzene, a type of molecular "Ferris wheel" (**Figure 13.32**). This compound, originally obtained from the reaction of hexaiodobenzene and diferrocenylzinc, has six ferrocenyl groups as substituents on a benzene ring.[29] The highly crowded nature of hexaferrocenylbenzene is illustrated by the alternating up/down arrangement of ferrocenes around the benzene ring, with the benzene itself adopting a chair conformation with alternating C—C distances of 142.7 and 141.1 pm.

Binuclear metallocenes—with two atoms, rather than one in the center of a sandwich structure—are also known. Perhaps the best known of these metallocenes is decamethyldizincocene, $(\eta^5\text{-}C_5Me_5)_2Zn_2$ (**Figure 13.33**), which was prepared from decamethylzincocene,

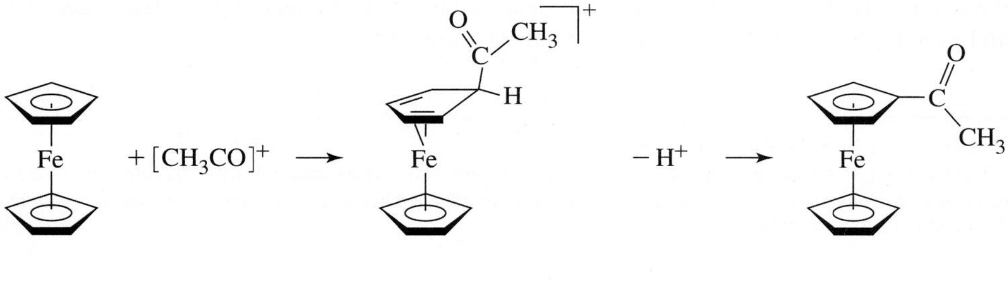

FIGURE 13.31 Electrophilic Acyl Substitution in Ferrocene.

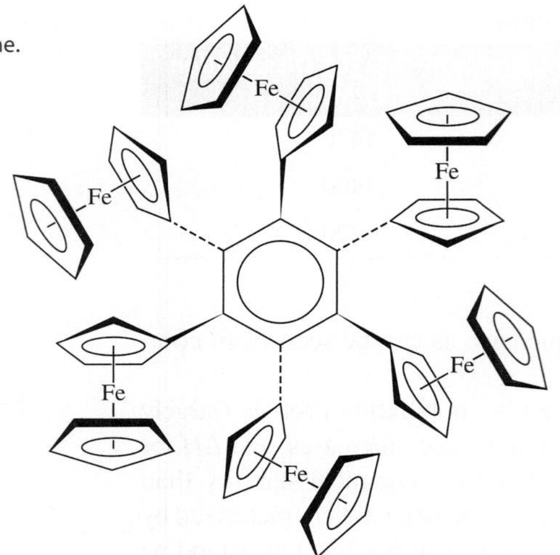

FIGURE 13.32
Hexaferrocenylbenzene.

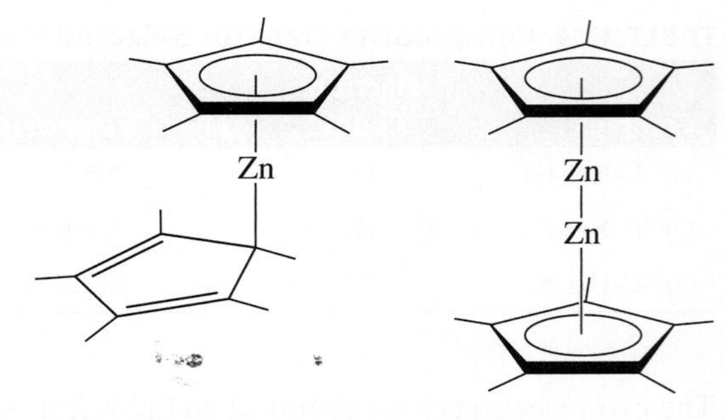

FIGURE 13.33 Decamethylzincocene and Decamethyldizincocene.

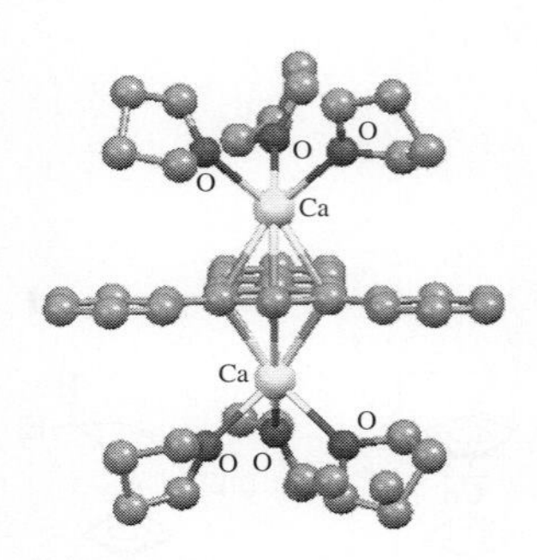

FIGURE 13.34 An Inverse Sandwich Compound, $[(thf)_3Ca\{\mu\text{-}C_6H_3\text{-}1,3,5\text{-}Ph_3\}Ca(thf)_3]$. (Molecular structure drawing created from CIF data, with hydrogen atoms omitted for clarity.)

$(\eta^5\text{-}C_5Me_5)_2Zn$, and diethylzinc.[30] Particularly notable is $(\eta^5\text{-}C_5Me_5)_2Zn_2$, the first example of a stable molecule with a zinc–zinc bond; moreover, its zinc atoms are in the exceptionally rare +1 oxidation state. Among Group 12 elements, mercury by far exhibits this oxidation state most often, with the best known example the Hg_2^{2+} ion*; cadmium and zinc compounds almost always have these metals in the +2 oxidation state. The metallocene $(\eta^5\text{-}C_5Me_5)_2Zn_2$ has parallel C_5Me_5 rings and a zinc–zinc distance of 230.5 pm, consistent with a single bond. Since this compound was first reported, renewed interest in molecules containing zinc–zinc bonds has yielded a variety of interesting compounds.[31]

A variation on the theme of metallocenes and related sandwich compounds is provided by the "inverse" sandwich in **Figure 13.34**, with calcium(I) ions on the outside and the cyclic pi ligand 1,3,5-triphenylbenzene in between. This compound was most efficiently prepared by reacting 1,3,5-triphenylbenzene with activated calcium in THF solvent using catalytic amounts of 1-bromo-2,4,6-triphenylbenzene.[32] Although the product of this reaction is highly sensitive to moisture and air and is pyrophoric, it represents a rare example of a +1 oxidation state among the alkaline earths.

Complexes Containing Cyclopentadienyl and CO Ligands

Many complexes are known containing both Cp and CO ligands. These include "half-sandwich" compounds such as $(\eta^5\text{-}C_5H_5)Mn(CO)_3$ and dimeric and larger cluster molecules. Examples are in **Figure 13.35**. As for the binary CO complexes, complexes of the second- and third-row transition metals show a decreasing tendency of CO to act as a bridging ligand.

Many other linear and cyclic pi ligands are known. Examples of complexes containing some of these ligands are in **Figure 13.36**.** Depending on the ligand and the electron requirements of the metal (or metals), these ligands may be capable of bonding in a mono-hapto or polyhapto fashion, and they may bridge two or more metals. Particularly interesting are the cases in which cyclic ligands can bridge metals to give "triple-decker" and even higher order sandwich compounds (Figure 13.1).

*The classic mercury(I) or "mercurous" ion.

For interesting historical accounts of the discovery of the first two of these molecules, see the following articles by D. Seyferth: on uranocene: *Organometallics*, **2004, *23*, 3562; on dibenzenechromium: *Organometallics*, **2002**, *21*, 1520 and **2002**, *21*, 2800.

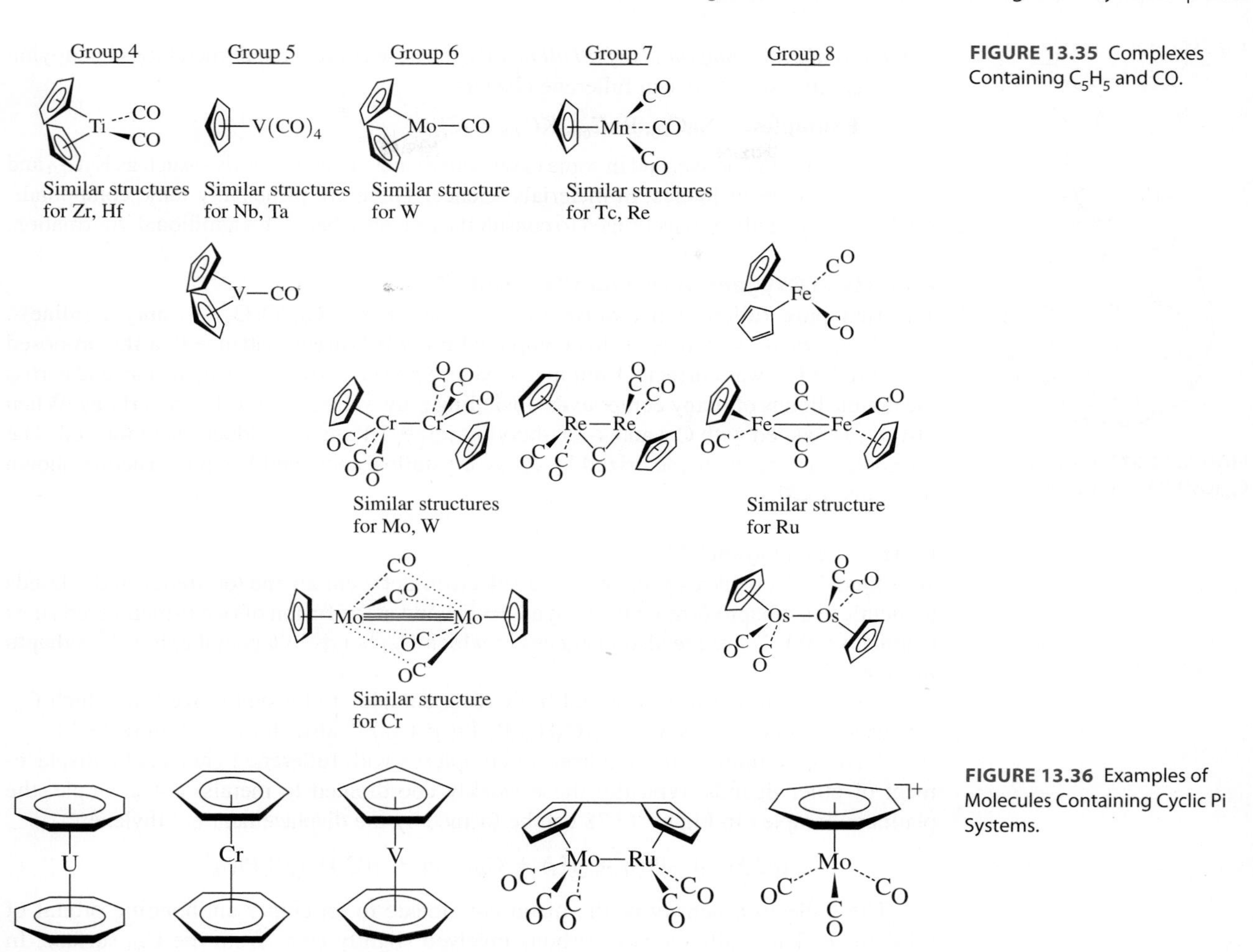

FIGURE 13.35 Complexes Containing C_5H_5 and CO.

FIGURE 13.36 Examples of Molecules Containing Cyclic Pi Systems.

13.5.3 Fullerene Complexes

As immense pi systems, fullerenes were recognized early as ligands to transition metals. Fullerene-metal compounds* have been prepared for a variety of metals. These compounds fall into several structural types:

- *Adducts to the oxygens of osmium tetroxide.*[33]

  Example: $C_{60}(OsO_4)$(4-*t*-butylpyridine)$_2$

- *Complexes in which the fullerene itself behaves as a ligand.*[34]

  Examples: $Fe(CO)_4(\eta^2\text{-}C_{60})$, $Mo(\eta^5\text{-}C_5H_5)_2(\eta^2\text{-}C_{60})$, $[(C_6H_5)_3P]_2Pt(\eta^2\text{-}C_{60})$

- *Fullerenes containing encapsulated (incarcerated) atoms, called* **incarfullerenes.** These may contain one, two, three, or four atoms, sometimes as small molecules, inside the fullerene structure.[35] Although examples of encapsulated nonmetals are known, most incarfullerenes contain metals.

  Examples: UC_{60}, LaC_{82}, Sc_2C_{74}, Sc_3C_{82}

*For a review of metal complexes of C_{60} during the early development of this field, see P. J. Fagan, J. C. Calabrese, B. Malone, *Acc. Chem. Res.*, **1992**, *25*, 134. A more recent review can be found in A. Hirsch and M. Brettreich, *Fullerenes*, Wiley-VCH, Weinheim, Germany, 2005, pp. 231–250.

- *Intercalation compounds of alkali metals.*[36] These contain alkali metal ions occupying interstitial sites between fullerene clusters.

  Examples: NaC_{60}, RbC_{60}, KC_{70}, K_3C_{60}

  These are conductive, and in some cases superconductive, materials—such as K_3C_{60} and Rb_3C_{60}—that are of interest in materials science. These are principally ionic compounds. The interested reader is encouraged to consult the reference here[37] for additional information.

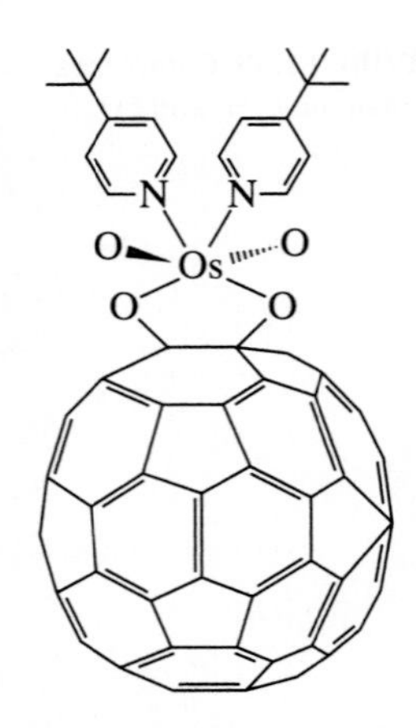

FIGURE 13.37 Structure of $C_{60}(OsO_4)$(4-*t*-butylpyridine)$_2$.

Adducts to Oxygens of Osmium Tetroxide[38]

The first pure fullerene derivative to be prepared was $C_{60}(OsO_4)$(4-*t*-butylpyridine)$_2$. The X-ray crystal structure of this compound provided direct evidence that the proposed structure for C_{60} was correct. Osmium tetroxide, a powerful oxidizing agent, can add across the double bonds of many compounds, including polycyclic aromatic hydrocarbons. When OsO_4 was reacted with C_{60} and 4-*tert*-butylpyridine, 1:1 and 2:1 adducts were formed. The 1:1 adduct has been characterized by X-ray crystallography, and has the structure shown in Figure 13.37.

Fullerenes as Ligands[39]

As a ligand, C_{60} behaves primarily as an electron-deficient alkene (or arene), and it bonds to metals in a dihapto fashion through a C—C bond at the fusion of two 6-membered rings (Figure 13.38). There are also instances in which C_{60} bonds in a pentahapto or hexahapto fashion.

Dihapto bonding was observed in the first complex to be synthesized, in which C_{60} acts as a ligand toward a metal, $[(C_6H_5)_3P]_2Pt(\eta^2\text{-}C_{60})$,[40] also shown in Figure 13.38.

A common route to the synthesis of complexes with fullerene ligands is by displacement of other ligands, typically those weakly coordinated to metals. For example, the platinum complex in Figure 13.38 can be formed by the displacement of ethylene:

$$[(C_6H_5)_3P]_2Pt(\eta^2\text{-}C_2H_4) + C_{60} \longrightarrow [(C_6H_5)_3P]_2Pt(\eta^2\text{-}C_{60})$$

The *d* electron density of the metal can donate to an empty antibonding orbital of a fullerene. This pulls the two carbons involved slightly away from the C_{60} surface. In addition, the distance between these carbons is elongated slightly as a consequence of this interaction, which populates an orbital that is antibonding with respect to the C—C bond. This increase in C—C bond distance is analogous to the elongation that occurs when ethylene and other alkenes bond to metals (Section 13.5.1). In some cases, more than one metal can become attached to a fullerene surface. A spectacular example is $[(Et_3P)_2Pt]_6C_{60}$[41] in Figure 13.39. In this structure, the six $(Et_3P)_2Pt$ units are arranged octahedrally around the C_{60}.

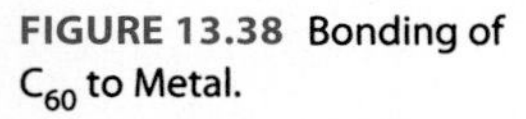

FIGURE 13.38 Bonding of C_{60} to Metal.

Reprinted by permission. Bonding of C_{60} to Metal by Gary O. Spessard and Gary L. Miessler. Reprinted by permission.

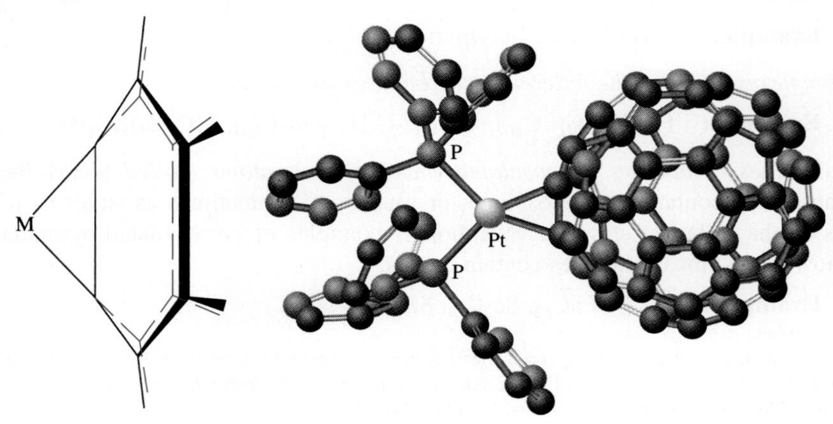

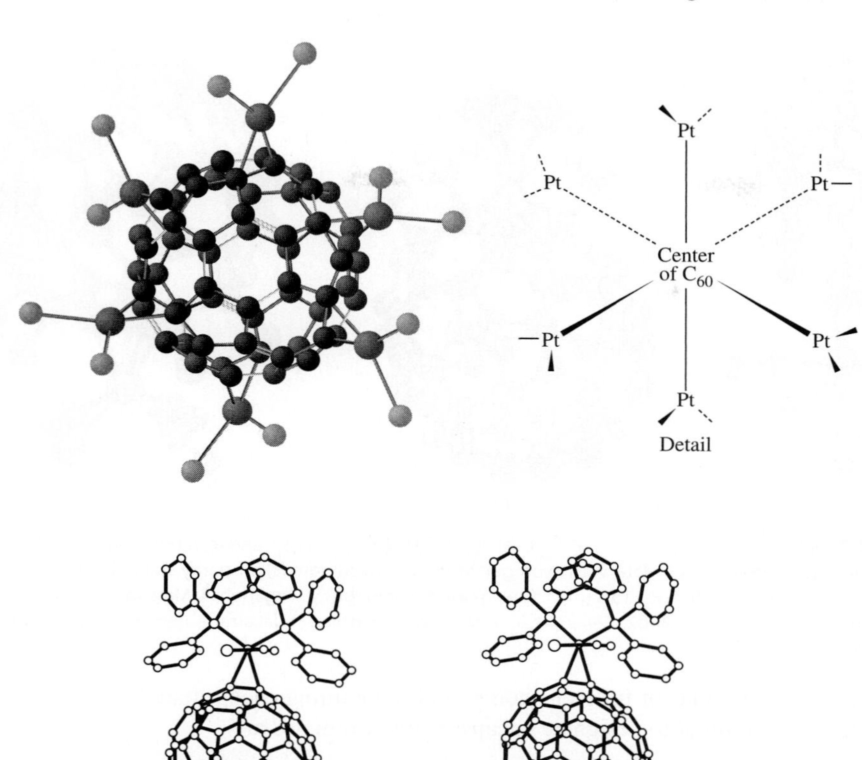

FIGURE 13.39 Structure of $[(Et_3P)_2Pt]_6C_{60}$.
Structure of $[(Et_3P)_2Pt]_6C_{60}$ by Gary O. Spessard and Gary L. Miessler. Reprinted by permission.

FIGURE 13.40 Stereoscopic View of $(\eta^2\text{-}C_{70})$ Ir(CO)Cl$(PPh_3)_2$.
Reproduced with permission from A. L Balch, V. J. Catalano, J. E. Lee, M. M. Olmstead, S. R. Parkin, *J. Am. Chem. Soc., 113,* 8953,. Copyright 1991. American Chemical Society.

Complexes of other fullerenes have also been prepared. An example is $(\eta^2\text{-}C_{70})$Ir(CO)Cl$(PPh_3)_2$ (**Figure 13.40**). As in the case of the known C_{60} complexes, bonding to the metal occurs at the fusion of two 6-membered rings.

C_{60} bonds to transition metals primarily in a dihapto fashion, but at least one example of a hexahapto structure has been reported. The coordination mode of the C_{60} in the triruthenium cluster in **Figure 13.41a** is best described as $\eta^2, \eta^2, \eta^2\text{-}C_{60}$; the C — C bonds bridged by the ruthenium atoms are slightly shorter than the other C — C bonds in the 6-membered ring.

Hybrids of a fullerene and a ferrocene have been reported in which an ion is sandwiched between a $\eta^5\text{-}C_5H_5$ ring and a η^5-fullerene (Figure 13.41b). The fullerenes used, $C_{60}(CH_3)_5$ and $C_{70}(CH_3)_3$, have methyl groups that apparently help stabilize these compounds. The methyl groups are bonded to carbons adjacent to the 5-membered ring to which the iron bonds. This pentamethylfullerene has also been used to form pentahapto complexes with a variety of transition metals (Figure 13.41c).

Complexes with Encapsulated Metals*

These complexes are "cage" organometallic complexes in which the metal is completely surrounded by the fullerene. Typically, complexes containing encapsulated metals are prepared by laser-induced vapor phase reactions between carbon and the metals. These compounds contain central metal cations surrounded by a fulleride, a fullerene that has been reduced.

---

*A recent listing of fullerenes and their encapsulated atoms and molecules, with references, can be found in F. Langa and J.-F. Nierengarten, *Fullerenes: Principles and Applications*, RSC Publishing, Cambridge, UK, 2007, pp. 8–9.

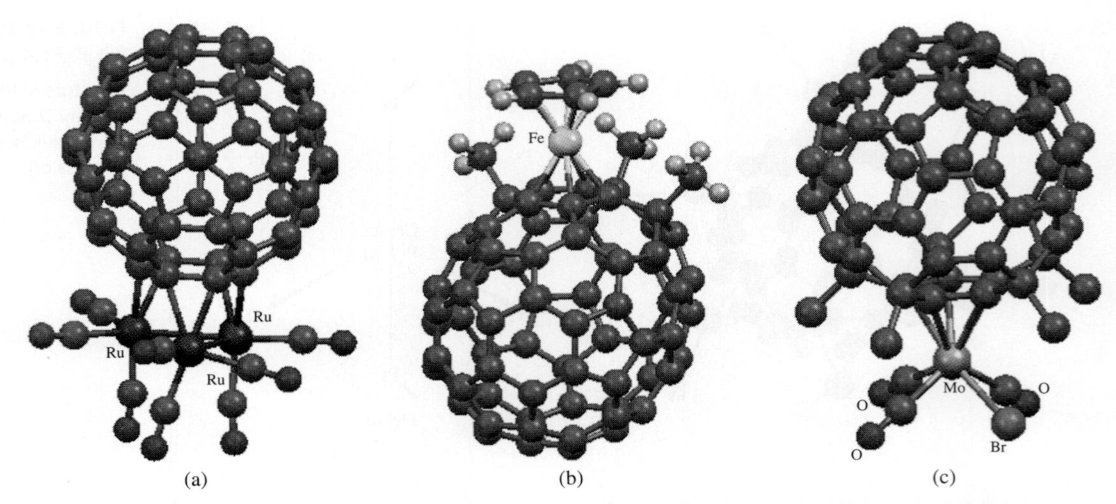

FIGURE 13.41 (a) $Ru_3(CO)_9(\mu_3\text{-}\eta^2, \eta^2, \eta^2\text{-}C_{60})$, (b) $Fe(\eta^5\text{-}C_5H_5)(\eta^5\text{-}C_{70}(CH_3)_3)$, and (c) $MoBr(CO)_3(\eta^5\text{-}C_{60}Me_5)$. (Structures generated with CIF data, with hydrogen atoms removed from (c) for clarity.) (a) H.-F. Hsu, J. R. Shapley, *J. Am. Chem. Soc.*, **1996**, *118*, 9192. (b) M. Sawamura, Y. Kuninobu, M. Toganoh, Y. Matsuo, M. Yamakana, E. Nakamura, *J. Am. Chem. Soc.*, **2002**, *124*, 9354. (c) Y. Matsuo, A. Iwashita, E. Nakamura, *Organometallics*, **2008**, *27*, 4611.

Chemical formulas of fullerene compounds containing encapsulated metals are written with the @ symbol to designate encapsulation: examples are

$U@C_{60}$	Contains U surrounded by C_{60}
$Sc_3@C_{82}$	Contains three atoms of Sc surrounded[42] by C_{82}

This designation indicates structure only and does not include ion charges. For example, $La@C_{82}$ is believed to contain La^{3+} surrounded by the $C_{82}{}^{3-}$. Small molecules and ions can also be encapsulated in fullerenes. An example is $Sc_3N@C_{78}$, which contains a triangular Sc_3N inside the C_{78} cage (**Figure 13.42**).[43]

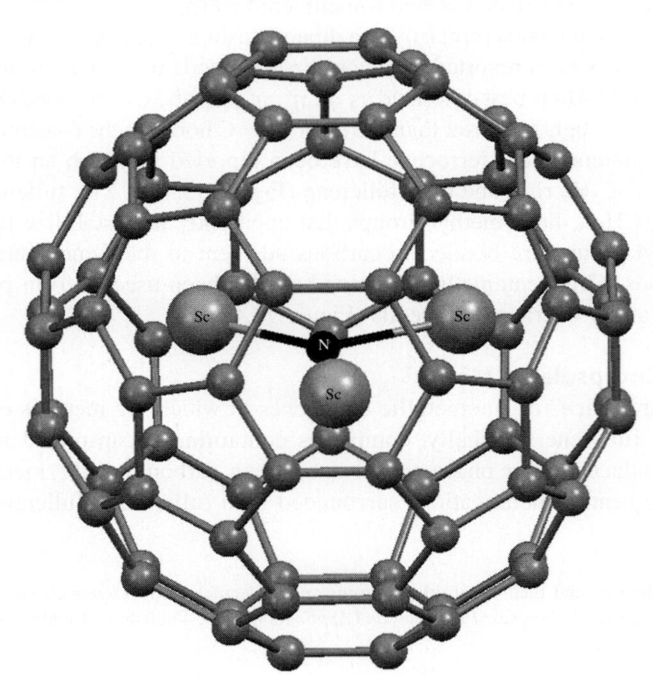

FIGURE 13.42 $Sc_3N@C_{78}$. At the low temperature used for the X-ray study, the Sc_3N is planar with angles of 130.3°, 113.8°, and 115.9°, and each Sc bonds loosely to a C—C bond that is part of the two six membered rings; however, at higher temperatures, the Sc_3N cluster moves freely inside the cage. (Structure generated with CIF data by Victor G. Young, Jr.)

13.6 Complexes Containing M—C, M═C, and M≡C Bonds

Complexes containing direct metal–carbon single, double, and triple bonds have been studied extensively. **Table 13.5** gives examples of the most important types of ligands in these complexes.

13.6.1 Alkyl and Related Complexes

Some of the earliest known organometallic complexes were those having σ bonds between main group metal atoms and alkyl groups. Examples include Grignard reagents, having magnesium–alkyl bonds, and alkyl complexes with alkali metals, such as methyllithium.

Stable transition-metal alkyls were initially synthesized in the first decade of the twentieth century; many such complexes are now known. The metal–ligand bonding in these complexes may be viewed as primarily involving covalent sharing of electrons between the metal and the carbon in a sigma fashion:

M ← CR_3 (R═H, alkyl, aryl)

sp^3 orbital

In terms of electron counting, the alkyl ligand may be considered a 2-electron donor $:CR_3^-$ (method A) or a 1-electron donor $\cdot CR_3$ (method B). Significant ionic contribution to the bonding may occur in complexes of highly electropositive elements, such as the alkali metals and alkaline earths.

TABLE 13.5 Complexes Containing M—C, M═C, and M≡C Bonds

Ligand	Formula	Example
Alkyl	$—CR_3$	$W(CH_3)_6$
Carbene (alkylidene)	$═CR_2$	$(OC)_5Cr═C(OCH_3)(C_6H_5)$
Carbyne (alkylidyne)	$≡CR$	$X—Cr(CO)_4≡C—C_6H_5$
Carbide (carbon)	$≡C$	$Cl_2(PR_3)_2Ru≡C$
Cumulene	$═C(═C)_nRR'$	$Cl—Ir(P(CH_3)_3)_2═C═C═C═C(C_6H_5)_2$

Many synthetic routes to transition-metal alkyl complexes have been developed. Two of the most important of these methods are:

1. Reaction of a transition-metal halide with organolithium, organomagnesium, or organoaluminum reagent

$$ZrCl_4 + 4\,PhCH_2MgCl \longrightarrow Zr(CH_2Ph)_4\,(Ph = \text{phenyl}) + 4\,MgCl_2$$

2. Reaction of a metal carbonyl anion with an alkyl halide

$$Na[Mn(CO)_5]^- + CH_3I \longrightarrow CH_3Mn(CO)_5 + NaI$$

Although many complexes contain alkyl ligands, transition-metal complexes that contain alkyl groups as the only ligands, are relatively rare. Examples include $Ti(CH_3)_4$, $W(CH_3)_6$, and $Cr[CH_2Si(CH_3)_3]_4$. Alkyl complexes have a tendency to be kinetically unstable;* their stability is enhanced by structural crowding, which protects the coordination sites of the metal by blocking pathways to decomposition. The 6-coordinate $W(CH_3)_6$ can be melted at 30 °C without decomposition, whereas the 4-coordinate $Ti(CH_3)_4$ is subject to decomposition at approximately –40 °C.[44] In an unusual use of alkyls, diethylzinc has been used to treat books and documents for their long-term preservation by neutralizing the acid in the paper. Many alkyl complexes are important in catalytic processes discussed in Chapter 14.

Other ligands have direct metal–carbon σ bonds (**Table 13.6**). In addition, there are many examples of **metallacycles**, complexes in which organic ligands attach to metals at two positions, thereby incorporating the metals into organic rings.[45] The reaction below is an example of a metallacycle synthesis. Metallacycles are important intermediates in catalytic processes (Chapter 14).

$$PtCl_2(PR_3)_2 + Li\text{–}CH_2CH_2CH_2CH_2\text{–}Li \longrightarrow (R_3P)_2Pt(C_4H_8) + 2\,LiCl$$

Metallacyclopentane

TABLE 13.6 Other Ligands Forming Sigma Bonds to Metals

Ligand	Formula	Example
Aryl	$-C_6H_5$	$Cp_2Ta(H)_2(C_6H_5)$
Alkenyl (vinyl)	$>C=C<$	$Cl-Pt(PR_3)_2-C(H)=CH_2$
Alkynyl	$-C\equiv C-$	$Cl-Pt(PR_3)_2-C\equiv CPh$

*An interesting historical perspective on alkyl complexes is in G. Wilkinson, *Science*, **1974**, *185*, 109.

13.6.2 Carbene Complexes

Carbene complexes contain metal–carbon double bonds.* First synthesized in 1964 by Fischer,[46] carbene complexes are known for the majority of transition metals and for a wide range of carbene ligands, including the simple carbene, $:CH_2$. The majority of such complexes contain one or two highly electronegative heteroatoms—such as O, N, or S—directly attached to the carbene carbon. These are designated as *Fischer-type carbene complexes*. Other carbene complexes contain only carbon and/or hydrogen attached to the carbene carbon. First synthesized several years after the initial Fischer carbene complexes,[47] these have been studied extensively by Schrock and several others. They are sometimes designated as *Schrock-type carbene complexes*, commonly referred to as *alkylidenes*. Distinctions between Fischer- and Schrock-type carbene complexes are summarized in **Table 13.7**. We will focus primarily on Fischer-type carbene complexes.

The formal double bond in carbene complexes may be compared with the double bond in alkenes. In the case of a carbene complex, the metal must use a *d* orbital to form the π bond with carbon (**Figure 13.43**).

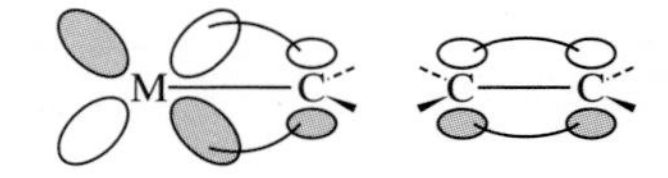

FIGURE 13.43 Bonding in Carbene Complexes and in Alkenes.

Carbene complexes having a highly electronegative atom—such as O, N, or S—attached to the carbene carbon tend to be more stable than complexes lacking such an atom. For example, $Cr(CO)_5[C(OCH_3)C_6H_5]$, with an oxygen on the carbene carbon, is much more stable than $Cr(CO)_5[C(H)C_6H_5]$. The stability of the complex is enhanced if the highly electronegative atom can participate in the π bonding, with the result a delocalized, 3-atom π system involving a *d* orbital on the metal and *p* orbitals on the carbon and on the electronegative atom. Such a delocalized 3-atom system provides more stability to the bonding π electron pair than would a simple metal-to-carbon π bond. An example of such a π system is in **Figure 13.44**.

The methoxycarbene complex $Cr(CO)_5[C(OCH_3)C_6H_5]$ illustrates the bonding just described.[48] To synthesize this complex, we can begin with $Cr(CO)_6$. As in organic chemistry, highly nucleophilic reagents can attack the carbonyl carbon. For example, phenyllithium can react with $Cr(CO)_6$ to give the anion $[C_6H_5C(O)Cr(CO)_5]^-$, which has two important resonance structures:

$$Li^+{:}C_6H_5^- + O{\equiv}C{-}Cr(CO)_5 \longrightarrow C_6H_5{-}\overset{\overset{\displaystyle O}{\|}}{C}{-}Cr^-(CO)_5 \longleftrightarrow C_6H_5{-}\overset{\overset{\displaystyle O^-}{|}}{C}{=}Cr(CO)_5 + Li^+$$

$$C_6H_5{-}\overset{\overset{\displaystyle O}{|\!\!\text{\textbackslash}^-}}{C}{\overset{-}{=}}Cr(CO)_5$$

TABLE 13.7 Fischer- and Schrock-Type Carbene Complexes

Characteristic	Fischer-Type Carbene Complex	Schrock-Type Carbene Complex
Typical metal [oxidation state]	Middle to late transition metal [Fe(0), Mo(0), Cr(0)]	Early transition metal [Ti(IV), Ta(V)]
Substituents attached to $C_{carbene}$	At least one highly electronegative heteroatom (such as O, N, or S)	H or alkyl
Typical other ligands in complex	Good π acceptors	Good σ or π donors
Electron count	18	10–18

*IUPAC has recommended that the term "alkylidene" be used to describe all complexes containing metal–carbon double bonds and that "carbene" be restricted to free $:CR_2$. For a detailed description of the distinction between these two terms—and between "carbyne" and "alkylidyne," discussed later in this chapter—see W. A. Nugent and J. M. Mayer, *Metal–Ligand Multiple Bonds*, Wiley InterScience, New York, 1988, pp. 11–16.

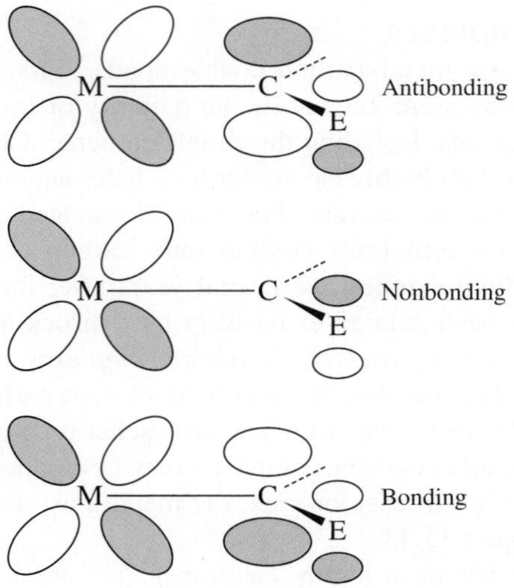

FIGURE 13.44 Delocalized Pi Bonding in Carbene Complexes. E designates a highly electronegative heteroatom, such as O, N, or S.

M—C, E: Antibonding

M—C, E: Nonbonding

M—C, E: Bonding

Alkylation by a source of CH_3^+, such as $[(CH_3)_3O][BF_4]$ or CH_3I, gives the methoxycarbene complex:

$$C_6H_5-\overset{\overset{\displaystyle O^-}{|}}{C}=Cr(CO)_5 + [(CH_3)_3O][BF_4] \longrightarrow C_6H_5-\overset{\overset{\displaystyle OCH_3}{|}}{C}=Cr(CO)_5 + BF_4^- + (CH_3)_2O$$

Evidence for double bonding between chromium and carbon is provided by X-ray crystallography, which measures this distance at 204 pm, compared with a typical Cr—C single-bond distance of approximately 220 pm.

One interesting aspect of this complex is that it exhibits a temperature-dependent proton NMR. At room temperature, a single resonance is found for the methyl protons; however, as the temperature is lowered, this peak first broadens and then splits into two peaks. How can this behavior be explained?

A single proton resonance, corresponding to a single magnetic environment, is expected for the carbene complex as illustrated, with a double bond between chromium and carbon, and a single bond (permitting rapid rotation about the bond) between carbon and oxygen. The room-temperature NMR is therefore as expected. However, the splitting of this peak at lower temperature into two peaks suggests two different proton environments.[49] Two environments are possible if rotation is hindered about the C—O bond. A resonance structure can be drawn showing the possibility of some double bonding between C and O; were such double bonding significant, *cis* and *trans* isomers (**Figure 13.45**) might be observable at low temperatures.

FIGURE 13.45 Resonance Structures, and *cis* and *trans* Isomers for $Cr(CO)_5[C(OCH_3)C_6H_5]$.

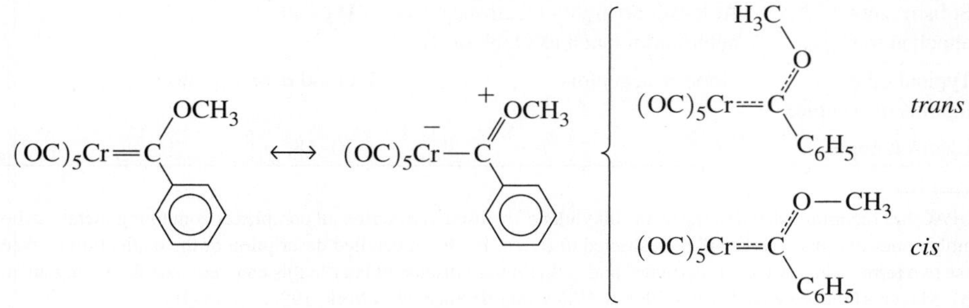

$(OC)_5Cr=C(OCH_3)(C_6H_5) \longleftrightarrow (OC)_5\overset{-}{Cr}-C(\overset{+}{O}CH_3)(C_6H_5)$

trans: $(OC)_5Cr{\cdots}C(O{\cdots}CH_3)(C_6H_5)$, with H_3C on O

cis: $(OC)_5Cr{\cdots}C(O-CH_3)(C_6H_5)$

Evidence for double-bond character in the C—O bond is provided by crystal structure data, which show a C—O bond distance of 133 pm, compared with a typical C—O single-bond distance of 143 pm.[50] The double bonding between C and O, although weak (typical C=O bonds are much shorter, approximately 116 pm), is sufficient to slow down rotation about the bond so that at low temperatures, proton NMR detects the *cis* and *trans* methyl protons separately. At higher temperature, there is sufficient energy to cause rapid rotation about the C—O bond so that the NMR spectrometer detects only an average signal, which is observed as a single peak.

X-ray crystallographic data show double-bond character in both the Cr—C and C—O bonds. This supports the statement that π bonding in complexes of this type (containing a highly electronegative atom, in this case oxygen) may be considered delocalized over three atoms. Although not absolutely essential for all carbene complexes, the delocalization of π electron density over three (or more) atoms provides an additional measure of stability.[51]

13.6.3 Carbyne (Alkylidyne) Complexes

Carbyne complexes have metal–carbon triple bonds; they are formally analogous to alkynes.* Many carbyne complexes are now known; examples of carbyne ligands include the following:

$$M{\equiv}C{-}R$$

where R = aryl, alkyl, H, $SiMe_3$, NEt_2, PMe_2, SPh, or Cl. Carbyne complexes were first synthesized fortuitously as products of the reactions of carbene complexes with Lewis acids.[52] The methoxycarbene complex $Cr(CO)_5[C(OCH_3)C_6H_5]$ was found to react with the Lewis acids BX_3 (X = Cl, Br, or I). First, the Lewis acid attacks the oxygen, the basic site on the carbene:

$$(CO)_5Cr{=}C(OCH_3)(C_6H_5) + BX_3 \longrightarrow [(CO)_5Cr{\equiv}C{-}C_6H_5]^+X^- + X_2BOCH_3$$

The intermediate loses CO, with the halide coordinating *trans* to the carbyne:

$$[(CO)_5Cr{\equiv}C{-}C_6H_5]^+X^- \longrightarrow X{-}Cr(CO)_4{\equiv}C{-}C_6H_5 + CO$$

The best evidence for the carbyne nature of the complex is provided by X-ray crystallography, which gives a Cr—C bond distance of 168 pm (for X = Cl), considerably shorter than the 204 pm for the parent carbene complex. The Cr≡C—C angle is, as expected, 180° for this complex; however, slight deviations from linearity are observed for many complexes in crystalline form, in part a consequence of crystal packing effects.

Bonding in carbyne complexes may be viewed as a combination of a σ bond plus two π bonds (**Figure 13.46**). The carbyne ligand has a lone pair of electrons in an *sp* hybrid on carbon; this lone pair can donate to a suitable orbital on Cr to form a σ bond. In addition, the carbon has two *p* orbitals that can accept electron density from *d* orbitals on Cr to form π bonds. Thus, the overall function of the carbyne ligand is as both a σ donor and π acceptor. (For electron counting purposes, a $:CR^+$ ligand can be considered a 2-electron donor; it is usually more convenient to count neutral CR as a 3-electron donor.)

---

*IUPAC has recommended that "alkylidyne" be used to designate complexes containing metal–carbon triple bonds.

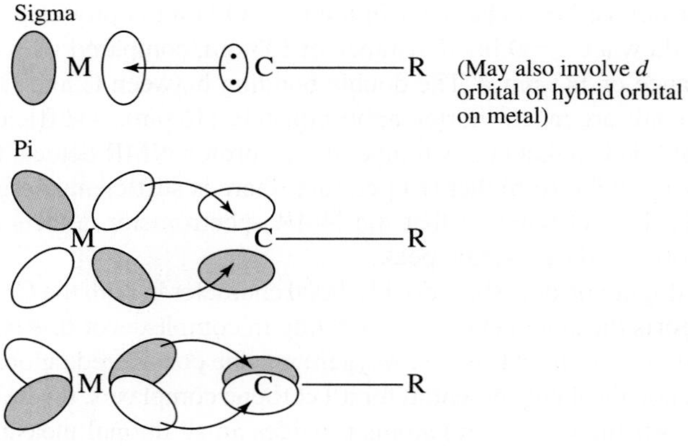

FIGURE 13.46 Bonding in Carbyne Complexes.

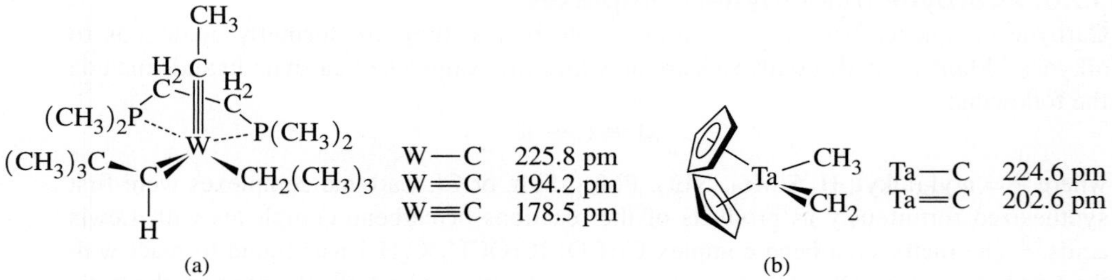

FIGURE 13.47 Complexes Containing Alkyl, Carbene, and Carbyne Ligands. Data from (a) M. R. Churchill, W. J. Young, *Inorg. Chem.*, **1979**, *18*, 2454. (b) L. J. Guggenberger, R. R. Schrock, *J. Am. Chem. Soc.*, **1975**, *97*, 6578.

Carbyne complexes can be synthesized in a variety of ways in addition to Lewis acid attack on carbene complexes. Synthetic routes for carbyne complexes and the reactions of these complexes have been reviewed.[53]

In some cases, molecules have been synthesized containing two or three of the types of ligands discussed in this section (alkyl, carbene, and carbyne). Such molecules provide an opportunity to make direct comparisons of lengths of metal–carbon single, double, and triple bonds (**Figure 13.47**).

EXERCISE 13.10

Are the compounds shown in Figure 13.47 18-electron species?

13.6.4 Carbide and Cumulene Complexes

The simplest possible carbon-containing ligand, a single carbon atom, is known, and the number of complexes of such **carbide**, **carbido**, or simply **carbon** ligands has grown considerably in the past decade. Although it is tempting, by extending the series $-CR_3$, $=CR_2$, $\equiv CR$, to assign a quadruple bond to carbon in complexes containing a metal bonded to a single carbon atom, such metal–ligand bonding is better described as $M\equiv C^-$. The first neutral carbide complex was a trigonal-bipyramidal ruthenium complex (**Figure 13.48**).[54] The Ru—C distance in this complex is perhaps longer than might be expected, 165.0 pm—only slightly shorter than the comparable distance in the structurally similar ruthenium carbyne complex, also shown.

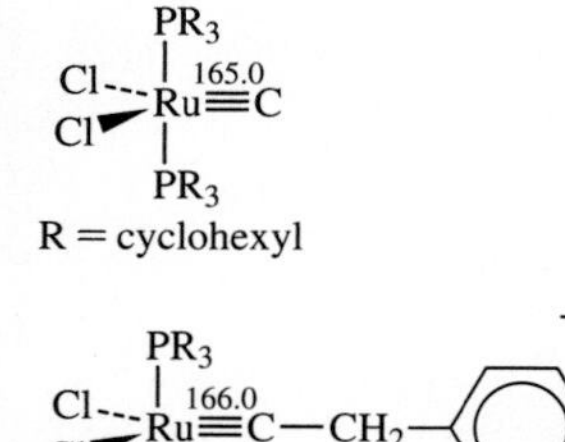

FIGURE 13.48 Carbide and Carbyne Complexes.

Calculations have indicated that bonds between transition metals and terminal carbon atoms are quite strong, with bond dissociation enthalpies comparable to those of

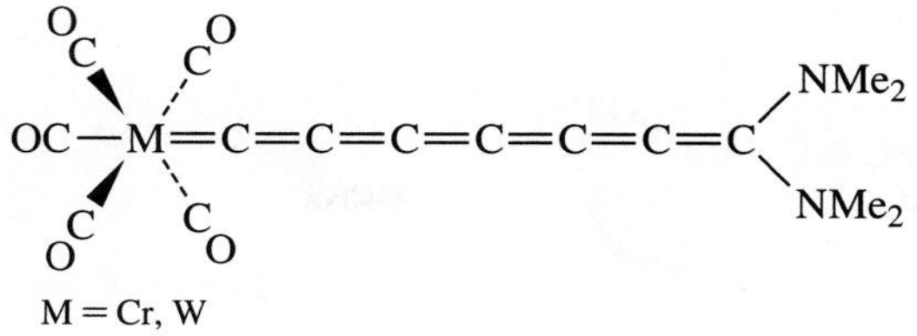

FIGURE 13.49 A Metallacumulene Complex.

transition-metal complexes with M≡N and M=O bonds.[55] In addition, the frontier orbitals of the carbide complex in Figure 13.48 (where R = methyl) have many similarities to those of CO, suggesting that such complexes may potentially show similar coordination chemistry to the carbonyl ligand.[56] Carbido complexes similar to the one shown in Figure 13.48 have also been used in the synthesis of CS, CSe, and CTe complexes.[57]

Ligands with chains of carbon atoms that have cumulated (consecutive) double bonds, designated *cumulenylidene* ligands, are also known. Such **metallacumulene** complexes have drawn interest because of possible applications as 1-dimensional molecular wires and for use in nanoscale optical devices.[58] In recent years, complexes with 2- and 3-carbon chains have also been developed as effective catalysts. Metallacumulene complexes with carbon chains containing three or more atoms have recently been reviewed.[59]

The longest cumulenylidene ligand reported to date is the heptahexaenylidene complex shown in **Figure 13.49**.[60] As in the case of extended organic pi systems, the difference in energy between the HOMO and the LUMO decreases as the length of the cumulene ligand increases.[61]

13.6.5 Carbon Wires: Polyyne and Polyene Bridges

Unsaturated carbon and hydrocarbon bridges have been studied in recent years in connection with their potential to serve as wires joining metal centers in molecular electronics. The most widely studied types of these bridges have been the **polyynediyl** bridges with alternating single and triple bonds and **polyenediyl** bridges with alternating single and double bonds. The bond conjugation (extended π system) is needed to enable electronic communication between the metal atoms at the ends of the bridges; saturated bridges or sections of bridges inhibit such communication. Examples of structures containing polyyne and polyene bridges are in **Figure 13.50**. These and related types of bridging structures having two or more carbon atoms have been reviewed.[62]

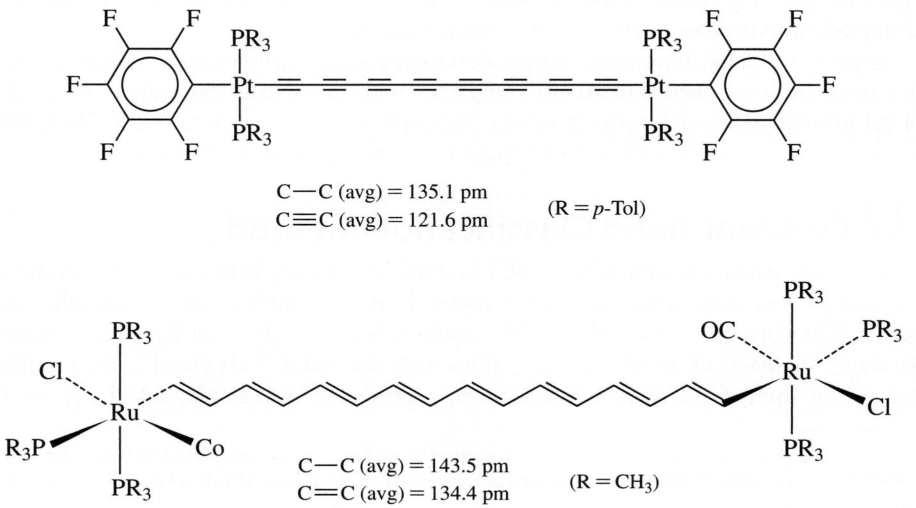

FIGURE 13.50 Molecules with Polyyne and Polyene Bridges.

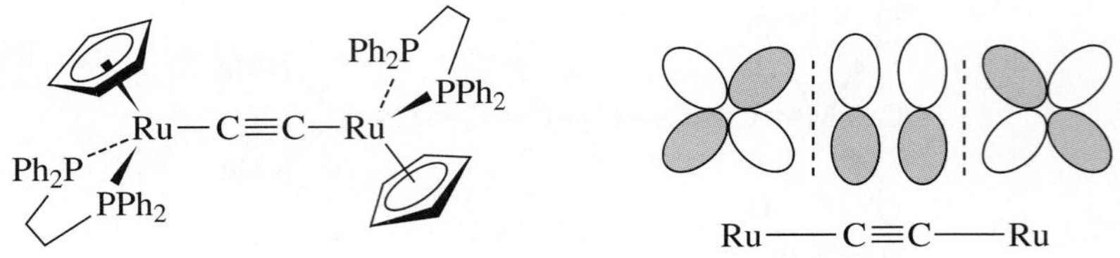

FIGURE 13.51 Conjugated π System in C_2-Bridged Complex.

Although bridges as long as 28 carbon atoms between transition metal atoms have been prepared, these still fall short of known purely organic molecular wires with lengths in the 100 atom range. However, transition metal-based wires can be more stable with respect to thermal decomposition than comparable organic molecules and show promise for higher temperature applications.[63] The transition metal-containing end groups appear to provide more stability than organic groups.

Crystallographic measurements of the carbon–carbon distances in the polyynediyl complexes show alternating long and short distances, as in the upper example in Figure 13.50, and support designating these molecules as having alternating single and triple bonds ($-C\equiv C-C\equiv C-$) rather than cumulated double bonds ($=C=C=C=C=$). The long and short bond distances are similar to comparable distances in organic molecules having alternating single and triple bonds.[64]

The π systems in these bridging systems, including the *d* orbitals of the transition metals at the ends, can be considered similarly to the linear π systems discussed in Section 13.4.4. Consequently, as chain length increases, the energy separation between occupied and unoccupied π orbitals becomes smaller, the energy necessary to excite electrons decreases, and absorption bands shift from the ultraviolet (for short π systems) toward the visible. This phenomenon has been observed for a series of diplatinum complexes similar to the one in Figure 13.50 but with *p*-tolyl end groups.[65] The positions of absorption maxima for the π–π^* transitions in these complexes shift progressively from the UV range (< 400 nm) for yellow complexes with carbon bridges of 12 or fewer atoms to approximately 490 nm for a red complex with a C_{28} bridge.

On a small scale, a conjugated π system is illustrated by the C_2-bridged diruthenium complex in **Figure 13.51**. Calculations have identified the highest two occupied molecular orbitals in this complex as bonding between the bridging carbons and antibonding with respect to the ruthenium–carbon bonds; the atomic orbitals contributing to one of these molecular orbitals are shown. One- and two-electron oxidations have each resulted in lengthening of the C—C bonds and shortening of the Ru—C bonds, an expected consequence of removing electrons from these orbitals.[66] Electrochemical studies have also been conducted on molecules having longer carbon bridges.

Among the most interesting recent developments in carbon-based molecular wires have been carbon-linked diplatinum complexes with the chains "insulated" by double-helical arrangements of lengthy bridging diphosphine ligands such as $Ph_2P(CH_2)_{20}PPh_2$ and a rotaxane structure with a macrocyclic ring circling the carbon chain.[67,68]

13.7 Covalent Bond Classification Method

The Covalent Bond Classification (CBC) Method* is an insightful approach specifically designed for covalent molecules and is particularly applicable to organometallic compounds. The conceptual core of the CBC method is to classify each ligand as a *neutral* species on the basis of its orbital interactions with the metal. This classification requires considering three fundamental interactions, represented by the symbols L, X, and Z

*A library of leading references and teaching materials is available at covalentbondclass.org. An extension of the CBC Method to three-center two-electron bonds is provided in J.C. Green, M.L.H. Green, G. Parkin, *Chem. Commun.*, **2012**, *48*, 11481.

TABLE 13.8 Fundamental Metal–Ligand Interactions in the CBC Method

Ligand Function	Orbital Requirement	Designations and Simple Examples
L	Ligand features an orbital that donates an electron pair to the metal	L: CO, PR_3; L_2: η^4-C_4H_4; L_3: η^6-C_6H_6
X	Ligand features an orbital that donates one electron to the metal	Cl, Br, I, H, R (alkyl)
Z	Ligand features an orbital that accepts a pair of electrons from the metal	BR_3, AlR_3, BX_3

(**Table 13.8**). A ligand designated with an *L-function* interacts with a metal by donating a pair of electrons; these ligands function as Lewis bases via formation of dative covalent bonds. Some ligands donate pairs via multiple orbitals, and are designated with multiple L functions; for example, η^6-C_6H_6 is generally classified as an L_3 ligand, on the basis of the three donor pairs (Figure 13.22). An *X-function* ligand interacts via a singly occupied orbital with a singly occupied metal orbital forming a normal covalent bond. A ligand designated with a *Z-function* features a metal–ligand interaction in which the ligand accepts a pair of electrons from the metal; these ligands function as Lewis acids.

The π molecular orbitals of C_3H_5, *cyclo*-C_4H_4, and *cyclo*-C_5H_5 indicate that ligands can engage metals via multiple different functions *simultaneously*. On the basis of the Section 13.4.4 energy-level diagrams, these ligands are classified as LX, LX_2, and L_2X, respectively. For example, when *cyclo*-C_5H_5 binds to a metal to afford a η^5-C_5H_5 binding mode, it utilizes two filled π molecular orbitals (L_2), and one half-filled π molecular orbital (X). Classifications of many ligands, instructions on dealing with bridging ligands, and details on more ligand function combinations are available.[69]

Once each ligand has been classified, an organometallic complex can be communicated via a general formula $[ML_lX_xZ_z]^{Q\pm}$, where l, x, and z indicate the number of L, X, and Z functions, respectively, and Q designates the charge. This process is shown in **Figure 13.52**. The CBC approach uncovers similarities between complexes that may seem

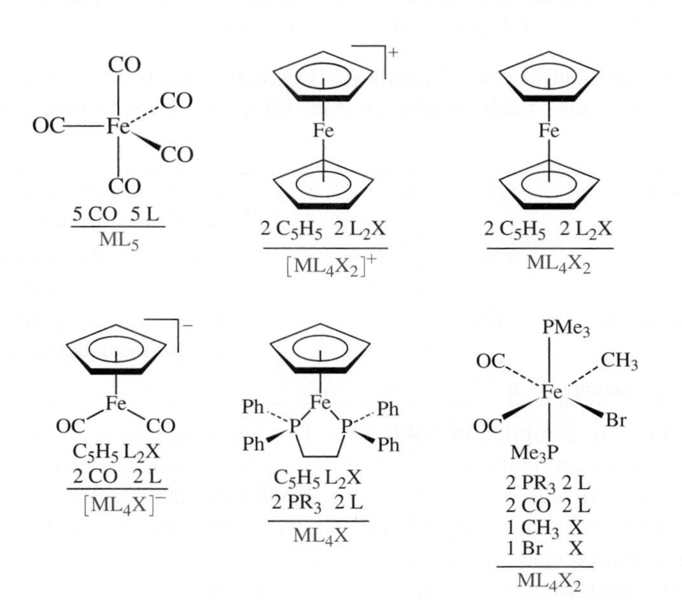

FIGURE 13.52 Covalent Bond Classification of Iron Complexes.

quite different. For example, ferrocene, $Fe(C_5H_5)_2$, and $Fe(CO)_2(PMe_3)_2(CH_3)Br$ are members of the same class (ML_4X_2).

For charged complexes, reduction of the $[ML_lX_xZ_z]$ formula to its *equivalent neutral class* allows comparison to neutral complexes. The reduction strategy requires adjusting ligand classifications as if the charge is localized on the ligands. Introductory transformations are provided in the margin, and more are available.[69] For a cation with an L function, we envision the complex containing an L^+ ligand. If a two-electron donor L ligand is rendered positively charged, it loses one electron and becomes like an X function, hence the equivalence shown in the margin. Similarly, for anionic complexes with X ligands, we envision the complex containing an X^- ligand. If a one-electron X ligand is rendered negatively charged, it gains one electron and becomes like an L ligand. These transformations are best applied sequentially, in the order shown. For example, when a cation contains both X and L functions, eliminate charge first with L functions, only using X functions if necessary. As examples, the charged complexes in Figure 13.52 are reduced to their equivalent neutral class in **Figure 13.53**. In $[Fe(CO)_2(C_5H_5)]^-$, the formula for the charged complex contains both L and X functions, but the negative charge is assigned to X. In $[Fe(C_5H_5)_2]^+$, the positive charge is assigned to L and not X.

Cations	Anions
$L^+ \rightarrow X$	$X^- \rightarrow L$
$X^+ \rightarrow Z$	$L^- \rightarrow LX$

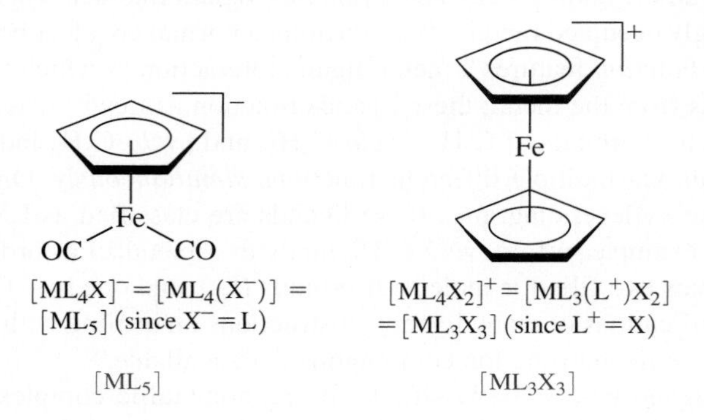

$[ML_5]$ $[ML_3X_3]$

FIGURE 13.53 Conversion to Equivalent Neutral Class.

EXERCISE 13.11

Determine the equivalent neutral class CBC formulas of $Cr(\eta^6\text{-}C_6H_6)_2$, $[Mo(CO)_3(\eta^5\text{-}C_5H_5)]^-$, $WH_2(\eta^5\text{-}C_5H_5)_2$, and $[FeCl_4]^{2-}$.

The equivalent neutral class formula ($[ML_lX_xZ_z]$) permits comparison of these iron complexes, and provides ready access to other information about these complexes. For example,

Electron-count	EN	$m + 2l + x$	m is number of valence electrons for the neutral metal ($m = 8$ for Fe)
Valence number	VN	$x + 2z$	Valence number is the number of electrons the metal uses in bonding
Ligand bond number	LBN	$l + x + z$	The number of ligand functions involved with the metal
d^n metal configuration	n	$m - \text{VN}$	

These values are compiled in **Table 13.9**. The EN values are equal to the values that are obtained by conventional electron counting methods, and the d^n configurations and valences are consistent with commonly assigned oxidation states for these complexes (ML_5, Fe(0); ML_4X, Fe(I); ML_4X_2, Fe(II); ML_3X_3, Fe(III)), although it should be recognized that valences and oxidation states are not always equivalent.[70] Ligand bond numbers are often equivalent to the coordination numbers that are generally associated with organometallic

TABLE 13.9 Information Available from the Equivalent Neutral Class Formula

Formula	Complexes	EN	LBN	VN	d^n
ML_5	$Fe(CO)_5$, $[Fe(CO)_2(\eta^5\text{-}C_5H_5)]^-$	18	5	0	8
ML_4X	$Fe(Ph_2PCH_2CH_2Ph_2)(\eta^5\text{-}C_5H_5)$	17	5	1	7
ML_4X_2	$Fe(C_5H_5)_2$, $Fe(CO)_2(PMe_3)_2(CH_3)Br$	18	6	2	6
ML_3X_3	$[Fe(\eta^5\text{-}C_5H_5)_2]^+$	17	6	3	5

complexes. For example, iron is traditionally considered to be 6-coordinate in $Fe(C_5H_5)_2$ (and not 10-coordinate), with each η^5-C_5H_5 ligand assigned to three coordinate sites.

An insightful outcome of the CBC method is MLX plots, matrices that communicate the abundances of known complexes of a given metal for all valence and electron number combinations. These plots are available for all transition metals[*,69] and that for iron is provided below.[**] The white squares indicate classes for which no iron complexes are known; iron complexes with these equivalent neutral class formulas, as well as those that comprise less than 1% of known examples, suggest opportunities for further research. It is interesting that ML_4X_2 (71%), ML_5 (20%), and ML_3X_4 (7%) constitute the vast majority of known iron complexes, and all of these satisfy the 18-electron rule. Although the ferricinium cation, $[Fe(\eta^5\text{-}C_5H_5)_2]^+$, has been known for decades, this complex represents a rare class for iron, as does $Fe(Ph_2PCH_2CH_2Ph_2)(\eta^5\text{-}C_5H_5)$.

Fe	Electron Number								
Valence	10	11	12	13	14	15	16	17	18
0	ML		ML_2		ML_3		ML_4 <1%		ML_5 20%
1		MLX		ML_2X		ML_3X		ML_4X <1%	
2	MX_2		MLX_2		ML_2X_2 <1%		ML_3X_2 <1%		ML_4X_2 71%
3		MX_3		MLX_3		ML_2X_3		ML_3X_3 <1%	
4	MX_2Z		MX_4 <1%		MLX_4		ML_2X_4		ML_3X_4 7%
5		MX_3Z		MX_5		MLX_5		ML_2X_5	
6	MX_2Z_2		MX_4Z		MX_6		MLX_6		ML_2X_6 <1%
7		MX_3Z_2		MX_5Z		MX_7		MLX_7	
8	MX_2Z_3		MX_4Z_2		MX_6Z		MX_8		MLX_8 <1%

[*]A library of leading references and teaching materials is available at covalentbondclass.org.

[**]MLX plot adapted from that in G. Parkin, "Classification of Organotransition Metal Compounds" in *Comprehensive Organometallic Chemistry III—From Fundamentals to Applications*, R. H. Crabtree, D. M. P. Mingos (eds.), Elsevier, 2007, Volume 1, p. 34, used with permission.

13.8 Spectral Analysis and Characterization of Organometallic Complexes

One of the most challenging aspects of organometallic research is the characterization of new products. Assuming that pure products can be isolated by chromatography, recrystallization, or other techniques, determining the structure can present a challenge. Many complexes can be crystallized and characterized by X-ray crystallography; however, not all organometallic complexes can be crystallized, and not all that crystallize lend themselves to structural solution by X-ray techniques. Furthermore, it is frequently desirable to use more convenient techniques than X-ray crystallography (although, in some cases, an X-ray structural determination is the only way to identify a compound conclusively). Infrared spectroscopy and NMR spectrometry are often the most useful. Mass spectrometry, elemental analysis, conductivity measurements, and other methods may be valuable in characterizing products of organometallic reactions. We will consider primarily IR and NMR as techniques used in the characterization of organometallic complexes.

13.8.1 Infrared Spectra

The number of IR bands, as discussed in Chapter 4, depends on molecular symmetry; consequently, by determining the number of such bands for a particular ligand (such as CO), we may be able to decide among several alternative geometries for a compound or at least reduce the number of possibilities. In addition, the position of the IR band can indicate the function of a ligand (e.g., terminal vs. bridging modes) and, in the case of π-acceptor ligands, can describe the electronic environment of the metal.

Number of Infrared Bands

In Section 4.4.2, a method is described for using molecular symmetry to determine the number of IR-active stretching vibrations. IR-active vibrational modes must result in a change in the dipole moment of the molecule. In symmetry terms, the equivalent statement is that IR-active vibrational modes must have irreducible representations of the same symmetry as the Cartesian coordinates x, y, or z (or a linear combination of these coordinates). The procedure developed in Chapter 4 is used in the following examples of carbonyl complexes. Identical reasoning applies to other linear monodentate ligands, such as CN^- and NO. We will begin by considering several simple cases.

Monocarbonyl Complexes

These complexes have a single C—O stretching mode and consequently show a single band in the IR.

Dicarbonyl Complexes

Two geometries, linear and bent, must be considered:

O—C—M—C—O

```
      M
     / \
    C   C
   /     \
  O       O
```

In the case of two CO ligands arranged linearly, only an antisymmetric vibration of the ligands is IR active; a symmetric vibrational mode produces no change in dipole moment and hence is inactive. However, if two CO ligands are oriented in a nonlinear fashion, both symmetric and antisymmetric vibrations result in changes in dipole moment, and both are IR active:

Symmetric Stretch

O⟵C—M—C⟶O

No change in dipole moment: IR inactive

M with two C—O ligands at an angle (O atoms moving outward)

Change in dipole moment: IR active

Antisymmetric Stretch

O⟵C—M—C⟵O

Change in dipole moment: IR active

M with two C—O ligands at an angle (one O moving outward, the other inward)

Change in dipole moment: IR active

Therefore, an IR spectrum is a convenient tool for determining the structure of molecules with two CO ligands: a single band indicates linear orientation of the CO ligands, and two bands indicate nonlinear orientation.

For molecules containing two CO ligands on the same metal atom, the relative intensities of the IR bands can be used to determine the approximate angle between the CO ligands, using the equation

$$\frac{I_{symmetric}}{I_{antisymmetric}} = cotan^2\left(\frac{\phi}{2}\right)$$

where the angle between the ligands is ϕ. For example, for two CO ligands at 90°, $cotan^2(45°) = 1$. For this angle, two IR bands of equal intensity would be observed. For an angle greater than 90°, the ratio is less than 1; the IR band due to symmetric stretching is less intense than the band due to antisymmetric stretching. If ϕ is less than 90°, the IR band for symmetric stretching is the more intense. (For C—O stretching vibrations, the symmetric band occurs at higher energy than the corresponding antisymmetric band.) This calculation is approximate and requires integrated values of intensities of absorption bands (rather than the more easily determined intensity at the wavelength of maximum absorption).

Complexes containing three or more carbonyls

Here, the predictions are not quite so simple. The exact number of carbonyl bands can be determined according to the symmetry approach of Chapter 4. For convenient reference, the numbers of bands expected for a variety of CO complexes are in **Table 13.10**. In carbonyl complexes, the number of C—O stretching bands cannot exceed the number of CO ligands. The alternative is possible in some cases (more CO groups than IR bands), when vibrational modes are not IR active (do not cause a change in dipole moment). Carbonyl complexes of T_d and O_h symmetry have a single carbonyl band in the IR spectrum.

Additional points relating to the number of IR bands are noteworthy. First, although we can predict the number of IR-active bands by the methods of group theory, fewer bands are sometimes observed. In some cases, bands may overlap to such a degree as to be indistinguishable; alternatively, one or more bands may be of very low intensity and not readily observed. In some cases, isomers may be present, and it may be difficult to determine which IR absorptions belong to which isomer.

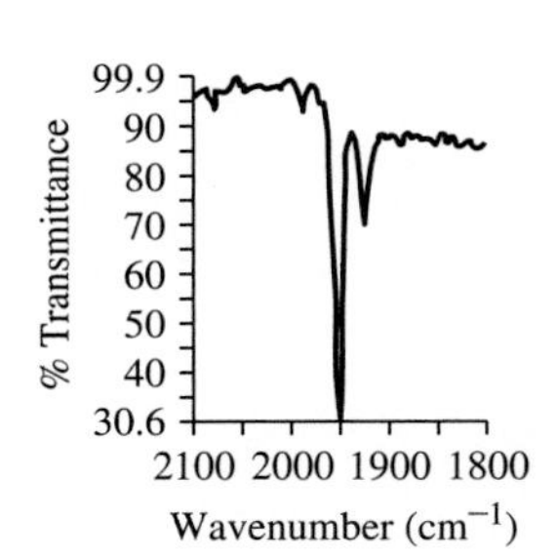

EXERCISE 13.12

The complex $Mo(CO)_3(NCC_2H_5)_3$ has the infrared spectrum in the margin. Is this complex more likely the *fac* or *mer* isomer?

TABLE 13.10 Carbonyl Stretching Bands

		Coordination Number		
Number of CO Ligands		**4**	**5**	**6**
3		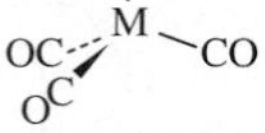	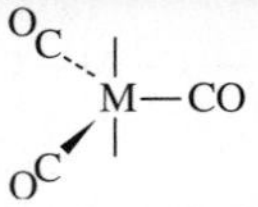	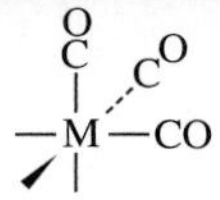
	IR bands:	2	1	2
			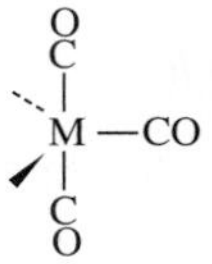	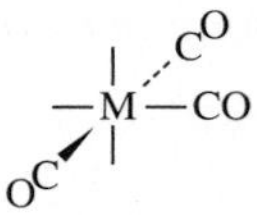
	IR bands:		3	3
			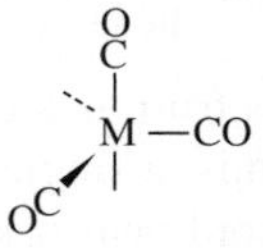	
	IR bands:		3	
4		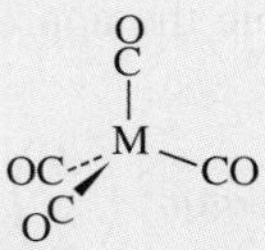		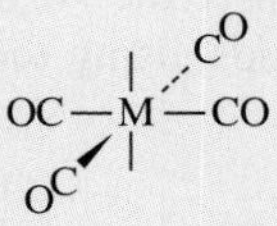
	IR bands:	1	4	1
			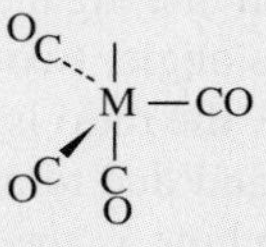	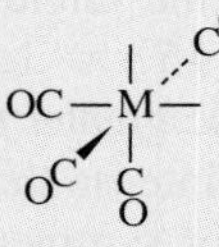
	IR bands:		3	4
5			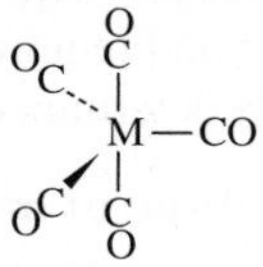	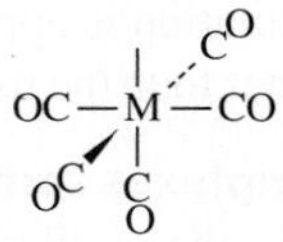
	IR bands:		2	3
6				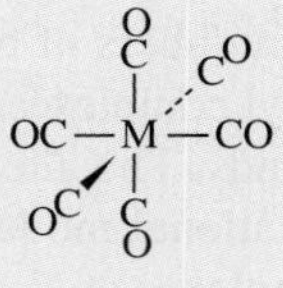
	IR bands:			1

Positions of IR Bands

We have already encountered two examples in which the position of the carbonyl stretching band provides useful information. In the case of the isoelectronic species $[Mn(CO)_6]^+$, $Cr(CO)_6$, and $[V(CO)_6]^-$, an increase in negative charge on the complex causes a significant reduction in the energy of the C—O band as a consequence of additional π back-bonding from the metal to the ligands (Section 13.4.1). The bonding mode is also reflected in the infrared spectrum, with energy decreasing in the order

terminal CO > doubly bridging CO > triply bridging CO

TABLE 13.11 Examples of Carbonyl Stretching Bands: Molybdenum Complexes

Complex	ν(CO), cm^{-1}
fac-$Mo(CO)_3(PF_3)_3$	2090, 2055
fac-$Mo(CO)_3(PCl_3)_3$	2040, 1991
fac-$Mo(CO)_3(PClPh_2)_3$	1977, 1885
fac-$Mo(CO)_3(PMe_3)_3$	1945, 1854

Data from F. A. Cotton, *Inorg. Chem.*, *3*, 702, **1964**.

The positions of infrared bands are also a function of other ligands present. For example, consider the data in **Table 13.11**. Going down the series of complexes in this table, the σ-donor ability of the phosphine ligands increases and the π-acceptor ability decreases. PF_3 is the weakest donor (as a consequence of the highly electronegative fluorines) and the strongest acceptor; conversely, PMe_3 is the strongest donor and the weakest acceptor. As a result, the molybdenum in $Mo(CO)_3(PMe_3)_3$ carries the greatest electron density; it is the most able to donate electron density to the π^* orbitals of the CO ligands. Consequently, the CO ligands in $Mo(CO)_3(PMe_3)_3$ have the weakest C—O bonds and the lowest energy stretching bands. Many comparable series are known.

The position of the carbonyl bands provide important clues to the electronic environment of the metal. The greater the electron density on the metal (and the greater the negative charge), the greater the back-bonding to CO, and the lower the energy of the carbonyl stretching vibrations. Similar correlations between the metal environment and IR spectra can be shown for other ligands. NO, for example, has an IR spectrum that is strongly correlated with the environment in a manner similar to that of CO. In combination with information on the number of IR bands, the positions of such bands for CO and other ligands are extremely useful in characterizing organometallic compounds.

13.8.2 NMR Spectra

NMR spectroscopy is also a valuable tool in characterizing organometallic complexes. The advent of high-field NMR instruments using superconducting magnets has revolutionized the study of these compounds. Convenient NMR spectra can now be taken using many metal nuclei as well as the more traditional nuclei such as ^{1}H, ^{13}C, ^{19}F, and ^{31}P; the combined spectral data of several nuclei make it possible to identify many compounds.

As in organic chemistry, chemical shifts, splitting patterns, and coupling constants are useful in characterizing the environments of individual atoms in organometallic compounds. The reader may find it useful to review the basic theory of NMR as presented in an organic chemistry text. More detailed discussions of NMR, especially relating to ^{13}C, are presented elsewhere.[71]

^{13}C NMR

Although the isotope ^{13}C has a low natural abundance (approximately 1.1%) and low sensitivity for the NMR experiment (about 1.6% as sensitive as ^{1}H), Fourier transform techniques make it possible to obtain useful ^{13}C spectra for most organometallic species of reasonable stability. Nevertheless, the time necessary to obtain a ^{13}C spectrum may still be an experimental difficulty for compounds present in very small amounts or of low solubility. Some useful features of ^{13}C spectra include:

1. An opportunity to observe organic ligands that do not contain hydrogen, such as CO and $F_3C—C{\equiv}C—CF_3$.
2. Direct observation of the carbon skeleton of organic ligands.

3. **3.** ^{13}C chemical shifts are more widely dispersed than ^{1}H shifts. This often makes it easy to distinguish between ligands in compounds containing several different organic ligands.

^{13}C NMR spectroscopy is a valuable tool for observing intramolecular rearrangement processes.[72]

Approximate ranges of chemical shifts for ^{13}C spectra of some categories of organometallic complexes are in **Table 13.12**. Several features of these data are noteworthy.

1. Terminal carbonyl peaks are frequently in the range δ 195 to 225 ppm; CO groups are usually easy to distinguish from other ligands.
2. The ^{13}CO chemical shift is correlated with the strength of the C—O bond; in general, the stronger the C—O bond, the lower (more upfield) the chemical shift.[73]
3. Bridging carbonyls have slightly greater chemical shifts than terminal carbonyls and consequently may lend themselves to easy identification. (However, IR is usually a better tool than NMR for distinguishing between bridging and terminal carbonyls.)
4. Cyclopentadienyl ligands have a wide range of chemical shifts in paramagnetic compounds and a much narrower range in diamagnetic compounds. Other organic ligands may also have fairly wide ranges in ^{13}C chemical shifts.*

^{1}H NMR

The ^{1}H spectra of organometallic compounds also provide useful structural information. For example, most protons bonded directly to metals (in hydride complexes, Section 13.4.3) are very strongly shielded, with chemical shifts commonly in the approximate range of –5 to –20 ppm relative to $Si(CH_3)_4$. Such protons are typically easy to detect because few other protons commonly appear in this region.

Protons in methyl complexes ($M—CH_3$) typically have chemical shifts between 1 and 4 ppm, similar to their positions in organic molecules. Cyclic π ligands, such as

TABLE 13.12 **^{13}C Chemical Shifts for Organometallic Compounds**

Ligand	^{13}C Chemical Shift (Range) [a]			
$M—CH_3$	–28.9 to 23.5			
$M{=}C\langle$	190 to 400			
$M{\equiv}C—$	235 to 401			
$M{\equiv}C$ (terminal)	470 to 556			
$M—CO$	177 to 275			
Neutral binary CO	183 to 223			
$M—(\eta^5\text{-}C_5H_5)$	68.2 to 121.3 (diamagnetic) –790 to 1430 (paramagnetic)			
$Fe(\eta^5\text{-}C_5H_5)_2$	69.2			
$M—(\eta^3\text{-}C_3H_5)$	C_2: 91 to 129	C_1 and C_3: 46 to 79		
$M—C_6H_5$	M—C: 130 to 193	*ortho*: 132 to 141	*meta*: 127 to 130	*para*: 121 to 131

[a]Parts per million (ppm) relative to $Si(CH_3)_4$.

*Extensive tables of chemical shifts and coupling constants can be found in B. E. Mann, "^{13}C NMR Chemical Shifts and Coupling Constants of Organometallic Compounds," in *Adv. Organomet. Chem.*, **1974**, *12*, 135.

TABLE 13.13 Examples of ^1H Chemical Shifts for Organometallic Compounds

Complex	^1H Chemical Shift [a]
$Mn(CO)_5\mathbf{H}$	−7.5
$W(\mathbf{CH_3})_6$	1.80
$Ni(\eta^2\text{-}C_2\mathbf{H}_4)_3$	3.06
$(\eta^5\text{-}C_5\mathbf{H}_5)_2Fe$	4.04
$(\eta^6\text{-}C_6\mathbf{H}_6)_2Cr$	4.12
$(\eta^5\text{-}C_5H_5)_2Ta(CH_3)(=\mathbf{CH_2})$	10.22

NOTE: [a]Parts per million relative to $Si(CH_3)_4$.

η^5-C_5H_5 and η^6-C_6H_6, most commonly have ^1H chemical shifts between 4 and 7 ppm and, because of the relatively large number of protons involved, may lend themselves to easy identification.* Protons in other types of organic ligands also have characteristic chemical shifts (**Table 13.13**).

As in organic chemistry, integration of NMR peaks of organometallic complexes can provide the ratio of atoms in different environments. For example, the area of a ^1H peak is usually proportional to the number of nuclei giving rise to that peak. However, for ^{13}C, this calculation is less reliable. Relaxation times of different carbon atoms in organometallic complexes vary widely; this may lead to inaccuracy in correlating peak area with the number of atoms (the correlation between area and number of atoms is dependent on rapid relaxation). Adding paramagnetic reagents may speed up relaxation and improve the validity of integration data. One paramagnetic compound often used is $Cr(acac)_3$ [acac = acetylacetonate = $H_3CC(O)CHC(O)CH_3^-$].**

Molecular Rearrangement Processes

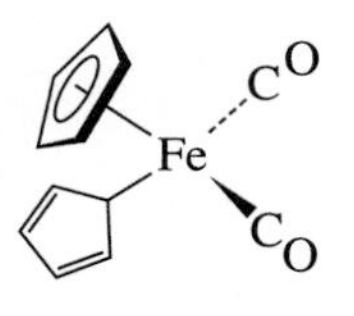

$(C_5H_5)_2Fe(CO)_2$ has interesting NMR behavior. This compound contains both η^1- and η^5-C_5H_5 ligands. The ^1H NMR spectrum at room temperature shows two singlets of equal area. A singlet would be expected for the five equivalent protons of the η^5-C_5H_5 ring but is surprising for the η^1-C_5H_5 ring, because the protons are not all equivalent. At lower temperatures, the peak at 4.5 ppm (η^5-C_5H_5) remains constant, but the other peak at 5.7 ppm spreads and then splits into new peaks near 3.5 and between 5.9 and 6.4 ppm—all consistent with a η^1-C_5H_5 ligand. A "ring whizzer" mechanism,[74] **Figure 13.54**, has been proposed by which the five ring positions of the monohapto ring interchange via 1,2-metal shifts so rapidly at 30 °C that the NMR spectrometer can detect only the average signal for the ring.[75] At lower temperatures, this process is slower, and the different resonances for the protons of η^1-C_5H_5 become apparent, also shown in Figure 13.54. More detailed discussions of NMR spectra of organometallic compounds, including nuclei not mentioned here, are given by Elschenbroich.[76]

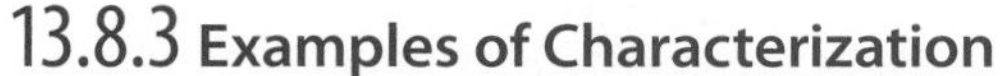

13.8.3 Examples of Characterization

We conclude this chapter with two examples of how spectral data may be used in the characterization of organometallic compounds. Further examples can be found in the problems at the end of this chapter and in Chapter 14.

---

*These are ranges for diamagnetic complexes. Paramagnetic complexes may have much larger chemical shifts, sometimes several hundred parts per million relative to tetramethylsilane.

**For a discussion of the problems associated with integration in ^{13}C NMR, see J. K. M. Saunders and B. K. Hunter, *Modern NMR Spectroscopy*, W. B. Saunders, New York, 1992.

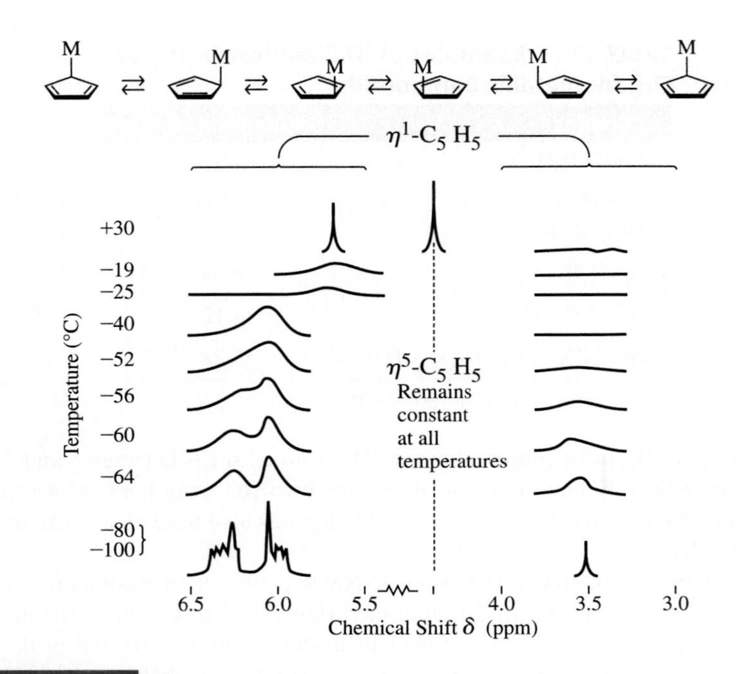

FIGURE 13.54 Ring-Whizzer Mechanism and Variable Temperature NMR Spectra of $(C_5H_5)_2Fe(CO)_2$. The central peak at 4.5 ppm due to the η^5-C_5H_5 ligand remains constant throughout; it is not shown except in the highest temperature spectrum to simplify the figure. (Reproduced with permission from M. H. Bennett, Jr., F. A. Cotton, A. Davison, J. W. Faller, S. J. Lippard, S. M. Morehouse, *J. Am. Chem. Soc.*, *88*, 4371. Copyright 1966. American Chemical Society.)

EXAMPLE 13.3

$[(C_5H_5)Mo(CO)_3]_2$ reacts with tetramethylthiuramdisulfide (tds) in refluxing toluene to give a molybdenum-containing product having the following characteristics:

H_3C, CH_3, N—C, S, S—S, C—N
tds

^{1}H NMR: Two singlets, at δ 5.48 (relative area = 5) and δ 3.18 (relative area = 6). (For comparison, $[(C_5H_5)Mo(CO)_3]_2$ has a single ^{1}H NMR peak at δ 5.30.)

IR: Strong bands at 1950 and 1860 cm^{-1}.

Mass spectrum: A pattern similar to the Mo isotope pattern with the most intense peak at $m/e = 339$. (The most abundant Mo isotope is ^{98}Mo.)

What is the most likely identity of this product?

The ^{1}H NMR singlet at δ 5.48 suggests retention of the C_5H_5 ligand (the chemical shift is a close match for the starting material). The peak at δ 3.18 is most likely due to CH_3 groups originating from the tds. The 5:6 ratio of hydrogens suggests a 1:2 ratio of C_5H_5 ligands to CH_3 groups.

IR shows two bands in the carbonyl region, indicating at least two CO ligands in the product.

The mass spectrum makes it possible to pin down the molecular formula. Subtracting the molecular fragments believed to be present from the total mass:

Total mass:	339
Mass of Mo (from mass spectrum pattern)	−98
Mass of C_5H_5	−65
Mass of two CO ligands	−56
Remaining mass	120

CO, CO, Mo, S, S, C, $N(CH_3)_2$

120 is exactly half the mass of tds; it corresponds to the mass of $S_2CN(CH_3)_2$, the dimethyldithiocarbamate ligand. Therefore, the likely formula of the product is $(C_5H_5)Mo(CO)_2[S_2CN(CH_3)_2]$. This formula has the necessary 5:6 ratio of protons in two magnetic environments and should give rise to two C—O stretching vibrations

(because the carbonyls would not be expected to be oriented at 180° angles with respect to each other in such a molecule).

In practice, additional information is likely to help characterize reaction products. For example, additional examination of the infrared spectrum in this case shows a moderately intense band at 1526 cm^{-1}, a common location for C—N stretching bands in dithiocarbamate complexes. Analysis of the fragmentation pattern of mass spectra may also provide useful information on molecular fragments.

EXAMPLE 13.4

When a toluene solution containing **I** and excess triphenylphosphine is heated to reflux, first compound **II** is formed, and then compound **III**. **II** has infrared bands at 2038, 1958, and 1906 cm^{-1}; **III** at 1944 and 1860 cm^{-1}. ^1H and ^{13}C NMR data [δ values (relative area)] are as follows:

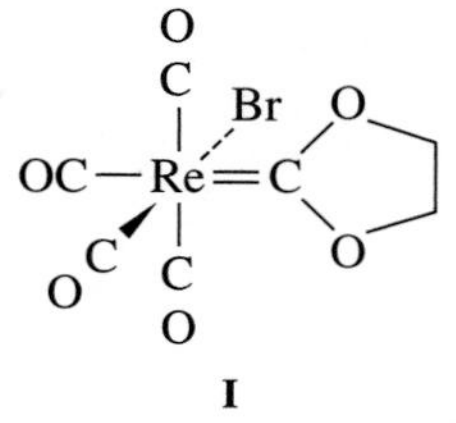

I	II	III
^1H: 4.83 singlet	7.62, 7.41 multiplets (15)	7.70, 7.32 multiplets (15)
	4.19 multiplet (4)	3.39 singlet (2)
^{13}C: 224.31	231.02	237.19
187.21	194.98	201.85
185.39	189.92	193.83
184.01	188.98	127.75–134.08 (several peaks)
73.33	129.03–134.71 (several peaks)	68.80
	72.26	

Additional useful information: the ^{13}C signal of **I** at δ 224.31 is similar to the chemical shift of carbene carbons in similar compounds; the peaks between δ 184 and 202 correspond to carbonyls; and the peak at δ 73.33 is typical of CH_2CH_2 bridges in dioxycarbene complexes.

Identify **II** and **III**.

This is a good example of the utility of ^{13}C NMR. Both **II** and **III** have peaks with similar chemical shifts to the peak at δ 224.31 for **I**, suggesting that the carbene ligand is retained in the reaction. Similarly, **II** and **III** have peaks near δ 73.33, a further indication that the carbene ligand remains intact.

The ^{13}C peaks in the range δ 184 to 202 can be assigned to carbonyl groups. **II** and **III** show new peaks in the range δ 129 to 135. The most likely explanation is that the chemical reaction involves replacement of carbonyls by triphenylphosphines and that the new peaks in the 129 to 135 range are due to the phenyl carbons of the phosphines.

^1H NMR data are consistent with replacement of CO ligands by phosphines. In both **II** and **III**, integration of the $-CH_2CH_2-$ peaks (δ 4.19 and 3.39, respectively) and the phenyl peaks (δ 7.32 to 7.70) give the expected ratios for replacement of one and two COs.

IR data are in agreement with these conclusions. In **II**, the three bands in the carbonyl region are consistent with the presence of three COs, either in a *mer* or a *fac* arrangement.* In **III**, the two C—O stretches correspond to two carbonyls *cis* to each other.

*In an octahedral complex of formula *fac*-$ML_3(CO)_3$ (having C_{3v} symmetry), only two carbonyl stretching bands are expected if all ligands L are identical. However, in this case, there are three different ligands in addition to CO; the point group is C_1, and three bands are expected.

The chemical formulas of these products can be written as follows:

II: $ReBr(CO)_3(\overline{COCH_2CH_2O})(PPh_3)$

III: *cis*-$ReBr(CO)_2(\overline{COCH_2CH_2O})(PPh_3)_2$

II **III**

EXERCISE 13.13

Using ^{13}C NMR data, determine whether **II** is more likely the *fac* or *mer* isomer.[77]

References

1. W. C. Zeise, *Ann. Phys. Chem.*, **1831**, *21*, 497–541. A translation of excerpts from this paper can be found in G. B. Kauffman, ed., *Classics in Coordination Chemistry*, Part 2, Dover, New York, 1976, pp. 21–37. A review of the history of the anion of Zeise's salt, including some earlier references, has been published: D. Seyferth, *Organometallics*, **2001**, *20*, 2.
2. R. A. Love, T. F. Koetzle, G. J. B. Williams, L. C. Andrews, R. Bau, *Inorg. Chem.*, **1975**, *14*, 2653.
3. L. Mond, *J. Chem. Soc.*, **1890**, *57*, 749.
4. V. Grignard, *Ann. Chim.*, **1901**, *24*, 433. An English translation of most of this paper is in P. R. Jones and E. Southwick, *J. Chem. Educ.*, **1970**, *47*, 290.
5. T. J. Kealy P. L. Pauson, *Nature*, **1951**, *168*, 1039.
6. J. D. Dunitz, L. E. Orgel, R. A. Rich, *Acta Crystallogr.*, **1956**, *9*, 373.
7. E. A. V. Ebsworth, D. W. H. Rankin, and S. Cradock, *Structural Methods in Inorganic Chemistry*, 2nd ed., Blackwell Scientific, Oxford UK, 1991.
8. P. R. Mitchell R. V. Parish, *J. Chem. Educ.*, **1969**, *46*, 311. See also W. B. Jensen, *J. Chem. Educ.*, **2005**, *82*, 28.
9. A. S. Goldman K. Krogh-Jespersen, *J. Am. Chem. Soc.*, **1996**, *118*, 12159.
10. D. S. Ginley M. S. Wrighton, *J. Am. Chem. Soc.*, **1975**, *97*, 3533; R. J. Klingler, W. Butler M. D. Curtis, *J. Am. Chem. Soc.*, **1975**, *97*, 3535.
11. A. L. Sargent M. B. Hall, *J. Am. Chem. Soc.*, **1989**, *111*, 1563 and references therein.
12. L. Mond, C. Langer, F. Quincke, *J. Chem. Soc.*,**1890**, *57*, 749; reprinted in *J. Organomet. Chem.*, **1990**, *383*, 1.
13. C. P. Horwitz D. F. Shriver, *Adv. Organomet. Chem.*, **1984**, *23*, 219.
14. W. Petz, *Coord. Chem. Rev.*, **2008**, *252*, 1689 and references cited therein.
15. Y. Mutoh, N. Kozono, K. Ikenage, Y. Ishii, *Coord. Chem. Rev.*, **2012**, *256*, 589.
16. P. V. Broadhurst, *Polyhedron*, **1985**, *4*, 1801.
17. J. Jiang, S. A. Koch, *Inorg. Chem.*, **2002**, *41*, 158.
18. J. Jiang, S. A. Koch, *Angew. Chem., Int. Ed.*, **2001**, *40*, 2629; T. B. Rauchfuss, S. M. Contakes, S. C. N. Hsu, M. A. Reynolds S. R. Wilson, *J. Am. Chem. Soc.*, **2001**, *123*, 6933; S. M. Contakes, S. C. N. Hsu, T. B. Rauchfuss, S. R. Wilson, *Inorg. Chem.*, **2002**, *41*, 1670.
19. J. R. Dethlefsen, A. Døssing, E. D. Hedegård, *Inorg. Chem.*, **2010**, *49*, 8769.
20. T. J. Crevier, S. Lovell, J. M. Mayer, A. L.Rheingold, I. A. Guzei, *J. Am. Chem. Soc.*, **1998**, *120*, 6607.
21. G. J. Kubas, *Comments Inorg. Chem.*, **1988**, *7*, 17; R. H. Crabtree, *Acc. Chem. Res.*, **1990**, *23*, 95; G. J. Kubas, *Acc. Chem. Res.*, **1988**, *21*, 120.
22. S. C. Abrahams, A. P. Ginsberg, K. Knox, *Inorg. Chem.*, **1964**, *3*, 558.
23. G. J. Kubas, R. R. Ryan, B. I. Swanson, P. J. Vergamini, H. J. Wasserman, *J. Am. Chem. Soc.*, **1984**, *106*, 451.
24. J. K. Burdett, O. Eisenstein, and S. A. Jackson, "Transition Metal Dihydrogen Complexes: Theoretical Studies," in A. Dedieu, ed., *Transition Metal Hydrides*, VCH, New York, 1992, pp. 149–184; N. K. Szymczak, D. R. Tyler, *Coord. Chem. Rev.*, **2008**, *252*, 212.
25. F. A. Cotton, *Chemical Applications of Group Theory*, 3rd ed., Wiley-Interscience, 1990, pp. 142–159.
26. A. Haaland J. E. Nilsson, *Acta Chem. Scand.*, **1968**, *22*, 2653; A. Haaland, *Acc. Chem. Res.*, **1979**, *12*, 415.
27. P. Seiler J. Dunitz, *Acta Crystallogr., Sect. B*, **1982**, *38*, 1741.
28. J. C. Giordan, J. H. Moore, J. A. Tossell, *Acc. Chem. Res.*, **1986**, *19*, 281; E. Rühl, A. P. Hitchcock, *J. Am. Chem. Soc.*, **1989**, *111*, 5069.
29. Y. Yu, A. D. Bond, P. W. Leonard, U. J. Lorenz, T. V. Timofeeva, K. P. C. Vollhardt, G. D. Whitener, A. A. Yakovenko, *Chem. Commun.*, **2006**, 2572.
30. I. Resa, E. Carmona, E. Gutierrez-Puebla, A. Monge, *Science*, **2004**, *305*, 1136; A. Grirrane,

I. Resa, A. Rodriguez, E. Carmona, E. Alvarez, E. Gutierrez-Puebla, A. Monge, A. Galindo, D. del Rio, R. A. Anderson, *J. Am. Chem. Soc.*, **2007**, *129*, 693.
31. D. L. Kays, S. Aldridge, *Angew. Chem., Int. Ed.*, **2009**, *48*, 4109.
32. S. Krieck, H. Görls, L. Yu, M. Reiher, M. Westerhausen, *J. Am. Chem. Soc.*, **2009**, *131*, 2977; S. Krieck, H. Görls, M. Westerhausen, *J. Am. Chem. Soc.*, **2010**, *132*, 12492.
33. J. M. Hawkins, A. Meyer, T. A. Lewis, S. D. Loren, F. J. Hollander, *Science*, **1991**, 252, 312.
34. P. J. Fagan, J. C. Calabrese, and B. Malone, "The Chemical Nature of C_{60} as Revealed by the Synthesis of Metal Complexes," in G. S. Hammond and V. J. Kuck, eds., Fullerenes, ACS Symposium Series 481, American Chemical Society, Washington, DC, 1992, pp. 177–186; R. E. Douthwaite, M. L. H. Green, A. H. H. Stephens, J. F. C. Turner, *Chem. Commun*, **1993**, 1522; P. J. Fagan, J. C. Calabrese, B. Malone, *Science*, **1991**, *252*, 1160.
35. J. R. Heath, S. C. O'Brien, Q. Zhang, Y. Liu, R. F. Curl, H. W. Kroto, R. E. Smalley, *J. Am. Chem. Soc.*, **1985**, *107*, 7779; H. Shinohara, H. Yamaguchi, N. Hayashi, H. Sato, M. Ohkohchi, Y. Ando, Y. Saito, *J. Phys. Chem.*, **1993**, *97*, 4259.
36. R. C. Haddon, A. F. Hebard, M. J. Rosseinsky, D. W. Murphy, S. H. Glarum, T. T. M. Palstra, A. P. Ramirez, S. J. Duclos, R. M. Fleming, T. Siegrist, and R. Tycko, "Conductivity and Superconductivity in Alkali Metal Doped C_{60}," in Hammond and Kuck, Fullerenes, pp. 71–89.
37. R. C. Haddon, *Acc. Chem. Res.*, **1992**, *25*, 127.
38. J. M. Hawkins, *Acc. Chem. Res.*, **1992**, *25*, 150 and references therein.
39. P. J. Fagan, J. C. Calabrese, B. Malone, *Acc. Chem. Res.*, **1992**, *25*, 134; D. Soto and R. Salcedo, *Molecules*, **2012**, *17*, 7151.
40. P. J. Fagan, J. C. Calabrese, B. Malone, *Science*, **1991**, *252*, 1160.
41. P. J. Fagan, J. C. Calabrese, B. Malone, *J. Am. Chem. Soc.*, **1991**, *113*, 9408. See also P. V. Broadhurst, *Polyhedron*, **1985**, *4*, 1801.
42. H. Shinohara, H. Yamaguchi, N. Mayashi, H. Sato, M. Ohkohchi, Y. Ando, Y. Saito, *J. Phys. Chem.*, **1993**, *97*, 4259.
43. M. M. Olmstead, A. de Bettencourt-Dias, J. C. Duchamp, S. Stevenson, D. Marciu, H. C. Dorn, A. L. Balch, *Angew. Chem., Int. Ed.*, **2001**, *40*, 1223.
44. A. J. Shortland, G. Wilkinson, *J. Chem. Soc., Dalton Trans.*, **1973**, 872.
45. B. Blom, H. Clayton, M. Kilkenny, J. R. Moss, *Adv. Organomet. Chem.*, **2006**, *54*, 149.
46. E. O. Fischer, A. Maasbol, *Angew. Chem., Int. Ed.*, **1964**, *3*, 580.
47. R. R. Schrock, *J. Am. Chem. Soc.*, **1974**, *96*, 6796.
48. E. O. Fischer, *Adv. Organomet. Chem.*, **1976**, *14*, 1.
49. C. G. Kreiter, E. O. Fischer, *Angew. Chem., Int. Ed.*, **1969**, *8*, 761.
50. O. S. Mills A. D. Redhouse, *J. Chem. Soc. A*, **1968**, 642.
51. K. H. Dötz, H. Fischer, P. Hoffmann, F. R. Kreissl, U. Schubert, and K. Weiss, *Transition Metal Carbene Complexes*, Verlag Chemie, Weinheim, Germany, 1983, pp. 120–122.
52. E. O. Fischer, G. Kreis, C. G. Kreiter, J. Müller, G. Huttner, H. Lorentz, *Angew. Chem., Int. Ed.*, **1973**, *12*, 564.
53. H. P. Kim and R. J. Angelici, "Transition Metal Complexes with Terminal Carbyne Ligands," in *Adv. Organomet. Chem.*, **1987**, *27*, 51; H. Fischer, P. Hoffmann, F. R. Kreissl, R. R. Schrock, U. Schubert, and K. Weiss, *Carbyne Complexes*, VCH, Weinheim, Germany, 1988.
54. R. G. Carlson, M. A. Gile, J. A. Heppert, M. H. Mason, D. R. Powell, D. Vander Velde, J. M. Vilain, *J. Am. Chem. Soc.*, **2002**, *124*, 1580.
55. J. B. Gary, C. Buda, M. J. A. Johnson, B. D. Dunietz, *Organometallics*, **2008**, *27*, 814.
56. **A.** Krapp, G. Frenking, *J. Am. Chem. Soc.*, **2008**, *130*, 16646.
57. S. R. Caskey, M. H. Stewart, J. E. Kivela, J. R. Sootsman, M. J. A. Johnson, J. W. Kampf, *J. Am. Chem. Soc.*, **2005**, *127*, 16750; Y. Mutoh, N. Kozono, M. Araki, N. Twuchida, K. Takano, Y. Ishii, *Organometallics*, **2010**, *29*, 519.
58. M. I. Bruce, *Coord. Chem. Rev.*, **2004**, *248*, 1603.
59. V. Cadierno J. Gimeno, *Chem. Rev.*, **2009**, *109*, 3512.
60. M. Dede, M. Drexler, H. Fischer, *Organometallics*, **2007**, *26*, 4294.
61. C. Coletti, A. Marrone, N. Re, *Acc. Chem. Res.*, **2012**, *45*, 139.
62. P. Aguirre-Etcheverry D. O'Hare, *Chem. Rev.*, **2010**, *110*, 4839.
63. Q. L. Zheng, J. C. Bohling, T. B. Peters, A. C. Frisch, F. Hampel, J. A. Gladysz, *Chem–Eur. J.*, **2006**, *12*, 6486.
64. S. Szafert, J. A. Gladysz, *Chem. Rev.*, **2006**, *106*, PR1.
65. Q. Zheng J. A. Gladysz, *J. Am. Chem. Soc.*, **2005**, *127*, 10508.
66. M. I. Bruce, K. Costuas, B. G. Ellis, J.-F. Halet, P. J. Low, B. Moubaraki, K. S. Murray, N. Ouddaï, G. J. Perkins, B. W. Skelton, A. H. White, *Organometallics*, **2007**, *26*, 3735.
67. L. de Quadras, F. Hampel, J. A. Gladysz, *Dalton Trans.*, **2006**, 2929.
68. N. Wiesbach, Z. Baranov, A. Gauthier, J. H. Reibenspies, J. A. Gladysz, *Chem. Commun.*, **2012**, *48*, 7562.
69. G. Parkin, "Classification of Organotransition Metal Compounds" in *Comprehensive Organometallic Chemistry III—From Fundamentals to Applications*, R. H. Crabtree, D. M. P, Mingos (eds.), Elsevier, 2007, Volume 1, pp. 1–57; M. L. H. Green, *J. Organomet. Chem.*, **1995**, *500*, 127.
70. G. Parkin, *J. Chem. Educ.*, **2006**, *83*, 791.
71. B. E. Mann, "^{13}C NMR Chemical Shifts and Coupling Constants of Organometallic Compounds," in *Adv. Organomet. Chem.*, **1974**, *12*, 135; P. W. Jolly and R. Mynott, "The Application of ^{13}C NMR Spectroscopy to Organo-Transition Metal Complexes," in *Adv. Organomet. Chem.*, **1981**, *19*, 257; E. Breitmaier and W. Voelter,

Carbon 13 NMR Spectroscopy, VCH, New York, 1987; E. A. V. Ebsworth, D. W. H. Rankin, and S. Cradock, *Structural Methods in Inorganic Chemistry*, 2nd ed., Blackwell, Oxford, 1991, pp. 414–425.
72. Breitmaier and Voelter, *Carbon 13 NMR Spectroscopy*. pp. 127–133, 166–167, 172–178.
73. P. C. Lauterbur, R. B. King, *J. Am. Chem. Soc.*, **1965**, *87*, 3266.
74. C. H. Campbell, M. L. H. Green, *J. Chem. Soc., A*, **1970**, 1318.
75. M. J. Bennett, Jr., F. A. Cotton, A. Davison, J. W. Faller, S. J. Lippard, S. M. Morehouse, *J. Am. Chem. Soc.*, **1966**, *88*, 4371. For a summary of early developments in the use of NMR to observe molecular rearrangement processes, see F. A. Cotton, *Inorg. Chem.*, **2002**, *41*, 643.
76. C. Elschenbroich, *Organometallics*, 3rd ed., Wiley-VCH, Wiesbaden, Germany, 2005.
77. G. L. Miessler, S. Kim, R. A. Jacobson, R. A. Angelici, *Inorg. Chem.*, **1987**, *26*, 1690.

General References

Much information on organometallic compounds is included in two general inorganic references, N. N. Greenwood and A. Earnshaw, *Chemistry of the Elements*, 2nd ed., Butterworth Heinemann, Oxford, 1997, and F. A. Cotton, G. Wilkinson, C. A. Murillo, and M. Bochman, *Advanced Inorganic Chemistry*, 6th ed., Wiley InterScience, New York, 1999. G. O. Spessard and G. L. Miessler, *Organometallic Chemistry*, Oxford University Press, New York, 2010, C. Elschenbroich, *Organometallics*, 3rd ed., Wiley-VCH, Wiesbaden, Germany, 2005, J. F. Hartwig, *Organotransition Metal Chemistry, From Bonding to Catalysis*, University Science Books, Mill Valley, CA, 2010 provide extensive discussion, with numerous references, of many additional types of organometallic compounds in addition to those discussed in this chapter. The most comprehensive references on organometallic chemistry are the multiple-volume sets edited by G. Wilkinson and F. G. A. Stone, *Comprehensive Organometallic Chemistry*, Pergamon Press, Oxford, 1982, by E. W. Abel, F. G. A. Stone, and G. Wilkinson, *Comprehensive Organometallic Chemistry II*, Pergamon Press, Oxford, 1995, and by R. H. Crabtree and D. M. P. Mingos, *Comprehensive Organometallic Chemistry III*, Elsevier, 2007. Each of these sets has an extensive listing of references on organometallic compounds that have been structurally characterized by X-ray, electron, or neutron diffraction. A useful reference to literature sources on the synthesis, properties, and reactions of specific organometallic compounds is J. Buckingham and J. E. Macintyre, editors, *Dictionary of Organometallic Compounds*, Chapman and Hall, London, 1984, to which supplementary volumes have also been published. The series *Advances in Organometallic Chemistry*, Academic Press, San Diego, provides valuable review articles.

Problems

13.1 Which of the following obey the 18-electron rule?
- **a.** $Fe(CO)_5$
- **b.** $[Rh(bipy)_2Cl]^+$
- **c.** $(\eta^5\text{-}Cp^*)Re({=}O)_3$, where $Cp^* = C_5(CH_3)_5$
- **d.** $Re(PPh_3)_2Cl_2N$
- **e.** $Os(CO)({\equiv}CPh)(PPh_3)_2Cl$
- **f.** The CE complexes in Table 13.3

13.2 Which of the following square-planar complexes have 16-electron valence configurations?
- **a.** $Ir(CO)Cl(PPh_3)_2$
- **b.** $RhCl(PPh_3)_3$
- **c.** $[Ni(CN)_4]^{2-}$
- **d.** *cis*-$PtCl_2(NH_3)_2$

13.3 On the basis of the 18-electron rule, identify the first-row transition metal for each of the following:
- **a.** $[M(CO)_7]^+$
- **b.** $H_3CM(CO)_5$
- **c.** $M(CO)_2(CS)(PPh_3)Br$
- **d.** $[(\eta^3\text{-}C_3H_3)(\eta^5\text{-}C_5H_5)M(CO)]^-$
- **e.** $(OC)_5M{=}C(OCH_3)(C_6H_5)$
- **f.** $[(\eta^4\text{-}C_4H_4)(\eta^5\text{-}C_5H_5)M]^+$
- **g.** $(\eta^3\text{-}C_3H_5)(\eta^5\text{-}C_5H_5)M(CH_3)(NO)$ (has linearly coordinated NO)
- **h.** $[M(CO)_4I(diphos)]^-$ (diphos = 1,2-bis(diphenylphosphino)ethane)

13.4 Determine the metal–metal bond order consistent with the 18-electron rule for the following:
- **a.** $[(\eta^5\text{-}C_5H_5)Fe(CO)_2]_2$
- **b.** $[(\eta^5\text{-}C_5H_5)Mo(CO)_2]_2^{2-}$

13.5 Identify the most likely second-row transition metal for each of the following:
- **a.** $[M(CO)_3(NO)]^-$
- **b.** $[M(PF_3)_2(NO)_2]^+$ (contains linear M—N—O)
- **c.** $[M(CO)_4(\mu_2\text{-}H)]_3$
- **d.** $M(CO)(PMe_3)_2Cl$ (square-planar complex)

13.6 On the basis of the 18-electron rule, determine the expected charge on the following:
- **a.** $[Co(CO)_3]^z$
- **b.** $[Ni(CO)_3(NO)]^z$ (contains linear M—N—O)
- **c.** $[Ru(CO)_4(GeMe_3)]^z$
- **d.** $[(\eta^3\text{-}C_3H_5)V(CNCH_3)_5]^z$
- **e.** $[(\eta^5\text{-}C_5H_5)Fe(CO)_3]^z$
- **f.** $[(\eta^5\text{-}C_5H_5)_3Ni_3(\mu_3\text{-}CO)_2]^z$

13.7 Determine the unknown quantity:

a. $[(\eta^5\text{-}C_5H_5)W(CO)_x]_2$ (has W—W single bond)

b.

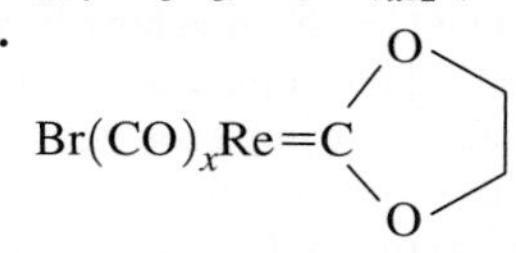

c. $[(CO)_3Ni\text{—}Co(CO)_3]^z$

d. $[Ni(NO)_3(SiMe_3)]^z$ (contains linear M—N—O)

e. $[(\eta^5\text{-}C_5H_5)Mn(CO)_x]_2$ (has Mn=Mn bond)

13.8 Determine the unknown quantity:

a. The hapticity of the lower ring in the mixed superphane (See S. Gath, R. Gleiter, F. Rominger, C. Bleiholder, *Organometallics*, **2007**, *26*, 644. Structure generated with CIF data, with hydrogen atoms omitted for clarity.)

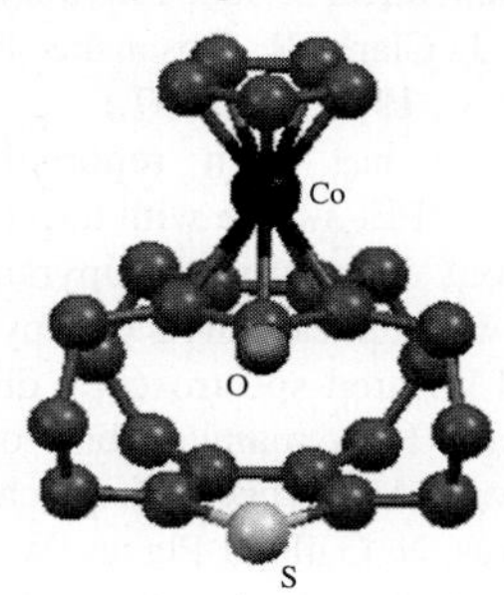

b. The *third* row 16-electron transition metal **M** on the left and the *first* row 16-electron transition metal **M′** on the right (See M. Tamm, A. Kunst, T. Bannenberg, S. Randoll, P. G. Jones, *Organometallics* **2007**, *26*, 417.)

13.9 Figure 13.35 shows examples of complexes containing both C_5H_5 and CO ligands for group 4–8 transition metals. What complexes containing both ligands would be consistent with the 18-electron rule for first row transition metals in groups 9–11? Consider complexes containing one or two metal atoms.

13.10 Nickel tetracarbonyl, $Ni(CO)_4$, is an 18-electron species. Using a qualitative molecular orbital diagram, explain the stability of this 18-electron molecule. (See A. W. Ehlers, S. Dapprich, S. V. Vyboishchikov, G. Frenking, *Organometallics*, **1996**, *15*, 105.)

13.11 The Re—O stretching vibration in $Re(^{16}O)I(HC{\equiv}CH)_2$ is at 975 cm^{-1}. Predict the position of the Re—O stretching band in $Re(^{18}O)I(HC{\equiv}CH)_2$. (See J. M. Mayer, D. L. Thorn, T. H. Tulip, *J. Am. Chem. Soc.*, **1985**, *107*, 7454.)

13.12 The compound $W(O)Cl_2(CO)(PMePh_2)_2$ has ν (CO) at 2006 cm^{-1}. Would you predict ν (CO) for $W(S)Cl_2(CO)(PMePh_2)_2$ to be at higher or lower energy? Explain briefly. (See J. C. Bryan, S. J. Geib, A. L. Rheingold, J. M. Mayer, *J. Am. Chem. Soc.*, **1987**, *109*, 2826.)

13.13 The vanadium–carbon distance in $V(CO)_6$ is 200 pm but only 193 pm in $[V(CO)_6]^-$. Explain.

13.14 Describe, using sketches, how the following ligands can act as both σ donors and π acceptors:

a. CN^- **b.** $P(CH_3)_3$ **c.** SCN^-

13.15 **a.** Account for the following trend in IR frequencies:

$[Cr(CN)_5(NO)]^{4-}$ ν (NO) = 1515 cm^{-1}

$[Mn(CN)_5(NO)]^{3-}$ ν (NO) = 1725 cm^{-1}

$[Fe(CN)_5(NO)]^{2-}$ ν (NO) = 1939 cm^{-1}

b. The ion $[RuCl(NO)_2(PPh_3)_2]^+$ has N—O stretching bands at 1687 and 1845 cm^{-1}. The C—O stretching bands of dicarbonyl complexes typically are much closer in energy. Explain.

13.16 Sketch the π molecular orbitals for the following:

a. CO_2

b. 1,3,5-Hexatriene

c. Cyclobutadiene, C_4H_4

d. Cyclo-C_7H_7

13.17 For the hypothetical molecule $(\eta^4\text{-}C_4H_4)Mo(CO)_4$:

a. Assuming C_{4v} geometry, predict the number of IR-active C—O bands.

b. Sketch the π molecular orbitals of cyclobutadiene. For each, indicate which *s*, *p*, and *d* orbitals of Mo are of suitable symmetry for interaction. (Hint: Assign the *z* axis to be collinear with the C_4 axis.)

13.18 Using the D_{5h} character table in Appendix C:

a. Assign symmetry labels (labels of irreducible representations) for the group orbitals shown in Figure 13.27.

b. Assign symmetry labels for the atomic orbitals of Fe in a D_{5h} environment.

c. Verify that the orbital interactions for ferrocene shown in Figure 13.28 are between atomic orbitals of Fe and group orbitals of matching symmetry.

13.19 Dibenzenechromium, $(\eta^4\text{-}C_6H_6)_2Cr$, is a sandwich compound having two parallel benzene rings in an eclipsed conformation. For this molecule,

a. Sketch the π orbitals of benzene.

b. Sketch the group orbitals, using π orbitals of the two benzene rings.

c. For each of the 12 group orbitals, identify the Cr orbital(s) of suitable symmetry for interaction.

d. Sketch an energy-level diagram of the molecular orbitals.

13.20 Although the bonding in the proposed ferrocene isomer $(\eta^4\text{-}C_4H_4)(\eta^6\text{-}C_6H_6)Fe$ has been studied theoretically, synthesis of this molecule has not been reported (C. M. Brett, B. E. Bursten, *Polyhedron*, **2004**, *23*, 2993). For this molecule:

a. What is its point group?

b. Generate a (reducible) representation using as a basis the p orbitals of the C_4H_4 and C_6H_6 rings that can participate in π interactions.

c. Reduce this representation to its irreducible components, and match these with the appropriate orbitals on Fe.

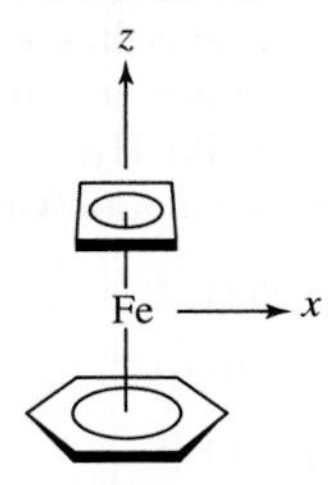

13.21 Predict the number of IR-active C—O stretching vibrations for $W(CO)_3(\eta^6\text{-}C_6H_6)$ assuming C_{3v} geometry.

13.22 In this chapter, the assertion was made that highly symmetric binary carbonyls of T_d and O_h symmetry should show only a single C—O stretching band in the infrared. Check this assertion by analyzing the C—O vibrations of $Ni(CO)_4$ and $Cr(CO)_6$ by the symmetry method described in Chapter 4.

13.23 $Mn_2(CO)_{10}$ and $Re_2(CO)_{10}$ have D_{4d} symmetry. How many IR-active carbonyl stretching bands would you predict for these compounds?

13.24 When tungsten hexacarbonyl is refluxed in butyronitrile, C_3H_7CN, product **X** is formed first. Continued reflux converts **X** to **Y**, and very long reflux (several days) converts **Y** to **Z**. However, even refluxing for several weeks does not convert compound **Z** into another product. In addition, in each reaction step, a colorless gas is liberated. The following infrared bands are observed (cm^{-1}):

X:		Y:		Z:	
	2077		2017		1910
	1975		1898		1792
	1938		1842		

a. Propose structures of **X**, **Y**, and **Z**. Where more than one isomer is possible, determine the isomer that is the best match for the infrared data. (Note: a weak band in **Y** may be obscured by other bands.)

b. Account for the trend in the position of the infrared bands in the sequence **X** → **Y** → **Z**.

c. Suggest why **Z** does not react further when refluxed in butyronitrile. (See G. J. Kubas, *Inorg. Chem.* **1983**, *22*, 692.)

13.25 The infrared spectra of *trans*- and *cis*-$[Fe(CO)_2(CN)_4]^{2-}$ and of $[Fe(CO)(CN)_5]^{3-}$ are shown below.

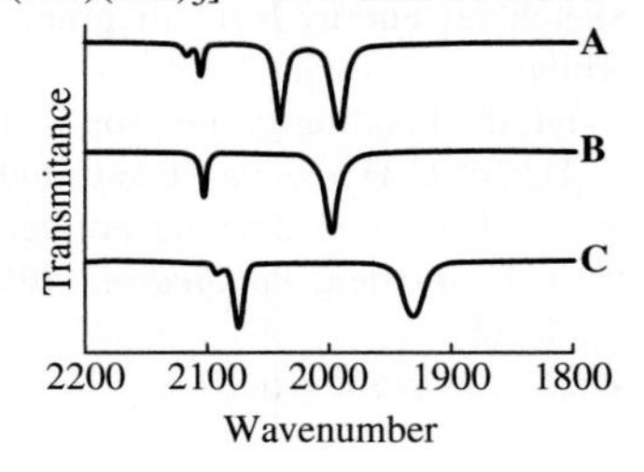

a. Which stretching bands occur at lower energy, those for the CO ligand or those for the CN^- ligand? Explain.

b. How many C—O and C—N stretching bands would you predict for each of these complexes on the basis of their symmetry? Match the complexes with their spectra. (Reproduced with permission from S. M. Contakes, S. C. N. Hsu, T. B. Rauchfuss, S. R. Wilson, *Inorg. Chem.*, *41*, 1670. Copyright 2002. American Chemical Society.)

13.26 Samples of $Fe(CO)(PF_3)_4$ show two carbonyl stretching bands, at 2038 and 2009 cm^{-1}.

a. How is it possible for this compound to exhibit two carbonyl bands?

b. $Fe(CO)_5$ has carbonyl bands at 2025 and 2000 cm^{-1}. Would you place PF_3 above or below CO in the spectrochemical series? Explain briefly. (See H. Mahnke, R. J. Clark, R. Rosanske, R. K. Sheline, *J. Chem. Phys.*, **1974**, *60*, 2997.)

13.27 Evidence has been reported for two isomers of $Ru(CO)_2(PEt_2)_3$, one with the carbonyls occupying axial positions in a trigonal-bipyramidal structure and the other with the carbonyls occupying equatorial positions. Could infrared spectroscopy distinguish between these isomers? How many carbon–oxygen stretching vibrations would be expected for each? (See M. Ogasawara, F. Maseras, N. Gallego-Planas, W. E. Streib, O. Eisenstein, K. G. Caulton, *Inorg. Chem.*, **1996**, *35*, 7468.)

13.28 Account for the observation that $[Co(CO)_3(PPh_3)_2]^+$ has only a single carbonyl stretching frequency.

13.29 In addition to the hexacarbonyl complexes shown in Section 13.4.1, the ion $[Ir(CO)_6]^{3+}$ has been reported. Predict the position of the carbonyl stretching vibration in this complex. (See C. Bach, H. Willner, C. Wang, S. J. Rettig, J. Trotter, F. Aubke, *Angew. Chem., Int. Ed.*, **1996**, *35*, 1974.)

13.30 Pathways to a variety of homoleptic transition-metal carbonyl cations have now been developed.

a. Three such cations are $[Hg(CO)_2]^{2+}$, $[Pt(CO)_4]^{2+}$ and $[Os(CO)_6]^{2+}$. Predict which of these has the lowest energy carbon–oxygen stretching vibration in the infrared. (See H. Willner, F. Aubke, *Angew. Chem., Int. Ed.*, **1997**, *36*, 2402.)

b. The cation $[\{Pt(CO)_3\}_2]^{2+}$ is believed to have the structure shown, with D_{2d} symmetry. Predict the number of carbon–oxygen stretching bands observable in the infrared for this ion, and predict the approximate region in the spectrum where these bands might be observed.

13.31 The Raman spectrum of $Mo(CO)_6$ in pyridine has bands at 2119 and 2015 cm^{-1}. Microwave irradiation of this solution yielded three products with the following Raman bands:

Compound **J**: 2071, 1981 cm^{-1}
Compound **K**: 1892 cm^{-1}
Compound **L**: 1600 cm^{-1}

a. Determine the irreducible representations matching the Raman-active bands of $Mo(CO)_6$.

b. On the basis of the Raman data, propose structures of **J**, **K**, and **L**. (See T. M. Barnard, N. E. Leadbeater, *Chem. Commun.*, **2006**, 3615.)

13.32 One of the first thionitrosyl complexes to be reported was $(\eta^5\text{-}C_5H_5)Cr(CO)_2(NS)$. This compound has carbonyl bands at 1962 and 2033 cm^{-1}. The corresponding bands for $(\eta^5\text{-}C_5H_5)Cr(CO)_2(NO)$ are at 1955 and 2028 cm^{-1}. On the basis of the IR evidence, is NS behaving as a stronger or weaker π acceptor in these compounds? Explain briefly. (See T. J. Greenough, B. W. S. Kolthammer, P. Legzdins, J. Trotter, *Chem. Commun.*, **1978**, 1036.)

13.33 The ^{14}N and ^{15}N derivatives of $TpOs(NS)Cl_2$ [Tp = hydrotris(1-pyrazolyl)borate, a tridentate ligand] have been prepared. The ^{14}N derivative has a nitrogen–sulfur stretch at 1284 cm^{-1}. Predict the N—S stretch for the ^{15}N derivative. (See T. J. Crevier, S. Lovell, J. M. Mayer, A. L. Rheingold, I. A. Guzei, *J. Am. Chem. Soc.*, **1998**, *120*, 6607.)

13.34 The compound $[Ru(CO)_6][Sb_2F_{11}]_2$ has a strong infrared band at 2199 cm^{-1}. The spectrum of $[Ru(^{13}CO)_6][Sb_2F_{11}]_2$ has also been reported. Would you expect the band observed at 2199 cm^{-1} for the ^{12}C compound to be shifted to higher or lower energy for the analogous ^{13}C compound? (See C. Wang, B. Bley, G. Balzer-Jöllenbeck, A. R. Lewis, S. C. Siu, H. Willner, F. Aubke, *Chem. Commun.*, **1995**, 2071.)

13.35 Predict the products of the following reactions:

a. $Mo(CO)_6 + Ph_2P{-}CH_2{-}PPh_2 \xrightarrow{\Delta}$

b. $(\eta^5\text{-}C_5H_5)(\eta^1\text{-}C_3H_5)Fe(CO)_2 \xrightarrow{h\nu}$

c. $(\eta^5\text{-}C_5Me_5)Rh(CO)_2 \xrightarrow{\Delta}$ (dimeric product, contains one CO per metal)

d. $V(CO)_6 + NO \longrightarrow$

e. $W(CO)_5[C(C_6H_5)(OC_2H_5)] + BF_3 \longrightarrow$

f. $[(\eta^5\text{-}C_5H_5)Fe(CO)_2]_2 + Al(C_2H_5)_3 \longrightarrow$

13.36 Complexes of formula $Rh(CO)(phosphine)_2Cl$ have the C—O stretching bands shown below. Match the infrared bands with the appropriate phosphine.
Phosphines: $P(p\text{-}C_6H_4F)_3$, $P(p\text{-}C_6H_4Me)_3$, $P(t\text{-}C_4H_9)_3$, $P(C_6F_5)_3$; $\nu(CO)$, cm^{-1}: 1923, 1965, 1984, 2004

13.37 For each of the following sets, which complex would be expected to have the highest C—O stretching frequency?

a.	$Fe(CO)_5$	$Fe(CO)_4(PF_3)$	$Fe(CO)_4(PCl_3)$	$Fe(CO)_4(PMe_3)$
b.	$[Re(CO)_6]^+$	$W(CO)_6$	$[Ta(CO)_6]^-$	
c.	$Mo(CO)_3(PCl_3)_3$	$Mo(CO)_3(PCl_2Ph)_3$	$Mo(CO)_3(PPh_3)_3$	$Mo(CO)_3py_3$ (py = pyridine)

13.38 When $Cr(CO)_5(PH_3)$ and $Cr(CO)_5(NH_3)$ are oxidized by one electron:

a. What is the change in LFSE?

b. Does the C—O distance increase or decrease? Explain briefly.

c. When $Cr(CO)_5(PH_3)$ is oxidized, what is the effect on the Cr—P distance? Explain briefly.

d. When $Cr(CO)_5(NH_3)$ is oxidized, what is the effect on the Cr—N distance? Explain briefly. (T. Leyssens, D. Peeters, A. G. Orpen, J. N. Harvey, *New J. Chem.*, **2005**, *29*, 1424.)

13.39 Arrange the following complexes in order of the expected frequency of their $\nu(CO)$ bands. (See M. F. Ernst, D. M. Roddick, *Inorg. Chem.*, **1989**, *28*, 1624.)

$Mo(CO)_4(F_2PCH_2CH_2PF_2)$

$Mo(CO)_4[(C_6F_5)_2PCH_2CH_2P(C_6F_5)_2]$

$Mo(CO)_4(Et_2PCH_2CH_2PEt_2)$ (Et = C_2H_5)

$Mo(CO)_4(Ph_2PCH_2CH_2PPh_2)$ (Ph = C_6H_5)

$Mo(CO)_4[(C_2F_5)PCH_2CH_2P(C_2H_5)]$

13.40 Free N_2 has a stretching vibration (not observable by IR; why?) at 2331 cm^{-1}. Would you expect the stretching vibration for coordinated N_2 to be at higher or lower energy? Explain briefly.

13.41 The ^1H NMR spectrum of the carbene complex shown below shows two peaks of equal intensity at 40 °C. At –40 °C, the NMR shows four peaks, two of a lower intensity and two of a higher intensity. The solution may be warmed and cooled repeatedly without changing the NMR properties at these temperatures. Account for this NMR behavior.

$(OC)_5Cr{=}C(OCH_3)(CH_3)$

13.42 The ^1H NMR spectrum of $(C_5H_5)_2Fe(CO)_2$ shows two peaks of equal area at room temperature but has four resonances of relative intensity 5:2:2:1 at low temperatures. Explain. (See C. H. Campbell, M. L. H. Green, *J. Chem. Soc., A*, **1970**, 1318.)

13.43 Of the compounds $Cr(CO)_5(PF_3)$ and $Cr(CO)_5(PCl_3)$, which would you expect to have

a. The shorter C—O bonds?

b. The higher energy Cr—C stretching bands in the infrared spectrum?

13.44 Select the best choice for each of the following:

a. Higher N—O stretching frequency:
$[Fe(NO)(mnt)_2]^-$ $[Fe(NO)(mnt)_2]^{2-}$

$mnt^{2-} = [S_2C_2(CN)_2]^{2-}$

b. Longest N—N bond:
N_2
$(CO)_5Cr{:}N{\equiv}N$
$(CO)_5Cr{:}N{\equiv}N{:}Cr(CO)_5$

c. Shorter Ta—C distance in $(\eta^5\text{-}C_5H_5)_2Ta(CH_2)(CH_3)$:
Ta—CH_2
Ta—CH_3

d. Shortest Cr—C distance:
$Cr(CO)_6$
Cr—CO in *trans*-$Cr(CO)_4I(CCH_3)$
Cr—CCH_3 in *trans*-$Cr(CO)_4I(CCH_3)$

e. Lowest C—O stretching frequency:
$Ni(CO)_4$
$[Co(CO)_4]^-$
$[Fe(CO)_4]^{2-}$

13.45 The dimanganese complex shown can be oxidized reversibly by 1–3 electrons. X-ray crystal structures for the neutral complex and the ions having charges of 1+ and 2+ have provided the following bond distances for the atoms in the MnCCMn chain:

Complex	Mn—C(Å)	C—C(Å)
MCCM	1.872	1.271
[MCCM]$^+$	1.800	1.291
[MCCM]$^{2+}$	1.733	1.325

Account for the trends in Mn—C and C—C distances. Be sure to take into account the relevant orbitals. (See S. Kheradmandan, K. Venkatesan, O. Blacque, H. W. Schmalle, H. Berke, *Chem–Eur. J.*, **2004**, *10*, 4872.)

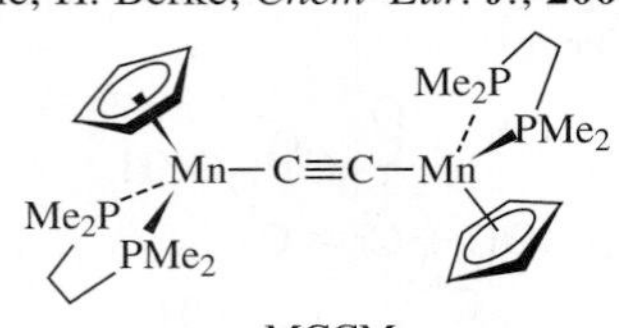

MCCM

13.46 Use the Covalent Bond Classification Method to classify these complexes with a $[ML_lX_xZ_z]^{Q\pm}$ formula. With charged species, reduce the formula to its equivalent neutral class.

a. $W(CH_3)_6$
b. $[W(\eta^5\text{-}C_5H_5)_2H_3]^+$
c. $[Fe(CO)_4]^{2-}$
d. $Mo(\eta^5\text{-}C_5H_5)(CO)_3H$
e. $Fe(CO)_4I_2$
f. $[Co(\eta^5\text{-}C_5H_5)_2]^+$

13.47 Use the Covalent Bond Classification Method to classify these hypothesized complexes with a $[ML_lX_xZ_z]^{Q\pm}$ formula. With charged species, reduce the formula to its equivalent neutral class. On the basis on the MLX plot for iron in Section 13.7, speculate on the likelihood that the following iron complexes have been prepared.

a. $[FeCl_3(PPh_3)_3]^{3+}$ **c.** $[FeCl_2(PPh_3)_2]^+$
b. $[FeCl_3(PPh_3)_2]^+$ **d.** $FeCl_2(PPh_3)_2$

13.48 A solution of blue $Mo(CO)_2(PEt_3)_2Br_2$ was treated with a tenfold excess of 2-butyne to give **X**, a dark green product. **X** had bands in the ^{1}H NMR at δ 0.90 (relative area = 3), 1.63 (2), and 3.16 (1). The peak at 3.16 was a singlet at room temperature but split into two peaks at temperatures below –20 °C. ^{31}P NMR showed only a single resonance. IR showed a single strong band at 1950 cm^{-1}. Molecular weight determinations suggest that **X** has a molecular weight of 580 ± 15. Suggest a structure for **X**, and account for as much of the data as possible. (See P. B. Winston, S. J. Nieter Burgmayer, J. L. Templeton, *Organometallics*, **1983**, *2*, 168.)

13.49 Photolysis at –78 °C of $[(\eta^5\text{-}C_5H_5)Fe(CO)_2]_2$ results in the loss of a colorless gas and the formation of an iron-containing product having a single carbonyl band at 1785 cm^{-1} and containing 14.7 percent oxygen by mass. Suggest a structure for the product.

13.50 Nickel carbonyl reacts with cyclopentadiene to produce a red diamagnetic compound of formula $NiC_{10}H_{12}$. The ^{1}H NMR spectrum of this compound shows four different types of hydrogen; integration gives relative areas of 5:4:2:1, with the most intense peak in the aromatic region. Suggest a structure of $NiC_{10}H_{12}$ that is consistent with this NMR spectrum

13.51 The carbonyl carbon–molybdenum–carbon angle in $Cp(CO)_2Mo[\mu\text{-}S_2C_2(CF_3)_2]_2MoCp$ ($Cp = \eta^5\text{-}C_5H_5$) is 76.05°. Calculate the ratio of intensities $I_{symmetric}/I_{antisymmetric}$ expected for the C—O stretching bands of this compound. (See K. Roesselet, K. E. Doan, S. D. Johnson, P. Nicholls, G. L. Miessler, R. Kroeker, S. H. Wheeler, *Organometallics*, **1987**, *6*, 480.)

13.52 The complex $(\pi\text{-}C_4BNH_6)Cr(CO)_3$ has recently been reported, the first example of the 1,2-dihydro-1,2-azaborine ligand. It has strong absorptions at 1898 and 1975 cm^{-1}, in comparison with 1892 and 1972 cm^{-1} for $(\pi\text{-}C_6H_6)Cr(CO)_3$. The C_4BNH_6 complex has carbon–carbon distances of 1.393, 1.421, and 1.374 Å, in order, in the ring.

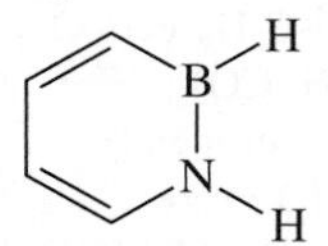

1,2-dihydro-1,2-azaborine

a. Is the C_4BNH_6 ligand a (slightly) stronger or weaker acceptor than C_6H_6? Explain.

b. Account for the differences in the C—C distances in the ring. (See A. J. V. Marwitz, M. H. Matus, L. N. Zakharov, D. A. Dixon, S.-Y. Liu, *Angew. Chem., Int. Ed.*, **2009**, *48*, 973.)

13.53 Reaction of Ir complex **A** with C_{60} gave a black solid residue **B** with the following spectral characteristics: mass spectrum: M$^+$ = 1056; ^{1}H NMR: δ 7.65 ppm (multiplet, 2H), 7.48 (multiplet, 2H), 6.89 (triplet, 1H), and 5.97 (doublet, 2H); IR: ν(CO) = 1998 cm^{-1}.

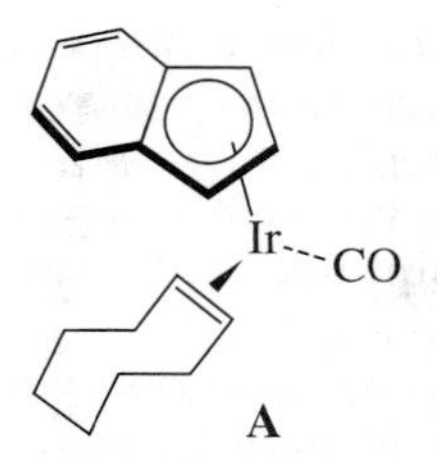

a. Propose a structure for **B**.

b. The carbonyl stretch of **A** was reported at 1954 cm^{-1}. How does the electron density at Ir change in going from **A** to **B**?

c. When **B** was treated with PPh_3, a new complex **C** formed rapidly, along with some C_{60}. What is a likely structure of **C**? (See R. S. Koefod, M. F. Hudgens, J. R. Shapley, *J. Am. Chem. Soc.*, **1991**, *113*, 8957.)

13.54 Reaction of $Cr(CO)_3(CH_3CN)_3$ with $NaCp^N$ (shown) in THF, followed by reaction with the tetrameric complex $[Cu(PPh_3)Cl]_4$, yields a yellow product having the following characteristics: IR (in THF): strong absorbances at 1906, 1808, 1773 cm^{-1}; ^1H NMR: δ 7.50 – 7.27 (multiplet, 15 H), 4.64 (apparent triplet, 2H), 4.52 (apparent triplet, 2H), 2.44 – 2.39 (multiplet, 4H), 2.19 (singlet, 6H); Elemental analysis: 60.29 % C, 4.82 % H, 2.40 % N by mass.

$NaCp^N$: Na^+ [cyclopentadienide–CH_2CH_2–NMe_2]

Propose a structure of this product. (See P. J. Fischer, A. P. Heerboth, Z. R. Herm, B. E. Kucera, *Organometallics*, **2007**, *26*, 6669.)

13.55 Although it is a simple molecule, diisocyanomethane, $H_2C(NC)_2$, was not isolated and characterized by X-ray crystallography until rather recently. This compound has been used in organometallic synthesis as follows: $(\eta^5\text{-}C_5H_5)Mn(CO)_3$ was dissolved in tetrahydrofuran and photolyzed, liberating some CO and forming compound **Q**. At –40 °C, a solution of $H_2C(NC)_2$ in dichloromethane was added to a solution of **Q**. Column chromatography of the resulting solution led to the isolation of compound **R**, which had the following characteristics: ^1H NMR (in CD_2Cl_2): δ 4.71 (relative area 5) 5.01, (2); ^{13}C NMR: δ 50.1, 83.4, 162.0, 210.5, 228.1; IR: 2147, 2086, 2010, 1903 cm^{-1}. Suggest structures for **Q** and **R**. (See J. Buschmann, R. Bartolmäs, D. Lentz, P. Luger, I. Neubert, M. Röttger, *Angew. Chem., Intl. Ed.*, **1997**, *36*, 2372.)

13.56 In solution, the cyclopentadienyl tricarbonyl dimers $[CpMo(CO)_3]_2$ and $[CpW(CO)_3]_2$ react to form the heterobimetallic complex $Cp(CO)_3Mo{-}W(CO)_3Cp$. However, the reaction does not go to completion; a mixture results in which the $[CpMo(CO)_3]_2$ and $[CpW(CO)_3]_2$ are in equilibrium with the mixed metal compound. The abundance of the three organometallic complexes is governed statistically by the number of $CpMo(CO)_3$ and $CpW(CO)_3$ fragments present. If 0.00100 mmol of $[CpMo(CO)_3]_2$ and 0.00200 mmol of $[CpW(CO)_3]_2$ are dissolved in toluene until equilibrium is achieved, calculate the amounts of the three organometallic complexes in the equilibrium solution. (See T. Madach, H. Vahrenkamp, *Z. Naturforsch.*, **1979**, *34b*, 573.)

13.57 Addition of $BH_3 \cdot THF$ to a slurry of $[K(15\text{-crown-}5)_2]_2\,[Ti(CO)_6]$ in tetrahydrofuran at –60 °C yielded an air-sensitive red solution from which an anion **Z** was isolated. **Z** had strong infrared peaks at 1945 and 1811 cm^{-1}. (Note: $[Ti(CO)_4(\eta^5\text{-}C_5H_5)]^-$ has bands at 1921 and 1779 cm^{-1}.) Other peaks are observed at 2495, 2132, and 2058 cm^{-1}; these are in the spectral region where B—H stretches commonly occur. The ^1H NMR spectrum at –95 °C showed broad singlets of relative intensity 3:1; these peaks became more complex at higher temperature. Propose a formula and structure for **Z**. (See P. J. Fischer, V. G. Young, Jr., J. E. Ellis, *Angew. Chem., Int. Ed.*, **2000**, *39*, 189.)

13.58 When a solution containing $(C_5Me_5)_2Os_2Br_4$ and $LiAlH_4$ is stirred at low temperature in diethyl ether, followed by addition of a small amount of ethanol, a white product is formed that can be isolated by sublimation. Analysis of the white product, an 18-electron complex, provided the following information:

Elemental analysis: 36.6 % C, 6.07 % H

^1H NMR: δ 2.02 (s, relative area-3), –11.00 (s, relative area-1)

^{13}C NMR: δ 2.02 (s), 94.2 (s)

IR: the region where Os—H stretches most likely occur is shown.

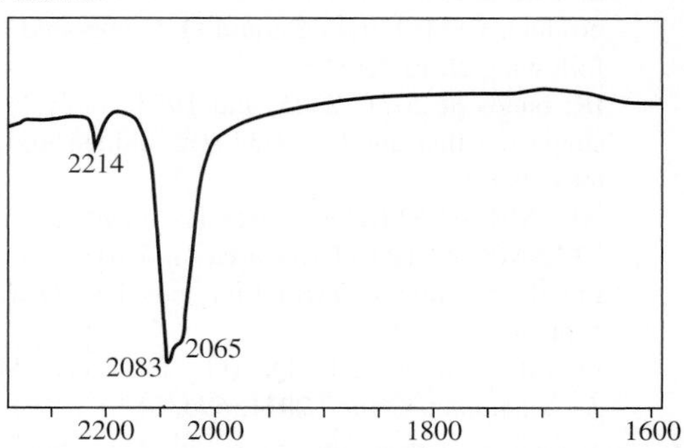

Propose the structure of the product, and account for as much of the data as possible. (See C. L. Gross, G. S. Girolami, *Organometallics*, **2007**, *26*, 160.)

13.59 Although thousands of examples of carbonyl complexes are known, the chemistry of the fluoroborylene (BF) ligand is still in its infancy.

a. Prepare a molecular orbital energy level diagram of BF, showing how the atomic orbitals of B and F interact.

b. On the basis of your answer to part **a**, would you expect BF to be a stronger or weaker π-acceptor ligand than CO? Explain.

c. The first example of a transition metal complex containing a bridging BF has recently been reported; its structure is shown below. (D. Vidovic, S. Aldridge, *Angew. Chem., Int. Ed.*, **2009**, *48*, 3669). The long-known dimeric Ru complex $[(\eta^5\text{-}C_5H_5)Ru(CO)_2]_2$ has C—O vibration bands at 1939 and 1971 cm^{-1}. Do you predict that the molecule shown below is likely to have C—O bands at higher or lower energy than $[(\eta^5\text{-}C_5H_5)Ru(CO)_2]_2$? Explain your reasoning.

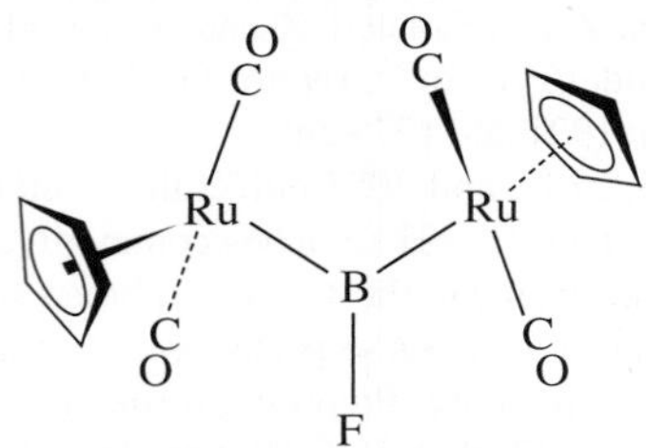

13.60 When hexaiodobenzene reacts with diferrocenylzinc, $[(\eta^5\text{-}C_5H_5)FeC_5H_4]_2Zn$, the most interesting product is obtained in, alas, only 4% yield. This compound contains no elements other than those that are in ferrocene. Moreover, its formula has a C/Fe atom ratio exactly ten percent higher than in ferrocene and an H/Fe atom ratio ten percent lower than in ferrocene. Predict (and sketch) this product. Why is it interesting? (Y. Yu, A. D. Bond, P. W. Leonard, U. J. Lorenz, T. V. Timofeeva, K. P. C. Vollhardt, G. D. Whitener, A. A. Yakovenko, *Chem. Commun.*, **2006**, 2572.)

13.61 When butadiene was added to a suspension of $Fe_2(CO)_9$ in *n*-hexane, a reaction occurred which, on isolation of product, yielded orange liquid **Q**. Compound **Q** had the following characteristics:

IR: bands at 2071, 2005, and 1975 cm^{-1}; calculations suggested that another band was hidden under one of these bands.

^1H NMR: δ3.52 (relative area 1), δ 2.90 (1), δ 2.76 (1)

^{13}C NMR: δ 212 (relative area 4), δ 70 (1), δ 36 (1)

Finally, reaction of **Q** with PPh_3 gave $Fe(CO)_4(PPh_3)$ (Ph = phenyl)

Identify and sketch **Q**. (G. J. Reiss, M. Finze, *J. Organomet. Chem.*, **2011**, *696*, 512.)

Use molecular modeling software for the following problems:

13.62 **a.** Generate and display the π and π^* orbitals of the cyclopentadienyl group, C_5H_5. Compare the results with the diagrams in Figure 13.22. Identify the nodes that cut through the plane of the atoms.

b. Generate and display the molecular orbitals of ferrocene. Identify the molecular orbitals that result from interactions of π orbitals of the C_5H_5 ligands with iron. Locate orbitals that show interactions of metal d_{xz} and d_{yz} with the ligands, and compare with the diagrams preceding Figure 13.27.

c. Compare the relative energies of the molecular orbitals with the molecular orbitals shown in Figure 13.28. How do the relative energies of the orbitals in the box in Figure 13.28 compare with the results of your calculations?

d. The d_{z^2} orbital is often described as essentially nonbonding in ferrocene. Do your results support this designation?

13.63 Generate and display the molecular orbitals of the anion of Zeise's salt, shown in Figure 13.4. Identify the molecular orbitals that show how the π and π^* orbitals of ethylene interact with orbitals on platinum.

13.64 Generate and display the molecular orbitals of $Cr(CO)_6$. Identify the t_{2g} and e_g^* orbitals (Figure 13.8). Verify that there are three equivalent and degenerate t_{2g} orbitals and that there are two degenerate e_g^* orbitals. Why do the e_g^* orbitals have different shapes?

13.65 Suppose a transition metal is bonded to a *cyclo*-C_3H_3 ligand as shown below.

a. Sketch each of the π molecular orbitals of $\eta^3\text{-}C_3H_3$.

b. For each of these orbitals, determine which *s*, *p*, and *d* orbitals of the metal are of appropriate symmetry for interaction.

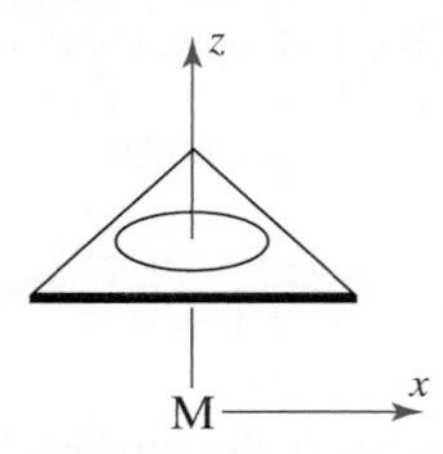

c. Using molecular modeling software, generate the molecular orbitals of *cyclo*-C_3H_3, and compare the π orbitals with your sketches in **a**.

d. Generate and display the molecular orbitals of the theoretical sandwich complex $[(\eta^3\text{-}C_3H_3)_2Ni]^{2-}$. Observe the orbitals showing overlap between lobes of the rings and the central Ni, and compare these with the orbitals you indicated for metal–ring interactions in **b**.

13.66 $(\eta^4\text{-}C_4H_4)_2Ni$ has apparently not been synthesized. However, calculations predict reasonable stability and eclipsed geometry for this molecule, with the rings parallel. (See Q. Li, J. Guan, *J. Phys. Chem. A*, **2003**, *107*, 8584.)

a. Sketch the group orbitals, using the π orbitals of the cyclobutadiene rings.

b. For each of the group orbitals, identify the Ni orbitals of suitable symmetry for interaction.

c. Using the *p* orbitals of the ligands as a basis, construct a (reducible) representation. Reduce it to its irreducible components, and match these with appropriate orbitals on the central Ni.

d. Generate and display the molecular orbitals of $(\eta^4\text{-}C_4H_4)_2Ni$. Observe the orbitals showing overlap between lobes of the rings and the central Ni, and compare with your results from **c**.

Chapter 14

Organometallic Reactions and Catalysis

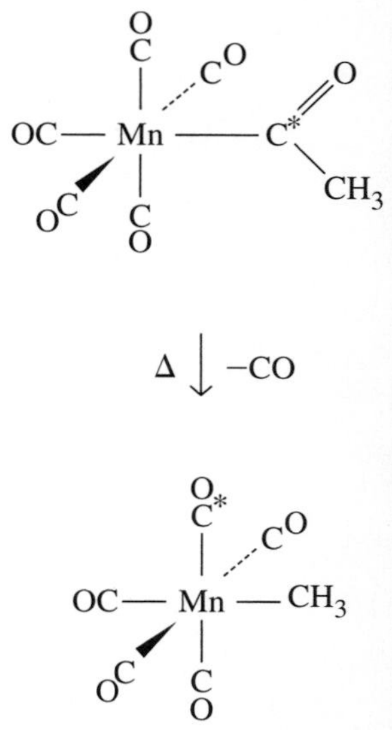

Organometallic compounds undergo a rich variety of reactions that can sometimes be combined into useful catalytic cycles. In this chapter, we will discuss organometallic reactions of particular importance for synthetic and catalytic processes.

14.1 Reactions Involving Gain or Loss of Ligands

Many reactions of organometallic compounds involve a change in metal coordination number by a gain or loss of ligands. If the oxidation state of the metal is retained, these reactions are considered addition or dissociation reactions; if the metal oxidation state is changed, they are termed oxidative additions or reductive eliminations. In classifying these reactions, it is often necessary to determine oxidation states of the metals; the donor pair method (Chapter 13) can be used.

Type of Reaction	Change in Coordination Number	Change in Formal Oxidation State of Metal
Addition	Increase	None
Dissociation	Decrease	None
Oxidative addition	Increase	Increase
Reductive elimination	Decrease	Decrease

We will first consider ligand dissociation reactions. These reactions provide an avenue to replace ligands such as carbon monoxide and phosphines.

14.1.1 Ligand Dissociation and Substitution

CO Dissociation

Chapter 13 introduced carbonyl dissociation reactions, in which CO may be lost thermally or photochemically. Such a reaction may result in rearrangement of the remaining molecule or replacement of CO by another ligand:

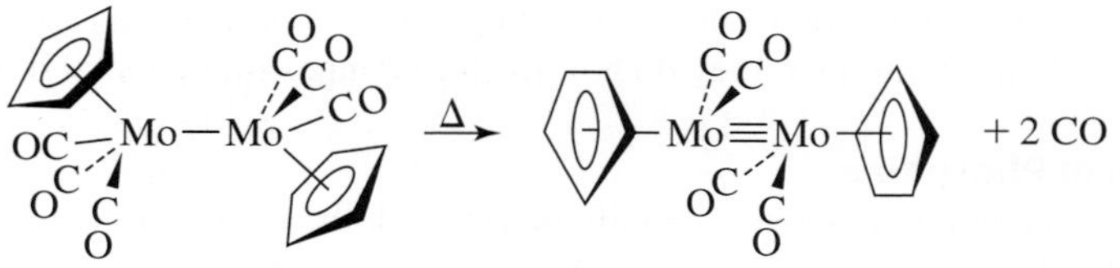

$$Fe(CO)_5 + P(CH_3)_3 \xrightarrow{\Delta} Fe(CO)_4(P(CH_3)_3) + CO$$

The second type of reaction shown above, involving ligand replacement, is an important way to introduce new ligands into complexes. Most thermal reactions involving replacement of CO by another ligand, L, have rates that are independent of the concentration of L; they are first order with respect to the metal complex. This behavior is consistent with a **dissociative** mechanism involving slow loss of CO, followed by rapid reaction with L:

$$\underset{18\,e^-}{Ni(CO)_4} \xrightarrow{k_1} \underset{16\,e^-}{Ni(CO)_3} + CO \quad \text{(slow)} \quad \text{loss of CO from 18-electron complex}$$

$$\underset{16\,e^-}{Ni(CO)_3} + L \longrightarrow \underset{18\,e^-}{Ni(CO)_3L} \quad \text{(fast)} \quad \text{addition of L to 16-electron intermediate}$$

The first step is rate limiting, and has the rate law Rate $= k_1[Ni(CO)_4]$. Some ligand replacement reactions show more complicated kinetics. Study of the reaction

$$Mo(CO)_6 + L \xrightarrow{\Delta} Mo(CO)_5L + CO \ (L = \text{phosphine})$$

has shown that, for some phosphines, the rate law is:

$$\text{Rate} = k_1[Mo(CO)_6] + k_2[Mo(CO)_6][L]$$

The two terms in the rate law imply parallel pathways for the formation of $Mo(CO)_5L$. The first term is consistent with a dissociative mechanism:

$$Mo(CO)_6 \xrightarrow{k_1} Mo(CO)_5 + CO \quad \text{(slow)}$$

$$Mo(CO)_5 + L \longrightarrow Mo(CO)_5L \quad \text{(fast)}$$

$$\text{Rate}_1 = k_1[Mo(CO)_6]$$

There is strong evidence that solvent is involved in the first-order mechanism for the replacement of CO; however, because the solvent is in great excess, it does not appear in the above rate law; this pathway exhibits pseudo–first order kinetics.[1] The second term is consistent with an **associative** process, involving a bimolecular reaction of $Mo(CO)_6$ and L to form a transition state that loses CO:

$$Mo(CO)_6 + L \xrightarrow{k_2} [Mo(CO)_6\text{---}L] \qquad \text{association of } Mo(CO)_6 \text{ and L}$$

$$[Mo(CO)_6\text{----}L] \longrightarrow Mo(CO)_5L + CO \qquad \text{loss of CO from transition state}$$

Formation of the transition state is the rate-limiting step in this mechanism; the rate law for this pathway is

$$\text{Rate}_2 = k_2[Mo(CO)_6][L]$$

The overall rate of formation of $Mo(CO)_5L$ is the sum of the rates of the unimolecular and bimolecular mechanisms, $\text{Rate}_1 + \text{Rate}_2$.

Although most CO substitution reactions proceed by dissociative mechanisms, the likelihood of an associative path increases for reactions involving (1) addition of highly nucleophilic ligands and (2) reactants containing metals of larger atomic radii that can more readily expand their coordination sphere. While ligand dissociation and association involve changes in coordination number, they do not involve changes in the metal oxidation state.*

Dissociation of Phosphine

Ligands other than carbon monoxide can dissociate, with the ease of dissociation a function of the strength of metal–ligand bonding. The metal–ligand bond strength depends on an interplay of electronic effects (for example, the match of energies of the metal and ligand

*Assuming that no oxidation-reduction reaction occurs between the ligand and the metal.

orbitals) and steric effects (for example, the degree to which crowding of ligands around the metal can reduce the strength of metal-ligand orbital overlap). These steric effects have been investigated for many ligands but especially for neutral donor ligands such as phosphines. Tolman defined the **cone angle** as the apex angle, θ, of a cone that encompasses the van der Waals radii of the outermost atoms of a ligand (**Figure 14.1a**).[2] An emerging strategy for classification of ligand steric bulk is determination of the **percent buried volume** ($\%V_{bur}$) from crystallographic data.[3] The $\%V_{bur}$ is defined as the percentage of the potential coordination sphere around the metal occupied by the ligand (Figure 14.1b). An excellent correlation between $\%V_{bur}$ and cone angles has been reported for tertiary phosphines, and the percent buried volume approach quantifies ligand steric bulk for ligands where Tolman's approach has limited efficacy.[4] Cone angles and $\%V_{bur}$ values of selected phosphines are in **Table 14.1**.*

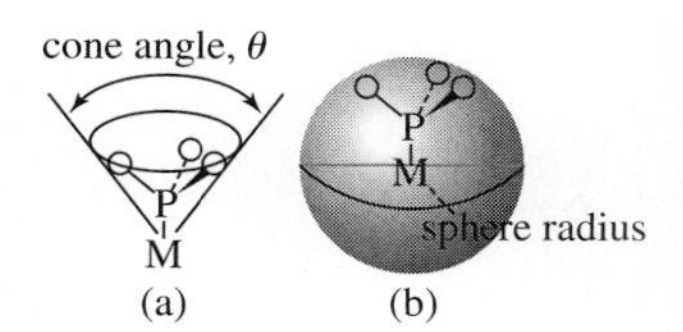

FIGURE 14.1 (a) Cone angle and (b) Percent Buried Volume. In both cases, the M—P distance is defined as 228 pm.

The presence of bulky ligands results in crowding around the metal. This can lead to more rapid ligand dissociation as a consequence of slight elongation of metal–ligand bonds to relieve steric hindrance at the metal center. The relative importance of ligand bulk and electronic effects in influencing phosphine dissocation rates can be difficult to predict and must be considered on a case by case basis. For example, the activity of the ruthenium olefin metathesis catalyst in **Figure 14.2** depends on the rate of phosphine dissocation, which is the initiation step of the catalytic cycle (Section 14.3.6).[5] The rate constants (k_1) were determined for a variety of PR_3 ligands (**Table 14.2**).

TABLE 14.1 Ligand Cone Angles and $\%V_{bur}$ for Tertiary Phosphines

Ligand	Cone Angle θ	$\%V_{bur}$	Ligand	Cone Angle θ	$\%V_{bur}$
PH_3	87°		$P(CH_3)(C_6H_5)_2$	136°	
PF_3	104°		$P(CF_3)_3$	137°	
$P(OCH_3)_3$	107° (128°)[a]	26.4[b]	$P(C_6H_5)_3$	145°	29.6
$P(OC_2H_5)_3$	109°		$P(cyclo\text{-}C_6H_{11})_3$	170°	31.8
$P(CH_3)_3$	118°	22.2	$P(t\text{-}C_4H_9)_3$	182°	36.7
PCl_3	124°		$P(C_6F_5)_3$	184°	37.3
$P(OC_6H_5)_3$	128° (155°)[a]	30.7	$P(o\text{-}C_6H_4CH_3)_3$	194°	41.4
$P(C_2H_5)_3$	132°	27.8	$P(C_6H_2Me_3)_3$ (2,4,6-Me)	212°	47.6

[a]Cone angles suggested by $\%V_{bur}$ analysis. The Tolman cone angles for these phosphites may be underestimates. [b]All of these $\%V_{bur}$ values were determined from structural data of the non-coordinated molecules, except this value, which was determined using crystallographic data from $(P(OCH_3)_3)AuCl$.

FIGURE 14.2 Phosphine dissocation equilibrium that influences the rate of olefin metathesis reactions.

*More recent approaches to measuring the bulk of phosphines and other ligands have included analysis of data in the Cambridge Structural Database (K. A. Bunten, L. Chen, A. L. Fernandez, A. J. Poë, *Coord. Chem. Rev.*, **2002**, *233–234*, 41) and calculations of molecular electrostatic potentials (C. H. Suresh, *Inorg. Chem.*, **2006**, *45*, 4982).

TABLE 14.2 Rates of Phosphine Dissociation and Steric/Electronic Effects

Phosphine	Rate Constant (k_1)(s^{-1}) at 353 K	Cone Angle (°)
PCy_3	0.13	170
$P(n\text{-}Bu)_3$	8.1×10^{-4}	130
$P(C_6H_5)_2(OMe)$	1.7	132
$P(p\text{-}CF_3C_6H_4)_3$	48	145
$P(p\text{-}ClC_6H_4)_3$	17.9	145
$P(p\text{-}FC_6H_4)_3$	8.5	145
$P(C_6H_5)_3$	7.5	145
$P(p\text{-}CH_3C_6H_5)_3$	4.1	145
$P(p\text{-}CH_3OC_6H_5)_3$	1.8	145

The donor capabilities of PCy_3 and $P(n\text{-}Bu)_3$ were deemed sufficiently similar that the 160-fold increase in dissociation rate for the PCy_3 complex was attributed to the greater steric bulk (larger cone angle) of PCy_3. An electronic effect is suggested upon comparison of the rate of phosphine dissociation of PCy_3 relative to $P(C_6H_5)_2(OMe)$; the latter phosphine dissociates *faster* despite its significantly smaller cone angle. The electron withdrawing substituents of $P(C_6H_5)_2(OMe)$ render this phosphine a weaker donor (resulting in a weaker Ru—P bond) relative to PCy_3. This electronic effect on dissociation is further illustrated in Figure 14.2 upon comparison of phosphine dissociation rates for a series of *para*-substituted triphenylphosphines, with identical cone angles (145°). For these phosphines, the dissocation rates increase as the electron-withdrawing capability of the *para* substituent increases, making the phosphine a weaker donor, resulting in weaker Ru—P bonds.

A classic example of ligand steric bulk influencing ligand substitution reaction rates is

$$cis\text{-}Mo(CO)_4L_2 + CO \longrightarrow Mo(CO)_5L + L \qquad (L = \text{phosphine or phosphite})$$

The rate of this reaction, which is first order in *cis*-$Mo(CO)_4L_2$, increases with increasing ligand bulk, as shown in **Figure 14.3**; the larger the cone angle, the more rapidly the phosphine or phosphite is lost.[6] The overall effect is substantial; for example, the rate for the most bulky ligand shown is more than 64000 times greater than that for the least bulky ligand.

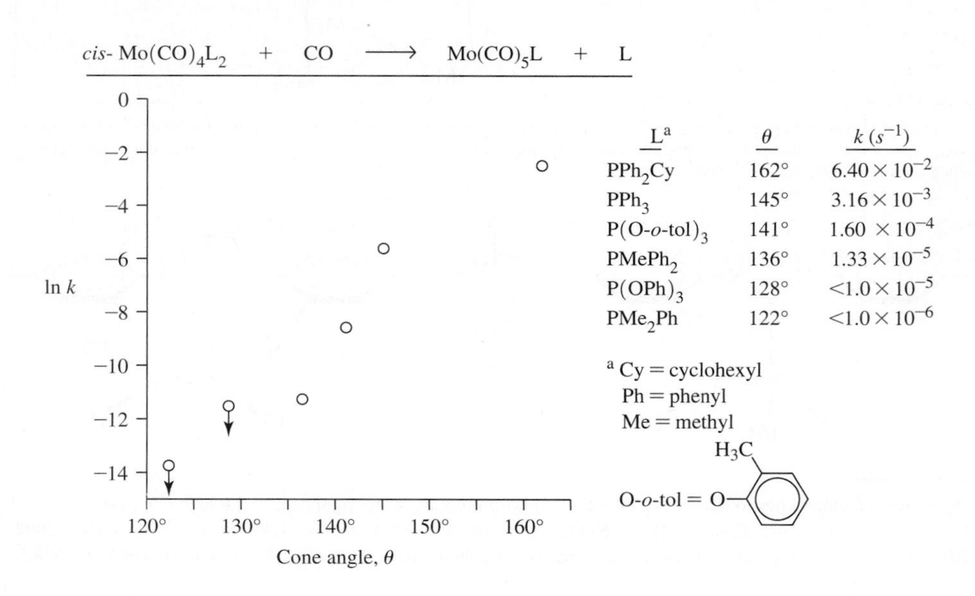

FIGURE 14.3 Reaction Rate Constant versus Cone Angle for Phosphine Dissociation.

14.1.2 Oxidative Addition and C—H Bond Activation

These reactions involve an increase in both the oxidation state and the coordination number of the metal. *Oxidative addition (OA)* reactions are essential steps in many catalytic processes. The reverse reaction, designated *reductive elimination (RE)*, is also very important. These reactions are described schematically by the following:

$$L_nM + X—Y \underset{RE}{\overset{OA}{\rightleftharpoons}} L_nM\langle^{X}_{Y}$$

Type of Reaction	Change in Coordination Number	Change in Formal Oxidation State of Metal	Change in Electron Count
Oxidative addition	Increase by 2	Increase by 2	Increase by 2
Reductive elimination	Decrease by 2	Decrease by 2	Decrease by 2

Oxidative addition reactions of square-planar d^8 complexes are ubiquitous; we will use one such complex, *trans*-$Ir(CO)Cl(PEt_3)_2$, to illustrate these reactions (**Figure 14.4**).

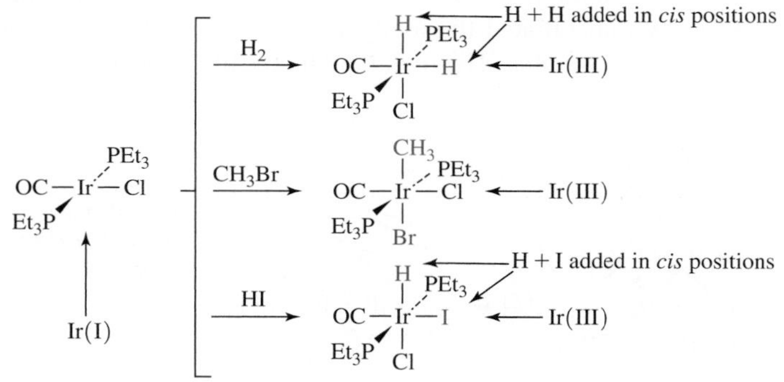

FIGURE 14.4 Examples of Oxidative Addition Reactions.

In Figure 14.4, the iridium oxidation state increases from (I) to (III), and the coordination number increases from 4 to 6. The new ligands add in a *cis* or *trans* fashion, with their orientation a function of the mechanistic pathway. The expansion of the coordination number of the metal brings the newly added ligands into close proximity to the original ligands; this can enable reactions to occur between them. Such reactions are encountered frequently in the mechanisms of catalytic cycles.

Oxidative addition and reductive elimination reactions play key roles in C—H activation reactions, where a strong C—H bond is cleaved by a transition-metal complex.* These are important reactions because they permit unfunctionalized hydrocarbons to be transformed into complex molecules. Bergman reported the following classic C—H reductive elimination/oxidative addition sequence.[7]

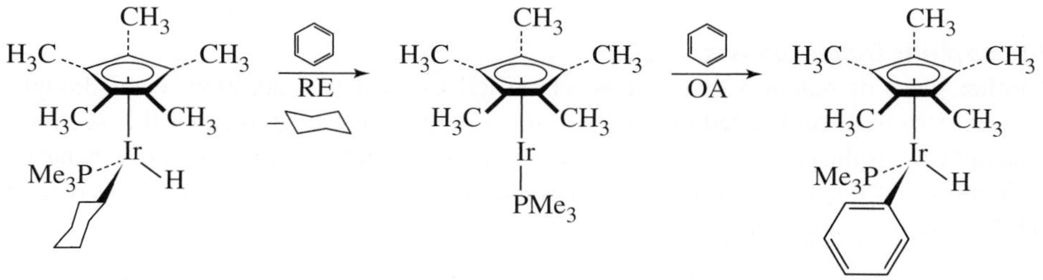

*C—H functionization is the topic in a special issue of *Acc. Chem. Res.*, **2012**, *45*, 777–958.

The first step involves reductive elimination of cyclohexane from a six-coordinate Ir(III) complex (ML_3X_3 via the CBC method in Section 13.7) to afford a four-coordinate Ir(I) intermediate (ML_3X). Like most reductive eliminations, this first step involves a decrease by two in both the oxidation state and coordination number of the metal. The second step results in oxidation from Ir(I) to Ir(III) as benzene oxidatively adds to the iridium center, resulting in activation of a C(sp^2)—H bond. Like most oxidative additions, this second step involves an increase by two in both the oxidation state and the coordination number of the metal.

Cyclometallations

These are reactions that incorporate metals into organic rings. *Orthometallations*, oxidative additions in which an *ortho* position of an aromatic ring becomes attached to the metal, are quite common. The first **Figure 14.5** example features oxidative addition in which the *ortho* carbon to the phosphorus of a triphenylphosphine ligand and the hydrogen originally in the *ortho* position add to iridium, resulting in a four-membered ring involving the iridium center. Orthometallation is not the favored pathway in the second reaction even though the platinum reactant also has triphenylphosphine ligands. A five-membered ring is formed instead by oxidative addition of a C—H bond of the naphthalene substituent, concurrently with reductive elimination of methane. In these reactions, the reactants are kinetic products that are converted to cyclometallated thermodynamic products. These reactions are also classified as C—H bond activations. The metal facilitates the cleavage of C—H bonds.

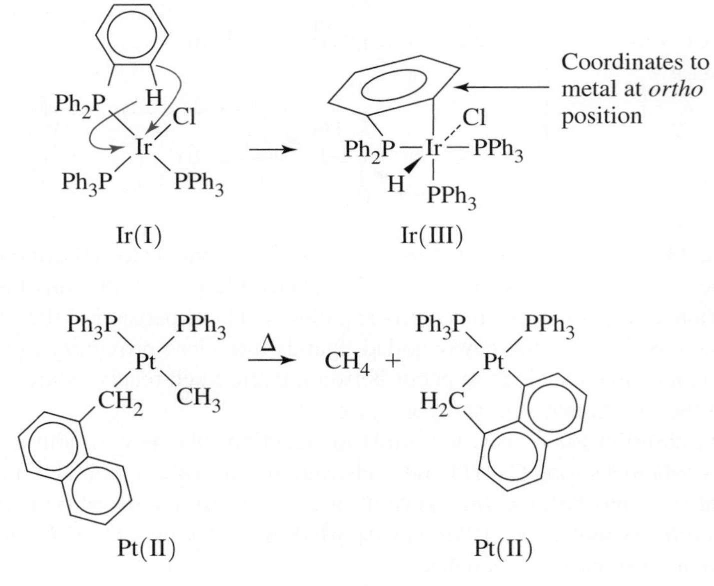

FIGURE 14.5 Cyclometallation Reactions. The first reaction is an orthometallation where the iridium coordinates to the carbon *ortho* to the phosphorus atom.

Nucleophilic Displacement

Another class of reaction that can be classified as oxidative addition, even though it does not strictly meet the requirements of the scheme at the beginning of this section, is nucleophilic displacement. Negatively charged organometallic complexes often behave as nucleophiles in displacement reactions. For example, $[(\eta^5\text{-}C_5H_5)Mo(CO)_3]^-$ can displace iodide from methyl iodide:

$$[(\eta^5\text{-}C_5H_5)Mo(CO)_3]^- + CH_3I \longrightarrow (\eta^5\text{-}C_5H_5)(CH_3)Mo(CO)_3 + I^-$$

This reaction results in formal oxidation of the metal (from Mo(0) to Mo(II)) and the coordination number increases by 1 (by the CBC method an $[ML_5X]^-$ complex (equivalent neutral class is ML_6) changes to an ML_5X_2 complex) as the iodide is expelled.

14.1.3 Reductive Elimination and Pd-Catalyzed Cross-Coupling

Reductive elimination is the reverse of oxidative addition. To illustrate this distinction, consider the following equilibrium:

$$\underset{\text{Ta(III)}}{(\eta^5\text{-}C_5H_5)_2TaH} + H_2 \underset{RE}{\overset{OA}{\rightleftharpoons}} \underset{\text{Ta(V)}}{(\eta^5\text{-}C_5H_5)_2TaH_3}$$

The forward reaction involves formal oxidation of the metal, accompanied by an increase in coordination number; it is an OA. The reverse reaction is an example of RE, which involves a decrease in both oxidation number and coordination number.

Reductive elimination can result in the formation of a wide range of bonds including H—H, C—H, C—C, and C—X (X = halide, amide, alkoxide, thiolate, and phosphide). Reductive eliminations from palladium are of great interest because they comprise an important step in Pd-catalyzed cross-coupling; **Figure 14.6** provides a generic catalytic cycle. Step 1 is oxidative addition of RX (R = alkyl, aryl) to a Pd(0) reactant, resulting in a square planar Pd(II) complex. If this product is *cis* (as shown), rapid *cis–trans* isomerization (step 2) is likely to avoid having a strong donor R *trans* to another good σ donor (L).* Step 3 is *transmetallation*, a ligand substitution reaction, that introduces R′ (R′ = alkyl, aryl) to the Pd(II) center.** Reductive elimination of a C—C bond (step 5) requires the *cis* orientation of R and R′, so a *cis–trans* isomerization is necessary (step 4) if the transmetallation product is *trans* (as shown).

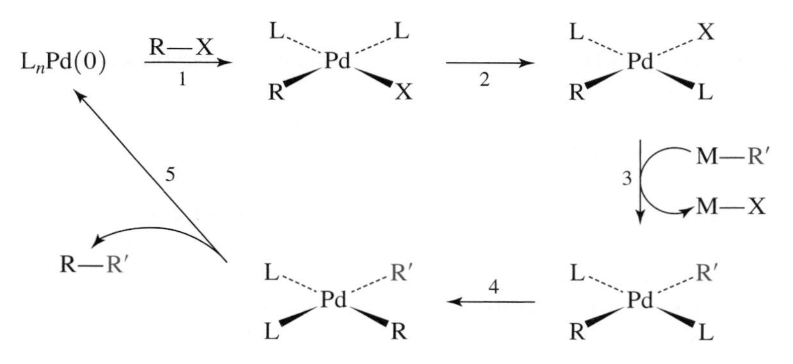

FIGURE 14.6 A generic Pd-catalyzed cross-coupling cycle.

The reductive elimination reaction (step 5) generally proceeds faster as the steric bulk imposed by L increases. The greater the steric bulk of the ligands, the greater the relief in hindrance upon the decrease in coordination number, and the faster the rate. A series of C—S bond-forming reductive elimination reactions with bidentate ligands (**Figure 14.7**) with different **bite angles** illustrates this trend.[8] Bite angles are steric parameters (similar to cone angle) specific to bidentate ligands.† As the bite angle increases, the bidentate ligand imposes increasing steric hindrance when coordinated. The rate of dialkylsulfide reductive elimination increases with the bite angle (**Table 14.3**).

---

*This is a consequence of the *trans* effect (Section 12.7).

**This scheme is extremely general. The identity of M distinguishes alternate Pd-catalyzed cross-coupling reactions, with varying mechanisms and R′substituents. The cross-coupling mechanisms associated with Stille, Suzuki, Sonogashira, and Negishi are associated with transmetallation reactions that employ M = Sn, B, Cu, and Zn, respectively.

†For the definition of bite angle, see C. P. Casey, G. T. Whiteker, *Isr. J. Chem.*, **1990**, *30*, 299.

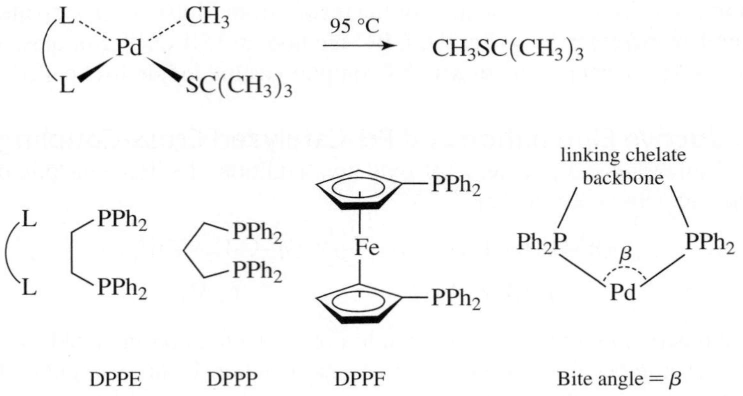

FIGURE 14.7 Reductive elimination of C—S bonds via Pd(II) complexes with bidentate phosphines with varying bite angles.

TABLE 14.3 Rates of Reductive Elimination with $(L_2)Pd(CH_3)(SC(CH_3)_3)$[a]

Phosphine (L_2)	Rate Constant (k_{obs}) at 95 °C (h^{-1})	Bite Angle (°)[b]
DPPE	0.069	85.8
DPPP	0.35	90.6
DPPF	1.4	99.1

[a]These reactions were carried out in the presence of excess triphenylphosphine to trap the Pd(0) formed.

[b]Bite angles data from T. Hayashi, M. Konishi, Y. Kobori, M. Kumada, T. Higuchi, K. Hirotsu, *J. Am. Chem. Soc.*, **1984**, *106*, 158.

The electronic properties of the ligands also have a large impact on the rates of reductive elimination from Pd(II) complexes for the formation of carbon–carbon and carbon–heteroatom bonds.[9]

Palladium-catalyzed cross-coupling reactions to form carbon–carbon bonds have become extremely important tools in organic synthesis during the past several decades. In recognition of their work as pioneers in this field, R. F. Heck, E.-I. Negishi, and A. Suzuki were awarded the Nobel Prize in Chemistry in 2010.* **Table 14.4** provides representative examples of the types of Pd-catalyzed cross-coupling reactions associated with these Nobel recipients and two other major chemists in this field, K. Sonogashira and J. K. Stille. The historical development of Pd-catalyzed cross-coupling reactions, together with numerous examples of the work of these and other contributors to this field, has been reviewed.[10]

TABLE 14.4 Types of Palladium-Catalyzed Cross-Coupling Reactions

Name	First Reported	M	Example of Catalyst Precursor
Heck	1968	Hg	Li_2PdCl_4
Sonogashira	1975	Cu	$PdCl_2(PPh_3)_2$
Negishi	1977	Zn	$PdCl_2(PPh_3)_2$
Stille	1978	Sn	$PhCH_2Pd(PPh_3)_2Cl$
Suzuki[a]	1979	B	$Pd(PPh_3)_4$

[a] Also referred to as Suzuki–Miyaura cross-coupling.

*The recipients' Nobel lectures and Powerpoint slides can be found at nobelprize.org/nobel_prizes/chemistry/laureates/2010/. In addition, the lectures of Negishi and Suzuki have been published in E.-I. Negishi, *Angew. Chem., Int. Ed.*, **2011**, *50*, 6738 and A. Suzuki, *Angew. Chem., Int. Ed.*, **2011**, *50*, 6722.

14.1.4 Sigma Bond Metathesis

All the C—H bond activation examples given in Sections 14.1.2 and 14.1.3 exhibit the classic characteristics of oxidative addition and involve metal centers with eight valence electrons (Ir(I), Pd(II), Fe(0)). These metals have relatively low oxidation states and are predisposed toward being able to oxidize C—H bonds.* Does this mean that complexes with metals in their formally highest oxidation states are inherently unable to effect C—H bond activation? It is fascinating that some d^0 transition-metal complexes are able to activate C—H bonds without any change in oxidation state. For example, consider the following classic reaction:

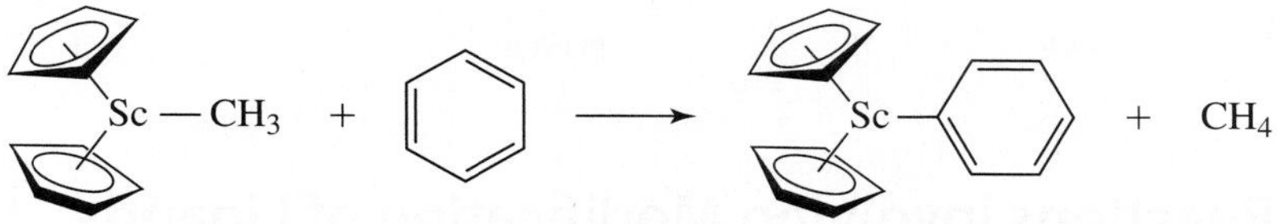

$$Cp_2Sc-CH_3 + C_6H_6 \longrightarrow Cp_2Sc-C_6H_5 + CH_4$$

In this reaction, a C—H bond in methane is made, and a C—H bond of benzene is cleaved, but the oxidation state of the scandium center remains +3.[11] This class of reaction, which is not limited to early transition metals, is called **sigma-bond metathesis**.** In this mechanism, the metal is first postulated to coordinate the bond to be activated in an η^2 fashion, followed by formation of a four-centered transition state that leads to an exchange of ligands at the metal (**Figure 14.8**).[12]

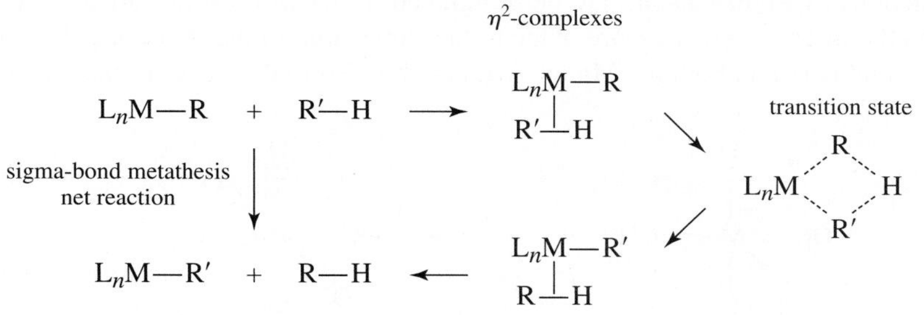

η^2-complexes

$L_nM—R$ + $R'—H$ ⟶ $L_nM—R$ / $R'—H$ ↘ transition state L_nM (R, H, R') ↙ $L_nM—R'$ / $R—H$ ⟶ $L_nM—R'$ + $R—H$

sigma-bond metathesis net reaction

FIGURE 14.8 General Sigma-Bond Metathesis Mechanism.

14.1.5 Application of Pincer Ligands

A significant current focus in organometallic chemistry is the design of **pincer ligands** that provide constrained metal geometries and afford complexes with frontier orbitals of proper energy and symmetry to permit useful stoichiometric and catalytic reactivity.[13] Pincer ligands (see examples in **Figure 14.9**) generally bind to metals via three coordination sites; the anionic or neutral donor site in the backbone attaches, and the phosphine donors "pinch" the metal center, affording a robust chelate. Pincer complexes have permitted novel C—H activation reactions, and their capacity to support bond activation reactions continues to be explored.† A pincer complex supported the first σ-methane complex to be characterized in solution, providing further evidence for the validity of the type of interaction postulated to occur in sigma-bond metathesis prior to formation of the four-centered transition state (Figure 14.8). Protonation of (PONOP)Rh(CH_3) at low temperature afforded the methane complex $[(PONOP)Rh(\sigma\text{-}CH_4)]^+$ shown in the margin.[14]

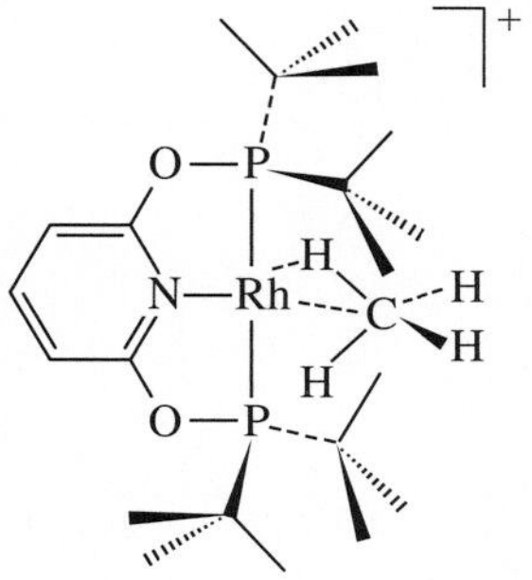

*The scope of this chapter does not permit a detailed discussion of proposed mechanisms for oxidative addition reactions. The reader is encouraged to consult references at the end of Chapter 13 and 14 for mechanistic details.

**A challenge in many cases of bond activation is to distinguish between a mechanism of sequential oxidative addition/reductive elimination or sigma-bond metathesis.

†For two representative examples see M. T. Whited, Y. Zhu, S. D. Timpa, C.-H. Chen, B. M. Foxman, O. V. Ozerov, R. H. Grubbs, *Organometallics*, **2009**, *28*, 4560; B. C. Bailey, H. Fan, J. C. Huffman, M.-H. Baik, D. J. Mindiola, *J. Am. Chem.* Soc., **2007**, *129*, 8781.

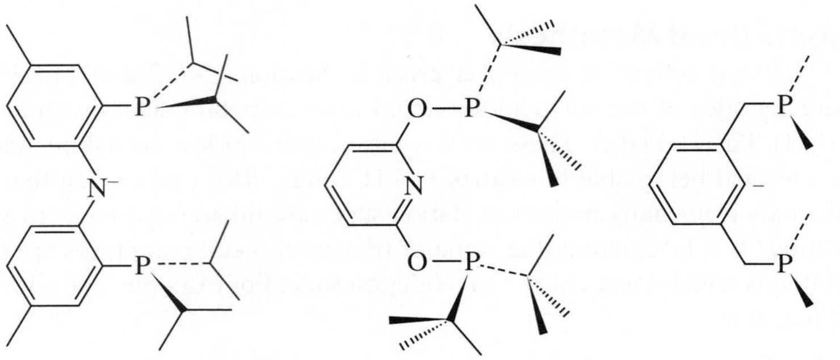

FIGURE 14.9 Pincer Ligands.

14.2 Reactions Involving Modification of Ligands

Many cases are known in which a ligand or molecular fragment appears to insert into a metal–ligand bond. Although some of these reactions are believed to occur by direct, single-step insertion, many "insertion" reactions do not involve a direct insertion step. The most studied of these reactions are carbonyl insertions.

14.2.1 Insertion

The reactions in **Figure 14.10** may be designated as 1,1 insertions, indicating that both bonds to the inserted molecule are made to the same atom in that molecule. For example, in the second reaction, both the Mn and CH_3 are bonded to the sulfur of the inserted SO_2.

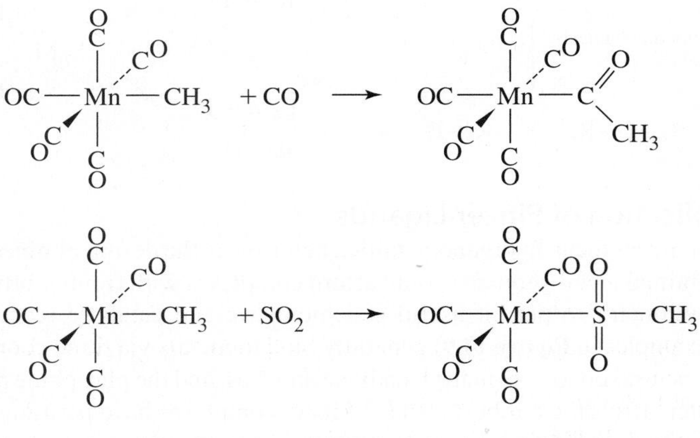

FIGURE 14.10 Examples of 1,1 Insertion Reactions.

$$CH_3Mn(CO)_5 + CO \longrightarrow (OC)_5Mn{-}C({=}O)CH_3$$

$$CH_3Mn(CO)_5 + SO_2 \longrightarrow (OC)_5Mn{-}S({=}O)_2{-}CH_3$$

In 1,2 insertions, bonds to the inserted molecule are made to adjacent atoms. For example, in the reaction of $HCo(CO)_4$ with tetrafluoroethylene (**Figure 14.11**), the product has the $Co(CO)_4$ group attached to one carbon and the H attached to the neighboring carbon.

14.2.2 Carbonyl Insertion (Alkyl Migration)

Carbonyl insertion, the reaction of CO with an alkyl complex to give an acyl [—C(=O)R] product, has been well-studied. The reaction of $CH_3Mn(CO)_5$ with CO is an excellent example:

$$H_3C-Mn(CO)_5 + CO \longrightarrow CH_3\overset{\overset{\displaystyle O}{\|}}{C}-Mn(CO)_5$$

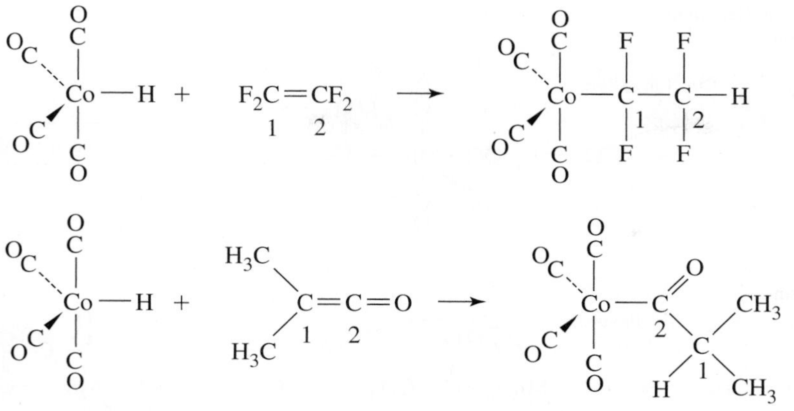

$$(OC)_4Co{-}H \; + \; \underset{1\quad 2}{F_2C{=}CF_2} \longrightarrow (OC)_4Co{-}\overset{1}{C}F_2{-}\overset{2}{C}F_2{-}H$$

$$(OC)_4Co{-}H \; + \; (H_3C)_2\overset{1}{C}{=}\overset{2}{C}{=}O \longrightarrow (OC)_4Co{-}\overset{2}{C}({=}O){-}\overset{1}{C}H(CH_3)_2$$

FIGURE 14.11 Examples of 1,2 Insertion Reactions.

The insertion of CO into a metal–carbon bond in alkyl complexes is of particular interest for its potential applications to organic synthesis and catalysis (Section 14.3), and its mechanism deserves careful consideration.

From the net equation, we might expect that the CO inserts directly into the $Mn{-}CH_3$ bond. However, other mechanisms are possible that would give the overall reaction stoichiometry and involve steps other than the insertion of an incoming CO. Three plausible mechanisms have been suggested:

Mechanism 1: CO Insertion

Direct insertion of CO into a metal–carbon bond.

Mechanism 2: CO Migration

Migration of a CO ligand to give intramolecular CO insertion. This would yield a 5-coordinate intermediate, with a vacant site available for attachment of an incoming CO.

Mechanism 3: Alkyl Migration

In this case, the alkyl group would migrate, rather than the CO, and attach itself to a CO *cis* to the alkyl. This would also give a 5-coordinate intermediate with a vacant site available for an incoming CO.

These mechanisms are described schematically in **Figure 14.12**. In both mechanisms 2 and 3, the intramolecular migration is considered to occur to one of the migrating group's nearest neighbors, located in *cis* positions.

Experimental evidence used to evaluate these mechanisms include the following:[15]

1. Reaction of $CH_3Mn(CO)_5$ with ^{13}CO gives a product with the labeled CO in carbonyl ligands only; *none* is found in the acyl position.
2. The reverse reaction

$$CH_3\overset{\overset{\displaystyle O}{\|}}{C}{-}Mn(CO)_5 \longrightarrow H_3C{-}Mn(CO)_5 + CO$$

occurs readily on heating $CH_3C({=}O)Mn(CO)_5$. When this reaction is carried out with ^{13}C in the acyl position, the product $CH_3Mn(CO)_5$ has the labeled CO entirely *cis* to CH_3. No labeled CO is lost in this reaction.
3. The reverse reaction, when carried out with ^{13}C in a carbonyl ligand *cis* to the acyl group, gives a product that has a 2:1 ratio of *cis* to *trans* product (*cis* and *trans* refer to the position of labeled CO relative to CH_3 in the product). Some labeled CO also dissociates from the Mn in this reaction.

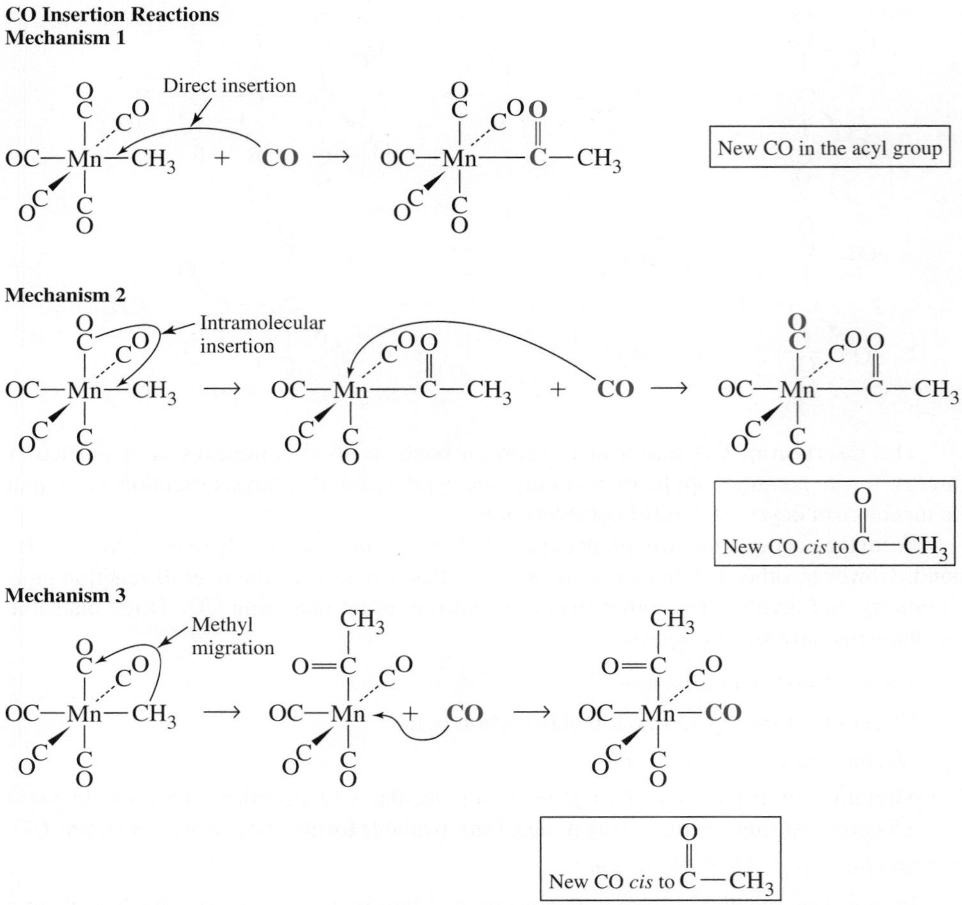

FIGURE 14.12 Possible Mechanisms for CO Insertion Reactions. Acyl groups are shown as $-\overset{O}{\overset{\|}{C}}-CH_3$ for clarity; the actual geometry around acyl carbons is trigonal.

The mechanisms can now be evaluated. Mechanism 1 is definitely ruled out by the first experiment. Direct insertion of ^{13}CO must result in ^{13}C in the acyl ligand, but none is found. Mechanisms 2 and 3 are both compatible with the results of this experiment.

The principle of microscopic reversibility requires that any reversible reaction must have identical pathways for the forward and reverse reactions, simply proceeding in opposite directions. If the forward reaction is carbonyl migration (mechanism 2), the reverse reaction must proceed by loss of a CO ligand, followed by migration of CO from the acyl ligand to the empty site. Because this migration is unlikely to occur to a *trans* position, all the product should be *cis*. If the mechanism is alkyl migration (mechanism 3), the reverse reaction must proceed by loss of a CO ligand, followed by migration of the methyl group of the acyl ligand to the vacant site. Again, all the product should be *cis*. Both mechanisms 2 and 3 would transfer labeled CO in the acyl group to a *cis* position and are therefore consistent with the experimental data for the second experiment (**Figure 14.13**).

EXERCISE 14.1

Show that heating of $CH_3-\overset{O}{\overset{\|}{^{13}C}}-Mn(CO)_5$ would not be expected to give the *cis* product by mechanism 1.

The third experiment differentiates conclusively between mechanisms 2 and 3. The CO migration of mechanism 2, with ^{13}CO *cis* to the acyl ligand, requires migration of

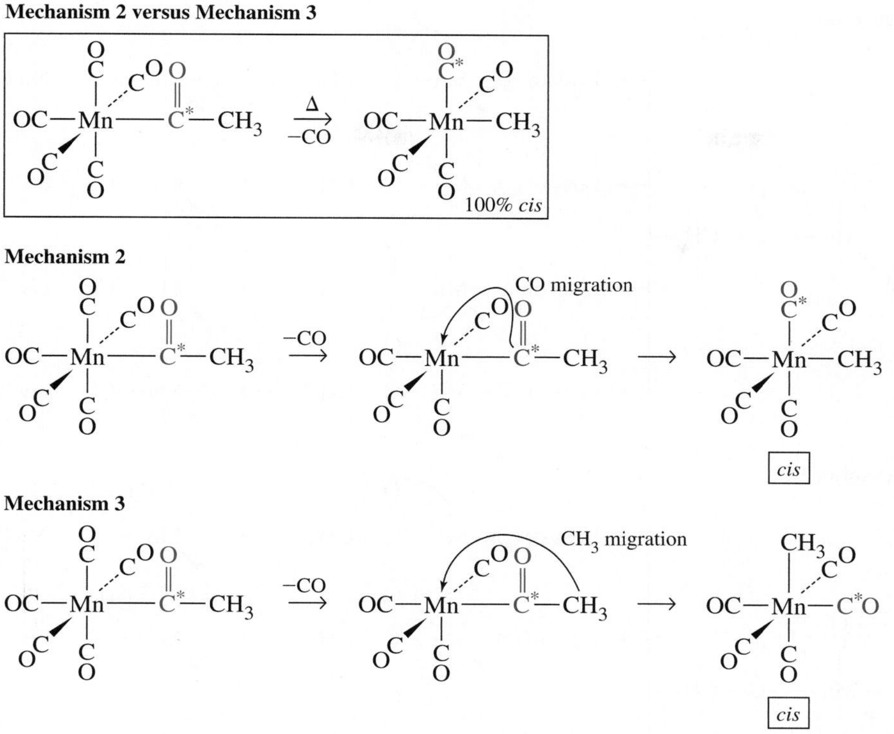

Mechanism 2 versus Mechanism 3

Mechanism 2

CO migration

Mechanism 3

CH_3 migration

FIGURE 14.13 Mechanisms of Reverse Reactions for CO Migration and Alkyl Insertion (1). C* indicates the location of ^{13}C.

CO from the acyl ligand to the vacant site. As a result, 25% of the product should have no ^{13}CO label and 75% should have the labeled CO *cis* to the alkyl (**Figure 14.14**). On the other hand, alkyl migration (mechanism 3) should yield 25% with no label, 50% with the label *cis* to the alkyl, and 25% with the label *trans* to the alkyl. Because this is the ratio of *cis* to *trans* found in the experiment, the evidence supports mechanism 3, the accepted reaction pathway.

We conclude that a reaction that initially appears to involve CO insertion does not involve CO insertion at all! It is common for reactions to differ substantially from how they might at first appear; the "carbonyl insertion" reaction may be more complicated than described here. It is extremely important to keep an open mind about alternative mechanisms. No mechanism can be proved; it is always possible to suggest alternatives consistent with the known data. For example, in the previous discussion of mechanisms 2 and 3, it was assumed that the intermediate was a square pyramid and that no rearrangement to other geometries (such as trigonal-bipyramidal) occurred. Other labeling studies, involving reactions of labeled $CH_3Mn(CO)_5$ with phosphines, have supported a square-pyramidal intermediate.[16]

EXERCISE 14.2

Predict the product distribution for the reaction of *cis*-$CH_3Mn(CO)_4(^{13}CO)$ with PR_3 (R = C_2H_5).

14.2.3 Examples of 1,2 Insertions

Two examples of 1,2 insertions are in Figure 14.11. An important application of 1,2 insertions of alkenes into metal–alkyl bonds is in the formation of polymers. One such process is the Cossee–Arlman mechanism,[17] proposed for the Ziegler–Natta polymerization of

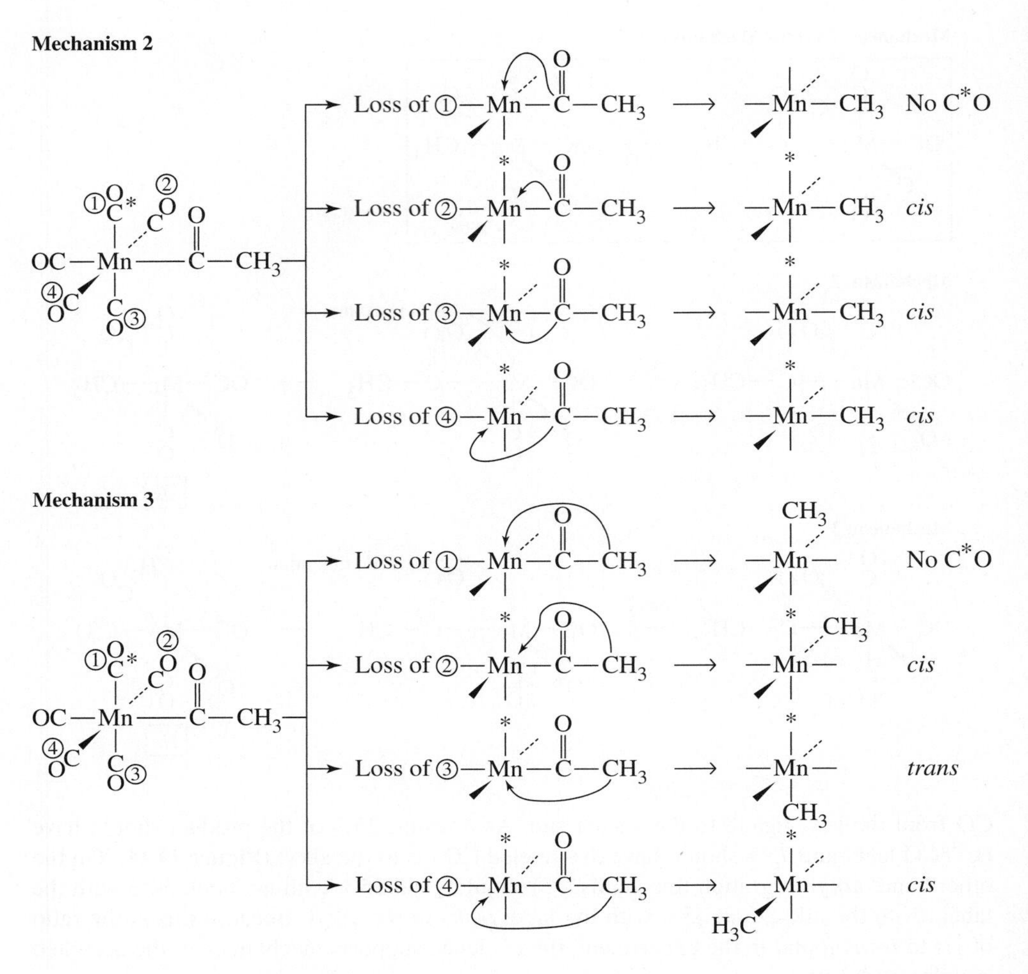

FIGURE 14.14 Mechanisms of Reverse Reactions for CO Migration and Alkyl Insertion (2). C* indicates the location of ^{13}C.

alkenes (Section 14.4.1). According to this mechanism, a polymer chain can grow, as a consequence of repeated 1,2 insertions, into a vacant coordination site:

$$M-CH_2R + \underset{1}{H_2C}=\underset{2}{CHR'} \longrightarrow M(-CH_2R)(\underset{1}{H_2C}=\underset{2}{CHR'}) \xrightarrow{\text{1,2 insertion}} M-\underset{1}{CH_2}-\underset{2}{CHR'}-CH_2R$$

(□ under the first M: vacant site; □ under the product M: site available for another $H_2C=CHR'$)

14.2.4 Hydride Elimination

Hydride elimination reactions are characterized by hydrogen atom transfer from a ligand to a metal. The most common type of hydride elimination is ***β* elimination**, with a proton in a β position* on an alkyl ligand transferred to the metal by way of an intermediate in which the metal, the α and β carbons, and the hydride are coplanar. An example of β elimination—the reverse of 1,2-insertion—is in **Figure 14.15**. Beta eliminations are important in many catalytic processes.

*The Greek letter α is used to designate the carbon atom directly attached to the metal, β is used for the next carbon atom, and so forth.

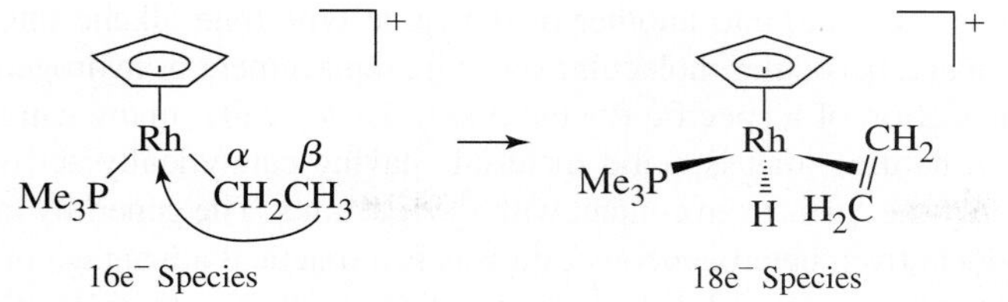

FIGURE 14.15 Example of β Elimination.

EXERCISE 14.3

Show that the reverse of the reaction shown in Figure 14.15 would be 1,2 insertion.

Because only complexes that have β hydrogens can undergo these reactions, alkyl complexes that lack β hydrogens tend to be more stable than those that have such hydrogens (although the former may undergo other types of reactions). Furthermore, coordinatively saturated complexes—complexes in which all coordination sites are filled—containing β hydrogens are in general more stable than complexes having empty coordination sites; the β-elimination mechanism requires transfer of a hydrogen to an empty coordination site. Finally, other types of elimination reactions are also known, such as the elimination of hydrogen from α and γ positions.[18]

14.2.5 Abstraction

Abstraction reactions are elimination reactions in which the coordination number of the metal does not change. In general, they involve removal of a substituent from a ligand by the action of an external reagent, such as a Lewis acid. Two types of abstractions, α and β, are illustrated in **Figure 14.16**; they involve removal of substituents from the α and β positions (with respect to the metal), respectively, of ligands. Alpha abstraction is used to synthesize carbyne complexes (Section 13.6.3).

14.3 Organometallic Catalysts

Organometallic reactions are of great interest industrially. The commercial interest in catalysis has been spurred by the fundamental problem of how to convert relatively inexpensive feedstocks (e.g., coal, petroleum, and water) into molecules of greater commercial value. This frequently involves conversion of simple molecules into more complex molecules (e.g., ethylene into acetaldehyde, methanol into acetic acid, or organic monomers into polymers),

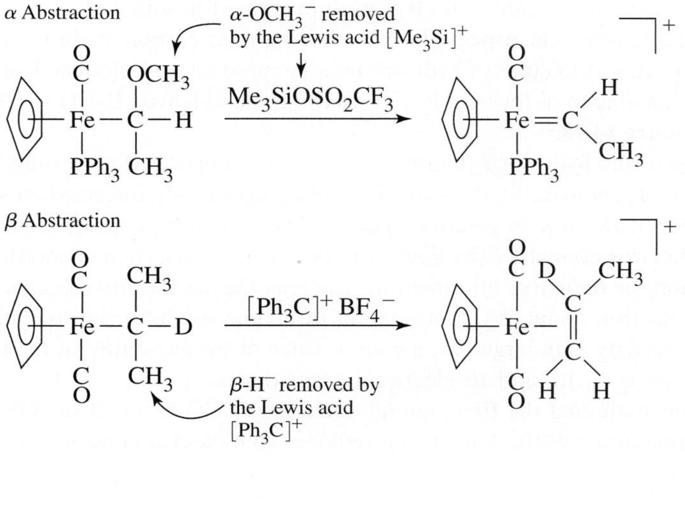

FIGURE 14.16 Abstraction Reactions.

conversion of one molecule into another of the same type (one alkene into another), or a selective reaction at a particular molecular site (e.g., replacement of hydrogen by deuterium, selective hydrogenation of a specific double bond). Historically, many catalysts have been heterogeneous in nature—that is, solid materials having catalytically active sites on their surfaces, with only the surfaces in contact with the reactants. The generally facile separation of reaction products from heterogeneous catalysts is a practical advantage of this approach.

Homogeneous catalysts, soluble in the reaction medium, are molecular species that are easier to study and modify for specific applications than heterogeneous catalysts. Transition-metal complexes are attractive for homogeneous catalysts: they exhibit a variety of oxidation states; an immense variety of ligands can become attached to them, including ligands that can engage in pi, sigma, and more complex interactions; the metals can change coordination numbers as they gain and lose ligands; and they can have different coordination numbers and geometries. Appropriate design of catalyst molecules may provide high selectivity in the processes catalyzed; it is not surprising that development of highly selective homogeneous catalysts is of considerable industrial interest.

In the examples that follow, the reader will find it useful to identify the catalysts, the species regenerated in each complete reaction cycle. In addition, the individual steps in these cycles provide examples of the various types of organometallic reactions introduced earlier in this chapter.

14.3.1 Catalytic Deuteration

If deuterium gas (D_2) is bubbled through a benzene solution of $(\eta^5\text{-}C_5H_5)_2TaH_3$ at an elevated temperature, the hydrogen atoms of benzene are slowly replaced by deuterium; eventually, perdeuterobenzene, C_6D_6, an NMR solvent, can be obtained.[19] Replacement of hydrogen by deuterium occurs in a series of alternating RE and OA steps (**Figure 14.17**).

The initial step is the reductive elimination of H_2 from the 18-electron $(\eta^5\text{-}C_5H_5)_2TaH_3$ to give the 16-electron $(\eta^5\text{-}C_5H_5)_2TaH$. $(\eta^5\text{-}C_5H_5)_2TaH$ reacts with benzene in the second step (oxidative addition) to give an 18-electron species containing a phenyl group σ-bonded to the metal. This species can undergo a second loss of H_2 to give another 16-electron species, $(\eta^5\text{-}C_5H_5)_2Ta—C_6H_5$. $(\eta^5\text{-}C_5H_5)_2Ta—C_6H_5$ subsequently adds D_2 (another oxidative addition) to form an 18-electron species (Step 4), which in the last step eliminates C_6H_5D. Repetition of this sequence in the presence of excess D_2 eventually leads to C_6D_6. In each subsequent cycle, the catalytic species $(\eta^5\text{-}C_5H_5)_2TaD$ is regenerated.

14.3.2 Hydroformylation

The hydroformylation, or oxo, process was introduced in 1938 and is the oldest homogeneous catalytic process in commercial use. It is used to convert terminal alkenes into aldehydes and other organic products, especially those having their carbon chain increased by one. Approximately 10 million tons of hydroformylation products are produced annually.[20] The conversion of an alkene of formula $R_2C{=}CH_2$ into an aldehyde $R_2CH—CH_2—CHO$ is outlined in **Figure 14.18**.[21]

Each step of the hydroformylation cycle may be categorized according to its characteristic type of organometallic reaction. The cobalt-containing intermediates in this cycle alternate between 18- and 16-electron species. The 18-electron species react to formally reduce their electron count by 2 (by ligand dissociation, 1,2 insertion of coordinated alkene, alkyl migration, or reductive elimination), whereas the 16-electron species can increase their formal electron count (by coordination of alkene or CO or by oxidative addition). The catalytic activity is in large part a consequence of the capability of the metal to react by way of a variety of 18- and 16-electron intermediates.

It is problematic that the first step (dissociation of CO from $HCo(CO)_4$) is inhibited by high CO pressure, yet the fourth step requires CO; careful control of this pressure is

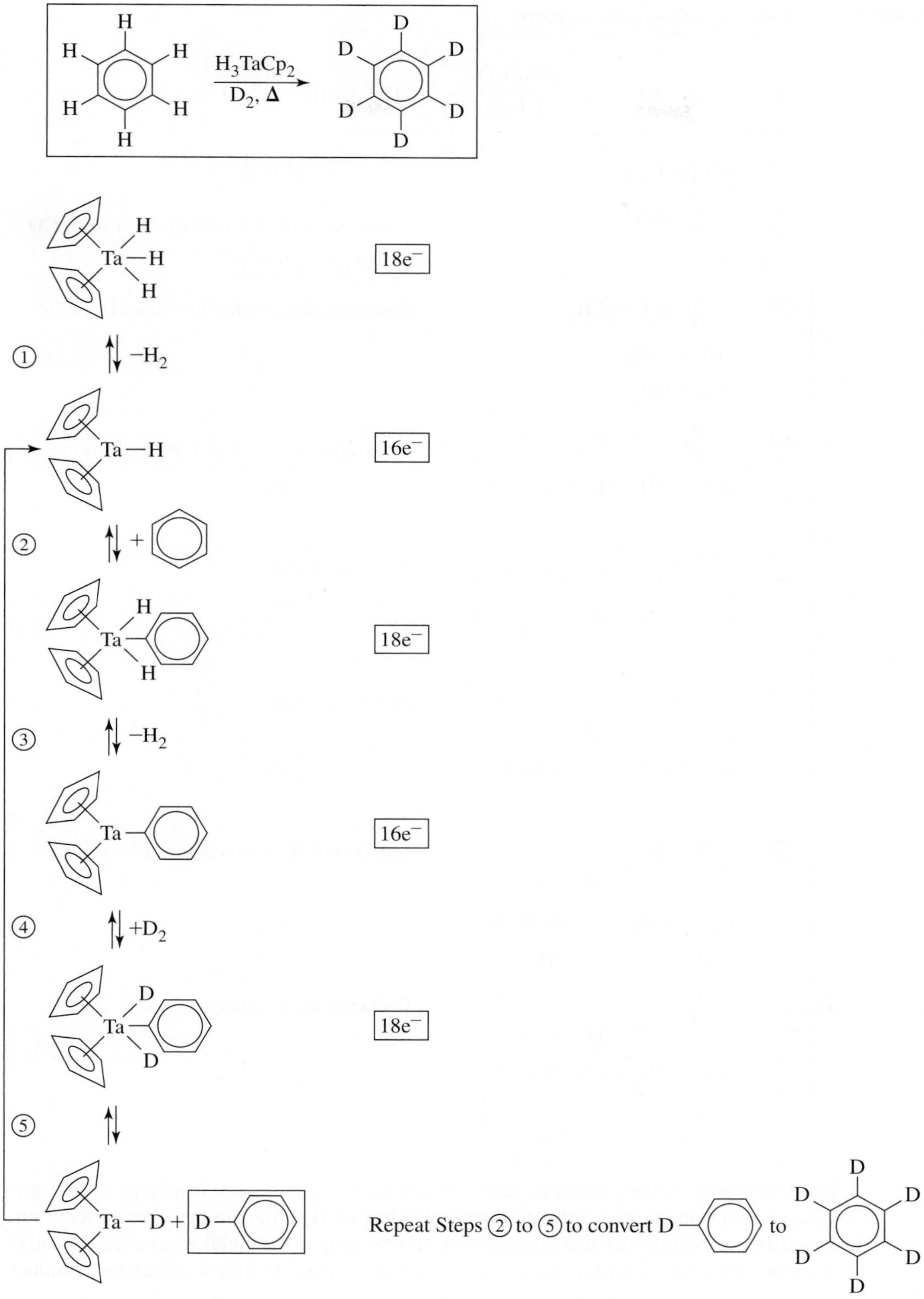

H
H H
H H
H
H_3TaCp_2
D_2, Δ
D
D D
D D
D

Ta—H H H
$18e^-$
① $-H_2$
Ta—H
$16e^-$
② +
Ta H H
$18e^-$
③ $-H_2$
Ta
$16e^-$
④ $+D_2$
Ta D D
$18e^-$
⑤
Ta—D + D—
Repeat Steps ② to ⑤ to convert D— to
D D D D D D

FIGURE 14.17 Catalytic Deuteration.

FIGURE 14.18 Hydroformylation Process.

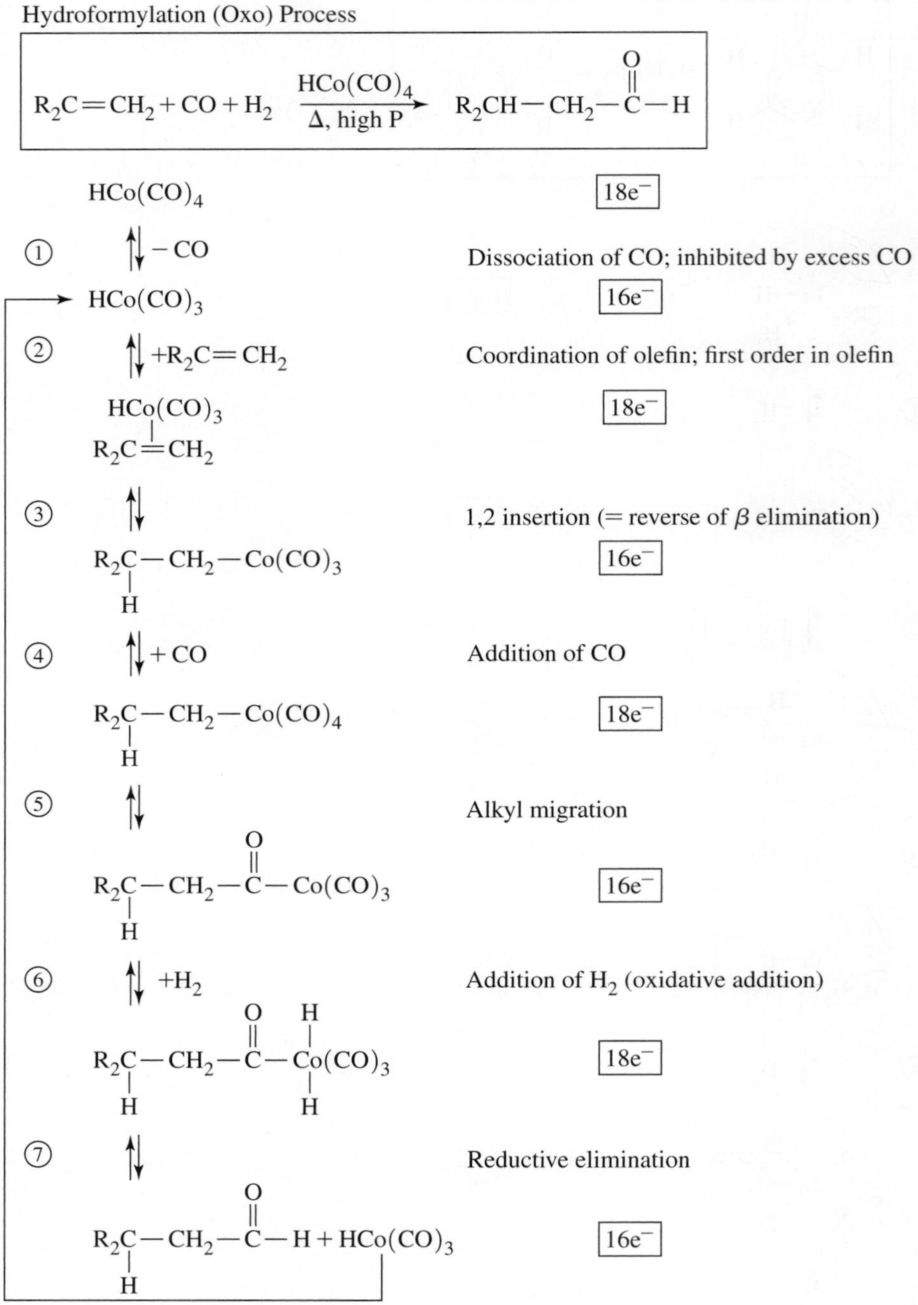

Hydroformylation (Oxo) Process

$$R_2C{=}CH_2 + CO + H_2 \xrightarrow[\Delta,\ \text{high P}]{HCo(CO)_4} R_2CH{-}CH_2{-}\overset{\overset{\displaystyle O}{\|}}{C}{-}H$$

$HCo(CO)_4$ — $18e^-$

① $\rightleftharpoons$ $-CO$ — Dissociation of CO; inhibited by excess CO

$HCo(CO)_3$ — $16e^-$

② $\rightleftharpoons$ $+R_2C{=}CH_2$ — Coordination of olefin; first order in olefin

$HCo(CO)_3$ coordinated to $R_2C{=}CH_2$ — $18e^-$

③ $\rightleftharpoons$ — 1,2 insertion (= reverse of β elimination)

$R_2C(H){-}CH_2{-}Co(CO)_3$ — $16e^-$

④ $\rightleftharpoons$ $+CO$ — Addition of CO

$R_2C(H){-}CH_2{-}Co(CO)_4$ — $18e^-$

⑤ $\rightleftharpoons$ — Alkyl migration

$R_2C(H){-}CH_2{-}C({=}O){-}Co(CO)_3$ — $16e^-$

⑥ $\rightleftharpoons$ $+H_2$ — Addition of H_2 (oxidative addition)

$R_2C(H){-}CH_2{-}C({=}O){-}CoH_2(CO)_3$ — $18e^-$

⑦ $\rightleftharpoons$ — Reductive elimination

$R_2C(H){-}CH_2{-}C({=}O){-}H + HCo(CO)_3$ — $16e^-$ (returns $HCo(CO)_3$ to step 2)

necessary for optimum yields and rates.* The second step is first order in alkene; it is the rate-determining step. In Step 3, the product shown has a CH_2 group, rather than a CR_2 group, bonded to the metal; such a preference for CH_2 attaching to the metal is enhanced by bulky R groups. However, coordination of the CR_2 can also occur, leading to a branched product.

*For more information on reaction conditions, see G. W. Parshall and S. D. Ittel, *Homogeneous Catalysis*, 2nd ed., John Wiley & Sons, New York, 1992, pp. 106–111.

Because linear products are generally more valuable, a challenge in the development of hydroformylation has been to design processes with high linear to branched ratios.

Detailed calculations of the geometries and energies of the steps in hydroformylation have been reported.[22] The calculated geometry of the product of the third reaction step is more interesting than that in Figure 14.18; it has an **agostic hydrogen**, a hydrogen that is bonded to the α carbon and also is strongly attracted to the metal, forming a bridge between the two, as shown in the margin. Agostic interactions, involving a hydrogen atom weakly bonded to a metal and having a weakened, elongated bond with a carbon have been found in a variety of organometallic complexes and have been proposed in intermediates in reactions of hydrogen at metal centers.[23]

R_2C — CH_2 — Co — CO, CO, CO; H — Co; Agostic hydrogen

Step 6 involves oxidative addition of H_2; however, high H_2 pressure can lead to addition of H_2 to the 16-electron intermediate from Step 3, which would then eliminate an alkane:

$$\underset{16\ e^-}{R_2CH{-}CH_2{-}Co(CO)_3} + H_2 \longrightarrow \underset{18\ e^-}{R_2CH{-}CH_2{-}Co(H)_2(CO)_3} \quad \text{oxidative addition}$$

$$\underset{18\ e^-}{R_2CH{-}CH_2{-}Co(H)_2(CO)_3} \longrightarrow R_2CH{-}CH_3 + \underset{16\ e^-}{HCo(CO)_3} \quad \text{reductive elimination}$$

Careful control of the experimental conditions is necessary to maximize yield of the desired products.[24,25] The catalyst in this mechanism is the 16-electron $HCo(CO)_3$.

The main industrial application of hydroformylation is in the production of butanal from propene ($CH_3CH{=}CH_2 \longrightarrow CH_3CH_2CH_2CHO$). Subsequent hydrogenation gives butanol, an important solvent. Other aldehydes are also produced industrially by hydroformylation, using either cobalt catalysts such as the one in Figure 14.18 or rhodium-based catalysts.

EXERCISE 14.4

Show how $(CH_3)_2CHCH_2CHO$ can be prepared from $(CH_3)_2C{=}CH_2$ by the hydroformylation process.

A shortcoming of the cobalt carbonyl-based hydroformylation process is that it produces only about 80 percent of the much more valuable linear aldehydes, with the remainder having branched chains. Modifying the catalyst by replacing one of the CO ligands of the starting complex by PBu_3 (Bu = *n*-butyl) to give $HCo(CO)_3(PBu_3)$ increases the selectivity of the process to give an approximately 9:1 ratio of linear to branched aldehydes.[26] The effect of phosphines on the thermodynamics of this cobalt-catalyzed hydroformylation system has been explored.[27] Finally, replacing the cobalt with rhodium yields far more active catalysts (much less catalyst needs to be present) that can function with higher linear to branched selectivity at significantly lower temperatures and pressures than cobalt-based catalysts.[28] A proposed mechanism for an example of such a catalytic process using $HRh(CO)_2(PPh_3)_2$ is in **Figure 14.19**.*

EXERCISE 14.5

Classify each step of the mechanism in Figure 14.19 according to its reaction type.

More recently, higher linear to branched ratios for rhodium-catalyzed hydroformylation have been achieved using bidentate ligands, which broaden the range of steric and electronic options available. For example, the BISBI ligand has enabled a linear

PPh_2
PPh_2
BISBI

*For a more detailed outline of the various cobalt- and rhodium-based hydroformylation catalysts, and for additional related references, see G. O. Spessard and G. L. Miessler, *Organometallic Chemistry*, Oxford University Press, New York, 2010, pp. 322–339.

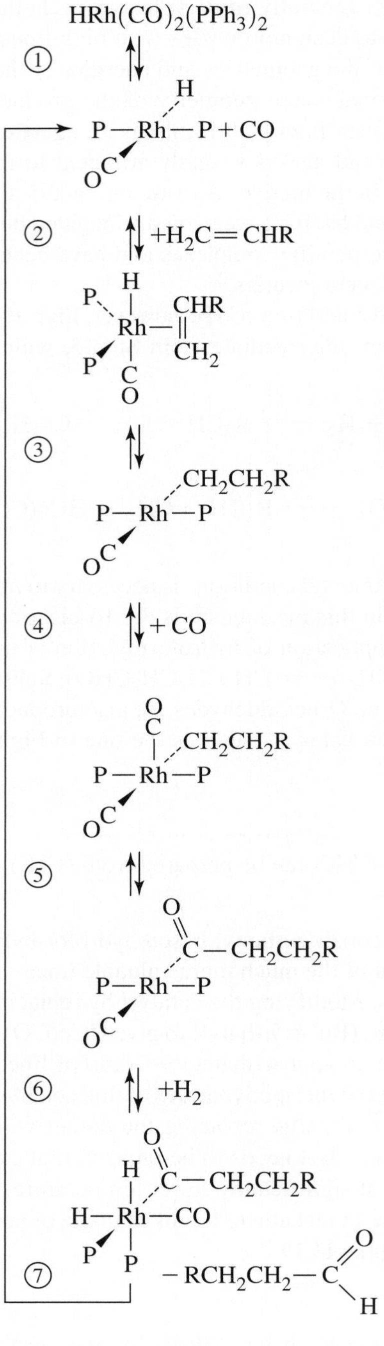

FIGURE 14.19 Hydroformylation using $HRh(CO)_2(PPh_3)_2$. P = PPh_3. Data from C. K. Brown, G. Wilkinson, *J. Chem. Soc., A*, **1970**, 2753.

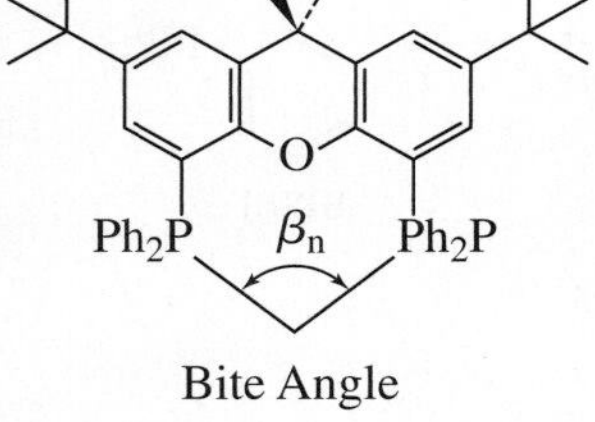

to branched ratio of 66:1 to be achieved.[29] In a series of oxygen-bridged diphosphine ligands such as the one in the margin, the linear to branched ratio was found to increase with the bite angle.[30] The rational design of phosphine ligands to improve hydroformylation catalysts has been reviewed.[31] Application of bidentate phosphine–phosphite ligands to increase rhodium hydroformylation catalyst performance is a current area of interest.[32]

14.3.3 Monsanto Acetic Acid Process

The synthesis of acetic acid from methanol and CO is a process that has been used with commercial success by Monsanto. The mechanism of this process is complex; a proposed outline is in **Figure 14.20**. The individual steps are the characteristic types of organometallic reactions described previously; the intermediates are 18- or 16-electron species having the capability to lose or gain, respectively, 2 electrons. Solvent molecules may occupy empty coordination sites in the 4- and 5-coordinate 16-electron intermediates. The first step, oxidative addition of CH_3I to $[RhI_2(CO)_2]^-$, is rate determining.*

The final step involving rhodium is reductive elimination of $IC(=O)CH_3$. Acetic acid is formed by hydrolysis of this compound. The catalytic species, $[Rh(CO)_2I_2]^-$, which may contain solvent in the empty coordination sites, is regenerated.

In addition to rhodium-based catalysts, iridium-based catalysts have also been developed for carbonylation of methanol. The iridium system, known as the *Cativa process*, follows a cycle similar to the rhodium system in Figure 14.20, beginning with oxidative addition of

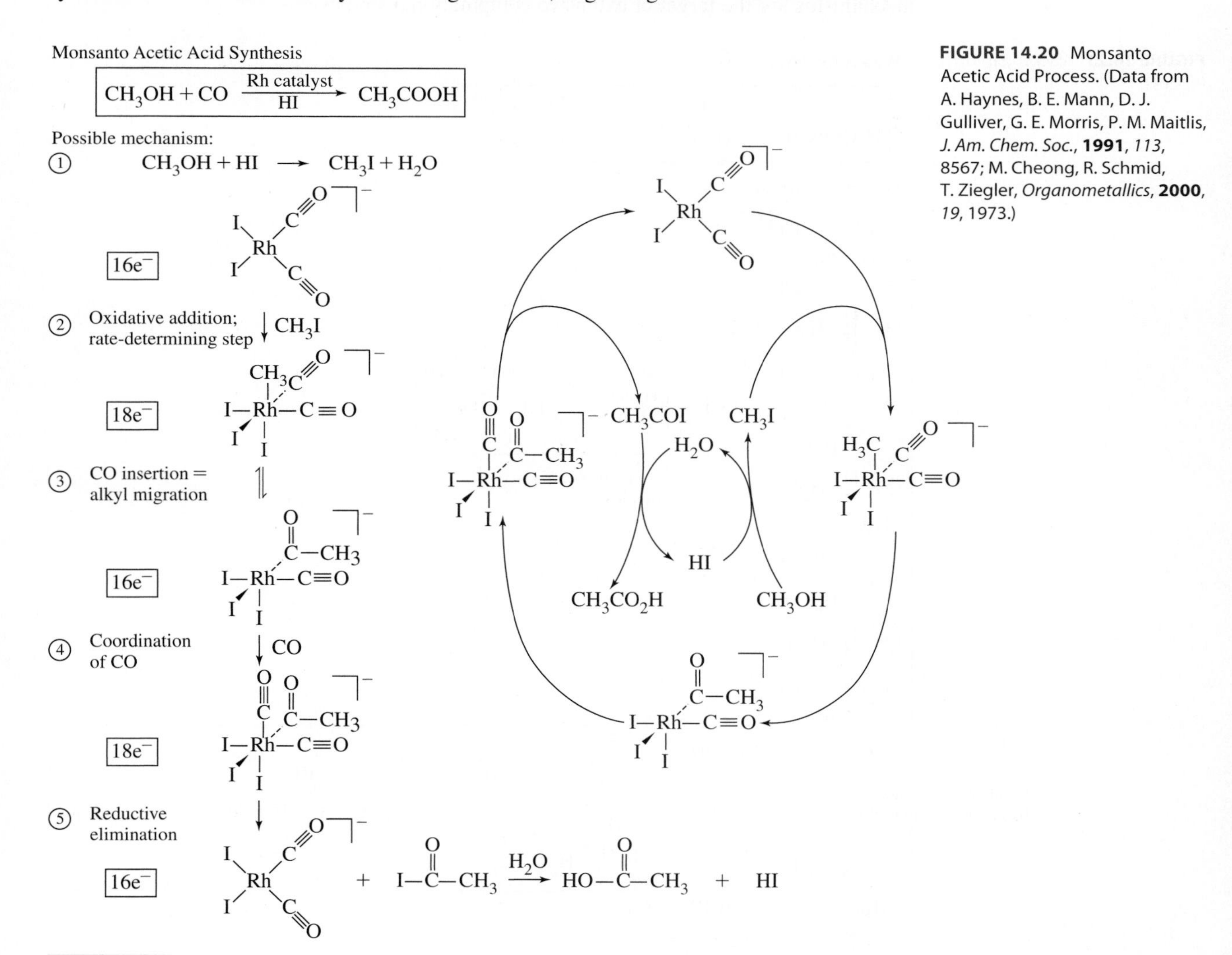

FIGURE 14.20 Monsanto Acetic Acid Process. (Data from A. Haynes, B. E. Mann, D. J. Gulliver, G. E. Morris, P. M. Maitlis, *J. Am. Chem. Soc.*, **1991**, *113*, 8567; M. Cheong, R. Schmid, T. Ziegler, *Organometallics*, **2000**, *19*, 1973.)

*A discussion of the mechanism of this reaction can be found in D. Forster, T. W. Deklava, *J. Chem. Educ.*, **1986**, *63*, 204, and references therein.

CH_3I to $[Ir(CO)_2I_2]^-$. The first step in the iridium system is several hundred times faster than in the Monsanto process; the second step, involving alkyl migration, is much slower, and it is rate determining for the Cativa process.[33] In addition to the catalytic cycle involving the anion $[Ir(CO)_2I_2]^-$, an alternative neutral cycle involving $Ir(CO)_3I$ or $Ir(CO)_2I$ has been reported.[34] Two excellent reviews of developments in catalytic methanol carbonylation are available.[35]

14.3.4 Wacker (Smidt) Process

The Wacker or Smidt process, used to synthesize acetaldehyde from ethylene, involves a catalytic cycle that uses $PdCl_4^{2-}$. A fiftieth-anniversary retrospective by a coauthor of the original report[36] on this process was published in 2009.[37] **Figure 14.21** outlines a proposed cycle; the fourth step is more complex than that shown and involves overall oxidation of the organic fragment and reduction of the Pd center. Regenerating the catalyst requires stoichiometric oxidation (by $CuCl_2$ below). The cycle may proceed by different mechanisms depending on whether the Cl^- and $CuCl_2$ concentrations are high or low.[38] These mechanisms are the target of extensive computational studies.[39]

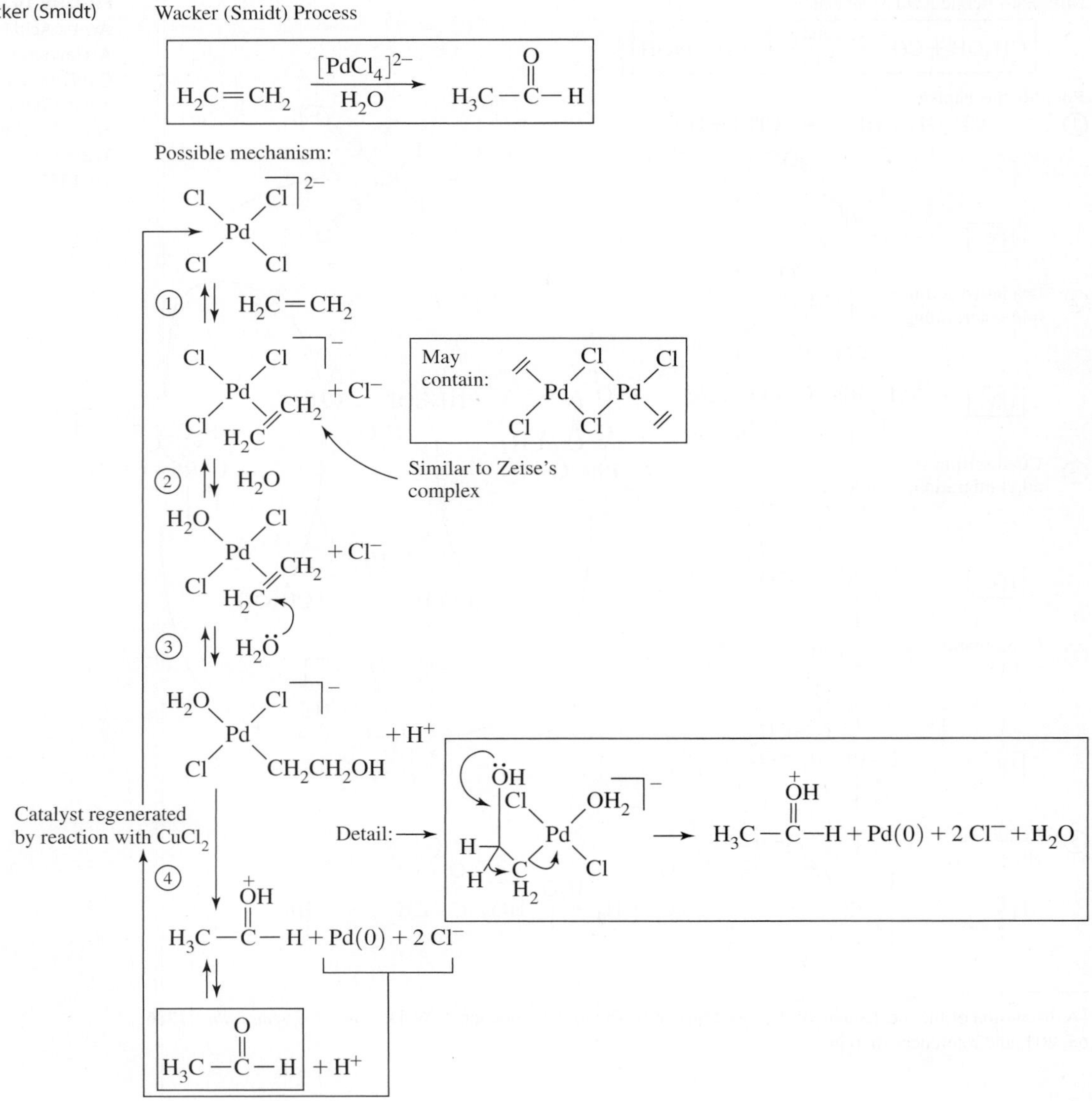

FIGURE 14.21 Wacker (Smidt) Process.

An important feature of this process is that it uses the ability of palladium to form complexes with the reactant ethylene, with the modification of ethylene occurring while it is attached to the metal. The palladium tunes the chemical behavior of ethylene to enable reactions to occur that would not be possible for free ethylene. The first ethylene complex with palladium in Figure 14.21 is isoelectronic with Zeise's complex, $[PtCl_3(\eta^2\text{-}H_2C{=}CH_2)]^-$, (Figure 13.4).

14.3.5 Hydrogenation by Wilkinson's Catalyst

Wilkinson's catalyst, $RhCl(PPh_3)_3$, participates in the same types of reactions as expected for 4-coordinate organometallic compounds; for example, many reactions bear similarities to Vaska's catalyst, *trans*-$IrCl(CO)(PPh_3)_2$. $RhCl(PPh_3)_3$ participates in a variety of catalytic and noncatalytic processes. The bulky phosphine ligands play an important role in making the complex selective—for example, they limit coordination of Rh to unhindered positions on alkenes. **Figure 14.22** outlines catalytic alkene hydrogenation.[40]

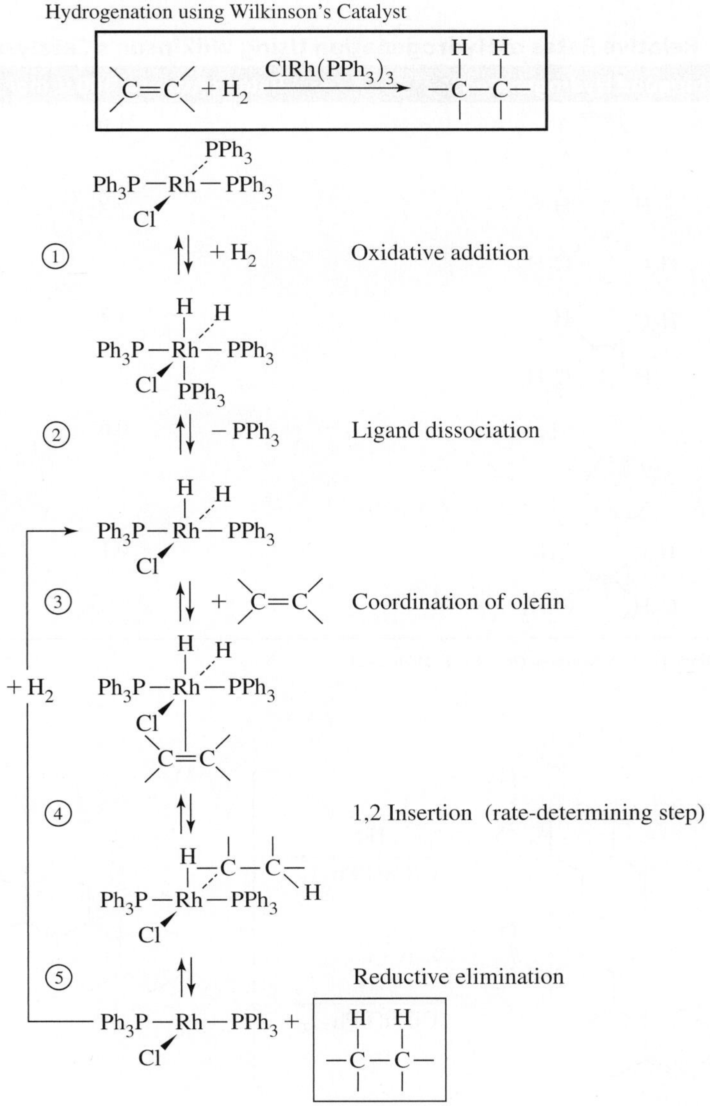

FIGURE 14.22 Catalytic Hydrogenation Involving Wilkinson's Catalyst.

The first two steps give the catalytic species $RhCl(H)_2(PPh_3)_2$, which has a vacant coordination site. If a C=C double bond is not too sterically hindered, it can coordinate to this site and gain the two hydrogens coordinated to Rh, resulting in hydrogenation of the double bond. Variation in the steric bulk around the double bond impacts the hydrogenation rates (**Table 14.5**) when Wilkinson's catalyst is used.

In molecules containing several double bonds, the least hindered double bonds are reduced with significantly lower activation barriers. The most hindered positions cannot coordinate effectively to Rh, largely because of the presence of the bulky phosphines, and hence do not react as rapidly. Consequently, Wilkinson's catalyst is useful for selective hydrogenations of C=C bonds that are not sterically hindered. Examples are in **Figure 14.23**.

Because the selectivity of Wilkinson's catalyst is largely a consequence of the bulky triphenylphosphine ligands, the selectivity can be fine-tuned by using phosphines having different cone angles than PPh_3.

TABLE 14.5 Relative Rates of Hydrogenation Using Wilkinson's Catalyst at 25 °C

Compound Hydrogenated	Rate Constant × 100 ($L\ mol^{-1}s^{-1}$)
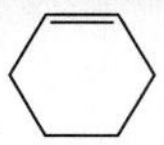	31.6
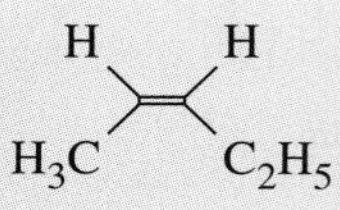 H H H_3C C_2H_5	9.9
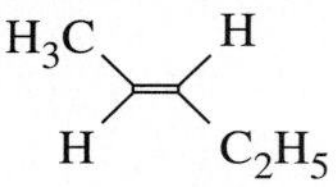 H_3C H H C_2H_5	1.8
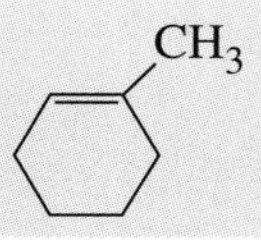 CH_3	0.6
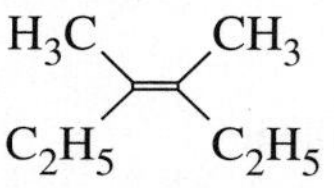 H_3C CH_3 C_2H_5 C_2H_5	< 0.1

Data from A. J. Birch, D. H. Williamson, *Org. React.*, **1976**, *24*, 1.

FIGURE 14.23 Selective Hydrogenation by Wilkinson's Catalyst.

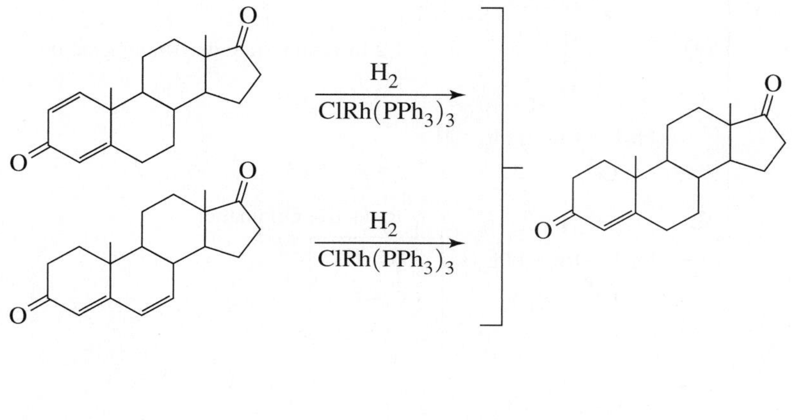

14.3.6 Olefin Metathesis

Olefin metathesis, first discovered in the 1950s, involves the formal exchange of :CR_2 fragments (R = H or alkyl) between alkenes, also known as olefins. For example, metathesis between molecules of formula $H_2C{=}CH_2$ and $HRC{=}CHR$ would yield two $H_2C{=}CHR$ molecules:

$$\begin{matrix} H_2 & HR \\ C & C \\ \| & \| \\ C & C \\ H_2 & HR \end{matrix} \quad \rightleftharpoons \quad \begin{matrix} H_2C{=}CHR \\ \\ H_2C{=}CHR \end{matrix}$$

New double bonds are formed between the top and bottom two carbons in the diagram; the original double bonds are severed.*

EXAMPLE 14.1

Predict the possible products of metathesis of the following olefins. Be sure to consider that two molecules of the same structure can also metathesize (undergo self-metathesis).

a. Between propene and 1-butene.

b. Between ethylene and cyclohexene.

Example **b** is an example of **ring-opening metathesis (ROM)**, in which metathesis opens a ring of a cyclic alkene. The reverse of this process is called, appropriately, **ring-closing metathesis (RCM)**. An example of ring-closing metathesis is in Figure 14.33.

EXERCISE 14.6 Predict the products of metathesis:

a. Between two molecules of propene

b. Between propene and cyclopentene

Metathesis, which is reversible and can be catalyzed by organometallic complexes, is important industrially. In recognition of the fundamental research to shed light on the pathway of metathesis, the 2005 Nobel Prize in Chemistry was awarded to three leaders in its development, Y. Chauvin, R. Grubbs, and R. Schrock. Their Nobel lectures provide an excellent background on the experiments leading to an understanding of metathesis.[41]

In the early stages of metathesis research, several mechanisms were proposed. We will consider three of the most important: an alkyl group transfer mechanism, a diolefin ("pairwise") mechanism, and a mechanism involving carbene complexes ("non-pairwise"). These are shown schematically in **Figure 14.24**.

Alkyl Exchange

This proposed mechanism was investigated by reacting two 2-butenes, $H_3C{-}CH{=}CH{-}CH_3$ and the fully deuterated $D_3C{-}CD{=}CD{-}CD_3$. If alkyl transfer could occur, $-CH_3$ and $-CD_3$ groups would be expected to exchange, giving mixtures of H and D atoms on each side of the double bond ($=CH{-}CD_3$ and $=CD{-}CH_3$). The result (**Figure 14.25**) was that the halves of the molecule on each side of the double bond exchanged as a whole; there was no evidence for exchange of alkyl groups.[42]

*Discussions of the history of the metathesis reaction written by two of its discoverers can be found in R. L. Banks, *Chemtech*, **1986**, *16*, 112 and H. Eleuterio, *Chemtech*, **1991**, *21*, 92.

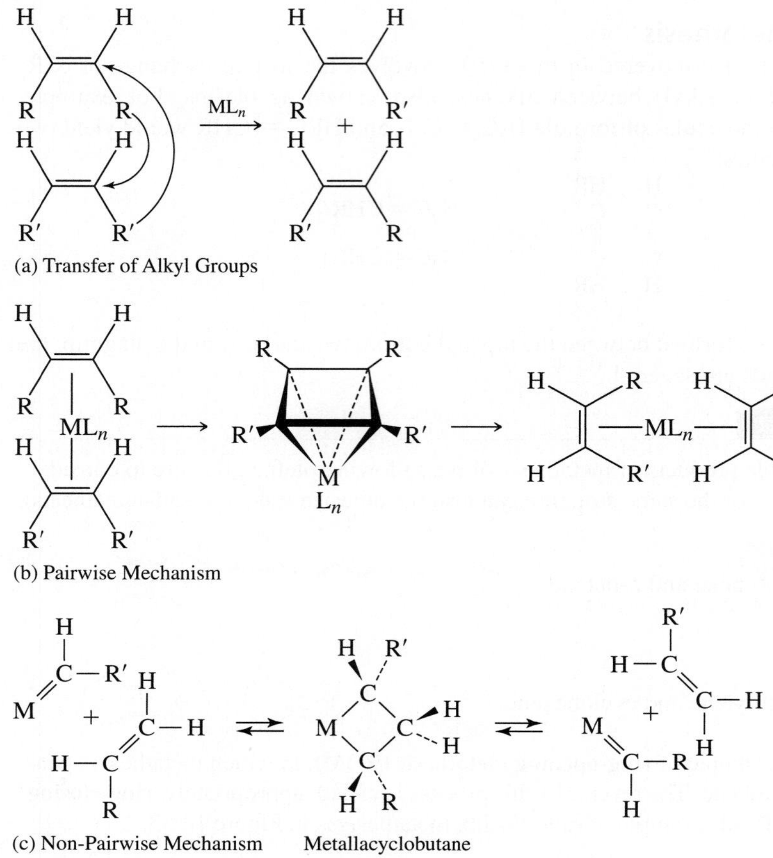

FIGURE 14.24 Proposed Metathesis Mechanisms (a) Transfer of Alkyl Groups (b) Pairwise Mechanism (c) Non-Pairwise Mechanism.

(a) Transfer of Alkyl Groups

(b) Pairwise Mechanism

(c) Non-Pairwise Mechanism Metallacyclobutane

$$H_3C{-}CH{=}CH{-}CH_3 + D_3C{-}CD{=}CD{-}CD_3 \longrightarrow H_3C{-}CH{=}CD{-}CD_3$$

The *only* product

FIGURE 14.25 Experiment to Test the Alkyl Exchange Mechanism.

Disinguishing between the other two mechanisms was more challanging; each could account for the results in Figure 14.25.

Diolefin (Pairwise) Mechanism

Bradshaw proposed that "the dismutation of olefins should proceed via a quasicyclobutane intermediate."[43] In this mechanism, the two alkenes would first coordinate to a transition metal, forming a quasicyclobutane (**Figure 14.26**), after which the metal-complex intermediate would break apart to form the new alkenes. Because formation of the intermediate would involve two alkenes attaching to the metal in pairs, this has been called both the *diolefin* and *pairwise mechanism*.

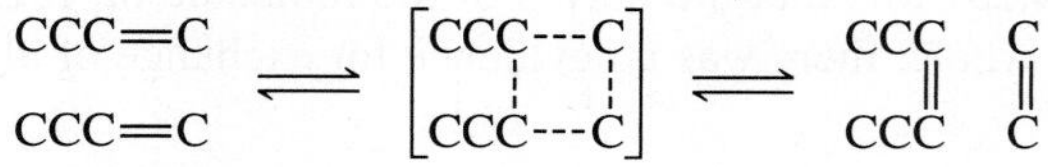

FIGURE 14.26 Quasicyclobutane Intermediate.

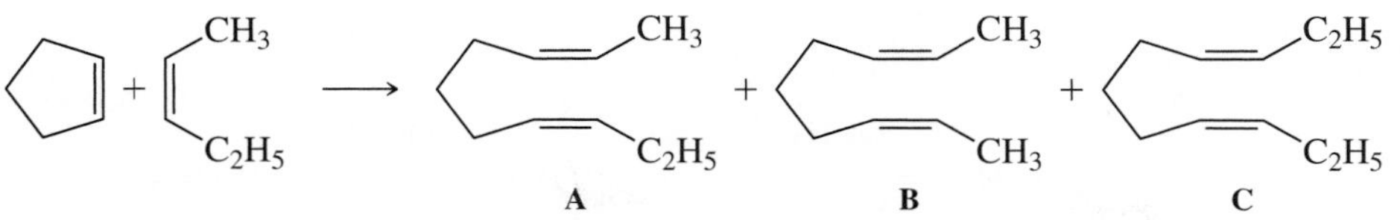

FIGURE 14.27 Hérisson and Chauvin's Experiment. Metathesis of cyclopentene and 2-pentene.

Hérisson and Chauvin subsequently reported the experiment shown in **Figure 14.27**. Metathesis of cyclopentene and 2-pentene, according to the diolefin mechanism, would form the first product shown, **A**, with terminal methyl and ethyl groups. This product could undergo further metathesis with $CH_3-CH=CH-C_2H_5$ to form products **B** and **C** (in addition to *trans* isomers and other products). Because the metathesis reactions are equilibria, the ultimate distribution of products would be expected to be in the proportion expected statistically on the basis of equal numbers of terminal methyl and ethyl groups in 2-pentene.[44] Analysis of products should initially show a higher proportion of the first product, and the statistical distribution of products would come later. In Hérisson and Chauvin's work, however, the statistical distribution of products was found even if the reaction was quenched before equilibrium was achieved, arguing against the pairwise mechanism.

EXERCISE 14.7

Show how the pairwise mechanism could form the last two products by metathesis involving the first product.

Carbene (Non-Pairwise) Mechanism

Hérisson and Chauvin proposed that metathesis reactions are catalyzed by carbene (alkylidene) complexes that react with alkenes via the formation of a cyclic intermediate, a metallacyclobutane, as shown in Figure 14.24c. In this mechanism, a metal carbene complex first reacts with an alkene to form the metallacyclobutane. This intermediate can either revert to reactants or form new products; because all steps in the process are in equilibria, an equilibrium mixture of alkenes results. This non-pairwise mechanism would enable the statistical mixture of products to form from the start by the action of catalytic amounts of the necessary carbene complexes, with both R and R′ groups, as shown in Figure 14.24c.

A variety of experimental evidence showed results consistent with a non-pairwise mechanism involving carbene complexes and a metallacyclobutadiene intermediate. For example, in the **Figure 14.28** reaction, the gaseous products $H_2C=CH_2$, $D_2C=CD_2$, and $H_2C=CD_2$ formed in the expected statistical ratio early in the reaction, with no indication that the mixed product $H_2C=CD_2$, the first product by the pairwise mechanism, forms before the others.[45]

The first direct demonstration that a carbene complex could engage in metathesis involved the carbene complex in **Figure 14.29**, which was able to replace its carbene ligand with a terminal alkene in a reaction consistent with the non-pairwise mechanism.[46]

Although alternatives to the non-pairwise mechanism were proposed, the preponderance of evidence has strongly supported the role of carbene complexes as catalysts for olefin metathesis. This mechanism, now known as the **Chauvin mechanism**, is believed to be the pathway of the majority of transition metal–catalyzed olefin metathesis reactions.*

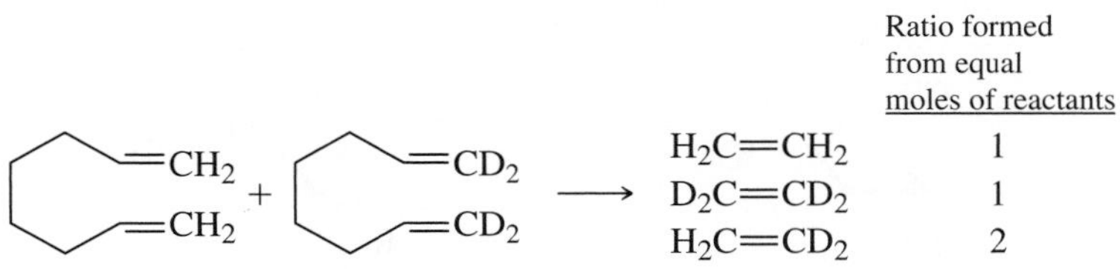

Product	Ratio formed from equal moles of reactants
$H_2C=CH_2$	1
$D_2C=CD_2$	1
$H_2C=CD_2$	2

FIGURE 14.28 A Mechanistic Test.

*For a short outline of the development of olefin metathesis, see C. P. Casey, *J. Chem. Educ.*, **2006**, *83*, 192.

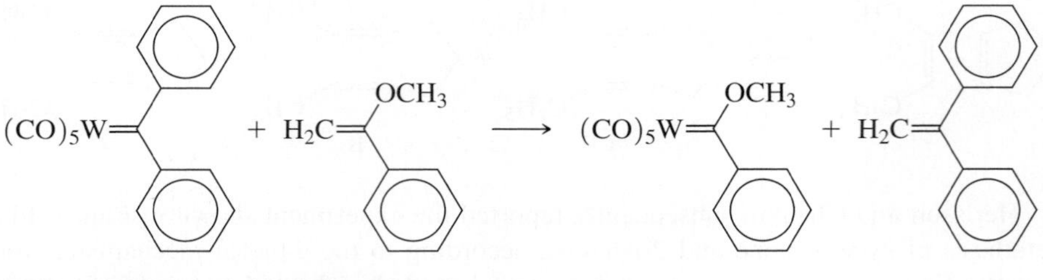

FIGURE 14.29 Metathesis of a Carbene Complex.

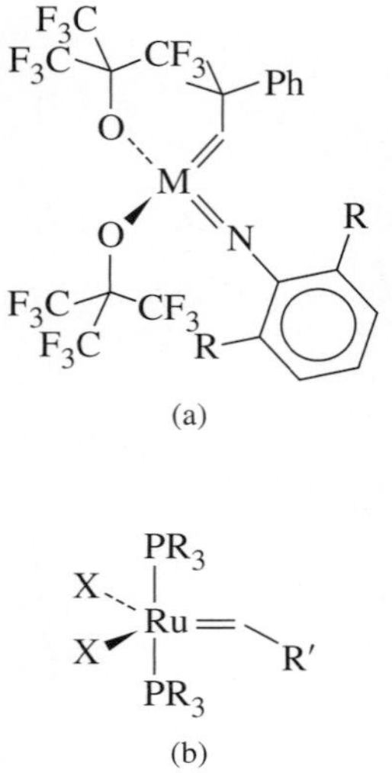

The most thoroughly studied catalysts that effect alkene metathesis are of two types (**Figure 14.30**). Schrock metathesis catalysts are the most effective of all metathesis catalysts but in general are highly sensitive to oxygen and water. These catalysts are available commercially; the catalyst having M = Mo and R = isopropyl is called Schrock's catalyst. A reaction utilizing this catalyst is a critical step of the synthesis of the natural product dactylol (**Figure 14.31**).[47]

The Figure 14.31 reaction is an example of **ring-closing metathesis (RCM)**, in which the metathesis of two double bonds leads to ring formation. Like ordinary metathesis, ring-closing metathesis is believed to occur by way of a metallacyclobutane intermediate; this intermediate is responsible for joining the originally separate carbons into a ring.

Grubbs metathesis catalysts in general have less catalytic activity than Schrock catalysts but are less sensitive to oxygen and water. The catalyst having R = cyclohexyl, X = Cl, and R′ = phenyl is marketed as Grubbs's catalyst. One requirement of these catalysts is the presence of bulky phosphine ligands. This bulkiness facilitates phosphine dissociation, a key step in the proposed catalyst mechanism (**Figure 14.32**).[48]

Although much research in homogeneous metathesis catalysis has focused on complexes resembling those in Figure 14.30, other avenues have also been pursued. The introduction of catalysts that contain ruthenium and *N*-heterocyclic carbene ligands results in so-called second generation Grubbs's catalysts.[49] The *N*-heterocyclic carbene ligands exceed trialkylphosphines in steric requirements and are more strongly electron donating;[50] both features support improved catalytic activity. An example of such a catalyst, and a

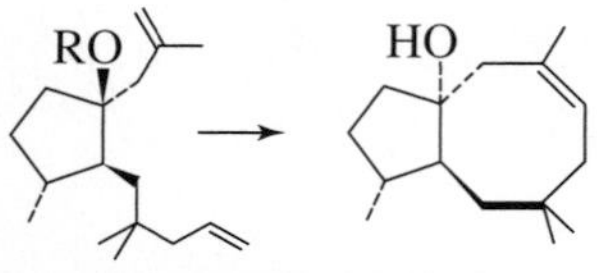

FIGURE 14.30 Metathesis Catalysts. (a) Schrock catalyst (M = Mo, W). (b) Grubbs catalyst (X = Cl, Br).

FIGURE 14.31 Ring-Closing Metathesis (RCM)

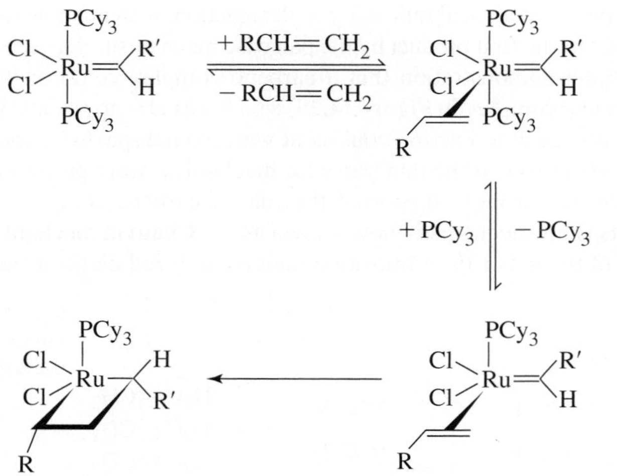

FIGURE 14.32 Proposed Mechanism for Formation of Metallacyclobutane from Ruthenium Catalyst.

ring-closing metathesis process that it catalyzes, are in **Figure 14.33**.[51] Modified versions of Grubbs's catalysts without any phosphine ligands are more thermally stable than other Grubbs's catalysts.[52]

The Figure 14.33 process has also been performed using Schrock's catalyst and Grubbs's catalyst. As shown in **Table 14.6**, the *N*-heterocyclic catalyst compares favorably with Schrock's catalyst and is far superior to Grubbs's catalyst—at least for this reaction.[51]

An interesting variation on olefin metathesis is the use of carbene complexes to catalyze alkene polymerization, also via a metallacyclobutane intermediate. An example is the use of W(CH—*t*-Bu)(OCD_2—*t*-Bu)$_2Br_2$ as a catalyst in the ring-opening polymerization of norbornene in the presence of $GaBr_3$, (**Figure 14.34**).[53] NMR data are consistent with the proposed structure of the metallacyclobutane, as well as the polymer growing off the carbene carbon.

Alkynes can also undergo metathesis reactions catalyzed by transition-metal carbyne complexes. The intermediates in these reactions are believed to be metallacyclobutadiene species, formed from the addition of an alkyne across a metal–carbon triple bond of the carbyne (**Figure 14.35**). The structures of a variety of metallacyclobutadiene complexes have been determined, and some have been shown to catalyze alkyne metathesis.

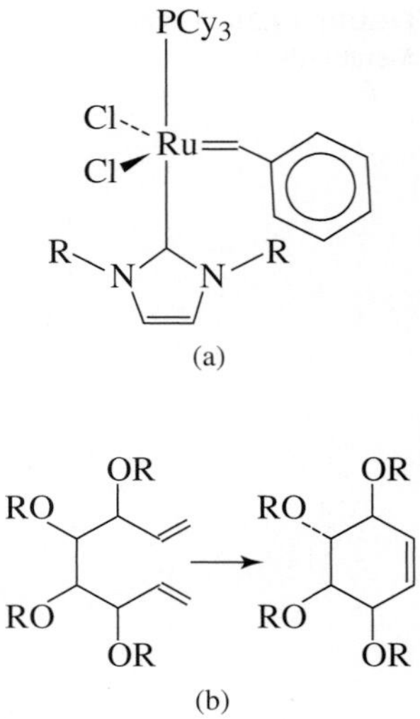

PCy$_3$ Cl Cl Ru R N N R (a)

OR RO RO OR → RO OR RO OR (b)

FIGURE 14.33 Ring-Closing Metathesis Catalyzed by the *N*-Heterocyclic Carbene Complex. (a) Catalyst (R = mesityl) (b) Ring-closing reaction (R = benzyl)

TABLE 14.6 Relative Activity of Metathesis Catalysts

Catalyst	Reaction Time (*h*)	Yield (%)
Schrock's catalyst	1	92
Grubbs's catalyst	60	32
Catalyst in Figure 14.33	2	89

Data from L. Ackermann, D. El Tom, A. Fürstner, *Tetrahedron*, **2000**, *56*, 2195.

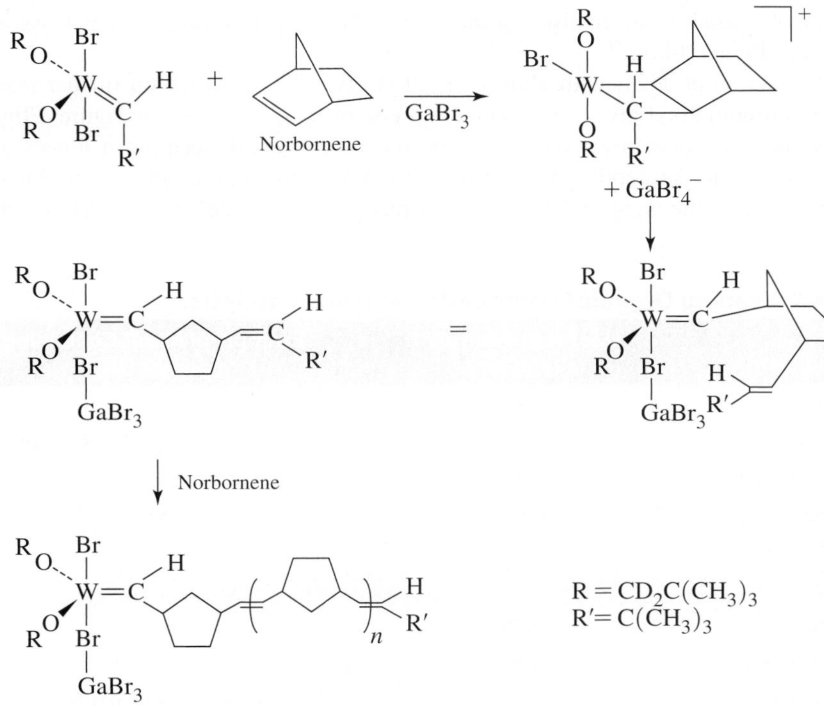

Norbornene GaBr$_3$ + GaBr$_4^-$ Norbornene

R = CD$_2$C(CH$_3$)$_3$
R′= C(CH$_3$)$_3$

FIGURE 14.34 Polymerization of Norbornene using a Carbene Catalyst.

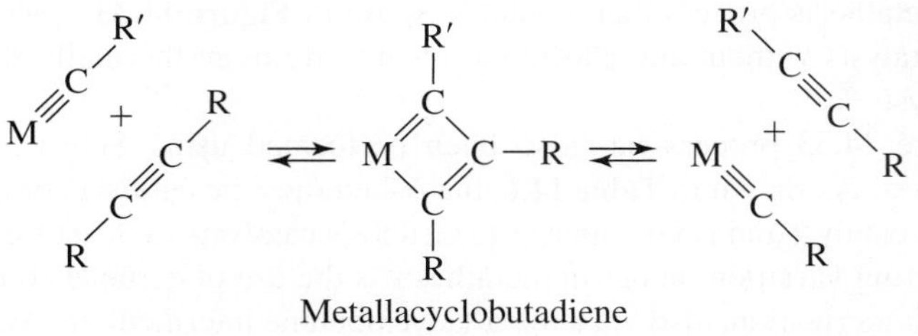

FIGURE 14.35 Alkyne Metathesis.

14.4 Heterogeneous Catalysts

In addition to homogeneous catalytic processes, heterogeneous processes, involving solid catalytic species, are very important, although the exact nature of the reactions occurring on the surface of the catalyst may be difficult to ascertain. The vast majority of the organic chemicals produced in greatest quantities in the United States are produced commercially by processes that involve metal catalysts; most of these processes involve heterogeneous catalysis. Selected examples from 2010 are given in **Table 14.7**.[54]

In many cases, the methods of preparing the catalysts and information on their function are proprietary. Nevertheless, it is important to mention several of these processes as important applications of organometallic reactions.

14.4.1 Ziegler–Natta Polymerizations

In 1955, Ziegler and coworkers reported that solutions of $TiCl_4$ in hydrocarbon solvents in the presence of $Al(C_2H_5)_3$ gave heterogeneous mixtures capable of polymerizing ethylene.[55] Subsequently, many other heterogeneous processes were developed for polymerizing alkenes, using aluminum alkyls in combination with transition-metal complexes. An outline of a possible mechanism for the Ziegler–Natta process proposed by Cossee and Arlman is in **Figure 14.36**.[56]

First, reaction of $TiCl_4$ with aluminum alkyl gives $TiCl_3$, which on further reaction with the aluminum alkyl gives a titanium alkyl complex, as shown in the figure. Ethylene (or propylene) can then insert into the titanium–carbon bond, forming a longer alkyl. This alkyl is further susceptible to insertion of ethylene to lengthen the chain. Although the mechanism of the Ziegler–Natta process has proved difficult to understand, direct

TABLE 14.7 Leading Organic Compounds and Metal Catalysts

Compound	U.S. Production 2010 ($\times 10^9$ kg)	Example of Metal-Containing Catalyst Used
Ethylene	23.97	
Propylene	14.08	$TiCl_3$ or $TiCl_4$ + AlR_3(R = alkyl)
1,2-Dichloroethane	8.81	$FeCl_3$, $AlCl_3$
Benzene	6.05	Pt on Al_2O_3 support
Ethylbenzene	4.24	$AlCl_3$
Styrene	4.10	ZnO, Cr_2O_3
Cumene	3.48	
Ethylene oxide	2.66	Ag
1,3-Butadiene	1.58	Fe_2O_3 or other metal oxide
Vinyl acetate	1.39	Pd salts

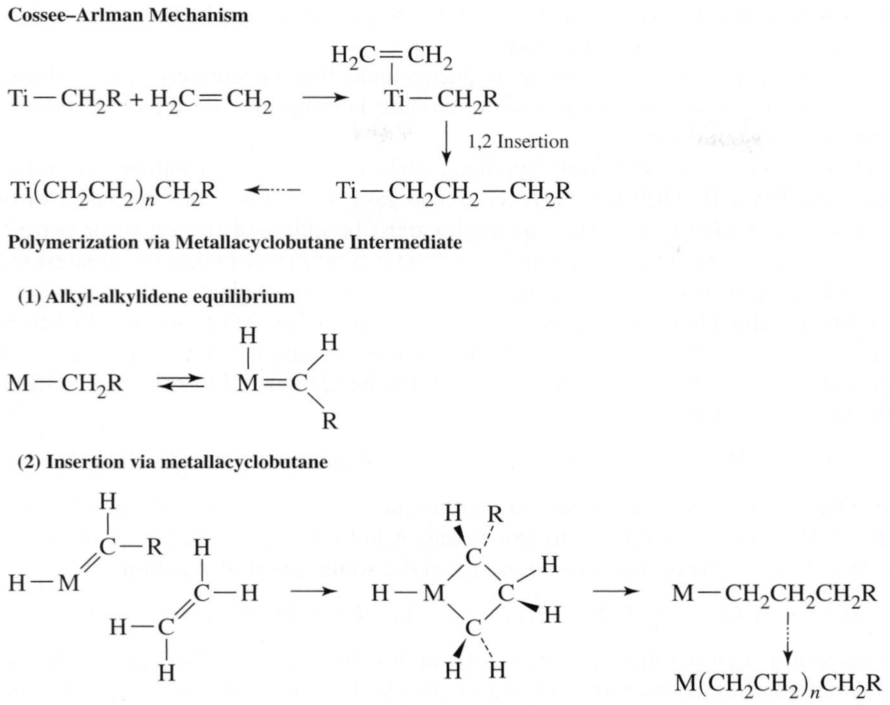

Cossee–Arlman Mechanism

$$\mathrm{Ti{-}CH_2R + H_2C{=}CH_2 \longrightarrow \underset{\displaystyle Ti{-}CH_2R}{\overset{\displaystyle H_2C{=}CH_2}{|}}}$$

$$\downarrow \text{ 1,2 Insertion}$$

$$\mathrm{Ti(CH_2CH_2)_nCH_2R \longleftarrow Ti{-}CH_2CH_2{-}CH_2R}$$

Polymerization via Metallacyclobutane Intermediate

(1) Alkyl-alkylidene equilibrium

$$\mathrm{M{-}CH_2R \rightleftarrows H{-}M{=}C(H)(R)}$$

(2) Insertion via metallacyclobutane

$$\mathrm{H{-}M{=}C(H){-}R + H_2C{=}CH_2 \longrightarrow H{-}M(CHR)(CH_2)C(H)(H) \longrightarrow M{-}CH_2CH_2CH_2R}$$

$$\downarrow$$

$$\mathrm{M(CH_2CH_2)_nCH_2R}$$

FIGURE 14.36 Ziegler–Natta Polymerization.

insertions of multiply bonded organics into titanium–carbon bonds have been demonstrated, supporting the Cossee–Arlman mechanism.[57]

However, an alternative mechanism, involving a metallacyclobutane intermediate, has also been proposed.[58] This mechanism, also shown in Figure 14.36, involves the initial formation of alkylidene from a metal alkyl complex, followed by addition of ethylene to give the metallacyclobutane, which then yields a product having ethylene inserted into the original metal–carbon bond. Distinguishing between these mechanisms has been extremely challenging, but experiments by Grubbs have strongly supported the Cossee–Arlman mechanism as the likely pathway for polymerization in most cases.[59] In at least one example, however, there is strong evidence for ethene polymerization involving a metallacycle intermediate.[60]

14.4.2 Water Gas Reaction

This reaction occurs at elevated temperatures and pressures between water (steam) and natural sources of carbon, such as coal or coke:

$$H_2O + C \longrightarrow H_2 + CO$$

The products of this reaction, an equimolar mixture of H_2 and CO, called *synthesis gas* or *syn gas* (some CO_2 may be produced as a by-product), can be used with metallic heterogeneous catalysts in the synthesis of a variety of useful organic products. For example, the **Fischer–Tropsch process**, developed by German chemists in the early 1900s, uses transition-metal catalysts to prepare hydrocarbons, alcohols, alkenes, and other products from synthesis gas.[61] For example:

$$H_2 + CO \longrightarrow \text{Alkanes} \qquad \text{Co catalyst}$$

$$3H_2 + CO \longrightarrow CH_4 + H_2O \qquad \text{Ni catalyst}$$

$$2H_2 + CO \longrightarrow CH_3OH \qquad \text{Co or Zn/Cu catalyst}$$

Various heterogeneous catalysts are used industrially—for example, transition metals on Al_2O_3 and mixed transition-metal oxides.

Most of these processes have been conducted under heterogenous conditions. However, there has been considerable interest in developing homogenous systems to catalyze the Fischer–Tropsch conversion.

These processes for obtaining synthetic fuels were used by a number of countries during World War II. They are, however, uneconomical in most cases because hydrogen and carbon monoxide in sufficient quantities must be obtained from coal or petroleum sources. Currently, South Africa, which has large coal reserves, makes the greatest use of Fischer–Tropsch reactions in the synthesis of fuels in its Sasol plants.

In **steam reforming**, natural gas—consisting chiefly of methane—is mixed with steam at high temperatures and pressures over a heterogeneous catalyst to generate carbon monoxide and hydrogen, resulting in the reverse of the Ni-catalyzed Fischer-Tropsch reaction on the preceding page:

$$CH_4 + H_2O \longrightarrow CO + 3H_2 \qquad \text{Ni catalyst, 700 °C to 1000 °C}$$

Other alkanes also react with steam to give mixtures of CO and H_2. Steam reforming is the principal industrial source of hydrogen gas. Additional hydrogen can be produced by recycling the CO to react further with steam in the **water gas shift reaction**:

$$CO + H_2O \longrightarrow CO_2 + H_2 \qquad \text{Fe}-\text{Cr or Zn}-\text{Cu catalyst, 400 °C}$$

This reaction is favored thermodynamically: at 400 °C, $\Delta G° = -14.0$ kJ/mol. Removal of CO_2 by chemical means from the product can yield hydrogen of greater than 99 percent purity. This reaction has been studied extensively with the objective of being able to catalyze formation of H_2 *homo*geneously.[62] An example is in **Figure 14.37**.[63] However, these processes have not yet proved efficient enough for commercial use.

In general, these processes, when performed using heterogeneous catalysts, require significantly elevated temperatures and pressures. There is interest in developing homogeneous catalysts that can perform the same functions under much milder conditions.

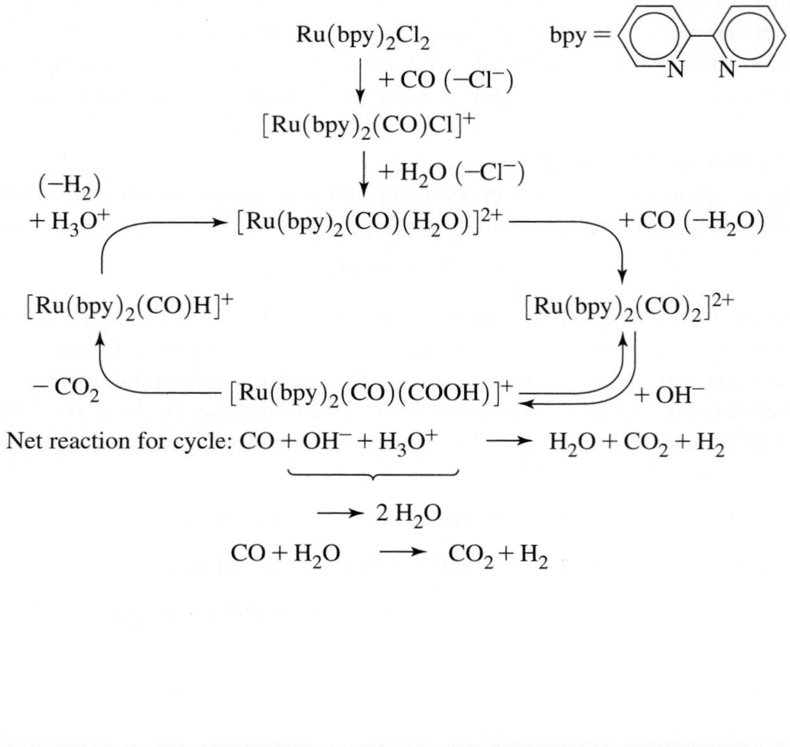

FIGURE 14.37 Homogeneous Catalysis of Water Gas Shift Reaction.
(Adapted with permission from H. Ishida, K. Tanaka, M. Morimoto, T. Tanaka, *Organometallics* **1986**, *5*, 724. Copyright 1986 American Chemical Society.)

$Ru(bpy)_2Cl_2$ bpy =

$\downarrow + CO\ (-Cl^-)$

$[Ru(bpy)_2(CO)Cl]^+$

$\downarrow + H_2O\ (-Cl^-)$

$(-H_2)$ $+ H_3O^+$ $\longrightarrow$ $[Ru(bpy)_2(CO)(H_2O)]^{2+}$ $\longrightarrow$ $+ CO\ (-H_2O)$

$[Ru(bpy)_2(CO)H]^+$ $[Ru(bpy)_2(CO)_2]^{2+}$

$-CO_2$ $[Ru(bpy)_2(CO)(COOH)]^+$ $\rightleftharpoons$ $+ OH^-$

Net reaction for cycle: $CO + OH^- + H_3O^+ \longrightarrow H_2O + CO_2 + H_2$

$\longrightarrow 2\,H_2O$

$CO + H_2O \longrightarrow CO_2 + H_2$

References

1. W. D. Covey, T. L. Brown, *Inorg. Chem.*, **1973**, *12*, 2820.
2. C. A. Tolman, *J. Am. Chem. Soc.*, **1970**, *92*, 2956; *Chem. Rev.*, **1977**, *77*, 313. K. A. Bunten, L. Chen, A. L. Fernandez, A. J. Poë, *Coord. Chem. Rev.*, **2002**, *233–234*, 41.
3. H. Clavier, S. P. Nolan, *Chem. Commun.*, **2010**, *46*, 841. This reference also provides percent buried volumes for a variety of N-heterocyclic carbenes.
4. A. Poater, B. Cosenza, A. Correa, S. Giudice, F. Ragone, V. Scarano, L. Cavallo, *Eur. J. Inorg. Chem.*, **2009**, 1759.
5. J. A. Love, M. S. Sanford, M. W. Day, R. H. Grubbs, *J. Am. Chem. Soc.*, **2003**, *125*, 10103.
6. D. J. Darensbourg, A. H. Graves, *Inorg. Chem.*, **1979**, *18*, 1257.
7. J. M. Buchanan, J. M. Stryker, R. G. Bergman, *J. Am. Chem. Soc.*, **1986**, *108*, 1537.
8. G. Mann, D. Baranano, J. F. Hartwig, A. L. Rheingold, I. A. Guzei, *J. Am. Chem. Soc.*, **1998**, *120*, 9205.
9. J. F. Hartwig, *Inorg. Chem.*, **2007**, *46*, 1936.
10. C. C. C. Johansson Seechurn, M. O. Kitching, T. J. Colacot, V. Snieckus, *Angew. Chem., Int. Ed.*, **2012**, *51*, 5062.
11. M. E. Thompson, S. M. Baxter, A. R. Bulls, B. J. Burger, M. C. Nolan, B. D. Santarsiero, W. P. Schaefer, J. E. Bercaw, *J. Am. Chem. Soc.*, **1987**, *109*, 203.
12. Z. Lin, *Coord. Chem. Rev.*, **2007**, *251*, 2280.
13. D. Morales-Morales and C. M. Jensen, *The Chemistry of Pincer Compounds*, 2007, Elsevier.
14. W. H. Bernskoetter, C. K. Schauer, K. I. Goldberg, M. Brookhart, *Science*, **2009**, *326*, 553.
15. T. C. Flood, J. E. Jensen, J. A. Statler, *J. Am. Chem. Soc.*, **1981**, *103*, 4410 and references therein.
16. T. C. Flood, J. E. Jensen, J. A. Statler, *J. Am. Chem. Soc.*, **1981**, *103*, 4410 and references therein.
17. P. Cossee, *J. Catal.*, **1964**, *3*, 80; E. J. Arlman, P. Cossee, *J. Catal.*, **1964**, *3*, 99.
18. J. D. Fellmann, R. R. Schrock, D. D. Traficante, *Organometallics*, **1982**, *1*, 481; J. F. Hartwig, *Organotransition Metal Chemistry, From Bonding to Catalysis*, University Science Books, Mill Valley, CA, 2010, and references therein.
19. J. W. Lauher, R. Hoffmann, *J. Am. Chem. Soc.*, **1976**, *98*, 1729 and references therein.
20. K. Weissermel and H.-J. Arpe, *Industrial Organic Chemistry*, Wiley-VCH, Weinheim, 2003.
21. R. F. Heck, D. S. Breslow, *J. Am. Chem. Soc.*, **1961**, *83*, 4023; see also F. Heck, *Adv. Organomet. Chem.*, **1966**, *4*, 243.
22. C-F. Huo, Y.-W. Li, M. Beller, H. Jiao, *Organometallics*, **2003**, *22*, 4665.
23. M. Brookhart, M. L. H. Green, G. Parkin, *Proc. Nat. Acad. Sci. USA*, **2007**, *104*, 6908.
24. T. Ziegler and L. Versluis, "The Tricarbonylhydridocobalt-Based Hydroformylation Reaction," in W. R. Moser and D. W. Slocum, eds., *Homogeneous Transition Metal-Catalyzed Reactions*, American Chemical Society, Washington, DC, 1992, pp. 75–93.
25. F. Hebrard, P. Kalak, *Chem. Rev.*, **2009**, *109*, 4272.
26. L. H. Slaugh, R. D. Mullineaux, *J. Organomet. Chem.*, **1968**, *13*, 469.
27. R. J. Klingler, M. J. Chen, J. W. Rathke, K. W. Kramarz, *Organometallics*, **2007**, *26*, 352.
28. J. A. Osborne, J. F. Young, G. Wilkinson, *Chem. Commun.*, **1965**, 17; C. K. Brown, G. Wilkinson, *J. Chem. Soc., A*, **1970**, 2753.
29. C. P. Casey, G. T. Whiteker, M. G. Melville, L. M. Petrovich, J. A. Garvey, Jr., D. R. Powell, *J. Am. Chem. Soc.*, **1992**, *114*, 5535.
30. L. A. van der Veen, P. H. Keeven, G. C. Schoemaker, J. N. H. Reek, P. C. J. Kramer, P. W. N. M. van Leeuwen, M. Lutz, A. L. Spek, *Organometallics*, **2000**, *19*, 872 and references cited therein.
31. J. A. Gillespie, D. L. Dodds, P. C. J. Kramer, *Dalton Trans.*, **2010**, *39*, 2751.
32. A. Gual, C. Godard, S. Castillón, C. Claver, *Tetrahedron: Asymmetry*, **2010**, *21*, 1135.
33. M. Cheong, R. Schmid, T. Ziegler, *Organometallics*, **2000**, *19*, 1973 and references therein.
34. J. Forster, *J. Chem. Soc., Dalton Trans.*, **1979**, 1639; A. Haynes, et al., *J. Am. Chem. Soc.*, **2004**, *126*, 2847.
35. A. Haynes, *Adv. Catal.*, **2010**, *53*, 1. C. M. Thomas, G. Süss-Fink, *Coord. Chem. Rev.*, **2003**, *243*, 125.
36. J. Smidt, W. Hafner, R. Jira, J. Sedlmeier, R. Sieber, H. Kojer, R. Rüttinger, *Angew. Chem.*, **1959**, *72*, 176.
37. R. Jira, *Angew. Chem., Int. Ed.*, **2009**, *48*, 9034.
38. J. A. Keith, P. M. Henry, *Angew. Chem., Int. Ed.*, **2009**, *48*, 9038.
39. G. Kovacs, A. Stirling, A. Lledos, G. Ujaque, *Chem. Eur. J.*, **2012**, *18*, 5612; J. A. Keith, R. J. Nielsen, J. Oxgaard, W. A. Goddard, III, *Organometallics*, **2009**, *28*, 1618; J. A. Keith, R. J. Nielsen, J. Oxgaard, W. A. Goddard, III, *J. Am. Chem. Soc.*, **2007**, *129*, 12342.
40. B. R. James, *Adv. Organomet. Chem.*, **1979**, *17*, 319; see also J. P. Collman, L. S. Hegedus, J. R. Norton, and R. G. Finke, *Principles and Applications of Organotransition Metal Chemistry*, University Science Books, Mill Valley, CA, 1987, pp. 531–535 and references therein.
41. R. H. Grubbs, *Angew. Chem, Int. Ed.*, **2006**, *45*, 3760; R. R. Schrock, *Adv. Synth. Cat.*, **2007**, *349*, 41; Y. Chauvin, *Adv. Synth. Cat.*, **2008**, *349*, 27. These Nobel lectures can also be found on the Nobel Prize Web site at http://nobelprize.org/nobel_prizes/chemistry/laureates/2005/.
42. N. Calderon, E. A. Ofstead, J. P. Ward, W. A. Judy, K. W. Scott, *J. Am. Chem. Soc.*, **1968**, *90*, 4133.
43. C. P. C. Bradshaw, E. J. Howman, L. Turner, *J. Catal.*, **1967**, *7*, 269.
44. J.-L. Hérisson, Y. Chauvin, *Makromol. Chem.*, **1971**, *141*, 161.
45. R. H. Grubbs, P. L. Burk, D. D. Carr, *J. Am. Chem. Soc.*, **1975**, *97*, 3265.

46. T. J. Katz, J. McGinness, *J. Am. Chem. Soc.*, **1975**, *97*, 1592 and **1977**, *99*, 1903.

47. A. Fürstner, K. Langemann, *J. Org. Chem.*, **1996**, *61*, 8746.

48. E. L. Dias, S. T. Nguyen, R. H. Grubbs, *J. Am. Chem. Soc.*, **1997**, *119*, 3887.

49. M. S. Sanford, J. A. Love, R. H. Grubbs, *J. Am. Chem. Soc.*, **2001**, *123*, 6543.

50. J. Huang, H.-J. Schanz, E. D. Stevens, S. P. Nolan, *Organometallics*, **1999**, *18*, 2370.

51. L. Ackermann, D. El Tom, A. Fürstner, *Tetrahedron*, **2000**, *56*, 2195.

52. S. B. Garber, J. S. Kingsbury, B. L. Gray, A. H. Hoveyda, *J. Am. Chem. Soc.*, **2000**, *122*, 8168.

53. J. Kress, J. A. Osborn, R. M. E. Greene, K. J. Ivin, J. J. Rooney, *J. Am. Chem. Soc.*, **1987**, *109*, 899.

54. W. A. Nugent and J. M. Mayer, *Metal–Ligand Multiple Bonds*, Wiley Interscience, New York, 1988, p. 311 and references therein; U. H. W. Bunz, L. Kloppenburg, *Angew. Chem., Int. Ed.*, **1999**, *38*, 478.

55. R. Chang and W. Tikkanen, *The Top Fifty Industrial Chemicals*, Random House, New York, 1988; *Chem. Eng. News*, July 2, 2011, p. 57.

56. K. Ziegler, E. Holzkamp, H. Breiland, H. Martin, *Angew. Chem.*, **1955**, *67*, 541.

57. J. Cossee, *J. Catal.*, **1964**, *3*, 80; E. J. Arlman, *J. Catal.*, **1964**, *3*, 89; E. J. Arlman, J. Cossee, *J. Catal.*, **1964**, *3*, 99.

58. J. J. Eisch, A. M. Piotrowski, S. K. Brownstein, E. J. Gabe, F. L. Lee, *J. Am. Chem. Soc.*, **1985**, *107*, 7219.

59. K. J. Ivin, J. J. Rooney, C. D. Stewart, M. L. H. Green, *Chem. Commun.*, **1978**, 604.

60. L. Clauson, J. Sato, S. L. Buchwald, M. L. Steigerwald, R. H. Grubbs, *J. Am. Chem. Soc.*, **1985**, *107*, 3377. For a brief review of experiments used to distinguish between the two mechanisms, see G. O. Spessard and G. L. Miessler, *Organometallic Chemistry*, Prentice Hall, Upper Saddle River, NJ, 1997, pp. 357–369.

61. W. H. Turner, R. R. Schrock, *J. Am. Chem. Soc.*, **1982**, *104*, 2331.

62. E. Fischer, H. Tropsch, *Brennst. Chem.*, **1923**, *4*, 276.

63. M. M. Taqui Khan, S. B. Halligudi, S. Shukla, *Angew. Chem., Int. Ed.*, **1988**, *27*, 1735 and R. Ziessel, *Angew. Chem., Int. Ed.*, **1992**, *30*, 844.

64. J. P. Collins, R. Ruppert, J. P. Sauvage, *Nouv. J. Chim.*, **1985**, *9*, 395.

General References

J. F. Hartwig, *Organotransition Metal Chemistry, From Bonding to Catalysis*, University Science Books, Mill Valley, CA, 2010, provides a detailed discussion, with numerous references, of many of the reactions and catalytic processes described in this chapter, as well as a variety of other types of organometallic reactions. In addition to providing extensive information on the structural and bonding properties of organometallic compounds, G. Wilkinson, F. G. A. Stone, and E. W. Abel, editors of *Comprehensive Organometallic Chemistry*, Pergamon Press, Oxford, UK, 1982, and E. W. Abel, F. G. A. Stone, and G. Wilkinson, editors of *Comprehensive Organometallic Chemistry II*, Pergamon Press, Oxford, 1995, give the most comprehensive information on organometallic reactions, with numerous references to the original literature. Two recent sources of general information and recent references on catalytic processes are P. W. N. M. van Leeuwen's, *Homogeneous Catalysis: Understanding the Art*, Kluwer Academic Publishers, Dordrecht, the Netherlands, 2004, and J. Hagen's *Industrial Catalysis: A Practical Approach*, 2nd ed., Wiley-VCH, Weinheim, Germany, 2006. S. T. Oyama and G. A. Somorjai, "Homogeneous, Heterogeneous, and Enzymatic Catalysis" in *J. Chem. Educ.*, **1988**, *65*, 765, gives examples of the types and amounts of catalysts used in a variety of industrial processes. The references listed at the end of Chapter 13 are also useful in connection with this chapter.

Problems

14.1 Predict the transition metal-containing products of the following reactions:

a. $[Mn(CO)_5]^- + H_2C{=}CH{-}CH_2Cl \longrightarrow$ initial product $\xrightarrow{-CO}$ final product

b. *trans*-$Ir(CO)Cl(PPh_3)_2 + CH_3I \longrightarrow$

c. $Ir(PPh_3)_3Cl \xrightarrow{\Delta}$

d. $(\eta^5\text{-}C_5H_5)Fe(CO)_2(CH_3) + PPh_3 \longrightarrow$

e. $(\eta^5\text{-}C_5H_5)Mo(CO)_3[C({=}O)CH_3] \xrightarrow{\Delta}$

f. $H_3C{-}Mn(CO)_5 + SO_2 \longrightarrow$ (no gases are evolved)

14.2 Predict the transition metal–containing products of the following reactions:

a. $H_3C{-}Mn(CO)_5 + P(CH_3)(C_6H_5)_2 \longrightarrow$ (no gases are evolved)

b. $[Mn(CO)_5]^- + (\eta^5\text{-}C_5H_5)Fe(CO)_2Br \xrightarrow{\Delta}$

c. *trans*-$Ir(CO)Cl(PPh_3)_2 + H_2 \longrightarrow$

d. $W(CO)_6 + C_6H_5Li \longrightarrow$

e. *cis*-$Re(CH_3)(PEt_3)(CO)_4 + {}^{13}CO \longrightarrow$ (show all expected products, percentage of each)

f. *fac*-$Mn(CO)_3(CH_3)(PMe_3)_2 + {}^{13}CO \longrightarrow$ (show all expected products, percentage of each)

14.3 Predict the transition metal–containing products of the following reactions:

a. *cis*-$Mn(CO)_4(^{13}CO)(COCH_3) \xrightarrow{\Delta}$ (show all expected products, percentage of each)

b. $C_6H_5CH_2{-}Mn(CO)_5 \xrightarrow{h\nu} CO +$

c. $V(CO)_6 + NO \longrightarrow$

d. $Cr(CO)_6 + Na/NH_3 \longrightarrow$

e. $Fe(CO)_5 + NaC_5H_5 \longrightarrow$

f. $[Fe(CO)_4]^{2-} + CH_3I \longrightarrow$

g. $H_3C{-}Rh(PPh_3)_3 \xrightarrow{\Delta} CH_4 +$

14.4 Heating $[(\eta^5\text{-}C_5H_5)Fe(CO)_3]^+$ with NaH in solution gives **A**, which has the empirical formula $C_7H_6O_2Fe$. **A** reacts rapidly at room temperature to eliminate a colorless gas **B**, forming a purple-brown solid **C** having the empirical formula $C_7H_5O_2Fe$. Treatment of **C** with iodine generates a brown solid **D** with the empirical formula $C_7H_5O_2FeI$, which on treatment with TlC_5H_5 gives a solid **E** with the formula $C_{12}H_{10}O_2Fe$. **E**, on heating, gives off a colorless gas, leaving an orange solid **F** with the formula $C_{10}H_{10}Fe$. Propose structural formulas for **A** through **F**.

14.5 $Na[(\eta^5\text{-}C_5H_5)Fe(CO)_2]$ reacts with $ClCH_2CH_2SCH_3$ to give **A**, a monomeric and diamagnetic substance of stoichiometry $C_{10}H_{12}FeO_2S$ having two strong IR bands at 1980 and 1940 cm^{-1}. Heating of **A** gives **B**, a monomeric, diamagnetic substance having strong IR bands at 1920 and 1630 cm^{-1}. Identify **A** and **B**.

14.6 The reaction of $V(CO)_5(NO)$ with $P(OCH_3)_3$ to give $V(CO)_4[P(OCH_3)_3](NO)$ has the rate law

$$\frac{-d[V(CO)_5(NO)]}{dt} = k_1\,[V(CO)_5(NO)] + k_2[P(OCH_3)_3][V(CO)_5(NO)]$$

a. Suggest a mechanism for this reaction consistent with the rate law.

b. One possible mechanism consistent with the last term in the rate includes a transition state of formula $V(CO)_5[P(OCH_3)_3](NO)$. Would this necessarily be a 20-electron species? Explain.

14.7 The rate law for the reaction
$H_2 + Co_2(CO)_8 \longrightarrow 2\,HCo(CO)_4$ is

$$\text{Rate} = \frac{k[Co_2(CO)_8][H_2]}{[CO]}$$

Propose a mechanism consistent with this rate law.

14.8 Which of the following *trans* complexes would you expect to react most rapidly with CO? Which would you expect to react least rapidly? Briefly explain your choices.

$Cr(CO)_4(PPh_3)_2$
$Cr(CO)_4(PPh_3)(PBu_3)$ (Bu = *n*-butyl)
$Cr(CO)_4(PPh_3)[P(OMe)_3]$
$Cr(CO)_4(PPh_3)[P(OPh)_3]$

(See M. J. Wovkulich, J. D. Atwood, *Organometallics*, **1982**, *1*, 1316.)

14.9 The equilibrium constants for the ligand dissociation reaction $NiL_4 \rightleftharpoons NiL_3 + L$ have been determined for a variety of phosphines. (See C. A. Tolman, W. C. Seidel, L. W. Gosser, *J. Am. Chem. Soc.*, **1974**, *96*, 53). For L = PMe_3, PEt_3, $PMePh_2$, and PPh_3, arrange these equilibria in order of the expected magnitudes of their equilibrium constants (from largest *K* to smallest).

14.10 In addition to pioneering the cone angle concept, C. A. Tolman proposed a parameter $\mathcal{X}$ (chi) as a measure of the electronic effect of phosphine and related ligands, based on infrared spectra of complexes containing these ligands (C. A. Tolman, *J. Am. Chem. Soc.*, **1970**, *92*, 2953).

a. What was the general formula of the complexes used by Tolman?

b. How was $\mathcal{X}$ defined?

c. What types of ligands had high values of $\mathcal{X}$?

d. To what extent does this approach distinguish between the sigma donor and pi acceptor nature of the ligands studied?

14.11 The nickel(II) pincer complexes $[N(o\text{-}C_6H_4PR_2)_2]NiX$ (**1**: R = Ph, X = H; **2**: R = Ph, X = Me) undergo different reactions with CO on the basis of the R and X groups, but the phosphine arms remain attached to the Ni center (See L.-C. Liang, Y.-T. Hung, Y.-L. Huang, P.-Y. Lee, W.-C. Chen, *Organometallics*, **2012**, *31*, 700 for structural details of these complexes.)

a. When R = Ph and X = H, the addition of CO results in the disappearance of a ^1H NMR resonance at –18 ppm, with the corresponding new signal (that also integrates to one H) at 8.61 ppm. Two $\upsilon(CO)$ stretching bands appear in the IR spectrum at 1993 and 1924 cm^{-1}. Propose a structure for this product and a mechanism for its formation.

b. When R = Ph and X = Me, the addition of CO results in the appearance of an intermediate with a single $\upsilon(CO)$ infrared band at 1621 cm^{-1}. This intermediate subsequently provides another nickel complex with three $\upsilon(CO)$ infrared bands (2002, 1943, 1695 cm^{-1}). Propose the structures of the intermediate and final product, and the reaction mechanisms involved in these reactions.

c. Why do the reactions proceed differently when X = H or X = Me?

14.12 The *N,N*-dibenzylcyclam ligand supports a novel double-cyclometallation reaction of zirconium(IV) (R = ^{t}Bu, Me, CH_2Ph) in the presence of 4 atm H_2 as depicted in Scheme 1 of R. F. Munhá, J. Ballmann, L. F. Veiros, B. O. Patrick, M. D. Fryzuk, *Organometallics*, **2012**, *31*, 4937.

a. Propose a mechanism for this reaction, which proceeds most slowly for R = CH_2Ph, that is consistent with the rate being independent of H_2 pressure.

b. Explain the reasoning for your selection. Why might the reaction be slowest with R = CH_2Ph)?

14.13 The complex shown below loses carbon monoxide on heating. Would you expect this carbon monoxide to be ^{12}CO, ^{13}CO, or a mixture of both? Why?

$$\mathrm{OC{-}Mn(^{13}C(=O)CH_3)(CO)(PPh_3)_2(CO)}$$

14.14 a. Predict the products of the following reaction, showing clearly the structure of each:

$$\mathrm{(OC)_3Mn(CH_3)(PMe_3)_2} + {}^{13}\mathrm{CO} \longrightarrow$$

b. Each product of this reaction has a new, rather strong IR band that is distinctly different in energy from any bands in the reactants. Account for this band, and predict its approximate location (in cm^{-1}) in the IR spectrum.

14.15 Give structural formulas for **A** through **D**:

$$(\eta^5\text{-}C_5H_5)_2Fe_2(CO)_4 \xrightarrow{\text{Na/Hg}} \mathbf{A} \xrightarrow{\text{Br}_2}$$

$$\mathbf{B} \xrightarrow{\text{LiAlH}_4} \mathbf{C} \xrightarrow{\text{PhNa}} \mathbf{A} + \mathbf{D} \text{ (a hydrocarbon)}$$

$\nu_{CO} = 1961, 1942, 1790\ cm^{-1}$ for $(\eta^5\text{-}C_5H_5)_2Fe_2(CO)_4$

A has strong IR bands at 1880 and 1830 cm^{-1}; **C** has a ^1H NMR spectrum consisting of two singlets of relative intensity 1:5 at approximately δ –12 ppm and δ 5 ppm, respectively. (Hint: Metal hydrides often have protons with negative chemical shifts.)

14.16 $Re(CO)_5Br$ reacts with the ion $Br{-}CH_2CH_2{-}O^-$ to give compound **Y** + Br^-.

a. What is the most likely site of attack of this ion on $Re(CO)_5Br$? (Hint: Consider the hardness (see Chapter 6) of the Lewis base.)

b. Using the following information, propose a structural formula of **Y**, and account for each of the following:
Y obeys the 18-electron rule.
No gas is evolved in the reaction.
^{13}C NMR indicates that there are five distinct magnetic environments for carbon in **Y**.
Addition of a solution of Ag^+ to a solution of **Y** gives a white precipitate.
(See M. M. Singh, R. J. Angelici, *Inorg. Chem.*, **1984**, *23*, 2699.)

14.17 What is a *nitrenium* ligand? How does it compare with an *N*-heterocyclic carbene? Which of these types of ligand is viewed as a stronger sigma donor, and what experimental evidence supports this view? What role do pincer ligands play in nitrenium complexes? (See Y. Tulchinsky, M. A. Iron, M. Botoshansky, M. Gandelman, *Nature Chem.*, **2011**, *3*, 525.)

14.18 The carbene complex **I** shown below undergoes the following reactions. Propose structural formulas for the reaction products.

$$\mathrm{(OC)_4BrRe{=}C(OCH_2CH_2O)}$$

a. When a toluene solution containing **I** and excess triphenylphosphine is heated to reflux, compound **II** is formed first, and then compound **III**. **II** has infrared bands at 2038, 1958, and 1906 cm^{-1} and **III** at 1944 and 1860 cm^{-1}. ^1H NMR data δ values (relative area) are as follows:
II: 7.62, 7.41 multiplets (15)
4.19 multiplet (4)
III: 7.70, 7.32 multiplets (15)
3.39 singlet (2)

b. When a solution of **I** in toluene is heated to reflux with 1,1-bis(diphenylphosphino)methane, a colorless product **IV** is formed that has the following properties:
IR: 2036, 1959, 1914 cm^{-1}
Elemental analysis (accurate to $\pm 0.3\%$): 35.87% C, 2.73% H

c. **I** reacts rapidly with the dimethyldithiocarbamate ion $S_2CN(CH_3)_2^-$ in solution to form $Re(CO)_5Br$ + **V**, a product that does not contain a metal atom. This product has no infrared bands between 1700 and 2300 cm^{-1}. However, it does show moderately intense bands at 1500 and 977 cm^{-1}. The ^1H NMR spectrum of **V** shows bands at δ 3.91 (triplet), 3.60 (triplet), 3.57 (singlet), and 3.41 (singlet).
(See G. L. Miessler, S. Kim, R. A. Jacobson, R. J. Angelici, *Inorg. Chem.*, **1987**, *26*, 1690.)

14.19 The complex **I** in the preceding problem can be synthesized from $Re(CO)_5Br$ and 2-bromoethanol in ethylene oxide solution with solid NaBr present. Suggest a mechanism for the formation of the carbene ligand.

14.20 $BrCH_2CH_2CH_2Mn(CO)_5$ is formed by reaction of $[Mn(CO)_5]^-$ with 1,3-dibromopropane. However, the reaction does not stop here; the product reacts with additional $[Mn(CO)_5]^-$ to yield a carbene complex. Propose a structure for this complex, and suggest a mechanism for its formation.

14.21 An acyl metal carbonyl ($R{-}C({=}O)M(CO)_x$) is generally easier to protonate than either a metal carbonyl or an organic ketone, such as acetone. Suggest an explanation.

14.22 Show how transition-metal complexes could be used to effect the following syntheses:
a. Acetaldehyde from ethylene
b. $CH_3CH_2COOCH_3$ from CH_3CH_2Cl
c. $CH_3CH_2CH_2CH_2CHO$ from $CH_3CH_2CH{=}CH_2$
d. $PhCH_2CH_2CH_2CHO$ from an alkene (Ph = phenyl)

e.

AcO H → D_2 → AcO H D D

f. $C_6D_5CH_3$ from toluene, $C_6H_5CH_3$.

14.23 The complex $Rh(H)(CO)_2(PPh_3)_2$ can be used in the catalytic synthesis of *n*-pentanal from an alkene having one less carbon. Propose a mechanism for this process. Give an appropriate designation for each type of reaction step (such as oxidative addition or alkyl migration), and identify the catalytic species.

14.24 It is possible to synthesize the following aldehyde from an appropriate 5-carbon alkene by using an appropriate transition-metal catalyst:

$$H_3C-CH_2-\underset{}{\overset{CH_3}{\overset{|}{C}H}}-CH_2-\overset{O}{\overset{\|}{C}}-H$$

Show how this synthesis could be effected catalytically. Identify the catalytic species.

14.25 Predict the products if the following compounds undergo metathesis:

a. cyclooctene + 1-butene ⟶

b. cyclopentene + C_2H_5CH=CHCH_3 (cis)

c. 1-butene + 2-butene

d. 1,7-octadiene

14.26 The complex $(CO)_5$**M**$=C(C_6H_5)_2$ [**M** = third row transition metal] can catalyze ROMP of 1–methylcyclobutene. Identify M, show the initial steps in this process, and clearly identify the structure of the polymer.

14.27 One of the classical experiments in the development of olefin metathesis was the "double cross" metathesis in which a mixture of cyclooctene, 2-butene, and 4-octene underwent metathesis.

a. What products would be expected from this metathesis?

b. How would the formation of these products differ in the pairwise and non-pairwise mechanisms? (See T. J. Katz, J. McGinness, *J. Am. Chem. Soc.*, **1975**, *97*, 1592.)

14.28 The complex $(\eta^5\text{-}C_5H_5)_2Zr(CH_3)_2$ reacts with the highly electrophilic borane $HB(C_6F_5)_2$ to form a product having stoichiometry $(CH_2)[HB(C_6F_5)_2]_2[(C_5H_5)_2Zr]$, and the product is a rare example of pentacoordinate carbon.

a. Suggest a structure for this product.

b. An isomer of this product, $[(C_5H_5)_2ZrH]^+$ $[CH_2\{B(C_6F_5)_2\}_2(\mu\text{–}H)]^-$, has been proposed as a potential Ziegler–Natta catalyst. Suggest a mechanism by which this isomer might serve as such a catalyst.

(See R. E. von H. Spence, D. J. Parks, W. E. Piers, M. MacDonald, M. J. Zaworotko, S. J. Rettig, *Angew. Chem., Int. Ed.*, **1995**, *34*, 1230.)

14.29 Terminal tungsten nitride complexes containing W≡N bonds have been reported to catalyze nitrile–alkyne cross-metathesis (NACM) by a mechanism that has parallels with the non-pairwise mechanism for olefin metathesis (A. M. Geyer, E. S. Weidner, J. B. Gary, R. L. Gdula, N. C. Kuhlmann, M. J. A. Johnson, B. D. Dunietz, J. W. Kampf, *J. Am. Chem. Soc.*, **2008**, *130*, 8984.) When *p*-methoxybenzonitrile and 3-hexyne were mixed in the presence of catalyst candidate $N{\equiv}W(OC(CF_3)_2Me)_3(DME)$, two principal products attributable to metathesis were observed.

a. Outline the NACM mechanism, as described in the reference.

b. Account for the formation of the two products, and suggest which was formed first.

14.30 At low temperature and pressure, a gas-phase reaction can occur between iron atoms and toluene. The product, a rather unstable sandwich compound, reacts with ethylene to give compound **X**. Compound **X** decomposes at room temperature to liberate ethylene; at –20 °C it reacts with $P(OCH_3)_3$ to give Fe(toluene)$[P(OCH_3)_3]_2$. Suggest a structure for compound **X**. (See U. Zenneck, W. Frank, *Angew. Chem., Int. Ed.*, **1986**, *25*, 831.)

14.31 The reaction of $RhCl_3 \cdot 3H_2O$ with tri-*o*-tolylphosphine in ethanol at 25 °C gives a blue-green complex **I** ($C_{42}H_{42}P_2Cl_2Rh$) that has ν(Rh—Cl) at 351 cm^{-1} and μ_{eff} = 2.3 BM. At a higher temperature, a diamagnetic yellow complex **II** that has an Rh:Cl ratio of 1:1 is formed that has an intense band near 920 cm^{-1}. Addition of NaSCN to **II** replaces Cl with SCN to give a product **III** having the following ^1H NMR spectrum:

Chemical Shift	Relative Area	Type
6.9–7.5	12	Aromatic
3.50	1	Doublet of 1:2:1 triplets
2.84	3	Singlet
2.40	3	Singlet

Treatment of **II** with NaCN gives a phosphine ligand **IV** with the empirical formula $C_{21}H_{19}P$ and a molecular weight of 604. **IV** has an absorption band at 965 cm^{-1} and the following ^1H NMR spectrum:

Chemical Shift	Relative Area	Type
7.64	1	Singlet
6.9–7.5	12	Aromatic
2.37	6	Singlet

Determine the structural formulas of compounds **I** through **IV**, and account for as much of the data as possible. (See M. A. Bennett, P. A. Longstaff, *J. Am. Chem. Soc.*, **1969**, *91*, 6266.)

14.32 Ring-closing metathesis (RCM) is not restricted to alkenes; similar reactions are also known for alkynes. The tungsten alkylidyne complex $W(\equiv CCMe_3)(OCMe_3)_3$ has been used to catalyze such reactions. Predict the structures of the cyclic products for metathesis of

a. $[MeC\equiv C(CH_2)_2OOC(CH_2)]_2$

b. $MeC\equiv C(CH_2)_8COO(CH_2)_9C\equiv CMe$

(See A Fürstner, G. Seidel, *Angew. Chem., Int. Ed.*, **1998**, *37*, 1734.)

14.33 More than two decades elapsed between the development of a homogeneous catalytic process for producing acetic acid in the late 1960s and convincing experimental evidence for the key intermediate $[CH_3Rh(CO)_2I_3]^-$ (Figure 14.20). Describe the evidence presented for this intermediate. (A. Haynes, B. E. Mann, D. J. Gulliver, G. E. Morris, P. M. Maitlis, *J. Am. Chem. Soc.*, **1991**, *113*, 8567.)

14.34 The compound $Fe(CO)_4I_2$ reacts with cyanide in methanol solution to form complex **A**, which has intense IR bands at 2096 and 2121 cm^{-1} and less intense bands at 2140 and 2162 cm^{-1}. Reaction of **A** with additional cyanide yields **B**. Ion **B** also has two pairs of infrared bands, a more intense pair at 1967 and 2022 cm^{-1} and a less intense pair at 2080 and 2106 cm^{-1}. Neither **A** nor **B** contains iodine. Propose structures of **A** and **B**. (See J. Jiang, S. A. Koch, *Inorg. Chem.*, **2002**, *41*, 158.)

14.35 Cation **1** reacts with the ion $HB(sec\text{-}C_4H_9)_3^-$, a potential source of hydride, to form **2**. The following data are reported for **2**:

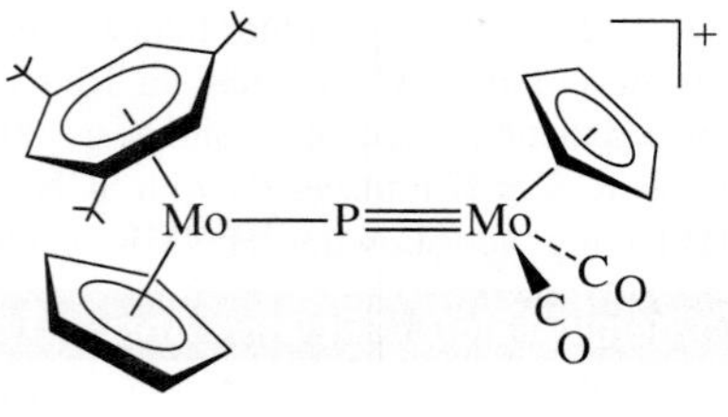

Mo—P≡Mo CO CO +

1

IR: strong bands at 1920, 1857 cm^{-1}.

^{1}H NMR: chemical shift (relative area):

5.46 (2)
5.28 (5)
5.15 (3)
4.22 (2)
1.31 (27)

In addition, a small peak is believed to be hidden under other peaks.

^{13}C NMR: resonance at 236.9 ppm, seven additional peaks and clusters of peaks between 32.4 and 115.7 ppm. Propose a structure for **2**, and account for as much of the experimental data as possible. (See I. Amor, D. García-Vivó, M. E. García, M. A. Ruiz, D. Sáez, H. Hamidov, J. C. Jeffery, *Organometallics*, **2007**, *26*, 466.)

14.36 Sometimes an organometallic complex can act as the intermediate in the synthesis of organic compounds. The heterocyclic compound shown below was reacted with $LiBH(C_2H_5)_3$ to form **X**, a halogen-free heterocycle. When **X** was treated with $Cr(CO)_3(CH_3CN)_3$, followed by reaction with HF·pyridine, transition-metal complex **Y**, which had strong absorptions at 1898 and 1975 cm^{-1}, formed. Addition of triphenylphosphine gave a new transition-metal complex plus heterocycle **Z**. The ^{1}H NMR spectrum of **Z** showed four single peaks of equal intensity between δ 6.4 and 7.8 ppm, a quartet at 4.9 ppm, and a triplet at 8.44 ppm. Propose structures for **X**, **Y**, and **Z**. (See A. J. V. Marwitz, M. H. Matus, L. N. Zakharov, D. A. Dixon, S-H. Liu, *Angew. Chem., Int. Ed.*, **2009**, *48*, 973.)

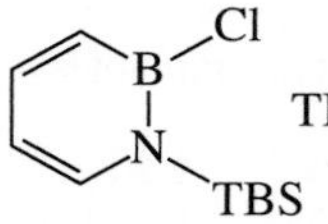

Cl B N TBS

TBS = *tert*-butyldimethylsilylallylamine

14.37 When hexaiodobenzene, C_6I_6, reacts with diferrocenylzinc, $[(\eta^5\text{-}C_5H_5)FeC_5H_4]_2Zn$, one of the products has a C to Fe atom ratio exactly 10 percent higher than in ferrocene and an H to Fe atom ratio 10 percent lower than in ferrocene. The product contains no elements other than those that occur in ferrocene. Suggest a structure of this product. (See Y. Yu, A. D. Bond, P. W. Leonard, U. J. Lorenz, T. V. Timofeeva, K. P. C. Vollhardt, G. D. Whitener, A. A. Yakovenko, *Chem. Commun.*, **2006**, 2572.)

Chapter 15

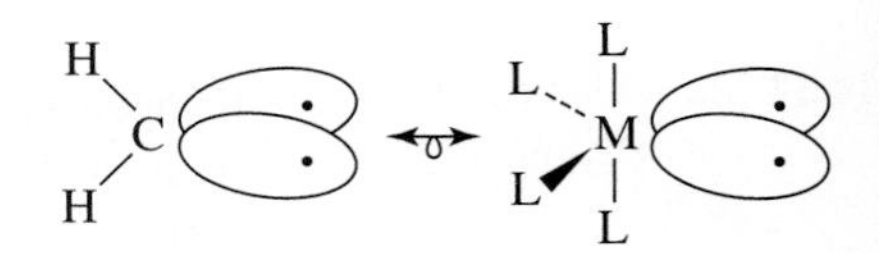

Parallels between Main Group and Organometallic Chemistry

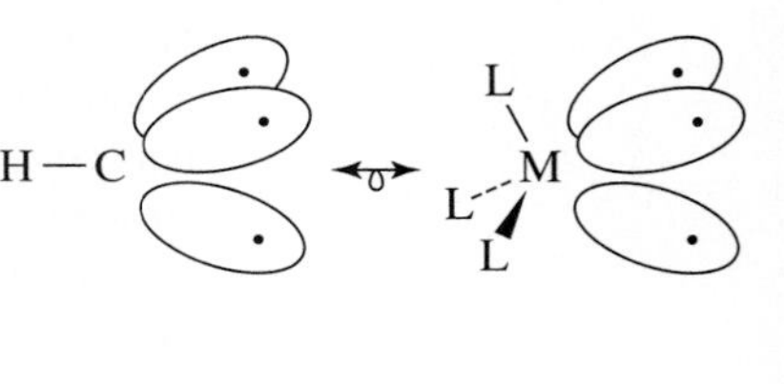

It is common to treat organic and inorganic chemistry as separate topics and, within inorganic chemistry, to consider separately the chemistry of main group compounds and organometallic compounds, as we have generally done so far in this text. However, valuable insights can be gained by examining parallels between these different classifications of compounds. Such an examination leads to a more thorough understanding of the compounds being compared and may suggest new chemical compounds or reactions. The objective of this chapter is to consider several of these parallels, especially between main group and organometallic compounds.

15.1 Main Group Parallels with Binary Carbonyl Complexes

Comparisons within main group chemistry have already been discussed in earlier chapters. These included the similarities and differences between borazine and benzene, the relative instability of silanes in comparison with alkanes, and differences in bonding in homonuclear and heteronuclear diatomic species, such as the isoelectronic N_2 and CO. In general, these parallels have centered around isoelectronic species. Similarities also occur between main group and transition-metal species that are electronically equivalent, species that require the same number of electrons to achieve a filled valence configuration.[1] For example, a halogen atom, one electron short of a valence shell octet, may be considered electronically equivalent to $Mn(CO)_5$, a 17-electron species one electron short of an 18-electron configuration. In this section, we will discuss briefly some parallels between main group atoms and ions and electronically equivalent binary carbonyl complexes.

Much chemistry of main group and metal carbonyl species can be rationalized from the way in which these species can achieve closed-shell (octet or 18-electron) configurations. These methods of achieving more stable configurations will be illustrated for the following electronically equivalent species:

	Examples of Electronically Equivalent Species	
Electrons Short of a Filled Shell	Main Group	Metal Carbonyl
1	Cl, Br, I	$Mn(CO)_5$, $Co(CO)_4$
2	S	$Fe(CO)_4$, $Os(CO)_4$
3	P	$Co(CO)_3$, $Ir(CO)_3$

Halogen atoms, one electron short of a valence shell octet, exhibit chemical similarities with 17-electron organometallic species; some of the most striking are the parallels between halogen atoms and $Co(CO)_4$ (**Table 15.1**). Both can reach filled-shell electron configurations by acquiring an electron or by dimerization. The neutral dimers can add across multiple carbon–carbon bonds and undergo disproportionation by Lewis bases. Anions of both electronically equivalent species have a 1– charge and can combine with H^+ to form acids; both HX (X = Cl, Br, or I) and $HCo(CO)_4$ are strong acids in aqueous solution. Both types of anions form precipitates with heavy metal ions such as Ag^+ in aqueous solution. The parallels between 7-electron halogen atoms and 17-electron binary carbonyl species are sufficiently strong to justify extending the label *pseudohalogen* to these carbonyls.

Similarly, 6-electron main group species show chemical similarities with 16-electron organometallic species. As for the halogens and 17-electron organometallic complexes, many of these similarities can be accounted for on the basis of the ways in which the species can acquire or share electrons to achieve filled-shell configurations. Some similarities between sulfur and the electronically equivalent $Fe(CO)_4$ are listed in **Table 15.2**.

The concept of electronically equivalent groups can be extended to 5-electron main group elements [Group 15] and 15-electron organometallic species. For example, phosphorus and $Ir(CO)_3$ both form tetrahedral tetramers (**Figure 15.1**). The 15-electron $Co(CO)_3$, which is isoelectronic with $Ir(CO)_3$, can replace one or more phosphorus atoms in the P_4 tetrahedron, as also shown in this figure.

The parallels between electronically equivalent main group and organometallic species summarize a considerable amount of their chemistry. The limitations of these parallels should also be recognized, however. For example, main group compounds having expanded shells (central atoms exceeding an electron count of 8) may not have organometallic analogs; organometallic analogs of IF_7 and XeF_4 are not known.

TABLE 15.1 Parallels between Cl and $Co(CO)_4$

Characteristic	Examples	Examples
Ion of 1– charge	Cl^-	$[Co(CO)_4]^-$
Neutral dimeric species	Cl_2	$[Co(CO)_4]_2$
Hydrohalic acid	HCl (strong acid in aqueous solution)	$HCo(CO)_4$ (strong acid in aqueous solution)[a]
Formation of interhalogen compounds	$Br_2 + Cl_2 \rightleftharpoons 2\,BrCl$	$I_2 + [Co(CO)_4]_2 \longrightarrow 2\,ICo(CO)_4$
Formation of heavy-metal salts of low solubility in water	AgCl	$AgCo(CO)_4$
Addition to unsaturated species	$Cl_2 + H_2C{=}CH_2 \longrightarrow H{-}CHCl{-}CHCl{-}H$	$[Co(CO)_4]_2 + F_2C{=}CF_2 \longrightarrow (CO)_4Co{-}CF_2{-}CF_2{-}Co(CO)_4$
Disproportion-ation by Lewis bases	$Cl_2 + N(CH_3)_3 \longrightarrow [ClN(CH_3)_3]Cl$	$[Co(CO)_4]_2 + C_5H_{10}NH$ (Piperidine) $\longrightarrow [(CO)_4Co(HNC_5H_{10})][Co(CO)_4]$

[a] However, $HCo(CO)_4$ is only slightly soluble in water.

TABLE 15.2 Parallels between Sulfur and $Fe(CO)_4$

Characteristic	Examples	
Ion of 2– charge	S^{2-}	$[Fe(CO)_4]^{2-}$
Neutral compound	S_8	$Fe_2(CO)_9$, $[Fe(CO)_4]_3$
Hydride	H_2S: $pK_1 = 7.24$, $pK_2 = 14.92$[a]	$H_2Fe(CO)_4$: $pK_1 = 4.44$[a], $pK_2 = 14$
Phosphine adduct	Ph_3PS	$Ph_3PFe(CO)_4$
Polymeric mercury compound	–S–Hg–S–Hg–S–Hg–S–	–$Fe(CO)_4$–Hg–$Fe(CO)_4$–Hg–$Fe(CO)_4$–
Compound with ethylene	H_2C—CH_2 bridged by S Ethylene sulfide	$(CO)_4Fe$ bonded to $H_2C{=}CH_2$ π complex

[a] pK values in aqueous solution at 25° C.

Organometallic complexes of ligands significantly weaker than CO in the spectrochemical series may not follow the 18-electron rule and may behave quite differently from electronically equivalent main group species. In addition, the reaction chemistry of organometallic compounds can be very different from that of main group compounds. For example, loss of ligands such as CO is far more common in organometallic chemistry than ligand dissociations in main group chemistry. Therefore, as in any scheme based on as simple a framework as electron counting, the concept of electronically equivalent groups has its limitations. However, it serves as valuable background for a versatile way to seek parallels between main group and organometallic chemistry: the concept of isolobal groups.

15.2 The Isolobal Analogy

An important contribution to the understanding of parallels between organic and inorganic chemistry has been the concept of isolobal molecular fragments, described elaborately by Roald Hoffmann in his 1981 Nobel lecture.[2] Hoffmann defined molecular fragments to be isolobal

> if the number, symmetry properties, approximate energy and shape of the frontier orbitals and the number of electrons in them are similar—not identical, but similar.

To illustrate this definition, we will compare fragments of methane with fragments of an octahedral transition-metal complex, ML_6. For simplicity, we will consider only σ bonding between the metal and the ligands.* The fragments to be discussed are in **Figure 15.2**.

The parent compounds have filled valence shell electron configurations, an octet for CH_4, and 18 electrons for ML_6 ($Cr(CO)_6$ is an example of such an ML_6 compound). Within the hybridization model, methane uses sp^3 hybrid orbitals in bonding, with 8 electrons occupying bonding pairs formed from interactions between the hybrids and 1s orbitals on hydrogen. The metal in ML_6 uses d^2sp^3 hybrids in bonding to the ligands (the d_{z^2} and $d_{x^2-y^2}$

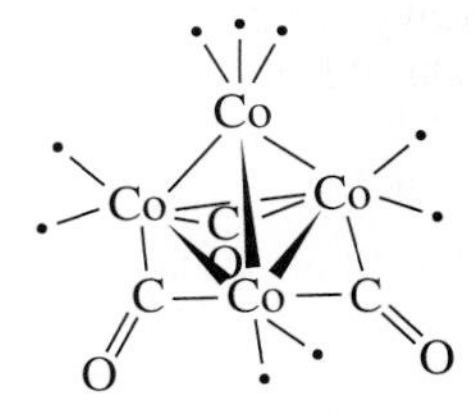

FIGURE 15.1 P_4, $[Ir(CO)_3]_4$, $P_3[Co(CO)_3]$, and $Co_4(CO)_{12}$.

*The model can be refined further to include π interactions between d_{xy}, d_{xz}, and d_{yz} orbitals with ligands having suitable donor and/or acceptor orbitals.

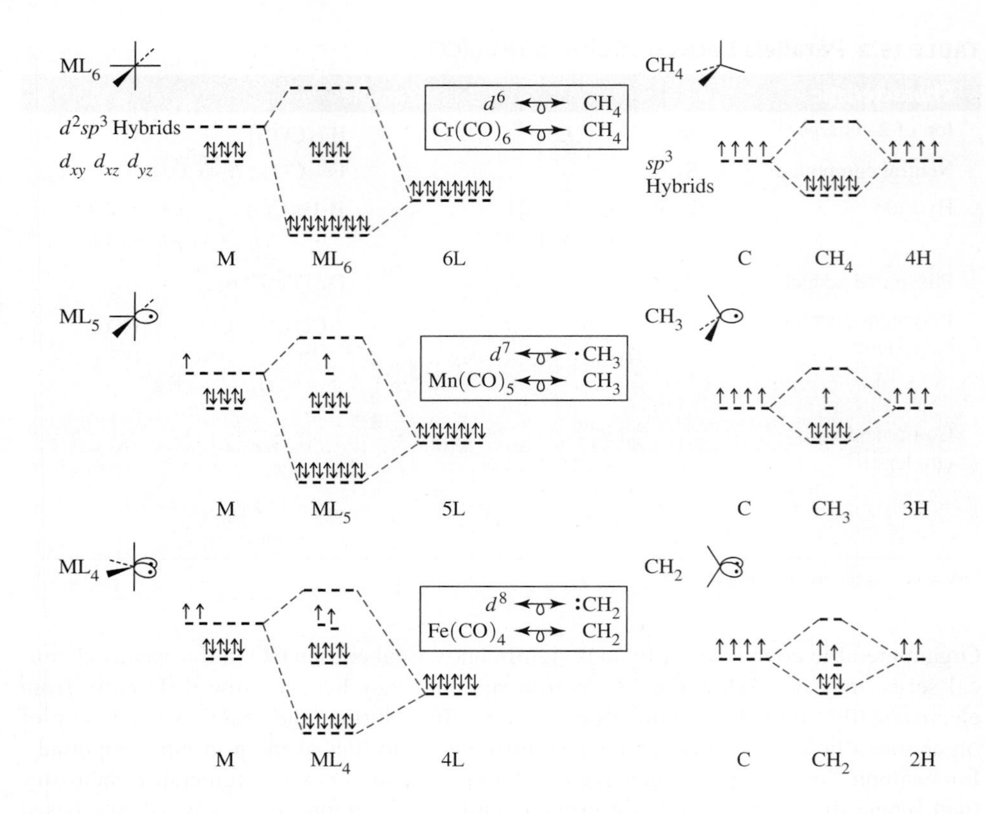

FIGURE 15.2 Orbitals of Octahedral and Tetrahedral Fragments.

contribute to these hybrids for σ interactions), with 12 electrons occupying bonding orbitals and 6 essentially nonbonding electrons occupying d_{xy}, d_{xz}, and d_{yz} orbitals.

Molecular fragments containing fewer ligands than the parent polyhedra can now be described. For the purpose of the analogy, these fragments will be assumed to preserve the geometry of the remaining ligands.

In the 7-electron fragment CH_3, three of the carbon sp^3 orbitals are involved in σ bonding with the hydrogens. The fourth hybrid is singly occupied and at higher energy than the σ-bonding pairs of CH_3 (Figure 15.2). This situation is similar to the 17-electron fragment $Mn(CO)_5$. The sigma interactions between the ligands and Mn in this fragment may be considered to involve five metal d^2sp^3 hybrid orbitals. The sixth hybrid is singly occupied and at higher energy than the five σ-bonding orbitals.

As Figure 15.2 shows, each of these fragments has a single electron in a hybrid orbital at the vacant site of the parent polyhedron. These orbitals are sufficiently similar to meet Hoffmann's isolobal definition. Using Hoffmann's symbol $\leftarrow\!\!\circ\!\!\rightarrow$ to designate groups as isolobal, we may write

$$CH_3 \leftarrow\!\!\circ\!\!\rightarrow ML_5$$

Similarly, 6-electron CH_2 and 16-electron ML_4 are isolobal. These fragments are 2 electrons short of a filled-shell octet or 18-electron configuration, so they are electronically equivalent; each has two singly occupied hybrid orbitals at otherwise vacant sites. Absence of a third ligand similarly gives a pair of isolobal fragments, 5-electron CH and 15-electron ML_3.

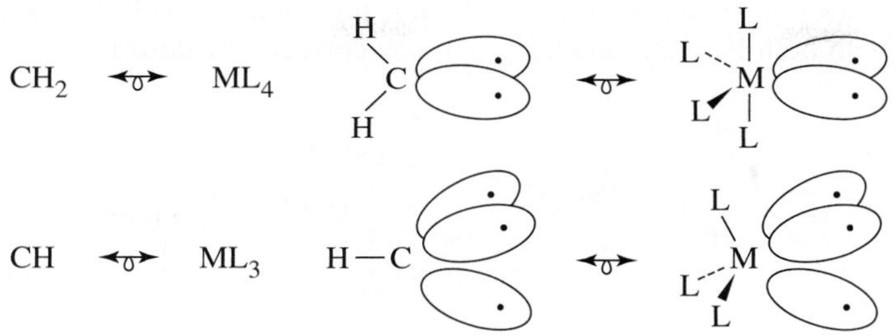

$CH_2 \leftarrow\!\!\circ\!\!\rightarrow ML_4$

$CH \leftarrow\!\!\circ\!\!\rightarrow ML_3$

To summarize,

	Organic	Inorganic	Organometallic Example	Vertices Missing from Parent Polyhedron	Electrons Short of Filled Shell
Parent	**CH_4**	**ML_6**	**$Cr(CO)_6$**	**0**	**0**
Fragments	CH_3	ML_5	$Mn(CO)_5$	1	1
	CH_2	ML_4	$Fe(CO)_4$	2	2
	CH	ML_3	$Co(CO)_3$	3	3

These fragments can be combined into molecules; two CH_3 fragments form ethane, and two $Mn(CO)_5$ fragments form $(OC)_5Mn{-}Mn(CO)_5$. Furthermore, these organic and organometallic fragments can be combined to afford $H_3C{-}Mn(CO)_5$.

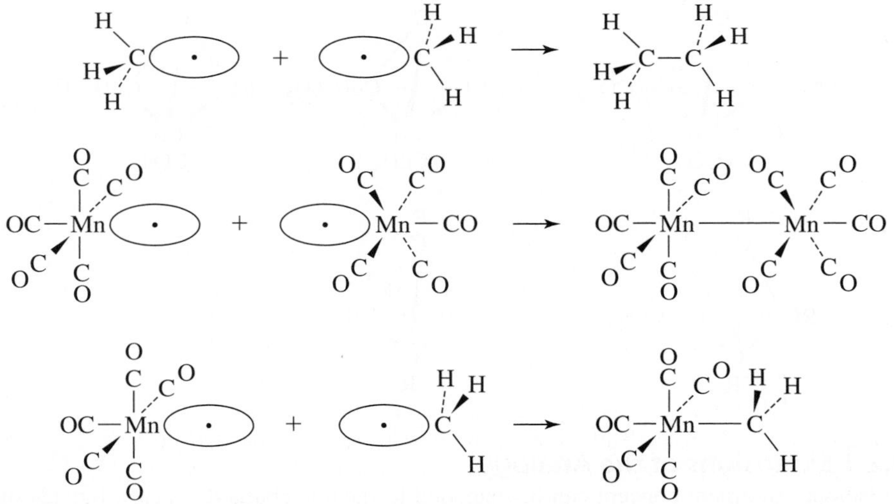

The organic and organometallic parallels are not always this complete. For example, although two 6-electron CH_2 fragments form ethylene, $H_2C{=}CH_2$, the neutral dimer of the isolobal $Fe(CO)_4$ is known only as a transient species, obtained photochemically[3] from $Fe_2(CO)_9$.* However, both CH_2 and $Fe(CO)_4$ form three-membered rings, cyclopropane and $Fe_3(CO)_{12}$. Although cyclopropane is a trimer of three CH_2 fragments, $Fe_3(CO)_{12}$ has two bridging carbonyls and is therefore not a perfect trimer of $Fe(CO)_4$. The isoelectronic

*It is interesting that the dianion, $[Fe_2(CO)_8]^{2-}$, is known. It features two trigonal bipyramidal Fe centers and a single bond between the iron atoms (J. M. Cassidy, K. H. Whitmire, G. J. Long, *J. Organomet. Chem.*, **1992**, *427*, 355). Metal–metal bonds are discussed in Section 15.3.

$Os_3(CO)_{12}$, on the other hand, is a trimeric combination of three $Os(CO)_4$ fragments, which are isolobal with both $Fe(CO)_4$ and CH_2, and can correctly be described as $[Os(CO)_4]_3$.

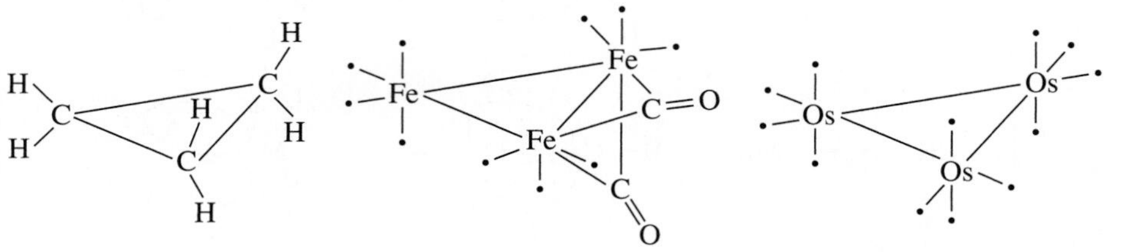

C_3H_6 $\quad$ $Fe_3(CO)_{12}$ $\quad$ $Os_3(CO)_{12}$

• = terminal carbonyl

The isolobal species $Ir(CO)_3$, $Co(CO)_3$, CR (R = H, alkyl, aryl), and P may also be combined in several different ways. As mentioned previously (Figure 15.1), $Ir(CO)_3$, a 15-electron fragment, forms $[Ir(CO)_3]_4$, which has T_d symmetry (Figure 15.1). The isoelectronic complex $Co_4(CO)_{12}$ has a nearly tetrahedral array of cobalt atoms but has three bridging carbonyls and hence C_{3v} symmetry (Figure 15.1). Compounds are also known that have a central tetrahedral structure, with one or more $Co(CO)_3$ fragments replaced by the isolobal CR fragment, as shown in **Figure 15.3**. This is similar to the replacement of phosphorus atoms in the P_4 tetrahedron by $Co(CO)_3$ fragments; P is isolobal with CR.

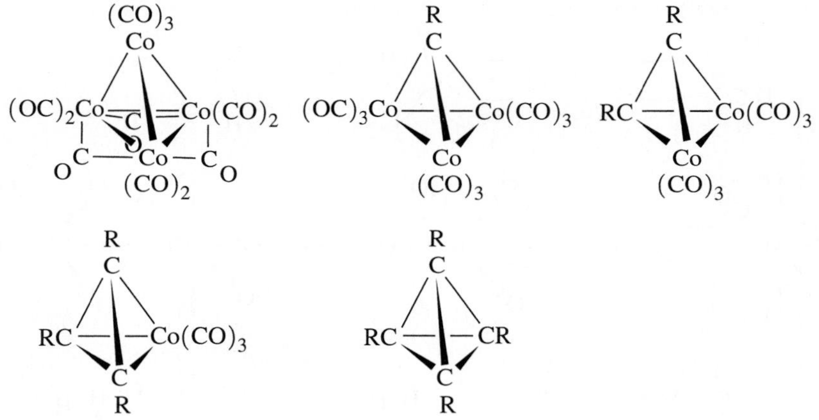

FIGURE 15.3 Structures Resulting from Combinations of Isolobal $Co(CO)_3$ and CR ($R = CH_3$).

15.2.1 Extensions of the Analogy

The isolobal fragment concept can be extended to include charged species, ligands other than CO, and organometallic fragments not based on octahedral geometries. Some ways of extending the isolobal parallels are summarized as follows:

1. The isolobal definition may be extended to isoelectronic fragments having the same coordination number. For example, because

$$Mn(CO)_5 \overset{\sigma}{\longleftrightarrow} CH_3, \quad \begin{matrix} Re(CO)_5 \\ [Fe(CO)_5]^+ \\ [Cr(CO)_5]^- \end{matrix} \overset{\sigma}{\longleftrightarrow} CH_3$$

(17-electron fragment) (7-electron fragment) (17-electron fragments) (7-electron fragment)

2. Gain or loss of electrons from two isolobal fragments yields isolobal fragments. For example, because

			$Cr(CO)_5$		
$Mn(CO)_5$	$\leftarrow\!\circ\!\rightarrow$	CH_3,	$Mo(CO)_5$	$\leftarrow\!\circ\!\rightarrow$	CH_3^+
			$W(CO)_5$		
(17-electron fragment)		(7-electron fragment)	(16-electron fragments)		(6-electron fragment)

$Fe(CO)_5$		
$Ru(CO)_5$	$\leftarrow\!\circ\!\rightarrow$	CH_3^-
$Os(CO)_5$		
(18-electron fragments)		(8-electron fragment)

All of these examples are one ligand short of the parent complex. $Fe(CO)_5$ is isolobal with CH_3^-, for example, because both have filled electron shells and both are one vertex short of the parent polyhedron. By contrast, $Fe(CO)_5$ and CH_4 are not isolobal. Both have filled electron shells (18 and 8 electrons, respectively), but CH_4 has all vertices of the tetrahedron occupied, whereas the $Fe(CO)_5$ fragment has an empty vertex on the basis of the parent octahedron.*

3. Other 2-electron donors are treated similarly to CO:**

$$Mn(CO)_5 \leftarrow\!\circ\!\rightarrow Mn(PR_3)_5 \leftarrow\!\circ\!\rightarrow [MnCl_5]^{5-} \leftarrow\!\circ\!\rightarrow Mn(NCR)_5 \leftarrow\!\circ\!\rightarrow CH_3$$

4. Ligands η^5-C_5H_5 and η^6-C_6H_6 are considered to occupy three coordination sites and to be 6-electron donors:†

$(\eta^5\text{-}C_5H_5)Fe(CO)_2$ $(\eta^6\text{-}C_6H_6)Mn(CO)_2$	$\leftarrow\!\circ\!\rightarrow$	$[Fe(CO)_5]^+$	$\leftarrow\!\circ\!\rightarrow$	$Mn(CO)_5$	(17-electron fragments)
$(\eta^5\text{-}C_5H_5)Mn(CO)_2$ $(\eta^6\text{-}C_6H_6)Cr(CO)_2$	$\leftarrow\!\circ\!\rightarrow$	$[Mn(CO)_5]^+$	$\leftarrow\!\circ\!\rightarrow$	$Cr(CO)_5$	(16-electron fragments)

5. Octahedral fragments of formula ML_n (where M has a d^x configuration) are isolobal with square-planar fragments of formula ML_{n-2} (where M has a d^{x+2} configuration and L is a 2-electron donor):

Octahedral Fragments		Square-planar Fragments
ML_n		ML_{n-2}
$Cr(CO)_5$ d^6	$\leftarrow\!\circ\!\rightarrow$	$[PtCl_3]^-$ d^8
$Fe(CO)_4$ d^8	$\leftarrow\!\circ\!\rightarrow$	$Pt(PR_3)_2$ d^{10}

---

*Although the parent geometry for the $Fe(CO)_5$ fragment is the octahedron within an isolobal context, it is important to mention that 18-electron $Fe(CO)_5$ exists as a stable trigonal bipyramidal complex.

**Hoffmann uses electron-counting method A, in which chloride is considered a negatively charged, 2-electron donor.

†η^5-C_5H_5 is considered the 6-electron donor $C_5H_5^-$.

The fifth of these isolobal analogy extensions is sophisticated and deserves further explanation. We will consider two examples, the parallels between d^6 ML_5 (octahedral) and d^8 ML_3 (square planar) fragments and the parallels between d^8 ML_4 (octahedral) and d^{10} ML_2 (square planar) fragments. The ML_3 and ML_2 fragments of a square planar parent structure are shown in **Figure 15.4**, with dsp^2 hybrid orbitals used for bond formation. The fragments of a square planar ML_4 molecule are compared with the fragments of an octahedral ML_6 molecule (Figure 15.2).

A square-planar d^8 ML_3 fragment, such as $[PtCl_3]^-$, has an empty lobe of a nonbonding hybrid orbital as its LUMO. This is comparable to the LUMO of a d^6 ML_5 fragment of an octahedron (e.g., $Cr(CO)_5$):

$[PtCl_3]^-$ ←o→ $Cr(CO)_5$

A d^8 fragment such as $[PtCl_3]^-$ would therefore be isolobal with $Cr(CO)_5$ and other ML_5 fragments, provided the empty lobe in each case had suitable energy.*

A d^{10} ML_2 fragment such as $Pt(PR_3)_2$ would have two valence electrons more than d^8 $PtCl_2$ (Figure 15.4). These electrons are considered to occupy two nonbonding hybrid orbitals. This situation is very comparable to the $Fe(CO)_4$ fragment (Figure 15.2); each complex has two singly occupied lobes:

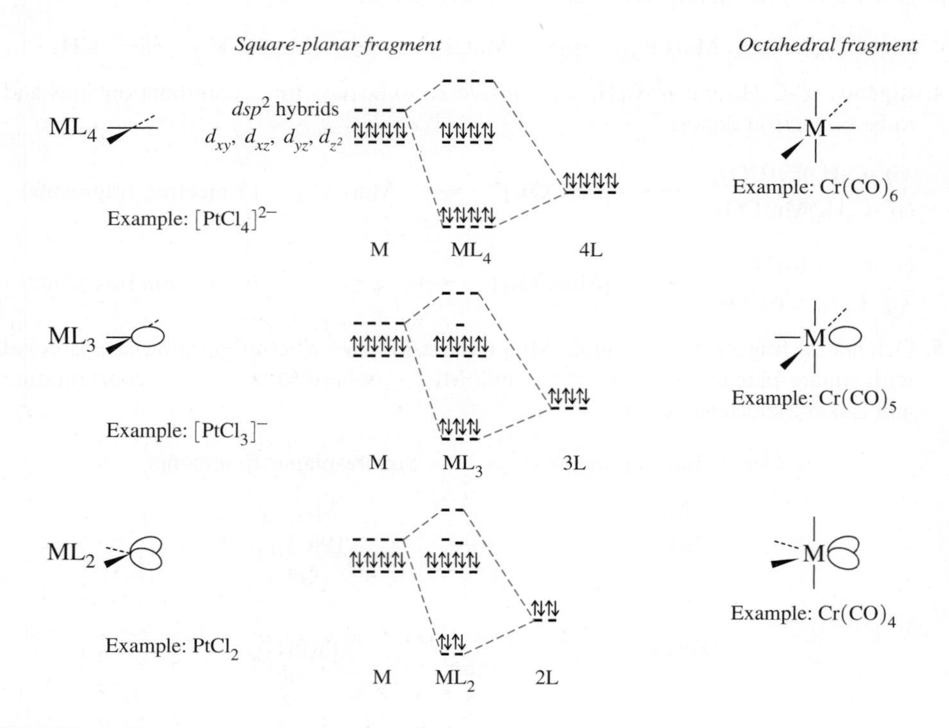

FIGURE 15.4 Comparison of Square-Planar Fragments with Octahedral Fragments.**

*The highest occupied orbitals of ML_3 have similar energies. For a detailed analysis of the energies and symmetries of ML_5, ML_3, and other fragments, see M. Elian, R. Hoffmann, *Inorg. Chem.*, **1975**, *14*, 1058, and T. A. Albright, R. Hoffmann, J. C. Thibeault, D. L. Thorn, *J. Am. Chem. Soc.*, **1979**, *101*, 3801.

**The $Cr(CO)_6$ diagram is provided in Figure 15.2. The $Cr(CO)_5$ and $Cr(CO)_4$ diagrams are identical to those of $Mn(CO)_5$ and $Fe(CO)_4$, respectively, upon removal of one electron from the $Mn(CO)_5$ diagram and two electrons from the $Fe(CO)_4$ diagram.

$Pt(PR_3)_2$ $\longleftrightarrow$ $Fe(CO)_4$

Isolobal fragments containing CO and η^5-C_5H_5 ligands are given in **Table 15.3**.

EXAMPLE 15.1

Propose examples of organometallic fragments isolobal with CH_2^+.

Here we will limit ourselves to the ligand CO and first-row transition metals. Other ligands and metals may be used with equally valid results.

CH_2^+ is two ligands and 3 electrons short of its parent compound (CH_4). The corresponding octahedral fragment will therefore be a 15-electron species with the formula ML_4, two ligands and 3 electrons short of its parent, $M(CO)_6$. If L = CO, the four carbon monoxides contribute 8 electrons, requiring that the metal contribute the remaining 7. The first-row d^7 metal is Mn. The overall result is $CH_2^+ \longleftrightarrow Mn(CO)_4$.

Other octahedral isolobal fragments can be found by changing the metal and the charge on the complex. A positive charge compensates for a metal with one more electron, and a negative charge compensates for a metal with one fewer electron:

d^7: $Mn(CO)_4$ $\quad$ d^8: $[Fe(CO)_4]^+$ $\quad$ d^6: $[Cr(CO)_4]^-$

EXERCISE 15.1 For the following, propose examples of isolobal organometallic fragments other than those just cited and in Table 15.3:

a. A fragment isolobal with CH_2^+
b. A fragment isolobal with CH^-
c. Three fragments isolobal with CH_3

EXERCISE 15.2 Find organic fragments isolobal with each of the following:

a. $Ni(\eta^5$-$C_5H_5)$ $\quad$ **b.** $Cr(CO)_2(\eta^6$-$C_6H_6)$ $\quad$ **c.** $[Fe(CO)_2(PPh_3)]^-$

TABLE 15.3 Examples of Isolobal Fragments

Neutral hydrocarbons	CH_4	CH_3	CH_2	CH	C
Isolobal organometallic fragments	$Cr(CO)_6$	$Mn(CO)_5$	$Fe(CO)_4$	$Co(CO)_3$	$Ni(CO)_2$
(Cp = η^5-C_5H_5)	$[Mn(CO)_6]^+$	$[Fe(CO)_5]^+$	$[Co(CO)_4]^+$	$[Ni(CO)_3]^+$	$[Cu(CO)_2]^+$
	$CpMn(CO)_3$	$CpFe(CO)_2$	CpCo(CO)	CpNi	
Anionic hydrocarbon fragments obtained by loss of H^+	CH_3^-	CH_2^-	CH^-		
Isolobal organometallic fragments	$Fe(CO)_5$	$Co(CO)_4$	$Ni(CO)_3$		
Cationic hydrocarbon fragments obtained by gain of H^+		CH_4^+	CH_3^+	CH_2^+	CH^+
Isolobal organometallic fragments		$V(CO)_6$	$Cr(CO)_5$	$Mn(CO)_4$	$Fe(CO)_3$

These analogies are not limited to octahedral and square-planar organometallic fragments; similar arguments can be used to derive fragments of different polyhedra. For example, $Co(CO)_4$, a 17-electron fragment of a trigonal bipyramid, is isolobal with $Mn(CO)_5$, a 17-electron fragment of an octahedron:

Examples of electron configurations of isolobal fragments of polyhedra having five through nine vertices are given in **Table 15.4**.

TABLE 15.4 Isolobal Relationships for Fragments of Polyhedra

Organic Fragment	Coordination Number of Transition Metal for Parent Polyhedron					Valence Electrons of Fragment
	5	6	7	8	9	
CH_3	d^9-ML_4	d^7-ML_5	d^5-ML_6	d^3-ML_7	d^1-ML_8	17
CH_2	d^{10}-ML_3	d^8-ML_4	d^6-ML_5	d^4-ML_6	d^2-ML_7	16
CH		d^9-ML_3	d^7-ML_4	d^5-ML_5	d^3-ML_6	15

The interested reader is encouraged to refer to Hoffmann's Nobel lecture for further information on how the isolobal analogy can be extended to other ligands and geometries. A newer concept, called *autogenic isolobality*, classifies species as analogous on the basis of vacant orbitals in the electronic configuration that have the same bonding potential.[4]

15.2.2 Examples of Applications of the Analogy

The isolobal analogy can be extended to any molecular fragment having frontier orbitals of suitable size, shape, symmetry, and energy. Realization of these analogies inspires research by suggesting target molecules that can appear unorthodox on first inspection. For example, the 5-electron fragment CH is isolobal with P and other Group 15 atoms. This suggests that phosphorus-containing analogs to organometallic complexes containing cyclic π ligands such as C_5H_5 might exist that are isolobal with the ubiquitous metallocenes, $[(C_5H_5)_2M]^n$. Indeed, not only can P_5^-, the analog to the cyclopentadienide ion $C_5H_5^-$, be prepared in solution,[5] but sandwich compounds containing P_5 rings, such as those shown in **Figure 15.5**, have been synthesized. The first of these, $(\eta^5$-$C_5Me_5)Fe(\eta^5$-$P_5)$, was prepared from the reaction of $[(\eta^5$-$C_5Me_5)Fe(CO)_2]_2$ with white phosphorus (P_4).[6]

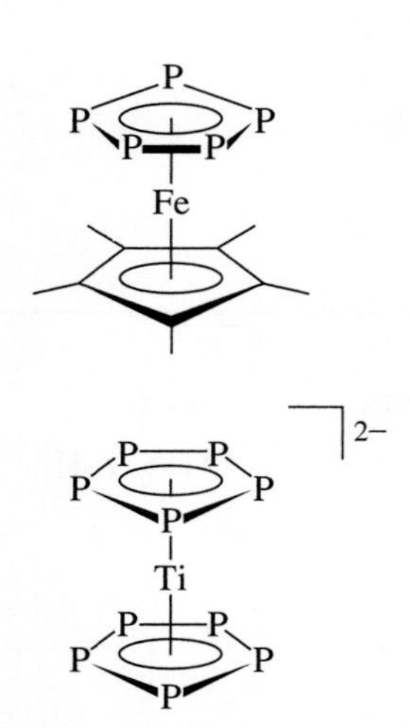

FIGURE 15.5 Metallocenes Containing P_5 Rings.

Perhaps the most interesting of all the phosphorus analogs of metallocenes is the carbon-free metallocene, $[(\eta^5$-$P_5)_2Ti]^{2-}$. This complex, prepared by the reaction of $[Ti(naphthalene)_2]^{2-}$ with P_4, contains parallel, eclipsed P_5 rings.* The P_5 ligand in this and other complexes is a weaker donor, but substantially stronger acceptor, than the cyclopentadienyl ligand.

Another example, $AuPPh_3$, a 13-electron fragment, has a single electron in a hybrid orbital pointing away from the phosphine.[7] This electron is in an orbital of similar symmetry but higher energy than the singly occupied hybrid in the $Mn(CO)_5$ fragment.

---

*E. Urnezius, W. W. Brennessel, C. J. Cramer, J. E. Ellis, P. von Ragué Schleyer, *Science*, **2002**, *295*, 832. For a discussion of the development of complexes containing ligands composed of group 15 atoms, see M. Scheer, *Dalton Trans.*, **2008**, 4372.

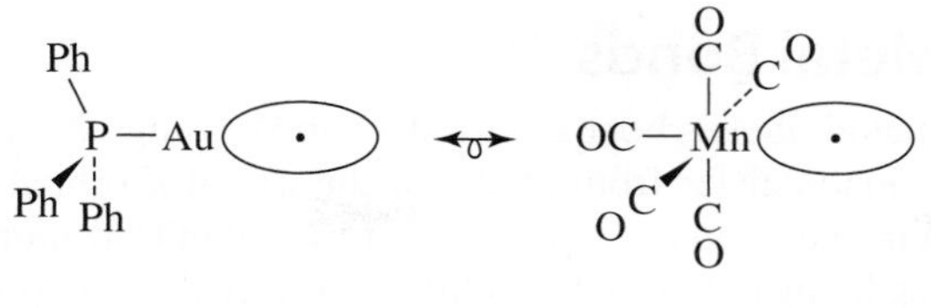

Nevertheless, $AuPPh_3$ can combine with the isolobal $Mn(CO)_5$ and CH_3 to form $(OC)_5Mn—AuPPh_3$ and $H_3C—AuPPh_3$.

A hydrogen atom, with a single electron in its 1*s* orbital, can be viewed as isolobal with species such as CH_3, $Mn(CO)_5$, and $AuPPh_3$. From this perspective, H_2 can be viewed as analogous to CH_4 and $HMn(CO)_5$. $AuPPh_3$ and H often show surprisingly similar behavior, such as their ability to bridge the following triosmium clusters.[8,9]

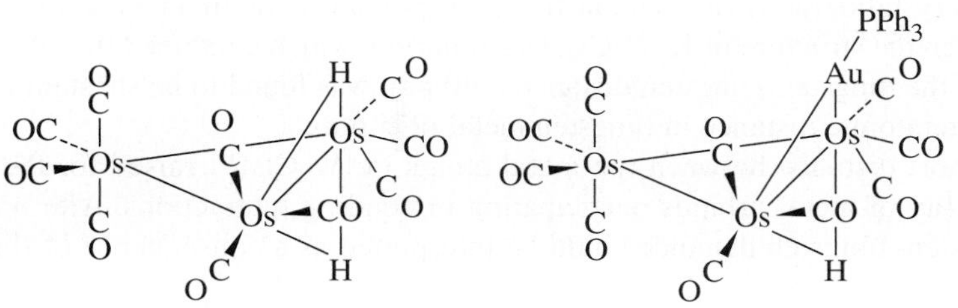

As further examples of the predictive power of this analogy in metal carbonyl chemistry, $HCo(CO)_4$ and $Co(CO)_4(AuPPh_3)$, as well as $HMn(CO)_5$ and $Mn(CO)_5(AuPPh_3)$ are known. Apparent limitations of the analogy between $AuPPh_3$ and H include $[HTi(CO)_6]^-$ and $HV(CO)_6$. These hydrides have not been reported, despite the existence of their isolobal "cousins" $[Ti(CO)_6(AuPEt_3)]^-$[10] and $V(CO)_6(AuPPh_3)$,[11] respectively, which have been prepared despite their high coordination number. It has been pointed out that metal-hydride bonds are often weaker than metal bonds to $AuPR_3$ ligands.[4]

Isolobal analogies find tremendous utility in suggesting new compounds. A survey of the current literature uncovers many examples of this analogy guiding research efforts.

EXAMPLE 15.2

CH_3 is isolobal with 17-electron $Zn(\eta^5\text{-}C_5Me_5)$ (extension 4 of the analogy).

CH_2 is isolobal with 16-electron $Ir(PPh_3)_2(CS)Cl$ (start with 16-electron $[Ir(CO)_4]^+$ from the parent octahedron, and substitute PPh_3, Cl^-, and CS, respectively, for CO, extension 2 of the analogy).

Recognition of these fragments as isolobal has been exploited in the syntheses of organometallic compounds composed of fragments isolobal with fragments of known compounds (examples in **Figure 15.6**).[12]

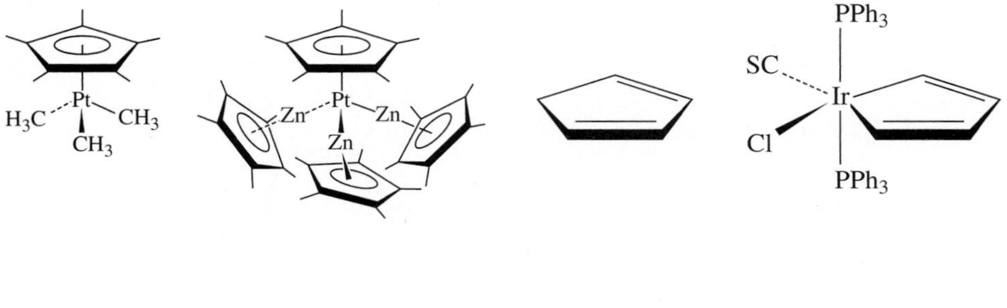

FIGURE 15.6 Compounds Composed of Isolobal Fragments.

15.3 Metal–Metal Bonds

The formation of metal–metal bonds was described using the isolobal analogy in Section 15.2. These bonds differ from others in the use of *d* orbitals on both atoms. In addition to the usual σ and π bonds, δ bonds are possible in transition-metal compounds. Furthermore, bridging by ligands and the ability to form cluster compounds make for great variety in structures containing metal–metal bonds. Examples of compounds with carbon–carbon, other main group, and metal–metal single, double, and triple bonds, together with a metal–metal quadruple bond, are shown in **Figure 15.7**.

For approximately a century, compounds containing two or more metal atoms have been known. The first of these to be correctly characterized contained ligands that bridged the metals involved; X-ray crystallographic studies eventually showed that the metal atoms were too far apart to be likely participants in direct metal–metal orbital interactions.

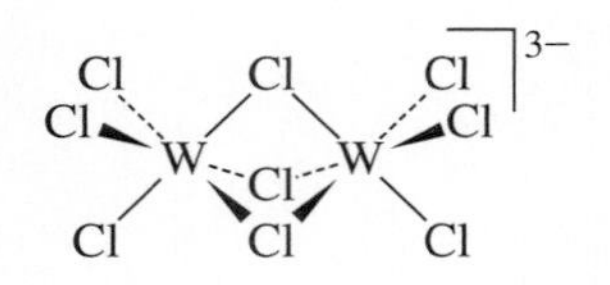

X-ray crystallography first demonstrated the possibility of direct metal–metal bonding in 1935 when the structure of $K_3W_2Cl_9$, was reported, which contained the $[W_2Cl_9]^{3-}$ ion. In this ion, the tungsten–tungsten distance (240 pm) was found to be substantially shorter than the interatomic distance in tungsten metal (275 pm).

The short distance between the metal atoms in $[W_2Cl_9]^{3-}$ raised for the first time the possibility of metal orbitals participating in bonding interactions. Alternatively, the short tungsten–tungsten distance could be interpreted as a consequence of the bridging

FIGURE 15.7 Single, Double, Triple, and Quadruple Bonds.

Organic	Inorganic	
$H_3C—CH_3$	F—F	$(OC)_5Mn—Mn(CO)_5$
$H_2C{=}CH_2$	O=O	$[(C_5H_5)Fe(NO)]_2$ (Fe=Fe)
H—C≡C—H	N≡N	$[Cl_4Os{\equiv}OsCl_4]^{2-}$
		$[Cl_4Re{\equiv}ReCl_4]^{2-}$

ligands. Questions regarding the electronic structure and metal–metal bonding in the series $[W_2Cl_9]^{3-}$, $[W_2Cl_9]^{2-}$, and $[W_2Cl_9]^{-}$ have been revisited via modern computational techniques.[13] This work provides evidence for a tungsten–tungsten triple bond in $[W_2Cl_9]^{3-}$.

The maturation of the chemistry of metal–metal-bonded species was spurred by the crystal structures of $[Re_3Cl_{12}]^{3-}$ and $[Re_2Cl_8]^{2-}$.[14] $[Re_3Cl_{12}]^{3-}$, originally believed to be monomeric $ReCl_4^-$, was shown in 1963 to be a trimeric cyclic ion having very short rhenium–rhenium distances (248 pm). In the following year, during a study on the synthesis of triruthenium complexes, the dimeric $[Re_2Cl_8]^{2-}$ was synthesized. This ion had a remarkably short metal–metal distance (224 pm) and was the first complex identified as having a quadruple bond:

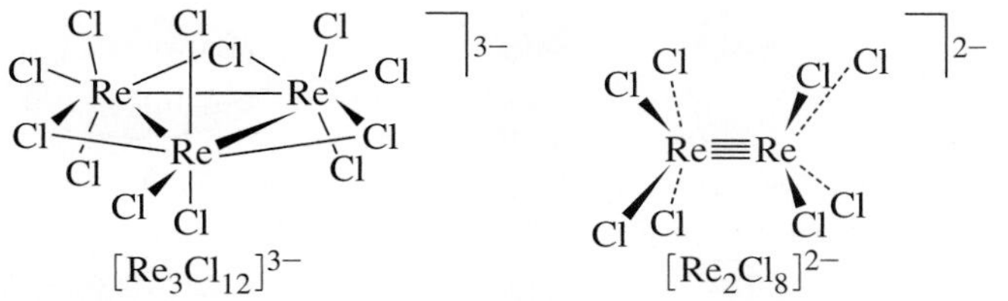

$[Re_3Cl_{12}]^{3-}$ $[Re_2Cl_8]^{2-}$

During the succeeding decades, thousands of transition-metal cluster compounds have been synthesized, including hundreds containing quadruple bonds and some that may contain quintuple bonds. Therefore, we need to consider how metal atoms can bond to each other and, in particular, how high bond orders between metals can be achieved.

15.3.1 Multiple Metal–Metal Bonds

Quadruple Bonds

Transition metals may form single, double, triple, or quadruple bonds (or bonds of fractional order) with other metal atoms. How are quadruple bonds possible? In main group chemistry, atomic orbitals in general can interact in a σ or π fashion, with the highest possible bond order of 3 a combination of one σ bond and two π bonds. When two transition-metal atoms interact, the most important interactions are between their outermost *d* orbitals. These *d* orbitals can combine to form not only σ and π orbitals, but also δ (delta) orbitals, as shown in **Figure 15.8**. If the *z* axis is chosen as the internuclear axis, the strongest interaction (involving greatest overlap) is the sigma interaction between the d_{z^2} orbitals. Next in effectiveness of overlap are the d_{xz} and d_{yz} orbitals, which form π orbitals as a result of interactions in two regions in space. The last, and weakest, of these interactions are between the d_{xy} and $d_{x^2-y^2}$ orbitals, which interact in four regions in the formation of δ molecular orbitals.

The relative energies of the resulting molecular orbitals are shown schematically in **Figure 15.9**. In the absence of ligands, an M_2 fragment would have five bonding orbitals resulting from *d–d* interactions, with molecular orbitals increasing in energy in the order σ, π, δ, δ^*, π^*, σ^*, as shown on the left side of Figure 15.9. In $[Re_2Cl_8]^{2-}$, our example of quadruple bonding, the configuration is eclipsed and exhibits D_{4h} symmetry. For convenience, we can choose the Re—Cl bonds to be oriented in the *xz* and *yz* planes. The ligand orbitals interact most strongly with the metal orbitals pointing toward them, in this case the δ and δ^* orbitals originating primarily from the $d_{x^2-y^2}$ atomic orbitals.* The

*Analysis of the symmetry of this ion shows that the *s*, p_x, and p_y orbitals are also involved.

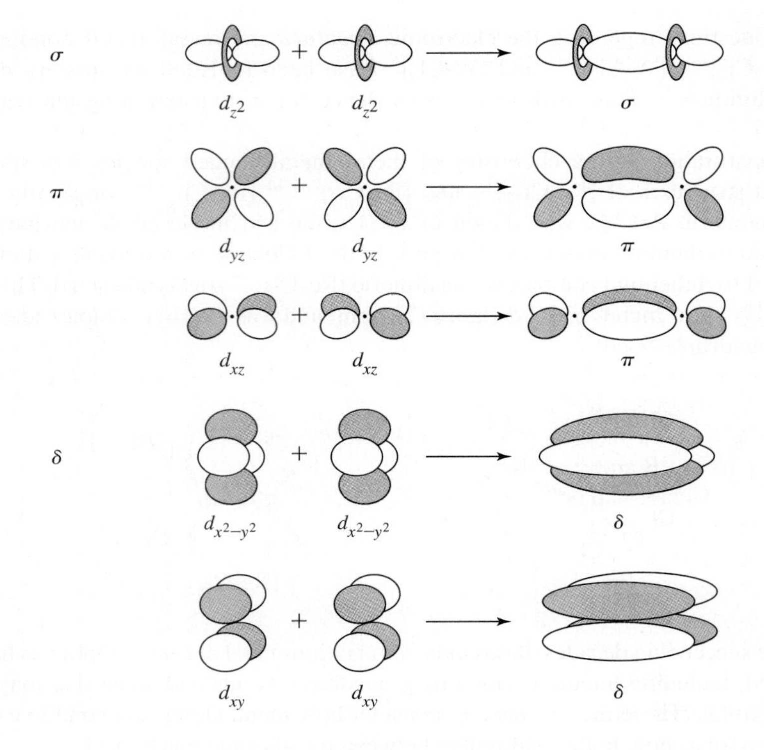

FIGURE 15.8 Bonding Interactions between Metal *d* Orbitals.

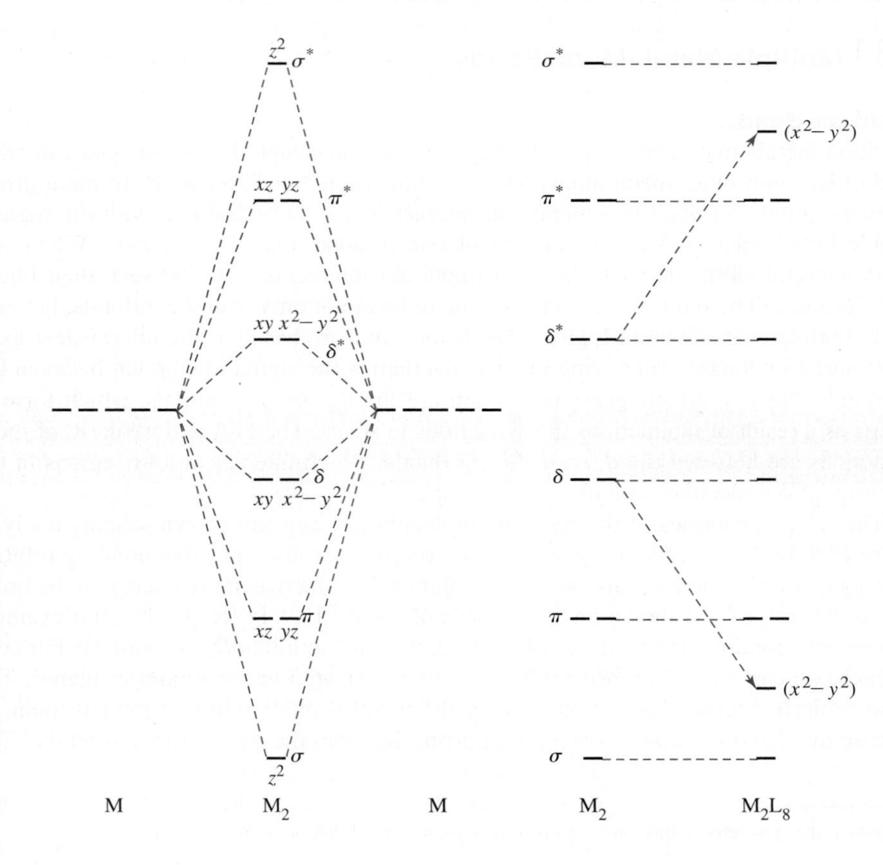

FIGURE 15.9 Relative Energies of Orbitals Formed from *d*-Orbital Interactions. In M_2L_8 complexes, a pair of electrons from the ligands occupies the bonding orbital with high metal $d_{x^2-y^2}$ contribution.

consequence of these interactions is new molecular orbitals, as shown on the right side of Figure 15.9. The relative energies of these orbitals depend on the specific metal–ligand interactions.

In $[Re_2Cl_8]^{2-}$, each formal Re(III) center has four *d* electrons. If the eight *d* electrons for this ion are placed into the four lowest energy orbitals shown in Figure 15.9 (not including the low-energy orbital arising from the $d_{x^2-y^2}$ interactions, occupied by ligand electrons), the total bond order is 4, corresponding to—in increasing energy—one σ bond, two π bonds, and one δ bond.* Although the δ bond is weakest, it is strong enough to maintain this ion in its eclipsed conformation. The weakness of the δ bond is illustrated by the small separation in energy of the δ and δ^* orbitals. This energy difference typically corresponds to the energy of visible light, with the consequence that most quadruply bonded complexes are vividly colored; $[Re_2Cl_8]^{2-}$ is royal blue, and $[Mo_2Cl_8]^{4-}$ is bright red. By comparison, main group compounds having filled π and empty π^* orbitals are often colorless (e.g., N_2 and CO), because the energy difference between these orbitals is commonly in the ultraviolet part of the spectrum.

Additional metal valence electrons populate δ^* orbitals and reduce the bond order. For example, $[Os_2Cl_8]^{2-}$, an osmium(III) species with a total of 10 *d* electrons, has a triple bond. The δ bond order in this ion is zero; in the absence of such a bond, the eclipsed geometry—as found in quadruply bonded complexes, such as $[Re_2Cl_8]^{2-}$—is absent. X-ray crystallographic analysis has shown $[Os_2Cl_8]^{2-}$ to be very nearly staggered (D_{4d}) as would be expected from VSEPR considerations. The eclipsed geometry in $[Re_2Cl_8]^{2-}$ is evidence of the importance of δ bonding in this complex.

$[Os_2Cl_8]^{2-}$

Similarly, fewer than 8 valence electrons would also give a bond order less than 4. Examples of such complexes are shown in **Figure 15.10**.

It is important to remember that the relative energies of these orbitals depend on the metal–ligand interactions and the metal geometry, and consequently some complexes vary from the pattern shown in Figure 15.9. For example, computational studies on $[W_2Cl_9]^{3-}$ suggest a bond order 3 on the basis of the six metal valence electrons (each metal is formally d^3) occupying one σ and two δ bonding orbitals.[13] The electronic structure is different than that in Figure 15.9 in these "face-shared" $[M_2Cl_9]^{z-}$ systems.

Metal–metal multiple bonding can have dramatic effects on bond distances, as measured by X-ray crystallography. One way of describing the shortening of interatomic distances by multiple bonds is by comparing the bond distances in multiple bonds to the distances for single bonds. The ratios of these distances is sometimes called the **formal shortness ratio**. Values of this ratio are compared here for main group triple bonds and for some of the shortest of the measured transition-metal quadruple bonds:

Multiple Bond Distance/Single Bond Distance			
Bond	**Ratio**	**Bond**	**Ratio**
C≡C	0.783	Cr≣Cr	0.767
N≡N	0.786	Mo≣Mo	0.807
		Re≣Re	0.848

The ratios found for several quadruply bonded chromium complexes are the smallest ratios found to date for any compounds. Moderate variations in bond distances have been observed. Mo—Mo quadruple bonds, for example, have been found in the range 206 to 224 pm, with nearly 95 percent between 206 and 217 pm.[15]

*More elaborate calculations have yielded an "effective" bond order of 3.2 for $[Re_2Cl_8]^{2-}$, recommended as an alternative label to a "weak" quadruple bond. See L. Gagliardi, B. O. Roos, *Inorg. Chem.*, **2003**, *42*, 1599.

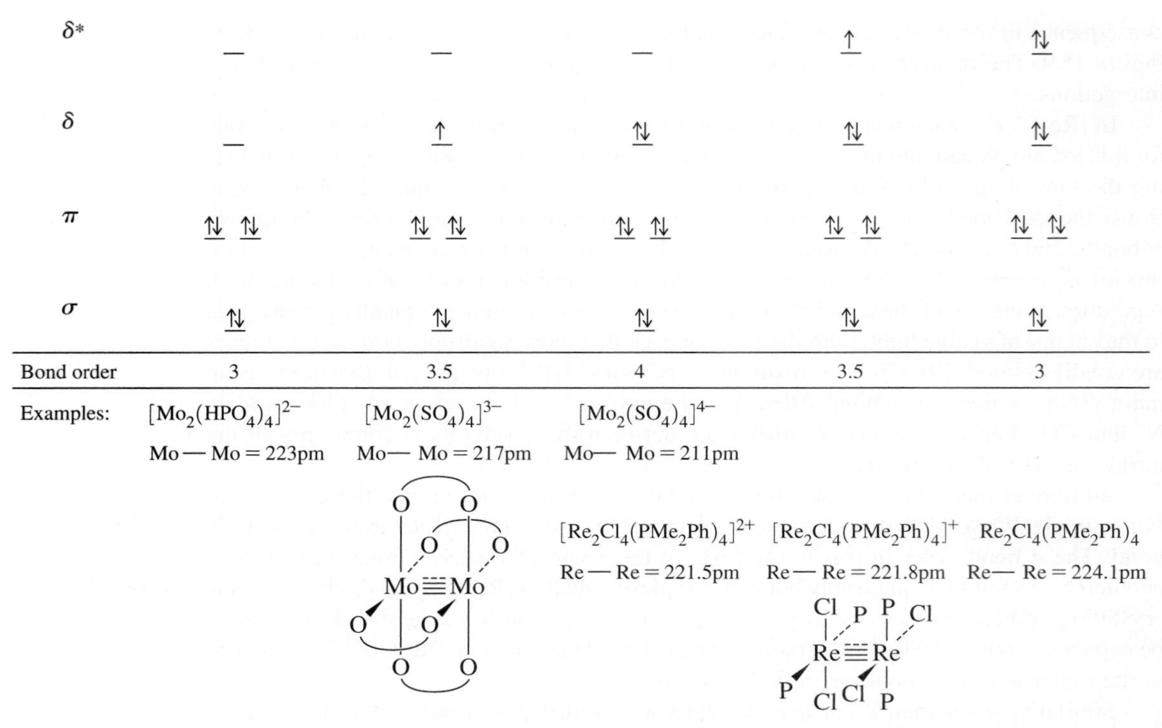

FIGURE 15.10 Bond Order and Electron Count in Dimetal Clusters. (Data from A. Bino, F. A. Cotton, *Inorg. Chem.*, **1979**, *18*, 3562; and F. A. Cotton, *Chem. Soc. Rev.*, **1983**, *12*, 35.)

The effect of δ and δ^* orbital populations on bond distances can be surprisingly small. For example, removal of δ^* electrons on oxidation of $Re_2Cl_4(PMe_2Ph_4)_4$ gives only very slight shortening of the Re—Re distances, as shown in **Table 15.5**.[16]

A possible explanation for the small change in bond distance is that, with increasing oxidation state of the metal, the *d* orbitals contract. This contraction may cause overlap of *d* orbitals involved in π bonding to become less effective. Thus, as δ^* electrons are removed, the overlap between orbitals responsible for metal–metal bonding becomes slightly weaker; the two factors—lowering the population of an antibonding orbital and decreasing the bonding overlap due to the increase in the Re oxidation state—very nearly offset each other. Another possible explanation is that coulombic repulsion between the metal atoms increases as the metal centers become relatively more positive upon oxidation.

TABLE 15.5 Effect of Oxidation on Re—Re Bond Distances in Re_2 Complexes

Complex	Number of *d* Electrons	Formal Re—Re Bond Order	Formal Oxidation State of Re	Re—Re Distance (pm)
$Re_2Cl_4(PMe_2Ph)_4$	10	3	2	224.1
$[Re_2Cl_4(PMe_2Ph)_4]^+$	9	3.5	2.5	221.8
$[Re_2Cl_4(PMe_2Ph)_4]^{2+}$	8	4	3	221.5

Quintuple Bonds

Can there be such a thing as a quintuple bond? Figure 15.8 shows five possible interactions between *d* orbitals, including two δ interactions—so it is reasonable to propose a compound having bonding electron pairs in orbitals arising from all five. In 2005 Power reported a compound with "fivefold" bonding.[17] This dimeric chromium(I) complex, shown in **Figure 15.11(a)** and on a larger scale in Figure 1.3, has ligands that are sufficiently bulky to protect the metals. In addition, the low chromium oxidation state affords each metal with the five electrons necessary to fully occupy molecular orbitals from the five sets of *d* orbital interactions. It is interesting that the Cr—Cr distance of 183.5 pm gives a formal shortness ratio of 0.774, not even as small as the lowest ratio (0.767) reported for quadruple bonds. Although Power's complex may in fact have one σ, two π, and two δ interactions, this does not imply bonding that has the full strength of the five interactions in Figure 15.8. The *trans*-bent geometry and apparent interactions between the chromium atoms and the aromatic rings make the bonding in this compound more complex; nevertheless, calculations support that the σ, π, and δ interactions between *d* orbitals do occur.[18]

Since the initial report of fivefold bonding in 2005, the record for shortest Cr—Cr bond has been reduced to 180.3 pm in 2007 (**Figure 15.11(b)**),[19] to 174.0 pm in 2008 (**Figure 15.11(c)**),[20] and to 172.9 pm in 2009 (**Figure 15.11(d)**, the shortest metal–metal bond observed to date).[21] The Figure 15.11(d) complex features a guanidinate ligand designed to apply steric pressure to further push the chromium atoms together. The formal shortness ratio of this complex is 0.729, the smallest such value reported to date; the apparently very strong Cr—Cr bonding can be rationalized with electron pairs occupying one σ, two π, and two δ orbitals, but the authors also emphasize the importance of the ligand design in enforcing a shorter bond.[21] In contrast to Power's complex, Tsai[20] and Kempe's[21] complexes do not have the complication of *trans*-bent geometry, and more direct *d* orbital

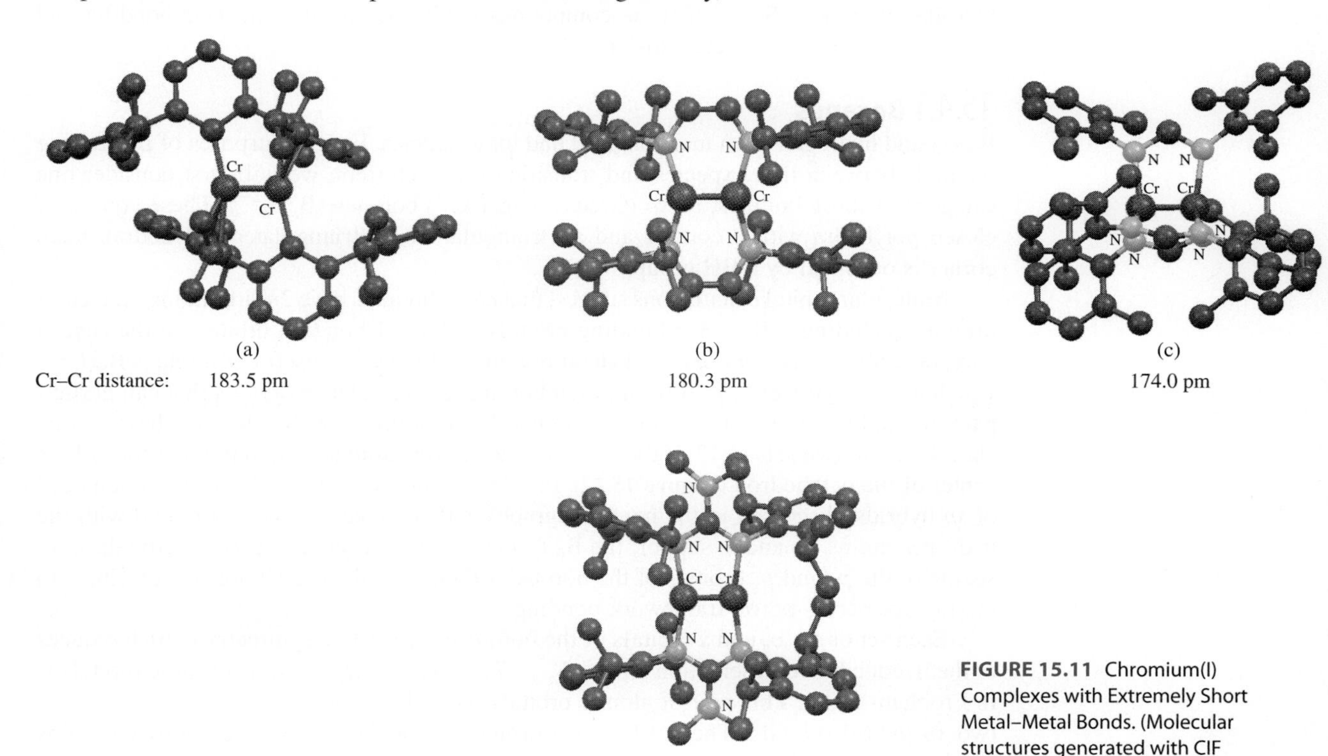

FIGURE 15.11 Chromium(I) Complexes with Extremely Short Metal–Metal Bonds. (Molecular structures generated with CIF data, with hydrogen atoms omitted for clarity.)

interactions may be one factor in making the bonding between the metals more efficient. Complexes containing Mo—Mo quintuple bonds (201.9 and 201.6 pm, and shortness ratios of 0.776 and 0.775, respectively) have been synthesized, and computations are consistent with the general orbital interactions in Figure 15.9.[22]

The bond order in complexes with very short metal–metal bonds has been controversial. Various calculations have given bond orders as low as 3.3 for "fivefold" bonded chromium(I) complexes. Factors such as the influence of bridging ligands on metal–metal distances, nonlinearity in Power's original complex that may reduce orbital interactions, and more complex interactions beyond the scope of this text are factors that merit consideration. The reader is encouraged to consult the sources cited[23] for discussions of the numerous factors involved in Cr—Cr multiple bonds in these dimers. Explorations of the reactivity of these new bonds are underway, with a particular emphasis on small molecule activation chemistry.[24]

15.4 Cluster Compounds

Cluster compounds have been included in previous sections of this chapter and earlier chapters. Transition-metal cluster chemistry has developed rapidly since the 1980s. Beginning with simple dimeric molecules, such as $Co_2(CO)_8$ and $Fe_2(CO)_9$,* chemists have developed syntheses of far more complex clusters, some with interesting and unusual structures and chemical properties. Large clusters have been studied with the objective of developing catalysts that may improve on the properties of heterogeneous catalysts; the surface of a cluster may mimic the behavior of the surface of a solid catalyst.

Before discussing transition-metal clusters, it is useful to consider some fundamental aspects of boron cluster chemistry. As mentioned in Chapter 8, boron forms numerous hydrides (boranes). Some of these compounds exhibit similarities in their bonding and structures to transition-metal clusters.

15.4.1 Boranes

Boron and hydrogen form many neutral and ionic species. For the purposes of illustrating parallels between these species and transition-metal clusters, we will first consider one category of these boranes, *closo* (Greek, "cagelike") boranes ($B_nH_n^{2-}$). These consist of closed polyhedra with n corners and all triangular faces (triangulated polyhedra). Each corner is occupied by a BH group.

Molecular orbital calculations suggest that *closo* boranes have $2n$+1 bonding molecular orbitals, including n B—H σ bonding orbitals and $n + 1$ bonding orbitals in the central core, described as **framework** or **skeletal** bonding orbitals.[25] A useful example is $B_6H_6^{2-}$, which has O_h symmetry. In this ion, each boron has four valence orbitals that can participate in bonding, giving a total of 24 boron valence orbitals for the cluster. These can be classified into two sets of 12. If the z axis of each boron atom is chosen to point toward the center of the octahedron (**Figure 15.12**), the six p_z and six s orbitals form sets (composed of sp hybrids, discussed in the next paragraph) with suitable symmetry to bond with the hydrogen atoms (6) and to support the B_6 framework (6). A second set of 12 orbitals, consisting of the p_x and p_y orbitals of the borons, is then exclusively available to contribute to orbitals for boron–boron framework bonding.

Each set of six p_z and s orbitals of the borons has the same symmetry (which reduces to the irreducible representation $A_{1g} + E_g + T_{1u}$; a symmetry analysis of these orbitals is in Problem 15.22). This pair of atomic orbitals on each boron can be considered to form two sp hybrid orbitals. These 12 hybrid orbitals point out toward the hydrogen atoms

*Some chemists define clusters as having at least three metal atoms.

(6, with $A_{1g} + E_g + T_{1u}$ collective symmetry) and in toward the center of the cluster (6, with collective $A_{1g} + E_g + T_{1u}$ symmetry). The A_{1g} and T_{1u} framework orbitals composed of these *sp* hybrids that point toward the center are shown in **Figure 15.13**. The orbitals under consideration to support the B_6 core are composed of the six aforementioned inward-pointing *sp* hybrids and 12 unhybridized boron 2*p* orbitals (p_x and p_y). The six

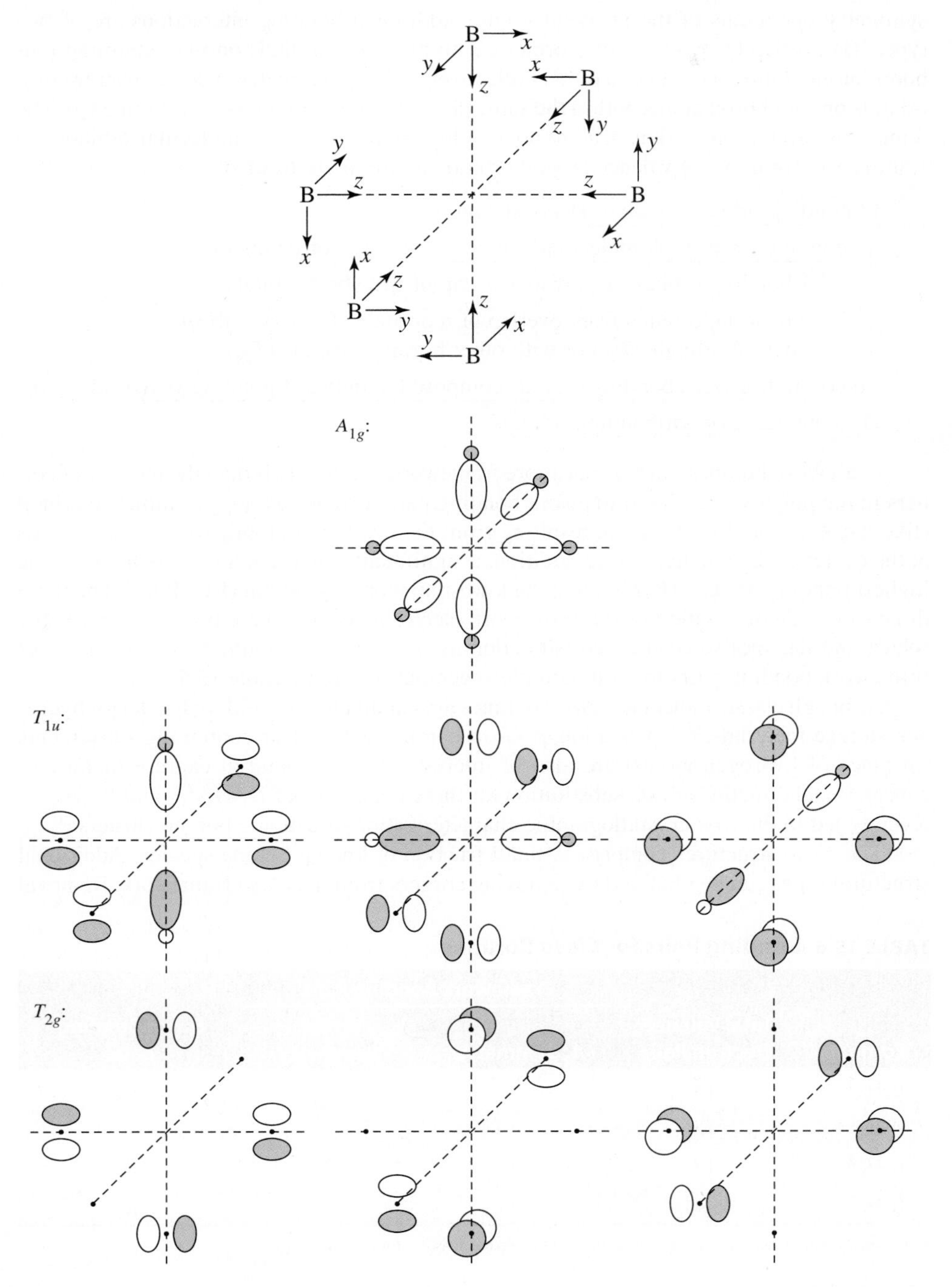

FIGURE 15.12 Coordinate System for Bonding in $B_6H_6^{2-}$.

FIGURE 15.13 Bonding in $B_6H_6^{2-}$.* (Data from "Some Bonding Considerations," in B. F. G. Johnson, ed., Transition Metal Clusters, p. 232.)

*Note that these T_{1u} orbitals also feature unhybridized 2*p* orbitals. The two E_g orbitals composed of the inward-pointing *sp* hybrids are not shown; these are nonbonding with respect to supporting the B_6 core.

hybrids directed away from the cluster framework (not shown) form the basis of orbitals responsible for bonding with the six hydrogen 1*s* orbitals.

Seven orbital combinations of the inward-pointing *sp* hybrids and unhybridized 2*p* orbitals lead to bonding interactions supporting the B_6 core (Figure 15.13). Constructive overlap of all six hybrid orbitals at the center of the octahedron yields a framework bonding orbital of A_{1g} symmetry; this orbital is completely symmetric with respect to all symmetry operations of the O_h point group. Additional bonding interactions are of two types: (1) overlap of two *sp* hybrid orbitals with parallel *p* orbitals on the remaining four boron atoms (three such interactions, collectively of T_{1u} symmetry) and (2) overlap of *p* orbitals on four boron atoms within the same plane (three interactions, T_{2g} symmetry). The remaining orbital interactions lead to nonbonding and antibonding molecular orbitals. To summarize, from the 24 valence atomic orbitals of boron are formed:

13 bonding orbitals ($= 2n + 1$), consisting of

- 7 bonding framework molecular orbitals ($= n + 1$), consisting of
  - 1 bonding orbital (A_{1g}) from overlap of *sp* hybrid orbitals
  - 6 bonding orbitals from overlap of *p* orbitals of boron with *sp* hybrid orbitals (T_{1u}) or with other boron *p* orbitals (T_{2g})
- 6 boron–hydrogen bonding orbitals composed of outward-pointing *sp* hybrids ($= n$)*

11 nonbonding or antibonding orbitals

In all *closo* boranes, there is one more framework bonding pair than the number of corners in the polyhedron. One framework bonding pair occupies a totally symmetric orbital (like the A_{1g} orbital in $B_6H_6^{2-}$), resulting from the overlap of atomic (or hybrid) orbitals at the center of the polyhedron. In addition, a significant gap in energy exists between the highest bonding orbital (HOMO) and the lowest nonbonding orbital (LUMO).[26] The three dimensional delocalization of the framework electrons renders these boranes remarkably robust and has motivated their classification as *spherically aromatic*.[27] The numbers of framework bonding pairs for common *closo* geometries are in **Table 15.6**.

Although classic molecules, *closo* boranes are candidates as building blocks for hydrogen storage materials.[28] Application of *closo* boranes with sulfur-containing substituents (in place of hydrogen atoms) are also of interest for boron neutron capture therapy of cancer,[29] and dimethylsulfide substitution for three hydrogens of $B_{12}H_{12}^{2-}$ and $B_{10}H_{10}^{2-}$ recently led to the first crystallographic characterization of cationic borane clusters.[30]

The *closo* structures comprise a small fraction of known borane species. Additional structural types can be obtained by removing corners from the *closo* framework. Removal

TABLE 15.6 Bonding Pairs for *Closo* Boranes

Formula	Total Valence Electron Pairs	Framework Bonding Pairs: A_1 Symmetry[a]	Framework Bonding Pairs: Other Symmetry	B—H Bonding pairs
$B_6H_6^{2-}$	13	1	6	6
$B_7H_7^{2-}$	15	1	7	7
$B_8H_8^{2-}$	17	1	8	8
$B_nH_n^{2-}$	2*n*+1	1	*n*	*n*

[a]Complete symmetry designation depends on the point group (such as A_{1g} for O_h symmetry).

*A *closo* borane features $4n + 2$ electrons (26 in the case of $B_6H_6^{2-}$) so all the bonding orbitals are filled.

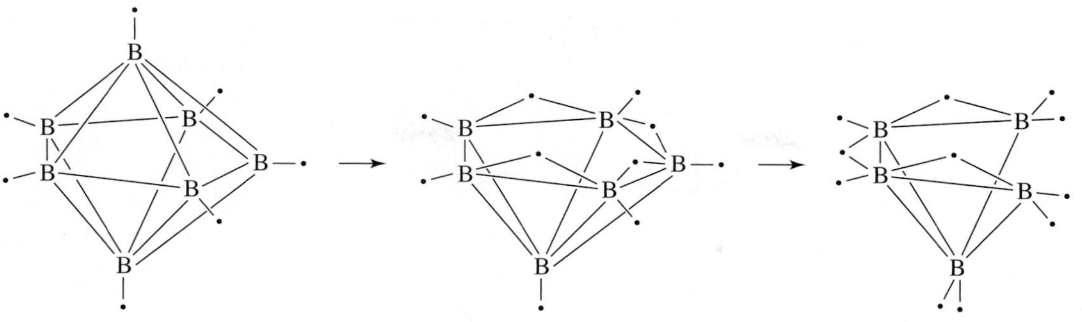

closso-$B_7H_7^{2-}$ nido-B_6H_{10} arachno-B_5H_{11}

FIGURE 15.14 *Closo, Nido,* and *Arachno* Borane Structures (small shaded circle in figure) = hydrogen.

of one corner yields a ***nido*** (nestlike) structure, removal of two corners an ***arachno*** (weblike) structure, removal of three corners a ***hypho*** (netlike) structure, and removal of four corners a ***klado*** (branched) structure.* Examples of three related *closo*, *nido*, and *arachno* borane structures are shown in **Figure 15.14**, and the structures for these boranes having 6 to 12 boron atoms are shown in **Figure 15.15**. Note that boranes resulting from formal removal of corners from *closo* boranes feature hydrogens that bridge boron atoms along edges; they may also have BH_2 groups.

Valence electron counts provide a convenient basis to classify these structural types. Various schemes for relating electron counts to structures have been proposed; most are based on rules formulated by Wade.[31] Wade's classification scheme is summarized in **Table 15.7**. It is remarkable that the pairs of framework bonding electrons solely depend on the number of corners of the parent polyhedron. The challenge is identifying the parent polyhedron on the basis of the borane formula. An example of a counting scheme to conveniently deduce the number of framework bonding pairs (and therefore the parent polyhedron) is presented later in this section.

In *closo* boranes, the total number of valence electron pairs is equal to the sum of the number of vertices in the polyhedron (each vertex has a boron–hydrogen bonding pair) and the number of framework bond pairs. For example, in $B_6H_6^{2-}$ there are 26 valence electrons, or 13 pairs ($= 2n + 1$, as mentioned previously). Six of these pairs are involved in bonding to the hydrogens (one per boron), and seven pairs are involved in framework bonding. The polyhedron of the *closo* structure is the parent polyhedron for the other structural types. **Table 15.8** summarizes electron counts and classifications for several boranes. The determination of the number of framework electron pairs for the other boranes is a bit counterintuitive on the basis of how the electrons are classified (see example).

EXAMPLE 15.3

Consider $B_{11}H_{13}^{2-}$ with $33 + 13 + 2 = 48$ valence electrons.

1. Each boron atom possesses at least one terminal hydrogen atom, and a B—H fragment contributes two electrons to framework bonding; one boron valence electron participates in the covalent bond to hydrogen. The framework contribution is $11 \times 2 = 22$ electrons.

**Hypho-* and *klado-* structures appear to be known only as derivatives. Additional details on naming boron hydrides and related compounds can be found in the IUPAC publication, G. J. Leigh, ed., *Nomenclature of Inorganic Chemistry: Recommendations 1990*, Blackwell Scientific Publications, Cambridge, MA, 1990, pp. 207–237.

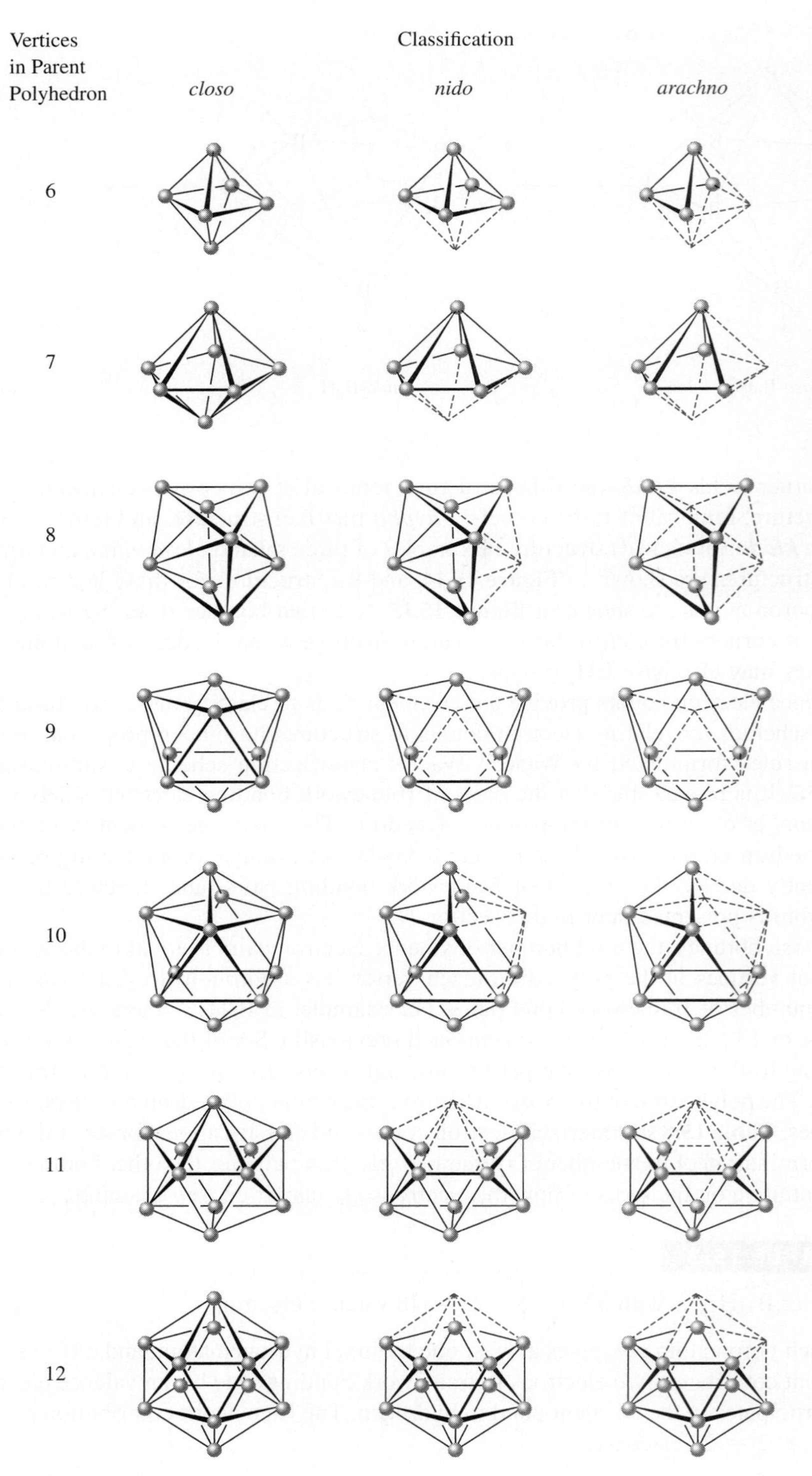

FIGURE 15.15 Structures of *Closo*, *Nido*, and *Arachno* Boranes Having 6 to 12 Borons. Terminal and bridging hydrogen atoms are not shown.

(Positions of hydrogen atoms can be found in N. N. Greenwood and A. Earnshaw, *Chemistry of the Elements, 2nd ed.*, Butterworth Heinemann, Oxford, 1997, pp. 153–154, 175, and 178.)

TABLE 15.7 Classification of Cluster Structures

Structure Type	Corners Occupied	Pairs of Framework Bonding Electrons	Empty Corners
Closo	n corners of n-cornered polyhedron	$n+1$	0
Nido	$(n-1)$ corners of n-cornered polyhedron	$n+1$	1
Arachno	$(n-2)$ corners of n-cornered polyhedron	$n+1$	2
Hypho	$(n-3)$ corners of n-cornered polyhedron	$n+1$	3
Klado	$(n-4)$ corners of n-cornered polyhedron	$n+1$	4

TABLE 15.8 Examples of Electron Counting in Boranes

Vertices in Parent Polyhedron	Classification	Boron Atoms in Cluster	Valence Electrons	Framework Electron Pairs	Examples	Formally Derived from
6	*Closo*	6	26	7	$B_6H_6^{2-}$	$B_6H_6^{2-}$
	Nido	5	24	7	B_5H_9	$B_5H_5^{4-}$
	Arachno	4	22	7	B_4H_{10}	$B_4H_4^{6-}$
7	*Closo*	7	30	8	$B_7H_7^{2-}$	$B_7H_7^{2-}$
	Nido	6	28	8	B_6H_{10}	$B_6H_6^{4-}$
	Arachno	5	26	8	B_5H_{11}	$B_5H_5^{6-}$
12	*Closo*	12	50	13	$B_{12}H_{12}^{2-}$	$B_{12}H_{12}^{2-}$
	Nido	11	48	13	$B_{11}H_{13}^{2-}$	$B_{11}H_{11}^{4-}$
	Arachno	10	46	13	$B_{10}H_{15}^{2-}$	$B_{10}H_{10}^{6-}$

2. Each hydrogen atom not yet accounted for (in this case 2 H atoms, $13 - 11 = 2$) contributes one electron to framework bonding regardless of whether these additional hydrogen atoms bridge two boron atoms or participate in a B—H bond as part of a BH_2 fragment. This contribution for $B_{11}H_{13}^{2-}$ is 2 electrons for a running total of $22 + 2 = 24$ framework electrons.

3. The charge of $B_{11}H_{13}^{2-}$ is –2; the contribution of the charge to the framework electron count is 2 electrons.

4. The total number of framework electrons in $B_{11}H_{13}^{2-}$ is $24 + 2 = 26$ electrons or 13 pairs, and the parent polyhedron has $13 - 1 = 12$ corners (icosahedron).

5. The parent polyhedron has 12 corners, so $B_{11}H_{13}^{2-}$, with one corner missing (with one less boron than 12), is a *nido* borane.

A Method for Classifying Boranes

The structures of boranes can be classified according to the following scheme:

closo boranes have the formula $B_nH_n^{2-}$

nido boranes are formally derived from $B_nH_n^{4-}$ ions

arachno boranes are formally derived from $B_nH_n^{6-}$ ions
hypho boranes are formally derived from $B_nH_n^{8-}$ ions
klado boranes are formally derived from $B_nH_n^{10-}$ ions

Other borane formulas can be related to these formulas by formally subtracting H^+ ions from the formula to make the number of B and H atoms equal, and then examining the resulting charge. For example, to classify $B_9H_{14}^-$, we can formally consider it to be derived from $B_9H_9^{6-}$:

$$B_9H_{14}^- - 5\ H^+ = B_9H_9^{6-}$$

The classification for this borane is therefore *arachno*.

EXAMPLE 15.4

Classify the following boranes by structural type.

$B_{10}H_{14}$

$B_{10}H_{14} - 4\ H^+ = B_{10}H_{10}^{4-}$ The classification is *nido*.

$B_2H_7^-$

$B_2H_7^- - 5\ H^+ = B_2H_2^{6-}$ The classification is *arachno*.

B_8H_{16}

$B_8H_{16} - 8\ H^+ = B_8H_8^{8-}$ The classification is *hypho*.

EXERCISE 15.3 Classify the following boranes by structural type:

a. $B_{11}H_{13}^{2-}$ **b.** $B_5H_8^-$ **c.** $B_7H_7^{2-}$ **d.** $B_{10}H_{18}$

15.4.2 Heteroboranes

The electron-counting schemes can be extended to isoelectronic species such as the carboranes, also known as *carbaboranes*. The CH^+ unit is isoelectronic with BH; many compounds are known in which one or more BH groups have been replaced by CH^+ (or by C, which also has the same number of electrons as BH). For example, replacement of two BH groups by CH^+ in *closo*-$B_6H_6^{2-}$ yields *closo*-$C_2B_4H_6$, a neutral compound. *Closo*, *nido*, and *arachno* carboranes are all known, most commonly containing two carbon atoms; examples are shown in **Figure 15.16**. Chemical formulas corresponding to these designations are given in **Table 15.9**. The utility of carboranes as pharmacophores in drug design (molecules that facilitate recognition and binding by biological macromolecules) motivates medicinal applications.[32] Unlike boranes, which are generally quite moisture sensitive, some carboranes are kinetically stable to hydrolysis, permitting their use *in vivo*.

Carboranes may be classified by structural type using the same method described previously for boranes. Because a carbon atom has the same number of valence electrons

TABLE 15.9 Examples of Formulas of Boranes and Carboranes

Type	Borane	Example	Carborane	Example
Closo	$B_nH_n^{2-}$	$B_{12}H_{12}^{2-}$	$C_2B_{n-2}H_n$	$C_2B_{10}H_{12}$
Nido	B_nH_{n+4}[a]	$B_{10}H_{14}$	$C_2B_{n-2}H_{n+2}$	$C_2B_8H_{12}$
Arachno	B_nH_{n+6}[a]	B_9H_{15}	$C_2B_{n-2}H_{n+4}$	$C_2B_7H_{13}$

[a]*Nido* boranes may also have the formulas $B_nH_{n+3}^-$ and $B_nH_{n+2}^{2-}$; *arachno* boranes may also have the formulas $B_nH_{n+5}^-$ and $B_nH_{n+4}^{2-}$.

$C_2B_4H_6$

closo

$C_2B_4H_8$

nido

$C_2B_8H_{10}^{4-}$

arachno
(Has one terminal H on each B and C atom)

FIGURE 15.16 Examples of Carboranes.

as a boron atom plus a hydrogen atom, formally each C should be converted to BH in the classification scheme. For example, for a carborane having the formula $C_2B_8H_{10}$,

$$C_2B_8H_{10} \longrightarrow B_{10}H_{12}$$

$$B_{10}H_{12} - 2\,H^+ = B_{10}H_{10}{}^{2-}$$

the classification of the carborane $C_2B_8H_{10}$ is therefore *closo*.

EXAMPLE 15.5

Classify the following carboranes by structural type:

$C_2B_9H_{12}{}^-$

$$C_2B_9H_{12}{}^- \longrightarrow B_{11}H_{14}{}^-$$
$$B_{11}H_{14}{}^- - 3H^+ = B_{11}H_{11}{}^{4-}$$ The classification is *nido*.

$C_2B_7H_{13}$

$$C_2B_7H_{13} \longrightarrow B_9H_{15}$$
$$B_9H_{15} - 6H^+ = B_9H_9{}^{6-}$$ The classification is *arachno*.

$C_4B_2H_6$

$$C_4B_2H_6 \longrightarrow B_6H_{10}$$
$$B_6H_{10} - 4H^+ = B_6H_6{}^{4-}$$ The classification is *nido*.

EXERCISE 15.4 Classify the following carboranes by structural type:

a. $C_3B_3H_7$ **b.** $C_2B_5H_7$ **c.** $C_2B_7H_{12}{}^-$

Many derivatives of boranes containing other main group atoms, designated *heteroatoms*, are also known. These *heteroboranes* may be classified by formally converting the heteroatom to a BH_x group having the same number of valence electrons, and then proceeding as in previous examples. For some of the more common heteroatoms, the substitutions are

Heteroatom	Replace with
C, Si, Ge, Sn	BH
N, P, As	BH_2
S, Se	BH_3

EXAMPLE 15.6

Classify the following heteroboranes by structural type:

SB_9H_{11}

$$SB_9H_{11} \longrightarrow B_{10}H_{14}$$
$$B_{10}H_{14} - 4H^+ = B_{10}H_{10}{}^{4-}$$ The classification is *nido*.

$CPB_{10}H_{11}$

$$CPB_{10}H_{11} \longrightarrow PB_{11}H_{12} \longrightarrow B_{12}H_{14}$$
$$B_{12}H_{14} - 2H^+ = B_{12}H_{12}{}^{2-}$$ The classification is *closo*.

EXERCISE 15.5 Classify the following heteroboranes by structural type:

a. SB_9H_9 **b.** $GeC_2B_9H_{11}$ **c.** $SB_9H_{12}{}^-$

Although it is reasonable that the set of electron-counting rules developed for boranes can be applied to similar compounds such as carboranes, how broadly can these rules can be applied? Can Wade's rules be used effectively on compounds similar to boranes or carboranes but with metal atoms incorporated into the cluster framework? Can the rules be extended even further to describe the bonding in polyhedral metal clusters?

15.4.3 Metallaboranes and Metallacarboranes

The CH group of a carborane is isolobal with 15-electron fragments of an octahedron such as $Co(CO)_3$. Similarly, BH, which has 4 valence electrons, is isolobal with 14-electron fragments such as $Fe(CO)_3$ and $Co(\eta^5\text{-}C_5H_5)$. These organometallic fragments have been found in borane and carborane derivatives in which the organometallic fragments substitute for the isolobal main group fragments. For example, organometallic derivatives of B_5H_9 have been synthesized (**Figure 15.17**). Calculations on the iron derivative support the view that $Fe(CO)_3$ in this compound bonds in a fashion isolobal with BH.[33] In both fragments, the orbitals involved in framework bonding within the cluster are similar (**Figure 15.18**). In BH, the orbitals participating in framework bonding are (a) an sp_z hybrid pointing toward the center of the polyhedron (similar to the orbitals participating in the bonding of A_{1g} symmetry in $B_6H_6^{2-}$ in Figure 15.13) and (b) p_x and p_y orbitals tangential to the surface of the cluster. In $Fe(CO)_3$, an $sp_zd_{z^2}$ hybrid points toward the center, and *pd* hybrid orbitals are oriented tangentially to the cluster surface.

There are many metallaboranes and metallacarboranes. Selected examples with *closo* structures are in **Table 15.10**.

Anionic boranes and carboranes can bind to metals by resembling cyclic organic ligands. For example, *nido* carboranes of formula $C_2B_9H_{11}^{2-}$ have *p* orbital lobes pointing toward the "missing" site of the icosahedron (remember that the *nido* structure corresponds to a *closo* structure, in this case the 12-vertex icosahedron, with one vertex missing). This arrangement of *p* orbitals can be compared with the *p* orbitals of the cyclopentadienyl ring (**Figure 15.19**).

Although the comparison between these ligands is not exact, the similarity is sufficient that $C_2B_9H_{11}^{2-}$ can bond to iron to form a carborane analog of ferrocene, $[Fe(\eta^5\text{-}C_2B_9H_{11})_2]^{2-}$. A mixed-ligand sandwich compound containing one

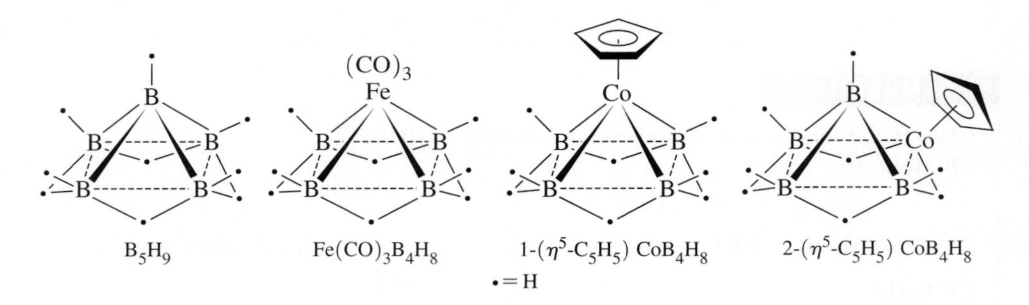

FIGURE 15.17 Organometallic Derivatives of B_5H_9.

FIGURE 15.18 Orbitals of Isolobal Fragments BH and $Fe(CO)_3$.

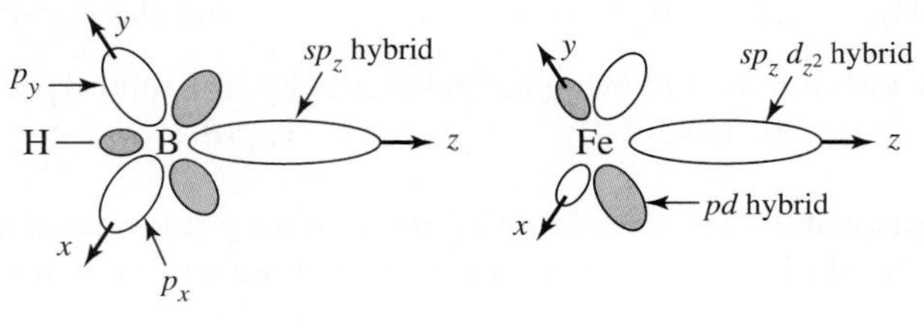

TABLE 15.10 Metallaboranes and Metallacarboranes with *Closo* Structures

Number of Skeletal Atoms	Shape		Examples	
6	Octahedron		$B_4H_6(CoCp)_2$	$C_2B_3H_5Fe(CO)_3$
7	Pentagonal bipyramid		$C_2B_4H_6Ni(PPh_3)_2$	$C_2B_3H_5(CoCp)_2$
8	Dodecahedron		$C_2B_4H_4[(CH_3)_2Sn]CoCp$	
9	Capped square antiprism		$C_2B_6H_8Pt(PMe_3)_2$	$C_2B_5H_7(CoCp)_2$
10	Bicapped square antiprism		$[B_9H_9NiCp]^-$	$CB_7H_8(CoCp)(NiCp)$
11	Octadecahedron		$[CB_9H_{10}CoCp]^-$	$C_2B_8H_{10}IrH(PPh_3)_2$
12	Icosahedron		$C_2B_7H_9(CoCp)_3$	$C_2B_9H_{11}Ru(CO)_3$

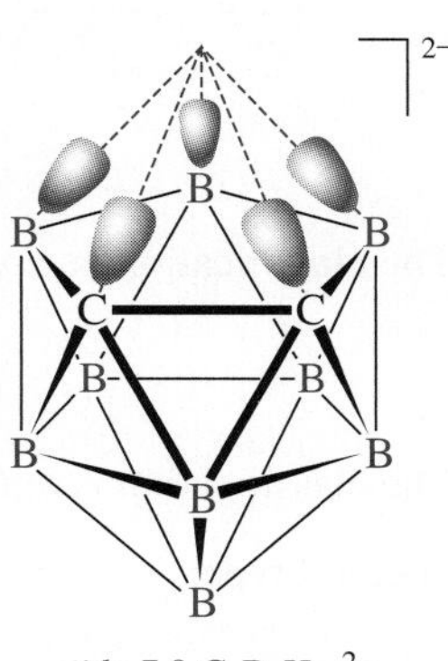

nido-7,8-$C_2B_9H_{11}^{2-}$

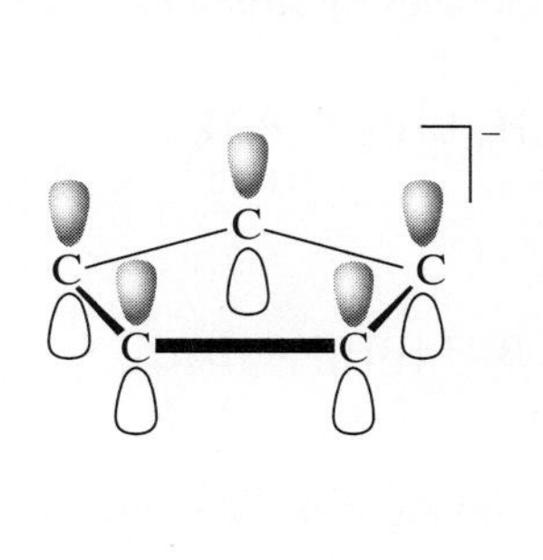

η^5-$C_5H_5^-$

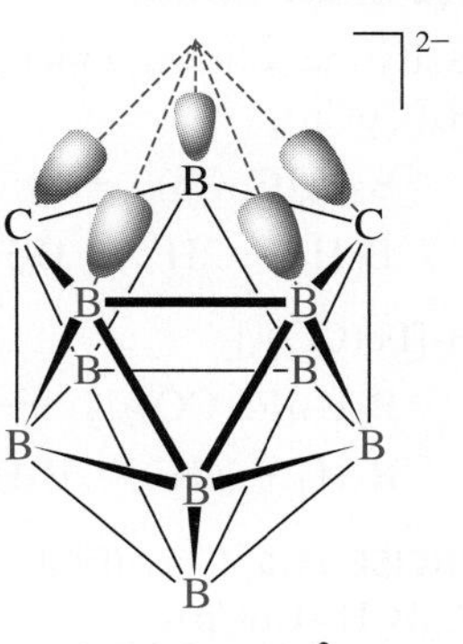

nido-7,9-$C_2B_9H_{11}^{2-}$

FIGURE 15.19 Comparison of Isomers of $C_2B_9H_{11}^{2-}$ with $C_5H_5^-$.

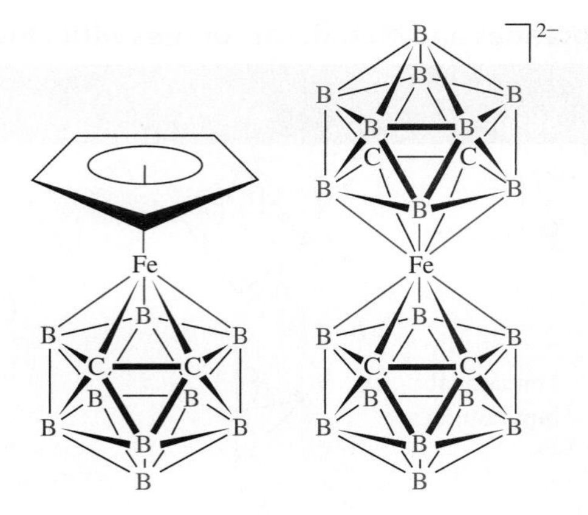

FIGURE 15.20 Carborane Analogs of Ferrocene.

carborane and one cyclopentadienyl ligand, $[Fe(\eta^5\text{-}C_2B_9H_{11})(\eta^5\text{-}C_5H_5)]$, has also been synthesized (**Figure 15.20**).[34] Many transition-metal complexes featuring borane and carborane ligands are known.[35]

Metallaboranes and metallacarboranes can be classified structurally by using a procedure similar to the method for boranes and their main group derivatives. To classify borane derivatives with transition-metal–containing fragments, it is convenient to determine how many electrons the metal-containing fragment needs to satisfy the requirement of the 18-electron rule. This fragment can be considered equivalent to a BH_x fragment needing the same number of electrons to satisfy the octet rule. For example, a 14-electron fragment such as $Co(\eta^5\text{-}C_5H_5)$ is 4 electrons short of 18; this fragment may be considered the equivalent of the 4-electron fragment BH, which is four electrons short of an octet. Shown here are examples of organometallic fragments and their corresponding isolobal BH_x fragments:

Valence Electrons in Organometallic Fragment	Example	Replace with
13	$Mn(CO)_3$	B
14	$Co(\eta^5\text{-}Cp)$	BH
15	$Co(CO)_3$	BH_2
16	$Fe(CO)_4$	BH_3

EXAMPLE 15.7

Classify the following metallaboranes by structural type:

$B_4H_6(CoCp)_2$

$$B_4H_6(CoCp)_2 \longrightarrow B_4H_6(BH)_2 = B_6H_8$$

$$B_6H_8 - 2H^+ = B_6H_6^{\ 2-}$$ The classification is *closo*.

$B_3H_7[Fe(CO)_3]_2$

$$B_3H_7[Fe(CO)_3]_2 \longrightarrow B_3H_7[BH]_2 = B_5H_9$$

$$B_5H_9 - 4H^+ = B_5H_5^{\ 4-}$$ The classification is *nido*.

EXERCISE 15.6 Classify the following metallaboranes by structural type:

a. $C_2B_7H_9(CoCp)_3$

b. $C_2B_4H_6Ni(PPh_3)_2$

15.4.4 Carbonyl Clusters

Many carbonyl clusters have structures similar to boranes. To what extent is the approach used to describe bonding in boranes applicable to bonding in carbonyl clusters and other clusters?

According to Wade, the valence electrons in a cluster can be assigned to framework and metal–ligand bonding.[36]

Total number of valence electrons in cluster	=	number of electrons involved in framework bonding	+	number of electrons involved in metal–ligand bonding

As we have seen, the number of electrons involved in framework bonding in boranes is related to the classification of the structure as *closo*, *nido*, *arachno*, *hypho*, or *klado*. Rearranging this equation gives

Number of electrons involved in framework bonding	=	total number of valence electrons in cluster	−	number of electrons involved in metal–ligand bonding

For a *closo* borane, one electron pair is assigned to one boron–hydrogen bond on each boron. The remaining valence electron pairs are framework bonding pairs. For a *closo* transition-metal carbonyl complex Wade suggests that 6 electron pairs per metal are involved either in metal–carbonyl bonding (to all carbonyls on a metal) or are nonbonding and therefore unavailable for participation in framework bonding. A *closo* metal–carbonyl cluster has 5 more electron pairs per framework atom, or 10 more electrons, than the corresponding borane. A metal–carbonyl analog of *closo*-$B_6H_6^{2-}$, which has 26 valence electrons, would therefore need a total of 86 valence electrons to adopt a *closo* structure. An 86-electron cluster that satisfies this requirement is $Co_6(CO)_{16}$, which has an octahedral framework similar to $B_6H_6^{2-}$. As in the case of boranes, *nido* structures correspond to *closo* geometries from which one vertex is empty, *arachno* structures lack two vertices, and so on.

A simpler way to compare electron counts in boranes and transition-metal clusters is to consider the different numbers of valence orbitals available to the framework atoms. Transition metals, with nine valence orbitals (one *s*, three *p*, and five *d* orbitals), have five more orbitals available for bonding than boron, which has only four valence orbitals; these five extra orbitals, when filled as a consequence of bonding within the framework and with surrounding ligands, give an increased electron count of 10 electrons per framework atom. Consequently, a useful rule of thumb is to increase the electron requirement of the cluster by 10 per framework atom when replacing a boron with a transition-metal atom. In the example cited previously, replacing the six borons in *closo*-$B_6H_6^{2-}$ with six cobalts should, therefore, increase the electron count from 26 to 86 for a comparable *closo* cobalt cluster. $Co_6(CO)_{16}$, an 86-electron cluster, meets this requirement.

The valence electron counts corresponding to the various structural classifications for main group and transition-metal clusters are summarized in **Table 15.11**.[37] In this table, n designates the number of framework atoms.

Examples of *closo, nido*, and *arachno* borane and transition-metal clusters are given in **Table 15.12**. The clusters $[Fe_4C(CO)_{12}]^{2-}$ and $Os_5C(CO)_{15}$ have metal atoms that form the same shapes as the boron atoms in B_4H_{10} and B_5H_9, respectively, but the lone carbon atoms play a vital role in establishing the expected valence electron count; these clusters are discussed in Section 15.4.5. Transition-metal clusters formally containing seven metal–metal framework bonding pairs are among the most common. Examples illustrating the structural diversity of these clusters are in **Table 15.13** and **Figure 15.21**.

TABLE 15.11 Electron Counting in Main Group and Transition-Metal Clusters

Structure Type	Main Group Cluster	Transition-Metal Cluster
Closo	$4n + 2$	$14n + 2$
Nido	$4n + 4$	$14n + 4$
Arachno	$4n + 6$	$14n + 6$
Hypho	$4n + 8$	$14n + 8$

TABLE 15.12 *Closo, Nido,* and *Arachno* Borane and Transition-Metal Clusters

Atoms in Cluster	Vertices in Parent Polyhedron	Framework Electron Pairs	Valence Electrons (Boranes)				Valence Electrons (Transition-Metal Clusters)			
			Closo	*Nido*	*Arachno*	Examples	*Closo*	*Nido*	*Arachno*	Examples
4	4	5	18				58			
	5	6		20		$B_4H_7^-$		60		$Co_4(CO)_{12}$
	6	7			22	B_4H_{10}			62	$[Fe_4C(CO)_{12}]^{2-}$
5	5	6	22			$C_2B_3H_5$	72			$Os_5(CO)_{16}$
	6	7		24		B_5H_9		74		$Os_5C(CO)_{15}$
	7	8			26	B_5H_{11}			76	$[Ni_5(CO)_{12}]^{2-}$
6	6	7	26			$B_6H_6^{2-}$	86			$Co_6(CO)_{16}$
	7	8		28		B_6H_{10}		88		$Os_6(CO)_{17}[P(OMe)_3]_3$
	8	9			30	B_6H_{12}			90	

TABLE 15.13 Clusters Formally Containing Seven Metal–Metal Framework Bond Pairs

Number of Framework Atoms	Structure Type	Shape	Examples
7	Capped *closo*[a]	Capped octahedron	$[Rh_7(CO)_{16}]^{3-}$
			$Os_7(CO)_{21}$
6	*Closo*	Octahedron	$Rh_6(CO)_{16}$
			$Ru_6C(CO)_{17}$
6	Capped *nido*[a]	Capped square pyramid	$H_2Os_6(CO)_{18}$
5	*Nido*	Square pyramid	$Ru_5C(CO)_{15}$
4	*Arachno*	Butterfly	$[Fe_4(CO)_{13}H]^{-}$[b]

Source: Data from "Some Bonding Considerations," in B. F. G. Johnson, ed., *Transition Metal Clusters,* p. 232.

[a]A capped *closo* cluster has a valence electron count equivalent to neutral B_nH_n. A capped *nido* cluster has the same electron count as a *closo* cluster.

[b]This complex has an electron count matching a *nido* structure, but it adapts the butterfly structure expected for *arachno*. This is one of the many examples in which the structure of metal clusters is not predicted accurately by Wade's rules. Limitations of Wade's rules are discussed in R. N. Grimes, "Metallacarboranes and Metallaboranes," in G. Wilkinson, F. G. A. Stone, and W. Abel, eds., *Comprehensive Organometallic Chemistry*, Vol. 1, Pergamon Press, Elmsford, NY, 1982, p. 473.

The predictions of structures of transition-metal carbonyl clusters using Wade's rules are not infallible. For example, the clusters $M_4(CO)_{12}$ (M = Co, Rh, Ir) have 60 valence electrons and are predicted to be *nido* complexes ($14n + 4$ valence electrons). A *nido* structure would correspond to a trigonal bipyramid parent structure with one position vacant. X-ray crystallographic studies, however, have shown these complexes to have tetrahedral metal cores.

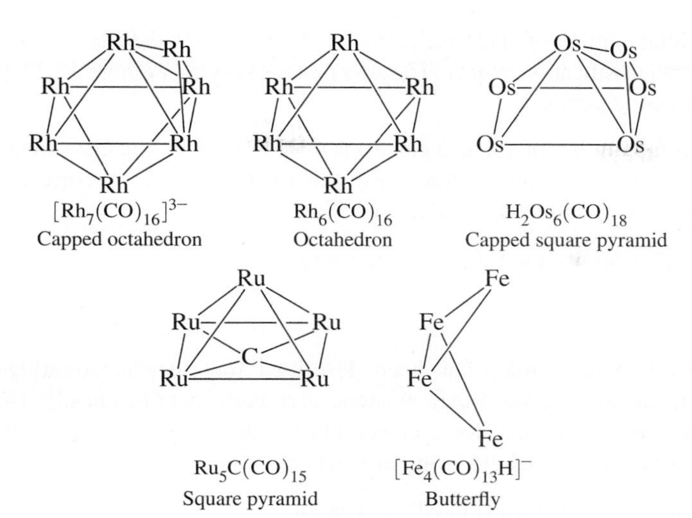

FIGURE 15.21 Metal Cores for Clusters Containing Seven Skeletal Bond Pairs.

Ionic clusters of main group elements can also be classified by a similar approach to that used for other clusters. Many such clusters are known;[38] they are sometimes called **Zintl ions**. Examples are shown in **Figure 15.22**. In some cases Zintl ions have additional groups outside the central core, as shown in the final ion in Figure 15.22, $\{Sn_9[Si(SiMe_3)_3]_3\}^-$.[39]

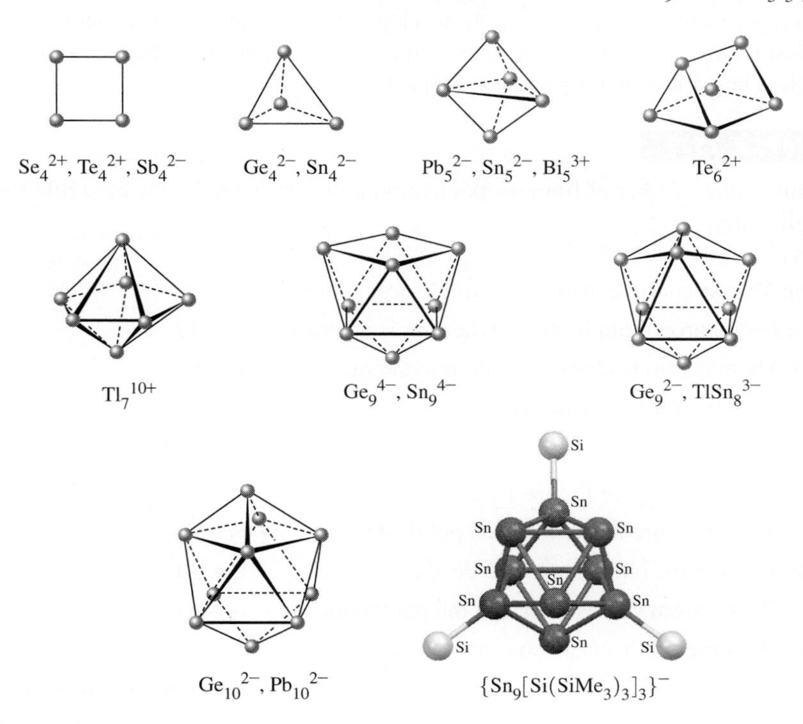

FIGURE 15.22 Ionic Clusters of Main Group Elements (Zintl Ions). (Molecular structures generated with CIF data.)

EXAMPLE 15.8

Classify the following main-group clusters:

a. Pb_5^{2-}: Total valence-electron count = 22 (including each of the 4 valence electrons per Pb, plus 2 electrons for the charge). Because $n = 5$, the total electron count = $4n + 2$; the classification is *closo*. (See Table 15.11.)

b. Sn_9^{4-}: Total number of valence electrons = 40. For $n = 9$, the electron count = $4n + 4$; the classification is *nido*. The structure, as shown in Figure 15.22, has one missing vertex.

c. Sb_4^{2-}: Total number of valence electrons = 22 = $4n + 6$. The classification is *arachno*. The square structure of this ion (Figure 15.22) corresponds to an octahedron with two vertices missing.

EXERCISE 15.7 Classify the following main-group clusters:

a. Ge_9^{2-}

b. Bi_5^{3+}

An extension of Wade's rules has been described for electron counting in boranes, heteroboranes, metallaboranes, other clusters, and even metallocenes.[40] This approach, called the ***mno* rule**, states that for a closed cluster structure to be stable, there must be $m + n + o$ framework skeletal electron pairs, where

m = number of condensed (linked) polyhedra

n = total number of vertices

o = number of single-atom bridges between two polyhedra

A fourth term, p, must be added for structures with missing vertices:

p = number of missing vertices (e.g., $p = 1$ for *nido*, $p = 2$ for *arachno*.)

This approach has been particularly developed for application to macropolyhedral structures, clusters involving linked polyhedra, and many examples have been described.[41] This approach is best illustrated by some examples.

EXAMPLE 15.9

Determine the number of framework electron pairs predicted by the *mno* rule for the following:

$B_{12}H_{12}^{2-}$ (see Figure 15.15)

m: This structure consists of a single polyhedron.	$m = 1$
n: Each boron atom in the polyhedron is a vertex.	$n = 12$
o: There are no bridges between polyhedra.	$o = 0$
p: The structure is *closo*, so $p = 0$.	
	$m + n + o =$ **13** electron pairs

$(\eta^5\text{-}C_2B_9H_{11})_2Fe^{2-}$ (see Figure 15.20)

m: This structure has two linked polyhedra.	$m = 2$
n: All carbons, borons, and the Fe are vertices.	$n = 23$
o: The Fe atom serves to bridge the polyhedra.	$o = 1$
p: The structure is *closo*, so $p = 0$.	
	$m + n + o =$ **26** electron pairs

Ferrocene, $(\eta^5\text{-}C_5H_5)_2Fe$ (see Figure 13.5)

m: The structure may be viewed as two linked polyhedra (pentagonal pyramids).	$m = 2$
n: Each atom in the structure is a vertex.	$n = 11$
o: The iron atom bridges the polyhedra.	$o = 1$

p: The structure is <u>not</u> *closo*; the top or bottom may be viewed as a pentagonal bipyramid lacking one vertex; the classification is *nido* for *each* linked polyhedron.

$p = 2$ (one open face per polyhedron)

$m + n + o + p =$ **16** electron pairs

EXERCISE 15.8 Determine the number of framework electron pairs predicted by the *mno* rule for the following:

a. $(\eta^5\text{-}C_5H_5)Fe(\eta^5\text{-}C_2B_9H_{11})$ (see Figure 15.20)

b. *nido*-7,8-$C_2B_9H_{11}^{2-}$ (see Figure 15.19)

15.4.5 Carbon-Centered Clusters

Many compounds have been synthesized, often fortuitously, in which one or more atoms have been partially or completely encapsulated within metal clusters. The most common have been the carbon-centered clusters, also called carbide or carbido clusters, with carbon exhibiting coordination numbers and geometries not found in classic organic molecules. Examples of these unusual coordination geometries are shown in **Figure 15.23**.

Encapsulated atoms contribute their valence electrons to the total electron count. For example, carbon contributes its 4 valence electrons in $Ru_6C(CO)_{17}$ to give a total of 86 electrons, corresponding to a *closo* electron count (Table 15.12).

How can carbon, with only four valence orbitals, form bonds to more than four surrounding transition-metal atoms? $Ru_6C(CO)_{17}$, with a central core of O_h symmetry, is a useful example. The 2*s* orbital of carbon has A_{1g} symmetry and the 2*p* orbitals have T_{1u} symmetry in the O_h point group. The octahedral Ru_6 core has framework bonding orbitals of the same symmetry as in $B_6H_6^{2-}$ described earlier in this chapter (see Figure 15.13): a centrally directed A_{1g} group orbital and two sets of orbitals, oriented tangentially to the core, of T_{1u} and T_{2g} symmetry. Therefore, there are two ways in which the symmetry match is correct for interactions between the carbon and the Ru_6 core: the interactions of A_{1g} and T_{1u} symmetry shown in **Figure 15.24** (the T_{2g} orbitals participate in Ru—Ru bonding but not in bonding with the central carbon). The net result is the formation of four C—Ru bonding orbitals, occupied by electron pairs in the cluster, and four unoccupied antibonding orbitals.[42]

EXERCISE 15.9 Classify the following clusters by structural type:

a. $[Re_7C(CO)_{21}]^{3-}$

b. $[Fe_4N(CO)_{12}]^{-}$

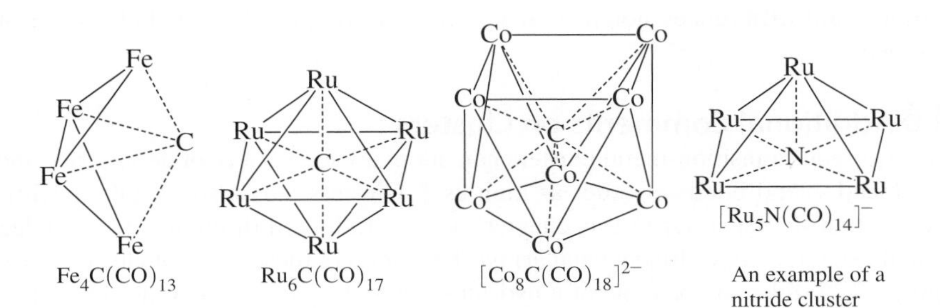

FIGURE 15.23 Carbide Clusters. CO ligands have been omitted for clarity.

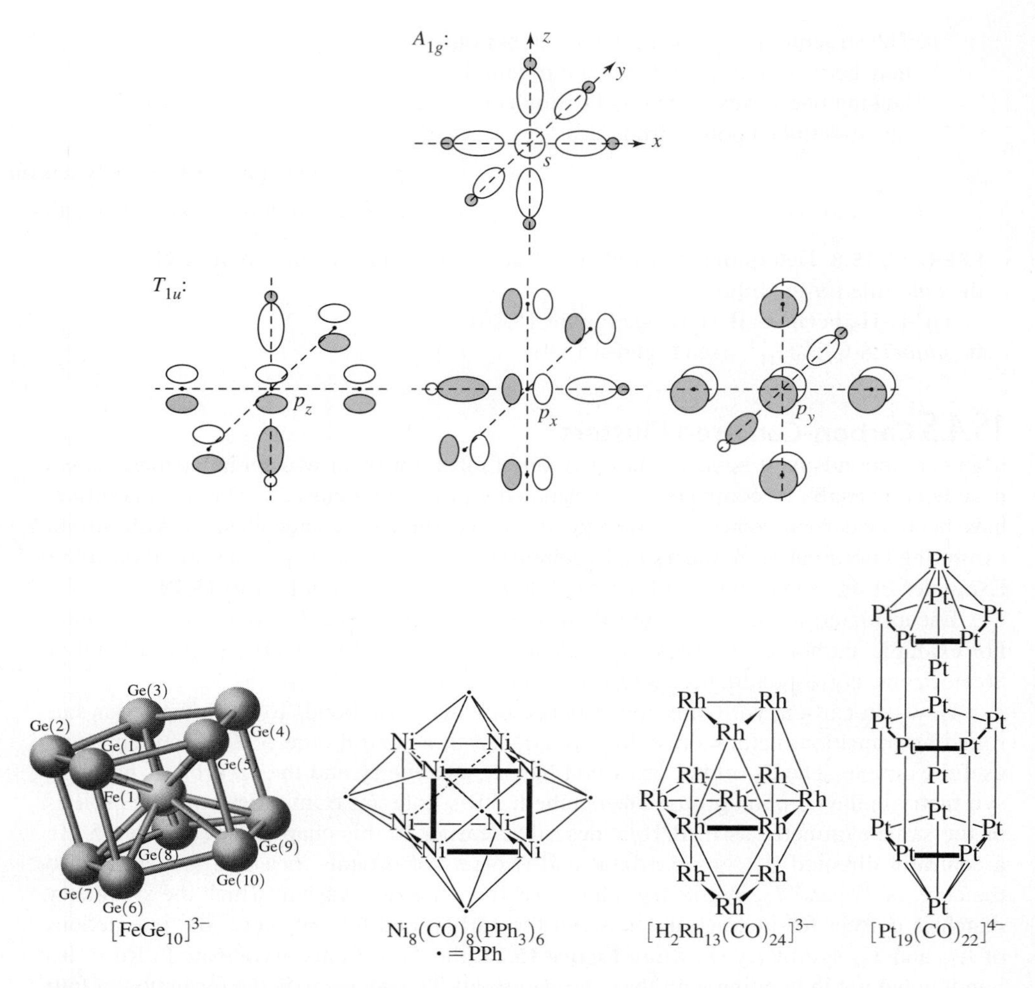

FIGURE 15.24 Bonding Interactions between Central Carbon and Octahedral Ru_6.

FIGURE 15.25 Examples of Large Clusters. CO and hydride ligands have been omitted to show the metal–metal bonding more clearly.

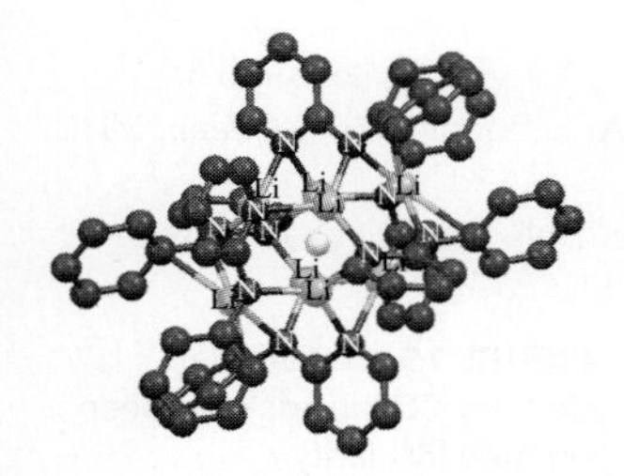

FIGURE 15.26 A Hydride Ion in a Cage of Eight Lithium Ions. (Molecular structures generated with CIF data, with hydrogen atoms omitted for clarity.)

A particularly interesting example of encapsulation is that of an iron atom inside a pentagonal prismatic Zintl ion, $[FeGe_{10}]^{3-}$ (**Figure 15.25**). This ion is formed from the potassium salt of the Zintl species $Ge_9{}^{4-}$ and $Fe(2,6\text{-}Mes_2C_6H_3)_2$ (mes = mesityl) in the presence of 2,2,2-crypt; the crypt and the solvent, ethylenediamine, are also present in the crystal. $[FeGe_{10}]^{3-}$ has been compared with the $[Ti(\eta^5\text{-}P_5)_2]^{2-}$ ion (see Section 15.2.2), with significant differences noted in structure and bonding despite their similarity in appearance.[43]

15.4.6 Additional Comments on Clusters

As we have seen, transition-metal clusters can adopt a wide variety of geometries and can involve metal–metal bonds of order as high as 5. Clusters may also include much larger polyhedra than shown so far in this chapter; polyhedra linked through vertices, edges, or faces; and extended three-dimensional arrays. Examples of these types of clusters are given in Figure 15.25. Even an example of a hydride-centered cluster, with a hydride ion within a cage of eight lithium ions, has been reported (**Figure 15.26**).[44]

References

1. J. E. Ellis, *J. Chem. Educ.*, **1976**, *53*, 2.
2. R. Hoffmann, *Angew. Chem., Int. Ed.*, **1982**, *21*, 711; see also H-J. Krause, *Z. Chem.*, **1988**, *28*, 129.
3. M. Poliakoff, J. J. Turner, *J. Chem. Soc. A*, **1971**, 2403.
4. H. G. Raubenheimer, H. Schmidbauer, *Organometallics*, **2012**, *31*, 2507.
5. M. Baudler, S. Akpapoglou, D. Ouzounis, F. Wasgestian, B. Meinigke, H. Budzikiewicz, H. Münster, *Angew. Chem., Int. Ed.*, **1988**, *27*, 280.
6. O. J. Scherer, T. Brück, *Angew. Chem., Int. Ed.*, **1987**, *26*, 59.
7. D. G. Evans, D. M. P. Mingos, *J. Organomet. Chem.*, **1982**, *232*, 171.
8. A. G. Orpen, A. V. Rivera, E. G. Bryan, D. Pippard, G. Sheldrick, K. D. Rouse, *Chem. Commun.*, **1978**, 723.
9. B. F. G. Johnson, D. A. Kaner, J. Lewis, P. R. Raithby, *J. Organomet. Chem.*, **1981**, *215*, C33.
10. P. J. Fischer, V. G. Young., Jr., J. E. Ellis, *Chem. Commun.*, **1997**, 1249.
11. A. Davison, J. E. Ellis, *J. Organomet. Chem.*, **1971**, *36*, 113.
12. (a) $Cp^*Pt(CH_3)_3$: S. Roth, V. Ramamoorthy, P. R. Sharp, *Inorg. Chem.*, **1990**, *29*, 3345. (b) $Cp^*Pt(ZnCp^*)_3$: T. Bolermann, K. Freitag, C. Gemel, R. W. Seidel, R. A. Fischer, *Organometallics,* **2011**, *30*, 4123. (c) $Ir[C_4H_4]Cl(CS)(PPh_3)_2$: G. R. Clark, P. M. Johns, W. R. Roper, L. J. Wright, *Organometallics*, **2008**, *27*, 451.
13. (a) G. Cavigliasso, T. Lovell, R. Stranger, *Dalton Trans.*, **2006**, 2017. (b) G. Cavigliasso, P. Comba, R. Stranger, *Inorg. Chem.*, **2004**, *43*, 6734.
14. F. A. Cotton, *Chem. Soc. Rev.*, **1975**, *4*, 27.
15. F. A. Cotton, L. M. Daniels, E. A. Hillard, C. A. Murillo, *Inorg. Chem.*, **2002**, 41, 2466. See also F. A. Cotton, C. A. Murillo, R. A. Walton, Eds., *Multiple Bonds between Metal Atoms*, Springer Science and Business Media, New York, 2005, pp. 71–74.
16. F. A. Cotton, *Chem. Soc. Rev.*, **1983**, *12*, 35.
17. T. Nguyen, A. D. Sutton, M. Brynda, J. C. Fettinger, G. J. Long, P. P. Power, *Science*, **2005**, *310*, 844.
18. G. Frenking, *Science,* **2005,** *310,* 796; U. Radius, F. Breher, *Angew., Int. Ed.,* **2006,** *45,* 3006.
19. K. A. Kreisel, G. P. A. Yap, O. Dmitrenko, C. R. Landis, K. H. Theopold, *J. Am. Chem. Soc.*, **2007**, *129*, 14162.
20. Y.-C. Tsai, C.-W. Hsu, J.-S. K. Yu, G.-H. Lee, Y. Wang, T.-S. Kuo, *Angew. Chem., Int. Ed.*, **2008**, *47*, 7250.
21. A. Noor, R. Kempe, *Chem. Rec.*, **2010**, *10*, 413; A. Noor, G. Glatz, R. Müller, M. Kaupp, S. Demeshko, R. Kempe, *Z. Anorg. Allg. Chem*, **2009**, *635*, 1149.
22. Y.-C. Tsai, H.-Z. Chen, C.-C. Chang, J.-S. K. Yu, G.-H. Lee, Y. Wang; T.-S. Kuo, *J. Am. Chem. Soc.*, **2009**, *131*, 12534.
23. G. Merino, K. J. Donald, J. S. D'Acchioli, R. Hoffmann, *J. Am. Chem. Soc.*, **2007**, *129*, 15295; M. Brynda, L. Gagliardi, B. O. Roos, *Chem. Phys. Lett.*, **2009**, *471*, 1; D. B. DuPré, *J. Chem. Phys. A*, **2009**, *113*, 1559; G. La Macchia, G. Li Manni, T. K. Todorova, M. Brynda, F. Aquilante, B. O. Moss, L. Gagliardi, *Inorg. Chem.*, **2010**, *49*, 5216.
24. J. Shen, G. A. P. Yap, J.-P. Werner, and K. H. Theopold, *Chem. Commun.*, **2011**, *47*, 12191; R. Kempe, C. Schwarzmaier, A. Noor, G. Glatz, M. Zabel, A. Y. Timoshkin, B. M. Cossairt, C. C. Cummins, M. Scheer, *Angew. Chem., Int. Ed.*, **2011**, *50*, 7283.
25. K. Wade, *Electron Deficient Compounds*, Thomas Nelson & Sons, London, 1971.
26. K. Wade, "Some Bonding Considerations," in B. F. G. Johnson, ed., *Transition Metal Clusters*, John Wiley & Sons, New York, 1980, p. 217.
27. Z. Chen, R. B. King, *Chem. Rev.*, **2005**, *105*, 3613.
28. K. Srinivasu, S. K. Ghosh, *J. Phys. Chem. C*, **2011**, *115*, 1450; S. Li., M. Willis; P. Jena, *J. Phys. Chem. C*, **2010**, *114*, 16849.
29. E. L. Crossley, H. Y. V. Ching, J. A. Ioppolo, and L. M. Rendina, *Bioinorganic Medicinal Chemistry*, Wiley-VCH: Weinheim, 2011, p. 298; A. H. Soloway, W. Tjarks, B. A. Barnum, F. -G. Rong, R. F. Barth, I. M. Codongi, J. G. Wilson, *Chem. Rev.*, **1998**, *98*, 1515.
30. E. J. M. Hamilton, H. T. Leung, R. G. Kultyshev, X. Chen, E. A. Meyers, S. G. Shore, *Inorg. Chem.*, **2012**, *51*, 2374.
31. K. Wade, *Adv. Inorg. Chem. Radiochem.*, **1976**, *18*, 1–66.
32. F. Issa, M. Kassiou, L. M. Rendina, *Chem. Rev.*, **2011**, *111*, 5701.
33. R. L. DeKock, T. P. Fehlner, *Polyhedron*, **1982**, *1*, 521.
34. M. F. Hawthorne, D. C. Young, P. A. Wegner, *J. Am. Chem. Soc.*, **1965**, *87*, 1818.
35. K. P. Callahan, M. F. Hawthorne, *Adv. Organomet. Chem.*, **1976**, *14*, 145.
36. K. Wade, *Adv. Inorg. Chem. Radiochem.*, **1980**, *18*, 1.
37. D. M. P. Mingos, *Acc. Chem. Res.*, **1984**, *17*, 311.
38. J. D. Corbett, *Angew. Chem., Int. Ed.*, **2000**, *39*, 670.
39. C. Schrenk, M. Neumaier, A. Schnepf, *Inorg. Chem.*, **2012**, *51*, 3989.
40. E. D. Jemmis, M. M. Balakrishnarajan, P. D. Pancharatna, *J. Am. Chem. Soc.*, **2001**, *123*, 4313.
41. E. D. Jemmis, M. M. Balakrishnarajan, P. D. Pancharatna, *Chem. Rev.*, **2002**, *102*, 93.
42. G. A. Olah, G. K. S. Prakash, R. E. Williams, L. D. Field, and K. Wade, *Hypercarbon Chemistry*, John Wiley & Sons, New York, 1987, pp. 123–133.
43. B. Zhou, M. S. Denning, D. L. Kays, J. M. Goicoechea, *J. Am. Chem. Soc.*, **2009**, *131*, 2802.
44. D. R. Armstrong, W. Clegg, R. P. Davies, S. T. Liddle, D. J. Linton, P. R. Raithby, R. Snaith, A. E. H. Wheatley, *Angew. Chem., Int. Ed.*, **1999**, *38*, 3367.

General References

A classic reference on parallels between main-group and organometallic chemistry is Roald Hoffmann's 1982 Nobel lecture, "Building Bridges between Inorganic and Organic Chemistry," in *Angew. Chem., Int. Ed.*, **1982**, *21*, 711–724, which describes in detail the isolobal analogy. Another very useful paper is John Ellis's "The Teaching of Organometallic Chemistry to Undergraduates," in *J. Chem. Educ.*, **1976**, *53*, 2–6. K. Wade's, *Electron Deficient Compounds*, Thomas Nelson, New York, 1971, provides detailed descriptions of bonding in boranes and related compounds. Metallacarboranes have been reviewed extensively by R. N. Grimes in E. W. Abel, F. G. A. Stone, and G. Wilkinson, editors, *Comprehensive Organometallic Chemistry II*, Pergamon Press, Oxford, 1995, Vol. 1, Chapter 9, pp. 373–430. Topics related to multiple bonds between metal atoms are discussed in detail in F. A. Cotton and R. A. Walton's, *Multiple Bonds between Metal Atoms*, 3rd ed., Springer Science and Business Media, Inc., New York, 2005. A useful review of Zintl ions and related clusters is provided in S. Scharfe, F. Kraus, S. Stegmaier, A. Schier, and T. F. Fässler, "Zintl Ions, Cage Compounds, and Intermetalloid Clusters of Group 14 and Group 15 Elements," *Angew. Chem. Int. Ed.*, **2011**, *50*, 3630–3670. Two articles in *Chemical and Engineering News* are recommended for further discussion of applications of cluster chemistry: E. L. Muetterties, "Metal Clusters," Aug. 20, 1982, pp. 28–41, and F. A. Cotton and M. H. Chisholm, "Bonds between Metal Atoms," June 28, 1982, pp. 40–46.

Problems

15.1 Predict the following products:

a. $Mn_2(CO)_{10} + Br_2 \longrightarrow$

b. $HCCl_3$ + excess $[Co(CO)_4]^- \longrightarrow$

c. $Co_2(CO)_8 + (SCN)_2 \longrightarrow$

d. $Co_2(CO)_8 + C_6H_5{-}C{\equiv}C{-}C_6H_5$ (product has a single Co—Co bond)

e. $Mn_2(CO)_{10} + [(\eta^5\text{-}C_5H_5)Fe(CO)_2]_2 \longrightarrow$

15.2 Propose organic fragments that are isolobal with

a. $Tc(CO)_5$

b. $[Re(CO)_4]^-$

c. $[Co(CN)_5]^{3-}$

d. $[CpFe(\eta^6\text{-}C_6H_6)]^+$

e. $[Mn(CO)_5]^+$

f. $Os_2(CO)_8$ (Find an organic molecule isolobal with this dimeric molecule.)

15.3 Propose two organometallic fragments not mentioned in this chapter that are isolobal with

a. CH_3

b. CH

c. CH_3^+

d. CH_3^-

e. $(\eta^5\text{-}C_5H_5)Fe(CO)_2$

f. $Sn(CH_3)_2$

15.4 Propose an organometallic molecule that is isolobal with each of the following:

a. Ethylene

b. P_4

c. Cyclobutane

d. S_8

15.5 Hydrides such as $NaBH_4$ and $LiAlH_4$ have been reacted with the complexes $[(C_5Me_5)Fe(C_6H_6)]^+$, $[(C_5H_5)Fe(CO)_3]^+$, and $[(C_5H_5)Fe(CO)_2(PPh_3)]^+$. (See P. Michaud, C. Lapinte, D. Astruc, *Ann. N. Y. Acad. Sci.*, **1983**, *415*, 97.)

a. Show that these complexes are isolobal.

b. Predict the products of the reactions of these complexes with hydride reagents.

15.6 Hoffmann has described the following molecules to be composed of isolobal fragments. Subdivide the molecules into fragments, and show that the fragments are isolobal.

a.

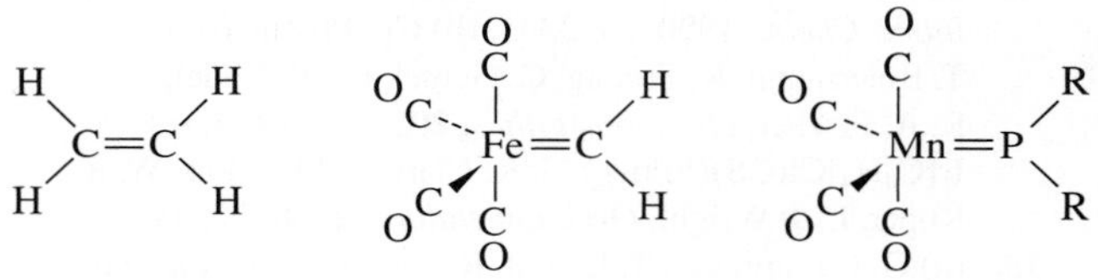

b.

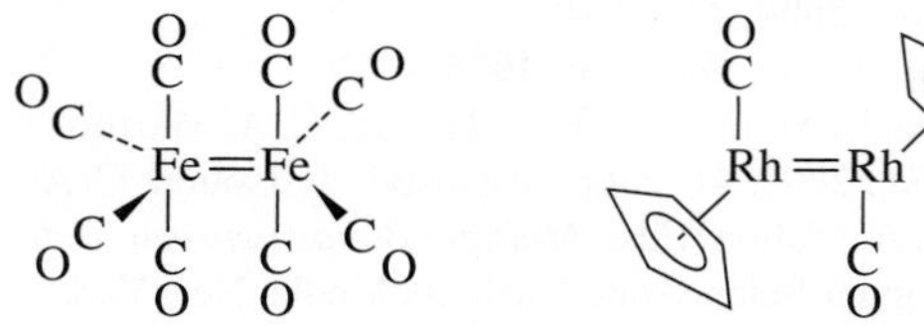

c.

15.7 Verify that the following compounds are composed of isolobal fragments:

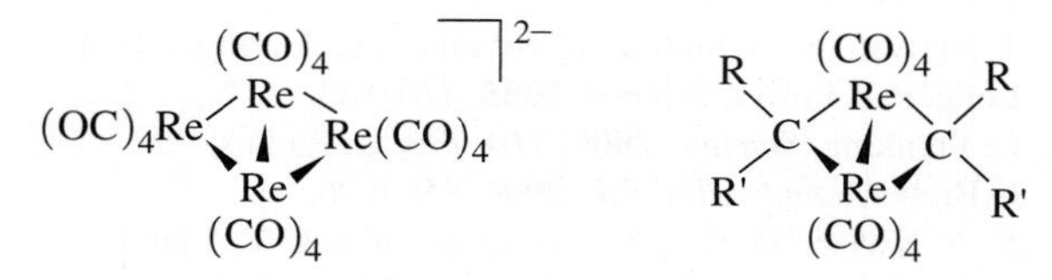

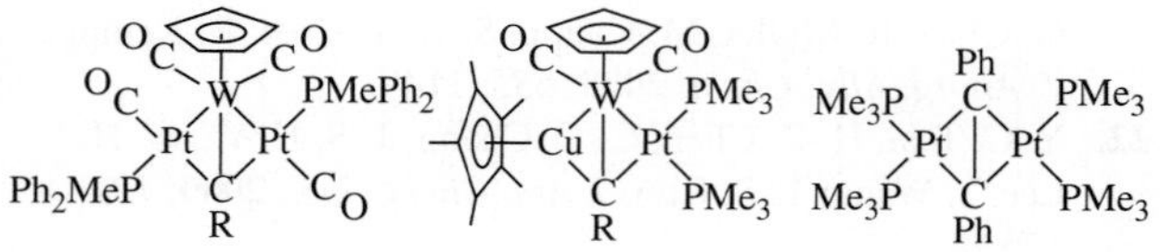

(See G. A. Carriedo, J. A. K. Howard, F. G. A. Stone, *J. Organomet. Chem.*, **1983**, *250*, C28.)

15.8 Calculations reported on the fragments $Mn(CO)_5$, $Mn(CO)_3$, $Cu(PH_3)$, and $Au(PH_3)$ have shown that the energies of their singly occupied hybrid orbitals are in the order $Au(PH_3) > Cu(PH_3) > Mn(CO)_3 > Mn(CO)_5$.

a. In the compound $(OC)_5Mn—Au(PH_3)$, would you expect the electrons in the Mn—Au bond to be polarized toward Mn or Au? Why? (Hint: sketch an energy-level diagram for the molecular orbital formed between Mn and Au.)

b. The $Cu(PPh_3)$ fragment bonds to C_5H_5 in a manner similar to the isolobal $Mn(CO)_3$ fragment. However, the geometry of the corresponding $Au(PPh_3)$ complex is significantly different:

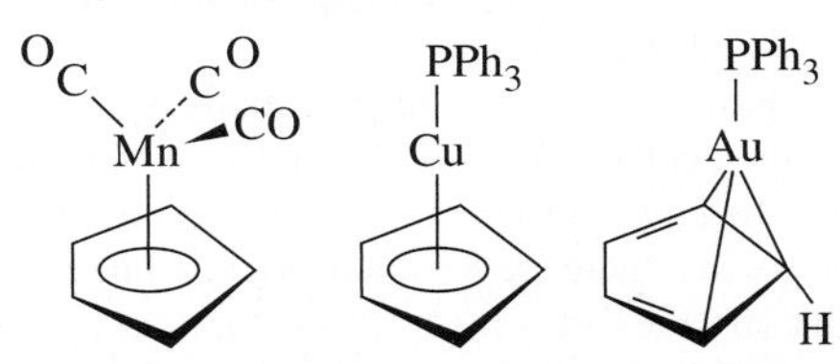

Suggest an explanation. (See D. G. Evans, D. M. P. Mingos, *J. Organomet. Chem.*, **1982**, *232*, 171.)

15.9 **a.** A tin atom can bridge two $Fe_2(CO)_8$ groups in a structure similar to that of spiropentane. Show that these two molecules are composed of isolobal fragments.

$(OC)_4Fe$ / Sn / $(OC)_4Fe$ — $Fe(CO)_4$ / $Fe(CO)_4$ H_2C / C / H_2C — CH_2 / CH_2

b. Tin can also bridge two $Mn(CO)_2(\eta^5\text{-}C_5Me_5)$ fragments; the compound formed has Mn—Sn—Mn arranged linearly. Explain this linear arrangement. (Hint: find a hydrocarbon isolobal with this compound. See W. A. Herrmann, *Angew. Chem., Int. Ed.*, **1986**, *25*, 56.)

15.10 The $AuPPh_3$ fragment is isolobal with the hydrogen atom. Furthermore, analogs of the unstable CH_5^+, CH_6^{2+}, and CH_7^{3+} ions can be prepared using $AuPPh_3$ instead of H. Predict the structures of the $AuPPh_3$ analogs of these ions, and suggest a reason for their stability. (Hint: see G. A. Olah, G. Rasul, *Acc. Chem. Res.*, **1997**, *50*, 245, and references therein.)

15.11 The isolobal fragments $Fe(CO)_3$ and CpCo are predicted to form a variety of analogous carbonyl complexes, for example,

$Fe_2(CO)_6(\mu_2\text{-}CO)_3$ and $Cp_2Co_2(\mu_2\text{-}CO)_3$

$Fe_2(CO)_6(\mu_2\text{-}CO)$ and $Cp_2Co_2(\mu_2\text{-}CO)$

In each case the cobalt compound is predicted to have a substantially shorter metal–metal bond than the analogous iron compound, even though the bond orders of the related compounds are the same. Suggest two reasons for this phenomenon. (See H. Wang, Y. Xie, R. B. King, H. F. Schaefer III, *J. Am. Chem. Soc.*, **2005**, *127*, 11646.)

15.12 The molecular orbitals of the known complex $[Ti(\eta^5\text{-}P_5)_2]^{2-}$, the fragment $[Ti(\eta^5\text{-}P_5)]^-$, and *cyclo*-$P_5^-$ have been calculated.

a. Compare the orbital interactions between *d* orbitals of Ti and P_5^- orbitals with the corresponding interactions between *d* orbitals of Fe and cyclopentadiene (Figure 13.28).

b. Of the species $[Ti(\eta^5\text{-}P_5)_2]^{2-}$, $[Ti(\eta^5\text{-}P_5)]^-$, and P_5^- which has the longest P—P distance? Why?

(See Z.-Z. Liu, W.-Q. Tian, J.-K. Feng, G. Zhang, W.-Q. Li, *J. Phys. Chem. A*, **2005**, *109*, 5645.)

15.13 The C_2 unit can form bridges between transition metals. For example, $(CO)_5Mn(\mu_2\text{-}C_2)Mn(CO)_5$ has been reported. (See P. Belanzoni, N. Re, A. Sgamellotti, C. Floriani, *J. Chem. Soc., Dalton Trans.*, **1997**, 4773.)

a. Show how two $Mn(CO)_5$ fragments can interact in a sigma fashion with a bridging C_2 ligand.

b. Show how pi interactions can occur between C_2 and the $Mn(CO)_5$ fragments, with the $Mn(CO)_5$, using orbitals *not* involved in the sigma interactions.

c. Compare your results with those reported in the literature.

15.14 The complex $Mo_2(NMe_2)_6$ (Me = methyl) contains a metal–metal triple bond. Would you predict this molecule to be more likely eclipsed or staggered? Explain.

15.15 In $[Re_2Cl_8]^{2-}$, the $d_{x^2-y^2}$ orbitals of rhenium interact strongly with the ligands. For one $ReCl_4$ unit in this ion, sketch the four group orbitals of the chloride ligands (assume one σ donor orbital per Cl). Identify the group orbital of suitable symetry to interact with the $d_{x^2-y^2}$ orbital of rhenium.

15.16 $[Tc_2Cl_8]^{2-}$ has a higher bond order than $[Tc_2Cl_8]^{3-}$; however, the Tc—Tc bond distance in $[Tc_2Cl_8]^{2-}$ is longer. Suggest an explanation. (Hint: see F. A. Cotton, *Chem. Soc. Rev.*, **1983**, *122*, 35, or F. A. Cotton and R. A Walton, *Multiple Bonds between Metal Atoms*, Clarendon Press, Oxford, 1993, pp. 122–123.)

15.17 Formal bond orders can sometimes be misleading. For example, $[Re_2Cl_9]^-$, which has a metal–metal bond order of 3.0, has a longer Re—Re bond (270.4 pm) than $[Re_2Cl_9]^{2-}$, (247.3 pm), which has a bond order of 2.5. Account for the shorter bond in $[Re_2Cl_9]^{2-}$. (Hint: see G. A. Heath, J. E. McGrady, R. G. Raptis, A. C. Willis, *Inorg. Chem.*, **1996**, *35*, 6838.)

15.18 The molybdenum compounds $[Mo_2(DTolF)_3]_2(\mu\text{-}OH)_2$, **1**, and $[Mo_2(DTolF)_3]_2(\mu\text{-}O)_2$, **2**, have similar core structures, as shown (DTolF = $[(p\text{-tolyl})NC(H)N(p\text{-tolyl})]^-$). The molybdenum–molybdenum distance in **2** (214.0 pm) is slightly greater than the comparable distance in **1** (210.7 pm). Suggest an explanation for this difference. Hydrogens and the toluene rings are omitted to show the bonding to the Mo more clearly. (F. A. Cotton, L. M. Daniels, I. Guimet, R. W. Henning, G. T. Jordan IV, C. Lin, C. A. Murillo, A. J. Schultz, *J. Am. Chem. Soc.*, **1998**, *120*, 12531.)

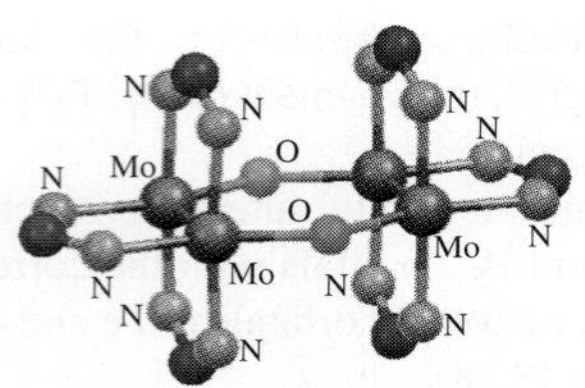

1 $[Mo_2(DTolF)_3]_2(\mu\text{–}OH)_2$

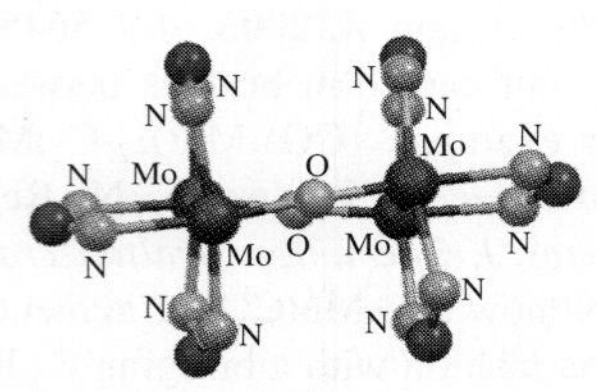

2 $[Mo_2(DTolF)_3]_2(\mu\text{–}O)_2$

15.19 The diiron complex $Fe_2(DPhF)_3$ (DPhF = diphenylformamidinate) has seven unpaired electrons. Calculations indicate that the order of energies of the molecular orbitals resulting from *d* orbital interactions is different from Figure 15.9: $\sigma\ \pi\ \pi^*\ \sigma^*\ \delta\ \delta^*$ in order of increasing energy.

a. What are the formal oxidation states of the iron atoms?

b. Account for the seven unpaired electrons.

(C. M. Zall, D. Zherebetskyy, A. L. Dzubak, E. Bill, L. Gagliardi, C. C. Lu, *Inorg. Chem.*, **2012**, *51*, 728.)

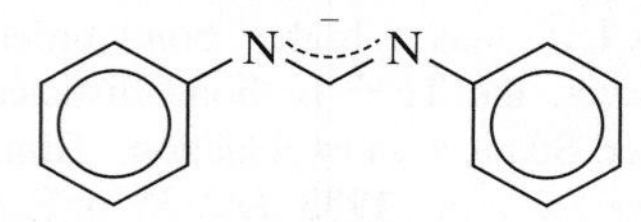

DPhF

15.20 Application of a "sterically fine-tuned" guanidinate ligand was considered essential to enforce the shortest Cr—Cr bond length known (A. Noor; G. Glatz; R. Müller; M. Kaupp; S. Demeshko; R. Kempe, *Z. Anorg. Allg. Chem.*, **2009**, *635*, 1149). Use the sketches in this reference to discuss the steric and electronic features of aminopyridinates, amidinates, and guanidinates that result in varying Cr—Cr bond lengths. How does the magnetic susceptibility measurement for the dichromium complex featured in this reference support an electronic structure consistent with a quintuple bond?

15.21 One objective of exploring the reactivity of dichromium complexes with proposed fivefold bonding is to obtain additional data to support these high bond orders. In this regard, $[^{H}L^{iPr}Cr]_2$ ($^{H}L^{iPr}$ = Ar—N=C(H)—(H)C=N—Ar, with Ar = 2,6-diisopropylphenyl) (Figure 15.11(b)) reacts with internal alkynes to afford 1:1 adducts (J. Shen, G. P. A. Yap, J.-P. Werner, K.H. Theopold, *Chem., Commun.*, **2011**, *47*, 12191). Does the product of the reaction with $CH_3C{\equiv}CCH_3$ support a quintuple bond in the starting dichromium complex? Provide arguments on the basis of crystallographic data and calculated bond orders.

15.22 Using the coordinate system of Figure 15.12, for $B_6H_6^{\ 2-}$, do the following:

a. Show that the p_z orbitals of the borons collectively have the same symmetry as the *s* orbitals (generate a representation based on the six p_z orbitals of the borons, and do the same for the six *s* orbitals).

b. Show that these representations reduce to $A_{1g} + E_g + T_{1u}$.

c. Show that the p_x and p_y orbitals of the borons form molecular orbitals of T_{2g} and T_{1u} symmetry.

15.23 For the *closo* cluster $B_7H_7^{\ 2-}$, which has D_{5h} symmetry, verify that there are 8 framework bonding electron pairs.

15.24 Classify the following as *closo*, *nido*, or *arachno*:

a. $C_2B_3H_7$

b. B_6H_{12}

c. $B_{11}H_{11}^{\ 2-}$

d. $C_3B_5H_7$

e. $CB_{10}H_{13}^{\ -}$

f. $B_{10}H_{14}^{\ 2-}$

15.25 Classify the following as *closo*, *nido*, or *arachno*:

a. $SB_{10}H_{10}^{\ 2-}$

b. $NCB_{10}H_{11}$

c. $SiC_2B_4H_{10}$

d. $As_2C_2B_7H_9$

e. $PCB_9H_{11}^{\ -}$

15.26 Classify the following as *closo* or *nido*:

a. $B_3H_8Mn(CO)_3$

b. $B_4H_6(CoCp)_2$

c. $C_2B_7H_{11}CoCp$

d. $B_5H_{10}FeCp$

e. $C_2B_9H_{11}Ru(CO)_3$

15.27 Tetrahedral $Ni(CO)_4$ can be used as the reference structure for the fragments $Ni(CO)_3$, $Ni(CO)_2$, and Ni(CO).

a. Determine the formulas of the hydrocarbon fragments of methane for each of these nickel carbonyl fragments.

Determine *x* for the *closo* clusters containing Ni(CO) and $Ni(CO)_2$ fragments:

b. $[Bi_3Ni_4(CO)_6]^x$

c. $[Bi_xNi_4(CO)_6]^{2-}$ (J. M. Goicoechea, M. W. Hull, S. C. Sevov, *J. Am. Chem. Soc.*, **2007**, *129*, 7885.)

15.28 Attempts to construct metallaboranes via cluster expansion reactions with metal carbonyl complexes sometimes result in unexpected products. For example, reaction of $B_3H_7(Cp^*RuH)_2$ and $Mo(CO)_3(CH_3CN)_3$ results in $B_2H_6(Cp^*RuCO)_2$ without any incorporation of the $Mo(CO)_3$ fragment (K. Geetharani; S. K. Bose; B. Varghese; S. Ghosh, *Chem. Eur. J.* **2010**, *16*, 11357). Classify these metalloboranes as *closo*, *nido*, or *arachno*. Does $B_2H_6(Cp^*RuCO)_2$ react similarly with the group 7 carbonyls $Mn_2(CO)_{10}$ and $Re_2(CO)_{10}$? Sketch the products of these reactions (K. Geetharani; S. K. Bose, S. Sahoo, B. Varghese; S. M. Mobin, S. Ghosh, *Inorg. Chem.*, **2011**, *50*, 5824).

15.29 This chapter deals separately with metal–metal bonds and metallaboranes, but reaction of $[Cp^*MoCl_4]$ and $LiBH_4$, followed by thermolysis with tellurium powder in toluene affords a metallaborane *with* metal–metal bonds, $[(Cp^*Mo)_4B_4H_4(\mu_4\text{-}BH)_3]$ (A. Thakur, S. Sahoo, S. Ghosh, *Inorg. Chem.*, **2011**, *50*, 7940). What structural feature apparently hinders BH capping of some of the cubane faces? The ^{11}B NMR spectrum of $[(Cp^*Mo)_4B_4H_4(\mu_4\text{-}BH)_3]$ is consistent with the presence of seven boron atoms; it featured three resonances with a 2:2:3 ratio. Sketch the abbreviated cubane structure provided in Chart 1 of this reference and label each boron atom with its chemical shift. The NMR data also indicates two Cp* environments. What differentiates the MoCp* fragments leading to this spectral feature?

15.30 The reaction of $[Fe(\eta^6\text{-arene})]^{2+}$ and $Tl_2[nido\text{-}7,8\text{-}C_2B_9H_{11}]$ results in arene displacement and formation of $[1\text{-}(\eta^6\text{-arene})\text{-}closo\text{-}1,2,3\text{-}FeC_2B_9H_{11}]$ where the number of methyl substituents on the arene vary from 1 to 6 (B. Štíbr, M. Bakardjiev, J. Holub, A. Růžička, Z. Padělková, P. Štěpnička, *Inorg. Chem.*, **2011**, *50*, 3097). Explain why product yields were lowest for the pentamethylbenzene and hexamethylbenzene complexes. Provide four unique trends (spectroscopic and electrochemical) that were correlated to increasing methyl substitution at the arene.

15.31 Classify the following as *closo*, *nido*, or *arachno*:
a. Ge_9^{4-}
b. $InBi_3^{2-}$
c. Bi_8^{2+}

15.32 Complexes with Zintl ions as ligands possess intriguing structures. An early study reported that reaction of K_3E_7 (E = P, As, Sb) with sources of $M(CO)_3$ (M = Cr, Mo, W), in the presence of cryptand [2.2.2] results in $[K(2.2.2)]_3[M(CO)_3(E_7)]$ (S. Charles, B. W. Eichhorn, A. L. Rheingold, S. G. Bott, *J. Am. Chem. Soc.*, **1994**, *116*, 8077). Why do these reactions require cryptand[2.2.2] to be successful? Discuss the relative π-donor abilities of these Zintl ions to the $Cr(CO)_3$ fragment (Hint: use IR ν(CO) data). What trend is observed in the electronic spectra for the $[K(2.2.2)]_3[Cr(CO)_3(E_7)]$ series? Briefly explain why this trend seems reasonable on the basis of how it is assigned. These trianions can be protonated to afford dianions. Sketch $[Cr(CO)_3(HSb_7)]^{2-}$ to clearly indicate the protonated site.

15.33 Zintl ions are being explored for small molecule activation reactions. The protonated Group 15 Zintl ion, $[HP_7]^{2-}$ effects hydrophosphination of carbodiimides, RN=C=NR (R = 2,6-diisopropylphenyl, ^{i}Pr, Cy) to afford amidine-functionalized Zintl ions $[P_7C(NHR)(NR)]^{2-}$ (R. S. P. Turbervill, J. M. Goicoechea, *Chem. Commun.*, **2012**, *48*, 1470; R. S. P. Turbervill, J. M. Goicoechea, *Organometallics*, **2012**, *31*, 2452). Provide two observations that support the hypothesis that the protonation step is an intramolecular process. Draw the Lewis structure of $[HP_7]^{2-}$ that indicates the negatively charged phosphorus atoms, and then use the arrow-pushing formalism to describe the proposed hydrophosphination mechanism for a general RN=C=NR.

15.34 Determine the number of skeletal electron pairs predicted by the *mno* rule for the following:
a. *arachno*-B_5H_{11}
b. $1\text{-}(\eta^5\text{-}C_5H_5)CoB_4H_{10}$ (see Figure 15.17)
c. $(\eta^5\text{-}C_5H_5)Fe(\eta^5\text{-}C_2B_9H_{11})$ (see Figure 15.20)

15.35 Assign the point groups of the following structures from this chapter:
a. The isolobal symbol
b. The metallocenes with P_5 rings shown in Figure 15.5
c. $[Re_2Cl_8]^{2-}$ and $[Os_2Cl_8]^{2-}$
d. A T_{2g} orbital of $B_6H_6^{2-}$ shown in Figure 15.13
e. The *nido* borane in the bottom center of Figure 15.15
f. The carborane analogs of ferrocene shown in Figure 15.20
g. Te_6^{2+} and Ge_9^{4-} in Figure 15.22

15.36 Assign the point groups:
a. The core of the iron(III) cluster $[Fe_8O_4(sao)_8(py)_4]\cdot 4py$ (sao = salicylaldoxime; py = pyridine), which has an Fe_4O_4 cubane structure inside an Fe_4 tetrahedron. (See I. A. Gass, C. J. Milios, A. G. Whittaker, F. P. A. Fabiani, S. Parsons, M. Murrie, S. P. Perlepes, E. K. Brechin, *Inorg. Chem.*, **2006**, *45*, 5281.)

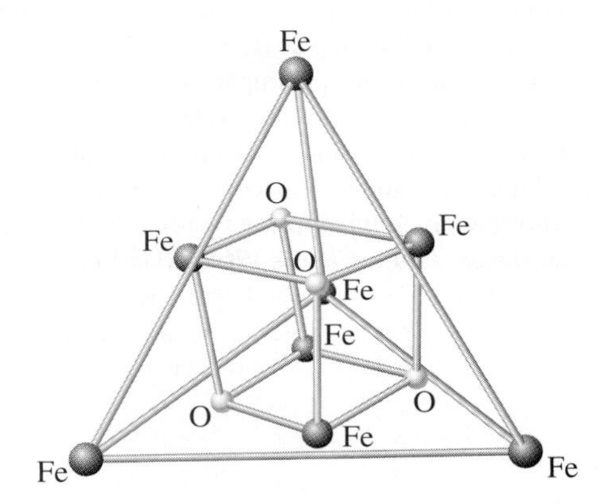

b. The $[As@Ni_{12}@As_{20}]^{3-}$ ion, which contains an As atom at the center of a Ni_{12} icosahedron, which in turn is surrounded by an As_{20} dodecahedron. (See B. W. Eichhorn, *Science*, **2003**, *300*, 778.)

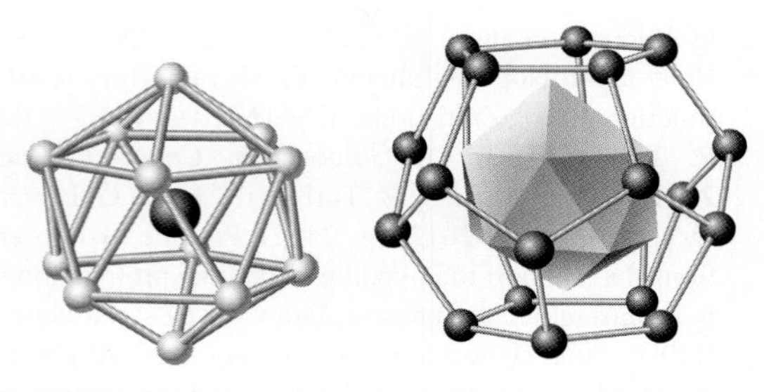

The following problems require the use of molecular modeling software.

15.37 The ligand P_5^- has been reported to act as a stronger acceptor than η^5-$C_5H_5^-$.

a. Using the group orbital approach as for ferrocene (see Figure 13.27), sketch the group orbitals arising from the P_5 rings. Then show how the group orbitals can interact with appropriate orbitals on a central transition metal.

b. On the basis of your diagram, suggest why the P_5^- ligand might act as a stronger acceptor than η^5-$C_5H_5^-$.

c. Using molecular modeling software, calculate and display the molecular orbitals of $[(\eta^5\text{-}P_5)_2Ti]^{2-}$. Compare your results with those described in the literature for this complex. (See E. Urnezius, W. W. Brennessel, C. J. Cramer, J. E. Ellis, P. von Ragué Schleyer, *Science*, **2002**, *295*, 832; a comparison of the molecular orbitals of η^5-$C_5H_5^-$ and P_5^- can be found in H-J. Zhai, L-S. Wang, A. E. Kuznetsov, A. I. Boldyrev, *J. Phys. Chem. A*, **2002**, *106*, 5600.)

15.38 Construct the ions $Re_2Cl_8^{2-}$ and $Os_2Cl_8^{2-}$ (see Figure 15.7), and calculate and display their molecular orbitals. Compare these orbitals with Figure 15.8, focusing on the metal–metal bonds and on *d*-orbital interactions with the ligands, and classify orbitals involved in metal–metal bonding as σ, π, or δ. How do your results compare with the quadruple and triple bonds reported respectively for these ions? (Depending on the software used, you may need to fix atoms in position before doing orbital calculations, especially for $Re_2Cl_8^{2-}$. Structural information on this ion can be found in F. A. Cotton, C. B. Harris, *Inorg. Chem.*, **1965**, *4*, 330.)

15.39 The maximum bond order for a diatomic chemical species has been calculated to be six, for example in Cr_2. What types of orbital interactions would be possible in this molecule? Calculate and display the molecular orbitals for Cr_2, and classify them as σ, π, or δ and as bonding or antibonding. Identify the atomic orbitals used in each molecular orbital. Are your results consistent with a bond order of six? (See G. Frenking, R. Tonner, *Nature*, **2007**, *446*, 276 and references cited therein.)

15.40 Draw *closo*-$B_6H_6^{2-}$ (see Figure 15.12), and calculate its molecular orbitals. Identify and display the following:

a. The orbital having A_{1g} symmetry resulting from overlap of *p* or hybrid orbitals pointing toward the center of the octahedron (see Figure 15.13).

b. Three orbitals having T_{1u} symmetry involving pi interactions between sets of four boron *p* orbitals.

c. Three orbitals having T_{2g} symmetry involving overlap of sets of four boron *p* orbitals within the same plane.

15.41 Draw the core carbon-centered cluster of $Ru_6C(CO)_{17}$ (see Figure 15.23), and calculate its molecular orbitals. Identify and display the following:

a. The orbital having A_{1g} symmetry resulting from overlap of *p* or hybrid orbitals pointing toward the center of the octahedron with the 2 *s*orbital of carbon (see Figure 15.24).

b. Three orbitals having T_{1u} symmetry involving pi interactions between sets of four ruthenium *p* or *d* orbitals with 2*p* orbitals on carbon.

c. Three orbitals having T_{2g} symmetry involving overlap of sets of four boron *p* or *d* orbitals within the same plane.

15.42 The Zintl ion cluster $[CoGe_{10}]^{3-}$ has pentagonal prismatic geometry. Calculate the molecular orbitals of this ion.

a. Identify the key molecular orbitals involved in bonding to cobalt.

b. Identify the key molecular orbitals involved in bonding between the germaniums in the rings.

c. Compare the bonding in this cluster with the bonding in ferrocene (Section 13.5.2).

(See J.-Q. Wang, S. Stegmaier, T. F. Fässler, *Angew. Chem., Int. Ed.*, **2009**, *48*, 1.)

Answers to Exercises

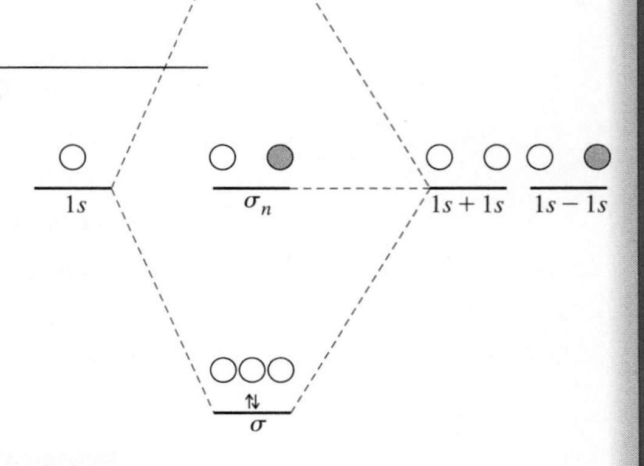

Chapter 2

2.1 $E = R_H\left(\frac{1}{2^2} - \frac{1}{3^2}\right) = R_H\left(\frac{5}{36}\right) = 2.179 \times 10^{-18}\ \text{J}\left(\frac{5}{36}\right) = 3.026 \times 10^{-19}\ \text{J}$

$$= 1.097 \times 10^7\ \text{m}^{-1}\left(\frac{5}{36}\right) = 1.524 \times 10^6\ \text{m}^{-1} \times \frac{\text{m}}{100\ \text{cm}} = 1.524 \times 10^4\ \text{cm}^{-1}$$

2.2 The nodal surfaces require $2z^2 - x^2 - y^2 = 0$, so the angular nodal surface for a d_{z^2} orbital is the conical surface where $2z^2 = x^2 + y^2$.

2.3 The angular nodal surfaces for a d_{xz} orbital are the planes where $xz = 0$, which means that either x or z must be zero. The yz and xy planes satisfy this requirement.

2.4 ↑↓ ↑ ↓ One pair with ↑ spin, one pair with ↓ spin, one exchange possibility for each; energy contribution $2\Pi_e$. One pair (first orbital), energy contribution Π_c. Total: $2\Pi_e + \Pi_c$

⇅ ⇅ __ $2\Pi_e + 2\Pi_c$

Π_c

⇅ ↑ ↓ $2\Pi_e + \Pi_c$

Π_e

⇅ ↑ ↑ $3\Pi_e + \Pi_c$

2.5 **a.** If the three $2p$ electrons all have the same spin, as in ↑1 ↑2 ↑3, there are three exchange possibilities (1 and 2, 1 and 3, or 2 and 3) and no pairs. Overall, the total energy is $3\Pi_e$. If there is one unpaired electron, as in ↑↓ ↑ __, there is one electron with ↓ spin, and no possibility of exchange; two electrons with ↑ spin, with one exchange possibility; and one pair. Overall; the total energy is $\Pi_e + \Pi_c$. Because Π_e is negative and Π_c is positive, the configuration with three unpaired electrons has a much lower energy.

b. If the three $2p$ electrons avoid pairing, but do not all have mutually parallel spins, as in ↑ ↑ ↓, only the two ↑ electrons can exchange. The total energy is Π_e. The energy of this state is intermediate between those in part a. It is $2\Pi_e$ higher than ↑ ↑ ↑ and lower than ↑↓ ↑ __ by Π_c.

2.6

Tin	Total	$5p$	$5s$	$4d$
Z	**50**	**50**	**50**	**50**
$(1s^2)$	2	2	2	2
$(2s^22p^6)$	8	8	8	8
$(3s^23p^6)$	8	8	8	8
$(3d^{10})$	10	10	10	10
$(4s^24p^6)$	8	8×0.85	8×0.85	8
$(4d^{10})$	10	10×0.85	10×0.85	9×0.35
$(5s^25p^2)$	4	3×0.35	3×0.35	
Z^*		5.65	5.65	10.85

2.7

Uranium	Total	$7s$	$5f$	$6d$
Z	**92**	**92**	**92**	**92**
$(1s^2)$	2	2	2	2
$(2s^22p^6)$	8	8	8	8
$(3s^23p^6)$	8	8	8	8
$(3d^{10})$	10	10	10	10
$(4s^24p^6)$	8	8	8	8
$(4d^{10})$	10	10	10	10
$(4f^{14})$	14	14	14	14
$(5s^25p^6)$	8	8	8	8
$(5d^{10})$	10	10	10	10
$(5f^3)$	3	3	2×0.35	3
$(6s^26p^6)$	8	8×0.85		8
$(6d^1)$	1	1×0.85		
$(7s^2)$	2	1×0.35		
Z^*		3.00	13.30	3.00

2.8 With 4 electrons, the electron configurations of B^+, Be, and Li^- are all $1s^2 2s^2$. Because the effective nuclear charge is greater for each $1s^2 2s^2$ configuration than the $1s^2 2s^1$ configuration of the preceding element, more energy is necessary to remove an electron from the species with $1s^2 2s^2$ configurations. For 5 electrons (C^+, B, and Be^-) the configurations are all $1s^2 2s^2 2p^1$. Because the $2p$ orbitals are significantly higher in energy than the $2s$ orbitals, in each case it is much easier to remove an electron from a $1s^2 2s^2 2p^1$ configuration than from the $1s^2 2s^2$ configuration of the preceding element. When a sixth electron is present (a $1s^2 2s^2 2p^2$ configuration), more energy is required to remove an electron because the electron being removed must overcome greater effective nuclear charge than for the preceding 5-electron species.

Chapter 3

3.1 POF_3: The octet rule results in single P—F and P—O bonds; formal charge arguments result in a double bond for P═O. The actual distance is 143 pm, considerably shorter than a regular P—O bond (164 pm).

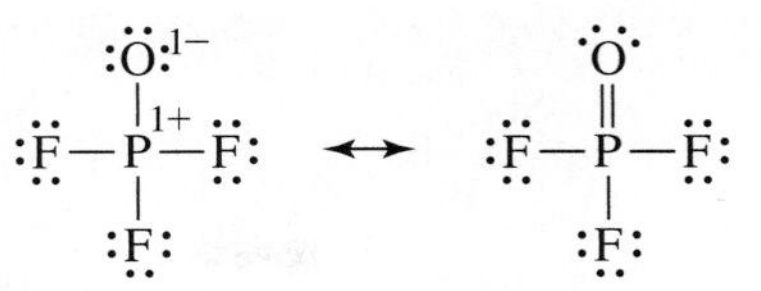

SOF_4: This is a distorted trigonal bipyramidal structure, with an S=O double bond and S—F single bonds required by formal charge arguments. The short S=O bond length of 141 pm is in agreement.

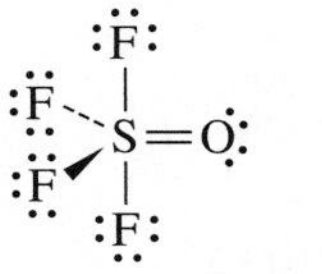

SO_3F^-: This is basically a tetrahedral structure, with two double bonds to oxygen atoms and single bonds to fluorine and the third oxygen. The S—O bond order is then 1.67 and the bond length is 143 pm, shorter than the 149 pm of SO_4^{2-}, which has a bond order of 1.5.

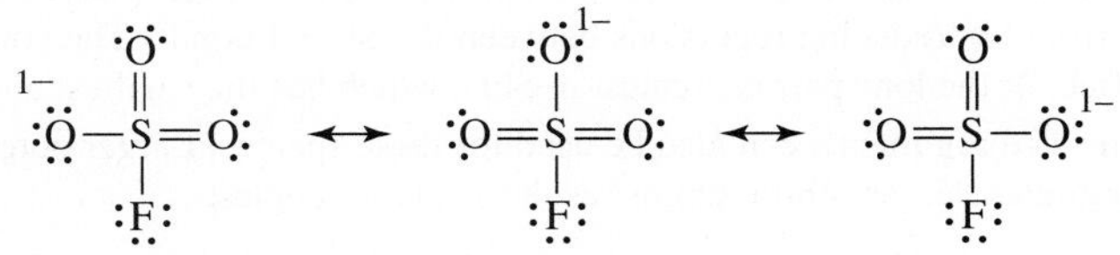

3.2

NH_2^-	NH_4^+	I_3^-	PCl_6^-
H—N—H < 109.5° because of lone pair repulsion	tetrahedral	linear	octahedral

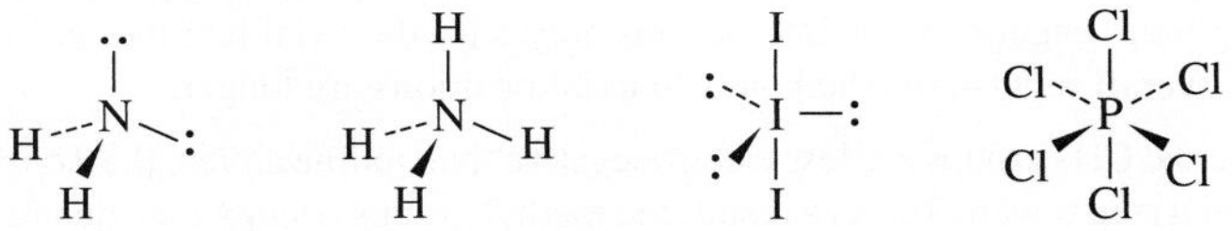

3.3

	$XeOF_2$	$ClOF_3$	$SOCl_2$
Steric number:	5	5	4
Angles:	F—Xe—O near 90°	F—Cl—F < 90° F—Cl—O > 90°	Cl—S—Cl = < 109.5° (exp 96°) Cl—S—O = < 109.5° (exp 106°)

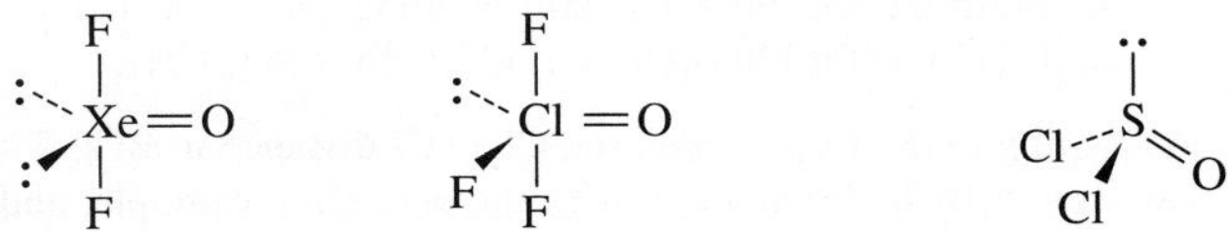

3.4 a. 92.2° 96.9° 99.7° (Calc)

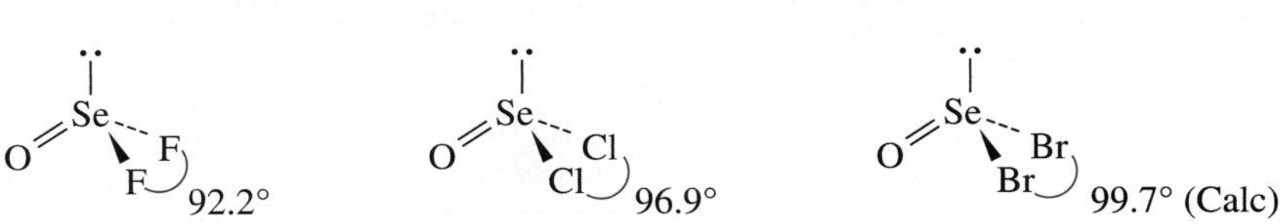

In $OSeF_2$, the fluorines exhibit the strongest attraction for electrons in the Se—F bonds. This reduces electron–electron repulsions near the Se atom, enhancing the ability of the lone pair and double bond to squeeze the rest of the molecule together.

b. $SbCl_3$ 97.1° $SbBr_3$ 98.2° SbI_3 99°

Cl, the most electronegative of the halogens in this series, pulls shared electrons the most strongly away from Sb, reducing electron density near Sb. The consequence is that the lone pair exerts the strongest influence on shape in $SbCl_3$.

c.

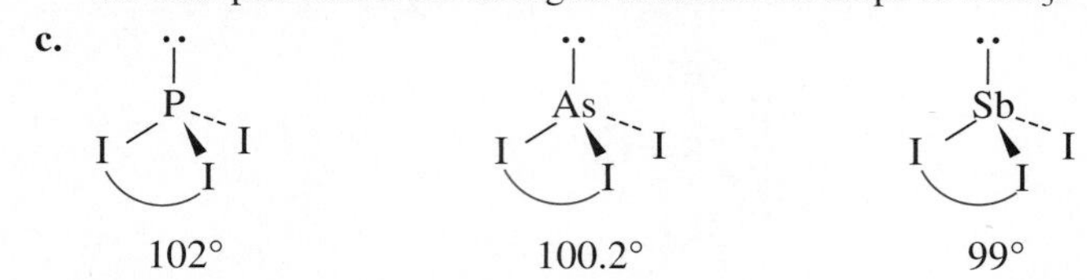

Phosphorus is the most electronegative of the central atoms. Consequently, it exerts the strongest pull on shared electrons, concentrating these electrons near P and increasing bonding pair–bonding pair repulsions, hence the largest angle in PI_3. Sb, the least electronegative central atom, has the opposite effect: Shared electrons are attracted away from Sb, reducing repulsions between the Sb—I bonds. The consequence is that the effect of the lone pair is greatest in SbI_3, which has the smallest angle.

Atomic size arguments can also be used for these species. Larger outer atoms result in larger angles; larger central atoms result in smaller angles.

3.5 **a.** Because fluorine is more electronegative than chlorine, electrons in N—F bonds are pulled more strongly away from nitrogen than electrons in N—Cl bonds. As a result, in NF_3 the bonding pair–bonding pair repulsions near the nitrogen are weaker. Repulsion by the lone pair then leads to smaller F—N—F angles than Cl—N—Cl angles.

b. In SOF_4 the bonding pairs to axial fluorine atoms are repelled at approximately 90° angles by two bonding pairs and the double bond to oxygen. By contrast, equatorial bonding pairs engage in 90° interactions only with the axial bonding pairs. Because of greater overall repulsions, the bonds to axial positions are longer.

c. Because the CH_3 groups are less electronegative than iodine atoms, the Te—C bonds are more polarized toward Te. As a result, the methyl groups occupy the equatorial positions, where there is less crowding and the methyl groups are farther from the lone pair.

d. Because the OCH_3 group is more electronegative than the CH_3 group, OCH_3 pulls electrons more strongly from sulfur, reducing electron–electron repulsions around sulfur. This enables the oxygen atoms in $FSO_2(OCH_3)$ to separate more from each other to minimize the bonding pair–bonding pair repulsions, giving a larger O—S—O angle (124.4° for $FSO_2(OCH_3)$, 123.1° for $FSO_2(CH_3)$).

3.6 According to the LCP model, the Cl···Cl distance in BCl_4^- should be approximately the same as in BCl_3. By analogy with the preceding example, and using the tetrahedral bond angle of 109.5° for BCl_4^-:

$$x = \text{B—Cl bond distance} = \frac{150.5 \text{ pm}}{\sin 54.75°} = 184 \text{ pm}$$

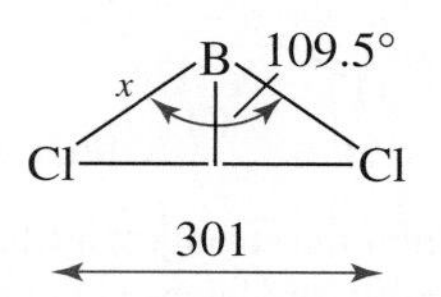

Chapter 4

4.1 S_2 is made up of C_2 followed by $\sigma_\perp$, which is shown in the figure below to be the same as i.

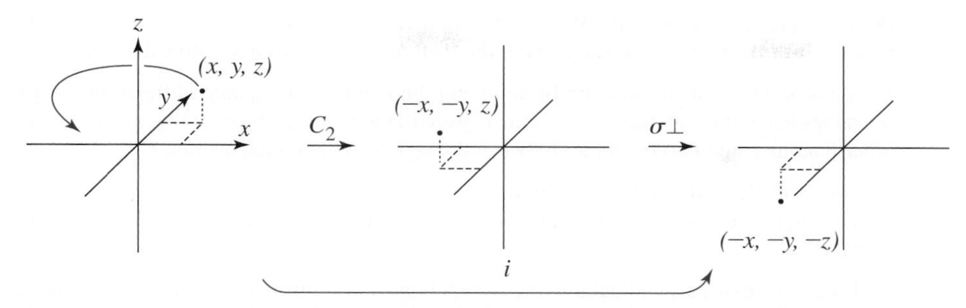

S_1 is made up of C_1 followed by $\sigma_\perp$, which is shown in the figure below to be the same as σ.

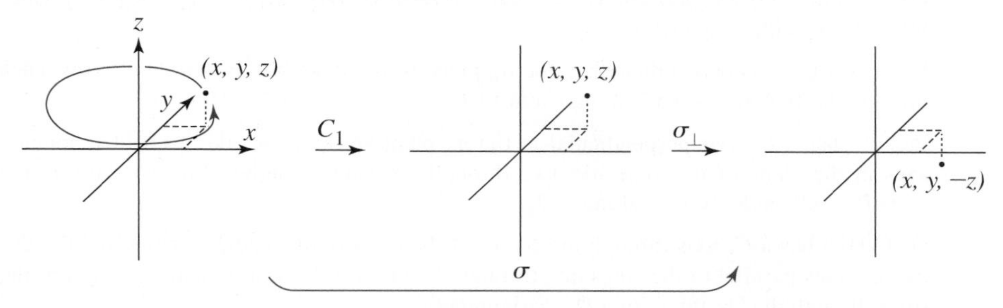

4.2 NH_3 has a threefold axis through the N perpendicular to the plane of the three hydrogen atoms and three mirror planes, each including the N and one H. $2C_3$, $3\sigma_v$

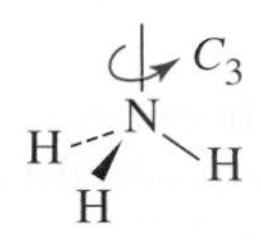

Cyclohexane in the boat conformation has a C_2 axis perpendicular to the plane of the lower four carbon atoms and two mirror planes that include this axis and are perpendicular to each other. C_2, $2\sigma_v$

Cyclohexane in the chair conformation has a C_3 axis perpendicular to the average plane of the ring, three perpendicular C_2 axes passing between carbon atoms, and three mirror planes passing through opposite carbon atoms and perpendicular to the average plane of the ring. It also contains a center of inversion and an S_6 axis collinear with the C_3 axis. A model is very useful in analysis of this molecule. $2C_3$, $3C_2$, $3\sigma_d$, i, $2S_6$

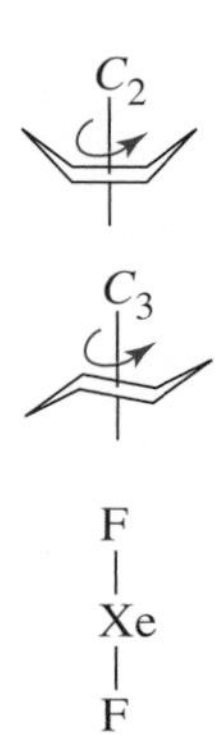

XeF_2 is a linear molecule, with a C_∞ axis through the three nuclei, an infinite number of perpendicular C_2 axes, a horizontal mirror plane (which is also an inversion center), and an infinite number of mirror planes that include the C_∞ axis. C_∞, $\infty\ C_2$, $i = \sigma_h$, $\infty\ \sigma_v$.

4.3 Several of the molecules below have symmetry elements in addition to those used to assign the point group. See the character tables in Appendix C for the complete list for each point group.

N_2F_2 has a mirror plane through all the atoms, which is the σ_h plane, perpendicular to the C_2 axis through the N═N bond. There are no other symmetry elements, so it is C_{2h}.

$B(OH)_3$ also has a σ_h mirror plane, the plane of the molecule, perpendicular to the C_3 axis through the B atom. Again, there are no others, so it is C_{3h}.

H_2O has a C_2 axis in the plane of the drawing, through the O atom and between the two H atoms. It also has two mirror planes, one in the plane of the drawing and the other perpendicular to it; overall, C_{2v}.

PCl_3 has a C_3 axis through the P atom and equidistant from the three Cl atoms. Like NH_3, it also has three σ_v planes, each through the P atom and one of the Cl atoms; overall, C_{3v}.

BrF_5 has one C_4 axis through the Br atom and the F atom in the plane of the drawing, two σ_v planes (each through the Br atom, the F atom in the plane of the drawing, and two of the other F atoms), and two σ_d planes between the equatorial F atoms; overall, C_{4v}.

HF, CO, and HCN all are linear, with the infinite rotation axis through the center of all the atoms. There are also an infinite number of σ_v planes, all of which contain the C_∞ axis; overall, $C_{\infty v}$.

N_2H_4 has a C_2 axis perpendicular to the N—N bond and splitting the angle between the two lone pairs. There are no other symmetry elements, so it is C_2.

$P(C_6H_5)_3$ has only a C_3 axis, much like that in NH_3 or $B(OH)_3$. The twist of the phenyl rings prevents any other symmetry; C_3.

BF_3 has a C_3 axis perpendicular to the σ_h plane of the molecule and three C_2 axes, each through the B atom and an F atom; overall, D_{3h}.

$PtCl_4^{2-}$ has a C_4 axis perpendicular to the σ_h plane of the molecule. It also has four C_2 axes in the plane of the molecule, two through opposite Cl atoms and two splitting the Cl—Pt—Cl angles, thus making it D_{4h}.

$Os(C_5H_5)_2$ has a C_5 axis through the center of the two cyclopentadienyl rings and the Os, five C_2 axes parallel to the rings and through the Os atom, and a σ_h plane parallel to the rings through the Os atom, for a D_{5h} assignment.

Benzene has a C_6 axis perpendicular to the σ_h plane of the ring and six C_2 axes in the plane of the ring, three through two C atoms each and three between the atoms. These are sufficient to make it D_{6h}.

F_2, N_2, and H—C≡C—H are all linear, each with a C_∞ axis through the atoms. There are also an infinite number of C_2 axes perpendicular to the C_∞ axes and a σ_h plane perpendicular to the C_∞ axes, sufficient to make them $D_{\infty h}$.

Allene, $H_2C{=}C{=}CH_2$, has a C_2 axis through the three carbon atoms and two C_2 axes perpendicular to the line of the carbon atoms, both at 45° angles to the planes of the H atoms. Two σ_d mirror planes through each H—C—H combination complete the assignment of D_{2d}.

$Ni(C_4H_4)_2$ has a C_4 axis through the centers of the C_4H_4 rings and the Ni, four C_2 axes perpendicular to the C_4 through the Ni, and four σ_d planes, each including two opposite carbon atoms of the same ring and the Ni; overall, D_{4d}.

$Fe(C_5H_5)_2$ has a C_5 axis through the centers of the rings and the Fe, five C_2 axes perpendicular to the carbon atoms and through the Fe, and five σ_d planes that include the C_5 axis; overall, D_{5d}.

$[Ru(en)_3]^{2+}$ has a C_3 axis perpendicular to the drawing through the Ru and three C_2 axes in the plane of the paper, each intersecting an en ring at the midpoint and passing through the Ru; overall, D_3.

4.4 **a.**
$$\begin{bmatrix} 5 & 1 & 3 \\ 4 & 2 & 2 \\ 1 & 2 & 3 \end{bmatrix} \times \begin{bmatrix} 2 & 1 & 1 \\ 1 & 2 & 3 \\ 5 & 4 & 3 \end{bmatrix}$$

$$= \begin{bmatrix} (5 \times 2) + (1 \times 1) + (3 \times 5) & (5 \times 1) + (1 \times 2) + (3 \times 4) & (5 \times 1) + (1 \times 3) + (3 \times 3) \\ (4 \times 2) + (2 \times 1) + (2 \times 5) & (4 \times 1) + (2 \times 2) + (2 \times 4) & (4 \times 1) + (2 \times 3) + (2 \times 3) \\ (1 \times 2) + (2 \times 1) + (3 \times 5) & (1 \times 1) + (2 \times 2) + (3 \times 4) & (1 \times 1) + (2 \times 3) + (3 \times 3) \end{bmatrix}$$

$$= \begin{bmatrix} 26 & 19 & 17 \\ 20 & 16 & 16 \\ 19 & 17 & 16 \end{bmatrix}$$

b.
$$\begin{bmatrix} 1 & -1 & -2 \\ 0 & 1 & -1 \\ 1 & 0 & 0 \end{bmatrix} \times \begin{bmatrix} 2 \\ 1 \\ 3 \end{bmatrix} = \begin{bmatrix} (1 \times 2) - (1 \times 1) - (2 \times 3) \\ (0 \times 2) + (1 \times 1) - (1 \times 3) \\ (1 \times 2) + (0 \times 1) + (0 \times 3) \end{bmatrix} = \begin{bmatrix} -5 \\ -2 \\ 2 \end{bmatrix}$$

c.
$$[1 \quad 2 \quad 3] \times \begin{bmatrix} 1 & -1 & -2 \\ 2 & 1 & -1 \\ 3 & 2 & 1 \end{bmatrix}$$

$$= [(1 \times 1) + (2 \times 2) + (3 \times 3) \qquad 1 \times (-1) + (2 \times 1) + (3 \times 2) \qquad 1 \times (-2) + 2 \times (-1) + (3 \times 1)]$$
$$= [14 \quad 7 \quad -1]$$

4.5 E: The new coordinates are

$$\begin{aligned} x' &= \text{new } x = x \\ y' &= \text{new } y = y \\ z' &= \text{new } z = z \end{aligned} \qquad \begin{bmatrix} 1 & 0 & 0 \\ 0 & 1 & 0 \\ 0 & 0 & 1 \end{bmatrix} = \text{transformation matrix for } E$$

In matrix notation,

$$\begin{bmatrix} x' \\ y' \\ z' \end{bmatrix} = \begin{bmatrix} 1 & 0 & 0 \\ 0 & 1 & 0 \\ 0 & 0 & 1 \end{bmatrix} \begin{bmatrix} x \\ y \\ z \end{bmatrix} = \begin{bmatrix} x \\ y \\ z \end{bmatrix} \quad \text{or} \quad \begin{bmatrix} x' \\ y' \\ z' \end{bmatrix} = \begin{bmatrix} x \\ y \\ z \end{bmatrix}$$

$\sigma_v'(yz)$: Reflect a point with coordinates (x, y, z) through the yz plane.

$$\begin{aligned} x' &= \text{new } x = -x \\ y' &= \text{new } y = y \\ z' &= \text{new } z = z \end{aligned} \qquad \begin{bmatrix} -1 & 0 & 0 \\ 0 & 1 & 0 \\ 0 & 0 & 1 \end{bmatrix} = \text{transformation matrix for } \sigma_v' \ (yz)$$

In matrix notation,

$$\begin{bmatrix} x' \\ y' \\ z' \end{bmatrix} = \begin{bmatrix} -1 & 0 & 0 \\ 0 & 1 & 0 \\ 0 & 0 & 1 \end{bmatrix} \begin{bmatrix} x \\ y \\ z \end{bmatrix} = \begin{bmatrix} -x \\ y \\ z \end{bmatrix} \quad \text{or} \quad \begin{bmatrix} x' \\ y' \\ z' \end{bmatrix} = \begin{bmatrix} -x \\ y \\ z \end{bmatrix}$$

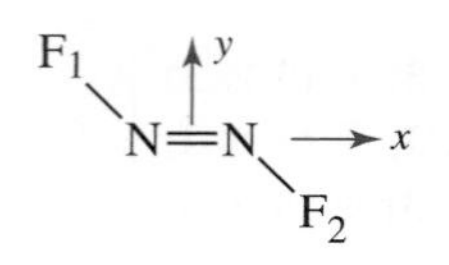

4.6 Representation Flow Chart: N_2F_2 (C_{2h})

Symmetry Operations

after E | after C_2 | after i | after σ_h

Matrix Representations (Reducible)

$$E: \begin{bmatrix} 1 & 0 & 0 \\ 0 & 1 & 0 \\ 0 & 0 & 1 \end{bmatrix} \quad C_2: \begin{bmatrix} -1 & 0 & 0 \\ 0 & -1 & 0 \\ 0 & 0 & 1 \end{bmatrix} \quad i: \begin{bmatrix} -1 & 0 & 0 \\ 0 & -1 & 0 \\ 0 & 0 & -1 \end{bmatrix} \quad \sigma_h: \begin{bmatrix} 1 & 0 & 0 \\ 0 & 1 & 0 \\ 0 & 0 & -1 \end{bmatrix}$$

Characters of Matrix Representations

$$3 \qquad -1 \qquad -3 \qquad 1$$

Block Diagonalized Matrices

$$\begin{bmatrix} [1] & 0 & 0 \\ 0 & [1] & 0 \\ 0 & 0 & [1] \end{bmatrix} \quad \begin{bmatrix} [-1] & 0 & 0 \\ 0 & [-1] & 0 \\ 0 & 0 & [1] \end{bmatrix} \quad \begin{bmatrix} [-1] & 0 & 0 \\ 0 & [-1] & 0 \\ 0 & 0 & [-1] \end{bmatrix} \quad \begin{bmatrix} [1] & 0 & 0 \\ 0 & [1] & 0 \\ 0 & 0 & [-1] \end{bmatrix}$$

Irreducible Representations

	E	C_2	i	σ_h	Coordinate Used
	1	–1	–1	1	x
	1	–1	–1	1	y (x and y give the same irreducible representation)
	1	1	–1	–1	z
Γ	3	–1	–3	1	

4.7 Chiral molecules may have only proper rotations. The C_1, C_n, and D_n groups—along with the rare T, O, and I groups—meet this condition.

4.8

$$\sigma_{xz}: \begin{bmatrix} 1 & 0 & 0 & 0 & 0 & 0 & 0 & 0 & 0 \\ 0 & -1 & 0 & 0 & 0 & 0 & 0 & 0 & 0 \\ 0 & 0 & 1 & 0 & 0 & 0 & 0 & 0 & 0 \\ 0 & 0 & 0 & 1 & 0 & 0 & 0 & 0 & 0 \\ 0 & 0 & 0 & 0 & -1 & 0 & 0 & 0 & 0 \\ 0 & 0 & 0 & 0 & 0 & 1 & 0 & 0 & 0 \\ 0 & 0 & 0 & 0 & 0 & 0 & 1 & 0 & 0 \\ 0 & 0 & 0 & 0 & 0 & 0 & 0 & -1 & 0 \\ 0 & 0 & 0 & 0 & 0 & 0 & 0 & 0 & 1 \end{bmatrix} \quad \sigma_{yz}: \begin{bmatrix} -1 & 0 & 0 & 0 & 0 & 0 & 0 & 0 & 0 \\ 0 & 1 & 0 & 0 & 0 & 0 & 0 & 0 & 0 \\ 0 & 0 & 1 & 0 & 0 & 0 & 0 & 0 & 0 \\ 0 & 0 & 0 & 0 & 0 & 0 & -1 & 0 & 0 \\ 0 & 0 & 0 & 0 & 0 & 0 & 0 & 1 & 0 \\ 0 & 0 & 0 & 0 & 0 & 0 & 0 & 0 & 1 \\ 0 & 0 & 0 & -1 & 0 & 0 & 0 & 0 & 0 \\ 0 & 0 & 0 & 0 & 1 & 0 & 0 & 0 & 0 \\ 0 & 0 & 0 & 0 & 0 & 1 & 0 & 0 & 0 \end{bmatrix}$$

4.9 The reducible representation (D_{2h} symmetry) is:

D_{2h}	E	$C_2(z)$	$C_2(y)$	$C_2(x)$	i	$\sigma(xy)$	$\sigma(xz)$	$\sigma(yz)$
Γ	18	0	0	–2	0	6	2	0

This reduces to: $3\,A_g + 3\,B_{1g} + 2\,B_{2g} + B_{3g} + A_u + 2\,B_{1u} + 3\,B_{2u} + 3\,B_{3u}$

Translational modes (matching x, y, and z): $B_{1u} + B_{2u} + B_{3u}$

Rotational modes (matching R_x, R_y, and R_z): $B_{1g} + B_{2g} + B_{3g}$

Vibrational modes (all that remain): $3\,A_g + 2\,B_{1g} + B_{2g} + A_u + B_{1u} + 2\,B_{2u} + 2\,B_{3u}$

4.10 **a.** $\Gamma_1 = A_1 + T_2$:

T_d	E	$8C_3$	$3C_2$	$6S_4$	$6\sigma_d$
Γ_1	4	1	0	0	2
A_1	1	1	1	1	1
A_2	1	1	1	–1	–1
E	2	–1	2	0	0
T_1	3	0	–1	1	–1
T_2	3	0	–1	–1	1

For A_1: $\frac{1}{24} \times [(4 \times 1) + 8(1 \times 1) + 3(0 \times 1) + 6(0 \times 1) + 6(2 \times 1)] = 1$

For A_2: $\frac{1}{24} \times [(4 \times 1) + 8(1 \times 1) + 3(0 \times 1) + 6(0 \times (-1)) + 6(2 \times (-1))] = 0$

For E: $\frac{1}{24} \times [(4 \times 2) + 8(1 \times (-1)) + 3(0 \times 2) + 6(0 \times 0) + 6(2 \times 0)] = 0$

For T_1: $\frac{1}{24} \times [(4 \times 3) + 8(1 \times 0) + 3(0 \times (-1)) + 6(0 \times 1) + 6(2 \times (-1))] = 0$

For T_2: $\frac{1}{24} \times [(4 \times 3) + 8(1 \times 0) + 3(0 \times (-1)) + 6(0 \times (-1)) + 6(2 \times 1)] = 1$

Adding the characters of A_1 and T_2 for each operation confirms the result.

b. $\Gamma_2 = A_1 + B_1 + E$:

D_{2d}	E	$2S_4$	C_2	$2C_2'$	$2\sigma_d$
Γ_2	4	0	0	2	0
A_1	1	1	1	1	1
A_2	1	1	1	–1	–1
B_1	1	–1	1	1	–1
B_2	1	–1	1	–1	1
E	2	0	–2	0	0

For A_1: $\frac{1}{8} \times [(4 \times 1) + 2(0 \times 1) + (0 \times 1) + 2(2 \times 1) + 2(0 \times 1)] = 1$

For A_2: $\frac{1}{8} \times [(4 \times 1) + 2(0 \times 1) + (0 \times 1) + 2(2 \times (-1)) + 2(0 \times (-1))] = 0$

For B_1: $\frac{1}{8} \times [(4 \times 1) + 2(0 \times (-1)) + (0 \times 1) + 2(2 \times 1) + 2(0 \times (-1))] = 1$

For B_2: $\frac{1}{8} \times [(4 \times 1) + 2(0 \times (-1)) + (0 \times 1) + 2(2 \times (-1)) + 2(0 \times 1)] = 0$

For E: $\frac{1}{8} \times [(4 \times 2) + 2(0 \times 0) + (0 \times (-2)) + 2(2 \times 0) + 2(0 \times 0)] = 1$

Adding the characters of A_1, B_1, and E for each operation confirms the result.

c. $\Gamma_3 = A_2 + B_1 + B_2 + 2E$:

C_{4v}	E	$2C_4$	C_2	$2\sigma_v$	$2\sigma_d$
Γ_3	7	–1	–1	–1	–1
A_1	1	1	1	1	1
A_2	1	1	1	–1	–1
B_1	1	–1	1	1	–1
B_2	1	–1	1	–1	1
E	2	0	–2	0	0

For A_1: $\frac{1}{8} \times [(7 \times 1) + 2((-1) \times 1) + ((-1) \times 1) + 2((-1) \times 1) + 2((-1) \times 1)] = 0$

For A_2: $\frac{1}{8} \times [(7 \times 1) + 2((-1) \times 1) + ((-1) \times 1) + 2((-1) \times (-1)) + 2((-1) \times (-1))] = 1$

For B_1: $\frac{1}{8} \times [(7 \times 1) + 2((-1) \times (-1)) + ((-1) \times 1) + 2((-1) \times 1) + 2((-1) \times (-1))] = 1$

For B_2: $\frac{1}{8} \times [(7 \times 1) + 2((-1) \times (-1)) + ((-1) \times 1) + 2((-1) \times (-1)) + 2((-1) \times 1)] = 1$

For E: $\frac{1}{8} \times [(7 \times 2) + 2((-1) \times 0) + ((-1) \times (-2)) + 2((-1) \times 0) + 2((-1) \times 0)] = 2$

4.11 The A_{2u} (matching symmetry of z) and both E_u (matching symmetry of x and y together) vibrational modes are IR active.

4.12 Vibrational analysis for NH_3:

C_{3v}	E	$2C_3$	$3\sigma_v$	
Γ_3	12	0	2	
A_1	1	1	1	z
A_2	1	1	–1	R_z
E	2	–1	0	(x, y) (R_x, R_y)

a. A_1: $\frac{1}{6}[(12 \times 1) + 2(0 \times 1) + 3(2 \times 1)] = 3$

A_2: $\frac{1}{6}[(12 \times 1) + 2(0 \times 1) + 3(2 \times (-1))] = 1$

E: $\frac{1}{6}[(12 \times 2) + 2(0 \times (-1)) + 3(2 \times 0)] = 4$

$\Gamma = 3A_1 + A_2 + 4E$

b. Translation: $A_1 + E$, based on the x, y, and z entries in the table

Rotation: $A_2 + E$, based on the R_x, R_y, and R_z entries in the table

Vibration: $2A_1 + 2E$ remaining from the total; the A_1 vibrations are symmetric stretch and symmetric bend. The E vibrations are asymmetric.

c. There are three translational modes, three rotational modes, and six vibrational modes, for a total of 12. With 4 atoms in the molecule, $3N = 12$, so there are $3N$ degrees of freedom in the ammonia molecule.

d. All the vibrational modes are IR active (all have x, y, or z symmetry).

4.13 Taking only the C—O stretching modes for $Mn(CO)_5Cl$ (only the vectors between the C and O atoms):

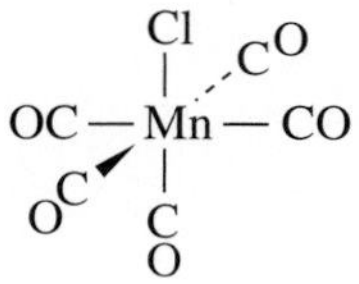

C_{4v}	E	$2C_4$	C_2	$2\sigma_v$	$2\sigma_d$	
Γ	5	1	1	3	1	
A_1	1	1	1	1	1	z
A_2	1	1	1	–1	–1	R_z
B_1	1	–1	1	1	–1	
B_2	1	–1	1	–1	1	
E	2	0	–2	0	0	(x, y) (R_x, R_y)

$\Gamma = 2A_1 + B_1 + E$

$Mn(CO)_5Cl$ should have four IR-active stretching modes, two from A_1 and two from E. The E modes are a degenerate pair; they give rise to a single infrared band. The B_1 mode is IR inactive.

4.14 Using as basis the I=O bonds, the following representation is obtained in the D_{5h} point group:

D_{5h}	E	$2C_5$	$2C_5^{\,2}$	$5C_2$	σ_h	$2S_5$	$2S_5^{\,2}$	$5\sigma_v$
Γ	2	2	2	0	0	0	0	2

This reduces to:

A_1'	1	1	1	1	1	1	1	1	$x^2 + y^2, z^2$
A_2''	1	1	1	−1	−1	−1	−1	1	z

The A_1' vibration matches the symmetry of $x^2 + y^2$ and z^2; it is Raman active. The A_2'' does not match xy, xz, yz, or a squared term; it matches z and is therefore active in the IR but not in the Raman spectrum. Therefore, the single Raman band is consistent with *trans* orientation.

Chapter 5

5.1 p_x and d_{xz} p_z and d_{z^2} s and $d_{x^2-y^2}$

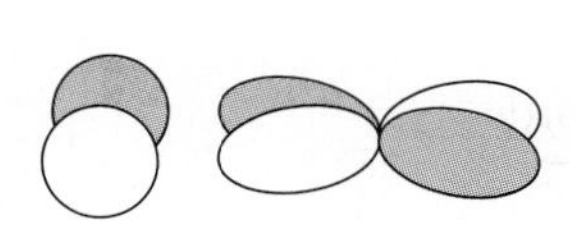

π interaction

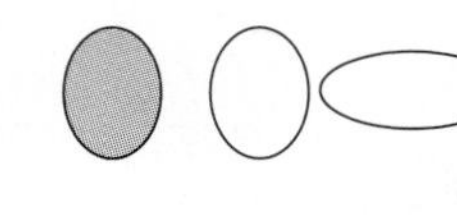

σ interaction

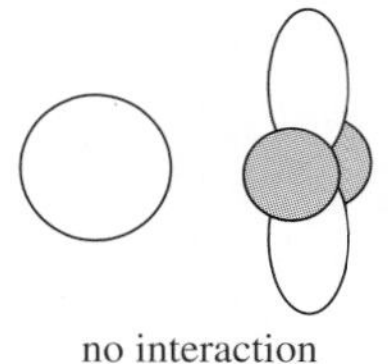

no interaction

5.2 In Figure 5.3, (a), σ^* is σ_u, σ is σ_g, π^* is π_g, π is π_u, δ^* is δ_u, and δ is δ_g.

5.3 Bonding in the OH^- ion.

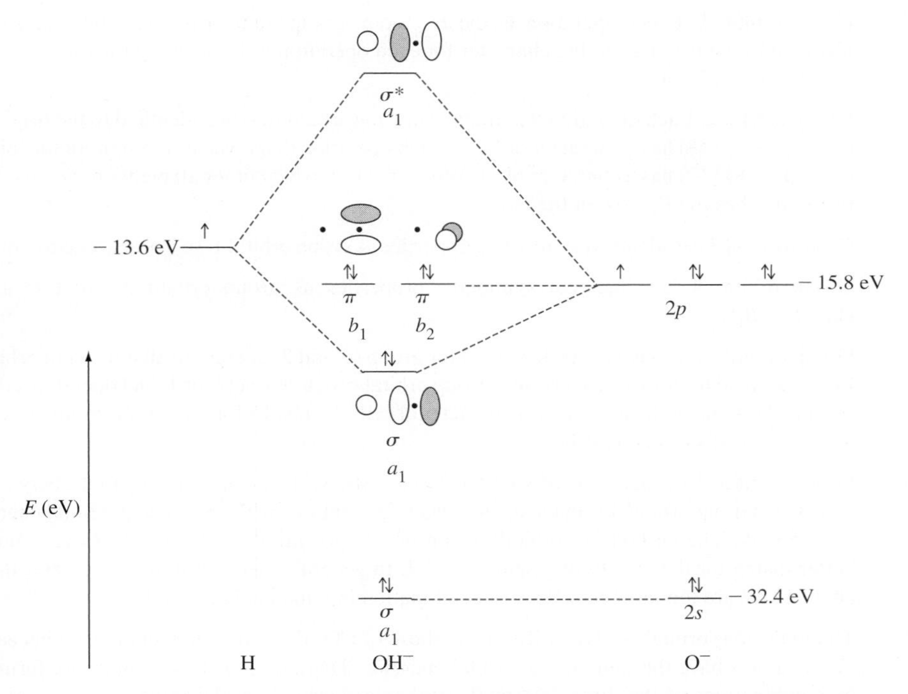

The energy match of the H 1*s* and the O 2*p* orbitals is fairly good, but that of the H 1s with the O 2*s* is poor. Therefore, molecular orbitals are formed between the H 1*s* and the O $2p_z$, as shown above (z is the axis through the nuclei). All the other O orbitals are nonbonding, either because of poor energy match or lack of useful overlap.

5.4 H_3^+ energy levels.

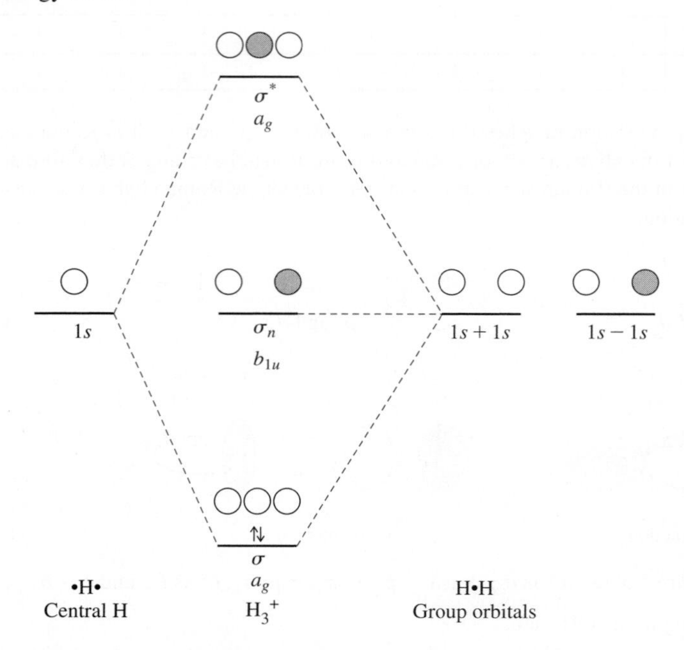

5.5 Group orbital 1: Every operation in the D_{2h} point group transforms the orbital into one identical to the original, so the character for each operation is 1, matching the top (A_g) row in the character table.

Group orbital 2: Each operation that transforms this orbital into one identical to the original (⊙ · ⊙ → ⊙ · ⊙) has a character of 1, and each operation that reverses the signs of the lobes (⊙ · ⊙ → ⊙ · ⊙) has a character of –1. When this is carried out for all eight operations, the result matches the B_{1u} row of the table.

Group orbital 3 has identical symmetry properties as group orbital 1, so it is also classified A_g.

Group orbital 4 has identical symmetry properties as group orbital 2, so it is also classified B_{1u}.

Group orbitals 5 through 8: As described for group orbital 2, operations that result in orbital lobes identical to those of the original group orbital give a character of 1, and operations that reverse the signs of the lobes give a character of −1. The results for these four group orbitals match the labels in Figure 5.18.

5.6 Group orbital 2 is made up of oxygen 2*s* orbitals, with an orbital potential energy of −32.4 eV. Group orbital 4 is made up of oxygen $2p_z$ orbitals, with an orbital potential energy of −15.9 eV. The carbon $2p_z$ orbital has an orbital potential energy of −10.7 eV, a much better match for the energy of group orbital 4. In general, energy differences greater than about 12 eV are too large for effective combination into molecular orbitals.

5.7 The molecular orbitals of N_3^- differ from those of CO_2 described in Section 5.4.2 because all the atoms have the same initial orbital energies. Therefore, the best orbitals are formed by combinations of the three 2*s* orbitals or three 2*p* orbitals of the same type (*x*, *y*, or *z*). The resulting pattern of orbitals is shown in the diagram. Note that the σ_n orbital is slightly higher in energy than the π_n orbitals as a consequence of an antibonding interaction with the 2*s* orbital of the central nitrogen. For symmetry labels on the orbitals, see Figure 5.25.

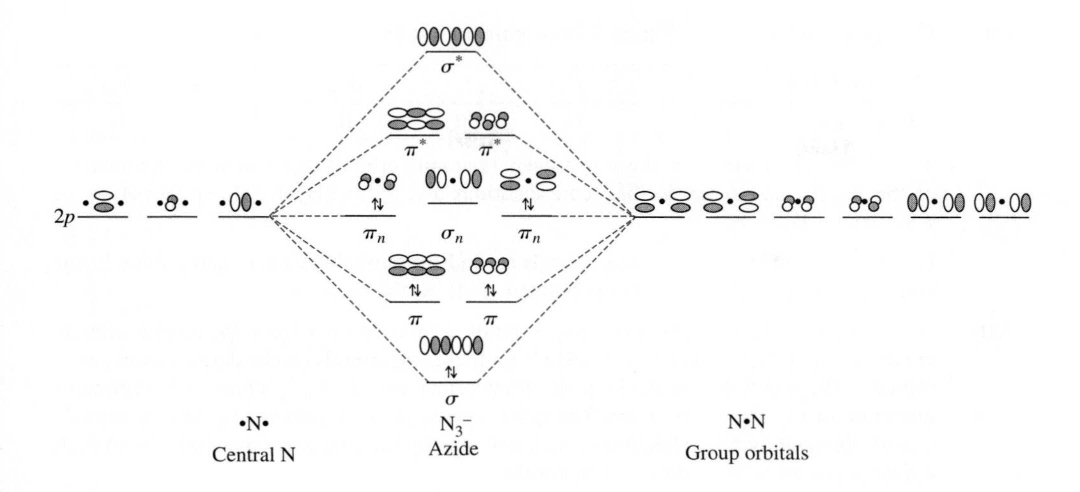

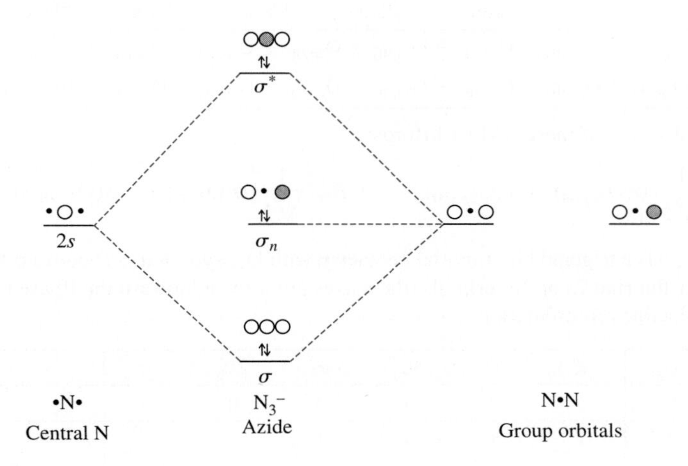

5.8 BeH_2 molecular orbitals.

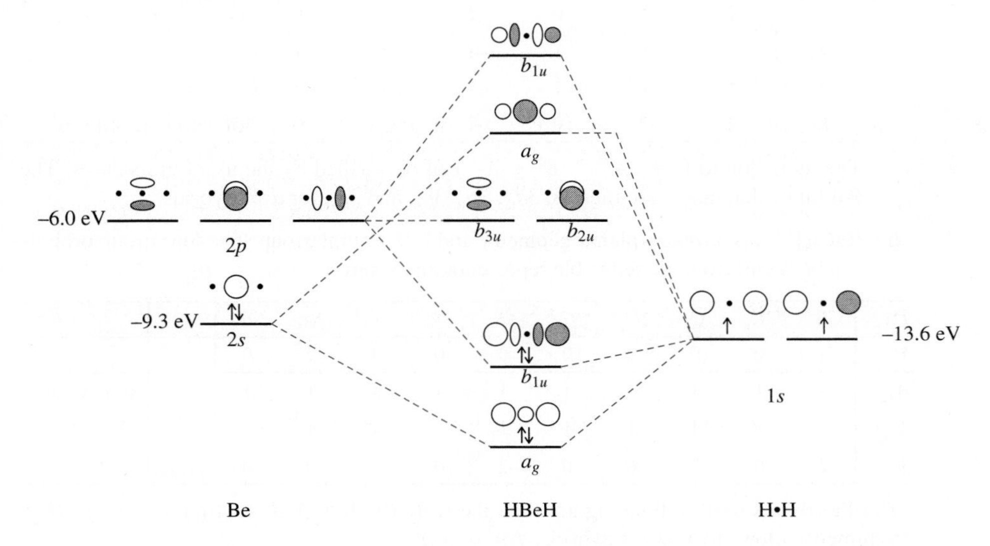

5.9 Using H_b as defined in the Figure 5.29 coordinate system:

Original Orbital	E	C_3	C_3^2	$\sigma_{v(a)}$	$\sigma_{v(b)}$	$\sigma_{v(c)}$
H_b becomes...	H_b	H_c	H_a	H_c	H_b	H_a

The linear combinations of these hydrogen 1*s* atomic orbitals that match the symmetries of the A_1, A_2, and E irreducible representations are: $A_1 = 2H_a + 2H_b + 2H_c$; $A_2 = 0$; $E = 2H_b - H_c - H_a$

Because these hydrogen 1*s* atomic orbitals are indistinguishable by symmetry, these linear combinations are identical to those obtained via H_a as the basis.

5.10 On the basis of Figure 5.18, the group orbitals composed of oxygen $2p_y$ atomic orbitals exhibit B_{2u} and B_{3g} symmetry. The SALCs for these two orbitals can be derived from group orbital 5 (Figure 5.18), by tracking the transformations of $O_{2p_y(A)}$ upon each symmetry operation of the D_{2h} point group. The linear combinations for these O_{2p_y} atomic orbitals can be obtained by multiplication of each outcome by the characters associated with each operation, followed by addition of the results.

Original Orbital	E	$C_{2(z)}$	$C_{2(y)}$	$C_{2(x)}$	i	$\sigma_{(xy)}$	$\sigma_{(xz)}$	$\sigma_{(yz)}$
$O_{2p_y(A)}$ becomes...	$O_{2p_y(A)}$	$-O_{2p_y(A)}$	$O_{2p_y(B)}$	$-O_{2p_y(B)}$	$-O_{2p_y(B)}$	$O_{2p_y(B)}$	$-O_{2p_y(A)}$	$O_{2p_y(A)}$
B_{2u}	$O_{2p_y(A)} + O_{2p_y(A)} + O_{2p_y(B)} + O_{2p_y(B)} + O_{2p_y(B)} + O_{2p_y(B)} + O_{2p_y(A)} + O_{2p_y(A)} = 4(O_{2p_y(A)}) + 4(O_{2p_y(B)})$							
B_{3g}	$O_{2p_y(A)} + O_{2p_y(A)} - O_{2p_y(B)} - O_{2p_y(B)} - O_{2p_y(B)} - O_{2p_y(B)} + O_{2p_y(A)} + O_{2p_y(A)} = 4(O_{2p_y(A)}) - 4(O_{2p_y(B)})$							

Normalization of these SALCs affords:

$$B_{2u}: \frac{1}{\sqrt{2}}[\psi(O_{2p_y(A)}) + \psi(O_{2p_y(B)})] \qquad B_{3g}: \frac{1}{\sqrt{2}}[\psi(O_{2p_y(A)}) - \psi(O_{2p_y(B)})]$$

5.11 **a.** PF_5 has a trigonal bipyramidal geometry with D_{3h} symmetry. The group orbitals for the five fluorine 2*s* or $2p_y$ orbitals (the *y* axes are directed toward the P) are found from the reducible representation.

D_{3h}	E	$2C_3$	$3C_2$	σ_h	$2S_3$	$3\sigma_v$		
Γ	5	2	1	3	0	3		
A_1'	1	1	1	1	1	1		$x^2 + y^2, z^2$
A_2'	1	1	−1	1	1	−1	R_z	
E'	2	−1	0	2	−1	0	(x, y)	$(x^2 - y^2, xy)$
A_1''	1	1	1	−1	−1	−1		
A_2''	1	1	−1	−1	−1	1	z	
E''	2	−1	0	−2	1	0	(R_x, R_y)	(xz, yz)

The reduction to $\Gamma = 2A_1' + E' + A_2''$ can be verified by the usual procedures. The P orbitals that match are then 3*s*, $3d_{z^2}$, $3p_x$, $3p_y$, and $3p_z$, for dsp^3 hybrids.

b. $[PtCl_4]^{2-}$ has a square-planar geometry and a D_{4h} point group. The four group orbitals can be found from the reducible representation, where $\Gamma = A_{1g} + B_{1g} + E_u$.

D_{4h}	E	$2C_4$	C_2	$2C_2'$	$2C_2''$	i	$2S_4$	σ_h	$2\sigma_v$	$2\sigma_d$		
Γ	4	0	0	2	0	0	0	4	2	0		
A_{1g}	1	1	1	1	1	1	1	1	1	1		$x^2 + y^2, z^2$
B_{1g}	1	−1	1	1	−1	1	−1	1	1	−1		$x^2 - y^2$
E_u	2	0	−2	0	0	−2	0	2	0	0	(x, y)	

The Pt orbitals used in bonding are then the *s*, d_{z^2} (both A_{1g}), $d_{x^2-y^2}(B_{1g})$, p_x, and p_y (E_u). Symmetry allows two sets of hybrids, dsp^2 or d^2p^2.

5.12 $SOCl_2$ has only a mirror plane and belongs to group C_s. Using s orbitals on O and the two Cls, we can obtain the reducible representation and its irreducible components shown in the table.

C_s	E	σ_h		
Γ	3	1		
A'	1	1	x, y, R_z	x^2, y^2, z^2, xy
A''	1	−1	z, R_x, R_y	yz, xz

$\Gamma = 2A' + A''$

The sulfur orbitals used in σ bonding are the $3p_x$, $3p_y$, and $3p_z$. The $3s$ could be involved, but the nonplanar shape of the molecule requires that all three p orbitals be used.

Chapter 6

6.1 $pK_{ion} = 34.4$ (Table 6.2) $\qquad K_{ion} = [CH_3CHN^+][CH_2CN^-] = 10^{-34.4}$

Because $[CH_3CHN^+] = [CH_2CN^-]$: $\qquad [CH_3CHN^+]^2 = 10^{-34.4}$

$$[CH_3CHN^+] = 10^{-17.2} = 6.3 \times 10^{-18} \text{ mol L}^{-1}$$

6.2 If C_6H_6 is the stronger Brønsted–Lowry acid, it can more effectively donate a hydrogen ion than n-butane, favoring the forward reaction. Alternatively, the conjugate base of n-butane, n-$C_4H_9^-$, is the stronger Brønsted–Lowry base and therefore has a stronger ability to accept hydrogen ions to yield its neutral species. From either viewpoint, the product side is favored.

6.3 The reaction of interest (3) is: $HC_2H_3O_2(aq) + H_2O(l) \longrightarrow H_3O^+(aq) + C_2H_3O_2^-(aq)$.

One approach is to apply Hess's law. Reaction 3 is the sum of the following two reactions:

(1) $\qquad 2\ H_2O(l) \longrightarrow H_3O^+(aq) + OH^-(aq)$

$$\Delta H_1^\circ = 55.9 \text{ kJ/mol}; \ \Delta S_1^\circ = -80.4 \text{ J/mol K}$$

(2) $\qquad HC_2H_3O_2(aq) + OH^-(aq) \longrightarrow H_2O(l) + C_2H_3O_2^-(aq)$

$$\Delta H_2^\circ = -56.3 \text{ kJ/mol}; \ \Delta S_2^\circ = -12.0 \text{ J/mol K}$$

Addition of these ΔH° and ΔS° values provide the desired thermodynamic parameters.

$$\Delta H_1^\circ + \Delta H_2^\circ = -0.4 \text{ kJ/mol}$$

$$\Delta S_1^\circ + \Delta S_2^\circ = -92.4 \text{ J/mol K}$$

Another approach is examination of the temperature dependence on the ionization equilibrium constant over a narrow range of temperatures.

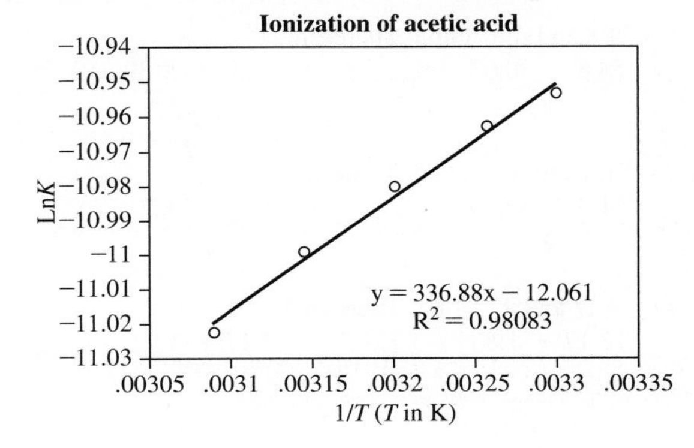

On the basis of $\ln K = -\frac{\Delta H^{\circ}}{R}\left(\frac{1}{T}\right) + \frac{\Delta S^{\circ}}{R}$, ΔH° and ΔS° for reaction 3 can be approximated.

$$-\frac{\Delta H_3^{\circ}}{R} = 336.88 \text{ K}$$

$$\Delta H_3^{\circ} = -2.8 \text{ kJ/ mol}$$

$$\frac{\Delta S_3^{\circ}}{R} = -12.061$$

$$\Delta S_3^{\circ} = -100 \text{ J/ mol K}$$

The values determined via the graphical approach are more negative than the values derived from Hess's law. The difference between the ΔS_3° values is small (about 8%) relative to the difference between the ΔH_3° values (the Hess's law value is roughly 14% of the graphically obtained value). If the data for 323 K are removed (further narrowing the temperature range examined), the equation of the best-fit line provides $\Delta H_3^{\circ} = -2.5$ kJ/mol and $\Delta S_3^{\circ} = -99$ J/mol K, a modest improvement relative to the Hess's law values.

6.4 Boron trifluoride has electron-withdrawing fluorine atoms; trimethyl boron has electron-donating methyl groups. As a result, boron trifluoride should be a stronger acid, with a more positive boron atom, toward ammonia. There are no significant steric interactions with ammonia.

With bulkier bases, the same results should be seen, because BF_3 is less bulky and would have less steric interference, regardless of the structure of the base. In this case, BF_3 is the stronger acid toward both sterically hindered and sterically unhindered bases.

6.5 **a.** Cu^{2+} is a borderline soft acid and will react more readily with NH_3 than with the harder OH^-, with the product $[Cu(NH_3)_4]^{2+}$ in a solution containing significant amounts of both NH_3 and OH^-. In the same fashion, it will react more readily with S^{2-} than with O^{2-}, forming CuS in basic solutions of sulfide.

b. On the other hand, Fe^{3+} is a hard acid and will react more readily with OH^- and O^{2-}. The product in basic ammonia is $Fe(OH)_3$ (approximately; the product is an ill-defined hydrated Fe(III) oxide and hydroxide mixture). In basic sulfide solution, the same $Fe(OH)_3$ product is formed. (There may also be some reduction of Fe(III) to Fe(II) and precipitation of FeS.)

c. Silver ion is a soft acid and is likely to combine more readily with PH_3 than with NH_3.

d. CO is a relatively soft base, Fe^{3+} is a very hard acid, Fe^{2+} is a borderline acid, and Fe is a soft acid. Therefore, CO is more likely to combine effectively with Fe(0) than with Fe(II) or Fe(III).

6.6. **a.** Al^{3+} has $I = 119.99$ and $A = 28.45$. Therefore,

$$\chi = \frac{119.99 + 28.45}{2} = 74.22 \qquad \eta = \frac{119.99 - 28.45}{2} = 45.77$$

Fe^{3+} has $I = 54.8$ and $A = 30.65$. Therefore,

$$\chi = \frac{54.8 + 30.65}{2} = 42.7 \qquad \eta = \frac{54.8 - 30.655}{2} = 12.1$$

Co^{3+} has $I = 51.3$ and $A = 33.50$. Therefore,

$$\chi = \frac{51.3 + 33.50}{2} = 42.4 \quad \eta = \frac{51.3 - 33.50}{2} = 8.9$$

b. OH^- has $I = 13.17$ and $A = 1.83$. Therefore,

$$\chi = \frac{13.17 + 1.83}{2} = 7.50 \quad \eta = \frac{13.17 - 1.83}{2} = 5.67$$

Cl^- has $I = 13.01$ and $A = 3.62$. Therefore,

$$\chi = \frac{13.01 + 3.62}{2} = 8.31 \quad \eta = \frac{13.01 - 3.62}{2} = 4.70$$

NO_2^- has $I > 10.1$ and $A = 2.30$. Therefore,

$$\chi = \frac{> 10.1 + 2.3}{2} > 6.2 \quad \eta = \frac{> 10.1 - 2.3}{2} > 3.9$$

c. H_2O has $I = 12.6$ and $A = -6.4$. Therefore,

$$\chi = \frac{12.6 + (-6.4)}{2} = 3.1 \quad \eta = \frac{12.6 - (-6.4)}{2} = 9.5$$

NH_3 has $I = 10.7$ and $A = -5.6$. Therefore,

$$\chi = \frac{10.7 + (-5.6)}{2} = 2.6 \quad \eta = \frac{10.7 - (-5.6)}{2} = 8.2$$

PH_3 has $I = 10.0$ and $A = -1.9$. Therefore,

$$\chi = \frac{10.0 + (-1.9)}{2} = 4.0 \quad \eta = \frac{10.0 - (-1.9)}{2} = 6.0$$

6.7 **a.**

Acid	E_A	C_A	Base	E_B	C_B	$\Delta H\,(E)$	$\Delta H\,(C)$	$\Delta H\,(total)$
BF_3	9.88	1.62	NH_3	1.36	3.46	−13.44	−5.60	−19.04
			CH_3NH_2	1.30	5.88	−12.84	−9.53	−22.37
			$(CH_3)_2NH$	1.09	8.73	−10.77	−14.14	−24.91
			$(CH_3)_3N$	0.808	11.54	−7.98	−18.70	−26.68

b.

Base	E_B	C_B	Acid	E_A	C_A	$\Delta H\,(E)$	$\Delta H\,(C)$	$\Delta H\,(total)$
Py	1.17	6.40	Me_3B	6.14	1.70	−7.18	−10.88	−18.06
			Me_3Al	16.9	1.43	−19.77	−9.15	−28.92
			Me_3Ga	13.3	0.881	−15.56	−5.64	−21.20

The amine series shows a steady increase in C_B and decrease in E_B as methyl groups are added. The methyl groups push electrons onto the N, and the lone-pair electrons are then made more available to the acid BF_3. As a result, covalent bonds to BF_3 are more likely with more methyl groups. The lone pairs of the molecules with fewer methyl groups are more tightly held, with more ionic bonding and a larger E_B and $\Delta H\,(E)$. The covalent effect is stronger and has a larger change in ΔH, so it determines the order in the total.

The B, Al, Ga series is less regular. Possible arguments for Al being the strongest in E_A include the following:

1. The larger central atom leads to more electrostatic bonding and less covalent bonding. This would make the order B $<$ Al $<$ Ga for electrostatic bonding, rather than the B $<$ Ga $<$ Al that is calculated.
2. The *d* electrons in Ga shield the outer electrons, so they are held less tightly. This results in less electrostatic attraction for the outer electrons and those of the pyridine, which makes them less likely to form either an electrostatic (ionic) or covalent bond.

Chapter 7

7.1 **a.** Atom or ion in the center of the cell = 1
8 atoms or ions at the corners of the cell, each $\frac{1}{8}$ within the cell = 1 Total = 2 atoms or ions per unit cell

b. Eight atoms at the corners of the cell, four of which are $\frac{1}{12}$ $\left(\frac{1}{2} \times \frac{1}{6}\right)$ in the cell and four of which are $\frac{1}{6}$ $\left(\frac{1}{2} \times \frac{1}{3}\right)$ in the cell $= \frac{4}{12} + \frac{4}{6} = 1$ atom per unit cell.

7.2 If the edge of the unit cell has a length of a, the face diagonal is $\sqrt{2}a$ and the body diagonal is $\sqrt{3}a$, from the Pythagorean theorem. The diagonal through a body-centered unit cell also has a length of $4r$, where r is the radius of each of the atoms (r for each of the corner atoms and $2r$ for the body-centered atom). Therefore,

$$4r = \sqrt{3}a, \quad \text{or} \quad a = 2.31r$$

7.3 A simple cubic array of anions with radius r_- has a body diagonal of length $2\sqrt{3}r_-$. With a cation of the ideal size in the body center, this distance is also $2r_+ + 2r_-$. Setting the two equal to each other and solving, $r_+ = 0.732r_-$, or a radius ratio of $r_+/r_- = 0.732$. Any ratio between 0.732 and 1.00 (the ideal for CN = 12) should fit the fluorite structure. $CaCl_2$ has r_+/r_- between $126/167 = 0.754$ (CN = 8) and $114/167 = 0.683$ (CN = 6). $CaBr_2$ has r_+/r_- between $126/182 = 0.692$ (CN = 8) and $114/182 = 0.626$ (CN = 6). $CaCl_2$ is on the edge of the CN = 6, CN = 8 boundary, but $CaBr_2$ is clearly in the CN = 6 region. Both crystallize with CN = 6 in structures that are similar to rutile (TiO_2).

7.4 Using the Born Mayer equation from Section 7.2.1:

$$U = \frac{NMZ_+Z_-}{r_0}\left[\frac{e^2}{4\pi\varepsilon_0}\right]\left(1 - \frac{\rho}{r_0}\right)$$

with $r_0 = r_+ + r_- = 116 + 167 = 283$ pm, $M = 1.74756$, $N = 6.02 \times 10^{23}$, $Z_+ = 1, Z_- = -1, \rho = 30$ pm, and $\dfrac{e^2}{4\pi\varepsilon_0} = 2.3071 \times 10^{-28}$ J m. Changing r_0 to meters in the first fraction, the result is -766 kJ mol^{-1}.

7.5

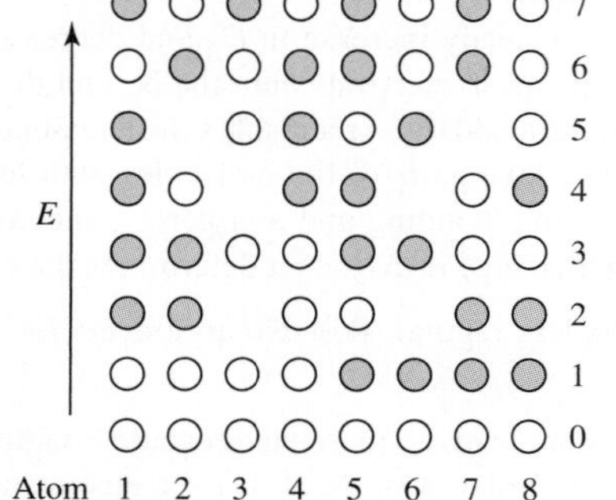

The diagram is approximate because some of the nodes are near, but not exactly over, an atomic nucleus. Using a much larger number of atoms would show the alternating signs more clearly.

7.6 Consider the bottom tetrahedron to have four oxygen atoms, or SiO_4. Proceeding clockwise, the next tetrahedron duplicates one oxygen (shared with the first tetrahedron), adding SiO_3. The last tetrahedron shares oxygens (one each) with the first two tetrahedra, adding SiO_2, for a total of three silicons and nine oxygens. Using the charges Si^{+4} and O^{2-} gives the formula $Si_3O_9{}^{6-}$.

Chapter 8

8.1 $H_2O + 2e^- \longrightarrow OH^- + H^-$

The half reactions to be added are: $\Delta G^\circ = -nFE^\circ$

$$H_2O + e^- \longrightarrow OH^- + \frac{1}{2}H_2 \quad E_1^\circ = -0.828\text{ V} \quad -(1\text{ mol})F(-0.828\text{ V}) = 0.828\text{ mol } F\text{ V}$$

$$\frac{1}{2}(H_2 + 2e^- \longrightarrow 2H^-) \quad E_2^\circ = -2.25\text{ V} \quad -(1\text{ mol})F(-2.25\text{ V}) = 2.25\text{ mol } F\text{ V}$$

$$H_2O + 2e^- \longrightarrow OH^- + H^- \quad E_3^\circ = -(2\text{ mol})\,F\text{ V} = 3.08\,F\text{ V}; \qquad E_3^\circ = -1.54\text{ V}$$

8.2 $2\,H_2O_2 \longrightarrow 2\,H_2O + O_2$

The half reactions involved in disproportionation are:

Reduction:	$H_2O_2 + 2e^- + 2\,H^+ \longrightarrow 2\,H_2O$	$E_1^\circ = 1.763\text{ V}$
Oxidation:	$H_2O_2 \longrightarrow O_2 + 2\,e^- + 2\,H^+$	$E_2^\circ = -0.695\text{ V}$
Overall:	$2\,H_2O_2 \longrightarrow 2\,H_2O + O_2$	$E_3^\circ = 1.068\text{ V}$

Because the voltage for the disproportionation reaction, E_3°, is positive, ΔG° must be negative, since $\Delta G^\circ = -nFE_3^\circ$.

8.3 The points for N_2 (0, 0); N_2O (1, 1.77); NH_3OH^+ (−1, 1.87) and $N_2H_4{}^+$ (−2, 0.46) are derived in the text. For all the reductions of adjacent species in the nitrogen Latimer diagram, the change in nitrogen oxidation state is −1.

$N_2H_4{}^+ \xrightarrow{1.275} NH_4{}^+$

$nE^\circ = (-1)(1.275\text{ V}) = -1.275\text{ V}$

$NH_4{}^+$ is 1.275 V lower than $N_2H_4{}^+$.

The Frost diagram point for $NH_4{}^+$ is (−3, −0.815 V).

$NO \xrightarrow{1.59} N_2O$

$nE^\circ = (-1)(1.59\text{ V}) = -1.59\text{ V}$

N_2O is 1.59 V lower than NO.

The Frost diagram point for NO is (2, 3.36 V).

$HNO_2 \xrightarrow{0.996} NO$

$nE^\circ = (-1)(0.996\text{ V}) = -0.996\text{ V}$

NO is 0.996 V lower than HNO_2.

The Frost diagram point for HNO_2 is (3, 4.36 V).

$N_2O_4 \xrightarrow{1.07} HNO_2$

$nE^\circ = (-1)(1.07\text{ V}) = -1.07\text{ V}$

HNO_2 is 1.07 V lower than N_2O_4.

The Frost diagram point for N_2O_4 is (4, 5.42 V).

$NO_3{}^- \xrightarrow{0.803} N_2O_4$

$nE^\circ = (-1)(0.803\text{ V}) = -0.803\text{ V}$

N_2O_4 is 0.803 V lower than $NO_3{}^-$.

The Frost diagram point for $NO_3{}^-$ is (5, 6.22 V).

8.4 The Latimer diagram for hydrogen in acidic solution is in Section 8.1.4. The substance with the element in the zero oxidation state (H_2) is the reference point (0,0) in the Frost diagram. Two other points need to be derived. The oxidation state change for each reduction of adjacent species in the Latimer diagram is −1.

$$H^+ \xrightarrow{0} H_2$$

$$nE^\circ = (-1)(0 \text{ V}) = 0 \text{ V}$$

The Frost diagram point for H^+ is (1, 0 V).

$$H_2 \xrightarrow{-2.25} H^-$$

$$nE^\circ = (-1)(-2.25 \text{ V}) = 2.25 \text{ V}$$

H^- is 2.25 V above H_2; the point for H^- is (−1, 2.25 V).

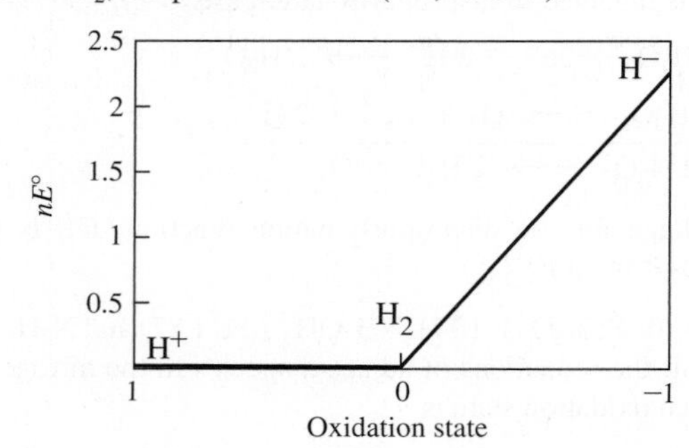

The Latimer diagram for hydrogen in basic solution is in Section 8.1.4. The substance with the element in the zero oxidation state (H_2) is the reference point (0,0) in the Frost diagram. Two other points need to be derived. The oxidation state change for each reduction of adjacent species in the Latimer diagram is −1.

$$H_2 \xrightarrow{-2.25} H^-$$

$$nE^\circ = (-1)(-2.25 \text{ V}) = 2.25 \text{ V}$$

H^- is 2.25 V above H_2; the point for H^- is (−1, 2.25 V).

$$H_2O \xrightarrow{-0.828} H_2$$

$$nE^\circ = (-1)(-0.828 \text{ V}) = 0.828 \text{ V}$$

H_2 is 0.828 V above H_2O; the point for H_2O is (1, −0.828 V).

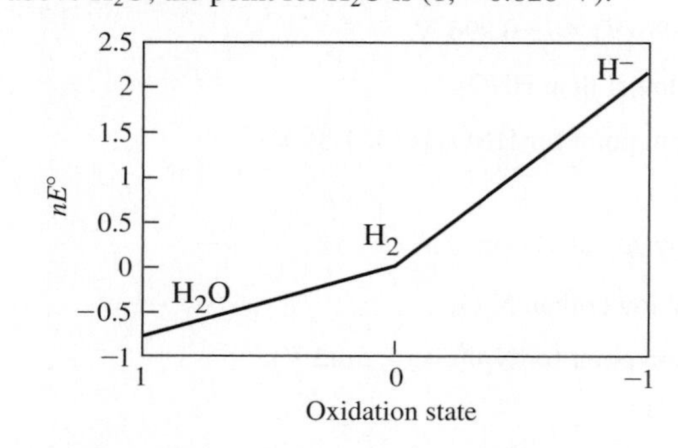

8.5 The fraction remaining for first-order decay = 0.5^n, where n = number of half-lives.

$$3.5 \times 10^{-2} = 0.5^n \qquad n \log 0.5 = \log(3.5 \times 10^{-2})$$

$$n = \frac{\log(3.5 \times 10^{-2})}{\log 0.5} = 4.836 \text{ half-lives}$$

$$\text{Age} = n(5730 \text{ years}) = 2.77 \times 10^4 \text{ years}$$

8.6 The balanced reaction of interest is $2\ HNO_2 + 11H^+ + 10e^- \longrightarrow N_2H_5^+ + 4\ H_2O$. If the coordinates of a nitrogen Frost diagram are available (see Exercise 8.3), these data can be used for rapid determination of this potential. The coordinates for HNO_2 and $N_2H_5^+$ are (3, 4.36 V) and (−2, 0.46 V), respectively. The slope of the line connecting these points is the desired potential; the slope is $\frac{0.46 - 4.36}{-2 - 3} = 0.78$ V. This number is reasonable since the slope of the line connecting these species is less steep than that connecting HNO_2 and NO; this half-reaction has a potential of 0.996 V.

Without the Frost diagram, the Latimer diagram data can be used through application of Hess' law. The sum of the free energies of the component half-reactions affords the free energy of the sum of these half-reactions. The balanced half-reaction ($2\ HNO_2 + 11H^+ + 10e^- \longrightarrow N_2H_5^+ + 4\ H_2O$) requires that we start with 2 moles of HNO_2. In this strategy n equals the number of moles of electrons in each balanced half-reaction, since $\Delta G^o = -nFE^o$.

	ΔG^o
$2\ HNO_2 + 2\ H^+ + 2e^- \longrightarrow 2\ NO + 2\ H_2O$	$-(2)(0.996)F$
$2\ NO + 2\ H^+ + 2e^- \longrightarrow N_2O + H_2O$	$-(2)(1.59)F$
$N_2O + 2\ H^+ + 2e^- \longrightarrow N_2 + H_2O$	$-(2)(1.77)F$
$N_2 + 4\ H^+ + H_2O + 2e^- \longrightarrow 2\ NH_3OH^+$	$-(2)(-1.87)F$
$2\ NH_3OH^+ + 1\ H^+ + 2e^- \longrightarrow N_2H_5^+ + 2\ H_2O$	$-(2)(1.41)F$
Sum $= 2\ HNO_2 + 11H^+ + 10e^- \longrightarrow N_2H_5^+ + 4\ H_2O$	$-(7.792)F$

$\Delta G^o = -nFE^o = -(7.792)F = -(10)FE^o$ (10 moles of electrons in balanced half-reaction)

$E^o = 0.779$ V

8.7 Since the oxidation state of nitrogen in N_2O is +1, in this proposed reaction NH_4^+ is the reducing agent, and NO_3^- is the oxidizing agent, with the common half-reaction species N_2O. The half-reactions are:

(1) $2\ NH_4^+ + H_2O \longrightarrow N_2O + 10\ H^+ + 8e^-$

(2) $2\ NO_3^- + 10\ H^+ + 8e^- \longrightarrow N_2O + 5\ H_2O$

The potentials for these reactions can be found as in Exercise 8.6. Because the nitrogen Frost diagram has been determined in Exercise 8.3, this is the most efficient strategy. The points of interest are NO_3^-: (5, 6.22 V); N_2O: (1, 1.77 V), and NH_4^+: (−3, −0.815 V).

The slope from NO_3^- to N_2O is $\frac{1.77 - 6.22}{1 - 5} = 1.11$ V (potential for $NO_3^- \longrightarrow N_2O$).

The slope from N_2O to NH_4^+ is $\frac{(-0.815) - 1.77}{(-3) - 1} = 0.646$ V (potential for $N_2O \longrightarrow NH_4^+$). Therefore, the potential for $NH_4^+ \longrightarrow N_2O = -0.646$ V.

$E^o = 1.11 + (-0.646) = 0.464$ V for $2NH_4^+ + 2NO_3^- \longrightarrow 2N_2O + 4H_2O$.

$$\Delta G^o = -nFE^o = -(8 \text{ moles } e^-)(96485\frac{\text{C}}{\text{mol } e^-})(0.464 \text{ V}) = -360 \text{ kJ}$$

On the basis of its positive potential and $\Delta G^o < 0$, this reaction proceeds spontaneously. Heating solid NH_4NO_3 results in evolution of N_2O, but this reaction must be carefully carried out to prevent detonation.

8.8

ClO_4^-/ClO_3^-:	$ClO_4^- + 2H^+ + 2e^- \longrightarrow ClO_3^- + H_2O$
$ClO_3^-/HClO_2$:	$ClO_3^- + 3H^+ + 2e^- \longrightarrow HClO_2 + H_2O$
$HClO_2/HClO$:	$HClO_2 + 2H^+ + 2e^- \longrightarrow HClO + H_2O$
$HClO/Cl_2$:	$2HClO + 2H^+ + 2e^- \longrightarrow Cl_2 + 2H_2O$
Cl_2/Cl^-:	$Cl_2 + 2e^- \longrightarrow 2Cl^-$

Chapter 9

9.1
- **a.** Triamminetrichlorochromium(III)
- **b.** Dichloroethylenediamineplatinum(II)
- **c.** Bis(oxalato)platinate(II) or bis(oxalato)platinate(2−)
- **d.** Pentaaquabromochromium(III) or pentaaquabromochromium(2+)
- **e.** Tetrachloroethylenediaminecuprate(II) or tetrachloroethylenediaminecuprate(2−)
- **f.** Tetrahydroxoferrate(III) or tetrahydroxoferrate(1−)

9.2
- **a.** Tris(acetylacetonato)iron(III)

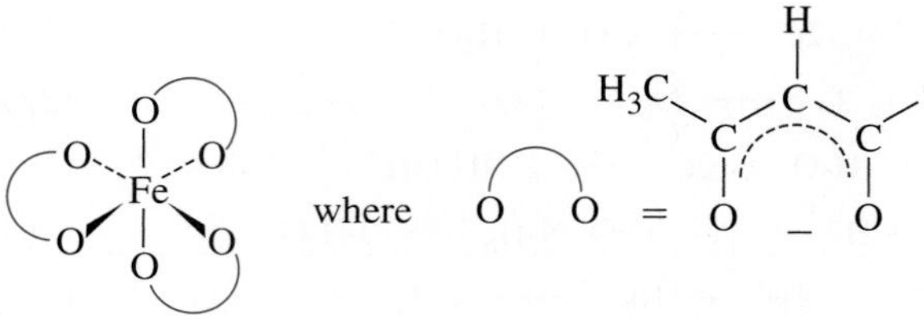

Δ isomer; Λ is also possible (Section 9.3.5)

- **b.** Hexabromoplatinate(2−)

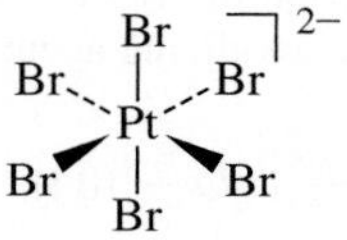

- **c.** Diamminetetrabromocobaltate(III)

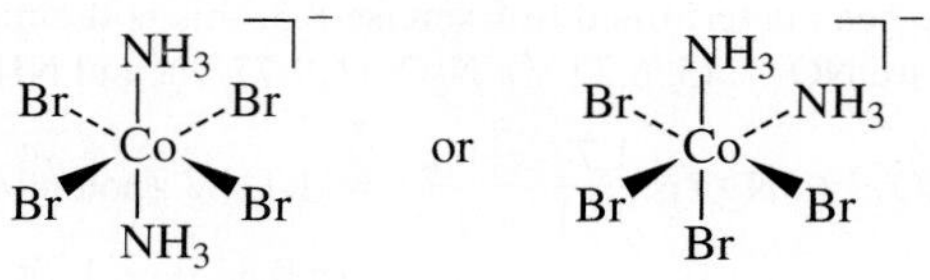

- **d.** Tris(ethylenediamine)copper(II)

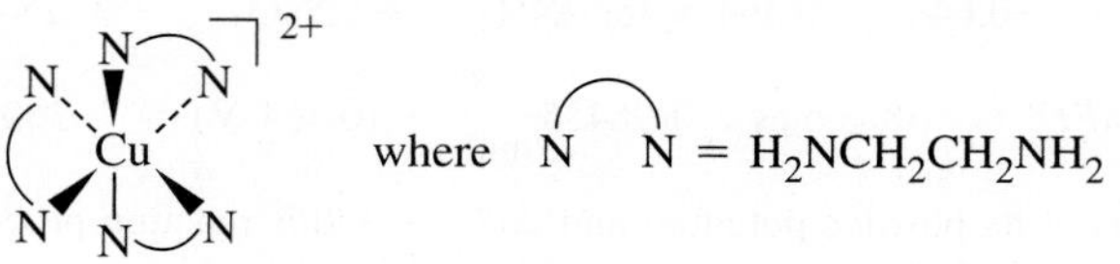

Δ isomer; Λ is also possible (Section 9.3.5)

e. Hexacarbonylmanganese(I)

f. Tetrachlororuthenate(1−)

9.3 Ma_2b_2cd has eight isomers, including two pairs of enantiomers, according to the method of Section 9.3.4.

M<aa><bb><cd> M<aa><bc><bd> M<ac><ad><bb> M<ab><ab><cd>

M<ab><ad><bc> M<ab><ac><bd>

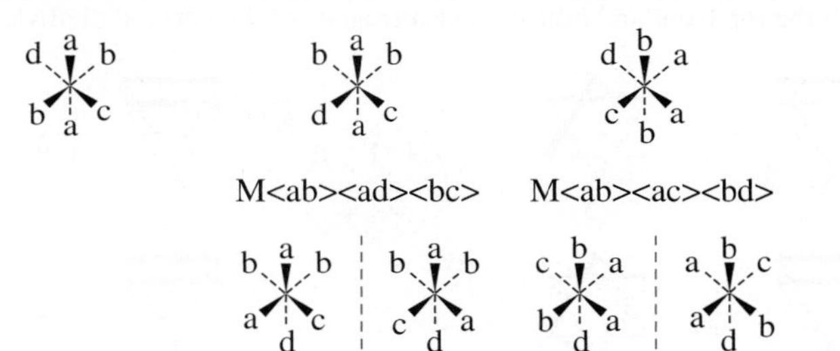

9.4 M(AA)bcde has six geometric isomers, each with enantiomers, for a total of 12 isomers.

M<Ab><Ac><de>	M<Ab><Ad><ce>	M<Ab><Ae><cd>
M<Ac><Ad><be>	M<Ac><Ae><bd>	M<Ad><Ae><bc>

9.5 This is a Λ configuration:

9.6 This can be analyzed more easily if it is flipped over, rotating about a horizontal axis in the plane of the paper. The result on the right can be checked for the two rings on the lower front as they relate to the ring across the back. One is Λ and one is Δ:

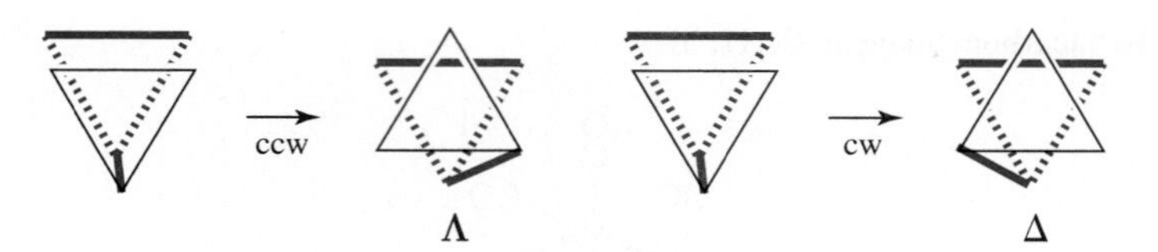

Rotating the complex to bring the other upper ring (top to back right) into the horizontal position, we find that one of the front rings is in the same plane, and the other is Δ. Overall, ΔΔΛ.

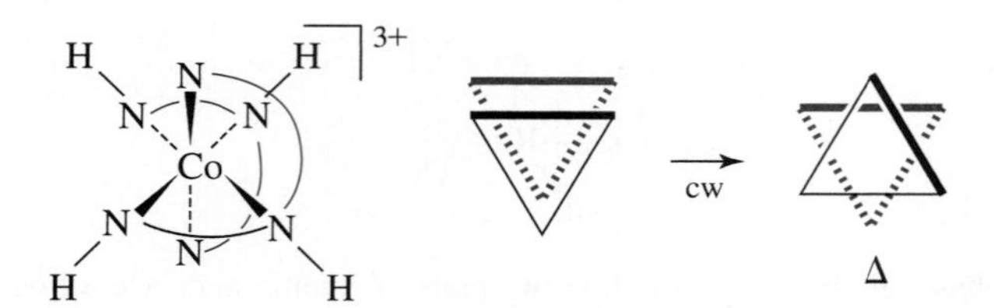

The other isomer is a meridional form that has two Λ and two Δ combinations. The first set has rings as originally shown. The second set is shown after a clockwise C_4 rotation about the axis, through the top front and bottom rear nitrogens of the vertical diethylenetriamine.

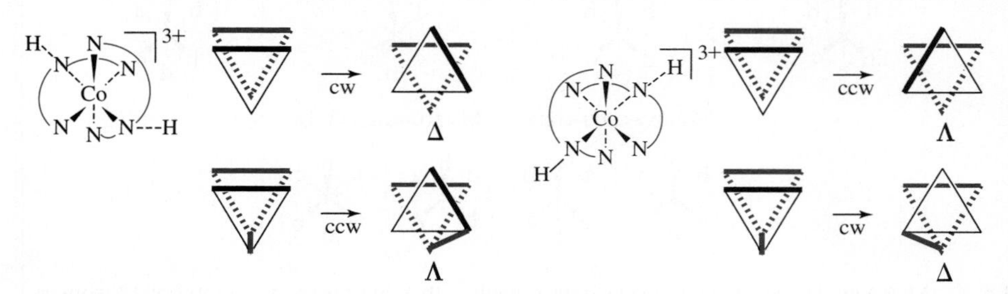

9.7 In the thiocyanate ligand, the harder, less polarizable terminal atom is nitrogen. This end of the ligand tends to interact more strongly with more polar, harder solvents, leading to stronger solvation. Consequently, formation of a bond between the metal and the less solvated sulfur at the opposite end of SCN^- is favored. The more polarizable, softer sulfur tends to interact more effectively with less polar solvents, thereby favoring bond formation between the metal and the nitrogen end of SCN^-.

Chapter 10

10.1 Nitrogen has three electrons in the $2p$ levels, with $m_l = -1, 0, +1$ and all with $m_s = +\frac{1}{2}$.

$$M_S = \frac{1}{2} + \frac{1}{2} + \frac{1}{2} = \frac{3}{2}, \quad M_L = -1 + 0 + 1 = 0, \quad \text{so} \quad S = \frac{3}{2} \text{ and } L = 0.$$

10.2 $S = n/2$, with n the number of unpaired electrons.

$$4\,S(S + 1) = 4(n/2)(n/2 + 1) = n^2 + 2n = n(n + 2), \text{ so } \sqrt{4S(S + 1)} = \sqrt{n(n + 2)}$$

10.3 Fe has the electron configuration $4s^2\,3d^6$, with four unpaired electrons. From the equations in Exercise 10.2,

$$\mu = \sqrt{4(4 + 2)} = 4.9\,\mu_B \quad \text{(Bohr magnetons)}$$

Fe^{2+} has the configuration $3d^6$ (the $4s$ electrons are lost first) with four unpaired electrons:

$$\mu = \sqrt{4(4 + 2)} = 4.9\,\mu_B$$

Cr has the electron configuration $4s^1\,3d^5$ with six unpaired electrons.

$$\mu = \sqrt{6(6 + 2)} = 6.9\ \mu_B$$

Cr^{3+} has the electron configuration $3d^3$ with three unpaired electrons.

$$\mu = \sqrt{3(3 + 2)} = 3.9\ \mu_B$$

Cu has the electron configuration $4s^1\,3d^{10}$ with one unpaired electron.

$$\mu = \sqrt{1(1 + 2)} = 1.7\ \mu_B$$

Cu^{2+} has the electron configuration $3d^9$ with one unpaired electron.

$$\mu = \sqrt{1(1 + 2)} = 1.7\ \mu_B$$

10.4 E: All six vectors remain unchanged; character = 6.

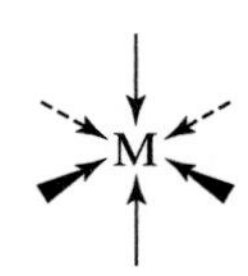

$8C_3$ and $6C_2$: All vectors are rotated into another position; character = 0.

$6C_4$: Rotation leaves only the vectors along the C_4 axis unchanged; character = 2.

$3C_2(=C_4^{\,2})$: Rotation leaves only the vectors along the C_2 axis unchanged; character = 2.

i: All vectors are switched to opposite sides; character = 0.

$6S_4$: Rotation moves four vectors, reflection switches the other two; character = 0.

$8S_6$: All vectors are moved; character = 0.

$3\sigma_h$: Reflection keeps all four vectors in the plane of reflection unchanged, switches the other two; character = 4.

$6\sigma_d$: This type of reflection has only two atoms in the reflection plane, and all others are moved; character = 2.

By the method of Section 4.4.2, the representation reduces to $A_{1g} + T_{1u} + E_g$. It can easily be verified—by summing the columns for these three representations, as provided in the chapter—that the total in each column matches the characters shown above.

10.5 The representations of the octahedral π orbitals can be found by using the x and z coordinates of the ligands in Figure 10.7. The characters for the symmetry operations are in the row labeled Γ_π in the table that precedes this exercise. In the C_2, C_4, and σ operations, some of the vectors do not move, but they are balanced by vectors whose direction is reversed. Only the E and $C_2 = C_4^{\,2}$ operations have nonzero totals, so they are the only ones used in finding the irreducible representations.

O_h	E	$8C_3$	$6C_2$	$6C_4$	$3C_2(=C_4^{\,2})$	i	$6S_4$	$8S_6$	$3\sigma_h$	$6\sigma_d$	
Γ_π	12	0	0	0	−4	0	0	0	0	0	
T_{1g}	3	0	−1	1	−1	3	1	0	−1	−1	
T_{2g}	3	0	1	−1	−1	3	−1	0	−1	1	(d_{xy}, d_{xz}, d_{yz})
T_{1u}	3	0	−1	1	−1	−3	−1	0	1	1	(p_x, p_y, p_z)
T_{2u}	3	0	1	−1	−1	−3	1	0	1	−1	

T_{1g}, T_{2g}, T_{1u}, and T_{2u} each total $= \frac{1}{48}[(12 \times 3) + 3(-4)(-1)] = 1$.

All other representations total 0.

10.6 A high-spin d^5 ion has three exchange possibilities in the t_{2g} level (1-2, 1-3, 2-3), and one exchange possibility in the e_g level (4-5), for a total exchange energy of $4\Pi_e$.

A low-spin d^5 ion has one unpaired electron. The set of three with the same spin has three exchange possibilities (1-2, 1-3, 2-3) and the set of two has one exchange possibility (4-5), for a total of four and a total exchange energy of $4\Pi_e$, the same as in the high spin case.

10.7 A low-spin d^6 ion has all six electrons in the lower t_{2g} levels, each at $-2/5\Delta_o$, so the total LFSE $= 6(-2/5\Delta_o) = -12/5\Delta_o$.

A high-spin d^6 ion has four electrons in the t_{2g} levels at $-2/5\Delta_o$ and two in the e_g levels at $3/5\Delta_o$. LFSE $= 4(-2/5\Delta_o) + 2(3/5\Delta_o) = -2/5\Delta_o$.

10.8 Using p_y orbitals of the four ligands, the reducible representation has four unchanged vectors for the E and σ_h operations and two for the C_2' and σ_v operations (the vectors along the C_2' axis and contained in the σ_v plane). All other operations result in changed positions and a character of 0. The p_x and p_z orbitals are similar, except that the p_x changes direction with the C_2' and σ_v operations, and the p_z changes direction with the C_2' and σ_v operations.

D_{4h}	E	$2C_4$	C_2	$2C_2'$	$2C_2''$	i	$2S_4$	σ_h	$2\sigma_v$	$2\sigma_d$	
Γ_{p_y}	4	0	0	2	0	0	0	4	2	0	σ
Γ_{p_x}	4	0	0	−2	0	0	0	4	−2	0	$\pi_{\parallel}$
Γ_{p_z}	4	0	0	−2	0	0	0	−4	2	0	$\pi_{\perp}$

Only the nonzero operations need to be used in finding the irreducible representations.

For Γ_{p_y}: A_{1g}, and B_{1g} each total $\frac{1}{16}[(4 \times 1) + 2(2)(1) + 1(4)(1) + 2(2)(1)] = 1$

E_u totals $\frac{1}{16}[(4 \times 2) + 2(2)(0) + 1(4)(2) + 2(2)(0)] = 1$

For Γ_{p_x}: A_{2g}, B_{2g} each total $\frac{1}{16}[(4 \times 1) + 2(-1)(-2) + 1(1)(4) + 2(-1)(-2)] = 1$

E_u totals $\frac{1}{16}[(4 \times 2) + 2(-2)(0) + 1(4)(2) + 2(-2)(0)] = 1$

For Γ_{p_z}: A_{2u}, B_{2u} each total $\frac{1}{16}[(4 \times 1) + 2(-1)(-2) + 1(-1)(-4) + 2(1)(2)] = 1$

E_g totals $\frac{1}{16}[(4 \times 2) + 2(-2)(0) + 1(-4)(-2) + 2(2)(0)] = 1$

All others total 0.

10.9 Energy changes:

d_{xy} total for 7, 8, 9, 10 $= 1.33e_\sigma$

d_{xz} total for 7, 8, 9, 10 $= 1.33e_\sigma$

d_{yz} total for 7, 8, 9, 10 $= 1.33e_\sigma$

d_{z^2} total for 7, 8, 9, 10 $= 0$ $\qquad \Delta_o = 3e_\sigma,\ \Delta_t = 1.33e_\sigma,\ \Delta_t = \frac{4}{9}\Delta_o$

$d_{x^2-y^2}$ total for 7, 8, 9, 10 $= 0$

Ligands are lowered by e_σ each.

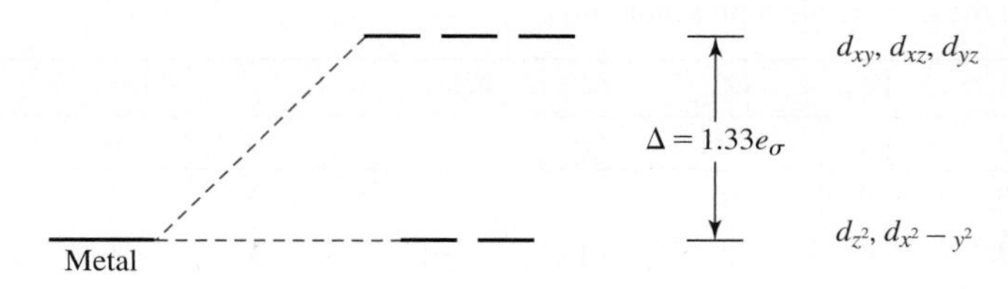

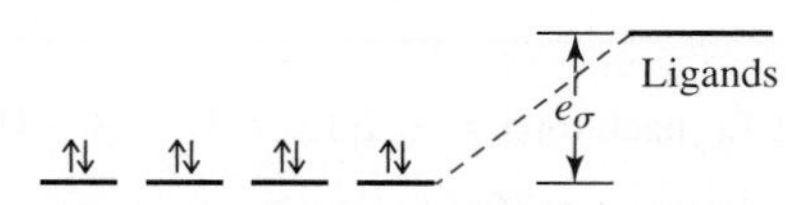

10.10 Adding π bonding to the results of Exercise 10.9 results in these changes in energy:

d_{xy}, d_{xz}, d_{yz} each total $0.89e_\pi$ for 7, 8, 9, 10

d_{z^2} and $d_{x^2-y^2}$ each total $2.67e_\pi$ for 7, 8, 9, 10

Ligands decrease by $2e_\pi$ each.

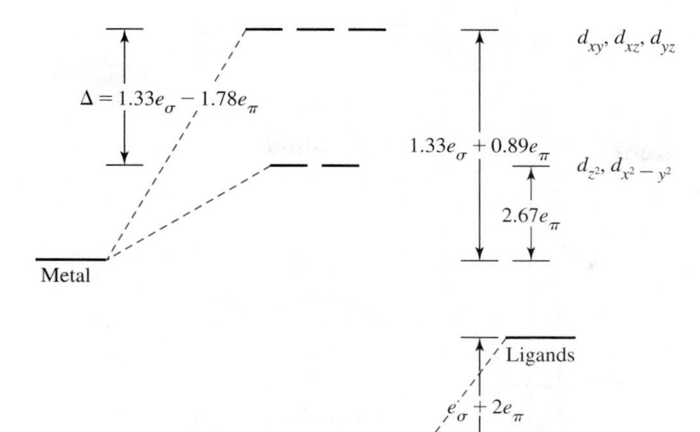

10.11 Square planar (positions 2, 3, 4, 5)

a. Energy changes for σ only:

d_{z^2} total for 2, 3, 4, 5 $= e_\sigma$

$d_{x^2-y^2}$ total for 2, 3, 4, 5 $= 3e_\sigma$

d_{xy}, d_{xz}, d_{yz} total for 2, 3, 4, 5 $= 0$

Ligands decrease by e_σ each.

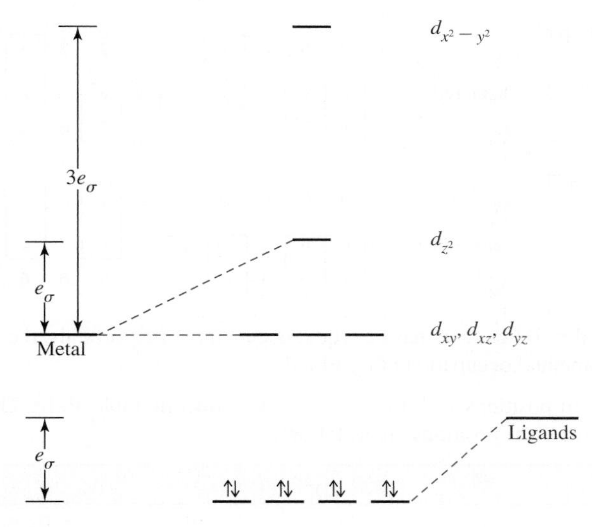

b. Adding π:

$d_{z^2}, d_{x^2-y^2}$ total for 2, 3, 4, 5 $= 0$

d_{xz}, d_{yz} total for 2, 3, 4, 5 $= 2e_\pi$

d_{xy} total for 2, 3, 4, 5 $= 4e_\pi$

Ligand π^* orbitals increase by $2e_\pi$ each.

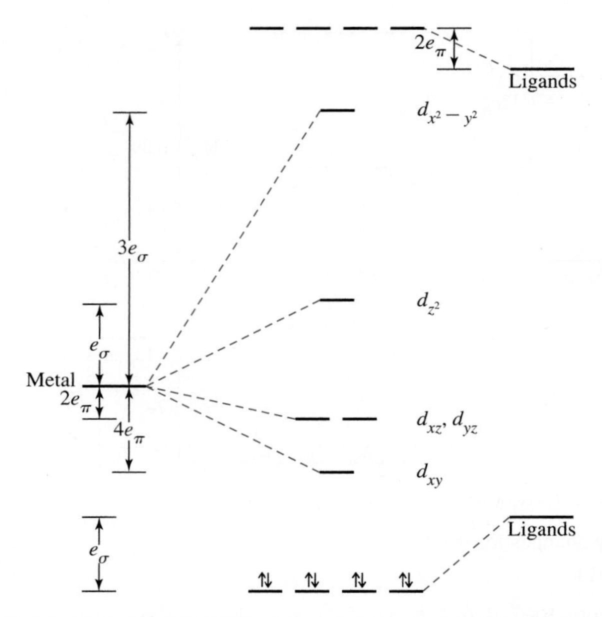

10.12 See Table 10.5 for the complete high-spin and low-spin configurations.

Number of electrons		1	2	3	4	5	6	7	8	9	10
High spin	e_g	0	0	0	1	2	2	2	2	3	4
	Jahn-Teller	W	W		S		W	W		S	
	t_{2g}	1	2	3	3	3	4	5	6	6	6
Low spin	e_g	0	0	0	0	0	0	1	2	3	4
	Jahn-Teller	W	W		W	W		S		S	
	t_{2g}	1	2	3	4	5	6	6	6	6	6

The weak Jahn–Teller cases have unequal occupation of t_{2g} orbitals; the strong Jahn–Teller cases have unequal occupation of e_g orbitals.

10.13 Ligands are in positions 1, 2, 6, 11, and 12 as shown in Table 10.10. The angular overlap parameters for these positions are as follows:

Position	z^2	$x^2 - y^2$	xy	xz	yz
1	1	0	0	0	0
2	$\frac{1}{4}$	$\frac{3}{4}$	0	0	0
6	1	0	0	0	0
11	$\frac{1}{4}$	$\frac{3}{16}$	$\frac{9}{16}$	0	0
12	$\frac{1}{4}$	$\frac{3}{16}$	$\frac{9}{16}$	0	0
Total	$2\frac{3}{4}$	$1\frac{1}{8}$	$1\frac{1}{8}$	0	0

The results are consistent with Figure 10.32. The highest energy molecular orbital involves the d_{z^2} orbital and has symmetry label A_1' in the D_{3h} point group; it is predicted by the angular overlap approach to have energy 2.75 e_σ. The degenerate pair involving the $d_{x^2-y^2}$ and d_{xy} orbitals (symmetry label E') has energy 1.125 e_σ, and the lowest energy degenerate pair (symmetry label E''), consisting of the d_{xz} and d_{yz} orbitals, does not interact with the ligand orbitals; its energy is 0 e_σ.

Chapter 11

11.1 Microstate table for d^2.

M_L \ M_S	-1	0	$+1$
$+4$		$2^+\ 2^-$	
$+3$	$2^-\ 1^-$	$2^+\ 1^-$ $2^-\ 1^+$	$2^+\ 1^+$
$+2$	$2^-\ 0^-$	$2^+\ 0^-$ $2^-\ 0^+$ $1^+\ 1^-$	$2^+\ 0^+$
$+1$	$2^-{-1}^-$ $1^-\ 0^-$	$2^+\ {-1}^-$ $2^-\ {-1}^+$ $1^+\ 0^-$ $1^-\ 0^+$	$2^+{-1}^+$ $1^+\ 0^+$
0	$-2^-\ 2^-$ $-1^-\ 1^-$	$-2^+\ 2^-$ $-1^+\ 1^-$ $0^+\ 0^-$ $-1^-\ 1^+$ $-2^-\ 2^+$	$-2^+\ 2^+$ $-1^+\ 1^+$
-1	$-1^-\ 0^-$ $-2^-\ 1^-$	$-1^+\ 0^-$ $-1^-\ 0^+$ $-2^-\ 1^+$ $-2^+\ 1^-$	$-1^+\ 0^+$ $-2^+\ 1^+$
-2	$-2^-\ 0^-$	$-1^+\ {-1}^-$ $-2^+\ 0^-$ $-2^-\ 0^+$	$-2^+\ 0^+$
-3	$-2^-\ {-1}^-$	$-2^+\ {-1}^-$ $-2^-\ {-1}^+$	$-2^+\ {-1}^+$
-4		$-2^+\ {-2}^-$	

11.2 ^2D $L = 2, S = \frac{1}{2}$ $\quad$ ^1P $L = 1, S = 0$ $\quad$ ^2S $L = 0, S = \frac{1}{2}$

^2D: $M_L = -2, -1, 0, 1, 2$; $M_S = -\frac{1}{2}, \frac{1}{2}$

^1P: $M_L = -1, 0, 1$; $M_S = 0$

^2S: $M_L = 0$; $M_S = -\frac{1}{2}, \frac{1}{2}$

M_L \ M_S	$-\frac{1}{2}$	$+\frac{1}{2}$
$+2$	x	x
$+1$	x	x
0	x	x
-1	x	x
-2	x	x

M_L \ M_S	0
$+1$	x
0	x
-1	x

M_L \ M_S	$-\frac{1}{2}$	$+\frac{1}{2}$
0	x	x

11.3 $L = 4, S = 0, J = 4$ $\quad$ ^1G

$L = 3, S = 1, J = 4, 3, 2$ $\quad$ ^3F

$L = 2, S = 0, J = 2$ $\quad$ ^1D

$L = 1, S = 1, J = 2, 1, 0$ $\quad$ ^3P

$L = 0, S = 0, J = 0$ $\quad$ ^1S

Following Hund's rules:

1. The highest spin (S) is 1, so the ground state is ^3F or ^3P.
2. The highest L in Step 1 is $L = 3$, so ^3F is the ground state.

11.4 The J values for each term are shown in the solution to Exercise 11.3. The full set of term symbols for a d^2 configuration is $^1G_4$, $^3F_4$, $^3F_3$, $^3F_2$, $^1D_2$, $^3P_2$, $^3P_1$, $^3P_0$, $^1S_0$. Hund's third rule predicts which J value corresponds to the lowest energy state. The d orbitals are less than half-filled, so the minimum J value for ^3F, $J = 2$, is the ground state. Overall, it is $^3F_2$.

11.5 High-spin d^6

1. ↑ ↑ (upper level); ↑↓ ↑ ↑ (lower level)
2. Spin multiplicity $= 4 + 1 = 5$, $S = 2$
3. Maximum possible value of $M_L = 2 + 2 + 1 + 0 - 1 - 2 = 2$, therefore, D term, $L = 2$.
4. ^5D

Low-spin d^6

1. — — (upper level); ↑↓ ↑↓ ↑↓ (lower level)
2. Spin multiplicity $= 0 + 1 = 1$, $S = 0$
3. Maximum value of $M_L = 2 + 2 + 1 + 1 + 0 + 0 = 6$; therefore, I term, $L = 6$.
4. ^1I

11.6.
a. The e_g level is asymmetrically occupied, so this is an E state.
b. The e_g and t_{2g} levels are symmetrically occupied, making this an A state.
c. The t_{2g} levels are asymmetrically occupied; this is a T state.

11.7 $[Fe(H_2O)_6]^{2+}$ is a high-spin d^6 complex. The weak field (left) part of the Tanabe–Sugano diagram for d^6 shows that the only excited state with the same spin multiplicity (5) as the ground state is the ^5E. The transition is therefore $^5T_2 \longrightarrow {}^5E$. The excited state $t_{2g}{}^3\, e_g{}^3$ is subject to Jahn–Teller distortion; consequently, as in the d^1 complex $[Ti(H_2O)_6]^{3+}$, the absorption band is split.

11.8 First assigning transitions, which are to the left of the crossover point of $^4A_2$ and $^4T_1$:

$^4T_1 \longrightarrow {}^4T_2$	not seen in this range ν_1	
$^4T_1 \longrightarrow {}^4A_2$	$16{,}000\ \text{cm}^{-1} = \nu_2$	
$^4T_1 \longrightarrow {}^4T_1$	$20{,}000\ \text{cm}^{-1} = \nu_3$	$\nu_3 / \nu_2 = 1.25$

From the Tanabe-Sugano diagram, $\nu_3/\nu_2 = 1.25$ at $\Delta/B = 10$; $\nu_3 = 25$ and $\nu_2 = 20$; and $\nu_3/\nu_2 = 1.25$ B and Δ can then be calculated:

$$\nu_2 = 20 \qquad B = E/\nu_2 = 16{,}000/20 = 800\ \text{cm}^{-1}$$

$$\nu_3 = 25 \qquad B = E/\nu_2 = 20{,}000/25 = 800\ \text{cm}^{-1}$$

$$\Delta/B = 10,\ \Delta = 10 \times 800 = 8{,}000\ \text{cm}^{-1}$$

11.9

$VO_4{}^{3-}$, vanadate	$CrO_4{}^{2-}$, chromate	$MnO_4{}^{-}$, permanganate
colorless	yellow	purple

As the charge on the nucleus increases, the vacant metal d orbitals are pulled to lower energies. The difference between the oxygen donor orbitals and the metal d orbitals grows smaller, and less energy is required for the LMCT transition. Permanganate, which absorbs yellow (the complement of purple) requires the least energy.

Chapter 12

12.1

$$ML_5X + Y \underset{k_{-1}}{\overset{k_1}{\rightleftharpoons}} ML_5XY$$

$$ML_5XY \xrightarrow{k_2} ML_5Y + X$$

Applying the stationary-state approach to ML_5XY,

$$\frac{d\,[ML_5XY]}{dt} = k_1\,[ML_5X][Y] - k_{-1}[ML_5XY] - k_2[ML_5XY] = 0$$

$$\text{and } [ML_5XY] = \frac{k_1[ML_5X][Y]}{k_{-1} + k_2}$$

From the second equation,

$$\frac{d\,[ML_5Y]}{dt} = k_2\,[ML_5XY]$$

Combining the two,

$$\frac{d[ML_5Y]}{dt} = \frac{k_1k_2[ML_5X][Y]}{k_{-1} + k_2} = k[ML_5X][Y]$$

12.2

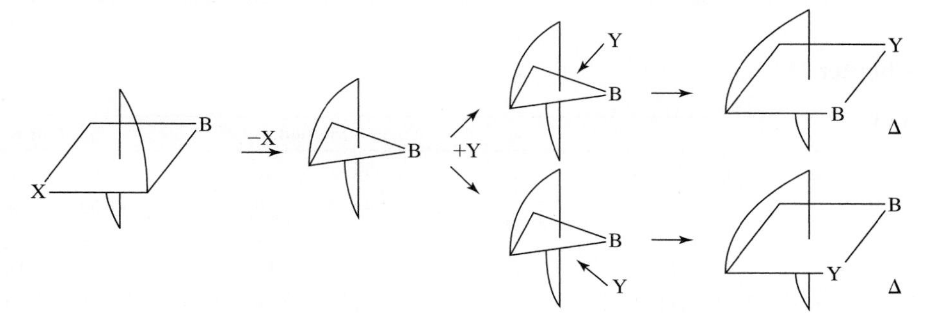

12.3 $[PtCl_4]^{2-} + NO_2^- \longrightarrow$ (a) $\qquad$ (a) $+ NH_3 \longrightarrow$ (b)

(a) $= [PtCl_3(NO_2)]^{2-}$ $\qquad$ (b) $=$ *trans*-$[PtCl_2(NO_2)(NH_3)]^-$

NO_2^- is a better *trans* director than Cl^-.

$[PtCl_3(NH_3)]^- + NO_2^- \longrightarrow$ (c) $\qquad$ (c) $+ NO_2^- \longrightarrow$ (d)

(c) $=$ *cis*-$[PtCl_2(NO_2)(NH_3)]^-$ $\qquad$ (d) $=$ *trans*-$[PtCl(NO_2)_2(NH_3)]^-$

Cl^- has a larger *trans* effect than NH_3, and NO_2^- has a larger *trans* effect than either Cl^- or NH_3. In the first step, Cl^- is a better leaving group than NH_3.

$[PtCl(NH_3)_3]^+ + NO_2^- \longrightarrow$ (e) $\qquad$ (e) $+ NO_2^- \longrightarrow$ (f)

(e) $=$ *trans*-$[PtCl(NO_2)(NH_3)_2]$ $\qquad$ (f) $=$ *trans*-$[Pt(NO_2)_2(NH_3)_2]$

Cl^- has a larger *trans* effect than NH_3, and NO_2^- has a larger *trans* effect than either Cl^- or NH_3.

$[PtCl_4]^{2-} + I^- \longrightarrow$ (g) $\qquad$ (g) $+ I^- \longrightarrow$ (h)

(g) $= [PtCl_3I]^{2-}$ $\qquad$ (h) $=$ *trans*-$[PtCl_2I_2]^{2-}$

I^- has a larger *trans* effect than Cl^-.

$[PtI_4]^{2-} + Cl^- \longrightarrow$ (i) $\qquad$ (i) $+ Cl^- \longrightarrow$ (j)

(i) $= [PtClI_3]^{2-}$ $\qquad$ (j) $=$ *cis*-$[PtCl_2I_2]^{2-}$

I^- has a larger *trans* effect than Cl^- and replacement of Cl^- in the second step would give no net change.

12.4 **a.** $[PtCl_4]^{2-} + 2\ NH_3 \longrightarrow$ *cis*-$PtCl_2(NH_3)_2 + 2\ Cl^-$

Because chloride is both a stronger *trans* director and a better leaving group, the major product is the *cis* isomer.

b. *cis*-$PtCl_2(NH_3)_2 + 2$ py $\longrightarrow$ *cis*-$[Ptpy_2(NH_3)_2]^{2+} + 2\ Cl^-$

As shown in (h) of Figure 12.13, the chloride is again a better leaving group, so the major product is the *cis* isomer.

c. *cis*-$[Ptpy_2(NH_3)_2]^{2+} + 2\ Cl^- \longrightarrow$ *trans*-$PtCl_2(NH_3)py$ + py + NH_3

The major product is the *trans* isomer, as shown in (e) and (f) of Figure 12.13. Chloride is the strongest *trans* director, so the *trans* isomer is formed regardless of whether py or NH_3 is the first ligand replaced.

d. *trans*-$PtCl_2(NH_3)py + NO_2^- \longrightarrow Pt < Cl(NO_2)> < (NH_3)py> + Cl^-$
(Cl^- is *trans* to NO_2^-; NH_3 is *trans* to py.)

Chloride is the strongest *trans* director (and the best leaving group), so one of them is replaced.

This sequence has been used to form the complex with four different groups. With some modification, it has been used with oxidation by Cl_2 and another substitution to form complexes with six different groups, with the geometry predictable to the last stage.

Chapter 13

13.1

		Method A		Method B	
a.	$[Fe(CO)_4]^{2-}$	Fe^{2-}	10	Fe	8
		4 CO	8	4 CO	8
				2–	2
			18		18
b.	$[(\eta^5\text{-}C_5H_5)_2\ Co]^+$	Co^{3+}	6	Co	9
		2 Cp^-	12	2 Cp	10
				1+	−1
			18		18
c.	$(\eta^3\text{-}C_5H_5)(\eta^5\text{-}C_5H_5)Fe(CO)$	Fe^{2+}	6	Fe	8
		η^3-Cp^-	4	η^3-Cp	3
		η^5-Cp^-	6	η^5-Cp	5
		CO	2	CO	2
			18		18
d.	$Co_2(CO)_8$	Co	9	Co	9
		4 CO	8	4 CO	8
		bridging CO	1	bridging CO	1
			18		18

13.2

		Method A		Method B	
a.	$[M(CO)_3PPh_3]^-$	3 CO	6	3 CO	6
		PPh_3	2	PPh_3	2
				1–	1
			8		9

Need 10 electrons for M^- or 9 electrons for M, so the metal is Co.

(continues)

		Method A		Method B	
b.	$HM(CO)_5$	5 CO	10	5 CO	10
		H^-	2	H	1
			12		11
	Need 6 electrons for M^+ or 7 electrons for M, so the metal is Mn.				
c.	$(\eta^4\text{-}C_8H_8)M(CO)_3$	3 CO	6	3 CO	6
		$\eta^4\text{-}C_8H_8$	4	$\eta^4\text{-}C_8H_8$	4
			10		10
	Need 8 electrons for M, so the metal is Fe.				
d.	$[(\eta^5\text{-}C_5H_5)M(CO)_3]_2$	3 CO	6	3 CO	6
		$\eta^5\text{-}C_5H_5{}^-$	6	$\eta^5\text{-}C_5H_5$	5
		M—M	1	M—M	1
			13		12
	Need 5 electrons for M^+ or 6 electrons for M, so the metal is Cr.				

13.3

	Method A		Method B	
$[Ni(CN)_4]^{2-}$	Ni(II)	8	Ni	10
	$4\ CN^-$	8	4 CN	4
			2–	2
		16		16
$PtCl_2en$	Pt(II)	8	Pt	10
	$2\ Cl^-$	4	2 Cl	2
	en	4	en	4
		16		16
$RhCl(PPh_3)_3$	Rh(I)	8	Rh	9
	Cl^-	2	Cl	1
	$3\ PPh_3$	6	$3\ PPh_3$	6
		16		16
$IrCl(CO)(PPh_3)_2$	Ir(I)	8	Ir	9
	Cl^-	2	Cl	1
	CO	2	CO	2
	$2\ PPh_3$	4	$2\ PPh_3$	4
		16		16

13.4 The N_2 sigma and pi levels are very close together in energy (see Chapter 5) and they are distributed evenly over both atoms. The CO levels are farther apart in energy and concentrated on C. Therefore, the geometric overlap for CO is better, and CO has better σ-donor and π-acceptor qualities than N_2.

13.5 The greater the negative charge on the complex, the greater the degree of pi acceptance by the CO ligands. This increases the population of the π^* orbitals of CO, weakening the carbon–oxygen bond. Consequently, $[V(CO)_6]^-$ has the longest C—O bond and $[Mn(CO)_6]^+$ has the shortest.

13.6 The two methods of electron counting are equivalent for these examples.

$M(CO)_4$	(M = Ni, Pd)	M	10
		4 CO	8
			18
$M(CO)_5$	(M = Fe, Ru, Os)	M	8
		5 CO	10
			18
$M(CO)_6$	(M = Cr, Mo, W)	M	6
		6 CO	12
			18
$Co_2(CO)_8$ (solution)		Co	9
(for each Co)		4 CO	8
		Co—Co	1
			18
$Co_2(CO)_8$ (solid)		Co	9
(for each Co)		3 CO	6
		2 μ_2-CO	2
		Co—Co	1
			18
$Fe_2(CO)_9$		Fe	8
(for each Fe)		3 CO	6
		3 μ_2-CO	3
		Fe—Fe	1
			18
$M_2(CO)_{10}$	(M = Mn, Te, Re)	M	7
(for each M)		5 CO	10
		M—M	1
			18
$Fe_3(CO)_{12}$	Fe on left	Fe	8
		4 CO	8
		2 Fe—Fe	2
			18
	Other Fe	Fe	8
		3 CO	6
		2 μ_2-CO	2
		2 Fe—Fe	2
			18
$M_3(CO)_{12}$	(Ru, Os)	M	8
(for each M)		4 CO	8
		2 M—M	2
			18
$M_4(CO)_{12}$	(M = Co, Rh), M on top	M	9
		3 CO	6
		3 M—M	3
			18

Other M	M	9
	2 CO	4
	2 μ_2-CO	2
	3 M—M	3
		18
$Ir_4(CO)_{12}$	Ir	9
(for each Ir)	3 CO	6
	3 Ir—Ir	3
		18

13.7 PMe_3 is a stronger σ donor and weaker π acceptor than CO. Therefore, the Mo in $Mo(PMe_3)_5H_2$ has a greater concentration of electrons and a greater tendency to back-bond to the hydrogens by donating to the σ^* orbital of H_2. This donation is strong enough to rupture the H—H bond, converting H_2 into two hydride ligands.

13.8

	Method A		Method B	
a. $(\eta^5\text{-}C_5H_5)(cis\text{-}\eta^4\text{-}C_4H_6)M(PMe_3)_2(H)$	$\eta^5\text{-}C_5H_5^-$	6	$\eta^5\text{-}C_5H_5$	5
	$\eta^4\text{-}C_4H_6$	4	$\eta^4\text{-}C_4H_6$	4
	2 PMe_3	4	2 PMe_3	4
	H^-	2	H	1
		16		14
M^{2+} needs 2 electrons, M needs 4. Zr fits.				
b. $(\eta^5\text{-}C_5H_5)M(C_2H_4)_2$	$\eta^5\text{-}C_5H_5^-$	6	$\eta^5\text{-}C_5H_5$	5
	2 C_2H_4	4	2 C_2H_4	4
		10		9
M^+ needs 8 electrons, M needs 9; Co fits.				

13.9 2-Node group orbitals.

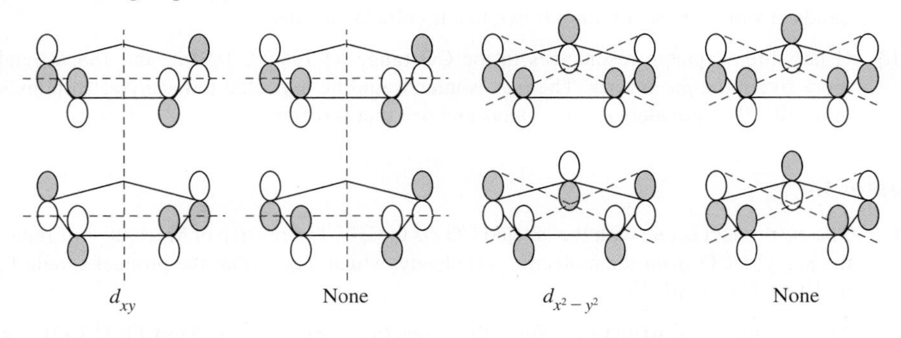

1-Node group orbitals.

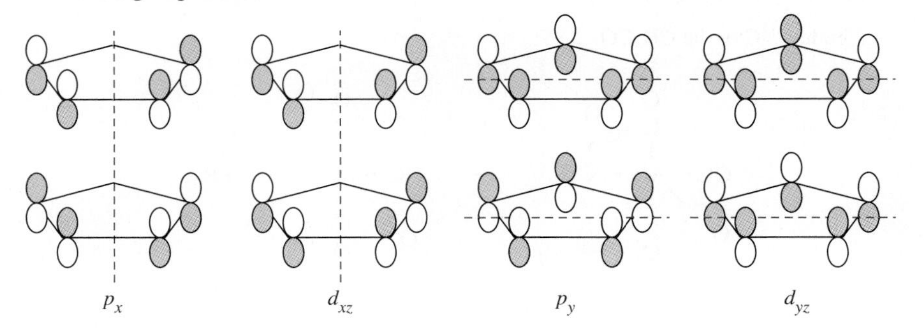

13.10

		Method A		Method B	
a.	$[((CH_3)_3CCH)((CH_3)_3CCH_2)(CCH_3)(C_2H_4(P(CH_3)_2)_2)W]$				
		W^+	5	W	6
		$(CH_3)_3)CCH$	2	$(CH_3)_3CCH$	2
		$(CH_3)_3CCH_2^-$	2	$(CH_3)_3CCH_2$	1
		CCH_3	3	CCH_3	3
		$C_2H_4(P(CH_3)_2)_2$	4	$C_2H_4(P(CH_3)_2)_2$	4
			16		16
b.	$Ta(Cp)_2(CH_3)(CH_2)$	Ta^{3+}	2	Ta	5
		$2\ Cp^-$	12	2 Cp	10
		CH_3^-	2	CH_3	1
		CH_2	2	CH_2	2
			18		18

13.11 $Cr(\eta^6\text{-}C_6H_6)_2$ — $\eta^6\text{-}C_6H_6 = L_3$ so $Cr(\eta^6\text{-}C_6H_6)_2 = ML_6$

$[Mo(CO)_3(\eta^5\text{-}C_5H_5)]^-$ — $\eta^5\text{-}C_5H_5 = L_2X$. The CBC formula would be $[ML_5X]^-$ as an ion. However, to determine the equivalent neutral class formula, the extra electron is counted as converting an X^- ligand to an L ligand; this gives the formula ML_6.

$WH_2(\eta^5\text{-}C_5H_5)_2$ — H is an X ligand, and $\eta^5\text{-}C_5H_5$ is an L_2X ligand. Therefore the CBC formula is ML_4X_4.

$[FeCl_4]^{2-}$ — Cl is an X ligand. As an ion, $[FeCl_4]^{2-}$ would be $[MX_4]^{2-}$. Each added negative charge is considered to convert an X^- ligand to an L, giving the equivalent neutral class formula ML_2X_2.

13.12 With just two bands, this is more likely the *fac* isomer. The *mer* isomer should show three bands; it would show two only if two bands coincide in energy.

13.13 II has three separate resonances in the CO range (at 194.98, 189.92, and 188.98) and is more likely the *fac* isomer. The *mer* isomer would be expected to have two carbonyls of magnetically equivalent environments and one that is different.

Chapter 14

14.1 The *cis* product is one with the labeled CO *cis* to CH_3. The reverse of Mechanism 1 removes the acetyl ^{13}CO from the molecule completely, which means that the product should have no ^{13}CO label at all.

14.2 The product distribution for the reaction of *cis*-$CH_3Mn(CO)_4(^{13}CO)$ with PR_3 ($R = C_2H_5$, $*C = {}^{13}C$):

25% has ^{13}C in the CH_3CO.

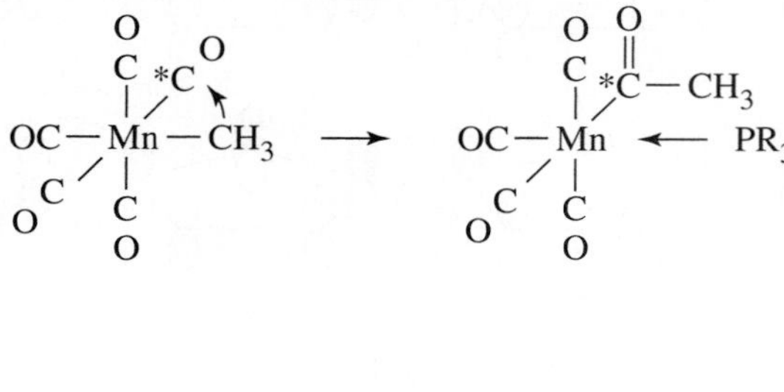

$(OC)_4Mn(*CO)(CH_3) \longrightarrow (OC)_4Mn(*C(=O)\text{—}CH_3) \longleftarrow PR_3$

25% has ^{13}CO *trans* to the CH_3CO.

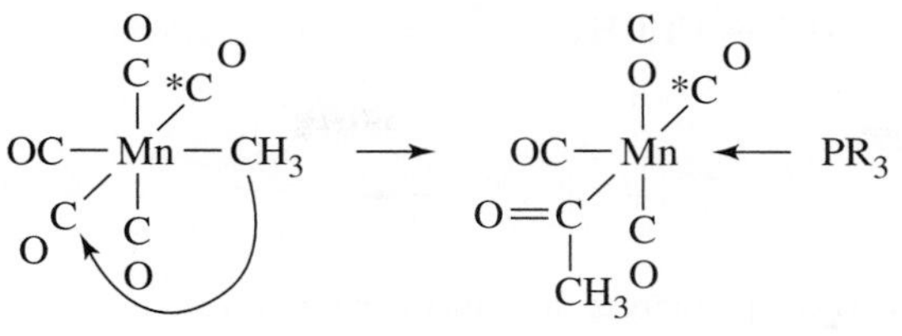

50% has ^{13}CO *cis* to the CH_3CO.

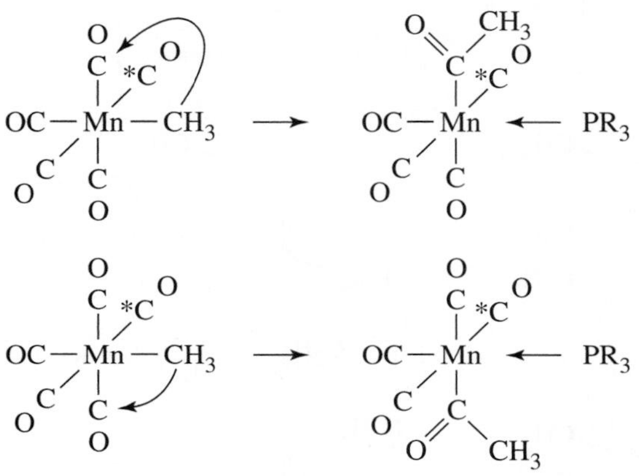

All the products have PEt_3 *cis* to the CH_3CO.

14.3 The reverse of the reaction has a pi-bonded ethylene and a hydride bonded to Rh rearranging, with Rh going to carbon 1 of the ethylene and the hydrogen going to carbon 2 of the ethylene, a 1,2 insertion of Rh and H into the double bond.

14.4 The hydroformylation process for the preparation of $(CH_3)_3CH{-}CH_2{-}CHO$ from $(CH_3)_2C{=}CH_2$ is exactly that of Figure 14.18, with $R = CH_3$.

14.5
1. Ligand dissociation
2. Coordination of olefin
3. 1,2 insertion
4. Ligand coordination
5. Alkyl migration (carbonyl insertion)
6. Oxidative addition
7. Reductive elimination

14.6. **a.** Metathesis between two molecules of propene, $H_2C{=}CHCH_3$:

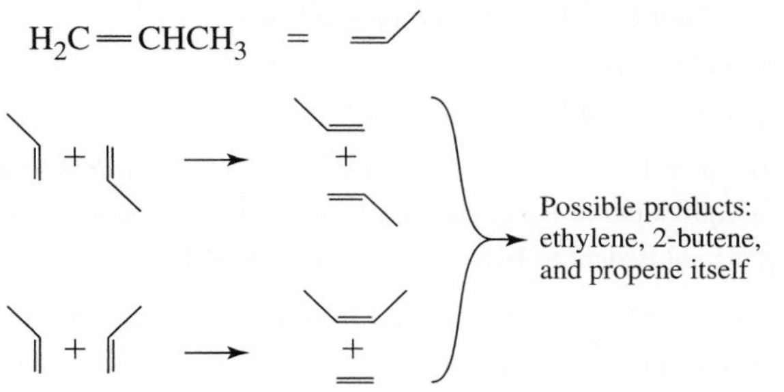

$H_2C{=}CHCH_3$ =

Possible products: ethylene, 2-butene, and propene itself

b. Metathesis between propene and cyclopentene:

propene = $H_2C{=}CHCH_3$ = [structure] cyclopentene = [structure]

[propene] + [cyclopentene] → [product] = 1,6-octadiene

1,6-octadiene can participate in further metathesis reactions. For example,

14.7

CH_3, C_2H_5, H_3C, C_2H_5 → CH_3, H_3C = CH_3, CH_3 + C_2H_5, C_2H_5

C_2H_5, CH_3, CH_3, C_2H_5 → C_2H_5, C_2H_5 = C_2H_5, C_2H_5 + H_3C, CH_3

Chapter 15

15.1 There are many possible answers. Examples include the following:

a. $Re(CO)_4$	$(\eta^5\text{-}C_5H_5)Fe(CO)$	
b. $Pt(CO)_3$	$[(\eta^5\text{-}C_5H_5)Co]^{2-}$	
c. $Re(CO)_5$	$[(\eta^5\text{-}C_5H_5)Mn(CO)_2]^-$	$(\eta^6\text{-}C_6H_6)Mn(CO)_2$

15.2
a. This is a 15-electron species with three vacant positions, isolobal with CH.
b. This is a 16-electron species with one vacant position, isolobal with CH_3^+.
c. This is a 15-electron species with three vacant positions, isolobal with CH.

15.3
a. $B_{11}H_{13}^{2-}$ is derived from $B_{11}H_{11}^{4-}$, a *nido* species.
b. $B_5H_8^-$ is derived from $B_5H_5^{4-}$, a *nido* species.
c. $B_7H_7^{2-}$ is a *closo* species.
d. $B_{10}H_{18}$ is derived from $B_{10}H_{10}^{8-}$, a *hypho* species.

15.4
a. $C_3B_3H_7$ is equivalent to B_6H_{10}, derived from $B_6H_6^{4-}$, a *nido* species.
b. $C_2B_5H_7$ is equivalent to B_7H_9, derived from $B_7H_7^{2-}$, a *closo* species.
c. $C_2B_7H_{12}^-$ is equivalent to $B_9H_{14}^-$, derived from $B_9H_9^{6-}$, an *arachno* species.

15.5
a. SB_9H_9 is equivalent to $B_{10}H_{12}$, derived from $B_{10}H_{10}^{2-}$, a *closo* species.
b. $GeC_2B_9H_{11}$ is equivalent to $B_{12}H_{14}$, derived from $B_{12}H_{12}^{2-}$, a *closo* species.
c. $SB_9H_{12}^-$ is equivalent to $B_{10}H_{15}^-$, derived from $B_{10}H_{10}^{6-}$, an *arachno* species.

15.6 **a.** $C_2B_7H_9(CoCp)_3$ is equivalent to $B_9H_{11}(CoCp)_3$ or $B_{12}H_{14}$, derived from $B_{12}H_{12}^{2-}$, a *closo* species.

b. $C_2B_4H_6Ni(PPh_3)_2$ is equivalent to $B_6H_8Ni(PPh_3)_2$ or B_7H_9, derived from $B_7H_7^{2-}$, a *closo* species.

15.7 **a.** Ge_9^{2-} The total valence electron count is 38, four for each Ge plus two for the charge; $n = 9$, so the electron count is $4n + 2$, and the structure is *closo*.

b. Bi_5^{3+} The valence electron count is 22, five for each Bi minus three for the charge. Because $n = 5$, the total count again is $4n + 2$, and the structure is *closo*.

15.8 **a.** $(\eta^5\text{-}C_5H_5)(\eta^5\text{-}C_2B_9H_{11})Fe$

$m = 2$, two linked polyhedra
$n = 17$; all Fe, B, and C atoms are vertices
$o = 1$; one atom, Fe, bridges the polyhedra
$p = 1$; the top polyhedron is not complete
$m + n + o + p = 21$ electron pairs

b. *nido*−7,8−$C_2B_9H_{11}^{2-}$

$m = 1$ single polyhedron
$n = 11$; each B and C atom is a vertex
$o = 0$; no bridges between polyhedra
$p = 1$; *nido* classification
$m + n + o + p = 13$ electron pairs

15.9 **a.** $[Re_7C(CO)_{21}]^{3-}$

7 Re	49
C	4
21 Co	42
3−	3
Total	98

A 98-electron, seven-metal cluster is two electrons short of a *closo* configuration. The structure would be expected to be a capped *closo*, such as the 98-electron species $[Rh_7(CO)_{16}]^{3-}$ and $Os_7(CO)_{21}$ in Table 15.13.

b. $[Fe_4N(CO)_{12}]^-$

4 Fe	32
N	5
12 CO	24
1−	1
Total	62

A 62-electron, four-metal cluster is classified as *arachno*.

Appendix C

Character Tables

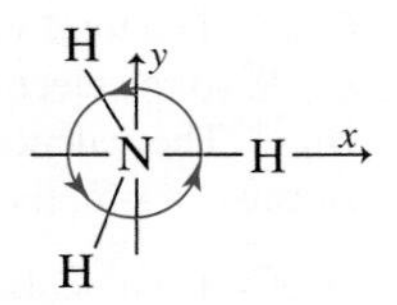

NOTE: $i = \sqrt{-1}$; $\varepsilon^* = \varepsilon$ with $-i$ substituted for i.

C_{3v}	E	$2C_3$	$3\sigma_v$	
A_1	1	1	1	z
A_2	1	1	−1	R_z
E	2	−1	0	$(x, y), (R_x, R_y)$

1. Groups of Low Symmetry

C_1	E
A	1

C_s	E	σ_h		
A'	1	1	x, y, R_z	x^2, y^2, z^2, xy
A''	1	−1	z, R_x, R_y	yz, xz

C_i	E	i		
A_g	1	1	R_x, R_y, R_z	$x^2, y^2, z^2, xy, xz, yz$
A_u	1	−1	x, y, z	

2. C_n, C_{nv}, and C_{nh} Groups

C_n Groups

C_2	E	C_2		
A	1	1	z, R_z	x^2, y^2, z^2, xy
B	1	−1	x, y, R_x, R_y	yz, xz

C_3	E	C_3	C_3^2		
A	1	1	1	z, R_z	$x^2 + y^2, z^2$
E	$\begin{Bmatrix} 1 \\ 1 \end{Bmatrix}$	$\begin{matrix} \varepsilon \\ \varepsilon^* \end{matrix}$	$\begin{matrix} \varepsilon^* \\ \varepsilon \end{matrix}$	$(x, y), (R_x, R_y)$	$(x^2 - y^2, xy), (yz, xz)$

$\varepsilon = e^{(2\pi i)/3}$

C_4	E	C_4	C_2	C_4^3		
A	1	1	1	1	z, R_z	$x^2 + y^2, z^2$
B	1	−1	1	−1		$x^2 - y^2, xy$
E	$\begin{Bmatrix} 1 \\ 1 \end{Bmatrix}$	$\begin{matrix} i \\ -i \end{matrix}$	$\begin{matrix} -1 \\ -1 \end{matrix}$	$\begin{matrix} -i \\ i \end{matrix}$	$(x, y), (R_x, R_y)$	(yz, xz)

C_5	E	C_5	C_5^2	C_5^3	C_5^4		
A	1	1	1	1	1	z, R_z	$x^2 + y^2, z^2$
E_1	1	ε	ε^2	ε^{2*}	ε^*	$(x, y), (R_x, R_y)$	(yz, xz)
	1	ε^*	ε^{2*}	ε^2	ε		
E_2	1	ε^2	ε^*	ε	ε^{2*}		$(x^2 - y^2, xy)$
	1	ε^{2*}	ε	ε^*	ε^2		

$\varepsilon = e^{(2\pi i)/5}$

C_6	E	C_6	C_3	C_2	C_3^2	C_6^5		
A	1	1	1	1	1	1	z, R_z	$x^2 + y^2, z^2$
B	1	-1	1	-1	1	-1		
E_1	1	ε	$-\varepsilon^*$	-1	$-\varepsilon$	ε^*	$(x, y), (R_x, R_y)$	(yz, xz)
	1	ε^*	$-\varepsilon$	-1	$-\varepsilon^*$	ε		
E_2	1	$-\varepsilon^*$	$-\varepsilon$	1	$-\varepsilon^*$	$-\varepsilon$		$(x^2 - y^2, xy)$
	1	$-\varepsilon$	$-\varepsilon^*$	1	$-\varepsilon$	$-\varepsilon^*$		

$\varepsilon = e^{(\pi i)/3}$

C_7	E	C_7	C_7^2	C_7^3	C_7^4	C_7^5	C_7^6		
A	1	1	1	1	1	1	1	z, R_z	$x^2 + y^2, z^2$
E_1	1	ε	ε^2	ε^3	ε^{3*}	ε^{2*}	ε^*	$(x, y),$ (R_x, R_y)	(yz, xz)
	1	ε^*	ε^{2*}	ε^{3*}	ε^3	ε^2	ε		
E_2	1	ε^2	ε^{3*}	ε^*	ε	ε^3	ε^{2*}		$(x^2 - y^2, xy)$
	1	ε^{2*}	ε^3	ε	ε^*	ε^{3*}	ε^2		
E_3	1	ε^3	ε^*	ε^2	ε^{2*}	ε	ε^{3*}		
	1	ε^{3*}	ε	ε^{2*}	ε^2	ε^*	ε^3		

$\varepsilon = e^{(2\pi i)/7}$

C_8	E	C_8	C_4	C_2	C_4^3	C_8^3	C_8^5	C_8^7		
A	1	1	1	1	1	1	1	1	z, R_z	$x^2 + y^2, z^2$
B	1	-1	1	1	1	-1	-1	-1		
E_1	1	ε	i	-1	$-i$	$-\varepsilon^*$	$-\varepsilon$	ε^*	$(x, y),$ (R_x, R_y)	(yz, xz)
	1	ε^*	$-i$	-1	i	$-\varepsilon$	$-\varepsilon^*$	ε		
E_2	1	i	-1	1	-1	$-i$	i	$-i$		$(x^2 - y^2, xy)$
	1	$-i$	-1	1	-1	i	$-i$	i		
E_3	1	$-\varepsilon$	i	-1	$-i$	ε^*	ε	$-\varepsilon^*$		
	1	$-\varepsilon^*$	$-i$	-1	i	ε	ε^*	$-\varepsilon$		

$\varepsilon = e^{(\pi i)/4}$

C_{nv} Groups

C_{2v}	E	C_2	$\sigma_v(xz)$	$\sigma_v'(yz)$		
A_1	1	1	1	1	z	x^2, y^2, z^2
A_2	1	1	-1	-1	R_z	xy
B_1	1	-1	1	-1	x, R_y	xz
B_2	1	-1	-1	1	y, R_x	yz

C_{3v}	E	$2C_3$	$3\sigma_v$		
A_1	1	1	1	z	$x^2 + y^2, z^2$
A_2	1	1	-1	R_z	
E	2	-1	0	$(x, y), (R_x, R_y)$	$(x^2 - y^2, xy), (xz, yz)$

C_{4v}	E	$2C_4$	C_2	$2\sigma_v$	$2\sigma_d$		
A_1	1	1	1	1	1	z	$x^2 + y^2, z^2$
A_2	1	1	1	−1	−1	R_z	
B_1	1	−1	1	1	−1		$x^2 - y^2$
B_2	1	−1	1	−1	1		xy
E	2	0	−2	0	0	$(x, y), (R_x, R_y)$	(xz, yz)

C_{5v}	E	$2C_5$	$2C_5^2$	$5\sigma_v$		
A_1	1	1	1	1	z	$x^2 + y^2, z^2$
A_2	1	1	1	−1	R_z	
E_1	2	2 cos 72°	2 cos 144°	0	$(x, y), (R_x, R_y)$	(xz, yz)
E_2	2	2 cos 144°	2 cos 72°	0		$(x^2 - y^2, xy)$

C_{6v}	E	$2C_6$	$2C_3$	C_2	$3\sigma_v$	$3\sigma_d$		
A_1	1	1	1	1	1	1	z	$x^2 + y^2, z^2$
A_2	1	1	1	1	−1	−1	R_z	
B_1	1	−1	1	−1	1	−1		
B_2	1	−1	1	−1	−1	1		
E_1	2	1	−1	−2	0	0	$(x, y), (R_x, R_y)$	(xz, yz)
E_2	2	−1	−1	2	0	0		$(x^2 - y^2, xy)$

C_{nh} Groups

C_{2h}	E	C_2	i	σ_h		
A_g	1	1	1	1	R_z	x^2, y^2, z^2, xy
B_g	1	−1	1	−1	R_x, R_y	xz, yz
A_u	1	1	−1	−1	z	
B_u	1	−1	−1	1	x, y	

C_{3h}	E	C_3	C_3^2	σ_h	S_3	S_3^5		
A'	1	1	1	1	1	1	R_z	$x^2 + y^2, z^2$
E'	$\begin{Bmatrix} 1 \\ 1 \end{Bmatrix}$	ε ε^*	ε^* ε	1 1	ε ε^*	ε^* ε	(x, y)	$(x^2 - y^2, xy)$
A''	1	1	1	−1	−1	−1	z	
E''	$\begin{Bmatrix} 1 \\ 1 \end{Bmatrix}$	ε ε^*	ε^* ε	−1 −1	$-\varepsilon$ $-\varepsilon^*$	$-\varepsilon^*$ $-\varepsilon$	(R_x, R_y)	(xz, yz)

$\varepsilon = e^{(2\pi i)/3}$

C_{4h}	E	C_4	C_2	C_4^3	i	S_4^3	σ_h	S_4		
A_g	1	1	1	1	1	1	1	1	R_z	$x^2 + y^2, z^2$
B_g	1	−1	1	−1	1	−1	1	−1		$x^2 - y^2, xy$
E_g	$\begin{Bmatrix} 1 \\ 1 \end{Bmatrix}$	i $-i$	−1 −1	$-i$ i	1 1	i $-i$	−1 −1	$-i$ i	(R_x, R_y)	(xz, yz)
A_u	1	1	1	1	−1	−1	−1	−1	z	
B_u	1	−1	1	−1	−1	1	−1	1		
E_u	$\begin{Bmatrix} 1 \\ 1 \end{Bmatrix}$	i $-i$	−1 −1	$-i$ i	−1 −1	$-i$ i	1 1	i $-i$	(x, y)	

C_{5h}	E	C_5	C_5^2	C_5^3	C_5^4	σ_h	S_5	S_5^7	S_5^3	S_5^9		
A'	1	1	1	1	1	1	1	1	1	1	R_z	$x^2 + y^2, z^2$
E_1'	1	ε	ε^2	ε^{2*}	ε^*	1	ε	ε^2	ε^{2*}	ε^*	(x, y)	
	1	ε^*	ε^{2*}	ε^2	ε	1	ε^*	ε^{2*}	ε^2	ε		
E_2'	1	ε^2	ε^*	ε	ε^{2*}	1	ε^2	ε^*	ε	ε^{2*}		$(x^2 - y^2, xy)$
	1	ε^{2*}	ε	ε^*	ε^2	1	ε^{2*}	ε	ε^*	ε^2		
A''	1	1	1	1	1	−1	−1	−1	−1	−1	z	
E_1''	1	ε	ε^2	ε^{2*}	ε^*	−1	$-\varepsilon$	$-\varepsilon^2$	$-\varepsilon^{2*}$	$-\varepsilon^*$	(R_x, R_y)	(xz, yz)
	1	ε^*	ε^{2*}	ε^2	ε	−1	$-\varepsilon^*$	$-\varepsilon^{2*}$	$-\varepsilon^2$	$-\varepsilon$		
E_2''	1	ε^2	ε^*	ε	ε^{2*}	−1	$-\varepsilon^2$	$-\varepsilon^*$	$-\varepsilon$	$-\varepsilon^{2*}$		
	1	ε^{2*}	ε	ε^*	ε^2	−1	$-\varepsilon^{2*}$	$-\varepsilon$	$-\varepsilon^*$	$-\varepsilon^2$		

$\varepsilon = e^{(2\pi i)/5}$

C_{6h}	E	C_6	C_3	C_2	C_3^2	C_6^5	i	S_3^5	S_6^5	σ_h	S_6	S_3		
A_g	1	1	1	1	1	1	1	1	1	1	1	1	R_z	$x^2 + y^2, z^2$
B_g	1	−1	1	−1	1	−1	1	−1	1	−1	1	−1		
E_{1g}	1	ε	$-\varepsilon^*$	−1	$-\varepsilon$	ε^*	1	ε	$-\varepsilon^*$	−1	$-\varepsilon$	ε^*	(R_x, R_y)	(xz, yz)
	1	ε^*	$-\varepsilon$	−1	$-\varepsilon^*$	ε	1	ε^*	$-\varepsilon$	−1	$-\varepsilon^*$	ε		
E_{2g}	1	$-\varepsilon^*$	$-\varepsilon$	1	$-\varepsilon^*$	$-\varepsilon$	1	$-\varepsilon^*$	$-\varepsilon$	1	$-\varepsilon^*$	$-\varepsilon$		$(x^2 - y^2, xy)$
	1	$-\varepsilon$	$-\varepsilon^*$	1	$-\varepsilon$	$-\varepsilon^*$	1	$-\varepsilon$	$-\varepsilon^*$	1	$-\varepsilon$	$-\varepsilon^*$		
A_u	1	1	1	1	1	1	−1	−1	−1	−1	−1	−1	z	
B_u	1	−1	1	−1	1	−1	−1	1	−1	1	−1	1		
E_{1u}	1	ε	$-\varepsilon^*$	−1	$-\varepsilon$	ε^*	−1	$-\varepsilon$	ε^*	1	ε	$-\varepsilon^*$	(x, y)	
	1	ε^*	$-\varepsilon$	−1	$-\varepsilon^*$	ε	−1	$-\varepsilon^*$	ε	1	ε^*	$-\varepsilon$		
E_{2u}	1	$-\varepsilon^*$	$-\varepsilon$	1	$-\varepsilon^*$	$-\varepsilon$	−1	ε^*	ε	−1	ε^*	ε		
	1	$-\varepsilon$	$-\varepsilon^*$	1	$-\varepsilon$	$-\varepsilon^*$	−1	ε	ε^*	−1	ε	ε^*		

$\varepsilon = e^{(\pi i)/3}$

3. D_n, D_{nd}, and D_{nh} Groups

D_n Groups

D_2	E	$C_2(z)$	$C_2(y)$	$C_2(x)$		
A	1	1	1	1		x^2, y^2, z^2
B_1	1	1	−1	−1	z, R_z	xy
B_2	1	−1	1	−1	y, R_y	xz
B_3	1	−1	−1	1	x, R_x	yz

D_3	E	$2C_3$	$3C_2$		
A_1	1	1	1		$x^2 + y^2, z^2$
A_2	1	1	−1	z, R_z	
E	2	−1	0	$(x, y), (R_x, R_y)$	$(x^2 - y^2, xy), (xz, yz)$

D_4	E	$2C_4$	$C_2(=C_4^2)$	$2C_2'$	$2C_2''$		
A_1	1	1	1	1	1		$x^2 + y^2, z^2$
A_2	1	1	1	−1	−1	z, R_z	
B_1	1	−1	1	1	−1		$x^2 - y^2$
B_2	1	−1	1	−1	1		xy
E	2	0	−2	0	0	$(x, y), (R_x, R_y)$	(xz, yz)

D_5	E	$2C_5$	$2C_5^2$	$5C_2$		
A_1	1	1	1	1		$x^2 + y^2, z^2$
A_2	1	1	1	−1	z, R_z	
E_1	2	2 cos 72°	2 cos 144°	0	$(x, y), (R_x, R_y)$	(xz, yz)
E_2	2	2 cos 144°	2 cos 72°	0		$(x^2 - y^2, xy)$

D_6	E	$2C_6$	$2C_3$	C_2	$3C_2'$	$3C_2''$		
A_1	1	1	1	1	1	1		$x^2 + y^2, z^2$
A_2	1	1	1	1	−1	−1	z, R_z	
B_1	1	−1	1	−1	1	−1		
B_2	1	−1	1	−1	−1	1		
E_1	2	1	−1	−2	0	0	$(x, y), (R_x, R_y)$	(xz, yz)
E_2	2	−1	−1	2	0	0		$(x^2 - y^2, xy)$

D_{nd} Groups

D_{2d}	E	$2S_4$	C_2	$2C_2'$	$2\sigma_d$		
A_1	1	1	1	1	1		$x^2 + y^2, z^2$
A_2	1	1	1	−1	−1	R_z	
B_1	1	−1	1	1	−1		$x^2 - y^2$
B_2	1	−1	1	−1	1	z	xy
E	2	0	−2	0	0	$(x, y), (R_x, R_y)$	(xz, yz)

D_{3d}	E	$2C_3$	$3C_2$	i	$2S_6$	$3\sigma_d$		
A_{1g}	1	1	1	1	1	1		$x^2 + y^2, z^2$
A_{2g}	1	1	−1	1	1	−1	R_z	
E_g	2	−1	0	2	−1	0	(R_x, R_y)	$(x^2 - y^2, xy)$ (xz, yz)
A_{1u}	1	1	1	−1	−1	−1		
A_{2u}	1	1	−1	−1	−1	1	z	
E_u	2	−1	0	−2	1	0	(x, y)	

D_{4d}	E	$2S_8$	$2C_4$	$2S_8^3$	C_2	$4C_2'$	$4\sigma_d$		
A_1	1	1	1	1	1	1	1		$x^2 + y^2, z^2$
A_2	1	1	1	1	1	−1	−1	R_z	
B_1	1	−1	1	−1	1	1	−1		
B_2	1	−1	1	−1	1	−1	1	z	
E_1	2	$\sqrt{2}$	0	$-\sqrt{2}$	−2	0	0	(x, y)	
E_2	2	0	−2	0	2	0	0		$(x^2 - y^2, xy)$
E_3	2	$-\sqrt{2}$	0	$\sqrt{2}$	−2	0	0	(R_x, R_y)	(xz, yz)

D_{5d}	E	$2C_5$	$2C_5^2$	$5C_2$	i	$2S_{10}^3$	$2S_{10}$	$5\sigma_d$		
A_{1g}	1	1	1	1	1	1	1	1		$x^2 + y^2, z^2$
A_{2g}	1	1	1	−1	1	1	1	−1	R_z	
E_{1g}	2	2 cos 72°	2 cos 144°	0	2	2 cos 72°	2 cos 144°	0	(R_x, R_y)	(xz, yz)
E_{2g}	2	2 cos 144°	2 cos 72°	0	2	2 cos 144°	2 cos 72°	0		$(x^2 - y^2, xy)$
A_{1u}	1	1	1	1	−1	−1	−1	−1		
A_{2u}	1	1	1	−1	−1	−1	−1	1	z	
E_{1u}	2	2 cos 72°	2 cos 144°	0	−2	−2 cos 72°	−2 cos 144°	0	(x, y)	
E_{2u}	2	2 cos 144°	2 cos 72°	0	−2	−2 cos 144°	−2 cos 72°	0		

D_{6d}	E	$2S_{12}$	$2C_6$	$2S_4$	$2C_3$	$2S_{12}^5$	C_2	$6C_2'$	$6\sigma_d$		
A_1	1	1	1	1	1	1	1	1	1		$x^2 + y^2, z^2$
A_2	1	1	1	1	1	1	1	−1	−1	R_z	
B_1	1	−1	1	−1	1	−1	1	1	−1		
B_2	1	−1	1	−1	1	−1	1	−1	1	z	
E_1	2	$\sqrt{3}$	1	0	−1	$-\sqrt{3}$	−2	0	0	(x, y)	
E_2	2	1	−1	−2	−1	1	2	0	0		$(x^2 - y^2, xy)$
E_3	2	0	−2	0	2	0	−2	0	0		
E_4	2	−1	−1	2	−1	−1	2	0	0		
E_5	2	$-\sqrt{3}$	1	0	−1	$\sqrt{3}$	−2	0	0	(R_x, R_y)	(xz, yz)

D_{nh} Groups

D_{2h}	E	$C_2(z)$	$C_2(y)$	$C_2(x)$	i	$\sigma(xy)$	$\sigma(xz)$	$\sigma(yz)$		
A_g	1	1	1	1	1	1	1	1		x^2, y^2, z^2
B_{1g}	1	1	−1	−1	1	1	−1	−1	R_z	xy
B_{2g}	1	−1	1	−1	1	−1	1	−1	R_y	xz
B_{3g}	1	−1	−1	1	1	−1	−1	1	R_x	yz
A_u	1	1	1	1	−1	−1	−1	−1		
B_{1u}	1	1	−1	−1	−1	−1	1	1	z	
B_{2u}	1	−1	1	−1	−1	1	−1	1	y	
B_{3u}	1	−1	−1	1	−1	1	1	−1	x	

D_{3h}	E	$2C_3$	$3C_2$	σ_h	$2S_3$	$3\sigma_v$		
A_1'	1	1	1	1	1	1		$x^2 + y^2, z^2$
A_2'	1	1	−1	1	1	−1	R_z	
E'	2	−1	0	2	−1	0	(x, y)	$(x^2 - y^2, xy)$
A_1''	1	1	1	−1	−1	−1		
A_2''	1	1	−1	−1	−1	1	z	
E''	2	−1	0	−2	1	0	(R_x, R_y)	(xz, yz)

D_{4h}	E	$2C_4$	C_2	$2C_2'$	$2C_2''$	i	$2S_4$	σ_h	$2\sigma_v$	$2\sigma_d$		
A_{1g}	1	1	1	1	1	1	1	1	1	1		$x^2 + y^2, z^2$
A_{2g}	1	1	1	−1	−1	1	1	1	−1	−1	R_z	
B_{1g}	1	−1	1	1	−1	1	−1	1	1	−1		$x^2 - y^2$
B_{2g}	1	−1	1	−1	1	1	−1	1	−1	1		xy
E_g	2	0	−2	0	0	2	0	−2	0	0	(R_x, R_y)	(xz, yz)
A_{1u}	1	1	1	1	1	−1	−1	−1	−1	−1		
A_{2u}	1	1	1	−1	−1	−1	−1	−1	1	1	z	
B_{1u}	1	−1	1	1	−1	−1	1	−1	−1	1		
B_{2u}	1	−1	1	−1	1	−1	1	−1	1	−1		
E_u	2	0	−2	0	0	−2	0	2	0	0	(x, y)	

D_{5h}	E	$2C_5$	$2C_5^2$	$5C_2$	σ_h	$2S_5$	$2S_5^3$	$5\sigma_v$		
A_1'	1	1	1	1	1	1	1	1		$x^2 + y^2, z^2$
A_2'	1	1	1	−1	1	1	1	−1	R_z	
E_1'	2	2 cos 72°	2 cos 144°	0	2	2 cos 72°	2 cos 144°	0	(x, y)	
E_2'	2	2 cos 144°	2 cos 72°	0	2	2 cos 144°	2 cos 72°	0		$(x^2 - y^2, xy)$
A_1''	1	1	1	1	−1	−1	−1	−1		
A_2''	1	1	1	−1	−1	−1	−1	1	z	
E_1''	2	2 cos 72°	2 cos 144°	0	−2	−2 cos 72°	−2 cos 144°	0	(R_x, R_y)	(xz, yz)
E_2''	2	2 cos 144°	2 cos 72°	0	−2	−2 cos 144°	−2 cos 72°	0		

D_{6h}	E	$2C_6$	$2C_3$	C_2	$3C_2'$	$3C_2''$	i	$2S_3$	$2S_6$	σ_h	$3\sigma_d$	$3\sigma_v$		
A_{1g}	1	1	1	1	1	1	1	1	1	1	1	1		$x^2 + y^2, z^2$
A_{2g}	1	1	1	1	−1	−1	1	1	1	1	−1	−1	R_z	
B_{1g}	1	−1	1	−1	1	−1	1	−1	1	−1	1	−1		
B_{2g}	1	−1	1	−1	−1	1	1	−1	1	−1	−1	1		
E_{1g}	2	1	−1	−2	0	0	2	1	−1	−2	0	0	(R_x, R_y)	(xz, yz)
E_{2g}	2	−1	−1	2	0	0	2	−1	−1	2	0	0		$(x^2 - y^2, xy)$
A_{1u}	1	1	1	1	1	1	−1	−1	−1	−1	−1	−1		
A_{2u}	1	1	1	1	−1	−1	−1	−1	−1	−1	1	1	z	
B_{1u}	1	−1	1	−1	1	−1	−1	1	−1	1	−1	1		
B_{2u}	1	−1	1	−1	−1	1	−1	1	−1	1	1	−1		
E_{1u}	2	1	−1	−2	0	0	−2	−1	1	2	0	0	(x, y)	
E_{2u}	2	−1	−1	2	0	0	−2	1	1	−2	0	0		

D_{8h}	E	$2C_8$	$2C_8^3$	$2C_4$	C_2	$4C_2'$	$4C_2''$	i	$2S_8$	$2S_8^3$	$2S_4$	σ_h	$4\sigma_d$	$4\sigma_v$		
A_{1g}	1	1	1	1	1	1	1	1	1	1	1	1	1	1		$x^2 + y^2, z^2$
A_{2g}	1	1	1	1	1	−1	−1	1	1	1	1	1	−1	−1	R_z	
B_{1g}	1	−1	−1	1	1	1	−1	1	−1	−1	1	1	1	−1		
B_{2g}	1	−1	−1	1	1	−1	1	1	−1	−1	1	1	−1	1		
E_{1g}	2	$\sqrt{2}$	$-\sqrt{2}$	0	−2	0	0	2	$\sqrt{2}$	$-\sqrt{2}$	0	−2	0	0	(R_x, R_y)	(xz, yz)
E_{2g}	2	0	0	−2	2	0	0	2	0	0	−2	2	0	0		$(x^2 - y^2, xy)$
E_{3g}	2	$-\sqrt{2}$	$\sqrt{2}$	0	−2	0	0	2	$-\sqrt{2}$	$\sqrt{2}$	0	−2	0	0		
A_{1u}	1	1	1	1	1	1	1	−1	−1	−1	−1	−1	−1	−1		
A_{2u}	1	1	1	1	1	−1	−1	−1	−1	−1	−1	−1	1	1	z	
B_{1u}	1	−1	−1	1	1	1	−1	−1	1	1	−1	−1	−1	1		
B_{2u}	1	−1	−1	1	1	−1	1	−1	1	1	−1	−1	1	−1		
E_{1u}	2	$\sqrt{2}$	$-\sqrt{2}$	0	−2	0	0	−2	$-\sqrt{2}$	$\sqrt{2}$	0	2	0	0	(x, y)	
E_{2u}	2	0	0	−2	2	0	0	−2	0	0	2	−2	0	0		
E_{3u}	2	$-\sqrt{2}$	$\sqrt{2}$	0	−2	0	0	−2	$\sqrt{2}$	$-\sqrt{2}$	0	2	0	0		

4. Linear Groups

$C_{\infty v}$	E	$2C_{\infty}^{\phi}$	...	$\infty\sigma_v$		
$A_1 \equiv \Sigma^+$	1	1	...	1	z	x^2+y^2, z^2
$A_2 \equiv \Sigma^-$	1	1	...	-1	R_z	
$E_1 \equiv \Pi$	2	$2\cos\phi$	...	0	(x, y), (R_x, R_y)	(xz, yz)
$E_2 \equiv \Delta$	2	$2\cos 2\phi$	...	0		(x^2-y^2, xy)
$E_3 \equiv \Phi$	2	$2\cos 3\phi$	...	0		
...	...	...	...	...		

$D_{\infty h}$	E	$2C_{\infty}^{\phi}$	...	$\infty\sigma_v$	i	$2S_{\infty}^{\phi}$	...	∞C_2		
$A_{1g} \equiv \Sigma_g^+$	1	1	...	1	1	1	...	1		x^2+y^2, z^2
$A_{2g} \equiv \Sigma_g^-$	1	1	...	-1	1	1	...	-1	R_z	
$E_{1g} \equiv \Pi_g$	2	$2\cos\phi$	...	0	2	$-2\cos\phi$	...	0	(R_x, R_y)	(xz, yz)
$E_{2g} \equiv \Delta_g$	2	$2\cos 2\phi$	...	0	2	$2\cos 2\phi$	...	0		(x^2-y^2, xy)
...	...	...	...	...	...	...	...	...		
$A_{1u} \equiv \Sigma_u^+$	1	1	...	1	-1	-1	...	-1	z	
$A_{2u} \equiv \Sigma_u^-$	1	1	...	-1	-1	-1	...	1		
$E_{1u} \equiv \Pi_u$	2	$2\cos\phi$	...	0	-2	$2\cos\phi$	...	0	(x, y)	
$E_{2u} \equiv \Delta_u$	2	$2\cos 2\phi$	...	0	-2	$-2\cos 2\phi$	...	0		
...	...	...	...	...	...	...	...	...		

5. S_{2n} Groups

S_4	E	S_4	C_2	S_4^3		
A	1	1	1	1	R_z	x^2+y^2, z^2
B	1	-1	1	-1	z	x^2-y^2, xy
E	$\left\{\begin{matrix}1 \\ 1\end{matrix}\right.$	$\begin{matrix}i \\ -i\end{matrix}$	$\begin{matrix}-1 \\ -1\end{matrix}$	$\left.\begin{matrix}-i \\ i\end{matrix}\right\}$	(x, y), (R_x, R_y)	(xz, yz)

S_6	E	C_3	C_3^2	i	S_6^5	S_6		
A_g	1	1	1	1	1	1	R_z	x^2+y^2, z^2
E_g	$\left\{\begin{matrix}1 \\ 1\end{matrix}\right.$	$\begin{matrix}\varepsilon \\ \varepsilon^*\end{matrix}$	$\begin{matrix}\varepsilon^* \\ \varepsilon\end{matrix}$	$\begin{matrix}1 \\ 1\end{matrix}$	$\begin{matrix}\varepsilon \\ \varepsilon^*\end{matrix}$	$\left.\begin{matrix}\varepsilon^* \\ \varepsilon\end{matrix}\right\}$	(R_x, R_y)	(x^2-y^2, xy) (xz, yz)
A_u	1	1	1	-1	-1	-1	z	
E_u	$\left\{\begin{matrix}1 \\ 1\end{matrix}\right.$	$\begin{matrix}\varepsilon \\ \varepsilon^*\end{matrix}$	$\begin{matrix}\varepsilon^* \\ \varepsilon\end{matrix}$	$\begin{matrix}-1 \\ -1\end{matrix}$	$\begin{matrix}-\varepsilon \\ -\varepsilon^*\end{matrix}$	$\left.\begin{matrix}-\varepsilon^* \\ -\varepsilon\end{matrix}\right\}$	(x, y)	

$\varepsilon = e^{(2\pi i)/3}$

S_8	E	S_8	C_4	S_8^3	C_2	S_8^5	C_4^3	S_8^7		
A	1	1	1	1	1	1	1	1	R_z	x^2+y^2, z^2
B	1	-1	1	-1	1	-1	1	-1	z	
E_1	$\left\{\begin{matrix}1 \\ 1\end{matrix}\right.$	$\begin{matrix}\varepsilon \\ \varepsilon^*\end{matrix}$	$\begin{matrix}i \\ -i\end{matrix}$	$\begin{matrix}-\varepsilon^* \\ -\varepsilon\end{matrix}$	$\begin{matrix}-1 \\ -1\end{matrix}$	$\begin{matrix}-\varepsilon \\ -\varepsilon^*\end{matrix}$	$\begin{matrix}-i \\ i\end{matrix}$	$\left.\begin{matrix}\varepsilon^* \\ \varepsilon\end{matrix}\right\}$	(x, y), (R_x, R_y)	
E_2	$\left\{\begin{matrix}1 \\ 1\end{matrix}\right.$	$\begin{matrix}i \\ -i\end{matrix}$	$\begin{matrix}-1 \\ -1\end{matrix}$	$\begin{matrix}-i \\ i\end{matrix}$	$\begin{matrix}1 \\ 1\end{matrix}$	$\begin{matrix}i \\ -i\end{matrix}$	$\begin{matrix}-1 \\ -1\end{matrix}$	$\left.\begin{matrix}-i \\ i\end{matrix}\right\}$		(x^2-y^2, xy)
E_3	$\left\{\begin{matrix}1 \\ 1\end{matrix}\right.$	$\begin{matrix}-\varepsilon^* \\ -\varepsilon\end{matrix}$	$\begin{matrix}-i \\ i\end{matrix}$	$\begin{matrix}\varepsilon \\ \varepsilon^*\end{matrix}$	$\begin{matrix}-1 \\ -1\end{matrix}$	$\begin{matrix}\varepsilon^* \\ \varepsilon\end{matrix}$	$\begin{matrix}i \\ -i\end{matrix}$	$\left.\begin{matrix}-\varepsilon \\ -\varepsilon^*\end{matrix}\right\}$		(xz, yz)

$\varepsilon = e^{(\pi i)/4}$

6. Tetrahedral, Octahedral, and Icosahedral Groups

T	E	$4C_3$	$4C_3^2$	$3C_2$		
A	1	1	1	1		$x^2 + y^2 + z^2$
E	$\begin{Bmatrix} 1 & \varepsilon & \varepsilon^* & 1 \\ 1 & \varepsilon^* & \varepsilon & 1 \end{Bmatrix}$					$(2z^2 - x^2 - y^2, x^2 - y^2)$
T	3	0	0	−1	$(R_x, R_y, R_z), (x, y, z)$	(xy, xz, yz)

$\varepsilon = e^{(2\pi i)/3}$

T_d	E	$8C_3$	$3C_2$	$6S_4$	$6\sigma_d$		
A_1	1	1	1	1	1		$x^2 + y^2 + z^2$
A_2	1	1	1	−1	−1		
E	2	−1	2	0	0		$(2z^2 - x^2 - y^2, x^2 - y^2)$
T_1	3	0	−1	1	−1	(R_x, R_y, R_z)	
T_2	3	0	−1	−1	1	(x, y, z)	(xy, xz, yz)

T_h	E	$4C_3$	$4C_3^2$	$3C_2$	i	$4S_6$	$4S_6^5$	$3\sigma_h$		
A_g	1	1	1	1	1	1	1	1		$x^2 + y^2 + z^2$
A_u	1	1	1	1	−1	−1	−1	−1		
E_g	$\begin{Bmatrix} 1 & \varepsilon & \varepsilon^* & 1 & 1 & \varepsilon & \varepsilon^* & 1 \\ 1 & \varepsilon^* & \varepsilon & 1 & 1 & \varepsilon^* & \varepsilon & 1 \end{Bmatrix}$									$(2z^2 - x^2 - y^2, x^2 - y^2)$
E_u	$\begin{Bmatrix} 1 & \varepsilon & \varepsilon^* & 1 & -1 & -\varepsilon & -\varepsilon^* & -1 \\ 1 & \varepsilon^* & \varepsilon & 1 & -1 & -\varepsilon^* & -\varepsilon & -1 \end{Bmatrix}$									
T_g	3	0	0	−1	3	0	0	−1	(R_x, R_y, R_z)	(xy, xz, yz)
T_u	3	0	0	−1	−3	0	0	1	(x, y, z)	

$\varepsilon = e^{(2\pi i)/3}$

O	E	$6C_4$	$3C_2(=C_4^2)$	$8C_3$	$6C_2$		
A_1	1	1	1	1	1		$x^2 + y^2 + z^2$
A_2	1	−1	1	1	−1		
E	2	0	2	−1	0		$(2z^2 - x^2 - y^2, x^2 - y^2)$
T_1	3	1	−1	0	−1	$(R_x, R_y, R_z), (x, y, z)$	
T_2	3	−1	−1	0	1		(xy, xz, yz)

O_h	E	$8C_3$	$6C_2$	$6C_4$	$3C_2(=C_4^2)$	i	$6S_4$	$8S_6$	$3\sigma_h$	$6\sigma_d$		
A_{1g}	1	1	1	1	1	1	1	1	1	1		$x^2 + y^2 + z^2$
A_{2g}	1	1	−1	−1	1	1	−1	1	1	−1		
E_g	2	−1	0	0	2	2	0	−1	2	0		$(2z^2 - x^2 - y^2, x^2 - y^2)$
T_{1g}	3	0	−1	1	−1	3	1	0	−1	−1	(R_x, R_y, R_z)	
T_{2g}	3	0	1	−1	−1	3	−1	0	−1	1		(xy, xz, yz)
A_{1u}	1	1	1	1	1	−1	−1	−1	−1	−1		
A_{2u}	1	1	−1	−1	1	−1	1	−1	−1	1		
E_u	2	−1	0	0	2	−2	0	1	−2	0		
T_{1u}	3	0	−1	1	−1	−3	−1	0	1	1	(x, y, z)	
T_{2u}	3	0	1	−1	−1	−3	1	0	1	−1		

I	E	$12C_5$	$12C_5^2$	$20C_3$	$15C_2$		
A	1	1	1	1	1		$x^2 + y^2 + z^2$
T_1	3	$\frac{1}{2}(1 + \sqrt{5})$	$\frac{1}{2}(1 - \sqrt{5})$	0	−1	$(x, y, z), (R_x, R_y, R_z)$	
T_2	3	$\frac{1}{2}(1 - \sqrt{5})$	$\frac{1}{2}(1 + \sqrt{5})$	0	−1		
G	4	−1	−1	1	0		
H	5	0	0	−1	1		$(xy, xz, yz, x^2 - y^2, 2z^2 - x^2 - y^2)$

I_h	E	$12C_5$	$12C_5^2$	$20C_3$	$15C_2$	i	$12S_{10}$	$12S_{10}^3$	$20S_6$	15σ		
A_g	1	1	1	1	1	1	1	1	1	1		$x^2 + y^2 + z^2$
T_{1g}	3	$\frac{1}{2}(1 + \sqrt{5})$	$\frac{1}{2}(1 - \sqrt{5})$	0	−1	3	$\frac{1}{2}(1 - \sqrt{5})$	$\frac{1}{2}(1 + \sqrt{5})$	0	−1	(R_x, R_y, R_z)	
T_{2g}	3	$\frac{1}{2}(1 - \sqrt{5})$	$\frac{1}{2}(1 + \sqrt{5})$	0	−1	3	$\frac{1}{2}(1 + \sqrt{5})$	$\frac{1}{2}(1 - \sqrt{5})$	0	−1		
G_g	4	−1	−1	1	0	4	−1	−1	1	0		
H_g	5	0	0	−1	1	5	0	0	−1	1		$(2z^2 - x^2 - y^2, x^2 - y^2, xy, xz, yz)$
A_u	1	1	1	1	1	−1	−1	−1	−1	−1		
T_{1u}	3	$\frac{1}{2}(1 + \sqrt{5})$	$\frac{1}{2}(1 - \sqrt{5})$	0	−1	−3	$-\frac{1}{2}(1 - \sqrt{5})$	$-\frac{1}{2}(1 + \sqrt{5})$	0	1	(x, y, z)	
T_{2u}	3	$\frac{1}{2}(1 - \sqrt{5})$	$\frac{1}{2}(1 + \sqrt{5})$	0	−1	−3	$-\frac{1}{2}(1 + \sqrt{5})$	$-\frac{1}{2}(1 - \sqrt{5})$	0	1		
G_u	4	−1	−1	1	0	−4	1	1	−1	0		
H_u	5	0	0	−1	1	−5	0	0	1	−1		

Index

C

I

P

Q

R

S

Electron Configurations of the Elements

Element	*Z*	*Configuration*
H	1	$1s^1$
He	2	$1s^2$
Li	3	[He]$2s^1$
Be	4	[He]$2s^2$
B	5	[He]$2s^22p^1$
C	6	[He]$2s^22p^2$
N	7	[He]$2s^22p^3$
O	8	[He]$2s^22p^4$
F	9	[He]$2s^22p^5$
Ne	10	[He]$2s^22p^6$
Na	11	[Ne]$3s^1$
Mg	12	[Ne]$3s^2$
Al	13	[Ne]$3s^23p^1$
Si	14	[Ne]$3s^23p^2$
P	15	[Ne]$3s^23p^3$
S	16	[Ne]$3s^23p^4$
Cl	17	[Ne]$3s^23p^5$
Ar	18	[Ne]$3s^23p^6$
K	19	[Ar]$4s^1$
Ca	20	[Ar]$4s^2$
Sc	21	[Ar]$4s^23d^1$
Ti	22	[Ar]$4s^23d^2$
V	23	[Ar]$4s^23d^3$
Cr	24	*[Ar]$4s^13d^5$
Mn	25	[Ar]$4s^23d^5$
Fe	26	[Ar]$4s^23d^6$
Co	27	[Ar]$4s^23d^7$
Ni	28	[Ar]$4s^23d^8$
Cu	29	*[Ar]$4s^13d^{10}$
Zn	30	[Ar]$4s^23d^{10}$
Ga	31	[Ar]$4s^23d^{10}4p^1$
Ge	32	[Ar]$4s^23d^{10}4p^2$
As	33	[Ar]$4s^23d^{10}4p^3$
Se	34	[Ar]$4s^23d^{10}4p^4$
Br	35	[Ar]$4s^23d^{10}4p^5$
Kr	36	[Ar]$4s^23d^{10}4p^6$
Rb	37	[Kr]$5s^1$
Sr	38	[Kr]$5s^2$
Y	39	[Kr]$5s^24d^1$
Zr	40	[Kr]$5s^24d^2$
Nb	41	*[Kr]$5s^14d^4$
Mo	42	*[Kr]$5s^14d^5$
Tc	43	[Kr]$5s^24d^5$
Ru	44	*[Kr]$5s^14d^7$
Rh	45	*[Kr]$5s^14d^8$
Pd	46	*[Kr]$4d^{10}$
Ag	47	*[Kr]$5s^14d^{10}$
Cd	48	[Kr]$5s^24d^{10}$
In	49	[Kr]$5s^24d^{10}5p^1$
Sn	50	[Kr]$5s^24d^{10}5p^2$
Sb	51	[Kr]$5s^24d^{10}5p^3$
Te	52	[Kr]$5s^24d^{10}5p^4$
I	53	[Kr]$5s^24d^{10}5p^5$
Xe	54	[Kr]$5s^24d^{10}5p^6$
Cs	55	[Xe]$6s^1$
Ba	56	[Xe]$6s^2$

Element	*Z*	*Configuration*
La	57	*[Xe]$6s^25d^1$
Ce	58	*[Xe]$6s^24f^15d^1$
Pr	59	[Xe]$6s^24f^3$
Nd	60	[Xe]$6s^24f^4$
Pm	61	[Xe]$6s^24f^5$
Sm	62	[Xe]$6s^24f^6$
Eu	63	[Xe]$6s^24f^7$
Gd	64	*[Xe]$6s^24f^75d^1$
Tb	65	[Xe]$6s^24f^9$
Dy	66	[Xe]$6s^24f^{10}$
Ho	67	[Xe]$6s^24f^{11}$
Er	68	[Xe]$6s^24f^{12}$
Tm	69	[Xe]$6s^24f^{13}$
Yb	70	[Xe]$6s^24f^{14}$
Lu	71	[Xe]$6s^24f^{14}5d^1$
Hf	72	[Xe]$6s^24f^{14}5d^2$
Ta	73	[Xe]$6s^24f^{14}5d^3$
W	74	[Xe]$6s^24f^{14}5d^4$
Re	75	[Xe]$6s^24f^{14}5d^5$
Os	76	[Xe]$6s^24f^{14}5d^6$
Ir	77	[Xe]$6s^24f^{14}5d^7$
Pt	78	*[Xe]$6s^14f^{14}5d^9$
Au	79	*[Xe]$6s^14f^{14}5d^{10}$
Hg	80	[Xe]$6s^24f^{14}5d^{10}$
Tl	81	[Xe]$6s^24f^{14}5d^{10}6p^1$
Pb	82	[Xe]$6s^24f^{14}5d^{10}6p^2$
Bi	83	[Xe]$6s^24f^{14}5d^{10}6p^3$
Po	84	[Xe]$6s^24f^{14}5d^{10}6p^4$
At	85	[Xe]$6s^24f^{14}5d^{10}6p^5$
Rn	86	[Xe]$6s^24f^{14}5d^{10}6p^6$
Fr	87	[Rn]$7s^1$
Ra	88	[Rn]$7s^2$
Ac	89	*[Rn]$7s^26d^1$
Th	90	*[Rn]$7s^26d^2$
Pa	91	*[Rn]$7s^25f^26d^1$
U	92	*[Rn]$7s^25f^36d^1$
Np	93	*[Rn]$7s^25f^46d^1$
Pu	94	[Rn]$7s^25f^6$
Am	95	[Rn]$7s^25f^7$
Cm	96	*[Rn]$7s^25f^76d^1$
Bk	97	[Rn]$7s^25f^9$
Cf	98	[Rn]$7s^25f^{10}$
Es	99	[Rn]$7s^25f^{11}$
Fm	100	[Rn]$7s^25f^{12}$
Md	101	[Rn]$7s^25f^{13}$
No	102	[Rn]$7s^25f^{14}$
Lr	103	*[Rn]$7s^25f^{14}7p^1$
Rf	104	[Rn]$7s^25f^{14}6d^2$
Db	105	[Rn]$7s^25f^{14}6d^3$
Sg	106	[Rn]$7s^25f^{14}6d^4$
Bh	107	[Rn]$7s^25f^{14}6d^5$
Hs	108	[Rn]$7s^25f^{14}6d^6$
Mt	109	[Rn]$7s^25f^{14}6d^7$
Ds	110	*[Rn]$7s^15f^{14}6d^9$
Rg	111	*[Rn]$7s^15f^{14}6d^{10}$
Cn	112	[Rn]$7s^25f^{14}6d^{10}$
Fl	114	[Rn]$7s^25f^{14}6d^{10}7p^2$
Lv	116	[Rn]$7s^25f^{14}6d^{10}7p^4$

*Elements with configurations that do not follow the simple order of orbital filling.

Evidence has been reported for elements having atomic numbers 113, 115, 117 and 118, but these have not been authenticated by the IUPAC. Configurations for elements 103–118 are predicted, not experimental.

Source: Data from Actinide configurations are from J. J. Katz, G. T. Seaborg, and L. R. Morss, *The Chemistry of the Actinide Elements*, 2nd ed., Chapman and Hall, New York and London, 1986.